ABOUT THE AUTHOR

Mahmood Nahvi is emeritus professor of electrical engineering at California Polytechnic State University (Cal Poly) in San Luis Obispo, California. He earned B.Sc., M.Sc., and Ph.D. degrees in electrical engineering, and has more than 50 years of educational, academic, research, and industrial experience in this field. He has taught undergraduate and graduate courses on various subjects including electric circuits, electronics, signals and systems, electromagnetics, random signals and noise, digital signal and image processing, and communications and control. Prior to joining the faculty at Cal Poly, he held positions as a guest professor and research scientist at the University of Stuttgart in Germany, working on signal processing in space communication systems, satellite microwave ranging, and synthetic aperture radar projects for the European Space Agency. In addition, he was a founding faculty member, professor, and department chair at Sharif University of Technology (former Aria-Mehr) in Teheran, Iran; a visiting professor at MIT; and a research scientist at UCLA, the Laboratory for Neural Control at the National Institutes of Health, and the Max-Planck Institute for Biological Cybernetics in Tübingen, Germany. He is a senior member of the Institute of Electrical and Electronics Engineers (IEEE).

Dr. Nahvi's areas of special interest and expertise include network theory, control theory, adaptation and learning in synthetic and living systems, communication and control in living systems, neural networks for pattern recognition and motor control, satellite communication, digital signal and image processing, and engineering education. In the area of engineering education, he has developed computer modules for electric circuits, signals, and systems that improve the teaching and learning of the fundamentals of electrical engineering. He is coauthor of *Electric Circuits* and *Electromagnetics* in Schaum's Outline Series published by McGraw-Hill.

This book is dedicated to the memory of
Janab Nahvi
humanist, scholar, and teacher.

BRIEF CONTENTS

CONTENTS

PREFACE

The subject of signals and systems is a requirement in undergraduate electrical and computer engineering programs. The subject provides the student a window through which he or she can look into and examine the field. In addition, it provides the necessary background for more specialized subjects, including communication, control, and signal processing. Several other engineering majors offer similar courses in the same subject matter.

This book is designed to serve as the primary textbook for a course in signals and systems at the junior undergraduate level. It is to be used mainly in the electrical, electronics, and computer engineering programs but is also appropriate for other engineering majors. It may be used in a one- or two-semester or two-quarter sequence according to the criteria of the curriculum and depending on an appropriate selection of material which meets the needs and backgrounds of students.

This book treats the continuous- and discrete-time domains separately in two parts. Part One (Chapters 1–9) covers continuous-time signals and systems; Part Two (Chapters 10–18) covers discrete-time signals and systems. Both parts stand alone and can be used independently of each other. This allows instructors to use the text for instruction on either domain separately, if desired. The book may also be used for courses that teach the two domains simultaneously in an integrated way, as the chapters in Parts One and Two provide parallel presentations of each subject. The parallelism of the chapters on the continuous- and discrete-time domains facilitates the integration of the two parts and allows for flexibility of use in various curricula. The chapter topics and the parallelism between the time-domain treatments are listed in the table below.

Part One, Continuous-Time Domain		Part Two, Discrete-Time Domain	
Chapter	Topic	Chapter	Topic
1	Introduction to Signals	10	Time-Domain Sampling and Reconstruction
2	Sinusoids	11	Discrete-Time Signals
3	Systems, Linearity, and Time Invariance	12	Linear Time-Invariant Discrete-Time Systems
4	Superposition, Convolution, and Correlation	13	Discrete Convolution
5	Differential Equations and LTI Systems	14	LTI Difference Equations
6	The Laplace Transform and Its Applications	15	The z-Transform and Its Applications
7	Fourier Series	16	Discrete-Time Fourier Transform
8	Fourier Transform	17	Discrete Fourier Transforms
9	System Function, the Frequency Response, and Analog Filters	18	System Function, the Frequency Response, and Digital Filters

Whether the subject of signals and systems in the continuous- and discrete-time domains is taught separately or in integrated form, the present organization of the book provides both pedagogical and practical advantages. A considerable part of the subject matter in signals and systems is on analysis techniques (such as solution methods in the time and frequency domains) which, although conceptually similar, use different tools.

Introducing the tools and applying them separately simplifies the structure of the course. Another advantage of the present organization is that the analyses of signals and systems in the continuous- and discrete-time domains can stand on their own (both conceptually and in terms of analysis tools). Each domain may be taught without requiring the other. Thus, for programs that are designed to offer a DSP course, the discrete-time part of the book will satisfy the prerequisites of such a course.

Each part begins with the introduction of signals and their models in the time domain. It then defines systems, linearity, and time invariance, along with examples. Time-domain solution methods, such as convolution and differential/difference equations, are presented next, followed by the transform domains. These are brought together in capstone chapters on the system function and frequency response. Chapter 10 on sampling provides a bridge between the continuous- and discrete-time domains.

Each chapter is made of sections and no subsections. Each section addresses a single discussion item, starting with the introduction of a topic, mathematical tools used to address that topic, the application of those tools, and one or two examples. To a large extent, therefore, each section is a learning unit and can provide the student with a concluding marker in learning the subject. In that sense the sections are modular and convenient for instruction. The modular organization of the book provides a direct approach and an effective tool for learning the fundamentals of signals and systems. As a vehicle for lectures, 5 to 10 essential sections may be covered in an hour, while others may be assigned as outside reading or homework.

Reference to other sections, figures, formulas, and other chapters is kept to a minimum. This provides easy and direct access to material, a feature much preferred by students and instructors. The modular structure of the chapters and sections also makes the book a convenient tool for instructional needs in a wide range of teaching scenarios at various levels of complexity.

Illustrative examples, end-of-chapter problems, and supplementary problems with solutions comprise other important components of the book. The book contains a total of nearly 475 examples, 175 problems with solutions, and 750 end-of-chapter problems. The examples and problems are of two types: (1) mathematical analyses and exercises that represent abstractions of engineering subjects and (2) contextual problems, such as those seen in electric circuits and devices, communication, and control. For the EE and CPE student these subjects provide a context to convey and develop fundamental concepts in signals and systems.

Examples from familiar signals and tangible systems in engineering can illustrate the utility of the relevant mathematical analysis. They can make the subject more attractive and generate motivation. In accordance with the above pedagogy, the book assumes that the reader is familiar with the operation of basic circuits and devices (such as passive *RLC* circuits and active circuits including dependent sources and operational amplifier models) and uses these to illustrate and reinforce the mathematical concepts. It also assumes familiarity with elementary trigonometric functions, complex numbers, differentiation/integration, and matrices. The Appendix at the end of the book can be used to refresh the reader's memory on electric circuits.

ORGANIZATION OF CHAPTERS

The detailed outline of the first part, covering signals and systems in the continuous-time domain, is as follows.

Chapter 1 introduces various signal types (such as those that are natural, societal, or human-made) and their models. It shows that, as functions of time, signals are specified by a wide set of parameters and characteristics (e.g., rate of change, time course, periodicity, and fine, coarse, and nested-loops structures). Time averages are discussed, along with some simple operations on signals.

Chapter 2 is on sinusoids and contains a review of basic trigonometry. The examples in this chapter employ simple sinusoids in illustrating some topics of practical interest such as phase and group delay, power, and more.

Chapter 3 introduces the definitions of linearity and time invariance. Examples teach the student how to test for these properties. This initial chapter is not intended to cover all properties of LTI systems, but only as much as is needed at this stage in such a course on signals and systems. More exposure will be provided throughout other parts of the book.

Chapter 4 discusses the time-domain solution of LTI systems by convolution. Convolution of a system's unit-impulse response with the input produces the system's response to that input. The chapter starts with convolution as a method of obtaining the response of a linear system. It uses the linearity and superposition properties to develop the convolution sum and integral. It then illustrates their evaluation by numerical, analytical, and graphical methods. The filtering property of convolution is explained next. The chapter also briefly touches on deconvolution. This latter concept is brought up in future chapters on solutions in the frequency domain.

Chapter 5 presents the time-domain solution of LTI systems by an examination of their describing differential equations in classical form. Parallels are drawn between the homogeneous and particular components of the total solution and the familiar components in the response of physical systems; that is, the natural and forced parts of the response from system analysis and design. The homogeneous and particular components of the total solution are then also related to the zero-input and zero-state responses. An example of a numerical computation of a response is provided at the end of the chapter.

Chapter 6 analyzes the solution of LTI systems by the Laplace transform in the frequency domain. Both the unilateral and bilateral forms of the transform are considered. The first half of the chapter focuses on the unilateral version, its inverse evaluation by the partial fraction expansion method, and some applications. The residue method of finding the inverse is also presented. The second half addresses the bilateral Laplace transform and its inverse. Comprehensive examples demonstrate how to obtain the response of an LTI system by the frequency domain approach and observe parallels with those in the time domain.

Chapters 7 and 8 are on Fourier analysis in the continuous-time domain. Chapter 7 discusses the Fourier series expansion of periodic signals, in both trigonometric and exponential forms, and visualizes methods of extending the expansion to nonperiodic signals, which is a topic presented in detail in Chapter 8. Introduction of the impulse function in the frequency domain provides a unified method of Fourier analysis for a large class of signals and systems. The convolution property of the Fourier transform enables system analysis in the frequency domain. The frequency variable f (or ω) in that analysis is more reminiscent of the actual real-world physical frequency than the complex frequency shown by s in the method of Laplace transforms.

Chapter 9 envisions a multiangle capstone perspective of the analysis methods presented up to this point. It introduces the system function, poles and zeros, the frequency response, and Bode plots. It then explains their relationships to each other and to the time-domain characteristics of a system. A vectorial interpretation of the system function and frequency response is included in order to help provide a qualitative understanding of the system's characteristics. Modeling a system by its dominant poles is then discussed in sufficient detail and illustrated by examples for first- and second-order systems. Interconnections between systems and the concept of feedback are then covered. Finally, the chapter concludes with a brief review of the effect of feedback on system behavior, along with an example of controller design.

The detailed outline of the second part of the book, covering signals and systems in the discrete-time domain, is as follows.

Chapter 10 is on the time-domain sampling of continuous-time signals and their reconstruction. It uses Fourier transform techniques and properties developed for continuous-time signals in Chapter 8 in order to derive the minimum sampling rate and a method for the error-free reconstruction of low-pass signals. The continuous-time signals used in the examples of this chapter are mostly built around 1 Hz to coincide, without loss of generality, with the *normalized* frequency encountered in the Fourier analysis of discrete-time signals in the second part of the book. The effects of sampling rate, the reconstruction filter, nearly low-pass signals, and the aliasing phenomenon are discussed. The chapter also extends the presentation to sampling and reconstruction of complex low-pass and bandpass signals.

Chapters 11 and 12 introduce discrete-time signals and LTI systems in a way that parallels the discussions in Chapters 1 and 3.

The discrete convolution and difference equations are discussed in Chapters 13 and 14. In these chapters, as in Part One, in addition to developing quantitative analytical techniques the text also aims to develop the student's intuitive sense of signals and systems.

Chapter 15 is on the z-transform and parallels the Laplace transform in providing a frequency domain analysis of discrete-time signals and systems. However, the chapter starts with the bilateral transform and its applications to signals and systems and then proceeds to the unilateral transform. The z-transform is normally defined on its own. However, it is related to the Laplace transform and can be derived from it. The relationship between these two transforms is explained in the chapter.

Chapters 16 and 17 discuss the discrete-time Fourier transform (DTFT) and the discrete Fourier transform (DFT). As is true of the z-transform, they can be defined on their own or be derived as extensions of the Fourier transform for continuous-time signals. Primary emphasis is given to the DTFT and DFT as stand-alone operations with a secondary reminder of their relationship to the Fourier transform for continuous-time signals. Having introduced the DTFT as an analysis tool, Chapter 16 introduces the concepts of decimation, interpolation, and sampling rate conversion. These concepts have a special place in discrete-time signal processing.

Chapter 18 is the discrete-time counterpart to Chapter 9. It encapsulates the system function, poles and zeros, and the frequency response. It includes an introduction to digital filters with relevant examples.

The book includes an appendix on the basics of electric circuit analysis.

PROJECTS

As a concept, projects cannot only reinforce a theory learned but also motivate it. Ideally, they have the most impact on learning when most of their formulation and solution steps are left to the student. With these ideas in mind, each chapter includes one or more projects germane to the subject of that chapter. These projects present self-contained theory and procedures that lead the student toward expected results and conclusions. Most projects are designed to be carried out in a laboratory with basic measurement instruments. They can also be implemented by using a simulation package such as Matlab. It is, however, recommended that they be done in a real-time laboratory environment whenever possible. For example, despite its simplicity, a simple passive *RLC* circuit can demonstrate many features of first- and second-order systems. Similarly, time and frequency responses, system function, oscillations, and the stability of systems can best be explored using an actual op-amp circuit.

PEDAGOGY

The book is designed with the following pedagogical features in mind.

1. One learns from being exposed to examples, each of which addresses a single, not-too-complicated question. The examples should be easy to grasp, relevant, and applicable to new scenarios.

2. One learns by doing, whether using paper and pencil, computer tools, projects, or laboratory experiments employing hardware. This leads students to search, explore, and seek new solutions. This point, along with the previous one, helps them develop their own methods of generalization, concept formation, and modeling.

3. One learns from exposure to a problem from several angles. This allows for the analysis of a case at various levels of complexity.

4. One needs to develop a qualitative and intuitive understanding of the principles behind, applications of, and solutions for the particular problem at hand. This is to supplement the quantitative and algorithmic method of solving the problem.

5. One benefits a great deal from gradual learning; starting from what has already been learned, one builds upon this foundation using familiar tools. In order to discuss a complex concept, one starts with the discussion and use of a simpler one upon which the former is based. An example would be introducing and using mathematical entities such as the frequency-domain variables s and z initially as complex numbers in exponential functions. The student first becomes familiar with the role the new variables play in the analysis of signals and systems before moving on to the Laplace and z-transforms. Another example would be the frequency

response, a concept that can be developed within the existing realm of sinusoids and as an experimentally measurable characteristic of a system, as opposed to the more mathematical formulation of evaluating the system function on the imaginary axis of the complex plane. Yet another example would be the convolution integral, which can initially be introduced as a weighted averaging process.

ACKNOWLEDGMENTS

The author is indebted to many faculty colleagues who reviewed the manuscript and provided valuable comments and suggestions. Reza Nahvi also contributed to the book significantly during its various stages. I also wish to acknowledge the expertise and assistance of the project team at McGraw-Hill Higher Education, especially global brand manager Raghothaman Srinivasan, executive marketing manager Curt Reynolds, developmental editor Katie Neubauer, and senior project manager Lisa Bruflodt. I thank them all for making this book possible.

Permissions were granted for the use of material from the following sources, whose details are given within the text:

Experimental Brain Research, Elsevier Inc.

Experimental Neurology, Springer Inc.

Squire, L. et al ed. *Fundamental Neuroscience*, Elsevier Inc. 2008.

Data from the following sources were used in constructing several figures, details for which are given within the text.

National Geophysical Data Center (NGDC), www.ngdc.noaa.gov.

Dr. Pieter Tans, NOAA/ESRL (www.esrl.noaa.gov/gmd/ccgg/trends/) and Dr. Ralph Keeling, Scripps Institution of Oceanography (scrippsco2.ucsd.edu).

Northern California Earthquake Data Center and Berkeley Seismological Laboratory, University of California, Berkeley, www.ncedc.org.

Bureau of Labor Statistics, www.bls.gov.

ONLINE RESOURCES

INSTRUCTOR AND STUDENT WEBSITE

Available at www.mhhe.com/nahvi 1e are a number of additional instructor and student resources to accompany the text. These include solutions for end-of-chapter problems and lecture PowerPoints. The site also features COSMOS, a complete online solutions manual organization system that allows instructors to create custom homework, quizzes, and tests using end-of-chapter problems from the text.

This text is available as an eBook at www.CourseSmart.com. At CourseSmart your students can take advantage of significant savings off the cost of a print textbook, reduce their impact on the environment, and gain access to powerful web tools for learning. CourseSmart eBooks can be viewed online or downloaded to a computer.

The eBooks allow students to do full text searches, add highlighting and notes, and share notes with classmates. CourseSmart has the largest selection of eBooks available anywhere. Visit www.CourseSmart.com to learn more and to try a sample chapter.

Craft your teaching resources to match the way you teach! With McGraw-Hill Create, www.mcgrawhillcreate.com, you can easily rearrange chapters, combine material from other content sources, and quickly upload content you have written like your course syllabus or teaching notes. Find the content you need in Create by searching through thousands of leading McGraw-Hill textbooks. Arrange your book to fit your teaching style. Create even allows you to personalize your book's appearance by selecting the cover and adding your name, school, and course information. Order a Create book and you'll receive a complimentary print review copy in three to five business days or a complimentary electronic review copy (eComp) via e-mail in minutes. Go to www.mcgrawhillcreate.com today and register to experience how McGraw-Hill Create empowers you to teach *your* students *your* way.

Chapter

1

Introduction to Signals

Contents

Introduction and Summary

A major part of this book is about signals and signal processing. In the conventional sense, signals are elements of communication, control, sensing, and actuation processes. They convey data, messages, and information from the source to the receiver and carry commands to influence the behavior of other systems. Radio, television, telephone, and computer communication systems use time-varying electromagnetic fields as signals. Command, control, and communication centers also use electromagnetic signals. Living systems employ sensory signals such as acoustic, visual, tactile, olfactory, or chemical. They also send signals by motion of their body parts such as the arms, hands, and face. The presence or unexpected absence of such signals is then detected by other living systems with whom communication is made. Neurons of the nervous system communicate with other neurons and control activity of muscles by electrical signals. Another group of signals of interest are those that represent variations of economic and societal phenomena (e.g., historical unemployment rate, stock market prices, and indexes such as the Dow Jones Industrial Average, median prices of houses, the federal funds interest rate, etc.). Still another group of signals of interest represent natural phenomena (pressure, temperature, and humidity recorded by weather stations, number of sunspots, etc.).

Signals, Information, and Meaning

As an element of communication and control processes, a signal is strongly related to other concepts such as data, codes, protocols, messages, information, and meaning. However, our discussion of signals and signal processing will be, to a large degree, confined outside of the context of such facets attached to a signal.

Signals and Waveforms

In this book a signal is a time-varying waveform. It may be an information-carrying element of a communication process that transmits a message. It may be the unwanted disturbance that interferes with communication and control processes, distorts the message, or introduces errors. It may represent observations of a physical system and our characterizations of it regardless of its influence (or lack thereof) on other systems.

We are interested in signals used in fields such as electrical communication, speech, computer and electronics, electromechanical systems, control systems, geophysical systems, and biomedical systems. Such signals represent variations of physical phenomena such as air pressure, electric field, light intensity and color in a visual field, vibrations in a liquid or solid, and so on. These signals are waveforms that depend on one or more variables such as time or space. (For example, a speech signal is a function of time but can also vary as a function of another variable such as space, if it is multiply recorded at several locations or if the microphone is moved around relative to the speaker. Geophysical signals are another set of such examples. Weather data collected at various stations at various times are still another such set.) The words *signals* and *waveforms* are, therefore, often used interchangeably.

Signals and Functions

We represent signals by mathematical functions. To this end, we often use the words *signals, waveforms*, and *functions* synonymously. Some simple elementary functions used in the mathematical representation of signals are steps, impulses, pulses, sinosoids, and exponentials. These are briefly described in section 1.14 of this chapter. Sinusoids are of special interest in signal analysis. They are treated in detail in the next chapter.

The chapter aims at achieving two interrelated goals. First, it presents the reader with a qualitative landscape of signals of common interest by giving actual examples such as natural, societal, financial, voice and speech, communication, and bioelectric signals. Second, in order to prepare the reader for the analytical conversation carried on throughout the book, it introduces, in detail, signal notations and elementary mathematical functions of interest (such as step, impulse, exponential, sinusoid, sinc, pulse, windows), and their basic properties such as the time average, even and odd parts, causality, and periodicity. It then introduces time transformation and scaling, which are parts of many mathematical operations on signals. Random signals are briefly introduced to broaden the scope of applications and projects. The Matlab programs in this chapter focus on generating and plotting signals and functions.

1.1 Discrete Versus Continuous; Digital Versus Analog

Discrete Versus Continuous

Some quantities appear to be analog in nature. They change in a continuous way. Geometrical and physical quantities are considered continuous. Some other examples are the following: time; muscle force; the intensity of sound, light, or color; the intensity of wind, ocean waves, or pain; the motion of the sun, moon, and planets; water flow from a spring; the growth of a tree; the radius of a circle; voltage, current, and field strength; the distance between two points; the size of a foot; the circumference of a waist measured using a rope.

Some quantities appear to be discrete and change in a discrete way. Quantities expressed by numbers are discrete. Some examples are the following: the number of fingers on our hands, teeth in our mouths, trees in an orchard, oranges on a tree, members of a tribe; the number of planets and stars in the sky; the price of a loaf of bread; the rings on a tree; the distance between two cities; the size of a shoe; voltage, current, and weight measurements; a credit card monthly balance; waist size measured using a pinched belt.

On some occasions a discrete quantity is treated (modeled) as a continuous variable. This may be for modeling convenience (e.g., the amount of hair on a person's head) or an effect of perception (e.g., the construction of a carpet, see Plate 1.1). Similarly, a continuous quantity may be expressed in discrete form. The number of colors appearing on a computer monitor is an example of this. Once a continuous variable is measured and expressed by a number it becomes discrete. Most computations are done in the discrete domain.

(*a*) Front side of a segment of a carpet appears as having a continuous structure.

(*b*) The back side shows the discrete structure.

Plate 1.1 An example of continuous versus discrete representation of a signal can be observed by comparing the continuous front *(a)* with the discrete back *(b)* of a hand-woven carpet. The carpet has a discrete structure that characterizes the carpet by the number of knots per unit length, measured on the back side. The pattern shown in this plate is repeated several times throughout the carpet (not shown). The pattern is deterministic and is provided to the weaver for exact implementation. The weaver, however, introduces unintentional randomness seen in the product as slight variations. These variations are not observed by an untrained person but are detected by a specialist or through magnification.

Digital Versus Analog

A continuous-time signal is converted into a discrete signal by sampling. The samples, however, are analog because they assume a continuous range of values. We can convert the sample values into a discrete set by assigning the value of each sample to one of n predetermined levels. The result is a digital signal. For instance, in the case of a binary discrete signal, there are only two predetermined levels into which the analog samples are forced: 0 and 1. Changes between these levels occur at the arrival time of a clock signal. Because of finite wordlength, which determines the resolution in the magnitude value of discrete-time functions, these are often called digital signals and discrete systems are called digital systems.

1.2 Deterministic Versus Random

Signals are said to be deterministic or probabilistic (random). Once it appears, a deterministic signal does not provide any new information, unless some of its properties change with time. A signal that could be predicted from its past values causes less surprise and carries less information. The only information available from the 60 Hz sinusoidal signal of a power line is its presence.[1] In contrast, a code that reduces the correlation between consecutive segments of the signal increases the information content of the signal.[2]

Some signals originating from natural, living, or societal systems vary with time in an exact and regular way, making them predictable. An example would be the rising sun. Or take the regularity of an electrocardiogram signal (EKG) that conveys health information. The appearance of an irregularity is taken as a sign of disease. As a third example, consider an advertisement for a candy brand touting the consistency of the product. In this case, the information intended to be conveyed is predictability. Within the above category we may also include signals that vary somewhat in a regular and complicated (but not random) manner. An example would be the positions of the planets in the sky, perceived and determined by an observer of the sky 5,000 years ago, and their application in predicting future events and fortune telling by astronomers, astrologers, and seers, or in decision making by rulers or elected officials in the past or current times.

In contrast, a signal may contain some stochastic (random) characteristic contained within quasi-deterministic features and, depending on the degree of randomness in the signal, still be considered predictable probabilistically within some statistical error rate.[3] The combination of regularity and randomness in natural or societal signals is to be expected. Such signals are the collective result of many interacting elements in physical systems. The signals, therefore, reflect the regularity of the physical structure and the irregularity of the message. The apparent randomness may also be due to our lack of

---

[1]The information provided by the above signal is only one *bit*. However, the information is normally very valuable and important.

[2]By some definitions, signals that appear most random contain the most information.

[3]Sometimes the signal might appear to be totally unpredictable (e.g., appearance and time of a shooting star).

knowledge about the system responsible for generating the signal or an inability to incorporate such knowledge in a model.

1.3   Examples of Natural and Societal Signals

Sunspot Numbers

Of interest in electrical communication (as well as in other fields) is the level of solar activities, signaled by the number of sunspots as a function of time. Figure 1.1(*a*) shows the annual mean number of sunspot records from AD 1700 to 2010 with the abscissa indicating days. A clear feature of the record is the pattern of its variation, which exhibits 28 cycles of activity with an average period of 11 years during the past 310 years. Each cycle has its own duration, peak, and valley values. One can also observe some waxing and waning of the peak values, suggesting stratification of the record into three centennial groups of cycles (one segment consisting of cycles from 1770 to about 1810, the second segment from 1810 to 1900, and the third from 1900 to 2010). In relation to the signal of Figure 1.1(*a*) one may define several variables exhibiting random behavior. Examples of such random variables are the number of sunspots (daily, monthly, and yearly numbers), period of cyclic variations, peaks and valleys, and rise and fall times within each cycle. A first step in the analysis of signals such as that in Figure 1.1(*a*) would be to estimate the mean and variance of the variables. To acquire more insight one would also find the correlation and interdependencies between them. Toward that goal one would use a more detailed set of sunspot data such as average daily measurements. These provide a better source for analysis of cyclic variation and fine structures within a cycle. Figure 1.1(*b*) plots such a set of data for the period of January 1, 1818, through September 30, 2010. In this figure and its subsets shown in Figure 1.1(*c*), (*d*), and (*e*), the numbers on the abscissas indicate the day, counting from the beginning of the time period for that figure. The dates of the beginning and end of the time period are shown at the left and right sides of the abscissa, respectively. The data of Figure 1.1(*b*) show many days with average sunspot numbers above 200 and even some above 300 per day. Daily averages also provide a better tool for analysis of cyclic variation and fine structure within a cycle. Bursts of activities lasting 10 to 20 days are observed in Figure 1.1(*c*), which plots the data for the years January 1, 1996, through December 31, 2009, covering the most recent cycle. Two extreme examples of yearlong daily measurements during the most recent cycle of sunspot activities are shown in Figure 1.1(*d*) and (*e*). Figure 1.1(*d*) (for the year 2001, which was an active one) indicates high levels of activity with the occurrence of strong periodic bursts. In contrast, Figure 1.1(*e*) (for the year 2009) shows weak levels of activity but still occurring in the form of bursts.

Atmospheric CO_2 Content

Carbon dioxide (CO_2) is one of the greenhouse gases associated with thermal changes of the atmosphere and is a signal for it. Monitoring atmospheric CO_2 and trends in its temporal and spatial variations is of potential importance to every person. Scientific work on atmospheric carbon dioxide uses long-term historical information as well as

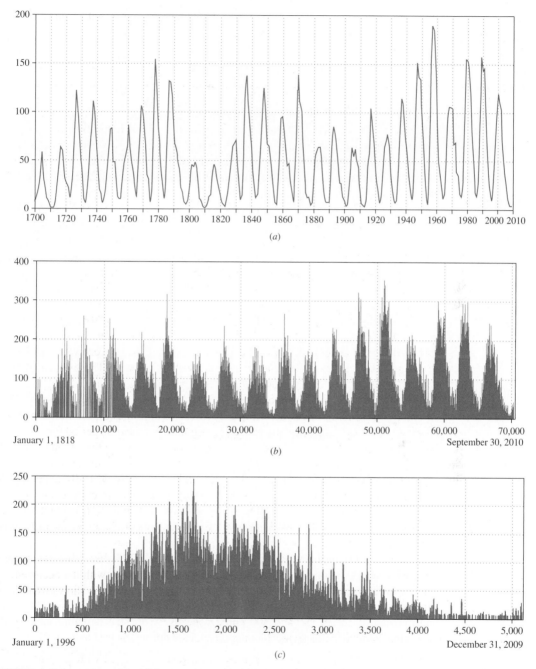

FIGURE 1.1 Sunspot numbers. *(a)* Mean annual values (AD 1700–2010); *(b)*, *(c)*, *(d)*, and *(e)* daily values for selected time intervals during 1818 to 2010.

Source: National Oceanic and Atmospheric Association's (NOAA) National Geophysical Data Center (NGDC) at www.ngdc.noaa.gov.

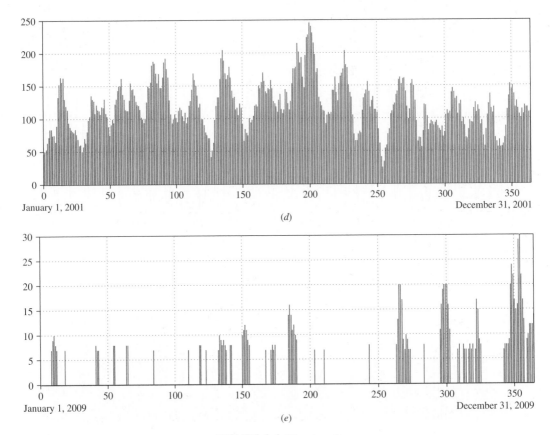

FIGURE 1.1 (*Continued*)

contemporary observations. Historical records indicate that trends in atmospheric CO_2 are associated with glacial cycles. During the last 400,000 years, the CO_2 content of the atmosphere fluctuated between a value below 200 to nearly 300 ppmv (parts per million volume). The data are obtained by analyzing gas contents of the air bubbles entrapped in polar ice sheets.

An example of the historical data is shown in Figure 1.2(*a*) which plots the results of measuring the CO_2 content of air bubbles in the ice cores of Vostock station in Antarctica. The air in these bubbles is from 400,000 to 5,000 years ago. The data for the plot of Figure 1.2(*a*) show long-term cyclic variations of 80 to 120,000 years with minima and maxima of nearly 180 and 300 ppm, respectively. Present-day atmospheric CO_2 shows much higher values which are unprecedented during the past half a million years. Contemporary measurements are done under controlled and calibrated conditions to avoid the influence of local sources (such as emissions) or environmental elements (such as trees) that absorb, trap, or remove CO_2 from the air. Figure 1.2(*b*) plots contemporary data for 1959 to 2010 from measurements at the Mauna Loa observatory station in Hawaii (chosen for its suitable location in terms of providing base measurements). The

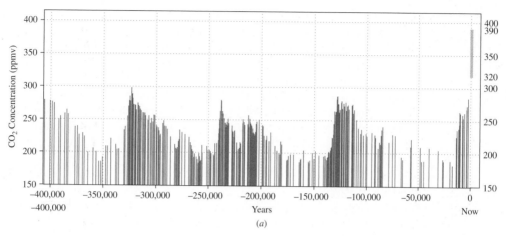

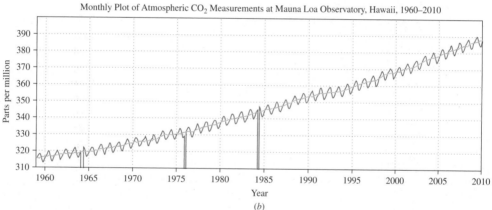

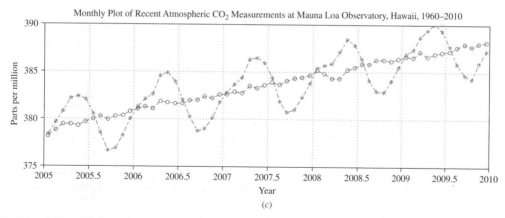

FIGURE 1.2 *(a)* Historical carbon dioxide record from Vostok ice cores. *(b)* and *(c)* Monthly carbon dioxide record from Mauna Loa observatory, Hawaii. The record for 1960–2010 is shown in *(b)*, while *(c)* shows the record for 2005–2010 at a higher resolution.

Sources: *(a)* Carbon Dioxide Information Analysis Center (CDIAC) of the U.S. Department of Energy at http://cdiac.ornl.gov/. *(b)* and *(c)* Dr. Pieter Tans, NOAA/ESRL (www.esrl.noaa.gov/gmd/ccgg/trends/).

measurements exhibit seasonal variations with a trend of steady growth. Since 1959 the annual average of CO_2 content has grown from about 315 to nearly 390 ppmv. For comparison, the contemporary values are also shown as a bar on the right-side end of the plot on Figure 1.1(a) using the same vertical scale as for the historical data. It is seen that while during the last 400,000 years atmospheric CO_2 remained within the range of 200 to 300 ppmv, it has grown from about 315 to nearly 390 ppmv during the last 50 years. The recent data for 2005 to 2010 and its trend are shown in Figure 1.2(c) on an expanded time scale to better visualize the change. The plot of Figure 1.2(c) shows seasonal variations of 6 to 10 ppmv riding on a steady growth of 2 ppmv per year.

Seismic Signals

Seismic signals are recordings of earth vibrations caused by factors as diverse as earthquakes, explosions, big falling objects (a giant tree, a building, etc.), the galloping of a horse, the passing of a car nearby, and the like. Seismic signals are of interest for several reasons among which are oil and gas exploration, treaty verification and monitoring, and, most importantly, earthquake studies and predictions. In the case of earthquakes, the source of vibrations is a sudden movement of the ground, such as slippage at an area of broken earth crust called a fault. The energy released by the slippage produces waves that propagate throughout the earth similar to the radiation of electromagnetic waves from an antenna. Earthquake studies and predictions are of great interest, especially for those living in earthquake areas close to faults such as the San Andreas fault. The San Andreas fault is the meeting boundary between two plates of the crust of the earth: the Pacific plate on the west and the North American plate (the continental United States). The fault runs from southern to northern California, for a length of about 800 miles. It is as narrow as a few hundred feet in some locations and as wide as about a mile in others. The Pacific plate has been moving slowly in the northwest direction with respect to the North American plate for more than a million years. The current rate of this movement is an average of 2 inches per year. Once in a while the fault may snap, making the two plates suddenly slip and causing the sudden release of energy and strong vibrations in the ground we know as an earthquake. Both the gradual and sudden movement of the earth can change features on the ground. See Figure 1.3.

A transducer, which senses the earth's movement and converts it to an electric signal, records the results (in analog form on paper and magnetic tape or as a digital signal by electronic device). Such a transducer is called a *seismograph*. The earthquake record is called a *seismogram*. Traditional seismographs record the signal on paper covering a slowly rolling cylinder that completes one turn every 15 minutes. This produces a trace that begins at the left side of the seismogram and records 15 minutes of data, at which time it reaches the right side of the paper. The trace then returns to the left side and starts at a location slightly below the previous trace, as is done by the sweep mechanism in a television cathode ray tube. The time of each trace is shown on the left side. Because of the three-dimensional nature of seismic wave propagation in the earth, a set of seismograms recorded at several locations, or several seismograms each recording the vibrations at a different location or direction, are needed to produce a three-dimensional picture.

FIGURE 1.3 Sharp twisting of Wallace Creek in Carrizo Plane, central California, due to earth plate movements. The stream channel crosses the San Andreas fault at the twisted segment. The sharp twist (made of two bends of 90°) indicates the narrowness of the fault at that location. The gradual movements of the earth at the fault (at an average rate of 2 inches per year) along with the sudden movements due to earthquakes have twisted the channel and shifted its downstream path by about 100 meters toward the northwest (approximate direct measurement by the author). It is estimated that during the past 1,500 years earthquakes have occurred along the San Andreas fault about once every 150 years. The earthquakes may also rupture the ground and produce larger abrupt shifts in the surface, as in the Tejon earthquake of 1857 (the largest earthquake in California with a magnitude of 7.9, an average slip of 4.5 meters, and a maximum displacement of about 9 meters in the Carrizo Plane area).

The motion may last from tens of seconds to several minutes. The seismogram reflects the duration of the motion, its intensity, and the physical properties of the ground. An example is shown in Figure 1.4.

Release of energy in an earthquake shakes and displaces the rocks and earth elements and causes them to vibrate back and forth. The frequency of the vibrations varies from nearly 0.1 to 30 hertz (Hz). The vibrations propagate and travel as seismic waves on the surface and also within the body of the earth (both within the crust and the core). Surface waves have a lower frequency and travel slower than body waves. Propagations can be in the form of longitudinal waves, where the direction of propagation is the same as the vibration of earth materials. (Visualize the forward motion of a running tiger in relation

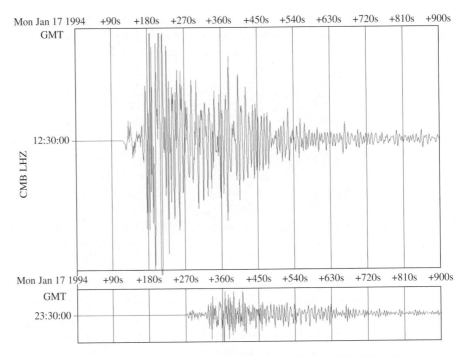

FIGURE 1.4 The seismogram of the Northridge earthquake of January 17, 1994, with a magnitude of 6.7. The main shock of the earthquake occurred at 4:30:55 a.m. local time (PST), 12:30:55 Coordinated Universal Time (UTC). The center of the earthquake was located at 1 mile south-southwest of Northridge and 20 miles west-northwest of Los Angeles. The seismogram in this figure shows vertical motion of the ground. It is plotted from the digital seismic data (sampled at the rate of 1 sample per second) which was recorded by Berkeley Digital Seismic Network (BDSN) at an earthquake station in Berkeley, California. The station is at a distance of 525 km from the earthquake location. The seismogram shows arrival of the main shock 63 seconds after the earthquake occurred in Northridge, indicating a travel speed of 8,333 m/s through the earth. An aftershock of the earthquake arriving at 23.32.51 UTC is shown in the lower trace. The guidelines of the BDSN web page were used to construct the plot of the seismogram in this figure.
Source: www.ncedc.org/bdsn/make_seismogram.html.

to the motion of its limbs.) The propagation can also be in the form of transverse waves, where the direction is perpendicular to the vibration of earth materials. (Visualize the forward motion of a snake in relation to the sideways motion of its body.) Considering the above characteristics, four main types of seismic waves are recognized and used in earthquake studies. These are the following:

1. A P- (primary) wave travels through the crust or the core of the earth and is longitudinal. It compresses the earth in the direction of its travel and is, therefore, called a *compression wave*. It propagates faster than other wave types, traveling at an average speed of 7 km per second (slower in the crust and faster within the core). It is, therefore, the first to be felt or registered as a seismogram. It is the first wave to shake the buildings when an earthquake occurs, hence called the *primary wave*.

2. An S- (secondary) wave also travels through the earth's body but is a transverse wave. Therefore, it is called a *sheer wave*. S-waves are about half as fast as P-waves. They shake the ground in a direction that is perpendicular to the direction of wave travel.

3. A Love wave is a transverse surface wave. It shakes the ground horizontally in a direction perpendicular to the direction of its travel.

4. A Rayleigh wave is also a surface wave, but not a transverse one. It shakes the ground in an elliptical motion with no component perpendicular to the direction of travel.

P- and S-waves are called *body waves*. They arrive first and result in vibrations at high frequencies (above 1 Hz). Love and Rayleigh waves are called *L-waves*. They arrive last and result in vibrations at low frequencies (below 1 Hz). The above collection of waves appear one way or another in the total waveform. Their contribution to the overall spectrum of the earthquake wave is influenced by the earthquake magnitude.

Societal Signals

Some signals originating from societal systems also vary in a somewhat regular but possibly complicated way, appearing as random. The apparent randomness may be due to our lack of knowledge about the system responsible for generating the signal, or an inability to incorporate such knowledge in a model.[4] Examples of such signals are indexes reflecting contraction and expansion of the economy, industrial averages, stock market prices, the unemployment rate, and housing prices.

Figure 1.5 shows the rate of unemployment in the United States and its relationship with recession.[5] The latter are shown by vertical bands on the top section of the plot. Note the periodicity in the unemployment rate, its rise associated with recession, and the shape of its decline after recovery from recession.

Another example of a societal signal containing regularity and randomness is the Dow Jones Industrial Average (DJIA) taken every minute, at the end of the day, or as a weekly average. The signal exhibits some features such as trends and oscillations that generally vary with the time period analyzed. There is great interest in predicting the signal's behavior within the next minute, day, month, or years. Analysis methods here have a lot in common with signal analysis in other fields such as communication, signal detection, and estimation.

---

[4]A correct understanding of the system that generates the signal or reshapes it is not essential for predicting or modeling it. An astronomer can adequately predict the movements of the planets based on the earth-center theory which is considered an incorrect theory. James Clerk Maxwell, in formulating the propagation of electromagnetic signals, assumed the existence of a medium for such propagation even though it presently is considered nonexistent.

[5]A period characterized by negative growth in gross domestic product (GDP) for two or more consecutive quarters of the year is called a *recession*.

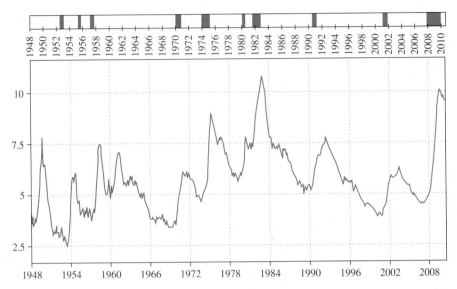

FIGURE 1.5 Plot of monthly unemployment rate in the United States as a percentage of the labor force from January 1948 to October 2010. The vertical bands on the top display periods of recession. Note the association between recessions and the rise in umemployment.
Source: Bureau of Labor Statistics, http://data.bls.gov.

1.4 Voice and Speech Signals

Voice and speech signals are vocal expressions. They are generally perceived through the auditory system, although other senses such as visual and tactile lip reading could also serve as a transmission channel. In the present context, by voice and speech signals we mean variations of a scalar that is plotted in analog form versus time. The scalar may be air pressure, electric voltage, or vibrations in a liquid or solid. A voice signal is the representation and reproduction of air pressure variation emanating (also called radiating) from a speaker's lips (or throat) and registered by a transducer such as a microphone in the form of a time-varying voltage. Time is a continuous variable. The air pressure is also a continuous variable. The auditory waveform can also vary as a function of an additional variable such as space, if it is multiply recorded at several locations or if the microphone is moved around relative to the speaker. This representation is raw and not coded in that it contains features both relevant and irrelevant to its information content. It can be directly converted back into the original signal by the use of a transducer such as an earphone without any decoding. In a broad sense, speech processing is the collection of operations done on speech for the purpose of encoding, communication, recognition, data reduction and storage, text-to-speech conversion, and so on. Speech processing is done in the analog and digital domains. The digital processing of speech has vastly expanded speech technologies.

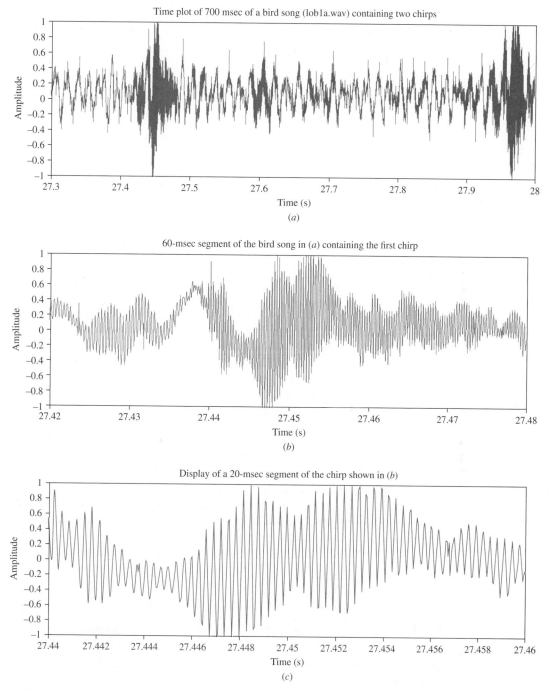

FIGURE 1.6 The waveform of a bird song (lob1a.wav, 22,050 samples per sec., 8 bits per sample, mono). The plot in *(a)* is 700 msec long and contains two *chirps*. The plots in *(b)*, *(c)*, and *(d)* reveal the fine structure of the *chirp*.

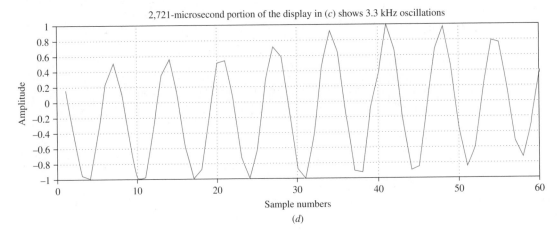

FIGURE 1.6 (*Continued*)

Speech signals contain a considerable degree of structural regularity. A 700-msec sample of a bird song is shown in Figure 1.6(*a*) . The sample contains two chirp sounds. The first chirp, lasting 60 msec, is displayed with an expanded time axis in Figure 1.6(*b*). It reveals some regularity in the form of oscillations, as seen in the finer structure of the chirp. The oscillations carry the chirp sound in their modulated amplitude. The expanded plot of the chirp in Figure 1.6(*c*) shows amplitude modulation of the carrier. To find the frequency of this carrier, a portion of Figure 1.6(*c*), lasting 2,721 microseconds (60 samples), is displayed in Figure 1.6(*d*). It contains 9 cycles. The period of oscillations is, therefore, $T = 302$ μsec, which corresponds to a frequency of $f \approx 3.3$ kHz.

A similar effect is observed in the signal of Figure 1.7(*a*), which shows the recording of the spoken phrase *a cup of hot tea*. The waveform shows five distinct bursts corresponding to the utterance of the five elements of the phrase. The waveform for *hot* shown in (*b*) lasts 150 msec and exhibits some regularity, in the form of a wavelet, in its fine structure. The fine structure of (*b*) is shown in (*c*) and (*d*).

1.5 Communication Signals

Communication signals such as those in radio, television, telephone, computer, microwaves, and radar systems are made of electromagnetic fields and waves. They are distinguished from each other by their frequency, bandwidth, and method of carrying information. Here are some examples.

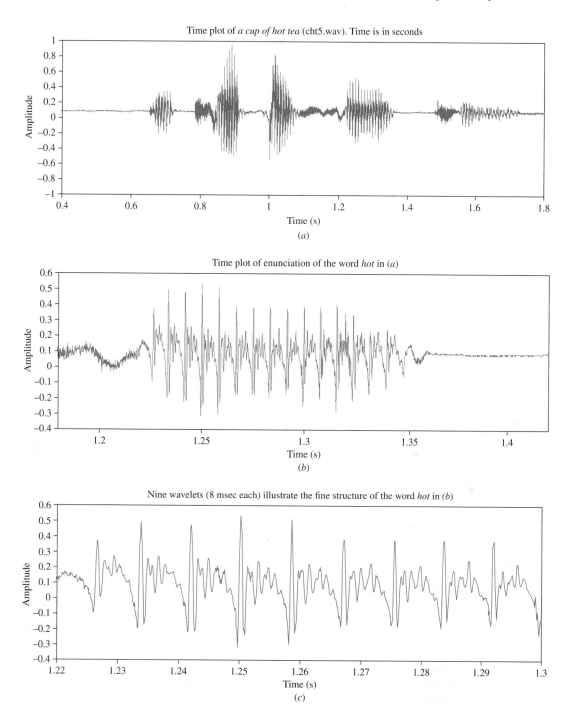

FIGURE 1.7 *(a)* The waveform of the utterance *a cup of hot tea* (cht5.wav, 22,050 samples per sec, 8 bits per sample, mono). The plot in *(b)* shows the waveform for *hot*. The fine structure of *(b)* is revealed in *(c)* and *(d)*.

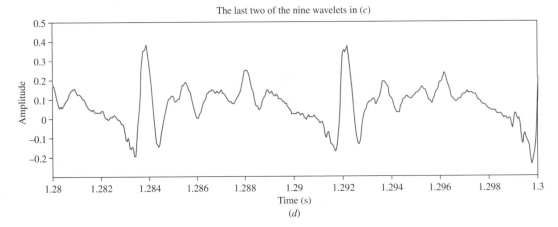

FIGURE 1.7 (*Continued*)

E*xample* 1.1

An energy signal

A signal picked up by a microwave antenna is modeled by the sinusoidal function $v(t) = A\cos(\omega t + \theta)$, where $\omega = 2\pi f$ and $f = 100$ MHz. The signal lasts only T seconds. Find the total energy in the signal as a function of A and T, and compute its value when $A = 1\ mV$ and $T = 1\ \mu$sec.

If the signal contains an integer number of cycles, we have

$$W = \int_0^T v^2(t)dt = A^2 \int_0^T \cos^2(\omega t + \theta)dt = \frac{A^2 T}{2}\ \text{Joules}$$

For $A = 1\ mV$ and $T = 1\ \mu$sec, $W = (10^{-12})/2\ J = 0.5$ pJ.

E*xample* 1.2

Dual-tone multi-frequency signaling (DTMF)

Touch-tone telephones dial using DTMF signaling. In this DTMF system each decimal digit (0 to 9) is identified by a signal that is the sum of two pure tones, one from low-band and another from high-band frequencies. The low-band frequencies are 697, 770, 852, 941 and the high-band frequencies are 1,209, 1,336, 1,477, 1,633, all in Hz. Frequency assignments are shown in Figure 1.8. For example, the signal for the digit 1 is the sum of the 697 and 1,209 Hz tones. Telephone key pads also include more buttons: the star $*$ and the # (called the pound key) which provide the capability to communicate with automated systems and computers. The combination of a low-band and the high-band tone of 1,633 Hz (the column on the right side of Figure 1.8) provide four additional signals called A, B, C, D tones, which are used for communication or control purposes.

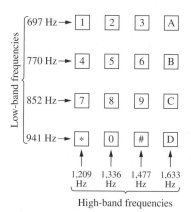

FIGURE 1.8 Dual-tone multi-frequency signaling (DTMF).

Phase-shift keyed (PSK) signals

a. A binary stream of data is represented by a two-level (e.g., 0 or 1) signal (to be called symbol). The signal modulates an RF carrier, at the frequency f_c, in the form of two-state phase-shift keying (2PSK), producing the signal $s_i(t) = \cos(\omega_c t + \theta_i)$, $i = 1, 2$, where θ_i is 0° or 180°. The modulated signal, therefore, shifts its phase by 180° whenever the data switch between the two levels. Each data bit maps into one signal (called a symbol) with a PSK phase of 0° or 180°. The bit and symbol rates are equal.

b. Now assume that each two successive bits in the binary stream of data are mapped into phase values of $\theta_i = 0$, $\pm\pi/2$, π (or $\pm\pi/4$, $\pm3\pi/4$). Each pair of data bits 00, 01, 10, and 11 produces one symbol and a signal with its own phase out of the four phase values: $s_i(t) = \cos(\omega_c t + \theta_i)$, $i = 1, 2, 3, 4$, where the phase θ_i assumes one of the four permitted values. This is called quadrature- (or quaternary-) phase-shift-keyed (QPSK) signaling (also 4PSK). The bit rate in QPSK is, therefore, twice the rate of 2PSK with the same carrier. If M values of phase are permitted, the modulation is called M-ary PSK (MPSK). The bit rate is then M times the symbol rate.

Signal Constellation

The QPSK signal can be generated by modulating the phase of a single sinusoidal carrier $\cos \omega t$ according to the permitted phase values θ_i. It can also be produced by modulating the amplitudes of two quadrature carriers by x_i and y_i, and be expressed by

$$s_i(t) = x_i \cos \omega_c t - y_i \sin \omega_c t, \quad i = 1, 2, 3, 4$$

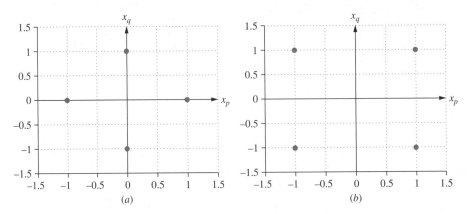

FIGURE 1.9 Two QPSK constellations.

where,

(x_i, y_i)
$= (1, 0), (0, -1), (-1, 0), (0, -1)$ for the case when $\theta_i = 0, \pm\pi/2, \pi$.

$(\sqrt{2}x_i, \sqrt{2}y_i)$
$= (1, 0), (0, -1), (-1, 0), (0, -1)$ for the case when $\theta_i = \pm\pi/4, \pm3\pi/4$.

The pair (x_i, y_i) are normally represented on an x-y plane (with the x-axis labeled in-phase and the y-axis labeled quadrature-phase) and called the signal constellation. See Figure 1.9.

Multilevel amplitude-modulation signals

Suppose the amplitude of a single analog pulse could be discretized to M levels with an acceptable error. Discretization maps the pulse height into a binary number of length n bits, where $M = 2^n$. For $M = 2$, we have $n = 1$, a one-bit binary number (0 or 1). For $M = 4$, we obtain $n = 2$ with four binary numbers, each of them being two bits long (00, 01, 10, 11). For $N = 8$, we have a three-bit binary number (word length $= 3$ bits). The bit rate is three times the symbol rate. Using the reverse of the above process a sequence of n data bits can be represented by a single pulse (a symbol) with an amplitude $M = 2^n$, which can modulate the amplitude of a sinusoidal carrier. The bit rate is n times the symbol rate.

Quadrature amplitude modulation signals (QAMs)

Communication signals that transmit digital data are generated by modulating some features of a sinusoid (called the carrier) by the binary data to be transmitted. Examples 1.3 and 1.4 introduced two types of signals generated by PSK and multilevel pulse amplitude modulation. Another type of communication signal is generated by a method called quadrature amplitude modulation (QAM), which combines PSK with

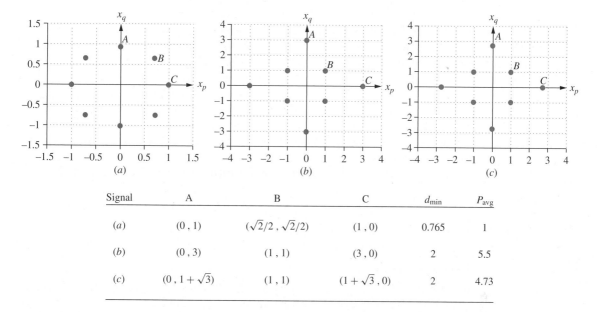

FIGURE 1.10 Three different constellations of eight-point QAM signals. They are differentiated from each other by their average power (P_{avg}) and the minimum distance (d_{min}) between their subsignals. The table above gives P_{avg}, d_{min}, and the coordinates of three subsignals in each constellation. The coordinates of the A, B, and C signals are shown in the table.

Signal	A	B	C	d_{min}	P_{avg}
(a)	(0 , 1)	($\sqrt{2}/2$, $\sqrt{2}/2$)	(1 , 0)	0.765	1
(b)	(0 , 3)	(1 , 1)	(3 , 0)	2	5.5
(c)	(0 , $1+\sqrt{3}$)	(1 , 1)	($1+\sqrt{3}$, 0)	2	4.73

pulse amplitude modulation. The QAM signal may be written as

$$s_i(t) = x_i \cos \omega_c t + y_i \sin \omega_c t = r \cos(\omega_c t + \theta)$$

The signal is represented in two-dimensional space by the location of various permitted pairs (x_i, y_i). The horizontal axis represents the x_i coordinate of such points and is called the *in-phase axis*. The vertical axis represents the y_i coordinate and is called the *quadrature axis*. The resulting set of points on the diagram is called the *signal constellation*. For example, a modem that recognizes 1 of 16 levels of a sinusoidal signal, and can distinguish 16 different phases, has a signal constellation of 256 elements and is called 256 QAM. It can carry 128 times more binary information than a 2PSK modem. Three 8-symbol constellations are shown in Figure 1.10.

1.6 Physiologic Signals

Physiologic signals are those which are obtained from a living tissue or organism and reflect the physiological state of that organism. As such, they are used in diagnosis, prosthesis, and research. Some physiologic signals require a transducer in order to be converted to electrical signals for recording, processing, and use. Examples are measurements of blood pressure, temperature, and blood oxygen content. Some others are electric potentials generated by the living tissue and recorded by placing two electrodes between two points in the body. These potentials are normally called electrophysio-logic signals.[6] Familiar examples of electrophysiologic signals are those recorded from the skin surface: electrocardiogram signals (EKG), electroencephalogram signals (EEG), surface electromyogram signals (surface EMG), and electro-oculogram signals (to measure eye movement). Other examples of electrophysiologic signals are electric potentials recorded by electrodes inserted under the skin or into a muscle (subdermal EMG), from the surface of the cortex (electrocorticogram, ECG), from the collection of neurons of the central nervous system (field potentials, gross potentials, or evoked potentials), from extracellular space near individual neurons (action potentials), or from inside a neuron (intracellular potentials). The above signals are of interest in clinical diagnosis and prosthetic devices, as well as in research. Here we discuss briefly the salient features of some of them.

1.7   Electrocardiogram Signals

EKG Signals

Electrocardiogram (EKG) is the most familiar of physiologic signals. It is of special interest because it relates to the operation of the heart, one of the most important functions in the living body. EKG is the recording of electric potentials associated with contraction of the heart muscles and picked up by electrodes on the surface of the body. The first noticeable temporal feature of an electrocardiogram is its periodicity which is produced by the ventricular cycle of contraction (heartbeat). This is called the *rate* and is measured in units of beats per minutes (BPM). An example of an EKG recording is shown in Figure 1.11. Each division on the electrocardiogram chart paper is 1 mm, with the grid conveniently designed for easy reading of the amplitude and beat rate. The vertical scale is 0.1 mV per division. The time scale is 40 msec per division (the distance between two thick lines being 200 msec). The time scale allows a direct reading of the rate off the trace without much computation. The distance between two consecutive heavy black lines (a box composed of 5 divisions) is 200 msec = 1/300 minutes, which corresponds to a rate of 300 BPM. A distance of two boxes represents 2/300 minutes (i.e., a rate of 150 BPM) and three boxes mean 3/300 minutes (i.e., a rate of 100 BPM).[7] The rate in the trace in Figure 1.11(*a*) is 70 BPM.

Standard EKG

Standard EKG is composed of 12 trace recordings (called leads). They are produced by two sets of recording electrodes: three limb electrodes (left and right arms, and the left leg) and six chest electrodes (which cover the projection of the normal position of the heart on the chest). The limb electrodes provide three bipolar (or differential)

---

[6]They are also called *bioelectric signals* as they are biologically based. Those recorded from the nervous system are sometimes referred to as *neuroelectric signals* also.

[7]For details see *Rapid Interpretation of EKG* by Dale Dubin, Cover Pub. Co., ©2000. 6th ed.; and *Introduction to ECG Interpretation* by Frank G. Yanowitz, http://library.med.utah.edu/kw/ecg/.

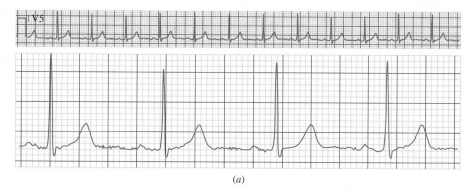

(a)

FIGURE 1.11 *(a)* Twelve consecutive cycles of continuous EKG recording (lead V_5 recorded by the electrode on the left side of the chest). Each division on the chart paper is 1 mm. One division on the time axis is 40 msec (a total of about 3 seconds and 4 beats on the lower trace), resulting in a rate of about 70 beats per minute (BPM). In practice, the BPM number is read directly off the chart (see text). One division on the vertical axis represents 100 μV, resulting in a base-to-peak value of 1.5 mV. Note the remarkable reproduction of the waveform in each cycle. The schematic of one cycle of EKG and its noted components are shown in Figure 1.11(*b*).

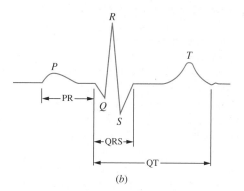

(b)

FIGURE 1.11 *(b)* Schematic representation of one cycle of an EKG and its noted components, recognized and understood as the manifestations of the physiological activity of the heart:

    P-wave (depolarization of the atria)

    QRS complex (depolarization of the ventricular muscle)

    ST-T wave (repolarization of the ventricular muscle)

recordings between each two electrodes, with the third electrode acting as the reference ground. These are called leads I, II, and II. The limb electrodes also provide three more leads (unipolar recordings from each electrode with the other two connected as the common reference). These are called *augmented voltages*: AVF (left foot), AVR (right arm), and AVL (left arm). The leads obtained from the chest electrodes are labeled V_1 through V_6. Collectively, the trace recordings from the 12 leads are referred to as the electrocardiogram. An example is shown in Figure 1.12(*a*).

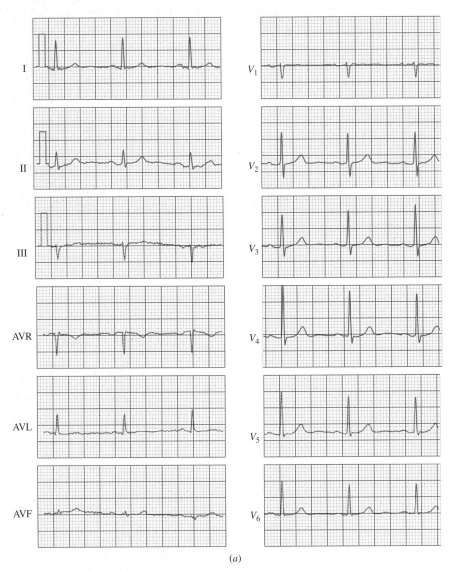

(a)

FIGURE 1.12 *(a)* The standard EKG chart contains 12 traces (12 leads) obtained from electric potential recordings by nine electrodes (three limb and six chest electrodes). In general, the traces are not recorded simultaneously. An example is shown along with the summary interpretation. Amplitude scale is 0.1 mV per division and time scale is 40 msec per division. *(b)* Interpretation of the above EKG is by a specialist.

Clinic ID:	Rate:	69	BPM	Interpretation:
Name:	Req. Physician:	PR:	162	msec	Sinus Rhythm
ID:	Technician:	QT/QTc:	374/391	msec	P:QRS - 1:1, Normal P axis, H Rate 69
Sex:	History:	QRSD:	82	msec	WITHIN NORMAL LIMITS
BP:	Medication:	P Axis:	37		
Weight:	Date of Report:	QRS Axis:	−1		
Height:	Reviewed By:	T Axis:	40		
DOB:	Review Date:				
Comments: Unconfirmed Report					

(*b*)

FIGURE 1.12 (*Continued*)

Interpretation of EKG

The temporal and spatial features obtained from the ensemble of 12 EKG leads are associated with the functioning of the heart and are used as diagnostic tools for it. In interpreting an EKG, the following time intervals and durations (along with the amplitudes and timings of the above components) are used:

PR interval,   QRT duration,   QP interval,   PP interval,   RR interval

An interpretation of the EKG of Figure 1.12(*a*) by a specialist is shown in Figure 1.12(*b*).

1.8 Electromyogram Signals

Electromyogram (EMG) signal is the recording of electric potentials arising from the activity of the skeletal muscles. EMG activity can be recorded by a pair of surface disk electrodes placed over the skin on the muscle (a better representation of the entire muscle activity, representing the activity of a larger number of motor units and, therefore, the entire muscle) or by a subdermal bipolar needle electrode inserted under the skin or inside the muscle. Three mechanisms can contribute to an increase in EMG activity. These are (1) more motor units of the muscle being recruited, (2) the rate of motor units involved being increased, and (3) stronger synchronization between the motor units involved.

Characteristics

EMG is the collective electric potentials of individual motor-unit activations involved in muscular contraction. This manifests itself in the form of a burst of electrical activity composed of spike-shaped wavelets. Traditionally, an EMG signal has been characterized by its amplitude, duration, and shape which are used in simple analysis. Within an EMG signal one observes bursts of spikes with multiple peaks. These spikes and their parameters (such as shape, duration, sharpness, and peak amplitude) are used for a more refined analysis of the EMG. The bandwidth of 2 Hz to 2 kHz often suffices for a general EMG recording. Some EMG recorders accommodate higher bandwidths (up to 20 kH), which would keep more details from the spike parameters. The amplitude of the EMG depends on the activity of the muscle and is normally on the order of several hundred microvolts.

Applications

The study of the EMG and its relation to movement is of interest in several areas. These are (1) clinical diagnosis, (2) kinesiological studies, (3) research on initiation and control of movement in living systems, and (4) development of movement-assisting devices and prostheses. In the last case, being of much practical application, the EMG of some muscles controls an external device that has replaced a limb that does not function or is lost.

Samples of EMG activities recorded during a series of experiments on the extension of the right-upper limb in the sagittal plane are shown in Figure 1.13. Extensions were made at normal [Figure 1.13(*a*)] or fast speeds [Figure 1.13(*b*)] and under four cases of different load conditions (the subject holding no load in his hand, or a load of 0.5, 1, and 3 kg). EMG activities from the biceps and triceps brachii were each recorded differentially by a pair of surface disk electrodes placed 2.5 cm apart on each muscle. The reference electrode was placed on the forehead. Simultaneously, the motion trajectories of 10 points on the arm were recorded by fast movies and photographically sampled by stroboscopic flashing on still film snapshots. The strobe light was synchronized with the initiation of the movement during the experiment and flashed at the rate of 50 Hz while the shutter of a still camera was kept open to take in the sequence of motion.

1.9 Electroencephalogram Signals

The potential difference measured between an electrode placed on the scalp and some other point such as the neck or ear has a time-varying magnitude of 20–100 μV and is called an electroencephalogram (EEG). The EEG is the electric field potential generated by the sum of the electrical activities of the population of neurons within a distance of several mm from the recording electrode filtered through the dura, skull, and skin. The filter has smoothing effects in time and space. However, two electrodes placed only 1 or 2 mm apart from each other may record distinctly different (but correlated) electrical activities. The number of electrodes and their placement on the scalp varies from under 10 to more than 20. The 10–20 system of electrode placement recommended by the International Federation of Societies for Electroencephalography and Clinical Neurophysiology (or a simpler version of it) is mostly used. An EEG signal is spontaneous or evoked. Its temporal characteristics depend on the recording location on the scalp and the physiological state (e.g., the subject being awake or asleep, eyes opened or closed, etc.). The basic variables of an EEG signal are amplitude and frequency. The frequency band ranges from less than 4 Hz (called *delta*) to more than 13 Hz (called *beta*). The intermediate bands are 4 to less than 8 Hz (called *theta*) and 8 to 13 Hz (called *alpha*).

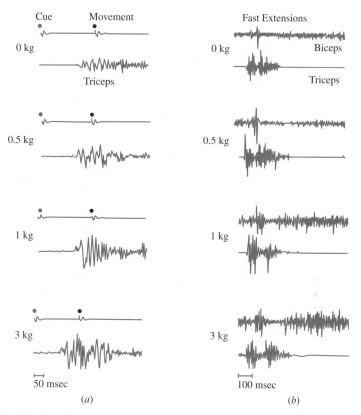

FIGURE 1.13 *(a)* EMG activities recorded from the triceps brachii in normal extensions of the arm under four load conditions are shown. The amplitude and duration of the EMG activity increase with the load. The EMG signal contains spikes whose number and amplitude also increase with the load. The upper trace in each case shows the cue to begin the extension and its start. The extension starts about 250 msec after the acoustic cue is given. EMG activity begins to manifest itself about 100 msec before the actual movement appears. Each trace is 500 msec long. *(b)* EMG activities recorded from the biceps and triceps brachii in fast extensions of human arm, again under four load conditions are shown. The triphasic activity pattern of triceps-biceps-triceps is observed in all cases. In addition, as the load increases (from 0 to 3 kg), a second phase of biceps activity becomes more pronounced. Each trace is 1 sec long. (Nahvi, 1983, unpublished.)

These basic variables of the EEG change with the behavioral state and the mental and physical aspects of the subject. These changes are well understood and established in the study of EEGs. Their association with physiological, behavioral, and mental states (e.g., eyes opened or closed, fear, relaxation, and sleep) is also well known. In a more detailed description of EEGs, one recognizes complex wave patterns (modeled as combinations of sinusoids and simple patterns) and waveforms described by special features (such as spikes). Examples of such patterns are shown in Figure 1.14.

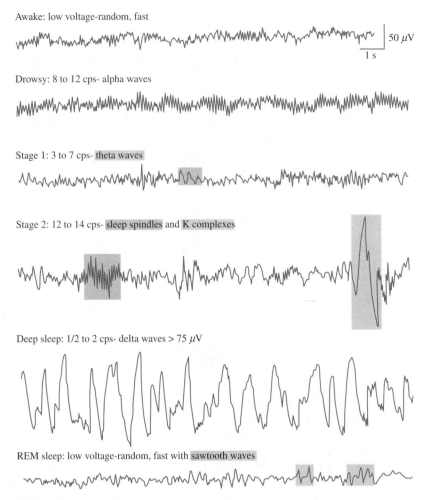

FIGURE 1.14 Examples of EEG patterns during various stages of sleep.
Source: This figure was published in *Fundamental Neuroscience*, Larry Squire, Darwin Berg, et al., page 963, Copyright Elsevier (2008).

1.10 Electrocorticogram Signals

Evoked Gross Potentials

An electrode made of a silver wire about 1 mm in diameter with a round tip (called the gross electrode) and placed on the cortex (with the reference electrode placed on the neck) records voltages of about 500 μV. Such recordings are referred to as an electrocorticogram (ECG). These voltages are field potentials resulting from the electrical activities of population of nearby neuronal elements. They are generated spontaneously or evoked in response to external (acoustic, visual, or tactile) stimuli. A gross evoked cortical response has a positive deflection (historically shown by a downward direction).

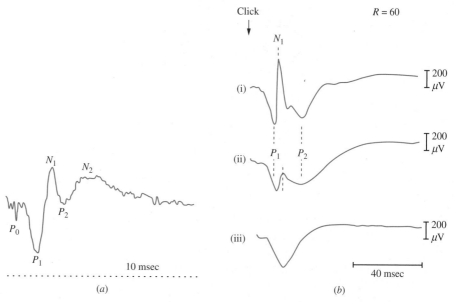

FIGURE 1.15 Cortical gross evoked potentials. *(a)* Average of 60 responses to acoustic clicks (one per second) recorded from the surface of the cerebellum of unanesthetized cats by a macro-electrode. A sharp negative deflection ($N1$) interrupts the positive component of the response, creating two positive deflections ($P1$ and $P2$), followed by a slow negative deflection ($N2$). The waveform changes from location to location but the general shape is the same. The shape of the evoked response and its peaks are associated with the elements of the electrical activity of the cellular neural network such as the incoming volley through input pathways, postsynaptic potentials at the neuronal elements, and the evoked action potentials of individual neurons. (Nahvi, 1965, unpublished.) *(b)* Average click evoked potentials from three locations on the surface of the cerebellum show the range of variations in the triphasic response characteristic.

Source: With kind permission from Springer Science+Business Media: *Experimental Brain Research.* "Firing Patterns induced by Sound in Single Units of the Cerebellar Cortex," Volume 8, Issue 4, 1969, pages 327-45, M.J. Navhi and R.J. Shofer.

It starts with a latency that varies from several msec to tens of msec and lasts from 10 to 30 msec. The positive component may be interrupted by a negative deflection. Examples of cortical evoked potentials in response to acoustic clicks are shown in Figure 1.15.

Using a smaller recording electrode allows a better discrimination of localized electrical activity of the neurons of the brain. A macroelectrode with a tip-size of about 0.1 mm records more localized extracellular field potentials (still called *gross potential* in contrast to single-unit action potentials, to be discussed shortly). A bipolar macro-electrode can be constructed to measure the potential difference between two adjacent (e.g., 100 or 200 microns apart) points or between a single point and a common reference point. Figure 1.16 shows examples of electrical potentials recorded from the coronal-precruciate cortex of an awake cat conditioned to twitch its nose in response to a hiss conditional stimulus (CS). Recent advances have enhanced the possibility of using similar cortical potentials to program and operate prosthetic devices for control of movement in persons with impaired motor skills.

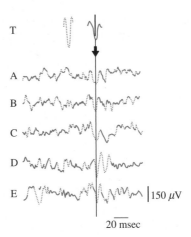

FIGURE 1.16 Cortical potentials associated with a nose twitch recorded from the coronal-precruciate cortex. The recording electrodes were made of teflon-coated stainless-steel wire of 0.1 mm diameter, insulated except at the tip, and implanted chronically and approximately 1.5 mm below the surface of the cortex. The five traces shown in this figure are representative samples of cortical recordings. Each trace (160 msec shown) contains a CS-evoked response (aligned at the vertical line) preceding a conditioned nose twitch movement. Nose twitch was monitored by TV film and concurrent subdermal EMG recordings from the activity of the levator oris muscles. The evoked response was successfully detected by cross-correlation with a template matched to the signal. The template (20 msec long) is shown on the top of the figure.

Source: With kind permission from Springer Science+Business Media: *Experimental Brain Research*, "Application of Optimum Linear Filter Theory to the Detection of Cortical Signals Preceding Facial Movement in Cat," Volume 16, Issue 5, 1973, pages 455-65, C.D. Woody and M.J. Nahvi.

1.11   Neuroelectric Signals from Single Neurons

Action Potentials

Neurons are processing elements of the brain. A neuron is composed of a set of dendrites that receive signals from other cells (as few as several and as numerous as several hundred thousand inputs), a cell body (5 to 70 microns across) that integrates and processes the information, and an axon that sends the output signal to other cells. A thin membrane separates the inside from the outside of the neuron. The liquid inside the neuron contains ions that are positively or negatively charged. Ion pumps in the membrane maintain different concentrations of positive and negative ions on the two sides of the membrane. This creates a negative electrical potential of -70 to -80 mV across the membrane (with the inside being more negative with respect to the outside) called the resting potential. An incoming signal from sensory receptors or other neurons may be inhibitory, causing the neuron to become hyperpolarized (resting potential becoming more negative), or excitatory, causing the neuron to become depolarized (resting potential becoming less negative). If depolarization reaches a certain threshold level, the electrical conductance of the membrane at the axon hillock is dramatically reduced for a brief period, triggering a rush of positive ions (such as Na^+ ions) from outside to the cell interior. The movement of positive ions creates a positive current flow and reduces the negative magnitude of the resting potential toward zero for a brief period of time. The cell is said to have fired. These events take about a millisecond, after which the cell membrane recovers and intracellular potential returns to the former resting potential value. The effect of the above sequence of events is a spike in voltage change called an *action potential*. It is an all-or-none event. The depolarization and the action potential propagate along the cell's axon away from the cell body and reach synapses on other cells, thereby transmitting information to other neurons. In summary, action potentials are all-or-none events that transmit information by the timing of their occurrence.

The action potentials are recorded by fine electrodes (called microelectrodes) with a tip size on the order of microns or less. Action potentials can be recorded on the outside or inside of neurons. In processing neuroelectric data, an action potential can be treated as a discrete-time unitary signal which carries information in its inter-spike time interval or its rate of spike. It can be spontaneous or evoked by external factors such as an acoustic, visual, or tactile stimulus.

Recordings of Extracellular Action Potentials

Microelectrodes with fine tips can pick up single-unit action potentials from neuronal elements in the extracellular space. Metal microelectrodes can be made of tungston wires which are insulated except for a few microns at the tip. Glass microelectrodes are made of glass pipets filled with KCl solutions with 3-mol density. Extracellular recordings of action potentials vary from 100 μV to 10 mV or higher. Their shape and magnitude vary depending on the recording distance from the neuronal element and the electrical properties of the recording electrode, but they are always brief and unitary in the form of spikes. An example is shown in Figure 1.17.

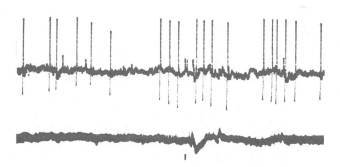

FIGURE 1.17 Single-unit recording of extracellular action potentials using a glass microelectrode from the auditory area of the cerebellum (upper trace) and simultaneous recording of surface gross potential (lower trace) evoked by an acoustic click (shown by the bar below the gross potental recording).

Source: With kind permission from Springer Science+Business Media: *Experimental Brain Research*, "Firing Patterns Induced by Sound in Single Units of the Cerebellar Cortex," Volume 8, Issue 4, 1969, pages 327-45, M.J. Nahvi and R.J. Shofer.

Recordings of Intracellular Potentials

The membrane of a neuron may be penetrated by a glass pipet microelectrode with a tip-size of 1 micron or less to record electrical potentials from inside the cell. The intracellular electrical potential recording is composed of (1) a resting potential of -70 to -80 mV and (2) postsynaptic potentials generated by the arrival of electric signals from other cells. Such signals generate electric potentials inside the cell which can be positive (excitatory postsynaptic potentials, EPSP, making the interior of the cell less negative and thus depolarizing it) or negative (inhibitory postsynaptic potentials, IPSP, making the interior of the cell more negative and thus hyperpolarizing it). Postsynaptic potentials are in the form of decaying exponentials with peak magnitudes of 0.1 to 10 mV and time constants of 5 msec to several minutes. They add algebraically to the resting potential. Upon reaching a depolarization threshold (-65 mV at the axon hillock and -35 mV at the cell body), the cell fires with one or more action potentials that travel through the axon and carry information to other cells. Examples of intracellular recordings are shown in Figure 1.18.

1.12 Applications of Electrophysiologic Signals

Understanding the physiological correlates of neuroelectric signals in the functioning of the central nervous system and methods for their analysis has found many applications in medicine and engineering. Advances in communication and computer technologies (both hardware and software) have provided vast opportunities for using biological signals in general (and electrophysiologic signals in particular) for many purposes and tasks. EEG signals (as well as EKG, EMG, and signals generated by eye movement) provide signal sources for controlling prostheses (prosthetic limbs, assistive devices, and robots), computer recognition of unuttered speech (synthetic generation of live speech) and real-time monitoring, analysis, and decision making in preventive care (such as for heart attacks). In such applications, several measures need to be defined, evaluated, and investigated. One important measure is the probability (or rate) of error. For example, (1) By how much may the error rate in an assistive device be reduced through use of multiple signals? (2) How accurately can one predict the occurrence of a signal intended for movement before the movement has been initiated? (3) Can one reduce the error rate by increasing the number of recording electrodes? (4) Can one record the single-unit activity of neurons and use them in daily tasks?

Besides current applications, studies of neuroelectric signals (especially at the neuronal and subneuronal levels) suggest applications in neurocomputing and new methods of mass storage of data, a field with much future promise.

1.13 Characterization and Decomposition of Signals

Complex signals are difficult to characterize unless we find ways to decompose and expand them into simpler components that can be represented by familiar mathematical functions such as DC levels, sinusoids, pulses, waveforms having specific features, and

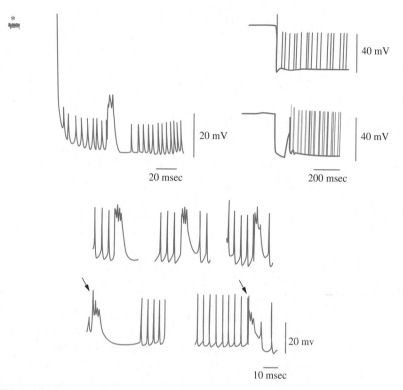

FIGURE 1.18 Recordings of intracellular potentials. *(a)* Top traces. Recordings of electrical potentials at penetration time from three morphologically identified neurons of the cerebellar cortex of an unanesthetized awake cat. The downward negative voltage shift shows the intracellular penetration of the electrode. The intracellular recording also shows the presence of both undershoot and overshoot action potentials. *(b)* Right traces. Sample recordings of action potentials. Two types of action potentials are recognized: one type is a simple spike with short duration and slower rise time; the other is a complex one with faster rise time but longer duration. The two types of action potentials seen in these recordings are attributed to Purkinje cells in the cerebellum.

Source: Reprinted from *Experimental Neurology*, Volume 67, Issue 2, Mahmood J. Nahvi, Charles D. Woody, Eric Tzebelikos, Charles E. Ribak, "Electrophysiologic Characterization of Morphologically Identified Neurons in the Cerebellar Cortex of Awake Cats," pages 368–376, (1980), with permission from Elsevier.

so on. One such expansion method was proposed by Jean Baptiste Joseph Fourier in the early 19th century[8] and later generalized by others. Fourier discovered that any periodic signal can be represented by a collection of sinusoidal functions, leading to the Fourier series and transform. The power of Fourier analysis is in the fact that sinusoids are eigenfunctions of an important class of system models called *linear time-invariant (LTI)* systems. Linear systems are convenient and rather accurate models of many physical

---

[8] *The Analytical Theory of Heat* by Joseph Fourier, Dover Phoenix Editions, 2003, originally published as *Theorie Analytic de la Chaleur* in 1822. Fourier (1768–1830) submitted his paper on this subject to the Academy of Science, Paris, in 1807.

systems within the range of interest. A sinusoidal signal entering a linear time-varying system appears at the output as a sinusoid. The system can change only the amplitude and phase of the signal. This is called the frequency response. The frequency response allows us to predict the output of an LTI system for any input, as long as the input is expressed in the form of a sum of sinusoids. The Fourier expansion of the input signal provides that expression. In Fourier analysis, sinusoids are the building blocks. In wavelet analysis, the building blocks are wavelets.

Almost every communication, control, and measurement process is vulnerable to noise. Noise interferes with a signal's mission and potentially degrades its performance. If it were not for noise, a large amount of information could be encoded in the amplitude of a single pulse and transmitted to a receiver. Below we mention important characteristics that notably influence the functioning of signals in combating the effect of noise.

Duration and amplitude. Mathematical models allow a function to have an infinite range and domain. Real signals, however, have finite duration and amplitude. These two characteristics translate into energy which is an indication of a signal's strength. (See Example 1.1 in section 1.5.) Signals with more energy are less vulnerable to noise and impairment.

Rate of change. Signals which change more often can carry more information with less error. The faster the signal switches from one form to another, the more data it transmits.

Size of the alphabet. The change in a signal may be digital (using one of N letters of an alphabet every T seconds). The alphabet may be binary (a voltage being high or low; the man in a picture either standing or sitting; a variable being $+1$ or -1; a flag being up or down). It may be tertiary (high, medium, or low; standing, sitting, or lying down; $+1$, 0, or -1; up, down, or at half mast). It may contain 10 decimal digits or be made of the 26 letters of the English language. The larger the size of the alphabet, the more informative the signal is. The information carried by a single letter of a binary alphabet is called a *bit*.

Frequency. The change in the signal may also be analog and continuous in amplitude, frequency, and phase (analog modulation). Similar to the digital case, analog high-frequency signals can transmit more data. The examples given in section 1.5 combine the case of digital and analog signals and show how the size of the alphabet and a high-frequency carrier increase the data rate.

Bandwidth. Signals are rarely made of a pure sinusoid. A signal may contain a range of frequencies called the bandwidth. Higher bandwidths increase the rate of information transmission and reduce the error rate. The bandwidth will be discussed in Chapters 7 and 8 on Fourier analysis.

Sampling rate. Many discrete signals are generated by sampling a continuous-time signal. For the discrete signal to contain the same information, the sampling should be done fast enough to capture the changes in the signal. The miminum rate is called the Nyquist rate. In practice, the reconstruction process is not perfect and a higher sampling rate is needed. When the sampling is done below the Nyquist rate, not only is some information lost, but the remaining information is also distorted.

Coarse and fine structure. Several structural levels can be observed in the signal examples of Figures 1.1, 1.2, and 1.3. Similar aspects are also seen in other signals (natural or synthetic) such as amplitude-modulated signals, radar signals, neuroelectric signals, seismic signals, and many more. Several tools such as filtering, Fourier methods, and wavelets can be employed to analyze these structures.

Other domains. Mathematical modeling of signals is not limited to waveforms and functions in the time domain. Signals may also be represented in the frequency or other domains (e.g., by an alphabet or a dictionary). For most of our discussions in this book we employ signals in the time and frequency domains.

1.14 Mathematical Representations of Signals

The following three classes of mathematical functions are used to represent signals. They are as follows:

1.   Periodic functions, such as sinusoids.
2.   Nonperiodic functions (also called aperiodic), such as impulse, step, ramp, exponentials, pulses, and sinc function.
3.   Random functions.

In this section we discuss functions that belong to the first two classes. There are several aspects that distinguish signals from these functions. For one thing, the domain of these functions is $-\infty < t < \infty$ while actual signals have a finite duration. The other thing is that actual signals, by necessity, include unpredictability while these functions are deterministic. The differences, however, don't undermine the value of these representations in the analysis and design of linear systems, as should become clear to the student throughout this book.

Periodic Functions

A function $f(t)$ is periodic with period T if

$$f(t) = f(t + T) \quad \text{for all t,} \quad -\infty < t < \infty$$

Note that nT, where n is a positive integer, is also a period of $f(t)$. The function repeats itself every T seconds. Therefore, it is sufficient to know its value during one period. This information may be provided by an analytical expression, a graph, or in tabular form. A well-known periodic function is the sinusoid $V_0 \cos(\omega t + \theta)$. Periodic functions may be very complex. However, as we will see in Chapter 7 on Fourier series, they can always be represented by a sum of sinusoids. Therefore, sinusoids play an important role in the analysis of signals and linear systems. Sinusoids are treated in detail in Chapter 2.

Sum of Periodic Functions

The sum of two periodic functions with respective periods T_1 and T_2 is periodic if a common period $T = n_1 T_1 = n_2 T_2$, where n_1 and n_2 are integers, can be found. This requires that $T_1/T_2 = n_2/n_1$ be a rational number. Otherwise, the sum is not periodic and usually called *almost periodic*. A similar conclusion applies to the product of two periodic functions.

Example **1.7**

For each time function given below, determine if it is periodic or aperiodic. Specify its period if periodic.

a. $\cos(3t + 45°) + \sin(\sqrt{2}t - 120°)$ b. $\cos(5t) + \cos \pi (t - 0.5)$
c. $\cos(\pi t + 10°) - \cos(2\pi t + \pi/3)$ d. $\sin(6.28t - 2\pi/3) + \cos 0.2(t - 0.5)$

Solution
a. Aperiodic   b. Aperiodic
c. Periodic, period $= 2$ d. Periodic, period $= 50\pi$

Example **1.8**

Repeat Example 1.7 for the time functions given below.

a. $\cos(3t + 45°) \sin(\sqrt{2}t - 120°)$ b. $\cos(5t) \cos \pi (t - 0.5)$
c. $\cos(\pi t + 10°) \cos(2\pi t + \pi/3)$ d. $\sin(6.28t - 2\pi/3) \cos 0.2(t - 0.5)$

Solution
a. Aperiodic   b. Aperiodic
c. Periodic, period $= 2$ d. Periodic, period $= 25\pi$

Elementary Functions

The unit-impulse, unit-step, unit-ramp, sinusoidal, exponential, and *sinc* functions, as well as pulses and windows, are often used as the elementary building blocks for modeling signals and they will be discussed here. Finite-duration pulses and windows will be discused in detail in Chapters 7 and 8 on Fourier analysis.

Unit Impulse

The unit-impulse function $\delta(t)$[9] (also called *Dirac's delta* function) is represented by a double-stem up arrow as shown in Figure 1.19. A delay of τ produces $\delta(t - \tau)$, while an advance of τ produces $\delta(t + \tau)$.

FIGURE 1.19 Representation of the unit-impulse function $\delta(t)$. A delayed unit-impulse $\delta(t - \tau)$ and an advanced unit-impulse $\delta(t + \tau)$ are also shown.

---

[9]In the strict mathematical sense, $\delta(t)$ is not a conventional function but a generalization of functions. It, therefore, has sometimes been called a *symbol* rather than a function, as emphasized by Ruel Churchill in his lectures (see also his *Operational Mathematics*, (New York: McGraw-Hill, 1958), pp. 26–28).

The unit-impulse function is defined in less rigorous ways than a function. One such definition is

$$\delta(t) = \begin{cases} 0, & t \neq 0 \\ \infty, & t = 0 \end{cases} \quad \text{and} \quad \int_{-\infty}^{\infty} \delta(t)dt = 1$$

Another definition involves the limit of a narrow pulse $\Delta_T(t)$ with unit area, as the pulse-width approaches zero, but the area remains equal to 1. For example, let

$$\Delta_T(t) = \begin{cases} \frac{1}{T}, & -\frac{T}{2} < t < \frac{T}{2} \\ 0, & \text{elsewhere} \end{cases}$$

Then,

$$\delta(t) = \lim_{T \to 0} \Delta_T(t)$$

Note that $\int_{-\infty}^{\infty} \Delta_T(t)dt = 1$. The initial pulse need not be rectangular, or even of finite duration, as long as it narrows down and keeps its area at 1. Examples are pulses described by exponential, Gaussian, and sinc functions listed below and shown in Figure 1.20. These functions will be described in detail shortly.

Rectangular pulses, Figure 1.20(a):

$$x_1(t) = \begin{cases} \frac{1}{T}, & -\frac{T}{2} < t < \frac{T}{2} \\ 0, & \text{elsewhere} \end{cases} \qquad -\infty < t < \infty \qquad \lim_{T \to 0} x_1(t) = \delta(t)$$

Exponential pulses, Figure 1.20(b):

$$x_2(t) = \frac{1}{2\tau}e^{-\frac{|t|}{\tau}} \qquad -\infty < t < \infty \qquad \lim_{\tau \to 0} x_2(t) = \delta(t)$$

Gaussian pulses, Figure 1.20(c):

$$x_3(t) = \frac{1}{\sigma\sqrt{2\pi}}e^{-\frac{x^2}{2\sigma^2}} \qquad -\infty < t < \infty \qquad \lim_{\sigma \to 0} x_3(t) = \delta(t)$$

Sinc pulses, Figure 1.20(d):

$$x_4(t) = \frac{\sin\left(\frac{\pi t}{T}\right)}{\frac{\pi t}{T}} \qquad -\infty < t < \infty \qquad \lim_{T \to 0} x_4(t) = \delta(t)$$

Sieving Property of Impulse

A more rigorous definition of $\delta(t)$, based on generalized functions and distribution, is given by its "sieving" property

$$\int_{-\infty}^{\infty} f(t)\delta(t - t_0)dt = f(t_0)$$

where $f(t)$ is a well-behaved function and continuous at t_0.

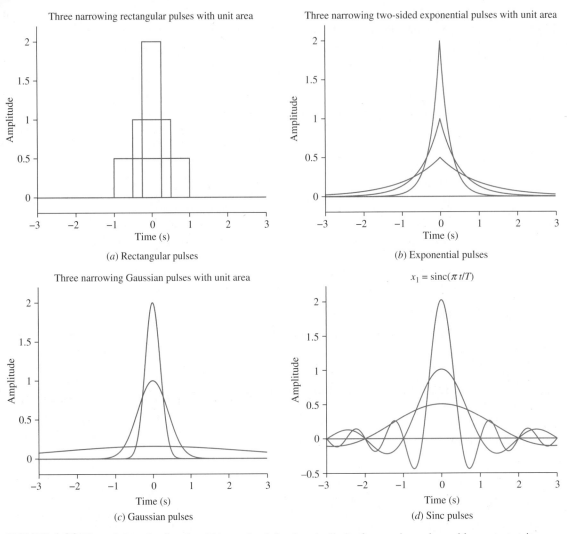

FIGURE 1.20 The unit-impulse function $\delta(t)$ may be defined as the limit of narrowing pulses with constant unit area.

Example 1.9

Let $x(t)$ be an even rectangular pulse at the origin

$$x(t) = \begin{cases} \frac{1}{T}, & -\frac{T}{2} < t < \frac{T}{2} \\ 0, & \text{elsewhere} \end{cases}$$

Shift it by t_0 and multiply by a well-behaved function $f(t)$, which is continuous at t_0, and integrate. Show that

$$\lim_{T \to 0} \int_{-\infty}^{\infty} f(t)x(t - t_0)dt = f(t_0)$$

Solution

$$I = \int_{-\infty}^{\infty} f(t)x(t-t_0)dt = \frac{1}{T} \int_{t_0-T/2}^{t_0+T/2} f(t)dt$$

For small T:

$$I \approx \frac{f(t_0 - T/2) + f(t_0 + T/2)}{2}$$

As the limit:

$$I = \lim_{T \to 0} \frac{f(t_0 - T/2) + f(t_0 + T/2)}{2} = f(t_0)$$

Unit Step

The unit-step function $u(t)$ is defined by

$$u(t) = \begin{cases} 1, & \text{for } t \geq 0 \\ 0, & \text{for } t < 0 \end{cases}$$

A delay of τ produces $u(t-\tau)$, while an advance of τ produces $u(t+\tau)$. See Figure 1.21.

Unit Ramp

The unit-ramp function $r(t)$ is defined by

$$r(t) = \begin{cases} t, & \text{for } t \geq 0 \\ 0, & \text{for } t < 0 \end{cases}$$

A delay of τ produces $r(t-\tau)$, while an advance of τ produces $r(t+\tau)$. See Figure 1.22.

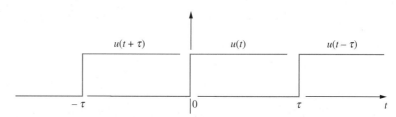

FIGURE 1.21 The unit-step function $u(t)$, a delayed unit-step $u(t-\tau)$, and an advanced unit-step $u(t+\tau)$.

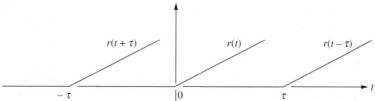

FIGURE 1.22 The unit-ramp function $r(t)$, a delayed unit-ramp $r(t-\tau)$, and an advanced unit-ramp $r(t+\tau)$.

In addition to impulse, step, and ramp functions, other user-defined functions may become desirable, especially when employing computer software. An example of such a user-defined function is $Rect(t) = u(t) - u(t - 1)$. Such additional functions will not be employed in this book.

Sinusoid

The sinusoidal function $v(t) = V_0 \cos(\omega t)$ is periodic. The period is $T = 2\pi/\omega$. The frequency is $f = 1/T = \omega/2\pi$. A delay of τ produces $v(t - \tau) = V_0 \cos(\omega t - \omega\tau) = V_0 \cos(\omega t - \theta)$, where $\theta = \omega\tau$ is the phase lag. Similarly, an advance of τ produces the phase lead $\theta = \omega\tau$. See Figure 1.23. For more details on sinusoids see Chapter 2.

Exponential

An exponential function grows or decays at a rate proportional to its value: $dv/dt = \alpha v$, where α is a positive number when exponentially growing, and a negative number when decaying. It is easy to show that $v(t) = V_0 e^{\alpha t}$, where V_0 is its initial value (i.e., at $t = 0$). Decaying exponentials are of special interest as they are often encountered in the analysis of stable systems. A causal decaying exponential is shown in Figure 1.24. It is also written as $v(t) = V_0 e^{-t/\tau} u(t)$, where τ is called the exponential's time constant. The maximum value of the function is V_0 at $t = 0$, which also produces the (maximum) decay rate of V_0/τ at that time. This is the downward slope of the line tangent to the exponential function at $t = 0$. Consequently, such a line intersects the time axis at $t = \tau$. At $t = \tau$, the decaying exponential function is reduced to $V_0 e^{-1} \approx 0.368 \, V_0$ and at $t = 2\tau$ to 0.135 times its initial value. At $t = 5\tau$, the reduction is more than 99%, which for

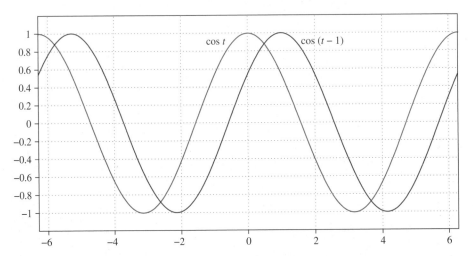

FIGURE 1.23 The sinusoidal function $\cos t$ is periodic with period $T = 2\pi \approx 6.2832$ seconds. A delay of τ seconds produces an equivalent amount, in radians, of phase lag (e.g., a one-second delay creates a phase lag of 1 radian ≈ 57 degrees). Similarly, an advance of τ seconds produces a phase lead of τ radians (or, equivalently, a phase lag of $2\pi - \tau$ radians).

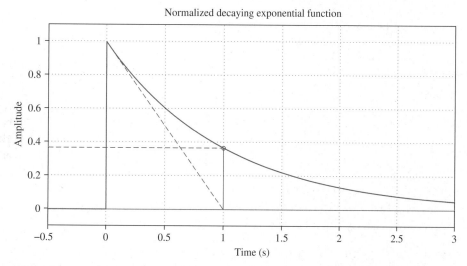

FIGURE 1.24 A decaying exponential e^{-t} with a time constant of $\tau = 1$ sec. The tangent to the function at $t = 0$ intersects the time axis at $t = 1$. The value of the function at $t = 1$ is ≈ 0.368.

most practical purposes may be considered reaching a zero value. The two parameters (V_0 and τ) of an exponential function may be computed from two measurements as described in Example 1.10.

Determining a time constant

A signal is known to be exponentially decaying toward its zero final value. Derive its time constant from the ratio of two measurements taken T seconds apart.

Solution

An exponentially decaying signal with zero final value can be represented by $v(t) = Ae^{-t/\tau}$, where A is the value of the signal at $t = 0$ and τ is its time constant. Let $V_1 = Ae^{-t_1/\tau}$ and $V_2 = Ae^{-t_2/\tau}$ be the measured values at t_1 and t_2, respectively. The time constant can be derived from a knowledge of V_1/V_2 and $T = t_2 - t_1$ by the following:

$$\frac{V_1}{V_2} = \frac{Ae^{-t_1/\tau}}{Ae^{-t_2/\tau}} = e^{(t_2 - t_1)/\tau} = e^{T/\tau}, \quad \text{and} \quad \ln\left(\frac{V_1}{V_2}\right) = T/\tau, \quad \tau = \frac{T}{\ln\left(\frac{V_1}{V_2}\right)}$$

For example, let $V_1 = 0.6$, $V_2 = 0.2$, and $T = 1$ s, as shown in Figure 1.25. Then $\tau = T/\ln(\frac{V_1}{V_2}) = \frac{1}{\ln 3} = 0.91$ s. Having found the time constant, the function can now be expressed by $v(t) = V_1 e^{-t/\tau}$ (taking the time of the first measurement as the time origin). If the time origin is already set, then the parameter A (the value of the exponential at $t = 0$) can be found from $V_1 = Ae^{-t_1/\tau}$ or $A = V_1 e^{-t_1/\tau}$.

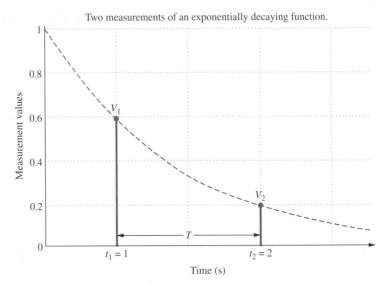

FIGURE 1.25 Finding the time constant of a decaying exponential from two measurements.

 *xample* **1.11**

Response of an *RC* circuit to step and impulse voltages

A 1-Ω resistor and a 1-F capacitor are connected in series with a voltage source $v_1(t)$ as in Figure 1.26.

a. With $v_1(t) = u(t)$ (V), the capacitor current and voltage signals are shown to be $i(t) = e^{-t}u(t)$ (A) and $v_2(t) = (1 - e^{-t})u(t)$ (V), respectively. Plot $i(t)$ and $v_2(t)$ for $-\tau < t < 5\tau$, where τ is their common time constant. For each signal determine initial and final values.

b. Repeat with $v_1(t) = \delta(t)$.

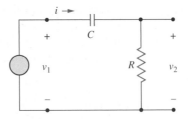

FIGURE 1.26 An *RC* circuit in series with a voltage source.

Solution

a. The unit-step voltage source is shown in Figure 1.27[a(i)]. The capacitor current decays exponentially from 1 A to a zero value with a time constant of 1 second. It is plotted in Figure 1.27[a(ii)] for $-1 < t < 5$. At $t = 0^+$ (in the

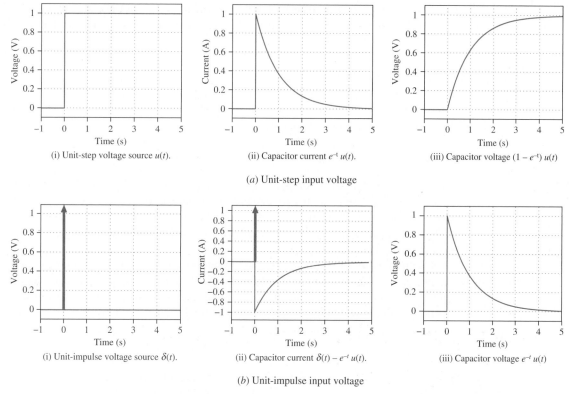

(i) Unit-step voltage source $u(t)$.

(ii) Capacitor current $e^{-t} u(t)$.

(iii) Capacitor voltage $(1 - e^{-t}) u(t)$

(a) Unit-step input voltage

(i) Unit-impulse voltage source $\delta(t)$.

(ii) Capacitor current $\delta(t) - e^{-t} u(t)$.

(iii) Capacitor voltage $e^{-t} u(t)$

(b) Unit-impulse input voltage

FIGURE 1.27 Capacitor current and voltage in response to unit-step **(a)** and unit-impulse **(b)** voltage sources (see the circuit of Figure 1.26).

direction of increasing t), the slope of the tangent to the exponential is 1 A/s. Therefore, for small values of t such as $t < 5$ ms, the exponential can be approximated by its tangent line $i = 1 - t/1{,}000$ A. For example, at $t = 5$ ms, the linear approximation results in a value of $i_{\text{approximate}} = 1 - 5/1{,}000 = 0.995$ A, while the exact value of the current is $i_{\text{exact}} = e^{-0.005} = 0.9950125$ A. The percentage error is

$$\epsilon\% = 100 \times \frac{i_{\text{exact}} - i_{\text{approximate}}}{i_{\text{exact}}} = 100 \times \frac{0.9950125 - 0.995}{0.9950125}$$

$$\approx 0.001\% \text{ (or one in } 100{,}000)$$

This is an extremely small error, prompting an approximation of capacitor current by the flat line of a unit-step function. (For further elaboration see problem 98.)

b. The unit impulse is derivative of unit step. The responses of the circuit to the unit-impulse voltage are derivatives of its responses to unit-step voltage. These are given in Figure 1.27(b).

Complex Exponential

The function e^{st}, with $s = \sigma + j\omega$ a complex number, is called a complex exponential function. It is the only mathematical function that retains its functional form under the derivative and integration operations.

$$\frac{d}{dt}\{e^{st}\} = se^{st} \quad \text{and} \quad \int e^{s\tau}d\tau = \frac{1}{s}e^{st}$$

Taking the derivative of a complex exponential function changes only its magnitude by a factor s. Similarly, taking its integral multiplies the magnitude by $1/s$. This invariance property remains true under the class of operations called linear time-invariant (LTI). Hence, the complex exponential function is called the characteristic function (also eigenfunction) of the LTI systems.

Complex exponential functions can represent sinusoids, including those with time-varying amplitudes, resulting in a simplification of analysis. For example, consider the function $\overline{V}(t) = \overline{V}e^{st}$ with the complex magnitude $\overline{V} = V_0e^{j\theta}$. Then a sinusoidal function of time with exponentially varying amplitude can be written as the real part (or the imaginary part) of a complex exponential.

$$v(t) = V_0e^{\sigma t}\cos(\omega + \theta)t = \mathcal{RE}\{\overline{V}(t)\}$$

$$u(t) = V_0e^{\sigma t}\sin(\omega + \theta)t = \mathcal{IM}\{\overline{V}(t)\}$$

where

$$\overline{V}(t) = V_0e^{\sigma t}e^{j(\omega+\theta)t} = v(t) + ju(t)$$

Throughout an analysis, the signal $\overline{V}(t)$ may be used instead of $v(t)$ or $u(t)$. At the end of the analysis the result will be converted back to the real or imaginary parts. Due to the above properties the complex exponential function occupies a unique place in the analysis of signals and systems, thereby called analysis in the s-domain.

Sinusoid with Exponentially Varying Amplitude

The function $v(t) = V_0e^{\sigma t}\cos(\omega t + \theta)$ is a sinusoidal function with exponential growth (if $\sigma > 0$) or decay (if $\sigma < 0$). It is the real part of the complex exponential $\overline{V}(t) = V_0e^{\sigma t}e^{j(\omega+\theta)t}$, where V_0, σ, ω, and θ are constant real numbers. An example is shown in Figure 1.28(a) for $V_0 = 1, \sigma = -2, \omega = 10\pi$, and $\theta = -\pi/2$. For $\sigma = 0$, the amplitude of the sinusoid is constant. For $\sigma > 0$, the amplitude of the sinusoid grows with time; see Figure 1.28(b). The complex exponential function and its real part are completely specified by the four parameters V_0, θ, σ, and ω. These parameters are combined in pairs as the complex amplitude $\mathbf{V} = V_0e^{j\theta}$ (also called the phasor $\mathbf{V} = V_0\angle\theta$) and the complex frequency $s = \sigma + j\omega$. Exponential and sinusoidal functions are important building blocks in the analysis and design of signals and linear systems and are encountered frequently.

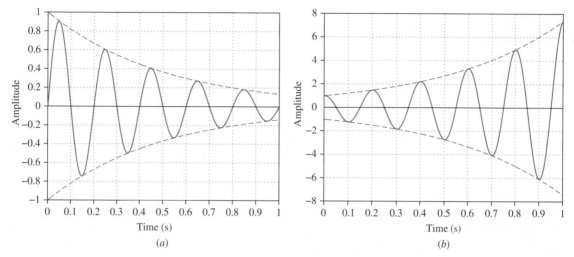

FIGURE 1.28 *(a)* An exponentially decaying sinusoid $V_0 e^{\sigma t} \sin(2\pi f t)$ and its envelope. $V_0 = 1$, $f = 5$ Hz, $\sigma = -2\,s^{-1}$ (or, equivalently, a time constant of $\tau = -1/\sigma = 0.5$ sec.). *(b)* An exponentially growing cosine function $e^{2t}\cos(10\pi t)$ and its envelope.

The Sinc Function

The unit-sinc function is given by $\text{sinc}(t) = \frac{\sin t}{t}$, $-\infty < t < \infty$, See Figure 1.29(a). The function contains positive and negative lobes whose amplitudes diminish with t. The main lobe is centered at $t = 0$ and has a maximum of sinc(0)=1. Zero-crossings occur at $t = \pm k\pi$, where k is a nonzero integer. The regularity of the zero-crossings produces positive and negative lobes which create a sort of periodicity.

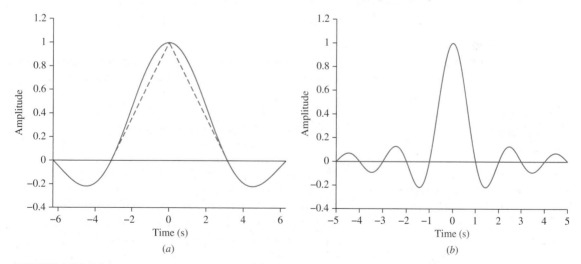

FIGURE 1.29 *(a)* A unit-*sinc* function $\text{sinc}(t) = \frac{\sin(t)}{t}$. Zero-crossings are at $t = \pm k\pi$. The area under the sinc function is equal to the area of the dashed triangle within the main lobe. *(b)* A *sinc* function $\text{sinc}(\pi t) = \frac{\sin(\pi t)}{\pi t}$. Zero-crossings are at $t = \pm k$ (Example 1.12).

Example
1.12

Write the mathematical expression for a sinc function with a maximum value of 1 and a main lobe that is 2 sec wide.

Solution

The general expression for a sinc function is $x(t) = \sin(at)/(bt)$. To find a we note that the first pair of zero-crossings occurs at ± 1. The other crossings occur at $t = \pm k$, where k is a nonzero integer. The numerator of the *sinc* function is, therefore, $\sin(\pi t)$. To find b we note that $x(0) = a/b = 1$ which results in $b = \pi$. The mathematical expression for the desired unit-sinc function is $\mathrm{sinc}(\pi t) = \frac{\sin \pi t}{\pi t}$. Figure 1.29($b$) plots this unit-sinc function.

Example
1.13

Determine the maximum value and zero-crossings of the sinc function $x(t) = \frac{\sin(2\pi f_0 t)}{\pi t}$ in terms of f_0. Determine f_0 so that the main lobe of the function is 1 ms wide.

Solution

The maximum of the function occurs at $t = 0$. The value of the maximum is

$$\lim_{t \to 0} \frac{\sin(2\pi f_0 t)}{\pi t} = \frac{\frac{d}{dt}[\sin(2\pi f_0 t)]}{\frac{d}{dt}(\pi t)}\bigg|_{t=0} = 2 f_0 \cos(2\pi f_0 t)\big|_{t=0} = 2 f_0$$

Zero-crossings occur at $2\pi f_0 t = \pm k\pi$ or $t = \pm\frac{k}{2 f_0}$, k being an integer. For $f_0 = 1$ kHz the main lobe is 1 ms wide.

Gaussian Function

The normalized Gaussian function is

$$e^{-\pi t^2}, \quad -\infty < t < \infty$$

It is an even function of t and positive with a peak value of 1 at $t = 0$. As $t \to 0$, the Gaussian function diminishes to zero faster than the exponential function does.

Example
1.14

Show that the area under the normalized Gaussian function is equal to 1:

$$\int_{-\infty}^{\infty} e^{-\pi t^2} dt = 1$$

Solution

Start with the square of the integral and transform it to the polar coordinate system:

$$\left[\int_{-\infty}^{\infty} e^{-\pi t^2} dt\right]^2 = \left[\int_{-\infty}^{\infty} e^{-\pi x^2} dx\right]\left[\int_{-\infty}^{\infty} e^{-\pi y^2} dy\right] = \int_{-\infty}^{\infty}\int_{-\infty}^{\infty} e^{-\pi(x^2+y^2)} dx\, dy$$

$$= \int_{r=0}^{\infty}\int_{\theta=0}^{2\pi} r e^{-\pi r^2} d\theta\, dr = \int_{0}^{\infty} 2\pi r e^{-\pi r^2} dr$$

where $x^2 + y^2 = r^2$ and $dxdy = rd\theta dr$. To evaluate the last integral let $\pi r^2 = t$ and $2\pi r dr = dt$. Then,

$$\int_0^\infty 2\pi r e^{-\pi r^2} dr = \int_0^\infty e^{-t} dt = 1$$

E*xample*
1.15

The following function is called the Gaussian density function with a mean value of μ and variance $= \sigma^2$.

$$f(x) = \frac{1}{\sigma\sqrt{2\pi}} e^{-\frac{(x-\mu)^2}{2\sigma^2}}$$

Show that a) $\int_{-\infty}^\infty f(x)dx = 1$ for any set of μ and σ, b) $\int_{-\infty}^\infty xf(x)dx = \mu$, and c) $\int_{-\infty}^\infty (x-\mu)^2 f(x)dx = \sigma^2$.

Solution

Let $x - \mu = \sigma\sqrt{2\pi}t$ and $dx = \sigma\sqrt{2\pi}dt$.

a. $\displaystyle\int_{-\infty}^\infty f(x)dx = \int_{-\infty}^\infty \frac{1}{\sigma\sqrt{2\pi}} e^{-t^2}\sigma\sqrt{2\pi}dt = \int_{-\infty}^\infty e^{-\pi t^2} dt = 1$

b. $\displaystyle\int_{-\infty}^\infty xf(x)dx = \int_{-\infty}^\infty (\mu + \sigma\sqrt{2\pi}t)e^{-\pi t^2} dt$

$$= \mu \int_{-\infty}^\infty e^{-\pi t^2} dt + \sigma\sqrt{2\pi} \int_{-\infty}^\infty te^{-\pi t^2} dt = \mu$$

c. $\displaystyle\int_{-\infty}^\infty (x-\mu)^2 f(x)dx = \int_{-\infty}^\infty 2\pi\sigma^2 t^2 e^{-\pi t^2} dt$

$$= \sigma^2 \left[-te^{-\pi t^2}\right]_{-\infty}^\infty + \sigma^2 \int_{-\infty}^\infty e^{-\pi t^2} dt = \sigma^2$$

Pulses and Windows

A pulse is a signal with its energy mostly concentrated during a finite duration of time. The exponential, sinc, and Gaussian functions introduced previously exemplify one class of such pulses. Another class consists of strictly time-limited pulses with zero value outside their durations:

$$x(t) = \begin{cases} f(t), & -\frac{\tau}{2} < t < \frac{\tau}{2} \\ 0, & \text{elsewhere} \end{cases}$$

Time-limited pulses are of much practical interest. They constitute the basic building blocks in analog electrical instruments such as function generators. They are also the focus of signal design for transmission of digital data. Finite-duration pulses also serve

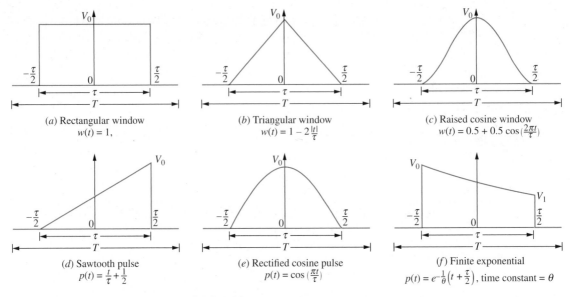

FIGURE 1.30 Windows and pulses with $V_0 = 1$.

as windows. A window is a finite-duration pulse through which we look at a data stream or a function such as an impulse response. The first row in Figure 1.30 shows three such windows (rectangular, triangular, and raised cosine). The second row shows three other familiar pulses from which a variety of pulse shapes and periodic signals may be constructed. All pulses are τ sec wide and expressed in the interval $-\tau/2 < t < \tau/2$. Elsewhere, the pulses have zero values.

1.15   Time Averages

Power and Energy

The square of $f(t)$ is called its power. Its energy from t_0 to $t_0 + T$ is

$$\text{Energy} = \int_{t_0}^{t_0+T} f^2(t)\,dt$$

If the integration over the infinite range of time results in a finite number, then $f(t)$ is called an energy signal. A finite-duration pulse has finite energy. However, not all energy functions are necessarily of finite length. An energy signal may have infinite duration. Examples are the sinc function, pulses, and functions whose amplitudes are modulated by an exponential decay. Clearly, energy signals are not periodic.

The average of power from t_0 to $t_0 + T$ is

$$\text{Average power} = \frac{1}{T} \int_{t_0}^{t_0+T} f^2(t)\,dt$$

For a periodic function with a period T, the integral may be taken over any time duration nT, where n is a positive integer. If a function has infinite energy it might have finite power, in which case it is called a power function. Periodic functions are power functions. Many random functions are also modeled as power functions. Nonperiodic functions can have finite power, too.

Average (DC) and Effective (RMS) Values

The average (also called the DC value) of $f(t)$ from t_0 to $t_0 + T$ is

$$f_{\text{avg}} = \frac{1}{T} \int_{t_0}^{t_0+T} f(t)dt$$

The rms (which stands for root-mean-square) or effective value of $f(t)$ from t_0 to t_0+T is

$$f_{\text{rms}} = \sqrt{\frac{1}{T} \int_{t_0}^{t_0+T} f^2(t)dt}$$

From the above definition, the average power in $f(t)$ is equal to the square of its rms value; that is, $P_{\text{avg}} = f_{\text{rms}}^2$. The rms value of a function, therefore, is a measure of its average power and is often used to identify signals and noise powers and their effects. Because of this, the rms value is also called the *effective* value (i:e., f_{eff} and f_{rms} are equivalent).

Calculating Averages

Averaging is a linear operation and also invariant with respect to time shift. Multiplying a function by a constant k, multiplies its DC and rms values by k and its average power by k^2. These properties can be used to simplify evaluation of the DC and rms values and average power.

Example **1.16**

a. Find the DC value and average power of the unit-sawtooth pulse

$$x(t) = \begin{cases} t, & 0 \le t < 1 \\ 0, & \text{elsewhere} \end{cases}$$

taken over a period of T.

b. Find the DC value and average power of the even triangular pulse

$$y(t) = \begin{cases} 1 - 2\frac{|t|}{\tau}, & -\frac{\tau}{2} \le t < \frac{\tau}{2}. \\ 0, & \text{elsewhere} \end{cases}$$

[Figure 1.30(b)] taken over a period of T.

Solution

a. The area under $x(t)$ is $\frac{1}{2}$ and its DC value during a period T is $x_{DC} = \frac{1}{2T}$. The average power is

$$\frac{1}{T}\int_0^1 t^2 dt = \frac{1}{(3T)}$$

b. The area under $y(t)$ is $\frac{\tau}{2}$ and its DC value is $y_{DC} = \frac{\tau}{2T}$. To evaluate the average power shift $y(t)$ by $\tau/2$, calculate the energy in the left-half side of the pulse and then double it:

$$P_{Avg} = \frac{2}{T}\int_0^{\tau/2}\left(\frac{2}{\tau}t\right)^2 dt$$

To evaluate the integral, let $\frac{2}{\tau}t \equiv \theta$ and use the result of part (a) to find

$$P_{Avg} = \frac{2}{T}\frac{\tau}{2}\int_0^1 \theta^2 d\theta = \frac{\tau}{3T}$$

Note that the DC value and average power in (b) may be directly derived from those in (a).

Example

1.17

Find the DC values, average powers, and the rms values of the pulses of Figure 1.30 taken over a period of T. In all cases a pulse width is τ and its height is $V_0 = 1$.

Solution

Pulse Shape	Equation $\left(-\dfrac{\tau}{2} < t < \dfrac{\tau}{2}\right)$	DC Value During T	P_{Avg} During T	RMS Value During T
Rectangular	1	$\dfrac{\tau}{T}$	$\dfrac{\tau}{T}$	$\sqrt{\dfrac{\tau}{T}}$
Triangular	$1 - 2\dfrac{|t|}{\tau}$	$\dfrac{\tau}{2T}$	$\dfrac{\tau}{3T}$	$\sqrt{\dfrac{\tau}{3T}}$
Raised cosine	$0.5 + 0.5\cos\left(\dfrac{2\pi t}{\tau}\right)$	$\dfrac{1}{2T}$	$\dfrac{3\tau}{8T}$	$\sqrt{\dfrac{3\tau}{8T}}$
Sawtooth	$\dfrac{t}{\tau} + \dfrac{1}{2}$	$\dfrac{\tau}{2T}$	$\dfrac{\tau}{3T}$	$\sqrt{\dfrac{\tau}{3T}}$
Rectified cosine	$\cos\left(\dfrac{\pi t}{\tau}\right)$	$\dfrac{2\tau}{\pi T}$	$\dfrac{\tau}{2T}$	$\sqrt{\dfrac{\tau}{2T}}$
Finite exponential	$e^{-\frac{1}{\theta}\left(t + \frac{\tau}{2}\right)}$	$\dfrac{\theta}{T}\left[1 - e^{-\frac{\tau}{\theta}}\right]$	$\dfrac{\theta}{2T}\left[1 - e^{-\frac{2\tau}{\theta}}\right]$	$\sqrt{\dfrac{\theta}{2T}\left[1 - e^{-\frac{2\tau}{\theta}}\right]}$

1.16 Operations on Signals

Signal processing involves analytical and computational operations on signals done for a variety of purposes; examples are enhancing a signal, reducing noise and disturbances, compensating for some deficiencies in a system's performance, recording and storing data, coding, filtering, analyzing, detecting, synthesizing, designing, and predicting a signal. Signals include speech, audio, images, communication, geophysical, sonar, radar, medical, societal, commercial, and financial. Presenting signal by components that are suitable for processing (such as in harmonics analysis) is a main step in signal processing. In addition, recognizing classes of waveforms and wavelets specific to a signal (such as those mentioned for seismic, physiologic, and speech signals) provides additional processing avenues tailored to that signal. Signal processing operations are, therefore, done in the time domain (i.e., on the signals that are expressed as functions of time) and in the frequency domain (on the frequency representations of signals). Examples of such operations of common interest are integration, differentiation, correlation, convolution, spectral analysis, and filtering. The last four operations involve integration of weighted signal and, therefore, have much in common. These operations are briefly introduced below and will be discussed in detail in future chapters.

1.17 Time Transformation

Some elementary operations are widely applied in signal processing. Among these are time shift (delay or advance), time reversal, and time scaling. These elementary operations provide basic mathematical tools and building blocks for higher-level operations such as correlation, convolution, and filtering. In addition, some elementary operations are also associated with physical phenomena such as time delay or space displacement. A graphical visualization by signal plots is often helpful, especially when time reversal and time shift are combined. These operations are described below.

Time Shift

For notational convenience, let τ be a positive constant. In a signal modeled by the function $x(t)$ change t to $t - \tau$ and obtain $x(t - \tau)$. This shifts the plot of $x(t)$ to the right by τ units and, therefore, is called shifting to the right. When the variable t represents actual time, the above operation represents a time delay as in an echo of a radar signal bouncing from a target or an audio signal bouncing from a wall. Conversely, a shift to the left by τ units converts $x(t)$ to $x(t + \tau)$ and is called an *advance*. In summary, to produce a delay (shift to the right) or an advance (shift to the left) in $x(t)$ change t to $t \mp \tau$, respectively. Time shift was briefly illustrated for elementary functions in section 1.14 and Figures 1.19, 1.21, 1.22, and 1.23.

E*xample*

1.18

The effect of 1-unit shifts (to the left and right) on a sawtooth pulse is shown in Figure 1.31. The left side is a shift to the left and the right side is shift to the right.

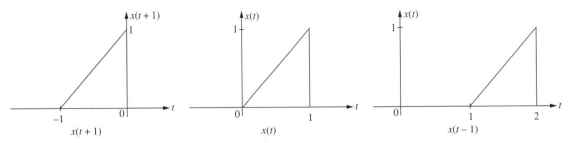

FIGURE 1.31 Time-shifting a sawtooth pulse.

E*xample*

1.19

The effect of 1-unit shifts on a unit-step function is given below. The left side shows a shift to the left and the right side shows a shift to the right.

$$u(t + 1) = \begin{cases} 1, & t \geq -1 \\ 0, & t < -1, \end{cases} \quad u(t) = \begin{cases} 1, & t \geq 0 \\ 0, & t < 0, \end{cases} \quad u(t - 1) = \begin{cases} 1, & t \geq 1 \\ 0, & t < 1 \end{cases}$$

E*xample*

1.20

The effect of 1-unit shifts (to the left and right) on a causal exponential function is shown below. The left side is a shift to the left and the right side is a shift to the right.

$$x(t + 1) = e^{-(t+1)}u(t + 1), \quad x(t) = e^{-t}u(t), \quad x(t - 1) = e^{-(t-1)}u(t - 1)$$

E*xample*

1.21

The effect of a 2-second delay on a causal exponential function with a 1-second time constant is shown by

$$x(t) = 2^{-t}u(t), \quad x(t - 2) = 2^{-(t-2)}u(t - 2) = 4 \times 2^{-t}u(t - 2)$$

The effect of the delay may also be shown in the following way:

$$x(t - 2) = \begin{cases} 2^{-(t-2)}, & (t - 2) \geq 0 \\ 0, & (t - 2) < 0 \end{cases} = \begin{cases} 4 \times 2^{-t}, & t \geq 2 \\ 0, & t < 2 \end{cases}$$

The last presentation expresses the effect of $u(t - 2)$ verbally; that is, by noting that $x(t - 2)$ is zero for $t < 2$ and equal to 4×2^{-t} for $t \geq 2$ and it appears to be more convenient and intuitive.

Time Reversal

Change t to $-t$ in $x(t)$ and you will have a time reversal (also called a time inversion). For example, reversing time in the causal exponential function $x(t) = 2^{-t}u(t)$ will produce

$$x(-t) = 2^{t}u(-t) = \begin{cases} 2^{t}, & t \le 0 \\ 0, & t > 0 \end{cases}$$

Time reversal flips the plot of $x(t)$ around the origin producing its mirror image with respect to the ordinate (vertical axis). Obviously, when the variable t represents actual time, its reversal is not realizable in real time but only *off-line* and on the recorded data (e.g., by reversing the motion of a tape recorder and running it backward) or by constructing a mathematical function that models the time-reversed signal.

***E**xample*
1.22

The effect of time reversal on a sawtooth pulse is shown in Figure 1.32.

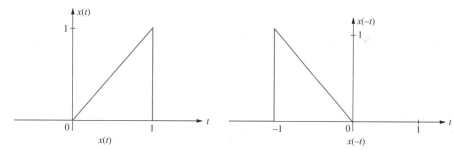

FIGURE 1.32 Time reversal of a sawtooth pulse.

***E**xample*
1.23

The effect of time reversal on a delayed unit-step function is given below.

$$x(t) = u(t - 1) = \begin{cases} 1, & t \ge 1 \\ 0, & t < 1, \end{cases} \qquad x(-t) = u(-t - 1) = \begin{cases} 1, & t \le -1 \\ 0, & t > -1 \end{cases}$$

***E**xample*
1.24

The effect of time reversal on three exponential functions is given below.

a. $x(t) = e^{-2t}u(t)$ $\qquad\qquad$ $x(-t) = e^{2t}u(-t)$
b. $x(t) = e^{-2t}u(t - 2)$ $\qquad\qquad$ $x(-t) = e^{2t}u(-t - 2)$
c. $x(t) = e^{-2t}u(-t + 1)$ $\qquad\qquad$ $x(-t) = e^{2t}u(t + 1)$

Combining Time Shift and Reversal

Mathematically, in the equation that describes $x(t)$ change t to $-(t - t_0)$ and you have combined time reversal and a time shift of t_0 (for both positive and negative values of t_0).[10] When using signal plots, note that the time-reversal operation is pivoted around $t = 0$.

The effect of time reversal on a sawtooth pulse followed by unit shifts to the right and left is shown in Figure 1.33. The figure on the top is the original pulse $x(t)$. The second row shows its reversal $x(-t)$. The reversal of $x(t)$ followed by 1-unit shift to the right is shown in the third row. The fourth row is $x(-t + 1)$, the reversal of $x(t)$ followed by a 1-unit shift to the right.

(a) $x(t) = \begin{cases} t, & 0 \le t < 1 \\ 0, & \text{elsewhere} \end{cases}$

(b) $x(-t) = \begin{cases} -t, & -1 < t \le 0 \\ 0, & \text{elsewhere} \end{cases}$

(c) $x(-t + 1) = \begin{cases} -t + 1, & 0 < t \le 1 \\ 0, & \text{elsewhere} \end{cases}$

(d) $x(-t - 1) = \begin{cases} -t - 1, & -2 < t \le -1 \\ 0, & \text{elsewhere} \end{cases}$

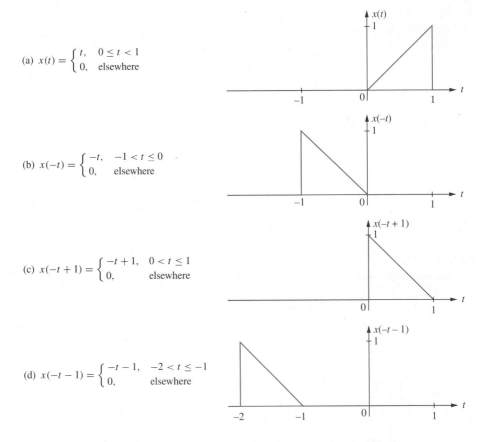

FIGURE 1.33 Combination of time reversal and unit shifts on a sawtooth pulse.

[10]Time reversal in conjunction with time shift is used in superposition of responses due to several inputs arriving at various times. The method is called *convolution* and will be discussed in detail in Chapter 4.

Example

1.26

This example examines the order of operations on a unit-step function when combining time reversal and shift.

a. Time reversal followed by time shift:

unit-step function:	$u(t)$
time reversal:	$u(-t)$
time reversal followed by unit shift to left:	$u(-t-1)$
time reversal followed by unit shift to right:	$u(-t+1)$

b. Time shift followed by time reversal:

unit shift to right:	$u(t-1)$
unit shift to right followed by time reversal:	$u(-t-1)$
unit shift to left:	$u(t+1)$
unit shift to left followed by time reversal:	$u(-t+1)$

The validity of the above results may be checked by applying the definition of the unit-step function or by applying the graphical effects of time shift and reversal. Note that changing the order of *time shift* and *time reversal* would lead to the same result only if at the same time *delay* and *advance* are also interchanged. As an example, $u(-t-1)$ may be obtained by a delay/reverse sequence; first introduce a 1-unit delay in $u(t)$ (by changing t to $t-1$) to produce $u(t-1)$, then reverse $u(t-1)$ (by changing t to $-t$) to obtain $u(-t-1)$. Conversely, $u(-t-1)$ may be obtained by a reverse/advance sequence; first reverse $u(t)$ to obtain $u(-t)$, then introduce a 1-unit advance in $u(-t)$ (by changing t to $t+1$) to produce $u(-(t+1)) = u(-t-1)$.

Example
1.27

The results of time reversal and unit shifts (to the left or right) on two exponential functions $x(t)$ and $y(t)$ are given below. In this example $y(t) = x(-t)$. Note that $x(-t+1) = y(t-1)$ and $x(-t-1) = y(t+1)$. This indicates, as in Example 1.26, that changing the order of *time shift* and *time reversal* would lead to the same result only if, at the same time, *delay* and *advance* are also interchanged.

	function	$\Longrightarrow$ unit shift to left	$\Longrightarrow$ time reversal
(a)	$x(t) = e^{-t}u(t)$	$\Longrightarrow x(t+1) = e^{-(t+1)}u(t+1)$	$\Longrightarrow x(-t+1) = e^{(t-1)}u(-t+1)$
(b)	$y(t) = e^{t}u(-t)$	$\Longrightarrow y(t+1) = e^{(t+1)}u(-t-1)$	
(c)	$x(t) = e^{-t}u(t)$	$\Longrightarrow x(t-1) = e^{-(t-1)}u(t-1)$	$\Longrightarrow x(-t-1) = e^{(t+1)}u(-t-1)$
(d)	$y(t) = e^{t}u(-t)$	$\Longrightarrow y(t-1) = e^{(t-1)}u(-t+1)$	

Time Scaling

Change t to at in $x(t)$, where a is a constant called the scale factor, and you will have time scaling. For example, a causal exponential function and its time-scaled (by a factor of 2) version are

$$x(t) = 2^{-t}u(t)$$

$$x(2t) = 2^{-2t}u(2t) = 4^{-t}u(t) = \begin{cases} 4^{-t}, & t \geq 0 \\ 0, & t < 0 \end{cases}$$

Scaling time by a factor greater than 1 compresses the time axis in the plot of $x(t)$. Conversely, a factor smaller than 1 stretches the time axis.

E*xample*

1.28

A sawtooth pulse $x(t)$ with a base of 1 second starts at $t = 1$. The pulse $x(2t)$ is a sawtooth with a base of 0.5 second and starts at $t = 0.5$. See Figure 1.34.

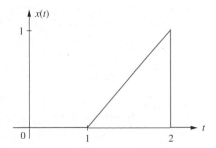 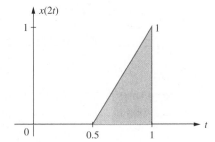

FIGURE 1.34 Time compression of a sawtooth pulse.

When the variable t represents actual time, its scaling is not realizable in real time but only *off-line*, on the recorded data (e.g., by running a tape recorder at a faster or slower speed), or through the mathematical functions that model the signal. Time transformation is used in multirate signal processing, decimation, and interpolation, all of which require storage of the signal.

General Form of Time Transformation

A time transformation that changes t to $at + b$, with a and b positive or negative numbers, can represent a combination of time shift, reversal, and scaling.

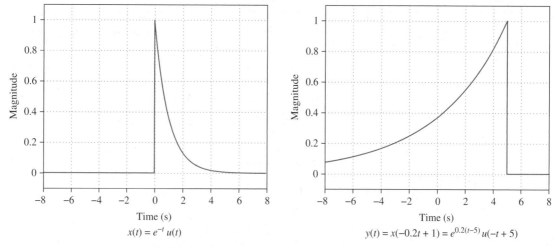

$$x(t) = e^{-t} u(t)$$

$$y(t) = x(-0.2t + 1) = e^{0.2(t-5)} u(-t + 5)$$

FIGURE 1.35 Time transformation from t to $1 - 0.2t$.

a. Find $y(t) = x(-0.2t + 1)$ from $x(t) = e^{-t}u(t)$.

b. Plot $x(t)$ and $y(t)$. Describe how to obtain the plot of $y(t)$ from the plot of $x(t)$.

Solution

a. $y(t) = e^{-(-0.2t+1)}u(-0.2t + 1) = e^{0.2(t-5)}u(-t + 5)$.

b. See Figure 1.35. The plot of $y(t)$ is produced by reversing the plot of $x(t)$, shifting it by 5 units to the right, and increasing its time constant by a factor of 5.

Summary of Time Transformation

The general form of time transformation means changing t to $at + b$ where a and b are positive or negative constants. In this section we have considered the following special cases individually:

> Time shift: $a = 1, b \neq 0$
>
> Time reversal: $a = -1, b = 0$
>
> Combined time shift and reversal: $a = -1, b \neq 0$
>
> Time scaling: $0 < a \neq 1, b = 0$

For implementation, we replace t by $at + b$ in the expression that describes a function. The expression may be verbal, analytical, or a computer program.

1.18 Even and Odd Functions

A function $f(t)$ is even if $f(t) = f(-t)$ for all t. An example of such a function is $\cos(\omega t)$. A function $f(t)$ is odd if $f(t) = -f(-t)$ for all t. An example of such a function is $\sin(\omega t)$. Any function $f(t)$ may be represented by the sum of two components such that $f(t) = f_e(t) + f_o(t)$, where $f_e(t)$ is an even function, called the even part of $f(t)$, and $f_o(t)$ is an odd function, called the odd part of $f(t)$. For example,

$$f(t) = \cos(\omega t - \theta) = f_e(t) + f_o(t), \quad \text{where}$$
$$f_e(t) = \cos\theta\,\cos(\omega t)$$
$$f_o(t) = \sin\theta\,\sin(\omega t)$$

Another example is

$$f(t) = \sin(\omega t - \theta) = f_e(t) + f_o(t), \quad \text{where}$$
$$f_e(t) = -\sin\theta\,\cos\omega t$$
$$f_o(t) = \cos\theta\,\sin\omega t$$

The even and odd parts of a function are uniquely determined from the following:

$$f_e(t) = \frac{f(t) + f(-t)}{2}$$
$$f_o(t) = \frac{f(t) - f(-t)}{2}$$

*E*xample
1.30

The even and odd parts of $f(t) = 2u(t)$ are

$$f_e(t) = \frac{f(t) + f(-t)}{2} = \frac{2u(t) + 2u(-t)}{2} = 1$$
$$f_o(t) = \frac{f(t) - f(-t)}{2} = \frac{2u(t) - 2u(-t)}{2} \triangleq \mathrm{sgn(t)}$$

where sgn(t) (pronounced as *signum*, standing for *sign of*) is defined by

$$\mathrm{sgn(t)} = \begin{cases} 1, & \text{for } t > 0 \\ -1, & \text{for } t < 0 \end{cases}$$

The unit-step can then be written as

$$u(t) = \frac{1}{2}(1 + \mathrm{sgn(t)})$$

*E*xample
1.31

Find the even and odd parts of the square pulse:

$$f(t) = \begin{cases} 2, & \text{for } 0 < t < 1 \\ 0, & \text{elsewhere} \end{cases}$$

Solution

Apply the rule to find

$$f_e(t) = \begin{cases} 1, & \text{for } -1 < t < 1 \\ 0, & \text{elsewhere} \end{cases}$$

$$f_o(t) = \begin{cases} -1, & \text{for } -1 < t < 0 \\ 1, & \text{for } 0 < t < 1 \\ 0, & \text{elsewhere} \end{cases}$$

Example 1.32

Find the even and odd parts of a sawtooth pulse:

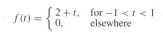

$$f(t) = \begin{cases} 2+t, & \text{for } -1 < t < 1 \\ 0, & \text{elsewhere} \end{cases}$$

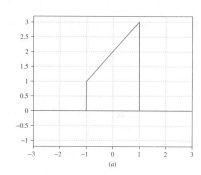

Solution

$$f_e(t) = \begin{cases} \frac{(2+t)+(2-t)}{2} = 2, & \text{for } -1 < t < 1 \\ 0, & \text{elsewhere} \end{cases}$$

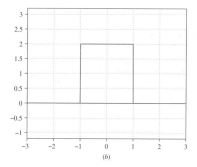

$$f_o(t) = \begin{cases} \frac{(2+t)-(2-t)}{2} = t, & \text{for } -1 < t < 1 \\ 0, & \text{elsewhere} \end{cases}$$

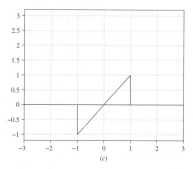

FIGURE 1.36 *(a)* A sawtooth pulse, *(b)* even part, *(c)* odd part.

Even and Odd Parts of Causal Signals

A function $f(t)$ is causal if $f(t) = 0$, $t < 0$. The even and odd parts of a causal function are related to each other and a causal function may be obtained from its even part or from its odd part alone. Any one of the three functions can specify the others by the equations given below:

$$f(t) = 2f_e(t)u(t) = 2f_o(t)u(t)$$

Example 1.33

Show that a causal function $f(t)$ may be obtained from its even part or from its odd part alone by $f(t) = 2f_e(t)u(t) = 2f_o(t)u(t)$.

Solution

$$f_e(t) = \frac{f(t) + f(-t)}{2}$$

$$\implies \begin{cases} \text{for } t < 0 \quad f(t) = 0, & \text{therefore, } f_e(t) = \tfrac{1}{2}f(-t), \quad t < 0 \\ \text{for } t > 0 \quad f(-t) = 0, & \text{therefore, } f_e(t) = \tfrac{1}{2}f(t), \qquad t > 0 \end{cases}$$

$$f_o(t) = \frac{f(t) - f(-t)}{2}$$

$$\implies \begin{cases} \text{for } t < 0 \quad f(t) = 0, & \text{therefore, } f_o(t) = -\tfrac{1}{2}f(-t), \; t < 0 \\ \text{for } t > 0 \quad f(-t) = 0, & \text{therefore, } f_o(t) = \tfrac{1}{2}f(t), \qquad t > 0 \end{cases}$$

In summary,

$$f_e(t) = \begin{cases} \tfrac{1}{2}f(-t), & \text{for } t < 0 \\ \tfrac{1}{2}f(t), & \text{for } t > 0 \end{cases}$$

$$f_o(t) = \begin{cases} -\tfrac{1}{2}f(-t), & \text{for } t < 0 \\ \tfrac{1}{2}f(t), & \text{for } t > 0 \end{cases}$$

Note that

$$f_e(t) = -f_o(t) = \frac{1}{2}f(-t), \text{ for } t < 0$$

$$f_e(t) = f_o(t) = \frac{1}{2}f(t), \text{ for } t > 0$$

Checking the results,

$$f_e(t) + f_o(t) = \begin{cases} 0, & \text{for } t < 0 \\ 2f_e(t) = 2f_o(t) = f(t), & \text{for } t > 0 \end{cases}$$

In conclusion, for a causal $f(t)$

$$f(t) = 2f_e(t)u(t) = 2f_o(t)u(t)$$

$$f_e(t) = \frac{f(t) + f(-t)}{2} = f_o(t)\text{sgn}(t)$$

$$f_o(t) = \frac{f(t) - f(-t)}{2} = f_e(t)\text{sgn}(t)$$

Simply put, in the case of a causal function

$$\begin{cases} \text{for } t > 0, & f_e(t) = f_o(t) = \frac{1}{2}f(t) \\ \text{for } t < 0, & f_e(t) = -f_o(t) \text{ and } f(t) = 0 \end{cases}$$

The even and odd parts of a causal function $f(t)$ are equal for $t > 0$; To find them, scale $f(t)$ down by a factor of 2. For $t < 0$, flip them around the origin (and also vertically for the odd part).

Example
1.34

Find the even and odd parts of $f(t) = e^{-\alpha t}u(t)$ and verify the statement in Example 1.33.

Solution

$$f_e(t) = \frac{e^{-\alpha t}u(t) + e^{\alpha t}u(-t)}{2} = 0.5e^{-\alpha|t|}, \qquad\qquad 2f_e(t)u(t) = e^{-\alpha t}u(t) = f(t)$$

$$f_o(t) = \frac{e^{-\alpha t}u(t) - e^{\alpha t}u(-t)}{2} = 0.5e^{-\alpha|t|}\text{sgn}(t), \qquad 2f_o(t)u(t) = e^{-\alpha t}u(t) = f(t)$$

1.19   Integration and Smoothing

The integral of a function $x(t)$ over a time period from t_1 to t_2 is given by

$$\int_{t_1}^{t_2} x(t)dt$$

The value of the integral is the area under the plot of x as the dependent variable versus t as the independent variable. For a discrete-time function $x(n)$, integration over the range $n = k_1$ to k_2 becomes summation of its element values within that range.

$$\sum_{n=k_1}^{k_2} x(n)$$

In either case integration removes discontinuities in the signal, reduces the variations, and results in smoothing.

Example
1.35

An example of the smoothing property of integration is shown in Figure 1.37. The function

$$x(t) = \begin{cases} 1, & 0 \le t < 0.5 \\ -1, & 2.5 \le t < 3, \\ 0, & \text{elsewhere} \end{cases}$$

which is shown in Figure 1.37(a), is discontinuous at four locations $t = 0,\ 0.5,\ 2.5,$ and 3. For example, at $t = 0$ the function experiences a unit jump: $x(0^+) - x(0^-) = 1$. Its integral $y(t) = \int_{-\infty}^{t} x(t)dt$, shown in Figure 1.37(b), becomes a continuous function. The discontinuities in $x(t)$ manifest themselves as discontinuities in the derivative of $y(t)$. At the locations of former discontinuities the derivative of $y(t)$ will depend on the direction from which one approaches the location. When approaching $t = 0$ from the left direction, the derivative of $y(t)$ is zero while from the right direction the derivative is 1. Integrating once more, $z(t) = \int_{-\infty}^{t} y(t)dt$, removes the discontinuities in the derivative of $y(t)$ [see Figure 1.37(c)].

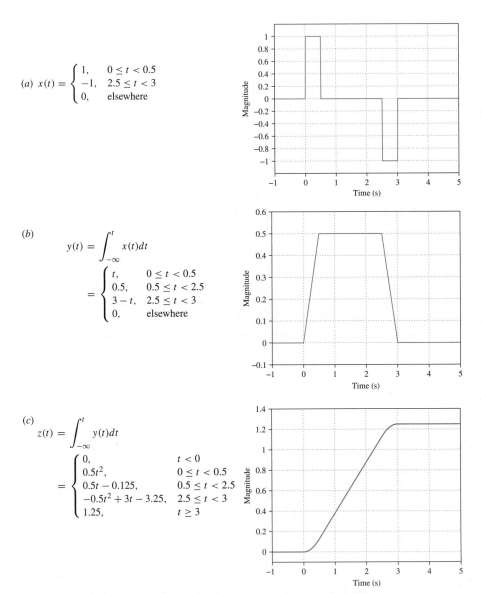

(a) $x(t) = \begin{cases} 1, & 0 \le t < 0.5 \\ -1, & 2.5 \le t < 3 \\ 0, & \text{elsewhere} \end{cases}$

(b)
$$y(t) = \int_{-\infty}^{t} x(t)dt$$

$$= \begin{cases} t, & 0 \le t < 0.5 \\ 0.5, & 0.5 \le t < 2.5 \\ 3 - t, & 2.5 \le t < 3 \\ 0, & \text{elsewhere} \end{cases}$$

(c)
$$z(t) = \int_{-\infty}^{t} y(t)dt$$

$$= \begin{cases} 0, & t < 0 \\ 0.5t^2, & 0 \le t < 0.5 \\ 0.5t - 0.125, & 0.5 \le t < 2.5 \\ -0.5t^2 + 3t - 3.25, & 2.5 \le t < 3 \\ 1.25, & t \ge 3 \end{cases}$$

FIGURE 1.37 Smoothing effect of integration.

Another illustrative example of the smoothing property of integration is its effect on high-frequency components of a signal. Take the integral of a cosine function at the frequency ω:

$$\int_{-\infty}^{t} \cos(\omega t)dt = \frac{1}{\omega}\sin(\omega t)$$

After integration, the amplitude of the sinusoid is reduced by the factor ω, reducing its power by ω^2. Integrating random functions and noise also reduces their variance and thus their power.

Practical Integrators

In practice real-time integration of signals is made by electrical, pneumatic, or hydraulic instruments. The following three examples discuss performance of three analog circuits commonly used for integration.

An ideal integrator circuit

The op amp circuit of Figure 1.38(*a*) is an ideal integrator (considering the op amp to be an ideal one). The input to the circuit is the voltage v_1 and the output is v_2, both functions of t. To find the input–output relationship, we first note that the ideal op amp draws zero current at the input terminals (nodes A and B) and when operating as an element of a linear circuit it brings the inverting and noninverting terminals to the same voltage level [$v_A = v_B = 0$ in Figure 1.38(*a*)]. The input voltage generates the resistor current $i = v_1/R$, all of which passes through the capacitor. The capacitor integrates the current and produces the output voltage

$$v_2(t) = -\frac{1}{C}\int_{-\infty}^{t} i(\tau)d\tau = -\frac{1}{RC}\int_{-\infty}^{t} v_1(\tau)d\tau$$

The factor $1/(RC)$ is the gain of the integrator circuit and is dimensionless. The negative sign in front of the above integral indicates that the circuit inverts the integration results. The output voltage is the integral of the input multiplied by the number $-1/(RC)$, where R is in ohms and C is in farads. For example, with $R = 1$ MΩ and $C = 1$ μF, the magnitude gain of the integrator circuit of Figure 1.38(*a*) is $1/(RC) = 1/(10^6 \times 10^{-6}) = 1$. The input voltage $v_1(t)$ produces an output voltage $v_2(t) = -\int_{\infty}^{t} v_1(t)$. With $v_1(t) = e^{st}$ and an integrator circuit with a unity magnitude gain, the output is $v_2(t) = -\frac{1}{s}v_1(t)$. The gain of the exponential signal passing through the integration is $-1/s$, which diminishes with increasing s. In the sinusoidal case of $v_1(t) = \cos \omega t$, the input and output voltages may be written as $v_1(t) = \mathcal{RE}\{e^{j\omega t}\}$ and $v_2(t) = \mathcal{RE}\{\frac{j}{\omega}e^{j\omega t}\}$, respectively. The gain of the integration operation is the complex number $\frac{j}{\omega}$, which means a magnitude change of $\frac{1}{\omega}$ and a $90°$ phase advance.

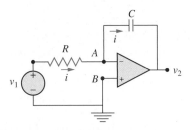

FIGURE 1.38 *(a)* An ideal integrator circuit. The op amp makes the inverting input a virtual ground ($v_A = 0$), thus the resistor current becomes $i = v/R$. All of the current goes through the capacitor (the op amp does't draw any current), establishing an output voltage proportional to the integral of the input voltage, with a negative sign.

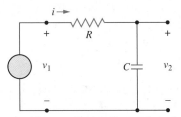

FIGURE 1.38 *(b)* An *RC* integration circuit. The capacitor voltage in a series *RC* circuit is the integral of the current, not the input voltage. It can be shown that in this circuit $v_2(t)$ is the integral of the product of the input voltage and an exponential weighting function, see Example 1.38. In addition, $v_2(t)$ is also modified when a load is connected to the output of the circuit.

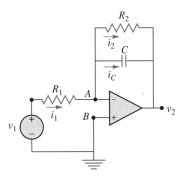

FIGURE 1.38 *(c)* A leaky integrator circuit using an ideal op amp. The circuit operates similar to the ideal integrator of Figure 1.38*(a)*, except that the resistor in parallel with the capacitor provides a path for the capacitor charges to leak through it, thereby reducing the output voltage v_2. The operation can also be qualitatively analyzed in terms of the effect of the frequency of the input signal; From an *AC* point of view, the input current $i_1 = v_1/R_1$ gets divided between i_2 and i_C, with the current in each element being inversely proportional to the impedance of that element. At higher frequencies, the capacitor has a lower impedance, thus absorbing a bigger share of the input current. This leaves less current to pass through R_2 and reduces v_2. The reduction in $v_2(t)$ accentuates as the frequency goes up.

E*xample*

1.38

RC integrator circuit

The passive *RC* circuit of Figure 1.38(*b*) [with $v_1(t)$ as the input and $v_2(t)$ as the output] implements an integration with an exponential weighting factor. In that figure, the input–output differential equation and its solution are

$$\frac{d}{dt}v_2(t) + \alpha v_2(t) = \alpha v_1(t), \quad \text{where } \alpha = \frac{1}{RC}$$

$$v_2(t) = \alpha \int_{-\infty}^{t} e^{-\alpha(t-\tau)} v_1(\tau) d\tau$$

In the above, $v_1(\tau)$ is weighted by $\alpha e^{-\alpha(t-\tau)}$, where $(t - \tau)$ is the distance from the past. The above integral is called a *convolution integral*.

E*xample*

1.39

Leaky integrator in the time domain

The integration equations obtained for the *RC* integrator of Example 1.38, however, may change when another device is connected to the output. The circuit of Figure 1.38(*c*) contains an op amp that overcomes this limitation. To find $v_2(t)$ as a weighted integral of $v_1(t)$, we first obtain the differential equation relating $v_2(t)$ to $v_1(t)$. From that circuit we have:

$$i_1(t) = \frac{v_1(t)}{R_1}, \quad i_2(t) = -\frac{v_2(t)}{R_2}, \quad i_C(t) = -C\frac{dv_2}{dt}$$

$$i_C(t) + i_2(t) = i_1(t)$$

$$\frac{d}{dt}v_2(t) + \frac{1}{R_2 C}v_2(t) = -\frac{1}{R_1 C}v_1(t)$$

The time-domain solution to the above differential equation is

$$v_2(t) = -\frac{1}{R_1 C}\int_{-\infty}^{t} e^{-\frac{(t-\tau)}{R_2 C}} v_1(\tau) d\tau$$

As in Example 1.38 the above is also a convolution integral. $v_1(\tau)$ is weighted by $\frac{1}{R_1 C} e^{-\frac{(t-\tau)}{R_2 C}}$, where $(t - \tau)$ is the distance from the past.

E*xample*

1.40

Leaky integrator in the *s*-domain

For a direct quantitative analysis of the operation of the leaky integrator in the *s*-domain, let $v_1(t) = V_1 e^{st}$. The input current $i_1 = v_1/R_1 = (V_1/R_1)e^{st} \equiv I_1 e^{st}$ gets divided between R_2 and C as $i_2(t) = I_2 e^{st}$ and $i_C(t) = I_C e^{st}$. [Note that $i_2(t)$ and $i_C(t)$ are necessarily exponentials.]

Kirchhoff's current law requires $I_2 e^{st} + I_C e^{st} = I_1 e^{st}$ or $I_2 + I_C = I_1$.

Kirchhoff's voltage law requires $R_2 I_2 e^{st} = \frac{I_C}{Cs} e^{st}$ or $R_2 I_2 = \frac{I_C}{Cs}$.

From the above set of equations we find

$$I_2 = \frac{I_1}{1 + R_2Cs}, \quad i_2(t) = \frac{I_1}{1 + R_2Cs}e^{st}, \quad \text{and } v_2(t) = -R_2i_2(t) = -\frac{R_2}{R_1}\frac{V_1}{1 + R_2Cs}e^{st}$$

With increasing s the magnitude of the output voltage gets reduced.

Time-Domain Solution Versus s-Domain

We have analyzed the operation of the leaky integrator by two methods, to be called the time- and s-domain methods. In future chapters the relationship between these two solutions will become clear. Here we verify that the s-domain solution obtained for the leaky integrator is in agreement with the solution obtained in the time domain. For this, after substituting for $v_1(t) = V_1e^{st}$, we evaluate the convolution integral for the time-domain solution. The result is the same as the solution obtained by direct application of $v_1(t) = V_1e^{st}$ in the s-domain.

$$v_1(t) = V_1e^{st}$$

$$v_2(t) = -\frac{1}{R_1C}\int_{-\infty}^{t} e^{-\frac{(t-\tau)}{R_2C}}v_1(\tau)d\tau$$

$$= -\frac{V_1}{R_1C}e^{-\frac{t}{R_2C}}\int_{-\infty}^{t} e^{\frac{\tau}{R_2C}}e^{s\tau}d\tau$$

$$= -\frac{V_1}{R_1C}e^{-\frac{t}{R_2C}}\int_{-\infty}^{t} e^{(s+\frac{1}{R_2C})\tau}d\tau$$

$$= -\frac{V_1}{R_1C}\frac{1}{\left(s + \frac{1}{R_2C}\right)}e^{-\frac{t}{R_2C}}e^{(s+\frac{1}{R_2C})t}$$

$$= -\frac{R_2}{R_1}\frac{V_1}{1 + R_2Cs}e^{st} \quad \text{(Q.E.D)}$$

*E*xample 1.41

Summary of responses of integrators to impulse, step, and pulse functions

The response of a unity-gain ideal integrator to a unit-impulse input is a unit-step function and its response to a unit-step is a unit ramp. A rectangular pulse input creates a steplike function that reaches its DC steady-state level linearly within the time duration of the input pulse. A unit-step input results in a unit ramp that grows linearly with time. The above sets of inputs are expressed mathematically in column 1 of Figure 1.39 with their plots shown in column 2. Column 3 shows the corresponding response of a unity-gain ideal integrator and column 4 shows the response of a leaky integrator with a time constant of 200 μsec. The leaky integrator doesn't sustain a DC level in its response to the impulse. Its response to a step input becomes an exponential with a finite final value rather than a ramp. This becomes a desired feature when accumulation of a small DC level would threaten to saturate the op amp in the circuit.

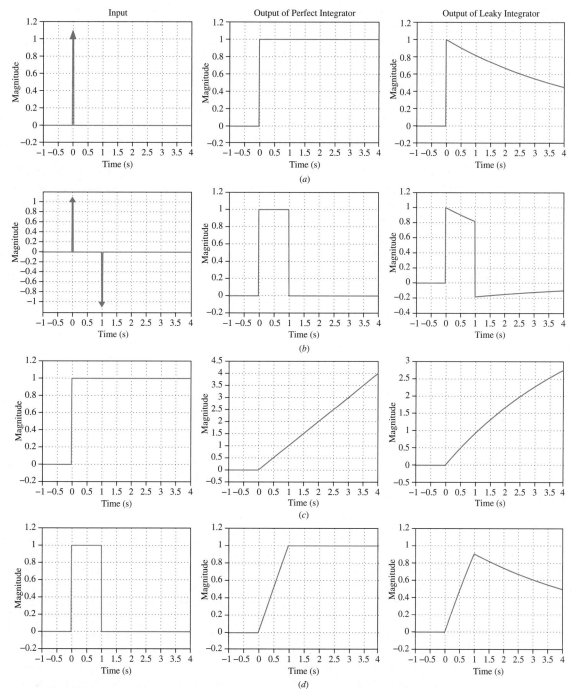

FIGURE 1.39 Responses of perfect and leaky integrators to impulse, step, and pulse voltage inputs. The inputs are *(a)* $\delta(t)$, *(b)* $\delta(t) - \delta(t-1)$, *(c)* $u(t)$, *(d)* $u(t) - u(t-1)$.

1.20 **Weighted Integration**

In integrating a function $x(\tau)$ from $\tau = -\infty$ to t, one sums up the past values of x with equal weight. In some applications, contributions from $x(\tau)$ to the integral may be weighted differently (for example, the recent past may be more valuable than the remote past). In such cases under the integral $x(\tau)$ is multiplied by a weighting function $h(t, \tau)$ which depends on τ and t.

$$y(t) = \int_{-\infty}^{t} x(\tau)h(t, \tau)d\tau$$

In the above weighted integral, τ represents the input's past time and t is the upper limit of integration (i.e., the present time of the output). In many cases of interest in signals and systems, the weighting factor h is a function of $(t - \tau)$. The RC and leaky integrators of Figure 1.38(b) and (c) are two examples of weighted integration. The following are some more examples.

Weighted integration

Consider an integration operation on $x(\tau)$ from $-\infty < \tau < t$, where the contribution to $y(t)$ from x at time τ diminishes exponentially with its distance from the present time t with a time constant of $(1/\alpha)$ sec. Find $y(t)$ for $x(\tau) = \cos(\pi\tau)$, given that the time constant of the weighting function is i) 0.5 sec, ii) 2 sec, and iii) ∞.

Solution
With a weighting function of $h(t, \tau) = e^{-\alpha(t-\tau)}$ the integral becomes

$$y(t) = \int_{-\infty}^{t} \cos(\pi\tau)e^{-\alpha(t-\tau)}d\tau = e^{-\alpha t}\int_{-\infty}^{t} e^{\alpha\tau}\cos(\pi\tau)d\tau$$

However,

$$\int_{-\infty}^{t} e^{\alpha\tau}\cos(\pi\tau)d\tau = \frac{e^{\alpha t}}{\alpha^2 + \pi^2}[\alpha\cos(\pi t) + \pi\sin(\pi t)], \quad (\alpha \geq 0)$$

Therefore,

$$y(t) = \frac{1}{\alpha^2 + \pi^2}[\alpha\cos(\pi t) + \pi\sin(\pi t)] = \frac{1}{\sqrt{\alpha^2 + \pi^2}}\sin(\pi t + \theta), \text{ where } \tan\theta = \frac{\alpha}{\pi}$$

- **(i)** $\alpha = 2$, $y(t) = 0.26851\sin(\pi t + 32.5°)$
- **(ii)** $\alpha = 0.5$, $y(t) = 0.31435\sin(\pi t + 9°)$
- **(iii)** $\alpha = 0$, $y(t) = 0.31831\sin(\pi t)$

The shorter the time constant of the weighting function, the more pronounced its effect is. Implementation of the above integral is illustrated in Figure 1.40 for the 2-second time constant.

The time axis in Figure 1.40(a–c) represents τ. Figure 1.40(a) shows $x(\tau) = \cos\pi\tau$. Figure 1.40(b) shows $h(t - \tau) = e^{-\frac{(t-\tau)}{2}}$ at $t = 2$, which is the weighting function for

(a) The function

$x = \cos(\pi\tau)$

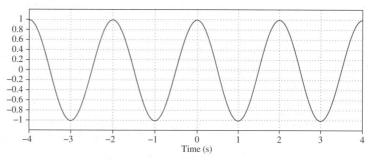

(b) The weight factor

$h(t - \tau) = e^{-\frac{(t-\tau)}{2}}$, at $t = 2$

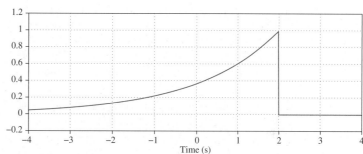

(c) The integrand

$x(\tau)h(t - \tau)$, at $t = 2$

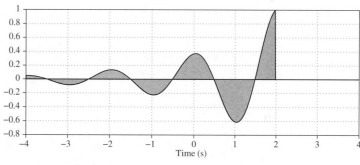

(d) The integral

$$y(t) = \int_{-\infty}^{t} \cos(\pi\tau)e^{-\frac{(t-\tau)}{2}}d\tau$$

$$= 0.31435\sin(\pi t + 9°),$$

$$-\infty < t < \infty$$

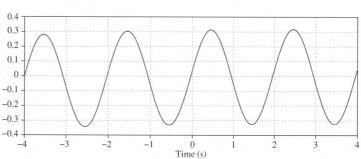

FIGURE 1.40 Integration of a 0.5-Hz cosine function multiplied by an exponential weighting factor with a 2-second time constant. (The operation is identical with convolution of the two functions to be discussed in Chapter 4.)

evaluating the integral at $t = 2$. The product $x(\tau)h(t - \tau)$ is shown in (c) for $t = 2$. The net area under the curve is the value of the integral at $t = 2$. It is seen that the $x(\tau)$s of the past gradually contributes less to the value of the integral due to the attenuating effect of multiplication by h. The decaying weighting function makes the memory of the past gradually fade away. The result of the integral as a function of t is shown in (d).

*E**xample***

1.43

In Example 1.42 we observed the effect of the time constant of the exponential weighting function on the integration outcome. Here we generalize that observation and show that the effect is a function of the ratio of the time constant to the period of the sinusoidal input. Let

$$y(t) = \int_{-\infty}^{t} x(\tau)h(t, \tau)d\tau$$

$$x(\tau) = \cos(2\pi f\tau)$$

$$h(t, \tau) = e^{-\alpha(t-\tau)}$$

a.   Find $y(t)$ and express it in terms of $k = \dfrac{\text{time constant of the weighting function}}{\text{period of the sinusoial input}}$. Describe the effect of k on the integration outcome.

b.   Obtain $y(t)$ for $f = 1$Hz and the following time constants: (i) 0.05 sec, (ii) 0.25 sec, (iii) 0.5 sec, (iv) 1 sec, (v) 2 sec, and (vi) ∞.

Solution

a.

$$y(t) = \int_{-\infty}^{t} x(\tau)h(t, \tau)d\tau = \int_{-\infty}^{t} \cos(2\pi f\tau)e^{-\alpha(t-\tau)}d\tau, \quad \alpha \geq 0$$

$$= e^{-\alpha t} \int_{-\infty}^{t} e^{\alpha\tau} \cos(2\pi f\tau)d\tau$$

$$= e^{-\alpha t} \left\{ \frac{e^{\alpha\tau}}{\alpha^2 + 4\pi^2 f^2} \left[\alpha \cos(2\pi f\tau) + \pi \sin(2\pi f\tau)\right] \right\}_{-\infty}^{t}$$

$$= \frac{1}{\alpha^2 + 4\pi^2 f^2} \left[\alpha \cos(2\pi ft) + 2\pi f \sin(2\pi ft)\right]$$

$$= \frac{1}{\sqrt{\alpha^2 + 4\pi^2 f^2}} \sin(2\pi ft + \theta), \quad \text{where } \theta = \tan^{-1}\left(\frac{\alpha}{2\pi f}\right).$$

In terms of the ratio $k = $ (time constant of the weighting function)/ (period of the sinusoial input) $= f/\alpha$:

$$y(t) = \frac{1}{2\pi f} \frac{k}{\sqrt{k^2 + \frac{1}{4\pi^2}}} \sin(2\pi ft + \theta), \quad \text{where } \theta = \tan^{-1}\left(\frac{1}{2\pi k}\right)$$

The exponentially decaying weighting function attenuates the amplitude and introduces a phase advance in the integration outcome, both of which depend on k. A shorter k has a more pronounced effect on both the amplitude and phase of the outcome. Here are some examples:

(i) Short time constant $\qquad$ $2\pi k \ll 1,$ $\qquad$ $y(t) \approx \frac{k}{f} \sin(2\pi ft + 90^0)$
$$= \frac{1}{\alpha} \sin(2\pi ft + 90^0)$$

(ii) Medium time constant $\qquad$ $2\pi k = 1,$ $\qquad$ $y(t) = \frac{1}{2\sqrt{2}\pi f} \sin(2\pi ft + 45^0)$

(iii) Long time constant $\qquad$ $2\pi k \gg 1,$ $\qquad$ $y(t) \approx \frac{1}{2\pi f} \sin(2\pi ft)$

b. For $f = 1$ Hz the period of the sinusoid is 1 second and $k =$ time constant. The integration outputs for the various time constants are then

(i) time constant $= 0.05$ $\qquad$ $k = 0.05$ $\qquad$ $y(t) = 0.04770 \sin(2\pi t + 72.5^\circ)$
(ii) time constant $= 0.25$ $\qquad$ $k = 0.25$ $\qquad$ $y(t) = 0.13425 \sin(2\pi t + 32.5^\circ)$
(iii) time constant $= 0.5$ $\qquad$ $k = 0.5$ $\qquad$ $y(t) = 0.15166 \sin(2\pi t + 17.6^\circ)$
(iv) time constant $= 1$ $\qquad$ $k = 1$ $\qquad$ $y(t) = 0.15717 \sin(2\pi t + 9^\circ)$
(v) time constant $= 2$ $\qquad$ $k = 2$ $\qquad$ $y(t) = 0.15865 \sin(2\pi t + 4.5^\circ)$
(vi) time constant $= \infty$ $\qquad$ $k = \infty$ $\qquad$ $y(t) = 0.15915 \sin(2\pi t)$

The last line above is simply

$$y(t) = \int_{-\infty}^{t} \cos(2\pi\tau)d\tau = \frac{1}{2\pi} \sin(2\pi t) = 0.15915 \sin(2\pi t)$$

1.21 Window Averaging

A special case of weighted integration is the finite duration integration, also called window averaging, as it takes the average of data seen through a window of T seconds length. We already have seen an example of window averaging in the plots of sunspot numbers from 1700 to 2010, where annual averages shown in Figure 1.1(a) are obtained from monthly and daily averages used in the plots of Figure 1.1(b) to (d). The averaging reduces the daily variations and smooths the data. In the case of sunspot numbers, the data is discrete and the windows are uniform, nonoverlapping, and 365 days long.

In another class of window averaging, the window slides along the time axis and has overlap with its neighbor. For example, the finite integral

$$y(t) = \frac{1}{T} \int_{t-T}^{t} x(\tau)d\tau$$

takes the average of $x(t)$ seen through a uniform window of length T and assigns the result as $y(t)$. In the above integral, the variable t sweeps from $-\infty$ to ∞. The window, therefore, moves along with t and $y(t)$ is called the *moving average*. For a discrete-time function the moving average under a uniform window of length N becomes

$$y(n) = \sum_{k=n-N}^{n} x(k)$$

Example
1.44

Moving averages

Moving averages are commonly used to reduce fluctuations in data and to discern trends. For this purpose, the effects of several features of a window need to be considered. The most important features are the window's width and shape. Some other factors such as magnitude scaling and the DC level in the original signal also affect the windowing outcome. An example is given in Figure 1.41, which shows a deterministic signal to which a random fluctuation is added

$$x(t) = [a\cos(\pi t) + b\cos(2\pi t) + c\tan^{-1}(t+6) + d\tan^{-1}(t-5)] + n(t)$$

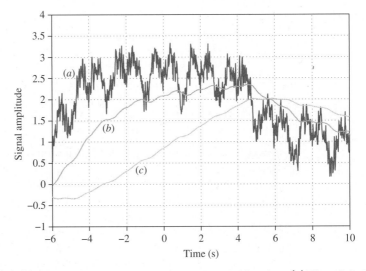

FIGURE 1.41 Moving averages are superimposed on the function. *(a)* The original function $x(t)$. *(b)* Averaged by an exponential window (time constant = 2). *(c)* Averaged by a uniform window. Both windows are 4 seconds wide.

The component $n(t)$ is a filtered Gaussian random variable. The coefficients $a, b, c,$ and d are constants. The signal is then window averaged by two different windows that are both 4 seconds wide. One is a uniform window and the other is an exponential one (with a 2-second time constant). In the latter case, recent signal values are considered more relevant and given more weight by the exponential weighting function. Superimposed on the figures are trends extracted from the data by window averaging. The rapid signal fluctuations due to the sinusoidal and random components are reduced. The slowly varying components have become prominent. The appropriate choice of window parameters is required to extract desired trends in the signal.

1.22 Differentiation and Differences

The derivative of a function $x(t)$ at a given point indicates the rate of change of the dependent variable x with respect to the independent variable t. Mathematically it is defined as

$$\frac{dx}{dt} = \lim_{\Delta t \to 0} \frac{\Delta x}{\Delta t}$$

The derivative of $x(t)$, when it exists, is the slope of its tangent line (tangent of the angle of the line with the horizontal axis). The faster the function changes with time, the bigger the derivative numerically.

Derivative of a steplike function

An example of interest is the derivative of a function that grows linearly with time from zero to one in T seconds. See Figure 1.42.

$$x(t) = \begin{cases} 0, & t < 0 \\ \frac{t}{T}, & 0 < t < T \\ 1, & t > T \end{cases} \quad \text{and} \quad \frac{dx}{dt} = \begin{cases} 0, & t < 0 \\ \frac{1}{T}, & 0 < t < T \\ 0, & t > T \end{cases}$$

The derivative of $x(t)$ is a rectangular pulse (width $= T$, height $= \frac{1}{T}$). Three such functions are shown in Figure 1.42(a) for $T = 1, 2$, and 3 seconds, with their derivatives given in Figure 1.42(b). As T becomes smaller, $x(t)$ approaches the unit step and its derivative approaches the unit impulse. At the limit,

$$\lim_{\Delta t \to 0} x(t) = u(t) \quad \text{and} \quad \lim_{\Delta t \to 0} \frac{dx}{dt} = \delta(t)$$

The unique derivative property of exponential function

The derivative of the exponential function is an exponential function with the same time constant but an amplitude change.

$$x(t) = e^{at}, \quad \frac{dx}{dt} = ae^{at} = ax(t), \quad \frac{dx}{dt} - ax = 0$$

This unique property belongs to exponential functions only. The constant a may be a real or complex number. Of special interest are the derivatives of complex exponential and sinusoidal functions:

$$\frac{d}{dt}e^{j\omega t} = j\omega e^{j\omega t}, \quad \frac{d}{dt}\sin \omega t = \omega \cos \omega t, \quad \frac{d}{dt}\cos \omega t = -\omega \sin \omega t$$

Another example of is the two-sided exponential function

$$x(t) = e^{-|at|}, \quad \frac{dx}{dt} = -ae^{-|at|}\text{sgn}(t), \quad \text{where sgn}(t) = \begin{cases} 1, & t > 0 \\ -1, & t < 0 \end{cases}.$$

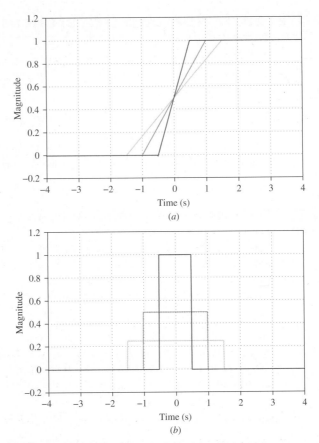

FIGURE 1.42 Three steplike time functions *(a)* and their derivatives *(b)*.

The derivative is made of exponential functions scaled by $\pm a$. At $t = 0$, the sign of the derivative depends on the direction from which the function approaches the origin, resulting in a biphasic character. As a becomes larger, the time constant becomes smaller, $x(t)$ becomes narrower, and its derivative becomes sharper. Accordingly, the derivative operator amplifies the high-frequency components of a signal compared to its low frequencies. It functions as a high-pass filter.

Practical Differentiator Circuits

Three commonly used analog differentiator circuits are described in the next three examples.

Ideal differentiator

The op amp circuit of Figure 1.43(*a*) is a perfect differentiator. By an approach similar to that of the ideal integrator, the input–output relationship in Figure 1.43(*a*) is found to be $v_2 = -RC \frac{dv_1}{dt}$.

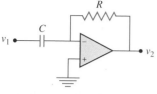

(*a*) Ideal differentiator

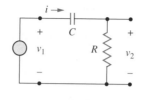

(*b*) *CR* differentiator

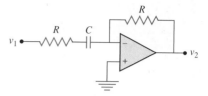

(*c*) Differentiator with high-frequency limiting capacity

FIGURE 1.43 Three differentiation circuits.

Example

Capacitor-resistor (*CR*) differentiator

The *CR* circuit of Figure 1.43(*b*) (with v_1 as the input and v_2 as the output) approximates a differentiator. To analyze the operation of the circuit we note that the output voltage v_2 is proportional to the current through the circuit, while the current is proportional to the derivative of the capacitor voltage. Consequently, v_2 is proportional to the derivative of the capacitor voltage which is $v_1 - v_2$. The input–output differential equation of the circuit is obtained from

$$i = C\frac{d}{dt}(v_1 - v_2)$$

$$v_2 = Ri = RC\frac{d}{dt}(v_1 - v_2).$$

$$\frac{dv_2}{dt} + \alpha v_2 = \frac{dv_1}{dt}, \quad \text{where} \ \alpha = \frac{1}{RC}$$

For a complex exponential input $v_1 = e^{j\omega t}$ we obtain $v_2 = \frac{j\omega}{\alpha + j\omega}e^{j\omega t}$.

The above equation approximates the derivative operation only at low frequencies where $\omega << \alpha$. At high frequencies, where $\omega >> \alpha$, the gain approaches unity. The output of the circuit in response to the exponential input $v_1 = e^{\omega t}$ is given below for three ranges of frequencies:

Low frequencies, $\omega << \alpha$ and $v_2 \approx \frac{j\omega}{\alpha}e^{j\omega t}$. It approximates a differetiator with a gain of *RC*.

Mid-range frequencies, $\omega = \alpha$ and $v_2 = 0.707e^{j(\omega t + \pi/4)}$.

High frequencies, $\omega >> \alpha$ and $v_2 \approx e^{j\omega t}$. It approximates a unity-gain transfer of the input to the output.

The above conclusions may also be derived by a direct examination of the circuit at various frequencies: At low frequencies the capacitor represents a large impedance compared to the resistor and receives the major share of input voltage, $v_C \approx v_1$. The capacitor current and the output voltage become $i = Cdv_C/dt \approx Cdv_1/dt$ and $v_2 = Ri \approx Cdv_1/dt$, respectively. By a similar argument, at high frequencies the capacitor functions almost as a short-circuited element and nearly all of the input voltage gets transferred to the output, $v_2 \approx v_1$.

Differentiator with high-frequency limiting

The relationship between v_2 and v_1 in the op amp circuit of Figure 1.43(c) is found by applying circuit laws and assuming an ideal op amp:

$$\frac{dv_2}{dt} + \alpha v_2 = -\frac{dv_1}{dt}, \quad \text{where} \quad \alpha = \frac{1}{RC}$$

Except for a sign inversion, it is similar to that of the CR circuit of Figure 1.46(b). However, unlike the passive CR circuit of Figure 1.43(b), the performance of the active circuit of Figure 1.43(c) is not affected by loading.

1.23 Concluding Remarks

The goal of this chapter has been to familiarize the student with some signal notations that are used later in the book. Actual signals convey information, are often complex, and are analyzed probabilistically. Examples of such signals were given at the start of this chapter. However, in this book we employ deterministic functions that, despite their simplicity, provide convenient and useful means for the analysis of signals and systems. The utility of these functions is primarily due to our interest in systems being modeled as linear time-invariant (LTI). For example, a great deal can be derived from knowledge of the step response of a system that lends itself to an LTI model. After describing mathematical functions of interest (such as step, impulse, exponential, sinusoids, sinc, pulses, and windows), the chapter introduced several basic operations on signals such as time transformation, integration, and differentiation.

1.24 Problems

Solved Problems

Note. To get started on Matlab or to find out about Matlab commands use the *Help* button on the menu bar of its command window.

1. **Plotting a function.** Digital computer software packages sample continuous-time functions and represent the samples as vectors or matrixes. This problem illustrates the use of basic commands for sampling and plotting functions.

   a. **Basic plot.** The Matlab program given below generates two vectors: one is t (the sequence of time samples) and the other is y [the sequence of the function $y(t) = \sin(2\pi t)$ evaluated at those time samples]. The program then plots y vs. t (with time as the abscissa and y as the ordinate).

   ```
   % CH1_PR1_a.m
   t=0:0.05:1;
   y=sin(2*pi*t);
   plot(t,y)
   ```

   The first line in the above list is a comment; for example, the file name. Comment lines start with the symbol %, which tells the computer to ignore that line. The second line in the above list generates a vector representing time. It determines sampling instances. Starting from $t = 0$ and ending at $t = 1$ second, with samples taken every 0.05 second, it produces a total of 21 points. The third line evaluates the sinusoid at the sampling instances and creates the vector *y* containing 21 samples. The last line plots the vector *y* *versus* the vector *t*. The plot appears as a continuous-time function. Run the above program under Matlab and observe the outcome.

b. **Using stem, stairs, and figure commands.** The following commands create plots as stem or stairs.

```
stem(y);        % plot y as stem.
stairs(t,y);    % plot y as stairs.
```

The command *figure* creates a window for each plot. Explore the effect of including the command *figure* by running the program given below, first with and then without the command *figure*, and then with and without the numbers associated with each figure.

```
t=0:0.05:1;
y=sin(2*pi*t);
figure(1); plot(t,y);
figure(2); stem(t,y);
figure(3); stairs(t,y);
```

c. **Subplot, customizing and saving plots.** The command

```
subplot(a,b,c);
```

places the current plot in an $a \times b$ matrix at a location determined by *c*.

Customizing a plot. The plot may be customized by the following commands:

```
axis([f])       % Specifies range of axes.
xlabel('f')     % Places a label on the abscissa.
ylabel('f')     % Places a label on the ordinate.
title('f')      % Gives the plot a title.
grid            % Puts a grid on the plot.
```

In addition, the color of the plot, its line width, and the shape may be specified.

The following commands save the current plot as a postscript file in the working directory.

```
print -dps filename.ps
print -dpsc filename.eps
```

A plot may also be saved as an *.m file using the menu bar of the plot window. The above features are illustrated by the program below.

```
t=0:0.05:2;
y=sin(2*pi*t);
subplot(3,1,1)
plot(t,y,'r','LineWidth',2 ) % Plots in red color with a line-width=2.
axis([0 2 -1.5 1.5]);        % The range of abscissa is from 0 to 2
                             % that of ordinate is from -1.5 to 1.5.
grid
title('Subplot, stem, stairs, customize and save.');
xlabel('time (s)'); ylabel('volts');
subplot(3,1,2)
stem(t,y,'b');               % plots in blue color
axis([0 2 -1.5 1.5]);
grid
xlabel('time (s)'); ylabel('volts')
subplot(3,1,3)
stairs(t,y,'m');             % plots in magenta color
axis([0 2 -1.5 1.5]);
grid
xlabel('time (s)'); ylabel('volts')
print -dps CH1_PR1_c.ps
```

d. **Multiple plots on the same graph.** A single command may plot more than one pair of vectors in one window. The following program plots in a single window three instantaneous voltages of a 3-phase 120-V, 60-Hz system:

$$v_1(t) = 170\sin(2\pi ft), \quad v_2(t) = 170\sin(2\pi ft - 120°), \quad v_3(t) = 170\sin(2\pi ft - 240°)$$

The plots are then saved as a postscript file.

```
f=60;                                % Sets the frequency at 60 Hz.
t=0:.0001:.05;                       % Time vector, t=0 to t=0.05 every 0.0001 second.
v1=170*sin(2*pi*f*t);                % Generates the vector of voltage values, v1.
v2=170*sin(2*pi*f*t-2*pi/3);         % Generates the vector of voltage values, v2.
v3=170*sin(2*pi*f*t-4*pi/3);         % Generates the vector of voltage values, v2.
plot(t,v1,'r',t,v2,'b',t,v3,'m')     % Plots v1, v2, and v3 versus t (in red, blue, and magenta).
xlabel('time (s)'); ylabel('volts');
title('Phase voltages in a 3-phase 120 V 60-Hz system.');
grid                                 % Places a grid on the plot.
print -dps CH1_PR1_d.ps              % Saves the plot under the name CH1_PR1_d.ps
```

Alternatively, one may apply the command *hold on* which holds the current window for adding more plots. For example, in the above program replace the single *plot* command by the following set of commands which holds the plot to includes three separate commands with individually controlled line widths and colors.

```
plot(t,v1,'r','LineWidth',2)    % Plots v1 versus t in red color with 2-units thickness.
hold on                         % Holds the current graph for more plots
plot(t,v2,'b','LineWidth',2)    % Plots v2 versus t in blue color with 2-units thickness.
plot(t,v3,'m','LineWidth',2)    % Plots v3 versus t in magneta color with 2-units thickness.
hold off
```

2. **Impulse and step functions.** Consider a finite-duration pulse $x_a(t)$ made of a single cycle of a sinusoid:

$$x_a(t) = \begin{cases} \frac{a^2}{2\pi}\sin(at), & 0 < t < \frac{2\pi}{a} \\ 0, & \text{elsewhere} \end{cases}$$

a. Sketch $x_a(t)$ for $a = 2n\pi$, $n = 1, 3, 5$. Find the area under each half cycle of $x(t)$.

b. Let $y_a(t) = \int_{-\infty}^{t} x_a(t)dt$. Find $y_a(t)$ and sketch it for $a = 2n\pi$, $n = 1, 3, 5$. Show that $y_a(t)$ is a finite-duration positive pulse. Find its duration, maximum value, and area.

c. Let $z_a(t) = \int_{-\infty}^{t} y_a(t)dt$. Find $z_a(t)$ and sketch it for $a = 2n\pi$, $n = 1, 3, 5$. Find its DC steady-state value and the transition time to reach the steady-state value.

d. Argue that

$$\text{(i)} \quad \lim_{a\to\infty} y_a(t) = \delta(t)$$

$$\text{(ii)} \quad \lim_{a\to\infty} z_a(t) = u(t)$$

Solution

a. $x_a(t) = \begin{cases} \frac{a^2}{2\pi}\sin(at), & 0 < t < \frac{2\pi}{a} \\ 0, & \text{elsewhere.} \end{cases}$ For $a = 2\pi n$ we have $x_a(t) = \begin{cases} 2\pi n^2 \sin(2\pi nt), & 0 < t < 1/n \\ 0, & \text{elsewhere} \end{cases}$

See Figure 1.44(a). The area under a half cycle of $x_a(t)$ is $\int_{-\infty}^{1/(2n)} 2\pi n^2 \sin(2\pi nt)dt = 2n$.

b. $y_a(t) = \int_{-\infty}^{t} x_a(t)dt = \int_{-\infty}^{t} 2\pi n^2 \sin(2\pi nt)\,[u(t) - u(t - 1/n)]\,dt = \begin{cases} n\,[1 - \cos(2\pi nt)], & 0 < t < 1/n \\ 0, & \text{elsewhere.} \end{cases}$

$y_a(t)$ is a positive finite-duration pulse. Its duration is $1/n$. Its maximum value is $2n$ which occurs at $t = 1/(2n)$. See Figure 1.44(b). The area under the pulse is

$$\int_0^{1/n} y_a(t)dt = \int_0^{1/n} n\,[1 - \cos(2\pi nt)]\,dt = 1$$

c. $$z_a(t) = \int_{-\infty}^{t} y_a(t)dt = \int_{-\infty}^{t} n[1 - \cos(2\pi nt)][u(t) - u(t - 1/n)] = \begin{cases} 0, & t < 0 \\ n(t - \frac{\sin(2\pi nt)}{2\pi n}), & 0 < t < 1/n \\ 1, & t > 1/n \end{cases}$$

The DC steady-state value is 1. The transition time from zero to 1 is $1/n$. See Figure 1.26(c).

d. As $a = 2\pi n \to \infty$, the pulse $y_a(t)$ becomes narrower and taller, but its area remains equal to 1. Similarly, $z_a(t)$ reaches the DC steady state in a shorter time. At the limit, therefore,

(i) $\lim_{a \to \infty} y_a(t) = \delta(t)$

(ii) $\lim_{a \to \infty} z_a(t) = u(t)$

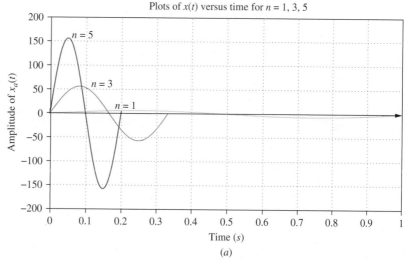

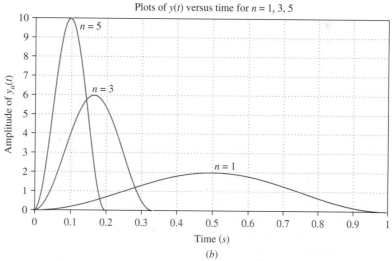

FIGURE 1.44

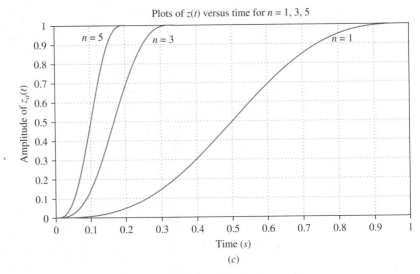

(c)

FIGURE 1.44 *(Continued)*

3.   a.   A pair of input–output waveforms for a physical system with an attenuating property is shown in Figure 1.45(a). Model the waveforms by sinusoids. Designate the input by $x(t) = A_1 \sin(2\pi f t + \theta_1) + B_1$ and the output by $y(t) = A_2 \sin(2\pi f t + \theta_2) + B_2$, where $A_i > 0$ and $-\pi < \theta_i < \pi$, $i = 1,\ 2$. Find the period T, frequency f, A_1, θ_1, A_2, and θ_2. Determine the system's attenuation $20\log(A_1/A_2)$ (in decibel) and its phase shift in degrees. Find the phase lag and the minimum time delay of the output with reference to the input. Express $y(t)$ in terms of $x(t)$.

   b.   Repeat for Figure 1.45(b).

Solution

Modeling a plot by a sinusoid.

   a.   Both traces in Figure 1.45(a) have a 0 DC value resulting in $B_1 = B_2 = 0$. For each trace the period is given by twice the time between its consecutive zero crossings or by twice the time between consecutive peaks. For both traces the period is measured to be $T = 0.4$ sec, corresponding to a frequency of $f = 1/T = 2.5$ Hz. The trace with a peak-to-peak value of 6 V is the input $x(t)$ with an amplitude $A_1 = 3$. Its phase θ_1 may be determined from two perspectives: From one point of view, $x(t)$ is a sinusoid that has been shifted to the left by $T/8$, corresponding to $\theta = 2\pi/8 = \pi/4$. From another point of view, the sinusoid has been shifted to the right $7T/8$, corresponding to $\theta = -7\pi/4$. The second point of view results in a phase outside the range specified by the problem and doesn't constitute an answer in the present setting. The input can be written as $x(t) = 3\sin(5\pi t + \pi/4)$.

   Similarly, the mathematical expression for the output trace in Figure 1.45(a) is found to be $y(t) = 2\sin(5\pi t - \pi/2)$. The system's attenuation is $20\log(3/2) = 3.52$ dB. Because the system is causal, the output lags the input by $\theta = \theta_1 - \theta_2 = \pi/4 + \pi/2 = 3\pi/4$. At the frequency of $f = 2.5$ Hz the phase lag of θ corresponds to a minimum time delay of $t_0 = \theta/(2\pi f) = (3\pi/4)/(5\pi) = 0.15$ sec. (The minimum time delay indicated directly from the plots is three divisions of 0.05 sec each, also yielding $t_0 = 3 \times 0.5 = 0.15$ sec). The particular output shown in Figure 1.45 (a) may be expressed in terms of the input by $y(t) = \frac{2}{3}x(t - t_0)$. (This will not necessarily be valid at other frequencies or for other input–output pairs.)

   b.   By a parallel analysis done on Figure 1.45(b) we obtain: period $T = 0.5$ sec, frequency $f = 2$ Hz, $x(t) = 5\sin(4\pi t + 0.2\pi)$, $y(t) = 3\sin(4\pi t - 0.4\pi)$, attenuation$= 20\log(5/3) = 4.44$ dB, phase lag $\theta = 0.6\pi$, and

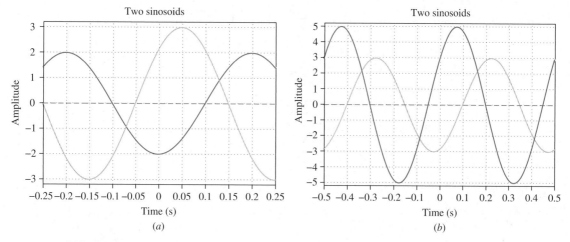

FIGURE 1.45 Modeling an input–output pair of systems in the form of sinusoidal functions.

a minimum time delay $t_0 = \theta/(2\pi f) = 0.6\pi/(4\pi) = 0.15$ sec.

Note. The knowledge of the phase lag of the sinusoidal output with respect to the input is not enough to specify the time delay between them unambiguously. The actual time delay may be any value given by $\frac{\theta}{2\pi f} + kT$, $k = 0, 1, \ldots, k$ being a nonnegative integer. For an unambiguous determination of time delay one needs additional information.

4. Modeling a plot by an exponential.

    a. A voltage signal that decays to a zero steady-state level is ploted in Figure 1.46. Model it by a single exponential function. Obtain the time constant of the exponential from the fact that during an interval that lasts τ seconds the signal is reduced by a factor of $e^{-1} = 0.368$. Alternatively, obtain the time constant from two measurements of the signal, v_1 and v_2, taken at two instances t_1 and t_2, T seconds apart, and compare the time constants obtained by the two approaches. When using the alternative approach, what are the preferred measurement instances t_1 and t_2 for better accuracy?

    b. How does one find the time constant if the signal decays toward a known nonzero final value? If the final value is not known, what additional information is needed to determine the complete mathematical expression for the signal? Show whether or not in that situation a third measurement of the signal amplitude at t_3 may be enough and, if so, how.

Solution

    a. The general form of a decaying exponential function with a single time constant is $A + Be^{-t/\tau}$, where $A + B$ is the initial value at $t = 0$, A is the final value at $t = \infty$, and τ is the time constant. From Figure 1.46, the initial value is 2 and the final value is 0. This gives us $A = 0$, $B = 2$, and $v(t) = 2e^{-\frac{t}{\tau}}u(t)$. The time constant is the amount of time it takes $x(t)$ to decay by a factor of $e^{-1} = 0.368$; for example, down from 2 at $t = 0$ to 0.736 at $t = \tau$. From the plot we observe $x(0.5) \approx 0.74$ which suggests $\tau = 0.5$ sec.

    The time constant may also be obtained from two measurements of magnitudes v_1 and v_2 made T seconds apart. The equation for computing τ from v_1 and v_2, and the time interval T between them, is (see Example 1.10)

$$\tau = \frac{T}{\ln\left(\frac{v_1}{v_2}\right)}$$

Measurements of the amplitudes v_1 and v_2 are more accurate during the steep segment of the function, while determinations of t_1 and t_2 are more accurate during the less steep segment. With those considerations in mind,

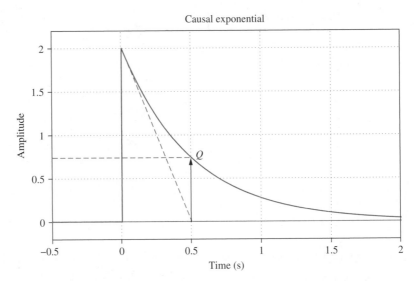

FIGURE 1.46 Determining the time constant of an exponential function.

we choose the following set of measurements from the plot in Figure 1.46: $v_1 = 2$ at $t_1 = 0$, $v_2 = 0.75$ at $t_2 = 0.5$, and $T = t_2 - t_1 = 0.5$.

The time constant of the exponential is computed to be

$$\tau = \frac{T}{\ln\left(\frac{x_1}{x_2}\right)} = \frac{0.5}{\ln\left(\frac{2}{0.75}\right)} = 0.51 \ \text{s}$$

b. If the final value A of the exponential is not zero, then v_1 and v_2 used in the above equation would be replaced by $v_1 - A$ and $v_2 - A$; in other words, the deviations of the value of the function from the final value at the two measurement instances. If the final value A is not known, a third measurement at t_3 would add a third equation to the set from which the three unknowns A, B, and τ, may be obtained.

5. A plot made of the sum of two exponentials. The function $v(t)$ plotted in Figure 1.47 is known to be the sum of two exponentials. Specify its parameters. The early part of the function is shown by the fast trace in (b). Assume the final value is zero.

Solution

We observe that $v(0) = v(\infty) = 0$. Therefore, we start with a sum of two exponential functions having time constants τ_1 and τ_2, with equal but opposing initial values $\pm B$ at $t = 0$, and zero final values at $t = \infty$.

$$v(t) = B\left(e^{-\frac{t}{\tau_1}} - e^{-\frac{t}{\tau_2}}\right)$$

The exponential with time constant τ_1 contributes positive values to the function and the exponential with time constant τ_2 provides negative values.

Both exponential components decay from their respective initial values of $\pm B$ toward a zero final value. The three parameters τ_1, τ_2, and B may be calculated from three measurements of the function at three instances of time

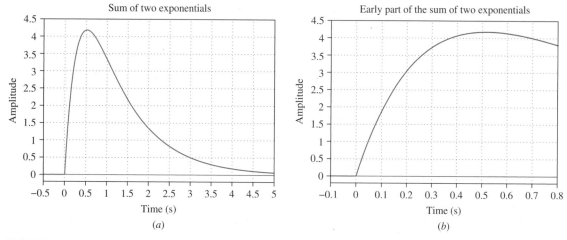

FIGURE 1.47 *(a)* A plot of the sum of two exponentials. *(b)* Fast trace of the early part of the plot.

t_1, t_2, and t_3. The measurements provide the following three independent equations:

$$v_1 = B\left(e^{-\frac{t_1}{\tau_1}} - e^{-\frac{t_1}{\tau_2}}\right)$$

$$v_2 = B\left(e^{-\frac{t_2}{\tau_1}} - e^{-\frac{t_2}{\tau_2}}\right)$$

$$v_3 = B\left(e^{-\frac{t_3}{\tau_1}} - e^{-\frac{t_3}{\tau_2}}\right)$$

The equations can be solved for B, τ_1, and τ_2 using computational and numerical methods. To obtain a solution of the above equations we choose $t_2 = 2t_1$ and $t_3 = 3t_1$. Then,

$$v_1 = B(x_1 - x_2)$$
$$v_2 = B\left(x_1^2 - x_2^2\right) = B(x_1 - x_2)(x_1 + x_2)$$
$$v_3 = B\left(x_1^3 - x_2^3\right) = B(x_1 - x_2)\left(x_1^2 + x_1 x_2 + x_2^2\right)$$

where $x_1 = e^{-\frac{t_1}{\tau_1}}$ and $x_2 = e^{-\frac{t_1}{\tau_2}}$. We then find

$$(x_1 + x_2) = v_2/v_1$$
$$\left(x_1^2 + x_1 x_2 + x_2^2\right) = v_3/v_1$$

The solution is

$$x^2 - \frac{v_2}{v_1}x + \left(\frac{v_2}{v_1}\right)^2 - \frac{v_2}{v_1} = 0$$

$$x_{1,2} = \frac{v_2}{2v_1}\left[1 \pm \sqrt{\frac{4v_1 v_3}{v_2^2} - 3}\right], \quad \tau_1 = \frac{-t_1}{\ln x_1}, \quad \text{and} \quad \tau_2 = \frac{-t_1}{\ln x_2}$$

(i) From the plots of Figure 1.47 the following three measurements and the resulting parameters were obtained:

$t_1 = 0.5$	$t_2 = 1$	$t_3 = 1.5$
$v_1 = 4.15$	$v_2 = 3.35$	$v_3 = 2.15$

$\Longrightarrow$

$x_1 = 0.575$	$x_2 = 0.232$	
$\tau_1 = 0.905$	$\tau_2 = 0.342$	$B = 12$

(ii) For better accuracy, the plots were enlarged and the following three measurements with the resulting parameters were obtained:

$t_1 = 0.5$	$t_2 = 1$	$t_3 = 1.5$
$v_1 = 4.166$	$v_2 = 3.324$	$v_3 = 2.166$

$\Longrightarrow$

$x_1 = 0.605$	$x_2 = 0.193$	
$\tau_1 = 0.995$	$\tau_2 = 0.304$	$B = 10.112$

(iii) To compare the above results with the actual parameters used by the computer to plot $v(t)$, the sample values at t_1, t_2, and t_3 were picked from the computer plots, resulting in the following:

$t_1 = 0.5$	$t_2 = 1$	$t_3 = 1.5$
$v_1 = 4.176$	$v_2 = 3.322$	$v_3 = 2.164$

$\Longrightarrow$

$x_1 = 0.60652$	$x_2 = 0.18898$	
$\tau_1 = 1$	$\tau_2 = 0.3$	$B = 10$

6. Model the plot of Figure 1.47 as a sum of two exponentials. The function that generated the plot may or not have been the sum of two exponentials.

Solution

As in problem 5, we observe that $v(0) = v(\infty) = 0$. The desired model should have the following form:

$$v(t) = v^+(t) - v^-(t), \quad \text{where } v^+(t) = Be^{-\frac{t}{\tau_1}} \text{ and } v^-(t) = Be^{-\frac{t}{\tau_1}}$$

The model has three parameters τ_1, τ_2, and B, which may be found from three measurements as done in problem 5. However, because the plot may not represent the sum of two exponentials, the resulting model parameters may depend on measurement times t_1, t_2, and t_3, possibly making the model inaccurate. The method described below can provide an approximate solution.

The early part of the plot changes noticeably faster than the later part, suggesting a shorter time constant for $v^-(t)$. During the later part of the plot, $v^-(t)$ has noticeably diminished and the plot is dominated mainly by $v^+(t)$ whose time constant τ_1 and initial value B may be estimated first. Two measurements v_1 and v_2, taken at t_1 and t_2, during the later part of the function yield

$$v_1 = B\left(e^{-\frac{t_1}{\tau_1}} - e^{-\frac{t_1}{\tau_2}}\right) \approx Be^{-\frac{t_1}{\tau_1}}$$

$$v_2 = B\left(e^{-\frac{t_2}{\tau_1}} - e^{-\frac{t_2}{\tau_2}}\right) \approx Be^{-\frac{t_2}{\tau_1}}$$

$$\tau_1 \approx \frac{t_2 - t_1}{\text{Ln}\left(\frac{v_1}{v_2}\right)} \quad \text{and} \quad B \approx v_2 e^{\left(\frac{t_2}{\tau_1}\right)}$$

Note that in addition to the time constant τ_1, the above two measurements will also provide us with an estimate of B. The above approach is applied to the later part of the plot of Figure 1.47(a) with the following results:

$$t_1 = 2, \quad v_1 = 1.353$$
$$t_2 = 4, \quad v_2 = 0.206$$
$$\tau_1 = \frac{t_2 - t_1}{\ln\left(\frac{v_1}{v_2}\right)} = \frac{2}{\ln\left(\frac{1.353}{0.206}\right)} = 1.06258$$
$$B \approx v_2 e^{\frac{t_2}{\tau_1}} = 0.206 e^{\frac{4}{1.062}} = 9.5$$

To find τ_2 we obtain a third measurement $v_3 = 1.833$ at $t_3 = 0.1$ (during the early part of the plot in Figure 1.47(b)) which yields the following:

$$v_3 = B\left(e^{-\frac{t_3}{\tau_1}} - e^{-\frac{t_3}{\tau_2}}\right)$$
$$1.833 = 9.5\left(e^{-\frac{0.1}{1.0625}} - e^{-\frac{0.1}{\tau_2}}\right)$$
$$e^{-\frac{0.1}{\tau_2}} = 0.7215, \quad \tau_2 = 0.306$$

7. Modeling a plot by a decaying sinusoid. Model the function plotted in Figure 1.48 by an exponentially decaying sinusoid:

$$x(t) = A + Be^{-t/\tau}\sin(2\pi f t + \theta)$$

Determine A, B, τ, f, and θ.

Solution

We begin with A, τ, and f which are easier to determine. From Figure 1.48(a), the final steady-state value is seen to be $A = 1$, and the time constant of the decay is computed to be $\tau = 0.4$ sec. (Use the value of the first two peaks of the function with a time separation of 0.3 sec to compute τ as done in problem 4.) The frequency f is the inverse of the period. The period T is twice the time between two adjacent crossings of $x(t)$ at $x(t) = 1$. From Figure 1.48(b) we obtain $T/2 = 0.15$ sec and $f = 1/(2 \times 0.15) = 10/3$ Hz.

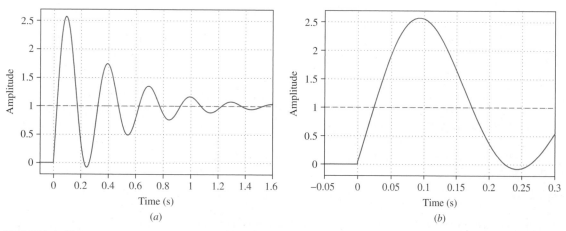

FIGURE 1.48 (*a*) Plot of a function to be modeled by an exponentially decaying sinusoid. (*b*) Fast trace of the early part of the function.

The phase angle θ is found by observing that the first crossing of the function at $x(t) = 1$ occurs at $t = 0.025$. The resulting equation is $\sin(2\pi f \times 0.025 + \theta) = 0$ or $\theta = -30°$. Finally, the parameter B is found from $x(t)|_{t=0} = 1 + B\sin(\theta) = 0$ or $B = 2$. In summary, the model for the plot of Figure 1.48 is

$$x(t) = \left[1 + 2e^{-2.5t}\sin\left(\frac{20\pi t}{3} - \frac{\pi}{6}\right)\right]u(t)$$

$$= \left\{1 + 2e^{-2.5t}\sin\left[\frac{20\pi}{3}\left(t - \frac{1}{40}\right)\right]\right\}u(t)$$

$$= \left[1 + 2e^{-2.5t}\sin\omega(t - \tau)\right]u(t), \quad \omega = \frac{20\pi}{3}, \quad \tau = 25 \text{ msec}$$

8. Time averages and time transformations. Consider the single sawtooth pulse function

$$x(t) = \begin{cases} 0, & t < 0 \\ t, & 0 < t < 1 \\ 0, & t > 1 \end{cases}$$

a. Find a mathematical expression for $x(-t)$.
b. Find a mathematical expression for $x(-t - 1)$.
c. Find the average of $x(t)$ and its rms value during $t = 0$ to $t = 1$.
d. Consider periodic functions $y_1(t)$, $y_2(t)$, and $y_3(t)$ with period $T = 2$, where for one period

$$y_1(t) = x(t) + x(-t), \quad -1 < t < 1$$
$$y_2(t) = x(t) - x(-t), \quad -1 < t < 1$$
$$y_3(t) = x(t) + x(t - 1), \quad 0 < t < 2$$

Find their average and rms values.

Solution

a. In the expression for $x(t)$ replace t by $-t$

$$x(-t) = \begin{cases} 0, & -t < 0 \\ -t, & 0 < -t < 1 \\ 0, & -t > 1 \end{cases}$$

To clean up the above expressions, trade the inequality signs with minus signs and order the segments in form of increasing t:

$$x(-t) = \begin{cases} 0, & t < -1 \\ -t, & -1 < t < 0 \\ 0, & t > 0 \end{cases}$$

Alternatively, one may start with $x(t) = t[u(t) - u(t - 1)]$ and change t to $-t$ to obtain $x(-t) = t[u(-t - 1) - u(-t)]$. In either approach the plot of $x(t)$ flips around the origin.

b. In the expression for $x(t)$ replace t by $-t - 1$ to find

$$x(-t - 1) = \begin{cases} 0, & -t - 1 < 0 \\ -t - 1, & 0 < -t - 1 < 1 \\ 0, & -t - 1 > 1 \end{cases} = \begin{cases} 0, & t < -2 \\ -t - 1, & -2 < t < -1 \\ 0, & t > -1 \end{cases}$$

Alternatively, one may start with $x(t) = t[u(t) - u(t - 1)]$ and change t to $-t - 1$ resulting in $x(-t - 1) = (t + 1)[u(-t - 2) - u(-t - 1)]$. In either approach the plot of $x(t)$ flips around the origin and then shifts to the right by one unit.

c. Average of $x(t)$ during $t = 0$ to $t = 1$ is

$$\frac{1}{T} \int_0^T x(t)dt = \int_0^1 t\,dt = \frac{1}{2}.$$

The rms value of $x(t)$ during $t = 0$ to $t = 1$ is

$$\sqrt{\frac{1}{T} \int_0^T x^2(t)dt} = \sqrt{\int_0^1 t^2 dt} = \frac{\sqrt{3}}{3}.$$

d. See Figure 1.49. The average and rms values for $y_1(t)$, $y_2(t)$, and $y_3(t)$ are obtained from those for $x(t)$.

Function	Average	RMS
$y_1(t)$	1/2	$\sqrt{3}/3$
$y_2(t)$	0	$\sqrt{3}/3$
$y_3(t)$	1/2	$\sqrt{3}/3$

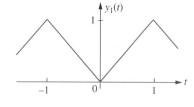 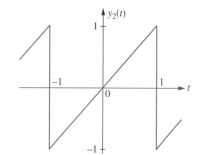 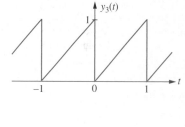

FIGURE 1.49

9. Time shift, time reversal, and their interchange. Given $u(t) = \begin{cases} 1, & t \geq 0 \\ 0, & t < 0 \end{cases}$

a. Express $u(-t - 3)$ in terms of t, plot it, and show how it may be constructed by *a delay followed by a time reversal* or, equivalently, by *a time reversal followed by an advance*.

b. Express $u(-t + 3)$ in terms of t, plot it, and show how it may be constructed by *a time reversal followed by a delay* or, equivalently, by *an advance followed by a time reversal*.

c. Conclude that a delay followed by a time reversal is equivalent to a time reversal followed by an advance. Similarly, a time reversal followed by a delay is equivalent to an advance followed by a time reversal.

Solution

a. See a_1 and a_2 in Figure 1.50(a).

b. See b_1 and b_2 in Figure 1.50(b).

The results shown in Figure 1.50 may be verified by (i) applying the definition of the unit-step function or by (ii) graphical effects of time shift and reversal. Note that changing the order of *time shift* and *time reversal* would lead to the same result only if at the same time *delay* and *advance* are also interchanged. As an example, $u(-t - 3)$ may be obtained by a delay/reversal sequence; first introduce a 3-unit delay in $u(t)$ (by changing t to $t - 3$) to produce $u(t - 3)$, then reverse $u(t - 3)$ (by changing t to $-t$) to obtain $u(-t - 3)$. Conversely, $u(-t - 3)$ may be obtained by a reverse/advance sequence; first reverse $u(t)$ to obtain $u(-t)$, then introduce a 3-unit advance in $u(-t)$ (by changing t to $t + 3$) to produce $u[-(t + 3)] = u(-t - 3)$.

(a_1) Delay $u(t-3) = \begin{cases} 1, & t \geq 3 \\ 0, & t < 3 \end{cases}$

and

Reversal $u(-t-3) = \begin{cases} 1, & t \leq -3 \\ 0, & t > 3 \end{cases}$

(a_2) Reversal $u(-t) = \begin{cases} 1, & t \leq 0 \\ 0, & 0, t > 0 \end{cases}$

and

Advance $u[-(t+3)] = u(-t-3) = \begin{cases} 1, & t \leq -3 \\ 0, & t > -3 \end{cases}$

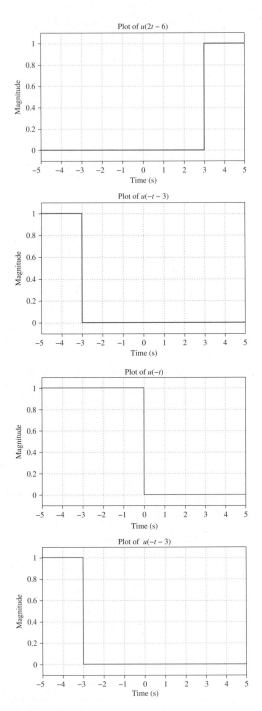

FIGURE 1.50 *(a)* Two ways to transform $u(t)$ to $u(-t-3)$.

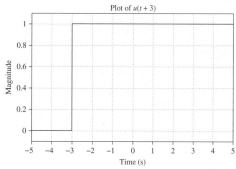

(b_1) Reversal $u(-t) = \begin{cases} 1, & t \le 0 \\ 0, & t > 0 \end{cases}$

and

Delay $u[-(t-3)] = u(-t+3) = \begin{cases} 1, & t \le 3 \\ 0, & t > 3 \end{cases}$

(b_2) Advance $u(t+3) = \begin{cases} 1, & t \ge -3 \\ 0, & t < -3 \end{cases}$

and

Reversal $u(-t+3) = \begin{cases} 1, & t \le 3 \\ 0, & t > 3 \end{cases}$

FIGURE 1.50 *(b)* Two ways to transform $u(t)$ to $u(-t+3)$.

10. The unit-step function is expressed as a function of time by $u(t) = \begin{cases} 1, & t \geq 0 \\ 0, & t < 0 \end{cases}$. In a similar way, express the following as functions of time and plot them:

a. $0.5u(t)$, $2u(t)$ b. $u(0.5t)$, $u(2t)$ c. $u(t+1)$, $u(t-2)$
d. $u(-t)$, $u(-t+1)$ e. $u(2t+1)$, $u(-3t+2)$ f. $u(t)u(2-t)$, $u(t)u(2+t)$

Solution

a. $0.5u(t) = \begin{cases} 0.5, & t \geq 0 \\ 0, & t < 0 \end{cases}$ $2u(t) = \begin{cases} 2, & t \geq 0 \\ 0, & t < 0 \end{cases}$

b. $u(0.5t) = \begin{cases} 1, & t \geq 0 \\ 0, & t < 0 \end{cases}$ $u(2t) = \begin{cases} 1, & t \geq 0 \\ 0, & t < 0 \end{cases}$

c. $u(t+1) = \begin{cases} 1, & t \geq -1 \\ 0, & t < -1 \end{cases}$ $u(t-2) = \begin{cases} 1, & t \geq 2 \\ 0, & t < 2 \end{cases}$

d. $u(-t) = \begin{cases} 1, & t \leq 0 \\ 0, & t > 0 \end{cases}$ $u(-t+1) = \begin{cases} 1, & t \leq 1 \\ 0, & t > 1 \end{cases}$

e. $u(2t+1) = \begin{cases} 1, & t \geq -0.5 \\ 0, & \text{elsewhere} \end{cases}$ $u(-3t+2) = \begin{cases} 1, & t \leq \frac{2}{3} \\ 0, & \text{elsewhere} \end{cases}$

f. $u(t)u(2-t) = \begin{cases} 1, & 0 \leq t \leq 2 \\ 0, & \text{elsewhere} \end{cases}$ $u(t)u(2+t) = \begin{cases} 1, & t \geq 0 \\ 0, & \text{elsewhere} \end{cases}$

11. A unit pulse is defined by $d(t) = \begin{cases} 1, & 0 < t < 1 \\ 0, & \text{elsewhere} \end{cases}$. In a similar way, express the following as functions of time and plot them:

a. $d(0.5t)$
b. $d(2t)$
c. $d(t+1)$
d. $d(-t+1)$
e. $d(2t+1)$
f. $d(t)d(2-t)$

Solution
See below.

a. $d(0.5t) = \begin{cases} 1, & 0 < 0.5t < 1 \\ 0, & \text{elsewhere} \end{cases} = \begin{cases} 1, & 0 < t < 2 \\ 0, & \text{elsewhere} \end{cases}$

b. $d(2t) = \begin{cases} 1, & 0 < 2t < 1 \\ 0, & \text{elsewhere} \end{cases} = \begin{cases} 1, & 0 < t < 0.5 \\ 0, & \text{elsewhere} \end{cases}$

c. $d(t+1) = \begin{cases} 1, & 0 < (t+1) < 1 \\ 0, & \text{elsewhere} \end{cases} = \begin{cases} 1, & -1 < t < 0 \\ 0, & \text{elsewhere} \end{cases}$

d. $d(-t+1) = \begin{cases} 1, & 0 < (-t+1) < 1 \\ 0, & \text{elsewhere} \end{cases} = \begin{cases} 1, & 0 < t < 1 \\ 0, & \text{elsewhere} \end{cases}$

e. $d(2t+1) = \begin{cases} 1, & 0 < (2t+1) < 1 \\ 0, & \text{elsewhere} \end{cases} = \begin{cases} 1, & -0.5 < t < 0 \\ 0, & \text{elsewhere} \end{cases}$

f. $d(t)d(2-t) = \begin{cases} 1, & 0 < t < 1 \text{ and } 0 < (2-t) < 1 \\ 0, & \text{elsewhere} \end{cases} = \begin{cases} 1, & 0 < t < 1 \text{ and } 1 < t < 2 \\ 0, & \text{elsewhere} \end{cases} = 0 \text{ (all } t)$

A Matlab program for creating some of the above functions is given below.

```
t=[-2.5:.01:2.5]; T=length(t)
% Create d(t)=u(t)-u(t-1);
a=1; b=0;
for i=1:T
    if a*t(i)+b >0;
    if a*t(i)+b <1;
        x1(i)=1;
    else
        x1(i)=0;
        end
    end
end
%Create d(t+1);
a=1; b=1;
for i=1:T
    if a*t(i)+b >0;
    if a*t(i)+b <1;
        x4(i)=1;
    else
        x4(i)=0;
        end
    end
end
%Create  d(-t+1);
a=-1; b=1;
for i=1:T
    if a*t(i)+b <1;
    if a*t(i)+b >0;
        x5(i)=1;
    else
        x5(i)=0;
        end
    end
end
%Create  d(2t+1);
a=2; b=1;
for i=1:T
    if a*t(i)+b >0;
    if a*t(i)+b <1;
        x6(i)=1;
    else
```

```
        x6(i)=0;
      end
    end
  end
end
%Create d(2-t) and d(t)d(2-t);
a=-1; b=2;
for i=1:T
        if a*t(i)+b <1;
        if a*t(i)+b >0;
        x7(i)=1;
    else
        x7(i)=0;
      end
    end
end; x1x7=x1.*x7;
```

12. Shifts and reversals of a right-sided function. Given $x(t) = 2^{-t}u(t) = \begin{cases} 2^{-t}, & t \geq 0 \\ 0, & t < 0 \end{cases}$, express the following as functions of t:

a. $x(t-3)$
b. $x(t+4)$
c. $x(-t)$
d. $x(-t-3)$
e. $x(-t-4)$

Solution

a. Shifting $x(t)$ to the right by 3 units produces $x(t-3) = 2^{-(t-3)}u(t-3) = \begin{cases} 8 \times 2^{-t}, & t \geq 3 \\ 0, & t < 3 \end{cases}$

b. Shifting $x(t)$ to the left by 4 units produces $x(t+4) = 2^{-(t+4)}u(t+4) = \begin{cases} \frac{2^{-t}}{16}, & t \geq -4 \\ 0, & t < -4 \end{cases}$

c. Reversal of $x(t)$ around the point $t = 0$ produces $x(-t) = 2^{t}u(-t) = \begin{cases} 2^{t}, & t \leq 0 \\ 0, & t > 0 \end{cases}$

d. Reversal of $x(t-3)$ around the point $t = 0$ produces $x(-t-3) = 2^{(t+3)}u(-t-3) = \begin{cases} 8 \times 2^{t}, & t \leq -3 \\ 0, & t > -3 \end{cases}$

e. Reversal of $x(t+4)$ around the point $t = 0$ produces $x(-t+4) = 2^{(t-4)}u(-t+4) = \begin{cases} \frac{1}{16}2^{t}, & t \leq 4 \\ 0, & t > 4 \end{cases}$

13. A one-sided function is given by $x(t) = \begin{cases} 2^{-t}, & t \geq 0 \\ 0, & t < 0 \end{cases}$. In a similar way, express the following as functions of time:

a. $x(0.5t)$
b. $x(2t)$
c. $x(t+1)$
d. $x(t-2)$
e. $x(2t+1)$
f. $x(t)x(2-t)$

Solution

See below.

a. $x(0.5t) = \begin{cases} 2^{-0.5t}, & t \geq 0 \\ 0, & t < 0 \end{cases}$

b. $x(2t) = \begin{cases} 2^{-2t}, & t \geq 0 \\ 0, & t < 0 \end{cases}$

c. $x(t+1) = \begin{cases} 2^{-(t+1)}, & (t+1) \geq 0 \\ 0, & \text{elsewhere} \end{cases} = \begin{cases} 0.5 \times 2^{-t}, & t \geq -1 \\ 0, & \text{elsewhere} \end{cases}$

d. $x(t-2) = \begin{cases} 2^{-(t-2)}, & (t-2) \geq 0 \\ 0, & \text{elsewhere} \end{cases} = \begin{cases} 4 \times 2^{-t}, & t \geq 2 \\ 0, & \text{elsewhere} \end{cases}$

e. $x(2t+1) = \begin{cases} 2^{-(2t+1)}, & (2t+1) \geq 0 \\ 0, & \text{elsewhere} \end{cases} = \begin{cases} 0.5 \times 2^{-2t}, & t \geq -0.5 \\ 0, & \text{elsewhere} \end{cases}$

f. $x(2-t) = \begin{cases} 2^{-(2-t)}, & (2-t) \geq 0 \\ 0, & \text{elsewhere} \end{cases} = \begin{cases} 0.25 \times 2^{t}, & t \leq 2 \\ 0, & \text{elsewhere} \end{cases}$

g. $x(t)x(2-t) = \begin{cases} 2^{-t} \times 0.25 \times 2^{t}, & t \geq 0 \text{ and } t \leq 2 \\ 0, & \text{elsewhere} \end{cases} = \begin{cases} 0.25, & 0 \leq t \leq 2 \\ 0, & \text{elsewhere} \end{cases}$

14. **Time transformations.** Let $x(t) = e^{-t}[u(t) - u(t-1)]$.

   a. Express $y(t) = x(-t-1)$ in terms of exponential and step functions.

   b. Describe the sequence of time transformations which generates $y(t)$ from $x(t)$.

   c. Write computer code to plot $x(t)$ and $y(t)$ for $-2.5 < t < 2.5$ and save the plots as postscript graphs.

   d. Verify the sequence of time transformations described in part (b) by examining the plots obtained in part (c).

Solution

a. In the expression given for $x(t)$ replace t by $-t-1$ to find

$$y(t) = x(-t-1) = e^{-(-t-1)}[u(-t-1) - u(-t-1-1)] = e^{(t+1)}[u(-t-1) - u(-t-2)]$$

b. A time reversal changes $x(t)$ to $x(-t)$. Shifting the new function to the left by one unit changes t to $t+1$, thus producing $x[-(t+1)] = x(-t-1)$. The sequence of operations is *time reversal followed by a left-shift.* Alternatively, one could start with a right shift, changing $x(t)$ to $x(t-1)$, then perform a time reversal that changes t to $-t$, thus producing $x(-t-1)$. The sequence of operations is then *right-shift followed by a time reversal.*

c. The Matlab code and resulting plots are shown below. Note that the code determines $y(t)$ by directly implementing the expression given in part (a).

```
t=linspace(-2.5, 2.5, 500);   x=zeros(1,500 );  y=zeros(1,500 );
for i=250:350;
    x(i)=exp(-t(i));
end
for i=50:150;
    y(i)=exp(t(i)+1);
end
```

$x(t)$ and $y(t)$ generated by the above Matlab code are plotted in Figure 1.51(a) and (b), respectively.

d. Flip the plot in Figure 1.51(a) around the origin and shift it to the left by one unit and you will get the plot in (b).

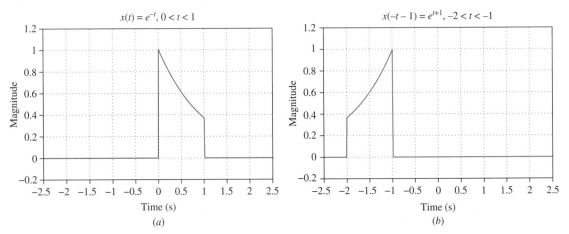

FIGURE 1.51 (a) A finite-duration exponential pulse $x(t) = e^{-t}[u(t) - u(t-1)]$. (b) A time reversal in $x(t)$ followed by a unit shift to the left produces $y(t) = x(-t-1) = e^{t+1}[u(-t-1) - u(-t-2)]$.

15. Even and odd parts of one-sided function.

   a. Let $y(t)$ be the causal part (the right-side segment) of a two-sided function $x(t)$. Can one relate the even and odd parts of $y(t)$ to the even and odd part of $x(t)$?

   b. Repeat for $z(t) = x(t)u(-t)$ [i.e., the left-sided function made of the left side of $x(t)$].

Solution

a.
$$y(t) = x(t)u(t) \quad y_e(t) = \frac{x(t)u(t) + x(-t)u(-t)}{2} \qquad x_e(t) = \frac{x(t) + x(-t)}{2}$$
$$y_o(t) = \frac{x(t)u(t) - x(-t)u(-t)}{2} \qquad x_o(t) = \frac{x(t) - x(-t)}{2}$$

b.
$$z(t) = x(t)u(-t) \quad z_e(t) = \frac{x(t)u(-t) + x(-t)u(t)}{2} \qquad z_e(t) = \frac{z(t) + z(-t)}{2}$$
$$z_o(t) = \frac{x(t)u(-t) - x(-t)u(t)}{2} \qquad z_o(t) = \frac{z(t) - z(-t)}{2}$$

Given the even and odd parts of $x(t)$, one can find $y(t)$ and $z(t)$ and, as a result, their even and odd parts. However, given either $y(t)$ or $z(t)$ (or, equivalently, the even or odd parts of each but not both), one cannot find the even and odd parts of $x(t)$. In such a case, one side of $x(t)$ will not be known.

16. Integration. A step function with non-zero rise-time may be obtained by integrating a finite-duration pulse. Consider

$$x(t) = \begin{cases} \sin 2\pi t, & 0 \le t < 0.5 \\ 0, & 0.5 \le t < 1.5 \\ \sin 2\pi t, & 1.5 \le t < 2 \\ 0, & \text{elsewhere} \end{cases}$$

which is shown in Figure 1.52(a). Find and plot $y(t) = \int_{-\infty}^{t} x(t)dt$ and $z(t) = \int_{-\infty}^{t} y(t)dt$.

Solution   See below and the plots in Figure 1.52.

$$t < 0 \begin{cases} x(t) = 0 \\ y(t) = \int_{-\infty}^{t} x(t)dt = 0 \\ z(t) = \int_{-\infty}^{t} y(t)dt = 0 \end{cases}$$

$$0 \le t < 0.5 \begin{cases} x(t) = \sin 2\pi t \\ y(t) = \int_{0}^{t} \sin 2\pi t dt = \frac{1}{2\pi}(1 - \cos 2\pi t) & y(0.5) = \frac{1}{\pi} \\ z(t) = \frac{1}{2\pi} \int_{0}^{t} (1 - \cos 2\pi t)dt = \frac{1}{2\pi}\left(t - \frac{\sin 2\pi t}{2\pi}\right) & z(0.5) = \frac{1}{4\pi} \end{cases}$$

$$0.5 \le t < 1.5 \begin{cases} x(t) = 0 \\ y(t) = y(0.5) + \int_{0.5}^{t} x(t)dt = \frac{1}{\pi} + 0 = \frac{1}{\pi} & y(1.5) = \frac{1}{\pi} \\ z(t) = z(0.5) + \int_{0.5}^{t} y(t)dt = \frac{1}{4\pi} + \int_{0.5}^{t} \frac{1}{\pi}dt = \frac{1}{\pi}\left(t - \frac{1}{4}\right) & z(1.5) = \frac{5}{4\pi} \end{cases}$$

$$1.5 \le t < 2 \begin{cases} x(t) = \sin 2\pi t \\ y(t) = y(1.5) + \int_{1.5}^{t} x(t)dt = \frac{1}{\pi} + \int_{1.5}^{t} \sin 2\pi t dt = \frac{1}{2\pi}(1 - \cos 2\pi t) & y(2) = 0 \\ z(t) = z(1.5) + \int_{1.5}^{t} y(t)dt = \frac{5}{4\pi} + \int_{1.5}^{t} \frac{1}{2\pi}(1 - \cos 2\pi t) & z(2) = \frac{3}{2\pi} \\ \quad = \frac{1}{2\pi}\left(1 + t - \frac{\sin 2\pi t}{2\pi}\right) \end{cases}$$

$$t \ge 2 \begin{cases} x(t) = 0 \\ y(t) = y(2) + \int_{2}^{t} x(t)dt = 0 \\ z(t) = z(2) + \int_{2}^{t} y(t)dt = \frac{3}{2\pi} \end{cases}$$

17. Random signals.

a. Write a Matlab program to produce a Gaussian random sequence of length k with an expected mean $= b$ and variance of $\sigma^2 = a^2$. The elements of the sequence are random variables which are independent from each other. Discretize the amplitudes with a resolution of $\Delta = 1/N$ and obtain the frequency distribution of its elements. Generate a sequence of 524,288 samples with mean $= 10$ and $\sigma = 5$. Call it $x_a(n)$. Find and plot the histogram of its amplitude with a bin-width of 0.1. Find and plot its covariance function and spectrum.

b. Repeat for a random signal $x_b(n)$ with uniform amplitude distribution between -5 to 25.

Solution

a. The Matlab command $randn(N, M)$ generates an N by M matrix of Gaussian (also called normal) random numbers with zero mean and unity variance. The histogram of amplitude values can be constructed using the $hist$ command. The Matlab file given below generates a sequence of 524,288 Gausian random variables (mean $= 10$, $\sigma = 5$) and obtains the amplitude distribution. Assuming a sampling rate of $F_s = 8$ kHz, the spectrum z versus frequency f is obtained by applying the command $[z, f] = spectrum(x, fs)$. The maximum frequency is $F_s/2 = 4$ kHz.

```
x=5*randn(1,1024*512)+10; figure(1); stairs(x); grid; axis([1  100  -5   25])
bin=-30:1/10:30; figure(2); hist(x,bin); h=hist(x,bin);  grid;
    axis([-30   30  0   4500 ])
```

(a)

$$x(t) = \begin{cases} \sin 2\pi t, & 0 \le t < 0.5 \\ 0, & 0.5 \le t < 1.5 \\ \sin 2\pi t, & 1.5 \le t < 2 \\ 0, & \text{elsewhere} \end{cases}$$

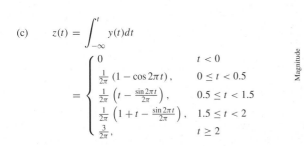

(b) $\displaystyle y(t) = \int_{-\infty}^{t} x(t)\,dt$

$$= \begin{cases} \frac{1}{2\pi}\left(1 - \cos 2\pi t\right), & 0 \le t < 0.5 \\ \frac{1}{\pi}, & 0.5 \le t < 1.5 \\ \frac{1}{2\pi}\left(1 - \cos 2\pi t\right), & 1.5 \le t < 2 \\ 0, & \text{elsewhere} \end{cases}$$

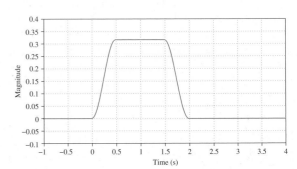

(c) $\displaystyle z(t) = \int_{-\infty}^{t} y(t)\,dt$

$$= \begin{cases} 0 & t < 0 \\ \frac{1}{2\pi}\left(1 - \cos 2\pi t\right), & 0 \le t < 0.5 \\ \frac{1}{2\pi}\left(t - \frac{\sin 2\pi t}{2\pi}\right), & 0.5 \le t < 1.5 \\ \frac{1}{2\pi}\left(1 + t - \frac{\sin 2\pi t}{2\pi}\right), & 1.5 \le t < 2 \\ \frac{3}{2\pi}, & t \ge 2 \end{cases}$$

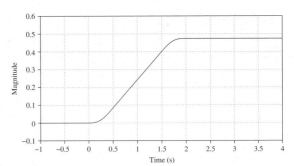

FIGURE 1.52

b. The Matlab command $rand(N, M)$ generates an N by M matrix of random numbers between 0 and 1 with uniform probability density. The following command generates the signal with uniform amplitude distribution from -5 to 25.

```
x=30*rand(1,1024*512)-5;
```

The rest is the same as in part (a).

Chapter Problems

18. Sketch the following functions made of unit impulses. They may be called difference operators:

a. forward difference operator: (i) 1st-order, $\delta(t + 1) - \delta(t)$ (ii) 2nd-order, $\delta(t + 2) - 2\delta(t + 1) + \delta(t)$

b. backward difference operator: (iii) 1st-order, $\delta(t) - \delta(t - 1)$ (iv) 2nd-order, $\delta(t) - 2\delta(t - 1) + \delta(t - 2)$

19. Sketch the following functions made of unit impulses:

a. $x(t) = 6\delta(t) - 4\delta(t-1) + 2\delta(t-2)$
b. $x_o(t) = -\delta(t+2) + 2\delta(t+1) - 2\delta(t-1) + \delta(t-2)$
c. $x_e(t) = \delta(t+2) - 2\delta(t+1) + 6\delta(t) - 2\delta(t-1) + \delta(t-2)$

Express $x(t)$ in terms of $x_o(t)$ and $x_e(t)$. Argue if $x_o(t)$ and $x_e(t)$ may be considered the odd and even parts of $x(t)$, respectively.

20. Sketch the following functions made of unit impulses. In each case determine if the function is even or odd. If it is neither, determine if by a time shift it may become even or odd.

a. $\delta(t) - \delta(t-1)$ b. $-2\delta(t) + 2\delta(t-1)$ c. $\delta(t+2) + \delta(t+1) - \delta(t-1) - \delta(t-2)$
d. $\delta(t+1) - \delta(t-1)$ e. $\delta(t+1) - 2\delta(t) + \delta(t-1)$ f. $\delta(t+2) - 2\delta(t+1) + 2\delta(t-1) - \delta(t-2)$
g. $-\delta(t+2) - \delta(t+1) + 4\delta(t) - 4\delta(t-1) + \delta(t-2) + \delta(t-3)$

21. Repeat problem 20 for the functions given below.

a. $\delta(t) - 2\delta(t-1) + \delta(t-2)$
b. $\delta(t) + \delta(t-1) - 3\delta(t-2) + \delta(t-3) + \delta(t-4)$
c. $\delta(t) + \delta(t-1) - \delta(t-2) - \delta(t-3)$
d. $\delta(t) - \delta(t-1) + \delta(t-4) - \delta(t-5) + \delta(t-8) - \delta(t-9)$
e. $\delta(t) - 3\delta(t-1) + 3\delta(t-2) - \delta(t-3)$
f. $\delta(t) - \delta(t-1) - \delta(t-4) + \delta(t-5) + \delta(t-8) - \delta(t-9)$
g. $\delta(t) - 2\delta(t-1) + \delta(t-2) + \delta(t-4) - 2\delta(t-5) + \delta(t-6) + \delta(t-8) - 2\delta(t-9) + \delta(t-10)$
h. $\delta(t) - 2\delta(t-1) + \delta(t-2) - \delta(t-4) + 2\delta(t-5) - \delta(t-6) + \delta(t-8) - 2\delta(t-9) + \delta(t-10)$

22. Repeat problem 20 for the functions given below. Assume $\left.\dfrac{\sin(k\pi)}{\pi k}\right|_{k=0} = 1.$

a. $\displaystyle\sum_{k=0}^{3}(k+1)\delta(t-k)$ b. $3\delta(t) - \displaystyle\sum_{k=1}^{3}\delta(t-k)$ c. $7\delta(t) - \displaystyle\sum_{k=-3}^{3}\delta(t-k)$

d. $\displaystyle\sum_{k=1}^{3}[\delta(t+k) - \delta(t-k)]$ e. $5\delta(t) - \displaystyle\sum_{k=1}^{5}\delta(t-k)$ f. $\displaystyle\sum_{k=-1}^{1}[d(t-3k) - d(t-3k-1)]$

g. $\displaystyle\sum_{k=-6}^{6}\dfrac{\sin(k\pi/2)}{k}\delta(t-k)$ h. $\displaystyle\sum_{k=-6}^{6}\dfrac{\sin(k\pi/5)}{k}\delta(t-k)$ i. $\displaystyle\sum_{k=-6}^{6}\dfrac{\sin(k\pi/8)}{k}\delta(t-k)$

23. Repeat problem 20 for the functions given below.

a. $e^{-t}\delta(t)$ b. $e^{-t/5}\delta(t-5)$ c. $\cos\left(\dfrac{\pi t}{6}\right)\delta(t-2)$

d. $\displaystyle\sum_{k=0}^{3}e^{-t}\delta(t-k)$ e. $\displaystyle\sum_{k=0}^{5}e^{-t/5}\delta(t-k)$ f. $\displaystyle\sum_{k=-2}^{2}\cos\left(\dfrac{\pi t}{6}\right)\delta(t-k)$

24. a. Find the values of the integrals given below.

(i) $\displaystyle\int_{-\infty}^{\infty}e^{-t}\delta(t)dt$ (ii) $\displaystyle\int_{-\infty}^{\infty}e^{-t/5}\delta(t-5)dt$ (iii) $\displaystyle\int_{-\infty}^{\infty}\cos\left(\dfrac{\pi t}{6}\right)\delta(t-1)dt$

b. Find the values of the summations given below.

(iv) $\displaystyle\sum_{k=0}^{3}\int_{-\infty}^{\infty}e^{-t}\delta(t-k)dt$ (v) $\displaystyle\sum_{k=0}^{5}\int_{-\infty}^{\infty}e^{-t/5}\delta(t-k)dt$ (vi) $\displaystyle\sum_{k=-2}^{2}\int_{-\infty}^{\infty}\cos\left(\dfrac{\pi t}{6}\right)\delta(t-k)dt$

25. Find the values of the integrals given below.

a. $1 - \displaystyle\int_{-\infty}^{t}[\delta(t) + \delta(t-1)]dt$ b. $1 - \displaystyle\int_{-\infty}^{t}[\delta(t) - \delta(t-1)]$ c. $-0.5 + \displaystyle\int_{-\infty}^{t}[\delta(t) - \delta(t-1) + \delta(t-2)]dt$

26. a. Sketch the following functions made of unit impulses and determine if each function is odd, even, or neither. Argue if $x_2(t)$ and $x_3(t)$ may be considered the odd and even parts of $x_1(t)$, respectively.

$$x_1(t) = 3d(t) - \sum_{k=1}^{3}(k+1)\delta(t-k)$$

$$x_2(t) = \sum_{k=1}^{3}[\delta(t+k) - \delta(t-k)]$$

$$x_3(t) = 7d(t) - \sum_{k=-3}^{3}(k+1)\delta(t-k)$$

   b. Find and sketch $y_i(t) = \int_{-\infty}^{t} x_i(\tau)d\tau$, $i = 1, 2, 3$, where the $x_i(t)$s are given in part (a). Show that $y_2(t)$ and $y_3(t)$ are the even and odd parts of $y_1(t)$, respectively.

27. Sketch the following functions made of unit steps:

   a. $u(t+1) - u(t-1)$

   b. $\sum_{k=1}^{N} u(t+k) - u(t-k)$, $N = 3, 5, 7$

   c. $u(t+0.5) - u(t-0.5)$

   d. $\sum_{\substack{k=1 \\ k \text{ odd}}}^{N} u(t+0.5k) - u(t-0.5k)$, $N = 5, 7, 9$

   e. $u(t) + u(t-1) - 2u(t-2)$

   f. $\sum_{k=0}^{N-1} u(t-k) - Nu(t-N)$, $N = 5, 7, 9$

   g. $u(t-1) - u(-t-1)$

   h. $u(t-1) - u(-t-1) + t[u(t+1) - u(t-1)]$

28. Find the time derivatives of the functions given in problem 27.

29. Let $x(t) = \sum_{k} a_k\delta(t-t_k)$ and $y(t) = \int_{-\infty}^{t} x(\tau)d\tau$, where a_k and t_k are constants. Determine the DC steady-state value of $y(t)$ and find condition in order for $y(t)$ to be a finite-duration pulse.

30. Let the area under a pulse and its energy (shown by A and E, respectively) be defined by

$$A = \int_{-\infty}^{\infty} x(t)dt$$

$$E = \int_{-\infty}^{\infty} |x|^2(t)dt$$

   a. Find A and E for $x_1(t)$, $ax_1(t)$, and $x_1(bt)$, where a and b are constants and $x_1(t)$ is a unit-sawtooth pulse given by

$$x_1(t) = \begin{cases} t, & 0 < t \leq 1 \\ 0, & \text{elsewhere} \end{cases}$$

   b. Repeat for $x_1(-t)$.
   c. Repeat for $x_1(t) + x_1(-t)$.

31. Repeat problem 30 for the unit-sawtooth pulse given by

$$x_2(t) = \begin{cases} 1-t, & 0 \leq t < 1 \\ 0, & \text{elsewhere} \end{cases}$$

32. Repeat problem 30 for a one-sided unit-exponential $x_3(t) = e^{-t}u(t)$.

33. Sketch the following functions made of unit steps.

a. $u(t) - u(t - 1)$ b. $-2u(t) + 2u(t - 1)$ c. $u(t + 2) + u(t + 1) - u(t - 1) - u(t - 2)$
d. $u(t + 1) - u(t - 1)$ e. $u(t + 1) - 2u(t) + u(t - 1)$ f. $u(t + 2) - 2u(t + 1) + 2u(t - 1) - u(t - 2)$
g. $-u(t + 2) - u(t + 1) + 4u(t) - 4u(t - 1) + u(t - 2) + u(t - 3)$.

Show that in all cases the functions are finite-duration pulses. Find their DC, rms, and average power values taken over the pulse duration.

34. For each function given in problem 33 determine if it is even or odd, and if it is neither, then determine if a time shift may make it even or odd.

35. Show that the functions given in problem 33 are integrals of the functions given in problem 20.

Hint: In the expressions for the functions in problem 33, replace u by δ and you will get the expressions for the functions in problem 20.

36. Repeat problems 33 through 34 for $y_i(t) = \int_{-\infty}^{t} x_i(\tau)d\tau$, where the $x_i(t)$s are as given in problem 28.

37. Repeat problems 33 through 34 for $y_i(t) = \int_{-\infty}^{t} x_i(\tau)d\tau$, where the $x_i(t)$s are as given in problem 30.

38. Let a unit square pulse be defined by $d(t) = u(t) - u(t - 1)$.

a. A staircaselike function with four steps is specified by $d(t) + 2d(t - 1) + 3d(t - 2) + 4d(t - 3)$. Sketch it and show that

$$\sum_{k=0}^{3} d(t - k) = \sum_{k=0}^{3} u(t - k) - 4u(t - 4)$$

b. Show that

$$\sum_{k=0}^{N-1} d(t - k) = \sum_{k=0}^{N-1} u(t - k) - Nu(t - N), \quad \text{where } N \geq 1 \text{ is an integer.}$$

39. Sketch the following functions made of unit pulses defined in problem 38.

a. $d(t)$ b. $-2d(t)$ c. $d(t + 1) + 2d(t) + d(t - 1)$
d. $d(t + 1) + d(t)$ e. $d(t + 1) - d(t)$ f. $d(t + 2) - d(t + 1) - d(t) + d(t - 1)$
g. $-d(t + 2) - d(t + 1) + 2d(t) - 2d(t - 1) - d(t - 2)$

Which of the above functions are integrals of the impulse functions of problem 20?

40. Sketch the following functions and express them in terms of steps.

$$x_a(t) = \begin{cases} -t, & -1 \leq t < 0 \\ t, & 0 \leq t < 1 \\ 0, & \text{elsewhere} \end{cases} \qquad x_b(t) = \begin{cases} t & 0, \leq t < 1 \\ 2 - t, & 1 \leq t < 2 \\ 0, & \text{elsewhere} \end{cases} \qquad x_c(t) = \begin{cases} t & 0, \leq t < 1 \\ t - 2, & 1 \leq t < 2 \\ 0, & \text{elsewhere} \end{cases}$$

$$x_d(t) = \begin{cases} -1 - t, & -1 \leq t < 0 \\ 1 - t, & 0 \leq t < 1 \\ 0, & \text{elsewhere} \end{cases} \qquad x_e(t) = \begin{cases} t & 0 \leq t < 1 \\ t - 1, & 1 \leq t < 2 \\ 1, & \text{elsewhere} \end{cases} \qquad x_f(t) = \begin{cases} 1 - t & -1, \leq t < 0 \\ t - 2 & <0 \leq t < 1 \\ 0 & \text{elsewhere} \end{cases}$$

41. Find the DC, rms, and average power value of each pulse in problem 40 taken over the pulse duration.

42. a. Sketch the finite-duration pulses specified below.

$$x_a(t) = \begin{cases} -1, & -2 \le t < 0 \\ 1, & 0 \le t < 2 \end{cases} \qquad x_b(t) = \begin{cases} 0, & t < 0 \\ t, & 0 \le t < 1 \\ 1, & 1 \le t < 2 \end{cases} \qquad x_c(t) = \begin{cases} -1, & -2 \le t < -1 \\ t, & -1 \le t < 1 \\ 1, & 1 \le t < 21 \end{cases}$$

b. In each case determine if the pulse is even or odd. If it is neither, determine if it may be made odd and/or even by a shift in time or level.

c. Find the average and rms value of each pulse taken over the pulse duration.

43. Express the pulses given in problem 42 in terms of steps and ramps.

44. a. Sketch the following functions and determine their average (DC level) values.

$$x_a(t) = \tan^{-1}(t) \qquad x_b(t) = \frac{1}{1 + e^{-t}} \qquad x_c(t) = \frac{1 - e^{-t}}{1 + e^{-t}}$$

b. In each case determine if the pulse is even or odd. If it is neither, determine if it may be made odd and/or even by a shift in time or level.

45. a. Write a Matlab program to plot the function $y(t) = u(t) - u(t - 1)$.

b. Repeat for the function specified in problem 33 g.

c. Repeat for $y(t) = \sum_k a_k u(t - t_k)$, where a_k and t_k are two known vectors of the same length.

46. a. Write a Matlab program to plot the function $y(t) = d(t) - d(t - 1)$, where $d(t) = u(t) - u(t - 1)$.

b. Repeat for the function specified in problem 39 g.

c. Repeat for $y(t) = \sum_k a_k d(t - t_k)$, where a_k and t_k are two vectors of the same length.

47. A unit ramp function is defined by $r(t) = \begin{cases} t, & t \ge 0 \\ 0, & t < 0 \end{cases}$.

a. Argue that the unit-step function $u(t)$ is the integral of a unit impulse, and that the unit-ramp function $r(t)$ is the integral of a unit step.

$$u(t) = \int_{-\infty}^{t} \delta(t)d\tau = \begin{cases} 1, & t > 0 \\ 0, & t < 0 \end{cases}$$

$$r(t) = \int_{-\infty}^{t} u(\tau)d\tau = \begin{cases} t, & t > 0 \\ 0, & t < 0 \end{cases}$$

b. Sketch the functions given below and graphically confirm the above relationships.

(i) $u(t)$ (ii) $u(-t)$ (iii) $u(t) - u(-t)$
(iv) $u(t - 1)$ (v) $u(-t - 1)$ (vi) $u(t - 1) - u(-t - 1)$
(vii) $r(t)$ (viii) $r(-t)$ (ix) $r(t) - r(-t)$
(x) $r(t - 1)$ (xi) $r(-t - 1)$ (xii) $r(t - 1) - r(-t - 1)$ (xiii) $r(t) - r(-t) + r(t - 1) - r(-t - 1)$

48. Consider the single sawtooth pulse function $x(t) = r(t) - r(t - 1) - u(t - 1)$ where $r(t)$ is the unit ramp.

a. Sketch $x(t)$, $x(t + 1)$, $x(t - 1)$, $x(-t)$, $x(-t + 1)$, and $x(-t - 1)$.

b. Write a computer program to plot $x(t)$ and $x(-t - 1)$ for $-3 < t < 3$. Save the plots as postscript graphs.

49. Consider a periodic signal

$$y(t) = \sum_{n=\infty}^{\infty} x(t + nT)$$

where $x(t)$ is one of the pulses given in problem 48, n is an integer, and T is the period. Define the DC value of $y(t)$ and its average power (shown by DC and P, respectively) by

$$DC = \frac{1}{T} \int_{t_0}^{t_0+T} y(t)\, dt, \quad P = \frac{1}{T} \int_{t_0}^{t_0+T} y^2(t)\, dt$$

Plot $y(t)$ and find the DC and P values for (i) $T = 4$, (ii) $T = 8$, (iii) $T = 12$.

50. Repeat problem 49 for

$$y(t) = \sum_{n=\infty}^{\infty}(-1)^n x(t+nT)$$

51. A single triangular waveform grows from 0 to 1 volt in 1 second and then is reduced to 0 in 2 seconds. Plot it and express it in terms of step functions.

52. The periodic rectangular pulse train $x(t)$ shown in Figure 1.53(a) is specified by a period $T = 10$ msec, a pulse duration $\tau = 2$ msec, and a base to peak value of $V_0 = 5$ V.

a. Find its frequency, duty cycle (defined by $\frac{\tau}{T}\%$), DC, and rms values.
b. Shift it by 1 msec to the right and subtract the result from $x(t)$ to find a new pulse train to be called $y(t)$. Plot $y(t)$ and find its DC and rms values.
c. Define $\zeta(t)$ (pronounced *zeta of* t) to be a 1-V, 1-msec wide rectangular pulse at the origin,

$$\zeta(t) = \begin{cases} 1, & 0 < t < 1 \text{ msec} \\ 0, & \text{elsewhere} \end{cases}.$$

Express $x(t)$ and $y(t)$ in terms of $\zeta(t)$. Then express their DC and rms values in terms of the area under $\zeta(t)$ and its energy.

53. Five periodic functions $x_i(t), i = 1, \ldots, 5$ shown in Figure 1.53(b)–(g), resepectively, are specified by $T = 10$ msec, $\tau = 1$ msec, and $V_0 = 5$ V.

a. Find the DC and rms value of each function.
b. Define

$$\zeta(t) = \begin{cases} 1, & 0 < t < 1 \text{ msec} \\ 0, & \text{elsewhere} \end{cases}$$

Express each function in Figure 1.53 in terms of $\zeta(t)$. Then determine their DC and rms values in terms of of the area under $\zeta(t)$ and its energy.
c. Investigate if any of the above functions can be obtained analytically by a linear combination of the others or by any combination of shift and/or reversal operations on the others.

54. Two periodic functions $x_1(t)$ and $x_2(t)$ [shown in Figure 1.54(a) and (b), respectively] are specified by $T = 10$ msec, $\tau = 1$ msec, and $V_0 = 5$ V.

a. Find the DC and rms values of each function.
b. Define $\eta(t)$ (pronounced *eta of* t) to be a single equilateral triangular pulse with a height of 1 and a base of τ:

$$\eta(t) = \begin{cases} 1 - 2\frac{|t|}{\tau}, & -\frac{\tau}{2} < t < \frac{\tau}{2} \\ 0, & \text{elsewhere} \end{cases}$$

Express $x_1(t)$ and $x_2(t)$ in terms of $\eta(t)$. Then determine their DC and rms values in terms of the area under $\eta(t)$ and its energy.
c. Investigate if $x_2(t)$ may be expressed analytically in terms of $x_1(t)$.

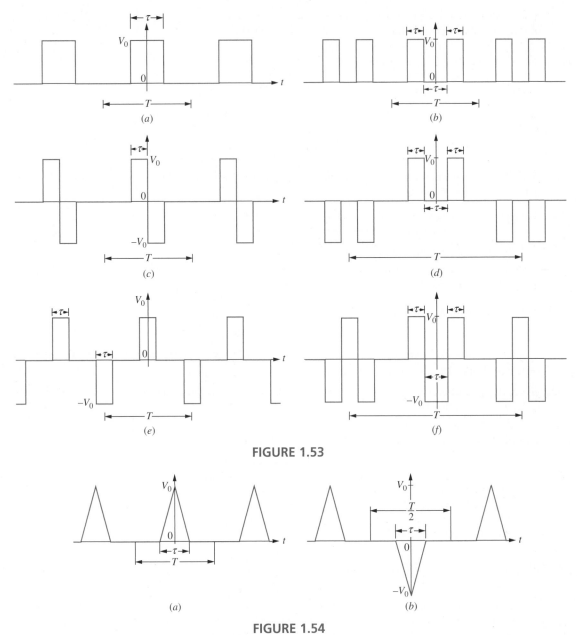

FIGURE 1.53

FIGURE 1.54

55. Three periodic functions $x_1(t)$, $x_2(t)$, and $x_3(t)$ shown in Figure 1.55 (a), (b), and (c), respectively, are specified by $T = 10$ msec, $\tau = 1$ msec, and $V_0 = 5$ V.

   a. Find the DC and rms values of each function.

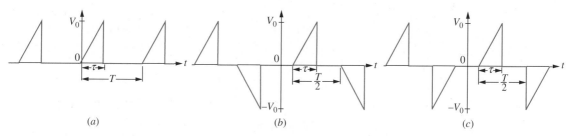

(a) (b) (c)

FIGURE 1.55

b. Define $\xi(t)$ (pronounced *cai of t*) to be a single right-angle triangular pulse with a height of 1 and a base of τ:

$$\xi(t) = \begin{cases} t, & -0 \le t < \tau \\ 0, & \text{elsewhere} \end{cases}$$

Express $x_1(t)$, $x_2(t)$, and $x_3(t)$ in terms of $\xi(t)$. Then determine their DC and rms values in terms of the area under $\xi(t)$ and its energy.

c. Investigate if any of the above functions can be obtained by a linear combination of the others or by any combination of shift and/or reversal operations on the others.

56. The periodic functions $x_1(t)$ and $x_2(t)$ shown in Figure 1.56 are called half-symmetric. Plot their time integrals

$$y_i(t) = \int_{-\infty}^{t} x_i(\tau)d\tau, \quad i = 1, 2$$

and compute their DC and rms values. Assume $y_1(-\infty) = y_2(-\infty) = 0$.

57. Repeat problem 56 for the periodic functions shown in:
   a. Figure 1.53(c), (d), (e), and (f).
   b. Figure 1.54(b).
   c. Figure 1.55(b) and (c).

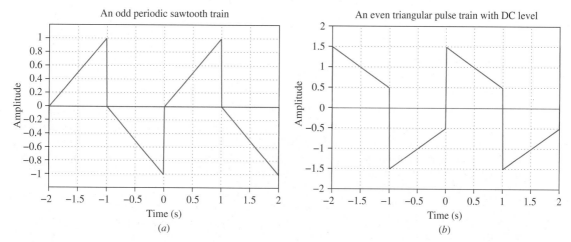

FIGURE 1.56 Two periodic functions with the same period T and half-symmetry property: $x(t) = -x(t + T/2)$.

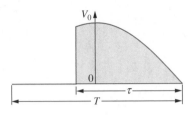

FIGURE 1.57

58. A finite-duration pulse is described by $x(t) = \cos(2\pi t/T)[u(t-T/4)-u(t-T/4)]$. Sketch the following functions and find their DC and rms values.

$$y_1(t) = \sum_{n=-\infty}^{\infty} V_0 x(t-nT), \quad y_2(t) = \sum_{n=-\infty}^{\infty} V_0 x(t-nT/2), \quad y_3(t) = \sum_{n=-\infty}^{\infty} (-1)^n V_0 x(t-nT)$$

59. The power drawn from a sinusoidal source is controlled by applying a threshold level λ to it in order to produce the signal:

$$x(t) = \begin{cases} \sin\omega t - \lambda, & \text{if } \sin\omega t > \lambda \\ 0, & \text{otherwise} \end{cases}$$

Sketch $x(t)$ and find its DC value and average power as a function of $\lambda > 0$.

60. a. Find the area and energy of a single pulse made of the partial sinusoid described by

$$x(t) = \begin{cases} V_0 \cos\left(\frac{\pi t}{T}\right), & \frac{T}{2}-\tau \leq t < \frac{T}{2} \\ 0, & \text{elsewhere} \end{cases}$$

b. Find the DC and rms values of the following periodic functions made of the pulse of part (a).

$$y_1(t) = \sum_{n=-\infty}^{\infty} x(t-nT), \quad y_2(t) = \sum_{n=-\infty}^{\infty} (-1)^n x(t-nT/2)$$

61. A finite-duration pulse is described by $x(t) = e^{-t}[u(t)-u(t-3)]$. Sketch the following functions and find their DC and rms values over a period of 6. Describe symmetry property of each function.

$$y_1(t) = x(t)+x(-t), \quad y_2(t) = x(t)-x(-t), \quad y_3(t) = x(t)-x(t-3),$$

62. A finite-duration sinc pulse is given by $x(t) = \frac{\sin(\pi t)}{\pi t}[u(t+2)-u(t-2)]$. Obtain the DC and rms values of the following pulses over the period of 10 seconds ($|t| < 5$).

a. $y_1(t) = x(t)$
b. $y_2(t) = x(t+2)+x(t)+x(t-2)$
c. $y_3(t) = x(t+1)+x(t)+x(t-1)$

63. a. Consider a finite-duration pulse $\eta(t)$ with area A and energy S. Find the area of the pulse $y(t) = a\eta(bt)$, where a and b are real constants, in terms of A, S, a, and b.
b. Let $\eta(t)$ be a unit rectangular pulse $\eta(t) = u(t)-u(t-1)$. Using the result of part (a) find the area and energy in the pulse shown in Figure 1.58(a) as a function of τ, τ_1, V_0, and V_1.
c. Let $\eta(t)$ be an equilateral triangular pulse specified by

$$\eta(t) = \begin{cases} 1-2\frac{|t|}{\tau}, & -\frac{\tau}{2} < t < \frac{\tau}{2} \\ 0, & \text{elsewhere} \end{cases}$$

Using the results of parts (a) and (b), find the area and energy of the pulse of Figure 1.58(b) in terms of τ, V_0, and V_1.

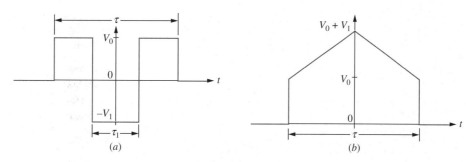

FIGURE 1.58

64. a. Find the area and energy of the following single pulse made of a segment of an exponential function:

$$x(t) = \begin{cases} e^{-t}, & 0 < t < 1 \\ 0, & \text{elsewhere} \end{cases}$$

b. Consider the periodic function $y_1(t) = \sum\limits_{k=-\infty}^{\infty} x(t-k)$, where k is an integer. Plot $y_1(t)$ and find its DC and rms values.

c. Repeat part (b) for

(i) $y_2(t) = \sum\limits_{k=-\infty}^{\infty} x(t-2k)$

(ii) $y_3(t) = \sum\limits_{k=-\infty}^{\infty} (-1)^k x(t-k)$

(iii) $y_4(t) = \sum\limits_{k=-\infty}^{\infty} x(t-2k) - x(t-2k-1)$

(iv) $y_5(t) = \sum\limits_{k=-\infty}^{\infty} x(t-2k) - (-1)^k x(t-2k-1)$

65. a. Find the area and energy of the following single pulse made of a segment of an exponential function:

$$x(t) = \begin{cases} e^{-t}, & 0 < t < 2 \\ 0, & \text{elsewhere} \end{cases}$$

b. Consider the periodic function $y_1(t) = \sum\limits_{k=-\infty}^{\infty} x(t-2k)$, where k is an integer. Plot $y(t)$ and find its DC and rms values.

c. Repeat part (b) for

(i) $y_2(t) = \sum\limits_{k=-\infty}^{\infty} (-1)^k x(t-2k)$

(ii) $y_3(t) = \sum\limits_{k=-\infty}^{\infty} x(t-4k)$

(iii) $y_4(t) = \sum\limits_{k=-\infty}^{\infty} (-1)^k x(t-4k)$

(iv) $y_5(t) = \sum\limits_{k=-\infty}^{\infty} x(2t-k)$

(v) $y_6(t) = \sum\limits_{k=-\infty}^{\infty} (-1)^k x(2t-k)$

66. a. Find the area and energy of the following single pulse made of a segment of an exponential function:

$$x(t) = \begin{cases} e^{-t} & 0 < t < 3 \\ 0 & \text{elsewhere} \end{cases}$$

b. Consider the periodic function $y_1(t) = \sum_{k=-\infty}^{\infty} x(t - 3k)$, where k is an integer. Plot $y(t)$ and find its DC and rms values.

c. Repeat part (b) for

(i) $y_2(t) = \sum_{k=-\infty}^{\infty} (-1)^k x(t - 3k)$ (ii) $y_3(t) = \sum_{k=-\infty}^{\infty} x(t - 6k)$

(iii) $y_4(t) = \sum_{k=-\infty}^{\infty} (-1)^k x(t - 6k)$ (iv) $y_5(t) = \sum_{k=-\infty}^{\infty} x(3t - k)$

(v) $y_6(t) = \sum_{k=-\infty}^{\infty} (-1)^k x(3t - k)$

67. a. Prove that $x(t) = x(-t)$ where

$$x(t) = \begin{cases} 1 + 2\frac{t}{\tau} & -\frac{\tau}{2} < t < 0 \\ 1 - 2\frac{t}{\tau} & 0 < t < \frac{\tau}{2} \\ 0 & \text{elsewhere} \end{cases}$$

Plot $x(t)$ and observe that it is an even function of t.

68. Let $x(t)$ be the even equilateral triangular pulse with a base τ and height 1 specified in problem 67.

a. Express $y(t) = x(t + \tau/2) + x(t) + x(t - \tau/2)$ section-wise in terms of t.

b. Find its rise time (the time it takes for the pulse to grow from 10% to 90% of its final value).

69. In the expression given for $x(t)$ in problem 67 replace t by $t - \tau/2$ and show that the new function is expressed by

$$x(t - \tau) = \begin{cases} 2\frac{t}{\tau}, & 0 < t < \frac{\tau}{2} \\ 2\left(1 - \frac{t}{\tau}\right), & \frac{\tau}{2} < t < \tau \\ 0, & \text{elsewhere} \end{cases}$$

Plot the new function and observe that it is the same as the plot obtained in problem 67, except for a shift to the right by the amount $\tau/2$.

70. Repeat problem 68 for

$$x(t) = \begin{cases} 0.5 + 0.5\cos\left(\frac{2\pi t}{\tau}\right), & -\frac{\tau}{2} < t < \frac{\tau}{2} \\ 0, & \text{elsewhere} \end{cases}$$

71. In the expression given for $x(t)$ in problem 70 replace t by $t - \tau/2$ and show that the new function is expressed by

$$x(t) = \begin{cases} 0.5 - 0.5\cos\left(\frac{2\pi t}{\tau}\right), & 0 < t < \tau \\ 0, & \text{elsewhere} \end{cases}$$

Plot the new function and observe that it is the same as the plot of $x(t)$ from problem 70, except for a shift to the right by the amount $\tau/2$.

72. Given the one-sided exponential function $x(t) = e^{-t}u(t)$, express the following as functions of time and plot them:

a. $x(t - 1)$, $x(t + 1)$

b. $x(-t)$, $x(-t + 1)$, $x(-t - 1)$

c. $x(4t)$, $x(0.25t + 1)$, $x(-0.25t + 0.5)$

73. The unit-step function $u(t)$ is defined by $u(t) = \begin{cases} 1, & t \geq 0 \\ 0, & t < 0 \end{cases}$. Plot

   a. $u(t)$, $u(t-2)$, $u(t+2)$
   b. $u(-t)$, $u(-t+2)$, $u(-t-2)$
   c. $u(2t)$, $u(-2t+6)$, $u(2t-6)$

74. Consider the single pulse specified by

$$x(t) = \begin{cases} e^{-t}, & 0 < t < 1 \\ 0, & \text{elsewhere} \end{cases}$$

   a. Plot $x(t)$, $x(t+1)$, $x(t-1)$, $x(-t)$, $x(-t+1)$, and $x(-t-1)$.
   b. By examining the graphs verify that the plot of $x(-t+1)$ may be found by flipping the plot of $x(t+1)$ around the vertical axis at $t = 0$. Similarly, the plot of $x(-t-1)$ is that of $x(t-1)$ flipped around the vertical line $t = 0$.
   c. Consider periodic functions $y_i(t)$, $i = 1, 2, 3, 4$ with periods $T = 2$, where, for one period,

$$\begin{aligned} y_1(t) &= x(t) + x(-t), & -1 < t < 1 \\ y_2(t) &= x(t) - x(-t), & -1 < t < 1 \\ y_3(t) &= x(t) - x(t-1), & 0 < t < 2 \\ y_3(t) &= x(t) + x(t-1), & 0 < t < 2 \end{aligned}$$

Plot $y_i(t)$, $i = 1, 2, 3, 4$ and find their average and rms values.

75. Repeat problem 74 for

$$x(t) = \begin{cases} t, & 0 < t < 1 \\ 0, & \text{elsewhere} \end{cases}$$

76. Consider the periodic functions $x_1(t)$ and $x_2(t)$ with period τ, where, for one period,

$$\begin{aligned} x_1(t) &= e^{-t/\tau}, & 0 < t < \tau \\ x_2(t) &= e^{-2t/\tau}, & 0 < t < \tau \end{aligned}$$

   a. Plot $x_1(t)$ and $x_2(t)$ and find their average and rms values.
   b. Find mathematical expressions for $dx_1(t)/dt$ and $dx_2(t)/dt$, and sketch them.

77. A 1-kHz periodic voltage $v(t)$ grows exponentially from -2.3 V to 2.3 V and decays back to -2.3 V, in both cases with a time constant of 0.5 msec.

   a. Find a mathematical expression for $v(t)$ and plot it. Find its rise and fall times, average, and rms values.
   b. Find a mathematical expression for $dv(t)/dt$ and sketch it.

78. A 1-kHz periodic voltage $v(t)$ grows exponentially from 1 to 4 V with a time constant of 0.15 msec, and decays back to 1 V with a time constant of 0.6 msec.

   a. Find a mathematical expression for $v(t)$ and plot it. Find its rise and fall times, average, and rms values.
   b. Let $x(t) = v(t) - V_0$, where V_0 is the average of $v(t)$. Find the rms value of $x(t)$ and relate it to V_0 and the rms value of $v(t)$.

79. Let $x(t) = \sin(2\pi f t)$. Define

$$y_1(t) = \begin{cases} x(t), & \text{when } x(t) > V_0 \\ 0, & \text{otherwise} \end{cases}$$

$$y_2(t) = \begin{cases} x(t), & \text{when } |x(t)| > V_0 \\ 0, & \text{otherwise} \end{cases}$$

where V_0 is a nonnegative constant.

a. Plot $y_1(t)$ and $y_2(t)$ and find their average and rms values as functions of V_0.

b. Find mathematical expressions for $dy_i(t)/dt$ $(i = 1, 2)$ and sketch them.

80. Consider a single cycle of a sinusoid $x(t)$ and its first and second integrals $y(t)$ and $z(t)$, respectively.

$$x(t) = \begin{cases} \sin t, & 0 < t < 2\pi \\ 0, & \text{elsewhere} \end{cases}$$

$$y(t) = \int_{-\infty}^{t} x(t)dt = \begin{cases} 1 - \cos t, & 0 < t < 2\pi \\ 0, & \text{elsewhere} \end{cases}$$

$$z(t) = \int_{-\infty}^{t} y(t)dt = \begin{cases} 0, & t < 0 \\ t - \sin t, & 0 < t < 2\pi \\ 2\pi, & t \geq 2\pi \end{cases}$$

a. Sketch $x(t)$, $y(t)$, and $z(t)$.

b. Show that the area under each half cycle of $x(t)$ is 2, its DC value is zero, and its rms value defined over the period $0 < t < 2\pi$ is $1/\sqrt{2}$.

c. Show that $y(t)$ is a finite-duration positive pulse that lasts 2π seconds with a maximum value of 2 and an area equal to 2π.

d. Show that $z(t)$ has a steady-state DC level of 2π and reaches it in 2π seconds.

81. A periodic waveform $f(t)$ with period T is made of isosceles triangular pulses (height $= 1$, base $= T$). It is passed through a level detector set at a positive-level threshold V_0, producing a rectangular output waveform $v(t)$ such that

$$v(t) = \begin{cases} 1, & f \geq V_0 \\ 0, & f < V_0 \end{cases}$$

Find the DC, rms, and average power values of $v(t)$ as functions of period T and the threshold level V_0.

82. The average power of a voltage signal is one Watt and its average value is zero. A DC level of V_0 is added to the signal. Find the new average power.

83. **On-off control.** A sinusoidal wave $\sin(2\pi f t)$ is gated by a periodic rectangular pulse with a period T and a duty cycle $(T_1/T) \times 100\%$. One cycle of the gated signal is given by

$$x(t) = \begin{cases} V_0 \sin(2\pi f t), & 0 \leq t \leq T_1 \\ 0, & T_1 \leq t \leq T \end{cases}$$

where $T_1 = k_1/f$, $T = k/f$, and k_1 and k are integers. Find the rms value of $x(t)$ as a function of $\alpha = k_1/k$.

84. **Light dimmer.** A sinusoidal signal $\sin(2\pi t/T)$ is sent through a gating device such that only an initial portion of each half cycle passes through. One period of the resulting signal is expressed by

$$v(t) = \begin{cases} \sin(2\pi t/T), & 0 \leq t \leq \tau \quad \text{and } T/2 \leq t \leq (T/2 + \tau) \\ 0, & \tau < t < T/2 \quad \text{and } (T/2 + \tau) < t < T \end{cases}$$

a. Let $\tau = \alpha T$ $(0 < \alpha < 0.5)$. Show that the rms of $v(t)$ as a function of α is given by

$$V_{rms}(\alpha) = \frac{2}{T} \int_0^{\alpha T} \sin(2\pi t/T)dt$$

Find $V_{rms}(\alpha)$ and plot it vs. α.

b. Observe that the plot is not linear. Devise and use a nonlinear scale for the α-axis such that the plot of $V_{rms}(\alpha)$ vs. α under the new scale becomes a linear one.

85. By using a computational/plotting software of your choice plot the function $x(t) = 10[e^{-t} - e^{-t/5}]u(t)$. Then use measurements off the graph to model it by the sum of two exponentials. Compare the model with $x(t)$.

86. In an electric device, a pair of input output voltage signals is given by $v_1(t) = 10\cos t$ and $v_2(t) = 2\sin(t + \pi/3)$, respectively.

a. Show that their phasor representations are $V_1 = 10\angle 0°$ and $V_2 = 2\angle -30°$.
b. Determine $H = V_2/V_1$.
c. By using a computational/plotting software of your choice, plot $v_1(t)$ and $v_2(t)$ on the same graph (sharing the same axes).
d. From measurements off the graph model the plots by $\hat{v}_1(t) = a_1 \cos(2\pi f t + \theta_1)$ and $\hat{v}_2(t) = a_2 \cos(2\pi f t + \theta_2)$. Then find the ratio of their phasors and compare with the original voltages.

87. Plot the exponential function $x(t) = 10e^{-t/5}\cos(10t)u(t)$. Then use measurements off the graph to model it. Compare the model with $x(t)$.

88. Let $x(t)$ be a continuous-time function that changes smoothly. Let $\zeta(t)$ (pronounced *zeta of t*) be the rectangular pulse

$$\zeta(t) = \begin{cases} 1, & 0 < t < T \\ 0, & \text{elsewhere} \end{cases}$$

$x(t)$ may be approximated by $\hat{x}(t)$ [made of the sum of rectangular pulses shifted in time and scaled by $x(nT)$]:

$$\hat{x}(t) = \sum_{n=-\infty}^{\infty} x(nT)\zeta(t - nT)$$

Apply the above approximation to the periodic functions listed below. Note that in all cases the period is 1 second.

$$x(t) = t, \qquad\qquad 0 < t \leq 1, \qquad x(t) = x(t \pm n), \ n \text{ an integer}$$
$$x(t) = \cos(2\pi t), \qquad\qquad \text{all t,}$$
$$x(t) = e^{-t}, \qquad\qquad 0 < t \leq 1, \qquad x(t) = x(t \pm n), \ n \text{ an integer}$$
$$x(t) = e^{-|t|}, \qquad -0.5 < t \leq 0.5, \quad x(t) = x(t \pm n), \ n \text{ an integer}$$

Define the approximation error by $\epsilon(t) = x(t) - \hat{x}(t)$. Find $\mathcal{E} = $ rms of $\epsilon(t)$:

$$\mathcal{E} = \sqrt{\int_{t_0}^{t_0+1} \epsilon^2(t)dt}$$

as a function of the sampling interval T and find its limit as $T \to 0$.

89. The unit-step function may be approximated by analytic functions. One such function is the *sigmoid* containing a parameter a and described by

$$x(t) = \frac{1}{1 + e^{-at}}$$

a. Verify that

$$x(-\infty) = 0, \quad x(\infty) = 1 \quad x(0) = 0.5 \quad \text{and} \quad \frac{dx}{dt}\Big|_{t=0} = a$$

b. By simulation method explore if
   (i) $\lim\limits_{a\to\infty} x(t) = u(t)$
   (ii) $\lim\limits_{a\to\infty} \dfrac{dx}{dt} = k\delta(t)$

90. It is desired to approximate the even rectangular pulse $x(t) = u(t+1) - u(t-1)$ by analytic functions. The following two families of functions, in which n is a positive integer, are proposed:

a. $\eta_n(t) = \dfrac{1}{1+t^{2n}}$

b. $\zeta_n(t) = \dfrac{1}{100} + \sum\limits_{k=1}^{n} \left[\dfrac{2}{k\pi} \sin\left(\dfrac{k\pi}{100}\right)\right] \cos\left(\dfrac{k\pi t}{100}\right)$

By simulation method explore if

a. $\lim\limits_{n\to\infty} \eta_n(t) = x(t)$

b. $\lim\limits_{n\to\infty} \zeta_n(t) = x(t)$

and examine how $\eta_n(t)$ and $\zeta_n(t)$ would or would not converge to $x(t)$ in the neighborhood of $t = \pm 1$.

91. a. Sketch the finite-duration pulse made of the half cycle of a cosine

$$x(t) = \begin{cases} \frac{1}{2}\cos t, & -\pi/2 < t < \pi/2 \\ 0, & \text{elsewhere} \end{cases}$$

and show that the area under the pulse is 1.

b. Sketch the family of finite-duration pulses $y_\omega(t) = \omega x(\omega t)$

$$y_\omega(t) = \begin{cases} \frac{\omega}{2}\cos\omega t, & -\pi/2 < \omega t < \pi/2 \\ 0, & \text{elsewhere} \end{cases}$$

for $\omega = n/\pi$, $n = 1, 2, 3, 4, 5$. Note that as ω increases, the pulse becomes narrower and taller. Show that the area under the pulse always remains equal to 1:

$$\int_{-\infty}^{\infty} y_\omega(t)dt = 1, \quad \text{for all } \omega$$

c. Argue that

$$\lim_{\omega\to\infty} \int_{-\infty}^{\infty} y_\omega(t)\phi(t)dt = \phi(0)$$

where $\phi(t)$ is a smooth function in the neighborhood of $t = 0$. Extend the argument to

$$\lim_{\omega\to\infty} \int_{-\infty}^{\infty} y_\omega(t - t_0)\phi(t)dt = \phi(t_0)$$

Argue that in that sense,

$$\lim_{\omega\to\infty} y_\omega(t) = \delta(t)$$

d. Find

$$\int_{-\infty}^{t} y_\omega(t)dt$$

and argue that

$$\lim_{\omega \to \infty} \int_{-\infty}^{t} y_\omega(t)dt = u(t)$$

e. Provide three other pulse functions that exhibit limit properties similar to those of $y_\omega(t)$ given in this problem.

92. a. Sketch the family of *sinc* functions $y_n(t) = \frac{\sin(\pi nt)}{\pi t}$ for $n = 1, 2, 3, 4, 5$. Note that as n increases, the main lobe becomes narrower and taller. Show that the net area under the function always remains equal to 1.

  b. Can you assert that

$$\lim_{n \to \infty} \int_{-\infty}^{\infty} y_n(t)\phi(t)dt = \phi(0)$$

where $\phi(t)$ is a smooth function in the neighborhood of $t = 0$? And, consequently, that

$$\lim_{n \to \infty} y_n(t) \stackrel{?}{=} \delta(t)$$

  c. Sketch

$$\int_{-\infty}^{t} y_n(t)dt$$

for $n = 1, 2, 3, 4, 5$ and argue that

$$\lim_{n \to \infty} \int_{-\infty}^{t} y_n(t)dt = u(t)$$

93. Consider a finite-duration pulse $x_a(t)$ made of a single cycle of a sinusoid

$$x_a(t) = \begin{cases} a^2 \sin at, & 0 < t < \frac{2\pi}{a} \\ 0, & \text{elsewhere} \end{cases}$$

  a. Sketch $x_a(t)$ for $a = 2n\pi$, $n = 1, 2, 3, 4, 5$. Find the area under each half cycle of $x(t)$ and its rms value defined over the duration of the pulse.

  b. Let

$$y_a(t) = \int_{-\infty}^{t} x_a(t)dt$$

Find $y_a(t)$ and sketch it for $a = 2n\pi$, $n = 1, 2, 3, 4, 5$. Show that $y_a(t)$ is a finite-duration positive pulse. Find its duration, its maximum value, and its area.

  c. Let

$$z_a(t) = \int_{-\infty}^{t} y_a(t)dt$$

Find $z_a(t)$ and sketch it for $a = 2n\pi$, $n = 1, 2, 3, 4, 5$. Find its DC steady-state value and the transition time to reach the steady-state value.

  d. Argue that

   (i) $\lim_{a \to \infty} y_a(t) = 2\pi \delta(t)$

   (ii) $\lim_{a \to \infty} z_a(t) = 2\pi u(t)$

94. Repeat problem 93 starting with the finite-duration odd pulse

$$x_a(t) = \begin{cases} -a^2 \sin at, & -\pi/a < t < \pi/a \\ 0, & \text{elsewhere} \end{cases}$$

and find the limits as $a \to \infty$.

95. Repeat problem 93 starting with the finite-duration odd pulse

$$x_b(t) = \begin{cases} b^2, & -1/b < t < 0 \\ -b^2, & 0 < t < 1/b \\ 0, & \text{elsewhere} \end{cases}$$

and find the limits as $b \to \infty$.

96. Repeat problem 93 starting with the finite-duration odd pulse

$$x_c(t) = \begin{cases} e^{ct}, & t < 0 \\ -e^{-ct}, & t > 0 \end{cases}$$

and find the limits as $c \to \infty$.

97. Consider the sinc function $x(t) = \frac{\sin(at)}{bt}$, $-\infty < t < \infty$.

a. Plot $x(t)$ and show that its maximum value is a/b and the area under it is π/b.

b. Consider the family of functions

$$y_k(t) = \begin{cases} x(t), & -k\pi/a < t < (k+1)\pi/a, \ k = 1, 2, 3, 4 \cdots \\ 0, & \text{elsewhere} \end{cases}$$

Show that $\lim_{a \to \infty} y_k(t) = \alpha_k \delta(t)$, where α_k are constants.

c. Consider the family of periodic functions

$$z(t) = x(t), \quad -\frac{3\pi}{2} < t < \frac{3\pi}{2}, \quad z(t) = z(t + T), \quad T = 3\pi, \quad \text{for all } t$$

Show that $z(t)$ is an even function and $\lim_{a \to \infty} z(t)$ is a train of impulses spaced every 3π seconds.

98. **Response of *RC* to step voltage and its approximation.** A 1-kΩ resistor and a 1,000-μF capacitor are connected in series with a voltage source $v_1(t)$ as in Figure 1.26. With $v_1(t) = u(t)$ (V), the capacitor current and voltage signals are shown to be $i(t) = e^{-t}u(t)$ (mA) and $v_2(t) = (1 - e^{-t})u(t)$ (V), respectively. Plot $i(t)$ and $v_2(t)$ for $-\tau < t < 5\tau$, where τ is their common time constant. For each signal determine initial and final values. An exponential function can be approximated as $e^x \approx 1 + x$ near $|x| \approx 0$ (as evidenced by the Taylor series expansion near zero of the exponential, or the behavior of its tangent at $x = 0$). Using that approximation, find the behavior of the current and voltage signals for $-0.001\tau < t < 0.005\tau$ and plot them. For the current signal find the maximum percentage difference between the exact and approximate values within the above range.

99. **Response of *RC* to narrow pulse voltage and its approximation.** Let the voltage source in the *RC* circuit of Figure 1.26 be a 1-kV, 1-ms pulse [i.e., $v_1(t) = 10^3[u(t) - u(t - 0.001)]$ V]. Obtain and plot $i(t)$ and $v_2(t)$ for the range of $-\tau < t < 5\tau$, and their linear approximations during $-0.001\tau < t < 0.005\tau$.

100. **Modem standard (QAM).** The 2400 bit/sec CCITT V.22 bis modem[11] uses the 16-symbol QAM signal constellation shown in Figure 1.59 at a bit rate of 2,400 bit/sec. Compute its average power (P_{avg}) and the minimum distance (d_{min}) between the subsignals. Give the coordinates of the four subsignals A, B, C, and D shown on the figure.

---

[11] International Telegraph and Telephone Consultative Committee (CCITT) has been renamed ITU-T. It is the standardization division for telecommunication of the International Telecommunication Union (ITU).

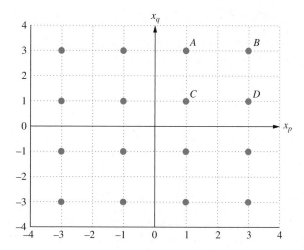

FIGURE 1.59 16-QAM constellation signaling for 2,400 bit/sec V.22 bis computer modem.

101. **Barker code.** A Barker code is a binary sequence $c_N(n)$, $n = 1, 2, \cdots N$ (shown by $+$ and $-$, or 1 and -1) with code length $N = 1, 2 \cdots 11, 13$, as listed in the table below.

N	$c_N(n)$
2	$1 -1$ or $-1\,1$
3	$1\,1 -1$
4	$1 -1\,1\,1$ or $1 -1 -1 -1$
5	$1\,1\,1 -1\,1$
7	$1\,1\,1 -1 -1\,1 -1$
11	$1\,1\,1 -1 -1 -1\,1 -1 -1\,1 -1$
13	$1\,1\,1\,1\,1 -1 -1\,1\,1 -1\,1 -1\,1$

To obtain an indicator for the pattern of dependency between 1s and -1s in the Barker code, for each code sequence $c(n)$ listed in the table above define $\rho(k)$ as the index of correlation between its elements:

$$\rho(k) = \sum_{n=1}^{N} c(n)c(n-k), \ k = 0, 1, 2, \ldots, N$$

For codes with lengths $N = 2, 3, 4, 5, 7, 11, 13$ find and plot $\rho(k)$ and determine the ratio $\rho(k)/\rho(0)$.

102. **Signal detection.** A deterministic periodic binary sequence $x(n)$ made of regularly alternating 1s and 0s is transmitted at the rate of 10^6 bits per second. For transmission it uses an on-off encoding scheme where the 1 bits are transmitted by $s(t)$ and the 0 bits are transmitted by silent periods each 1 μs long. The encoded signal is a periodic signal $x(t)$ given by

$$x(t) = \begin{cases} s(t), & 0 < t < 0.5 \ \mu s \\ 0, & 0.5 \ \mu s < t < 1 \ \mu s \end{cases}, \quad x(t) = x(t + 1\mu s) \text{ all t}$$

To detect the occurrence of a 1 or 0, the received signal $\hat{x}(t)$ [generally a corrupted version of $x(t)$ due to communication noise and interference] is continually matched with the template $s(t)$ and the results are observed every $T = 1/M \ \mu$s (i.e., at intervals $t = k/M \ \mu$s, $k = 0, 1, 2 \cdots, M - 1$). The integer M indicates the frequency (or number) of observations during transmission of each bit. The present problem illustrates the effect of $s(t)$ on separability of 1s and 0s at the receiver. For simplicity we assume no interference or noise; that is, $\hat{x}(t) = x(t)$. Assume a matching scheme where $x(t)$ is multiplied by a sliding $s(t)$-shaped window slid every $1/M$ µs. The results are integrated over 1 µs, resulting in observation values $y(k)$:

$$y(k) = \int_0^{1\ \mu s} x(t)s(t - k\mu s)dt, \quad k = 0, 1, 2 \cdots (M - 1)$$

a. Let $s(t)$ be a 1-volt rectangular 1 µs pulse. Find the maximum value y_{Max} and its location k_0. Find the distance between the peak value and its immediate neighbor $y_{Max} - y(k_0 - 1)$. Consider this to be a *separability distance*.

b. Repeat for a binary $s(t) = \pm 1$ made of a 7-segment Barker code and observations every $1/7$ µs. Compare with the corresponding value obtained in part (a).

c. Repeat for a binary $s(t) = \pm 1$ made of an 11-segment Barker code and observations every $1/11$ µs. Compare with corresponding values obtained in parts (a) and (b).

d. Repeat for a binary $s(t) = \pm 1$ made of a 13-segment Barker code and observations every $1/13$ µs. Compare with corresponding values obtained in parts (a), (b), and (c).

e. Discuss the gain in *separability distance* versus the increase in eventual transmission rate as you increase the length of the Barker code used.

103. A ranging signal for space communication and control uses a binary code c_n of order n with a period $T_n = 2^n$. The signal is a periodic binary function produced by assigning an amplitude of +1 to the high values of the code and 0 to the low values. The $(n + 1)$-th code $c^{(n+1)}$ is constructed by cascading c_n with its logical complement. Alternatively, the code of order n may be generated through the recursive equation

$$C^{n+1}(t) = c^n(t) - c^n(t - Tn)$$

starting from $c_0(t) = -1$. Examples of the signal for the first five codes ($n = 0, 1, 2, 3, 4$) are as follows:

$c_0 = 0$

$c_1 = 01$

$c_2 = 0110$

$c_3 = 01101001$

$c_4 = 0110100110010110$

Find the DC and rms values of the codes for order $n = 4, 8$, and 10.

104. Signal encoding. Consider a binary sequence $x(n)$ made of $+$ and $-$ (or, equivalently, $+1$ and -1) which occur randomly with equal probability of 50% and independently of each other. By encoding, we convert $x(n)$ into a time signal $x(t)$. In this problem you compare two such schemes called *on-off* and *polar encoding*.

a. In on-off encoding, the 1 bits are transmitted by $r_a(t) = R_a \sin(2\pi f t)$, $0 < t < T$ (shown by the phasor $R_a \angle 0°$) with an energy content of $T R_a^2/2$, and the 0 bits are transmitted by no pulse (a silent period T seconds long shown by the phasor $0 \angle 0°$, with zero energy content). Define the *distance* between the encoded 1s and 0s to be $D_a = T R_a^2/2$. The resulting encoded time signal is $x_a(t)$. Find the average power P_a in $x_a(t)$.

b. In polar encoding, the 1 and 0 bits are transmitted by $r_b(t) = R_b \sin(2\pi f t)$ and $-r_b(t)$, respectively (and shown by the phasors $R_b \angle 0°$ and $R_b \angle 180°$, each with the same energy content $T R_b^2/2$). However, because of the 180° phase shift, the *distance* between 1 and 0 will be $D_b = T R_b^2$. The resulting encoded time signal is $x_b(t)$. Find the average power P_b in $x_b(t)$.

c. Find the relationship between R_a and R_b if distances between 1s and 0s are to be the same. Then compare P_a with P_b.

105. **Morse code.** The international Morse code[12] converts symbols (letters, numbers, and some other characters) to a sequences of short (dots, pronounced *dit*) and long (dashes, pronounced *dah*) tone bursts. In terms of timing, a *dit* takes one unit of time (also called an element), a *dah* takes three units of time, and a pause between a *dit* and a *dah* takes one dit of time. The pause between letters lasts three units, and that between words lasts seven units of time. The code is given in the table below.

Symbol	Morse Code	Symbol	Morse Code	Symbol	Morse Code
A	. —	U	. . —	0	— — — — —
B	— . . .	V	. . . —	1	. — — — —
C	— . — .	W	. — —	2	. . — — —
D	— . .	X	— . . —	3	. . . — —
E	.	Y	— . — —	4	. . . . —
F	. . — .	Z	— — . .	5	. . . . .
G	— — .			6	— . . . .
H	. . . .			7	— — . . .
I	. .			8	— — — . .
J	. — — —			9	— — — — .
K	— . —			Period	. — . — . —
L	. — . .			Comma	— — . . — —
M	— —			?	. . — — . .
N	— .			Hyphen	— . . . . —
O	— — —			Apostrophe	. — — — — .
P	. — — .			Colon	— — — . . .
Q	— — . —			Quotation	. — . . — .
R	. — .			/	— . . — .
S	. . .			@	. — — . — .
T	—				

a. The standard word for measuring transmission speed is the word *paris*. Find the Morse code for it, count the number of *dits*, *dahs*, and spaces, and determine how many elements it takes.

b. Find the Morse code sequence for the phrase *"An introduction to signals and systems, @ Cal Poly State University."*

c. Count the number of *dits*, *dahs*, and spaces, and find their percentage in the code sequence of part (b).

d. Consider a tone burst $r(t) = \sin(2\pi f t)$, $0 < t < T$, lasting T seconds with an rms value of $\sqrt{2}/2$ V, and assume each *dit* is transmitted by an $r(t)$. This is called continuous wave (CW) operation. Find the total energy in the signal representing the code sequence found in (a), its duration, and its average power during transmission.

e. Assume a CW operator using a bencher iambic key transmits 30 WPM (words per minute, based on the standard of 50 time-elements per word). Determine how long it takes the operator to transmit the phrase in (b), and the average power needed during transmission.

---

[12]The original code was invented by Samuel Morse in 1838 and was used for the first time in 1844 to send a message over a telegraph line between Baltimore and Washington. The international Morse code, devised in 1851 and currently in use with some modifications, simplified the code. Proficiency in communication using Morse code, which has been a requirement for obtaining the amater Radio license, was phased out by 2010.

106. **Mapping a code.** Let a Morse code sequence (made of *dits*, *dahs*, and pauses) be mapped into a sequence of tertiary numbers made of 1, −1, and 0.

a. Map the Morse code sequence obtained in problem 105(b) into a new sequence based on the "tertiary code" described above.
b. Count the number of 1s, −1s, and 0s in the tertiary sequence and find their percentage in the code sequence.
c. Assume the 1 and −1 bits in the tertiary code sequence are transmitted by $r(t) = \sin(2\pi f t)$ and $-r(t) = -\sin(2\pi f t) = \sin(2\pi f t - 180°)$ of duration T (also called *polar* or phase-shift keying), respectively. A 0 bit is signaled by a silent period of T seconds. Find the total energy in the signal representing the code sequence, its duration, and its average power during transmission.

107. **ASCII code.** A popular code for converting symbols to 8-bit binary numbers is the ASCII (American Standard Code for Information Interchange) code.

a. Obtain an ASCII code table, then find the ASCII code sequence for the phrase *"An introduction to signals and systems, @ Cal Poly State University."*
b. Count the number of 1s and 0s and find their percentage in the code sequence.
c. Assume each 1 is transmitted by $r(t) = \sin(2\pi f t)$ and each 0 bit is signaled by a silent period of T seconds. Find the total energy in the signal representing the code sequence, its duration, and its average power during transmission.

1.25 Project 1: Binary Signal in Noise

Summary

This project formulates detection of a signal in noise and errors associated with it when the signal is corrupted by additive noise.

The Signal

In a binary communication system the high bit is transmitted as the sinc voltage function $\zeta(t) = \sin(2\pi t)/(2\pi t)$. For example, the data sequence $z(k) = \{0, 1, 0, 0, 1, 0, 1, 1, 0, 0, 1, 0, 0\}$ is represented by $s(t) = \zeta(t) + \zeta(t-3) + \zeta(t-5) + \zeta(t-6) + \zeta(t-9)$. See Figure 1.60. Note that in this scheme all zero crossings occur at $t = k$, where k is an integer. At $t = k$ the signal value is, therefore, either zero (for a low bit) or one (for a high bit). The presence of a high bit at $t = k$ can be detected from the value of $s(t)$ at such times, as the interferences from neighboring bits are zero. To detect the presence or absence of a high bit, the signal is passed through a threshold level λ and its value at $t = k$ is examined. Detection is unambiguous and error-free if the signal is not corrupted by noise (e.g., no noise is added to the signal) and if $\zeta(t)$ extends over the range $-\infty < t < \infty$.

Detection of Signal in Noise

With additive noise $n(t)$, the received waveform is $x(t) = s(t) + n(t)$. At $t = k$

$$x(k) = \begin{cases} 1 + n(k), & \text{if a high bit was sent} \\ n(k), & \text{if a low bit was sent} \end{cases}$$

and the descision rule becomes

$$\begin{cases} x(k) > \lambda & \text{the transmitted bit is assumed high} \\ x(k) < \lambda & \text{the transmitted bit is assumed low} \end{cases}$$

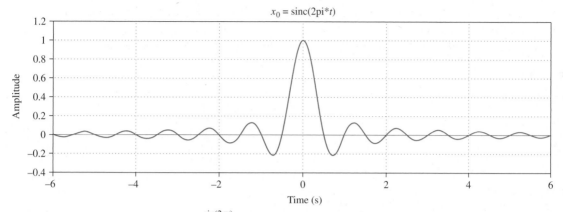

(a) Plot of $\zeta(t) = \dfrac{\sin(2\pi t)}{2\pi t}$ representing a high bit in a binary communication system.

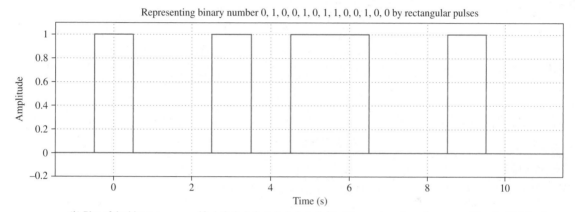

(b) Plot of the binary sequence {0, 1, 0, 0, 1, 0, 1, 1, 0, 0, 1, 0, 0} with a rectangular pulse representing the high level.

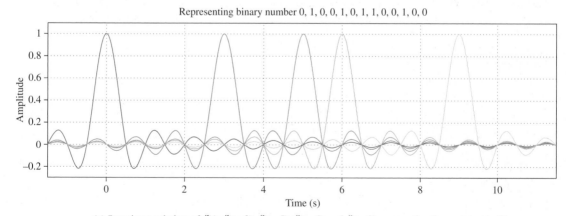

(c) Superimposed plots of $\zeta(t)$, $\zeta(t-3)$, $\zeta(t-5)$, $\zeta(t-6)$, and $\zeta(t-9)$ representing the sequence in (b).

FIGURE 1.60 Representation of a binary number by sinc signals. The signal at $t = k$ is either zero or 1.

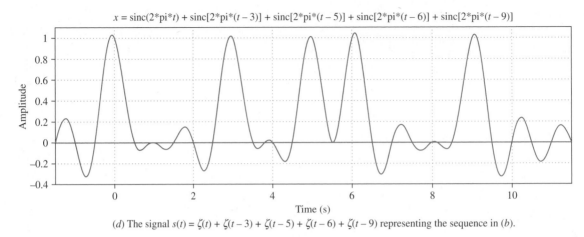

(d) The signal $s(t) = \zeta(t) + \zeta(t-3) + \zeta(t-5) + \zeta(t-6) + \zeta(t-9)$ representing the sequence in (b).

FIGURE 1.60 (*Continued*)

Under this rule, a high bit may be diagnosed as a low (to be called a miss) or a low bit may be diagnosed as a high (to be called a false alarm). To compute error rates, we assume that the noise has a random value equally likely from -1 to 1 volt and independent from neighboring values. Mathematically, $n(t)$ is a random variable (R.V.) with uniform probability distribution between -1 and 1 volt and zero correlation between neighboring values: $p(n) = 0.5$, $-1 < n < 1$ [see Figure 1.61(a)]. For a low bit, $x(k) = n(k)$; $x(k)$ is an R.V. with the same probability distribution function as noise. Then the probability of a false alarm is $P(x > \lambda) = 0.5(1 - \lambda)$, where P indicates *"the probability of."* For a high bit, $x(k) = 1 + n(k)$; $x(k)$ is an R.V. with a uniform probability distribution function between 0 and 2 [see Figure 1.61(b)]. Then the probability of a false alarm is $P(x > \lambda) = 0.5(2 - \lambda)$. From Figure 1.61(a) and (b) we can readily deduce that for $0 < \lambda < 1$.

$$\text{Probability of detection} \equiv P_d = 1 - 0.5\lambda$$

$$\text{Probability of false alarm} \equiv P_f = 0.5(1 - \lambda)$$

The choice of threshold level follows some optimization criteria; for example, minimizing a cost function such as $c(\lambda) = 1 - p_d + \alpha p_f$ where $\alpha > 0$ is the relative cost of a false alarm compared to a miss. In the present case, $p_m = 1 - p_d$ and p_f are given in Figure 1.61(c) and (d).

Simulation

To simulate the project, proceed as follows:

(i) Generate a binary sequence $z(k)$ of length N in which the high and low bits occur randomly and equally likely.

(ii) Obtain $s(t)$ from $z(k)$.

(iii) Generate a random sequence $n(t)$ with values between -1 and 1 with uniform probability.

(iv) Obtain $x(t) = s(t) + n(t)$ and $x(k)$ (with k an integer).

(v) Pass $x(k)$ through a level detector λ to obtain the binary sequence $y(k)$ such that $y(k) = \begin{cases} 1, & x(k) \geq \lambda \\ 0, & x(k) < \lambda \end{cases}$. Do this for $-1 < \lambda < 2$ in steps of size 0.2 and plot $y(k)$.

(vi) For each λ compare $y(k)$ with $z(k)$ to find the total number of misses (N_m) and false detections (N_f). (If N is large enough, you may estimate $1 - p_d$ and p_f by N_m/N and N_f/N, respectively.)

(a) Probability distribution of the amplitude of $x(k) = n(k)$ is uniform with a mean of 0 when the signal is absent (low bit). A false detection occurs when $x(k) = n(k) > \lambda$, in which case $p_f = 0.5(1 - \lambda)$. In this figure, $\lambda = 0.5$ results in $p_f = 0.25$ (shaded area)

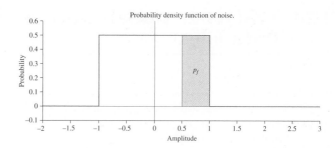

(b) Probability distribution of the amplitude of $x(k) = s(k) + n(k)$ is uniform with a mean of 1 when the signal is present (high bit). The signal is detected when $x(k) = s(k) + n(k) > \lambda$, in which case $p_d = 0.5(2 - \lambda)$. In this figure, $\lambda = 0.5$ results in $p_d = 0.75$ (shaded area)

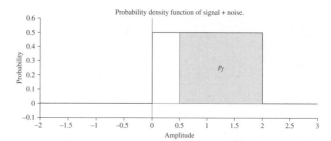

(c) Probability of a false alarm as a function of threshold level. In the present case,

$$p_f = \begin{cases} 1, & \lambda < -1 \\ 0.5(1 - \lambda), & -1 < \lambda < 1 \\ 0, & \lambda > 1 \end{cases}$$

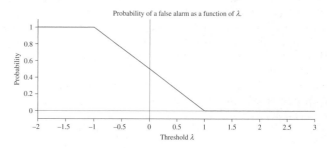

(d) Probability of a miss as a function of threshold level. In the present case,

$$p_m = 1 - p_d = \begin{cases} 0, & \lambda < 0 \\ 0.5\lambda, & 0 < \lambda < 2 \\ 1, & \lambda > 2 \end{cases}$$

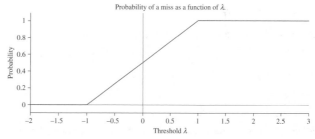

FIGURE 1.61

(vii) Let each miss cost 1 cent and each false detection α. The total cost is, therefore, $C(\alpha) = N_m + \alpha N_f$. For each λ find $c(\alpha)$, $\alpha = 0.1, 0.2, 0.5, 1, 2, 5, 10$. Plot the family of curves showing $c(\alpha, \lambda)$ versus λ with α as a parameter. For each α determine the λ that produces the smallest cost.

Discussion and Conclusions

Discuss the advantages, potentials, and limitations of the modulation method studied in this project.

1.26 Project 2: Signals Stored as Computer Data Files

In this project you will access the data used to produce Figures 1.1, 1.2, 1.4, 1.6, and 1.7 in this chapter. The Internet sources for Figures 1.1, 1.2, and 1.4 are given along each figure. The data for Figures 1.6 and 1.7 are available on the book's website. Using a software package or a computer tool of your choice, retrieve the data and reproduce the above figures. Then store them as ASCII files.

Chapter

2

Sinusoids

Contents

Introduction and Summary

Sinusoidal functions form the cornerstone of the analysis of signals and systems. Many signals, whether natural or synthetic, continuous or discrete, contain repetitive features. A few examples of natural phenomena that exhibit repetitive features include: day and night; seasonal variations of weather attributes such as temperature, humidity, and rainfall; the motion of heavenly objects (sun, moon, planets, and stars); ocean waves; the flappings of the wings of a hummingbird the galloping of a horse; the sounds of a person's heartbeat, breathing, blood pressure in the veins, and swinging of hands when walking; eruption of geysers; radiation from a quasar; oscillations of a cesium clock; and the yearly number of sunspots. Wheels, windmills, rotating machines, the pendulum of a grandfather clock, the motion and sound of a pneumatic drill, the whistle of a train, the sound of a train's steam engine, the *click-clack* sound of a train on the rail, the sound of a car engine, the roar of a jet engine, radio and radar signals, and AC power are some human-made examples. At the root of many such phenomena we detect oscillatory components with one or more frequencies. The building block that describes them is the sinusoidal function, which is the main theme of this chapter. Sinusoids have been used for millennia to build mathematical models of natural signals and systems. In ancient astronomy, the sine of an angle became a prime tool in modeling cosmic systems. It was noticed that the combination of sinusoids can account for some of their complicated motions. The art of computing sinusoids became precise and refined.

In the early part of the 18th century, mathematicians noted that a finite sum of harmonically related sinusoids is itself a periodic function. In 1807, Fourier suggested that a periodic function may be represented by an infinite series of sinusoids, later to be called its Fourier series expansion. The observation that a sinusoidal excitation evokes a sinusoidal response from a class of systems called linear time-invariant (LTI) has made sinusoids and the Fourier method of waveform analysis a very important tool for linear systems analysis.

This chapter provides a basic set of definitions and operational tools used in working with sinusoids. Both time- and phasor-domain representations are used to illustrate operations such as combinations of sinusoids, their power, and spectra. Examples draw on familiar situations and contexts and provide an additional intuitive and qualitative understanding of the subject. Simulations and computational solutions are included to broaden the scope of the examples and to facilitate explorations of the concepts. Phase shift and delay in passing through a filter are illustrated by employing sinusoids. Several examples of approximating a periodic function are discussed through the solved problems. The project and some problems at the end of the chapter highlight the power of the computational tools and the need for their use in working with signals.

2.1 Sine and Cosine

Refer to the right triangle of Figure 2.1(a). The sine of the angle θ is the ratio of the opposite side to the hypotenuse.

$$\sin\theta = \frac{AB}{AC}$$

Similarly, the cosine of θ is the ratio of the neighboring side to the hypotenuse.

$$\cos\theta = \frac{CB}{CA}$$

Note that $\sin^2\theta + \cos^2\theta = 1$. If the hypotenuse is equal to one, the sine and cosine of an angle θ are equal to the ordinate and abscissa, respectively, of point A, a point on the unit circle located at the angle θ. (See Figure 2.1b.)

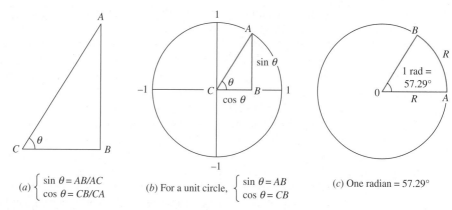

$$(a) \begin{cases} \sin\theta = AB/AC \\ \cos\theta = CB/CA \end{cases}$$

(b) For a unit circle, $\begin{cases} \sin\theta = AB \\ \cos\theta = CB \end{cases}$

(c) One radian = 57.29°

FIGURE 2.1 *(a)* Definitions of $\sin\theta = AB/AC$ and $\cos\theta = CB/CA$. *(b)* In a right triangle with hypotenuse equal to one, $\sin\theta = AB$ (the side opposite of the angle) and $\cos\theta = CB$ (the side adjacent to the angle). *(c)* One radian is the size of the angle whose circular arc is equal in length to the radius of the circle.

2.2 Angles

Angles are expressed in degrees or radians. If the circumference of a circle is divided into 360 equal arc segments, each segment will cover one degree. The degree symbol is shown by a small circle on the upper-right side of the angle: $\theta°$. A right angle is 90°.

Another unit is the radian, which stands for *radial angle*. One radian is the size of the angle whose circular arc is equal in length to the radius of the circle. (See Figure 2.1c.) Since the circumference of a circle is $2\pi R$, it contains 2π radians. Accordingly, 1 rad = $360°/(2\pi) = 360°/(2 \times 3.141593\cdots) = 57.29577951°$.

Table 2.1 shows the sine and cosine of seven angles expressed in degrees and radians.[1]

TABLE 2.1 Sine and Cosine of Seven Angles

θ in degrees	0°	1°	30°	45°	60°	89°	90°
θ in radians	0	$\dfrac{1}{180}\pi$	$\dfrac{\pi}{6}$	$\dfrac{\pi}{4}$	$\dfrac{\pi}{3}$	$\dfrac{89}{180}\pi$	$\dfrac{\pi}{2}$
$\sin\theta$	0	$0.01745 \approx \dfrac{1}{180}\pi$	1/2	$\sqrt{2}/2$	$\sqrt{3}/2$	$0.99985 \approx 1$	1
$\cos\theta$	1	$0.99985 \approx 1$	$\sqrt{3}/2$	$\sqrt{2}/2$	1/2	$0.01745 \approx \dfrac{1}{180}\pi$	0

---

[1] Another unit is the *grad* which rhymes with Brad. Divide a right angle into 100 parts and each will be one grad. The grad unit is almost obsolete and seldom used.

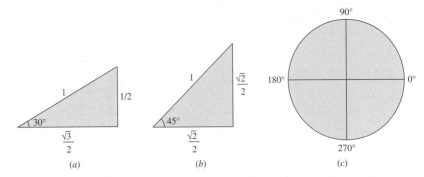

FIGURE 2.2 *(a)* In a right triangle, the side opposite of a 30° angle is half the length of the hypotenuse (sin 30° = 0.5). *(b)* In an isosceles right triangle with the hypotenuse equal to one, the two sides are sin 45° = cos 45° = $\sqrt{2}/2$ in length. *(c)* The circle is divided into four quadrants of 90° each.

Remember: (1) In a right triangle, the side opposite of a 30° angle is half the length of the hypotenuse (Figure 2.2a). (2) In an isosceles right triangle with the hypotenuse equal to one, the two sides are sin 45° = cos 45° = $\sqrt{2}/2$ in length (Figure 2.2b).

Reference and Direction

Sinusoids are used to model physical entities (such as electric fields) as functions of time and space. As such, angles are algebraic quantities that can be positive or negative. Therefore, in addition to specifying the unit, we need to determine a reference position and a direction from which the angles on a circle are measured. The horizontal coordinate axis in the Cartesian plane is the zero reference for the angle and the counterclockwise direction on the circle is its positive direction. Using the above notations, the circle is divided into four quadrants of 90° ($\pi/2$ radians) each. (See Figure 2.2c.) Angles may be measured and plotted using the graduated arc of a protractor.

Inverses of Trigonometric Functions

The inverse of $x = \sin\theta$ is shown by $\theta = \sin^{-1} x$ (also shown by $\theta = arc\sin x$, which reads as arc sine). The inverse expression says that θ is the angle whose sine value is x. Similar notations apply to the other trigonometric functions such as cosines and tangents. An inverse function has multiple answers. For example, the answer to $\theta = \sin^{-1} 0.5$ is $\theta = 30°$ or 150°. To produce a unique answer, one needs to specify to which quadrant the angle belongs.

2.3   Series Approximations

The sine and cosine may be approximated by their series expansions shown below.

$$\sin\theta = \theta - \frac{\theta^3}{3!} + \frac{\theta^5}{5!} - \frac{\theta^7}{7!} + \frac{\theta^9}{9!} - \cdots$$

$$\cos\theta = 1 - \frac{\theta^2}{2!} + \frac{\theta^4}{4!} - \frac{\theta^6}{6!} + \frac{\theta^8}{8!} - \cdots$$

θ is in radians. For small angles ($\theta << 1$ radian), the first few terms of the sine series are sufficient to give an accurate estimate of $\sin \theta$. As an example,

$$\sin 1° = \left(\frac{\pi}{180}\right) - \frac{1}{3!}\left(\frac{\pi}{180}\right)^3 + \frac{1}{5!}\left(\frac{\pi}{180}\right)^5 + \cdots$$

$$= 0.017453293 - 0.000005317 + 0.000000002 - \cdots$$

It is seen that the first term in the series is quite capable of estimating the sum; that is, $\sin\theta \approx \theta$, for $\theta << 1$, where θ is in radians.[2]

2.4 Trigonometric Identities and Relations

Some useful trigonometric relations are listed in Table 2.2. Also see Figure 2.3.

TABLE 2.2 Some Trigonometric Relationships

$\sin(-a) = -\sin a$	$\cos(-a) = \cos a$
$\sin(a + 90°) = \cos a$	$\cos(a + 90°) = -\sin a$
$\sin(a - 90°) = -\cos a$	$\cos(a - 90°) = \sin a$
$\sin(90° - a) = \cos a$	$\cos(90° - a) = \sin a$
If $a + b = 90°$, $\cos a = \sin b$, $\sin a = \cos b$	
$\sin(a + 180°) = -\sin a$	$\cos(a + 180°) = -\cos a$
$\sin(a + 360°) = \sin a$	$\cos(a + 360°) = \cos a$
$\sin(a + b) = \sin a \cos b + \cos a \sin b$	$\cos(a + b) = \cos a \cos b - \sin a \sin b$
$\sin a + \sin b = 2\sin\frac{1}{2}(a + b)\cos\frac{1}{2}(a - b)$	$\cos a + \cos b = 2\cos\frac{1}{2}(a + b)\cos\frac{1}{2}(a - b)$
$\sin a \sin b = \frac{1}{2}\cos(a - b) - \frac{1}{2}\cos(a + b)$	$\cos a \cos b = \frac{1}{2}\cos(a + b) + \frac{1}{2}\cos(a - b)$
$\sin a \cos b = \frac{1}{2}\sin(a + b) + \frac{1}{2}\sin(a - b)$	$\cos a \sin b = \frac{1}{2}\sin(a + b) - \frac{1}{2}\sin(a - b)$
$\sin 2a = 2\sin a \cos a$	$\cos 2a = \cos^2 a - \sin^2 a = 2\cos^2 a - 1 = 1 - 2\sin^2 a$
$\sin^2 a = (1 - \cos 2a)/2$	$\cos^2 a = (1 + \cos 2a)/2$
$\sin 3a = 3\sin a - 4\sin^3 a$	$\cos 3a = 4\cos^3 a - 3\cos a$
See problem 2.7 for $\sin(na)$ and $\cos(na)$.	
$A\sin a + B\cos a = C\sin(a + \theta) = C\cos(a - \phi)$	$C = \sqrt{A^2 + B^2}$, $\theta = \tan^{-1}(B/A)$, $\phi = \tan^{-1}(A/B)$

---

[2] An iterative algorithm by Al-Kāsi (died H.G. 832, AD 1428–1429) computes $\sin 1°$ to any desired accuracy. See A. Aaboe, "Al-Kāsi's Iteration Method for Determination of Sin 1°," *Scripta Mathematica*, 20 (1954), pp. 24–29.

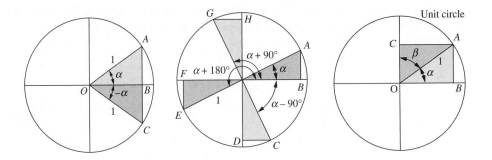

FIGURE 2.3 Visualization of some useful trigonometric relationships.

(a) symmetry	(b) 90° and 180° phase differences		(c) $\alpha + \beta = 90°$
$\begin{cases} \sin(-\alpha) = -\sin(\alpha) \\ \cos(-\alpha) = \cos(\alpha) \end{cases}$	$\begin{cases} \sin(\alpha + 90°) = \cos\alpha \\ \cos(\alpha + 90°) = -\sin\alpha \end{cases}$	$\begin{cases} \sin(\alpha + 180°) = -\sin\alpha \\ \cos(\alpha + 180°) = -\cos\alpha \end{cases}$	$\begin{cases} \sin(\alpha) = \cos(\beta) \\ \cos(\alpha) = \sin(\beta) \end{cases}$

2.5 Sinusoidal Waveforms

Look at the uniform rotary motion of a ceiling fan. The actual trajectory of the tip of a blade is a circle. This trajectory is projected on a plane perpendicular to the direction of view and perceived as the motion. If you view the fan from below at the center (a direction perpendicular to the plane of motion), each blade appears to move along a circular path

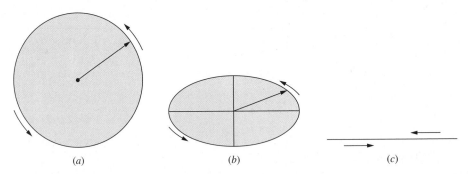

(a)   (b)   (c)

FIGURE 2.4 The trajectory of the tip of a blade of a uniformly rotating ceiling fan viewed from three directions.
(a) Viewed from below at the center (perpendicular to the plane of motion), the trajectory is a circular path traversed at a constant angular speed. *(b)* Viewed from below and off-center (at a slant), the trajectory appears as an ellipse. *(c)* Viewed from the side (a direction in the plane of motion), the tip appears to go back and forth on a straight line with its position proportional to the sine of the angle of rotation. The perceived trajectory is the projection of the circular rotation on the axis and is a sinusoidal function.

at a constant speed (Figure 2.4*a*). If you view it at a slant, the motion appears to follow an ellipse (Figure 2.4*b*). Finally, if you view the fan from the side (a direction in the plane of motion), the tip appears to move from one end of a segment of a straight line to the other end with a nonuniform speed (Figure 2.4*c*). The position of the tip along this path is proportional to the sine of the angle of rotation θ. It is modeled by a sinusoidal function.

A sinusoid is a function proportional to the sine of its argument. One way to visualize the sinusoidal function $\sin\theta$ is to project on the vertical axis the trajectory of the tip of a unity vector with angle θ as θ changes. Such a function is shown in Figure 2.5(*a*). Similarly, projection on the horizontal axis generates $\cos\theta$. (See Figure 2.5*b*.)

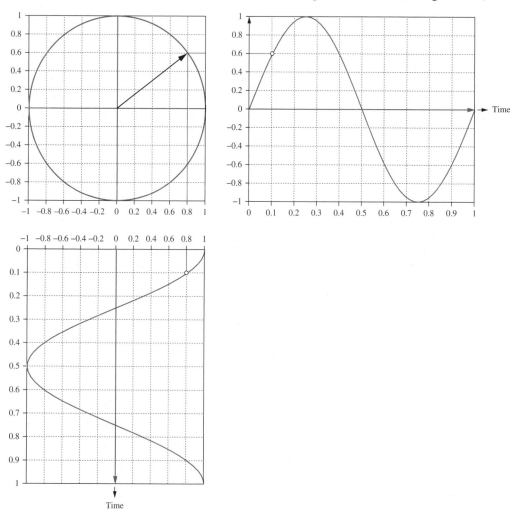

FIGURE 2.5 Projections of a vector on the x and y axes generate the sinusoidal functions sine and cosine. The sinusoidal function $v(t) = V\cos(\omega t + \theta)$, therefore, may be interpreted as the projection of the rotating vector $V\angle(\omega t + \theta)$ on the x-axis. V is the length of the vector and θ is its angle with the x-axis at $t = 0$.

If the vector rotates with the uniform angular speed of ω radians per second, then, after t seconds, the angle traversed by it is $\theta = \omega t$ and the sinusoid becomes a function of time. If the radius is V and the angle begins from zero, then the sinusoid is $v(t) = V \sin \omega t$. If at $t = 0$ the vector is at the angle θ_0, then $v(t) = V \sin(\omega t + \theta_0)$.

2.6 Sine or Cosine?

A sine function may equivalently be expressed in cosine form (and vice versa) as listed in Table 2.2:

$$\sin \theta = \cos(\theta - \pi/2)$$
$$\cos \theta = \sin(\theta + \pi/2)$$

Therefore, $v(t) = V_0 \sin(\omega t + \theta_0) = V_0 \cos(\omega t + \theta_1)$, where $\theta_1 = (\theta_0 - \pi/2)$. In this book, sine and cosine functions are both referred to as sinusoids. But when a sinusoid is represented by a phasor (to be discussed in section 2.8), the cosine form of the function is used.

2.7 Period and Frequency

The function $v(t)$ is periodic with period T if

$$v(t) = v(t + T) \quad \text{for all } t$$

The sinusoidal function $v(t) = V \sin(\omega t + \theta)$ is clearly periodic with period $T = 2\pi/\omega$ because

$$v(t + 2\pi/\omega) = V \sin(\omega t + 2\pi + \theta) = V \sin(\omega t + \theta) = v(t)$$

If it takes T seconds for the periodic function $v(t)$ to go through one full cycle of change, then the repetition rate is $1/T$ cycles per second. This quantity is called the frequency $f = 1/T$ and its unit is the Hertz (shown by Hz), where 1 Hz = 1 cycle per second. The angular frequency ω, period T, and frequency f are related by

$$\omega = 2\pi f = \frac{2\pi}{T}, \quad f = \frac{1}{T} = \frac{\omega}{2\pi}, \quad T = \frac{1}{f} = \frac{2\pi}{\omega}$$

Decimal multiples and submultiples of Hz are also used. For example,

mHz	for	millihertz $= 10^{-3}$ Hz
kHz	for	kilohertz $= 10^{3}$ Hz
MHz	for	megahertz $= 10^{6}$ Hz
GHz	for	gigahertz $= 10^{9}$ Hz
THz	for	terahertz $= 10^{12}$ Hz

The frequency spectrum of electromagnetic waves ranges from extremely low frequencies (ELF, 3–30 Hz) in the radio range to gamma rays (up to 10^{23} Hz).

2.8 Phasors

A sinusoid $V \sin(\omega t + \theta)$ or $V \cos(\omega t + \theta)$ is completely specified by its amplitude V, phase θ, and angular frequency ω (or $f = \omega/2\pi$). The amplitude and phase of a sinusoid may be combined into a two-dimensional (or complex) number $\mathbf{V}$ called the *complex amplitude*, which can be expressed in polar form as $\mathbf{V} = V e^{j\theta} = V\angle\theta$, or in Cartesian form as $\mathbf{V} = V \cos\theta + jV \sin\theta$. The complex amplitude may be shown by a vector or point in the complex plane (Figure 2.6). The real and imaginary parts of $\mathbf{V}$ are the abscissa and the ordinate of the tip of the vector, respectively. The quantity $\mathbf{V}$ is also called a phasor. In representing a sinusoid by a phasor, the cosine form $V \cos(\omega t + \theta)$ is normally meant.

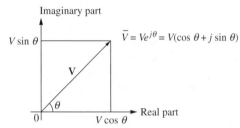

FIGURE 2.6 Phasor $\mathbf{V} = \mathbf{V}\angle\theta$ (shown by a vector) is the complex representation of the sinusoid $v(t) = V \cos(\omega t + \theta)$.

2.9 Lag and Lead

Consider two vectors $\mathbf{V_1}$ and $\mathbf{V_2}$ with phase angles θ_1 and θ_2, respectively (Figure 2.7). The phase difference between $\mathbf{V_1}$ and $\mathbf{V_2}$ is $\theta_0 = \theta_1 - \theta_2$. It is said that $\mathbf{V_2}$ lags $\mathbf{V_1}$ by θ_0. The word *lag* assumes more physical meaning if the angles vary with time. For example, if both change uniformly with an angular velocity ω, then the phase lag θ_0 translates into a time lag of $\tau_0 = \theta_0/\omega$ seconds. If θ_0 is negative, then $\mathbf{V_2}$ leads $\mathbf{V_1}$ by $\tau_0 = \theta_0/\omega$ seconds. More on this subject in the next section.

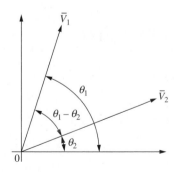

FIGURE 2.7 Lead and lag. Phasor $\mathbf{V_1}$ leads phasor $\mathbf{V_2}$ by $\theta_0 = \theta_1 - \theta_2$. Equivalently, $\mathbf{V_2}$ lags $\mathbf{V_1}$ by θ_0.

2.10 Time Shift and Phase Shift

If a function $f(t)$ is delayed by τ seconds, one obtains $f(t - \tau)$. The delay operation shifts the graph of $f(t)$ to the right by the amount τ. Likewise, an advance by τ seconds produces $f(t + \tau)$ and shifts the graph of $f(t)$ to the left by the amount τ.

In the case of sinusoidal functions, *time shift* translates into *phase shift*. A delay produces a phase lag and an advance produces a phase lead. Delay the function $\cos(\omega t)$ by τ seconds to get $cos\omega(t - \tau) = \cos(\omega t - \theta)$, where $\theta = \omega\tau$. The graph of $\cos(\omega t)$ is shifted to the right by τ, which corresponds to a *phase lag* of $\theta = \omega\tau = 2\pi f\tau$. See Figure 2.8(*a*). Similarly, a time advance by τ seconds means a shifted graph to the left. See Figure 2.8(*b*). In general, a time shift of τ corresponds to a phase shift of $\theta = \omega\tau = 2\pi f\tau$. For a given phase shift, the higher the frequency, the smaller the time shift.

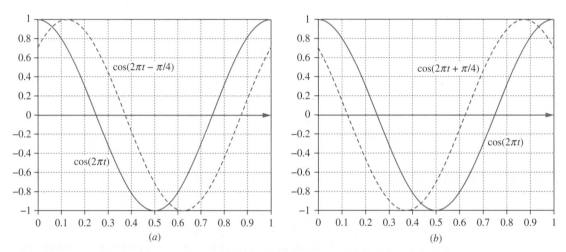

FIGURE 2.8 *(a)* **Phase lag**. A 125-msec delay in $\cos 2\pi t$ produces $\cos 2\pi(t - 0.125) = \cos(2\pi t - \pi/4)$, shown by the dashed curve, creating a 45° phase lag. *(b)* **Phase lead**. A 125-msec advance in $\cos 2\pi t$ produces $\cos 2\pi(t + 0.125) = \cos(2\pi t + \pi/4)$, shown by the dashed curve, creating a 45° phase lead.

2.11 Summing Phasors

Let $\mathbf{V_1} = V_1\angle\theta_1$ and $\mathbf{V_2} = V_2\angle\theta_2$ be two phasors. Their sum is

$$\mathbf{V} = \mathbf{V_1} + \mathbf{V_2} = V\angle\theta$$

where V and θ are found from

$$V\cos\theta = V_1\cos\theta_1 + V_2\cos\theta_2$$

$$V\sin\theta = V_1\sin\theta_1 + V_2\sin\theta_2$$

See Figure 2.9.

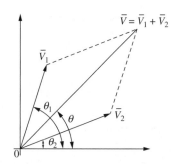

FIGURE 2.9 Sum of two phasors $\mathbf{V} = \mathbf{V}_1 + \mathbf{V}_2$.

Example

2.1

Use the phasor notation to find the amplitude and phase of $v(t) = 2\cos(\omega t + 25°) - 3\sin(\omega t - 30°)$.

Solution

Let

$$v_1 = 2\cos(\omega t + 25°) \qquad\qquad \Longrightarrow \mathbf{V}_1 = 2\angle 25°$$

$$v_2 = -3\sin(\omega t - 30°) = 3\cos(\omega t + 60°) \Longrightarrow \mathbf{V}_2 = 3\angle 60°$$

$$\mathbf{V} = \mathbf{V}_1 + \mathbf{V}_2 = 2\angle 25° + 3\angle 60°$$

$$= 3.3126 + j3.4433 = 4.7781\angle 46.1° \quad \Longrightarrow v(t) = 4.7781\cos(\omega t + 46.1°)$$

2.12 Combination of Sinusoids

The sum of two sinusoids with frequencies f_1 and f_2 is periodic only if the ratio of the two frequencies is a rational number; that is, if $f_1/f_2 = k_1/k_2$, where k_1 and k_2 are integers. The common divisor f found from $f_1 = k_1 f$ and $f_2 = k_2 f$ is then called the fundamental frequency and f_1 and f_2 are its harmonics. For example, $v(t) = \cos(\pi t) + \cos(5\pi t)$ is periodic because it contains the fundamental frequency 0.5 Hz and its harmonic at 2.5 Hz. In contrast, $v(t) = \cos(\pi t) + \cos t$ is not periodic. The same is true of $v(t) = 2\cos(\pi t)\cos t = \cos[(\pi - 1)t] + \cos[(\pi + 1)t]$. We conclude that a finite sum of sinusoidal waveforms that are harmonics of a fundamental frequency is a periodic waveform with period equal to the inverse of the fundamental frequency.

Beat Frequency

Consider the sum of two sinusoids $v(t) = \cos(\omega_1 t) + \cos(\omega_2 t)$, where ω_1 and ω_2 are not harmonics of a fundamental frequency and, therefore, $v(t)$ is not periodic. However, $v(t)$ may be written as

$$v(t) = \cos(\omega_1 t) + \cos(\omega_2 t) = 2\cos(\omega_0 t)\cos(\omega_d t)$$

$$\omega_0 = \frac{\omega_2 + \omega_1}{2}$$

$$\omega_d = \frac{\omega_2 - \omega_1}{2}$$

$v(t)$ is a sinusoid at frequency ω_0 whose amplitude is modulated by a sinusoid at frequency ω_d. If ω_1 and ω_2 are very close, the amplitude of v varies slowly. ω_d is then called the *beat frequency*.

2.13 Combination of Periodic Signals

The observations regarding a sum of two sinusoids can be generalized to the sum of two periodic functions $v_1(t)$ and $v_2(t)$ with periods T_1 and T_2, respectively. The sum $v_1(t) + v_2(t)$ is periodic if the ratio T_1/T_2 is a rational number. The period of the sum is then $T = n_1 T_1 = n_2 T_2$, where n_1 and n_2 are integers. In such a case, $v_1(t)$, $v_2(t)$, and their sum contain harmonics of the frequency $f = 1/T$.

2.14 Representation of a Sum of Sinusoids

A waveform, which is a sum of sinusoids, whether periodic or not, is completely specified by the complex amplitude (magnitude and phase) of its components and their frequencies. These may be given in tabular or graphical form.

Example 2.2

The waveform $v(t) = 1 + 3\cos(\pi t) + \cos(5\pi t + 30°) + 0.5\cos(20t + 90°)$ may be represented by the table below. The waveform is a sum of periodic functions and the table contains a finite number of sinusoids. However, the waveform $v(t)$ is itself not periodic because the elements of the table are not harmonics of a fundamental frequency.

f in Hz	Amplitude ∠ Angle
0	$1\angle 0°$
0.5	$3\angle 0°$
2.5	$1\angle 30°$
$10/\pi$	$0.5\angle 90°$

Example 2.3

The waveform $v(t) = 1 + 3\cos(\pi t) + \cos(5\pi t + 30°) + 0.5\cos(20\pi t + 90°)$ is represented by the table below.

f in Hz	Amplitude ∠ Angle
0	$1\angle 0°$
0.5	$3\angle 0°$
2.5	$1\angle 30°$
10	$0.5\angle 90°$

The table contains a finite number of harmonics of 0.5 Hz. The waveform, therefore, is periodic.

2.15   Power in a Sinusoid

The instantaneous power in a waveform $v(t)$ is defined by $p(t) = v^2(t)$. The average of $p(t)$ over the duration 0 to T is

$$P = <p(t)> = <v^2(t)> = \frac{1}{T}\int_0^T v^2(t)dt$$

The rms value of $v(t)$ over the duration 0 to T is $\sqrt{P}$. For a sinusoidal signal with amplitude V, the average power and rms value (shown by V_{rms}) defined over its period T are

$$P = \frac{1}{T}\int_0^T V^2\cos^2(\omega t + \theta)dt = \frac{V^2}{2}$$

$$V_{\text{rms}} = \sqrt{P} = \frac{V}{\sqrt{2}}$$

The average power and rms values are normally defined over one period of the sinusoid and, therefore, are independent of its frequency and phase angle.

The average power in a waveform that is made of harmonics is the sum of the average powers in the harmonics, because the integral of cross terms in the expression for instantaneous power is zero.

*E*xample
2.4

Find the average power in $v(t) = V_1\cos(n\omega t) + V_2\cos(m\omega t)$, where m and n are integers.

$$P = <v^2(t)> = <V_1^2\cos^2(n\omega t) + V_2^2\cos^2(m\omega t) + 2V_1V_2\cos(n\omega t)\,\cos(m\omega t)>$$

$$= V_1^2<\cos^2(n\omega t)> + V_2^2<\cos^2(m\omega t)> + 2V_1V_2<\cos(n\omega t)\,\cos(m\omega t)>$$

$$= \frac{V_1^2}{2} + \frac{V_2^2}{2}$$

This example shows the superposition of power and is valid if a common period can be found for all components. The following is a counterexample to the concept of universal application of superposition of power.

*E*xample
2.5

Find the average power in $v(t) = v_1(t) + v_2(t)$, where $v_1(t) = V_1\cos(\omega t)$ and $v_2(t) = V_2\cos(\sqrt{2}\omega t)$.

$$P = <v^2(t)> = <v_1^2(t)> + <v_2^2(t)> + 2<v_1(t)v_2(t)>$$

$$= \frac{V_1^2}{2} + \frac{V_2^2}{2} + 2V_1V_2<\cos(\omega t)\,\cos(\sqrt{2}\omega t)>$$

But

$$2\cos(\omega t)\,\cos(\sqrt{2}\omega t) = \cos(\sqrt{2}+1)\omega t + \cos(\sqrt{2}-1)\omega t$$

These two components do not have a common period and the average of their sum is not zero for any time interval. Therefore, superposition of power does not apply in this case.

2.16 One-Sided Line Spectrum

The table or graph showing the amplitude of a sinusoid becomes an especially useful tool when its components are harmonics, in which case it is called a one-sided line spectum. The spectrum is discrete and contains a finite or infinite number of lines.

 2.6

Plot the one-sided line spectrum of

$$x(t) = \sum_{n>0,\ n\ \text{odd}}^{\infty} \frac{1}{n}\cos(100n\pi t + \theta_n)$$

Compute the power in the fundamental frequency and the next five nonzero harmonics.

Solution
The spectrum is given by $\frac{1}{n}$, $n = 1, 3, 5, \ldots$. See Figure 2.10. The first seven nonzero amplitudes are

$$1, \ \frac{1}{3}, \ \frac{1}{5}, \ \frac{1}{7}, \ \frac{1}{9}, \ \frac{1}{11}, \ \frac{1}{13}$$

The power in the above harmonics is

$$\frac{1}{2}\left[1 + \frac{1}{9} + \frac{1}{25} + \frac{1}{49} + \frac{1}{81} + \frac{1}{121} + \frac{1}{169}\right] \approx 0.599$$

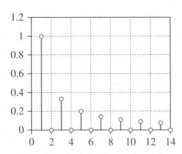

FIGURE 2.10 The one-sided line spectrum of $x(t)$ which is a sum of sinusoids

$$x(t) = \sum_{n>0,\ n\ \text{odd}}^{13} \frac{1}{n}\cos(100n\pi t + \theta_n)$$

The line spectrum is insensitive to the phase of the harmonics or their time shift and does not possess all the information needed to reconstruct the waveform. Two completely different waveforms may have identical spectra. Nevertheless, the spectrum is a useful characteristic of the waveform because it conveys information about the density of its average power.

Example
2.7

Consider the even periodic squarewave signal with period T, peak-to-peak value A, and zero DC level. One period of the signal is described by

$$v(t) = \begin{cases} -\frac{A}{2} & -\frac{T}{2} < t < -\frac{T}{4} \\ \frac{A}{2} & -\frac{T}{4} < t < \frac{T}{4} \\ -\frac{A}{2} & \frac{T}{4} < t < \frac{T}{2} \end{cases} \quad \text{and } v(t) = v(t + T)$$

As you will see in Chapter 7 on the Fourier series, the above square wave may be represented by an infinite series of the following form:

$$v(t) = \frac{2A}{\pi} \left\{ \cos \omega t - \frac{1}{3} \cos 3\omega t + \frac{1}{5} \cos 5\omega t - \frac{1}{7} \cos 7\omega t + \cdots \right\}$$

Compute the percentage of average power in the fundamental component and in the next five nonzero harmonics.

Solution
The one-sided line spectrum of $v(t)$ is given by $2A/(\pi n)$, n odd. The total power in $v(t)$ is $A^2/4$. The power in the n-th harmonic is $2A^2/(\pi n)^2$. The ratio of the power in each harmonic to the total power is

$$\frac{2A^2/(\pi n)^2}{A^2/4} = \frac{8}{(\pi n)^2}, \quad n \text{ odd}$$

The percentages are shown in Table 2.3, which may be called the relative power spectral density of the signal.

TABLE 2.3 Relative Power Spectral Density of a Square Wave

n	1	3	5	7	9	11	Higher harmonics
Percentage power	81.06	9.01	3.24	1.65	1.00	0.67	3.34

In summary, out of every 1,000 watts in the waveform, about 810 watts come from the fundamental component, 156 watts from the next five components, and 34 watts from the remaining higher components.

2.17 Complex Representation of Sinusoids and the Two-Sided Spectrum

In section 2.8 we introduced the phasor $\mathbf{V}$ as a vector drawn from the origin to the tip at point M in the xy plane. The vector is expressed in the rectangular coordinate system by $x = \rho \cos \theta$ and $y = \rho \sin \theta$, where ρ is the length of the vector and θ is its angle with the abscissa. The x and y coordinates may be written in a combined form as

$$\mathbf{V} \equiv \rho \cos \theta + j\rho \sin \theta \equiv \rho e^{j\theta}$$

where j is the imaginary number $j^2 = -1$. Of special interest are the following identities:

$$e^{j\theta} = \cos \theta + j \sin \theta$$

$$\cos \theta = \mathcal{RE}\left\{e^{j\theta}\right\} = \frac{e^{j\theta} + e^{-j\theta}}{2}$$

$$\sin \theta = \mathcal{IM}\left\{e^{j\theta}\right\} = \frac{e^{j\theta} - e^{-j\theta}}{2j}$$

A sinusoid, therefore, may be represented by the sum of two complex conjugate exponential functions of time:

$$V \cos(\omega t + \theta) = \frac{V}{2}e^{j\theta}e^{j\omega t} + \frac{V}{2}e^{-j\theta}e^{-j\omega t}$$

In the above representation, the term $e^{j\omega t}$ specifies the time dependency of the sinusoid and $\frac{V}{2}e^{j\theta}$ (called the complex amplitude) determines its magnitude and phase. The two-sided spectrum of the sinusoid represents the complex amplitude and its conjugate versus frequency as shown in Figure 2.11(a). The two-sided spectrum of a signal made of several sinusoids is the sum of the individual spectra. See Example 2.8. Unlike the one-sided line spectrum, the two-sided spectrum includes the phases of the harmonics and possesses all the information needed to reconstruct the waveform.

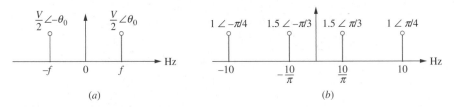

(a)   (b)

FIGURE 2.11 **(a)** Two-sided spectrum of $V \cos(2\pi f t + \theta)$. **(b)** Two-sided spectrum of $2 \cos(20\pi t + \pi/4) + 3 \cos(20t + \pi/3)$.

2.8

Plot the two-sided spectrum of $v(t) = 2 \cos(20\pi t + \pi/4) + 3 \cos(20t + \pi/3)$.

Solution

See Figure 2.11(b).

2.18 Problems

Solved Problems

1. Determine the phase relationship between $x_1(t) = \cos(t + 10°)$ and $x_2(t) = \cos(t - 30°)$ by choosing the correct answer:

a. x_1 leads x_2 by $10°$

b. x_1 leads x_2 by $40°$

c. x_2 lags x_1 by $30°$

d. none of the above

Solution

The phase of $x_1(t)$ with reference to $x_2(t)$ is $10° - (-30°) = 40°$. The correct answer is b.

2. For each time function given below determine if it is periodic or aperiodic, and specify the period if the function is periodic.

a. $\cos(3t + 45°) + \sin(\sqrt{2}t - 120°)$

c. $\cos(\pi t + 10°) - \cos(2\pi t + \pi/3)$

b. $\cos(5t) + \cos \pi(t - 0.5)$

d. $\sin(6.28t - 2\pi/3) + \cos 0.2(t - 0.5)$

Solution

A combination of two periodic functions with periods T_1 and T_2 is periodic if the ratio of the two periods is a rational number, $\frac{T_1}{T_2} = \frac{k_1}{k_2}$, where k_1 and k_1 are integers. In that case, the common period is $T = k_2 T_1 = k_1 T_2$. Expressed in terms of ω for sinusoids, the periodicity condition becomes $\frac{\omega_1}{\omega_2} = \frac{k_2}{k_1}$, with k_1 and k_1 integers.

a. $\frac{\omega_1}{\omega_2} = \frac{3}{\sqrt{2}} \neq \frac{k_2}{k_1}$, aperiodic

c. $\frac{\omega_1}{\omega_2} = \frac{\pi}{2\pi} = \frac{1}{2}$, periodic (period $= 2$)

b. $\frac{\omega_1}{\omega_2} = \frac{5}{\pi} \neq \frac{k_2}{k_1}$, aperiodic

d. $\frac{\omega_1}{\omega_2} = \frac{6.28}{0.2} = \frac{157}{5}$, periodic (period $= 50\pi$)

3. Repeat problem 2 for the time functions given below.

a. $\cos(3t + 45°) \sin(\sqrt{2}t - 120°)$

c. $\cos(\pi t + 10°) \cos(2\pi t + \pi/3)$

b. $\cos(5t) \cos \pi(t - 0.5)$

d. $\sin(6.28t - 2\pi/3) \cos 0.2(t - 0.5)$

Solution

a. Aperiodic, b. Aperiodic, c. Periodic (period $= 2$), d. Periodic (period $= 25\pi$).

4. Consider the sum of two sinusoids having the same frequency but different phases $x(t) = \cos \omega t + \cos(\omega t + \theta)$, where $0 \leq \theta \leq 2\pi$ is the phase.

a. Express it in the form of $x(t) = A \cos(\omega t + \phi)$, with $A \geq 0$ and $-\pi \leq \phi \leq \pi$. Determine A and ϕ as functions of θ and obtain their values for $\theta = k\pi/4, 0 \leq k \leq 8$; k is an integer. Plot A and ϕ versus θ.

b. Repeat part a, allowing A to assume any value and $0 \leq \phi \leq \pi$. In problem 5 you will be asked to write a Matlab program and execute it to plot the magnitude and phase of $x(t)$ as functions of θ for $0 < \theta < 2\pi$.

Solution

a. Expand $\cos(\omega t + \theta)$ in terms of $\sin \omega t$ and $\cos \omega t$. Collect terms and convert the sum into a single cosine form $A \cos(2\pi f t + \phi)$ as shown below.

$$x(t) = \cos \omega t + \cos(\omega t + \theta) = \cos \omega t + \cos \omega t \cos \theta - \sin \omega t \sin \theta$$
$$= (1 + \cos \theta) \cos \omega t - \sin \theta \sin \omega t = A \cos(\omega t + \phi),$$

where

$$A = \sqrt{(1 + \cos \theta)^2 + \sin^2 \theta} = \sqrt{2(1 + \cos \theta)} \text{ and } \phi = \tan^{-1}\left(\frac{\sin \theta}{1 + \cos \theta}\right).$$

Values of A and ϕ for $\theta = k\pi/4,\ 0 \le k \le 8$, are given in the table below.

θ	0	$\pi/4$	$\pi/2$	$3\pi/4$	π	$5\pi/4$	$3\pi/2$	$7\pi/4$	2π
A	2	1.848	$\sqrt{2}$	0.765	0	0.765	$\sqrt{2}$	1.848	2
ϕ	0	22.5°	45°	67.5°	0	−67.5°	−45°	−22.5°	0

b. $x(t) = (1 + \cos\theta)\cos\omega t - \sin\theta \sin\omega t = 2\cos^2(\theta/2)\cos\omega t - \sin\theta \sin\omega t = 2\cos(\theta/2)\cos(\omega t + \theta/2)$. The amplitude of $x(t)$ is $2\cos(\theta/2)$, which can be a positive or negative number. See the table below.

θ	0	$\pi/4$	$\pi/2$	$3\pi/4$	π	$5\pi/4$	$3\pi/2$	$7\pi/4$	2π
A	2	1.848	$\sqrt{2}$	0.765	0	−0.765	−$\sqrt{2}$	−1.848	−2
ϕ	0	22.5°	45°	67.5°	90°	112.5°	135°	157.5°	180°

5. The following Matlab program plots $x(t) = \cos(2\pi t) + \cos(2\pi t + \theta)$ for 11 values of θ from 0 to 2π.

```
theta=linspace(0,2*pi,11);
t=linspace(0,1,500);
for j=1:11;
    for i=1:500;
        x(i,j)=cos(2*pi*t(i))+cos(2*pi*t(i)+theta(j));
    end
end
for j=1:11;
    plot(t,x,'LineWidth',2)
    axis([0 1 -2.2 2.2]);
end
xlabel('Time(s)');
ylabel('x(t)');
title('Plot of x(t)=cos(2\pit)+cos(2\pit+\theta) for 11 values of \theta
from 0 to 2\pi.')
grid
```

Execute the program and examine the plots. Verify that the plots are in agreement with the results developed in problem 4.

6. a. Convert the sum of two sinusoids with different frequencies $x(t) = \cos\omega_1 t + \cos\omega_2 t$ into a product of two sinusoids.

b. Given $\omega_2 = (1 + k)\omega_1$, find and plot $x(t)$ for $k = 1/2,\ 1/4,\ 1/16$.

c. Interpret the plots in light of the approach of problem 4.

Solution

a. Using the identity

$$\cos a + \cos b = 2\cos\left(\frac{a-b}{2}\right)\cos\left(\frac{a+b}{2}\right)$$

we find $\cos\omega_1 t + \cos\omega_2 t = 2\cos(\omega_d t)\cos(\omega_s t)$, where $\omega_d = (\omega_1 - \omega_2)/2$ and $\omega_s = (\omega_1 + \omega_2)/2$.

b. Using the result of part a we find

$$x(t) = 2\cos(\omega_d t)\cos(\omega_s t), \quad \text{where } \omega_d = (k/2)\omega_1 \text{ and } \omega_s = (1 + k/2)\omega_1.$$

See Figure 2.12.

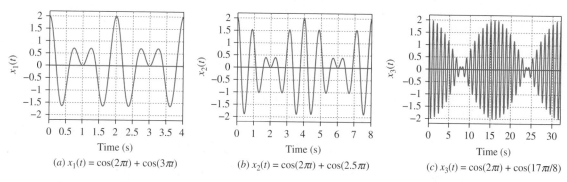

(a) $x_1(t) = \cos(2\pi t) + \cos(3\pi t)$ (b) $x_2(t) = \cos(2\pi t) + \cos(2.5\pi t)$ (c) $x_3(t) = \cos(2\pi t) + \cos(17\pi t/8)$

FIGURE 2.12 Plot of $x(t) = \cos\omega_1 t + \cos(1 + k)\omega_1 t$ for three values of k.

c. View $x(t)$ as the sum of $\cos(\omega_1 t) + \cos(\omega_1 t + \theta) = 2\cos(\theta/2)\cos(\omega t + \theta/2)$, where $\theta = k\omega_1 t$. $x(t)$ is a sinusoid whose amplitude depends on θ, which in the present case is a time-varying quantity. This is reflected in the envelopes of the plots of Figure 2.12. When the phase angle is an odd multiple of π, the amplitude becomes zero. In Figure 2.12 with $\omega = 2\pi$, the zero amplitudes occur at $t_M = M/(2k)$, where M is an odd number. In Figure 2.12(a), $k = 1/2$ results in $t_M = 1, 3, \ldots$. In Figure 2.12(b), $k = 1/2$ results in $t_M = 2, 6, \ldots$, and in Figure 2.12(c), $k = 1/16$ results in $t_M = 8, 24, \ldots$.

7. Find $\cos n\theta$ and $\sin n\theta$ in terms of $\cos\theta$ and $\sin\theta$. Then express $\cos n\theta$ in terms of $\cos\theta$ only.

Solution

Start with	$e^{j\theta} = x + jy$, where $x = \cos\theta$, $y = \sin\theta$, and $x^2 + y^2 = 1$
Raise it to the power of n	$e^{jn\theta} = (x + jy)^n$
But	$e^{jn\theta} = \cos n\theta + j\sin n\theta$
Therefore,	$\cos n\theta = \mathcal{RE}\{(x + jy)^n\}$ and $\sin n\theta = \mathcal{IM}\{(x + jy)^n\}$

Apply the binomial expansion

$$(a + b)^n = a^n + na^{n-1}b + \frac{n(n-1)}{2!}a^{n-2}b^2 + \frac{n(n-1)(n-2)}{3!}a^{n-3}b^3 + \cdots + \frac{n(n-1)}{2!}a^2 b^{n-2} + nab^{n-1} + b^n$$

to find the real and imaginary parts of $(x + jy)^n$ and set them equal to $\cos n\theta$ and $\sin n\theta$, respectively. Then use $x^2 + y^2 = 1$ to simplify. Here are some examples:

$$(x + jy)^2 = (x^2 - y^2) + j2xy, \qquad \cos 2\theta = 2x^2 - 1, \qquad \sin 2\theta = 2xy$$
$$(x + jy)^3 = (x^3 - 3xy^2) + j(3x^2 y - y^3), \qquad \cos 3\theta = 4x^3 - 3x, \qquad \sin 3\theta = 3y - 4y^3$$
$$(x + jy)^4 = (x^4 - 6x^2 y^2 + y^4) + j(4x^3 y - 4xy^3), \quad \cos 4\theta = 8x^4 - 8x^2 + 1, \quad \sin 4\theta = 4x^3 y - 4xy^3$$

$$\begin{aligned} \text{In summary,} \quad &\cos 2\theta = 2\cos^2\theta - 1 && \sin 2\theta = 2\sin\theta\cos\theta \\ &\cos 3\theta = 4\cos^3\theta - 3\cos\theta && \sin 3\theta = 3\sin\theta - 4\sin^3\theta \\ &\cos 4\theta = 8\cos^4\theta - 8\cos^2\theta + 1 && \sin 4\theta = 4\sin\theta\cos\theta - 4\sin^3\theta\cos\theta \end{aligned}$$

8. Phase shift and delay. The input signal to a device (to be called a filter) is a sinusoid $v_i(t) = V \cos \omega t$. The signal at the output terminal of the filter is $v_o(t) = V \cos(\omega t - \theta)$ for all V and ω. The filter doesn't change the frequency of the signal and has a unity amplitude gain but introduces a phase lag θ. The transformation is shown below.

$$v_i(t) = V \cos \omega t \implies v_o(t) = V \cos(\omega t - \theta)$$

Given $v_i(t) = (1 + 2 \cos \omega_0 t) \cos \omega_c t$, where $\omega_0 = 10^7$ rad/s and $\omega_c = 10^8$ rad/s, find $v_o(t)$ for the following three cases:

a. Linear phase $\theta = 10^{-9}\omega$.
b. Constant phase $\theta = 10^{-1}$.
c. Linear phase with a constant bias $\theta = (10^{-9}\omega + 10^{-1})$.

θ and ω are in radians and radians per second, respectively.

Solution
Assuming $\omega_l = \omega_c - \omega_0 = 0.9 \times 10^8$ and $\omega_h = \omega_c + \omega_0 = 1.1 \times 10^8$, the input signal may be expanded into its sinusoidal components:

$$v_i(t) = \cos \omega_l t + \cos \omega_h t + \cos \omega_c t$$

The output is

$$v_o(t) = \cos(\omega_l t - \theta_l) + \cos(\omega_h t - \theta_h) + \cos(\omega_c t - \theta_c)$$

a. In the case of a linear phase filter, $\theta_l = 10^{-9}\omega_l$, $\theta_h = 10^{-9}\omega_h$, and $\theta_c = 10^{-9}\omega_c$. The output is

$$v_o(t) = \cos(\omega_l t - 10^{-9}\omega_l) + \cos(\omega_h t - 10^{-9}\omega_h) + \cos(\omega_c t - 10^{-9}\omega_c)$$
$$= \cos \omega_l(t - 10^{-9}) + \cos \omega_h(t - 10^{-9}) + \cos \omega_c(t - 10^{-9})$$
$$= v_i(t - 10^{-9})$$

Because of the linear phase property, the output is a delayed version of the input. The delay is equal to the slope of the phase lag versus the ω curve, which is equal to 10^{-9} seconds.

b. In the case of a filter with constant phase, $\theta_l = \theta_h = \theta_c = 0.1$.

$$v_o(t) = \cos(\omega_l t - 0.1) + \cos(\omega_h t - 0.1) + \cos(\omega_c t - 0.1)$$
$$= \cos \omega_l(t - 1.1 \times 10^{-9}) + \cos \omega_h(t - 0.9 \times 10^{-9}) + \cos \omega_c(t - 10^{-9})$$

Because of the constant phase, the time delay depends on the frequency. Higher frequencies pass through the filter faster than lower frequencies, producing a distortion. The time delays are given below.

ω in rad/s	Delay in sec
$\omega_\ell = 0.9 \times 10^8$	1.11×10^{-9}
$\omega_c = 10^8$	10^{-9}
$\omega_h = 1.1 \times 10^8$	0.91×10^{-9}

c. In the case of a linear phase with a constant bias we have

$$v_o(t) = \cos(\omega_l t - \theta_\ell) + \cos(\omega_h t - \theta_h) + \cos(\omega_c t - \theta_c)$$

where

$$\theta_l = 10^{-9}\omega_l + 0.1 = 0.19$$

$$\theta_h = 10^{-9}\omega_h + 0.1 = 0.21$$

$$\theta_c = 10^{-9}\omega_c + 0.1 = 0.2$$

Substituting for θ we get

$$v_o(t) = \cos(\omega_l t - 0.19) + \cos(\omega_h t - 0.21) + \cos(\omega_c t - 0.2)$$

$$= \cos\omega_l(t - 2.11 \times 10^{-9}) + \cos\omega_h(t - 1.9 \times 10^{-9}) + \cos\omega_c(t - 2 \times 10^{-9})$$

As in case b. the three frequency components in $v_o(t)$ undergo three different delays as given below.

ω in rad/s	Delay in sec
$\omega_\ell = 0.9 \times 10^8$	2.11×10^{-9}
$\omega_c = 10^8$	2×10^{-9}
$\omega_h = 1.1 \times 10^8$	1.9×10^{-9}

In summary,

a. In a linear phase filter all input frequencies are delayed by the same amount, resulting in $y(t) = x(t - \tau)$.
b. In a filter with constant phase, higher frequencies are delayed by a lesser amount, thus creating a distortion.
c. In a biased linear phase filter, individual frequencies experience different delays but share a common group delay.

By way of simplification, one may consider that the total signal is delayed by an average of 2×10^{-9} seconds. However, in this case the nonuniformity in delays doesn't produce destructive distortion. By combining the low- and high-frequency components in $v_o(t)$, we find

$$v_o(t) = \left[1 + 2\cos\omega_0(t - 11 \times 10^{-9})\right]\cos\omega_c(t - 2 \times 10^{-9})$$

Note that the amplitude $1 + 2\cos\omega_0 t$ is delayed by 11 nanoseconds, becoming $1 + 2\cos\omega_0(t - 11 \times 10^{-9})$, but experiences no distortion. The carrier $\cos\omega_c t$ is delayed by 2 nanoseconds and becomes $\cos\omega_c(t - 2 \times 10^{-9})$. This phenomenon is generalized in the next problem.

9. **A simple case of group delay.** In section 2.10 we illustrated the relationship between phase shift and time shift for a single sinusoid, and problem 8 examined time delays experienced by a group of sinusoids in passing through a filter. In this problem we formulate a simple case of group delay. A more general treatment of group delay is found in Chapter 9.

Consider a linear filter that has unity gain within the passband $\omega_c \pm \Delta\omega$ and introduces a phase lag θ in a sinusoidal signal passing through it. In other words,

$$v_i(t) = V\cos(\omega t) \implies v_o(t) = V\cos(\omega t - \theta)$$

for all V and $(\omega_c - \Delta\omega) < \omega < (\omega_c + \Delta\omega)$. Within the above band the phase lag is $\theta = \theta_c + \tau_0(\omega - \omega_c)$. Let the input to the filter be an amplitude-modulated waveform

$$v_i(t) = [1 + 2A\cos(\omega_0 t)]\cos(\omega_c t)$$

where $\cos(\omega_c t)$ is the carrier and $1 + 2A\cos(\omega_0 t)$ is the modulating signal with $\omega_0 < \Delta\omega$. Find the output $v_o(t)$.

Solution

Expand the input signal into its sinusoidal components:

$$v_i(t) = A\cos(\omega_\ell t) + A\cos(\omega_h t) + \cos(\omega_c t)$$

where

$$\omega_\ell = \omega_c - \omega_0$$
$$\omega_h = \omega_c + \omega_0$$

The input signal, therefore, falls within the passband. The output is

$$v_o(t) = A\cos(\omega_\ell t - \theta_\ell) + A\cos(\omega_h t - \theta_h) + \cos(\omega_c t - \theta_c)$$

where

$$\theta_\ell = \theta_c + \tau_0(\omega_\ell - \omega_c) = \theta_c - \tau_0\omega_0$$
$$\theta_h = \theta_c + \tau_0(\omega_h - \omega_c) = \theta_c + \tau_0\omega_0$$

Combining the first two terms of $v_o(t)$ and noting that

$$\omega_c = \frac{\omega_h + \omega_\ell}{2}$$

$$\omega_0 = \frac{\omega_h - \omega_\ell}{2}$$

we get

$$v_o(t) = 2A\cos\left(\omega_0 t - \frac{\theta_h - \theta_\ell}{2}\right)\cos\left(\omega_c t - \frac{\theta_h + \theta_\ell}{2}\right) + \cos(\omega_c t - \theta_c)$$

Define $\tau_c = \theta_c/\omega_c$. Then

$$\theta_h + \theta_\ell = 2\theta_c = 2\tau_c\omega_c$$
$$\theta_h - \theta_\ell = 2\theta_0 = 2\tau_0\omega_0$$

Therefore,

$$v_o(t) = 2A\cos(\omega_0 t - \theta_0)\cos(\omega_c t - \theta_c) + \cos(\omega_c t - \theta_c)$$
$$= 2A\cos\omega_0(t - \tau_0)\cos\omega_c(t - \tau_c) + \cos(\omega_c t - \tau_c)$$
$$= [1 + 2A\cos\omega_0(t - \tau_0)]\cos\omega_c(t - \tau_c)$$

The modulating signal and the carrier are delayed by τ_0 and τ_c, respectively. In general, the modulating signal may contain many frequency components. As long as they remain within the band $2\Delta\omega$, all will be delayed by the same amount τ_0, which makes the modulating signal undistorted. The carrier $\cos\omega_c t$ is delayed by τ_c, a different amount. This, however, doesn't introduce any distortion in the modulating signal. As a group, the total signal is delayed by an average of τ_c.

10. Let $x(t)$ be the sum of a finite set of sinusoids in which the frequency and amplitude of the components are known and specified by the two vectors f and a, respectively.

$$x(t) = \sum_{k=1}^{N} a_i \sin(2\pi f_i t)$$

Write a Matlab program to sample $x(t)$ for $0 \le t \le T$ at the rate of f_s. Choose $T = 10$ seconds, $f_s = 10$ kHz, and

f	5	25	50	75	100	125	150	250	275	300	325	350	375	400
a	1	1	1	1	1	1	1	1	1	1	1	1	1	1

then, compute the average power by integrating $P_x = \dfrac{1}{T} \displaystyle\int_0^T x(t)^2 dt$ and compare with $P_{\text{theory}} = \dfrac{1}{2} \displaystyle\sum_{i=1}^{N} a_i^2$.

Solution

See the program below.

```
clear
f=[5 25 50 75 100 125 150 250 275 300 325 350 375 400];
a=[1 1  1  1  1   1   1   1   1   1   1   1   1   1 ];
t=0:.001:10;
M=length(t); N=length(f);
for i=1:N
    for j=1:M
        q(i,j)=a(i)*sin(2*pi*f(i)*t(j));
    end;
end;
x=sum(q,1);   % Constructing x(t)
Px=sum(x.*x)/M;    Pt=sum(v1.*v1)/2    % Finding power in x(t)
```

The average power is $P_x = 6.9993$ [obtained by averaging the integral of $x^2(t)$]. The expected theoretical value is $P_{\text{theory}} = 7$ (obtained from the amplitude vector a), both in watts.

11. Pass the signal $x(t)$ of problem 10 through a device that changes the magnitude and phase of a sinusoidal input, but not its frequency. An input $\sin(\omega t)$ produces the output $H\sin(\omega t + \theta)$, where H and θ (called the gain and phase of the filter, respectively) depend on the frequency of the sinusoid. Assume $\theta = 0$, but that H decreases from unity (at DC) to zero (at high frequencies) as given below.[3]

$$H(f) = \frac{1}{\sqrt{1 + (f/150)^{10}}}, \quad \text{where } f \text{ is in Hz.}$$

Call the output of the filter $y(t)$ and compute its average power. Finally, by examining $H(f)$, suggest a method to predict an approximate value for P_y without integration.

Solution

Continue the program in problem 10 with the following:

```
% Finding the vector H
for i=1:N;
    H=1./sqrt(1+(f/150).^10)
end;
b=a.*H;
```

---

[3] This is called an ideal 5th-order lowpass Butterworth filter with cutoff frequency at 150 Hz.

```
for i=1:N
    for j=1:M
        r(i,j)=b(i)*sin(2*pi*f(i)*t(j));
    end;
end;
y=sum(r,1);
```

The result is $P_y = 3.1762$. By approximation,

$$H(f) = \begin{cases} 1, & f < 150 \\ 0.707, & f = 150 \\ 0, & f > 150 \end{cases}, \quad \sum H^2 = 6.5,$$

and

$$P_y \approx \left(\frac{6.5}{14}\right) P_x = 3.25.$$

12. An amplifier with a 1-Ω output resistance feeds an 8-Ω loudspeaker. See Figure 2.13(a). The signal put out by the voltage source is

$$v_1(t) = 9[\sin(100t + 45°) + \cos(10,000t)]$$

a. Find the average power in the loudspeaker.
b. Place a 100-μF capacitor in parallel with the loudspeaker as shown in Figure 2.13(b). Find the average power in the loudspeaker and compare it with the result in part a.

Solution

a. The average power delivered by a sinusoidal voltage to a resistor is $V_{max}^2/(2R)$, where V_{max} is the amplitude of the sinusoid. By voltage division, the voltage at the terminals of the loudspeaker is

$$v_2(t) = 8 \; [\sin(100t + 45°) + \cos(10,000t)]$$

The average power in the loudspeaker is, therefore,

$$P = 8^2/16 + 8^2/16 = 4 + 4 = 8 \text{ W}$$

b. With the capacitor in place, the magnitude of the frequency response is

$$|V_2/V_1| = \frac{R}{\sqrt{1 + R^2 C^2 \omega^2}}$$

where $R = 8/9 \; \Omega$ is the Thevenin equivalent resistance of the circuit seen by the capacitor. See Figure 2.13(c). Substituting for the two ω components in $v_1(t)$ and applying superposition of power we find the power in the loudspeaker to be

$$P = \frac{64}{16} \times \frac{64}{81} \left[\frac{1}{1 + \frac{64}{81} 10^{-8} (100)^2} + \frac{1}{1 + \frac{64}{81} 10^{-8} (10,000)^2} \right] = 3.16 + 1.76 = 4.92 \text{ W}$$

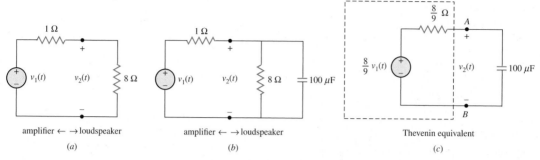

amplifier ← → loudspeaker

(a)

amplifier ← → loudspeaker

(b)

Thevenin equivalent

(c)

FIGURE 2.13 *(a)* An amplifier with a 1-Ω output resistance feeds an 8-Ω loudspeaker. *(b)* A 100-μF capacitor in parallel with the loudspeaker reduces its power at high frequencies. *(c)* Thevenin's equivalent of *(b)*.

Chapter Problems

13. The four corners of a quadrilateral in the xy plane are labeled in clockwise direction as $ABCD$. It is given that $AC = 1$, ABC and ADC are right angles, and BD is perpendicular to AC. The intersection of AC and BD is labeled O and the angle BAC called α. Show that $OC = \sin^2 \alpha$ and $BD = \sin 2\alpha$.

14. The expression $A \cos(2\pi ft + \theta)$, where $A > 0$ is the amplitude, f is the frequency (in Hz), and $-\pi < \theta < \pi$ is the phase in radians, represents a sinusoidal signal in cosine form. Determine the amplitude, frequency, and phase of the following signals when represented in the above form.

a. $3 \cos(3t + 45°)$ b. $2 \sin(\sqrt{2}t - 120°)$ c. $-4 \cos 5t$ d. $2 \cos \pi (t - 0.5)$
e. $2.5 \cos(\pi t + 10°)$ f. $- \cos(2\pi t + \pi/3)$ g. $-3 \sin(6.28t - 2\pi/3)$ h. $2 \cos 0.2(t - 1)$

15. Sketch and label the signals given in problem 14.

Problems 16–21

Identify the correct frequency of the given sinusoids.

16. The frequency of $\sin 10^6 \pi t$ is

   a. 2 MHz

   b. 2 MHz

   c. 500 kHz

   d. 250 kHz

   e. none of the above

17. The frequency of $\cos 10^6 t$ is most nearly

   a. 318 kHz

   b. 159 kHz

   c. 159 Hz

   d. 16 Hz

   e. none of the above

18. The frequency of $\sin(\pi t/6 + 3)$ is

   a. 1/6 Hz

   b. 1/12 Hz

   c. 1/3 Hz

   d. $2/\pi$ Hz

   e. none of the above

19. The frequency of $\sin(t/6) + 3$ is

   a. $\pi/3$ Hz

   b. $1/(12\pi)$ Hz

   c. 1/12 Hz

   d. $\pi/3$ Hz

   e. none of the above

20. The frequency of $\cos(3000\pi t + \theta)$ is

   a. 6 kHz

   b. 3 kHz

   c. 1.5 kHz

   d. 300 Hz

   e. none of the above

21. The frequency of $\cos 10^3 t$ is most nearly

   a. 159 kHz

   b. 159 Hz

   c. 80 Hz

   d. 50 Hz

   e. none of the above

Problems 22–31

Identify the correct period of the given signals.

22. The period of $\cos 10^6 \pi t$ is

   a. 1 ms

   b. 2 μs

   c. 1 μs

   d. 1 ns

   e. none of the above

23. The period of $3 \sin 10^6 t$ is most nearly

   a. 6 ms

   b. 6 μs

   c. 1 μs

   d. 6 ns

   e. 1 ns

24. The period of $2 + \cos(2 \times 10^5 \pi t)$ is

   a. 2 ms

   b. 20 μs

   c. 10 μs

   d. 2 ns

   e. none of the above

25. The period of $\sin 20\pi t + 4 \cos 2\pi t$ is

   a. 1 s

   b. 500 ms

   c. 50 ms

   d. 500 μs

   e. none of the above

26. The period of $\sin 5\pi t + \cos 3\pi t$ is

   a. 5 s

   b. 2.5 s

   c. 2 s

   d. 1.2 s

   e. none of the above

27. The period of $\sin t + \cos 2t$ is

   a. $3\pi/2$ s

   b. π s

   c. 1 s

   d. 1 ms

   e. none of the above

28. The period of $\sin 5t + \cos 3t$ is

   a. 5π s

   b. 2π s

   c. $(1/5 + 1/3)2\pi$ s

   d. $3\pi/5$ s

   e. none of the above

29. The period of $\sin 4t + 3 \cos 2t$ is

   a. 2π s

   b. π s

   c. $\pi/2$ s

   d. $\pi/3$ s

   e. none of the above

30. The period of $\sin 200t + \sin 201t$ is

   a. 2.1π s

   b. 2π s

   c. 1.99π s

   d. π s

   e. none of the above

31. The period of $\sin 200\pi t + \sin 201\pi t$ is

   a. 2.2 s

   b. 2.1 s

   c. 2 s

   d. 1.9 s

   e. none of the above

Problems 32–37

Identify the correct phase θ of the following sinusoids when expressed in the form of $\cos(\omega t + \theta)$.

32. The phase of $\cos(\pi t + 30°)$ is

a. 30 degrees

b. 3π radians

c. 3 radians

d. 60 degrees

e. none of the above

33. The phase of $\sin(2\pi t - 30°)$ is

a. $-30°$

b. $90°$

c. $-120°$

d. $120°$

e. none of the above

34. The phase of $\cos(\pi t + 1)$ is

a. 30 degrees

b. 3π radians

c. 1 radian

d. 60 degrees

e. none of the above

35. The phase of $\sin(\pi t + 1)$ is most nearly

a. -1 radian

b. -23 degrees

c. 1 radian

d. 23 radians

e. -33 degrees

36. The phase of $\cos \pi (t - 0.5)$ is

a. 90 degrees

b. $-\pi/2$ radians

c. $-\pi$ radians

d. -5 radians

e. none of the above

37. The phase of $\cos 500\pi (t + 10^{-3})$ is

a. 30 degrees

b. 0.5π radian

c. 0.5 radian

d. 2 radians

e. none of the above

Problems 38–45

Determine the correct phase relationship between the sinusoids given below.

38. $x_1 = \cos(t + 10°)$, $x_2 = \cos(t - 30°)$

a. x_1 leads x_2 by $10°$

b. x_1 lags x_2 by $130°$

c. x_1 leads x_2 by $40°$

d. x_2 lags x_1 by $130°$

e. none of the above

39. $x_1 = \cos(t + 30°)$, $x_2 = \cos(t + 55°)$

a. x_1 lags x_2 by $65°$

b. x_1 lags x_2 by $25°$

c. x_2 lags x_1 by $55°$

d. x_1 leads x_2 by $85°$

e. none of the above

40. $x_1 = \cos(t + 10°)$, $x_2 = \sin(t - 30°)$

a. x_1 leads x_2 by $10°$

b. x_1 leads x_2 by $40°$

c. x_2 lags x_1 by $130°$

d. x_1 lags x_1 by $130°$

e. none of the above

41. $x_1 = \cos(t + 30°)$, $x_2 = \sin(t + 55°)$

a. x_1 leads x_2 by $65°$

b. x_1 lags x_2 by $85°$

c. x_1 lags x_2 by $25°$

d. x_2 leads x_1 by $35°$

e. none of the above

42. $x_1 = \cos t$, $x_2 = \cos(t - 5 \, ms)$

a. x_1 leads x_2 by 1 degree

b. x_1 leads x_2 by $9/(2\pi)$ degrees

c. x_1 lags x_2 by 1 radian

d. x_1 and x_2 are almost in phase

e. none of the above

43. $x_1 = \cos t$, $x_2 = \cos(t - 1)$

a. x_1 leads x_2 by 55 degrees

b. x_1 leads x_2 by 1 radian

c. x_1 leads x_2 by 1 degree

d. x_1 and x_2 are almost in phase

e. x_2 lags x_1 by 0.1 radian

44. $x_1 = \cos 10^4 t$, $x_2 = \cos 10^4 (t - 10\mu s)$

    a. x_1 lags x_2 by 10 degrees

    b. x_1 and x_2 are almost in phase

    c. x_1 leads x_2 by 5 degrees

    d. x_2 lags x_1 by 10 degrees

    e. x_2 lags x_1 by 0.1 radian

45. $x_1 = \cos 10^5 t$, $x_2 = \cos 10^6 (t + 1\mu s)$

    a. x_1 leads x_2 by 1 degree

    b. x_1 leads x_2 by 1 radian

    c. x_1 and x_2 are almost in phase

    d. x_1 lags x_2 by 1 radian

    e. none of the above

46. Use phasors to convert each time function given below into the form $A\cos(2\pi f t + \theta)$.

    a. $3\cos(3t + 45°) + 2\sin(3t - 120°)$ b. $-4\cos(5t) + 2\cos 5(t - 0.5)$

    c. $2.5\cos(\pi t + 10°) - \cos(\pi t + \pi/3)$ d. $-3\sin(6.28t - 2\pi/3) + 2\cos 6.28(t - 0.5)$

47. For each time function given below determine if it is periodic or aperiodic, and specify its period if periodic.

    a. $\cos 5t + \cos \pi t$ b. $\cos(5t) + \cos 3.1416t$

    c. $\cos(\pi t + 10°) - \cos(2\pi t + \pi/3)$ d. $\sin(6.28t - 2\pi/3) + \cos 0.2(t - 1)$

    e. $\sin 5t + \cos(3t + \theta)$ f. $\sin t + \sin 2\pi t$

    g. $\sin \pi t + \sin 3.141592t$ h. $\cos(1.14t + \theta) + \sin 3.141592t$

48. For each time function given below determine if it is periodic or aperiodic, and specify its period if periodic.

    a. $\cos 2\pi t + \cos 6.28t$ b. $\cos 2\pi t + \cos 6.2816t$

    c. $\cos 2\pi t + \cos 6.28159t$ d. $\cos 6.2816t + \cos 6.28159t$

    e. $\cos \sqrt{2}t + \cos 1.41t$ f. $\cos 1.4142t + \cos 1.41t$

49. For each time function given below determine if it is periodic, aperiodic, or whether more information is needed. Specify its period if periodic.

    a. $\cos 3.14t + \sin 2\pi t$ b. $\cos 3.14t + \sin 3.1416t$

    c. $\cos \pi t - \sin(100\pi t + \pi/3)$ d. $\sin(6.28t - \pi/3) + \cos(t - 1)$

50. A sinusoidal voltage $v(t)$ has a frequency of 100 Hz, a zero DC value, and a peak value of 13 V which it reaches at $t = 1$ ms. Write its equation as a function of time in cosine form and plot it for $0 < t < 10$ msec.

51. For the following cases determine the phase lag of $x_2(t)$ with reference to $x_1(t)$.

    a. $x_1(t) = \cos(t + 30°)$ and $x_2(t) = \cos(t - 10°)$ b. $x_1(t) = \sin(t + 30°)$ and $x_2(t) = \cos(t - 10°)$

    c. $x_1(t) = \cos(t + 30°)$ and $x_2(t) = \sin(t - 10°)$ d. $x_1(t) = \sin(t + 30°)$ and $x_2(t) = \sin(t - 10°)$

52. Consider the sum of two sinusoids having the same frequency but different phases $x(t) = \cos \omega t + \cos(\omega t + \theta)$, where $0 \le \theta \le 2\pi$ is the phase. Problem 4 expressed it as $x(t) = A\cos(\omega t + \phi)$ and obtained the A and *phi* values for $\theta = k\pi/4$, $0 \le k \le 8$, k an integer. Write a Matlab program to plot $x(t)$ for $\omega = 2\pi$ and the above 9 values of θ. As in problem 4 consider the following two conditions for A and ϕ:

    a. $A \ge 0$, and $-\pi \le \phi \le \pi$.

    b. A any value, and $0 \le \phi \le \pi$.

53. The following three measurements are made off a sinusoidal signal $x(t) = X_0 \cos(2\pi f t + \theta)$, where the frequency is assumed to be 1 MHz.

t	0	0.1 μs	0.2 μs
$x(t)$	2.1213	0.4693	−1.3620

    a. Verify the above frequency assumption.

    b. Find the amplitude and phase of $x(t)$.

c. Knowing the frequency, what is the minimum number of samples from which the amplitude may be computed? Show how this minimum sample number is obtained.

54. For each set of measurements given in Table 2.4, model the signal by $x(t) = A\cos(2\pi f t + \theta)$ (i.e., find the amplitude, frequency, and phase of the sinusoid).

TABLE 2.4 (Problem 54)

		t	0	0.1 μs	0.2 μs
a.	$x_1(t)$		1.7321	1.8437	1.9263
b.	$x_2(t)$		1.6209	0.7615	-0.1725
c.	$x_3(t)$		0.1732	0.1889	0.1979
d.	$x_4(t)$		1.4494	2.0148	2.5306
e.	$x_5(t)$		1.4330	1.3831	1.3201
f.	$x_6(t)$		4.2092	4.9875	5.0184
g.	$x_7(t)$		-0.2499	-0.3516	-0.4500

55. Write the mathematical expression $A\cos(2\pi f t + \theta)$ for the sinusoidal signals whose measurements are recorded in Table 2.5 and then sketch them.

TABLE 2.5 (Problem 55)

	Signal	Period	x_{max}	x_{min}	$x(0)$	Slope at $t = 0$
a.	$x_1(t)$	10 ms	3	-1	2.7321	+
b.	$x_2(t)$	6.67 μs	3.5	0.9	2.85	-
c.	$x_3(t)$	8.33 μs	4	0	-0.3823	-
d.	$x_4(t)$	12.5 ms	-1	-4	-1.0511	+
e.	$x_5(t)$	10 ms	2.5	0.5	2.366	+
f.	$x_6(t)$	1334 μs	-2.5	-7.5	-3.1823	-
g.	$x_7(t)$	4 ms	1	-5	-1.0729	+

56. A sinusoidal voltage $v(t)$ has a frequency of 200 Hz, a zero DC value, and a peak value of 13 V, which it reaches at $t = 0.5$ ms. Write its equation as a function of time in a cosine form.

57. A low-frequency periodic signal $s(t)$ is modeled by a DC value added to the sum of the first N harmonics of a fundamental frequency f_s (in Hz) as given below:

$$s(t) = C_0 + \frac{1}{2}\sum_{n=1}^{N} C_n \cos(2\pi n f_s t)$$

The highest frequency in $s(t)$ is $f_0 = N f_s$. The signal $s(t)$ modulates a sinusoidal carrier $\cos(2\pi f_c t)$, $f_c > f_0$. The modulated waveform is $x(t) = s(t)\cos(2\pi f_c t)$. Let $f_s = 100$ kHz, $f_0 = 1$ MHz, and $f_c = 100$ MHz. The modulated waveform $x(t)$ is passed through an ideal bandpass filter with unity amplitude gain within the band of 99 MHz to 101 MHz and zero gain outside it. The phase lag introduced by the filter in a sinusoidal input at frequency f within the above band is $\theta = 0.2\pi \left(1 + 10^{-8} f\right)$ radian. Find the output of the filter $y(t)$ and discuss possible distortions.

58. Use the phasor notation to show that $V_1 \cos(\omega t + \theta_1) + V_2 \cos(\omega t + \theta_2) = V\cos(\omega t + \theta)$, where

$$\theta = \tan^{-1}\left\{\frac{V_1 \sin\theta_1 + V_2 \sin\theta_2}{V_1 \cos\theta_1 + V_2 \cos\theta_2}\right\} \text{ and } V = \sqrt{V_1^2 + V_2^2 + 2V_1 V_2 \cos(\theta_1 - \theta_2)}$$

59. Shift the periodic squarewave signal of Example 2.7 by a constant value and show that the shifted waveform may be represented by an infinite series of the following form:

$$v(t) = \sum_{n=1}^{\infty} a_n \cos\left(\frac{2\pi nt}{T} + \theta_n\right), \quad \text{where } |a_n| = \frac{2A}{\pi n} \quad \text{and } n \text{ odd.}$$

Show that the shift does not affect conclusions regarding power distribution obtained in Example 2.7.

60. The unit-ramp function $x(t) = t$ can be approximated during $-T/2 < t < T/2$ by the finite series

$$y(t) = \frac{T}{\pi} \sum_{n=1}^{N} \sin\left(\frac{2\pi nt}{T}\right)$$

One measure of approximation error is

$$\mathcal{E} = \frac{1}{T} \int_{-\frac{T}{2}}^{-\frac{T}{2}} |x(t) - y(t)|^2 dt.$$

Write a program to generate $x(t)$ and $y(t)$, plot them, and compute the error as defined above. Run the program for $N = 1, \ldots, 10$ and plot $\mathcal{E}$ versus N.

61. Motion of a free electron in a sinusoidal electric field. An electron has a negative electric charge of $e = 1.602 \times 10^{-19}$ C and a mass of $m = 9.109 \times 10^{-31}$ kg. When placed in an electric field of field strength $\mathcal{E}$, it experiences a force of $e \times \mathcal{E}$. Determine the span of the motion of an electron in vaccuum when subjected to an electric field $\mathcal{E} = 10^{-6} \cos(2\pi ft)$ V/m at frequencies of (a) 60 Hz, (b) 1 kHz, (c) 1 MHz, and (d) 1 GHz.

62. The following Matlab program is written to sweep from left to right in the xy plane. In each sweep it plots a sinusoid, then moves down an incremental value, similar to the motion of the electron beam in a cathod ray tube or the operation of the recording element in a seismograph.

```
t=[0:0.001:1];
hold on
for k=1:30,
    x=sin(2*pi*k*t)+10-2*k-0.3;
    axis([0 1 -50 10]);
    plot(t,x);
    xlabel('Time (s)'); ylabel('harmonics'); title(' harmonics of 1 Hz');
    grid;
end
hold off
```

a. Execute the program and examine the plot to verify that it agrees with expectation.
b. In successive steps replace the fourth line in the program with a new command line from the following list and note if the resulting plot agrees with your expectation.

```
x=sin(2*pi*k*t)+2*k+0.3;      axis([0 1 0 50]);
x=sin(2*pi*k*t)+50-2*k-0.3;   axis([0 1 0 50]);
x=sin(2*pi*k*t)+60-2*k-0.5;   axis([0 1 0 60]);
x=sin(2*pi*k*t)+90-2*k-1;     axis([0 1 30 90]);
x=sin(2*pi*k*t)+90-4*k-1;     y=cos(2*pi*k*t)+90-4*k-3; axis([0 1 30 90]);
plot(t,x,t,y);
```

Explore an alternative set of variables that enhance the wavy appearance of the two-dimensional plot.

2.19   Project: Trajectories, Wave Polarization, and Lissajous Patterns

Purpose

To investigate the trajectories of the sinusoidal motion of a point in the xy plane and obtain parameters of the motion from the patterns of the trajectories.

Introduction and Summary

The motion of a point M in the xy plane can be stated as a time-varying vector drawn from the origin to its tip at point M. The motion is described by two equations that specify the Cartesian coordinates of M as functions of time. The path traversed by the tip of the vector, called its trajectory, may be found by eliminating the variable t from those equations. Some parameters of the motion (such as the amplitude, phase, and frequency) may be deduced from the trajectory. In this project you will generate and examine several classes of trajectories where the x and y coordinate values vary sinusoidally with time. The project contains six sections.

   i. $x(t)$ and $y(t)$ have the same frequency.
   ii. $x(t)$ and $y(t)$ have slightly different frequencies.
   iii. The frequencies of $x(t)$ and $y(t)$ are harmonics of a principle frequency.
   iv. The frequencies of $x(t)$ and $y(t)$ are not harmonics of a principle frequency.
   v. The effect of the sampling rate.
   vi. Implementation through an electric circuit.

   The present project may be carried out by mathematical simulation or in real time by physical oscillators. Examples from both are included.

Section I. $x(t)$ and $y(t)$ have the same frequency. Consider the time-varying vector **E** in the xy plane drawn from the origin to point M whose projections on the x and y axes are given by

$$\begin{cases} x = E_x \cos \omega t \\ y = E_y \cos(\omega t + \theta) \end{cases}$$

Depending on the phase difference and amplitude ratio, three types of trajectories are observed: linear (zero phase or π), circular (equal amplitudes with $\pm\pi/2$ phase), and elliptical (all other values).

a. **Linear trajectory.** For $\theta = 0$ or π, the trajectory is a straight line. See Figure 2.14(a). Show that $y = \pm\frac{E_y}{E_x}x$. Rotate the xy coordinate system by an angle γ in the counterclockwise direction to have a new coordinate system $\alpha\beta$. Show that the relationship between the two coordinate systems is

$$\alpha = x \cos \gamma + y \sin \gamma$$
$$\beta = -x \sin \gamma + y \cos \gamma$$

   Find the equation of the trajectory in the new coordinate system. Determine the appropriate γ value to represent **E** by a one-dimensional vector that oscillates in time along the α-axis only.

b. **Circular trajectory.** For $E_y = E_x = E$ and $\theta = \pi/2$ or $-\pi/2$, the motion has a circular trajectory. See Figure 2.14(b). Show that $x^2 + y^2 = E^2$ and determine the direction of motion for the trajectory.

c. **Elliptical trajectory.** For the general case, the tip of the vector moves along an elliptical trajectory. Show that

$$\left(\frac{x}{E_x}\right)^2 + \left(\frac{y}{E_y}\right)^2 - \frac{2xy}{E_x E_y}\cos\theta = \sin^2\theta$$

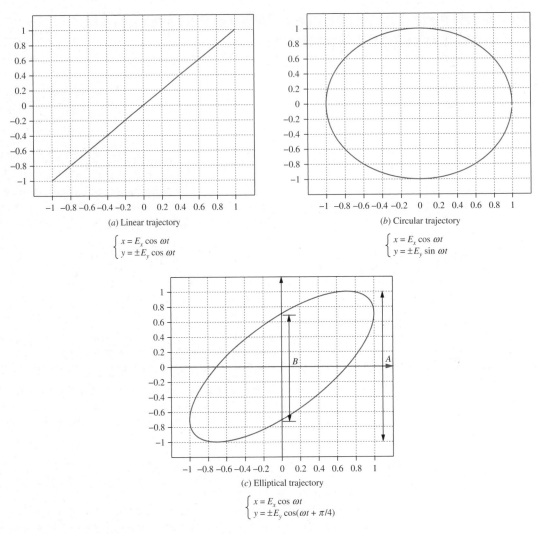

FIGURE 2.14 Trajectories of the motion of the tip of a rotating vector in the xy plane. The trajectories have the form of simple Lissajous patterns.

and determine the direction of motion for the trajectory. Show that the phase angle θ can be found from

$$\sin\theta = \pm(B/A)$$

where A and B are as shown on Figure 2.14(c). Rotate the xy coordinate system by an appropriate angle to align the x-axis along the major axis of the ellipse. Determine the rotation angle. Write the equation for the trajectory in the new coordinate system.

Simulation by Computer

Run the Matlab code given below to generate a linear trajectory.

```
f=1; w=2*pi*f; T=1/f; N=1; a=1; b=1; theta=0;
% Motion parameters.
t=linspace(0,N*T,100*N*T); x=a*cos(w*t); y=b*cos(w*t+theta); plot(x,y)
% Trajectory.
```

Change the parameters of the motion and the variables in the above code to explore how each one may or may not change the shape of the trajectory. From the plots of the trajectories obtain the amplitudes of the horizontal and vertical motions and the phase difference between them. Compare with the values used in the simulation.

Parallels with Electromagnetic Wave Polarization

The electric field vector $\mathbf{E}$ in an electromagnetic plane wave is a time-varying vector $\mathbf{E}$ that lies in the plane that is perpendicular to the direction of propagation. It has two components; namely, in the x and y directions. In the case of sinusoidal time variation, the electric field is an $\mathbf{E}$ vector with x and y components (each of which vary sinusoidally with time) as follows:

$$\begin{cases} x = E_x \cos \omega t \\ y = E_y \cos(\omega t + \theta) \end{cases}$$

This is the same vector we discussed at the beginning of this section with three possible tip trajectories. Each trajectory is associated with one type of wave polarization. The electromagnetic wave, therefore, is said to be polarized as any of the above types. In addition, the motion of the tip of the electric field vector can be in the clockwise or counterclockwise directions, labeled left or right, circular or elliptical.

Section II. $x(t)$ and $y(t)$ have slightly different frequencies.

$$\begin{cases} x(t) = a \cos \omega_1 t \\ y(t) = b \cos \omega_2 t \end{cases}$$

With ω_1 and ω_2 approximately the same (but not exactly equal), the difference in frequencies will appear as a time-varying phase difference, resulting in a trajectory that slowly moves between the above three patterns. Construct an example where

$$\begin{cases} x(t) = \cos 2\pi t \\ y(t) = b \cos 2.1\pi t \end{cases}$$

(corresponding to $f_1 = 1$ Hz and $f_2 = 1.05$ Hz, respectively). Modify the Matlab program given in section I to plot the Lissajous patterns for this example, allowing the plots to be generated for (a) 2 seconds, (b) 3 seconds, (c) 8 seconds, and (d) 150 seconds. Determine the time needed for a full cycle in each plot. Explore the effect for $f_2 = (1 + k) f_1$, with $k = 0.2, 0.1$, and 0.01.

Section III.   The frequencies of $x(t)$ and $y(t)$ are harmonics of a principle frequency. In this case, more complex patterns are generated whose shapes and parameters are associated with the ratio of the two frequencies. Four examples are shown in Figure 2.15 for $x = \cos \omega t$ and four different variations of $y(t)$.

Using Matlab, plot trajectories of y versus x with $x = \cos(at)$ and $y = \cos(bt)$, for $a/b = 3$ and $3/5$. Repeat for $x = \cos(at)$ and $y = \sin(bt)$. In each case eliminate t between y and x to obtain the equation relating them together and verify its representation by a plot obtained through Matlab. Determine the number of crossings of a horizontal line and a vertical line and relate them to the ratio a/b. Suggest a method to measure the frequency of a sinusoidal signal from Lissajous patterns.

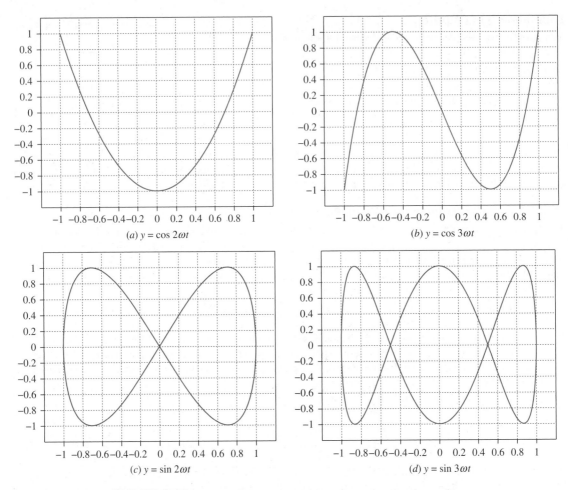

FIGURE 2.15 Four Lissajous patterns with $x = \cos \omega t$ and various $y(t)$.

Section IV. The frequencies of $x(t)$ and $y(t)$ are not harmonics of a principle frequency. In theory, the motion is not periodic and it takes $T = \infty$ for it to repeat. In practice (e.g., in simulation by Matlab or in a real-time experiment using two physical oscillators), the Lissajous pattern will eventually repeat itself. Run the following Matlab code, observe the trend, and compare it with the observations in section II. Repeat the procedure after replacing $\cos(\pi t)$ by $\cos(3.1t)$.

```
N=4; %Repeat for N=2, 8, 20, 50
t=linspace(0,N*pi,800);
x=cos(t);  y=cos(pi*t); plot(x,y);
```

Section V. Effect of low sampling rate. In plotting a trajectory, the x and y coordinates need to be sampled sufficiently fast. Otherwise the plot will exhibit the artifacts caused by a low number of samples. This effect is shown by the following two examples.

a. Generate three examples of Lissajous patterns derived from the same motion but with three different sampling rates. You may use the following code for this purpose.

```
N=2; % Repeat for N=20 and 50.
t=linspace(0,N*pi,100);
x= cos(2*t); y=sin(3*t); plot(y,x)
```

b. For the following set of samples (N) and time (T)

   i. $N = 20$, $T = 100,\ 50,\ 25$

   ii. $N = 50$, $T = 10,\ 20,\ 25$

   iii. $N = 100$, $T = 40,\ 50$

   iv. $N = 200$, $T = 2$

run the Matlab code given below and explain trends in the plots.

```
t=linspace(0,T,N);
x=cos(2*pi*t); y=cos(2*pi*t+pi/4); plot(x,y)
```

Section VI.   Implementation through an electric circuit. An oscilloscope whose horizontal and vertical deflections are controlled by the signals $x(t)$ and $y(t)$, respectively, eliminates t between them and displays the xy motion trajectory. For this purpose, an ordinary oscilloscope can be used if set in the X deflection mode.

a. Start with the situation described in section I of this project. Configure the circuit of Figure 2.16(a) with $R = 1\ \mathrm{k}\Omega$ and $C = 0.1\ \mu\mathrm{F}$. Set up the function generator to provide a 2-volt peak-to-peak sinusoidal voltage signal v_1 at 1590 Hz and connect it to the RC circuit as shown, and also to the horizontal axis of the scope and set it in X deflection mode. Connect the capacitor voltage v_2 to the vertical axis of the scope. The relationship between v_1 and v_2 can be readily found using terminal characteristics of R and C, and Kirchhoff's current and voltage laws. The result is

$$v_1(t) = V_1 \cos \omega t, \quad v_2(t) = V_2 \cos(\omega t + \theta), \quad V_2 = \frac{1}{\sqrt{1 + R^2 C^2 \omega^2}} V_1, \quad \text{and } \theta = -\tan^{-1}(RC\omega)$$

Show that at the above settings,

$$v_1(t) = \cos 10{,}000t \ \text{ and } \ v_2(t) = 0.707 \cos\left(10{,}000t - \frac{\pi}{4}\right)$$

The elliptic pattern shown in Figure 2.16(a) should appear on the scope. Find the equation of the trajectory on the screen as predicted from theory. Compute the phase angle between v_1 and v_2 from

$$\sin\theta = \pm(B/A)$$

where A and B are shown on the ellipse. The phase angle is expected to be 45°. The amplitudes of the horizontal and vertical signals may be changed through the horizontal and vertical gains of the scope. The phase angle between the horizontal and vertical signals can be changed by changing the frequency of the signal generator. Explore the effect of the above parameters on the shape of the trajectory.

To obtain a linear trajectory, replace the capacitor in the circuit by a resistor. The circuit and resulting trajectory are shown in Figure 2.16(b). Relate the slope of the trajectory to the resistor values and the horizontal and vertical gains of the oscilloscope.

To obtain a circular trajectory use the circuit configuration of Figure 2.16(c), which contains an isolation transformer.

b. Continue with real-time implementations of the situations described in sections II, III, IV, and V. For that purpose you will employ two sinusoidal signal generators directly connected to the horizontal and vertical channels.

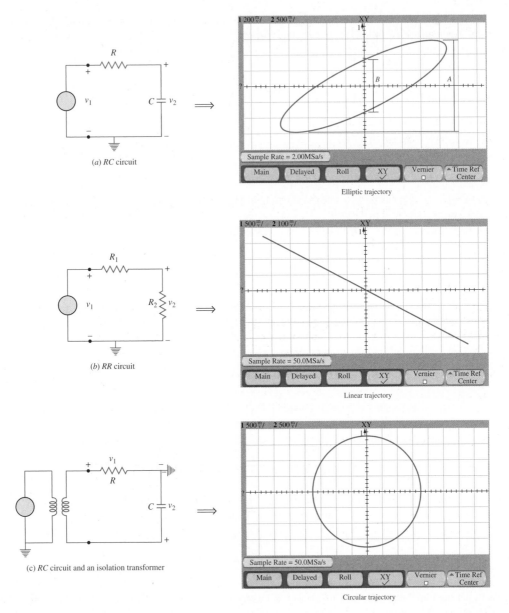

FIGURE 2.16 Three circuits for generating x and y coordinate signals and the resulting trajectories captured off an oscilloscope.

Conclusions

Describe your overall conclusions from the project. What applications may it have? Where can one draw the line that distinguishes the results obtained by the digital approach (using the computer) from the analog approach (using the funtion generator and oscilloscope)?

Chapter 3

Systems, Linearity, and Time Invariance

Contents

Introduction and Summary

In its common usage, the word *system* signifies a combination or collection of interconnected elements or parts performing certain functions or producing certain effects. The elements may be physical (as in electrical or mechanical systems), models of nonphysical entities (as in computational, economic, societal, environmental, and ecological systems), or a combination of physical and nonphysical models (as in digital communication systems, data communication systems, and systems of computer networks).

Systems and subsystems may be analyzed at various levels and from different viewpoints. One may analyze the operation of a system at the component level. As an example, we may analyze an amplifier circuit using Kirchhoff's current and voltage laws, along with the characteristics of elements such as energy sources, resistors, inductors, capacitors, switches, transistors, op-amps, and the like. At this level, the focus is more on the internal structure of the amplifier, for instance, how it functions, and how it should be designed to produce the desired performance. One is interested in not only what it does but also how it does it. At a higher level, one addresses the function and usage of the amplifier within a bigger system. We call this the system level. From this point of view, the internal structure of the amplifier is of less interest than its operation within the bigger system. Therefore, a description of the function and performance is more important than how the internal structure of the amplifier leads to that performance.

Clearly, the borderline between the component and system levels depends on what is called a *component* and what is called a *system*. For example, consider a digital signal processing (DSP) system, such as a stand-alone board or one embedded in a computer, that acquires analog signals, converts them to digital data, performs DSP operations on the data, and sends the results to a user. The system is made of several subsystems such as analog-to-digital and digital-to-analog (A/D and D/A) converters, interfaces, processor, and communication ports, each of which is constructed from components (integrated circuits, resistors, capacitors, op-amps, etc.). Each of the components may be considered a system on its own. At the same time, most probably, the DSP system under consideration may be functioning within a larger system such as a speech processing system, a telephone network, a digital communication system, a data transmission network, or a computer network. In that role, the DSP system is a component of a larger system.

What then constitutes the system view? How important is the distinction between the system view and the component view? Should analysis of a system be detached from the components of which it is composed? Is there an advantage in developing a unified approach to describing various systems regardless of whether the components and variables are electrical, mechanical, thermal, or societal? In the present discussion we will avoid such an abstract approach as much as possible. We will use mathematical models of systems, with occasional insights into their physical structure, derived mostly from electrical and electromechanical systems. For our purposes, a system is defined by an input space, an output space, and the relationship between them. The formulation of equations that describe the system is shown in the next section, followed by the system's classifications.

In this book we consider the class of engineering systems that interact with the outside world through signals and waveforms which are functions of time and space. This class covers a vast population of systems. We concentrate on a special subset called linear time-invariant (LTI) systems, knowledge of which greatly facilitates the analysis and design of a large number of engineering systems. Furthermore, the systems are single-input, single-output deterministic, and real-valued (described by linear differential or difference equations with real-valued coefficients). Unless otherwise specified, the input and output are also real-valued functions of time. The linearity and time-invariance properties are sunmmarized below.

Linearity

input	$\Longrightarrow$ output
$x_1(t)$	$\Longrightarrow y_1(t)$
$x_2(t)$	$\Longrightarrow y_2(t)$
$x_3(t) = ax_1(t) + bx_2(t) \Longrightarrow y_3(t) = ay_1(t) + by_2(t),$	
for all $x_1(t), x_2(t), a,$ and b	

Time Invariance

input	$\Longrightarrow$ output
$x(t)$	$\Longrightarrow y(t)$
$x(t - T) \Longrightarrow y(t - T),$	
for all $x(t)$ and T	

The chapter shows how to test for LTI properties and how to construct or predict the response to a new input by using superposition of known responses and applying the derivative and integral properties. After presenting several examples from various engineering fields, the chapter introduces a system's responses to the impulse, step, exponential, and sinusoidal inputs. While the technique of superposition will be formally presented in the next chapter, the examples and solved problems in this chapter are designed to illustrate the application of the fundamentals independent of any technique. Methods for the analysis and solutions of LTI systems (including Fourier techniques) are direct consequences of LTI properties and are summarized for expansion and discussion in later chapters.

In an introductory textbook on signals and systems such as this one, LTI models of simple and familiar devices and systems are used to illustrate the concepts. Real-life systems on the other hand, be they synthetic or natural, are neither LTI nor simple. However, the LTI modeling provides an initial step toward the analysis and design of such complex systems. Also, a brief discussion of complex systems exposes the reader to some prominent examples such as the human central nervous system and adaptive learning models. The former is too advanced to lend itself to the basic methods presented in such an introductory book, but the latter are not. Finally, the open-loop control project at the end of the chapter shows how to apply LTI properties to produce a desired output and, therefore, helps the reader develop basic skills in this area.

3.1 Formulation of Equations

Mathematically, a system is specified by the following three entities:

   i.   The collection of all possible inputs, designated as the input set $X = \{x_i\}$.

  ii.   The collection of all possible outputs, designated as the output set $Y = \{y_j\}$.

 iii.   The input-output transformation or the mapping function $X \Rightarrow Y$.

See Figure 3.1.

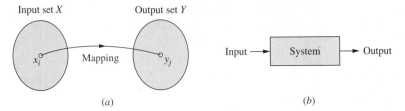

FIGURE 3.1 *(a)* A system has an input space, output space, and input-output mapping. *(b)* Block diagram representation.

Systems are, in general, made of interconnected elements. The performance of a system is then governed by two groups of relationships:

i.   Group one: The elements' individual characteristics (i.e., the relationships among each element's variables).

ii. Group two: Interconnection constraints (i.e., the relationships created by interconnecting the elements).

The above groups of relationships are combined and formulated as the system's dynamical equations in two ways.

1. **Input-output formulation.** The input-output formulation limits itself to the mapping function; the output is expressed in terms of input only. The internal variables don't appear in the system's dynamical equations. Once the input-output relationship is determined, one no longer needs to know the internal structure of the system or the physical laws governing the internal variables.

2. **State-variable formulation.** In an alternative formulation, the internal states of the system constitute the variables of its dynamical equations. This is also called the state-space description. In this formulation, the output is expressed in terms of the input and internal states. An example would be the voltage appearing at the output of an *RLC* circuit which is expressed in terms of (a) the input to the circuit, (b) the capacitors' voltages, and (c) the inductors' currents. These last two groups of variables constitute the state space of the system at any moment.

In formulating and solving dynamical equations, we observe great similarities and strong analogies between many physical systems in various areas such as the electrical, mechanical, and thermal arts. This is especially evident in the case of linear time-invariant systems, to be introduced later in this chapter.

3.2 Classifications of Systems

Input and output spaces in Figure 3.1 may be continuous or discrete. The sets may be finite or infinite, bounded or unbounded. The mapping may be one-to-one or not, deterministic or probabilistic, with or without memory. The mapping may be expressed by a table, an equation, or a graph. Based on such aspects, systems are classified in several different ways. Some classifications are listed below.

Continuous Versus Discrete, Analog Versus Digital Systems

A system is called *continuous-time* if its input and output signals may change at any instance of time (or at any location of space, in which case it should be called *continuous-space*). Most physical systems are continuous-time. A system is called *discrete-time* if the values of its inputs and outputs may change (or be observed) only at discrete instances of time, such as every T seconds.

If the system's input and output assume a continuous range of values the system is called *analog*. If they can assume only one of n predetermined levels the system is called discrete. In a binary system, signals belong to one of two states, each represented by a level; for example, 0 or 1. If the signal level may switch between only these levels and at discrete times, the system is called *digital*.

The above groupings are based on the properties of input and output signals and reflect the signal processing function of the system. In a different venue, a system may be concerned with discrete events, in which case it is called a *discrete-event* system. An example is a decision system that classifies members of a discrete set of events into another discrete set of decisions.

Static Versus Dynamic Systems

A system is said to be without memory if the output at any moment depends only on the input at that moment. Memoryless systems are sometimes called instantaneous, or static, because the output at each instant is influenced only by the input at that instant. The input effect is immediate and does not last. Past inputs have no influence on present or future outputs. Passive resistive circuits are examples of memoryless systems.

Systems with storage capability (energy or state) have memory, as past input values influence the present state of the system and its output. Electric circuits made of *RLC* elements store energy and are systems with memory. Mechanical systems with masses and springs (i.e., energy storage elements) have memory, too. Energy storage elements, however, are not the only devices that provide memory in physical systems. Introduction of a nonlinear active element in a resistive memoryless system may provide it with information storage capability, thus providing memory. Flip-flops, Schmitt triggers, and comparators with feedback are examples of systems that do not store energy, but rather information about past inputs. Systems that contain such elements are said to have memory. Systems with memory are also called *dynamic* and the equations that describe them are called the *dynamical equations* of the system.

Systems with Feedback

The output of a system may physically travel back through a path and be fed back (e.g., by simple addition) to the input of the system. This creates a feedback system. See Figure 3.2. An example is the reverberation and echo present in an acoustic system. Feedback paths may also be created to modify the behavior of a system; for example, to reduce the system's sensitivity to noise or increase its linearity range. In other situations

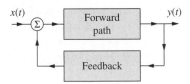

FIGURE 3.2 A system with feedback.

the feedback signal (be it the output or a state variable) may be used to design new commands.

3.3 Causality

The distinction between causal and noncausal systems becomes clear when input $x(t)$ and output $y(t)$ are functions of time only. A causal system is a system whose output is zero until an input arrives. Therefore, if $x(t) = 0$ for $t < t_0$, then $y(t) = 0$ for $t < t_0$. With time being the independent variable, a causal system may be realizable and a noncausal system is physically impossible. If the independent variable is something other than time (e.g., space), a noncausal system may be physically realizable.

3.4 Linearity, Time Invariance, and LTI Systems

The most important and useful properties of some systems are linearity and time invariance, which allow us to predict the output of a large class of inputs based on a single input-output pair, and often provide an analytical and closed-form formulation of the relationship between input and output.

Linearity

Let $x(t)$ and $y(t)$ designate the input and output of a system, respectively. In some systems, superposition of inputs results in superposition of outputs. If the property remains in effect for all possible inputs, the system is called *linear*. One consequence of linearity is proportionality: multiplying an input by a number, multiplies the output by the same number. Another consequence of linearity is the ability to predict the output in response to new inputs that may be represented by linear combinations of previously experienced inputs.

To test for linearity, let two arbitrary inputs x_1 and x_2 produce outputs y_1 and y_2, respectively. If the input $x = ax_1 + bx_2$ produces $y = ay_1 + by_2$ for all constants a and b, then the system is called linear.

$$
\begin{aligned}
x_1 &\implies y_1 \\
x_2 &\implies y_2 \\
x = ax_1 + bx_2 &\implies ay_1 + by_2
\end{aligned}
$$

Time Invariance

The characteristics of some systems are independent of time. As a result, a time shift in the input causes a time shift of the same value in the output, with no other effects. The system is called time invariant if the above property holds true for all inputs, at all times, and for all time shifts. This property allows the use of previously measured input-output pairs for future prediction. If some characteristics of a system change with time, a delay in an input may not result in a similar delay in the output, or may result in a longer delay. The system is then time variant.

To test for time invariance, let an arbitrary input $x(t)$ produce the output $y(t)$. If the input $x(t-T)$ produces the output $y(t-T)$ for all T, then the system is called time invariant.

$$x(t) \implies y(t)$$
$$x(t-T) \implies y(t-T)$$

Linearity and time invariance are two separate properties of systems. Systems that are both linear and time invariant (LTI) are of great interest because they may often be described sufficiently well by a certain single input-output pair. The output to any other input may then be predicted from such a description.

E*xample* 3.1

Is the system $y(t) = kx(t)$, with k a constant, linear? Is it time invariant?

Solution
Testing for linearity:

$$\begin{cases} x_1 & \implies y_1 = kx_1 \\ x_2 & \implies y_2 = kx_2 \\ x = ax_1 + bx_2 \implies y = kx = k(ax_1 + bx_2) = akx_1 + bkx_2 = ay_1 + by_2 \end{cases}$$

The system is linear because $y = ay_1 + by_2$ for all a, b, x_1, and x_2.
    Testing for time invariance:

$$\begin{cases} x(t) & \implies kx(t) = y(t) \\ x(t-T) \implies kx(t-T) = y(t-T) \end{cases}$$

The system is also time invariant because the shift property is valid for all T and x.

E*xample* 3.2

Is the system $y(t) = k_1 x(t) + k_0$, where k_0 and k_1 are nonzero constants, linear? Is it time invariant?

Solution
Testing for linearity:

$$\begin{cases} x_1 & \implies y_1 = k_1 x_1 + k_0 \\ x_2 & \implies y_2 = k_1 x_2 + k_0 \\ x = ax_1 + bx_2 \implies y = k_1 x + k_0 = k_1 [ax_1 + bx_2] + k_0 \neq ay_1 + by_2 \end{cases}$$

The system is not linear because $y \neq ay_1 + by_2$.
    Testing for time invariance:

$$\begin{cases} x(t) & \implies y(t) = k_1 x(t) + k_0 \\ x(t-T) \implies k_1 x(t-T) + k_0 = y(t-T) \end{cases}$$

The system is time invariant because the shift property is valid for all T and all x.

Examples 3.3 to 3.5

In the systems of Examples 3.3 to 3.5, $x(t)$ is the input and $y(t)$ is the output.

a. Determine if the system is (i) linear, (ii) time invariant.

b. Find the unit-impulse response $h(t)$ for the LTI systems.

$$y(t) = \int_{-\infty}^{t} x(\tau)d\tau$$

Solution

a. Testing for linearity:

$$y_1(t) = \int_{-\infty}^{t} x_1(\tau)d\tau \quad \text{and} \quad y_2(t) = \int_{-\infty}^{t} x_2(\tau)d\tau$$

$$x_3(t) = ax_1 + bx_2(t) \implies y_3(t) = \int_{-\infty}^{t} (ax_1 + bx_2)d\tau$$

$$y_3(t) = a\int_{-\infty}^{t} x_1 d\tau + b\int_{-\infty}^{t} x_2(\tau)d\tau = ay_1 + by_2(t)$$

for all x_1, x_2, a, and b. The system is linear.

b. Testing for time invariance:

$$x_1(t) = x(t - t_0) \implies y_1(t) = \int_{-\infty}^{t} x(\tau - t_0)d\tau = \int_{-\infty}^{t-t_0} x(\theta)d\theta = y(t - t_0)$$

for all $x(t)$ and t_0. The system is time invariant with

$$h(t) = \int_{-\infty}^{t} \delta(\tau)d\tau = u(t)$$

$$y(t) = t\int_{-\infty}^{t} x(\tau)d\tau$$

Solution

a. Testing for linearity:

$$y_1(t) = t\int_{-\infty}^{t} x_1(\tau)d\tau \quad \text{and} \quad y_2(t) = t\int_{-\infty}^{t} x_2(\tau)d\tau$$

$$x_3(t) = ax_1(t) + bx_2(t) \implies y_3(t) = t\int_{-\infty}^{t} [ax_1(\tau) + bx_2(\tau)]d\tau$$

$$y_3(t) = at\int_{-\infty}^{t} x_1(\tau)d\tau + bt\int_{-\infty}^{t} x_2(\tau)d\tau = ay_1(t) + by_2(t)$$

for all x_1, x_2, a, and b. The system is linear.

b. Testing for time invariance:

$$x_1(t) = x(t - t_0) \implies y_1(t) = t \int_{-\infty}^{t} x(\tau - t_0)d\tau = t \int_{-\infty}^{t-t_0} x(\theta)d\theta \neq y(t - t_0)$$

The system is not time invariant.

Note 1. To see that

$$y_1(t) \neq y(t - t_0), \text{ let } y_\alpha(t) = \int_{-\infty}^{t} x(\tau)d\tau$$

Then

$$y(t) = ty_\alpha(t), \ y(t - t_0) = (t - t_0)y_\alpha(t - t_0), \text{ and } y_1(t) = ty_\alpha(t - t_0) \neq y(t - t_0)$$

Time variance could also have been directly concluded from the presence of factor t in the expression for $y(t)$.

$$\frac{dy(t)}{dt} + y(t) = x(t)$$

Solution

a. Testing for linearity:

$$\frac{dy_1(t)}{dt} + y_1(t) = x_1(t) \qquad (1)$$

$$\frac{dy_2(t)}{dt} + y_2(t) = x_2(t) \qquad (2)$$

$$\frac{dy_3(t)}{dt} + y_3(t) = ax_1(t) + bx_2(t) \quad (3)$$

Multiply (Eq. 1) by a and (Eq. 2) by b and add them together to obtain

$$\frac{d}{dt}[ay_1(t) + by_2(t)] + [ay_1(t) + by_2(t)] = ax_1(t) + bx_2(t) \quad \text{(Eq. 4)}$$

Compare (Eq. 4) with (Eq. 3). You will find $y_3(t) = ay_1(t) + by_2(t)$ for all $x_1, x_2, a,$ and b. The system is linear.

b. Testing for time invariance:

$$x_1(t) = x(t - t_0) \implies \frac{dy_1(t)}{dt} + y_1(t) = x(t - t_0)$$

Compare with $\frac{dy(t)}{dt} + y(t) = x(t)$ to conclude $y_1(t) = y(t - t_0)$ for all $x(t)$ and t_0.

The system is also time invariant. The unit-impulse response of the system $h(t)$ is obtained from

$$\frac{dh(t)}{dt} + h(t) = \delta(t), \quad h(t) = e^{-t}u(t)$$

3.5 Derivative and Integral Properties of LTI Systems

The linearity and time-invariance properties lead to the derivative and integral properties of the input and output.

$$x(t) \implies y(t)$$
$$\frac{dx}{dt} \implies \frac{dy}{dt}$$
$$\int_{-\infty}^{t} x(\tau)d\tau \implies \int_{-\infty}^{t} y(\tau)d\tau$$

One result of these properties is that dynamic systems are represented by linear differential equations with constant coefficients

$$\frac{d^n y}{dt^n} + a_{n-1}\frac{d^{n-1}y}{dt^{n-1}} + \cdots + a_1\frac{dy}{dt} + a_0 y = b_m\frac{d^m x}{dt^m} + b_{m-1}\frac{d^{m-1}x}{dt^{m-1}} + \cdots + b_1\frac{dx}{dt} + b_0 x$$

where $d^k y/dt^k$ and $d^k x/dt^k$ are the k-th time derivatives of $y(t)$ and $x(t)$, respectively, and a_k and b_k are constants.

3.6 Examples from Electric Circuits

Electric circuits are interconnections of electrical elements and devices. The variables of interest in an electric circuit are currents through branches and voltages between connection nodes. These variables are under three types of constraints:

   i.   The terminal current-voltage characteristics of constituent elements.
   ii. Kirchhoff's current law.
   iii. Kirchhoff's voltage law.

These constraints provide equations that are necessary and sufficient to solve for all currents and voltages in the circuit. The equations are called *equilibrium* or *dynamical equations*. In a circuit made of ideal sources and linear elements (resistors, capacitors, inductors, and dependent sources), the dynamic equations are linear differential equations. Let the circuit contain ℓ (passive) elements with n loops and m nodes. To obtain element currents and voltages (a total of 2ℓ variables), we use the following sets of equations:

   1.   Elements' i-v characteristics (ℓ equations) are linear or nonlinear, see table below.
   2.   KVL (n equations): $\sum_k v_k = 0$
   3.   KCL ($m - 1$ equations): $\sum_k i_k = 0$

Element	*v-i*	*i-v*	Property
Resistor, R	$v = Ri$	$i = \dfrac{v}{R}$	linear
Inductor, L	$v = L\dfrac{di}{dt}$	$i = \displaystyle\int_{-\infty}^{t} v\, dt$	linear
Capacitor, C	$v = \displaystyle\int_{-\infty}^{t} i\, dt$	$i = C\dfrac{dv}{dt}$	linear
Diode	$v = 0$ when $i \geq 0$	$i = 0$ when $v < 0$	nonlinear
Switch	$v = 0$ when open	$i = 0$ when closed	nonlinear

The KVL and KCL equations are composed of linear combinations of the 2ℓ unknown voltages and currents, their derivatives, or their integrals. It may be shown that $\ell = n + m - 1$, and that the total number of independent equations is 2ℓ, equivalent to the total number of unknowns. The coefficients of the equations are derived from the element values of the circuits. If element values don't change with time, we obtain a set of linear differential equations with constant coefficients, making the systems time invariant. In these mathematical models, the input and output spaces are the collection of all continuous-time analog waveforms.

Input-Output Description

The set of node (or loop) equations may be reduced to a set of differential equations containing only the independent sources (i.e., inputs) and the desired unknowns (i.e., outputs). This is the input-output relationship of the system.

State-Space Description

The state of the circuit at any moment is completely known from the set of currents in the inductors and voltages at the terminals of the capacitors at that moment. These are called the *state variables* and their space is called the state space. The dynamical equations of the circuit may be formulated directly in terms of state variables. The equations are of first order.

The systems of Examples 3.6 to 3.12 below are all causal and time invariant. Otherwise, they may be static, dynamic, linear, or nonlinear.

Example
3.6

In the circuit of Figure 3.3 find the capacitor voltage v_c as a function of the source voltage v_s. Show that the relationship is linear and time invariant.

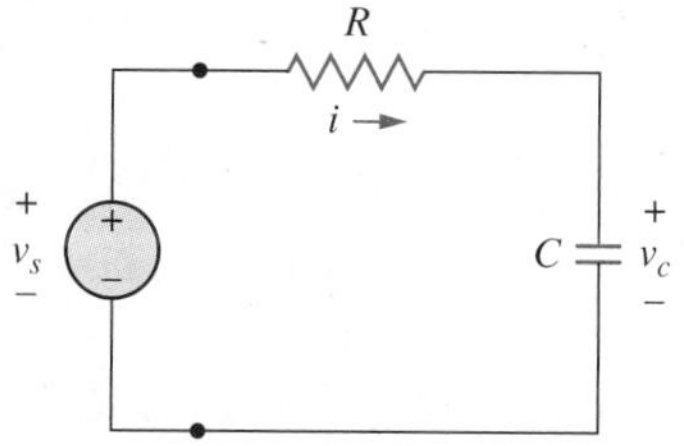

FIGURE 3.3 A linear *RC* circuit.

Solution

Applying KVL around the loop we have

$$v_s = Ri + v_c \quad \text{and} \quad i = C\frac{dv_c}{dt}$$

The input-output relationship is, therefore,

$$RC\frac{dv_c}{dt} + v_c = v_s$$

To test for linearity, let an input voltage v_{s1} produce the output voltage v_{c1}:

$$RC\frac{dv_{c1}}{dt} + v_{c1} = v_{s1} \qquad (1)$$

Let another input v_{s2} produce the output v_{c2}:

$$RC\frac{dv_{c2}}{dt} + v_{c2} = v_{s2} \qquad (2)$$

Multiplying both sides of (1) by a constant a and (2) by another constant b and adding the results, we have

$$RC\frac{d}{dt}(av_{c1} + bv_{c2}) + (av_{c1} + bv_{c2}) = av_{s1} + bv_{s2} \qquad (3)$$

From (3) we deduce that $v_c = av_{c1} + bv_{c2}$ is the capacitor voltage produced by the source $v_s = av_{s1} + bv_{s2}$. Since in the above derivation no limitations were imposed on a, b, v_{s1}, and v_{s2}, the system is linear and is described by a linear differential equation. The coefficients of the equation are constant and the system is time invariant.

Example 3.7

Voltage follower

In the circuit of Figure 3.4 (with an ideal op-amp, drawing zero current and having an infinite open-loop gain) the output is equal to the input $v_o = v_s$. The system isolates the load Z_ℓ from the signal source v_s, making the internal resistance of the signal source ineffective. The system is causal, memoryless, linear, and time invariant with unity gain.

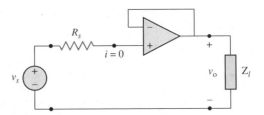

FIGURE 3.4 A voltage follower circuit.

E*xample*

3.8

Inverting amplifier

The circuit of Figure 3.5 (with an ideal op-amp, drawing zero current and having an infinite open-loop gain) is an amplifier. It is described by the following input-output relationship:

$$v_2 = -\frac{R_2}{R_1} v_1$$

(For the derivation apply KCL at the inverting terminal of the op-amp.)

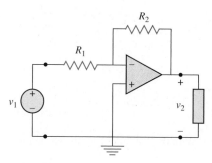

FIGURE 3.5 An inverting amplifier.

The system is causal, static, linear, and time invariant. With a real op-amp, the output of Figure 3.5 saturates at V_{SH} and $-V_{SL}$, making the system nonlinear. The circuit is then limited by the following relationships:

$$v_2 = \begin{cases} V_{SH}, & -\dfrac{R_2}{R_1} v_1 \geq V_{SH} \\[2mm] -\dfrac{R_2}{R_1} v_1, & -V_{SL} < -\dfrac{R_2}{R_1} v_1 < V_{SH} \\[2mm] -V_{SL}, & -\dfrac{R_2}{R_1} v_1 \leq -V_{SL} \end{cases}$$

E*xample*

3.9

Integrator

In the circuit of Figure 3.5 (with an ideal op-amp) replace the feedback resistor R_2 by a capacitor C and the circuit then becomes an integrator. It is described by the following input-output relationship:

$$v_2 = -\frac{1}{RC} \int_{-\infty}^{t} v_1 \, dt$$

The system is causal, dynamical, linear, and time invariant. With a real op-amp, the output saturates at V_{SH} and $-V_{SL}$, making the system nonlinear.

Multi-input system

The summing circuit of Figure 3.6 has n inputs v_i, $i = 1 \cdots n$, and one output v_{out}.

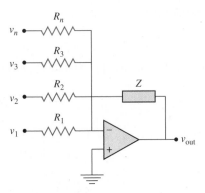

FIGURE 3.6 A summing circuit.

With the feedback element Z being a resistor R_f, the input-output relationship is

$$v_{\mathrm{out}} = -\left(\frac{R_f}{R_1} v_1 + \frac{R_f}{R_2} v_2 + \cdots + \frac{R_f}{R_n} v_n \right)$$

(For the derivation apply KCL at the inverting terminal of the op-amp.) The circuit is linear, time invariant, and memoryless. Replacing the feedback resistor R_f with a capacitor C, we obtain a summing integrator with the input-output relationship

$$v_{\mathrm{out}} = -\frac{1}{C} \int_{-\infty}^{t} \left(\frac{1}{R_1} v_1 + \frac{1}{R_2} v_2 + \cdots + \frac{1}{R_n} v_n \right) dt$$

The circuit is then dynamic.

Multi-output system

The circuit of Figure 3.7 has one input i_s, and two outputs $v(t)$ and $i(t)$. Let $R_1 = 3\ \Omega$, $R_2 = 2\ \Omega$, $L = 1$ H, and $C = 1$ F. Formulate the system's dynamical equations to find $v(t)$ and $i(t)$.

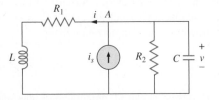

FIGURE 3.7 A multi-output system.

Solution

We write node and loop equations using inductor current i and capacitor voltage v.

$$\begin{cases} v' + \frac{v}{2} + i = i_s & \text{(KCL at node A)} \\ v - 3i - i' = 0 & \text{(KVL around the external loop)} \end{cases}$$

Moving the derivative terms to the left side and the rest of the terms to the right side we get

$$\begin{cases} v' = -0.5v - i + i_s & \text{(KCL)} \\ i' = v - 3i & \text{(KVL)} \end{cases}$$

The above equations may also be written in matrix form in terms of the input (i_s), state (v, i), and the output.

$$\begin{bmatrix} v' \\ i' \end{bmatrix} = \begin{bmatrix} -0.5 & -1 \\ 1 & -3 \end{bmatrix} \begin{bmatrix} v \\ i \end{bmatrix} + \begin{bmatrix} 1 \\ 0 \end{bmatrix} i_s$$

The input-output relationships are

$$v'' + 3.5v' + 2.5v = i'_s + 3i_s$$

$$i'' + 3.5i' + 2.5i = i_s$$

The system is causal, dynamic, linear, and time invariant.

Example **3.12**

State-variable description

The circuit of Figure 3.8 has one input x, one output y, and two state variables z_1 and z_2, which are the voltages at the output terminals of op-amps 1 and 2, respectively.

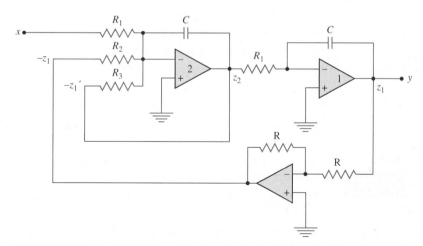

FIGURE 3.8 An op-amp circuit that solves a single-input single-output differential equation and provides access to state variables. See Example 3.12.

Given $R_1 = 1 \ M\Omega$, $R_2 = 500 \ k\Omega$, $R_3 = 333 \ k\Omega$, and $C = 1 \ \mu F$, the state and output equations become

$$z'_1 = -z_2$$
$$z'_2 = 2z_1 - 3z_2 - x$$
$$y = z_1$$

which may be written in matrix form as

$$\begin{bmatrix} z'_1 \\ z'_2 \end{bmatrix} = \begin{bmatrix} 0 & -1 \\ 2 & -3 \end{bmatrix} \begin{bmatrix} z_1 \\ z_2 \end{bmatrix} + \begin{bmatrix} 0 \\ -1 \end{bmatrix} x \quad \text{and} \quad y = \begin{bmatrix} 1 & 0 \end{bmatrix} \begin{bmatrix} z_1 \\ z_2 \end{bmatrix}$$

The circuit solves the following LTI differential equation

$$y'' + 3y' + 2y = x$$

Example 3.13   Multi-input multi-output system with state-variable description

An electronic circuit has two inputs (x_1, x_2) and two outputs (y_1, y_2), all functions of time. It is described by the following input-output relationships:

$$y'_1 = -y_2 - x_1$$
$$y'_2 = -3y_1 - 2y_2 - x_2$$

which may be written in matrix form as

$$\begin{bmatrix} y'_1 \\ y'_2 \end{bmatrix} = - \begin{bmatrix} 0 & 1 \\ 3 & 2 \end{bmatrix} \begin{bmatrix} y_1 \\ y_2 \end{bmatrix} - \begin{bmatrix} 1 & 0 \\ 0 & 1 \end{bmatrix} \begin{bmatrix} x_1 \\ x_2 \end{bmatrix}$$

The system is linear, time invariant, and dynamic. The project at the end of Chapter 5 synthesizes an analog circuit to solve a set of differential equations.

3.7 Examples from Other Fields

In this section we present three additional examples commonly encountered in basic engineering systems. The systems are all causal and time invariant. In addition, they may be static, dynamic, linear, or nonlinear.

Example 3.14   Motion of a pendulum

A simple rigid pendulum is made of a bob of mass m and a rigid bar of length l and negligible mass. See Figure 3.9(a).

Let the external torque applied to the pivot of the pendulum be τ. The damping force of the environment opposing the motion of the bob is proportional to its velocity, $-\beta \ v = -\beta \ \theta' l$, resulting in the damping torque $-\beta \theta' \ l^2$. The pivot has a coil that

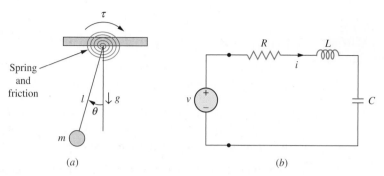

FIGURE 3.9 *(a)* A pendulum, *(b)* Its linearized model is represented by an RLC circuit.

creates an opposing torque of $-k_s\theta$. The torque produced by gravity is $-mgl\sin\theta$. The equation of motion (i.e., the relationship between the pendulum's displacement, θ, and the external torque) is found by applying Newton's law:

$$ml^2\theta'' + \beta\theta'l^2 + k_s\theta + mgl\sin\theta = \tau$$

$$\theta'' + \frac{\beta}{m}\theta' + \frac{k_s}{ml^2}\theta + \frac{g}{l}\sin\theta = \frac{1}{ml^2}\tau$$

This is a time invariant, dynamic, and nonlinear system. However, if θ is small (e.g., $\theta < 10°$), one can assume $\sin\theta \approx \theta$, and the system becomes linear:

$$\theta'' + \frac{\beta}{m}\theta' + \left[\frac{k_s}{ml^2} + \frac{g}{l}\right]\theta = \frac{1}{ml^2}\tau$$

The dynamical equation of this linear system is then the same as the dynamical equation of a series *RLC* circuit driven by a voltage source v (Figure 3.9*b*) with the differential equation

$$q'' + \frac{R}{L}q' + \frac{1}{LC}q = \frac{1}{L}v$$

where q is the charge on the capacitor at time t.

With no external torque, friction, spring, or stiffness in the pendulum, the equation of motion becomes

$$\theta'' + \frac{g}{l}\sin\theta = 0$$

For small displacements, $\sin\theta \approx \theta$, and we have a zero-input, second-order linear system

$$\theta'' + \frac{g}{l}\theta = 0$$

The systems described in Examples 3.15 and 3.16 represent more comprehensive cases of linear systems whose input and output are functions of time and space.

A public address sound system

In a public address sound system, let the air pressure at the microphone constitute the input and the air pressure at a given point in space constitute the output. Consider a single input-output system with a microphone (the input) and a loudspeaker (the output), both placed at fixed positions. The input and output spaces are collections of all real-valued time functions. This system may be partitioned into six subsystems, as shown in Figure 3.10, and described below.

1. The microphone that converts the incoming sound into electrical signals.
2. The voltage amplifier that amplifies the voltage signals.
3. The current amplifier that provides the power to drive the loudspeaker.
4. The loudspeaker that converts the electrical signal into sound.
5. The air through which sound propagates.
6. The echo from walls that may feed back into the microphone.

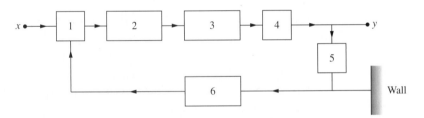

FIGURE 3.10 Elements of a public address system.

Each subsystem has an input and an output. Assume that within the operation range a sinusoidal input to the i-th subsystem is transformed into a sinusoidal output with a magnitude gain of A_i and a phase addition of θ_i, both of which are functions of frequency. We show this transfromation by $A_i \angle \theta_i$ and call it the transfer function for the i-th subsystem. The transfer function of the forward path (made of subsystems 1 through 4) is $A \angle \theta$, where $A = A_1 \cdot A_2 \cdot A_3 \cdot A_4$ and $\theta = \theta_1 + \theta_2 + \theta_3 + \theta_4$. Similarly, the transfer function of the feedback path (made of subsystems 5 and 6) is $B \angle \phi$. The total phase shift around the loop is $\theta + \phi$. If this loop phase equals $360°$, a very small input may produce a large output, which manifests itself by whistling. The system is then called unstable. The above system becomes multi-input–multi-output if it employs several microphones and loudspeakers.

Acoustic system in a room

In another system that models the acoustic properties of a room, the sound generated by a loudspeaker constitutes the input and the acoustic vibrations picked up by a microphone constitute the output. The input and output become functions of time and space. To characterize the system, we gather data by placing a sound source at a location s in the room to produce input $x(s, t)$ and a microphone at another location to measure the response $y(s, t)$. We then repeat the time-space response measurement for frequencies and locations of interest. The input-output inventory created by the above experiment will describe the acoustic properties of the room. The system then becomes multivariable, multi-input–multi-output, where the time t is a continuous variable and s may assume discrete or continuous values.

3.8 Response of LTI Systems to Impulse and Step Inputs

An LTI system may be completely specified by its response $h(t)$ to a unit impulse or, equivalently, its response to a step input. Methods for finding the unit-impulse response and its use in analyzing a system and predicting its output to a new input will be discussed in Chapters 4, 5, and 6. Here we provide three examples.

Example **3.17**

The response of an LTI system to a unit impulse is measured and modeled by $h(t) = e^{-t}u(t)$. Find its unit-step response $g(t)$.

Solution

$$g(t) = \int_{-\infty}^{t} h(\tau)d\tau = \int_{0}^{t} e^{-\tau}d\tau = (1 - e^{-t})u(t)$$

Example **3.18**

The response of an LTI system to a unit step is measured and modeled by $g(t) = (1 - e^{-2t})u(t)$. Find its unit-impulse response.

Solution

$$h(t) = \frac{dg}{dt} = \frac{d}{dt}\left[(1 - e^{-2t})u(t)\right] = 2e^{-2t}u(t)$$

Example **3.19**

Unit-impulse response in an *RC* circuit

In the circuit of Figure 3.3 a unit-impulse voltage is applied by the voltage source, $v_s = \delta(t)$. Find capacitor and resistor voltages v_c and v_R, respectively.

Direct Approach

A unit-impulse voltage at v_s produces an impulse current $i = \delta(t)/R$ in the circuit which charges the capacitor to an initial voltage $v_c(0^+) = 1/(RC)$. For $t > 0$, this initial voltage is discharged through the resistor with a time-constant RC. The unit-impulse responses are, therefore,

$$v_c = \frac{1}{RC}e^{-t/RC}u(t)$$

$$v_R = v_s - v_c = \delta(t) - \frac{1}{RC}e^{-t/RC}u(t)$$

Alternative Approach

A unit impulse may be considered the derivative of a unit step: $\frac{du(t)}{dt} = \delta(t)$. The capacitor and resistor voltages in response to a unit-step input voltage are

$$\text{Capacitor voltage} = (1 - e^{-t/RC})u(t)$$
$$\text{Resistor voltage} = e^{-t/RC}u(t)$$

Taking derivatives of the above voltages results in the unit-impulse responses found by the direct approach.

3.9 Response of LTI Systems to Exponential Inputs

The response of LTI systems[1] to an exponential input e^{st}, where s is a constant, is an exponential $H(s)e^{st}$. This may be derived directly from the linearity and time-invariance properties. To show this, and to find the scale factor $H(s)$, start with the unit-impulse response $h(n)$ (entry 1 in the table below) and follow steps 2 through 7, where you will have the desired result.

Step	Actions and Their Rationale	Input	$\Longrightarrow$	Output
1	From unit-impulse response we have:	$\delta(t)$	$\Longrightarrow$	$h(t)$
2	From time invariance we have:	$\delta(t - \tau)$	$\Longrightarrow$	$h(t - \tau)$
3	From linearity we have:	$e^{s\tau}\delta(t - \tau)$	$\Longrightarrow$	$e^{s\tau}h(t - \tau)$
4	From linearity we have:	$\int_{-\infty}^{\infty} e^{s\tau}\delta(t - \tau)d\tau$	$\Longrightarrow$	$\int_{-\infty}^{\infty} e^{s\tau}h(t - \tau)d\tau$
5	In entry #4 change τ to $\tau = t - \theta$ to obtain:	$\int_{-\infty}^{\infty} e^{s(t-\theta)}\delta(\theta)d\theta$	$\Longrightarrow$	$\int_{-\infty}^{\infty} e^{s(t-\theta)}h(\theta)d\theta$
6	From sifting property of $\delta(\theta)$ we get:	e^{st}	$\Longrightarrow$	$e^{st}\int_{-\infty}^{\infty} e^{-s\theta}h(\theta)d\theta$
7	Now let $\int_{-\infty}^{\infty} e^{-st}h(t)dt = H(s)$, then	e^{st}	$\Longrightarrow$	$H(s)e^{st}$

---

[1] We assume the LTI system under consideration has an impulse response.

Note that the above property is not derived from the Laplace transform operation. Exponential time functions are called characteristic functions (or eigenfunctions) of linear time-invariant systems. The input e^{st} produces the output $H(s)e^{st}$, which is an exponential with the same exponent. $H(s)$ and s are, in general, complex numbers. The system scales only the amplitude of the input by the factor H.

E*xample*
3.20

Response of an *RC* circuit to e^{st}

In the circuit of Figure 3.11, $v(t) = e^{st}$. Find $H(s)$.

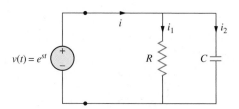

FIGURE 3.11 Response of a parallel *RC* circuit to e^{st} is $H(s)e^{st}$.

Solution

$$i_1 = \frac{v_s}{R} = \frac{e^{st}}{R}, \quad i_2 = C\frac{dv_s}{dt} = Cse^{st}$$

and

$$i = i_1 + i_2 = \left(\frac{1}{R} + Cs\right)e^{st} = H(s)e^{st}$$

From which $H(s) = \frac{1}{R} + Cs$

E*xample*
3.21

DC steady-state response

Given $h(t)$, the unit-impulse response of an LTI system, find its response to $x(t) = 1$.

Solution

$$e^{st} \Longrightarrow H(s)s^{st}, \quad \text{where } H(s) = \int_{-\infty}^{\infty} h(t)e^{-st}dt$$

DC input

$$\left. e^{st} \right|_{s=0} = 1 \Longrightarrow \left. H(s)e^{st} \right|_{s=0} = H(0) = \int_{-\infty}^{\infty} h(t)dt$$

Alternative Solution

$$u(t) \Longrightarrow g(t) = \int_{-\infty}^{t} h(\tau)d\tau.$$

DC input

$$u(t)\Big|_{t=\infty} = 1 \Longrightarrow g(t)\Big|_{t=\infty} = \int_{-\infty}^{\infty} h(t)dt$$

3.10   Response of LTI Systems to Sinusoids

In Section 3.9, let $s = j\omega$. Then we will have

Input	$\Longrightarrow$	Output
e^{st}	$\Longrightarrow$	$H(s)e^{st}$
$e^{j\omega t}$	$\Longrightarrow$	$H(\omega)e^{j\omega t}$
$\cos(\omega t) = \mathcal{RE}\left[e^{j\omega t}\right]$	$\Longrightarrow$	$\mathcal{RE}\left[H(\omega)e^{j\omega t}\right] = |H(\omega)|\cos(\omega t + \theta)$
Alternatively:		
$\cos(\omega t) = \dfrac{e^{j\omega t} + e^{-j\omega t}}{2}$	$\Longrightarrow$	$\dfrac{H(\omega)e^{j\omega t} + H^*(\omega)e^{-j\omega t}}{2} = |H(\omega)|\cos(\omega t + \theta)$

In the above table,

$$H(\omega) = H(s)\Big|_{s=j\omega} = |H(\omega)|\angle\theta \text{ and } H^*(\omega) = H(s)\Big|_{s=-j\omega} = |H(\omega)|\angle -\theta$$

We deduce that the response of an LTI system to a sinusoid is a sinusoid of the same frequency. The system generally changes the amplitude and phase of the input. The change depends on the frequency and is called the system's frequency response.

Note: The scale factors $H(s)$ and $H(\omega)$ are two different functions. For systems with real-valued elements, $H(s)$ is a real function of s (with s being a complex or real number). Therefore, $H(\omega)$ is a real function of $j\omega$, which makes it a complex function of ω. It is for convenience and simplicity that the same notation H is used to represent the scale factor when we switch from s to ω.

Example **3.22**

Response of an *RC* circuit to sinusoids

In the circuit of Figure 3.11 let $R = 1\ k\Omega$, $C = 1\ \mu F$, and $v(t) = \cos(1,000t) + \cos(2,000t)$. Find $i(t)$.

Solution

With $v(t)$ as the input and $i(t)$ as the output, the frequency response of the circuit is $H(\omega) = 10^{-6}(1,000 + j\omega)$. Let $v(t) = v_1 + v_2$, with $v_1(t) = \cos(1,000t)$ and $v_2(t) = \cos(2,000t)$. From linearity,

$$H_1 = 10^{-3}(1 + j) = \sqrt{2} \times 10^{-3}\angle 45°$$

$$H_2 = 10^{-3}(1 + 2j) = \sqrt{5} \times 10^{-3}\angle 63.4°$$

$$i_1 = \sqrt{2} \times 10^{-3}\cos(1,000t + 45°)$$

$$i_2 = \sqrt{5} \times 10^{-3}\cos(2,000t + 63.4°)$$

$$i = i_1 + i_2 = \sqrt{2} \times 10^{-3}\cos(1,000t + 45°) + \sqrt{5} \times 10^{-3}\cos(2,000t + 63.4°)$$

Note the amplitude and phase changes at $\omega = 1,000$ and $2,000$ rad/s.

3.11 Use of Superposition and Other Properties of LTI Systems

The following three examples illustrate the use of superposition in LTI systems. This subject will be treated in detail in Chapter 4.

*xample*
3.23

An input-output pair for an LTI system is shown in Figure 3.12, where $d(t)$ is the input and $p(t)$ is the output.

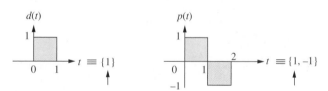

FIGURE 3.12 An input-output pair for the LTI system of Example 3.23.

a. Express $p(t)$ in terms of $d(t)$.

b. For convenience represent the unit rectangular pulse $d(t) = u(t) - u(t-1) \equiv \{1\}$. Consider an input $x(t) = \{1, -1, -1\}$. By superposition find $y(t)$, the system's output, and verify that it is equal to $x(t) - x(t-1)$.

c. Repeat part b for $x(t) = \{1, -1, -1, 1\}$.

d. Repeat part b for $x(t) = \{1, -1, -1, 1, 1 \cdots\}$.

e. Find the unit-step response of the system.

f. Write the input-output relationship.

Solution

a. $p(t) = d(t) - d(t-1)$

b. $x(t) = d(t) - d(t-1) - d(t-2)$

$y(t) = p(t) - p(t-1) - p(t-2) = \{1, -1\} - \{0, 1, -1\} - \{0, 0, 1, -1\}$

$\quad = \{\underset{\uparrow}{1}, -2, 0, 1\}$

$x(t) - x(t-1) = \{\underset{\uparrow}{1}, -1, -1\} - \{0, 1, -1, -1\} = \{\underset{\uparrow}{1}, -2, 0, 1\} = y(t)$

c. $x(t) = d(t) - d(t-1) - d(t-2) + d(t-3)$

$y(t) = p(t) - p(t-1) - p(t-2) + p(t-3) = \{\underset{\uparrow}{1}, -2, 0, 2, -1\}$

$x(t) - x(t-1) = \{\underset{\uparrow}{1}, -1, -1, 1\} - \{0, 1, -1, -1, 1\} = \{\underset{\uparrow}{1}, -2, 0, 2, -1\} = y(t)$

d. $x(t) = d(t) - d(t-1) - d(t-2) + \displaystyle\sum_{k=3}^{\infty} d(t-k)$

$y(t) = p(t) - p(t-1) - p(t-2) + \displaystyle\sum_{k=3}^{\infty} p(t-k) = \{\underset{\uparrow}{1}, -2, 0, 2\}$

$x(t) - x(t-1) = \{\underset{\uparrow}{1}, -1, -1, 1, 1, \cdots\} - \{0, 1, -1, -1, 1, 1, \cdots\}$

$\quad = \{\underset{\uparrow}{1}, -2, 0, 2\} = y(t)$

e. $x(t) = \displaystyle\sum_{k=0}^{\infty} d(t-k)$

$y(t) = \displaystyle\sum_{k=0}^{\infty} p(t-k) = \{\underset{\uparrow}{1}\}$

$x(t) - x(t-1) = \{\underset{\uparrow}{1}, 1, 1, 1, 1, \cdots\} - \{0, 1, 1, 1, 1, 1, \cdots\} = \{\underset{\uparrow}{1}\}$

f. $y(t) = x(t) - x(t-1)$

In summary,

$$
\begin{array}{ll}
x(t) & \Longrightarrow y(t) \\
\{\underset{\uparrow}{1}\} & \Longrightarrow \{\underset{\uparrow}{1}, -1\} \\
\{\underset{\uparrow}{1}, -1, -1\} & \Longrightarrow \{\underset{\uparrow}{1}, -2, 0, 1\} \\
\{\underset{\uparrow}{1}, -1, -1, 1\} & \Longrightarrow \{\underset{\uparrow}{1}, -2, 0, 2, -1\} \\
\{\underset{\uparrow}{1}, -1, -1, 1, 1, \cdots\} & \Longrightarrow \{\underset{\uparrow}{1}, -2, 0, 2\} \\
\{\underset{\uparrow}{1}, 1, 1, 1, 1, \cdots\} & \Longrightarrow \{\underset{\uparrow}{1}\}
\end{array}
$$

Example

3.24

Find $h(t)$, the unit-impulse response of the system in Example 3.23.

Solution

Take the derivatives of the given input-output pairs:

$$x(t) = u(t) - u(t-1) \Longrightarrow y(t) = u(t) - 2u(t-1) + u(t-2)$$

$$x'(t) = \delta(t) - \delta(t-1) \Longrightarrow y'(t) = \delta(t) - 2\delta(t-1) + \delta(t-2)$$

From examination of $x'(t)$ and $y'(t)$ we conclude that $h(t)$ is made of unit impulses. The first impulse in $y'(t)$ is due to the first impulse in $x'(t)$; therefore, $h(0) = \delta(t)$. From the second impulse $y'(1) = -2\delta(t-1)$ we derive $h(1) = -\delta(t)$ and from the third impulse $y'(2) = \delta(t-2)$ we conclude $h(2) = 0$ and $h(t) = 0$, $t > 2$. The unit-impulse response of the system is $h(t) = \delta(t) - \delta(t-1)$. It is verified as follows:

$$
\begin{aligned}
\delta(t) &\Longrightarrow \delta(t) - \delta(t-1) \\
\delta(t-1) &\Longrightarrow \delta(t-1) - \delta(t-2) \\
\delta(t) - \delta(t-1) &\Longrightarrow [\delta(t) - \delta(t-1)] - [\delta(t-1) - \delta(t-2)] \\
&= \delta(t) - 2\delta(t-1) + \delta(t-2)
\end{aligned}
$$

Example 3.25

The response of an LTI system to $u(t)$ is $g(t) = e^{-t}u(t)$. Find its responses to the following inputs:

a. $u(t) - u(t-1)$

b. $tu(t)$

c. $tu(t) - u(t-1)$

d. $\delta(t)$

Solution

a. $x(t) = u(t) - u(t-1)$

$$
y(t) = g(t) - g(t-1) = e^{-t}u(t) - e^{-(t-1)}u(t-1) = \begin{cases} 0, & t < 0 \\ e^{-t}, & 0 < t < 1 \\ -1.718e^{-t}, & t > 1 \end{cases}
$$

b. $x(t) = tu(t) = \displaystyle\int_{-\infty}^{t} u(\tau)d\tau$

$$
y(t) = \int_{-\infty}^{t} g(\tau)d\tau = \int_{0}^{t} e^{-\tau}d\tau = (1 - e^{-t})u(t)
$$

c. $x(t) = tu(t) - u(t-1)$

$$
y(t) = (1 - e^{-t})u(t) - e^{-(t-1)}u(t-1) = \begin{cases} 0, & t < 0 \\ 1 - e^{-t}, & 0 < t < 1 \\ 1 - 3.718e^{-t}, & t > 1 \end{cases}
$$

d. $x(t) = \delta(t) = \dfrac{du(t)}{dt}$

$$
y(t) = \frac{dg(t)}{dt} = \delta(t) - e^{-t}u(t)
$$

3.12 LTI Systems and Fourier Analysis

The relationship between sinusoidal signals and linear time-invariant systems has important consequences.[2] In almost all cases of interest, a sinusoidal input to an LTI system

---

[2] Sinusoids are called eigenfunctions of LTI systems, which are described by linear differential equations with constant real coefficients.

produces a sinusoidal output with the same frequency but possibly different amplitude and phase. Moreover, by superposition, if an input is made of a sum of sinusoids, the output will be a sum of sinusoids also, each weighted by the frequency response. Thus, by expanding a signal into its sinusoidal components, the Fourier method provides a powerful tool for the analysis and synthesis of signals and linear systems. In almost all cases, the frequency response of an LTI system contains all the input-output information. Using generalized harmonic analysis, inputs are transformed into the frequency domain. The frequency response then provides the change in phase and amplitude of each harmonic. Since the frequency response of an LTI system may be measured experimentally, Fourier analysis enables us to predict the output of an LTI system given any input and without modeling the internal structure of the system.

The above discussion illustrates the convenience of having the input in the form of a sinusoid. Fourier analysis is a method for representing waveforms as sums of sinusoids. It, therefore, provides a useful and powerful tool for the analysis and synthesis of signals and linear systems. Historically, the Fourier method started with the analysis and exploration of continuous-time signals made of a finite weighted sum of harmonic sinusoids. This led to the infinite Fourier series for periodic signals and then was extended to the Fourier transform. In this book we will follow the historical route, starting with the Fourier series (FS), its properties, and its applications in Chapter 7. In Chapter 8 we consider the Fourier transform (FT) and generalize it to periodic and nonperiodic signals.

3.13 Analysis and Solution Methods for LTI Systems

A linear time-invariant system is completely specified by its impulse response.[3] The input-output relationship for such a system is given in one of the following forms:

  i. Convolution integral of the input with the unit-impulse response.
  ii. Linear differential equations with constant coefficients.
  iii. System (transfer) function $H(s)$ or the frequency response $H(\omega)$.

In summary, the analysis of an LTI system may be done in the time or frequency domains. In the time domain the output is found by convolving the input with the impulse response of the system, or by solving the input-output differential equation (see Chapters 4 and 5). In the frequency domain, as will be seen in Chapters 6–9, the time functions will be transformed and the transform of the output is equal to the product of the transforms of the input signal and the system's impulse response. The different methods are generally convertible to each other. In fact, the transform of the impulse response is the frequency response of the system, and the set of integro-differential equations of LTI systems in the time domain become linear equations in the frequency domain.

---

[3]There are exceptions, however. The system $y(t) = x(t) + x(-t)$ is linear but does not have an impulse response.

3.14 Complex Systems

Previous examples showed simple systems that are analyzed by commonly used mathematical models, tools, and techniques. These examples are appropriate for the scope and aim of this book, but give only a limited, albeit clear and correct, view of engineering systems. Many systems of interest are much more complex than those discussed here. Despite their complexity, however, these systems can be analyzed and designed using classes of analytic methods and computer tools of the type discussed in this book. Below are three examples.

Multiple Input-Output and Multivariable Systems

Systems can have multiple inputs and outputs. The input and output are then given in the form of vectors and the input-output description assumes a matrix form.

$$
X = \begin{bmatrix} x_1(t) \\ x_2(t) \\ \vdots \\ x_i(t) \\ \vdots \\ x_m(t) \end{bmatrix} \implies Y = \begin{bmatrix} y_1(t) \\ y_2(t) \\ \vdots \\ y_j(t) \\ \vdots \\ y_n(t) \end{bmatrix}
$$

Probabilistic Systems

Some systems are deterministic. Their responses can be predicted from the inputs. However, in some systems the input-output relationship is probabilistic. We may predict only some output averages from the input, but a complete output specification is not possible. An example would be a binary digital communication channel shown in Figure 3.13. The input $x(n)$ and the output $y(n)$ are both binary numbers (0 or 1). Due to factors such as channel noise and interference, mapping from the input to the output is not totally predictable. An input $x(n) = 0$ may result in $y(n) = 0$ (with a probability of p) or $y(n) = 1$ (with a probability of $1 - p$). Similarly, an input $x(n) = 1$ may result in $y(n) = 1$ (with a probability of q) or $y(n) = 0$ (with a probability of $1 - q$). Probabilistic systems are distinguished from deterministic systems whose inputs are stochastic processes (normally called random signals and noise).

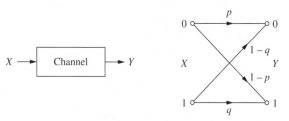

FIGURE 3.13 A probabilistic system.

Adaptive Systems

The properties of a system may change in order to adapt to new circumstances according to a predetermined rule or goal, or to seek another appropriate goal. Such adaptation generally requires feedback. In a simple adaptive system the output signal is sensed and used to modify the system parameters. The block diagram of an adaptive system is shown in Figure 3.14.

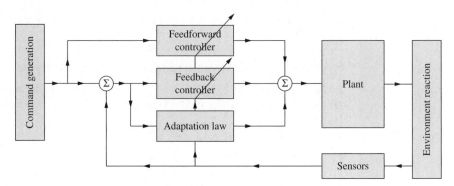

FIGURE 3.14 An adaptive system.

Some other examples of complex systems are the following: manufacturing systems, fuzzy systems, purposive and goal-seeking systems, self-organizing systems, decision systems, organizations as systems, computer networks, power systems, and qualitative analysis of systems. Powerful tools and techniques exist for analyzing and synthesizing such systems. These will not be discussed in this book.

3.15 Neuronal Systems

Still more complex are some systems for which no satisfactory analysis methods or models have been developed yet. Biological neuronal systems in general, and the central nervous system (CNS) in particular, are examples of such systems. The CNS receives, processes, and stores data. It controls the behavior of the living organism. It learns from experience and adapts to new situations. It is an accepted belief that the brain extracts features, generalizes, constructs models of the external world, forms ideas, reasons, makes decisions, and thinks. Other functions attributed to the brain are motivation, emotions, fear, pleasure, abstract thoughts, consciousness, and so on. For all those reasons, understanding how the brain works has been one of the greatest challenges of science and remains so at present. Despite the large amount of impressive data collected on the structure and the function of the nervous system, no satisfactory model exists to explain how these functions are performed. The existing methods of system analysis and modeling don't seem to promise a breakthrough, but they are the only tools presently available. In Chapter 1 we presented several classes of bioelectric signals associated with the activity of the CNS, along with a brief introduction to neurons. Here we describe some basic operations from which an input-output relationship in a neuronal unit may be constructed and models for it built.

The CNS contains individual nerve cells or neurons that are assembled and interconnected into subsystems and networks. There are about 10^{10} to 10^{11} neurons in the human CNS. They communicate with each other by electrical and chemical connections called synapses. A neuron receives electrical impulses from as few as less than ten to as many as several hundred thousand (e.g., a Purkinje cell in the cerebellum) and sends its output signal to an average of 1,000 other neurons. Moreover, rather than being a simple processing element of information, many neurons are composed of a complex collection of dendritic processing elements of their own. A thin membrane separates the inside from the outside of the neuron. By holding unequal concentrations of ions (higher concentration of positive ions outside and negative ions inside), it develops and maintains a negative intracellular voltage of 65 to 70 mV with respect to the extracellular space. Structurally, a neuron is made of a cell body, dendrites, and an axon. See Figure 3.15. The cell body varies across in size from a few microns to as much as 75 microns. It is the site of integration of the incoming signals and processing of the information. Dendrites are short neuronal processes that branch off the cell body like tree branches and receive signals from other neurons. A neuron has many dendrites but a single axon. The axon is a thin tube arising from the cell body at a location called the axon hillock. The axon may branch off and carry the electrical signal to several other neurons or even back to itself (called axon collateral). The axon varies in length from a few microns to more than

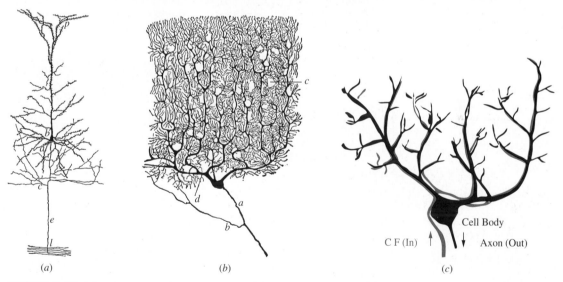

(a)   (b)   (c)

FIGURE 3.15 *(a)* A pyramidal cell of the cerebral cortex of rabbit. *(b)* A Purkinje cell of human cerebellum showing the flat dendritic tree. *(c)* A schematic representation of a Purkinje cell of the cerebellar cortex [dendrites, cell body, and climbing fibers (CF)]. Climbing fibers wrap around the cell and deliver one type of input to the cell. The other group of input fibers (called mossy fibers) are not shown. The output signal is carried by the cell axon.

Source (*a, b,* and *c*): From S. Ramon Cajal, *Histologie du System Nerveux de L'Homme & des Vertebres*, 1909 Paris. Part *(c)* of this figure is an adaptation from the original.

a meter. The axon carries the output signal of the neuron. Therefore, a neuron has multiple inputs but only a single output. A schematic is shown in Figure 3.15(*c*). The cells shown in Figure 3.16 were marked with intracellularly injected horseradish proxidate. The electrical signals from the cells were shown in Chapter 1.

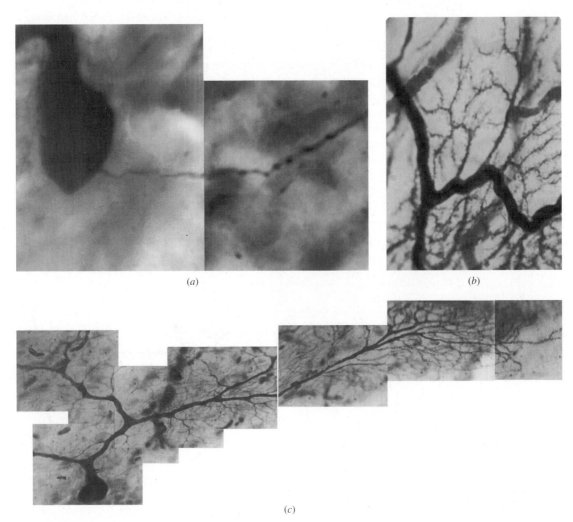

(*a*)

(*b*)

(*c*)

FIGURE 3.16 Purkinje cells from the cat's cerebellum. *(a)* The cell body of a Purkinje cell and its axon is shown on the top left. Purkinje cell bodies are typically about 50 to 75 microns across. *(b)* A dendritic segment from another cell shows fingerlike spines, which are the arrival sites of signals from other neurons. *(c)* On the bottom another Purkinje cell shows the cell body and its dendritic branches. The dendrites cover a span of about 1 mm.

Sources: From Nahvi, Woody, et al., "Electrophysiologic Characterization of Morphologically Identified Neurons in the Cerebellar Cortex of Awake Cats," *Experimental Neurology*, 67 (1980), pp. 368–76; and Nahvi and Woody, unpublished.

Signaling and Processing

Dendrites receive electrical signals in the form of narrow electric pulses (called action potentials, spikes, or nerve impulses) from other neurons. An action potential is a unitary all-or-none narrow electrical pulse that travels along the axon of the transmitting neuron. The sites where axon terminals from transmitting neurons impinge upon receiving neurons are called *synapses*. They are located on fingerlike projections (called spines) arising from the dendrites. See Figure 3.16(*b*). An action potential arriving at a synapse produces a postsynaptic potential inside the cell. Postsynaptic potentials can be positive [called excitatory postsynaptic potentials (EPSPs)] or negative [called inhibitory postsynaptic potentials (IPSPs)]. Their magnitudes vary from one to several mV. The postsynaptic potentials spread through the cell body as decaying exponentials with time and algebraically add to the intracellular potential. Thus, an EPSP reduces the negative electrical potential inside the neuronal unit (it is said to make the intracellular space depolarized) and an IPSP makes it more negative (making it hyperpolarized). It is said that the gain of the input pathways is modified by adaptation, due to conditioning and experience, and provides learning and memory. With sufficient depolarization, the intracelluar potential is reduced to a threshold level (e.g., -50 mV), causing a momentary change in the conductance of the membrane. This allows positive sodium ions to move from the outside to the interior, thus producing a momentary increase of the interior voltage from the negative resting potential to zero or even a positive level. This is the action potential. The phenomenon progresses along the axon away from the cell body and propagates the action potential. The action potential constitutes the output signal of the neuron.

Modeling

Based on the above observations, simple models for neurons are suggested and used in modeling biological and artificial neural networks. See Figure 3.17.

An incoming action potential produces an exponentially decaying postsynaptic potential. The gain of the input is modifiable based on a learning algorithm. The collection of input action potentials are grouped into excitatory and inhibitory inputs, producing EPSPs and IPSPs, respectively. The postsynaptic potentials and the resting potential of

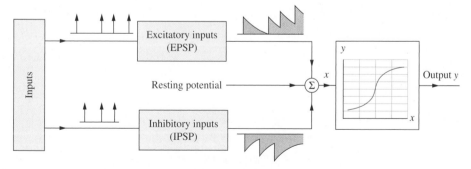

FIGURE 3.17 A model for a neuronal unit. $x(t)$ is the state of internal polarization (sum of postsynaptic and resting potentials). The output of the cell is $y(t)$, the rate of firing of spikelike action potentials.

the unit algebraically add to produce an internal variable labeled $x(t)$. Transformation of $x(t)$ to the output $y(t)$ (e.g., representing the rate of firing of the unit) is accomplished by a sigmoid transfer function (shown in the figure) or a piecewise linear transfer function with saturation levels. Artificial neural networks have been used in signal processing, pattern recognition, decision systems, and similar applications. Their impact on our understanding of the operation and modeling of the brain remains to be demonstrated.

3.16 Problems

Solved Problems

Problems 1–5

In the following systems $x(t)$ is the input and $y(t)$ is the output. (a) Determine if the systems are (i) linear, (ii) time invariant. (b) Find the unit-impulse response $h(t)$ for the LTI systems.

1. $y(t) = 2x^2(t) + 1$

Solution

a. Testing for linearity:

$$y_1 = 2x_1^2 + 1 \text{ and } y_2 = 2x_2^2 + 1$$
$$x_3 = ax_1 + bx_2 \implies y_3 = 2(ax_1 + bx_2)^2 + 1 = 2a^2x_1^2 + 2b^2x_2^2 + 2abx_1x_2 + 1 \neq ay_1 + by_2$$

The system is nonlinear.

b. Testing for time invariance:

$$x_1(t) = x(t - t_0) \implies y_1(t) = 2x^2(t - t_0) + 1 = y(t - t_0)$$

for all $x(t)$ and t_0. The system is time invariant.

2. $y(t) = x(5t)$

Solution

a. Testing for linearity:

$$y_1(t) = x_1(5t) \text{ and } y_2(t) = x_2(5t)$$
$$x_3(t) = ax_1(t) + bx_2(t) \implies y_3(t) = x_3(5t) = ax_1(5t) + bx_2(5t) = y_1(t) + y_2(t)$$

for all x_1, x_2, a, and b. The system is linear.

b. Testing for time invariance:

$$x_1(t) = x(t - t_0) \implies y_1(t) = x[5(t - t_0)] = x(5t - 5t_0) \neq y(t - t_0)$$

The system is not time invariant.

3. $y(t) = 2x(t) + 3 \int_{-\infty}^{t} x(\tau)d\tau$

Solution

Let $y_a(t) = 2x(t)$ be an LTI system with the unit-impulse response $h_a(t) = 2\delta(t)$ and

$$y_b(t) = 3 \int_{-\infty}^{t} x(\tau)d\tau$$

be an LTI system with the unit-impulse response $h_b(t) = 3u(t)$. Then, $y(t) = y_a(t) + y_b(t)$ is the LTI with unit-impulse response $h(t) = h_a(t) + h_b(t) = 2\delta(t) + 3u(t)$.

4. $y(t) = \int_{-\infty}^{\infty} x(\tau)h(t - \tau)d\tau$, where $h(t)$ is a continuous function.

Solution

a. To test for linearity, let $x(t) = ax_1(t) + bx_2(t)$. The output is

$$y(t) = \int_{-\infty}^{\infty} [ax_1(\tau) + bx_2(\tau)]h(t - \tau)d\tau = a \int_{-\infty}^{\infty} x_1(\tau)h(t - \tau)d\tau + b \int_{-\infty}^{\infty} x_2(\tau)h(t - \tau)d\tau = ay_1(t) + by_2(t)$$

for all x_1, x_2, a, and b. The system is linear.

b. To test for time invariance, let a new input be $x_1(t) = x(t - t_0)$. The new output is then

$$y_1(t) = \int_{-\infty}^{\infty} x(\tau - t_0)h(t - \tau)d\tau$$

By changing the variable τ to a new variable θ via $\tau - t_0 = \theta$, we get $\tau = \theta + t_0$, $d\tau = d\theta$, and

$$y_1(t) = \int_{-\infty}^{\infty} x(\theta)h[t - (\theta + t_0)]d\theta = \int_{-\infty}^{\infty} x(\theta)h(t - t_0 - \theta)d\theta = y(t - t_0)$$

The system is, therefore, also time invariant. To find the unit-impulse response, let $x(t) = \delta(t)$. Then

$$\int_{-\infty}^{\infty} \delta(\tau)h(t - \tau)d\tau = h(t - \tau)|_{\tau=0} = h(t)$$

5. $y(t) = \int_{-\infty}^{\infty} x(\tau)g(t + \tau)d\tau$

Solution

a. To test for linearity, let $x(t) = ax_1(t) + bx_2(t)$. The output is

$$y(t) = \int_{-\infty}^{\infty} [ax_1(\tau) + bx_2(\tau)]g(t + \tau)d\tau = a \int_{-\infty}^{\infty} x_1(\tau)g(t + \tau)d\tau + b \int_{-\infty}^{\infty} x_2(\tau)g(t + \tau)d\tau = ay_1(t) + by_2(t)$$

The system is linear. To test for time invariance, let a new input be $x_1 = x(t - t_0)$. The new output is

$$y_1(t) = \int_{-\infty}^{\infty} x(\tau - t_0)g(t + \tau)d\tau$$

By changing the variable τ to a new variable θ via $\tau - t_0 = \theta$, we get $\tau = \theta + t_0$, $d\tau = d\theta$, and

$$y_1(t) = \int_{-\infty}^{\infty} x(\theta)g[t + (\theta + t_0)]d\theta = \int_{-\infty}^{\infty} x(\theta)g[(t + \theta) + t_0]d\theta = y(t + t_0)$$

$x(t - t_0)$ produces $y(t + t_0)$. Shifting the input in one direction shifts the output in the opposite direction. The system is not time invariant.

6. The signal

$$x(t) = \sum_{k=0}^{4} \delta(t - k)$$

is the input to an LTI system. Find the output if the unit-impulse response of the system is (a) $h_1(t) = \delta(t) - \delta(t - 1)$, and (b) $h_2(t) = \delta(t) - 2\delta(t - 1) + \delta(t - 2)$.

Solution
Apply superposition.

$x(t) =$	$\delta(t)$	$\delta(t-1)$	$\delta(t-2)$	$\delta(t-3)$	$\delta(t-4)$	
(a)	$\delta(t)$	$-\delta(t-1)$				
		$\delta(t-1)$	$-\delta(t-2)$			
			$\delta(t-2)$	$-\delta(t-3)$		
				$\delta(t-3)$	$-\delta(t-4)$	
					$-\delta(t-4)$	$-\delta(t-5)$
$y_1(t) =$	$\delta(t)$	0	0	0	0	$-\delta(t-5)$

$x(t) =$	$\delta(t)$	$\delta(t-1)$	$\delta(t-2)$	$\delta(t-3)$	$\delta(t-4)$		
(b)	$\delta(t)$	$-2\delta(t-1)$	$+\delta(t-2)$				
		$\delta(t-1)$	$-2\delta(t-2)$	$+\delta(t-3)$			
			$\delta(t-2)$	$-2\delta(t-3)$	$+\delta(t-4)$		
				$\delta(t-3)$	$-2\delta(t-4)$	$+\delta(t-5)$	
					$\delta(t-4)$	$-2\delta(t-5)$	$+\delta(t-6)$
$y_2(t) =$	$\delta(t)$	$-\delta(t-1)$	0	0	0	$-\delta(t-5)$	$\delta(t-6)$

A simpler visualization of the above is shown below.

$$x = \{1, \ 1, \ 1, \ 1, \ 1\}$$
$$\uparrow$$

$h_1 = \{1, -1, 0, 0, 0, 0\}$	$h_2 = \{1, -2, 1, 0, 0, 0, 0\}$
$\quad\quad\uparrow$	$\quad\quad\uparrow$
$y_1 = \{1, -1, 0, 0, 0, 0\}+$	$y_2 = \{1, -2, 1, 0, 0, 0, 0\}+$
$\quad\quad\uparrow$	$\quad\quad\uparrow$
$\quad\ \{0, 1, -1, 0, 0, 0\}+$	$\quad\ \{0, 1, -2, 1, 0, 0, 0\}+$
$\quad\quad\quad\uparrow$	$\quad\quad\quad\uparrow$
$\quad\ \{0, 0, 1, -1, 0, 0\}+$	$\quad\ \{0, 0, \ 1, -2, 1, 0, 0\}+$
$\quad\quad\quad\quad\uparrow$	$\quad\quad\quad\quad\uparrow$
$\quad\ \{0, 0, 0, 1, -1, 0\}+$	$\quad\ \{0, 0, 0, 1, -2, 1, 0\}+$
$\quad\quad\quad\quad\quad\uparrow$	$\quad\quad\quad\quad\quad\uparrow$
$\quad\ \{0, 0, 0, 0, 1, -1\}$	$\quad\ \{0, 0, 0, 0, 1, -2, 1\}$
$\quad\quad\quad\quad\quad\quad\uparrow$	$\quad\quad\quad\quad\quad\quad\uparrow$
$\ = \{1, 0, 0, 0, 0, -1\}$	$\ = \{1, -1, 0, 0, 0, -1, 1\}$
$\quad\uparrow$	$\quad\uparrow$

7. The response of an LTI system to a narrow pulse $\zeta(t)$ (width $= 1\ \mu s$, height $= 10^6$) is $\eta(t) = e^{-t}u(t)$. Find its response to a unit-square pulse $d(t) = u(t) - u(t-1)$.

Solution

Let $\alpha = 10^{-6}$ be an incremental time. We express $x(t)$ in terms of $\zeta(t)$ and α. Then we use the linearity and time invariance properties of the system to express $y(t)$ in terms of $\eta(t)$ and α.

$$x(t) = \alpha \sum_{k=0}^{N} \zeta(t - k\alpha) \qquad y(t) = \alpha \sum_{k=0}^{N} \eta(t - k\alpha) = \alpha \sum_{k=0}^{N} e^{-(t-k\alpha)}, \quad \text{where } N = 10^6 - 1.$$

$k = 0,\ 0 < t < \alpha,$	$y(t) = \alpha e^{-t}$	$y(t) \approx \alpha e^{-\alpha} \approx 1 - e^{-\alpha}$
$k = 1,\ \alpha < t < 2\alpha,$	$y(t) = \alpha(1 + e^{\alpha})e^{-t}$	$y(t) \approx \alpha(1 + e^{-\alpha}) \approx 1 - e^{-2\alpha}$
$k = 2,\ 2\alpha < t < 3\alpha,$	$y(t) = \alpha(1 + e^{\alpha} + e^{2\alpha})e^{-t}$	$y(t) \approx \alpha(1 + e^{-\alpha} + e^{-2\alpha}) \approx 1 - e^{-3\alpha}$
$\cdots$	$\cdots$	$\cdots$
$k = \ell,\ \ell\alpha < t < (\ell+1)\alpha,$	$y(t) = \alpha(1 + e^{\alpha} + e^{2\alpha} + \cdots + e^{\ell\alpha})e^{-t}$	$y(t) \approx 1 - e^{-(\ell+1)\alpha}$
$\cdots$		
$k \geq N,\ t \geq N\alpha,$	$y(t) = \alpha(1 + e^{\alpha} + e^{2\alpha} + \cdots + e^{N\alpha})e^{-t}$	$\approx (e - 1) e^{-t}$

Note 1. Taylor series expansion of the exponential function is:

$$e^x = 1 + x + \frac{x^2}{2!} + \frac{x^3}{3!} + \cdots + \frac{x^n}{n!} + \cdots$$

Then, $1 - e^x \approx -x$ for $x << 1$.

Note 2. Sum of the finite geometrical series is

$$\sum_{k=0}^{N} x^k = \frac{1 - x^{N+1}}{1 - x}.$$

Then,

$$\sum_{k=0}^{N} x^k \approx \frac{1}{1 - x}$$

for $x < 1$ and N large.

Alternate Approach

$$x(t) \approx \int_{-\infty}^{t} \zeta(t)dt - \int_{-\infty}^{t} \zeta(t - 1)dt$$

$$y(t) \approx \int_{-\infty}^{t} \eta(t)dt - \int_{-\infty}^{t} \eta(t - 1)dt = \left(1 - e^{-t}\right)u(t) - \left(1 - e^{-(t-1)}\right)u(t - 1) = \begin{cases} 0, & t < 0 \\ 1 - e^{-t}, & 0 < t < 1 \\ (e - 1)e^{-t}, & t > 1 \end{cases}$$

In this alternate approach we have approximated $\zeta(t)$ by the unit impulse $\delta(t)$. The unit-impulse response of the system is, therefore, $h(t) \approx e^{-t}u(t)$. But $x(t) = u(t) - u(t - 1)$. Therefore,

$$y(t) = \int_{0}^{t} e^{-t}dt - \int_{1}^{t} e^{-(t-1)}dt = \begin{cases} 0, & t < 0 \\ 1 - e^{-t}, & 0 < t < 1 \\ (e - 1)e^{-t}, & t > 1 \end{cases}$$

8. An LTI system is specified by its unit-impulse response $h(t) = e^{-\frac{t}{1000}} u(t)$ V.

   a. Find its unit-step response $g(t)$.
   b. Find its response to the pulse $d(t) \equiv u(t) - u(t-1)$. Approximate the response by a piecewise linear curve and describe your method.
   c. Find its response to the pulse $d(t) \equiv 10^3[u(t) - u(t - 10^{-3})]$. Approximate the response by a piecewise linear curve and describe your method.

Solution

The unit-impulse response is a slowly decaying exponential that decreases from an initial value of 1 (at $t = 0$) to a final value of zero (at $t = \infty$), with a long time constant of 1,000 seconds. For example, at $t = 1$ its value is $h(1) = e^{-0.001} = 0.999$ (a reduction of only 0.1%). Because of the very slow decay, during its early part it can be considered to be a unit step.

   a. The unit-step response is

$$g(t) = \int_{-\infty}^{t} h(\tau)d\tau = \int_{-\infty}^{t} e^{-\frac{\tau}{1000}} u(\tau)d\tau = \int_{0}^{t} e^{-\frac{\tau}{1000}} d\tau = 1,000(1 - e^{-\frac{t}{1000}})u(t).$$

It is an exponential function that grows very slowly (with a time constant of 1,000 seconds) from an initial value of zero to a final value of 1,000. During its early part it can be considered to be $\approx t$ (a ramp with a slope of $1\ s^{-1}$ because $e^{-\alpha} \approx 1 - \alpha$ if $\alpha << 1$.) Overall, the unit-step response appears like a ramp that, after several hundred seconds, flattens to the level of 1,000. The system is a leaky integrator with a long time constant of 1,000 seconds.

   b. The response to the unit pulse $u(t) - u(t - 1)$ is

$$g(t)-g(t-1)=1,000(1-e^{-\frac{t}{1,000}})u(t)-1,000(1-e^{-\frac{(t-1)}{1,000}})u(t-1)=\begin{cases} 0, & t<0 \\ 1,000(1-e^{-\frac{t}{1,000}}), & 0<t<1 \\ 1,000e^{-\frac{t}{1,000}}(e^{\frac{1}{1,000}}-1)\approx e^{-\frac{t}{1,000}}, & t>1 \end{cases}$$

During the pulse period the response grows almost linearly from 0 to 1. At $t = 1$, the response is $1,000(1 - e^{-0.001}) = 0.999 \approx 1$ (note that $e^{-0.001} \approx 0.999$ and $e^{0.001} \approx 1.001$). From that value it slowly decays to zero with a time constant of 1,000 seconds. Once beyond the early part, the response to a 1-V, 1-s pulse soon becomes identical to the unit-impulse response. The 1-V, 1-s rectangular pulse approximates an impulse.

   c. The response to the 1-kV, 1-ms pulse is obtained by a way similar to part b. The response is $\approx e^{-\frac{t}{1,000}}$. This is even closer to the unit-impulse response than was the case in part b.

Chapter Problems

Note: Use the linearity and time-invariance properties of LTI systems to solve problems 9 through 14.

9. An input-output pair of an LTI system was shown in Figure 3.12. Represent the unit rectangular pulse by $d(t) = u(t) - u(t - 1) \equiv \{1\}$. The input-output pair is, therefore, given by $\{1\} \Longrightarrow \{1, -1\}$. Find the system's response to the following inputs:

   a. $x(t) = \{1, 1, 1, 1, 1\}$

   c. $x(t) = \{1, -1, 1, -1, 1\}$

   e. $x(t) = \{1, 0, -1, 1, 0, -1, 1, 0, -1\}$

   b. $x(t) = \{0, 1, -1, 1, 0\}$

   d. $x(t) = \{1, 0, 1, 0, 1\}$

   f. $x(t) = \{1, 0, 0, 0, 1\}$

10. Find the response of the system of problem 9 to the following inputs:

a. $x(t) = u(t)$

b. $x(t) = 1$

c. $x(t) = \sum_{n=0}^{\infty} (-1)^n d(t - n)$

d. $x(t) = \sum_{n=0}^{\infty} (n + 1)d(t - n)$

11. Following the notation of Figure 3.12, an input-output pair of an LTI system is given by

$$x(t) = \{\underset{\uparrow}{1}\} \implies y(t) = \{\underset{\uparrow}{1}, 1\}$$

Find its response to the following inputs:

a. $x(t) = \{\underset{\uparrow}{1}, 1, 1, 1, 1, 1, 1\}$

b. $x(t) = \{\underset{\uparrow}{1}, -1, 1, -1, 1, -1, 1\}$

12. Find the response of the system of problem 11 to the following inputs:

a. $x(t) = d(t) + 2d(t - 1) + 3d(t - 2) + 4d(t - 3) + 5d(t - 4)$
b. $x(t) = d(t) - 2d(t - 1) + 3d(t - 2) - 4d(t - 3) + 5d(t - 4)$

13. Find the response of the system of problem 11 to the following inputs:

a. $x(t) = u(t)$

b. $x(t) = 1$

c. $x(t) = \sum_{n=0}^{99} (n + 1)d(t - n)$

d. $x(t) = \sum_{n=0}^{99} (-1)^n (n + 1)d(t - n)$

14. An input-output pair of an LTI system is given by

$$x(t) = \{\underset{\uparrow}{1}, -1\} \implies y(t) = \{\underset{\uparrow}{1}, -1, 1\}$$

a. Find and sketch the system's output $y_2(t)$ for the input $x_2(t) = \{\underset{\uparrow}{1}, 1\}$.

b. Find the unit-step response of the system.

c. Give the input-output relationship.

15. Determine if the output of an LTI system due to an arbitrary input can be predicted from knowledge of the system's response to one of the following inputs:

a. $x(t) = \sin(2t)$, all t
b. $x(t) = e^{-t}$, all t
c. $x(t) = e^{-t} \sin(2t)$, all t
d. $x(t) = u(t)$

e. $x(t) = \delta(t)$
f. $x(t) = e^{-(1+j)t}$, all t
g. $x(t) = e^{-t} u(t - t_0)$
h. $x(t) = u(t - t_0)$

i. $x(t) = \sin(2t)u(t - t_0)$
j. $x(t) = e^{\sigma t} \sin(\omega t)u(t - t_0)$
k. $x(t) = \delta(t) + 1$
ℓ. $x(t) = u(t) - u(t - 1)$

Support your answer by reasoning.

16. In each of the following relationships $x(t)$ is the input and $y(t)$ is the output of a system. a and b are constants. Determine which system is (1) linear and/or (2) time invariant, and find the unit-impulse response if linear.

a. $y(t) = ax + b$
d. $y(t) = x(t) + x(-t)$
g. $y(t) = \sin x$
j. $y(t) = x(t)u(t)$

b. $y(t) = x(t - 1)$
e. $y(t) = x(2t)$
h. $x(t) = \sin y$
k. $y(t + 2) = x(t)$

c. $y(t) = |x(t)|$
f. $y(t) = x(2t) + 1$
i. $y(t) = x(2t + 1)$
ℓ. $y(t) = x(t + 2) + x(-t)$

17. Repeat problem 16 for the following systems

a. $y(t) = \int_0^t x(\tau)d\tau$

b. $y(t) = \int_{-\infty}^t x(\tau - 1)d\tau$

c. $y(t) = t\int_{-\infty}^{t-1} x(\tau)d\tau$

d. $y(t) = ax(t) + b\dfrac{dx(t)}{dt}$

e. $y(t) = x(t+2) - \dfrac{dx(t-1)}{dt}$

f. $y(t) = \dfrac{dx(3t)}{dt}$

g. $\dfrac{dy(t)}{dt} + y = 2\dfrac{d^2x(t)}{dt^2} + 3\dfrac{dx(t)}{dt} + x$

h. $y(t) = \int_{-\infty}^{\infty} x(\tau)x(t+\tau)d\tau$

i. $y(t) = \int_{t-T}^t x(\tau)d\tau$

j. $y(t) = 2x(t) + 3\int_{-\infty}^t x(\tau)d\tau$

k. $y = \dfrac{dx(t)}{dt} + \int_{-\infty}^{t-1} x(\tau)d\tau$

ℓ. $y(t) = x(t) + t\dfrac{dx(t)}{dt}$

m. $\dfrac{dy(t)}{dt} = \dfrac{dx(t)}{dt} + x$, $t > 0$, $y(0) = y_0$

n. $\dfrac{dy(t)}{dt} + y = x(t)u(t)$

o. $\dfrac{dy(t)}{dt} + y = tx(t)$

18. In each of the following relationships $x(t)$ is the input, $y(t)$ is the output, and $g(t)$ is a time function. Determine which system is (1) linear and/or (2) time invariant, and find the unit-impulse response if linear.

a. $y(t) = \int_{-\infty}^{\infty} x(\tau)g(t-\tau)d\tau$

c. $y(t) = g(t)\int_{-\infty}^t x(\tau)d\tau$

b. $y(t) = \int_0^{\infty} x(\tau)g(\tau - t)d\tau$

d. $y(t) = \int_{-\infty}^{\infty} x(\tau)g(t+\tau)d\tau$

19. a. Two systems are cascaded, with system 1 preceding system 2. System 1 is specified by $y = x + 1$ and system 2 by $z = \dfrac{dy}{dt}$. Determine if the combined system (with x as the input and z as the output) is (1) linear and (2) time invariant.

b. Switch the order of the connection so system 2 precedes system 1. Determine if the new combination is (1) linear and (2) time invariant.

20. The unit-impulse response of an LTI system is a unit-square pulse, $h(t) = u(t) - u(t-1)$. Find and sketch its response to a rectangular pulse of width $= T$ seconds and height $= 1/T$. Find the limit of the response as $T \to 0$.

21. The unit-impulse response of an LTI system is a unit-square pulse $d(t) = u(t) - u(t-1)$. Find and sketch its response to the input $x(t) = d(t) - d(t-1) + d(t-2)$.

22. Define $d(t) = u(t) - u(t-1)$. Find and sketch the responses of the systems specified below, where $x(t)$ is the input, $y(t)$ is the output, and $h(t)$ is the unit-impulse response.

a. $x(t) = h(t) = d(t)$

b. $x(t) = d(t) + d(t-10)$ and $h(t) = d(t/5)$

c. $x(t) = d(t) + d(t-2)$ and $h(t) = (1-t)d(t)$

23. The response of an LTI system to a narrow pulse (width $= 1\ \mu s$, height $= 100$) is $e^{-2t}u(t)$. Find its response to a unit-square pulse $d(t) = u(t) - u(t-1)$.

24. Find the time constant of the signal $x(t) = (0.5)^t$.

25. The response of an LTI system to a unit-square pulse $d(t) = u(t) - u(t-1)$ is $0.5^t u(t)$. Find its response to (a) $x_2(t) = d(0.5t)$ and (b) $x(t) = \sum_{n=0}^{\infty} (-1)^n x_2(t-n)$

26. Find the response of the system of problem 25 to $x(t) = u(t) - u(t-100)$.

27. The response of an LTI system to a unit-square pulse $d(t) = u(t) - u(t-1)$ is $0.5^t u(t)$. Find its response to $x(t) = 1$.

28. Draw the diagram of a linear electric circuit made of resistors, a capacitor, and op-amps, to produce the unit-impulse response $h(t) = ae^{-bt}u(t)$. Determine element values in terms of a and b.

29. The unit-impulse response of an LTI system is $h(t) = e^{-t}u(t)$. Find its response to the following:

a. An infinite train of unit impulses

$$x_a(t) = \sum_{n=-\infty}^{\infty} \delta(t-n)$$

b. An infinite train of alternating positive-negative unit impulses

$$x_b(t) = \sum_{n=-\infty}^{\infty} (-1)^n \delta(t-n)$$

c. An infinite train of alternating positive-negative unit-square pulses

$$x_c(t) = \sum_{n=-\infty}^{\infty} (-1)^n d(t-2n), \text{ where } d(t) = u(t) - u(t-1)$$

30. The unit-impulse response of an LTI system is $h(t) = (1-e^{-t})u(t)$. Find its response to $x(t) = 1.6u(t) - 0.6u(t-1)$.

31. Find and sketch the responses of the systems specified below. [$\delta(t)$ is the unit-impulse function.]

a. $x(t) = \sum_{n=-1}^{3} \delta(t-n)$ and $h(t) = \sum_{n=0}^{4} \delta(t-n)$

b. $x(t) = \delta(t) - \delta(t-1)$ and $h(t) = \sum_{n=0}^{4} \delta(t-n)$

c. $x(t) = \sum_{n=0}^{4} \delta(t-n)$ and $h(t) = \delta(t) - \delta(t-1)$

d. $x(t) = -\delta(t+1) + \delta(t-1) + 2\delta(t-2) + \delta(t-3)$ and $h(t) = \delta(t) + 2\delta(t-1) + \delta(t-2)$

32. The LTI systems in Table 3.1 are specified by their unit-impulse responses, where $d(t) \equiv u(t) - u(t-1)$ and n is an integer. In each case find and sketch the output to the given input. *Hint:* Use the linearity, superposition, integration, and differentiation properties of LTI systems.

TABLE 3.1 (Problem 32)

	Input	Unit-Impulse Response
a.	$d(t) - d(t-1)$	$d(t) - d(t-1)$
b.	$d(t)$	$e^{-\beta t}u(t)$
c.	$\sin(t)u(t)$	$\delta(t) + \delta(t-\pi)$
d.	$\sum_{n=0}^{\infty} (-1)^n \delta(t-n)$	$u(t)$
e.	$\sum_{n=0}^{\infty} (-1)^n \delta(t-n)$	$e^{-\beta t}u(t)$
f.	$\delta(t) + d(t)$	$e^{-t}u(t)$
g.	$\delta(t) + \delta(t-1)$	$\dfrac{\sin 2\pi t}{t}$

33. The LTI systems in Table 3.2 are specified by their unit-step responses. In each case find and sketch the output to the given input. Again, $d(t) \equiv u(t) - u(t-1)$ and n is an integer.

TABLE 3.2 (Problem 33)

Input	Unit-Step Response
a. $d(t)$	$e^{-\beta t} u(t)$
b. $d(t)$	$(1 - e^{-2t}) u(t)$
c. $\displaystyle\sum_{n=-\infty}^{\infty} (-1)^n d(t-n)$	$u(t)$
d. $\displaystyle\sum_{n=-\infty}^{\infty} (-1)^n d(t-n)$	$e^{-\beta t} u(t)$
e. $\cos(t) u(t - \pi/2)$	$u(t) + u(t - \pi)$
f. $\displaystyle\sum_{n=0}^{\infty} \delta(t-n)$	$u(t)$
g. $\delta(t) + d(t)$	$u(t)$

34. The response of an LTI system to $x(t) = u(t)$ is $y(t) = e^{-t} u(t)$. Find its responses to the following inputs:

a. $x(t) - x(t-1)$ b. $\displaystyle\int_{-\infty}^{t} x(t) dt$ c. $\displaystyle\int_{-\infty}^{t} x(t) dt - u(t-1)$ d. $\dfrac{dx(t)}{dt}$

35. Repeat problem 34 for the LTI system with the input-output pair

$$x(t) = e^{-2t} u(t) \Longrightarrow y(t) = \left(e^{-t} - e^{-2t}\right) u(t)$$

36. The response of an LTI system to $u(t)$ is $\left(2e^{-t} - e^{-5t}\right) u(t)$. Find its responses to

a. $\delta(t)$ b. 1   c. $\displaystyle\int_{-\infty}^{t} u(\tau) d\tau$ d. $\displaystyle\int_{t-1}^{t} u(\tau) d\tau$

37. The response of an LTI system to $u(t)$ is $e^{-2t} u(t)$. Find its responses to

a. $u(t) + u(t-2)$ b. $3u(t-1) + \displaystyle\int_{-\infty}^{t} u(\tau) d\tau$ c. $\dfrac{du(t-1)}{dt} + \displaystyle\int_{-\infty}^{t} u(\tau) d\tau$ d. $\dfrac{du(t)}{dt} + u(t)$

38. An LTI system is specified by its unit-impulse response $h(t) = e^{-10^6 t} u(t)$.
   a. Find its unit-step response and sketch it.
   b. Find its response to the unit-square pulse $d(t) \equiv u(t) - u(t-1)$ and sketch it.
   c. Describe a method to approximate the system's response in part b during 1 msec $< t <$ 1 sec.

39. The response of an LTI system to $u(t)$ is $2^{-t} u(t)$. Find its response to

a. $u(t-1) - u(t-2)$ b. $t \displaystyle\int_{-\infty}^{t} u(\tau) d\tau$ c. $\displaystyle\int_{-\infty}^{1} u(\tau) d\tau + u(t+1)$ d. $\dfrac{du(t)}{dt} - \dfrac{du(t-2)}{dt}$

40. The unit-impulse response of an LTI system is $h(t) = u(t) - u(t-1)$. Find and plot its output for the periodic input x(t), where, for one period T,

a. $x_1(t) = \begin{cases} 2 & 0 < t < 1 \\ 0 & 1 < t < 6, \end{cases}$ period $T = 6$

b. $x_2(t) = \begin{cases} 3 & 0 < t < 1 \\ -1 & 1 < t < 2, \end{cases}$ period $T = 2$

Problems 41--48

Identify the correct output of the given LTI systems. The unit-impulse response $h(t)$ and the input $x(t)$ are given.

41. $h(t) = e^{-2t}u(t)$ and $x(t) = 1$. The output is most nearly

a. e^{-2t}

b. 1

c. 0.5

d. e^{-t}

e. none of the above

42. $h(t) = e^{-t}u(t)$ and $x(t) = u(-t)$. The output for $t < 0$ is most nearly

a. -1

b. 1

c. e^t

d. e^{-t}

e. none of the above

43. $h(t) = e^{-t}u(t)$ and $x(t) = u(-t)$. The output for $t > 0$ is most nearly

a. -1

b. 1

c. e^t

d. e^{-t}

e. none of the above

44. $h(t) = (e^{-t} + e^{-2t})u(t)$ and $x(t) = u(-t)$. The output for $t < 0$ is most nearly

a. $A(e^{-t} + e^{-2t})$

b. $Be^{-t} + Ce^{-2t}, \ B \neq C$

c. 0

d. 1.5

e. none of the above

45. $h(t) = (e^{-t} - 2e^{-2t})u(t)$ and $x(t) = u(-t)$. The output for $t < 0$ is most nearly

a. $A(e^{-t} + e^{-2t})$

b. $Be^{-t} + Ce^{-2t}, \ B \neq C$

c. 0

d. 1

e. none of the above

46. $h(t) = e^{-2t}u(t)$ and $x(t) = u(t) - u(t-1)$. The output for $0 < t < 1$ is most nearly

a. 0

b. A

c. Be^{-2t}

d. $C(1 - e^{-2t})$

e. none of the above

47. $h(t) = x(t) = e^{-2t}[u(t) - u(t-1)]$. The output for $t > 2$ is most nearly

a. 0

b. A

c. Be^{-2t}

d. $C(1 - e^{-2t})$

e. none of the above

48. $h(t) = e^{-3t}u(t)$ and $x(t) = \delta(t) - \delta(t-1)$. The output for $t > 0$ is most nearly

a. 0

b. A

c. Be^{-3t}

d. $C(1 - e^{-3t})$

e. none of the above

Problems 49--52

Identify the correct output of the given LTI systems. The unit-step response $g(t)$ and the input $x(t)$ are given.

49. $g(t) = (1 - e^{-2t})u(t)$ and $x(t) = u(t) - u(t-1)$. The output during $0 < t < 1$ is most nearly

a. A

b. Be^{-2t}

c. $(1 - e^{-2t})$

d. $C + De^{-2t}, \ C \neq D$

e. none of the above

50. $g(t) = (1 - e^{-2t})u(t)$ and $x(t) = u(t) - u(t-1)$.
The output during $t > 1$ is most nearly

a. A

b. Be^{-2t}

c. $(1 - e^{-2t})$

d. $C + De^{-2t}$, $C \neq D$

e. none of the above

51. $g(t) = (1 - e^{-2t})u(t)$ and $x(t) = 2u(t) - u(t-1)$.
The output during $0 < t < 1$ is most nearly

a. A

b. Be^{-2t}

c. $C(1 - e^{-2t})$

d. $D + Ee^{-2t}$, $D \neq E$

e. none of the above

52. $g(t) = (1 - e^{-2t})u(t)$ and $x(t) = 2u(t) - u(t-1)$.
The output during $t > 1$ is most nearly

a. A

b. Be^{-2t}

c. $C(1 - e^{-2t})$

d. $D + Ee^{-2t}$, $D \neq E$

e. none of the above

53. The input-output relationship of an LTI system is

$$\frac{d^2 y}{dt^2} + 5\frac{dy}{dt} + 6y(t) = x(t)$$

a. Show that the $h(t)$ and $g(t)$ given below are the responses of the system to the unit-impulse and unit-step inputs, respectively.

$$h(t) = (e^{-2t} - e^{-3t})u(t)$$

$$g(t) = \left(\frac{1}{6} - \frac{1}{2}e^{-2t} + \frac{1}{3}e^{-3t}\right)u(t)$$

Problems 54–56

These problems are concerned with an LTI system that has an oscillatory step response. The project at the end of the chapter expands upon this class of systems.

54. The input-output relationship of an LTI system is

$$\frac{d^2 y}{dt^2} + 2\frac{dy}{dt} + 4y(t) = 4x(t)$$

a. Show that the $h(t)$ and $g(t)$ given below are the responses of the system to the unit-impulse and unit-step inputs, respectively.

$$h(t) = \frac{4}{\sqrt{3}}e^{-t}\sin\sqrt{3}t\,u(t)$$

$$g(t) = \left[1 - \sqrt{\frac{4}{3}}e^{-t}\cos(\sqrt{3}t - 30°)\right]u(t)$$

b. Plot the two complex numbers $s_{1,2} = -1 \pm j\sqrt{3}$ on the complex plane. Note that the complex numbers $s_{1,2}$ are the roots of the equation $s^2 + 2s + 4 = 0$. Draw a semicircle of radius 2 in the left half-plane and center it at the origin. Describe the correspondence between the system's time responses and the diagram on the complex plane.

55. The input-output relationship of an LTI system is

$$\frac{d^2y}{dt^2} + 2\sigma\frac{dy}{dt} + \omega_0^2 y(t) = \omega_0^2 x(t)$$

a. Show that the $h(t)$ and $g(t)$ given below are the responses of the system to the unit-impulse and unit-step inputs, respectively.

$$h(t) = \frac{\omega_0^2}{\omega_d} e^{-\sigma t} \sin\omega_d t\, u(t)$$

$$g(t) = \left[1 - \sqrt{\frac{\omega_0^2}{\omega_d^2}} e^{-\sigma t} \cos(\omega_d t - \phi)\right] u(t), \quad \text{where } \omega_0^2 = \sigma^2 + \omega_d^2 \text{ and } \phi = \tan^{-1}\left(\frac{\sigma}{\omega_d}\right)$$

b. Plot the two complex numbers $s_{1,2} = -\sigma \pm j\omega_d$ on the complex plane. Note that the complex numbers $s_{1,2}$ are the roots of the equation $s^2 + 2\sigma s + \omega_0^2 = 0$. Draw a semicircle of radius ω_0 in the left half-plane and center it at the origin. Show σ, ω_d, and ϕ on the complex plane. Describe the correspondence between the system's time responses and the diagram on the complex plane.

56. The input-output relationship of the LTI system in problem 55 is also written as

$$\frac{d^2y}{dt^2} + 2\xi\omega_0\frac{dy}{dt} + \omega_0^2 y(t) = \omega_0^2 x(t)$$

a. Show that the unit-impulse and unit-step responses, expressed in terms of ω_0 and ξ, are

$$h(t) = \frac{\omega_0}{\sqrt{1-\xi^2}} e^{-\xi\omega_0} \sin\omega_0\sqrt{1-\xi^2}t\, u(t)$$

$$g(t) = \left[1 - \frac{1}{\sqrt{1-\xi^2}} e^{-\xi\omega_0} \sin(\omega_0\sqrt{1-\xi^2}t + \theta)\right] u(t), \quad \text{where } \theta = \cos^{-1}\xi$$

b. Plot the two complex numbers $s_{1,2} = -\omega_0(\xi \pm j\sqrt{1-\xi^2})$ on the complex plane. Note that the complex numbers $s_{1,2}$ are roots of the equation $s^2 + 2\xi\omega_0 s + \omega_0^2 = 0$. Draw a semicircle of radius ω_0 in the left half-plane and center it at the origin. Show the angle θ.

c. Show that in an oscillatory step response,

$$\text{period of oscillations is } T = \frac{2\pi}{\omega_0\sqrt{1-\xi^2}} \quad \text{and the}$$

$$\text{relative overshoot is } \rho = e^{\frac{-\pi\xi}{\sqrt{1-\xi^2}}}$$

57. In the circuit of Figure 3.7, let $R_1 = 3\text{ k}\Omega$, $R_2 = 2\text{ k}\Omega$, $L = 1\text{ H}$, and $C = 1\text{ }\mu\text{F}$.

a. Show that the capacitor voltage v and inductor current i are related to the current source by

$$\begin{cases} v' = -500v - 10^6 i + 10^6 i_s \\ i' = v - 3{,}000i \end{cases}$$

or

$$\begin{bmatrix} v' \\ i' \end{bmatrix} = \begin{bmatrix} -500 & -10^6 \\ 1 & -3{,}000 \end{bmatrix} \begin{bmatrix} v \\ i \end{bmatrix} + \begin{bmatrix} 10^6 \\ 0 \end{bmatrix} i_s$$

b. Show that the above relationships may be reduced to

$$v'' + 3{,}500v' + 2.5 \times 10^6 v = 10^6(i_s' + 3{,}000i_s)$$

$$i'' + 3{,}500i' + 2.5 \times 10^6 i = 10^6 i_s$$

c. Argue that the system is causal, dynamic, linear, and time invariant.

58. State equations for a pendulum. In the pendulum of Example 3.14, let the applied external torque τ constitute the input, $y = \theta$ be the output, and $z_1 = \theta'$ and $z_2 = \theta$ be the two state variables. Show that the state and output equations of the pendulum system are

$$z_1' = -k_d z_1 - \frac{k_s}{ml^2} z_2 - \frac{g}{l} \sin z_2 + \frac{1}{ml^2} \tau$$

$$z_2' = z_1$$

$$y = z_2$$

Show that for small angles we have

$$\begin{bmatrix} z_1' \\ z_2' \end{bmatrix} = \begin{bmatrix} -k_d & -\frac{k_s}{m\ell^2} - \frac{g}{\ell} \\ 1 & 0 \end{bmatrix} \begin{bmatrix} z_1 \\ z_2 \end{bmatrix} + \tau \begin{bmatrix} \frac{1}{m\ell^2} \\ 0 \end{bmatrix}$$

which may be written in matrix form as the following first-order differential equations

$$\begin{cases} \mathbf{Z}' = \mathbf{AZ} + \mathbf{BX} \\ \mathbf{Y} = \mathbf{CZ} \end{cases}$$

where

$$\mathbf{Z} = \begin{bmatrix} z_1 \\ z_2 \end{bmatrix}, \quad \mathbf{X} = \begin{bmatrix} \tau \\ 0 \end{bmatrix}, \quad \mathbf{A} = \begin{bmatrix} -k_d & -\frac{k_s}{m\ell^2} - \frac{g}{\ell} \\ 1 & 0 \end{bmatrix}, \quad \mathbf{B} = \begin{bmatrix} \frac{1}{m\ell^2} & 0 \end{bmatrix}, \quad \text{and} \quad \mathbf{C} = \begin{bmatrix} 0 & 1 \end{bmatrix}$$

3.17 Project: Open-Loop Control

Open-Loop Procedure

An LTI system is completely specified by the following two parameters:

$$\xi = \text{damping ratio}$$

$$\omega_0 = \text{undamped natural frequency}$$

The unit-step response of the system is

$$g(t) = 1 - \frac{1}{\sqrt{1-\xi^2}} e^{-\sigma t} \cos(\omega_d t - \phi), \text{ where } \sigma = \omega_0 \xi, \quad \phi = \sin^{-1} \xi, \text{ and } \omega_0 = \sqrt{\omega_d^2 + \sigma^2}$$

a. Show that the response overshoot (before it reaches the DC steady state) is

$$\rho = e^{-\pi \frac{\xi}{\sqrt{1-\xi^2}}}$$

b. Show that for a system with 50% overshoot the damping ratio is $\xi = 0.21545$. Write the equation for the step response of such a system if the period of oscillation equals $T = 200$ s and then plot it. From the plot verify the percentage overshoot and period of oscillation assumed initially. See Figure 3.18(a).

c. To avoid oscillations and reach the final steady state in finite time, apply the input $x(t) = au(t) + (1-a)u(t-t_0)$. The output may be found by superposition. Determine a and t_0 so that the response reaches its DC steady state in 100 s. See Figure 3.18(b).

d. To reduce the transition time, apply the input $u(t) - ku(t-T_1) + ku(t-T_2)$, where the parameters k, T_1, and T_2 can be adjusted. Find relationships between the three parameters such that the system reaches the steady state in T_2 s. Find example pairs of T_1 and T_2 for $k = 1, 2$. An example is shown in Figure 3.18(c) for $k = 1$, $T_1 = 46$,

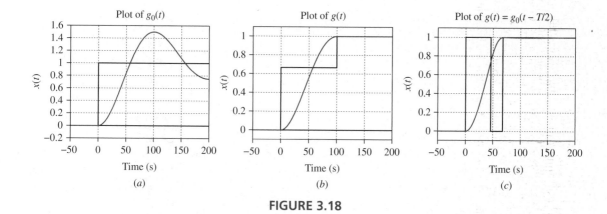

FIGURE 3.18

and $T_2 = 67.5$. Reproduce Figure 3.18(c). Then, given $k = 2$ and $T_1 = 50$, find T_2 and plot the response. *Hint:* To accomplish this part you need $g(T_2) = 1$ and $dg(t)/dt = 0$ at $t = T_2$.

e. Given the overshoot ρ and k, develop and plot a family of curves from which one may find T_1 and T_2. Keep the period of oscillation at $T = 200$ s and change the overshoot from 20% to 90% in steps of 10%.

Discussion and Conclusions

The procedure described in this project is for open-loop control. Do you see any practical application for it? How sensitive are the values of T_1 and T_2 to a changing k as it becomes larger? Does the procedure work for any inputs other than a step function?

Chapter 4

Superposition, Convolution, and Correlation

Contents

Introduction and Summary

Convolution is the general formulation of the superposition of responses of linear systems in the time domain. It is a cornerstone of linear systems analysis and filtering. It states that an LTI system is completely described by its unit-impulse response $h(t)$. The convolution of two continuous-time functions $x(t)$ and $h(t)$ is defined by the integral

$$y(t) = \int_{-\infty}^{\infty} x(\tau)h(t - \tau)d\tau \quad (1)$$

When both $x(t)$ and $h(t)$ are right-sided, the limits of the integral given above are reduced to 0 to t. The above integral produces the response of the LTI system having the unit-impulse response $h(t)$ to the input $x(t)$ (or vice versa). In addition, through its properties, the convolution operation can provide insight into the functioning of a system in terms of its subsystems. For example, the unit-impulse response of a cascade of several LTI systems can be obtained from convolution of the individual unit-impulse responses, and the cascading order may be changed without affecting the overall unit-impulse response of the ensemble. As another example, the unit-impulse response of several LTI systems in parallel is equal to the sum of the individual unit-impulse responses. Such properties provide useful tools not only for the analysis but also for the synthesis and design of systems (e.g., how to reduce design complexity by breaking down a system into simpler subsystems).

Convolution constitutes the core subject of this chapter. By way of several examples we first show the use of superposition to find the response of an LTI system to a new input given the response to other inputs. Expanding upon the superposition and time-invariance properties, we then derive the convolution sum and integral and present their properties. The graphical method of visualizing the convolution integral is illustrated next. The method is especially helpful in determining the limits of the integral and evaluating it when $x(t)$ and/or $h(t)$ are sectionwise continuous functions. Finally, the chapter briefly introduces the concepts of filtering by convolution, the matched filter, correlations, and deconvolution. An application of a matched filter is provided as a project at the end of the chapter.

4.1 Superposition of Responses

The linearity property leads to the superposition of responses due to individual inputs. If a complex input can be broken down into simpler components with known responses, the output may then be constructed from the responses to the input components.

E*xample*
4.1

An input-output pair $x_1(t) \rightarrow y_1(t)$ for an LTI system is given in Figure 4.1(a). Find the response of the system to inputs x_2 and x_3 of Figure 4.1(b) and 4.1(c) in terms of $y_1(t)$ and evaluate them.

Solution
Express $x_2(t)$ and $x_3(t)$ in terms of $x_1(t)$. Then construct their outputs using $y_1(t)$ as the building block. For brevity and convenience we use the following representations:

$$x_1(t) = \{\underset{\uparrow}{1}, -1, \ 0\}$$

$$x_2(t) = \{\underset{\uparrow}{1}, \ 0, -1\}$$

$$x_3(t) = \{\underset{\uparrow}{1}, \ 1, -2\}$$

1. $x_1(t) \Longrightarrow y_1(t) = \{\underset{\uparrow}{1}, -2, 1, 0\}$

2. $x_2(t) = x_1(t) + x_1(t-1)$

$y_2(t) = y_1(t) + y_1(t-1) = \{\underset{\uparrow}{1}, -2, 1, 0\} + \{\underset{\uparrow}{0}, 1, -2, 1\} = \{\underset{\uparrow}{1}, -1, -1, 1\}$

3. $x_3(t) = x_1(t) + 2x_1(t-1)$

$y_3(t) = y_1(t) + 2y_1(t-1) = \{\underset{\uparrow}{1}, -2, 1, 0\} + \{\underset{\uparrow}{0}, 2, -4, 2\} = \{\underset{\uparrow}{1}, 0, -3, 2\}$

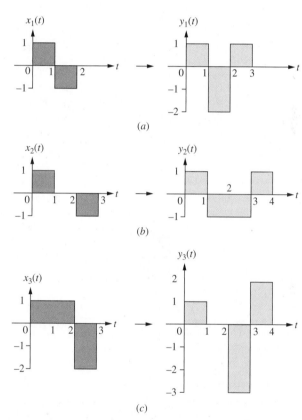

FIGURE 4.1 The outputs y_2 and y_3 (due to inputs x_2 and x_3) are found by knowing the input-output pair x_1, y_1.

Example

4.2

Given the input-output pair $x_1(t) \rightarrow y_1(t)$ for the LTI system shown in Figure 4.1(a), find the output due to $x_4(t) = \{\underset{\uparrow}{1}, -1, -1\}$.

Solution

Express $x_4(t)$ in terms of $x_1(t)$, then construct its response $y_4(t)$ in terms of $y_1(t)$ and evaluate it as done below.

$$x_4(t) = x_1(t) - \sum_{k=2}^{\infty} x_1(t - k)$$

$$y_4(t) = y_1(t) - \sum_{k=2}^{\infty} y_1(t - k) = \{\underset{\uparrow}{1}, -2, 1\}$$

$$-\{\underset{\uparrow}{0}, 0, 1, -2, 1\}$$

$$-\{\underset{\uparrow}{0}, 0, 0, 1, -2, 1\}$$

$$-\{\underset{\uparrow}{0}, 0, 0, 0, 1, -2, 1\}$$

$$-\{\underset{\uparrow}{0}, 0, 0, 0, 0, 1, -2, 1\}$$

$$\cdots \ \cdots \ \ \cdots \ \ \cdots \ \ \cdots \ \ \cdots$$

$$= \{\underset{\uparrow}{1}, -2, 0, 1, \ 0, 0, \ 0, 0\}, \quad (\text{See Figure 4.2.})$$

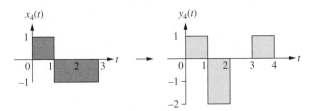

FIGURE 4.2 The output y_4 (due to input x_4) is found by knowing the input-output pair x_1, y_1 given in Figure 4.1(a).

Alternative Solution

From an examination of the given relationship $x_1(t) \rightarrow y_1(t)$ in Figure 4.1(a), one can deduce that the input-output relationship for the systems of Examples 4.1 and 4.2 is $y(t) = x(t) - x(t - 1)$. The relationship can be used to directly obtain the output given any input.

Example 4.3

The response of an LTI system to a unit pulse at the origin, $d(t) = u(t) - u(t - 1)$, is $\hat{h}(t) = 4d(t) + 3d(t - 1) + 2d(t - 2) + d(t - 3)$, a 4-step downstairs curve. as shown in Figure 4.3(a). Find the system's response to $x(t) = u(t) - u(t - 4)$.

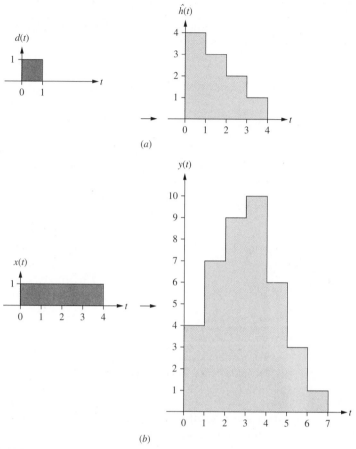

FIGURE 4.3 The output y_2 (due to input x_2 in [**b**]) is found by knowing the input-output pair $d(t)$, $\hat{h}(t)$ in **(a)**.

Solution

Since time is discrete, we designate it by the integer variable n. Then the unit pulse at the origin is represented by $d(n) = \{1\}$ and produces the output $\hat{h}(n) = \{4, 3, 2, 1\}$. Similarly, the input $x(n) = u(n) - u(n-4)$ may be expressed in terms of $d(n)$ by

$$x(n) = \sum_{k=0}^{3} d(n-k)$$

By superposition, the response to $x(n)$ is

$$y(n) = \sum_{k=0}^{3} \hat{h}(n-k)$$

Because $d(n)$, $\hat{h}(n)$, and $x(n)$ assume constant values during discrete time intervals, they may also be represented by sequences of numbers as shown in Table 4.1.

TABLE 4.1 Finding $y(n)$ by Superposition

$$d(n) = \{\ldots, 0, \underset{\uparrow}{1}, 0, 0, 0, 0, 0, 0, 0, 0, \ldots\}$$

$$\hat{h}(n) = \{\ldots, 0, \underset{\uparrow}{4}, 3, 2, 1, 0, 0, 0, 0, 0, \ldots\}$$

$$x(n) = d(n) + d(n-1) + d(n-2) + d(n-3)$$

$$= \{\ldots, 0, \underset{\uparrow}{1}, 1, 1, 1, 0, 0, 0, 0, 0, \ldots\}$$

$$y(n) = \hat{h}(n) + \hat{h}(n-1) + \hat{h}(n-2) + \hat{h}(n-3)$$

$$= \{\ldots, 0, \underset{\uparrow}{4}, 3, 2, 1, 0, 0, 0, 0, 0, \ldots\}$$

$$+ \{\ldots, 0, 0, \underset{\uparrow}{4}, 3, 2, 1, 0, 0, 0, 0, \ldots\}$$

$$+ \{\ldots, 0, 0, 0, \underset{\uparrow}{4}, 3, 2, 1, 0, 0, 0, \ldots\}$$

$$+ \{\ldots, 0, 0, 0, 0, \underset{\uparrow}{4}, 3, 2, 1, 0, 0, \ldots\}$$

$$= \{\ldots, 0, \underset{\uparrow}{4}, 7, 9, 10, 6, 3, 1, 0, 0, \ldots\}$$

The system's response to $x(t) = d(t) + d(t-1) + d(t-2) + d(t-3)$ is $y(t) = 4d(t) + 7d(t-1) + 9d(t-2) + 10d(t-3) + 6d(t-4) + 3d(t-5) + d(t-6)$, shown in Figure 4.3($b$).

Alternative Solution
From an examination of the given relationship $d(t) \rightarrow \hat{h}(t)$, we can deduce that the input-output relationship for the system is $y(t) = 4x(t) + 3x(t-1) + 2x(t-2) + x(t-3)$. The relationship can then be used to directly obtain the output given any input.

Example

4.4

A continuous-time waveform $x(t)$ that may change its value at unit-time steps is represented by a finite sequence made of four elements: $\{\ldots, 0, 0, \underset{\uparrow}{x_0}, x_1, x_2, x_3, 0, 0, 0 \ldots\}$ $\equiv \{\underset{\uparrow}{x_0}, x_1, x_2, x_3\}$. The response of the LTI system of Example 4.3 to a unit pulse may, therefore, be shown by $\hat{h}(t) = \{\underset{\uparrow}{4}, 3, 2, 1\}$. The response of that system to the above $x(t)$ is

$$y(t) = x_0\hat{h}(t) + x_1\hat{h}(t-1) + x_2\hat{h}(t-2) + x_3\hat{h}(t-3) = \sum_{n=0}^{3} x_n\hat{h}(t-n)$$

For example, let $x(t) = \{\underset{\uparrow}{1}, 2, 3, 4\}$. Then $y(t) = \hat{h}(t) + 2\hat{h}(t-1) + 3\hat{h}(t-2) + 4\hat{h}(t-3)$ can be constructed as shown below.

$$
\begin{array}{rrrrrrrrr}
y(t) & = & 4, & 3, & 2, & 1, & 0, & 0, & 0, \\
 & + & 0, & 8, & 6, & 4, & 2, & 0, & 0, \\
 & + & 0, & 0, & 12, & 9, & 6, & 3, & 0, \\
 & + & 0, & 0, & 0, & 16, & 12, & 8, & 4, \\
 & = & 4, & 11, & 20, & 30, & 20, & 11, & 4, \\
\end{array}
$$

Example 4.5

Pat has established a credit account with a bank that requires N monthly payments of $(a + \frac{1}{N})$ dollars each for every dollar of purchase. Statements are issued at the end of each month and payments are made in the following month. Pat's monthly charges in dollars from January (marked by ↑) to December are

$$x(n) = \{\underset{\uparrow}{50}, 100, 200, 150, 300, 250, 350, 300, 100, 150, 400, 500\}$$

with no purchase from then on. Starting with zero debt and for $N = 5$ and $a = 0.02$, calculate the monthly payments for as long as they last.

Solution

The monthly charges $x(n)$ (charges made to the account in the nth month) and the monthly payments due $y(n)$ (for payments due in the nth month) are listed in Table 4.2. The system along with its input and output functions are shown in Figure 4.4.

TABLE 4.2 Monthly Charges and Payments of Example 4.5

Monthly Charges	Payments Due																	
	Jan	Feb	Mar	Apr	May	June	July	Aug	Sep	Oct	Nov	Dec	Jan	Feb	Mar	Apr	May	June
Jan: $50	11	11	11	11	11													
Feb: $100		22	22	22	22	22												
Mar: $200			44	44	44	44	44											
Apr: $150				33	33	33	33	33										
May: $300					66	66	66	66	66									
June: $250						55	55	55	55	55								
July: $350							77	77	77	77	77							
Aug: $300								66	66	66	66	66						
Sep: $100									22	22	22	22	22					
Oct: $150										33	33	33	33	33				
Nov: $400											88	88	88	88	88			
Dec: $500												110	110	110	110	110		
Monthly Payments	0	11	33	77	110	176	220	275	297	286	253	286	319	253	231	198	110	0

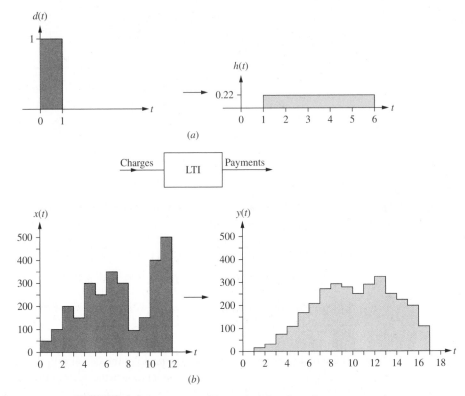

(a)

(b)

FIGURE 4.4 The system of Example 4.5 and two input-output pairs.

In the circuit of Figure 4.5(a), let v_s be made of a rectangular voltage pulse (starting at the origin) of magnitude k and duration T, superimposed on a unit step. See Figure 4.5(b). Find v_c and determine k so that the capacitor voltage reaches its DC value in T seconds.

Solution

Let the step response be shown by $g(t)$. It may be found that

$$g(t) = \left(1 - e^{-t/\tau}\right)u(t), \quad \tau = RC$$

The rectangular pulse of height k and width T may be written as $k\,[u(t) - u(t - T)]$. Using the linearity and time-invariance properties of the circuit, the capacitor voltage in response to the pulse is determined to be $k\,[g(t) - g(t - T)]$. The response to the pulse superimposed on a unit step is then

$$v_c(t) = k\,[g(t) - g(t - T)] + g(t) = \begin{cases} 0, & t < 0 \\ (k+1)\left(1 - e^{-t/\tau}\right), & 0 \le t < T \\ 1 + \left(ke^{T/\tau} - 1 - k\right)e^{-t/\tau}, & t \ge T \end{cases}$$

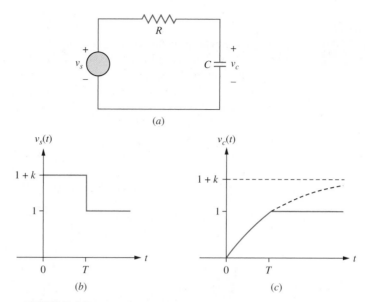

FIGURE 4.5 Charging a capacitor to a DC final value in finite time.

For the capacitor voltage to reach its DC value at T we need

$$v_c(t) = 1 + e^{-t/\tau}\left(ke^{T/\tau} - 1 - k\right) = 1, \quad \text{all } t \geq T$$

This requires

$$ke^{T/RC} - 1 - k = 0, \quad \text{or } k = \frac{1}{e^{T/RC} - 1}$$

An Alternate and More Elegant Solution

During $0 < t \leq T$ the circuit experiences a step input of size $(k + 1)$. In response, the capacitor voltage at $t = T$ is

$$v_c(T) = (1 + k)(1 - e^{-T/\tau})$$

Since for $t > T$ the input is $v_s = 1$, the response will remain constant if $v_c(T) = 1$.

$$v_c(T) = (1 + k)(1 - e^{-T/\tau}) = 1 \quad \text{or } k = \frac{1}{e^{T/\tau} - 1}$$

Note that k is a positive number. See Figure 4.5(c).

4.2 Convolution Sum

Superposition of a system's response to a sequence of pulses leads to the convolution sum as illustrated in the following examples.

Let $\zeta(t)$ (written and pronounced as *zeta*) designate a rectangular pulse of width T and unit height with its left edge at the origin. Then $\zeta(t - nT)$ represents such a pulse shifted to the right by the amount nT. Let the response of an LTI system to $\zeta(t)$ be $\psi(t)$ (written *psi* and pronounced *sie*). Consider an input $x(t)$ to the LTI system made of a sequence of rectangular pulses of width T and height x_n, each placed at $t = nT$. The input can be expressed in terms of $\zeta(t)$:

$$x(t) = \sum_{n=-\infty}^{\infty} x_n \zeta(t - nT), \quad x_n = \{\underset{\uparrow}{1}, \, 2, \, 0.5, \, -0.5, \, 0, \, 1\}$$

See Figure 4.6. Find the response of the system to $x(t)$.

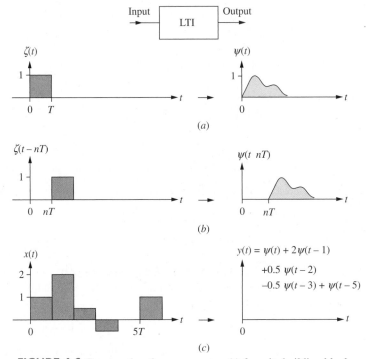

FIGURE 4.6 Constructing the response to $x(t)$ from its building blocks.

Solution
Using the linearity and time-invariance properties of the system, the output may be found by the superposition of responses to individual pulses in the input. The summation is called the *convolution sum*.

$$y(t) = \sum_{n=-\infty}^{\infty} x_n \psi(t - nT)$$

The summation operation and resulting output are shown in Table 4.3.

TABLE 4.3 Finding $y(t)$ from the Convolution Sum

$\zeta(t)$	$\Longrightarrow$	$\psi(t)$
$\zeta(t - nT)$	$\Longrightarrow$	$\psi(t - nT)$
$x_n \zeta(t - nT)$	$\Longrightarrow$	$x_n \psi(t - nT)$

$$x(t) = \zeta(t) + 2\zeta(t - T) + 0.5\zeta(t - 2T) - 0.5\zeta(t - 3T) + \zeta(t - 5T)$$
$$y(t) = \psi(t) + 2\psi(t - T) + 0.5\psi(t - 2T) - 0.5\psi(t - 3T) + \psi(t - 5T)$$

*E*xample
4.8

The response of an LTI system to a unit pulse $d(t) = u(t) - u(t - 1)$ is $e^{-t}u(t)$. Find the system's response to the input $x(t) = 4d(t) + 3d(t - 1) + 2d(t - 2) + d(t - 3)$.

Solution
Applying the linearity and time-invariance properties, we find $y(t)$ from Table 4.4.

TABLE 4.4 Finding $y(t)$ from the Convolution Sum

$t < 0$	$y(t) = 0$
$0 \le t < 1$	$y(t) = 4e^{-t}$
$1 \le t < 2$	$y(t) = 4e^{-t} + 3e^{-(t-1)} = (4 + 3e)\,e^{-t}$
$2 \le t < 3$	$y(t) = 4e^{-t} + 3e^{-(t-1)} + 2e^{-(t-2)} = (4 + 3e + 2e^2)e^{-t}$
$t \ge 3$	$y(t) = 4e^{-t} + 3e^{-(t-1)} + 2e^{-(t-2)} + e^{-(t-3)} = (4 + 3e + 2e^2 + e^3)e^{-t}$

Note the components in the amplitude of the output and how each input pulse contributes to the total output. The above system is not realizable from a finite number of lumped elements, delays, and differentiators.

*E*xample
4.9

Let $\xi(t)$ (written *xi* and pronounced *cai*) designate a rectangular pulse of width T and height $1/T$ with its left edge at the origin. Then $\xi(t - nT)$ represents such a pulse shifted to the right by the amount nT. Note that the area under every such pulse is unity. Let the response of an LTI system to $\xi(t)$ be $\eta(t)$ (written and pronounced as *eta*). See Figure 4.7. Find the response of the system to

$$\hat{x}(t) = \sum_{n=-\infty}^{\infty} T e^{-snT} \xi(t - nT)$$

and determine its limit as $T \to 0$.

Solution
By time invariance, the response of the system to $\xi(t - nT)$ is $\eta(t - nT)$ and, by superposition, the response to $\hat{x}(t)$ is

$$\hat{y}(t) = \sum_{n=-\infty}^{\infty} T e^{-snT} \eta(t - nT)$$

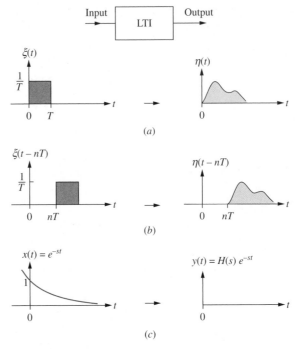

FIGURE 4.7 Formulating response to an exponential input.

As T approaches zero, $\xi(t)$ approaches the unit-impulse function, $\delta(t)$, and $\hat{x}(t)$ approaches e^{-st}. If

$$\lim_{T \to 0} \eta(t) = h(t)$$

then

$$y(t) = \lim_{T \to 0} \hat{y}(t) = \left[\int_{-\infty}^{\infty} h(t)e^{st}dt \right] e^{-st} = H(s)e^{-st}$$

Note that in passing through the system the exponential input e^{-st} is multiplied by the scale factor $H(s)$.

4.3 Convolution Integral

We extend Example 4.9 to derive a general expression for the convolution integral. Again, let $\xi(t)$ designate a rectangular pulse of width T and unit area with its left edge at the origin. Then $\xi(t - nT)$ represents such a pulse shifted to the right by the amount nT. Let the response of the LTI system to $\xi(t)$ be $\eta(t)$. Let the input $x(t)$ to the LTI system be approximated by a stepwise function made of narrow rectangular pulses of width T and height $x(nT)$ placed at $t = nT$. This stepwise approximation may be expressed by

$$\hat{x}(t) = \sum_{k=-\infty}^{\infty} Tx(nT)\xi(t - nT)$$

See Figure 4.8. The output is then found by superposition as shown in Table 4.5.

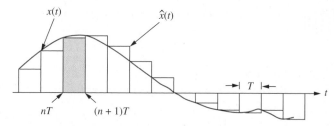

FIGURE 4.8 Stepwise approximation to a continuous-time function. Each step is T seconds wide and has a height of $x(nT)$. The step that starts at $t = nT$ can be expressed by $Tx(nT)\xi(t - nT)$.

TABLE 4.5 Developing the Convolution Integral

$\xi(t)$	$\Rightarrow \; \eta(t)$
$\xi(t - nT)$	$\Rightarrow \; \eta(t - nT)$
$Tx(nT)\xi(t - nT)$	$\Rightarrow \; Tx(nT)\eta(t - nT)$
$\displaystyle \hat{x}(t) = \sum_{n=-\infty}^{\infty} Tx(nT)\xi(t - nT)$	$\Rightarrow \; \displaystyle \hat{y}(t) = \sum_{n=-\infty}^{\infty} Tx(nT)\eta(t - nT)$

In the limit $T \to 0$,

$$\lim_{T \to 0} nT = \tau$$

$$\lim_{T \to 0} T = d\tau$$

$$\lim_{T \to 0} \xi(t) = \delta(t)$$

$$\lim_{T \to 0} \overline{x}(t) = x(t)$$

$$\lim_{T \to 0} \overline{y}(t) = \int_{-\infty}^{\infty} x(\tau)h(t - \tau)d\tau$$

The convolution sum becomes the convolution integral

$$y(t) = \int_{-\infty}^{\infty} x(\tau)h(t - \tau)d\tau$$

Special Case

The convolution of two right-sided functions $x(t)u(t)$ and $h(t)u(t)$ becomes

$$\int_{-\infty}^{\infty} [x(\tau)u(\tau)][h(t-\tau)u(t-\tau)]d\tau = \int_{0}^{t} x(\tau)h(t-\tau)d\tau. \text{ See Example 4.10.}$$

Example 4.10

Given $x(t) = e^{-\alpha t}u(t)$ and $h(t) = e^{-\beta t}u(t)$, find

$$y(t) = \int_{-\infty}^{\infty} x(\tau)h(t-\tau)d\tau$$

Solution

Because $x(\tau) = 0$ for $\tau < 0$, the lower limit of the integral is set to 0.

$$y(t) = \int_{0}^{\infty} x(\tau)h(t-\tau)d\tau = \int_{0}^{\infty} e^{-\alpha\tau}e^{-\beta(t-\tau)}u(t-\tau)d\tau$$

But $u(t-\tau) = 0$ for $\tau > t$. Therefore, the upper limit of the integral is reduced to t

$$y(t) = \int_{0}^{t} e^{-\alpha\tau}e^{-\beta(t-\tau)}d\tau = e^{-\beta t}\int_{0}^{t} e^{-(\alpha-\beta)\tau}d\tau = \frac{1}{\beta-\alpha}\left(e^{-\alpha t} - e^{-\beta t}\right)u(t)$$

Example 4.11

Given $x(t) = u(t)$ and $h(t) = \sin(\omega t)u(t)$, find $y(t) = x(t) \star h(t)$.

Solution

$$y(t) = \int_{-\infty}^{\infty} x(t-\tau)h(\tau)d\tau = \int_{0}^{t} x(t-\tau)h(\tau)d\tau$$

$$= \int_{0}^{t} \sin(\omega\tau)d\tau = \frac{-1}{\omega}\left[\cos(\omega\tau)\right]_{\tau=0}^{\tau=t} = \frac{1}{\omega}\left[1 - \cos(\omega t)\right]u(t)$$

Summary

The convolution of two functions $x(t)$ and $h(t)$ produces a new function $y(t)$ defined by

$$y(t) = \int_{-\infty}^{\infty} x(\tau)h(t-\tau)d\tau$$

where $-\infty < t < \infty$. It will be shown that the convolution operation between $x(t)$ and $h(t)$ is commutative:

$$\int_{-\infty}^{\infty} x(\tau)h(t-\tau)d\tau = \int_{-\infty}^{\infty} x(t-\tau)h(\tau)d\tau$$

The operation involves time reversal of one of the functions, its time shift, multiplication by the other function, and integration. The amount of shift constitutes the independent variable in the convolution result. For simplicity and convenience, the convolution integral of $x(t)$ and $h(t)$ is also shown by $y(t) = x(t) \star h(t) = h(t) \star x(t)$.

4.4 Graphical Convolution

The convolution integral of (1) may be evaluated by plotting the function $x(\tau)h(t - \tau)$ [or $x(t - \tau)h(\tau)$] versus τ and then evaluating the area under the curve either geometrically (from the graph) or analytically (from the integral). This is called the graphical method. It is especially helpful in determining the upper and lower limits of integration when the functions are specified graphically or by piecewise equations.[1] The process is summarized by the flip-shift-multiply-integrate recipe as described below:

Step 1. Flip $h(\tau)$ around $\tau = 0$ to obtain $h(-\tau)$.

Step 2. Shift $h(-\tau)$ by the amount t to obtain $h(t - \tau)$. A positive value of t shifts $h(-\tau)$ to the right. A negative value of t shifts $h(-\tau)$ to the left. Start with $t = -\infty$.

Step 3. Multiply $h(t - \tau)$ by $x(\tau)$ to obtain $x(\tau)h(t - \tau)$.

Step 4. Integrate $x(\tau)h(t - \tau)$ over the range of $-\infty < \tau < \infty$. Call the sum $y(t)$ for the given t.

Step 5. Go back to step 2. Increment the shift in $h(t - \tau)$ and continue the cycle until $y(t)$ is found for all values of t.

Needless to say, the flip and shift may be performed on $x(t)$.

Example 4.12

Find the convolution $y(t) = x(t) \star h(t)$, where $x(t) = e^{-t}u(t)$ and $h(t) = u(t)$.

Solution

$$y(t) = \int_{-\infty}^{\infty} x(\tau)h(t - \tau)d\tau = \int_{-\infty}^{\infty} e^{-\tau}u(\tau)u(t - \tau)d\tau$$

Because $u(\tau)$ is zero for $\tau < 0$, its presence in the integrand brings up the lower limit of the integral to zero. Similarly, since $u(t - \tau) = 0$ for $\tau > t$, the upper limit of the integral becomes t. Therefore,

$$y(t) = \int_{0}^{t} e^{-\tau}d\tau = \left(1 - e^{-t}\right)u(t)$$

The above convolution is graphically visualized in Figure 4.9.

---

[1]The method is identical to the graphical method for evaluating the discrete convolution.

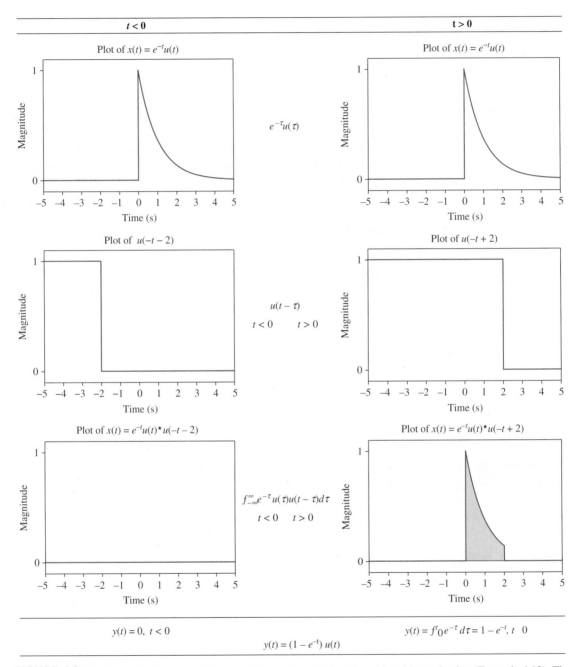

FIGURE 4.9 Graphical visualization of the convolution $y(t) = [e^{-t}u(t)] \star u(t)$ and its evaluation (Example 4.12). The column on the left is for $t < 0$ and that on the right for $t > 0$.

*E**xample***

4.13

The unit-impulse response of an LTI system is $h(t) = \alpha e^{-\beta t} u(t)$. Find its response to the unit pulse $x(t) = u(t) - u(t-1)$.

$$y(t) = x(t) \star h(t) = \int_{-\infty}^{\infty} h(\tau) x(t - \tau) d\tau$$

Solution

First, sketch graphs of $h(\tau)$ and $x(\tau)$. Then flip $x(\tau)$ around the vertical axis to get $x(-\tau)$. See Figure 4.10(a).

Next, shift the pulse $x(-\tau)$ to the left to get $x(t - \tau)$. For $t < 0$ the product $h(\tau)x(t - \tau)$ is zero and so is $y(t)$. See Figure 4.10(b).

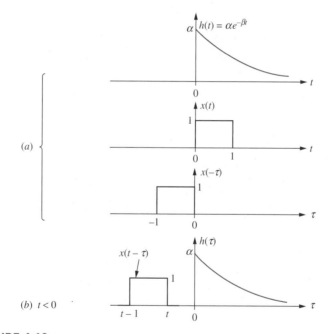

FIGURE 4.10 Convolution of the unit pulse with a causal exponential function (Example 4.13).

$$h(t) = \alpha e^{-\beta t} u(t)$$

$$x(t) = u(t) - u(t - 1),$$

$$y(t) = x(t) \star h(t)$$

$$= \begin{cases} 0, & t < 0 \\ \frac{\alpha}{\beta}(1 - e^{-\beta t}), & 0 \leq t < 1 \\ \frac{\alpha}{\beta}(e^{\beta} - 1)e^{-\beta t}, & t \geq 1 \end{cases}$$

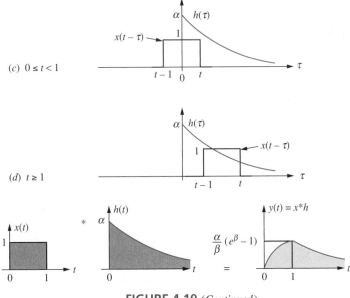

FIGURE 4.10 (*Continued*)

Then, shift the pulse $x(-\tau)$ to the right. The product $h(\tau)x(t - \tau)$ is either zero or $\alpha e^{-\beta t}$.

For $0 \leq t < 1$ (Figure 4.10c), the integration is done from $\tau = 0$ to $\tau = t$ and

$$y(t) = \int_0^t \alpha e^{-\beta \tau} d\tau = \frac{\alpha}{\beta} \left(1 - e^{-\beta t}\right), \quad 0 \leq t < 1$$

For $t \geq 1$ (Figure 4.10d), the integration is from $t - 1$ to t and

$$y(t) = \int_{t-1}^t \alpha e^{-\beta \tau} d\tau = \frac{-\alpha}{\beta} \left(e^{-\beta t} - e^{-\beta(t-1)}\right) = \frac{\alpha}{\beta}(e^{\beta} - 1)e^{-\beta t}, \quad t \geq 1$$

*E**xample**
4.14*

Find the convolution of the unit pulse $x(t) = u(t) - u(t - 1)$ with itself.

Solution

Using the graphical method we find the values of $y(t)$ at the break points $t = 0, 1,$ and 2. Because $x(t)$ is either zero or one, the product $x(\tau)x(t - \tau)$ is either zero or one and its integral between break points is a straight line. It is therefore sufficient to compute $y(t)$ at the break points $t = 0, 1,$ and 2, and then connect them by straight lines. The result is a triangular pulse that is twice as wide as the rectangular pulse (see Figure 4.11).

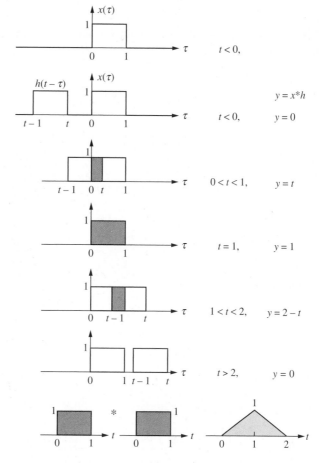

FIGURE 4.11 Convolution of the unit pulse with itself results in a triangular pulse (Example 4.14).

$$
y(t) = \begin{cases} 0, & t < 0 \\ t, & 0 \le t < 1 \\ 2 - t, & 1 \le t < 2 \\ 0, & t \ge 2 \end{cases}
$$

*E*xample
4.15

(1) Find the convolution of $x(t)$ with the unit impulse $\delta(t)$. (2) Repeat for the unit impulse located at t_0, $\delta(t - t_0)$.

Solution

1. $x(t) \star \delta(t) = \displaystyle\int_{-\infty}^{\infty} x(\tau)\delta(t - \tau)d\tau = x(t)$

2. $x(t) \star \delta(t - t_0) = \displaystyle\int_{-\infty}^{\infty} x(\tau)\delta(t - \tau + t_0)d\tau = x(t - t_0)$

Example

4.16

Given $f(t) = \delta(t - t_0) + \delta(t + t_0)$ and $x(t) = \frac{\sin t}{t}$, find and plot $y(t) = x(t) \star f(t)$ for $t_0 = 7\pi/2$.

Solution

$$y(t) = x(t) \star f(t) = \frac{\sin t}{t} \star [\delta(t - t_0) + \delta(t + t_0)] = y_1(t) + y_2(t)$$

$$y_1(t) = \frac{\sin t}{t} \star \delta(t - t_0) = \frac{\sin(t - t_0)}{t - t_0}$$

$$y_2(t) = \frac{\sin t}{t} \star \delta(t + t_0) = \frac{\sin(t + t_0)}{t + t_0}$$

For $t_0 = 7\pi/2$ we obtain

$$y(t) = \frac{\sin(t - 7\pi/2)}{t - 7\pi/2} + \frac{\sin(t + 7\pi/2)}{t + 7\pi/2}$$

See Figure 4.12.

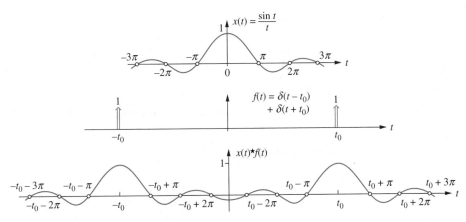

FIGURE 4.12 (For Example 4.16)

4.5 Properties of Convolution

Convolution is an operation with several algebraic properties. Three such properties with important consequences are the commutative, associative, and distributive laws. In addition, we note that convolution is a linear and time-invariant operation. (For the proof see problem 4 in Chapter 3.) It, therefore, exhibits the additional properties of LTI systems that were presented in Chapter 3. Here we briefly summarize these properties.

Commutative Property

$x(t)$ and $h(t)$ in the convolution integral (1) may switch places and the result remains the same:

$$x(t) \star h(t) = h(t) \star x(t)$$

To verify this property, change the integration variable τ in (1) to a new variable θ so that $\tau = t - \theta$ and $d\tau = -d\theta$. Substituting for τ and $t - \tau$ we have

$$y(t) = \int_{\tau=-\infty}^{\infty} x(\tau)h(t-\tau)d\tau = \int_{\theta=-\infty}^{\infty} x(t-\theta)h(\theta)d\theta$$

This property indicates that the output of an LTI system with the unit-impulse response $h(t)$ to an input $x(t)$ is the same as the output of another system with the unit-impulse response $x(t)$ to an input $h(t)$. The commutative property, in conjunction with some others (such as the differentiation property, to be presented shortly by Example 4.20), can facilitate and simplify the analysis of LTI systems.

Example 4.17

The unit-impulse response of an LTI system is $\cos(2\pi f t)u(t)$, where f is an integer. Find its response to the pulse $u(t) - u(t-1)$.

Solution

The above response is the same as that of an LTI system with an input $\cos(2\pi f t)u(t)$, where f is an integer, and the unit-impulse response $u(t) - u(t-1)$. The input-output relationship for the second system is

$$y(t) = \int_{-\infty}^{t} [x(\tau) - x(\tau-1)]\,d\tau$$

For $x(t) = \cos(2\pi f t)u(t)$ we get

$$y(t) = \frac{\sin(2\pi f t)}{2\pi f}\left[u(t) - u(t-1)\right]$$

Distributive Property

According to this property,

$$x(t) \star [h_1(t) + h_2(t)] = x(t) \star h_1(t) + x(t) \star h_2(t)$$

This is the linearity property of the convolution integral. It expresses the parallel operation of two LTI systems with unit-impulse responses $h_1(t)$ and $h_2(t)$ and their equivalence to an LTI system whose unit-impulse response is $h(t) = h_1(t) + h_2(t)$.

Example 4.18

Use the result of Example 4.17 to find $y(t) = h(t) \star x(t)$, where $h(t) = u(t) - 2u(t-1) + u(t-2)$, $x(t) = \cos(2\pi f t)u(t)$, and f is an integer.

Solution

Let $h_1(t) = u(t) - u(t-1)$. Then, $h(t) = h_1(t) - h_1(t-1)$ and $y(t) = y_1(t) - y_1(t-1)$. From Example 4.17 we have

$$y_1(t) = \frac{\sin(2\pi ft)}{2\pi f} \big[u(t) - u(t-1) \big]$$

Therefore,

$$y(t) = \frac{\sin(2\pi ft)}{2\pi f} \big[u(t) - 2u(t-1) + u(t-2) \big]$$

Alternative Solution

Let

$$h_0(t) = u(t), \quad y_0(t) = h_0(t) \star x(t) = \frac{\sin(2\pi ft)}{2\pi f} u(t)$$

Then,

$$h(t) = h_0(t) - 2h_0(t-1) + h_0(t-2),$$

$$y(t) = y_0(t) - 2y_0(t-1) + y_0(t-2)$$

$$= \frac{\sin(2\pi ft)}{2\pi f} \big[u(t) - 2u(t-1) + u(t-2) \big]$$

Associative Property

According to this property,

$$x(t) \star [h_1(t) \star h_2(t)] = [x(t) \star h_1(t)] \star h_2(t)$$

The proof can be derived directly from the definition of convolution. The associative property provides an alternative method for obtaining the output of the cascade of two LTI systems with unit-impulse responses $h_1(t)$ and $h_2(t)$. Either convolve the input $x(t)$ with the result of the convolution of $h_1(t)$ and $h_2(t)$, or convolve the input with $h_1(t)$ first and then convolve the result with $h_2(t)$. Alternatively, one can interchange the order of $h_1(t)$ and $h_2(t)$ and the output remains the same.

E*xample*

4.19

Find $y(t) = x(t) \star h_1(t) \star h_2(t)$, where $x(t) = \cos(\omega_0 t)$, $h_1(t) = ae^{-at}u(t)$, $h_2(t) = \delta(t) - h_1(t)$ and $\omega_0 = a$.

Solution

First obtain

$$z_1(t) = x(t) \star h_1(t) = \int_{-\infty}^{t} ae^{-a(t-\tau)} \cos(\omega_0 \tau) d\tau = (\sqrt{2}/2) \cos(\omega_0 t - \pi/4)$$

Then find

$$y(t) = z_1(t) \star h_2(t) = z(t) \star \delta(t) - z(t) \star h_1(t)$$

$$= (\sqrt{2}/2) \cos(\omega_0 t - \pi/4) - 0.5 \cos(\omega_0 t - \pi/2) = 0.5 \cos(\omega_0 t)$$

Alternative Solution

First obtain

$$z_2(t) = x(t) \star h_2(t) = \cos(\omega_0 t) - \int_{-\infty}^{t} ae^{-a(t-\tau)} \cos(\omega_0 \tau) d\tau$$
$$= \cos(\omega_0 t) - (\sqrt{2}/2) \cos(\omega_0 t - \pi/4)$$

Then find

$$y(t) = z_2(t) \star h_1(t)$$
$$= (\sqrt{2}/2) \cos(\omega_0 t - \pi/4) - 0.5 \cos(\omega_0 t - \pi/2) = 0.5 \cos(\omega_0 t)$$

Comment

The above solution is performed in the time domain. A frequency domain solution of this problem provides an easier, more intuitive, and elegant approach.

Other Properties of Convolution

In addition to the algebraic properties given above, other properties arising from the LTI nature of the convolution integral (such as the differentiation, integration, and shift properties) may help in obtaining the output of an LTI system and provide insight into the systems' synthesis and design. For example, consider an LTI system with the unit-impulse response $h(t)$ and an input $x(t)$ that produces the output $y(t)$. Then from the differentiation property of LTI systems,

$$\frac{dx(t)}{dt} \quad \Longrightarrow \quad \frac{dy(t)}{dt}$$

Example
4.20

Find $y(t) = h(t) \star x(t)$, where $h(t) = \cos(2\pi f t)u(t)$ and $x(t)$ is the triangular pulse

$$x(t) = \begin{cases} t, & 0 \le t < 1 \\ 2 - t, & 1 \le t < 2 \\ 0, & \text{elsewhere} \end{cases}$$

shown in Figure 4.13(a). Assume f to be an integer.

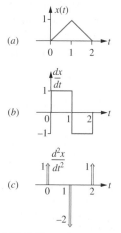

(a)

(b)

(c)

FIGURE 4.13 (For Example 4.20)

Solution

Take the first two derivatives of $x(t)$. The second derivative is composed of impulses. Convolve $h(t)$ with the second derivative of $x(t)$ and integrate the result twice.

$$\frac{dx}{dt} = u(t) - 2u(t-1) + u(t-2) \qquad\qquad \text{See Figure 4.13(b).}$$

$$\frac{d^2x}{dt^2} = \delta(t) - 2\delta(t-1) + \delta(t-2) \qquad\qquad \text{See Figure 4.13(c).}$$

$$\frac{d^2y}{dt^2} = h(t) \star \frac{d^2x}{dt^2} = h(t) \star \big[\delta(t) - 2\delta(t-1) + \delta(t-2)\big]$$

$$= h(t) - 2h(t-1) + h(t-2)$$

$$= \cos(2\pi f t)u(t) - 2\cos\big[2\pi f(t-1)\big]u(t-1) + \cos\big[2\pi f(t-2)\big]u(t-2)$$

Since f is an integer, the phase delays become multiples of 2π and so

$$\frac{d^2y}{dt^2} = \cos(2\pi f t)\big[u(t) - 2u(t-1) + u(t-2)\big]$$

$$\frac{dy}{dt} = \frac{\sin(2\pi f t)}{2\pi f}\big[u(t) - 2u(t-1) + u(t-2)\big]$$

$$y(t) = \frac{1 - \cos(2\pi f t)}{(2\pi f)^2}\big[u(t) - 2u(t-1) + u(t-2)\big]$$

4.6 Filtering by Convolution

Filtering is a frequency selection operation. A filter allows signals within a frequency band (passband) to pass and attenuates or blocks those within another band (stopband). This may be performed by matching the signal with a template; for example, through convolution or correlation. The method is called *filtering in the time domain*. Here we illustrate the filtering operation in the time domain by two examples.

E*xample*
4.21

A signal is composed of the sum of N sinusoids:

$$x(t) = \sum_{k=1}^{N} A_k \sin(\omega_k t)$$

To extract a possible component at the frequency ω_0, we convolve $x(t)$ with $h(t) = \sin \omega_0 t$. The convolution outcome is

$$y(t) = \lim_{T \to \infty} \frac{1}{2T} \int_{-T}^{T} h(\tau)x(t-\tau)d\tau = \sum_{k=1}^{N} \lim_{T \to \infty} \frac{1}{2T} \int_{-T}^{T} A_k \sin \omega_0 \tau \sin \omega_k(t-\tau)d\tau$$

To evaluate the integral, we convert the product of sinusoids in the integrand to the sum of two cosines

$$\sin \omega_0 \tau \sin \omega_k(t - \tau) = \frac{1}{2}\left\{\cos[(\omega_0 + \omega_k)\tau - \omega_k t] - \cos[(\omega_0 - \omega_k)\tau + \omega_k t]\right\}$$

and examine the integrals. If $\omega_k \neq \omega_0$ for $k = 1, \ldots, N$, then all the integrals under the summation become zero, resulting in $y(t) = 0$. On the other hand, if one of the frequency components of $x(t)$ (e.g., ω_j) is equal to ω_0, then

$$y(t) = \lim_{T \to \infty} \frac{1}{2T} \int_{-T}^{T} \frac{A_j}{2} \sin \omega_0 \tau \sin \omega_0(t - \tau)d\tau = -\frac{A_j}{2} \cos(\omega_0 t)$$

Example **4.22**

Consider a 0.5-Hz periodic square wave $x(t)$ with a DC value of 0.5 and base-to-peak value of 0 to 1, where for one period

$$x(t) = \begin{cases} 1, & 0 \leq t < 1 \\ 0, & 1 \leq t < 2 \end{cases}$$

See Figure 4.14(a). To filter out frequencies other than 0.5 Hz, convolve $x(t)$ with a signal made of a single cycle of a 0.5-Hz sinusoid:

$$h(t) = \begin{cases} \sin(\pi t), & 0 \leq t < 2 \\ 0, & \text{elsewhere} \end{cases}$$

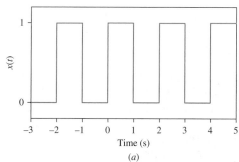
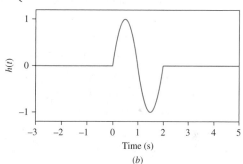
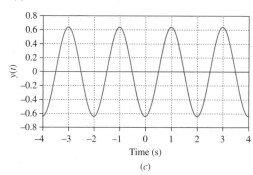

FIGURE 4.14 Convolving a 0.5-Hz square wave with a single cycle of a 0.5-Hz sinusoid extracts the principal harmonic of the square wave. *(a)* A periodic 0.5-Hz square wave $x(t)$. *(b)* A single cycle of a 0.5-Hz sinusoid $h(t) = \sin(\pi t)[u(t) - u(t-2)]$. *(c)* The convolution outcome $y(t) = h(t) \star x(t) = -\frac{2}{\pi}\cos(\pi t)$.

See Figure 4.14(*b*). The convolution output is

$$y(t) = \int_{-\infty}^{\infty} h(\tau)x(t-\tau)d\tau = \int_{0}^{2} \sin(\pi\tau)x(t-\tau)d\tau$$

$$= \begin{cases} \text{For } 0 < t < 1 \ y(t) = \int_{0}^{t} \sin(\pi\tau)d\tau + \int_{t+1}^{2}\sin(\pi\tau)d\tau = -\frac{2}{\pi}\cos(\pi t) \\ \text{For } 1 < t < 2 \ y(t) = \int (t-1)0^{t}\sin(\pi\tau)d\tau = -\frac{2}{\pi}\cos(\pi t) \end{cases}$$

$$= -\frac{2}{\pi}\cos(\pi t) \ \text{ all } t$$

The convolution outcome, $y(t)$, is a 0.5-Hz sinusoid, as in Figure 4.14(*c*). Convolution has filtered the 0.5-Hz component (called the *principal harmonic*) of the square wave signal. The step-by-step evaluation of the above integral is given in problem 6.

The filtering operations in Examples 4.21 and 4.22 were done by convolution. They used representations of the signals in the time domain and, therefore, are called time-domain filterings. Filtering may also be performed by representing the signal through its frequency components and then either selecting the desired or rejecting the undesired ones. This method is called *filtering in the frequency domain*. A signal's representation in the frequency domain provides a parallel path for signal analysis, which computationally can become much more efficient than in the time domain. Frequency-domain representation of signals is introduced in Chapters 7 and 8.

4.7 **Matched Filter**

The convolution $\eta(t)^\star\eta(T-t)$ concentrates the energy of the signal at $t = T$ and is called a matched filter. If the signal is corrupted by noise, the filter maximizes the ratio of the signal to the noise at the output. In that sense, the filter is optimum. This section presents the concept in detail. Further practice and exploration of the matched filter operation is given in the project at the end of the present chapter.

Let $\eta(t)$ be a continuous-time signal with a finite duration (from 0 to T). Outside that period $\eta(t) = 0$. Consider an LTI system with the unit-impulse response $h(t) = \eta(T-t)$. The unit-impulse response is constructed by time-reversing $\eta(t)$ and shifting it to the right by T units. Pass the signal through the system (where the filter is matched to the signal; hence, a *matched* filter) and examine the filter's output.

The output is

$$y(t) = \eta(t)^\star h(t) = \int_{0}^{t} \eta(\tau)h(t-\tau)d\tau$$

But,

$$h(t) = \eta(T-t)$$

Therefore,

$$y(t) = \int_{0}^{t} \eta(\tau)\eta[T-(t-\tau)]d\tau$$

At, $t = T$

$$y(T) = \int_0^T |\eta(\tau)|^2 \tau$$

The filter's output at $t = T$ is the energy in the signal. It may be shown that it is the maximum value of $y(t)$.

Example 4.23

Let $\eta(t)$ be a finite-duration continuous-time signal with amplitude ± 1. Switching between the ± 1 levels may occur only at multiples of θ seconds, where θ is the bit duration.

a. Consider the $\eta(t)$ shown in Figure 4.15(a) with seven bits, each bit being 1 second long ($\theta = 1$). Take an LTI system whose unit-impulse response is constructed by time-reversing $\eta(t)$ and shifting it to the right by $T = 7\theta$. The unit-impulse response obtained by these operations will be $h(t) = \eta(T - t)$, where $T = 7\theta$. Find the outcome of the convolution $y(t) = \eta(t) \star h(t)$.

b. Repeat for the 13-bit long signal
$$\eta(t) = \{\underset{\uparrow}{1}, 1, 1, 1, 1, -1, -1, 1, 1, -1, 1, -1, 1\}.$$

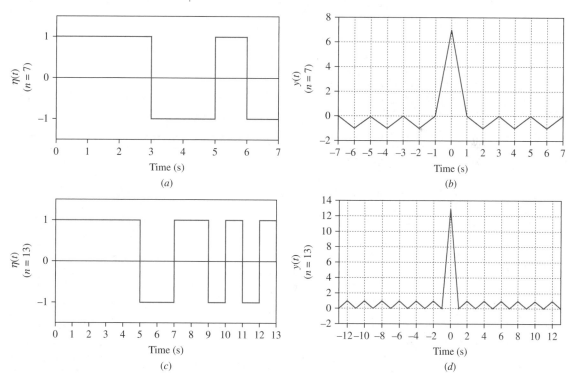

FIGURE 4.15 *(a)* A seven-segment signal. *(b)* Filter's output matched to the seven-segment signal. *(c)* A 13-segment signal. *(d)* Filter's output matched to the 13-segment signal.

Solution

The signal can be specified by its samples $\eta(n\theta)$ and for simplicity, be represented in discrete form as $\eta(n) = \{1, 1, 1, -1, -1, 1, -1\}$. Similarly, the unit-impulse response may be represented by $h(n) = \{-1, 1, -1, -1, 1, 1, 1\}$. Since $\eta(t)$ and $h(t)$ are constants during any interval $n\theta < t < (n+1)\theta$, the filter's output will be piecewise linear (see Example 4.14). We, therefore, first evaluate samples of the filter output $y(t)$ at $t = n\theta$, $-\infty < n < \infty$.

$$y(t) = \sum_{\tau=-\infty}^{\infty} \eta(\tau)h(t-\tau)d\tau = \begin{cases} \int_0^t \eta(\tau)h(t-\tau)d\tau, & 0 < t < T \\ \int_{t-T}^T \eta(\tau)h(t-\tau)d\tau, & T < t < 2T \\ 0, & \text{elsewhere} \end{cases}$$

$$y(n\theta) = \sum_{k=-\infty}^{\infty} \eta(k\theta)h(n\theta - k\theta)\theta$$

$$y(n) = \sum_{k=-\infty}^{\infty} \eta(k)h(n-k) = \begin{cases} \sum_0^n \eta(k)h(n-k), & 0 < n < 7 \\ \sum_{n-7}^7 \eta(k)h(n-k), & 7 < n < 14 \\ 0, & \text{elsewhere} \end{cases}$$

$y(n)$ is obtained by a convolution sum. Flip $h(k)$ around the origin and shift it by n units, then multiply term by term by $\eta(k)$ and add up. You will have $y(n)$. The operation is shown in Table 4.6.

TABLE 4.6 Calculation of $y(n) = \sum \eta(k)h(n-k)$ for $\eta(n) = \{1, 1, 1, -1, -1, 1, -1\}$

			−7	−6	−5	−4	−3	−2	−1	0	1	2	3	4	5	6	
	$\eta(k)$	$\Rightarrow$	0	0	0	0	0	0	0	1	1	1	−1	−1	1	−1	
	$h(k)$	$\rightarrow$	0	0	0	0	0	0	0	−1	1	−1	−1	1	1	1	
0	$h(-k)$	$\rightarrow$	1	1	1	−1	−1	1	−1	0	0	0	0	0	0	0	$y(0) = 0$
1	$h(1-k)$	$\rightarrow$	0	1	1	1	−1	−1	1	−1	0	0	0	0	0	0	$y(1) = -1$
2	$h(2-k)$	$\rightarrow$	0	0	1	1	1	−1	−1	1	−1	0	0	0	0	0	$y(2) = 0$
3	$h(3-k)$	$\rightarrow$	0	0	0	1	1	1	−1	−1	1	−1	0	0	0	0	$y(3) = -1$
4	$h(4-k)$	$\rightarrow$	0	0	0	0	1	1	1	−1	−1	1	−1	0	0	0	$y(4) = 0$
5	$h(5-k)$	$\rightarrow$	0	0	0	0	0	1	1	1	−1	−1	1	−1	0	0	$y(5) = -1$
6	$h(6-k)$	$\rightarrow$	0	0	0	0	0	0	1	1	1	−1	−1	1	−1	0	$y(6) = 0$
7	$h(7-k)$	$\rightarrow$	0	0	0	0	0	0	0	1	1	1	−1	−1	1	−1	$y(7) = 7$
8	$h(8-k)$	$\rightarrow$	0	0	0	0	0	0	0	0	1	1	1	−1	−1	1	$y(8) = 0$
9	$h(9-k)$	$\rightarrow$	0	0	0	0	0	0	0	0	0	1	1	1	−1	−1	$y(9) = -1$
10	$h(10-k)$	$\rightarrow$	0	0	0	0	0	0	0	0	0	0	1	1	1	−1	$y(10) = 0$
11	$h(11-k)$	$\rightarrow$	0	0	0	0	0	0	0	0	0	0	0	1	1	1	$y(11) = -1$
12	$h(12-k)$	$\rightarrow$	0	0	0	0	0	0	0	0	0	0	0	0	1	1	$y(12) = 0$
13	$h(13-k)$	$\rightarrow$	0	0	0	0	0	0	0	0	0	0	0	0	0	1	$y(13) = -1$
14	$h(14-k)$	$\rightarrow$	0	0	0	0	0	0	0	0	0	0	0	0	0	0	$y(10) = 0$

The leftmost column also shows, at rows 6, 7, 8, the markers $\uparrow$ n $\downarrow$.

Three sample calculations are given below:

$$y(3) = (1)(-1) + (1)(1) + (1)(-1) = -1$$
$$y(7) = (1)(1) + (1)(1) + (1)(1) + (-1)(-1) + (-1)(-1) + (1)(1) + (-1)(-1) = 7$$
$$y(10) = (-1)(1) + (-1)(1) + (1)(1) + (-1)(-1) = 0$$

Because $\eta(t)$ is a constant during any interval $nT < t < (n+1)T$, the filter's output is piecewise linear. Note that the filter output is at a maximum when $n = 7$. The filter has integrated the energy of the signal and placed it at the output at $t = T$. At other times, the filter's output is low. See Figure 4.15(b).

The 13-bit long signal $\eta(t) = \{1, 1, 1, 1, 1, -1, -1, 1, 1, -1, 1, -1, 1\}$ and the result of passing it through a filter matched to it are shown in Figure 4.15(c) and d, respectively. The maximum output is 13 and it occurs at $t = T$.

Repeat the above procedure for a code of length 15,

$$\eta(t) = \{1, 1, 1, -1, -1, -1, -1, 1, -1, 1, -1, -1, 1, 1, -1\}.$$

4.8 Deconvolution

Convolution of $x(t)$ and $h(t)$ produces $y(t)$. Can one reverse the operation, obtaining $h(t)$ from $x(t)$ and $y(t)$? In other words, can one extract the unit-impulse response of an LTI system from an input-output pair? The answer is yes, and one does so through an operation called deconvolution. It reverses the effect of the convolution and, therefore, is closely related to the inverse of the system. Here we illustrate the deconvolution concept by an example in which the functions involved change their values at unit-time steps, resulting in a convolution sum. More discussion and examples on these topics will be provided in Chapters 9 and 18.

Example 4.24

An input-output pair x_1, y_1 for a causal LTI system is given by

$$x_1 = \{1, -1\} \quad \to \quad y_1 = \{1, -1, 1\}$$

in Figure 4.16(a). Find the response of the system to the unit pulse $d(t) = u(t) - u(t - 1)$ and call it $\hat{h}(t)$. Then use $\hat{h}(t)$ to find the response to the input $x_2 = u(t) - u(t - 2)$.

Solution
Since time is discrete, we designate it by the integer variable n. Then the unit pulse $d(t) = u(t) - u(t - 1)$ is represented by $d(n)$ as shown in Figure 4.15(b). Let $\hat{h}(n)$ be the response to $d(n)$. Then,

$$x_2(n) = d(n) + d(n - 1)$$
$$y_2(n) = \hat{h}(n) + \hat{h}(n - 1)$$

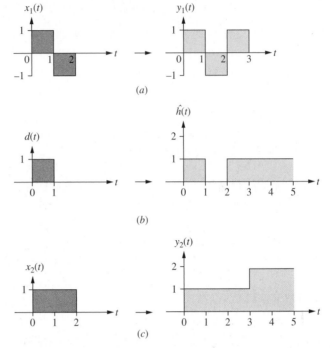

FIGURE 4.16 Deconvolution of the given input-output pair in *(a)* provides the response to the unit pulse *(b)* which is used to construct the reponse to x_2 *(c)*.

From the known pair $x_1(n)$ and $y_1(n)$ we search for $\hat{h}(n)$.

$$x_1(n) = d(n) - d(n-1)$$

$$y_1(n) = \hat{h}(n) - \hat{h}(n-1)$$

$$\hat{h}(n) = y_1(n) + \hat{h}(n-1), \ n \geq 0, \ \text{and} \ \hat{h}(n) = 0, \ n < 0$$

By successive application of the above rule, $\hat{h}(n)$ may be found to be as shown in Table 4.7.

TABLE 4.7 Finding $\hat{h}(n)$ by Successive Calculations

n	$x_1(n)$	$y_1(n)$	$\hat{h}(n-1)$	$\hat{h}(n) = y_1(n) + \hat{h}(n-1)$
-1	0	0	0	0
0	1	1	0	1
1	-1	-1	1	0
2	0	1	0	1
3	0	0	1	1
4	0	0	1	1
⋮	⋮	⋮	⋮	⋮

Now we find $y_2(n)$ as shown in Table 4.8.

TABLE 4.8 Finding $y(n)$ by Successive Calculations

n	$x_2(n) = d(n) + d(n-1)$	$y_2(n) = \hat{h}(n) + \hat{h}(n-1)$
-1	0	0
0	1	1
1	1	1
2	0	1
3	0	2
4	0	2
$\vdots$	$\vdots$	$\vdots$

The result is $y_2(t) = \{\underset{\uparrow}{1}, 1, 1, 2, 2, \cdots\}$.

Alternative Solution

In this example, $x_2(t)$ may be obtained by superposition of time-shifted $x_1(t)$ functions, allowing a direct construction of $y_2(t)$ from $y_1(t)$.

$$x_2(t) = \sum_{k=0}^{\infty}(k+1)x_1(t-k) \implies y_2(t) = \sum_{k=0}^{\infty}(k+1)y_1(t-k) = \{\underset{\uparrow}{1}, 1, 1, 2, 2, \cdots\}$$

4.9 Autocorrelation

The autocorrelation function, $r_{xx}(\tau)$, of an energy signal $x(t)$ is defined by

$$r_{xx}(\tau) = \int_{-\infty}^{\infty} x(t)x(t-\tau)dt, \quad -\infty < \tau < \infty$$

For a power signal the autocorrelation is given by

$$r_{xx}(\tau) = \lim_{T \to \infty} \frac{1}{2T} \int_{-T}^{T} x(t)x(t-\tau)dt, \quad -\infty < \tau < \infty$$

It is seen that autocorrelation is an even function: $r_{xx}(\tau) = r_{xx}(-\tau)$.

Example
4.25

Find the autocorrelation of the energy signal $x(t) = e^{-t}u(t)$.

Solution

$$r_{xx}(\tau) = \int_{-\infty}^{\infty} x(t)x(t-\tau)dt$$

$$= \int_{-\infty}^{\infty} \left[e^{-t}u(t)\right]\left[e^{-(t-\tau)}u(t-\tau)\right]dt = e^{\tau}\int_{0}^{\infty} e^{-2t}u(t-\tau)dt$$

$$\tau < 0, \quad r_{xx}(\tau) = e^{\tau}\int_{0}^{\infty} e^{-2t}dt = \frac{1}{2}e^{\tau}$$

$$\tau > 0, \quad r_{xx}(\tau) = e^{\tau}\int_{\tau}^{\infty} e^{-2t}dt = \frac{1}{2}e^{-\tau}$$

Hence,

$$r_{xx}(\tau) = \frac{1}{2}e^{-|\tau|} \quad \text{for } -\infty < \tau < \infty$$

Example
4.26

Find the autocorrelation of the power signal $x(t) = \cos 2\pi t$.

Solution

$$r_{xx}(\tau) = \lim_{T\to\infty}\frac{1}{2T}\int_{-T}^{T} \cos 2\pi t \cos 2\pi(t-\tau)dt$$

$$= \lim_{T\to\infty}\frac{1}{4T}\int_{-T}^{T} [\cos 2\pi(2t-\tau) + \cos 2\pi\tau]\,dt = \frac{1}{2}\cos(2\pi\tau)$$

Example
4.27

An early correlator

A tape recorder running at the speed of v cm/sec with two playback heads placed at a distance of d cm from each other produces a signal $x(t)$ and its delayed version $x(t-\tau)$, where $\tau = d/v$. By employing a multiplier and an integrator one can build a device to estimate the autocorrelation of $x(t)$ defined by

$$r_{xx}(\tau) = \lim_{T\to\infty}\frac{1}{2T}\int_{-T}^{T} x(t)x(t-\tau)dt$$

The physical settings for d may need to be adjusted for each desired time shift τ with the data played back at each new setting. By practical necessity, the duration of integration will be finite, resulting in an estimation:

$$r_{xx}(\tau) \approx \frac{1}{2T}\int_{-T}^{T} x(t)x(t-\tau)dt$$

Similar practical limitations apply when the correlation is obtained by numerical computation. In that case, the limiting factor is computation time. Another factor will be the finite length of the data available for computation. In the latter case, T (and τ) in the above integral will be chosen such that $x(t)$ and $x(t-\tau)$ exist during the integration period.

A parallel formulation is employed to find the autocorrelation of a discrete-time signal. It uses summation rather than integration:

$$r_{xx}(k) \approx \frac{1}{(2M+1)} \sum_{n=-M}^{M} x(n)x(n-k)$$

k is the amount of shift and $2M+1$ is the data length over which the summation operation is performed. If the mean value of $x(n)$ is excluded from summation, the result is called the *covariance function* of the signal. An estimate of the covariance function is given by

$$\text{Cov}_{xx}(k) \approx \frac{1}{(2M+1)} \sum_{n=-M}^{M} [x(n) - \overline{X}][x(n-k) - \overline{X}]$$

where

$$\overline{X} = \frac{1}{(2M+1)} \sum_{n=-M}^{M} x(n)$$

is an estimate of the mean value of $x(n)$. If finite data of size N is available, an unbiased estimate of the covariance function is obtained from

$$\text{Cov}_{xx}(k) \approx \frac{1}{(N-k)} \sum_{n=0}^{N-k} [x(n) - \overline{X}][x(n+k) - \overline{X}], \quad \text{Cov}_{xx}(k) = \text{Cov}_{xx}(-k)$$

Examples 4.28 and 4.29 illustrate two such cases.

Example **4.28**

Obtain and plot the covariance function for the number of annual sunspots.

Solution
The time series of annual sunspot numbers for the years 1700–2010 (Figure 4.17a, also Figure 1.1 in Chapter 1) has 305 data points. An estimate of the covariance as a function of the shift $k \geq 0$ is given by

$$\text{Cov}_{xx}(k) \approx \frac{1}{(305-k)} \sum_{n=0}^{305-k} [x(n) - \overline{X}][x(n+k) - \overline{X}]$$

The estimation error increases as k in the above summation is increased. $\text{Cov}_{xx}(k)$ is computed for $0 \leq k \leq 150$ and plotted in Figure 4.17(b). As expected, the plot

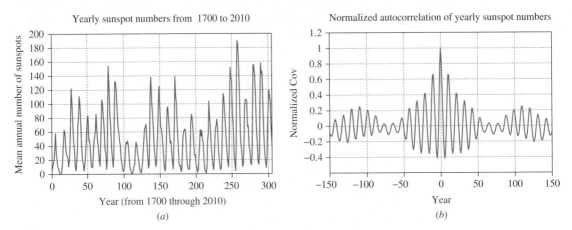

FIGURE 4.17 *(a)* Annual sunspot numbers for the years 1700–2010 and *(b)* unbiased estimate of its covariance function.
Source: 4.17 *(a)* NOAA's National Geophysical Data Center (NGDC) at www.ngdc.noaa.gov.

of Figure 4.17(*b*) is an even function. It shows periodicity of the data with a period of about 10 years. The waxing and waning of the numbers over a 100-year period is also reflected in the envelope of the covariance function.

Note: Problem 7 contains the Matlab code for the numerical evaluations and plots in this example.

Find the covariance function of monthly and yearly rates of unemployment in the United States. The unemployment rate is defined by the U.S. Labor Department as a percentage of the labor force aged 16 years and above. The data is available at www.bls.gov/data/.

Solution

Monthly unemployment rates from January 1968 through December 2010 (a total of 516 months) and an unbiased estimate of their covariance function are plotted in Figure 4.18(*a*). Yearly unemployment rates from 1940 through 2009 (a total of 70 years) and an unbiased estimate of their covariance function are plotted in Figure 4.18(*b*). In both cases the main lobes in the covariance functions indicate a correlation between successive data points (monthly and yearly) which gradually diminishes with the time lapse between them. The periodicity of annual rates (1940–2010) is reflected as periodicity in the covariance function.

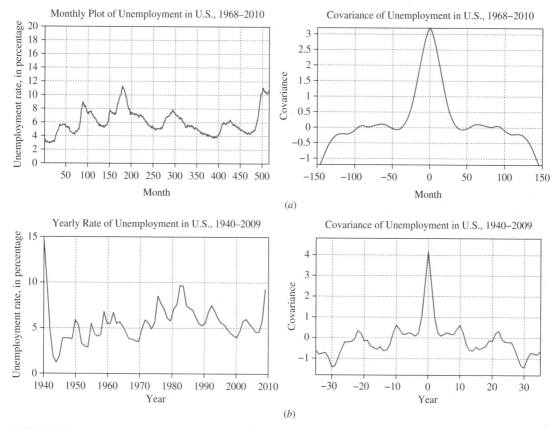

FIGURE 4.18 *(a)* Monthly U.S. unemployment rate for January 1968 through December 2010 and its covariance. *(b)* Annual U.S. unemployment rate for 1940 through 2009 and its covariance.
Source: Bureau of Labor Statistics, http://data.bls.gov.

Note: Problem 8 contains the Matlab code for the numerical evaluations and plots in this example.

4.10 Cross-Correlation

The cross-correlation function $r_{xy}(\tau)$ between two energy signals $x(t)$ and $y(t)$ is defined by

$$r_{xy}(\tau) = \int_{-\infty}^{\infty} x(t)y(t-\tau)dt, \quad -\infty < \tau < \infty$$

For power signals we scale the integral:

$$r_{xy}(\tau) = \lim_{T\to\infty} \frac{1}{2T} \int_{-T}^{T} x(t)y(t-\tau)dt, \quad -\infty < \tau < \infty$$

It is seen that $r_{xy}(\tau) = r_{yx}(-\tau)$. As in the autocorrelation operation, one of the signals is shifted, multiplied by the other, and then integrated.

Example 4.30

Find the cross-correlations $r_{xy}(\tau)$ and $r_{yx}(\tau)$, where $x(t) = \cos t$ and $y(t) = e^{-t}u(t)$. Show that $r_{xy}(\tau) = r_{yx}(-\tau)$.

Solution

$$r_{xy}(\tau) = \int_{-\infty}^{\infty} x(t)y(t-\tau)dt = \int_{-\infty}^{\infty} \cos t \left[e^{-(t-\tau)}u(t-\tau) \right] dt$$

$$= e^{\tau} \int_{\tau}^{\infty} e^{-t} \cos t \, dt = \frac{1}{2}(\cos \tau - \sin \tau)$$

$$r_{yx}(\tau) = \int_{-\infty}^{\infty} y(t)x(t-\tau)dt = \int_{-\infty}^{\infty} \left[e^{-t}u(t) \right] \cos(t-\tau)dt$$

$$= \int_{0}^{\infty} e^{-t} \cos(t-\tau)dt = \frac{1}{2}(\cos \tau + \sin \tau)$$

$$r_{xy}(\tau) = r_{yx}(-\tau)$$

Correlation operations enhance similarities within a signal or between two signals. The autocorrelation exposes periodicities in signals that are weak, are masked by other features, or are embedded in noise and disturbances. This is because shifting a periodic signal by multiples of its period reverts the signal to the original form and creates a periodic component in the correlation function (with the same period). Similarly, the cross-correlation between two signals serves as a method for matching a signal with a template. It is, therefore, used to detect signals of interest embedded in noise and hence the method is called *correlation detection*.

Example 4.31

Let a signal be composed of the sum of N sinusoids:

$$x(t) = \sum_{k=1}^{N} \sin(\omega_k t)$$

Each ω_k is a fixed but unknown frequency. To find out if the signal contains a certain frequency ω_0, we cross-correlate $x(t)$ with $h(t) = \sin \omega_0 t$ at zero shift.

$$r_{xh}(0) = \lim_{T \to \infty} \frac{1}{2T} \int_{-T}^{T} x(t) \sin(\omega_0 t)dt = \lim_{T \to \infty} \frac{1}{2T} \sum_{k=1}^{N} \int_{-T}^{T} \sin(\omega_0 t) \sin(\omega_k t)dt$$

$$= \lim_{T \to \infty} \frac{1}{4T} \sum_{k=1}^{N} \int_{-T}^{T} \left[\cos(\omega_0 - \omega_k)t - \cos(\omega_0 + \omega_k)t \right] dt \approx \begin{cases} \frac{1}{2} & \text{if } \omega_k = \omega_0 \\ 0 & \text{if } \omega_k \neq \omega_0 \end{cases}$$

A nonzero output indicates the presence of the ω_0 component in $x(t)$.

Example

4.32

Cross-covariance between unemployment rates in the United States and in California

Monthly unemployment rates in California and the rest of the United States from 1976 through 2010 are plotted in Figure 4.19(*a*) Obtain an unbiased estimate of their cross-covariance function and plot it.

Solution

An unbiased estimate of the cross-covariance function for a finite data of size N is obtained from

$$\text{Xcov}_{xy}(k) \approx \frac{1}{(N-k)} \sum_{n=0}^{N-k} \left[x(n) - \overline{X} \right] \left[x(n-k) - \overline{X} \right],$$

$$\text{Xcov}_{xy}(k) = \text{Xcov}_{yx}(-k)$$

In the present case, $N = 420$. The cross-covariance functions $\text{Xcov}_{xy}(k)$ and $\text{Xcov}_{yx}(k)$, where x and y are monthly unemployment rates for California and the United States respectively, are shown in Figure 4.19(*b*), with a maximum shift of 150. The rates in California and in the United States are strongly correlated. Note that $\text{Xcov}_{xy}(k) = \text{Xcov}_{yx}(-k)$.

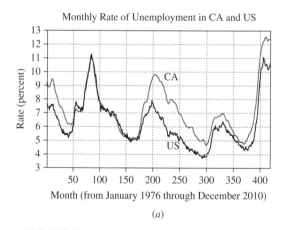

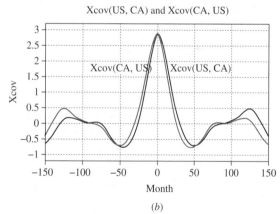

(*a*)   (*b*)

FIGURE 4.19 *(a)* Plot of monthly unemployment rates in California and in the United States from 1976 through 2010 and *(b)* their cross-covariance functions.

Source: 4.19 *(a)* Bureau of Labor Statistics, http://data.bls.gov.

Note: Problem 9 contains the Matlab code for the numerical evaluations and plots in this example.

4.11   **Correlation and Convolution**

Convolution and correlation are two related operations. The autocorrelation function of an energy signal $x(t)$ is defined by

$$r_{xx}(t) = \int_{-\infty}^{\infty} x(\tau)x(\tau - t)d\tau, \quad -\infty < \tau < \infty$$

In section 4.7 the filter matched to an energy signal $x(t)$ was defined by its impulse response $h(t) = x(T - t)$. Here we will show that the output of the matched filter is the autocorrelation of $x(t)$ shifted to the right by T seconds. The output of the matched filter to the input $x(t)$ is

$$y(t) = \int_{-\infty}^{\infty} x(t - \tau)h(\tau)d\tau = \int_{-\infty}^{\infty} x(t - \tau)x(T - \tau)d\tau = r_{xx}(t - T)$$

The matched filter takes the autocorrelation of the signal. A detector based on a matched filter (as in the project at the end of the present chapter) is, therefore, called a *correlation detector*.

For an energy signal that can be represented in discrete form as $\eta(n)$, the autocorrelation is given by

$$r_{\eta\eta}(n) = \sum_{k=-\infty}^{\infty} \eta(k)\eta(k - n)$$

Find the autocorrelation function of $\eta(n) = \{1, 1, 1, -1, -1, 1, -1\}$ and compare with the output of the matched filter in Example 4.23.

Solution

Shift $\eta(k)$ by n units (to the left or right) to find $\eta(k - n)$. Then multiply by $\eta(k)$ term by term and add up. You will then have $r_{\eta\eta}(n)$. The operation is shown in Table 4.9.
    Three sample calculations are given below:

$$r_{\eta\eta}(-4) = (1)(-1) + (1)(1) + (1)(-1) = -1$$

$$r_{\eta\eta}(0) = (1)(1) + (1)(1) + (1)(1) + (-1)(-1) + (-1)(-1)$$

$$+ (1)(1) + (-1)(-1) = 7$$

$$r_{\eta\eta}(3) = (-1)(1) + (-1)(1) + (1)(1) + (-1)(-1) = 0$$

TABLE 4.9 Calculation of $r_{\eta\eta}(n) = \sum \eta(k)\eta(k-n)$

			-7	-6	-5	-4	-3	-2	$\leftarrow$ -1	k $\rightarrow$ 0	1	2	3	4	5	6	
	$\eta(k)$	$\Rightarrow$	0	0	0	0	0	0	0	1	1	1	-1	-1	1	-1	
-7	$\eta(k+7)$	$\rightarrow$	1	1	1	-1	-1	1	-1	0	0	0	0	0	0	0	$r_{\eta\eta}(-7)=0$
-6	$\eta(k+6)$	$\rightarrow$	0	1	1	1	-1	-1	1	-1	0	0	0	0	0	0	$r_{\eta\eta}(-6)=-1$
-5	$\eta(k+5)$	$\rightarrow$	0	0	1	1	1	-1	-1	1	-1	0	0	0	0	0	$r_{\eta\eta}(-5)=0$
-4	$\eta(k+4)$	$\rightarrow$	0	0	0	1	1	1	-1	-1	1	-1	0	0	0	0	$r_{\eta\eta}(-4)=-1$
-3	$\eta(k+3)$	$\rightarrow$	0	0	0	0	1	1	1	-1	-1	1	-1	0	0	0	$r_{\eta\eta}(-3)=0$
-2	$\eta(k+2)$	$\rightarrow$	0	0	0	0	0	1	1	1	-1	-1	1	-1	0	0	$r_{\eta\eta}(-2)=-1$
$\uparrow$ -1	$\eta(k+1)$	$\rightarrow$	0	0	0	0	0	0	1	1	1	-1	-1	1	-1	0	$r_{\eta\eta}(-1)=0$
n 0	$\eta(k)$	$\rightarrow$	0	0	0	0	0	0	0	1	1	1	-1	-1	1	-1	$r_{\eta\eta}(0)=7$
$\downarrow$ 1	$\eta(k-1)$	$\rightarrow$	0	0	0	0	0	0	0	0	1	1	1	-1	-1	1	$r_{\eta\eta}(1)=0$
2	$\eta(k-2)$	$\rightarrow$	0	0	0	0	0	0	0	0	0	1	1	1	-1	-1	$r_{\eta\eta}(2)=-1$
3	$\eta(k-3)$	$\rightarrow$	0	0	0	0	0	0	0	0	0	0	1	1	1	-1	$r_{\eta\eta}(3)=0$
4	$\eta(k-4)$	$\rightarrow$	0	0	0	0	0	0	0	0	0	0	0	1	1	1	$r_{\eta\eta}(4)=-1$
5	$\eta(k-5)$	$\rightarrow$	0	0	0	0	0	0	0	0	0	0	0	0	1	1	$r_{\eta\eta}(5)=0$
6	$\eta(k-6)$	$\rightarrow$	0	0	0	0	0	0	0	0	0	0	0	0	0	1	$r_{\eta\eta}(6)=-1$
7	$\eta(k-7)$	$\rightarrow$	0	0	0	0	0	0	0	0	0	0	0	0	0	0	$r_{\eta\eta}(7)=0$

The autocorrelation function is at a maximum when $n = 0$; that is, with no shift, when the operation integrates the energy of the signal. Note the close correspondence, except for a shift of 7 units, between Tables 4.6 and 4.9, resulting in $y(n) = r_{\eta\eta}(n-7)$.

4.12 Concluding Remarks

In theory, convolution is a mathematical operation defined by equation (1) at the beginning of this chapter. It can be applied routinely, as it requires no more than an introductory course in calculus. The same observation applies to correlation operations. In practice, however, applying the convolution or correlation integrals to signals of common interest (especially those which are sectionwise continuous, discontinuous, or contain singularity functions) may become a challenge that can only be overcome by a conceptual and intuitive understanding of the operation. This chapter has attempted to present a balanced mix of theoretical and conceptual understandings of the topic and illustrate those through many examples. A point of departure from theory often is the fact that in many practical situations the signals involved are given in the form of tabular data and not as analytic functions. In such situations, computers become vital tools. The data needs to be read in, a computer program needs to evaluate the convolution/correlation integral, and the result needs to be plotted and stored. Several examples and solved problems that use Matlab programs are provided to help the student with an introductory experience in this area. The project included in this chapter offers further practice and explorations in signal detection by convolution and correlation.

4.13 Problems

Solved Problems

1. The unit-impulse response of an LTI system is $d(t) = u(t) - u(t-1)$. Find and plot its output for the periodic input $x(t)$, where, for one period T

a. $x_1(t) = \begin{cases} 2, & 0 \leq t < 1 \\ 0, & 1 \leq t < 6 \end{cases}$ Period $T = 6$ b. $x_2(t) = \begin{cases} 3, & 0 \leq t < 1 \\ -1, & 1 \leq t < 2 \end{cases}$ Period $T = 2$

Hint: Use the graphical method.

Solution

a. Express $x_1(t)$ in terms of $d(t)$ so that

$$x_1(t) = \sum_{k=-\infty}^{\infty} 2d(t - 6k)$$

Let

$$\zeta(t) = d(t) \star d(t) = \begin{cases} t, & 0 \leq t < 1 \\ 2 - t, & 1 \leq t < 2 \\ 0, & \text{elsewhere} \end{cases}$$

Then,

$$y_1(t) = \sum_{k=-\infty}^{\infty} 2\zeta(t - 6k) = \begin{cases} 2t, & 0 \leq t < 1 \\ 4 - 2t, & 1 \leq t < 2 \\ 0, & 2 \leq t < 6 \end{cases} \quad \text{Period } T = 6$$

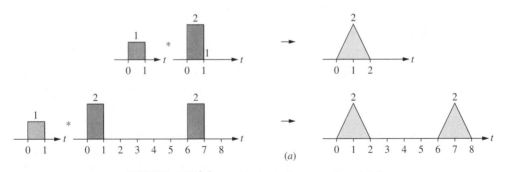

(a)

FIGURE 4.20(a) Graphical solution to problem 1 *(a)*.

b. Let

$$\eta(t) = 3d(t) - d(t-1) \text{ and } \xi(t) = \eta(t) \star d(t) = \begin{cases} 3t, & 0 \leq t < 1 \\ 7 - 4t, & 1 \leq t < 2 \\ t - 3, & 2 \leq t < 3 \\ 0, & \text{elsewhere} \end{cases}$$

Then,

$$x_2(t) = \sum_{k=-\infty}^{\infty} \eta(t-2k) \text{ and } y_2(t) = \sum_{k=-\infty}^{\infty} \xi(t-2k) = \begin{cases} -1 + 4t, & 0 < t < 1 \\ 7 - 4t, & 1 < t < 2 \end{cases} \quad \text{Period } T = 2. \text{ See Figure 4.20}(b).$$

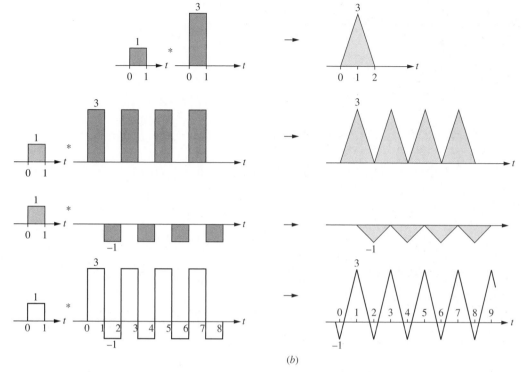

(b)

FIGURE 4.20(b) Graphical solution to problem 1 *(b)*.

2. Find $y(t) = x(t) \star h(t)$, where

a. $h(t) = e^{-3t}u(t)$ and $x(t) = 1$
b. $h(t) = e^{-3t}u(t)$ and $x(t) = u(t-2)$
c. $h(t) = e^{-3t}u(t)$ and $x(t) = e^{-t}u(-t)$
d. $h(t) = \delta(t-2) + u(t)$ and $x(t) = (\cos t)u(t)$

Solution

a. $h(t) = e^{-3t}u(t)$, $x(t) = 1$ $y(t) = \int_{-\infty}^{\infty} x(\tau)h(t-\tau)d\tau = \int_{-\infty}^{t} e^{-3(t-\tau)}d\tau = e^{-3t}\int_{-\infty}^{t} e^{3\tau}d\tau = \frac{1}{3}$

Alternatively,

$$y(t) = \int_{-\infty}^{\infty} h(\tau)x(t-\tau)d\tau = \int_{0}^{\infty} e^{-3\tau}d\tau = \frac{1}{3}$$

b. $h(t) = e^{-3t}u(t)$, $x(t) = u(t-2)$ $y(t) = \int_{-\infty}^{\infty} x(\tau)h(t-\tau)d\tau = \int_{2}^{t} e^{-3(t-\tau)}d\tau = \frac{1}{3}[1 - e^{-3(t-2)}]u(t-2)$

Alternatively,

$$y(t) = \int_{-\infty}^{\infty} h(\tau)x(t-\tau)d\tau = \int_{0}^{t-2} e^{-3\tau}d\tau = \frac{1}{3}[1 - e^{-3(t-2)}]u(t-2)$$

c. $h(t) = e^{-3t}u(t), \ x(t) = e^{-t}u(-t) \quad y(t) = \displaystyle\int_{-\infty}^{\infty} x(\tau)h(t-\tau)d\tau = \begin{cases} \int_{-\infty}^{t} e^{-3(t-\tau)}e^{-\tau}d\tau = \frac{1}{2}e^{-t} & t < 0 \\ \int_{-\infty}^{0} e^{-3(t-\tau)}e^{-\tau}d\tau = \frac{1}{2}e^{-3t} & t > 0 \end{cases}$

Alternatively,

$$y(t) = \int_{-\infty}^{\infty} h(\tau)x(t-\tau)d\tau = \begin{cases} \int_{0}^{\infty} e^{-3\tau}e^{-(t-\tau)}d\tau = \frac{1}{2}e^{-t} & t < 0 \\ \int_{t}^{\infty} e^{-3\tau}e^{-(t-\tau)}d\tau = \frac{1}{2}e^{-3t} & t > 0 \end{cases}$$

d. $\qquad h(t) = \delta(t-2) + u(t), \ x(t) = (\cos t)u(t) \qquad y(t) = x(t-2) + \displaystyle\int_{-\infty}^{t} x(\tau)d\tau$

$$= [\cos(t-2)]u(t-2) + \int_{0}^{t} (\cos \tau)d\tau$$

$$= [\cos(t-2)]u(t-2) + (\sin t)u(t)$$

3. The response of an LTI system to $u(t)$ is $g(t) = \left(2e^{-t} - e^{-5t}\right)u(t)$. Find its responses to

a. $x(t) = \delta(t)$ \qquad b. $x(t) = 1$ \qquad c. $x(t) = \displaystyle\int_{-\infty}^{t} u(\tau)d\tau$

Solution

a. $x(t) = \delta(t) \qquad \Longrightarrow y(t) = \dfrac{d}{dt}g(t) = (5e^{-5t} - 2e^{-t})u(t) + (2e^{-t} - e^{-5t})\Big|_{t=0} \delta(t)$

$$= (5e^{-5t} - 2e^{-t})u(t) + \delta(t)$$

b. $x(t) = 1 \qquad \Longrightarrow y(t) = \lim_{t \to \infty} g(t) = 0$

c. $x(t) = \displaystyle\int_{-\infty}^{t} u(\tau)d\tau \Longrightarrow y(t) = \int_{-\infty}^{t} g(\tau)d\tau = \left[\int_{0}^{t} (2e^{-\tau} - e^{-5\tau})d\tau\right]u(t)$

$$= (1.8 - 2e^{-t} + 0.2e^{-5t})u(t)$$

4. The response of an LTI system to $x(t) = u(t)$ is $y(t) = e^{-2t}u(t)$. Find its responses to

a. $x(t) + x(t-2)$ \qquad b. $3x(t-1) + \displaystyle\int_{-\infty}^{t} x(\tau)d\tau$ \qquad c. $\dfrac{dx(t-1)}{dt}$ \qquad d. $\dfrac{dx(t)}{dt} + x(t)$

Solution

a. $x(t) + x(t-2) \qquad \Longrightarrow y(t) = e^{-2t}u(t) + e^{-2(t-2)}u(t-2)$

b. $3x(t-1) + \displaystyle\int_{-\infty}^{t} x(\tau)d\tau \Longrightarrow y(t) = 3e^{-2(t-1)}u(t-1) + \int_{0}^{t} e^{-2\tau}d\tau$

$$= 3e^{-2(t-1)}u(t-1) + \frac{1}{2}\left(1 - e^{-2t}\right)u(t)$$

c. $\dfrac{dx(t-1)}{dt} \qquad \Longrightarrow y(t) = \dfrac{d}{dt}\left[e^{-2(t-1)}u(t-1)\right] = -2e^{-2(t-1)}u(t-1) + \delta(t-1)$

d. $\dfrac{dx(t)}{dt} + x(t) \qquad \Longrightarrow y(t) = \dfrac{d}{dt}\left[e^{-2t}u(t)\right] + e^{-2t}u(t)$

$$= -2e^{-2t}u(t) + e^{-2t}\Big|_{t=0} \delta(t) + e^{-2t}u(t) = \delta(t) - e^{-2t}u(t)$$

5. Evaluate and plot the convolution integral $y(t) = x(t) \star h(t)$, where

a. $x(t) = \sin(2\pi t)u(t)$ and $h(t) = u(t)$
b. $x(t) = e^{-t/50}u(t)$ and $h(t) = e^{-2t}u(t)$
c. $x(t) = e^{-t}u(-t)$ and $h(t) = e^{-2t}u(t)$
d. $x(t) = e^{t}u(-t)$ and $h(t) = e^{-2t}u(t)$
e. $x(t) = e^{-3t}u(-t)$ and $h(t) = e^{-t}u(t)$
f. $x(t) = e^{-|t|}$ and $h(t) = e^{-t/5}u(t)$

Provide a sample Matlab code for the above operations.

Hint: Evaluate $\left[e^{-\alpha t}u(t)\right] \star \left[e^{-\beta t}u(t)\right]$ and $\left[e^{-\alpha t}u(-t)\right] \star \left[e^{-\beta t}u(t)\right]$. In each case observe the condition(s) for the existence of the integral.

Note 1. Convolution of two right-sided exponential functions.

$$\left[e^{-\alpha t}u(t)\right] \star \left[e^{-\beta t}u(t)\right] = \int_{-\infty}^{\infty} \left[e^{-\alpha \tau}u(\tau)\right]\left[e^{-\beta(t-\tau)}u(t-\tau)\right]d\tau = \int_{0}^{t} e^{-\alpha\tau}e^{-\beta(t-\tau)}d\tau = \frac{e^{-\alpha t} - e^{-\beta t}}{\beta - \alpha}u(t)$$

Note 2. Convolution of a left-sided and a right-sided exponential function. It exists only if $\alpha + \beta > 0$.

$$\left[e^{\alpha t}u(-t)\right] \star \left[e^{-\beta t}u(t)\right] = \begin{cases} \int_{-\infty}^{t} e^{\alpha\tau}e^{-\beta(t-\tau)}d\tau = \frac{e^{\alpha t}}{\alpha+\beta}, & t \leq 0 \\ \\ \int_{-\infty}^{0} e^{\alpha\tau}e^{-\beta(t-\tau)}d\tau = \frac{e^{-\beta t}}{\alpha+\beta}, & t \geq 0 \end{cases}$$

$$= \frac{1}{\alpha+\beta}\left[e^{\alpha t}u(-t) + e^{-\beta t}u(t)\right], \quad \alpha + \beta > 0$$

Alternatively,

$$\left[e^{\alpha t}u(-t)\right] \star \left[e^{-\beta t}u(t)\right] = \begin{cases} \int_{0}^{\infty} e^{-\beta\tau}e^{\alpha(t-\tau)}d\tau = \frac{e^{\alpha t}}{\alpha+\beta}, & t \leq 0 \\ \\ \int_{t}^{\infty} e^{-\beta\tau}e^{\alpha(t-\tau)}d\tau = \frac{e^{-\beta t}}{\alpha+\beta}, & t \geq 0 \end{cases}$$

$$= \frac{1}{\alpha+\beta}\left[e^{\alpha t}u(-t) + e^{-\beta t}u(t)\right], \quad \alpha + \beta > 0$$

Note 3. Convolution of a two-sided and a right-sided exponential function. It exists only if $\alpha + \beta > 0$.

$$\left[e^{-\alpha|t|}\right] \star \left[e^{-\beta t}u(t)\right] = \left[e^{\alpha t}u(-t)\right] \star \left[e^{-\beta t}u(t)\right] + \left[e^{-\alpha t}u(t)\right] \star \left[e^{-\beta t}u(t)\right]$$

$$= \frac{1}{\alpha+\beta}\left[e^{\alpha t}u(-t) + e^{-\beta t}u(t)\right] + \frac{e^{-\alpha t} - e^{-\beta t}}{\beta - \alpha}u(t)$$

$$= \frac{e^{\alpha t}}{\alpha+\beta}u(-t) + \left(\frac{2\alpha e^{-\beta t}}{\alpha^2 - \beta^2} + \frac{e^{-\alpha t}}{\beta - \alpha}\right)u(t)$$

Solution

Answers to parts a to f and plots of $y(t)$ are given below.

a. $\sin(2\pi t)u(t) \star u(t) = 0.159(1 - \cos 2\pi t)u(t)$.
b. $e^{-t/50}u(t) \star e^{-2t}u(t) = 0.5051\left(e^{-t/50} - e^{-2t}\right)u(t)$.
c. $e^{-t}u(-t) \star e^{-2t}u(t) = e^{-t}u(-t) + e^{-2t}u(t)$.
d. $e^{t}u(-t) \star e^{-2t}u(t) = \frac{1}{3}[e^{t}u(-t) + e^{-2t}u(t)]$.

e. $e^{-3t}u(-t) \star e^{-t}u(t) = \infty$. (It doesn't exist. However, a plot is produced by Matlab for a finite data segment.)

f. $e^{-|t|} \star e^{-t/5}u(t) = e^{t}u(-t) \star e^{-t/5}u(t) + e^{-t}u(t) \star e^{-t/5}u(t)$.

Here is the sample Matlab program:

```
%Part a, x(t)=sin(2pit) u(t), h(t)=u(t)
t=[-1:0.01:3.5];
x=0*t;
h=0;
z=0*t;
for i=100:450
    x(i)=sin(2*pi*t(i));
    h(i)=1;
end
figure(1)
y=conv(x,h)/100;
ty=-2:0.01:7-0.01;
plot(ty,y,'LineWidth',2)
grid
xlabel('Time (s)');
ylabel('Magnitude' );
title('Plot of  convolution y(t)=sin(2\pit)*u(t).');
axis([-1  3.5  -.1  .4])
```

6. Evaluate the convolution integral of Example 4.22 for $t = 0$ to 2 at incremental steps of 0.25 second each.

Solution

See Figure 4.21.

(a)
$$\begin{cases} x(t) = \begin{cases} 1, & 0 < t < 1 \\ 0, & 1 < t < 2 \end{cases}, \; x(t) = x(t-2). \\[2em] h(t) = \begin{cases} \sin(\pi t), & 0 < t < 2 \\ 0, & \text{elsewhere} \end{cases} \\[2em] y(t) = \int h(\tau)x(t-\tau)d\tau \end{cases}$$

(a)

FIGURE 4.21 Details of the convolution of a single cycle of a sinusoid $h(t)$ with a square wave $x(t)$. **(a)** $h(t)$, $x(t)$, and their superimposed plots. Rows **(b)** to **(j)** show $h(\tau)$, $x(t - \tau)$, and their product for nine values of $0 \leq t \leq 2$, at increments of 0.25. In each case, the convolution outcome is the area of the shaded region under the curve $h(\tau)x(t - \tau)$. Its numerical value is given for each t.

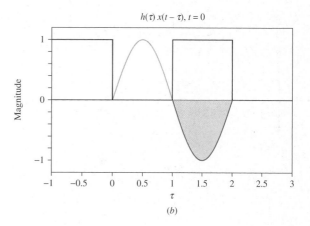

(b) $\begin{cases} t = 0, \\ y = \int h(\tau)x(0 - \tau)d\tau \\ \quad = \int_{1}^{2} \sin(\pi t)dt = -\dfrac{2}{\pi} \end{cases}$

$h(\tau)\,x(t - \tau),\, t = 0$

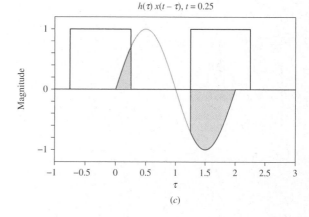

(c) $\begin{cases} t = 0.25, \\ y = \int h(\tau)x(0.25 - \tau)d\tau \\ \quad = \int_{0}^{0.25} \sin(\pi t)dt + \int_{1.25}^{2} \sin(\pi t)dt = -\dfrac{\sqrt{2}}{\pi} \end{cases}$

$h(\tau)\,x(t - \tau),\, t = 0.25$

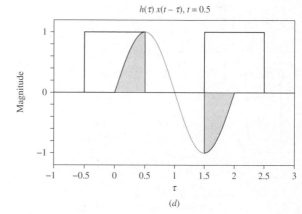

(d) $\begin{cases} t = 0.5, \\ y = \int h(\tau)x(0.5 - \tau)d\tau \\ \quad = \int_{0}^{0.5} \sin(\pi t)dt + \int_{1.5}^{2} \sin(\pi t)dt = 0 \end{cases}$

$h(\tau)\,x(t - \tau),\, t = 0.5$

FIGURE 4.21 (*Continued*)

(e) $\begin{cases} t = 0.75, \\ y = \displaystyle\int h(\tau)x(0.75 - \tau)d\tau \\ \quad = \displaystyle\int_0^{0.75} \sin(\pi t)dt + \int_{1.75}^2 \sin(\pi t)dt = \dfrac{\sqrt{2}}{\pi} \end{cases}$

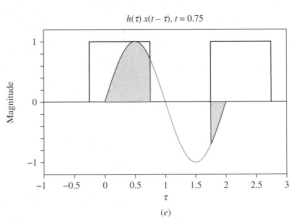

(e)

(f) $\begin{cases} t = 1, \\ y = \displaystyle\int h(\tau)x(1 - \tau)d\tau \\ \quad = \displaystyle\int_0^1 \sin(\pi t)dt = \dfrac{2}{\pi} \end{cases}$

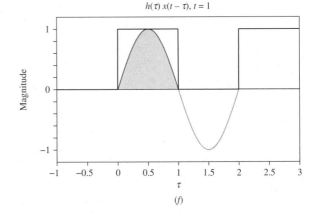

(f)

(g) $\begin{cases} t = 1.25, \\ y = \displaystyle\int h(\tau)x(1.25 - \tau)d\tau \\ \quad = \displaystyle\int_{0.25}^{1.25} \sin(\pi t)dt = \dfrac{\sqrt{2}}{\pi} \end{cases}$

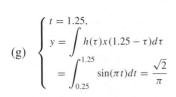

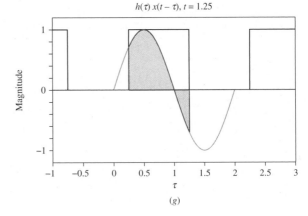

(g)

FIGURE 4.21 (*Continued*)

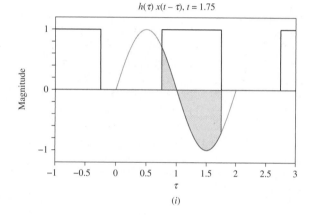

(h) $\begin{cases} t = 1.5, \\[2mm] y = \displaystyle\int h(\tau)x(1.5 - \tau)d\tau \\[4mm] = \displaystyle\int_{0.5}^{1.5} \sin(\pi t)dt = 0 \end{cases}$

$h(\tau) x(t - \tau), t = 1.5$

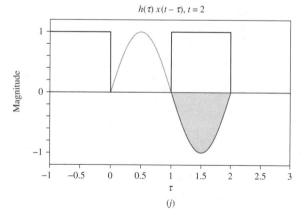

(i) $\begin{cases} t = 1.75, \\[2mm] y = \displaystyle\int h(\tau)x(1.75 - \tau)d\tau \\[4mm] = \displaystyle\int_{0.75}^{1.75} \sin(\pi t)dt = -\dfrac{\sqrt{2}}{\pi} \end{cases}$

$h(\tau) x(t - \tau), t = 1.75$

(j) $\begin{cases} t = 2, \\[2mm] y = \displaystyle\int h(\tau)x(2 - \tau)d\tau \\[4mm] = \displaystyle\int_{1}^{2} \sin(\pi t)dt = -\dfrac{2}{\pi} \end{cases}$

$h(\tau) x(t - \tau), t = 2$

FIGURE 4.21 (*Continued*)

7. Write Matlab code to read the annual sunspot numbers for the years 1700–2010, and compute and plot the covariance function for them. See Example 4.28.

Solution

Obtain the time series of annual sunspot numbers for the years 1700–2010 and place it in a data file. [Source: NOAA's National Geophysical Data Center (NGDC) at www.ngdc.noaa.gov] In this problem we have called it *yearly_plt_ascii.dat*. The following Matlab program loads the data, calculates its covariance, and plots it.

```
load yearly_plt_ascii.dat; A=yearly_plt_ascii; yearly_plt_sunspots=A;
save yearly_plt_sunspots -ascii;
m=1:1:length(A); figure(1); plot(m,A,'b','LineWidth',2); axis([1 length(A)
0 200]);
xlabel('Year (from 1700 through 2010)'); ylabel('mean annual number of
sunspots');
title('Sunspot numbers from 1700 to 2010.'); grid;
%
% Find autocorrelation of A and plot it.
maxlag=150; B=xcov(A,maxlag,'unbiased');
[c,lags]=xcov(A,150,'coeff');yearly_plt_sunspots_Xcorr=B;
save yearly_plt_sunspots_Xcorr -ascii;
figure(2); plot(lags,c,'m','LineWidth',2); axis([-maxlag maxlag -.6 1.2])
xlabel('Year '); ylabel('Normalized Cov'); title('Autocorrelation of
sunspot numbers.');
```

8. Write a Matlab program to load and plot the monthly (from January 1968 through December 2010) and yearly (from 1940 through 2009) rates of unemployment in the United States. Find and plot their autocovariance functions. The data is available at www.bls.gov/data/. See Example 4.29.

Solution

```
load BLS_Data_US_Unemployment_1968_2010.dat;
A=BLS_Data_US_Unemployment_1968_2010;
BLS_USUnemp_1968_2010=A; save BLS_USUnemp_1968_2010 -ascii;
m=1:1:length(A); figure(1); plot(m,A,'b','LineWidth',2);
axis([1 length(A) 2 13]); grid;
%
% Find autocorrelation.
maxlag=150; E=xcov(A,maxlag,'unbiased'); [c,lags]=xcov(A,150,'unbiased');
BLS_USUnemp_1968_2010_XCORR_a=E; save BLS_USUnemp_1968_2010_XCORR_a -ascii;
figure(2); plot(lags,E,'m','LineWidth',2); axis([-maxlag maxlag -1.2 3.2])
grid;
%
load aat_all.txt; x=aat_all; z=x(1:71,10); E=z';
t = linspace(1940,2009,71); figure(3); plot(t,E); grid;
%
% Find autocorrelation.
maxlag=35; F=xcov(E,maxlag,'unbiased'); [c,lags]=xcov(F,35,'unbiased');
BLS_USUnemp_1940_2009_XCORR_a=E; save BLS_USUnemp_1940_2009_XCORR_a -ascii;
figure(4); plot(lags,F,'m','LineWidth',2);
axis([-maxlag maxlag -1.8 4.8]); grid;
```

9. Write a Matlab program to load and plot the monthly unemployment rates in California and in the United States from 1976 through 2010. Find unbiased estimates of their cross-covariance functions and plot them. The data is available at www.bls.gov/data/. See Example 4.32.

Solution

```
load BLS_Data_US_Unemployment_1976_2010.dat;
B=BLS_Data_US_Unemployment_1976_2010;
BLS_USUnemp_1976_2010=B; save BLS_USUnemp_1976_2010 -ascii;
load BLS_Data_CA_Unemployment_1976_2010.dat;
C=BLS_Data_CA_Unemployment_1976_2010';
BLS_CAUnemp_1976_2010=C; save BLS_CAUnemp_1976_2010 -ascii;
%
m=1:1:length(C); figure(1); plot(m,C,'r','LineWidth',2); hold on
plot(m,B,'b','LineWidth',2); axis([1 length(C) 3 13])
grid; hold off;
%
% Find cross-correlation.
maxlag=150; D=xcov(B,C,maxlag,'unbiased');
[c,lags]=xcov(B,C,150,'unbiased');
BLS_USCAUnemp_1976_2010_XCORR_a=D;
save BLS_USCAUnemp_1976_2010_XCORR_a -ascii;
E=xcov(C,B,maxlag,'unbiased'); [c,lags]=xcov(C,B,150,'unbiased');
BLS_CAUSUnemp_1976_2010_XCORR_a=E; save
BLS_CAUSUnemp_1976_2010_XCORR_a -ascii;
%
figure(2): plot(lags,D,'b','LineWidth',2); hold on;
plot(lags,E,'r','LineWidth',2); axis([-maxlag maxlag -1.2 3.2])
grid;
```

10. **Convolution with a delayed unit-step function**. Find the convolution $y(t) = x(t) \star u(t - t_0)$ and model it by a combination of an integrator and a delay element.

Solution

$$y(t) = x(t) \star u(t - t_0) = \int_{-\infty}^{\infty} x(\tau)u(t - t_0 - \tau)d\tau$$

But $u(t - t_0 - \tau) = 0$ for $\tau > t - t_0$. Therefore,

$$y(t) = \int_{-\infty}^{t-t_0} x(\tau)d\tau = \int_{-\infty}^{t} x(\tau - t_0)d\tau$$

The convolution result may be interpreted as the output of an integrator followed by a delay element, or a delay element followed by an integrator $y(t) = x(t) \star u(t) \star \delta(t - t_0)$.

Chapter Problems

11. **Convolution with an impulse.**

a. Show that the convolution of a unit impulse located at t_0 with a function that is continuous at t_0 shifts the function by t_0:

$$\phi(t) \,\check{}\, \delta(t - t_0) = \int_{-\infty}^{\infty} \phi(t - \tau)\delta(\tau - t_0)d\tau = \phi(t - t_0)$$

b. The convolution of two periodic functions $\phi(t)$ and $\eta(t)$ that have the same period T is defined by

$$\phi(t)^*\eta(t) = \frac{1}{T}\int_{-\frac{T}{2}}^{\frac{T}{2}} \phi(t-\tau)\eta(\tau)d\tau$$

Show the convolution of a periodic function $\phi(t)$ with a periodic train of unit impulses (with the same period T) to be $\phi(t)$ itself with a scale factor $1/T$:

$$\phi(t) \star \sum_{k=-\infty}^{\infty} \delta(t-kT) = \frac{1}{T}\int_{-\frac{T}{2}}^{\frac{T}{2}} \left[\phi(t-\tau)\sum_{k=-\infty}^{\infty} \delta(\tau-kT)\right]d\tau = \frac{1}{T}\phi(t)$$

The convolution integrals in problems 12 through 17 address the LTI systems of some of the problems in Chapter 3 whose outputs you derived by superposition. Here, the convolution operation will be performed by evaluating the integral in the time domain.

12. The unit-impulse response of an LTI system is $\delta(t) - e^{-t}u(t)$. Find its responses to the following inputs:

    a. $u(t) - u(t-1)$ b. $tu(t)$ c. $tu(t) - u(t-1)$ d. $\delta(t) - \delta(t-1)$

13. Find $y(t) = d(t) \star h(t)$, where $d(t) = u(t) - u(t-1)$ and

    a. $h(t) = \delta(t) - \delta(t-1)$ b. $h(t) = \sum_{n=0}^{\infty}(-1)^n\delta(t-n)$ c. $h(t) = d(t)$ d. $h(t) = e^{-t}$

14. Find $y(t) = d(t) \star h(t)$, where $d(t) = u(t) - u(t-1)$ and $h(t) = e^{-10^6 t}u(t)$. Describe a method to approximate $y(t)$ during $0 \le t < 1$ sec.

15. Find $y(t) = x(t) \star h(t)$ for the sets of inputs and unit-impulse responses given in Table 3.1 of Chapter 3.

16. Repeat problem 15 for the set of inputs and unit-impulse responses given in Table 3.2 of Chapter 3.

17. Find $y(t) = x(t) \star h(t)$, where

    a. $h(t) = e^{-2t}u(t)$ and $x(t) = 1$ b. $h(t) = e^{-2t}u(t)$ and $x(t) = u(t+1)$
    c. $h(t) = e^{-2t}u(t)$ and $x(t) = e^{-t}u(-t)$ d. $h(t) = e^{-2t}u(t)$ and $x(t) = e^{3t}u(-t)$
    e. $h(t) = (e^{-2t} + e^{-3t})u(t)$ and $x(t) = e^{-t}u(-t)$ f. $h(t) = e^{-t}u(t)$ and $x(t) = e^{-2t}u(-t)$
    g. $h(t) = x(t) = e^{-2t}[u(t) - u(t-1)]$ h. $h(t) = (1 - e^{-2t})u(t)$ and $x(t) = u(t) - u(t-1)$

Evaluate the convolution integrals of problems 18 through 24 directly in the time domain. (They will be evaluated again in the frequency domain by the Laplace transform. See Chapter 6.)

18. Evaluate the convolution integral $y(t) = x(t) \star h(t)$, where

    a. $h(t) = \sin t\,[u(t) - u(t-2\pi)]$ and $x(t) = \sin tu(t)$ b. $h(t) = \delta(t-2\pi) + 2u(t)$ and $x(t) = \cos tu(t)$
    c. $h(t) = e^{-t}u(t)$ and $x(t) = \sin 2tu(t)$ d. $h(t) = e^{-t}u(t)$ and $x(t) = \sin 2tu(-t)$

19. Evaluate convolution integrals $y(t) = x(t) \star h(t)$ for the following cases:

    a. $x(t) = \sin t$ and $h(t) = e^{-0.1t}\sin 2tu(t)$ b. $x(t) = \sin tu(t)$ and $h(t) = e^{-0.1t}\sin 2tu(t)$

    c. $x(t) = \sum_{n=0}^{n=\infty} \delta(t-n)$ and $h(t) = \begin{cases} \sin \pi t, & 0 < t < 1 \\ 0, & \text{elsewhere} \end{cases}$ d. $x(t) = u(-t)$ and $h(t) = e^{-\alpha t}u(t)$

    e. $x(t) = \begin{cases} 1/T, & 0 < t < T \\ 0, & \text{elsewhere} \end{cases}$ and $h(t) = e^{-\alpha t}u(t)$ f. $x(t) = tu(t)$ and $h(t) = e^{-\alpha t}u(t)$

20. Let $x(t) = \frac{1}{T}[u(t) - u(t - T)]$. Show that

$$\lim_{T \to 0} [x(t) \star h(t)] = h(t)$$

21. Find $y(t) = x(t) \star h(t)$ for the following three cases:
   a. $x(t) = (\sin t)u(t)$ and $h(t) = e^{-\alpha t}(\sin t)u(t)$ for $\alpha = 0.1, \ 1, \ 10$
   b. $x(t) = (\cos t)u(t)$ and $h(t) = (\cos 2t)u(t)$
   c. $x(t) = (\cos t)u(t)$ and $h(t) = (\cos 1.1t)u(t)$

22. Directly evaluate the convolution integral $y(t) = x(t) \star h(t)$, where

   a. $x(t) = \cos t u(t)$ and $h(t) = \delta(t - 2\pi) - 2u(t)$
   b. $x(t) = e^{-t/100}u(t)$ and $h(t) = 2e^{-t}u(t)$
   c. $x(t) = e^{-|t|}$ and $h(t) = e^{-t/10}u(t)$
   d. $x(t) = (1 + \sin 2t)u(t)$ and $h(t) = e^{-2t}u(t)$
   e. $x(t) = e^{-2t}u(t)$ and $h(t) = (1 + e^{-t})u(t)$
   f. $x(t) = (\sin 3t)u(t)$ and $h(t) = (e^{-t} - 2e^{-2t})u(t)$
   g. $x(t) = e^{-t}u(t)$ and $h(t) = \delta(t) - e^{-t}u(t)$

23. The input to an LTI system is $x(t) = (\sin 3t)u(t)$. Find the output $y(t)$ and its steady-state value for
   a. $h(t) = (\sin t)u(t)$
   b. $h(t) = e^{-t}(\sin t)u(t)$.

24. Directly evaluate the convolution integral $y(t) = y_1(t) + y_2(t) = [x_1(t) + x_2(t)] \star h(t)$ for the cases listed below. $\delta'(t)$ is a unit doublet, the time derivative of a unit impulse. $y_1(t)$ is the contribution to the output by the input $x_1(t)$ and $y_2(t)$ is the contribution to the output by $x_2(t)$.

   a. $h(t) = (1 - 2t)e^{-t}u(t)$ $x_1(t) = \left(e^{-3t} - e^{-2t}\right)u(t)$ and $x_2(t) = \delta(t) + \delta'(t)$
   b. $h(t) = e^{-3t}u(t)$ $x_1(t) = e^{-2t}u(t)$ and $x_2(t) = \delta(t)$
   c. $h(t) = \frac{1}{8}\left(9e^{-t} - e^{-9t}\right)u(t)$ $x_1(t) = 9u(t)$ and $x_2(t) = \delta(t) + \delta'(t)$
   d. $h(t) = \left(e^{-2t} - e^{-3t}\right)u(t)$ $x_1(t) = 6u(t)$ and $x_2(t) = \delta(t)$
   e_1. $h(t) = e^{-3t}u(t)$ $x_1(t) = e^{-2t}u(t)$ and $x_2(t) = \delta(t)$
   e_2. "   " $x_2(t) = 0$
   e_3. "   " $x_2(t) = -\delta(t)$
   f_1. $h(t) = 0.5e^{-t/2}u(t)$ $x_1(t) = 3e^{-t}u(t)$ and $x_2(t) = 6\delta(t)$
   f_2. "   " $x_2(t) = 0$
   f_3. "   " $x_2(t) = -6\delta(t)$

25. The impulse response of an LTI system is $h(t) = (e^{-t} + e^{-2t})u(t)$ and the input is $x(t)u(t_0 - t)$. Show that the output for $t > t_0$ is $y(t) = Ae^{-t} + Be^{-2t}$, $t > t_0$. Find A and B in terms of $x(t)u(t_0 - t)$.

26. The input $x(t)$ and the output $y(t)$ in an LTI system are related by the differential equation

$$\frac{d^2y}{dt^2} + 2\frac{dy}{dt} + y = \frac{dx}{dt} + 2x$$

Show that

$$y(t) = \int_{-\infty}^{\infty} h(t - \tau)x(\tau)d\tau$$

where $h(t)$ is the system's unit-impulse response.

27. Consider $x(t) = 0.01 \cos t + e^{-|t|}$ and $h(t) = e^{-t}u(t)$. Find

$$y(t) = \int_{-\infty}^{\infty} x(\tau)h(t-\tau)d\tau, \quad -\infty < t < \infty$$

and compare with the cross-correlation between $x(t)$ and $h(-t)$.

28. Correlation. Consider $x(t) = 0.01 \cos t$, $y(t) = e^{-t}u(t)$, and $z(t) = x(t) + y(t)$. Find and plot

$$r_{xz}(\tau) = \int_{-\infty}^{\infty} x(t)z(t-\tau)dt, \quad -\infty < \tau < \infty$$

29. Barker code. The Barker code is a binary sequence $c_N(n)$, $n = 1, 2, \ldots, N$ (shown by $+$ and $-$, or 1 and -1), with code length $N = 1, 2 \cdots 11, 13$, as listed in the table below.

N	$c_N(n)$
2	$1 -1$ or $-1\ 1$
3	$1\ 1 -1$
4	$1 -1\ 1\ 1$ or $1 -1 -1 -1$
5	$1\ 1\ 1 -1\ 1$
7	$1\ 1\ 1 -1 -1\ 1 -1$
11	$1\ 1\ 1 -1 -1 -1\ 1 -1 -1\ 1 -1$
13	$1\ 1\ 1\ 1\ 1 -1 -1\ 1\ 1 -1\ 1 -1\ 1$

To obtain an indicator for the pattern of dependency between 1s and -1s in the Barker code, for each code sequence $c(n)$ listed in the table above define $\rho(k)$ as the index of correlation between the elements:

$$\rho(k) = \sum_{n=1}^{N} c(n)c(n-k), \quad k = 0, 1, 2, \ldots, N$$

For codes with lengths $N = 2, 3, 4, 5, 7, 11$, and 13, find and plot $\rho(k)$ and determine the ratio $\rho(0)/\rho(k)$.

30. Signal detection. A deterministic periodic binary sequence $x(n)$ made of regularly alternating 1s and 0s is transmitted at the rate of 10^6 bits per second. For transmission it uses an on-off encoding scheme where the 1 bits are transmitted by $s(t)$ and the 0 bits are transmitted by silent periods each 1 μs long. The encoded signal is a periodic signal $x(t)$ given by

$$x(t) = \begin{cases} s(t), & 0 < t < 0.5\ \mu s \\ 0, & 0.5\ \mu s < t < 1\ \mu s \end{cases}, x(t) = x(t + 1\mu s)\ \text{all}\ t$$

To detect the occurrence of a 1 or 0, the received signal $\hat{x}(t)$ [generally a corrupted version of $x(t)$ due to communication noise and interference] is continually matched with the template $s(t)$ and the results are observed every $T = 1/M\ \mu s$ (i.e., at intervals $t = k/M\ \mu s$, $k = 0, 1, 2 \ldots, M - 1$). The integer M indicates the frequency (or number) of observations during transmission of each bit. The present problem illustrates the effect of $s(t)$ on separability of 1s and 0s at the receiver. For simplicity we assume no interference or noise [i.e., $\hat{x}(t) = x(t)$]. Assume a matching scheme where $x(t)$ is multiplied by a sliding $s(t)$-shaped window that is slid every $1/M\ \mu s$. The results are integrated over 1 μs, resulting in observation values $y(k)$:

$$y(k) = \int_{0}^{1\ \mu s} x(t)s(t - k\mu s)dt, \quad k = 0, 1, 2 \ldots (M - 1)$$

a. Let $s(t)$ be a 1-volt rectangular 1-μs pulse. Find the maximum value y_{Max} and its location k_0. Find the distance between the peak value and its immediate neighbour $y_{\text{Max}} - y(k_0 - 1)$. Consider this to be a *separability distance*.

b. Repeat for a binary $s(t) = \pm 1$ made of a seven-segment Barker code and observations every $1/7$ μs. Compare with the corresponding value obtained in part a.

c. Repeat for a binary $s(t) = \pm 1$ made of an 11-segment Barker code and observations every $1/11$ μs. Compare with corresponding values obtained in parts a and b.

d. Repeat for a binary $s(t) = \pm 1$ made of a 13-segment Barker code and observations every $1/13$ μs. Compare with corresponding values obtained in parts a, b, and c.

e. Discuss the gain in *separability distance* versus the increase in eventual transmission rate as you increase the length of the Barker code used.

4.14   Project: Signal Detection by Matched Filter

Objectives and Summary

This project explores, by computer simulation, detection of known and deterministic signals of finite length embedded in noise. This is done by a matched filter followed by a decision rule. The data is made of signals of finite duration that occur randomly and are corrupted by additive white Gaussian noise (which is independent of the signal). The signal used in this project is Barker code of length 7, 13, or 15. In procedure 1, signals without noise are convolved with templates matched to them, and their properties examined. Procedure 2 examines the detection of signals embedded in noise. To determine signal occurrence, the data is convolved with an LTI system whose unit-impulse response is matched to the signal and the output is then passed through a threshold level. The performance is measured by obtaining the probability of correct detection and the rate of false alarm for several signal-to-noise ratios and for several threshold levels.

Procedure 1: Operation of a Matched Filter

In this procedure you will implement, by computer simulation, the operation of a matched filter and examine its properties.

a. Write Matlab code to implement the operation of the matched filter of section 4.7 with the seven-segment signal $\eta(n) = \{1, 1, 1, -1, -1, 1, -1\}$. Obtain the output and plot it. The signal and the filter's output are shown in Figure 4.15(*a*) and (*b*), respectively.

b. Repeat the above procedure for the 13-segment signal $\eta(t) = \{1, 1, 1, 1, 1, -1, -1, 1, 1, -1, 1, -1, 1\}$. The signal and the filter's output are shown in Figure 4.15(*c*) and (*d*), respectively.

c. Repeat the above procedure for the 15-segment signal $\eta(t) = \{1, 1, 1, -1, -1, -1, -1, 1, -1, 1, -1, -1, 1, 1, -1\}$.

Procedure 2: Filtering and Detection

a. **Signal and noise.** Design and run a detection experiment involving a binary signal embedded in noise. The data to be transmitted is binary. It assumes one of two states (e.g., *high* or *low*, 1 or 0, *plus* or *minus*, ...), which occur randomly. One of the states is transmitted by the seven-bit long waveform $\eta(t) = \{1, 1, 1, -1, -1, 1, -1\}$ of procedure 1a. The other state is transmitted by the absence of $\eta(t)$; that is, seven zeros. The result is a signal $s(t)$ which contains $\eta(t)$ repeatedly but randomly. The noise is an additive white Gaussian stochastic process. The input arriving at the receiver is the signal plus noise $x(t) = s(t) + w(t)$, within which $\eta(t)$ appears repeatedly but irregularly. To generate $s(t)$, pass a random signal through a threshold level and convolve it with $\eta(t)$. An example is shown in Figure 4.22.

b. **Filtering and detection.** The goal is to detect the occurrence of $\eta(t)$. Detection is done by first filtering the data through a matched filter with $h(t) = \eta(NT - t)$ and then passing the filter's output through a threshold level λ. In this operation, two types of error may occur: a miss or a false alarm. Both of these depend on the signal-to-noise ratio (SNR) and the threshold λ. An example is shown in Figure 4.23. By simulation of a long stretch of data,

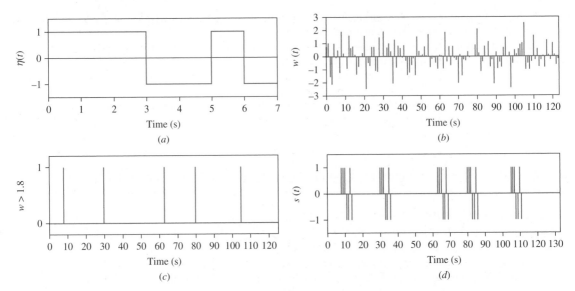

FIGURE 4.22 Generating $s(t)$. **(a)** A seven-segment $\eta(t)$. **(b)** A sample of a white Gaussian noise with zero mean and average power of 1 [shown by $N(0, 1)$]. **(c)** The noise is passed through a threshold level of 1.8 to generate a sequence of random events. **(d)** Convolution of $\eta(t)$ with the sequence of random events generates the signal $s(t)$.

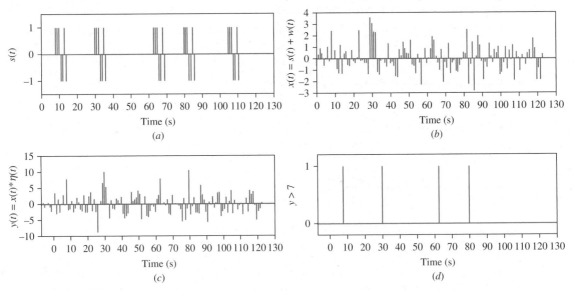

FIGURE 4.23 Detecting occurrences of $\eta(t)$. **(a)** A sample of the signal $s(t)$ containing 5 randomly occurring $\eta(t)$. **(b)** A sample of white Gaussian noise $N(0, 1)$ with zero mean and average power of 1. **(c)** The sum of the signal and noise in (a) and (b), $x(t) = s(t) + w(t)$. **(d)** Comparing filter's output with a threshold level $\lambda = 7$ in order to decide on the presence of $\eta(t)$. This figure shows one miss and no false alarm.

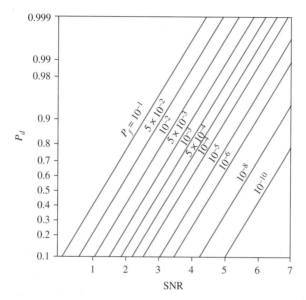

FIGURE 4.24 Theoretical relationship between P_d, P_f, and the signal-to-noise ratio (SNR) in an optimum linear filter for detection of signals in additive white Gaussian noise. The filter is matched to the signal. It maximizes the signal-to-noise ratio at the output. The threshold level provides a tradeoff between correct detections and false alarms.

Source: From Woody and Nahvi, "Application of Optimum Linear Filter Theory to Detection of Cortical Signals Preceding Facial Movement in Cat," *Experimental Brain Research*, 16 (1973), pp. 455–65.

measure the probability of detection and the rate of false alarm for two types of signals (using $\eta(t)$ in procedure 1a and 1b), six signal-to-noise ratios, and five threshold levels of your choice. Use the normalized frequency of occurrence of an event as an estimate of its probability. Compile your results in the form of the *receiver operating curve* (ROC). An example is shown in Figure 4.24. Discuss the detector's performance.

Note. The ROC of a matched filter is the plot of the relationship between P_d, P_f, and the signal-to-noise ratio (SNR) in an optimum linear filter for detection of signals containing noise. P_d is the probability of detection and P_f is the probability of false alarm. The filter is matched to the signal. It maximizes the SNR at the output. The threshold level provides a tradeoff between correct detections and false alarms.

c. **Effect of signal-to-noise ratio.** Define the signal-to-noise ratio to be the ratio of signal power to noise power at the output of the filter. Investigate the effect of SNR on P_d and P_f for five values of SNR of your choice.

Discussion and Conclusions

Discuss the detector's performance. Discuss the effect of the length of η on the detection outcome. How does one choose the threshold level λ? Discuss application of the above detection scheme in other fields which involves missing a signal or producing a false alarm.

Chapter 5

Differential Equations and LTI Systems

Contents

Introduction and Summary

Linear differential equations with constant coefficients are the main vehicles for describing a linear time-invariant (LTI) system. The input-output relationships for the majority of linear time-invariant systems, as well as their state-space dynamical equations, are expressed by linear differential equations with constant coefficients. Examples of these were given in Chapter 3. A linear differential equation with constant coefficients is written as:

$$\frac{d^n y}{dt^n} + a_{n-1}\frac{d^{n-1} y}{dt^{n-1}} + \cdots + a_1\frac{dy}{dt} + a_0 y = b_m\frac{d^m x}{dt^m} + b_{m-1}\frac{d^{m-1} x}{dt^{m-1}} + \cdots + b_1\frac{dx}{dt} + b_0 x \quad (1)$$

where $x(t)$ is the input and $y(t)$ is the output. Also, $d^k y/dt^k$ and $d^k x/dt^k$ are the kth time derivatives of $y(t)$ and $x(t)$, respectively, and a_k and b_k are constants. The differential equation is of order n. It is customary to arrange the output terms on the left side and the input terms on the right side of the equation. In this book differential equations are solved by one of the following methods:

1. Time-domain method: $\begin{cases} \text{homogeneous and particular solutions} \\ \text{convolution} \\ \text{zero-input and zero-state solutions} \\ \text{numerical solution} \end{cases}$

2. Transform method

In this chapter we discuss both a time-domain and a numerical method of finding the solution. Solutions using the Fourier or Laplace transforms are discussed in other chapters. This chapter starts with the classical method of finding the particular and homogeneous solutions. The method is simple and intuitive, as the components of the solution reflect the physical and measurable attributes of a system's response. These include things such as the forced and natural responses (especially the transient and steady-state responses) and its natural frequencies. The effect of an input discontinuity on the initial conditions and the response is then described. The impulse, step, exponential, and sinusoidal inputs are used in order to bring to the reader's attention the utility of tools such as the system function and the frequency response. After a detailed presentation of the classical method, the chapter briefly describes the zero-input and zero-state responses and the connection between the above methods and the convolution method of finding the response of an LTI system. The project at the end of the chapter is intended to help the student with second-order systems in more detail. It contains mathematical analyses and measurements that can be carried out in the laboratory or by simulation.

5.1 Formulation of Differential Equations

A system is an interconnection of its subsystems and elements. Each subsystem or element is described by relationships between certain variables of interest. The interconnection of the elements imposes additional constraints on these variables. The system is also connected to the outside world by input-output ports. The result is a set of equations that include internal and external variables. By eliminating some internal variables we obtain a set of equations that describe the system. When the elements are linear and their interconnections are also expressed by linear equations, the system becomes linear.[1] Similarly, the time invariance of the elements and their interconnections produce a time-invariant system. For the LTI systems of interest, the resulting equations are ordinary linear differential equations with constant coefficients. Some examples in sections 3.6 and 3.7 of Chapter 3 illustrated how the differential equations were formulated for LTI systems. The present section discusses more examples of formulating differential equations and their possible linearization. The examples cover electrical, mechanical, thermal, biological, and hydraulic systems. In the case of some nonlinear systems, a linear model may be constructed to represent the system within a limited range of operation.

The input-output differential equation of an LTI system may also be constructed from the input-output data, along with some reasonable assumptions about the system, or from knowledge of the system's response to a given input such as the impulse or step functions. In fact, any input-output pair (except exponential or sinusoidal) would work for that purpose. Mathematical tools for such formulations and modeling are provided by the Laplace transform and will be discussed in Chapter 8. Here we include 11 simple examples to illustrate the concept. For the sake of simplicity, the first 10 examples in this section are limited to first-order equations, but the methods apply to systems of higher order as well.

Example **5.1**

An electrical system

A 10-μF capacitor carrying an initial charge of 1 μC is connected at $t = 0$ in parallel with a 10-MΩ resistor. Formulate the differential equation that describes capacitor charge as a function of time.

Solution

Let q, v, and i represent the electric charge, the voltage, and the current coming out of the capacitor, respectively. All are functions of time. Then

From the capacitor side we have $i = -\dfrac{dq}{dt}$

From the resistor side we have $i = \dfrac{v}{R} = \dfrac{q}{RC} = 10^{-2}q$

By eliminating i we get $\dfrac{dq}{dt} + 10^{-2}q = 0$

It will be shown that the general solution to the above equation is $q = Ce^{-t/100}$. In the present case, given $q(0) = 10^{-6}$ we find $q(t) = 10^{-6}e^{-10^{-2}t}$. The charge on the capacitor dissipates exponentially with a time constant of 100 seconds.

---

[1] A linear system may also contain nonlinear elements.

A piecewise linear system

The symbol for a rectifier diode and the convention for its terminal variables are shown in Figure 5.1(*a*). An ideal diode is a voltage-controlled switch that remains open ($R = \infty$) as long as the voltage across its terminal is negative, and operates as a short ($R = 0$) when the current through it is positive. Its terminal characteristic may be expressed by

$$\begin{cases} i = 0, & \text{when } v < 0 \\ v = 0, & \text{when } i > 0 \end{cases}$$

In the circuit of Figure 5.1(*b*) the diode is ideal. Assume $R_1 = R_2 = 500\,\Omega$, $C = 2\,\mu F$ and $v_1(t) = u(t) - u(t - t_0)$, $t_0 = 10$ msec. Find differential equations that describe the capacitor voltage as a function of time.

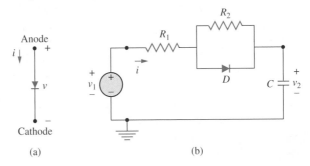

FIGURE 5.1

Solution

From KVL around the loop we have $\quad v_1 = Ri + v_2$

From the capacitor side we have $\quad i = C\frac{dv_2}{dt}$

By eliminating $i(t)$ we get $\quad RC\frac{dv_2}{dt} + v_2 = v_1$

During $t < 0$, $v_1 = 0$ and the circuit is at rest [all variables are zero, including $v_2(0) = 0$]. For $t > 0$, the diode can be in one of two states: on or off. Absent any a priori knowledge about the state of the diode, we assume a state and solve the problem. Then we test to see if the assumption holds true.

a. During $0 \leq t < t_0$, $v_1 = 1$, we assume the diode is on. Then $R = R_1 = 500\,\Omega$, $RC = 10^{-3}$ and

$$\frac{dv_2}{dt} + 1{,}000v_2 = 1{,}000, \quad 0 \leq t < t_0, \quad v_2(0) = 0$$

As will be seen later, the solution to the above differential equation is

$$v_2 = 1 - e^{-1{,}000t}, \quad 0 \leq t < t_0$$

To test the assumption, we note that during the above period $i = \frac{v_1 - v_2}{500} > 0$, which would turn the diode on. The assumption, then, was correct. In contrast, assuming the diode to be off would result in

$$R = R_1 + R_2 = 1,000, \quad RC = 2 \times 10^{-3}, \quad v_2 = 1 - e^{-500t},$$

and

$$i = \frac{v_1 - v_2}{1,000} = \frac{1}{1,000}e^{-500t} > 0$$

which would create a positive voltage across the diode. This contradicts the terminal characteristics of the diode indicating that the assumption was incorrect.

b. During $t \geq t_0$, $v_1 = 0$, we assume the diode is off. Then $R = R_1 + R_2 = 1,000 \ \Omega$, $RC = 2 \times 10^{-3}$ and

$$\frac{dv_2}{dt} + 500v_2 = 0, \quad t \geq t_0, \quad v_2(t_0) \approx 1$$

The solution to the above differential equation is

$$v_2 = e^{-500(t - t_0)}, \quad t \geq t_0$$

By a similar test we affirm that the assumption is correct.

Example 5.3

Mechanical system

A solid object with mass M slides on a lubricated horizontal surface with speed v under an external driving force. Formulate the differential equation that describes the speed in terms of the driving force.

Solution
According to Newton's law, the net force acting on the object is equal to the product of the mass with acceleration. In the present case the motion is one-dimensional. The net horizontal component of the force in the direction of motion of the object is $f - bv$, where f represents the external driving force (which produces the motion) and bv models the damping effect of the viscosity of the lubrication on the support surface (which opposes it).

$$f - bv = M\frac{dv}{dt}$$

Collecting the unknown terms on the left-hand side and dividing both sides by M we find

$$\frac{dv}{dt} + \frac{b}{M}v = \frac{f}{M}$$

Example 5.4

Thermal system

This example formulates the relationship between the temperature of a liquid and that of a solid body immersed in it. The solid body doesn't contain any energy-generating source. The parameters involved are the following:

θ_1 temperature of the liquid in $^\circ C$
θ_2 temperature of the solid object in $^\circ C$

A surface area of the solid object in m^2
M mass of the solid object in kg
C specific heat of the solid object in Calories/$(°C \times \text{kg})$
h surface heat transfer coefficient in Calories/$(°C \times \text{hr} \times m^2)$

Find the differential equation that describes θ_2 in terms of θ_1.

Solution
In accordance with the principle of conservation of energy, the heat exchange between the solid body and the surrounding liquid has to satisfy the condition below:

Energy transferred from the liquid = Energy stored in the solid body

Assuming that the temperature is uniform throughout the solid body, during a time period dt the above equation becomes

$$hA(\theta_1 - \theta_2)dt = MCd\theta_2$$

or

$$\frac{d\theta_2}{dt} + \frac{hA}{MC}\theta_2 = \frac{hA}{MC}\theta_1$$

The above is a first-order differential equation with θ_1 being the input variable and θ_2 the output variable. M, A, and C are constants, but h is generally a function of θ_1 and θ_2, making the equation a nonlinear one. In avoiding a mathematically rigorous linearization procedure, one may approximate h by its average value during a small perturbation.

Example
5.5

Hydraulic system

An open storage tank of cylindrical shape with cross-section area A receives liquid at a flow rate of q liters per second. It partly stores the incoming liquid and partly discharges it to atmospheric space through a sharp orifice.

a. Let the height of the liquid stored in the tank be h. Set up the differential equation that describes h in terms of input flow rate q.

b. Assume the liquid has been arriving in the tank at a constant rate of Q_0 for a long time. In the steady state the discharge becomes equal to the intake and the tank level reaches a steady-state height H_0, to be called the DC steady-state operating point. Linearize the equation around the operating point Q_0 and H_0 to obtain a linear differential equation describing perturbations (small variations in h produced by small variations in q).

Solution

a. From the application of the basic laws of hydraulics it can be shown that the flow rate through the orifice is $k\sqrt{h}$, where h is the height of the liquid stored in the tank and k is a constant whose value is determined from a combination of theory and measurements. From conservation of mass we have

Mass stored + Mass out = Mass in

Assuming an incompressible liquid we can substitute volumes for the masses.

Volume stored + Volume out = Volume in

During a period dt we have

$$\begin{cases} \text{volume stored} = A \times dh \\ \text{volume in} = q \times dt \\ \text{volume out} = k\sqrt{h} \times dt \end{cases}$$

which results in the following differential equation:

$$A\frac{dh}{dt} + k\sqrt{h} = q$$

where $q(t)$ is the known variable (the system's input) and $h(t)$ is the unknown (the system's response). A and k are constants, making the system described by the above equation time invariant. However, the term $\sqrt{h}$ causes it to become nonlinear. (It fails the linearity test because $\sqrt{h_1} + \sqrt{h_2} \neq \sqrt{h_1 + h_2}$.)

b.   The operating point satisfies the steady-state equation

$$k\sqrt{H_0} = Q_0$$

In its neighborhood the input and output variables can be expressed by $Q_0 + q$ and $H_0 + h$, respectively, where q and h are small perturbations. The input-output equation then becomes

$$A\frac{d(H_0 + h)}{dt} + k\sqrt{H_0 + h} = Q_0 + q$$

But

$$\frac{d(H_0 + h)}{dt} = \frac{dh}{dt}$$

and, since h is assumed to be small, the $\sqrt{H_0 + h}$ term can be approximated by the first two terms in its Taylor series:

$$\sqrt{H_0 + h} \approx \sqrt{H_0} + \frac{1}{2\sqrt{H_0}}h$$

The input-output equation is, therefore, approximated by

$$A\frac{dh}{dt} + k\sqrt{H_0} + \frac{k}{2\sqrt{H_0}}h = Q_0 + q$$

Noting that $k\sqrt{H_0} = Q_0$, the above equation is reduced to

$$A\frac{dh}{dt} + \frac{k}{2\sqrt{H_0}}h = q$$

This is a linear differential equation with constant coefficients.

Response of hydraulic system of Example 5.5 to a step perturbation

The coefficient k of the discharge orifice in Example 5.5 is given by

$$k = \sqrt{2g} \times a \times c$$

where g is the gravitational acceleration and a is the cross section of the discharge orifice. The coefficient c has no dimension and its value depends on the geometry of the orifice. It can be determined by a combination of theoretical analysis and experimental measurement. Assume that the area of the cross section of the tank is $A = 1$ m^2, the cross section of the discharge orifice is $a = 1$ cm^2, acceleration due to gravity is $g = 9.81$ m $\times$ s^{-2}, and $c = 0.65$. Assume that the liquid has been flowing into the tank at a constant rate of $Q_0 = 0.2$ liters/second for a long time so that the fluid height in the tank has reached a steady-state level H_0. (a) Fing H_0. (b) Find the response of the tank to an input perturbation in the form of a step increase of 10 cc/second. (c) Find the new steady-state level by using the linearized differential equation. Then verify that it is in agreement with the value found from the nonlinear equation.

Solution

In the following solution we will use the SI system of units.

$$k = \sqrt{2g} \times a \times c = \sqrt{2 \times 9.81} \times 10^{-4} \times 0.65 = 2.88 \times 10^{-4}$$

a. $\quad H_0 = \left(\dfrac{Q_0}{k}\right)^2 = \left(\dfrac{0.0002}{2.88 \times 10^{-4}}\right)^2 = 0.482$ m

b. $\quad A\dfrac{dh}{dt} + \alpha h = 10^{-5} u(t), \quad \alpha = \dfrac{k}{2\sqrt{H_0}} = \dfrac{2.88 \times 10^{-4}}{2\sqrt{0.482}} = 2.072 \times 10^{-4}$

$$h(t) = \dfrac{10^{-5}}{\alpha}\left(1 - e^{-\alpha t}\right) = 0.0483\left(1 - e^{-\alpha t}\right)$$

$$h(\infty) = 0.0483 \text{ m}, \quad H_0 + h(\infty) = 0.483 + 0.0482 = 0.5308 \text{ m}$$

c. $\quad H(\infty) = \left(\dfrac{Q_0 + q}{k}\right)^2 = \left(\dfrac{2.1 \times 10^4}{2.88 \times 10^4}\right)^2 = 0.532$ m

$$H(\infty) = H_0 + h(\infty)$$

Modeling a mixing process

A tank contains 200 liters of liquid with a salt concentration of 30 grams/liter. At $t = 0$, fresh liquid, whose concentration is 5 grams/liter, starts flowing in the tank at the rate of 10 liters/minute. At the same time, the tank discharges at the same rate of 10 liters/minute. Throughout the process the content of the tank is kept at a uniform concentration level by continuous stirring.

a.   Write the differential equation that describes salt concentration in the tank as a function of time. Show that the tank's concentration of salt decreases exponentially with time and determine its time constant τ.

b.   Find the concentration at $t = 3\tau$.

c.   Determine the time required to reduce the concentration of salt in the tank to the level of 6 grams/liter.

Solution

a.   Let the concentration of the salt in the tank at time t be represented by $c(t)$ (in grams/liter). The amount of salt entering the tank during a brief period dt is $10 \times 5dt = 50dt$ grams and the amount discharged is $10c \times dt$. The change in concentration is dc. The change in the total salt content of the tank is, therefore,

$$200 \times dc = 50 \times dt - 10c \times dt, \ t > 0, \ \text{ or } \ \frac{dc}{dt} = 0.25 - 0.05c, \ t > 0,$$

$$\text{with } c(0) = 30$$

The only solution that satisfies the above equation and the initial condition of $c(0) = 30$ is

$$c(t) = 5 + 25e^{-0.05t}, \ t > 0$$

The concentration of salt inside the tank starts at the level of 30 grams/liter at $t = 0$ and decreases exponentially to a final value of 5 grams/liter. The time constant of the process is $\tau = 20$ seconds. Theoretically, it requires $t = \infty$ to reach the final value, but, as seen next, after a time lapse of 3 to 5 time constants it reaches a level close enough to the final value.

b.   At $t = 3\tau$ seconds, $c = 5 + 25e^{-3} = 6.245$ grams/liter.

c.   The time required to reach $c = 6$ is the solution of the equation $5 + 25e^{-0.05t} = 6$, which is $t = 64.38$ seconds.

Example

5.8

Modeling a population's growth

Annual measurements of a certain population show an increase of 2.5% per year. That is, the number of people (shown by y_n) on the last day of the nth year is related to y_{n-1} (the number a year before) by $y_n = (1 + 0.025)y_{n-1}$. Assume growth is a continuous function of time and the rate of growth is proportional to the population number:

$$\frac{dy(t)}{dt} = \alpha y(t)$$

α a constant. Then, construct a simple model of population growth by a first-order differential equation.

Solution

The above proportionality assumption translates directly into the following first-order equation and its solution

$$\frac{dy(t)}{dt} - \alpha y(t) = 0, \quad y(t) = y_0 e^{\alpha t}, \quad \text{where } y_0 = y(0)$$

The annual rate of increase of 2.5% is produced by continuously compounding the population increase during the year. If the unit of time is chosen to be 1 year, then $e^\alpha = 1.025$ and $\alpha = \ln(1.025) = 0.02469261.$[2]

Modeling an interest rate problem

Pat has invested X_0 units of money in an account with an annual interest rate of $r\%$.

a. Find the account value (sum of principal X_0 and the accrued interest) at the end of the year if the interest is compounded (i) annually, (ii) semiannually, (iii) quarterly, (iv) monthly, (v) daily, and, finally, (vi) continuously.

b. Find the differential equation from which the account value can be obtained as a function of time when interest is compounded continuously.

c. At an interest rate of $r = 3\%$, how long does it take for Pat's money to double if the interest is compounded (i) continuously and (ii) monthly?

Solution

a. Let the account value at time t be shown by $x(t)$. At the end the first year,

(i)   annual compounding $x = X_0(1+r)$

(ii) semiannual compounding $x = X_0\left(1+\dfrac{r}{2}\right)^2$

(iii) quarterly compounding $x = X_0\left(1+\dfrac{r}{4}\right)^4$

(iv) monthly compounding $x = X_0\left(1+\dfrac{r}{12}\right)^{12}$

(v)   daily compounding $x = X_0\left(1+\dfrac{r}{360}\right)^{360}$

(vi) continuous compounding $x = \lim_{n\to\infty}\left[X_0\left(1+\dfrac{r}{n}\right)^n\right] = X_0 e^r$

b. With continuous compounding of interest, at the end of the first year $x = X_0 e^r$. At the end of the nth year Pat will have $x = X_0 e^{nr}$ or $x(t) = X_0 e^{rt}$, if t is specified in units of year. The differential equation whose solution is $X_0 e^{rt}$ is

$$\frac{dx}{dt} - rx(t) = 0, \quad \text{with } x(0) = X_0$$

---

[2] Note the parallels to the relationship between nominal and effective values of annual interest rates.

c. Pat's investment will double in $t = (\ln 2)/r = 23.105$ years, if interest is compounded continuously. If interest is compounded monthly, doubling time will be 23.134 years.

Modeling a system from measurements

The impulse response of a system is measured experimentally and a table of sampled data (h_i vs. t_i, $i = 1, 2, \dots, t > 0$) is obtained. Based on a preliminary examination, it is hypothesized that the data fits a decaying exponential function $h(t) = h_0 e^{-\alpha t}$. Test this hypothesis and find parameters for the exponential model $h(t)$.

Solution

To test the hypothesis, the data points are plotted in the form of $\ln(h)$ versus time (i.e., with a linear abscissa for time and a logarithmic ordinate for h). The plot is then examined visually and is seen to fit a straight line. The slope of the line becomes α because $\ln(h) = \ln(h_0) + \ln(e^{-\alpha t}) = \ln(h_0) - \alpha t$. As an example, assume the following two measurements:

$$\begin{cases} t_1 = 0, \ h_1 = 2, & \ln(h_1) = 0.693 \\ t_2 = 5, \ h_2 = 1.213, & \ln(h_2) = 0.193 \end{cases} \quad \alpha = \frac{\ln(h_1) - \ln(h_2)}{t_2 - t_1} = \frac{0.693 - 0.193}{5 - 0} = 0.1$$

The unit-impulse response is modeled by

$$h(t) = 2e^{-0.1t} u(t)$$

The only differential equation that can produce the above unit-impulse response is

$$\frac{dy(t)}{dt} + 0.1 y(t) = 2x(t)$$

A second-order electrical system

In the circuit of Figure 5.2, $R = 500 \ \Omega$, $L = 1$ mH, and $C = 10 \ \mu$F. The variables of interest are the elements' voltages and currents.

a. Write the terminal characteristics of the elements and the interconnection constraints.

b. Find differential equations that relate $i_R(t)$, $i_c(t)$, $i_L(t)$, and $v(t)$ to $v_s(t)$.

c. Summarize your observations.

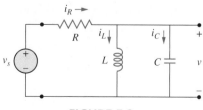

FIGURE 5.2

Solution

(1) Resistor characteristic:
$$i_R(t) = \frac{v_s(t) - v(t)}{R}$$

(2) Inductor characteristic:
$$i_L(t) = \frac{1}{L} \int_{-\infty}^{t} v(\tau)d\tau$$

(3) Capacitor characteristic:
$$i_C(t) = C\frac{dv(t)}{dt}$$

(4) Apply KCL at output node:
$$i_R(t) = i_C(t) + i_L(t)$$

(5) Substitute 1, 2, and 3 in 4:
$$\frac{v_s(t) - v(t)}{R} = \frac{1}{L} \int_{-\infty}^{t} v(\tau)d\tau + C\frac{dv(t)}{dt}$$

(6) Differentiate 5:
$$\frac{d^2v}{dt^2} + \frac{1}{RC}\frac{dv}{dt} + \frac{1}{LC}v = \frac{1}{RC}\frac{dv_s}{dt}$$

(7) $v(t)$: Plug in element values in 6:
$$\frac{d^2v}{dt^2} + 200\frac{dv}{dt} + 10^8 v = 200\frac{dv_s}{dt}$$

(8) $i_R(t)$: Find $v(t)$ from 1 and apply to 7:
$$\frac{d^2i_R}{dt^2} + 200\frac{di_R}{dt} + 10^8 i_R = \frac{1}{500}\frac{d^2v_s}{dt^2} + 2 \times 10^5 v_s$$

(9) $i_L(t)$: Integrate 7 and apply 2:
$$\frac{d^2i_L}{dt^2} + 200\frac{di_L}{dt} + 10^8 i_L = 2 \times 10^5 v_s$$

(10) $i_C(t)$: Differentiate 7 and apply 3:
$$\frac{d^2i_C}{dt^2} + 200\frac{di_C}{dt} + 10^8 i_C = \frac{1}{500}\frac{d^2v_s}{dt^2}$$

Observations

a.   KCL applies to the differential equations for $i_R(t)$, $i_L(t)$ and $i_C(t)$: [i.e., with regard to their right side $(8) = (9) + (10)$].

b.   The left sides of all differential equations that describe the system's variables have the same form. The implications of the above two observations will be discussed in sections 5.3 and 5.4.

5.2 Solution in the Time Domain by the Classical Method

Consider the nth-order differential equation
$$\frac{d^n y}{dt^n} + a_{n-1}\frac{d^{n-1}y}{dt^{n-1}} + \cdots + a_1\frac{dy}{dt} + a_0 y = f(t), \quad t \geq 0$$

where t is time. We assume the differential equation represents a real physical system,[3] and $f(t)$ represents the total contribution from the input and its derivatives.[4] By a solution we mean a function $y(t)$ that (i) satisfies the differential equation and (ii) is of such a nature that it and its $n-1$ derivatives at $t=0$ are equal to a set of prescribed initial conditions. Such a solution is also often called the complete solution. The following discussion elaborates on this point.

If the differential equation is valid for $-\infty < t < \infty$ and $f(t)$ is known for all times, then initial conditions at $t=-\infty$ are assumed to be zero. If the input is given only for $t \geq 0$, then additional information must be provided to summarize the effect of the unknown past on the state of the system at $t \geq 0$ and hence on the future output. For example, the information about the past history of the system may be supplied in the form of the value of y and its first $n-1$ derivatives at $t=0$. The initial conditions, therefore, contain the cummulative effect of all past inputs to the system.

From the above perspective, one expects that the response at $t \geq 0$ is affected by two factors: (i) the input after $t \geq 0$ and (ii) the initial conditions at $t=0$. The solution $y(t)$ is a function that satisfies the equation for $t \geq 0$ *and* meets the conditions for y and its $n-1$ derivatives at $t=0$. To achieve this goal, the solution is composed of two components. One component makes the left side of the differential equation equal to the right side for $t \geq 0$ and is called the particular solution, $y_p(t)$. The particular solution satisfies the differential equation for $t \geq 0$, but generally doesn't agree with the initial conditions by itself. It, therefore, is not the complete solution. To make it comply with the initial conditions as well, we add to it a second component called the homogeneous solution, $y_h(t)$. When substituted in the differential equation, the homogeneous solution makes the left side equal to zero. It is obtained by setting the left side of the differential equation equal to zero.

Example 5.12

Find the response of the first-order differential equation

$$y'(t) + y(t) = 1, \quad t \geq 0, \quad y(0) = 0$$

Solution

Try $y_p(t) = 1, \ t \geq 0$. Then

$$y_p(t) + y'_p(t) = 1, \ t \geq 0$$

The proposed $y_p(t)$ satisfies the equation, but because $y_p(0) = 1$, it contradicts the required initial condition $y(0) = 0$. It is not the solution to the equation. To remove the contradiction we add a second component, $y_h(t)$, and try $y(t) = y_p(t) + y_h(t)$.

---

[3] A consequence of this assumption is that if $f(t) = 0$ for $t < t_0$, then $y(t) = 0$ for $t < t_0$.

[4] For simplicity, we may represent the time derivative of a function by a *prime sign'* placed at its upper-right corner:

$$y' = \frac{dy}{dt}, \ y'' = \frac{d^2y}{dt^2}, \ y''' = \frac{d^3y}{dt^3}, \dots, y^{(n)} = \frac{d^ny}{dt^n}$$

The function of $y_h(t)$ is to contribute $y_h(0) = -1$ such that $y(0) = y_p(0) + y_h(0) = 1 - 1 = 0$, in agreement with the required initial condition. At the same time, $y_h(t)$ should contribute zero to the right side of the equation as shown below:

$$y(t) + y'(t) = \underbrace{y_p(t) + y'_p(t)}_{\Downarrow} + \underbrace{y_h(t) + y'_h(t)}_{\Downarrow} = 1$$
$$\qquad\qquad\quad 1 \qquad\qquad\quad 0$$

In other words, $y_h(t)$ is the solution to the so-called homogeneous equation

$$y_h(t) + y'_h(t) = 0$$

The only function that when added to its derivative sums up to zero is an exponential function $y_h(t) = Ae^{-st}$, where A is a constant and can assume any value. In the present example, $y_h(t) = Ae^{-t}$ and $y(t) = 1 + Ae^{-t}$. The parameter A will be adjusted to satisfy the initial condition. Here, $A = -1$ will set $y(0) = 0$. The complete solution is, therefore, $y(t) = 1 - e^{-t}$, $t \geq 0$. There is not enough information available to determine $y(t)$ for $t < 0$.

In summary, the solution of the linear time-invariant differential equation with constant coefficients is the sum of the particular solution and the homogeneous solution.

$$y(t) = y_p(t) + y_h(t)$$

In circuits and systems theory we refer to the input $x(t)$ as the forcing function, the particular solution as the forced response, and the homogeneous solution as the natural response. The exponent s is called the complex frequency and the roots of the characteristic equation (to be discussed shortly) are the natural frequencies. In the following sections we will expand upon ways to find the two components of the response and show how together they complete the solution. We start with the particular solution.

5.3   The Particular Solution

A function that satisfies the differential equation without necessarily meeting the initial conditions is called the particular solution. The particular solution depends on the forcing function $f(t)$ and may be found by inspection or a process of *guess and test* (i.e., substitution).[5] A short list of forcing functions and their particular solutions is given in Table 5.1. More details on the responses to the step, impulse, exponential, and sinusoidal forcing functions are found in sections 5.10 to 5.14.

The constants c_i in a particular solution are found by its substitution into the differential equation. See the examples in section 5.6.[6]

---

[5]This method is also known by the phrase *undetermined coefficients*.

[6]When the particular solution contains a large number of constants c_i, their determination by substitution becomes tedious. In such cases more efficient methods are employed.

TABLE 5.1 A Short Table of Forcing Functions and Their Particular Solutions

Forcing Function	$\Rightarrow$	Form of the Particular Solution
1	$\Rightarrow$	c
e^{st}	$\Rightarrow$	$\begin{cases} ce^{st}, & s \text{ is not a root of the characteristic equation.} \\ cte^{st}, & s \text{ is a root of the characteristic equation} \end{cases}$
		(Characteristic equation is described in the next section.)
$\sin \omega t$ or $\cos \omega t$	$\Rightarrow$	$c_1 \sin \omega t + c_2 \cos \omega t = A \cos(\omega t + \phi)$
t^k	$\Rightarrow$	$c_0 + c_1 t + \cdots + c_k t^k$
$t^k e^t$	$\Rightarrow$	$e^t(c_0 + c_1 t + \cdots + c_k t^k)$
$e^{\sigma t} \sin \omega t$ or $e^{\sigma t} \cos \omega t$	$\Rightarrow$	$e^{\sigma t}(c_1 \sin \omega t + c_2 \cos \omega t) = A e^{\sigma t} \cos(\omega t + \phi)$
$t \sin \omega t$ or $t \cos \omega t$	$\Rightarrow$	$c_1 \sin \omega t + c_2 \cos \omega t + c_3 t \sin \omega t + c_4 t \cos \omega t$

Example

5.13

The differential equations for the circuit of Figure 5.2 were found in Example 5.11. Use those results to find the particular responses in $i_R(t)$, $i_C(t)$, $i_L(t)$, and $v(t)$ to $v_s(t) = e^{st}u(t)$ and evaluate them for $s = 0$.

Solution

The particular solution to an exponential input e^{st} is an exponential function He^{st}. H is a function of s and is found by substituting e^{st} into the differential equation.

$$\frac{d^2 v}{dt^2} + 200 \frac{dv}{dt} + 10^8 v = 200 \frac{dv_s}{dt}, \qquad H_v(s) = \frac{200s}{s^2 + 200s + 10^8}, \qquad H_v(0) = 0$$

$$\frac{d^2 i_R}{dt^2} + 200 \frac{di_R}{dt} + 10^8 i_R = \frac{1}{500} \frac{d^2 v_s}{dt^2} + 2 \times 10^5 v_s, \quad H_R(s) = \frac{0.002s^2 + 2 \times 10^5}{s^2 + 200s + 10^8}, \quad H_R(0) = 2 \times 10^{-3}$$

$$\frac{d^2 i_L}{dt^2} + 200 \frac{di_L}{dt} + 10^8 i_L = 2 \times 10^5 v_s, \qquad H_L(s) = \frac{2 \times 10^5}{s^2 + 200s + 10^8}, \qquad H_L(0) = 2 \times 10^{-3}$$

$$\frac{d^2 i_C}{dt^2} + 200 \frac{di_C}{dt} + 10^8 i_C = \frac{1}{500} \frac{d^2 v_s}{dt^2}, \qquad H_C(s) = \frac{0.002s^2}{s^2 + 200s + 10^8}, \qquad H_C(0) = 0$$

Example

5.14

In the circuit of Figure 5.2 find the particular responses in $v(t)$ to an input $v_s(t) = \cos(\omega t)u(t)$ for $\omega = 9,900,\ 10,000,$ and $10,100$ rad/s.

Solution

Using the results of Example 5.13, we first find the particular solutions to exponential inputs e^{st} with $s = j10,000$ and $j(10,000 \pm 100)$.

input	$\Rightarrow$	$H(s)$		$\Rightarrow$	particular response
$e^{j9,900t}$		$H_v(s)\big|_{s=j9,900}$	$\approx 0.707\angle \pi/4$		$v_p(t) \approx 0.707 e^{j(9,900t + \pi/4)}$
$e^{j10,000t}$		$H_v(s)\big|_{s=j10,000}$	$= 1$		$v_p(t) = e^{j10,000t}$
$e^{j10,100t}$		$H_v(s)\big|_{s=j10,100}$	$\approx 0.707\angle -\pi/4$		$v_p(t) \approx 0.707 e^{j(10,100t - \pi/4)}$

The sinusoidal input $v_s(t) = \cos(\omega t)u(t)$ is the sum of two complex conjugate exponentials. Its particular solution is found by superposition of the individual exponentials. The results are

input $\quad\Longrightarrow\quad$ particular response

$\cos(9,900t)u(t) \qquad\qquad v_p(t) \approx 0.707\cos(9,900t + \pi/4)u(t)$

$\cos(10,000t)u(t) \qquad\qquad v_p(t) = \cos(10,000t)u(t)$

$\cos(10,100t)u(t) \qquad\qquad v_p(t) \approx 0.707\cos(9,900t - \pi/4)u(t)$

5.4 The Homogeneous Solution

The equation with zero input is called the homogeneous equation and its solution is called the homogeneous solution.

$$\frac{d^n y}{dt^n} + a_{n-1}\frac{d^{n-1}y}{dt^{n-1}} + \cdots + a_1\frac{dy}{dt} + a_0 y = 0, \quad t \geq 0$$

The only function that can satisfy the above equation is an exponential Ae^{st}. By substituting Ae^{st} into the differential equation we find

$$Ae^{st}(s^n + a_{n-1}s^{n-1} + \cdots + a_1 s + a_0) = 0$$

The above equation is satisfied if either $A = 0$ (which results in a zero homogeneous solution and is of no interest) or s is a root of the following equation (called the *characteristic equation*):

$$s^n + a_{n-1}s^{n-1} + \cdots + a_1 s + a_0 = 0$$

The characteristic equation of an nth-order differential equation has n roots s_k, $k = 1, 2, \ldots n$. Assume at first no multiple roots. Each exponential $A_k e^{s_k t}$ satisfies the homogeneous equation individually. However, for the n boundary conditions to be met, the homogeneous solution should contain all n exponentials. Therefore, the homogeneous solution is

$$y_h(t) = \sum_{k=1}^{n} A_k e^{s_k t}$$

$A_k e^{s_k t}$ is called a natural response and s_k is called a natural frequency as these have correlations with responses of physical systems.

 xample **5.15**

In the circuit of Figure 5.2 find the homogeneous responses in i_R, $i_C(t)$, $i_L(t)$, and $v(t)$.

Solution

The differential equations have the same characteristic equation and natural frequencies

$$s^2 + 200s + 10^8 = 0, \quad s_{1,2} = -100 \pm j\sqrt{10^8 - 10^4} \approx -100 \pm j10^4$$

The homogeneous solutions have the same functional form $y_h(t)$ shown below:

$$y_h(t) = e^{-100t}(\alpha e^{j10^4 t} + \beta e^{-j10^4 t}),$$

$$= e^{-100t}[A\cos(10^4 t) + B\sin(10^4 t)],$$

$$= Ce^{-100t}\cos(10^4 t + \theta),$$

In the above equation, α and β are complex conjugate constants. A, B, C, and θ are real-valued constants. The constants are to be determined by the input and the initial conditions for each variable.

5.5 Composing the Complete Solution

The complete solution needs to satisfy (1) the differential equation *and* (2) the initial conditions. You can easily see that the sum of the homogeneous and particular solutions can satisfy the differential equation. By setting coefficients A_k in the homogeneous solution to the appropriate values, the homogeneous solution makes it possible for the complete solution to satisfy the boundary conditions also. In this sense, the homogeneous solution complements the particular solution in meeting the boundary conditions and, therefore, is also called the complementary solution. The following recipe is, therefore, used to find the complete response:

1. From the characteristic equation find the natural frequencies and natural response with the unknown coefficients.
2. Find the forced response and form the complete response.
3. Determine the unknown coefficients of the natural response by matching the complete response to the initial conditions.

5.6 Examples of Complete Solutions

Example
5.16

Find the response of the first-order differential equation

$$y'(t) + 0.5y(t) = 1, \quad t \geq 0, \quad y(0) = 1$$

Solution
The forced response (from Table 5.1, or by inspection) is $y_p(t) = c$. By substituting $y_p(t)$ in the equation we find $c = 2$. However, we observe that $y_p(0) \neq 1$. Therefore, we need to add the natural response to meet the initial condition. To find the natural response we form the characteristic equation $s + 0.5 = 0$ and find its root at $s = -0.5$. The natural response is $y_h(t) = Ae^{-0.5t}$. The complete response is $y(t) = 2 + Ae^{-0.5t}$. The initial condition requires $y(0) = 2 + A = 1$, from which we find $A = -1$. The complete response is, therefore,

$$y(t) = 2 - e^{-0.5t}, \quad t \geq 0$$

xample

5.17

Find the solution to the second-order differential equation

$$y''(t) + 3y'(t) + 2y(t) = \sin 3t, \quad t \ge 0, \quad y(0) = 0, \quad y'(0) = 0$$

Solution

The characteristic equation is $s^2 + 3s + 2 = 0$. The natural frequencies are $s_{1,2} = -1, -2$. The natural response is $y_h(t) = A_1 e^{-t} + A_2 s^{-2t}$. The forced response is $y_p(t) = c_1 \sin 3t + c_2 \cos 3t$. By substituting $y_p(t)$ in the equation we find $c_1 = -7/130$ and $c_2 = -9/130$. The complete response is

$$y(t) = A_1 e^{-t} + A_2 s^{-2t} - \frac{1}{130}(7 \sin 3t + 9 \cos 3t)$$

Constants A_1 and A_2 are found from the initial conditions

$$y(0) = A_1 + A_2 - \frac{9}{130} = 0$$

$$y'(0) = -A_1 - 2A_2 - \frac{21}{130} = 0$$

These give $A_1 = 39/130$ and $A_2 = -30/130$. The complete response is, therefore,

$$y(t) = \frac{1}{130}\left(39e^{-t} - 30e^{-2t} - 7 \sin 3t - 9 \cos 3t\right), \quad t \ge 0$$

xample

5.18

Given

$$y''(t) + 2y'(t) + 5y(t) = x'(t) + x(t)$$

find $y(t)$ for $x(t) = \sin t, \quad t \ge 0, \quad y(0) = 0, \quad y'(0) = 0$.

Solution

The characteristic equation is $s^2 + 2s + 5 = 0$. The natural frequencies are $s_{1,2} = -1 \pm j2$. The natural response is

$$y_h(t) = A_1 e^{(-1+j2)t} + A_2 s^{(-1-j2)t}$$

Since $y_h(t)$ is a real function of t and the natural frequencies are complex conjugates of each other, the coefficients A_1 and A_2 must also be a complex conjugate pair: $A_1 = A_2^* = (C/2)e^{j\theta}$. The natural response may, therefore, be written as

$$y_h(t) = \frac{C}{2}e^{j\theta}e^{(-1+j2)t} + \frac{C}{2}e^{-j\theta}e^{(-1-j2)t}$$

$$= \frac{C}{2}e^{-t}\left[e^{j(2t+\theta)} + e^{-j(2t+\theta)}\right]$$

$$= Ce^{-t}\cos(2t + \theta)$$

The forced response is $y_p(t) = k \sin(t + \phi)$. By substituting $y_p(t)$ in the equation we find $k = 1/\sqrt{10} =$ and $\phi = \sin^{-1}(1/\sqrt{10}) = 18.44°$.[7] The complete response is

$$y(t) = y_h(t) + y_p(t) = Ce^{-t} \cos(2t + \theta) + \frac{1}{\sqrt{10}} \sin(t + 18.44°)$$

The two parameters C and θ are determined from the initial conditions.

$$y(0) = C \cos \theta + 0.1 = 0$$
$$y'(0) = -C \cos \theta - 2C \sin \theta + 0.3 = 0$$
$$C = \frac{\sqrt{5}}{10}, \quad \text{and} \quad \theta = \tan^{-1}(-2) = 116.56°$$

The complete response is then

$$y(t) = \frac{\sqrt{5}}{10} e^{-t} \cos(2t + 116.56°) + \frac{\sqrt{10}}{10} \sin(t + 18.44°)$$
$$= \frac{\sqrt{5}}{10} \left[e^{-t} \cos(2t + 116.56°) + \sqrt{2} \sin(t + 18.44°) \right], \quad t \geq 0$$

Alternative Solution

The homogeneous solution is $e^{-t}(C \cos 2t + D \sin 2t)$. The particular solution is $y_p(t) = A \cos t + B \sin t$. By substituting $y_p(t)$ in the equation, we find $A = 0.1$, $B = 0.3$, and $y_p(t) = 0.1 \cos t + 0.3 \sin t$.

The complete solution is

$$y(t) = y_h(t) + y_p(t) = e^{-t}(C \cos 2t + D \sin 2t) + 0.1 \cos t + 0.3 \sin t$$

The two parameters C and D are determined from the initial conditions.

$$y(0) = C + 0.1 = 0$$
$$y'(0) = -C + 2D + 0.3 = 0$$
$$C = -0.1 \quad \text{and} \quad D = -0.2$$

The complete solution is then

$$y(t) = \frac{1}{10} \left[\cos t + 3 \sin t - e^{-t}(\cos 2t + 2 \sin 2t) \right], \quad t \geq 0$$

Example
5.19

As previously mentioned in section 5.2, in circuits and systems theory we refer to the particular solution as the forced response and the homogeneous solution as the natural response. Exponent s is called the complex frequency and the roots of the characteristic equation are the natural frequencies. In some cases, the natural and forced responses, including a possible steady-state response and initial conditions, can be deduced directly from the properties of the system and its elements. The circuit

---

[7]The sinusoidal forced response may also be found more easily by employing the *system function* described in section 5.11.

of Figure 5.2 provides such an example. Find the responses of $i_R(t)$, $i_C(t)$, $i_L(t)$, and $v(t)$ to a unit-step input $v_s(t) = u(t)$.

Solution

The differential equations were obtained previously in Example 5.11. The particular solutions to $v_s(t) = u(t)$ and the homogeneous solutions were found in Examples 5.13 and 5.15, respectively. The particular solutions are specific to each variable but homogeneous solutions have the same functional form $e^{-at}(A \cos \omega t + B \sin \omega t)$, where $a = -100$ and $\omega = 10,000$. The total solution is the sum of the particular and homogeneous solutions. Since the functions and variables of the equation are all zero for $t < 0$, in the following table we have multiplied the total responses by $u(t)$. The initial conditions can be derived from the behavior of the differential equations in transition from $t = 0^-$ (where $v_s = 0$) to $t = 0^+$ (where $v_s = 1$). (In the present case, the initial conditions can also be derived directly from the circuit.) The results are summarized below.

1. $v(t) = e^{-at}(A_v \cos \omega t + B_v \sin \omega t)u(t)$, $v(0^+) = 0$, $v'(0^+) = 200$
2. $i_R(t) = [e^{-at}(A_R \cos \omega t + B_R \sin \omega t) + 2 \times 10^{-3}]u(t)$, $i_R(0^+) = 2 \times 10^{-3}$, $i'_R(0) = 2 \times 10^{-3}\delta(t)$
3. $i_L(t) = [e^{-at}(A_L \cos \omega t + B_L \sin \omega t) + 2 \times 10^{-3}]u(t)$, $i_L(0^+) = 0$, $i'_L(0^+) = 0$
4. $i_C(t) = e^{-at}(A_C \cos \omega t + B_C \sin \omega t)u(t)$, $i_C(0^+) = 2 \times 10^{-3}$, $i'_C(0) = 2 \times 10^{-3}\delta(t)$

The constants in (1) and (3) can be determined by application of the initial conditions. The results are

$$A_v = 0 \qquad \text{and} \quad B_v = 0.02 \text{ V/s}$$
$$A_L = -2 \times 10^{-3} \quad \text{and} \quad B_L = -2 \times 10^{-5} \text{ A/s}$$

Application of the initial conditions to (2) results in $A_R = 0$ but cannot determine B_R because it produces an identity with both sides containing an impulse function. A similar situation arises with (4), resulting in $A_C = 2 \times 10^{-3}$ but leaving B_C undetermined. The apparent difficulty can be overcome by first finding the solution to the following differential equation:

$$\frac{d^2y}{dt^2} + 200\frac{dy}{dt} + 10^8 y = u(t)$$

Then, by using the derivative and superposition properties of the LTI systems, construct the solutions to (2) and (4). The solution $y(t)$ to a unit-step input is given by the first line of the table below. Responses to the first two derivatives of the above input are obtained simply by taking the derivatives of the unit-step response. They are given by the second and third lines in the table, respectively.

Input	$\Longrightarrow$	Complete Solution
$v_s = u(t)$	$\Longrightarrow$	$y(t) = 10^{-8}\left[1 - e^{-100t}\left(\cos \omega t + 10^{-2} \sin \omega t\right)\right]u(t)$, $\omega = 10,000$
$\dfrac{dv_s}{dt}$	$\Longrightarrow$	$\dfrac{dy}{dt} = 10^{-4}e^{-100t}\left(\sin \omega t\right)u(t)$
$\dfrac{d^2 v_s}{dt^2}$	$\Longrightarrow$	$\dfrac{d^2 y}{dt^2} = e^{-100t}\left(\cos \omega t - 10^{-2} \sin \omega t\right)u(t)$

By applying superposition we can find the solutions for (2) and (4) in response to the input $v_s = u(t)$.

(2) $\dfrac{d^2 i_R}{dt^2} + 200\dfrac{di_R}{dt} + 10^8 i_R = \dfrac{1}{500}\dfrac{d^2 v_s}{dt^2} + 2 \times 10^5 v_s,$

$\qquad i_R(t) = \dfrac{1}{500}\dfrac{d^2 y}{dt^2} + 2 \times 10^5 y(t)$

$\qquad\qquad = 2 \times 10^{-3}\left(1 - 0.02e^{-100t}\sin\omega t\right)u(t)$

(4) $\dfrac{d^2 i_C}{dt^2} + 200\dfrac{di_C}{dt} + 10^8 i_C = \dfrac{1}{500}\dfrac{d^2 v_s}{dt^2},$

$\qquad i_C(t) = \dfrac{1}{500}\dfrac{d^2 y}{dt^2}$

$\qquad\qquad = 2 \times 10^{-3}e^{-100t}\left(\cos\omega t - 10^{-2}\sin\omega t\right)u(t)$

From the above responses one can easily verify that

(2) $i_R(0^+) = 2 \times 10^{-3}$ and $i_R'(0) = 2 \times 10^{-3}\delta(t)$
(4) $i_C(0^+) = 2 \times 10^{-3}$ and $i_C'(0) = 2 \times 10^{-3}\delta(t)$

5.7   Special Case: Multiple Roots

A root of an equation repeated p times is called a multiple root of the pth order. The contribution to the natural response from a multiple root s_k of order p is

$$(A_k + A_{k+1}t + A_{k+2}t^2 + \cdots + A_{k+p-1}t^{p-1})e^{-s_k t}$$

For example, the equation $s^2 + 2s + 1 - 0$ has a multiple root of order 2 at $s = -1$. The natural response due to it is $(A_1 + A_2 t)e^{-t}$.

 xample
5.20

Find the solution to the differential equation below.

$$y''' + 5y'' + 7y' + 3y = 0, \quad t \geq 0, \quad y(0) = 1, \quad y'(0) = 1, \quad y''(0) = 1$$

Solution
The characteristic equation is

$$s^3 + 5s^2 + 7s + 3 = 0$$

The roots are $s_{1,2} = -1$ and $s_3 = -3$. The solution contains the natural response only:

$$y(t) = (A_1 + A_2 t)e^{-t} + A_3 e^{-3t}$$
$$y'(t) = (-A_1 + A_2 - A_2 t)e^{-t} - 3A_3 e^{-3t}$$
$$y''(t) = (A_1 - 2A_2 + A_2 t)e^{-t} + 9A_3 e^{-3t}$$

From the initial conditions we see that

$$y(0) = A_1 + A_3 = 1$$
$$y'(0) = -A_1 + A_2 - 3A_3 = 1$$
$$y''(0) = A_1 - 2A_2 + 9A_3 = 1$$

Hence,

$$A_1 = 0, \quad A_2 = 4, \quad A_3 = 1, \quad \text{and} \quad y(t) = 4te^{-t} + e^{-3t}, \quad t \geq 0$$

5.8 When the Input Contains Natural Frequencies

An LTI system may be driven by a forcing function that has the same frequency as the natural frequency of the system; for example, $x(t) = e^{s_0 t}$, with s_0 being a root of the characteristic equation. In that case, the particular solution is $y_p(t) = Ate^{s_0 t}$. The factor A is obtained by substituting $y_p(t)$ in the equation.

 *Example* **5.21**

Find the solution of the differential equation

$$y'(t) + y(t) = e^{-t}, \quad t \geq 0, \text{ and } y(0) = 0$$

Solution
The homogeneous solution is Ae^{-t}. By substitution we find the particular solution to be te^{-t}. The complete solution is $y(t) = Ae^{-t} + te^{-t}$. From the initial condition we find $y(0) = A = 0$ and so $y(t) = te^{-t}, \ t \geq 0$.[8]

 Example **5.22**

Find the solution of the differential equation

$$y''(t) + y(t) = \cos t, \quad t \geq 0$$

given the following initial conditions:

$$\begin{cases} a) & y(0) = 0 \text{ and } y'(0) = 0 \\ b) & y(0) = 1 \text{ and } y'(0) = 0 \end{cases}$$

Solution
The particular solution and its derivatives are

$$y_p(t) = t(A \cos t + B \sin t)$$
$$y'(t) = A \cos t + B \sin t + t(-A \sin t + B \cos t)$$
$$y''(t) = (-A \sin t + B \cos t) + (-A \sin t + B \cos t) + t(-A \cos t - B \sin t)$$

---

[8]By inspection one can see that $y(t) = te^{-t}$ is the solution.

By substituting those into the differential equation we obtain

$$y''(t) + y(t) = (-A \sin t + B \cos t) + (-A \sin t + B \cos t)$$
$$+ t(-A \cos t - B \sin t) + t(A \cos t + B \sin t) = \cos t$$

which results in $A = 0$, $B = 0.5$, and $y_p(t) = 0.5t \sin t$. The homogeneous solution is $y_h(t) = C \cos t + D \sin t$. The complete solution is

$$y(t) = y_p(t) + y_h(t) = 0.5t \sin t + C \cos t + D \sin t$$

The initial conditions are $y(0) = C$ and $y'(0) = D$. The complete solutions for $t \geq 0$ are (a) $y(t) = 0.5t \sin t$ and (b) $0.5t \sin t + \cos t$.

5.9   When the Natural Response May Be Absent

The natural response would be absent if the forced response by itself meets the initial conditions. Example 5.22(*a*) was such a case. Four more examples are given below.

Example 5.23

Find the response of the first-order differential equation

$$y'(t) + 0.5y(t) = 1, \quad t \geq 0, \quad y(0) = 2$$

Solution
The forced response is $y_p(t) = 2$. The natural response is $y_h(t) = Ae^{-0.5t}$. The total response is $y(t) = 2 + Ae^{-0.5t}$. The initial condition requires $A = 0$. The complete response is, therefore,

$$y(t) = y_p(t) = 2, \quad t \geq 0$$

Example 5.24

Find the solution to the second-order differential equation

$$y''(t) + 3y'(t) + 2y(t) = \sin 3t, \quad t \geq 0, \quad y(0) = -\frac{9}{130}, \quad y'(0) = -\frac{21}{130}$$

Solution
This is the equation of Example 5.17, except for the different initial conditions. The particular solution was found to be

$$y_p(t) = -\frac{1}{130}(7 \sin 3t + 9 \cos 3t)$$

The above solution by itself meets the initial conditions: $y_p(0) = -\frac{9}{130}$ and $y'_p(0) = -\frac{21}{130}$. The natural response is, therefore, zero.

xample
5.25

Consider the differential equation

$$y'(t) + y(t) = \begin{cases} \cos t, & t \le t_0 \\ 0.5, & t > t_0 \end{cases}$$

Find t_0 so that $y(t) = 0.5$, $t > t_0$.

Solution

Before the switch the response is $y(t) = A \sin t + B \cos t$. The unknown coefficients A and B can be found by substitution:

$$y'(t) + y(t) = (A \cos t - B \sin t) + (A \sin t + B \cos t)$$
$$= (A + B) \cos t + (A - B) \sin t = \cos t$$

From which $A = B = 0.5$ and $y(t) = 0.5(\sin t + \cos t)$, $t \le t_0$. During the switch $y(t)$ remains continuous. After the switch the forced response is 0.5. For the natural response to disappear, the forced response alone must meet the initial condition, that is,

$$y(t_0) = 0.5(\sin t_0 + \cos t_0) = 0.5$$

For the above equation to be valid, the switch must act at the following instances:

$$\sin(t_0 + \pi 4) = \frac{\sqrt{2}}{2}, \qquad t_0 = k_0 \frac{\pi}{2} \pm 2k\pi, \quad k_0 = 0 \text{ or } 1, \text{ and } k \text{ an integer}$$
$$t_0 = \cdots - 7\pi/2, \ -2\pi, \ -3\pi/2, \ 0, \ \pi/2, \ 2\pi, \ 5\pi/2, \ldots$$

xample
5.26

Find the solution of the differential equation

$$y'(t) + y(t) = e^{-2t}, \quad t > 0, \text{ and } y(0) = -1$$

Solution

By inspection and substitution we find the particular solution to be $y_p(t) = -e^{-2t}$, which by itself satisfies the initial condition $y_p(0) = -1$. There is no natural component in the response.

5.10 Response to an Exponential Input

Consider the differential equation

$$\frac{d^n y}{dt^n} + a_{n-1} \frac{d^{n-1} y}{dt^{n-1}} + \cdots + a_1 \frac{dy}{dt} + a_0 y = e^{st}$$

where s is not a root of the characteristic equation. The particular response is found by substitution:

$$y_p(t) = \frac{1}{s^n + a_{n-1} s^{n-1} + \cdots + a_1 s + a_0} e^{st}$$

Regardless of the period during which the differential equation is specified, the particular response is proportional to the input. If the equation is valid for $-\infty < t < \infty$, then the complete response becomes proportional to the input as it is made of the particular response only. If the period of the equation is semi-infinite (e.g., $t \geq 0$) *and* the system is causal, then the total response will contain additional exponential components of natural frequencies whose strengths are found from the initial conditions.

5.11 The System Function

The response of the differential equation

$$\frac{d^n y}{dt^n} + a_{n-1}\frac{d^{n-1} y}{dt^{n-1}} + \cdots + a_1\frac{dy}{dt} + a_0 y = b_m\frac{d^m x}{dt^m} + b_{m-1}\frac{d^{m-1} x}{dt^{m-1}} + \cdots + b_1\frac{dx}{dt} + b_0 x$$

to an exponential input is an exponential

$$X_0 e^{st} \quad \Rightarrow \quad Y_0 e^{st}$$

where Y_0 is found by substitution:[9]

$$\left(s^n + a_{n-1}s^{n-1} + \cdots + a_1 s + a_0\right) Y_0 e^{st} = \left(b_m s^m + b_{n-1}s^{m-1} + \cdots + b_1 s + b_0\right) X_0 e^{st}$$

From the above equation, the amplitude of the output is $Y_0 = H(s)X_0$, where

$$H(s) = \frac{b_m s^m + b_{n-1}s^{m-1} + \cdots + b_1 s + b_0}{s^n + a_{n-1}s^{n-1} + \cdots + a_1 s + a_0}$$

It is seen that the output is proportional to the input. The proportionality factor, $H(s)$, is called the *system function*. It is easily obtained from the differential equation by employing the notation $s = \frac{d}{dt}$ to represent the derivative operator.[10] The system function not only gives the response amplitude to an exponential input, but also can provide the AC steady-state response and the response to any input expressed as a sum of exponentials or sinusoids. In addition, the denominator of the system function contains the natural frequencies of the system, which determine its natural response. In fact, the complete response may easily be written from the system function if the input and the initial conditions are given. This makes $H(s)$ a powerful tool in linear system analysis. The following example illustrates this point.

---

[9]The special case of s being a root of the characteristic equation was addressed in section 5.8.

[10]Using the D operator, the differential equation may be written as $D^n[y] = D^m[x]$, where D^n and D^m are linear operators acting on $y(t)$ and $x(t)$, respectively. They are constructed from sums of s^k, where s^k represents differentiation repeated k times.

$$D^n = s^n + a_{n-1}s^{n-1} + \cdots + a_1 s + a_0$$
$$D^m = b_m s^m + b_{m-1}s^{n-1} + \cdots + b_1 s + b_0$$

The system function is then

$$H(s) = \frac{D^m(s)}{D^n(s)} = \frac{b_m s^m + b_{m-1}s^{m-1} + \cdots + b_1 s + b_0}{s^n + a_{n-1}s^{n-1} + \cdots + a_1 s + a_0}$$

Example

5.27

Using the system function, find the solution to the differential equation

$$y''' + 5y'' + 7y' + 3y = 3x' + 2x$$

where $x = e^{-2t}$, $t \geq 0$, $y(0) = 0$, $y'(0) = 0$, and $y''(0) = 3$.

Solution

The system function is

$$H(s) = \frac{3s + 2}{s^3 + 5s^2 + 7s + 3}$$

The roots of the denominator polynomial are $s_{1,2} = -1$ and $s_3 = -3$. The natural response is

$$y_n(t) = (A_1 + A_2 t)e^{-t} + A_3 e^{-3t}$$

The system function evaluated at $s = -2$ (exponent of the input) is

$$H(s)\big|_{s=-2} = \frac{-3 \times 2 + 2}{-8 + 5 \times 4 - 7 \times 2 + 3} = -4$$

The forced response is

$$y_p(t) = H(s)|_{s=-2}\, e^{-2t} = -4e^{-2t}$$

The total response is

$$y(t) = y_h(t) + y_p(t) = (A_1 + A_2 t)e^{-t} + A_3 e^{-3t} - 4e^{-2t}$$

To find A_1, A_2, and A_3 we first find $y'(t)$ and $y''(t)$.

$$y'(t) = (-A_1 + A_2 - A_2 t)e^{-t} - 3A_3 e^{-3t} + 8e^{-2t}$$
$$y''(t) = (A_1 - 2A_2 + A_2 t)e^{-t} + 9A_3 e^{-3t} - 16e^{-2t}$$

From the initial conditions we get

$$y(0) = A_1 + A_3 - 4 = 0$$
$$y'(0) = -A_1 + A_2 - 3A_3 + 8 = 0$$
$$y''(0) = A_1 - 2A_2 + 9A_3 - 16 = 3$$

Solving for A_1, A_2, A_3 we have $A_1 = 2.25$, $A_2 = -0.5$, $A_3 = 1.75$. After substituting for A_1, A_2, A_3 we have the complete solution.

$$y(t) = (2.25 - 0.5t)e^{-t} + 1.75e^{-3t} - 4e^{-2t}, \quad t \geq 0$$

5.12 Sinusoidal Steady-State Response

Consider the differential equation with a sinusoidal input

$$\frac{d^n y}{dt^n} + a_{n-1}\frac{d^{n-1}y}{dt^{n-1}} + \cdots + a_1\frac{dy}{dt} + a_0 y = X_0 \cos(\omega t), \quad -\infty < t < \infty$$

where X_0 and ω are real constants. The input may be written as the sum of two exponentials with $s_1 = j\omega$ and $s_2 = -j\omega$:

$$X_0 \cos(\omega t) = \frac{X_0}{2}\left(e^{j\omega t} + e^{-j\omega t}\right)$$

From the discussion on exponential inputs and the linearity property we find the responses to the above components and their sum. The system function for $s = \pm j\omega$ may be written as

$$H(s)\big|_{s=\pm j\omega} = |H(\omega)|e^{\pm j\theta}$$

which is a real function of $j\omega$ (or a complex function of ω). For simplicity, in this book we use the following notation:

$$H(s)\big|_{s=j\omega} = H(\omega) = |H(\omega)|e^{j\theta}$$

Construction of sinusoidal steady-state responses from exponential inputs is given below.

Input	$\Rightarrow$	Response
$e^{j\omega t}$	$\Rightarrow$	$|H(\omega)|e^{j(\omega t+\theta)}$
$e^{-j\omega t}$	$\Rightarrow$	$|H(\omega)|e^{-j(\omega t+\theta)}$
$e^{j\omega t} + e^{-j\omega t}$	$\Rightarrow$	$2|H(\omega)|\cos(\omega t + \theta)$
$X_0 \cos \omega t$	$\Rightarrow$	$|H(\omega)|X_0 \cos(\omega t + \theta)$

The differential equation scales the magnitude of the sinusoidal input by the factor $|H(\omega)|$ and adds $\theta = \angle H(\omega)$ to its phase angle. The above result may also be derived directly from the real property of the system.

$$\cos \omega t = \mathcal{RE}\{e^{j\omega t}\} \;\Rightarrow\; \mathcal{RE}\{H(\omega)e^{j\omega t}\} = |H(\omega)|\cos(\omega t + \theta)$$

For more details on this subject see the discussion on the frequency response in Chapter 9.

5.13 Unit-Step Response

The solution of the following differential equation

$$\frac{d^n y}{dt^n} + a_{n-1}\frac{d^{n-1}y}{dt^{n-1}} + \cdots + a_1\frac{dy}{dt} + a_0 y = u(t)$$

is called the unit-step response. Assume a causal system $y(t) = 0$, $t < 0$. For $t > 0$, the response is made of the homogeneous solution and the particular solution.

$$y(t) = y_h(t) + y_p(t)$$

Let the roots of the characteristic equation be s_k, $k = 1, \ldots, n$, with none being a multiple root. The homogeneous solution is

$$y_h(t) = \sum_{k=1}^{n} A_k e^{s_k t}$$

The particular solution is a constant whose value may be found by substitution.

$$y_p(t) = \frac{1}{a_0}$$

The total solution is, therefore,

$$y(t) = \left[\sum_{k=1}^{n} A_k e^{s_k t} + \frac{1}{a_0} \right] u(t)$$

The constants A_k are found from the initial conditions of y and its $n - 1$ derivatives at $t = 0^+$. These remain at a zero value because the step jump in the right side of the differential equation produces a jump in the highest derivative on the left side and leaves the rest of the terms unchanged.[11]

$$y(0^+) = \frac{1}{a_0} + \sum_{k=1}^{n} A_k = 0$$

$$y'(0^+) = \sum_{k=1}^{n} A_k s_k = 0$$

$$y''(0^+) = \sum_{k=1}^{n} A_k s_k^2 = 0$$

$$\vdots$$

$$y^{(n-1)}(0^+) = \sum_{k=1}^{n} A_k s_k^{n-1} = 0$$

*E*xample 5.28

The circuit of Figure 5.3 is a second-order passive low-pass filter. Let $R = 1\ k\Omega$, $L = 1/(\sqrt{2}\pi)$ H and $C = 1/(\sqrt{2}\pi)\ \mu$F.

a. Find the differential equation that relates $v_2(t)$ to $v_1(t)$.

b. Find its unit-step response.

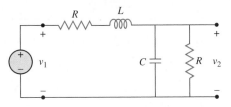

FIGURE 5.3 A second-order passive low-pass filter.

---

[11]For a broader discussion on the effect of discontinuity in the forcing function see section 5.15.

Solution

a.   Let i be the current flowing from left to right through the RL path. Then,

From KVL around the RL loop $Ri + L\dfrac{di}{dt} = v_1 - v_2$

From KCL at the output node $i = \dfrac{v_2}{R} + C\dfrac{dv_2}{dt}$

Eliminate i from the above equations to find

$$\frac{dv_2^2}{dt^2} + \left(\frac{1}{RC} + \frac{R}{L}\right)\frac{dv_2}{dt} + \frac{2}{LC}v_2 = \frac{1}{LC}v_1$$

Substitute for the element values to obtain

$$\frac{dv_2^2}{dt^2} + \sqrt{2}\omega_0\frac{dv_2}{dt} + \omega_0^2 v_2 = \frac{\omega_0^2}{2}v_1$$

where $\omega_0 \equiv \sqrt{\frac{2}{LC}} = 2{,}000\pi$.

b.   The characteristic equation and its roots are

$$s^2 + \sqrt{2}\omega_0 s + \omega_0^2 = 0, \quad s_{1,2} = -\omega_0\left(1 \pm j\right)/\sqrt{2}$$

The particular solution is a constant, and by substitution, its value is found to be 0.5. The total solution is

$$v_2(t) = \left[0.5 + e^{-\frac{\omega_0 t}{\sqrt{2}}}\left(A\cos\frac{\omega_0 t}{\sqrt{2}} + B\sin\frac{\omega_0 t}{\sqrt{2}}\right)\right]u(t)$$

The initial conditions are $v_2(0) = v_2'(0) = 0$. By using the initial conditions we find $A = -0.5$ and $B = -0.5/\omega_0$. The unit-step response is, therefore,

$$v_2(t) = 0.5\left[1 - e^{-\frac{\omega_0 t}{\sqrt{2}}}\left(\cos\frac{\omega_0 t}{\sqrt{2}} + \frac{1}{\omega_0}\sin\frac{\omega_0 t}{\sqrt{2}}\right)\right]u(t)$$

*E*xample
5.29

The active circuit of Figure 5.4 is called a Sallen-Key low-pass filter. Let $m = n = 1$, $R = 159\ \Omega$, $C = 1\ \mu F$, and $\beta \equiv \frac{R_1}{R_1 + R_2} = 0.63$.

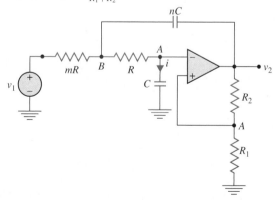

FIGURE 5.4 An equal-component Sallen-Key low-pass filter circuit.

a. Show that

$$\frac{dv_2^2}{dt^2} + \sqrt{2}\omega_0 \frac{dv_2}{dt} + \omega_0^2 v_2(t) = \frac{\omega_0^2}{\beta} v_1(t), \quad \text{where } \omega_0 = \frac{1}{RC} = 2,000\pi$$

b. Find its unit-step response.

Solution

a. The op-amp is operating in a noninverting configuration with a gain of $v_2/v_A = 1 + R_2/R_1 \equiv 1/\beta$. From the above observation and from application of KCL at nodes A and B we obtain the following three equations:

$$\text{Noninverting amplifier:} \quad v_A = \beta v_2 \tag{1}$$

$$\text{KCL at node A:} \quad \frac{v_A - v_B}{R} + C\frac{dv_A}{dt} = 0 \tag{2}$$

$$\text{KCL at node B:} \quad \frac{v_B - v_A}{R} + \frac{v_B - v_1}{R} + C\frac{d(v_B - v_2)}{dt} = 0 \tag{3}$$

Eliminate V_A and V_B between the above three equations to find

$$\frac{dv_2^2}{dt^2} + \frac{1}{RC}\left(3 - \frac{1}{\beta}\right)\frac{dv_2}{dt} + \frac{1}{R^2C^2}v_2(t) = \frac{1}{R^2C^2}\frac{1}{\beta}v_1(t)$$

Substitute for element values to obtain

$$\frac{dv_2^2}{dt^2} + \sqrt{2}\omega_0\frac{dv_2}{dt} + \omega_0^2 v_2(t) = \frac{\omega_0^2}{\beta}v_1(t)$$

where $\omega_0 = \frac{1}{RC} = 2,000\pi$.

b. From the linearity property we deduce that the unit-step response is the same as that of Example 5.28 but scaled by $2/\beta$.

5.14 Unit-Impulse Response

The solution of the following differential equation

$$\frac{d^n y}{dt^n} + a_{n-1}\frac{d^{n-1}y}{dt^{n-1}} + \cdots + a_1\frac{dy}{dt} + a_0 y = \delta(t)$$

is called the unit-impulse response and shown by $h(t)$. The unit-impulse response is the solution of the homogeneous equation for $t > 0$.

$$h(t) = \sum_{k=1}^{n} B_k e^{s_k t}$$

The constants B_k are found from the nonzero initial conditions of y and its $n - 1$ derivatives at $t = 0^+$. To find these conditions we examine the two sides of the equation during $0^- < t < 0^+$. During this period the right side of the equation contains a unit impulse $\delta(t)$. The left side should also contain an impulse of the same order and strength.

Such an impulse can appear only in the highest-order derivative of y:

$$\frac{d^n y}{dt^n}(t) = \delta(t), \quad \text{during } 0^- < t < 0^+$$

which means a unit jump in $y^{(n-1)}$ at $t = 0$. Since $y^{(n-1)}(0^-) = 0$, we conclude that $y^{(n-1)}(0^+) = 1$. All other lower-level derivatives remain zero through the transition from 0^- to 0^+. The equations that determine the constants B_k are

$$y(0^+) = \sum_{k=1}^{n} B_k = 0$$

$$y'(0^+) = \sum_{k=1}^{n} B_k s_k = 0$$

$$y''(0^+) = \sum_{k=1}^{n} B_k s_k^2 = 0$$

$$\vdots$$

$$y^{(n-1)}(0^+) = \sum_{k=1}^{n} B_k s_k^{n-1} = 1$$

Alternate Method

The unit-impulse response is the time derivative of the unit-step response:

$$h(t) = \frac{d}{dt} g(t) = \frac{d}{dt}\left[\left(\sum_{k=1}^{n} A_k e^{s_k t} + \frac{1}{a_0}\right) u(t)\right]$$

$$= \sum_{k=1}^{n} A_k s_k e^{s_k t} u(t) + \left(\sum_{k=1}^{n} A_k e^{s_k t} + \frac{1}{a_0}\right)\bigg|_{t=0} \delta(t)$$

But,

$$\left(\sum_{k=1}^{n} A_k e^{s_k t} + \frac{1}{a_0}\right)\bigg|_{t=0} = y(0^+) = 0.$$

Therefore,

$$h(t) = \sum_{k=1}^{n} A_k s_k e^{s_k t} u(t).$$

*E*xample
5.30

Find the unit-step and unit-impulse responses of the passive low-pass *RC* and high-pass *CR* filters.

Solution

The input-output differential equations and responses are derived and included in Figure 5.5. Note that, as expected, the unit-step responses of the two circuits add up to a unit step, and the sum of the unit-impulse responses is a unit impulse.

Circuit   Differential Equation   Unit-Step Response   Unit-Impulse Response

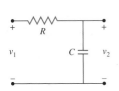

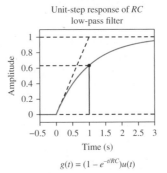

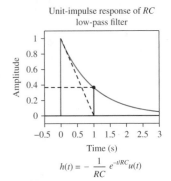

$$\frac{dv_2}{dt} + \frac{1}{RC} v_2 = \frac{1}{RC} v_1$$

$$g(t) = (1 - e^{-t/RC})u(t)$$

$$h(t) = -\frac{1}{RC} e^{-t/RC} u(t)$$

FIGURE 5.5 (a)

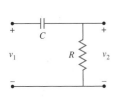

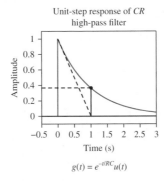

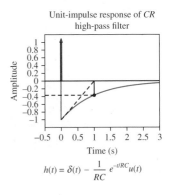

$$\frac{dv_2}{dt} + \frac{1}{RC} v_2 = \frac{dv_1}{dt}$$

$$g(t) = e^{-t/RC} u(t)$$

$$h(t) = \delta(t) - \frac{1}{RC} e^{-t/RC} u(t)$$

FIGURE 5.5 (b)

5.15 Effect of Discontinuity in the Forcing Function

Consider an LTI system described by the differential equation

$$\frac{d^n y}{dt^n} + a_{n-1}\frac{d^{n-1}y}{dt^{n-1}} + \cdots + a_1\frac{dy}{dt} + a_0 y = b_m\frac{d^m x}{dt^m} + b_{m-1}\frac{d^{m-1}x}{dt^{m-1}} + \cdots + b_1\frac{dx}{dt} + b_0 x$$

Let the input $x(t)$ contain a discontinuity (e.g., unit jump, unit impulse, etc.) at t_0. What is the effect of such a discontinuity on $y(t)$ and its derivatives at t_0? This question has direct relevance to the derivation of initial conditions at t_0^+ based on their values at t_0^-. For simplicity we assume $t_0 = 0$, but the conclusions are valid for any t_0. In this section we examine the effect of a unit jump in $x(t)$.

Unit Jump

A unit jump in the input at $t = 0$ means $x(0^+) = x(0^-) + 1$. Let the following quantities be known at $t = 0^-$:

$$y^{(n-1)}(0^-), \ldots, y''(0^-), \quad y'(0^-), \quad y(0^-)$$

It is not readily clear whether $y(t)$ and its derivatives are, or are not, continuous at $t = 0$. This depends on the structure of the differential equation. The unit jump in the input at $t = 0$ produces a sequence of progressively higher-order impulses (i.e., impulse, doublet, and their higher derivatives) in the right side of the equation, the highest order of which is an impulse of order m with strength b_m. In the functional sense, an impulse of order m corresponds to the mth derivative of a unit-step function, the Dirac δ (unit-impulse) function being an impulse of the first order and a unit doublet being an impulse of second order. Using the symbol $\delta^m(t)$ to show a unit inpulse of order m, the right side of the equation in the vicinity of $t = 0$ will be $b_m\delta^m(t)$. Since the differential equation is valid during the period $0^- \le t \le 0^+$, the left side should also contain an impulse of the same order and strength: $b_m\delta^m(t)$. Such an impulse can appear only in the highest derivative of y; that is,

$$y^{(n)}(t) = b_m\delta^m \quad \text{in the vicinity of } t = 0$$

However, this may or may not give us all of the information we would like to have about the post-jump conditions; in other words,

$$y^{(n-1)}(0^+), \ldots, y''(0^+), \quad y'(0^+), \quad y(0^+)$$

The analytic solution and simulation of systems represented by the differential equation are simplified if a finite jump in the input does not change the n initial conditions (y and its $n - 1$ derivatives). That happens when the right side of the differential equation is limited to $b_0 x(t)$ and has no derivative components.

Example 5.31

In the following differential equations find the initial conditions at $t = 0^+$ necessary for obtaining the complete response to

a.   a unit impulse $x(t) = \delta(t)$

b.   a unit step $x(t) = u(t)$

c.   a unit ramp $x(t) = tu(t)$. Note that $y(t)$ and all its derivatives are zero at $t = 0^-$.

  (i) $y' + 2y = x$
  (ii) $y' + 2y = 3x$
  (iii) $y' + 2y = x'$
  (iv) $y' + 7y = 5x'$
  (v) $y'' + 2y' + 3y = 2x$
  (vi) $y'' + 2y' + 3y = x' + x$
  (vii) $y'' + 3y' + 40y = 4x'$
  (viii) $y'' + 5y' + 4y = x' + 2x$
  (ix) $y''' + 5y'' + 6y' + y = 2x$
  (x) $y''' + 5y'' + 6y' + y = 3x' + 2x$

Solution

a. For $x(t) = \delta(t)$ we have

	Differential Equation	$y(0^+)$	$y'(0^+)$	$y''(0^+)$
(i)	$y' + 2y = x$	1		
(ii)	$y' + 2y = 3x$	3		
(iii)	$y' + 2y = x'$	$\delta(t)$		
(iv)	$y' + 7y = 5x'$	$5\delta(t)$		
(v)	$y'' + 2y' + 3y = 2x$	0	2	
(vi)	$y'' + 2y' + 3y = x' + x$	1	$\delta(t)$	
(vii)	$y'' + 3y' + 40y = 4x'$	4	$4\delta(t)$	
(viii)	$y'' + 5y' + 4y = x' + 2x$	1	$\delta(t)$	
(ix)	$y''' + 5y'' + 6y' + y = 2x$	0	0	2
(x)	$y''' + 5y'' + 6y' + y = 3x' + 2x$	0	3	$3\delta(t)$

b. For $x(t) = u(t)$ we have

	Differential Equation	$y(0^+)$	$y'(0^+)$	$y''(0^+)$
(i)	$y' + 2y = x$	0		
(ii)	$y' + 2y = 3x$	0		
(iii)	$y' + 2y = x'$	1		
(iv)	$y' + 7y = 5x'$	5		
(v)	$y'' + 2y' + 3y = 2x$	0	0	
(vi)	$y'' + 2y' + 3y = x' + x$	0	1	
(vii)	$y'' + 3y' + 40y = 4x'$	0	4	
(viii)	$y'' + 5y' + 4y = x' + 2x$	0	1	
(ix)	$y''' + 5y'' + 6y' + y = 2x$	0	0	0
(x)	$y''' + 5y'' + 6y' + y = 3x' + 2x$	0	0	3

c. For $x(t) = tu(t)$ we have

	Differential Equation	$y(0^+)$	$y'(0^+)$	$y''(0^+)$
(i)	$y' + 2y = x$	0		
(ii)	$y' + 2y = 3x$	0		
(iii)	$y' + 2y = x'$	0		
(iv)	$y' + 7y = 5x'$	0		
(v)	$y'' + 2y' + 3y = 2x$	0	0	
(vi)	$y'' + 2y' + 3y = x' + x$	0	0	
(vii)	$y'' + 3y' + 40y = 4x'$	0	0	
(viii)	$y'' + 5y' + 4y = x' + 2x$	0	0	
(ix)	$y''' + 5y'' + 6y' + y = 2x$	0	0	0
(x)	$y''' + 5y'' + 6y' + y = 3x' + 2x$	0	0	0

5.16 Solution by Convolution

A linear differential equation with constant coefficients represents an LTI relationship between the input and output. In the previous sections it was shown how to obtain the unit-impulse response, $h(t)$. One expects that, like any LTI system, such a differential equation is also completely characterized by its unit-impulse response. This is in fact true. From $h(t)$ one can construct the differential equation in its familiar form. The simple case of a first-order equation, given the unit-impulse response, was presented in Example 5.10. Formal methods of such modeling use the Laplace transform and frequency response methods. It is, therefore, not surprising that the solution of an LTI differential equation may be obtained by convolving its unit-impulse response with the input. In other words, the convolution integral satisfies the differential equation and thus, constitutes the complete solution. See Example 5.32 below. The use of convolution as an instrument to obtain the solution of a differential equation is illustrated by Examples 5.33 and 5.34.

Consider the differential equation

$$\frac{dy}{dt} + ay(t) = x(t)$$

The unit-impulse response of the equation is

$$h(t) = e^{-at}u(t)$$

Verify that the convolution integral $y(t) = x(t) \star h(t)$ satisfies the equation.

Solution

$$y(t) = \int_{-\infty}^{t} x(\tau)e^{-a(t-\tau)}d\tau = e^{-at}\int_{-\infty}^{t} x(\tau)e^{a\tau}d\tau$$

$$\frac{dy}{dt} = -ae^{-at}\int_{-\infty}^{t} x(\tau)e^{a\tau}d\tau + e^{-at}\frac{d}{dt}\int_{-\infty}^{t} x(\tau)e^{a\tau}d\tau$$

$$= -ae^{-at}\int_{-\infty}^{t} x(\tau)e^{a\tau}d\tau + e^{-at}x(t)e^{at}$$

By substitution,

$$\frac{dy}{dt} + ay(t) = x(t).$$

Given the unit-impulse response of an LTI differential equation, use convolution to obtain its solution to the input e^{st}. Compare the result with the response obtained by the direct method in sections 5.10 and 5.11 and conclude a relationship between the two methods.

Solution

Let $h(t)$ and $y(t)$ represent the unit-impulse response and the solution to $x(t) = e^{st}$, respectively. Then,

$$y(t) = x(t) \star h(t) = \int_{-\infty}^{\infty} h(\tau)e^{s(t-\tau)}d\tau = e^{st}\int_{-\infty}^{\infty} h(\tau)e^{-s\tau}d\tau = H(s)e^{st}$$

The above result confirms the previous observation that the response of an LTI system to an exponential is an exponential of the same form but magnitude-scaled by a factor $H(s)$, called the system function (see sections 5.10 and 5.11). The result obtained in the present example also suggests the following relationship:

$$H(s) = \underbrace{\int_{-\infty}^{\infty} h(\tau)e^{-s\tau}d\tau}_{\text{Example 5.33}} = \underbrace{\frac{b_m s^m + b_{n-1}s^{m-1} + \cdots + b_1 s + b_0}{s^n + a_{n-1}s^{n-1} + \cdots + a_1 s + a_0}}_{\text{Section 5.11}}$$

Consider the differential equation

$$\frac{d^2 y}{dt^2} + 4\frac{dy}{dt} + 3y = 2\frac{dx}{dt} + 8x$$

a. Verify that the response to $x(t) = e^{st}$ is $y(t) = H_a e^{st}$ and determine H_a.

b. Verify that the unit-impulse response is $h(t) = (3e^{-t} - e^{-3t})u(t)$.

c. Show that

$$H_b = \int_{-\infty}^{\infty} h(\tau)e^{-s\tau}d\tau = H_a(s)$$

Solution

a. Substitute e^{st} for $x(t)$ and collect terms.

$$\frac{d^2 y}{dt^2} + 4\frac{dy}{dt} + 3y = (2s + 8)e^{st}$$

The only solution to the above equation is $y(t) = H_a e^{st}$, which, by substitution, results in

$$(s^2 + 4s + 3)H_a e^{st} = (2s + 8)e^{st} \implies H_a = \frac{2s + 8}{s^2 + 4s + 3}$$

b. Substitute $h(t)$ and its derivatives in the left-hand side of the equation and verify that it equals the right-hand side with $x(t) = \delta(t)$.

$$h(t) = (3e^{-t} - e^{-3t})u(t)$$

$$\frac{dh}{dt} = (-3e^{-t} + 3e^{-3t})u(t) + 2\delta(t)$$

$$\frac{d^2 h}{dt^2} = (3e^{-t} - 9e^{-3t})u(t) + 2\delta'(t)$$

$$\frac{d^2 h}{dt^2} + 4\frac{dh}{dt} + 3h(t) = 8\delta(t) + 2\delta'(t)$$

c.

$$H_b = 3 \int_0^\infty e^{-\tau} e^{-s\tau} d\tau - \int_0^\infty e^{-3\tau} e^{-s\tau} d\tau$$

$$= \frac{-3}{s+1} \left[e^{-(s+1)\tau} \right]_{\tau=0}^\infty + \frac{1}{s+3} \left[e^{-(s+3)\tau} \right]_{\tau=0}^\infty$$

$$= \frac{3}{s+1} - \frac{1}{s+3} = \frac{2s+8}{s^2+4s+3}, \quad \mathcal{RE}(s) > -1, \quad H_b = H_a$$

Example 5.35

To prevent aliasing in sampling a speech signal $v_1(t)$, the signal is first passed through a first-order low-pass filter (called an antialiasing filter) such as the *RC* filter of Figure 5.5(*a*) made of a 530-Ω resistor and a 100-nF capacitor.

a. Show that the filter attenuates the power in frequencies above 3 kHz by a factor of two or more.

b. The *RC* filter may be replaced by a computer software program that performs convolution between $v_1(t)$ and the filter's unit-impulse response $h(t)$. Find $h(t)$ and construct the convolution integral. Then observe that the resulting convolution integral performs a weighted integration and averaging as discussed in Chapter 1.

Solution

The differential equation of the RC filter is

$$\frac{dv_2}{dt} + \frac{1}{RC} v_2 = \frac{1}{RC} v_1, \quad RC = 53 \times 10^{-6}$$

The response to a sinusoidal input $v_1(t) = \cos \omega t$ is $v_2(t) = |H(\omega)| \cos(\omega t + \theta)$ (see section 5.12). The ratio of output-to-input power is

$$|H(\omega)|^2 = \frac{1}{1 + \left(\frac{\omega}{\omega_0}\right)^2}, \quad \text{where } \omega_0 = 1/(RC) = 6{,}000\pi \quad \text{(corresponding to 3 kHz)}.$$

For $f \geq 3$ kH, the above ratio is less than 0.5. The unit-impulse response of the filter and the convolution integral are

$$h(t) = \frac{1}{RC} e^{-t/RC} u(t), \quad v_2(t) = \frac{1}{RC} e^{-t/RC} \int_{-\infty}^t v_1(\tau) e^{\tau/RC} d\tau$$

The convolution introduces an exponential weighting factor in $v_1(t)$ before integration. This is similar to section 1.20 in Chapter 1.

5.17 Zero-Input and Zero-State Responses

Consider an LTI system described by the differential equation

$$\frac{d^n y}{dt^n} + a_{n-1} \frac{d^{n-1} y}{dt^{n-1}} + \cdots + a_1 \frac{dy}{dt} + a_0 y = f(t), \quad t \geq 0$$

with nonzero initial conditions at $t = 0$. From the previous discussions, one expects that the complete response at $t \geq 0$ is affected by two factors: (i) the initial state at $t = 0$ and

(ii) the input during $t \geq 0$. Due to the linearity property of the differential equation, one expects the complete response to be the sum of the above two responses. One is due to the initial conditions only and with no input; hence, the zero-input response. The other is due to the input only and with zero initial conditions; hence, the zero-state response. Therefore,

$$y(t) = y_1(t) + y_2(t)$$

where $y_1(t)$ and $y_2(t)$ are zero-input and zero-state responses of the system, respectively.

Zero Input

The zero-input response $y_1(t)$ is the complete solution of the following equation

$$\frac{d^n y_1}{dt^n} + a_{n-1}\frac{d^{n-1}y_1}{dt^{n-1}} + \cdots + a_1\frac{dy_1}{dt} + a_0 y_1 = 0, \quad t \geq 0$$

with nonzero initial conditions

$$y_1^{(n-1)}(0), \ldots, y_1''(0), \quad y_1'(0), \quad y_1(0).$$

The solution is made of natural responses only

$$y_1(t) = \sum_{k=1}^{n} B_k e^{s_k t}$$

where the coefficients B_k are found from the following n equations:

$$y(0) = \sum_{k=1}^{n} B_k$$

$$y'(0) = \sum_{k=1}^{n} B_k s_k$$

$$y''(0) = \sum_{k=1}^{n} B_k s_k^2$$

$$\vdots$$

$$y^{(n-1)}(0) = \sum_{k=1}^{n} B_k s_k^{n-1}$$

Zero State

The zero-state response $y_2(t)$ is the complete solution of the following equation:

$$\frac{d^n y_2}{dt^n} + a_{n-1}\frac{d^{n-1}y_2}{dt^{n-1}} + \cdots + a_1\frac{dy_2}{dt} + a_0 y_2 = f(t), \quad t \geq 0$$

with zero initial conditions

$$y_2^{(n-1)}(0) = \cdots = y_2''(0) = y_2'(0) = y_2(0) = 0$$

The solution is the sum of the homogeneous and particular solutions

$$y_2(t) = y_p(t) + \sum_{k=1}^{n} C_k e^{s_k t}$$

where the coefficients C_k are found from the following equations:

$$y_p(0) + \sum_{k=1}^{n} C_k = 0$$

$$y'_p(0) + \sum_{k=1}^{n} C_k s_k = 0$$

$$y''_p(0) + \sum_{k=1}^{n} C_k s_k^2 = 0$$

$$\vdots$$

$$y_p^{(n-1)}(0) + \sum_{k=1}^{n} C_k s_k^{n-1} = 0$$

By linearity, the complete response is

$$y(t) = y_1(t) + y_2(t) = \sum_{k=1}^{n} (B_k + C_k) e^{s_k t} + y_p(t)$$

One may easily verify that $B_k + C_k = A_k$, $k = 1, \ldots, n$, where the constants A_k are the coefficients of the natural components found in section 5.5.

Example 5.36

Consider the system described by the differential equation

$$y''(t) + 5y'(t) + 6y(t) = x(t)$$

a. Find the zero-input response for $x(t) = 0$, $t \geq 0$, $y(0) = 2$, $y'(0) = -5$.

b. Find the zero-state response for $x(t) = 2e^{-t}$, $t \geq 0$, $y(0) = 0$, $y'(0) = 0$.

c. Find the complete response for $x(t) = 2e^{-t}$, $t \geq 0$, $y(0) = 2$, $y'(0) = -5$ and verify that the zero-input and zero-state responses add up to the complete response.

Solution

a. The zero-input response $y_1(t)$ is made of the natural response only. From the characteristic equation $s^2 + 5s + 6 = 0$ we find $y_1(t) = B_1 e^{-2t} + B_2 s^{-3t}$. The initial conditions require that

$$y_1(0) = B_1 + B_2 = 2$$
$$y'_1(0) = -2B_1 - 3B_2 = -5$$

from which $B_1 = B_2 = 1$ and so

$$y_1(t) = e^{-2t} + e^{-3t}, \quad t \geq 0$$

b.   The zero-state response is $y_2(t) = C_1 e^{-2t} + C_2 s^{-3t} + e^{-t}$, in which e^{-t} is the particular solution. To satisfy the initial conditions we need

$$y_2(0) = C_1 + C_2 + 1 = 0$$
$$y_2'(0) = -2C_1 - 3C_2 - 1 = 0$$

from which $C_1 = -2$, $C_2 = 1$ and so

$$y_2(t) = -2e^{-2t} + e^{-3t} + e^{-t}, \quad t \geq 0$$

c.   The complete response to $x(t) = 2e^{-t}$, $t > 0$, is $y(t) = A_1 e^{-2t} + A_2 s^{-3t} + e^{-t}$. However, the initial conditions require that

$$y(0) = A_1 + A_2 + 1 = 2$$
$$y'(0) = -2A_1 - 3A_2 - 1 = -5$$

from which $A_1 = -1$, $A_2 = 2$ and so

$$y(t) = -e^{-2t} + 2e^{-3t} + e^{-t}, \quad t \geq 0$$

Note that $A_1 = B_1 + C_1$ and $A_2 = B_2 + C_2$. Consequently, $y(t) = y_1(t) + y_2(t)$.

5.18 Zero-State Response and Convolution

The zero-state response of a differential equation may also be obtained by the convolution of its unit-impulse response with a function that is zero for $t < 0$ and equal to the input of the differential equation for $t > 0$. This is illustrated by the following example.

Find the zero-state response of the differential equation of Example 5.36 by convolution.

Solution
The differential equation is

$$y''(t) + 5y'(t) + 6y(t) = x(t)$$

The unit-impulse response is $h(t) = (e^{-2t} - e^{-3t})u(t)$ and $x(t) = e^{-t}$ for $t > 0$. The zero-state response is

$$y_2(t) = x(t) \star h(t) = \int_0^t e^{\tau - t} \left(e^{-2\tau} - e^{-3\tau} \right) d\tau = \left(e^{-t} - 2e^{-2t} + e^{-3t} \right) u(t)$$

The convolution produces the same solution as the zero-state response obtained in Example 5.36 by the classical method.

5.19   Properties of LTI Differential Equations

A linear differential equation with constant coefficients establishes a linear time-invariant relationship between the input $x(t)$ and the output $y(t)$. The most widely used properties of such a relationship are

1. Linearity
2. Conjugate symmetry
3. Derivative and integral

Table 5.2 summarizes some properties of linear time-invariant systems described by linear differential equations with real constant coefficients.

TABLE 5.2 Properties of Linear Differential Equations with Real Constant Coefficients

	Property	Input	$\Rightarrow$	Output
1	Basic pair	$x(t)$	$\Rightarrow$	$y(t)$
2	Linearity	$ax_1(t) + bx_2(t)$	$\Rightarrow$	$ay_1(t) + by_2(t)$
3	Complex function	$\mathcal{RE}[x(t)] + \mathcal{IM}[x(t)]$	$\Rightarrow$	$\mathcal{RE}[y(t)] + \mathcal{IM}[y(t)]$
4	Real part	$\mathcal{RE}[x(t)]$	$\Rightarrow$	$\mathcal{RE}[y(t)]$
5	Imaginary part	$\mathcal{IM}[x(t)]$	$\Rightarrow$	$\mathcal{IM}[y(t)]$
6	Time delay	$x(t - t_0)$	$\Rightarrow$	$y(t - t_0)$
7	Differentiation	$\dfrac{dx(t)}{dt}$	$\Rightarrow$	$\dfrac{dy(t)}{dt}$
8	Integration	$\displaystyle\int_{-\infty}^{t} x(t)\, dt$	$\Rightarrow$	$\displaystyle\int_{-\infty}^{t} y(t) dt$

5.20   Solution by Numerical Methods

A differential equation may be approximated by a discrete equation[12] by substituting for the derivatives of $y(t)$. This may be done in several ways. In the backward approximation method (also called the backward Euler algorithm),

$$\frac{dy(t)}{dt} \approx \frac{y(t) - y(t - \Delta t)}{\Delta t}$$

The time is incremented by Δt, while $t = n\Delta t$, $\infty < n < \infty$, $y(t)$, and its derivatives become functions of the discrete variable n as listed below.

$$y(n\Delta t) \Rightarrow y(n)$$

$$y'(n\Delta t) \Rightarrow \frac{y(n) - y(n - 1)}{\Delta t}$$

---

[12]This results in a difference equation.

$$y''(n\Delta t) \Rightarrow \frac{y(n) - 2y(n-1) + y(n-2)}{\Delta t^2}$$

$$y'''(n\Delta t) \Rightarrow \frac{y(n) - 3y(n-1) + 3y(n-2) - y(n-3)}{\Delta t^3}$$

$y(n)$ is calculated iteratively from the weighted finite sums of its own past values and the input's present and past values sampled at intervals of T.

This method may be implemented graphically, or on a computer. The method may also be applied to nonlinear and time-varying differential equations. The method is computationally inefficient and the solution is not in closed form. However, it may be used when input data is given in tabular form or flows in real time.

*E**xample**
5.38

Convert the following second-order differential equation to a difference equation using the backward Euler algorithm.

$$y''(t) + 2\zeta\omega_0 y'(t) + \omega_0^2 y(t) = x(t)$$

Solution

Let the time increment be T. Substituting for $y(t)$ and its derivatives and collecting terms of $y(n)$, $y(n-1)$, and $y(n-2)$, we obtain

$$\left(\frac{1}{T^2} + \frac{2\zeta\omega_0}{T} + \omega_0^2\right) y(n) - \left(\frac{2}{T^2} + \frac{2\zeta\omega_0}{T}\right) y(n-1) + \left(\frac{1}{T^2}\right) y(n-2) = x(n)$$

In simplified form,

$$y(n) = b_0 x(n) - a_1 y(n-1) - a_2 y(n-2)$$

where,

$$b_0 = \frac{T^2}{1 + 2\zeta\omega_0 T + \omega_0^2 T^2}$$

$$a_1 = -\frac{2 + 2\zeta\omega_0 T}{1 + 2\zeta\omega_0 T + \omega_0^2 T^2}$$

$$a_2 = \frac{1}{1 + 2\zeta\omega_0 T + \omega_0^2 T^2}$$

5.21 Concluding Remarks

This chapter has presented a detailed solution method for linear differential equations with constant coefficients. The method recognizes the correlation of the homogeneous and particular parts of the solution with familiar components of the responses in physical systems—the natural and forced responses. Many physical systems of interest are non-linear, which may seem to make the present chapter idealized and irrelevant. Such is not the case, however. Linearization of complex nonlinear systems may require extensive analytical manipulation of equations, a task that was daunting several decades ago and made practical applications less likely. The help provided by computers, however, has made the use of the smalll scale piecewise linear model approach a practical tool for analysis and design. Examples and problems in this chapter are designed primarily to clarify concepts, but they also provide brief insights into some broader aspects of system analysis such as modeling and system functions.

5.22 Problems

Solved Problems

Find $y(t)$, the solution to the differential equation in each of problems 1–6.

1. $\dfrac{d^2 y}{dt^2} + 2\dfrac{dy}{dt} + y = x(t) - \dfrac{dx}{dt}$, $t > 0$, with $y(0) = 1$, $\dfrac{dy}{dt}(0) = -1$, and $x(t) = e^{-3t} - e^{-2t}$, $t > 0$

Solution
The characteristic equation and natural frequencies are

$$s^2 + 2s + 1 = 0, \quad s_{1,2} = -1$$

The homogeneous solution is

$$y_h(t) = (A + Bt)e^{-t}$$

The system function and its values at s corresponding to the right-hand side of the differential equation are

$$H(s) = \frac{-s + 1}{s^2 + 2s + 1}, \quad H_1 = H(s)\big|_{s=-3} = 1, \quad \text{and } H_2 = H(s)\big|_{s=-2} = 3$$

The particular solution is

$$y_p(t) = H_1 e^{-3t} - H_2 e^{-2t} = e^{-3t} - 3e^{-2t}$$

The total solution is

$$y(t) = y_p(t) + y_h(t) = e^{-3t} - 3e^{-2t} + (A + Bt)e^{-t}$$

By applying the initial conditions we find A and B.

$$\begin{cases} y(0) = A + 1 - 3 = 1, & A = 3 \\ y'(0) = B - A - 3 + 6 = -1, & B = -1 \end{cases}$$

The solution to the differential equation is, therefore,

$$\boxed{y(t) = e^{-3t} - 3e^{-2t} + (3 - t)e^{-t}, \quad t > 0}$$

2. $\dfrac{d^2 y}{dt^2} + 200\dfrac{dy}{dt} + 10^8 y = 200\dfrac{dx}{dt}$, $t > 0$, with $y(0^+) = 0$, $\dfrac{dy}{dt}(0^+) = 200$, and $x(t) = 1$, $t > 0$

Solution

$$s^2 + 200s + 10^8 = 0, \quad s_{1,2} \approx -100 \pm j10^4, \quad y_h(t) = e^{-100t}\left(A\cos 10^4 t + B\sin 10^4 t\right), \quad y_p(t) = 0$$

By applying the initial conditions, we find A, B, and the total solution.

$$y(0) = A = 0, \quad y'(0) = 10^4 B = 200, \quad B = 0.02$$

$$\boxed{y(t) = 0.02e^{-100t}\sin 10^4 t, \ t > 0}$$

3. $\dfrac{d^2 y}{dt^2} + 200\dfrac{dy}{dt} + 10^8 y = 5x(t) + 100\dfrac{dx}{dt}$, where $x(t) = u(t)$

Solution

$$H(s) = \frac{100s + 5}{s^2 + 200s + 10^8}, \quad H_1 = H(s)\big|_{s=0} = 5\times 10^{-8}, \quad y_p(t) = 5\times 10^{-8}$$

$$y_h(t) = e^{-100t}\left(A\cos 10^4 t + B\sin 10^4 t\right)$$

$$y(t) = y_h(t) + y_p(t) = e^{-100t}\left(A\cos 10^4 t + B\sin 10^4 t\right) + 5\times 10^{-8}$$

The conditions at $t = 0^+$ are $y(0^+) = 0$ and $y'(0^+) = 100$. By applying the initial conditions we find A and B.

$$\begin{cases} y(0) = A + 5\times 10^{-8} = 0, & A = -5\times 10^{-8} \\ y'(0) = -100A + 10^4 B = 100, & B = 0.01 \end{cases}$$

The total solution is

$$\boxed{\begin{aligned} y(t) &= \left[e^{-100t}\left(-5\times 10^{-8}\cos 10^4 t + 0.01\sin 10^4 t\right) + 5\times 10^{-8}\right]u(t) \\ &\approx (5\times 10^{-8} + 0.01e^{-100t}\sin 10^4 t)u(t) \end{aligned}}$$

4. $\dfrac{d^2 y}{dt^2} + 650\dfrac{dy}{dt} + 30{,}000y = 10^4\left(3x(t) + \dfrac{dx}{dt}\right)$, where $x(t) = (\sin 100t)u(t)$

Solution

$$H(s) = \frac{10^4(s+3)}{s^2 + 650s + 30{,}000}, \quad H(s)\big|_{s=j100} = 10\frac{3 + 100j}{20 + 65j} = 14.71\angle 15.4°$$

The particular solution is $14.71\sin(100t + 15.4°)$.
The characteristic equation and its roots are $s^2 + 650s + 30{,}000 = 0$, $s_{1,2} = -600, -50$.
The homogeneous solution is $Ae^{-600t} + Be^{-50t}$.
The total solution is $14.71\sin(100t + 15.4°) + Ae^{-600t} + Be^{-50t}$.
Initial conditions are $y(0^+) = 0$ and $y'(0^+) = 0$, which result in $A = 2.933$ and $B = -6.836$.

$$\boxed{y(t) = \left[14.71\sin(100t + 15.4°) + 2.933e^{-600t} - 6.836e^{-50t}\right]u(t)}$$

An alternative method for finding $y_p(t)$ in problem 4

$$y_p(t) = C \cos \omega t + D \sin \omega t, \quad \omega = 100$$
$$y_p'(t) = 100 \left(-C \sin \omega t + D \cos \omega t \right)$$
$$y_p''(t) = -10^4 \left(C \cos \omega t + D \sin \omega t \right)$$

Substitute the above expressions in the differential equation

$$-10^4 \left(C \cos \omega t + D \sin \omega t \right) + 65{,}000 \left(-C \sin \omega t + D \cos \omega t \right) + 30{,}000 \left(C \cos \omega t + D \sin \omega t \right)$$
$$= 10^4 (3 \sin \omega t + 100 \cos \omega t)$$
$$- \left(C \cos \omega t + D \sin \omega t \right) + 6.5 \left(-C \sin \omega t + D \cos \omega t \right) + 3 \left(C \cos \omega t + D \sin \omega t \right)$$
$$= (3 \sin \omega t + 100 \cos \omega t)$$

By collecting all terms and equating same-kind coefficients we get

$$\begin{cases} 6.5D + 2C = 100 \\ 6.5C - 2D = -3 \end{cases}$$

from which we find $C = 3.902$ and $D = 14.184$.
The particular solution becomes $y_p(t) = 3.902 \cos 100t + 14.184 \sin 100t = 14.71 \sin(100t + 15.38°)$.

5. $\dfrac{d^2 y}{dt^2} + 3\dfrac{dy}{dt} + 2y = -\dfrac{dx}{dt}$, where $x(t) = (\sin 3t) u(t)$

Solution

The homogeneous solution is $y_h(t) = A e^{-t} + B e^{-2t}$.
The particular solution is

$$y_p(t) = C \cos \omega t + D \sin \omega t, \quad \omega = 3$$
$$y_p'(t) = 3 \left(-C \sin \omega t + D \cos \omega t \right)$$
$$y_p''(t) = -9 \left(C \cos \omega t + D \sin \omega t \right)$$

Substitute the above expressions in the differential equation to get

$$-9 \left(C \cos \omega t + D \sin \omega t \right) + 9 \left(-C \sin \omega t + D \cos \omega t \right) + 2 \left(C \cos \omega t + D \sin \omega t \right) = -3 \cos \omega t$$

By collecting terms on the left-hand side and equating with same-kind coefficients on the right-hand side we get

$$\begin{cases} 7C - 9D = 3 \\ 7D + 9C = 0 \end{cases} \text{, from which we find } C = 0.162 \text{ and } D = -0.208.$$

$y(t) = (0.162 \cos 3t - 0.208 \sin 3t + A e^{-t} + B e^{-2t}) u(t)$
Initial conditions are $y(0) = y'(0) = 0$, which give $y(0) = 0.162 + A + B = 0$ and $y'(0) = -0.208 \times 3 - A - 2B = 0$.

$$\begin{cases} A + B = -0.162 \\ A + 2B = -0.623 \end{cases} \text{, } A = 0.3, \quad B = -0.462$$

The total solution is, therefore,

$$\boxed{\begin{aligned} y(t) &= \left(0.162 \cos 3t - 0.208 \sin 3t + 0.3 e^{-t} - 0.462 e^{-2t} \right) u(t) \\ &= \left[0.263 \cos(3t + 52°) + 0.3 e^{-t} - 0.462 e^{-2t} \right] u(t) \end{aligned}}$$

6. $\dfrac{d^3 y}{dt^3} + \dfrac{d^2 y}{dt^2} + \dfrac{dy}{dt} + y = e^{-t} u(t)$

Solution

The characteristic equation and the natural frequencies are

$$s^3 + s^2 + s + 1 = (s+1)(s^2+1) = 0, \quad s_{1,2,3} = -1, \pm j$$

The homogeneous solution is $y_h(t) = Ae^{-t} + B\sin t + C\cos t$. The particular solution is $y_p(t) = Dte^{-t}$. By substitution we determine $D = 0.5$. The total solution for $t > 0$ is

$$y(t) = Ae^{-t} + B\sin t + C\cos t + 0.5te^{-t}$$
$$y'(t) = -Ae^{-t} + B\cos t - C\sin t + 0.5e^{-t} - 0.5te^{-t}$$
$$y''(t) = Ae^{-t} - B\sin t - C\cos t - e^{-t} + 0.5te^{-t}$$

From the initial conditions we find $y(0) = A + C = 0$, $y'(0) = -A + B + 0.5 = 0$, and $y''(0) = A - C - 1 = 0$. The constants are $A = 0.5$, $B = 0$, and $C = -0.5$.

The solution to the differential equation is

$$\boxed{y(t) = [0.5(1+t)e^{-t} - 0.5\cos t]u(t)}$$

7. In the circuit of Figure 5.2, $R = 500\ \Omega$, $L = 1$ mH, and $C = 10\ \mu$F

  a. Find the system function $H(s) = V/V_s$ and its natural frequencies.

  b. Let $v_s(t) = 1$, $t > 0$. Show that regardless of the initial state of the circuit at $t = 0$, the response at $t > 0$ is $v(t) = V_0 e^{-\alpha t}\cos(\omega t + \theta)$ and determine α and ω.

  c. Given $i_L(0^-) = v_C(0^-) = 0$ and $v_s(t) = 1$, $t > 0$, find V_0 and θ in part b.

  d. Given $v_s(t) = 1$, $t > 0$, specify $i_L(0^-)$ and $v_C(0^-)$ so that the response $v(t)$ becomes zero at $t > 0$.

Solution

  a. We first find the system function by applying KVL and LCL in the s domain. From voltage division:

$$H(s) = \frac{V}{V_s} = \frac{Z_{LC}}{R + Z_{LC}} = \frac{1}{RC}\frac{s}{s^2 + \frac{1}{RC}s + \frac{1}{LC}} = \frac{200s}{s^2 + 200s + 10^8}$$

  b. $v_s(t) = 1$ and $\dfrac{d^2v}{dt^2} + 200\dfrac{dv}{dt} + 10^8 v = 200\dfrac{dv_s}{dt} = 0$, for $t > 0$.

    The characteristic equation and its roots are $s^2 + 200s + 10^8 = 0$, $s_{1,2} = -100 \pm \sqrt{10^4 - 10^8} \approx -100 \pm 10^4 j$.

    The homogeneous solution is $v_h(t) = e^{-100t}(A\cos 10{,}000t + B\sin 10{,}000t) = V_0 e^{-100t}\cos(10{,}000t + \theta)$.

    The particular solution is zero and the complete solution is $v(t) = v_h(t) = V_0 e^{-100t}\cos(10{,}000t + \theta)$, $t > 0$.

  c. $i_L(0^+) = i_L(0^-) = 0$, $v_C(0^+) = v_C(0^-) = 0$, $i_C(0^+) = \dfrac{v_s(0^+)}{R} = \dfrac{1}{500} = C\dfrac{dv}{dt}\Big|_{t=0^+}$ and $v'(0^+) = \dfrac{10^5}{500} = 200$.

    The initial conditions for $v(t)$ in the differential equation are $v(0^+) = 0$, and $v'(0^+) = 200$. Applying the initial conditions we get:

$$v(0^+) = V_0\cos\theta = 0, \quad V_0 \neq 0, \quad \cos\theta = 0, \quad \theta = \frac{\pi}{2}$$
$$v'(0^+) = V_0[-100\cos\theta - 10{,}000\sin\theta] = 200, \quad V_0 = -0.02$$
$$v(t) = -0.02e^{-100t}\cos\left(10{,}000t + \frac{\pi}{2}\right) = 0.02e^{-100t}\sin 10{,}000t, \quad t \geq 0$$

  d. $i_L(0^-) = \dfrac{1}{500}$ and $v_C(0^-) = 0$

8. Find the sinusoidal steady-state response of the circuit of problem 7 to $v_s(t) = \sin(\omega t)$. Write the response in the form of $v(t) = V\sin(\omega t + \theta)$, where V is a nonnegative number. Determine V and θ as functions of ω. Write a

Matlab program to plot $20 \log_{10} V$ (in units of dB) and θ versus $\log_{10} \omega$. [The plots will be called the Bode diagram for the given $H(s)$.]

Solution

Evaluate $H(s)$ at $s = j\omega$ to find its magnitude and phase, which are the magnitude and phase of $v(t)$, too.

$$H(s)\big|_{s=j\omega} = \frac{200s}{s^2 + 200s + 10^8}\bigg|_{s=j\omega} = \frac{j200\omega}{10^8 - \omega^2 + j200\omega} = |H(\omega)|e^{j\theta(\omega)}$$

$$|H(\omega)| = \frac{200\omega}{\sqrt{(10^8 - \omega^2)^2 + (200\omega)^2}}$$

$$\theta(\omega) = 90° - \tan^{-1}\left(\frac{200\omega}{10^8 - \omega^2}\right) = \tan^{-1}\left(\frac{10^8 - \omega^2}{200\omega}\right)$$

$$20\log_{10} |H(\omega)| = 20\log_{10} 200 + 20\log_{10} \omega - 10\log_{10}\left[(10^8 - \omega^2)^2 + (200\omega)^2\right]$$

The following Matlab program produces the Bode diagram.

```
num=[200 0];
den=[1 200 10^8];
sys=tf(num,den)
grid
bode(sys)
```

Chapter Problems

9. Find $y(t)$, the solution of

$$\frac{dy}{dt} + y = x(t), \quad t \geq 0, x(t) = 1$$

for each of the following initial conditions:

a. $y(0) = -1$
b. $y(0) = -0.5$
c. $y(0) = 0$
d. $y(0) = 0.5$
e. $y(0) = 1$
f. $y(0) = 2$

Verify that the following Matlab program plots the solution functions.

```
t=linspace(0,5,1001);
y0=[-1 -0.5 0 0.5 1 2];
for i=1:6;
    y =1+(y0(i)-1)*exp(-t);
    plot(t,y,'r','LineWidth',2)
    title('Response of dy/dt+y=1 under various initial conditions');
    xlabel('Time (s)');
    ylabel('y(t)');
    axis([0  5 -1.5  2.5]);
    hold on
end
grid
```

```
hold off
print -dpsc CH5_Pr9.eps
```

10. Repeat problem 9 with $x(t) = \sin 2t$. Plot the solution functions using Matlab.

11. Repeat problem 10 with $x(t) = \cos 2t$.

12. Repeat problem 10 with $x(t) = \sin t$.

13. Find $y(t)$, the solution to each of the following differential equations:

a. $\dfrac{dy}{dt} + y = (\sin 2t)u(t)$ b. $\dfrac{d^2y}{dt^2} + y = (\sin 2t)u(t)$ c. $\dfrac{d^2y}{dt^2} + 2\dfrac{dy}{dt} + 2y = e^{-\frac{t}{100}}u(t)$

d. $\dfrac{dy}{dt} + y = (\sin t)u(t)$ e. $\dfrac{dy}{dt} + 2y = 5u(t)$ f. $\dfrac{dy}{dt} + 2y = e^{-3t}u(t)$

14. Find $y(t)$, the solution to each of the following differential equations:

a. $\dfrac{dy}{dt} + 3y = e^{-2t}$, $t > 0$, with $y(0) = 1$

b. $\dfrac{d^2y}{dt^2} + 10\dfrac{dy}{dt} + 9y = 0$, $t > 0$, with $y(0) = 1$, $\dfrac{dy}{dt}(0) = 0$

c. $\dfrac{d^2y}{dt^2} + 10\dfrac{dy}{dt} + 9y = 9$, $t > 0$, with $y(0) = 1$, $\dfrac{dy}{dt}(0) = 0$

d. $\dfrac{d^2y}{dt^2} + 5\dfrac{dy}{dt} + 6y = 6$, $t > 0$, with $y(0) = 0$, $\dfrac{dy}{dt}(0) = 1$

15. The input-output relationship in a system is

$$2\dfrac{dy(t)}{dt} + y(t) = 3x(t)$$

where x is the input and y is the output. Given $x(t) = e^{-t}$, $t > 0$, find $y(t)$ for the following three initial conditions:

a. $y(0) = 3$
b. $y(0) = 0$
c. $y(0) = -3$

16. Find $y(t)$, the solution to each of the following differential equations:

a. $\dfrac{d^2y}{dt^2} + \dfrac{dy}{dt} = u(t)$

b. $\dfrac{d^2y}{dt^2} + \dfrac{dy}{dt} = 1.6u(t) - 0.6u(t - 1)$

17. Find $y(t)$, the solution to each of the following differential equations:

a. $\dfrac{d^2y}{dt^2} + 3\dfrac{dy}{dt} + 2y(t) = u(t)$

b. $\dfrac{d^3y}{dt^3} + 13\dfrac{d^2y}{dt^2} + 32\dfrac{dy}{dt} + 20y(t) = u(t)$

18. Find $y(t)$, the solution to each of the following differential equations:

a. $\dfrac{d^2y}{dt^2} + 2\dfrac{dy}{dt} + y = x(t) - \dfrac{dx}{dt}$, $t > 0$, with $y(0) = 1$, $\dfrac{dy}{dt}(0) = -1$, and $x(t) = e^{-2t} - e^{-3t}$

b. $\dfrac{d^2y}{dt^2} + 200\dfrac{dy}{dt} + 10^8 y = 200\dfrac{dx}{dt}$, $t > 0$, with $y(0) = 0$, $\dfrac{dy}{dt}(0) = 200$, and $x(t) = t$

19. Find $y(t)$, the solution to each of the following differential equations:

a. $\dfrac{d^2y}{dt^2} + 200\dfrac{dy}{dt} + 10^8 y = 5x(t) + 100\dfrac{dx}{dt}$, where $x(t) = (t+1)u(t)$

b. $\dfrac{d^2y}{dt^2} + 650\dfrac{dy}{dt} + 30000y = 10^4\left(3x(t) + \dfrac{dx}{dt}\right)$, where $x(t) = (\sin 200t)u(t)$

c. $\dfrac{d^2y}{dt^2} + \dfrac{dy}{dt} = x(t) + 2\dfrac{dx}{dt}$, where $x(t) = e^{-2t}u(t)$

d. $\dfrac{d^2y}{dt^2} + 4\dfrac{dy}{dt} + 3y = \dfrac{dx}{dt}$, where $x(t) = (\sin 3t)u(t)$

20. Find $y(t)$, the solution to the following differential equation:

$$\dfrac{d^4y}{dt^4} + 2\dfrac{d^3y}{dt^3} + 2\dfrac{d^2y}{dt^2} + 2\dfrac{dy}{dt} + y = x(t), \text{ where}$$

a. $x(t) = e^{-t}u(t)$
b. $x(t) = (\cos t)u(t)$
c. $x(t) = e^{-t}, \ t \geq 0$, and $y'''(0) = y''(0) = y'(0) = 0$, and $y(0) = 1$
d. $x(t) = \sin t, \ t \geq 0$, and $y'''(0) = y''(0) = y'(0) = 0$, and $y(0) = 1$

21. In the circuit of Figure 5.2, $R = 1\ \Omega$, $L = 1$ H, and $C = 1$ F.

a. Show that

$$\dfrac{d^2v}{dt^2} + \dfrac{dv}{dt} + v = \dfrac{dv_s}{dt}$$

b. Given $v_s = 1$, $t > 0$, $i(0) = 0.5$ A, and $v(0) = 0.5$ V, find v for $t > 0$.

22. In the circuit of Figure 5.2, $R = 1\ \Omega$, $L = 1$ H, and $C = 1$ F.

a. Show that

$$\dfrac{d^2i}{dt^2} + \dfrac{di}{dt} + i = v_s(t)$$

where $i(t)$ is the current in the inductor.

b. Given $v_s(t) = 1$, $t > 0$, $i(0) = 0.5$ A, and $v(0) = 0.5$ V, find i for $t > 0$.

23. In the circuit of Figure 5.6, $R = 1\ \Omega$, $L = 1$ H, and $C = 2$ F.

a. Show that

$$2\dfrac{d^2v}{dt^2} + 2\dfrac{dv}{dt} + v = v_s + i_s + \dfrac{di_s}{dt}$$

b. Given $v_s = 1$, $i_s = e^{-t}$, $t > 0$, $i(0) = 1$ A, and $v(0) = 1$ V, find v for $t > 0$.

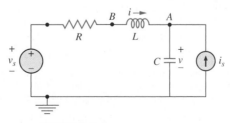

FIGURE 5.6 (For Problem 23)

24. In the circuit of Figure 5.7

a. Show that

$$\frac{dv}{dt} + \frac{1}{RC}v = \frac{1}{RC}v_s$$

b. With $RC = 1$ s and $v_s = u(t)$, find v.

c. With $RC = 0.1$ s, let $v_s = 2u(t) - u(t - T)$, where T is a constant. Find T so that v reaches its final steady-state value in T seconds.

d. With $RC = 0.1$ s, let $v_s = (1 + k)u(t) - ku(t - T)$, where k and T are constants. Determine the relationship between k and T so that v reaches its final steady-state value in T seconds.

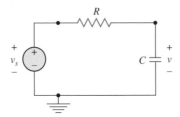

FIGURE 5.7 (For Problem 24)

25. a. In the circuit of Figure 5.8, show that

$$\frac{d^2v}{dt^2} + \frac{R}{L}\frac{dv}{dt} + \frac{1}{LC}v = \frac{1}{LC}v_s$$

b. With $R = 123.5\ \Omega$, $L = 8.207$ H, $C = 100\ \mu$F, and $v_s = u(t)$, show that $v(t)$ is expressed by

$$v(t) = [1 - Ae^{-\sigma t}\sin(\omega_d t + \phi)]u(t)$$

Find the constants in the above expression and verify that the period of its decaying oscillations is 184.3 ms.

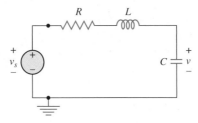

FIGURE 5.8 (For Problem 25)

26. **Open-loop control.** Oscillations in the step response of Figure 5.8 can be reduced or eliminated by modifying the input step voltage before it acts on the circuit. In this problem you will investigate the effects of three such modifications. In each case find the output voltage $v(t)$. Determine the time it takes to reach the steady-state value (exactly or approximately). The three cases are as follows:

a. $v_s = ku(t) + (1 - k)u(t - T)$, where $T = 90$ msec (Figure 5.9a). Find k so that v almost reaches its steady-state value in 90 ms.

b. $v_s = u(t) - u(t - T_1) + u(t - T_2)$, where $T_1 = 40$ msec and $T_2 = 60$ (Figure 5.9b).

c. $v_s = u(t) - 2u(t - T_1) + 2u(t - T_2)$, where $T_1 = 46$ msec and $T_2 = 56$ msec (Figure 5.9c).

Sample plots of the system's responses superimposed on the unit-step response are shown in Figure 5.9 for each case.

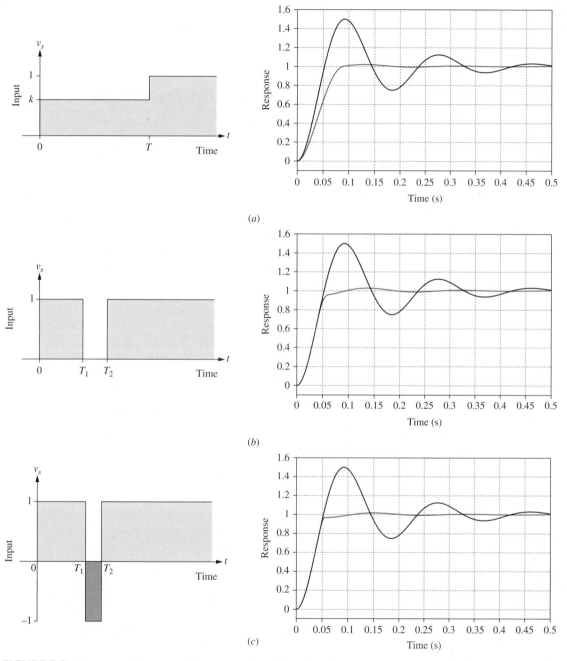

FIGURE 5.9 Three cases of input modification to improve transition time and reduce oscillations in the step response of a second-order system. For each case the responses to modified inputs are shown superimposed on the system's unit-step response. (For Problem 26.)

27. In the circuit of Figure 5.2, $R = 500\ \Omega$, $L = 10$ mH, and $C = 100\ \mu$F. Find $v(t)$ for (a) $v_s(t) = u(t)$ and (b) $v_s(t) = \delta(t)$.

28. Find the sinusoidal steady-state response of the circuit of problem 27 to $v_s(t) = \sin \omega t$. Follow the procedure of problem 8.

5.23 Project: System of Differential Equations

Summary

A two-input, two-output linear time-invariant system is described by

$$\frac{dy_1}{dt} + y_2 = x_1$$

$$\frac{dy_2}{dt} + 3y_2 - 2y_1 = x_2$$

where $x_1(t)$ and $x_2(t)$ are the input, and $y_1(t)$ and $y_2(t)$ are the output pairs.

a. Find the differential equations that separately describe $y_1(t)$ and $y_2(t)$ in terms of $x_1(t)$ and $x_2(t)$. Show that the characteristic equations and natural frequencies of the two differential equations you obtain are identical. They are, thus, called the natural frequencies of the system. Find the outputs of the above system to $x_1(t) = u(t)$ and $x_2(t) = (\cos t)u(t)$.

b. Write the differential equations in matrix form as given below and determine $\mathbf{A}$.

$$\begin{bmatrix} \frac{dy_1}{dt} \\ \frac{dy_2}{dt} \end{bmatrix} = \begin{bmatrix} 0 & -1 \\ 2 & -3 \end{bmatrix} \times \begin{bmatrix} y_1 \\ y_2 \end{bmatrix} + \begin{bmatrix} x_1 \\ x_2 \end{bmatrix} \quad \text{or} \quad \mathbf{Y}' = \mathbf{AY} + \mathbf{X}$$

Find the roots of the following equation and show that they are the same as those found in part a.

$$|s\mathbf{I} - \mathbf{A}| = 0$$

where $\mathbf{I}$ is the 2×2 identity matrix and $|s\mathbf{I} - \mathbf{A}|$ is the determinant.

c. Refer to the circuit of Figure 5.10 in which $x_1(t)$ and $x_2(t)$ are voltage sources with the polarities shown. Verify that the outputs of op-amps 1 and 2 can represent y_1 and y_2, respectively, and determine appropriate elements values.

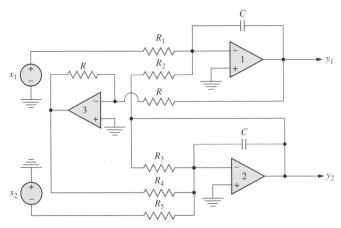

FIGURE 5.10

d. Remove op-amp number 3 from the circuit of Figure 5.10. Write the new input-output equations. Find the response to $x_1(t) = x_2(t) = u(t)$. Show that the system is unstable. Can you infer the unstable behavior of the system by merely examining the circuit?

e. You have been asked to configure a circuit for solving the system shown by

$$\mathbf{Y}' = \mathbf{AY} + \mathbf{BX}$$

where $\mathbf{A}$ and $\mathbf{B}$ are 2×2 matrices with nonzero elements. Suggest a circuit and discuss if there are any limiting factors or considerations.

Chapter **6**

The Laplace Transform and Its Applications

Contents

Introduction and Summary

A signal may be represented in various forms; for example, by its variations as a function of time, by its decomposition into components, or by a mathematical operation (transformation) that maps it from one domain to another. Each form may have a certain advantage over the others. Some transforms provide tools for converting a set of integro-differential equations into a set of simultaneous linear equations. This property, in conjunction with the use of matrix notations, greatly simplifies the analysis and design of linear systems and is especially useful in the state-space approach and multivariable systems. The Laplace transform belongs to a class of transforms, called linear integral transforms, that have such a property. The Fourier transform is another member of this class. In this chapter we present the Laplace transform and its applications. The Fourier series and transform are discussed in Chapters 7 and 8.

The Laplace transform converts linear time-invariant differential equations into algebraic equations that are generally easier to solve. In doing so, it incorporates the initial conditions into the solution. Partial differential equations and some time-varying differential equations are also more easily solved by application of the Laplace transform. Such features make the Laplace transform an attractive tool in operational mathematics. In addition, Laplace transform solutions can provide insights into the properties and behaviors of systems. Another application of the Laplace transform is the development and use of the system function, through which natural frequencies, poles and zeros, stability, feedback, and frequency response can be discussed.

This chapter starts with the definition of the Laplace transform in both its unilateral and bilateral forms. The former version is applied to the analysis of LTI systems subjected to causal (right-sided) inputs. The effect of nonzero initial conditions is explored, as are the zero-state and zero-input responses of the systems in the transform domain. The many examples based on causal functions provide a smooth transition to the bilateral version of the transform, where a broad class of applications is addressed. These include applications dealing with forcing functions that contain discontinuities and impulses.

The chapter also covers the inverse Laplace transform and expansion by partial fractions. Tables of useful Laplace transform pairs are given as a handy aid. The essentials of regions of convergence, properties, and theorems involving the Laplace transform are explained as well. Finally, the application of the transform to circuits and their describing dynamical equations serves to highlight the power and usefulness of this tool.

The organization of the chapter permits sequential, parallel, or mixed approaches to the study of both forms of the transform. The examples and problems serve to clarify the concepts presented, bridge the discussions on the unilateral and bilateral Laplace transforms, and bring to the reader's attention parallels between the time and transform domains.

6.1 Definition of the Laplace Transform

The Laplace transform of a function $x(t)$ is defined by

$$\text{Unilateral Laplace transform: } \mathcal{L}[x(t)] = X(s) = \int_0^\infty x(t)e^{-st}dt \quad (1-a)$$

$$\text{Bilateral Laplace transform: } \mathcal{L}[x(t)] = X(s) = \int_{-\infty}^\infty x(t)e^{-st}dt \quad (1-b)^1$$

For the above integrals to converge, the new variable s (in general, a complex number) has to be limited to a part of the s-plane called the region of convergence (ROC). It is seen that the Laplace transform produces a function of the new variable s. The mapping between $x(t)$ and $X(s)$ is one-to-one so that one can find $x(t)$ from $X(s)$.

Unilateral Transform

The Laplace transform has traditionally been used in its unilateral form. The unilateral Laplace transform provides a simple but powerful tool for analysis of causal LTI systems given the input for $t > 0$ and the initial conditions at $t = 0$. This is especially suitable for the case of physical systems such as those containing electrical and mechanical components. The effect of past inputs manifests itself in the form of initial conditions. The unilateral Laplace transform method is widely used in analysis and design of feedback and control systems. The fact that the integral in the unilateral Laplace transform excludes the negative half of $x(t)$ does not imply that the transform is defined only for causal functions or that the negative half of $x(t)$ is set to zero and its effect discarded. Through the differentiation property, the unilateral Laplace transform incorporates the initial conditions into the solution of the differential equation. We refer to the unilateral Laplace transform simply as the Laplace transform, requiring a knowledge of the time function for $t > 0$ only.

Bilateral Transform

If the conditions of the LTI system at $t = 0$ are not known, we need to know its past history (e.g., all past values of the input) in order to determine the future values of the output. In such a case the bilateral Laplace transform is used in which the lower bound of the integral in (1-a) is taken to be $-\infty$. The bilateral Laplace transform is also used in noncausal systems and will be discussed later on. Moreover, in many communication and

---

[1] In both cases the transform is shown by $X(s)$.

information applications we deal with signals and systems that are modeled for the whole range of time. Mathematical causality and initial conditions don't command a prominant feature anymore. There is a need to consider the past values of functions explicitly, not merely as initial conditions. The bilateral Laplace transform is a generalization of the Laplace transform that answers that need. It requires the knowledge of $x(t)$ from $-\infty$ to ∞.

Region of Convergence

The Laplace transform exists only in the region of convergence (ROC). The roots of the denominator of the Laplace transform of a function are called its poles. By definition, the region of convergence doesn't include any poles. The Laplace transform is analytic in that region. As will be seen, for the right-sided functions the ROC is $\mathcal{RE}[s] > a$, where, depending on the time function, a is a positive or negative number. It is a contiguous region on the right side of the pole with the largest real part. For the left-sided functions the ROC is $\mathcal{RE}[s] < b$, where b may be a positive or negative number. It is a contiguous region on the left side of the pole with the smallest real part. For the two-sided functions the ROC is $a < \mathcal{RE}[s] < b$, a vertical strip with no poles inside it.

Uniqueness Theorem

In a nutshell, the uniqueness theorem states that the Laplace transform pairs are unique. More precisely, if $X_1(s) = \mathcal{L}[x_1(t)]$ and $X_2(s) = \mathcal{L}[x_2(t)]$ are equal in a region of the s-plane, then $x_1(t)$ and $x_2(t)$ are functionally equal for $t > 0$. [2] Due to the uniqueness theorem, a time function $x(t)$ and its Laplace transform $X(s)$ form a transform *pair*. Knowledge of one would lead to the other. This property is used to determine the inverse Laplace transform of a function from a known transform. Tables of Laplace transform pairs can be comfortably used for transition from one domain to the other without fear of multiplicity errors.

Equivalence of the Unilateral and Bilateral Transforms

Obviously, the unilateral and bilateral transforms of a function $x(t)$ become the same if $x(t) = 0$ for $t < 0$. In other words, the Laplace transform of a function $x(t)u(t)$ is the same regardless of which definition is used. If $x(t)$ contains an impulse at $t = 0$, we set the lower limit of the unilateral integral to 0^-. This will result in the same expression for $X(s)$ regardless of which transform definition is used.

The unilateral Laplace transform (referred to simply as the Laplace transform), along with its inverse, properties, and applications are discussed in sections 6.2 through 6.19 of this chapter. The bilateral transform is considered in sections 6.20 through 6.25.

---

[2] Two time functions may be different at a set of isolated points with little practical overall effect; for example, their Laplace transform integrals will be equal in a region of the s-plane.

6.2 Linearity Property

Linearity is a basic property of the Laplace transform. It states that

$$\text{if:} \qquad x(t) \qquad \Longrightarrow X(s)$$
$$y(t) \qquad \Longrightarrow Y(s)$$

$$\text{then:} \qquad ax(t) + by(t) \Longrightarrow aX(s) + bY(s)$$

where a, b are any constants and x, y are any functions. This property, also called the linearity theorem, is due to the integration in Eq. (1-a) being a linear operation. The linearity property may be derived directly from the definition of the Laplace transform.

$$\mathcal{L}\left[ax_1(t) + bx_2(t)\right] = \int \left[ax_1(t) + bx_2(t)\right] e^{-st} dt$$

$$= a \int x_1(t)e^{-st} dt + b \int x_2(t)e^{-st} dt$$

$$= aX_1(s) + bX_2(s)$$

6.3 Examples of the Unilateral Laplace Transform

The unilateral Laplace transform of a function $x(t)$ is defined by

$$\mathcal{L}[x(t)] = X(s) = \int_0^\infty x(t)e^{-st} dt$$

The variable s is, in general, a complex number. For the transform to exist, the above integral should converge. This requirement generally limits s to a part of the s-plane called the region of convergence (ROC).

Example
6.1

Transform of a constant

The Laplace transform of $x(t) = 1, \quad t > 0$, is

$$X(s) = \int_0^\infty e^{-st} dt = \frac{-1}{s}\left[e^{-st}\right]_{t=0}^{t=\infty}$$

At $t = \infty$, the value of e^{-st} either goes to zero (when $\mathcal{RE}[s] > 0$, in which case the integral converges) or ∞ (when $\mathcal{RE}[s] \le 0$, in which case the integral doesn't converge). Therefore, the Laplace transform exists only if s is located in the region specified by $\mathcal{RE}[s] > 0$.

$$X(s) = -\frac{1}{s}\left(e^{-\infty} - 1\right) = \frac{1}{s}, \quad \mathcal{RE}[s] > 0$$

The ROC is the right-half plane starting from $\mathcal{RE}[s] = 0$ and stretching to ∞.

E*xample*

6.2

A rectangular pulse

A rectangular pulse and its Laplace transform are given below.

$$x(t) = \begin{cases} 1, & 0 < t < T \\ 0, & t > T \end{cases} \quad \text{and} \quad X(s) = \int_0^T e^{-st}\,dt = \frac{1 - e^{-sT}}{s}$$

The transform exists everywhere in the s-plane.

E*xample*

6.3

An exponential

The Laplace transform of $x(t) = e^{at}$, $t > 0$ is

$$X(s) = \int_0^\infty e^{at} e^{-st}\,dt = \frac{-1}{s-a}\left[e^{-(s-a)t}\right]_{t=0}^{t=\infty}$$

By an argument similar to that of Example 6.1, we note that the integral converges if $\mathcal{RE}(s-a) > 0$. The Laplace transform of e^{at} is then

$$X(s) = \frac{-1}{s-a}(0-1) = \frac{1}{s-a}, \quad \mathcal{RE}[s] > a$$

The ROC is the half-plane to the right of the line $\mathcal{RE}[s] > a$. For a decaying exponential $a < 0$ and the ROC includes the imaginary axis. For a growing exponential $a > 0$ and the ROC excludes the imaginary axis. It is observed that for $a = 0$ we get $x(t) = 1$ and

$$X(s) = \frac{1}{s}, \quad \mathcal{RE}[s] > 0$$

E*xample*

6.4

Two exponentials

Find the Laplace transform of $y(t) = e^{-t} - e^{-2t}$, $t > 0$, shown in Figure 6.1. Also see Example 6.10.

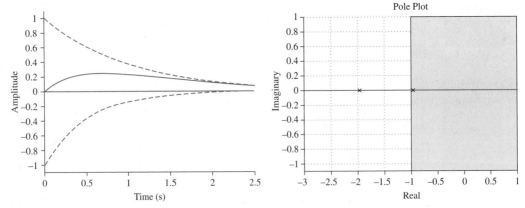

FIGURE 6.1 Sum of two exponentials $x(t) = e^{-t} - e^{-2t}, t > 0, \implies X(s) = \frac{1}{(s+1)(s+2)}, \mathcal{RE}[s] > -1$

Solution

Applying the linearity property we get

$$Y(s) = \frac{1}{s+1} - \frac{1}{s+2} = \frac{1}{(s+1)(s+2)} = \frac{1}{s^2 + 3s + 2} \quad \text{where } \mathcal{RE}[s] > -1$$

*E*xample 6.5 Trigonometric functions

Use the linearity property and transform of an exponential function to find the Laplace transforms of $\sin(\omega t)$ and $\cos(\omega t)$.

Solution

Transforms of trigonometric functions are derived in the following table.

x(t)	$\Longrightarrow$ X(s)	ROC
$e^{j\omega t}$	$\Longrightarrow \dfrac{1}{s - j\omega}$	$\mathcal{RE}[s] > 0$
$e^{-j\omega t}$	$\Longrightarrow \dfrac{1}{s + j\omega}$	$\mathcal{RE}[s] > 0$
$\sin(\omega t) = \dfrac{1}{2j}\left[e^{j\omega t} - e^{-j\omega t}\right]$	$\Longrightarrow \dfrac{1}{2j}\left[\dfrac{1}{s - j\omega} - \dfrac{1}{s + j\omega}\right] = \dfrac{\omega}{s^2 + \omega^2}$	$\mathcal{RE}[s] > 0$
$\cos(\omega t) = \dfrac{1}{2}\left[e^{j\omega t} + e^{-j\omega t}\right]$	$\Longrightarrow \dfrac{1}{2}\left[\dfrac{1}{s - j\omega} + \dfrac{1}{s + j\omega}\right] = \dfrac{s - a}{s^2 + \omega^2}$	$\mathcal{RE}[s] > 0$

*E*xample 6.6 Decaying or growing sinusoids

The Laplace transforms of $e^{at}\sin(\omega t)$, $t > 0$, $e^{at}\cos(\omega t)$, $t > 0$, and $e^{at}\cos(\omega t + \theta)$, $t > 0$, may be deduced from the transform of an exponential with the exponent $a \pm j\omega$ as derived below. See Figure 6.2.

x(t)	$\Longrightarrow$ X(s)	ROC
$e^{at}, \; t > 0$	$\Longrightarrow \dfrac{1}{s - a}$	$\mathcal{RE}[s] > a$
$e^{at}\sin(\omega t) = \dfrac{e^{(a+j\omega)t} - e^{(a-j\omega)t}}{2j}$	$\Longrightarrow \dfrac{1}{2j}\left[\dfrac{1}{s - a - j\omega} - \dfrac{1}{s - a + j\omega}\right] = \dfrac{\omega}{(s - a)^2 + \omega^2}$	$\mathcal{RE}[s] > a$
$e^{at}\cos(\omega t) = \dfrac{e^{(a+j\omega)t} + e^{(a-j\omega)t}}{2}$	$\Longrightarrow \dfrac{1}{2}\left[\dfrac{1}{s - a - j\omega} + \dfrac{1}{s - a + j\omega}\right] = \dfrac{s - a}{(s - a)^2 + \omega^2}$	$\mathcal{RE}[s] > a$

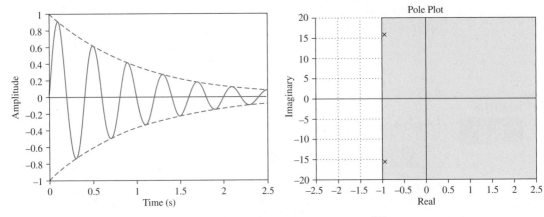

FIGURE 6.2 Exponential sinusoid $x(t) = e^{-t}\sin(5\pi t)$, $t > 0$, $\Longrightarrow X(s) = \frac{15.71}{s^2+2s+247.74}$, ROC $= \mathcal{RE}[s] > -1$

Hyperbolic functions

Use the linearity property and transform of an exponential function to find the Laplace transforms of $\sinh(at)$ and $\cosh(at)$.

Solution
Transforms of hyperbolic functions are derived in the following table.

x(t)	$\Longrightarrow$	X(s)	ROC
e^{at}	$\Longrightarrow$	$\dfrac{1}{s-a}$	$\mathcal{RE}[s] > a$
e^{-at}	$\Longrightarrow$	$\dfrac{1}{s+a}$	$\mathcal{RE}[s] > -a$
$\sinh(at) = \dfrac{1}{2}\left[e^{at} - e^{-at}\right]$	$\Longrightarrow$	$\dfrac{1}{2}\left[\dfrac{1}{s-a} - \dfrac{1}{s+a}\right] = \dfrac{a}{s^2-a^2}$	$\mathcal{RE}[s] > a$
$\cosh(at) = \dfrac{1}{2}\left[e^{at} + e^{-at}\right]$	$\Longrightarrow$	$\dfrac{1}{2}\left[\dfrac{1}{s-a} + \dfrac{1}{s+a}\right] = \dfrac{s}{s^2-a^2}$	$\mathcal{RE}[s] > a$

6.4 Differentiation and Integration Properties

Transforms of the derivative and integral of $x(t)$ are related to $X(s)$ by

$$\frac{dx(t)}{dt} \Longrightarrow sX(s) - x(0)$$

$$\int_0^t x(\tau)d\tau \Longrightarrow \frac{X(s)}{s}$$

Proof of the Derivative Property

First apply the definition of the Laplace transform to $dx(t)/dt$.

$$\mathcal{L}\left[\frac{dx(t)}{dt}\right] = \int_0^\infty \frac{dx(t)}{dt} e^{-st} dt$$

Then integrate by parts.

$$\int u dv = uv - \int v du \text{ where } u = e^{-st}, \quad du = -se^{-st}, \quad dv = \frac{dx(t)}{dt} dt, \text{ and } v = x(t)$$

$$\mathcal{L}\left[\frac{dx(t)}{dt}\right] = \left[e^{-st}x(t)\right]_0^\infty - \int_0^\infty x(t)(-se^{-st})dt = sX(s) - x(0)$$

The derivative property is of prime importance in the solution of LTI differential equations by the Laplace transform. It incorporates the initial conditions in the solution.

Proof of the Integral Property

First apply the definition of the Laplace transform to $\int_0^t x(\tau)d\tau$. The result is a double integral in τ and t.

$$\mathcal{L}\left[\int_{\tau=0}^t x(\tau)d\tau\right] = \int_{t=0}^\infty \int_{\tau=0}^t x(\tau)e^{-st} d\tau dt$$

Integration is done on a region that is defined by $t \geq 0$, $\tau \geq 0$ and $t \geq \tau$. Then, change the order and integrate over the same region to get

$$\mathcal{L}\left[\int_{\tau=0}^t x(\tau)d\tau\right] = \int_{\tau=0}^\infty x(\tau)\left(\int_{t=\tau}^\infty e^{-st}dt\right)d\tau = \frac{1}{s}\int_{\tau=0}^\infty x(\tau)e^{-s\tau}d\tau = \frac{X(s)}{s}$$

The use of integral property is illustrated in the following example.

Example **6.8**

A DC voltage source, V_1, gets connected to a capacitor C through a resistor R at $t = 0$. Capacitor voltage at $t = 0$ is V_0 volt. Use a Laplace transform to find current i flowing into the capacitor for $t > 0$.

Solution

From KVL around the loop for $t > 0$ we have the following equation in the time domain.

$$Ri(t) + \frac{1}{C}\int_0^t i(\tau)d\tau + V_0 = V_1$$

Taking Laplace transform of both sides we have

$$RI(s) + \frac{1}{C}\frac{I(s)}{s} + \frac{V_0}{s} = \frac{V_1}{s}$$

from which we find $I(s)$

$$I(s) = \frac{(V_1 - V_0)}{R}\frac{1}{s + \frac{1}{RC}}$$

The current in the time is

$$i(t) = \frac{(V_1 - V_0)}{R} e^{-\frac{t}{RC}}, \, t \geq 0$$

The derivative and integral properties provide circuit solutions in the transform domain. See section 6.18.

xample

6.9

Given $\frac{dy(t)}{dt} + 2y(t) = e^{-t}$, $t \geq 0$, $y(0) = 1$ find Y(s).

Solution

Take the transform of both sides of the differential equation

$$\mathcal{L}\left[\frac{dy(t)}{dt} + 2y(t)\right] = \frac{1}{s+1}, \quad \mathcal{RE}[s] > -1$$

By the linearity property of the Laplace transform

$$\mathcal{L}\left[\frac{dy(t)}{dt} + 2y(t)\right] = \mathcal{L}\left[\frac{dy(t)}{dt}\right] + 2\mathcal{L}[y(t)]$$

and by the derivative property

$$\mathcal{L}\left[\frac{dy(t)}{dt}\right] = sY(s) - y(0) = sY(s) - 1$$

Therefore,

$$\mathcal{L}\left[\frac{dy(t)}{dt} + 2y(t)\right] = sY(s) - 1 + 2Y(s) = \frac{1}{s+1}, \quad \mathcal{RE}[s] > -1$$

from which we find

$$Y(s) = \frac{1}{s+1}, \quad \mathcal{RE}[s] > -1$$

The only time function with the above unilateral Laplace transform is $y(t) = e^{-t}$, $t \geq 0$. The solution is made of the particular part only. The given initial condition $y(0) = 1$ has resulted in elimination of the homogeneous solution.

Repeated application of the derivative property results in the relationships between transforms of higher derivatives of $x(t)$ to $X(s)$, as shown in the list below.

1. $\mathcal{L}\left[\dfrac{dx(t)}{dt}\right] = sX(s) - x(0)$

2. $\mathcal{L}\left[\dfrac{d^{(2)}x(t)}{dt^2}\right] = s^2X(s) - sx(0) - x'(0)$

   $\vdots$ $\vdots$

n. $\mathcal{L}\left[\dfrac{d^{(n)}x(t)}{dt^n}\right] = s^nX(s) - s^{n-1}x(0) - s^{n-2}x'(0) \cdots - x^{(n-1)}(0)$

Repeated application of integral property is left as an exercise.

Title	SIGNALS & SYSTEMS
Condition	Good
Location	Books Row G 10 Bay 5 item 6185
Description	The cover and pages are in good condition! Any other included accessories are also in good condition showing some minor use. There is no dust jacket on this item. Supports Goodwill job training programs.
Source	EBOOKS
SKU	3ZHMCR001Q7L
	0073380709
ASIN	9780073380704
Code	elanal
Employee	
Date Added	6/1/2021 3:15:50 PM

3ZHMCR001Q7L

6.5 Multiplication by *t*

Multiplication of $x(t)$ by t is equivalent to taking the derivative of its Laplace transform with respect to s and then negating the result:

$$tx(t) \Longleftrightarrow -\frac{dX(s)}{ds}$$

To derive the above property take the derivative of both sides of the transform integral with respect to s.

$$\frac{dX(s)}{ds} = \frac{d}{ds}\int_0^\infty x(t)e^{-st}dt = \int_0^\infty x(t)\frac{d}{ds}[e^{-st}]dt$$

$$= \int_0^\infty -tx(t)e^{-st}dt = -\mathcal{L}[tx(t)]$$

The above property may be extended to multiplication by t^n resulting in the following summary table:

$tx(t)$	$\Longleftrightarrow$	$-\dfrac{dX(s)}{ds}$
$\vdots$	$\vdots$	
$t^n x(t)$	$\Longleftrightarrow$	$(-1)^n\dfrac{d^n X(s)}{ds^n}$

*E*xample
6.10

The following table uses the *t*-multiplication property to find the Laplace transform pairs shown.

1	$\Longleftrightarrow$	$\dfrac{1}{s}$	$\mathcal{RE}[s] > 0$
t	$\Longleftrightarrow$	$-\dfrac{d}{ds}\left[\dfrac{1}{s}\right] = \dfrac{1}{s^2}$	$\mathcal{RE}[s] > 0$
t^n	$\Longleftrightarrow$	$\dfrac{n!}{s^{n+1}}$	$\mathcal{RE}[s] > 0$
te^{-at}	$\Longleftrightarrow$	$-\dfrac{d}{ds}\left[\dfrac{1}{s+a}\right] = \dfrac{1}{(s+a)^2}$	$\mathcal{RE}[s] > -a$
$t^n e^{-at}$	$\Longleftrightarrow$	$\dfrac{n!}{(s+a)^{n+1}}$	$\mathcal{RE}[s] > -a$
$t\sin \omega t$	$\Longleftrightarrow$	$\dfrac{2\omega s}{(s^2+\omega^2)^2}$	$\mathcal{RE}[s] > 0$
$t\cos \omega t$	$\Longleftrightarrow$	$\dfrac{s^2-\omega^2}{(s^2+\omega^2)^2}$	$\mathcal{RE}[s] > 0$

6.6 Multiplication by e^{at}

Multiplication of a time function $x(t)$ by e^{at} shifts its Laplace transform by the amount a to the right.

$$x(t)e^{at} \Longrightarrow X(s - a)$$

This property, also called s-domain shift, is the counterpart of the time shift.

Decaying or growing sinusoids revisited

Use multiplication by e^{at} to find the Laplace transforms of $e^{at}\sin(\omega t)$, $t > 0$, $e^{at}\cos(\omega t)$, $t > 0$, and $e^{at}\cos(\omega t + \theta)$, $t > 0$.

Solution
We start with the transforms of $\sin(\omega t)$ and $\cos(\omega t)$, and apply the multiplication by exponential property to find

$\sin \omega t$	$\Longrightarrow$	$\dfrac{\omega}{s^2 + \omega^2}$	$\mathcal{RE}[s] > 0$
$\cos \omega t$	$\Longrightarrow$	$\dfrac{s}{s^2 + \omega^2}$	$\mathcal{RE}[s] > 0$
$e^{at}\sin \omega t$	$\Longrightarrow$	$\dfrac{\omega}{(s - a)^2 + \omega^2}$	$\mathcal{RE}[s] > a$
$e^{at}\cos \omega t$	$\Longrightarrow$	$\dfrac{s - a}{(s - a)^2 + \omega^2}$	$\mathcal{RE}[s] > a$
$e^{at}\cos(\omega t + \theta)$	$\Longrightarrow$	$\dfrac{(s - a)\cos\theta - \omega\sin\theta}{(s - a)^2 + \omega^2}$	$\mathcal{RE}[s] > a$

6.7 Time-Shift Property

Shifting of $x(t)u(t)$ to the right by T seconds results in the multiplication of $X(s)$ by e^{-sT}.

$$x(t - T)u(t - T) \Longrightarrow X(s)e^{-sT}$$

The shifted function $x(t - T)u(t - T)$, therefore, is zero at $0 < t < T$[3]

---

[3]Note that this property applies to right-shift (or delay) of $x(t)u(t)$, which is not the same as a right-shift of $x(t)$ if $x(t) \neq 0$, $t < 0$.

Example

6.12

Rectangular pulse revisited

Using the shift and linearity properties, find the Laplace transform of the rectangular pulse $x(t) = u(t) - u(t - T)$.

Solution

$$u(t) \implies \frac{1}{s}$$

$$u(t - T) \implies \frac{e^{-sT}}{s}$$

$$u(t) - u(t - T) \implies \frac{1}{s} - \frac{e^{-sT}}{s} = \frac{1 - e^{-sT}}{s}$$

Example

6.13

A 1-volt, 1-sec rectangular voltage source is connected to a 1-μF capacitor through a 1-MΩ resistor. The capacitor has zero initial charge. Find the Laplace transform of the current in the circuit.

Solution

Let the voltage be given by $u(t) - u(t - 1)$. The current and its transform are

$$i(t) = 10^{-6} \left[e^{-t} - e^{-(t-1)} u(t-1) \right]$$

$$= 10^{-6} \times \begin{cases} 0, & t < 0 \\ e^{-t}, & 0 \le t < 1 \\ (1 - e)e^{-t}, & t \ge 1 \end{cases} \implies I(s) = 10^{-6} \frac{1 - e^{-s}}{s + 1}$$

6.8 Scale Change

Changing the time scale by a positive factor α changes the s scale by a factor of $\frac{1}{\alpha}$ and multiplies the transform by $\frac{1}{\alpha}$.

$$\mathcal{L}[x(\alpha t)] = \int_0^\infty x(\alpha t)e^{-st}\,dt = \frac{1}{\alpha} \int_0^\infty x(\tau)e^{-(s/\alpha)\tau}\,d\tau = \frac{1}{a} X\left(\frac{s}{\alpha}\right), \quad \alpha > 0$$

In summary,

$$x(\alpha t) \implies \frac{1}{\alpha} X\left(\frac{s}{\alpha}\right), \quad \alpha > 0$$

6.9 Convolution Property

Convolution is an important and exceedingly useful property of the Laplace transform. It states that convolution in the time domain is equivalent to multiplication in the s-domain.

$$x(t) \star h(t) \iff X(s)H(s)$$

To show the above, we start with the convolution of two causal functions $x(t)$ and $h(t)$ given by

$$y(t) = x(t) \star h(t) = \int_0^t x(t - \tau)\, h(\tau)d\tau$$

By direct evaluation, the Laplace transform of $y(t)$ is

$$Y(s) = \int_0^\infty y(t)e^{-st}dt = \int_0^\infty \left[\int_0^t x(t - \tau)\, h(\tau)\, d\tau \right] e^{-st}\, dt$$

$$= \int_0^\infty \left[\int_0^\infty x(t - \tau)u(t - \tau)\, h(\tau)\, d\tau \right] e^{-st}\, dt$$

Interchanging the order of integration we get

$$Y(s) = \int_0^\infty h(\tau) \left[\int_0^\infty x(t - \tau)u(t - \tau)\, e^{-st}dt \right] d\tau$$

By the time-shift property,

$$\int_0^\infty x(t - \tau)u(t - \tau)\, e^{-st}dt = X(s)e^{-s\tau}$$

Therefore,

$$Y(s) = X(s) \int_0^\infty h(\tau)e^{-s\tau}\, d\tau = X(s)H(s)$$

Observanda

The convolution property is of great importance because it establishes the s-domain input-output relationship for LTI systems,

$$Y(s) = X(s)H(s)$$

where $Y(s)$, $X(s)$, and $H(s)$ are the Laplace transforms of the output, the input, and the impulse response, respectively. The ROC of $Y(s)$ is the intersection of the ROCs of $X(s)$ and $H(s)$. A null ROC indicates $y(t)$ doesn't exist.

Example
6.14

Convolution of two exponentials

Using the Laplace transform method find $y(t) = h(t) \star x(t)$, where $h(t) = e^{-t}u(t)$ and $x(t) = e^{-2t}u(t)$.

Solution

$$H(s) = \frac{1}{s + 1}, \quad \mathcal{RE}[s] > -1, \quad \text{and} \quad X(s) = \frac{1}{s + 2}, \quad \mathcal{RE}[s] > -2$$

Apply the convolution property to get

$$Y(s) = H(s)X(s) = \frac{1}{(s+1)(s+2)}, \quad \mathcal{RE}[s] > -1$$

To find $y(t)$ we expand $Y(s)$ into fractions that are transforms of known time functions. The method is called inverse Laplace transform and will be discussed in details in sections 6.13 to 6.17.

$$Y(s) = \frac{1}{(s+1)(s+2)} = \frac{1}{s+1} - \frac{1}{s+2}, \quad \mathcal{RE}[s] > -1 \text{ and } y(t) = e^{-t} - e^{-2t}, \ t > 0$$

Example **6.15**

The impulse response of an LTI system is $h(t) = e^{-t}\cos(2t)u(t)$. Using the Laplace transform method, find the system's response to $x(t) = -4e^{-3t}u(t)$.

Solution
The transform of the impulse response and the input are

$$h(t) = e^{-t}\cos(2t)u(t) \Longrightarrow H(s) = \frac{s+1}{(s+1)^2 + 4} \quad \mathcal{RE}[s] > -1$$

$$x(t) = -4e^{-3t}u(t) \Longrightarrow X(s) = \frac{-4}{s+3} \quad \mathcal{RE}[s] > -3$$

$$Y(s) = H(s)X(s) = \frac{s+1}{(s+1)^2 + 4} \cdot \frac{(-4)}{s+3} = -\frac{s+3}{(s+1)^2 + 4} + \frac{1}{s+3}, \quad \mathcal{RE}[s] > -1$$

The inverse of $Y(s)$ may be found from the known transform pairs.

$$y(t) = -e^{-t}(\cos 2t + \sin 2t)u(t) + e^{-3t}u(t) = \left[\sqrt{2}e^{-t}\cos\left(2t + \frac{3\pi}{4}\right) + e^{-3t} \right] u(t)$$

Note the natural and the forced components of the response.

6.10   Initial-Value and Final-Value Theorems

The initial-value theorem states that

$$\lim_{s \to \infty} \{sX(s)\} = x(0)$$

Strictly speaking, this applies only if $X(s)$ is a proper rational function or only if the proper part of $X(s)$ is used. Otherwise, one may employ impulse and other generalized functions to extend the above theorem. (A proper rational function is the ratio of two polynomials in which the power of the numerator is less than that of the denominator.)
    The final-value theorem states that

$$\lim_{s \to 0} \{sX(s)\} = x(\infty)$$

The final-value theorem is applicable only when all poles of $sX(s)$ have negative real parts. Otherwise, as $t \to 0$, the limit becomes infinite or indeterminate.

6.11 Lower Limit of Integration: $0^-, 0, 0^+$

The lower limit of integration in the unilateral Laplace transform is zero. But does that mean 0^-, 0, or 0^+? Unless the time function contains an impulse at the origin, there is no problem. However, if $x(t)$ includes a $\delta(t)$, one must decide whether to include or exclude the impulse from integration. Some authors specify the lower limit to be 0^+ to exclude the impulse. In that case, the effect of $\delta(t)$ will translate into initial conditions at $t = 0^+$. Some authors take $t = 0^-$ to be the lower limit to include the impulse directly in the transform. We keep the lower limit at $t = 0$ and accept $\mathcal{L}[\delta(t)] = 1$. Other forms of discontinuity of $x(t)$ at $t = 0$ (e.g., a finite jump) are accounted for in the derivative property by selecting $x(0^+)$. Many questions regarding the effect of discontinuities of the time function at the lower limit of the integral will vanish if the bilateral Laplace transform is used.

6.12 Laplace Transform of the Unit Impulse

A unit impulse located at $t = T > 0$ is represented by $\delta(t - T)$. Its Laplace transform is evaluated to be

$$\mathcal{L}[\delta(t - T)] = \int_0^\infty \delta(t - T)e^{-st}dt = e^{-sT}$$

For $T = 0$, the value of the integral seems to be unclear because of the presence of the impulse at $t = 0$.[4] Here we determine the transform of $\delta(t)$ from

$$\mathcal{L}[\delta(t)] = \lim_{T \to 0} e^{-sT} = 1$$

Alternate Method

$\delta(t)$ may be considered the limit of a narrowing rectangular pulse, $x_T(t)$, of duration T and height $1/T$ at the origin.

$$\delta(t) = \lim_{T \to 0} x_T(t)$$

$$x_T(t) = \begin{cases} \frac{1}{T}, & 0 < t < T \\ 0, & \text{elsewhere} \end{cases}$$

In the s-domain,

$$X_T(s) = \frac{1}{T} \int_0^T e^{-st}dt = \frac{1 - e^{-sT}}{Ts}$$

$$\mathcal{L}[\delta(t)] = \lim_{T \to 0} X_T(s) = 1$$

---

[4]To overcome the above ambiguity, one may define the lower limit of the unilateral Laplace transform to be at $t = 0^-$, in which case, $\mathcal{L}[\delta(t)] = 1$.

The above two derivations obtain the Laplace transform of $\delta(t)$ without getting involved with the question of the lower limit being 0^- or 0^+.

Generalization

The perceived ambiguity surrounding the Laplace transform of the unit-impulse function disappears if the bilateral Laplace transform is used and the unit impulse $\delta(t)$ is considered as a generalized function defined by

$$\int_{-\infty}^{\infty} \delta(t - t_0)\phi(t)dt = \phi(t_0)$$

In the above definition let $\phi(t) = e^{-st}$. Then

$$\mathcal{L}[\delta(t - t_0)] = e^{-st_0}$$
$$\mathcal{L}[\delta(t)] = 1$$

6.13   The Inverse Laplace Transform

The inverse transform is a time function $x(t)$ whose Laplace transform is given by $X(s)$, including the ROC. The inverse transform may be found from

$$x(t) = \mathcal{L}^{-1}[X(s)] = \frac{1}{2\pi j} \int_C X(s)e^{st}ds$$

where the integration path C is a vertical line defined by $s = \sigma_0 + j\omega$ in the ROC, from $\omega = -\infty$ to ∞. In the case of the unilateral Laplace transform [concerning $x(t)$ for $t > 0$] the path closes itself in the left half of the s-plane through a very large semicircle surrounding all poles of $X(s)$ as shown in Figure 6.3. Then the value of the integral over the semicircle becomes zero [because $t > 0$ and $\mathcal{RE}[s] < 0$ result in $e^{st} = 0$] and, according to Cauchy's theorem on contour integration of functions of complex variables, the above integral becomes equal to the sum of residues of the function $X(s)e^{st}$ at all poles (to the left of the ROC). Finding the inverse Laplace transform then becomes a matter of finding the residues.

Example **6.16**

Given the unilateral Laplace transform $X(s) = B(s)/(s^3 + 4s^2 + 9s + 10)$, what can be said about $x(t)$? Assume $B(s)$ is a polynomial of second order or less in s.

Solution

The roots of the denominator of $X(s)$ (called its poles) are at $s_{1,2} = -1 \pm j2$ and $s_3 = -2$. They are displayed in Figure 6.3. The ROC is $\mathcal{RE}[s] > -1$; that is, the

RHP to the right of the pole with the largest real part. The contour integral for evaluation of $x(t)$ circles the poles in a counterclockwise direction and generates $x(t)$ for $t \geq 0$, with a functional form determined by the poles:

$$x(t) = k_1 e^{-(1+2j)t} + k_1^* e^{-(1-2j)t} + k_2 e^{-2t}$$
$$= e^{-t}[C_1 \cos(2t) + C_2 \sin(2t)] + k_2 e^{-2t}$$
$$= Ce^{-t} \cos(2t + \theta) + k_2 e^{-2t}, \quad t > 0$$

To determine the parameters k_1 and k_2 $(C_1, C_2, C, \text{and } \theta)$ we need $B(s)$ or, equivalently, the zeros of $X(s)$.

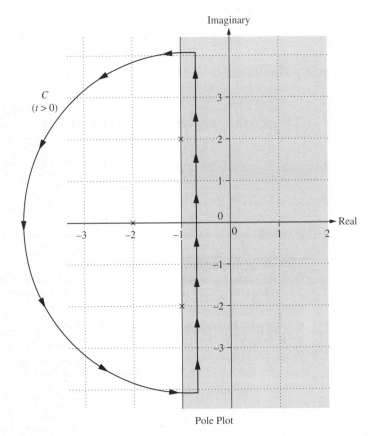

Pole Plot

FIGURE 6.3 A time function $x(t)$, $t > 0$, is found from its unilateral Laplace transform $X(s)$ by

$$x(t) = \mathcal{L}^{-1}[X(s)] = \frac{1}{2\pi j} \int_C X(s) e^{st} ds$$

where the path C goes from $-\infty < \mathcal{IM}[s] < \infty$ within the ROC.

For $t > 0$, the path may be closed by an infinitely large semicircle to the left, as shown in this figure, without affecting the value of the above integral. Then, according to Cauchy's theorem on contour integration of functions of complex variables, the value of the integral becomes equal to the sum of the residues of $X(s)e^{st}$ at its poles within the contour C; that is, at poles of $X(s)$ to the left of its ROC. The residues may be found by expanding $X(s)$ into its fractions.

The inverse Laplace transform can, however, be determined by first expanding $X(s)$ into simpler components and then using tables of transform pairs. One such method, which is in harmony with the residue method, is partial fraction expansion discussed next.

6.14 Partial Fraction Expansion; Simple Poles

Let $X(s)$ be a rational function with distinct poles (denominator polynomial has no repeated roots) s_i, $i = 1, 2, \ldots n$. In addition, let the degree of the numerator polynomial be less than that of the denominator polynomial. Then

$$X(s) = \frac{b_m s^m + b_{m-1} s^{m-1} + \cdots + b_0}{s^n + a_{n-1} s^{n-1} + \cdots + a_0}$$

$$= \frac{b_m s^m + b_{m-1} s^{m-1} + \cdots + b_0}{(s - s_1)(s - s_2) \cdots (s - s_n)}$$

$$= \frac{k_1}{s - s_1} + \frac{k_2}{s - s_2} + \cdots + \frac{k_n}{s - s_n}$$

where k_i, $i = 1, 2, \ldots, n$ are constants called residues of $X(s)$. They are found from

$$k_i = X(s)(s - s_i)\big|_{s=s_i}$$

From the table of Laplace transform pairs and based on the uniqueness theorem we find $x(t)$

$$x(t) = k_1 e^{s_1 t} + k_2 e^{s_2 t} + \cdots + k_n e^{s_n t}$$

 Example
6.17   Find $x(t)$ for $t > 0$, given

$$X(s) = \frac{s + 1}{s^2 + 3s}$$

Solution

$$X(s) = \frac{k_1}{s} + \frac{k_2}{s+3}$$

$$k_1 = X(s)s\big|_{s=0} = \frac{s+1}{s+3}\bigg|_{s=0} = \frac{1}{3}$$

$$k_2 = X(s)(s+3)\big|_{s=-3} = \frac{s+1}{s}\bigg|_{s=-3} = \frac{2}{3}$$

$$x(t) = \frac{1}{3} + \frac{2}{3}e^{-3t}, \quad t > 0$$

Example **6.18**

Find $x(t)$ for $t > 0$ given

$$X(s) = \frac{s^2 + 3s + 2}{(s+1)(s+2)(s+5)}$$

Solution

$$X(s) = \frac{k_1}{s+1} + \frac{k_2}{s+2} + \frac{k_3}{s+5}$$

$$k_1 = X(s)(s+1)\big|_{s=-1} = \frac{s^2 + 3s + 2}{(s+2)(s+5)}\bigg|_{s=-1} = 0$$

The result indicates that $X(s)$ has a zero at $s = -1$, which cancels the denominator term $s + 1$. Similarly, $k_2 = 0$.

$$X(s) = \frac{1}{s+5}, \quad x(t) = e^{-5t}, \quad t > 0$$

Example **6.19**

Find $x(t)$ for $t > 0$ given

$$X(s) = \frac{s^2 + 3s + 3}{(s+1)(s-2)(s+5)}$$

Solution

$$X(s) = \frac{k_1}{s+1} + \frac{k_2}{s-2} + \frac{k_3}{s+5}$$

$$k_1 = X(s)(s+1)\big|_{s=-1} = \frac{s^2 + 3s + 3}{(s-2)(s+5)}\bigg|_{s=-1} = -\frac{1}{12}$$

$$k_2 = X(s)(s-2)\big|_{s=2} = \frac{s^2 + 3s + 3}{(s+1)(s+5)}\bigg|_{s=2} = \frac{13}{21}$$

$$k_3 = X(s)(s+5)\big|_{s=-5} = \frac{s^2 + 3s + 3}{(s+1)(s-2)}\bigg|_{s=-5} = \frac{13}{28}$$

$$x(t) = -\frac{1}{12}e^{-t} + \frac{13}{21}e^{2t} + \frac{13}{28}e^{-5t}, \quad t > 0$$

Needless to say that in the above class of expansions, poles are not required to be real valued (i.e., be on the real axis in the s-plane). The method works as long as the poles are not repeated. They can be anywhere in the s-plane. When they are not on the real axis they occur in complex conjugate pairs. See Example 6.20.

Example 6.20

Find $x(t)$, $t > 0$, given

$$X(s) = \frac{s+1}{s(s^2+4)} = \frac{s+1}{s(s+2j)(s-2j)} = \frac{k_1}{s} + \frac{k_2}{s-2j} + \frac{k_3}{s+2j}$$

Solution

Contributions from the poles at $s = \pm j2$ may be combined in the s-domain and $X(s)$ be expanded in the form of

$$X(s) = \frac{s+1}{s(s^2+4)} = \frac{k_1}{s} + \frac{As+B}{s^2+4}$$

$$k_1 = X(s)s\big|_{s=0} = \frac{s+1}{s^2+4}\bigg|_{s=0} = 0.25$$

Coefficients A and B may be found by combining the fractions on the right side and matching with the original function.

$$\frac{0.25}{s} + \frac{As+B}{s^2+4} = \frac{0.25s^2+1+As^2+Bs}{s(s^2+4)} = \frac{s+1}{s(s^2+4)}$$

By matching numerator terms we get $0.25s^2 + 1 + As^2 + Bs = s + 1$ from which $A = -0.25$, and $B = 1$.

$$X(s) = \frac{0.25}{s} + \frac{-0.25s+1}{s^2+4}$$

$$x(t) = 0.25 - 0.25\cos(2t) + 0.5\sin(2t) = 0.25 + 0.56\cos(2t - 116.6°), \quad t > 0$$

Alternate Method

One may also evaluate contributions of the poles at $s = \pm j2$ to $x(t)$ (i.e., $k_2 e^{j2t} + k_3 e^{-j2t}$) and combine them in the time domain.

$$k_2 = X(s)(s - 2j)\big|_{s=2j} = \frac{(s+1)}{s(s+2j)}\bigg|_{s=2j} = 0.28\angle-116.6°$$

$$k_3 = k_2^* = 0.28\angle 116.6°$$

$$x(t) = 0.25 + 0.28 e^{j(2t-116.6°)} + 0.28 e^{-j(2t-116.6°)}$$

$$= 0.25 + 0.56\cos(2t - 116.6°), \quad t > 0$$

Example 6.21

The system function of an LTI system is

$$H(s) = \frac{2s+1}{s+1}$$

Find the system's response to the input

$$x(t) = e^{-t}\sin(2t)\,u(t)$$

Solution

The Laplace transform of $x(t)$ is

$$X(s) = \frac{2}{(s+1)^2 + 4}$$

The Laplace transform of $y(t)$ is

$$Y(s) = X(s)H(s) = \frac{2(2s+1)}{(s+1)(s^2+2s+5)}$$

$$= \frac{2(2s+1)}{(s+1)(s+1+2j)(s+1-2j)}$$

$$= \frac{k_1}{s} + \frac{k_2}{s+1-2j} + \frac{k_2^*}{s+1+2j}$$

k_1, k_2, and $y(t)$ are found from

$$k_1 = Y(s)(s+1)\big|_{s=-1} = -\frac{1}{2}$$

$$k_2 = Y(s)(s+1-2j)\big|_{s=-1+2j} = \frac{\sqrt{17}}{4} \angle -76°, \text{ and } k_2^* = \frac{\sqrt{17}}{4} \angle 76°$$

$$y(t) = -\frac{1}{2}e^{-t} + \frac{\sqrt{17}}{4}e^{-t}\left[e^{-j76°}e^{j2t} + e^{j76°}e^{-j2t}\right]$$

$$= \frac{1}{2}e^{-t}\left[-1 + \sqrt{17}\cos(2t - 76°)\right], \ t > 0$$

Alternate Method

$$Y(s) = \frac{2(2s+1)}{(s+1)(s^2+2s+5)} = \frac{k_1}{s+1} + \frac{As+B}{(s+1)^2 + 4}$$

where $k_1 = -\frac{1}{2}$ as before. By taking the common denominator of the terms on the right-hand side of the above equation and then matching the two sides we find $A = 0.5$ and $B = 4.5$. Then

$$Y(s) = -\frac{1}{2}\frac{1}{s+1} + \frac{1}{2}\frac{s+9}{(s+1)^2 + 4}$$

$$y(t) = \frac{1}{2}e^{-t}\left[-1 + \cos(2t) + 4\sin(2t)\right] = \frac{1}{2}e^{-t}\left[-1 + \sqrt{17}\cos(2t - 76°)\right], \ t > 0$$

Another Alternate Method to Find As + B

$$Y(s) = \frac{2(2s+1)}{(s+1)(s^2+2s+5)} = \frac{k_1}{s+1} + \frac{As+B}{(s+1)^2 + 4}$$

Multiply both sides by $(s^2 + 2s + 5)$ and set $s = -1 + j2$.

$$\frac{2(2s+1)}{s+1}\bigg|_{s=-1+j2} = A(-1 + j2) + B$$

By matching the real and imaginary parts of the two sides of the above equation we find $A = 0.5$ and $B = 4.5$. This method is more convenient and recommended.

6.15 Partial Fraction Expansion; Multiple-Order Poles

Example
6.22

A pole that occurs more than once is called a multiple-order pole. A partial fraction involving an nth-order pole contains additional terms up to the nth power of the repeated factor. This is illustrated in the following example.

Find $x(t)$ for $t > 0$, given

$$X(s) = \frac{s^2 + 3}{(s + 2)(s + 1)^2} = \frac{k_1}{s + 2} + \frac{k_2}{(s + 1)^2} + \frac{k_3}{s + 1}$$

Solution

$$X(s) = \frac{k_1}{s + 2} + \frac{k_2}{(s + 1)^2} + \frac{k_3}{s + 1}$$

$$k_1 = X(s)(s + 2)\big|_{s=-2} = 7 \quad \text{and} \quad k_2 = X(s)(s + 1)^2\big|_{s=-1} = 4$$

k_3 may be found by taking the common denominator of the fractions and matching it with the original function

$$\frac{7}{s + 2} + \frac{4}{(s + 1)^2} + \frac{k_3}{s + 1} = \frac{7(s + 1)^2 + 4(s + 1) + k_3(s + 1)(s + 2)}{(s + 2)(s + 1)^2} = \frac{s^2 + 3}{(s + 2)(s + 1)^2}$$

Matching terms of numerators on both sides results in $k_3 = -6$.

$$x(t) = 7e^{-2t} + 2(2t - 3)e^{-t}, \quad t > 0$$

Alternate Method

k_3 may also be found from

$$k_3 = \frac{d}{ds}\left[X(s)(s + 1)^2\right]_{s=-1} = \frac{d}{ds}\left[\frac{s^2 + 3}{s + 2}\right]_{s=-1} = -6$$

Generalization

The alternate method of finding residues may be generalized to the case of nth-order poles. This generalization is included here for the sake of completeness. In many cases the method of matching coefficients may prove to be easier, safer, and more practical. Let

$$X(s) = \frac{A(s)}{(s + a)^n B(s)}$$

be a proper rational fraction with an nth-order pole at $s = -a$. The partial fraction expansion of $X(s)$ is

$$X(s) = \frac{k_0}{(s + a)^n} + \frac{k_1}{(s + a)^{n-1}} + \cdots + \frac{k_i}{(s + a)^{n-i+1}} + \cdots + \frac{k_n}{s + a} + \cdots \text{etc.}$$

where,

$$k_0 = [X(s)(s + a)^n]_{s=-a}$$

$$k_1 = \frac{d}{ds}[X(s)(s + a)^{n-1}]_{s=-a}$$

$$k_i = \frac{1}{i!}\frac{d^i}{ds^i}[X(s)(s + a)^n]_{s=-a}$$

6.16 Summary of the Laplace Transform Properties and Theorems

See Table 6.1 for a summary of the Laplace transform properties and theorems.

TABLE 6.1 Some Properties of the Unilateral Laplace Transform

	Property	Time Domain	$\Longleftrightarrow$	s-Domain
1	Definition	$x(t) = \dfrac{1}{2\pi j} \oint_C X(s)e^{st}ds$	$\Longleftrightarrow$	$X(s) = \displaystyle\int_0^\infty x(t)e^{-st}dt$
2	Linearity	$ax(t) + by(t)$	$\Longleftrightarrow$	$aX(s) + bY(s)$
3	Time delay	$x(t-T)u(t-T)$	$\Longleftrightarrow$	$X(s)e^{-st}, \quad T > 0$
4	Multiplication by t	$tx(t)$	$\Longleftrightarrow$	$-\dfrac{dX(s)}{ds}$
5	Multiplication by t^2	$t^2x(t)$	$\Longleftrightarrow$	$\dfrac{d^2X(s)}{ds^2}$
6	Multiplication by t^n	$t^nx(t)$	$\Longleftrightarrow$	$(-1)^n\dfrac{d^nX(s)}{ds^n}$
7	Multiplication by e^{at}	$x(t)e^{at}$	$\Longleftrightarrow$	$X(s-a)$
8	Scale change	$x(at)$	$\Longleftrightarrow$	$\dfrac{1}{a}X\left(\dfrac{s}{a}\right), \quad a > 0$
9	Integration	$\displaystyle\int_0^t x(t)dt$	$\Longleftrightarrow$	$\dfrac{X(s)}{s}$
10	Differentiation	$\dfrac{dx}{dt}$	$\Longleftrightarrow$	$sX(s) - x(0)$
11	2nd-order differentiation	$\dfrac{d^2x}{dt^2}$	$\Longleftrightarrow$	$s^2X(s) - sx(0) - x'(0)$
12	nth-order differentiation	$\dfrac{d^nx}{dt^n}$	$\Longleftrightarrow$	$s^nX(s) - s^{n-1}x(0) - s^{n-2}x'(0)\cdots - x^{(n-1)}(0)$
13	Convolution	$\displaystyle\int_0^t y(\tau)x(t-\tau)d\tau$	$\Longleftrightarrow$	$X(s)Y(s)$

Initial value theorem	$\displaystyle\lim_{s\to\infty} \{sX(s)\} = x(0)$
Final value theorem	$\displaystyle\lim_{s\to 0} \{sX(s)\} = x(\infty)$
Zero s	$\displaystyle\int_0^\infty x(t)dt = X(0)$

6.17 A Table of Unilateral Laplace Transform Pairs

See Table 6.2 for a table of unilateral Laplace transform pairs. In this table $x(t)$ is defined for $t > 0$.

TABLE 6.2 Unilateral Laplace Transform Pairs

	$x(t) = \mathcal{L}^{-1}[X(s)]$	$\Longleftrightarrow$	$X(s) = \mathcal{L}[x(t)] = \int_0^\infty x(t)e^{-st}dt$	Region of Convergence
1	$\delta(t)$	$\Longleftrightarrow$	1	all s
2	1	$\Longleftrightarrow$	$\dfrac{1}{s}$	$\mathcal{RE}[s] > 0$
3	e^{at}	$\Longleftrightarrow$	$\dfrac{1}{s-a}$	$\mathcal{RE}[s] > a$
4	$\sin(\omega t)$	$\Longleftrightarrow$	$\dfrac{\omega}{s^2+\omega^2}$	$\mathcal{RE}[s] > 0$
5	$\cos(\omega t)$	$\Longleftrightarrow$	$\dfrac{s}{s^2+\omega^2}$	$\mathcal{RE}[s] > 0$
6	$e^{at}\sin(\omega t)$	$\Longleftrightarrow$	$\dfrac{\omega}{(s-a)^2+\omega^2}$	$\mathcal{RE}[s] > a$
7	$e^{at}\cos(\omega t)$	$\Longleftrightarrow$	$\dfrac{s-a}{(s-a)^2+\omega^2}$	$\mathcal{RE}[s] > a$
8	$\sin(\omega t + \theta)$	$\Longleftrightarrow$	$\dfrac{s\sin\theta + \omega\cos\theta}{s^2+\omega^2}$	$\mathcal{RE}[s] > 0$
9	$\cos(\omega t + \theta)$	$\Longleftrightarrow$	$\dfrac{s\cos\theta - \omega\sin\theta}{s^2+\omega^2}$	$\mathcal{RE}[s] > 0$
10	$e^{at}\sin(\omega t + \theta)$	$\Longleftrightarrow$	$\dfrac{(s-a)\sin\theta + \omega\cos\theta}{(s-a)^2+\omega^2}$	$\mathcal{RE}[s] > a$
11	$e^{at}\cos(\omega t + \theta)$	$\Longleftrightarrow$	$\dfrac{(s-a)\cos\theta - \omega\sin\theta}{(s-a)^2+\omega^2}$	$\mathcal{RE}[s] > a$
12	$\sinh(at)$	$\Longleftrightarrow$	$\dfrac{a}{s^2-a^2}$	$\mathcal{RE}[s] > a$
13	$\cosh(at)$	$\Longleftrightarrow$	$\dfrac{s}{s^2-a^2}$	$\mathcal{RE}[s] > a$
14	t^n	$\Longleftrightarrow$	$\dfrac{n!}{s^{n+1}}$	$\mathcal{RE}[s] > 0$
15	$t^n e^{at}$	$\Longleftrightarrow$	$\dfrac{n!}{(s-a)^{n+1}}$	$\mathcal{RE}[s] > a$

Note: The ROC for the unilateral Laplace transform is always the region of the s-plane to the right of that pole of $X(s)$ with largest real part. Hence, if σ_0 is the real part of such a pole then ROC $= \{s; \mathcal{RE}[s] > \sigma_0\}$.

6.18 Circuit Solution

Linear circuits may be analyzed by applying the Laplace transform technique. This can be done in one of two ways. In one approach, covered in this section, the current-voltage relationships of circuit elements are written in the s-domain by taking the Laplace transforms of their time-domain versions. In doing this, the initial voltages of the capacitors and initial currents of the inductors are incorporated in the s-domain terminal characteristics. By applying Kirchhoff's laws to the transform circuit, the Laplace transforms of the desired variables (and hence their time-domain counterparts) are obtained. In the second approach, covered in section 6.19, the Laplace transform is applied to the dynamical equations of the circuit.

The terminal characteristics of R, L, and C and their transforms, along with the equivalent circuit models, are summarized in Table 6.3. Models are shown in Figure 6.4. Note that $V(s)$ and $I(s)$ in Table 6.3 are not phasors. They are the Laplace transforms of the elements' terminal voltages and currents, respectively. In the special case where initial conditions are zero, the transform characteristics become the same as a generalized phasor.

TABLE 6.3 Laplace Transform Models of R, L, C

Element	Time Domain	⇔	Transform Domain	Figure
Resistor	$v(t) = Ri(t)$	⇔	$V(s) = RI(s)$	Figure 6.4(a)
Capacitor	$v(t) = v(0) + \dfrac{1}{C}\displaystyle\int_0^t i(t)dt$	⇔	$V(s) = \dfrac{v(0)}{s} + \dfrac{I(s)}{Cs}$	Figure 6.4(b)
Inductor	$i(t) = i(0) + \dfrac{1}{L}\displaystyle\int_0^t v(t)dt$	⇔	$I(s) = \dfrac{i(0)}{s} + \dfrac{V(s)}{Ls}$	Figure 6.4(c)

It should be remembered that the actual capacitor voltage is the inverse transform of $V(s)$. Similarly, the actual inductor current is the inverse transform of $I(s)$.

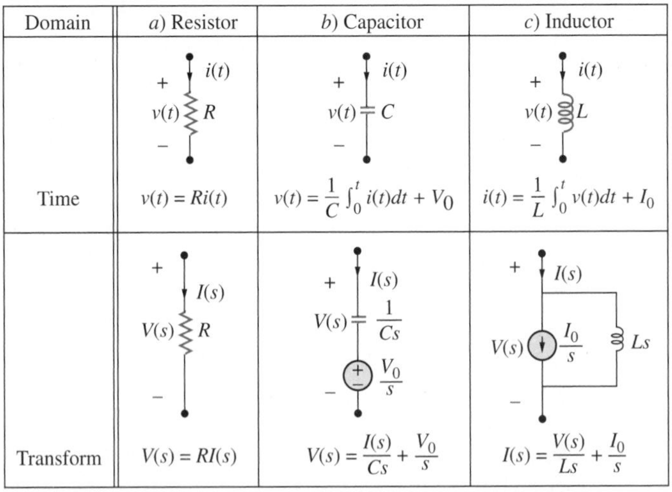

Domain	a) Resistor	b) Capacitor	c) Inductor
Time	$v(t) = Ri(t)$	$v(t) = \dfrac{1}{C}\int_0^t i(t)dt + V_0$	$i(t) = \dfrac{1}{L}\int_0^t v(t)dt + I_0$
Transform	$V(s) = RI(s)$	$V(s) = \dfrac{I(s)}{Cs} + \dfrac{V_0}{s}$	$I(s) = \dfrac{V(s)}{Ls} + \dfrac{I_0}{s}$

FIGURE 6.4 Laplace transform models of R, L, and C, including initial conditions.

Example 6.23

In Figure 6.5(a) $R_1 = 3\ \Omega$, $R_2 = 2\ \Omega$, $L = 1\ H$, and $C = 1\ F$. Find $v(t)$ and $i(t)$ given

$$i_g = \begin{cases} 1\ A, & t < 0 \\ \cos t, & t > 0 \end{cases}$$

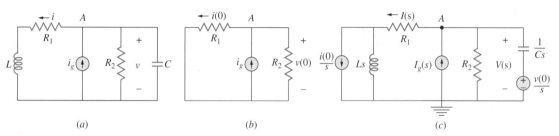

FIGURE 6.5 *(a)*An *RLC* circuit with $R_1 = 3\ \Omega$, $R_2 = 2\ \Omega$, $L = 1\ H$, $C = 1\ F$, and $i_g = u(-t) + \cos t\, u(t)$. *(b)* The equivalent DC circuit for $t < 0$. *(c)* The Laplace transform model for $t > 0$. See Example 6.23.

Solution

For $t < 0$ we have the DC circuit of Figure 6.5(b), which generates $i(0) = 0.4$ A and $v(0) = 1.2$ V. Applying the above conditions to $t > 0$ we construct the transform circuit of Figure 6.5(c) from which the dynamical equations of the circuit in terms of $V(s)$ and $I(s)$ are formulated.

$$\begin{cases} \text{KCL at node } A: & \left[V(s) - \dfrac{v(0)}{s} \right] s + \dfrac{V(s)}{2} + I(s) = I_g(s) \\[2mm] \text{KVL around the external loop:} & -V(s) + 3I(s) + \left[I(s) - \dfrac{i(0)}{s} \right] s = 0 \end{cases}$$

where $v(0) = 1.2$ V, $i(0) = 0.4$ A, and $I_g(s) = s/(s^2 + 1)$ is the Laplace transform of $i_g = \cos t$. Substituting for $v(0)$, $i(0)$, and $I_g(s)$ we obtain the following two equations:

$$\begin{cases} V(s)\left(s + \dfrac{1}{2} \right) + I(s) = \dfrac{1.2s^2 + s + 1.2}{s^2 + 1} \\[3mm] -V(s) + I(s)(s + 3) = 0.4 \end{cases}$$

Solving for $V(s)$ and $I(s)$ we obtain

$$V(s) = \frac{(s + 3)(1.2s^2 + s + 1.2) - 0.4(s^2 + 1)}{(s^2 + 1)(s^2 + 3.5s + 2.5)}$$

$$I(s) = \frac{(1.2s^2 + s + 1.2) + 0.4(s + 0.5)(s^2 + 1)}{(s^2 + 1)(s^2 + 3.5s + 2.5)}$$

Applying partial fraction expansions

$$V(s) = \frac{0.6667}{s+1} - \frac{0.0184}{s+2.5} + \frac{0.5513s + 0.621}{s^2 + 1}$$

$$I(s) = \frac{0.3332}{s+1} - \frac{0.0367}{s+2.5} + \frac{0.1034s + 0.2414}{s^2 + 1}$$

Taking the inverse transforms

$$v(t) = 0.6667e^{-t} - 0.0185e^{-2.5t} + 0.8304\cos(t - 48.4°), \quad t > 0$$

$$i(t) = 0.3332e^{-t} - 0.0367e^{-2.5t} + 0.2626\cos(t - 66.8°), \quad t > 0$$

6.19 Solution of Dynamical Equations

Linear circuits may also be analyzed by applying the Laplace transform to their dynamical equations. In this method, circuit laws are applied to the time-domain circuit to obtain linear differential equations, called dynamical equations, in terms of the desired variables. The Laplace transform is then used as a solution tool.

Again, let in the circuit of Figure 6.5(a) $R_1 = 3\ \Omega$, $R_2 = 2\ \Omega$, $L = 1\ H$, $C = 1\ F$, and

$$i_g = \begin{cases} 1\ A, & t < 0 \\ \cos t, & t \geq 0 \end{cases}$$

Write two equations involving $v(t)$, $i(t)$, their first-order derivatives, and the source i_g only. Solve using the Laplace transform technique.

Solution
For $t < 0$ we have the DC circuit of Figure 6.5(b) in which $i = 0.4$ A and $v = 1.2$ V. For $t \geq 0$ we use variables i and v to formulate the dynamical equations of the circuit. These variables determine the state of the system and are called state variables. The dynamical equations based on state variables are called state equations. From the circuit of Figure 6.5(a) we have

$$\begin{cases} \text{KCL at node } A: & v' + \frac{v}{2} + i = i_g \\ \text{KVL around the external loop:} & v - 3i - i' = 0 \end{cases}$$

Moving the derivative terms to the left side and the rest of the terms to the right side we get

$$\begin{cases} v' = -0.5v - i + i_g \\ i' = v - 3i \end{cases}$$

By taking the Laplace transforms of the above equations we have

$$\begin{cases} \mathcal{L}[v'] = sV(s) - v(0) \\ \mathcal{L}[i'] = sI(s) - i(0) \end{cases}$$

$$\begin{cases} sV(s) - v(0) = -0.5V(s) - I(s) + I_g(s) \\ sI(s) - i(0) = V(s) - 3I(s) \end{cases}$$

Note that the above includes the initial conditions. Substituting for $v(0)$, $i(0)$, and $I_g(s)$ and collecting terms

$$\begin{cases} V(s)(s + \frac{1}{2}) + I(s) = \dfrac{1.2s^2 + s + 1.2}{s^2 + 1} \\ -V(s) + I(s)(s + 3) = 0.4 \end{cases}$$

Solving for $V(s)$ and $I(s)$ we find

$$V(s) = \frac{(s + 3)(1.2s^2 + s + 1.2) - 0.4(s^2 + 1)}{(s^2 + 1)(s^2 + 3.5s + 2.5)}$$

$$I(s) = \frac{(1.2s^2 + s + 1.2) + 0.4(s + 0.5)(s^2 + 1)}{(s^2 + 1)(s^2 + 3.5s + 2.5)}$$

Note that these are the same expressions found from the transform circuit of Example 6.23. Taking the inverse transforms

$$v(t) = 0.6667e^{-t} - 0.0185e^{-2.5t} + 0.8304\cos(t - 48.4°), \quad t > 0$$

$$i(t) = 0.3332e^{-t} - 0.0367e^{-2.5t} + 0.2626\cos(t - 66.8°), \quad t > 0$$

Matrix Form

The state equations derived in Example 6.24 may be written in matrix form, with inputs, states, coefficients, and the outputs shown as

$$\mathbf{X}' = \mathbf{A} \times \mathbf{X} + \mathbf{B}$$

where $\mathbf{X}$ is the array of state variables, $\mathbf{A}$ is the coefficients matrix, and $\mathbf{B}$ is the array of inputs. The state equations for the circuit of Examples 6.24 are

$$\begin{cases} \text{KCL:} & v' = -0.5v - i + i_g \\ \text{KVL:} & i' = v - 3i \end{cases}$$

These may be written as

$$\begin{bmatrix} v' \\ i' \end{bmatrix} = \begin{bmatrix} -0.5 & -1 \\ 1 & -3 \end{bmatrix} \begin{bmatrix} v \\ i \end{bmatrix} + \begin{bmatrix} i_g \\ 0 \end{bmatrix}$$

Example

6.25

Consider the circuit of Example 6.23 once more. Without using the Laplace transform, find and solve two separate differential equations for $v(t)$ and $i(t)$. Identify the forced and natural components of the responses.

Solution

By using the generalized phasor (i.e., the s-operator) and generalized impedances we first write the state equations in the following form:

$$V(0.5 + s) + I = I_s$$
$$V - I(3 + s) = 0$$

and obtain system functions V/I_g and I/I_g.

$$\frac{V}{I_g} = \frac{s + 3}{s^2 + 3.5s + 2.5}$$

$$\frac{I}{I_g} = \frac{1}{s^2 + 3.5s + 2.5}$$

The differential equations are

$$v'' + 3.5v' + 2.5v = i'_g + 3i_g$$
$$i'' + 3.5i' + 2.5i = i_g$$

The above equations apply at all times. For $t < 0$ we have $i_g = 1$ A, from which we get $v = 1.2$ V and $i = 0.4$ A. These results agree with the DC steady-state response directly obtained from the resistive circuit. For $t \geq 0$, $i_g = \cos t$. Solutions to the differential equations contain a forced part (shown by the subscript $_f$) and a natural part (shown by the subscript $_n$). The forced and natural responses are the particular and homogeneous parts of the solution of the differential equation.

$$v(t) = v_f + v_n \quad \text{and} \quad i(t) = i_f + i_n$$

The complex amplitude of the forced response is found by evaluating system functions at the complex frequency of the current source, $s = j1$, and multiplying it by $I_g = 1$. We get $V = 0.8.3\angle - 48.4°$ and $I = 0.2626\angle - 66.8°$ which in time domain means

$$v_f = 0.83\cos(t - 48.4°) \quad \text{and} \quad i_f = 0.2626\cos(t - 66.8°)$$

The natural frequencies are found from the characteristic equation

$$s^2 + 3.5s + 2.5 = 0, \quad s = -1, \quad -2.5$$

The natural responses are

$$v_n(t) = Ae^{-t} + Be^{-2.5t} \quad \text{and} \quad i_n(t) = Ce^{-t} + De^{-2.5t}$$

Total responses at $t > 0$ are

$$v(t) = 0.83\cos(t - 48.4°) + Ae^{-t} + Be^{-2.5t}$$
$$i(t) = 0.2626\cos(t - 66.8°) + Ce^{-t} + De^{-2.5t}$$

To find $A, B, C,$ and D we need the initial conditions $i(0^+), i'(0^+), v(0^+),$ and $v'(0^+).$ In transition from $t = 0^-$ to $t = 0^+$ i and v do not change, remaining at $i(0^+) = 0.4$ A and $v(0^+) = 1.2$ V. To find $i'(0^+)$ and $v'(0^+)$ we apply the state equations at $t = 0^+.$

$$\begin{cases} v' = -0.5v - i + i_g \\ i' = v - 3i \end{cases} \Rightarrow \begin{cases} v'(0^+) = -0.5 \times 1.2 - 0.4 + 1 = 0 \\ i'(0^+) = 1.2 - 3 \times 0.4 = 0 \end{cases}$$

The initial conditions $i(0^+), i'(0^+), v(0^+),$ and $v'(0^+)$ can also be found directly from time analysis of the circuit around $t = 0.$ It should not be a surprise that $i'(0^+) = v'(0^+) = 0,$ as the value of the current source and its derivative didn't change at $t = 0.$ Applying the initial conditions to v we get

$$v(0^+) = 0.83 \cos 48.4° + A + B = 1.2$$
$$v'(0^+) = -0.83 \sin(-48.4°) - A - 2.5B = 0$$

from which we find $A = 0.6667,$ $B = -0.0185,$ and

$$v(t) = 0.6667e^{-t} - 0.0185e^{-2.5t} + 0.8304 \cos(t - 48.4°), \quad t > 0$$

Similarly,

$$i(t) = 0.3332e^{-t} - 0.0367e^{-2.5t} + 0.2626 \cos(t - 66.8°), \quad t > 0$$

6.20 Bilateral Laplace Transform

The bilateral Laplace transform of a function $x(t)$ is defined by

$$\mathcal{L}[x(t)] = X(s) = \int_{-\infty}^{\infty} x(t)e^{-st} dt$$

The range of the integral requires the knowledge of $x(t)$ from $-\infty$ to $\infty.$ Again, for the transform to exist, the above integral should converge, limiting s to a region of convergence (ROC) in the s-plane.

*E**xample*
6.26

Unit-step function

The bilateral Laplace transform of $x(t) = u(t)$ is

$$X(s) = \int_{-\infty}^{\infty} x(t)e^{-st} dt = \int_{0}^{\infty} e^{-st} dt = \frac{1}{s}, \quad \mathcal{RE}[s] > 0$$

It is readily seen that because $x(t) = 0$ for $t < 0,$ the unilateral and bilateral transforms of $x(t)$ are identical. $X(s)$ has a pole at the origin. The ROC is on the right of the pole (i.e., the right-half plane, RHP).

Example 6.27

Left-sided unit-step function

The bilateral Laplace transform of $x(t) = u(-t)$ is

$$X(s) = \int_{-\infty}^{0} e^{-st}dt = \frac{-1}{s}, \quad \mathcal{RE}[s] < 0$$

The ROC is the right-half plane, RHP (on the left of the pole at $s = 0$).

Example 6.28

Unit impulse

The bilateral Laplace transform of $\delta(t)$ is

$$\int_{-\infty}^{\infty} \delta(t)e^{-st}dt = 1 \quad \text{ROC} = \{s\}, \text{ all of the } s - \text{plane}$$

Example 6.29

Right-sided exponential

The bilateral Laplace transform of $x(t) = e^{at}u(t)$ is

$$X(s) = \int_{0}^{\infty} e^{at}e^{-st}dt = \frac{1}{s - a}, \quad \mathcal{RE}[s] > a$$

Again, because $x(t) = 0$ for $t < 0$, the unilateral and bilateral transforms are the same. $X(s)$ has a pole at $s = a$. The ROC is the portion of the s-plane to the right of the pole.

Example 6.30

Left-sided exponential

The bilateral Laplace transform of $x(t) = e^{bt}u(-t)$ is

$$X(s) = \int_{-\infty}^{0} e^{bt}e^{-st}dt = \frac{-1}{s - b}, \quad \mathcal{RE}[s] < b$$

The bilateral transform has a pole at $s = b$. The ROC is to the left of the pole.

Example 6.31

Two-sided exponential

The bilateral Laplace transform of $x(t) = e^{-c|t|}, \ c > 0, \ -\infty < t < \infty$ is

$$X(s) = \int_{-\infty}^{0} e^{ct}e^{-st}dt + \int_{0}^{\infty} e^{-ct}e^{-st}dt = -\frac{1}{s - c} + \frac{1}{s + c}$$

$$= \frac{-2c}{s^2 - c^2}, -c < \mathcal{RE}[s] < c$$

Table 6.4 summarizes transform pairs for various exponential functions.

TABLE 6.4 Bilateral Transforms of Exponential Functions and Their ROC

$x(t)$	$X(s)$	ROC	Existence of the Transform
$e^{at}u(t)$	$\dfrac{1}{s-a}$	$\mathcal{RE}[s] > a$	Transform always exists in the ROC
$e^{bt}u(-t)$	$\dfrac{-1}{s-b}$	$\mathcal{RE}[s] < b$	Transform always exists in the ROC
$e^{-c|t|}$	$\dfrac{-2c}{s^2-c^2}$	$-c < \mathcal{RE}[s] < c$	Transform exists only if $c > 0$
$e^{at}u(t) + e^{bt}u(-t)$	$\dfrac{a-b}{(s-a)(s-b)}$	$a < \mathcal{RE}[s] < b$	Transform exists only if $b > a$

Example

6.32

One-sided decaying or growing sinusoids

The bilateral Laplace transforms of $e^{at}\sin(\omega t)u(t)$ and $e^{at}\cos(\omega t)u(t)$ may be deduced from Example 6.29. Similarly, the transforms of $e^{bt}\sin(\omega t)u(-t)$ and $e^{bt}\cos(\omega t)u(-t)$ may be deduced from Example 6.30. These are derived in Table 6.5. The transforms of $e^{at}\sin(\omega t)u(t)$ and $e^{at}\cos(\omega t)u(t)$ each have two poles at $s_{1,2} = a \pm j\omega$. Their ROC are to the right of the poles. Similarly, the transforms of $e^{bt}\sin(\omega t)u(-t)$ and $e^{bt}\cos(\omega t)u(-t)$ have two poles at $s_{1,2} = b \pm j\omega$, with the ROC to the left.

TABLE 6.5 Bilateral Laplace Transforms of One-Sided Decaying or Growing Exponentials and Sinusoids and Their ROC

$x(t)$	$\Longrightarrow$	$X(s)$	ROC
$e^{at}u(t)$	$\Longrightarrow$	$\dfrac{1}{s-a}$	$\mathcal{RE}[s] > a$
$e^{bt}u(-t)$	$\Longrightarrow$	$\dfrac{-1}{s-b}$	$\mathcal{RE}[s] < b$
$e^{at}\sin(\omega t)u(t) = e^{at}\left[\dfrac{e^{j\omega t}-e^{-j\omega t}}{2j}\right]u(t)$	$\Longrightarrow$	$\dfrac{1}{2j}\left[\dfrac{1}{s-a-j\omega}-\dfrac{1}{s-a+j\omega}\right] = \dfrac{\omega}{(s-a)^2+\omega^2}$	$\mathcal{RE}[s] > a$
$e^{at}\cos(\omega t)u(t) = e^{at}\left[\dfrac{e^{j\omega t}+e^{-j\omega t}}{2}\right]u(t)$	$\Longrightarrow$	$\dfrac{1}{2}\left[\dfrac{1}{s-a-j\omega}+\dfrac{1}{s-a+j\omega}\right] = \dfrac{s-a}{(s-a)^2+\omega^2}$	$\mathcal{RE}[s] > a$
$e^{bt}\sin(\omega t)u(-t) = e^{bt}\left[\dfrac{e^{j\omega t}-e^{-j\omega t}}{2j}\right]u(-t)$	$\Longrightarrow$	$\dfrac{1}{2j}\left[\dfrac{-1}{s-b-j\omega}+\dfrac{1}{s-b+j\omega}\right] = -\dfrac{\omega}{(s-b)^2+\omega^2}$	$\mathcal{RE}[s] < b$
$e^{bt}\cos(\omega t)u(-t) = e^{bt}\left[\dfrac{e^{j\omega t}+e^{-j\omega t}}{2}\right]u(-t)$	$\Longrightarrow$	$\dfrac{1}{2}\left[\dfrac{-1}{s-b-j\omega}-\dfrac{1}{s-b+j\omega}\right] = -\dfrac{s-b}{(s-b)^2+\omega^2}$	$\mathcal{RE}[s] < b$

6.33 Two-sided decaying sinusoids

The bilateral transforms of $e^{-c|t|}\sin(\omega t)$ and $e^{-c|t|}\cos(\omega t)$, $c>0$, $-\infty < t < \infty$ are given in Table 6.6.

TABLE 6.6 Bilateral Transforms of Two-Sided Decaying Exponentials Sinusoids and Their ROC

$x(t)$ $\Longrightarrow$	$X(s)$	ROC
$e^{-c|t|}, c>0$ $\Longrightarrow$	$\dfrac{-2c}{s^2-c^2}$	$-c < \mathcal{RE}[s] < c$
$e^{-c|t|}\sin(\omega t)$ $\Longrightarrow$	$\dfrac{\omega}{(s+c)^2+\omega^2} - \dfrac{\omega}{(s-c)^2+\omega^2} = \dfrac{-4\omega cs}{(s^2-c^2)^2+2\omega^2(s^2+c^2)+\omega^4}$	$-c < \mathcal{RE}[s] < c$
$e^{-c|t|}\cos(\omega t)$ $\Longrightarrow$	$\dfrac{s+c}{(s+c)^2+\omega^2} - \dfrac{s-c}{(s-c)^2+\omega^2} = -2c\dfrac{(s^2-c^2)-\omega^2}{(s^2-c^2)+2\omega^2(s^2+c^2)+\omega^4}$	$-c < \mathcal{RE}[s] < c$

6.21 Region of Convergence of the Bilateral Laplace Transform

In section 6.2 we discussed the ROC for the unilateral transform. Similar considerations about the convergence of the integral apply to the ROC of the bilateral transform, resulting in more restrictions. The ROC can include the LHP, the RHP, or be limited to a vertical band in the s-plane. In this section we begin with some examples, then summarize the features of the ROC.

6.34

For each time function in Table 6.7(a), identify the correct expression for its bilateral Laplace transform [in the form of A, B, C, ..., L from Table 6.7(b)], and the correct ROC [in the form of 1, 2, 3, ..., 12, from Table 6.7(c). Enter your answers in Table 6.7(a)].

TABLE 6.7a

(Questions)
$x_1(t) = e^t u(t)$
$x_2(t) = e^{-t} u(t)$
$x_3(t) = e^t u(-t)$
$x_4(t) = e^{-t} u(-t)$
$x_5(t) = e^{2t} u(t)$
$x_6(t) = e^{-2t} u(t)$
$x_7(t) = e^{2t} u(-t)$
$x_8(t) = e^{-2t} u(-t)$
$x_9(t) = e^{3t} u(t)$
$x_{10}(t) = e^{-3t} u(t)$
$x_{11}(t) = e^{3t} u(-t)$
$x_{12}(t) = e^{-3t} u(-t)$

TABLE 6.7b

(Possible answers)	
A	$\frac{-1}{s+2}$
B	$\frac{1}{s-1}$
C	$\frac{-1}{s-1}$
D	$\frac{1}{s+3}$
E	$\frac{1}{s+1}$
F	$\frac{-1}{s+3}$
G	$\frac{-1}{s+1}$
H	$\frac{1}{s-2}$
I	$\frac{1}{s-3}$
J	$\frac{-1}{s-3}$
K	$\frac{1}{s+2}$
L	$\frac{-1}{s-2}$

TABLE 6.7c

(Possible answers)	
1	$\mathcal{RE}[s] > 2$
2	$\mathcal{RE}[s] < -2$
3	$\mathcal{RE}[s] > 3$
4	$\mathcal{RE}[s] > 1$
5	$\mathcal{RE}[s] < 3$
6	$\mathcal{RE}[s] > -2$
7	$\mathcal{RE}[s] < 2$
8	$\mathcal{RE}[s] < -1$
9	$\mathcal{RE}[s] < -3$
10	$\mathcal{RE}[s] > -3$
11	$\mathcal{RE}[s] < 1$
12	$\mathcal{RE}[s] > -1$

Solution

From Table 6.4, the answers are, respectively: B4, E12, C11, G8, H1, K6, L7, A2, I3, D10, J5, F9.

Example

6.35

Find the expression for the bilateral Laplace transform and the ROC of the following time functions which are constructed from $x_k(t)$, $k = 1, 2 \cdots 12$, given in Table 6.7(a).

1. $x_a(t) = x_3(t) + x_6(t) + x_{10}(t)$
2. $x_b(t) = x_1(t) + x_7(t) + x_{11}(t)$
3. $x_c(t) = x_2(t) + x_5(t) + x_{11}(t)$

Solution

Using the linearity property and the results of Table 6.4 we find

1. $X_a(s) = X_3(s) + X_6(s) + X_{10}(s) = \dfrac{-1}{s-1} + \dfrac{1}{s+2} + \dfrac{1}{s+3}$

   $= \dfrac{s^2 - 2s - 11}{s^3 + 4s^2 + s - 6}$ $\quad -2 < \mathcal{RE}[s] < 1$

2. $X_b(s) = X_1(s) + X_7(s) + X_{11}(s) = \dfrac{1}{s-1} - \dfrac{1}{s-2} - \dfrac{1}{s-3}$

   $= \dfrac{-s^2 + 2s + 1}{s^3 - 6s^2 + 11s - 6}$ $\quad 1 < \mathcal{RE}[s] < 2$

3. $X_c(s) = X_2(s) + X_5(s) + X_{11}(s) = \dfrac{1}{s+1} + \dfrac{1}{s-2} - \dfrac{1}{s-3}$

   $= \dfrac{s^2 - 6s + 5}{s^3 - 4s^2 + s + 6}$ $\quad 2 < \mathcal{RE}[s] < 3$

Example

6.36

The ROC of two-sided exponentials

Find the bilateral transforms of the following functions:

1. $x_1(t) = e^{-3|t|}$, $-\infty < t < \infty$
2. $x_2(t) = e^{3t}u(-t) + e^{-2t}u(t)$
3. $x_3(t) = e^{-2t}u(-t) + e^{-3t}u(t)$
4. $x_4(t) = e^{3t}u(-t) + e^{2t}u(t)$
5. $x_5(t) = e^{-3t}u(-t) + e^{2t}u(t)$
6. $x_6(t) = e^{-3t}u(-t) + e^{-2t}u(t)$

Solution

The transforms are derived below.

1. $X_1(s) = \displaystyle\int_{-\infty}^{0} e^{3t}e^{-st}dt + \int_{0}^{\infty} e^{-3t}e^{-st}dt = -\dfrac{1}{s-3} + \dfrac{1}{s+3}$

   $= \dfrac{-6}{s^2 - 9}, \quad -3 < \mathcal{RE}[s] < 3$

2. $X_2(s) = \int_{-\infty}^{0} e^{3t} e^{-st} dt + \int_{0}^{\infty} e^{-2t} e^{-st} dt = -\dfrac{1}{s-3} + \dfrac{1}{s+2}$

$\qquad = \dfrac{-5}{s^2 - s - 6}, \qquad -2 < \mathcal{RE}[s] < 3$

3. $X_3(s) = \int_{-\infty}^{0} e^{-2t} e^{-st} dt + \int_{0}^{\infty} e^{-3t} e^{-st} dt = -\dfrac{1}{s+2} + \dfrac{1}{s+3}$

$\qquad = \dfrac{-1}{s^2 + 5s + 6}, \qquad -3 < \mathcal{RE}[s] < -2$

4. $X_4(s) = \int_{-\infty}^{0} e^{3t} e^{-st} dt + \int_{0}^{\infty} e^{2t} e^{-st} dt = -\dfrac{1}{s-3} + \dfrac{1}{s-2}$

$\qquad = \dfrac{-1}{s^2 - 5s + 6}, \qquad 2 < \mathcal{RE}[s] < 3$

5. $X_5(s) = \int_{-\infty}^{0} e^{-3t} e^{-st} dt + \int_{0}^{\infty} e^{2t} e^{-st} dt = \infty,$

$\qquad$ transform doesn't exist No ROC

6. $X_6(s) = \int_{-\infty}^{0} e^{-3t} e^{-st} dt + \int_{0}^{\infty} e^{-2t} e^{-st} dt = \infty,$

$\qquad$ transform doesn't exist No ROC

Note that the ROC of $X(s)$, if it exists, is the single vertical band in the s-plane limited by its two poles. Transforms of $x_5(t)$ and $x_6(t)$ don't exist because the convergence of their integrals would have required $2 < \mathcal{RE}[s] < -3$ and $-2 < \mathcal{RE}[s] < -3$, respectively, which are not realizable.

Summary on ROC of bilateral transform
From the above examples we observe that

1.   The ROC, if it exists, is a contiguous region in the s-plane. It doesn't contain any poles.
2.   The ROC of a right-sided function is the area to the right of its pole with the greatest real part, $\mathcal{RE}[s] > a$.
3.   The ROC of a left-sided function is the area to the left of its pole with the smallest real part, $\mathcal{RE}[s] < b$.
4.   The ROC of a two-sided function, if it exists, is a vertical strip $a < \mathcal{RE}[s] < b$. The poles on the left of the ROC are due to the causal portion of $x(t)$ and the poles on the right are due to the anticausal portion.

6.22 Properties of the Bilateral Laplace Transform

The basic properties of the Laplace transform were presented in sections 6.2 and 6.4 to 6.11. These properties apply to the bilateral transform as well. However, in this case, because of the range of the integral: (a) the initial and final value theorems become mute,

(b) the differentiation property becomes simpler, and (c) the time-shift property will apply uniformly to shifts to the left as well as to the right. These properties are listed in Table 6.8 below.

TABLE 6.8 Properties of the Bilateral Laplace Transform

1	Definition	$\mathcal{L}[x(t)] \equiv X(s) = \displaystyle\int_{-\infty}^{\infty} x(t)e^{-st}dt$
2	Linearity	$\mathcal{L}[ax_1(t) + bx_2(t)] = aX_1(s) + bX_2(s)$
3	Time shift	$\mathcal{L}[x(t-T)] = X(s)e^{-sT}$
4	Multiplication by t	$\mathcal{L}[tx(t)] = -\dfrac{dX(s)}{ds}$
5	Multiplication by e^{at}	$\mathcal{L}[x(t)e^{at}] = X(s-a)$
6	Differentiation	$\mathcal{L}\left[\dfrac{dx(t)}{dt}\right] = sX(s)$
7	Integration	$\mathcal{L}\left[\displaystyle\int_{-\infty}^{t} x(t)dt\right] = \dfrac{X(s)}{s}$
8	Convolution	$\mathcal{L}[x(t) \star h(t)] = X(s)H(s)$
9	Uniqueness	$x(t) \Longleftrightarrow X(s)$

Time Shift

To show this property, apply the definition of the bilateral transform and change $(t - T)$ to τ as shown below.

$$\mathcal{L}[x(t-T)] = \int_{-\infty}^{\infty} x(t-T)e^{-st}dt = \int_{-\infty}^{\infty} x(\tau)e^{-s(\tau+T)}d\tau$$

$$= e^{-sT}\int_{-\infty}^{\infty} x(\tau)e^{-s\tau}d\tau = e^{-sT}X(s)$$

A time shift doesn't influence the ROC.

6.23 Inverse of the Bilateral Laplace Transform

The inverse of the Laplace transform is found from

$$x(t) = \mathcal{L}^{-1}[X(s)] = \frac{1}{2\pi j}\int_{C} X(s)e^{st}ds$$

where the integral is taken along a line (a contour) in the ROC. As in the case of the unilateral transform, the integration path goes from $-\infty < \mathcal{IM}[s] < \infty$ within the ROC. According to Cauchy's theorem on contour integration of functions of complex variables, the value of the above integral along a closed contour is equal to the sum of residues of the function $X(s)e^{st}$ at all poles inside the contour. Evaluating the integral then becomes a matter of finding the residues. The closed integration path may be traversed by an infinitely large semicircle without affecting the value of the above integral. To evaluate $x(t)$ for $t > 0$, we choose a closed contour which is a very large semicircle ($R = \infty$) in the LHP and on which $\mathcal{RE}[s] < 0$. The value of the integral over the semicircle then becomes zero and the inverse transform becomes equal to the sum of the residues of $X(s)e^{st}$ at its poles within the contour; that is, to the left of the ROC. Similarly, to evaluate $x(t)$ for $t < 0$, we traverse the closed contour through a very large semicircle in the RHP, on which $\mathcal{RE}[s] > 0$. The value of the integral over the semicircle then becomes zero and the inverse transform becomes equal to the sum of the residues of the function $X(s)e^{st}$ at its poles within the contour; that is, to the right of the ROC. The residues may be found by expanding $X(s)$ to its fractions. The sum of residues of the function $X(s)e^{st}$ at poles to the left of the ROC produces the causal portion of $x(t)$. The sum of poles to the right generates its anticausal portion. In practice, one expands $X(s)$ into its partial fractions and then uses tables of transform pairs.

xample

6.37

Given the bilateral Laplace transform

$$X(s) = \frac{B(s)}{(s + 2)(s + 1)(s - 1.5)(s - 3)}, \text{ with the ROC of } -1 < \mathcal{RE}[s] < 1.5$$

(see Figure 6.6), what can be said about $x(t)$?

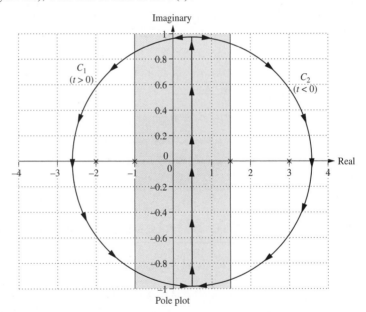

Pole plot

FIGURE 6.6 The ROC of $X(s)$ is the vertical band $-1 < \mathcal{RE}[s] < 1.5$ in the s-plane. The poles to the left of the ROC constitute the causal part of the time function and those on the right produce the anticausal part.

Solution

In this case the poles are at -2, -1, 1.5, and 3. The time function is two-sided. It is made of two left-sided and two right-sided exponentials. The two poles on the left of ROC constitute the causal part of $x(t)$ and those on its right produce its anticausal segment. The initial vale of each exponetial is set by the residue of $X(s)$ at the corresponding pole. The Laplace transform and time function are summarized below:

$$X(s) = \frac{B(s)}{(s+2)(s+1)(s-1.5)(s-3)}$$

$$= \frac{k_1}{s+2} + \frac{k_2}{s+1} + \frac{k_3}{s-1.5} + \frac{k_4}{s-3}, \quad \text{ROC is } -1 < \mathcal{RE}[s] < 1.5$$

$$x(t) = -\left[k_4 e^{3t} + k_3 e^{1.5t}\right] u(-t) + \left[k_2 e^{-t} + k_1 e^{-2t}\right] u(t)$$

Example
6.38

This example considers five different ROCs associated with the pole configuration in Figure 6.6 and evaluates their corresponding time functions. Given

$$H(s) = \frac{s^3 + 7s^2 + 1.5s - 34}{s^4 - 1.5s^3 - 7s^2 + 4.5s + 9}$$

a.   Find the poles of $H(s)$ and expand it to its partial fractions.

b.   Find all possible inverse bilateral Laplace transforms of $H(s)$.

Solution

a.   First find the roots of the denominator:

$$s^4 - 1.5s^3 - 7s^2 + 4.5s + 9 = 0$$

$$s = -2, \ -1, \ 1.5, \ 3$$

Then expand $H(s)$ to its partial fractions.

$$H(s) = \frac{s^3 + 7s^2 + 1.5s - 34}{(s+2)(s+1)(s-1.5)(s-3)} = \frac{k_1}{s+2} + \frac{k_2}{s+1} + \frac{k_3}{s-1.5} + \frac{k_4}{s-3}$$

$$k_1 = \left. \frac{s^3 + 7s^2 + 1.5s - 34}{(s+1)(s-1.5)(s-3)} \right|_{s=-2} = 0.9714$$

$$k_2 = \left. \frac{s^3 + 7s^2 + 1.5s - 34}{(s+2)(s-1.5)(s-3)} \right|_{s=-1} = -2.95$$

$$k_3 = \left. \frac{s^3 + 7s^2 + 1.5s - 34}{(s+2)(s+1)(s-3)} \right|_{s=1.5} = 0.9616$$

$$k_4 = \left. \frac{s^3 + 7s^2 + 1.5s - 34}{(s+2)(s+1)(s-1.5)} \right|_{s=3} = 2.0167$$

b.   The transform has four poles at $s = -2, \ -1, \ 1.5, \ 3$. The five possible ROCs, and the time functions associated with each, are listed below.

ROC1 is $\mathcal{RE}[s] < -2$ $\implies$ $h_1(t) = [-k_1 e^{-2t} - k_2 e^{-t} - k_3 e^{1.5t} - k_4 e^{3t}] u(-t)$

ROC2 is $-2 < \mathcal{RE}[s] < -1$ $\implies$ $h_2(t) = [-k_2 e^{-t} - k_3 e^{1.5t} - k_4 e^{3t}] u(-t) + k_1 e^{-2t} u(t)$

ROC3 is $-1 < \mathcal{RE}[s] < 1.5$ $\implies$ $h_3(t) = [-k_3 e^{1.5t} - k_4 e^{3t}] u(-t) + [k_1 e^{-2t} + k_2 e^{-t}] u(t)$

ROC4 is $1.5 < \mathcal{RE}[s] < 3$ $\implies$ $h_4(t) = -k_4 e^{3t} u(-t) + [k_1 e^{-2t} + k_2 e^{-t} + k_3 e^{1.5t}] u(t)$

ROC5 is $\mathcal{RE}[s] > 3$ $\implies$ $h_5(t) = [k_1 e^{-2t} + k_2 e^{-t} + k_3 e^{1.5t} + k_4 e^{3t}] u(t)$

Example

6.39

Using partial fraction expansion find the inverse of the following bilateral Laplace transforms:

$$\alpha) \quad X_\alpha(s) = \frac{s^2 - 2s - 11}{s^3 + 4s^2 + s - 6} \qquad -2 < \mathcal{RE}[s] < 1$$

$$\beta) \quad X_\beta(s) = \frac{-s^2 + 2s + 1}{s^3 - 6s^2 + 11s - 6} \qquad 1 < \mathcal{RE}[s] < 2$$

$$\gamma) \quad X_\gamma(s) = \frac{s^2 - 6s + 5}{s^3 - 4s^2 + s + 6} \qquad 2 < \mathcal{RE}[s] < 3$$

Solution

We first find the roots of the denominator of $X(s)$. (This may be done by an analytical method, a successive approximation technique, or using a software package such as Matlab.) We then expand each $X(s)$ into its partial fractions. The fractions containing the poles on the left of the ROC generate the causal part of $x(t)$. The poles on the right generate the anticausal part. See Figure 6.7 − α, β, γ.

$$\alpha) \quad X_\alpha(s) = \frac{s^2 - 2s - 11}{s^3 + 4s^2 + s - 6}, \quad -2 < \mathcal{RE}[s] < 1$$

$$s^3 + 4s^2 + s - 6 = 0, \ s_{1,2,3} = 1, -2, -3.$$

$$X_\alpha(s) = \frac{s^2 - 2s - 11}{((s-1)(s+2)(s+3)} = \frac{k_1}{s-1} + \frac{k_2}{s+2} + \frac{k_3}{s+3}$$

$$k_1 = (s-1)X_\alpha(s)\Big|_{s=1} = \frac{s^2 - 2s - 11}{(s+2)(s+3)}\Big|_{s=1} = -1$$

$$k_2 = (s+2)X_\alpha(s)\Big|_{s=-2} = \frac{s^2 - 2s - 11}{(s-1)(s+3)}\Big|_{s=-2} = 1$$

$$k_3 = (s+3)X_\alpha(s)\Big|_{s=-3} = \frac{s^2 - 2s - 11}{(s-1)(s+2)}\Big|_{s=-3} = 1$$

$$x_\alpha(t) = e^t u(-t) + e^{-2t}u(t) + e^{-3t}u(t)$$

Similarly,

$$\beta) \quad X_\beta(s) = \frac{-s^2 + 2s + 1}{s^3 - 6s^2 + 11s - 6} = \frac{1}{s-1} - \frac{1}{s-2} - \frac{1}{s-3}, \quad 1 < \mathcal{RE}[s] < 2$$

$$x_\beta(t) = e^t u(t) + e^{2t}u(-t) + e^{3t}u(-t)$$

and,

$$\gamma) \quad X_\gamma(s) = \frac{s^2 - 6s + 5}{s^3 - 4s^2 + s + 6} = \frac{1}{s+1} + \frac{1}{s-2} - \frac{1}{s-3}, \quad 2 < \mathcal{RE}[s] < 3$$

$$x_\gamma(t) = e^{-t}u(t) + e^{2t}u(t) + e^{3t}u(-t)$$

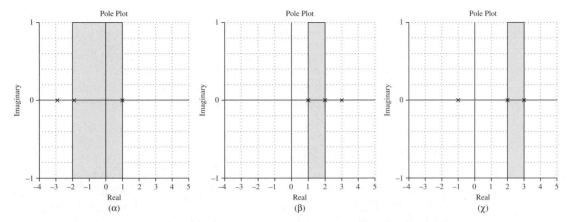

FIGURE 6.7 The ROCs for three transforms. For $t < 0$ the integration contour circles the RHP poles in a clockwise direction. For $t > 0$ it circles the LHP poles in a counterclockwise direction. See Example 6.39.

6.24 A Table of Bilateral Laplace Transform Pairs

See Table 6.9 for a list of bilateral Laplace transform pairs. In this table $x(t)$ is defined for $-\infty < t < \infty$.

TABLE 6.9 Bilateral Laplace Transform Pairs

	$x(t) = \mathcal{L}^{-1}[X(s)]$	$\Longleftrightarrow$	$X(s) = \int_{-\infty}^{\infty} x(t)e^{-st}dt$	Region of Convergence
1	$\delta(t)$	$\Longleftrightarrow$	1	all s
2	$u(t)$	$\Longleftrightarrow$	$\dfrac{1}{s}$	$\mathcal{RE}[s] > 0$
3	$u(-t)$	$\Longleftrightarrow$	$\dfrac{-1}{s}$	$\mathcal{RE}[s] < 0$
4	$e^{at}u(t)$	$\Longleftrightarrow$	$\dfrac{1}{s-a}$	$\mathcal{RE}[s] > a$
5	$e^{bt}u(-t)$	$\Longleftrightarrow$	$\dfrac{-1}{s-b}$	$\mathcal{RE}[s] < b$
6	$e^{-c|t|}$	$\Longleftrightarrow$	$\dfrac{-2c}{s^2-c^2}$	$-c < \mathcal{RE}[s] < c$
7	$e^{at}\sin(\omega t)u(t)$	$\Longleftrightarrow$	$\dfrac{\omega}{(s-a)^2+\omega^2}$	$\mathcal{RE}[s] > a$
8	$e^{at}\cos(\omega t)u(t)$	$\Longleftrightarrow$	$\dfrac{s-a}{(s-a)^2+\omega^2}$	$\mathcal{RE}[s] > a$
9	$e^{bt}\sin(\omega t)u(-t)$	$\Longleftrightarrow$	$-\dfrac{\omega}{(s-b)^2+\omega^2}$	$\mathcal{RE}[s] < b$
10	$e^{bt}\cos(\omega t)u(-t)$	$\Longleftrightarrow$	$-\dfrac{s-b}{(s-b)^2+\omega^2}$	$\mathcal{RE}[s] < b$

6.25 System Function

From the convolution property we see that $Y(s) = H(s)X(s)$, where $Y(s)$ and $X(s)$ are the Laplace transforms of an input-output pair to an LTI system and $H(s)$ is the Laplace transform of its unit-impulse response. We define $H(s)$ to be the *system function*. In this section we present some applications. More detail will be given in Chapter 9.

H(s) Is the Ratio Y(s)/X(s)

The system function is the ratio of the Laplace transform of the output to the Laplace transform of the input. This may be considered the first definition of the system function. Based on this definition, the system function can be obtained from input-output pair (of an LTI system) regardless of how that pair was obtained. For instance, the output to a given input may be measured experimentally and modeled as a mathematical function. The system function can then be obtained as the ratio of input-output Laplace tranforms. Example 6.40 illustrates the above method.

Example **6.40**

You are asked to identify a two-port electronic device and model it by an LTI system. For this purpose, a 1-volt rectangular voltage pulse lasting 1 ms is applied to the input of the device:

$$v_1(t) = \begin{cases} 1, & 0 < t < 1 \text{ msec.} \\ 0, & \text{elsewhere} \end{cases}$$

The output is recorded and modeled by

$$v_2(t) = \begin{cases} 0, & t < 0 \\ 1 - e^{-1,000t}, & 0 < t < 1 \text{ msec} \\ 1.718e^{-1,000t}, & t > 1 \text{ msec} \end{cases}$$

Assume the device is linear and time invariant and find its system function.

Solution

Laplace transforms will be used. The input is $v_1(t) = u(t) - u(t - 0.001)$ and the output may be represented as $v_2(t) = y(t) - y(t - 0.001)$, where $y(t) = (1 - e^{-1,000t})u(t)$. The system function is then found from

$$v_1(t) = u(t) - u(t - 0.001) \implies V_1(s) = \frac{(1 - e^{-0.001s})}{s}$$

$$y(t) = (1 - e^{-1000t})u(t) \implies Y(s) = \frac{1}{s} - \frac{1}{s + 1,000} = \frac{1,000}{s(s + 1,000)}$$

$$v_2(t) = y(t) - y(t - 0.001) \implies V_2(s) = (1 - e^{-0.001s})Y(s) = 1,000\frac{1 - e^{-0.001s}}{s(s + 1,000)}$$

$$H(s) = \frac{V_2(s)}{V_1(s)} \implies H(s) = 1,000\frac{1 - e^{-0.001s}}{s(s + 1,000)} \times \frac{s}{1 - e^{-0.001s}} = \frac{1,000}{s + 1,000}$$

Note: Because we are modeling the device by a causal LTI system, $y(t)$ is in fact the unit-step response of the system. Derivation of $H(s)$ can, therefore, be simplified by the following approach:

Input: $\qquad\qquad x(t) = u(t) \qquad\qquad \implies \quad X(s) = \dfrac{1}{s}$

Output: $\qquad\quad y(t) = (1 - e^{-1,000t})u(t) \quad \implies \quad Y(s) = \dfrac{1,000}{s(s + 1,000)}$

System function: $\quad H(s) = \dfrac{Y(s)}{X(s)} \qquad\qquad\quad \implies \quad H(s) = \dfrac{1,000}{s + 1,000}$

This is a consequence of $H(s)$ being the ratio of the transform of the output to that of the input.

$H(s)$ Is the Laplace Transform of $h(t)$

The system function is the Laplace transform of the unit-impulse response. This is a consequence of $H(s)$ being the ratio of the transform of the output to that of the input. The Laplace transform of a unit impulse is 1 which makes $H(s)$ the Laplace transform of $h(t)$.

Example 6.41

Find the system function of the LTI system with the unit-impulse response $h(t) = e^{-at}u(t)$.

Solution

From the table of Laplace transform pairs we find

$$H(s) = \frac{1}{s + a}, \quad \mathcal{RE}[s] > -a$$

Example 6.42

Find the system function of a finite integrator with unity gain and duration T seconds.

$$h(t) = \begin{cases} 1, & 0 \le t \le T \\ 0, & \text{elsewhere} \end{cases}$$

Solution

$$H(s) = \int_0^T e^{-st}dt = \frac{1 - e^{-sT}}{s}, \quad \text{all s}$$

Example 6.43

Find $H(s)$ of the system with the unit-impulse response $h(t) = \sqrt{2}e^{-t}\cos(t + 45°)u(t)$.

Solution

From the table of Laplace transform pairs we find

$$e^{-at}\cos(\omega t + \theta)u(t) \qquad\Longrightarrow\qquad \frac{(s+a)\cos\theta - \omega\sin\theta}{(s+a)^2 + \omega^2} \qquad \mathcal{RE}[s] > -a$$

$$h(t) = \sqrt{2}e^{-t}\cos(t + 45°)u(t) \quad\Longrightarrow\quad H(s) = \frac{s}{s^2 + 2s + 2} \qquad \mathcal{RE}[s] > -1$$

6.26 Comprehensive Examples

This section presents two examples for finding the responses of LTI systems by the time and transform-domain methods, and observes parallels between them.

 *xample*

6.44

Consider the LTI system specified by the unit-impulse response $h(t) = 5e^{-t}\sin(t)u(t)$. Find the system's response to $x(t) = e^{t}u(-t)$ by the following three methods:

1. Bilateral Laplace transform method.
2. Convolution of $x(t)$ with $h(t)$.
3. Time-domain solution of the differential equation.

Method 1. Laplace Transform

Multiply the bilateral Laplace transforms of $x(t)$ and $h(t)$ with each other, then take the inverse transform of the result to obtain $y(t)$.

$$x(t) = e^{t}u(-t) \qquad\qquad X(s) = \frac{-1}{s-1}, \qquad\qquad \mathcal{RE}[s] < 1$$

$$h(t) = 5e^{-t}\sin tu(t) \qquad\qquad H(s) = \frac{5}{s^2 + 2s + 2}, \qquad\qquad \mathcal{RE}[s] > -1$$

$$Y(s) = X(s)H(s) = \frac{-5}{(s-1)(s^2+2s+2)} = \frac{-1}{s-1} + \frac{s+3}{s^2+2s+2}, \qquad -1 < \mathcal{RE}[s] < 1$$

$$y(t) = e^{t}u(-t) + e^{-t}(\cos t + 2\sin t)u(t)$$

$$\qquad\quad = e^{t}u(-t) + \sqrt{5}e^{-t}\cos(t - 63°)u(t) \qquad\qquad -\infty < t < \infty$$

Method 2. Convolution

$$y(t) = x(t) \star h(t) = \int_{-\infty}^{\infty} x(t-\tau)h(\tau)d\tau$$

$$\mathbf{t \leq 0} \qquad y_\ell(t) = \int_0^\infty e^{(t-\tau)} \times 5e^{-\tau}\sin\tau d\tau$$

$$\qquad\qquad = 5e^{t}\int_0^\infty e^{-2\tau}\sin\tau d\tau = e^{t}$$

Note: $\displaystyle\int_{\tau_1}^{\tau_2} e^{-2\tau}\sin\tau\, d\tau = -\frac{1}{5}e^{-2\tau}(\cos\tau + 2\sin\tau)\Big|_{\tau_1}^{\tau_2}$

$t \geq 0 \quad y_r(t) = \displaystyle\int_0^\infty e^{(t-\tau)} \times 5e^{-\tau}\sin\tau\, d\tau$

$\qquad\qquad = 5e^t \displaystyle\int_t^\infty e^{-2\tau}\sin\tau\, d\tau = \sqrt{5}e^{-t}\cos(t - 63°)$

$\qquad y(t) = y_\ell(t) + y_r(t) = e^{-t}u(-t) + \sqrt{5}e^{-t}\cos(t - 63°)u(t) \quad -\infty < t < \infty$

Method 3. Time-Domain Solution
Solve the system's input-output differential equation.

$$\frac{d^2y}{dt^2} + 2\frac{dy}{dt} + 2y = 5x$$

$t \leq 0 \qquad \dfrac{d^2y_\ell}{dt^2} + 2\dfrac{dy_\ell}{dt} + 2y_\ell = 5e^t$

$\qquad\qquad y_\ell(t) = \dfrac{5}{s^2 + 2s + 2}\Big|_{s=1} e^t = e^t \quad$ particular solution only

$t \geq 0 \qquad \dfrac{d^2y_r}{dt^2} + 2\dfrac{dy_r}{dt} + 2y_r = 0 \qquad\qquad$ initial conditions: $y(0) = 1, \ y'(0) = 1$

$\qquad\qquad y_r(t) = \sqrt{5}e^{-t}\cos(t - 63°) \qquad\quad$ homogeneous solution only

$\qquad\qquad y(t) = y_\ell(t) + y_r(t) \qquad\qquad\qquad$ total solution for $-\infty < t < \infty$

$\qquad\qquad\quad = e^t u(-t) + \sqrt{5}e^{-t}\cos(t - 63°)u(t)$

Example
6.45

Consider the causal LTI system described by the differential equation

$$\frac{d^2y}{dt^2} + 4\frac{dy}{dt} + 5y = 2x$$

where $x(t)$ is the input and $y(t)$ is the output. Find the system's response to $x(t) = e^{-t}u(-t) + e^{-2t}u(t)$ by:

1. Bilateral Laplace transform.
2. Convolution of $x(t)$ with $h(t)$.
3. Time-domain solution of the differential equation.
4. Superposition of responses to the causal and anticausal inputs, transform-domain approach.
5. Superposition of responses to the causal and anticausal inputs, time-domain approach.

Method 1. Laplace Transform

Take the bilateral transform of the two sides of the differential equation.

$$\frac{d^2y}{dt^2} + 4\frac{dy}{dt} + 5y = 2x$$

$$s^2 Y(s) + 4s Y(s) + 5Y(s) = 2X(s)$$

$$Y(s) = \frac{2}{s^2 + 4s + 5} X(s)$$

$$x(t) = e^{-t}u(-t) + e^{-2t}u(t)$$

$$X(s) = \frac{-1}{s+1} + \frac{1}{s+2} = \frac{-1}{(s+1)(s+2)}, \qquad -2 < \mathcal{RE}[s] < -1$$

$$Y(s) = \frac{-2}{(s^2 + 4s + 5)(s+1)(s+2)} \qquad\qquad -2 < \mathcal{RE}[s] < -1$$

$$= \frac{-2}{(s+2+j)(s+2-j)(s+1)(s+2)}$$

$$= \frac{k_1}{s+2+j} + \frac{k_1^*}{s+2-j} + \frac{k_2}{s+1} + \frac{k_3}{s+2}$$

$$k_1 = \frac{-2}{(s+2-j)(s+1)(s+2)}\Big|_{s=-2-j} = \frac{\sqrt{2}}{2}\angle 135°$$

$$k_2 = \frac{-2}{(s^2+4s+5)(s+2)}\Big|_{s=-1} = -1$$

$$k_3 = \frac{-2}{(s^2+4s+5)(s+1)}\Big|_{s=-2} = 2$$

$$y(t) = e^{-t}u(-t) + 2e^{-2t}u(t) + \sqrt{2}e^{-2t}\cos(t - 135°)u(t) \qquad -\infty < t < \infty$$

Method 2. Convolution

The unit-impulse response of the differential equation is $h(t) = 2e^{-2t}\sin tu(t)$.

$$y(t) = x(t) \star h(t) = \int_{-\infty}^{\infty} x(t - \tau)h(\tau)d\tau$$

$$\mathbf{t \le 0} \qquad y_\ell(t) = 2\int_0^\infty e^{(t-\tau)} \times e^{-2\tau}\sin \tau d\tau = 2e^{-t}\int_0^\infty e^{-\tau}\sin \tau d\tau = e^t$$

Note: $\displaystyle \int_{\tau_1}^{\tau_2} e^{-\tau} \sin \tau d\tau = -\frac{1}{2} e^{-\tau} (\cos \tau + \sin \tau) \Big|_{\tau_1}^{\tau_2}$

t ≥ 0 $\displaystyle y_r(t) = 2 \int_t^{\infty} e^{-(t-\tau)} \times e^{-2\tau} \sin \tau d\tau + 2 \int_0^t e^{-2(t-\tau)} \times e^{-2\tau} \sin \tau d\tau$

$$= 2e^{-t} - \int_t^{\infty} e^{-\tau} \sin \tau d\tau + 2e^{-2t} \int_0^t \sin \tau d\tau$$

$$= e^{-2t}(\cos t + \sin t) + 2e^{-2t}(1 - \cos t)$$

$$= 2e^{-2t} + \sqrt{2} e^{-2t} \cos(t - 135°)$$

$$y(t) = y_\ell(t) + y_r(t)$$

$$= e^{-t} u(-t) + 2e^{-2t} u(t) + \sqrt{2} e^{-2t} \cos(t - 135°) u(t) \quad -\infty < t < \infty$$

Method 3. Time-Domain Solution

Solve the differential equation for $t < 0$ and $t > 0$ separately.

t ≤ 0 $\displaystyle \frac{d^2 y_\ell}{dt^2} + 4\frac{dy_\ell}{dt} + 5y_\ell = 2e^{-t}$

$\displaystyle y_\ell(t) = y_{\ell p}(t) = \frac{2}{s^2 + 4s + 5}\Big|_{s=-1} e^{-t} = e^{-t}$ total solution for **t ≤ 0**

t ≥ 0 $\displaystyle \frac{d^2 y_r}{dt^2} + 4\frac{dy_r}{dt} + 5y_r = 2e^{-2t}$ boundary conditions:

$$y(0) = 1, \quad y'(0) = -1$$

$\displaystyle y_{rp}(t) = \frac{2}{s^2 + 4s + 5}\Big|_{s=-2} e^{-2t} = 2e^{-2t}$ particular solution

$y_{rh}(t) = Ae^{-2t} \cos(t + \theta)$ homogeneous solution,

$\quad\quad = \sqrt{2} e^{-2t} \cos(t - 135°)$ apply boundary conditions

$y_r(t) = y_{rp}(t) + y_{rh}(t)$ total solution for **t ≥ 0**

$\quad\quad = 2e^{-2t} + \sqrt{2} e^{-2t} \cos(t - 135°)$

$y(t) = y_\ell(t) + y_r(t)$

$\quad\quad = e^{-t} u(-t) + 2e^{-2t} u(t) + \sqrt{2} e^{-2t} \cos(t - 135°) u(t)$

(total solution for $-\infty < $ **t** $ < \infty$)

Method 4. Superposition in the Transform Domain

Use the transform method to find the system's response to the causal and anticausal components of the input separately. Then add them together.

$$x(t) \;=\; x_1(t) + x_2(t)$$

$$x_1(t) = e^{-t}u(-t) \qquad\qquad\qquad X_1(s) = \frac{-1}{s+1}, \qquad\qquad \mathcal{RE}[s] < -1$$

$$Y_1(s) = \frac{2}{s^2+4s+5}X_1(s) = \frac{-2}{(s^2+4s+5)(s+1)} \qquad -2 < \mathcal{RE}[s] > -2$$

$$\qquad = \frac{-1}{s+1} + \frac{s+3}{s^2+4s+5}$$

$$y_1(t) = e^{-t}u(-t) + \sqrt{2}e^{-2t}\cos(t-45°)u(t) \qquad\qquad -\infty < t < \infty$$

$$x_2(t) = e^{-2t}u(t) \qquad\qquad\qquad X_2(s) = \frac{1}{s+2}, \qquad\qquad \mathcal{RE}[s] > -2$$

$$Y_2(s) = \frac{2}{s^2+4s+5}X_2(s) = \frac{2}{(s^2+4s+5)(s+2)} \qquad -2 < \mathcal{RE}[s] < -1$$

$$\qquad = \frac{2}{s+2} - 2\frac{s+2}{s^2+4s+5}$$

$$y_2(t) = 2e^{-2t}u(t) - 2e^{-2t}\cos(t)u(t) \qquad\qquad -\infty < t < \infty$$

$$y(t) \;=\; y_1(t) + y_2(t)$$

$$\qquad = e^{-t}u(-t) + \sqrt{2}e^{-2t}\cos(t-45°)u(t)$$

$$\qquad\quad + 2e^{-2t}u(t) - 2e^{-2t}\cos(t)u(t)$$

$$\qquad = e^{-t}u(-t) + 2e^{-2t}u(t) + \sqrt{2}e^{-2t}\cos(t-135°)u(t) \qquad -\infty < t < \infty$$

Method 5. Superposition in the Time Domain

Use the time-domain approach to find the system's response to the causal and anti-causal components of the input separately. Then add them together.

$$x(t) \;=\; x_1(t) + x_2(t)$$

$$x_1(t) = e^{-t}u(-t)$$

$$y_1(t) = y_{1p}(t) + y_{1h}(t) \qquad\qquad \text{particular plus homogeneous responses to } x_1(t)$$

$$\qquad = e^{-t}u(-t) + \sqrt{2}e^{-2t}\cos(t-45°)u(t) \qquad\qquad -\infty < t < \infty$$

$$x_2(t) = e^{-2t}u(t)$$

$$y_2(t) = y_{2p}(t) + y_{2h}(t) \qquad\qquad \text{particular plus homogeneous responses to } x_2(t)$$

$$\qquad = 2e^{-2t}u(t) - 2e^{-2t}\cos(t)u(t) \qquad\qquad -\infty < t < \infty$$

$$y(t) \;=\; y_1(t) + y_2(t) \qquad\qquad \text{total response to } x(t)$$

$$\qquad = e^{-t}u(-t) + 2e^{-2t}u(t) + \sqrt{2}e^{-2t}\cos(t-135°)u(t) \qquad -\infty < t < \infty$$

6.27 Concluding Remarks

The Laplace transform is a powerful tool for the analysis and synthesis of LTI systems. It converts a set of integro-differential equations into algebraic equations and, in this capability, is used to solve differential equations. This chapter started with the historical developments of the unilateral Laplace transform and its use in solving differential equations with given initial conditions. The discussion was kept at the basic level of solving linear differential equations with constant coefficients with one input and one output. Even within the framework of the unilateral transform, one needs to include singularity functions such as the unit-impulse function, especially those occurring at $t = 0$. Although most of the chapter discussed the unilateral transform, it did so in a way that its extension to the bilateral form does not present a rift but rather a way to overcome limitations in the power of the unilateral transform and clarify some apparent ambiguities in its application.

The importance of the Laplace transform in system analysis and design, however, is in its ability to decribe the behavior of an LTI system in the frequency domain. The vehicle for this task is the system function, which was introduced briefly in this chapter. The frequency domain analysis of LTI systems is the subject of Chapter 9.

6.28 Problems

Solved Problems

1. Find the Laplace transform of the functions given below. Specify the region of convergence.

a. $e^{3t}u(t)$ $\implies$ $\dfrac{1}{s-3}$, $\sigma > 3$

b. $\sqrt{2}\cos(3t + 45°)u(t)$ $\implies$ $\dfrac{s-3}{s^2+9}$ $\sigma > 0$

c. $e^{-3t}\sin(3t)u(t)$ $\implies$ $\dfrac{3}{(s+3)^2+9}$ $\sigma > 0$

d. $\dfrac{1}{b-a}\left[e^{-at} - e^{-bt}\right]u(t)$ $\implies$ $\dfrac{1}{b-a}\left[\dfrac{1}{s+a} - \dfrac{1}{s+b}\right] = \dfrac{1}{(s+a)(s+b)}$ $\sigma > -min(a,b)$

2. Find the inverse Laplace transform of

$$X(s) = \frac{s+2}{s^2+1}, \quad \sigma > 0$$

by the following two expansion methods and compare results:

(a) $X(s) = \dfrac{A}{s+j} + \dfrac{A^*}{s-j}$ (b) $X(s) = \dfrac{s}{s^2+1} + \dfrac{2}{s^2+1}$

Solution

a. $X(s) = \dfrac{A}{s+j} + \dfrac{A^*}{s-j}, \quad A = \dfrac{1}{2} + j = 1.118\angle 63°, \quad x(t) = 2.236\cos(t - 63°), \quad t > 0$

b. $X(s) = \dfrac{s}{s^2+1} + \dfrac{2}{s^2+1}, \quad x(t) = \cos t + 2\sin t = 2.236\cos(t - 63°), \quad t > 0$

3. Find the inverse Laplace transform of $X(s) = 1/(s+2)(s+1)^2, \quad \sigma > -1.$

Solution

$$X(s) = \frac{1}{(s+2)(s+1)^2} = \frac{k_1}{s+2} + \frac{k_2}{(s+1)^2} + \frac{k_3}{s+1}$$

$$k_1 = X(s)(s+2)\Big|_{s=-2} = \frac{1}{(s+1)^2}\Big|_{s=-2} = 1$$

$$k_2 = X(s)(s+1)^2\Big|_{s=-1} = \frac{1}{s+2}\Big|_{s=-1} = 1$$

$$k_3 = \frac{d}{ds}\left[X(s)(s+1)^2\right]\Big|_{s=-1} = \frac{d}{ds}\left[\frac{1}{s+2}\right]_{s=-1} = -\frac{1}{(s+2)^2} = -1$$

$$\text{Check:} \quad \frac{1}{s+2} + \frac{1}{(s+1)^2} - \frac{1}{s+1} = \frac{1}{(s+2)(s+1)^2}$$

$$x(t) = e^{-2t} + te^{-t} - e^{-t} = e^{-2t} + (t-1)e^{-t}, \ t > 0$$

4. Find the inverse of the Laplace transforms given below. Time functions are causal.

a. $X_1(s) = \dfrac{s+3}{s^2+9} = \dfrac{s}{s^2+9} + \dfrac{3}{s^2+9} \implies x_1(t) = [\cos(3t) + \sin(3t)]\,u(t) = \sqrt{2}\cos\left(3t - \dfrac{\pi}{4}\right)u(t)$

b. $X_2(s) = \dfrac{1}{(s-1)^2} = -\dfrac{d}{ds}\left[\dfrac{1}{s-1}\right] \implies x_2(t) = te^t u(t)$

5. Find the inverse Laplace transforms of

$$X(s) = \frac{3}{s^2 + 6s + 18} = \frac{3}{(s+3)^2 + 9}$$

for each of the following ROCs: (a) $\sigma < -3$, (b) $\sigma > -3$.

a. Let

$$X_1(s) = \frac{3}{s^2+9}, \qquad \sigma < -3 \qquad \implies \qquad x_1(t) = -\sin(3t)u(-t)$$

Then

$$X(s) = \frac{3}{(s+3)^2+9} = X_1(s+3), \qquad \sigma < -3 \qquad \implies \qquad x(t) = e^{-3t}x_1(t) = -e^{-3t}\sin(3t)u(-t)$$

b. Let

$$X_1(s) = \frac{3}{s^2+9}, \qquad \sigma > -3 \qquad \implies \qquad x_1(t) = \sin(3t)u(t)$$

Then

$$X(s) = \frac{3}{(s+3)^2+9} = X_1(s+3), \qquad \sigma > -3 \qquad \implies \qquad x(t) = e^{-3t}x_1(t) = e^{-3t}\sin(3t)u(t)$$

Alternate Solution

a. $\sigma < -3, \ X(s) = \dfrac{3}{(s+3)^2+9} = \dfrac{k_1}{s+3+3j} + \dfrac{k_1^*}{s+3-3j}$

$k_1 = X(s)(s+3+3j)\Big|_{s=-3-3j} = \dfrac{3}{s+3-3j}\Big|_{s=-3-3j} = \dfrac{j}{2}$

$x(t) = \left[-\dfrac{j}{2}e^{(-3-3j)t} + \dfrac{j}{2}e^{(-3+3j)t} \right] = -e^{-3t}\left[\dfrac{e^{j3t}-e^{-j3t}}{2j} \right] u(-t) = -e^{-3t}\sin(3t)u(-t)$

b. $\sigma > -3, \ X(s) = \dfrac{3}{(s+3)^2+9} = \dfrac{k_1}{s+3+3j} + \dfrac{k_1^*}{s+3-3j},$

$k_1 = X(s)(s+3+3j)\Big|_{s=-3-3j} = \dfrac{3}{s+3-3j}\Big|_{s=-3-3j} = \dfrac{j}{2}$

$x(t) = \left[\dfrac{j}{2}e^{(-3-3j)t} - \dfrac{j}{2}e^{(-3+3j)t} \right] = e^{-3t}\left[\dfrac{e^{j3t}-e^{-j3t}}{2j} \right] u(t) = e^{-3t}\sin(3t)u(t)$

6. Find the inverse Laplace transform of

$$X(s) = \dfrac{2s+7}{s^2+7s+12} = \dfrac{1}{s+3} + \dfrac{1}{s+4}$$

for each of the following ROCs: (a) $\sigma < -4$, (b) $-4 < \sigma < -3$, (c) $\sigma > -3$.

a. $\sigma < -4$ $x(t) = -\left[e^{-3t}+e^{-4t} \right]u(-t)$

b. $-4 < \sigma < -3$ $x(t) = -e^{-3t}u(-t) + e^{-4t}u(t)$

c. $\sigma > -3$ $x(t) = x(t) = \left[e^{-3t}+e^{-4t} \right]u(t)$

7. The input to an LTI system is $x(t) = (\sin 2t)u(t)$. Use the Laplace transform to find the output $y(t)$ and its steady-state part for (a) $h(t) = e^{-t}u(t)$, (b) $h(t) = (\sin t)u(t)$.

Solution

$$x(t) = (\sin 2t)u(t) \qquad X(s) = \dfrac{2}{s^2+4}, \ \sigma > 0$$

a. $h(t) = e^{-t}u(t)$ $H(s) = \dfrac{1}{s+1}, \ \sigma > -1$

$$Y(s) = X(s) \times H(s) = \dfrac{2}{(s^2+4)(s+1)} = \dfrac{0.4}{s+1} - 0.4\dfrac{s-1}{s^2+4}, \ \sigma > 0$$

$$y(t) = \left[\dfrac{2}{5}e^{-t} - 0.4\cos(2t) + 0.2\sin(2t) \right]u(t)$$

$$= \left[\dfrac{2}{5}e^{-t} - \dfrac{\sqrt{5}}{5}\cos(2t+26.6°) \right]u(t) = \left[\dfrac{2}{5}e^{-t} + \dfrac{\sqrt{5}}{5}\sin(2t-63.4°) \right]u(t)$$

b. $h(t) = (\sin t)u(t)$ $H(s) = \dfrac{1}{s^2+1}, \ \sigma > 0$

$$Y(s) = X(s) \times H(s) = \dfrac{2}{(s^2+4)(s^2+1)}, \ \sigma > 0$$

$$= \dfrac{As+B}{s^2+4} + \dfrac{Cs+D}{s^2+1} = \dfrac{(A+C)s^3 + (B+D)s^2 + (A+4C)s + B + 4D}{(s^2+4)(s^2+1)},$$

$$A = C = 0, \quad D = -B = \frac{2}{3}$$

$$Y(s) = -\frac{2}{3}\frac{1}{s^2+4} + \frac{2}{3}\frac{1}{s^2+1},$$

$$y(t) = \left[\frac{2}{3}\sin(t) - \frac{1}{3}\sin(2t)\right]u(t)$$

8. The unit-impulse response of an LTI system is $h(t) = e^{-t}\sin 2t\, u(t)$. Use the Laplace transform to find its response to (a) $x(t) = (\sin t)u(t)$, (b) $x(t) = \sin t$.

Solution

$$h(t) = e^{-t}(\sin 2t)u(t), \qquad H(s) = \frac{2}{(s+1)^2+4}, \quad \sigma > -1$$

a. $x(t) = (\sin t)u(t), \qquad X(s) = \frac{1}{s^2+1}, \quad \sigma > 0$

$$Y(s) = H(s) \times X(s) = \frac{2}{(s^2+2s+5)(s^2+1)}, \quad \sigma > 0$$

$$= \frac{As+B}{(s+1)^2+4} + \frac{Cs+D}{s^2+1}$$

$$= \frac{(A+C)s^3 + (B+2C+D)s^2 + (A+5C+2D)s + B + 5D}{(s^2+2s+5)(s^2+1)},$$

$$A = -C = 0.2, \quad B = 0, \quad D = 0.4$$

$$Y(s) = \frac{0.2s}{(s+1)^2+4} + \frac{-0.2s+0.4}{s^2+1}$$

$$= 0.2\frac{s+1}{(s+1)^2+4} - 0.1\frac{2}{(s+1)^2+4} - 0.2\frac{s}{s^2+1} + 0.4\frac{1}{s^2+1}$$

$$y(t) = \left[0.1e^{-t}(2\cos 2t - \sin t) + 0.4\sin t - 0.2\cos t\right]u(t)$$

b. $x(t) = \sin t, \qquad y(t) = y_{ss}(t) = [0.4\sin t - 0.2\cos t] = 0.447\,\sin(t - 26.6°), \text{ at } t >> 0.$

9. The unit-impulse response of an LTI system is $h(t) = e^{-t}\sin 2t\, u(t)$. Use the Laplace transform to find its response to $x(t) = \sin t$.

Solution

$$h(t) = e^{-t}(\sin 2t)u(t), \quad H(s) = \frac{2}{(s+1)^2+4}, \quad \sigma > -1$$

$$x(t) = \sin t, \quad X(s) = \frac{\delta(s-j) - \delta(s+j)}{2j}$$

$$Y(s) = H(s) \times X(s) = \frac{H(s)|_{s=j}\,\delta(s-j) - H(s)|_{s=-j}\,\delta(s+j)}{2j}$$

$$H(s)|_{s=j} = \frac{1}{2+j} = 0.447\,e^{-j\,26.6°}$$

$$H(s)|_{s=-j} = \frac{1}{2-j} = 0.447\,e^{j\,26.6°}$$

$$Y(s) = 0.447\,\frac{e^{-j\,26.6°}\,\delta(s-j) - e^{j\,26.6°}\,\delta(s+j)}{2j}$$

$$y(t) = 0.447\,\sin(t - 26.6°)$$

An Alternative Approach
Use the frequency response.

$$H(\omega) = \frac{2}{5 - \omega^2 + j\,2\omega} \qquad H(\omega)\Big|_{\omega=1} = \frac{2}{5 - 1 + j\,2} = \frac{1}{2 + j} = 0.447\,e^{-j\,26.6^\circ}$$

$$x(t) = \sin t \qquad y(t) = 0.447\,\sin(t - 26.6^\circ)$$

10. Use the Laplace transform to find $Y(s)$ and $y(t)$, $t > 0$ when

$$\frac{d^2 y}{dt^2} + 10\frac{dy}{dt} + 9y = 9, \quad t \geq 0$$

Initial conditions are $y(0) = 1$ and $y'(0) = 0$. Identify the zero-state and zero-input responses.

Solution

$$\mathcal{L}[y(t)] = Y(s)$$

$$\mathcal{L}\left[\frac{dy}{dt}\right] = sY(s) - y(0) = sY(s) - 1$$

$$\mathcal{L}\left[\frac{d^2 y}{dt^2}\right] = s^2 Y(s) - y(0)s - y'(0) = s^2 Y(s) - s$$

$$\mathcal{L}\left[\frac{d^2 y}{dt^2} + 10\frac{dy}{dt} + 9y\right] = s^2 Y(s) - s + 10[sY(s) - 1] + 9Y(s)$$

$$= (s^2 + 10s + 9)Y(s) - (s + 10) = \mathcal{L}[9] = \frac{9}{s}$$

$$Y(s) = \frac{9}{s(s^2 + 10s + 9)} + \frac{s + 10}{s^2 + 10s + 9}$$

$$\uparrow \qquad\qquad\qquad \uparrow$$

Zero-state response $+$ Zero-input response

$$Y(s) = \frac{1}{s} - \frac{s + 10}{s^2 + 10s + 9} + \frac{s + 10}{s^2 + 10s + 9} = \frac{1}{s}, \quad y(t) = 1, \quad t > 0$$

11. Use the Laplace transform to find $Y(s)$ and $y(t)$, $t > 0$ given

$$\frac{d^2 y}{dt^2} + 2\frac{dy}{dt} + y = x(t) - \frac{dx}{dt}, \quad y(0) = 1, \quad y'(0) = -1, \quad \text{and } x(t) = e^{-3t} - e^{-2t}, \quad t \geq 0$$

Solution

$$\left[s^2 Y(s) - sy(0) - y'(0)\right] + 2\left[sY(s) - y(0)\right] + Y(s) = X(s) - [sX(s) - x(0)], \quad x(0) = 1 - 1 = 0$$

$$Y(s)(s + 1)^2 - (s + 1) = X(s)(1 - s) = (1 - s)\left[\frac{1}{s + 3} - \frac{1}{s + 2}\right]$$

$$Y(s)(s + 1)^2 = (s + 1) + \frac{s - 1}{(s + 3)(s + 2)} = \frac{s^3 + 6s^2 + 12s + 5}{(s + 3)(s + 2)}$$

$$Y(s) = \frac{s^3 + 6s^2 + 12s + 5}{(s + 3)(s + 2)(s + 1)^2} = \frac{1}{s + 3} - \frac{3}{s + 2} + \frac{As + B}{(s + 1)^2}$$

$$A = \frac{d}{ds}\left[Y(s)(s + 1)^2\right]\Big|_{s=-1} = 3,$$

$$(As + B)\big|_{s=-1} = \left[Y(s)(s + 1)^2\right]\big|_{s=-1}, \ B = 2$$

$$Y(s) = \frac{1}{s+3} - \frac{3}{s+2} + \frac{3s+2}{(s+1)^2} = \frac{1}{s+3} - \frac{3}{s+2} + \frac{3}{s+1} - \frac{1}{(s+1)^2}$$

$$y(t) = e^{-3t} - 3e^{-2t} + (3 - t)e^{-t}, \ t \geq 0$$

12. a. Let $H(s)$ be the Laplace transform of an LTI system and $Y_{ss}(s)$ the Laplace transform of the steady-state component of its response to a sinusoidal input $x(t) = \sin(\omega t)u(t)$. Find $Y_{ss}(s)$ and show that $y_{ss}(t) = A \sin(\omega t + \theta)$, where $B = |H(\omega)|$ and $\theta = \angle H(\omega)$.

b. Apply to $H(s) = (s + 2)/(s^2 + 4s + 6)$ and $x(t) = 10 \sin(5t + 36°)u(t)$ to find $y_{ss}(t)$.

Solution

a.

$$Y(s) = H(s)X(s) = H(s)\frac{\omega}{s^2 + \omega^2} = \frac{A}{s - j\omega} + \frac{A^*}{s + j\omega} + \cdots,$$

where

$$A = H(s)\frac{\omega}{s + j\omega}\bigg|_{s=j\omega} = \frac{H(\omega)}{2j} = \frac{|H(\omega)|e^{j\theta}}{2j} \text{ and } A^* = -\frac{|H(\omega)|e^{-j\theta}}{2j}$$

$$Y_{ss}(s) = \frac{|H(\omega)|}{2j}\left[\frac{e^{j\theta}}{s - j\omega} - \frac{e^{-j\theta}}{s + j\omega}\right]$$

$$y_{ss}(t) = |H(\omega)|\frac{e^{j(\omega t + \theta)} - e^{j(\omega t + \theta)}}{2j} = |H(\omega)| \sin(\omega t + \theta)$$

b. $H(s)\big|_{s=j5} = 0.195\angle -65°, \ y_{ss}(t) = 1.95 \sin(5t + 36° - 65°) = 1.95 \sin(5t - 29°)$

Chapter Problems

I. Laplace Transform

(Problems 13–17)

13. Find the Laplace transform of the time functions given below.

a. $tu(t)$ b. $e^{-3t}u(t)$ c. $\sqrt{2}\cos(2t + 45°)u(t)$

d. $2^{-t}u(t)$ e. $e^{3t}\sin(2t)u(t)$ f. $\sqrt{2}e^{-3t}\cos(2t + 45°)u(t)$

g. $te^{at}u(t)$ h. $t \sin(at)u(t)$ i. $t \cos(at)u(t)$

14. Find the Laplace transform of each of the time functions below. Specify the region of convergence. Assume a and ω are positive quantities.

a. $u(t) - u(t - 1)$ b. $e^{-at}u(t - 1)$ c. $[1 - e^{-at}]u(t)$ d. $e^{at}\cos(\omega t)u(t)$

e. $e^{at}\cos(\omega t + \theta)u(t)$ f. $e^{at}\sin(\omega t + \theta)u(t)$ g. $te^{at}\cos(\omega t)u(t)$ h. $\frac{1}{b-a}\left[e^{-at} - e^{-bt}\right]u(t)$

15. Find the Laplace transform of

a. $x(t) = \begin{cases} t, & 0 \leq t < 1 \\ 0, & \text{elsewhere} \end{cases}$

b. $x(t) = \begin{cases} 1 + t, & -1 \leq t < 0 \\ 1 - t, & 0 \leq t < 1 \\ 0, & \text{elsewhere} \end{cases}$

16. Find the Laplace transform of each of the time functions below. Specify the region of convergence. Assume a and ω are positive quantities.

a. $u(-t) - u(-t - 1)$ b. $e^{at}u(-t + 1)$ c. $[1 - e^{at}]u(-t)$ d. $e^{at}\sin(\omega t)u(-t)$

e. $e^{-|at|}$, all t f. $e^{-|at|}\cos \omega t$, all t g. $e^{-|at|}\sin \omega t$, all t h. $e^{-|at|}\sin(\omega t + \theta)$, all t

17. A transformation from the time domain to another domain (called p-domain) is known to be linear. Given the transform pair

$$e^{-at}\sin bt \iff \frac{b}{(p + a)^2 + b^2}$$

If the above information about the transformation is sufficient, find the inverse transform of

a. $\dfrac{2}{p^2 + p + 0.5}$

b. $\dfrac{p + 1}{p^2 + 2p + 12}$

If it is not, determine what additional information is needed.

II. Inverse Laplace Transform

(Problems 18–26; note that $\sigma = \mathcal{RE}[s]$)

18. Find the inverse Laplace transform of

$$X(s) = \frac{s + 2}{s^2 + 2s + 2}, \quad \sigma > -1$$

by the following two expansion methods and compare results:

a. $X(s) = \dfrac{A}{s + 1 - j} + \dfrac{A^*}{s + 1 + j}$

b. $X(s) = \dfrac{s + 1}{(s + 1)^2 + 1} + \dfrac{1}{(s + 1)^2 + 1}$

19. Find the inverse of the Laplace transforms given below. All time functions are causal.

a. $\dfrac{e^{-s}}{s + 1}$ b. $\dfrac{s + 3}{s^2 + 9}$ c. $\dfrac{1 - e^{-\pi s/3}}{s^2 + 9}$

d. $\dfrac{2e^{-3s}}{s - 1}$ e. $\dfrac{2}{s^2}$ f. $\dfrac{-3}{s + 3}$

g. $\dfrac{1}{(s + 1)^2}$ h. $\dfrac{s}{(s + 1)^2}$ i. $\dfrac{1}{(s - 1)^2}$

20. Repeat problem 19 assuming the time functions to be anticausal.

21. Find the inverse transforms of the functions given below.

a. $\dfrac{1}{s}$, $\sigma > 0$ b. $\dfrac{e^{-(s+1)}}{s + a}$, $\sigma > -a$ c. $\dfrac{a}{s(s + a)}$, $\sigma > 0$

d. $\dfrac{s - a}{(s - a)^2 + \omega^2}$, $\sigma > -a$ e. $\dfrac{(s + a)^2 - \omega^2}{[(s + a)^2 + \omega^2]^2}$, $\sigma > -a$ f. $\dfrac{1}{(s + a)(s + b)}$, $\sigma > -min(a, b)$

22. Find the inverse Laplace transform of

a. $X(s) = \dfrac{5s^2 + 3s + 1}{(s^2 + 4)(s + 1)}$, $\sigma > 0$ b. $X(s) = \dfrac{5s^2 + 3s + 1}{(s^2 + 4)(s + 1)^2}$, $\sigma > 0$

23. Find the inverse Laplace transform of

$$X(s) = \frac{1}{(s+1)^2(s+2)}, \quad \sigma > -1$$

24. Find the inverse Laplace transforms of

a. $X(s) = \dfrac{3}{s^2 + 6s + 18}$ b. $X(s) = \dfrac{3}{s^2 + 6s + 9}$

For each of the following ROCs:

(i) $\sigma < -3$ (ii) $\sigma > -3$

25. Find the inverse Laplace transforms of

a. $X(s) = \dfrac{2s + 7}{s^2 + 7s + 12}$,   b. $X(s) = \dfrac{1}{(s+4)(s^2 + 6s + 18)}$, and   c. $X(s) = \dfrac{1}{(s+3)(s^2 + 8s + 25)}$

if

(i) $\sigma < -4$,   (ii) $-4 < \sigma < -3$, and   (iii) $\sigma > -3$.

26. Find all the time functions for which

a. $X(s) = \dfrac{1}{(s+2)(s^2 - 1)}$ b. $X(s) = \dfrac{s + 10}{(s+2)(s^2 - 1)}$ c. $X(s) = \dfrac{s - 10}{(s+2)(s^2 - 1)}$

d. $X(s) = \dfrac{s + 1.9}{(s+2)(s^2 - 1)}$ e. $X(s) = \dfrac{s + 0.9}{(s+2)(s^2 - 1)}$ f. $X(s) = \dfrac{s^2 - 0.81}{(s+2)(s^2 - 1)}$

III. LTI, Convolution, and the Laplace Transform

(Problems 27–33)

27. The unit-impulse response of an LTI system is $h(t) = \sin t \, [u(t) - u(t - 2\pi)]$.

a. Find $H(s)$.

b. Find the response of the system to $x(t) = \sin t u(t)$ by (i) using the Laplace transform method and (ii) using the convolution integral.

28. Find $y(t) = x(t) \star h(t)$ for the following cases:

a. $x(t) = (\sin t)u(t)$ and $h(t) = e^{-0.1t} \sin 2t u(t)$

b. $x(t) = \displaystyle\sum_{n=0}^{n=\infty} \delta(t - n)$ and $h(t) = \begin{cases} \sin \pi t, & 0 < t < 1 \\ 0, & \text{elsewhere} \end{cases}$

c. $x(t) = u(t)$ and $h(t) = e^{-\alpha t} u(t)$

d. $x(t) = tu(t)$ and $h(t) = e^{-\alpha t} u(t)$

e. $x(t) = \begin{cases} \frac{1}{T}, & 0 < t < T \\ 0, & \text{elsewhere} \end{cases}$ and $h(t) = e^{-\alpha t} u(t)$

29. Let $x(t) = \frac{1}{T}[u(t) - u(t - T)]$ and $h(t)$ be a well-behaved function. Find $Y(s) = X(s)H(s)$. Then, show that $\lim_{T \to 0}[Y(s)] = H(s)$ and conclude that $\lim_{T \to 0}[x(t) \star h(t)] = h(t)$.

30. Find $y(t) = x(t) \star h(t)$ for the following three cases:

a. $x(t) = (\sin t)u(t)$ and $h(t) = e^{-\alpha t}(\sin t)u(t)$ for $\alpha = 0.1, \ 1, \ 10$

b. $x(t) = (\cos t)u(t)$ and $h(t) = (\cos 2t)u(t)$

c. $x(t) = (\cos t)u(t)$ and $h(t) = (\cos 1.1t)u(t)$

31. The input to an LTI system is $x(t) = \sin 2t u(t)$. Find the output $y(t)$ and the steady-state part for

a. $h(t) = e^{-t}u(t)$
b. $h(t) = \sin t u(t)$

32. The unit-impulse response of an LTI system is $h(t) = \delta(t-2) + 2u(t)$ and its input is $x(t) = \cos t u(t)$.

a. Find $Y(s) = X(s)H(s)$ and its inverse. Identify the amplitude and phase of the output.
b. Find $y(t)$ by direct evaluation of the convolution integral $y(t) = x(t) \star h(t)$.

33. The impulse response of an LTI system is $h(t) = e^{-3t}u(t)$. Find $Y(s) = X(s)H(s)$ and $y(t)$ for the following inputs:

a. $x(t) = [\sin(2t) + \cos(2t)]u(t)$
b. $x(t) = [\sin(2t) + \cos(3t)]u(t)$

IV. Differential Equations and the Laplace Transform
(Problems 34–48)

34. Given

$$\frac{dy}{dt} + 3y = x(t) \ \text{ and } \ x(t) = e^{-2t}, \ t \geq 0$$

find $Y(s)$ and $y(t)$, $t \geq 0$, for a) $y(0) = 1$, b) $y(0) = 0$, c) $y(0) = -1$, d) $y(0) = y_0$.

35. Given

$$\frac{d^2y}{dt^2} + 2\frac{dy}{dt} + y = x(t) - \frac{dx}{dt}, \ \ y(0) = 1, \ \ y'(0) = 2, \ \text{ and } x(t) = e^{-3t} - e^{-2t}, \ t \geq 0$$

find $Y(s)$ and $y(t)$, $t \geq 0$.

36. a. An LTI system is given by

$$y'(t) + ay(t) = x(t)$$

Let $h(t)$ be the system's unit-impulse response. Show that the convolution integral

$$\int_{-\infty}^{\infty} x(\tau)h(t-\tau)d\tau$$

satisfies the differential equation.
b. Repeat for an LTI system described by an nth-order differential equation.

37. Find $Y(s)$ and $y(t)$, the solutions to each of the following differential equations:

a. $\dfrac{d^2y}{dt^2} + 2\dfrac{dy}{dt} + y = x(t) - \dfrac{dx}{dt}$, $t > 0$, with $y(0^+) = 1$, $\dfrac{dy}{dt}(0^+) = 0$, and $x(t) = e^{-3t} - e^{-2t}$, $t > 0$

b. $\dfrac{d^2y}{dt^2} + 200\dfrac{dy}{dt} + 10^8 y = 200\dfrac{dx}{dt}$, $t > 0$, with $y(0^+) = 0$, $\dfrac{dy}{dt}(0^+) = 200$, and $x(t) = -1$, $t > 0$

38. Find $Y(s)$ and $y(t)$, the solutions to each of the following differential equations:

a. $\dfrac{d^2y}{dt^2} + 200\dfrac{dy}{dt} + 10^8 y = 5x(t) + 100\dfrac{dx}{dt}$, where $x(t) = 2u(t)$

b. $\dfrac{d^2y}{dt^2} + 650\dfrac{dy}{dt} + 30000y = 10^4 \left[3x(t) + \dfrac{dx}{dt}\right]$, where $x(t) = (\sin 100t)u(t)$

c. $\dfrac{d^2 y}{dt^2} + \dfrac{dy}{dt} = x(t) + 2\dfrac{dx}{dt}$, where $x(t) = e^{-2t} u(t)$

d. $\dfrac{d^2 y}{dt^2} + 3\dfrac{dy}{dt} + 2y = -\dfrac{dx}{dt}$, where $x(t) = (\sin 3t) u(t)$

39. Find $Y(s)$ and $y(t)$, the solutions to each of the following differential equations:

a. $\dfrac{d^2 y}{dt^2} + 10\dfrac{dy}{dt} + 9y = 0$, $t > 0$, with $y(0) = 1$, $\dfrac{dy}{dt}(0) = 0$

b. $\dfrac{dy}{dt} + y = 5(\sin 2t) u(t)$

c. $\dfrac{d^2 y}{dt^2} + 10\dfrac{dy}{dt} + 9y = 9$, $t > 0$, with $y(0) = 1$, $\dfrac{dy}{dt}(0) = 0$

d. $\dfrac{dy}{dt} + 2y = t u(t)$

e. $\dfrac{dy}{dt} + y = 2(\sin t) u(t)$

f. $\dfrac{dy}{dt} + 2y = t e^{-t} u(t)$

40. Use the Laplace transform to find $Y(s)$ and $y(t)$ given

$$\dfrac{d^2 y}{dt^2} + 10\dfrac{dy}{dt} + 9y = 9, \quad t \geq 0, \quad y(0) = 0, \quad y'(0) = 0$$

41. In the circuit of Figure 6.8, $R = 1\ \Omega$, $L = \sqrt{2}\ H$, and $C = \sqrt{2}\ F$. Given $v_C(0) = 1$ V and $i_L(0) = 1$ A, find $v_C(t)$ and $i_L(t)$ for $t \geq 0$ if the applied voltage source for $t \geq 0$ is

a. $v(t) = 1$
b. $v(t) = 2$

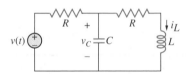

FIGURE 6.8 (For Problem 41)

42. A rectangular voltage source (0 to $1/T$ V, T msec long) charges a capacitor C through a resistor R. See Figure 6.9.

a. Find $i(t)$ and $v_C(t)$ by solving circuit equations in the time domain. Find their limits as $T \to 0$.
b. Find $i(t)$ and $v_C(t)$ by the Laplace transform method and find their limits as $T \to 0$.
c. Let the voltage source be a unit impulse. Find $i(t)$ and $v_C(t)$ and compare with results of parts (a) and (b).

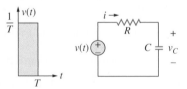

FIGURE 6.9 (For Problem 42)

43. In a parallel *RLC* circuit (with $R = 1/3\ \Omega$, $L = 1/2\ H$, and $C = 1\ \mu F$), $i_L(0) = 1$ and $v_C(0) = 1$. For $t \geq 0$ the circuit is fed by a 1-A parallel current source. Find $i_L(t)$ and $v_C(t)$ for $t \geq 0$ by using the Laplace transform method.

44. In the circuit of Figure 6.10, $R = 1\ k\Omega$, $L = 1\ mH$, and $C = 1\ \mu F$. The initial conditions are $i_2(0^+) = -i_1(0^+) = 1\ mA$, and $v_C(0^+) = 1\ V$. Find $i_1(t)$, $i_2(t)$, and $v_C(t)$ for $t \geq 0$ by using the Laplace transform method, given

   a. $v(t) = 1\ V$, $t > 0$
   b. $v(t) = 2\ V$, $t > 0$

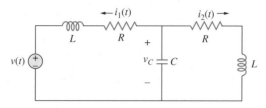

FIGURE 6.10 (For Problem 44)

45. In the circuit of Figure 6.11, $R_1 = 1/3\ \Omega$, $R_2 = 10\ m\Omega$, $L = 1/2\ H$, $C = 1\ \mu F$, $i(0^+) = 1\ A$, and $v(0^+) = 1\ V$. Find $i(t)$ and $v(t)$ for $t \geq 0$ by using the following steps:

   a. Formulate time-domain differential equations for i and v using (i) circuit laws, (ii) impedances (s-operator).
   b. Solve the differential equations using (i) time-domain approach, (ii) the Laplace transform approach.
   c. Formulate state equations and solve for i and v.

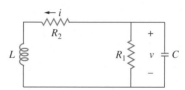

FIGURE 6.11 (For Problem 45)

46. In the circuit of Figure 6.12, $R = 1\ \Omega$, $L = 1\ H$, $C = 2\ F$ and

$$\begin{cases} v(t) = 1, & t \geq 0 \\ i(t) = e^{-t}, & t \geq 0 \end{cases} \text{ with initial conditions } \begin{cases} v_C(0) = 1\ V \\ i_L(0) = 1\ A \end{cases}$$

Find $i_L(t)$ and $v_C(t)$ for $t > 0$.

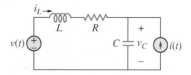

FIGURE 6.12 (For Problem 46)

47. The circuit of Figure 6.13 ($R = 5\,\Omega$, $L = 0.1\,H$, and $C = 20\,\mu F$) with zero energy is connected to a DC voltage source ($V = 10$ volts) at $t = 0$. Find the currents i_1 and i_2 for $t \geq 0$.

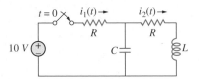

FIGURE 6.13 (For Problem 47)

48. A series RLC circuit ($R = 5\,\Omega$, $L = 1\,H$, and $C = 500\,\mu F$) with zero energy is connected to a DC voltage source ($V = 10$ volts) at $t = 0$. Find the current for $t \geq 0$.

49. The switch in the circuit of Figure 6.14 closes at $t = 0$. Using the Laplace transform method find steady states $v_c(t)$ and $i(t)$ for

a. $v(t) = \sin 2\pi f_0 t$, $f_0 = 159$ Hz.
b. $v(t)$, a periodic square pulse (50% duty cycle) which switches from 0 to 1 volt every $1/159$ sec.

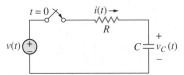

FIGURE 6.14 (For Problem 49)

50. In the circuit of Figure 6.15, $R_1 = 3\,k\Omega$, $R_2 = 2\,k\Omega$, $L = 1\,H$, and $C = 1\,\mu F$. Find $v(t)$ and $i(t)$ for

a. $i_g = 10 \cos 1{,}000t$ mA, all t
b. $i_g = \begin{cases} 10 \text{ mA}, & t < 0 \\ 10 \cos 1{,}000t \text{ mA}, & t \geq 0 \end{cases}$

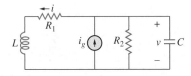

FIGURE 6.15 (For Problem 50)

V. System Function
(Problems 51–57)

51. In the circuit of Figure 6.16, $R = 1\,k\Omega$, $L = 1\,H$, and $C = 1{,}000\,\mu F$.

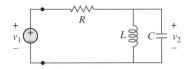

FIGURE 6.16 (For Problem 51)

Verify that the system function is $H(s) = \dfrac{V_2}{V_1} = \dfrac{s}{s^2 + s + 1{,}000}$.

52. In the series *RLC* circuit of Figure 6.17, $v_s(t)$ is the input and element voltages $v_1(t)$, $v_2(t)$, and $v_3(t)$ (across R, L, and C, respectively,) are the outputs.

a. Verify the three differential equations and system functions given below. Observe that the system functions have the same set of poles and relate that property to the natural response of the circuit.

Resistor:
$$\frac{dv_1^2}{dt^2} + \frac{R}{L}\frac{dv_1}{dt} + \frac{1}{LC}v_1 = \frac{R}{L}\frac{dv_s}{dt}, \quad H_1(s) = \frac{V_1}{V_s} = \frac{\frac{R}{L}s}{s^2 + \frac{R}{L}s + \frac{1}{LC}} = \frac{\frac{1}{Q}\left(\frac{s}{\omega_0}\right)}{\left(\frac{s}{\omega_0}\right)^2 + \frac{1}{Q}\left(\frac{s}{\omega_0}\right) + 1}$$

Inductor:
$$\frac{dv_2^2}{dt^2} + \frac{R}{L}\frac{dv_2}{dt} + \frac{1}{LC}v_2 = \frac{d^2v_s}{dt^2}, \quad H_2(s) = \frac{V_2}{V_s} = \frac{s^2}{s^2 + \frac{R}{L}s + \frac{1}{LC}} = \frac{\left(\frac{s}{\omega_0}\right)^2}{\left(\frac{s}{\omega_0}\right)^2 + \frac{1}{Q}\left(\frac{s}{\omega_0}\right) + 1}$$

Capacitor:
$$\frac{dv_3^2}{dt^2} + \frac{R}{L}\frac{dv_3}{dt} + \frac{1}{LC}v_3 = \frac{1}{LC}v_s, \quad H_3(s) = \frac{V_3}{V_s} = \frac{\frac{1}{LC}}{s^2 + \frac{R}{L}s + \frac{1}{LC}} = \frac{1}{\left(\frac{s}{\omega_0}\right)^2 + \frac{1}{Q}\left(\frac{s}{\omega_0}\right) + 1}$$

where $\omega_0 = \frac{1}{\sqrt{LC}}$ and $Q = \frac{L\omega_0}{R}$

b. For each of the following three cases obtain the location of poles and zeros of $H_1(s)$, $H_2(s)$, and $H_3(s)$ and find their step and impulse responses:

(i) $R > 2\sqrt{\frac{L}{C}}$,   (ii) $R = 2\sqrt{\frac{L}{C}}$,   (iii) $R < 2\sqrt{\frac{L}{C}}$

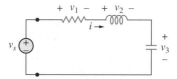

FIGURE 6.17 (For Problem 52)

53. In the series *RLC* circuit of Figure 6.18 connect a current source $i_s(t)$ in parallel with an element and consider it to be the input (three possible inputs). Let an element's current or voltage be the output (six possible outputs), for a total of 18 input-output pairs each represented by a system function. Discuss the common features of the collection of system functions and interpret them in terms of the circuit's natural response. Relate your observations to the results of problem 52.

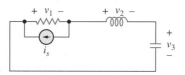

FIGURE 6.18 (For Problem 53)

54. In the parallel *RLC* circuit of Figure 6.19, $i_s(t)$ is the input and the elements' currents $i_1(t)$, $i_2(t)$, and $i_3(t)$ (in the resistor, inductor, and the capacitor, respectively) are the outputs.

a. Verify the three differential equations and system functions given below. Observe that the system functions have the same set of poles and relate that property to the natural response of the circuit.

Resistor: $\dfrac{di_1^2}{dt^2} + \dfrac{1}{RC}\dfrac{di_1}{dt} + \dfrac{1}{LC}i_1 = \dfrac{1}{RC}\dfrac{di_s}{dt}$, $H_1(s) = \dfrac{I_1}{I_s} = \dfrac{\frac{1}{RC}s}{s^2 + \frac{1}{RC}s + \frac{1}{LC}} = \dfrac{\frac{1}{Q}\left(\frac{s}{\omega_0}\right)}{\left(\frac{s}{\omega_0}\right)^2 + \frac{1}{Q}\left(\frac{s}{\omega_0}\right) + 1}$

Inductor: $\dfrac{di_2^2}{dt^2} + \dfrac{1}{RC}\dfrac{di_2}{dt} + \dfrac{1}{LC}i_2 = \dfrac{1}{LC}i_s$, $H_2(s) = \dfrac{I_2}{I_s} = \dfrac{\frac{1}{LC}}{s^2 + \frac{1}{RC}s + \frac{1}{LC}} = \dfrac{1}{\left(\frac{s}{\omega_0}\right)^2 + \frac{1}{Q}\left(\frac{s}{\omega_0}\right) + 1}$

Capacitor: $\dfrac{di_3^2}{dt^2} + \dfrac{1}{RC}\dfrac{di_3}{dt} + \dfrac{1}{LC}i_3 = \dfrac{d^2i_s}{dt^2}$ $H_3(s) = \dfrac{I_3}{I_s} = \dfrac{s^2}{s^2 + \frac{1}{RC}s + \frac{1}{LC}} = \dfrac{\left(\frac{s}{\omega_0}\right)^2}{\left(\frac{s}{\omega_0}\right)^2 + \frac{1}{Q}\left(\frac{s}{\omega_0}\right) + 1}$

where $\omega_0 = \dfrac{1}{\sqrt{LC}}$ and $Q = RC\omega_0$

b. For each of the following three cases obtain the location of poles and zeros of $H_1(s)$, $H_2(s)$, and $H_3(s)$ and find their step and impulse responses:

(i) $R < \dfrac{1}{2}\sqrt{\dfrac{L}{C}}$,   (ii) $R = \dfrac{1}{2}\sqrt{\dfrac{L}{C}}$,   (iii) $R > \dfrac{1}{2}\sqrt{\dfrac{L}{C}}$

FIGURE 6.19 (For Problem 54)

55. In a parallel *RLC* circuit place a voltage source $v_s(t)$ in series with an element (as in Figure 6.20) and consider it to be the input (three possible inputs). Let an element's current or voltage be the output (six possible outputs), for a total of 18 input-output pairs each represented by a system function. Discuss the common features of the collection of system functions and interpret them in terms of the circuit's natural response. Relate your observations to the results of problem 54.

FIGURE 6.20 (For Problem 55)

56. a. In the circuit of Figure 6.21, $R = 50\,\Omega$, $L = 1$ mH, and $C = 10\,\mu$F. Find the system function $H(s) = V_2/V_1$, its poles and zeros, and show them on the s-plane.

b. Let $v_1(t) = 1$, $t \geq 0$. Show that regardless of the initial state of the circuit at $t = 0$, the response at $t \geq 0$ is $v_2(t) = V_0 e^{-\alpha t}\cos(\omega t + \theta)$ and determine α and ω.

c. Given $i_L(0^-) = v(0) = 0$ and $v_1(t) = 1$, $t \geq 0$, find V_0 and θ in part b.

d. Given $v_1(t) = 1$, $t \geq 0$, specify $i(0)$ and $v(0)$ so that the response becomes zero at $t \geq 0$.

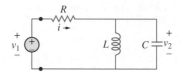

FIGURE 6.21 (For Problem 56)

57. The only poles of an LTI system are at $-1 \pm j$. The system has no zeros at finite values of s. Its steady-state response to a unit-step input is 5. Find (a) the system function $H(s)$, (b) the system's unit-impulse response, and (c) the dominant part of the response to the input $e^{-t/100}u(t)$ at $t > 10$.

6.29 Project: Pulse-Shaping Circuit

Summary and Objectives

This project is an exercise in using the Laplace transform in circuit analysis. The circuit under study is a passive RC circuit (made of a cascade of resistors and capacitors). It is shown in Figure 6.22. The input is $v_1(t)$ and the output is $v_2(t)$. The goal of the project is to find its system function, frequency, and time responses by analytic methods, simulation, and measurement and relate the results. Simulation may be done by Matlab or Spice. Laboratory measurements may use a function generator, an oscilloscope, and the circuit of Figure 6.22. Alternatively, the measurements can be made on the graphs obtained from simulation.

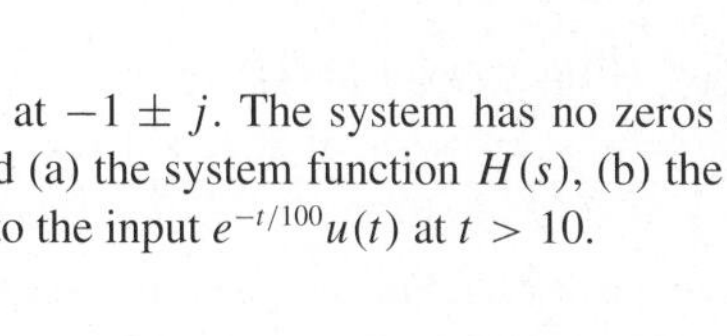

FIGURE 6.22 A two-terminal ladder RC circuit with open terminals. Element values are $R_1 = 470\ \Omega$, $R_2 = 4,700\ \Omega$, $C_1 = 10\ nF$, and $C_2 = 47\ nF$.

Analysis

This part of the project has four sections, which are to be done by mathematical analysis. You may simulate the circuit or use a computation package to obtain the plots.

System Function   In the circuit of Figure 6.22, write KCL at nodes A, B, and C using node voltages $\overline{V}_a$, $\overline{V}_b$, and $\overline{V}_c$ in the s-domain as variables.

Show that three node voltage equations in the s-domain are given in matrix form by

$$\begin{bmatrix} 2Y_{R1} + Y_{C1} & -Y_{R1} & 0 \\ -Y_{R1} & Y_{R1} + Y_{C1} + Y_{C2} & -Y_{C2} \\ 0 & -Y_{C2} & Y_{R2} + Y_{C2} \end{bmatrix} \begin{bmatrix} V_a \\ V_b \\ V_c \end{bmatrix} = Y_{R1} \begin{bmatrix} \overline{V}_1 \\ 0 \\ 0 \end{bmatrix}, \quad \text{where } Y_R = \frac{1}{R} \text{ and } Y_C = Cs$$

To simplify, you may divide both sides of the matrix equation by Y_{R1} and note that $Y_{R1} = 1/470$, $Y_{R2} = 0.1Y_{R1}$, $Y_{C1} = 10^{-8}s$, and $Y_{C2} = 4.7Y_{C1}$. Show that

$$H(s) = \frac{V_2}{V_1} = \frac{R_2 C_2 s}{R_1^2 C_1^2 R_2 C_2 s^3 + (R_1^2 C_1^2 + R_1^2 C_1 C_2 + 3R_1 C_1 R_2 C_2)s^2 + (3R_1 C_1 + 2R_1 C_2 + R_2 C_2) + 1}$$

$$= \frac{2.209 \times 10^{-4} s}{4.879 \times 10^{-15} s^3 + 3.241 \times 10^{-9} s^2 + 2.792 \times 10^{-4} s + 1}$$

Using Matlab find the roots of the characteristic equation and show that

$$H(s) = \frac{4.5227 \times 10^{10} s}{(s + 5.633 \times 10^5)(s + 9.719 \times 10^4)(s + 3.743 \times 10^3)}$$

Time Responses   Write the differential equation. Find and plot its response to (i) a unit impulse, (ii) a unit step, and (iii) a 1-V, 100-μs rectangular voltage pulse. Compute the 10% to 90% rise time, the 90% to 10% fall time, and the drop in the pulse response.

Frequency Response   Find $v_2(t) = V_2 \sin(\omega t + \theta)$, the steady-state response to a sinusoidal input $v_1(t) = \sin(\omega t)$ with a frequency range of 1 Hz to 100 MHz. Plot $20 \log V_2$ (in dB units) and θ (in degrees) versus $\log \omega$. You will use semilog graph paper having eight decades for the given frequency range, and 12 vertical divisions.

Measurement

    a. Record the unit-impulse response. Model it by a mathematical expression and compare with the result obtained in section 1(*b*).

    b. Record the unit-step response. Model it by a mathematical expression and compare with the result obtained in section 1(*b*).

    c. Record the response to a 100-μs, 1-V pulse. Model it by a mathematical expression and compare with the result obtained in section 1(*b*). Verify that the pulse response is a superposition of the responses to two steps.

    d. Measure and record the frequency response of the circuit, $H(\omega) = H(s)|_{s=j\omega}$, for 10 Hz $\leq f \leq$ 100 kHz and plot it. Use log scale for the frequency axis and uniform scale for the magnitude.

Conclusions

    a. By qualitative reasoning determine if the system is low pass, high pass, or bandpass.

    b. From the poles of the system [obtained from the denominator of $H(s)$] reason that the system can't oscillate and that the recorded step response is in agreement with expectations.

    c. Approximate the above system by a first-order model containing a single pole. Specify the location of the pole. Represent the first-order approximation by an *RC* circuit and compare it with Figure 6.22.

    d. Summarize your overall conclusions, especially the relationship between the system function, frequency response, impulse response, step response, and pulse response.

Chapter 7

Fourier Series

Contents

Introduction and Summary

Linearity is a property that provides for the ability to predict the output of a system to a new input if the new input can be expressed as a linear combination of functions for which the outputs are already known. The new output then becomes the linear combination of the known outputs weighted appropriately. Let the set of the known input-output pairs in a linear system be called the repertoire set. To be of greatest and most efficient use, the repertoire set should be as small as possible (or its representation be as simple as possible), while the population of inputs that it can describe be as large as one may encounter. (The time-invariance property will reduce the complexity of the repertoire and its description still more.) As we already have seen on several occasions in Chapters 3, 4, and 5, an exponential input e^{st}, with s allowed to be a complex number, produces the exponential output $H(s)e^{st}$. The function e^{st} is called a characteristic function (or an *eigenfunction*) of the LTI system. The function $H(s)$, therefore, describes the system. The question now is how large a class of input functions can be represented by e^{st} and how easily this can be done. For a periodic function with period T, the answer to the question is clear and simple: In the complex plane restrict s to the imaginary axis $j\omega$, leading to the set of exponential functions $e^{jn\omega_0 t}$, where $\omega_0 = 2\pi/T$ is called the fundamental frequency and n is an integer in the range $-\infty < n < \infty$. The description of a periodic function by the above set of exponentials is called its exponential Fourier series expansion, or simply its Fourier series. It is shown by

$$x(t) = \sum_{n=-\infty}^{\infty} X_n e^{jn\omega_0 t}$$

X_n is called the Fourier coefficient and $X_n e^{jn\omega_0 t}$ is called the nth harmonic of $x(t)$. The Fourier series representation of functions allows for their analysis in the frequency domain. It remains to be shown how conveniently the coefficients X_n can be obtained from $x(t)$ and how the above expansion facilitates the analysis of LTI systems.

This chapter covers material that has practical applications and, as such, places less emphasis on purely theoretical considerations (e.g., the uniform convergence of the Fourier series). Throughout the chapter, real-valued signals are assumed, unless specified otherwise. The chapter starts with approximating periodic signals, and then considers signal synthesis by a finite set of weighted sinusoids and the resulting error. It then introduces the trigonometric form of the series (useful in dealing with some problems of interest such as filtering), followed by its exponential form. For the sake of completeness, the generalized Fourier series and vectorial interpretation are briefly introduced. However, the exponential form of the Fourier series is used throughout the rest of the chapter. The properties and applicability of the series are then discussed. The role and importance of the Fourier series in the frequency domain analysis of signals and systems is illustrated within the context of real-life problems. Examples and problems are designed to teach analytical solutions, while bringing to the student's attention the need for and advantages of using a computer. Computation engines and simulation software for Fourier analysis provide more than just powerful tools for deriving analytical solutions. By removing the constraints imposed by heavy manual calculations, computer programs enable the user to expand his or her horizons, explore in detail many aspects of Fourier analysis, and examine these same aspects both quantitatively and qualitatively within an engineering context. Finally, the chapter uses a traditional route in extending the Fourier series to the Fourier transform (rather than the reverse). The project proposed at the end of the chapter is intended as an introduction to the use of computer methods in Fourier analysis and a practical bridge between the Fourier series and transform.

7.1 Signal Synthesis

With regard to the representation of periodic signals by the Fourier series, several questions come to mind. For example: (1) Does the infinite sum converge to the value of $x(t)$ for all instances of times? (2) How is the instantaneous error [defined by the square of the difference between $x(t)$ and its series expansion at any instance of time] distributed over time? (3) Can a finite sum of weighted exponentials approximate $x(t)$ in order to meet a desired average error taken over one period? If so, what is the optimum set of coefficients? To provide some insight into the above questions we start with two examples on synthesizing a signal so that a periodic function may be approximated by a finite sum. In this formulation we desire to approximate $x(t)$ by

$$y(t) = \sum_{n=1}^{N} X_n \phi_n(t)$$

over an interval $t_1 < t < t_2$. The instantaneous difference will be defined by $e(t) = y(t) - x(t)$ and its instantaneous power by $|e(t)|^2$. It can be shown that the average power of the difference

$$\mathcal{E} = \frac{1}{t_2 - t_1} \int_{t_1}^{t_2} |e(t)|^2 dt$$

is minimized if $X_n = \int_{t_1}^{t_2} x(t)\phi^*(t)dt$.

Example

7.1

Consider a periodic sawtooth waveform $x(t) = t$, $-1 < t < 1$, which repeats itself every 2 seconds. It will be shown (shortly) that the above waveform may be expressed by an infinite sum of sinusoidal components called its harmonics:

$$x(t) = \frac{2}{\pi} \left(\sin \pi t - \frac{1}{2} \sin 2\pi t + \frac{1}{3} \sin 3\pi t - \frac{1}{4} \sin 4\pi t + \frac{1}{5} \sin 5\pi t + \cdots \right)$$

In this example we will approximate $x(t)$ by the finite sum of its first N harmonics. The finite sum is

$$y(t) = \frac{2}{\pi} \sum_{n=1}^{N} \frac{(-1)^{n-1}}{n} \sin n\pi t$$

The average power of the difference (to be called the average error) is

$$\mathcal{E} = \frac{1}{2} \int_{-1}^{1} |y(t) - x(t)|^2 dt$$

a. Plot $y(t)$ and $|y(t) - x(t)|^2$ for $N = 1$ and 3 for the period $-2 < t < 2$. Overlay $x(t)$ on the plots and qualitatively discuss the salient features of the error.

b. Find $\mathcal{E}$ for $N = 1$ by analytical and numerical (computational) methods and verify that they are in agreement.

    c.   Find $\mathcal{E}$ for $N = 2, 3, 4$, and 5 by a numerical method and plot it versus N. Qualitatively and quantitatively, describe the decrease in $\mathcal{E}$ as the number of harmonics increases.

Solution

    a.   See Figure 7.1(a) and (b).

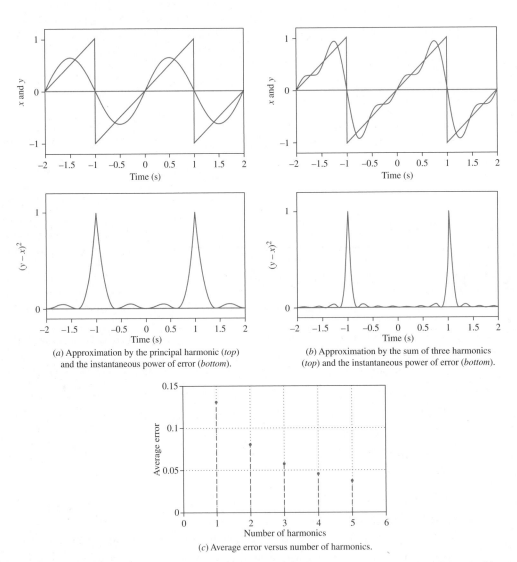

    (a) Approximation by the principal harmonic (*top*)
    and the instantaneous power of error (*bottom*).

    (b) Approximation by the sum of three harmonics
    (*top*) and the instantaneous power of error (*bottom*).

    (c) Average error versus number of harmonics.

FIGURE 7.1 Approximating a periodic sawtooth waveform by finite sums of its harmonics and the resulting average errors.

The difference between the sawtooth and its approximation is oscillatory, with a growing amplitude. It is not evenly distributed along time. It is zero at $t = -2, 0$, and 2, and 1 at the discontinuity points $t = -1$ and 1.

b. $\mathcal{E}_1 = \dfrac{1}{2} \displaystyle\int_{-1}^{1} \left(t - \dfrac{2}{\pi} \sin \pi t \right)^2 dt = 0.1307$

c. See Figure 7.1(c).

$N = 1, \quad y_1(t) = \dfrac{2}{\pi} \sin \pi t, \qquad\qquad\qquad\qquad \mathcal{E}_1 = 0.1307$

$N = 2, \quad y_2(t) = \dfrac{2}{\pi} \left(\sin \pi t - \dfrac{1}{2} \sin 2\pi t \right), \qquad\qquad \mathcal{E}_2 = 0.0800$

$N = 3, \quad y_3(t) = \dfrac{2}{\pi} \left(\sin \pi t - \dfrac{1}{2} \sin 2\pi t + \dfrac{1}{3} \sin 3\pi t \right), \ \ \mathcal{E}_3 = 0.0575$

$N = 4, \quad y_4(t) = \dfrac{2}{\pi} \left(\sin \pi t - \dfrac{1}{2} \sin 2\pi t + \dfrac{1}{3} \sin 3\pi t \right.$
$\qquad\qquad\qquad \left. - \dfrac{1}{4} \sin 4\pi t \right), \qquad\qquad\qquad\qquad \mathcal{E}_4 = 0.0448$

$N = 5, \quad y_5(t) = \dfrac{2}{\pi} \left(\sin \pi t - \dfrac{1}{2} \sin 2\pi t + \dfrac{1}{3} \sin 3\pi t \right.$
$\qquad\qquad\qquad \left. - \dfrac{1}{4} \sin 4\pi t + \dfrac{1}{5} \sin 5\pi t \right), \qquad\qquad \mathcal{E}_5 = 0.0367$

Example 7.2

Following the approach of Example 7.1 we will approximate a periodic square wave by the finite sum of its first N harmonics $y(t)$ and obtain the average error as defined in Example 7.1. Consider the periodic square wave $x(t)$ with a period of 2 seconds, which switches between ± 1 at equal time intervals. One cycle of the waveform is described by

$$x(t) = \begin{cases} -1, & -1 < t \le -0.5 \\ 1, & -0.5 < t \le 0.5 \\ -1, & 0.5 < t \le 1 \end{cases}$$

The waveform has a zero DC value and a peak-to-peak value of 2. Its Fourier series expansion is

$$x(t) = \frac{4}{\pi} \left(\cos \pi t - \frac{1}{3} \cos 3\pi t + \frac{1}{5} \cos 5\pi t - \frac{1}{7} \cos 7\pi t + \frac{1}{9} \cos 9\pi t \right)$$

You will approximate $x(t)$ by the finite sum of its first N harmonics, $y(t)$. The average error will be defined, as in Example 7.1, by

$$\mathcal{E} = \frac{1}{2} \int_{-1}^{1} |e(t)|^2 dt, \quad \text{where } e(t) = y(t) - x(t) \text{ is the instantaneous difference.}$$

a. Plot $y(t)$ and $|e(t)|^2$ for $N = 1$ and 3 for the period $-2 < t < 2$. Overlay $x(t)$ on the plots and qualitatively discuss the salient features of the error.

b. Find $\mathcal{E}$ for $N = 1$ by analytical and numerical (computational) methods and verify that they are in agreement.

c. Find $\mathcal{E}$ for $N = 2, 3, 4$ and 5 by a numerical method and plot it versus N. Qualitatively and quantitatively describe the decrease in $\mathcal{E}$ as the number of harmonics increases.

Solution

a. See Figure 7.2(a) and (b). The difference between the square wave and its approximation is oscillatory with maximum error at the discontinuity points.

b.
$$\mathcal{E}_1 = \int_{-0.5}^{0.5} \left(1 - \frac{4}{\pi} \cos \pi t \right)^2 dt = 0.1894$$

c. See Figure 7.2(c).

$$N = 1, \quad y_1(t) = \tfrac{4}{\pi} \cos \pi t, \qquad\qquad\qquad\qquad\qquad \mathcal{E}_1 = 0.1894$$

$$N = 2, \quad y_2(t) = \tfrac{4}{\pi} \left(\cos \pi t - \tfrac{1}{3} \cos 3\pi t \right), \qquad\qquad \mathcal{E}_2 = 0.0993$$

$$N = 3, \quad y_3(t) = \tfrac{4}{\pi} \left(\cos \pi t - \tfrac{1}{3} \cos 3\pi t + \tfrac{1}{5} \cos 5\pi t \right), \quad \mathcal{E}_3 = 0.0669$$

$$N = 4, \quad y_4(t) = \tfrac{4}{\pi} \left(\cos \pi t - \tfrac{1}{3} \cos 3\pi t + \tfrac{1}{5} \cos 5\pi t \right.$$
$$\left. - \tfrac{1}{7} \cos 7\pi t \right), \qquad\qquad\qquad\qquad \mathcal{E}_4 = 0.0504$$

$$N = 5, \quad y_5(t) = \tfrac{4}{\pi} \left(\cos \pi t - \tfrac{1}{3} \cos 3\pi t + \tfrac{1}{5} \cos 5\pi t \right.$$
$$\left. - \tfrac{1}{7} \cos 7\pi t + \tfrac{1}{9} \cos 9\pi t \right), \qquad\qquad \mathcal{E}_5 = 0.0404$$

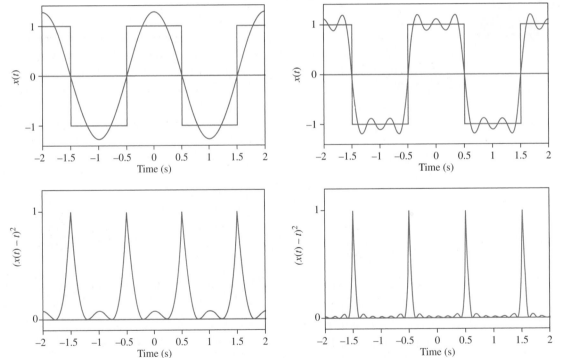

(a) Approximation by the principal harmonic (*top*) and the instantaneous power of error (*bottom*).

(b) Approximation by the sum of three harmonics (*top*) and the instantaneous power of error (*bottom*).

FIGURE 7.2 Approximating a periodic square wave by finite sums of its harmonics and the resulting errors.

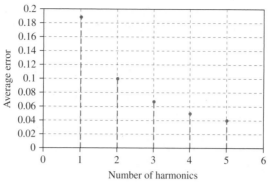

(*c*) Average error versus number of harmonics.

FIGURE 7.2 (*Continued*)

In both examples the error is greater at instances of discontinuity in the function. By increasing the number of harmonics one may reduce the average error but not the instantaneous values at discontinuity points.

7.2 Fourier Series Expansion

The question now is how large a class of input functions can be represented by e^{st} and how easily this can be done. For a periodic function with period T, the answer to the question is clear and simple: In the complex plane restrict s to the imaginary axis $j\omega$, leading to the set of exponential functions $e^{jn\omega_0 t}$, where $\omega_0 = 2\pi/T$ is called the fundamental frequency and n is an integer with $-\infty < n < \infty$. The description of a periodic function by the above set of exponentials is called its exponential Fourier series expansion, or simply its Fourier series. It is shown by

$$x(t) = \sum_{n=-\infty}^{\infty} X_n e^{jn\omega_0 t}$$

X_n is called the Fourier coefficient and $X_n e^{jn\omega_0 t}$ is called the nth harmonic of $x(t)$.

The coefficients X_n can be obtained from $x(t)$ using orthogonality of the set of exponential functions. Two functions $x(t)$ and $y(t)$ are said to be orthogonal to each other on an interval t_1 to t_2 if

$$\int_{t_0}^{t_1} x(t) y^*(t) dt = 0$$

The set of exponential functions $e^{jn\omega_0 t}$ have such a property on any interval of length $T = \frac{2\pi}{\omega_0}$. For those functions the integral yields

$$\int_{t_0}^{t_0+T} e^{jn\omega_0 t} e^{-jm\omega_0 t} dt = \int_{t_0}^{t_0+T} e^{j(n-m)\omega_0 t} dt = \begin{cases} 0, & n \neq m \\ T, & n = m \end{cases} = T\delta_{mn}$$

The orthogonality of the set of $e^{jn\omega_0 t}$, $-\infty < n < \infty$ on any interval T suggests the easy approach of finding X_n through the following integral:

$$\int_{t_0}^{t_0+T} x(t)e^{-jn\omega_0 t} = \int_{t_0}^{t_0+T} \left[\sum_{m=-\infty}^{\infty} X_m e^{jm\omega_0 t}\right] e^{-jn\omega_0 t} dt$$

Exchange the order by which the integration and the summation operations are performed to obtain

$$\sum_{m=-\infty}^{\infty} X_m \left[\int_{t_0}^{t_0+T} e^{j(m-n)\omega_0 t} dt\right] = \sum_{n=-\infty}^{\infty} T X_n \delta_{mn} = T X_n$$

Therefore,

$$x(t) = \sum_{n=-\infty}^{\infty} X_n e^{jn\omega_0 t}$$

$$X_n = \frac{1}{T} \int_{t_0}^{t_0+T} x(t)e^{-jn\omega_0 t} dt$$

It is seen that when $x(t)$ is a real-valued function of time, X_n and X_{-n} are complex conjugates, $X_n = X_{-n}^*$. The series, therefore, can be represented in trigonometric form (to be discussed in section 7.5).

7.3 The Generalized Fourier Series and Vectorial Representation of Signals

In the previous section we introduced a way to expand a periodic signal in the form of a sum of exponential functions. The expansion is called the Fourier series. The exponentials are eigenfunctions of LTI systems, making the Fourier series a very attractive tool for the analysis and design of LTI systems. In fact, our approach to the Fourier expansion was taken with that idea in mind.

Signals may also be represented by a set of functions $\phi_n(t)$ other than exponentials. The representation has the familiar form

$$x(t) = \sum_n X_n \phi_n(t)$$

The expansion is called a *generalized Fourier series* and the set of X_n are called generalized Fourier coefficients. To look into the applicability of such expansions, assume the functions $\phi_n(t)$ are orthonormal.

$$\int_{t_0}^{t_1} \phi_i(t)\phi_j^*(t)dt = \begin{cases} 1, & i = j \\ 0, & i \neq j \end{cases}$$

The Fourier coefficients X_n can then be found, as shown in section 7.2, by multiplying $x(t)$ with $\phi_n^*(t)$ and integrating, as was done in the previous section for the Fourier coefficients

$$X_n = \int_{t_0}^{t_1} x(t)\phi^*(t)dt$$

In view of the above expansion, one may think of the signal $x(t) = \sum_{n=1}^{N} X_n \phi_n(t)$ as the vector **x** in an orthogonal n-dimensional space. The vector **x** is specified by the ordered set X_n which constitutes its coordinate values. The coordinate system within which the signal is defined may also have infinite dimensions. The vectorial interpretation of signals is very useful in that it facilitates many analytical operations on signals. For the time being, the present chapter concentrates on the trigonometric and exponential Fourier series representations.

7.4 Dirichlet Conditions and Beyond

A rigorous proof of Fourier's theorem requires determination of conditions for convergence of the series to the original time function. One set of conditions was given by Dirichlet in 1829. The Dirichlet conditions are that

1. $x(t)$ be absolutely integrable over the period T:

$$\int_0^T |x(t)|dt < \infty$$

2. $x(t)$ have a finite number of discontinuities and a finite number of maxima and minima during the period T.

If the Dirichlet conditions are met by a function $x(t)$, then its Fourier series representation converges to $x(t)$ at all points where the function is continuous. At points where $x(t)$ is discontinuous, the series converges to the average of the limits of $x(t)$ from the left and right sides. For example, if $x(t)$ has a finite discontinuity at a point, the series converges to the midpoint of the discontinuity. (For more examples see section 7.13.) In almost all practical situations dealing with physical signals and systems, the Dirichlet conditions are satisfied and the Fourier series exists. [An example of a function for which the Dirichlet conditions do not apply is $x(t) = \sin(1/t)$.]

The Dirichlet conditions are sufficient to ensure validity of the series representation. They are, however, not necessary for the convergence of the series.[1] A periodic time function may violate the Dirichlet conditions and still have a valid Fourier series representation in the form described in section 7.1. As an example, a periodic impulse train in the time domain can be represented by the sum of harmonics of equal strength in the frequency domain. See section 7.12 on the Fourier series of impulse trains.

If a periodic function is square-integrable over one period

$$\int_0^T |x(t)|^2 dt < M, \quad \text{where } M \text{ is any large number}$$

---

[1] "The necessary conditions for convergence are not known" according to J. W. Nilsson, *Electric Circuits*, 3rd ed. (Upper Saddle River, NJ: Prentice Hall, 1990), p. 671.

then its Fourier series converges in the mean-squared sense. For more details on proof of Fourier's theorem and its convergence, see the references listed in footnotes 2 and 3.[2,3]

7.5 Trigonometric Fourier Series

A periodic signal with period T[4] may, in general, be represented by the infinite sum[5]

$$x(t) = \frac{1}{2}a_0 + a_1 \cos(\omega t) + a_2 \cos(2\omega t) \cdots + b_1 \sin(\omega t) + b_2 \sin(2\omega t) + \cdots$$

$$= \frac{1}{2}a_0 + \sum_{n=1}^{\infty} a_n \cos(n\omega t) + \sum_{n=1}^{\infty} b_n \sin(n\omega t) \tag{7.1}$$

$$a_n = \frac{2}{T} \int_0^T x(t) \cos(n\omega t) \, dt, \quad \text{for } n = 0, 1, 2, \dots \tag{7.2a}$$

$$b_n = \frac{2}{T} \int_0^T x(t) \, \sin(n\omega t) \, dt, \quad \text{for } n = 1, 2, \dots \tag{7.2b}$$

where $\omega = 2\pi/T$ is the fundamental frequency and $n\omega, n > 1$ are the harmonic frequencies (or simply, its harmonics). A first question pertains to the conditions for validity of the series representation. We are interested in the application of the Fourier series in the analysis of signals and linear systems, and are comforted in the knowledge that all periodic functions that model practical signals and systems may be represented by the Fourier series. Nonetheless, some conditions for convergence of the series are briefly discussed in section 7.4.

Derivation of the Trigonometric Fourier Coefficients

To derive a_n given in (7.2a) we multiply both sides of (7.1) by $\cos(m\omega t)$, integrate over one period, and then interchange the order of summation and integration.

$$\int_0^T x(t) \cos(m\omega t) \, dt = \frac{1}{2}a_0 \int_0^T \cos(m\omega t)dt + \sum_{n=1}^{\infty} \int_0^T a_n \cos(n\omega t) \cos(m\omega t)dt$$

$$+ \sum_{n=1}^{\infty} \int_0^T b_n \sin(n\omega t) \cos(m\omega t)dt$$

---

[2]H. S. Carslaw, *Introduction to Theory of Fourier's Series and Integrals* (London: Macmillan, 1930).

[3]L. Carlson, "On Convergence and Growth of Partial Sums by Fourier Series," *Acta Mathematica*, June 1966, pp. 135–157.

[4]$x(t) = x(t + T)$ for all t.

[5]Presented by Fourier to the French Academy in 1807. See J. B. J. Fourier, "Theory Analytique de la Chaleur," 1822. An English edition of the work is published by Dover, New York.

For $m = 0$: $a_0 = \dfrac{2}{T} \displaystyle\int_0^T x(t)\,dt$

For $m \neq 0$: $\displaystyle\int_0^T \sin(n\omega t)\cos(m\omega t)\,dt = 0$, all n and m

$$\int_0^T \sin(n\omega t)\sin(m\omega t)\,dt = \int_0^T \cos(n\omega t)\cos(m\omega t)\,dt = \begin{cases} 0, & n \neq m \\ \frac{T}{2}, & n = m \end{cases}$$

Therefore, all integral terms become zero except when $n = m$, which results in (7.2a). Similarly, to derive coefficients b_n given in (7.2b) multiply both sides of (7.1) by $\sin(m\omega t)$ and integrate over one period.

Note: The limits of the integrals for evaluating the Fourier coefficients are from t_0 to $t_0 + T$, with t_0 chosen for the convenience of integration as long as the integral is taken over one period.

Simplified Derivation of the Coefficients

The derivations of the trigonometric coefficients may be simplified by choosing a new variable $\tau = \omega t$ in the integrals of (7.2a) and (7.2b). Hence, $dt = \frac{1}{\omega}d\tau = \frac{T}{2\pi}d\tau$. The trigonometric coefficients are then found from

$$a_n = \frac{1}{\pi}\int_0^{2\pi} \chi(\tau)\cos(n\tau)\,d\tau$$

$$b_n = \frac{1}{\pi}\int_0^{2\pi} \chi(\tau)\sin(n\tau)\,d\tau$$

where $\chi(\tau) = x(\frac{\tau}{\omega})$. A plot of $\chi(\tau)$ versus τ is identical to a plot of $x(t)$ versus ωt. In other words, compress the time axis in the x plot by a factor of ω and you will get the plot of χ. In general, the limits of the integrals are from θ to $\theta + 2\pi$, with θ chosen to simplify the integration.

Sine and Cosine Series

The sine and cosine terms of (7.2a) and (7.2b) may be combined to form a series with sines or cosines only:

$$x(t) = \frac{1}{2}c_0 + \sum_{n=1}^{\infty} c_n \cos(n\omega t - \theta_n) = \frac{1}{2}c_0 + \sum_{n=1}^{\infty} c_n \sin(n\omega t + \phi_n)$$

where $c_0 = a_0$ and

$$c_n = \sqrt{a_n^2 + b_n^2}, \quad \theta_n = \tan^{-1}\left(\frac{b_n}{a_n}\right), \quad \text{and} \quad \phi_n = \tan^{-1}\left(\frac{a_n}{b_n}\right), \quad n \geq 1$$

Example
7.3

Rectangular pulse train

Rectangular pulses and pulse trains have a special place in Fourier analysis, signal processing, and filter design. In this example we derive the trigonometric Fourier coefficients of a rectangular pulse train. (We will revisit it in sections 7.6 and 7.12 when we consider the exponential form of the Fourier series.)

a. Find the coefficients of the trigonometric Fourier series for the periodic rectangular pulse train $x(t)$ with period T, pulse duration τ, baseline at 0, and pulse height V_0.

b. Write the first seven terms in $x(t)$ when the pulse train becomes a square.

c. A pulse with a period T and pulse duration τ is said to have a duty cycle of $\frac{\tau}{T} \times 100\%$. For example, a pulse with $\tau = T/100$ has 1% duty cycle. Write the first seven terms in $x(t)$ when $\tau = T/100$ (a narrow pulse train with a 1% duty cycle).

Solution

a. Choose the time origin such that the signal is an even function. During the period $-\frac{T}{2} < t < \frac{T}{2}$ the pulse is given by

$$x(t) = \begin{cases} V_0, & -\frac{\tau}{2} \leq t < \frac{\tau}{2} \\ 0, & \text{elsewhere} \end{cases}$$

The trigonometric Fourier coefficients are

$$a_n = \frac{2}{T} \int_{-\frac{\tau}{2}}^{\frac{\tau}{2}} V_0 \cos\left(\frac{2\pi n}{T}t\right) dt = \begin{cases} 2V_0 \frac{\sin\left(\pi n \frac{\tau}{T}\right)}{\pi n}, & n > 0 \\ 2\frac{V_0 \tau}{T} & n = 0 \end{cases}$$

Note that $a_0/2 = V_0 \tau / T$ is the DC value of the waveform. The coefficients b_n are zero because $x(t)$ is even, $x(t)\sin(2\pi n \frac{t}{T})$ is odd, and the integral in (7.2b) becomes zero.

b. In the case of a square pulse where $\tau = \frac{T}{2}$,

$$a_n = V_0 \frac{\sin \frac{\pi n}{2}}{\frac{\pi n}{2}}$$

$$x(t) = \frac{V_0}{2} + \frac{2V_0}{\pi}\left[\cos(\omega t) - \frac{1}{3}\cos(3\omega t) + \frac{1}{5}\cos(5\omega t) - \frac{1}{7}\cos(7\omega t) + \cdots\right]$$

$$= V_0 [0.5 + 0.637\cos(\omega t) - 0.212\cos(3\omega t) + 0.127\cos(5\omega t)$$

$$-0.091\cos(7\omega t) + \cdots]$$

where $\omega = \frac{2\pi}{T}$ is the fundamental angular frequency. The even rectangular pulse train with 50% duty cycle has odd harmonics only.

c. In the case of a narrow pulse with $\tau = T/100$, $a_n = 2V_0 \sin\left(\frac{\pi n}{100}\right)/(\pi n)$. For $n < 10$,

$$\sin\left(\frac{\pi n}{100}\right) \approx \frac{\pi n}{100}, \quad \Rightarrow \quad a_n \approx 2V_0 \frac{\frac{\pi n}{100}}{\pi n} = 0.02V_0$$

$$x(t) \approx \frac{V_0}{100} + \frac{2V_0}{100}[\cos(\omega t) + \cos(2\omega t) + \cos(3\omega t) + \cos(4\omega t) + \cdots]$$

For the narrow pulse, the coefficients a_n have become smaller and almost equal, with all harmonics being present. As n increases, the spectral lines do not diminish as fast as in (7.2b). See section 7.12 for a comprehensive picture.

*E*xample
7.4

Filtering

When a periodic waveform passes through an LTI system, its Fourier coefficients undergo different gains at different frequencies. Some harmonics may be attenuated more strongly than others and thus filtered out. In this example a 159.155-Hz square wave $v_1(t)$ with its base-to-peak voltage spanning zero to 1 volt is passed through a $1 \ k\Omega$ resistor in series with a $1 \ \mu F$ capacitor. See Figure 7.3(a). Find the frequency components of the capacitor voltage as a function of time. Plot the amplitudes of the DC value and the next nth harmonics versus n for $n = 1$ to 7, and call it the one-sided line spectra.

Solution

We first find the expansion of the input and pass it through the circuit that works as a filter. Choosing the time origin at the center of a pulse makes the input voltage an even function of time so that its Fourier series will contain cosine terms only. The Fourier coefficients of an input cosine series are

$$v_1(t) = \frac{1}{2} + \sum_{n=1}^{\infty} a_n \cos(n\omega_0 t)$$

$$\omega_0 = 2\pi \times 159.155 = 1,000 \text{ rad/s} \quad \text{and} \quad a_n = \frac{\sin(\frac{\pi n}{2})}{(\frac{\pi n}{2})}, \quad n \geq 1$$

The input voltage and its line spectrum are plotted in Figure 7.3(b) and (c), respectively. Now note that a sinusoidal component $\cos \omega t$ of $v_1(t)$ produces $|H|\cos(\omega t + \theta)$ at the terminals of the capacitor, where

$$|H(\omega)| = \frac{1}{\sqrt{1 + R^2 C^2 \omega^2}} \quad \text{and} \quad \theta(\omega) = -\tan^{-1}(RC\omega)$$

The capacitor voltage in response to the square wave input may be shown by the series

$$v_2(t) = \frac{1}{2} + \sum_{n=1}^{\infty} H_n \frac{2\sin(\frac{\pi n}{2})}{\pi n} \cos(n\omega_0 t + \theta_n)$$

where H_n is found from $|H(\omega)|$ by setting $\omega = n\omega_0$. Note that if $\omega_0 = 1/(RC)$, as in the present example, we obtain $RC\omega = n$ and

$$|H(\omega)| = H_n = \frac{1}{\sqrt{1 + n^2}} \quad \text{and} \quad \theta_n = -\tan^{-1}(n)$$

The sum of the DC value and the first four nonzero harmonics in $v_1(t)$ and $v_2(t)$ are listed below.

$$v_1(t) = 0.5 + 0.637\cos(\omega_0 t) - 0.212\cos(3\omega_0 t) + 0.127\cos(5\omega_0 t) - 0.091\cos(7\omega_0 t)$$

$$v_2(t) = 0.5 + 0.450\cos(\omega_0 t - 0.45°) - 0.067\cos(3\omega_0 t - 71.6°)$$
$$+ 0.025\cos(5\omega_0 t - 78.7°) - 0.013\cos(7\omega_0 t - 81.9°)$$

The output voltage and its line spectrum are plotted in Figure 7.3(d) and (e), respectively.

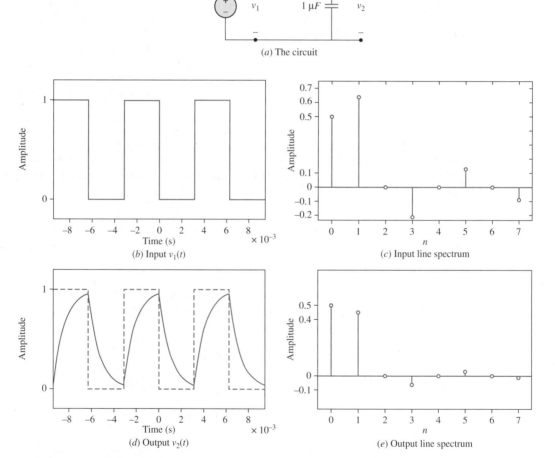

(a) The circuit

(b) Input $v_1(t)$

(c) Input line spectrum

(d) Output $v_2(t)$

(e) Output line spectrum

FIGURE 7.3 An *RC* circuit with a 1-msec time constant receives a rectangular pulse train voltage at 159.155 Hz as shown in *(b)*. The output voltage across the capacitor is shown in *(d)*. The one-sided spectra of the input and output voltages are shown in *(c)* and *(e)*, on the upper and lower traces, respectively. Note attenuation of spectral lines at the output.

7.6 **Exponential Fourier Series**

As introduced in section 7.1, the Fourier series expansion of a periodic signal $x(t)$ may, alternatively, be expressed by the infinite sum of exponential terms

$$x(t) = \sum_{n=-\infty}^{\infty} X_n e^{j2\pi nt/T} \quad (7.3)$$

$$X_n = \frac{1}{T} \int_T x(t) e^{-j2\pi nt/T} dt \quad (7.4)$$

where the integral is taken over a time period which is T seconds long. Coefficients X_n are called the exponential Fourier coefficients of $x(t)$ and the sum is the exponential Fourier series representation, or expansion, of $x(t)$. From now on we use the exponential form of the Fourier series unless specified otherwise. To derive X_n, first multiply both sides of (7.3) by $e^{-j2\pi mt/T}$ and integrate over one period. By exchanging the order of summation and integration, and observing that $e^{j2\pi nt/T}$ and $e^{-j2\pi mt/T}$ are orthogonal, all terms vanish except when $n = m$, from which (7.4) results. The four examples given below will be revisited again in more detail.

 Example **7.5**

Rectangular pulse train

Find the exponential Fourier coefficients of a periodic rectangular pulse train with period T and pulse width τ. See Figure 7.4. During $-\frac{T}{2} < t < \frac{T}{2}$ the pulse is given by

$$x(t) = \begin{cases} 1, & -\frac{\tau}{2} \leq t < \frac{\tau}{2} \\ 0, & \text{elsewhere} \end{cases}$$

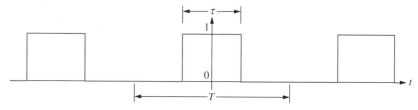

FIGURE 7.4 An even periodic rectangular pulse train.

Solution

$$X_n = \frac{1}{T} \int_{-\frac{T}{2}}^{\frac{T}{2}} x(t) e^{-j2\pi nt/T} dt = \frac{1}{T} \int_{-\frac{\tau}{2}}^{\frac{\tau}{2}} e^{-j2\pi nt/T} dt = \begin{cases} \frac{\sin\left(\pi n \frac{\tau}{T}\right)}{\pi n}, & n \neq 0 \\ \frac{\tau}{T}, & n = 0 \end{cases}$$

Example 7.6

Triangular pulse train

Find the exponential Fourier coefficients of a periodic triangular pulse train with period T. The base of the triangle is τ and the height is 1. See Figure 7.5.

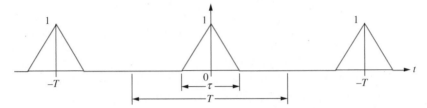

FIGURE 7.5 An even periodic triangular pulse train.

During the period $-\frac{T}{2} < t < \frac{T}{2}$ the pulse is given by

$$x(t) = \begin{cases} 1 + 2\frac{t}{\tau}, & -\frac{\tau}{2} \leq t < 0 \\ 1 - 2\frac{t}{\tau}, & 0 \leq t < \frac{\tau}{2} \\ 0, & \text{elsewhere} \end{cases}$$

solution

$$X_n = \frac{1}{T} \left[\int_{-\tau/2}^{0} \left(1 + 2\frac{t}{\tau} \right) e^{-j2\pi nt/T} dt + \int_{0}^{\tau/2} \left(1 - 2\frac{t}{\tau} \right) e^{-j2\pi nt/T} dt \right]$$

$$= \begin{cases} \frac{2T}{\tau} \frac{\sin^2\left(\frac{\pi n\tau}{2T} \right)}{(\pi n)^2}, & n \neq 0 \\ \frac{\tau}{2T}, & n = 0 \end{cases}$$

Example 7.7

Impulse train

Find the exponential Fourier coefficients of a periodic impulse train. See Figure 7.6.

$$x(t) = \sum_{n=-\infty}^{\infty} \delta(t - nT)$$

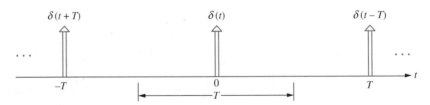

FIGURE 7.6 An impulse train.

Solution

$$X_n = \frac{1}{T} \int_{-T/2}^{T/2} \delta(t)e^{-j2\pi nt/T} dt = \frac{1}{T}, \quad -\infty < n < \infty$$

In summary,

$$\sum_{n=-\infty}^{\infty} \delta(t - nT) = \frac{1}{T} \sum_{n=-\infty}^{\infty} e^{j2\pi nt/T}$$

All harmonics are equally represented in the waveform.

7.7 Properties of Fourier Series

Some features of the time functions reflect upon the Fourier coefficients in ways that facilitate their derivation. These are called properties of Fourier series. Several such properties are listed in Table 7.1 and discussed in the following sections. Some other properties, not listed in Table 7.1, are more applicable when in the guise of Fourier transform (e.g., differentiation and integration).

TABLE 7.1 Some Properties of the Fourier Series of Real-Valued Time Function

Property	Time Domain	$\Longleftrightarrow$	Frequency Domain
Definition	$x(t) = \sum_{-\infty}^{\infty} X_n e^{j2\pi nt/T}$	$\Longleftrightarrow$	$X_n = \frac{1}{T} \int_T x(t)e^{-j2\pi nt/T} dt$
Superposition	$ax(t) + by(t)$	$\Longleftrightarrow$	$aX_n + bY_n$
Time reversal (real-valued functions)	$x(-t)$	$\Longleftrightarrow$	X_n^*
Time shift	$x(t - t_0)$	$\Longleftrightarrow$	$X_n e^{-j2\pi nt_0/T}$
Real-valued function	$x(t) = x^*(t)$	$\Longleftrightarrow$	$X_n = X_{-n}^*$
Even symmetry	$x(t) = x(-t)$	$\Longleftrightarrow$	$X_n = X_{-n}$
Odd symmetry	$x(t) = -x(-t)$	$\Longleftrightarrow$	$X_n = -X_{-n}$
Half-wave symmetry	$x(t) = -x(t + T/2)$	$\Longleftrightarrow$	$X_n = 0$ for n even
Even part of $x(t)$	$\frac{x(t)+x(-t)}{2}$	$\Longleftrightarrow$	$\mathcal{RE}\{X_n\}$
Odd part of $x(t)$	$\frac{x(t)-x(-t)}{2}$	$\Longleftrightarrow$	$j\mathcal{IM}\{X_n\}$

Theorems:	DC Value	$\frac{1}{T} \int_T x(t)dt = X_0$
	Zero Time	$x(0) = \sum_{-\infty}^{\infty} X_n$
	Parseval's Theorem	$\frac{1}{T} \int_T |x(t)|^2 dt = \sum_{-\infty}^{\infty} |X_n|^2$

7.8 Time Reversal and Shift

Time Reversal

Time reversal in a real-valued function changes the Fourier coefficients to their complex conjugate.

$$x(t) \Longleftrightarrow X_n$$

$$x(-t) \Longleftrightarrow X_n^*$$

To prove the above, let $y(t) = x(-t)$. Then

$$Y_n = \frac{1}{T} \int_{t=-T/2}^{T/2} x(-t)e^{-j2\pi nt/T}\, dt = \frac{1}{T} \int_{t=-T/2}^{T/2} x(t)e^{j2\pi nt/T}\, dt = X_n^*$$

Time Shift

A time shift, or delay in time, by t_0 seconds multiplies X_n by $e^{-j2\pi nt_0/T}$:

$$x(t - t_0) \qquad \Longleftrightarrow \qquad X_n e^{-j2\pi nt_0/T}$$

Proof. Let $y(t) = x(t - t_0)$. Then

$$Y_n = \frac{1}{T} \int_{-T/2}^{T/2} y(t)e^{-j2\pi nt/T}\, dt = \frac{1}{T} \int_{-T/2}^{T/2} x(t - t_0)e^{-j2\pi nt/T}\, dt$$

$$= \frac{1}{T} \int_{-T/2}^{T/2} x(\tau)e^{-j2\pi n(\tau+t_0)/T}\, dt = X_n e^{-j2\pi nt_0/T}$$

 Example **7.8**

Find, by applying the definition, the exponential Fourier coefficients of the periodic rectangular pulse train with a period T. During one period the pulse is shown by

$$x(t) = \begin{cases} 1, & 0 \le t < \tau \\ 0, & \tau \le t < T \end{cases}$$

Show that the same result is obtained by applying the shift property to the coefficients of the rectangular pulse train in Example 7.5.

Solution

$$X_n = \frac{1}{T} \int_0^\tau e^{-j2\pi nt/T}\, dt = \begin{cases} \frac{1}{j2\pi n}\left[1 - e^{-j2\pi n\tau/T}\right] = \frac{\sin\left(\pi n\frac{\tau}{T}\right)}{\pi n}e^{-j\pi n\tau/T}, & n \ne 0 \\ \frac{\tau}{T}, & n = 0 \end{cases}$$

The pulse train of this example is the same as that of Example 7.5 but delayed by $\tau/2$ units of time. The Fourier coefficients are, therefore, those of Example 7.5 multiplied by $e^{-j\pi n\tau/T}$.

Example

7.9

Find the exponential Fourier series coefficients for the periodic signals shown in Figure 7.7.

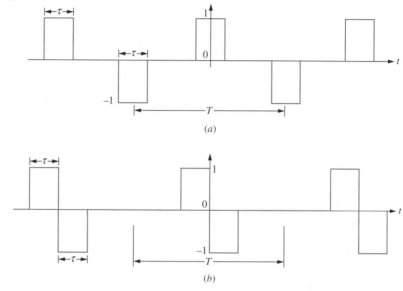

(a)

(b)

FIGURE 7.7

Solution

Consider the periodic rectangular even pulse train $x(t)$ of Example 7.5 with period T and pulse width τ. One period of the pulse and its Fourier coefficients are

$$x(t) = \begin{cases} 1, & -\tau/2 \le t < \tau/2 \\ 0, & \text{elsewhere} \end{cases} \iff X_n = \frac{\sin\left(\frac{\pi n \tau}{T}\right)}{\pi n}$$

For Figure 7.7(a), we have

$$x(t) - x(t - T/2) \iff X_n - X_n e^{-j\pi n} = \begin{cases} 2\frac{\sin\left(\frac{\pi n \tau}{T}\right)}{\pi n}, & n \text{ odd} \\ 0 & n \text{ even} \end{cases}$$

For Figure 7.7(b),

$$x(t + \tau/2) - x(t - \tau/2) \iff X_n e^{j\frac{\pi n \tau}{T}} - X_n e^{-j\frac{\pi n \tau}{T}} = 2j\frac{\sin^2\left(\frac{\pi n \tau}{T}\right)}{\pi n}$$

7.9 Conjugate Symmetry

If $x(t)$ is real, $x(t) = x^*(t)$, then $X_n = X^*_{-n}$. In other words, a real function $x(t)$ is completely specified from X_n, $n \ge 0$, which may be given in the form of its real and imaginary parts or by its magnitude and phase. This property can be observed throughout the examples in this chapter, as the time functions are all real valued.

7.10 Waveform Symmetry

Certain types of symmetry in a periodic waveform result in specific properties of its Fourier series. In this section consider three properties concerning real functions and their Fourier series: even symmetry, odd symmetry, and half-wave symmetry.

Even Symmetry

A function $x(t)$ is even if $x(t) = x(-t)$. The plot of a waveform represented by an even function is symmetrical with respect to the vertical axis. An example of an even function is $x(t) = 1 + x^2$. The cosine is an even function by its definition. Also, note that

$$\cos(t) = 1 - \frac{t^2}{2!} + \frac{t^4}{4!} - \frac{t^6}{6!} + \frac{t^8}{8!} + \dots.$$

which is made of even terms. The exponential Fourier coefficients of an even function are purely real numbers. In the trigonometric representation of the Fourier series for such a function, $b_n = 0$ and the series contains only cosine terms and possibly a constant if its average is nonzero.

Odd Symmetry

A function $x(t)$ is odd if $x(t) = -x(-t)$. The plot of a waveform represented by an odd function is symmetrical with respect to the origin. An example of an odd function is $x(t) = x + x^3$. The sine is an odd function by its definition. Also note that

$$\sin(t) = t - \frac{t^3}{3!} + \frac{t^5}{5!} - \frac{t^7}{7!} + \frac{t^9}{9!} - \dots.$$

which is made of odd terms. The exponential Fourier coefficients of an odd function are purely imaginary numbers. In the trigonometric representation of the Fourier series of such a function, $a_n = 0$ and the series contains sine terms only.

The product of two odd functions is an even function and the sum of two odd functions is an odd function. Adding a constant value to an odd function removes its odd property. Conversely, a waveform may become odd by removing its average value, in which case the series representing the original waveform is seen to contain sine terms and a constant value representing the average.

Half-Wave Symmetry

A function $x(t)$ is said to have half-wave symmetry if $x(t) = -x(t + T/2)$, where T is the period. If a waveform has half-wave symmetry, then $X_n = 0$ for n even, and its series contains only odd harmonics.

Even and Odd Parts of a Function

Any function $x(t)$ may be split into an even and an odd part: $x(t) = x_e(t) + x_o(t)$, where

$$x_e(t) = \frac{x(t) + x(-t)}{2}$$

$$x_o(t) = \frac{x(t) - x(-t)}{2}$$

If $x(t)$ is real, its even and odd parts are related to the trigonometric coefficients a_n and b_n.

Relationships between the Fourier series coefficients of a real-valued waveform and its even and odd parts are summarized as follows:

$x(t)$	$\Longleftrightarrow$	$X_n = \frac{1}{2}(a_n + jb_n)$
$x(-t)$	$\Longleftrightarrow$	$X_{-n} = \frac{1}{2}(a_n - jb_n)$
$x_e(t) = \frac{x(t)+x(-t)}{2}$	$\Longleftrightarrow$	$X_n = X_{-n} = a_n/2$, (cosine terms only)
$x_o(t) = \frac{x(t)-x(-t)}{2}$	$\Longleftrightarrow$	$X_n = -X_{-n} = jb_n/2$, (sine terms only)

Summary

The symmetry properties are summarized below:

$$\text{Even function} \quad x(t) = x(-t) \quad \Longleftrightarrow X_n = X_{-n}$$

$$\text{Odd function} \quad x(t) = -x(-t) \quad \Longleftrightarrow X_n = -X_{-n}$$

$$\text{Half-wave symmetry} \quad x(t) = -x(t + T/2) \Longleftrightarrow X_n = 0 \ (n \text{ even})$$

The symmetry conditions considered above may be used to simplify the Fourier representation of a waveform. When representing a waveform by a mathematical function, the time origin and the DC level may be chosen conveniently so that its Fourier series may contain sine or cosine terms only. The square pulse train (Figure 7.4 with $\tau = T/2$) can be odd, even, or neither, depending on placement of the time origin. Similarly, the sawtooth waveform [Figure 7.13(a) with $\tau = T$] becomes odd (and its series will contain only sine terms) if the DC value is subtracted. In summary, depending on the time origin, a rectangular pulse train becomes even, odd, or neither. Removing the DC level may make it an odd function.

7.11 Time Averages

The *average* (or DC) value of a waveform $x(t)$ over the time duration t_0 to $t_0 + T$ is

$$X_{\text{avg}} \equiv\, <x(t)> \equiv \frac{1}{T} \int_{t_0}^{t_0+T} x(t)\, dt$$

Similarly, the root-mean-square (rms) or effective (eff) value of the waveform $x(t)$ over the same time duration is

$$X_{\text{rms}} \equiv X_{\text{eff}} \equiv \left\{ \frac{1}{T} \int_{t_0}^{t_0+T} x^2(t)\, dt \right\}^{\frac{1}{2}}$$

One can see that $X_{\text{rms}}^2 = <x^2(t)>$.

The average and rms values of a periodic waveform may be obtained directly from the above definitions or from the Fourier series coefficients of the waveform. Let

$$x(t) = \frac{1}{2}a_0 + a_1 \cos(\omega t) + a_2 \cos(2\omega t) \cdots + b_1 \sin(\omega t) + b_2 \sin(2\omega t) \cdots$$

By applying the above definitions and considering that the sine and cosine functions have zero average values and are orthogonal, we get

$$X_{\text{avg}} = a_0/2$$

$$X_{\text{rms}} = \sqrt{\left(\frac{1}{2}a_0\right)^2 + \frac{1}{2}a_1^2 + \frac{1}{2}a_2^2 + \cdots + \frac{1}{2}b_1^2 + \frac{1}{2}b_2^2 + \cdots} = \sqrt{c_0^2 + \frac{1}{2}c_1^2 + \frac{1}{2}c_2^2 + \cdots}$$

where $c_0 = a_0/2$ and $c_n^2 = a_n^2 + b_n^2$, $n > 0$. In the exponential form, $X_{\text{avg}} = X_0$.

Parseval's Theorem

This theorem states that the (average) power in a periodic waveform $x(t)$ is the sum of powers in its harmonics.

$$P = \frac{1}{T} \int_{-T/2}^{T/2} |x(t)|^2 dt = \sum_{-\infty}^{\infty} |X_n|^2$$

This is simply a consequence of orthogonality of the harmonics. $|X_n|^2$ is called the power spectrum density of $x(t)$. The power contained in the harmonics $N_1 \le n \le N_2$ is

$$P = \sum_{n=N_1}^{N_2} 2|X_n|^2$$

7.12 Pulses and Impulses

Periodic trains such as rectangular, triangular, sawtooth, and sinc pulses have many applications in electrical engineering. Function, pulse, and signal generators, which can produce such pulse trains, constitute basic instruments of an analog electrical engineering laboratory. In addition, pulse signals are often used to transmit binary numbers. They are also encountered in finite-duration signal processing, discrete-time signal modeling, digital signal processing, speech processing and coding, and pattern recognition. Similarly, impulse trains are the mathematical instrument for sampling continuous-time signals. Because of their importance and repeated use, we will revisit and examine in detail the exponential Fourier series expansion of rectangular and triangular pulse and impulse trains. The sinc pulse train is analyzed in an exercise. More discussion on windows is found in Chapters 8, 16, and 17.

Rectangular Pulse Revisited

Consider the periodic rectangular pulse train $x(t)$ and its exponential Fourier coefficients (see Example 7.5):

$$x(t) = \begin{cases} 1, & -\frac{\tau}{2} < t < \frac{\tau}{2} \\ 0, & \text{elsewhere} \end{cases}, \quad x(t) = x(t+T)$$

$$X_n = \begin{cases} \dfrac{\sin\left(\pi n \frac{\tau}{T}\right)}{\pi n}, & n \neq 0 \\ \dfrac{\tau}{T}, & n = 0 \end{cases}$$

The coefficients X_n exhibit the following features (see Figure 7.8):

1. X_n is real and an even function of n, decreasing with n. The plot of X_n has an infinite number of alternating positive and negative lobes whose amplitudes decrease with n. The main lobe extends from $-n_0$ to n_0, where n_0 is the integer found by truncating T/τ. When T/τ is an integer, the amplitude of the harmonic represented by that integer is zero. The major part of the power in the signal is concentrated at low frequencies within the main lobe.

2. The DC value, or average of the pulse train, is $X(0) = \tau/T$.

3. For $\tau = \frac{T}{2}$, the waveform becomes a square pulse train and

$$X_n = \begin{cases} \dfrac{\sin\left(\frac{\pi n}{2}\right)}{\pi n}, & n \neq 0 \\ \dfrac{1}{2} & n = 0 \end{cases}$$

$$x(t) = \left\{ \cdots + \frac{1}{5\pi} e^{-j5\omega_0 t} - \frac{1}{3\pi} e^{-j3\omega_0 t} + \frac{1}{\pi} e^{-j\omega_0 t} + \frac{1}{2} + \frac{1}{\pi} e^{j\omega_0 t} \right.$$

$$\left. - \frac{1}{3\pi} e^{j\omega_0 t} + \frac{1}{5\pi} e^{j5\omega_0 t} \cdots \right\}, \quad \text{where } \omega_0 = \frac{2\pi}{T}$$

In this case the average power in $x(t)$ is

$$P = \frac{1}{T} \int_{-\frac{T}{4}}^{\frac{T}{4}} dt = \frac{1}{2}$$

The average power in the principal harmonic is $2|X_1|^2 = 2/\pi^2$. As the percentage of the total power, it is

$$\frac{2|X_1|^2}{P} \times 100 = \frac{4}{\pi^2} \times 100 \approx 40.5\%$$

4. As the pulse narrows its spectrum broadens.

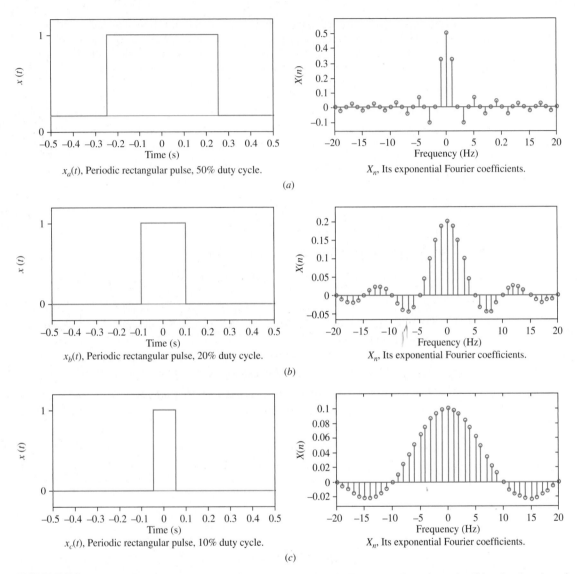

$x_a(t)$, Periodic rectangular pulse, 50% duty cycle.

X_n, Its exponential Fourier coefficients.

(a)

$x_b(t)$, Periodic rectangular pulse, 20% duty cycle.

X_n, Its exponential Fourier coefficients.

(b)

$x_c(t)$, Periodic rectangular pulse, 10% duty cycle.

X_n, Its exponential Fourier coefficients.

(c)

FIGURE 7.8 The time plots on the left show one cycle of a 1-Hz periodic rectangular pulse train with pulse duration of 0.5, 0.2, and 0.1 sec in *(a)*, *(b)*, and *(c)*, respectively. The time origin has been chosen so that the pulses are even functions of time. Their exponential Fourier coefficients are, therefore, real valued and shown in the right column, respectively. As the pulse narrows, its frequency spectrum broadens.

Triangular Pulse Revisited

The exponential Fourier coefficients of the periodic triangular pulse train with period T and base τ were obtained in Example 7.6.

$$x(t) = \begin{cases} 1 + 2\frac{t}{\tau}, & -\frac{\tau}{2} < t < 0 \\ 1 - 2\frac{t}{\tau}, & 0 < t < \frac{\tau}{2} \\ 0, & \text{elsewhere} \end{cases}, \quad x(t) = x(t+T)$$

$$X_n = \begin{cases} \frac{2T}{\tau} \frac{\sin^2\left(\frac{\pi n \tau}{2T}\right)}{(\pi n)^2}, & n \neq 0 \\ \frac{\tau}{2T}, & n = 0 \end{cases}$$

See Figure 7.9.

The Fourier coefficients exhibit the following features:

1. $X(n)$ is a real and even function of n. It is always nonnegative. It extends over the infinite frequency range.
2. The DC value of the transform is $X(0) = \tau/(2T)$, which is equal to the average of the area in one cycle.
3. Compared to the transform of the rectangular pulse, the mathematical expression of $X(f)$ for the triangular pulse indicates that the lobes attenuate faster with increased frequency.
4. For $\tau = T/2$, the Fourier series become

$$X_n = \begin{cases} 4\frac{\sin^2\left(\frac{\pi n}{4}\right)}{\pi^2 n^2}, & n \neq 0 \\ \frac{1}{4} & n = 0 \end{cases}$$

$$x(t) = \left\{ \cdots \frac{2}{9\pi^2}e^{-j3\omega_0 t} + \frac{1}{\pi^2}e^{-j2\omega_0 t} + \frac{2}{\pi^2}e^{-j\omega_0 t} + \frac{1}{4} + \frac{2}{\pi^2}e^{j\omega_0 t} + \frac{1}{\pi^2}e^{j2\omega_0 t} \right.$$

$$\left. + \frac{2}{9\pi^2}e^{j3\omega_0 t} \cdots \right\}, \text{ where } \omega_0 = \frac{2\pi}{T}$$

In this case the average power in $x(t)$ is

$$P = \frac{2}{T} \int_0^{\frac{T}{4}} \left(\frac{4t}{T}\right)^2 dt = \frac{1}{6}$$

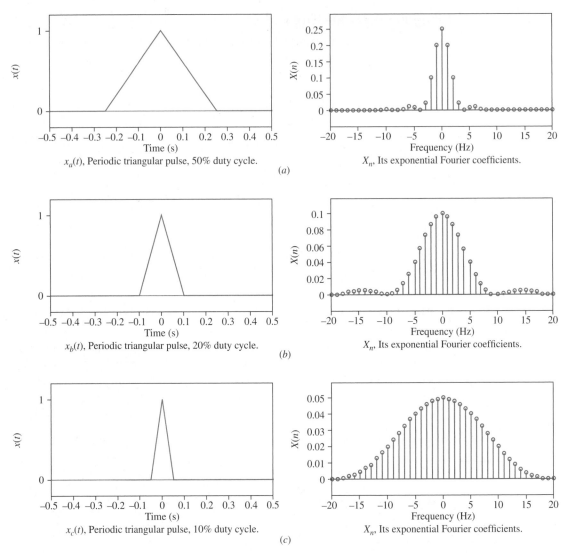

FIGURE 7.9 The time plots on the left show one cycle of a 1-Hz periodic triangular pulse train with pulse duration of 0.5, 0.2, and 0.1 sec in *(a)*, *(b)*, and *(c)*, respectively. The time origin has been set so that the pulses are even functions of time. Their exponential Fourier coefficients are, therefore, real valued and are shown in the right column, respectively. As the pulse narrows, its frequency spectrum broadens.

The average power in the principal harmonic is $2|X_1|^2 = \frac{8}{\pi^4}$. As the percentage of the total power, it is

$$\frac{2|X_1|^2}{P} \times 100 = \frac{48}{\pi^4} \times 100 \approx 49.3\%$$

Observation

Convolve a periodic square pulse $x(t)$ (spanning from zero to 1 volt with a period T and 50% duty cycle) with itself. The result is a periodic triangular pulse $y(t) = x(t) \star x(t)$ (same period T, 100% duty cycle, and height $= T/2$). The exponential Fourier coefficients of $x(t)$ and $y(t)$ are given below.

$$X_n = \frac{\sin(\frac{\pi n}{2})}{\pi n} \qquad Y_n = T \left[\frac{\sin(\frac{\pi n}{2})}{\pi n} \right]^2 = T \times X_n^2$$

Note the appearance of the multiplier T in Y_n. This is in agreement with the convolution property of the Fourier analysis, to be discussed in Chapter 8.

Impulse Train Revisited

A periodic unit-impulse train with period T and its exponential Fourier series coefficients are (see Example 7.7)

$$x(t) = \sum_{n=-\infty}^{\infty} \delta(t - nT)$$

$$X_n = \frac{1}{T}, \quad -\infty < n < \infty$$

$$\sum_{n=-\infty}^{\infty} \delta(t - nT) = \frac{1}{T} \sum_{n=-\infty}^{\infty} e^{j2\pi nt/T}$$

The Fourier coefficients are equal to $1/T$ for all n. Since the impulse is the limit of a narrow pulse with unit area, we derive the above result as the limit of the Fourier coefficients of a narrowing rectangular pulse train. We start with a periodic rectangular pulse $x(t)$ of unit height and duration τ. We express its Fourier coefficients X_n in terms of the duty cycle $\alpha = \tau/T$

$$X_n = \frac{\sin(\alpha \pi n)}{\pi n}$$

and find their limit when $\tau \to 0$.

First consider the case where the pulse duration τ becomes small but the height remains the same. For $\alpha n \ll 1$, we have $\sin(\alpha \pi n) \approx \alpha \pi n$ and $X_n = \alpha$. Therefore,

$$x(t) \approx 2\alpha \, [0.5 + \cos(\omega t) + \cos(2\omega t) + \cos(3\omega t) + \cos(4\omega t) + \cdots]$$

where $\omega = 2\pi/T$. The magnitude spectrum is diminished but remains flat for a wide frequency range. Compare with part c in Example 7.3.

Now consider the case where the pulse duration τ becomes small but the height is increased by a factor $1/\tau$ so that the pulse area remains the same. The exponential Fourier coefficients are then

$$X_n = \frac{\sin(\pi n \frac{\tau}{T})}{\pi n \tau} = \frac{\sin(\pi \alpha n)}{\pi n \tau}$$

where $\alpha = \tau/T$ is the duty cycle. For $\alpha n \ll 1$, we have $\sin(\pi \alpha n) \approx \pi \alpha n$ and $X_n = 1/T$. In the limit when $\tau \to 0$, the waveform becomes a train of unit impulses every T seconds and all its exponential Fourier coefficients become equal to $1/T$.

$$x(t) = \sum_{n=-\infty}^{\infty} \delta(t - nT) \iff X_n = \frac{1}{T}$$

From the above pair we find

$$\sum_{n=-\infty}^{\infty} \delta(t - nT) = \frac{1}{T} \sum_{n=-\infty}^{\infty} e^{j2\pi nt/T}$$

A similar result may be obtained from the limit behavior of other pulse trains that progressively narrows, becoming an impulse train.

7.13   Convergence of the Fourier Series

The Fourier series converges to $x(t)$ whenever $x(t)$ is continuous. If $x(t)$ has a finite discontinuity at t, its Fourier series converges to the mid-value between $x(t^-)$ and $x(t^+)$, which is the average of the limits from the left and right sides. In other words,

$$\lim_{N \to \infty} \sum_{n=-N}^{N} X_n e^{j2\pi nt/T} = \frac{x(t^-) + x(t^+)}{2}$$

This, however, does not mean that within the discontinuity region the difference between the sum and $x(t)$ (the actual value) or $\frac{x(t^-)+x(t^+)}{2}$ [the average value of $x(t^-)$ and $x(t^+)$] could be reduced to a desired value by increasing N. The convergence is not uniform.

7.14 Finite Sum

The Fourier series coefficients minimize the average power (or the rms) of difference between $x(t)$ and the finite sum of its harmonics. Let

$$\psi(t) = \sum_{n=-N}^{N} \Psi_n e^{j2\pi nt/T}$$

be a finite sum of harmonics of $x(t)$ with arbitrary coefficients Ψ_n. Let the energy of the difference between $x(t)$ and $\psi(t)$ over one period be defined by

$$\epsilon = \int_{t_0}^{t_0+T} |x(t) - \psi(t)|^2 \, dt$$

The above measure is a function of the coefficients $\{\Psi_n\}$ which constitute a set of variables. It may be shown that given N, ϵ is at a minimum when $\Psi_n = X_n$, in which case

$$\psi(t) = \sum_{n=-N}^{N} X_n e^{j2\pi nt/T}$$

To show this, we express ϵ in terms of coefficients Ψ

$$\epsilon = \int_{-T/2}^{T/2} \left| x(t) - \sum_{n=-N}^{N} \Psi_n e^{j2\pi nt/T} \right|^2 \, dt$$

To find the set of Ψ_n which minimizes ϵ, noting that $x(t)$ is a real-valued function, we get

$$\frac{d\epsilon}{d\Psi_n} = 0 \quad \text{for } -N \le n \le N$$

Noting that $x(t)$ is a real-valued function

$$\frac{d\epsilon}{d\Psi_n} = \frac{d}{d\Psi_n} \left[\int_{-T/2}^{T/2} \left| x(t) - \sum_{n=-N}^{N} \Psi_n e^{j2\pi kt/T} \right|^2 \, dt \right] = 0$$

which results in

$$\int_{-T/2}^{T/2} \left[x(t) - \sum_{k=-N}^{N} \Psi_k e^{j2\pi kt/T} \right] e^{j2\pi nt/T} \, dt = 0, \quad -N \le n \le N$$

Separating the two parts we have

$$\int_{-T/2}^{T/2} x(t) e^{j2\pi nt/T} \, dt = \int_{-T/2}^{T/2} e^{j2\pi nt/T} \left[\sum_{k=-N}^{N} \Psi_k e^{j2\pi kt/T} \right] \, dt$$

The aim is to find Ψ_n so that the right-side integral is equal to the left-side integral. Let the integral on the right side be called I_n. To evaluate I_n we exchange the order of integration and summation to get

$$I_n = \sum_{k=-N}^{N} \Psi_k \int_{-T/2}^{T/2} e^{j2\pi(n+k)t/T} dt = \sum_{k=-N}^{N} T\Psi_k \frac{\sin(n+k)\pi}{(n+k)\pi}$$

$$= \sum_{k=-N}^{N} T\Psi_k \delta(n+k) = T\Psi_{-n}$$

Therefore,

$$\int_{-T/2}^{T/2} x(t)e^{-j2\pi nt/T} dt = T\Psi_n$$

$$\Psi_n = \frac{1}{T} \int_{-T/2}^{T/2} x(t)e^{-j2\pi nt/T} dt$$

This shows that for the average power of difference between $x(t)$ and the finite sum to be at a minimum, the coefficients of the finite sum should be the coefficients of the Fourier series. The maximum of the difference, however, is not reduced.

7.15   Gibbs' Phenomenon

Let $x(t)$ be a periodic rectangular pulse train with period T and pulse width τ. The Fourier series expansion of $x(t)$ was obtained in sections 7.6 and 7.12. Consider the finite sum

$$\hat{x}(t) = \sum_{n=-M}^{M} X_n e^{j2\pi nt/T}$$

where X_n is the Fourier coefficient of $x(t)$. $\hat{x}(t)$ is obtained from a finite segment of the Fourier expansion, selected by a uniform window in the frequency domain. Examine the difference $x(t) - \hat{x}(t)$ as a function of M. As M increases, the difference approaches zero, except at the discontinuity points of $x(t)$ where it persists in the form of horns. See details in Figure 7.10. This is a special case of a broader effect called Gibbs' phenomenon. (In the case of Fourier series we already have examined the effect for a sawtooth and rectangular pulse train.) The mathematical reasoning behind Gibbs' phenomenon will be discussed in section 8.22 of Chapter 8 on the Fourier transform.

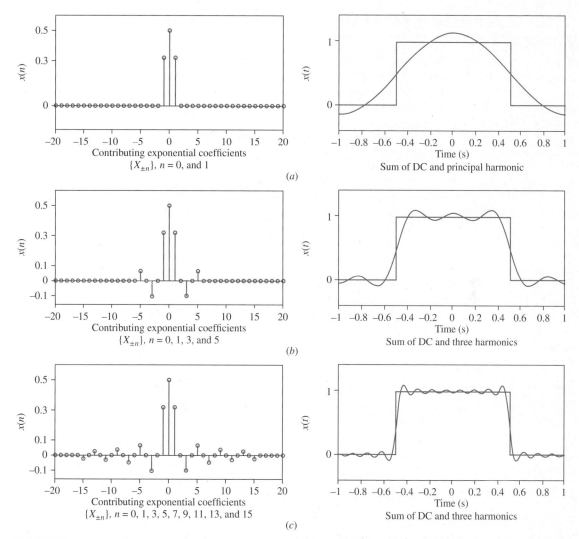

FIGURE 7.10 Gibbs' Phenomenon. The sum of a finite number of harmonics, N, of a rectangular pulse train converges toward the value of the pulse as the number of harmonics increases, except at the pulse edges where the horns persist (due to Gibbs' phenomenon) as illustrated in this figure. The left column shows frequency representations of finite numbers of harmonics of a square pulse [$N = 1$, 3, and 8, in *(a)*, *(b)*, and *(c)*, respectively]. The sum of harmonics is shown by the time functions in the right column. Increasing the number of harmonics from $N = 1$ (*top row*) to $N = 3$ (*middle row*) and 8 (*bottom row*) does not reduce the approximation error at the pulse edges. For a treatment of the theoretical foundation of Gibbs' phenomenon see Chapter 8.

7.16 Extension of Fourier Series to Transforms

Fourier series analysis may be extended to nonperiodic signals leading to the Fourier transform. For an intuitive approach to the Fourier transform, consider the periodic signal $x(t)$ with period T which satisfies the Dirichlet conditions. $x(t)$ and its Fourier coefficients are related by

$$X_n = \frac{1}{T} \int_{-\frac{T}{2}}^{\frac{T}{2}} x(t) e^{-j2\pi nt/T} dt, \quad -\infty < n < \infty$$

$$x(t) = \sum_{-\infty}^{\infty} X_n e^{j2\pi nt/T}$$

Now define a time-limited pulse $h(t)$ to represent one cycle of $x(t)$:

$$h(t) = \begin{cases} x(t), & -\frac{T}{2} < t < \frac{T}{2} \\ 0, & \text{elsewhere} \end{cases}$$

The Fourier coefficients X_n may be written in terms of $h(t)$.

$$X_n = \frac{1}{T} \int_{-\infty}^{\infty} h(t) e^{-j2\pi nt/T} dt, \quad -\infty < n < \infty$$

Define $f = n/T$ for $-\infty < n < \infty$. Note that f is a discrete variable with $\Delta f = 1/T$. Define

$$X(f) = T X_n = \int_{-\infty}^{\infty} h(t) e^{-j2\pi ft} dt$$

Therefore,

$$x(t) = \sum_{n=-\infty}^{\infty} X_n e^{j2\pi nt/T} = \sum_{f=-\infty}^{\infty} X(f) e^{j2\pi ft} \Delta f$$

Note that the summation is still taken over $f = \frac{n}{T}$, which is a discrete variable. If we allow T to go to ∞, then $x(t) = h(t)$ and $X_n = 0$. However, the variable $f = \frac{n}{T}$ and the product $X(f) = T X_n$ may converge to finite values. As $T \to \infty$, we have

$x(t) \to h(t)$ and $X(f) \to H(f)$,

$\Delta f \to df$, f becomes continuous and the sum becomes an integral, $\sum \to \int$

As a result,

$$H(f) = \int_{-\infty}^{\infty} h(t) e^{-j2\pi ft} dt$$

$$h(t) = \int_{-\infty}^{\infty} H(f) e^{j2\pi ft} df$$

The pair $h(t)$ and $H(f)$ are called a Fourier transform pair. Note that in the above discussion we required $h(t)$ to satisfy the Dirichlet conditions. The above concepts are illustrated in Figure 7.11.

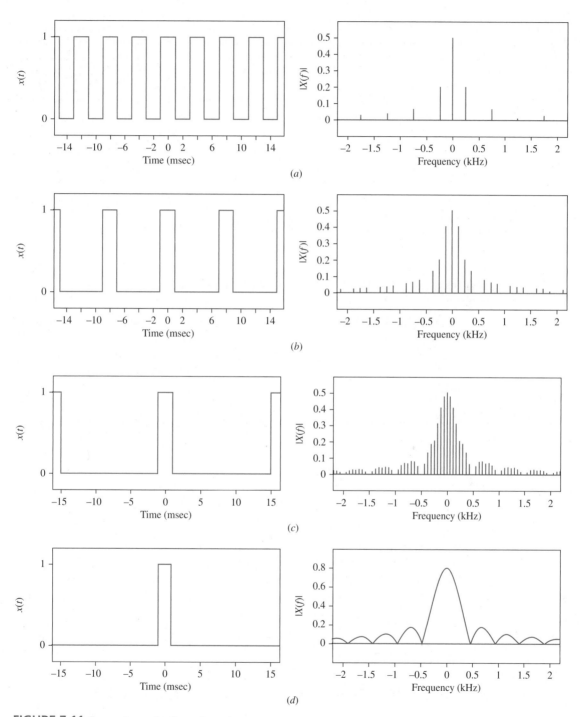

FIGURE 7.11 See next page for figure legend.

FIGURE 7.11 Four periodic pulse trains $x(t)$ are shown in the left column along with their representations in the frequency domain in the right column, in the form of discrete lines. The periods of the pulses are $T = 4$, 8, 16, and 400 msec in *(a)*, *(b)*, *(c)*, and *(d)*, respectively. In all cases the pulse width is $\tau = 2$ msec. The plots in the right column, with their abscissa in Hz, show the magnitude of exponential Fourier coefficients ($|X(f_n)|$, where $f_n = n/T$), also called spectral lines. The spectral lines represent X_n. They are located at $f = n/T$. As the period increases, the spectral lines move closer to each other. (They also become smaller in magnitude. In this figure, the plots of X_n are normalized to the same level.) The resolution in the frequency domain is the inverse of the period in the time domain. In this figure:

$$T = 4 \text{ msec} \quad \Delta f = 250 \text{ Hz}$$
$$T = 8 \text{ msec} \quad \Delta f = 125 \text{ Hz}$$
$$T = 16 \text{ msec} \quad \Delta f = 62.5 \text{ Hz}$$
$$T = 400 \text{ msec} \quad \Delta f = 2.5 \text{ Hz}$$

The envelope of the line spectrum $X(f_n) = TX(n/T)$, however, remains relatively unchanged. At $T = \infty$, the frequency f becomes continuous and the limit of the product of TX_n (Fourier series coefficients) becomes $X(f)$, which is the Fourier transform. See the next section.

7.17 Envelope of Fourier Coefficients

In this section we will show that the Fourier transform of one period of a periodic signal is T times the envelope of Fourier series coefficients of the periodic signal, with $f = n/T$. A periodic signal $x(t)$ with period T may be considered the result of convolving a time-limited pulse $h(t)$ with an infinite train of unit impulses arriving every T seconds [to be called $s(t)$]. See Figure 7.12.

$$x(t) = h(t) \star s(t) = h(t) \star \sum_{n=-\infty}^{\infty} \delta(t - nT) = \sum_{n=-\infty}^{\infty} h(t - nT)$$

where $h(t)$ is one cycle of $x(t)$:

$$h(t) = \begin{cases} x(t), & t_0 < t < t_0 + T \\ 0, & \text{elsewhere} \end{cases}$$

The convolution in time between $h(t)$ and $s(t)$ can be performed as a product in the frequency domain; that is, $X(f) = H(f)S(f)$, where $H(f)$ and $S(f)$ are Fourier transforms of $h(t)$ and $s(t)$, respectively. The transform of the impulse train is

$$S(f) = \frac{1}{T} \sum_{n=-\infty}^{\infty} \delta(f - nf_0)$$

where $f_0 = \frac{1}{T}$. Therefore,

$$X(f) = \frac{H(f)}{T} \sum_{n=-\infty}^{\infty} \delta(f - nf_0) = \frac{1}{T} \sum_{n=-\infty}^{\infty} H(nf_0)\delta(f - nf_0) = \sum_{n=-\infty}^{\infty} X_n \delta(f - nf_0)$$

It is seen that the Fourier coefficients of the periodic signal are samples of the Fourier transform of a single cycle taken every f_0 Hz and divided by T.

$$X_n = \frac{1}{T} H(f)\big|_{f=nf_0} = \frac{1}{T} H(f)\big|_{f=n/T}$$

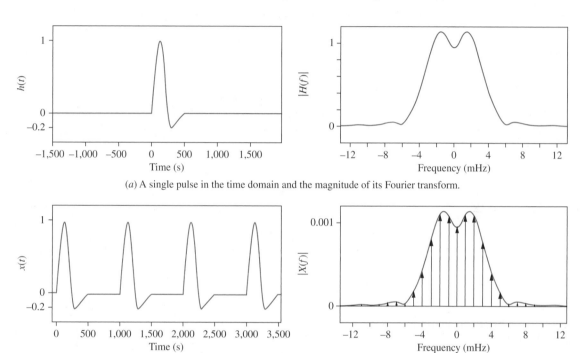

(*a*) A single pulse in the time domain and the magnitude of its Fourier transform.

(*b*) A periodic pulse train and the magnitude of its Fourier transform with its envelope.

FIGURE 7.12 A single pulse and the normalized magnitude of its Fourier transform are shown in *(a)*. The Fourier transform of the periodic pulse in *(b)* is made of impulses in the frequency domain. The envelope of the tip of the impulses is the Fourier transform of the single pulse multiplied by $1/T$. Time functions are in the left column and transforms in the right. The period of the time function in this figure is $T = 1000$ seconds. It results in the frequency resolution $\Delta f = 1/T = 1$ mHz.

7.18 Concluding Remarks

Through the Fourier series we express a periodic signal by an ensemble of orthogonal functions called its harmonics. By knowing the frequency behavior of an LTI system (to be called the frequency response and presented in Chapters 8 and 9), we can examine the effect of passing each harmonic, and then the entire ensemble, through the system. This chapter has emphasized the exponential form of the expansion because of its close relationship with the Fourier transform. In fact, as will be seen in the next chapter, by introducing impulse functions to represent periodic time functions in the frequency domain, we will develop a unified form of the Fourier transform that will be applicable to all signals. Computer software can be used to supplement analytical solutions. Such a use is strongly recommended.

7.19 Problems

Notations

The Fourier series expansion of a waveform with period T is

$$x(t) = \frac{1}{2}a_0 + \sum_{n=1}^{\infty} a_n \cos(n\omega t) + \sum_{n=1}^{\infty} b_n \sin(n\omega t), \text{ where } \omega = \frac{2\pi}{T},$$

$$a_0 = \frac{2}{T} \int_0^T x(t)dt, \quad a_n = \frac{2}{T} \int_0^T x(t) \cos(n\omega t)dt,$$

$$b_n = \frac{2}{T} \int_0^T x(t) \sin(n\omega t)dt, \ n = 1, 2, \ldots$$

in trigonometric form, and

$$x(t) = \sum_{n=-\infty}^{\infty} X_n e^{j2\pi nt/T}, \ X_n = \frac{1}{T} \int_0^T x(t)e^{-j2\pi nt/T}dt, \ -\infty \le n \le \infty, \ n \text{ an integer},$$

in exponential form.

Relation Between Fourier Series Coefficients in Exponential and Trigonometric Forms

Start with the exponential expansion

$$x(t) = \sum_{n=-\infty}^{\infty} X_n e^{\frac{j2\pi nt}{T}} = X_0 + \sum_{n=1}^{\infty} \left[X_n e^{\frac{j2\pi nt}{T}} + X_{-n} e^{-\frac{j2\pi nt}{T}} \right]$$

Note that X_{-n} is the complex conjugate of X_{-n} [when $x(t)$ is a real function of t as in our case]. Therefore,

$$X_n = \mathcal{RE}\{X_n\} + j\mathcal{IM}\{X_n\} \quad \text{and} \quad X_{-n} = \mathcal{RE}\{X_n\} - j\mathcal{IM}\{X_n\}$$

$$x(t) = X_0 + \sum_{n=1}^{\infty} [\mathcal{RE}\{X_n\} + j\mathcal{IM}\{X_n\}] e^{\frac{j2\pi nt}{T}} + \sum_{n=1}^{\infty} [\mathcal{RE}\{X_n\} - j\mathcal{IM}\{X_n\}] e^{-\frac{j2\pi nt}{T}}$$

$$= X_0 + \sum_{n=1}^{\infty} 2\mathcal{RE}\{X_n\} \cos\left(\frac{2\pi nt}{T}\right) + \sum_{n=1}^{\infty} 2j\mathcal{IM}\{X_n\} \sin\left(\frac{2\pi nt}{T}\right)$$

$$= \frac{a_0}{2} + \sum_{n=1}^{\infty} a_n \cos\left(\frac{2\pi nt}{T}\right) + \sum_{n=1}^{\infty} b_n \sin\left(\frac{2\pi nt}{T}\right)$$

$$a_0 = 2X_0, \ a_n = 2\mathcal{RE}\{X_n\} \quad \text{and} \quad b_n = 2j\mathcal{IM}\{X_n\}$$

Solved Problems

1. a. Find the exponential Fourier series expansion of the sawtooth waveform shown in Figure 7.13(a).

b. Describe how Fourier coefficients of the waveforms shown in Figure 7.13(b) through (f) may be found from the coefficients of the waveform in Figure 7.13(a).

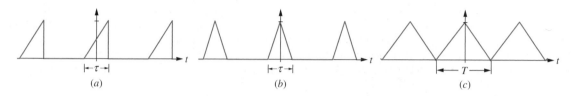

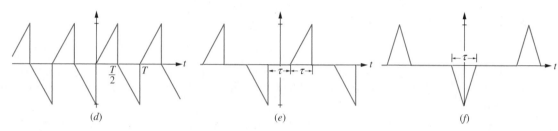

FIGURE 7.13

Solution

a. Let T be the period and $\pm V_0$ represent the peak-to-peak value of the waveforms.

$$X_n = \frac{1}{T}\int_{-\frac{\tau}{2}}^{\frac{\tau}{2}} x(t)e^{-\frac{j2\pi nt}{T}}\,dt = \frac{1}{T}\int_{-\frac{\tau}{2}}^{\frac{\tau}{2}} \left(\frac{V_0}{2} + \frac{V_0}{\tau}t\right)e^{-\frac{j2\pi nt}{T}}\,dt$$

$$= \frac{V_0}{2T}\int_{-\frac{\tau}{2}}^{\frac{\tau}{2}} e^{-\frac{j2\pi nt}{T}}\,dt + \frac{V_0}{T\tau}\int_{-\frac{\tau}{2}}^{\frac{\tau}{2}} te^{-\frac{j2\pi nt}{T}}\,dt$$

$$= \frac{V_0}{2}\frac{\sin(\frac{\pi n\tau}{T})}{\pi n} + \frac{V_0}{T\tau}\int_{-\frac{\tau}{2}}^{\frac{\tau}{2}} te^{-\alpha t}\,dt, \text{ where } \alpha = \frac{j2\pi n}{T}$$

But

$$\int te^{-\alpha t}\,dt = -\frac{1}{\alpha}e^{-\alpha t}\left(t + \frac{1}{\alpha}\right),$$

Therefore,

$$X_n = \frac{V_0}{2}\frac{\sin(\frac{\pi n\tau}{T})}{\pi n} - \frac{V_0}{\alpha T\tau}\left[e^{-\alpha t}\left(t + \frac{1}{\alpha}\right)\right]_{t=-\frac{\tau}{2}}^{\frac{\tau}{2}}$$

$$= \frac{V_0}{2}\frac{\sin(\frac{\pi n\tau}{T})}{\pi n} + j\frac{V_0}{2\pi n}\left[\cos\left(\frac{\pi n\tau}{T}\right) - \frac{\sin\left(\frac{\pi n\tau}{T}\right)}{\frac{\pi n\tau}{T}}\right]$$

In summary

$$X_n = \mathcal{RE}\{X_n\} + j\mathcal{IM}\{X_n\}$$

$$\mathcal{RE}\{X_n\} = \frac{V_0}{2} \frac{\sin(\frac{\pi n \tau}{T})}{\pi n}, \quad \mathcal{IM}\{X_n\} = \frac{V_0}{2\pi n}\left[\cos\left(\frac{\pi n \tau}{T}\right) - \frac{\sin\left(\frac{\pi n \tau}{T}\right)}{\frac{\pi n \tau}{T}}\right]$$

b. Let the waveform in Figure 7.13(a) be $x(t)$ and those in Figure 13(b) to (f) be called $y_b(t)$, $y_c(t)$, $y_d(t)$, $y_e(t)$, $y_b(t)$, and $y_f(t)$, respectively. To obtain Fourier coefficients of the waveforms shown in Figure 7.13(b), first construct $x(t + \tau/2) + x(-t + \tau/2)$, a triangular waveform with a base of 2τ. Find its Fourier coefficients by applying time shift, time-reversal, and superposition properties to X_n. Then, change τ to $\tau/2$ to find Y_b (see problem 2). Then again, in Y_b let $\tau = T$ to find Y_c.

For Figure 7.13(d) to (f) we have:

$$y_d(t) = x(t - \tau/2) - x(t - 3\tau/2), \text{ and } \tau = T/2.$$

$$y_e(t) = x(t - \tau) - x(t - 3\tau)$$

$$y_f(t) = y_b(t - T/2) - y_b(t)$$

2. Apply superposition, time-shift, and time-reversal properties of the Fourier series to the results of problem 1 to find the Fourier coefficients of the triangular pulse train shown in Figure 7.13(b) (with the base $= \tau$ and height $= V_0$).

Solution

Let $x(t)$ be the real-valued periodic sawtooth pulse train shown in Figure 7.13(a). Then, $x(t + \tau/2) + x(-t + \tau/2)$ is a triangular pulse train with the base $= 2\tau$ and height $= V_0$, similar to Figure 7.13(b) except for a base that is twice as wide. To find the Fourier coefficints of the waveform of Figure 7.13(b), take the following steps:

$$x(t) \Longleftrightarrow \mathcal{RE}\{X_n\} + j\mathcal{IM}\{X_n\}$$

Where,

$$\mathcal{RE}\{X_n\} = \frac{V_0}{2} \frac{\sin(\frac{\pi n \tau}{T})}{\pi n} \quad \text{and} \quad \mathcal{IM}\{X_n\} = \frac{V_0}{2\pi n}\left[\cos\left(\frac{\pi n \tau}{T}\right) - \frac{\sin\left(\frac{\pi n \tau}{T}\right)}{\frac{\pi n \tau}{T}}\right]$$

$$x(t + \tau/2) \Longleftrightarrow [\mathcal{RE}\{X_n\} + j\mathcal{IM}\{X_n\}] e^{j\pi n \tau/T}$$

$$x(-t + \tau/2) \Longleftrightarrow [\mathcal{RE}\{X_n\} - j\mathcal{IM}\{X_n\}] e^{-j\pi n \tau/T}$$

$$x(t + \tau/2) + x(-t + \tau/2) \Longleftrightarrow \mathcal{RE}\{X_n\} \left(e^{j\pi n \tau/T} + e^{-j\pi n \tau/T}\right) + j\mathcal{IM}\{X_n\}\left(e^{j\pi n \tau/T} - e^{-j\pi n \tau/T}\right)$$

$$= 2\mathcal{RE}\{X_n\}\cos(\pi n \tau/T) - 2\mathcal{IM}\{X_n\}\sin(\pi n \tau/T)$$

$$= \frac{V_0}{\pi n}\sin\left(\frac{\pi n \tau}{T}\right)\cos\left(\frac{\pi n \tau}{T}\right) - \frac{V_0}{\pi n}\left[\cos\left(\frac{\pi n \tau}{T}\right)\sin\left(\frac{\pi n \tau}{T}\right) - \frac{\sin^2\left(\frac{\pi n \tau}{T}\right)}{\frac{\pi n \tau}{T}}\right]$$

$$= \frac{V_0 T}{\tau}\left[\frac{\sin\left(\frac{\pi n \tau}{T}\right)}{\pi n}\right]^2$$

Finally, change τ to $\tau/2$ to find the Fourier coefficients of the waveform of Figure 7.13(b):

$$Y_b = \frac{2V_0 T}{\tau}\left[\frac{\sin\left(\frac{\pi n \tau}{2T}\right)}{\pi n}\right]^2$$

This is in agreement with the result obtained in Example 7.6.

3. Find the first 10 coefficients X_n, $n = 1, 2 \ldots, 10$ in the exponential Fourier series expansion of a 1-kHz periodic rectangular pulse train with a base-to-peak voltage of 0 to k volts and a pulse width $k/10$ msec for (a) $k = 1$, and (b) $k = 2$. Choose time reference so that the pulse train is represented by an even function.

Solution

$$X_n = \frac{k}{\pi n} \sin\left(\frac{\pi n k}{10}\right)$$

a. $k = 1$ $X_n = \frac{1}{\pi n} \sin\left(\frac{\pi n}{10}\right) = 10^{-4}\{1000, 984, 935, 858, 757, 637, 505, 368, 234, 109, 0\}$

b. $k = 2$ $X_n = \frac{2}{\pi n} \sin\left(\frac{\pi n}{5}\right) = 10^{-4}\{4000, 3742, 3027, 2018, 935, 0, -624, -865, -757, -416, 0\}$

4. Matlab file for Example 7.1 is given below. Add commands to obtain and save all plots in Figure 7.10. Run the program and reproduce the plots.

```
T=2; w=2*pi/T; N=1000; t=linspace(-T,T,2*N*T); z=0*t; tt=0;
for i=1:2*N*T/4;
tt(i)=t(i)+T;
end
for i=2*N*T/4+1:6*N*T/4;
tt(i)=t(i);
end
for i=6*N*T/4+1:2*N*T;
tt(i)=t(i)-T;
end
x1=(2/pi)*sin(w*t);
figure(1)
plot(t,x1,'r','LineWidth',2); hold on
plot(t,z,'b','LineWidth',1); plot(t,tt,'b','LineWidth',2); hold off
axis([-T/2 T/2 -1.2 1.2]); xlabel('Time(s)'); ylabel('x(t)');
title('Approximating a periodic sawtooth by the first harmonic.'); grid
print -dpsc  Ch7_Ex1_Fig1_a1.eps
e1=x1-tt; E1=e1.^(2); AE1=sum(E1)/(2*N*T);
%
x2=(2/pi)*(sin(w*t)-(1/2)*sin(2*w*t));
e2=x2-tt; E2=e2.^(2); AE2=sum(E2)/(2*N*T);
 %
x3=(2/pi)*(sin(w*t)-(1/2)*sin(2*w*t)+(1/3)*sin(3*w*t));
e3=x3-tt; E3=e3.^(2); AE3=sum(E3)/(2*N*T);
%
x4=(2/pi)*(sin(w*t)-(1/2)*sin(2*w*t)+(1/3)*sin(3*w*t)-(1/4)*sin(4*w*t));
e4=x4-tt; E4=e4.^(2); AE4=sum(E4)/(2*N*T);
 %
x5=(2/pi)*(sin(w*t)-(1/2)*sin(2*w*t)+(1/3)*sin(3*w*t)-(1/4)*sin(4*w*t)+(1/5)
*sin(5*w*t));
e5=x5-tt; E5=e5.^(2); AE5=sum(E5)/(2*N*T);
```

```
%
AE=[AE1 AE2 AE3 AE4 AE5]
```

5. Matlab file for Example 7.2 is given below. Add commands to obtain and save all plots in Figure 7.2. Run the program and reproduce the plots.

```
T=2; w=2*pi/T; N=1000; t=linspace(-T,T,2*N*T); z=0*t; p=0*t;
for i=1:2*N*T/8;
 p(i)=2;
end
for i=6*N*T/8+1:10*N*T/8 ;
 p(i)=2;
end
for i=14*N*T/8+1:2*N*T;
p(i)=2;
end
tt=p-1;
%
x1=  (4/pi)*cos(w*t) ;
x2=  (4/pi)*(cos(w*t)-(1/3)*cos(3*w*t)) ;
x3=  (4/pi)*(cos(w*t)-(1/3)*cos(3*w*t)+(1/5)*cos(5*w*t)) ;
x4=  (4/pi)*(cos(w*t)-(1/3)*cos(3*w*t)+(1/5)*cos(5*w*t)-(1/7)*cos(7*w*t)) ;
x5=  (4/pi)*(cos(w*t)-(1/3)*cos(3*w*t)+(1/5)*cos(5*w*t)-(1/7)*cos(7*w*t)
+(1/9)
*cos(9*w*t)) ;
%
e1=x1-tt; E1=e1.^(2); AE1=sum(E1)/(2*N*T);
e2=x2-tt; E2=e2.^(2); AE2=sum(E2)/(2*N*T);
e3=x3-tt; E3=e3.^(2); AE3=sum(E3)/(2*N*T);
e4=x4-tt; E4=e4.^(2); AE4=sum(E4)/(2*N*T);
e5=x5-tt; E5=e5.^(2); AE5=sum(E5)/(2*N*T);
AE=[AE1 AE2 AE3 AE4 AE5]
```

Chapter Problems

6. Find the trigonometric Fourier series for a 1-kHz square wave with a peak-to-peak voltage spanning -1 to 1 volts (0 DC value, 50% duty cycle). Choose the time origin so that the series contains (a) cosine terms only and (b) sine terms only.

7. The four periodic signals shown in Figure 7.14(*a*), (*b*), (*c*), and (*d*) are to be expanded in trigonometric Fourier series. For each signal determine if statements 1, 2, and 3 given below are true or false.

Statement 1. The time origin may be chosen such that $a_n = 0$, $n \geq 1$.

Statement 2. The time origin may be chosen such that $b_n = 0$, $n \geq 0$.

Statement 3. The baseline (zero level) may be chosen such that $a_0 = 0$.

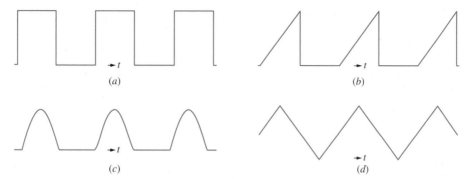

(a)   (b)

(c)   (d)

FIGURE 7.14 (For Problem 7)

8. Consider a 1-kHz periodic rectangular pulse signal. For each case given below, determine pulse width τ in msec such that the signal contains no harmonics at the listed frequencies (k is an integer).

    a. No harmonics at $2k$ kHz.

    b. No harmonics at $4k$ kHz.

    c. No harmonics at $8k$ kHz.

9. a. Represent the waveform $x(t) = \sin t + \sin(t + \pi/4)$ by its trigonometric Fourier series.

    b. Let $y(t) = x(t - t_0)$. Specify a value for t_0 so that the trigonometric representation of $y(t)$ contains (i) a_n or (ii) b_n only.

10. Shift $x(t) = A \sin \omega_0 t + B \sin(\omega_0 t - \theta)$ by t_0 seconds so that its trigonometric Fourier series contains (i) a_n or (ii) b_n only. In each case determine t_0 as a function of A, B, ω_0, and θ.

11. Describe the waveforms shown in Figure 7.15 in terms of a periodic rectangular waveform and obtain their exponential Fourier coefficients. Assume period T and a peak-to-peak value of ± 1.

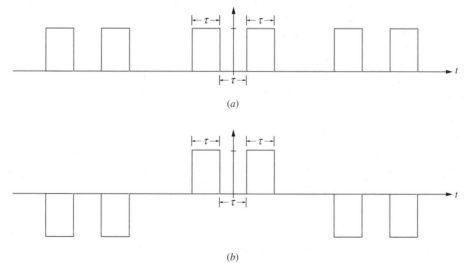

(a)

(b)

FIGURE 7.15 (For Problem 11)

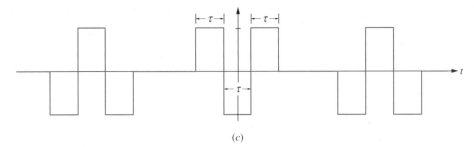

(c)

FIGURE 7.15 (*Continued*)

12. Find the trigonometric Fourier coefficients of a periodic rectangular voltage with $T = 100 \ \mu s$, $\tau = 40 \ \mu s$, and a peak-to-peak value of ± 1 volt.

13. Effect of pulse width. Find the exponential Fourier coefficients of a periodic rectangular waveform with period $T = 1$ ms and a peak-to-peak value of ± 1 volt for the following pulse width τ:

   a. 500 μs.
   b. 200 μs.
   c. 100 μs.
   d. 10 μs.
   e. 1 μs.

14. Effect of pulse width. (i) Find the exponential coefficients of Fourier expansion of a periodic rectangular waveform with $T = 1$ ms and a peak-to-peak value of $\pm V_0$ volt for the following cases:

   a. $\tau = 500 \ \mu s$ and $V_0 = 1$ volt.
   b. $\tau = 200 \ \mu s$ and $V_0 = 2.5$ volts.
   c. $\tau = 100 \ \mu s$ and $V_0 = 5$ volts.
   d. $\tau = 10 \ \mu s$ and $V_0 = 50$ volts.
   e. $\tau = 1 \ \mu s$ and $V_0 = 500$ volts.

(ii) In each case determine which harmonic is the first to be zero. Relate it to the pulse width τ.

15. Effect of period. Find the exponential coefficients of Fourier expansion of a periodic rectangular waveform with pulse duration of $\tau = 1 \ \mu s$ and a peak-to-peak value of ± 1 volt, for the period T being:

   a. 2 μs.
   b. 5 μs.
   c. 10 μs.
   d. 100 μs.
   e. 200 μs.

16. Effect of period. (i) Find the exponential Fourier coefficients of a periodic rectangular waveform with pulse width $\tau = 1 \ \mu s$ and a peak-to-peak value of $\pm V_0$ volt for the following cases:

   a. $T = 2 \ \mu s$ and $V_0 = 1$ volt.
   b. $T = 5 \ \mu s$ and $V_0 = 2.5$ volts.
   c. $T = 10 \ \mu s$ and $V_0 = 5$ volts.
   d. $T = 100 \ \mu s$ and $V_0 = 50$ volts.
   e. $T = 200 \ \mu s$ and $V_0 = 100$ volts.

(ii) In each case determine which harmonic is the first to be zero. Relate it to the pulse width τ.

17. Express, as a function of τ, the exponential Fourier coefficients of a periodic rectangular waveform with $T = 1$ s, a base-to-peak value of V_0 volts and pulse width $\tau = 1/V_0$. Find $\lim_{\tau \to 0} X_n$.

18. Find the exponential Fourier series for a 10-kHz periodic rectangular waveform with a base-to-peak voltage of 0 to 2 volts and a duty cycle of δ, changing from 10% to 90% in step of 10%. In all cases choose the time origin so that the series results in even functions.

19. Repeat problem 18 but this time with a base-to-peak value of $V_0 = 1/\tau$ so that the area of the pulse remains unity independent of the duty cycle.

20. Find the exponential Fourier series coefficients of a periodic even-square pulse signal (50% duty cycle, zero DC value). Then shift the signal by a half-period and add its series coefficients to those of the original signal. Verify that the sum is zero for all values of n.

21. Find the exponential Fourier series coefficients of a periodic square wave voltage having a 1 V base-to-peak value (0.5 V DC level, 50% duty cycle). Then shift the signal by a half-period and add its series coefficients to those of the original signal. Verify that the sum is zero for all n, except at $n = 0$ for which it is 1.

22. Find the exponential Fourier coefficients of a periodic signal $x(t)$ made of rectangular pulses (height $= V_0$, duration $= \tau$, period $= T$). Find and plot the envelope of X_n for

    a. $V_0 = 1$ V, $\tau = 1$ msec, and $T = 2$ msec.
    b. $V_0 = 5$ V, $\tau = 1$ msec, and $T = 10$ msec.
    c. $V_0 = 100$ V, $\tau = 1$ msec, and $T = 200$ msec.

23. Consider a rectangular pulse train with a period of 4 μsec and a base-to-peak value of 1 V. Find the exponential Fourier coefficients of the pulse train when the duty cycle is (a) 25%, (b) 50%, and (c) 75%. In each case evaluate the first seven coefficients X_n, $n = 0, \ldots, 6$.

24. Find the trigonomteric Fourier series coefficients of the periodic signals shown in Figure 7.13(a) to $-f$ for $T = 200$ μs, $\tau = 20$ μs, and a peak-to-peak value of 3 V. In each case anticipate the form of the series from the possible symmetry property of the signal.

25. Find the exponential Fourier coefficients X_n for the periodic rectangular waveform of Figure 7.15(a) with $T = 2$ msec, $\tau = 0.5$ msec, and $V_0 = 1$ V. Evaluate the magnitude and phase of the first three non-zero coefficients X_n, $n \geq 0$. Compute the DC power and average power in the first two nonzero harmonics and give their percentage share of the total power.

26. A periodic rectangular voltage pulse train (frequency $= 1$ kHz, amplitude $= 1$ V, pulse duration $= 1$ μs) is fed into a series RC circuit ($R = 100$ Ω, $C = 10$ nF). Find the DC value and amplitudes and phases of the first six harmonics appearing across the capacitor.

27. a. A 1-kHz periodic rectangular pulse train voltage is applied to a linear time-invariant circuit specified by the frequency response $H(\omega) = 1/(1 + j5 \times 10^{-6}\omega)$. Express the input and output voltages in Fourier series forms $\sum_n A \cos(n\omega_0 t + \alpha)$ and $\sum_n B \cos(n\omega_0 t + \beta)$, respectively.
    b. Find the DC and the first four harmonics of the input and output.

28. Find the Fourier series coefficients of the periodic signals shown in Figure 7.16(a) and (b). Assume period $T = 10$, $\tau = 3$, $\tau_1 = 1$ (all in msec), and $V_0 = V_1 = 1$ V. *Hint*: Use the superposition property of the Fourier series.

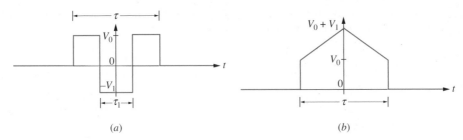

(a)   (b)

FIGURE 7.16 (For Problem 28)

29. Find the exponential Fourier series coefficients of a half-wave rectified sinusoid with a base-to-peak voltage of 0 to V_o volts and frequency f. See Figure 7.17(a). The time origin is chosen such that the waveform is represented by an even function.

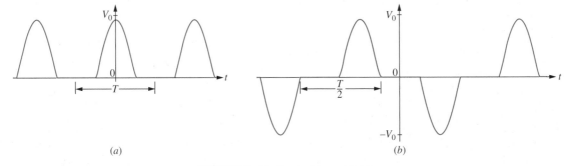

(a)   (b)

FIGURE 7.17 (For Problems 29–31)

30. Repeat problem 29 for (a) a full-wave rectified waveform and (b) the waveform of Figure 7.17(b).

31. Invert the half-wave rectified sinusoid (problem 29) and shift it by a half-period. Then determine its exponential Fourier coefficients and add them to the coefficients of the original signal. Verify that the sum is zero for all n except at $n = 1$, for which it is equal to half of the peak value of the sinusoid.

32. a. Find the exponential Fourier coefficients of a chopped sinusoid described by

$$v(t) = \begin{cases} V_0 \sin\left(\frac{2\pi t}{T} - \frac{\pi}{4}\right), & -\frac{T}{8} < t \le \frac{3T}{8} \\ 0, & \frac{3T}{8} < t \le \frac{7T}{8} \end{cases}, \text{ and } v(t) = v(t - T)$$

   b. Repeat for $|v(t)|$ where $v(t)$ is given in (a).

33. Use a computer to solve this problem.

   a. Compute numerical values of the first 10 exponential Fourier series coefficients of a half-wave rectified circle with a unity radius and period 2. Choose the time origin so that the waveform is represented by an even function.
   b. Repeat for a full-wave rectified circle.

34. Power in a signal is controlled by applying a threshold level λ to a sinusoidal source with a peak value of V_0 volts. The signal is described by

$$x(t) = \begin{cases} \sin \omega t - \lambda, & \text{if } \sin \omega t > \lambda \\ 0, & \text{otherwise} \end{cases}$$

Let $V_0 = \sqrt{2}$ and $\lambda = 1$. Use a computer to obtain numerical values of:

a. the average power in $x(t)$,
b. its exponential Fourier coefficients X_n, $n = 0, 1, \ldots, 9$,
c. the percentage of power in the principal harmonic.

35. A single pulse is made of the partial sinusoid and described by

$$x(t) = \begin{cases} V_0 \cos\left(\frac{\pi t}{T}\right), & \frac{T}{2} - \tau < t < \frac{T}{2} \\ 0, & \text{elsewhere} \end{cases}$$

See Figure 7.18.

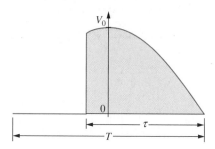

FIGURE 7.18 (For Problem 35)

Find exponential Fourier coefficients of the following periodic functions made of the pulse of part

a. $y_a(t) = \sum\limits_{n=-\infty}^{\infty} x(t - 2nT)$

b. $y_b(t) = \sum\limits_{n=-\infty}^{\infty} (-1)^n x(t - nT)$

Then, in each case find the percentage of power in the principal harmonic.

36. Verify even-symmetry property of the periodic waveforms shown in Figure 7.19 and find their trigonometric Fourier series.

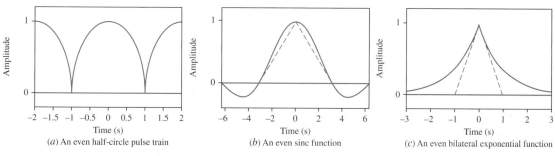

(a) An even half-circle pulse train   (b) An even sinc function   (c) An even bilateral exponential function

FIGURE 7.19 Waveforms with even symmetry: $x(t) = x(-t)$.

37. Verify odd-symmetry property of the periodic waveforms shown in Figure 7.20 and find their trigonometric Fourier series.

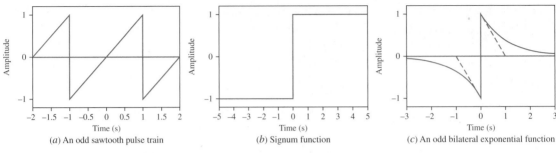

FIGURE 7.20 Waveforms with odd symmetry: $x(t) = -x(-t)$.

38. Verify half-wave symmetry property of the periodic waveforms shown in Figure 7.21 and find their trigonometric Fourier series.

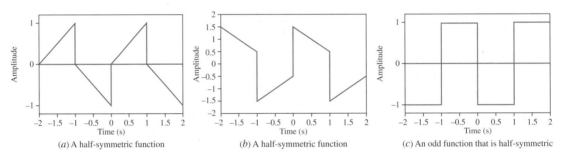

FIGURE 7.21 Waveforms with half-wave symmetry: $x(t) = -x(t + T/2)$.

39. Find the exponential Fourier coefficients of $x(t) = a(t) \cos 2\pi f_c t$, where $f_c = 100$ Hz, for the two cases of $a(t)$ given below. In each case sketch the two-sided spectrum, label important points, and determine whether $x(t)$ is band-limited.

a. $a(t) = \cos 2\pi f_1 t + \sin 2\pi f_2 t$, $f_1 = 10$ Hz and $f_2 = 20$ Hz.

b. $a(t) = 3 + 2\cos 2\pi f_1 t + \cos 2\pi f_2 t$, $f_1 = 15$ Hz and $f_2 = 30$ Hz.

40. The exponential Fourier series coefficients of a periodic signal $x(t)$ are

$$X_n = \begin{cases} 1.57, & n = 0 \\ 0, & n \text{ even and } \neq 0 \\ \frac{(-1)^{(n-1)/2}}{n}, & n \text{ odd} \end{cases}$$

a. Find the DC power ($n = 0$), the average power in the fundamental frequency ($n = 1$), and the average power in the third harmonic ($n = 3$).

b. Determine $x(t)$ as a function of $\omega_0 t$.

c. Find the total average power $< x^2(t) >$ and the number of harmonics that approximate the total power within a 1% error band.

41. An LTI system has a finite-duration unit-impulse response, lasting from $t = 0$ to τ. Given the exponential Fourier series coefficients of the system's response to a train of unit impulses arriving every τ seconds, can you find the Fourier series coefficients of its response to a train of unit impulses arriving every $T > \tau$ seconds? If the answer is positive, show how that could be done. If the answer is negative, specify what additional information is needed.

42. The exponential Fourier series coefficients of a periodic signal $x(t)$ are $X_n = (1/2)^n$, $-\infty < n < \infty$. Find the DC power ($n = 0$), average power in the fundamental frequency ($n = 1$), and the total average power $< x^2(t) >$ or an approximation of it within an error band of less than 1%.

43. An ideal low-pass filter passes frequencies below 1 kHz without any change in their magnitude or phase, and completely blocks frequencies above 1 kHz. The input to the filter is the periodic rectangular waveform shown in Figure 7.4 with $T = 8$ and $\tau = 2$ in msec. Find the exponential Fourier series coefficients of the input and output signals and their total average powers.

7.20 Project: Computer Explorations in Fourier Analysis

Summary

In this project you will generate periodic time series (sine, periodic rectangular, and triangular pulses, etc.) by sampling continuous-time signals and finding their Fourier transforms. Sampling interval ΔT is called resolution in time. Its inverse is $F_s = 1/\Delta T$ and is called the sampling rate. By applying high sampling rates the discrete-time signals approximate continuous-time signals. Computer-based software packages approximate the transform of a finite segment of the signal, and after frequency scaling, plot it as the signal's Fourier representation. The algorithm is called Discrete Fourier Transform (DFT) or Fast Fourier Transform (FFT). The resolution in frequency is $\Delta f = 1/T = 1/(N\Delta T)$, where N is the number of samples used in the DFT operation (i.e., T is the duration of the signal in time). By taking the DFT of long segments of the signal you approximate the Fourier transform of the original continuous-time signal.

Notations

The following notations are used; f is the frequency in Hz, F_s is the sampling rate (samples/sec), and N is the number of samples in a sequence.

Deliverable

The report will contain the following parts:

   a. Simulation codes and resulting graphs (time and frequency plots).
   b. Answers to questions and documentations of each section.
   c. Your interpretation of results and observations.
   d. Your conclusions.

Sinusoidal Signals

   a. Generate and display one cycle of a 100-Hz sine wave at the sampling rate of $F_s = 2$ kHz (20 points). Note that the time duration of the signal is $N/F_s = 10$ msec. This is equal to one period of the signal and generates one full cycle of the 100-Hz sinusoid. Next, evaluate the Fourier transform (DFT) of the signal. In Matlab the FFT command evaluates the DFT.
   b. Repeat for $N = 40$, $F_s = 2$ kHz, $f = 100$ Hz.
   c. Repeat for $N = 40$, $F_s = 4$ kHz, $f = 100$ Hz.

Document and interpret your results. Summarize your observations and conclusions and answer the following questions:

   a. How many lines do you see in the spectra of the signals?
   b. How are the spectral lines related to the frequency and amplitude of the sinusoidal signals?

c. Observe similarities and differences of frequency scales in transform plots. Conclude a relationship between sampling rate in the time domain and frequency range in the frequency domain.

Almost-Sinusoidal Signals

In this section you will generate three signals that are slightly longer than one cycle of a sinusoid. The periodic signals will contain harmonics other than the fundamental frequency of the sine wave.

a. Generate a 100-Hz sine wave at the sampling rate of $F_s = 2$ kHz. Choose $N = 21$, which produces slightly more than one cycle of the sinusoid. Find and display its transform.
b. Repeat for $N = 22$, $F_s = 2$ kHz, $f = 100$ Hz.
c. Repeat for $N = 42$, $F_s = 4$ kHz, $f = 100$ Hz.

As described in the summary, from the program's point of view, the length of signals constitutes one period of operation. The data length is $T = N \Delta T$, which is 10.5 msec in parts (a) and (c), and 11 msec in part (b). In either case, data length is slightly greater than the period of the 100-Hz sinusoids. Repeating the signals a, b, and c creates almost-sinusoidal signals with some harmonics. Summarize your observations and conclusions in order to provide answers to the following questions:

a. How many lines do you see in the spectra of the signals?
b. Determine if the position of dominant spectral lines is related to the frequency and amplitude of the sinusoid.
c. Where are the additional frequency components located?
d. How are they influenced by the number of additional points in the time signals?

Rectangular Pulses: Effect of Pulse Width

In the following sections and throughout the rest of this project, pulse width is shown by τ.

a. Generate one cycle of a 100-Hz rectangular pulse (period $T = 10$ msec, pulse width $\tau = 5$ msec) at the 20-kHz sampling rate (producing $N = 200$ points) and find its transform.
b. Repeat for $T = 10$ msec, $\tau = 2$ msec, $F_s = 20$ kHz.
c. Repeat for $T = 10$ msec, $\tau = 1$ msec, $F_s = 20$ kHz.

Rectangular Pulses: Effect of Period

a. Generate a rectangular pulse (period $= 2$ msec, pulse width $= 1$ msec) at a 100-kHz sampling rate. Find its transform.
b. Repeat for $T = 5$, $\tau = 1$, $F_s = 100$ kHz.
c. Repeat for $T = 10$, $\tau = 1$, $F_s = 100$ kHz.

Time-Frequency Relationship

The pulses you used were specified by the following three variables in the time domain:

a. Pulse width τ, indicating concentration of energy in time.
b. Period T, indicating one cycle of the signal in time.
c. Sampling interval $\Delta T = 1/F_s$, indicating resolution in time.

Their transforms in the frequency domain exhibit three aspects:

a. Band width B, indicating concentration of energy in the frequency domain.
b. Frequency span $F_0 = F_s/2$, indicating the frequency span; that is, the highest frequency in the signal.
c. Frequency increment Δf, indicating resolution in the frequency domain.

Note that the rectangular pulse signals contain low frequencies and the envelope of the Fourier coefficients of the rectangular pulse is approximately proportional to a finite section of the function $X(f) = \frac{\sin(\pi f \tau)}{\pi f}$ sampled at frequency intervals $\Delta f = \frac{1}{T}$. The bandwidth may be defined as the frequency of the first zero crossing of $X(f)$, $B = f_0 = \frac{1}{\tau}$. The frequency span is $F_0 = \frac{F_s}{2}$ and the frequency resolution is $\Delta f = \frac{1}{T}$. In the above procedures you changed T, τ, and F_s in the time domain and observed the effects on B, F_0, and Δf in the frequency domain. Verify that your worksheets exhibit such relationships between the above variables.

Triangular Pulses

Repeat the procedure of rectangular pulses for a train of triangular pulses. Compare the attenuation of harmonics in rectangular and triangular pulses. Find the percentage of the total energy residing in the first two harmonics of the triangular pulse and compare to that for the first two harmonics of the rectangular pulse.

Tone Burst

a. Modulate the amplitude of a 1-kHz sinusoidal signal by a rectangular pulse of width τ, sampled at the rate of $F_s = 10$ kHz. Take 50 msec of the data and obtain its transform. Display the signal in time and frequency domains. Show examples for $\tau = 5, 10, 20$, and 50 msec.
b. Repeat for Hanning windows of width $\tau = 5, 10, 20$, and 50 msec.

Observe and record the frequency shift and magnitude change of the spectra produced by amplitude modulation. Observe and record the effect of the Hanning window on the bandwidth of modulated signals.

Conclusions

i. Describe the observations on the spread in spectrum of almost-sinusoidal signals.
ii. For the periodic rectangular pulse waveforms qualitatively describe relationship between line spectrum, pulse width, and period.
iii. Repeat for triangular pulse.
iv. Describe your observations on the relationships between the Fourier series and Fourier transform.
v. Summarize your overall conclusions from this project.

Chapter 8

Fourier Transform

Contents

Introduction and Summary

In the analysis of LTI systems, starting with a collection of known input-output pairs made of

$$\xi_k(t) \implies \eta_k(t)$$

we would like to find the response to a new input. If the new input can be expressed as a weighted sum of the members of the known input set, the output will then be the weighted sum of the corresponding members of the output set

$$\sum_k X_k \xi_k(t) \implies \sum_k X_k \eta_k(t)$$

It is said that the input is expanded as a sum of signals $\xi_k(t)$ within its function space the same way that a vector is expanded as a sum of its components within its vector space. In Chapter 7 we noted that for weighting coefficients X_k to be easily calculated, we would like the members of the set of $\xi_k(t)$ to be orthogonal to each other. We chose the exponentials $e^{j2\pi ft}$ as the building blocks for expanding a periodic signal and called the resulting expansion the Fourier series of the signal. The choice of exponentials as the building blocks (in addition to their orthogonality property) was motivated by the fact that the exponentials are eigenfunctions of the linear systems. (The output has the same functional form as the input except for a scale factor; the scale factor is easily obtained as a function of the frequency.) The Fourier series, therefore, provides the tool necessary to obtain the output given a periodic input signal. Moreover, the harmonics generated by the Fourier series not only simplify the analysis, but also help us understand the operation intuitively as they are associated with the actual frequency experienced by our senses (auditory, visual, tactile, etc.) or by synthetic systems.

A similar reasoning applies to the case of nonperiodic signals, leading again to the choice of exponentials as building blocks. In Chapter 7 we examined the limiting form of the Fourier series coefficients when the period of the signal is increased to infinity. Based on that, we developed the following expression for $X(f)$:

$$X(f) = \int_{-\infty}^{-\infty} x(t) e^{-j2\pi ft} dt$$

$X(f)$ obtained from the above integral is called the Fourier transform of $x(t)$. Accordingly, a nonperiodic signal is transformed from the time domain to the frequency domain not in the form of a series but as a weighted integral transform. The transformation of $x(t)$ to $X(f)$ is more than a mere abstract mathematical operation that facilitates analysis and design. $X(f)$ reflects information about the *physical frequency content* of the signal. It suggests that a nonperiodic function has a continuous spectrum; that is, the signal contains a continuum of frequencies. The concept of a continuous spectrum may seem novel and unfamiliar at first but is easy to interpret. Instead of signal power being concentrated at a discrete set of frequencies, such power is spread over a range of frequencies. From $X(f)$ one can find the power within a frequency band of a signal, its interference with other signals, its vulnerability to noise, and so on. The Fourier transform has widespread applications in engineering, in general, and in communications, control, and signal processing, in particular. As in the case of the Fourier series, in addition to providing a powerful and exact tool for quantitative analysis, the Fourier transform adds an intuitive and qualitative component to understanding a process. It presents the outcome of theoretical work as a tangible and measurable quantity. Because of the above features, from its early inception it has been widely used and proven to be an essential tool and the main instrument in frequency-domain analysis of signals and systems.

We note the parallels and similarities between the Fourier and Laplace transforms. In the Laplace transform, we employ e^{st} and in the Fourier transform $e^{j\omega t}$; that is, we restrict s in the complex plane to the $j\omega$—axis. One may then question the need for the Fourier transform. The answer is the same as in the case of the Fourier series: an intimate and measurable connection between the new variable f in the Fourier transform and the physical frequency. Although the Laplace transform provides a very powerful analysis and design tool, it lacks such a property. The new variable s in the Laplace transform (which is sometimes called the complex frequency) is a mathematical entity not a transparent physical quantity.

This chapter starts with the Fourier transform of energy signals for which the integral converges to a known value at each frequency. We first explore the time-frequency relationship for some continuous-time signals of common interest in signal processing. The objective is to illustrate the practical correlates of the mathematical models of Fourier analysis, to visualize the time-frequency relationships, and to provide the student with the additional intuitive understanding often essential in applications. Single and repetitive pulses of finite duration, such as rectangular and triangular pulses [as well as pulses described by mathematical models of infinite time duration such as $\sin(t)/t$, exponential, Gaussian, etc.], will be transformed into the frequency domain and their time-frequency variations examined and interpreted.

With the introduction of singularities and generalized functions such as the *Dirac* δ (or impulse) function, the Fourier integral may be extended to power signals as well, resulting in the generalized Fourier transform. With this extension, the Fourier series is viewed as a function, allowing us to treat periodic, nonperiodic, and random signals by a unified formulation. This is a very useful and convenient tool when periodic and nonperiodic functions are encountered simultaneously.

The project proposed at the end of the chapter can be implemented by either using software or within the real time, in actual laboratory environment using equipment such as a function generator, a digital oscilloscope, or a spectrum analyzer. Assuming access to such a laboratory is available, the second approach is strongly recommended.

8.1 Fourier Transform of Energy Signals

The Fourier transform of a function $x(t)$ is defined as

$$\mathcal{FT}\{x(t)\} \equiv X(f) = \int_{-\infty}^{\infty} x(t)e^{-j2\pi f t}dt$$

This definition requires the integral to converge for all f. For most functions of interest in engineering (called "good functions" or "well-behaved functions"), the integral converges and the Fourier transform exists. One set of sufficient conditions that guarantees the convergence of the transform is called the *Dirichlet* conditions.

The first *Dirichlet* condition requires that $x(t)$ be absolutely integrable:

$$\int_{-\infty}^{\infty} |x(t)|\, dt < \infty$$

The second *Dirichlet* condition requires that $x(t)$ have a finite number of discontinuities and a finite number of maxima and minima during every finite interval.

Notation

In this book the time functions are real-valued (unless specified otherwise). $X(f)$ is then a real function of the variable $j2\pi f$ and, therefore, a complex function of f. For simplicity, however, we use the notation $X(f)$ rather than the more informative notation $X(j2\pi f)$.

In Cartesian form $X(f)$ is represented by its real and imaginary parts:

$$X(f) = \mathcal{RE}\{X(f)\} + j\mathcal{IM}\{X(f)\}$$

It may also be represented in polar form by its magnitude and phase:

$$X(f) = |X(f)|e^{j\angle X(f)} = |X(f)|\angle X(f),$$

where

$$|X(f)| = \sqrt{\mathcal{RE}\{X(f)\}^2 + \mathcal{IM}\{X(f)\}^2} \quad \text{and} \quad \angle X(f) = \tan^{-1}\frac{\mathcal{IM}\{X(f)\}}{\mathcal{RE}\{X(f)\}}$$

$X(f)$ is said to represent the frequency components of the time function $x(t)$, reminiscent of harmonic analysis of periodic signals by the Fourier series.

8.2   Inverse Fourier Transform

If the Fourier integral converges, then $x(t)$ is called the inverse Fourier transform of $X(f)$ and is obtained from the following integral:

$$x(t) = \int_{-\infty}^{\infty} X(f)e^{j2\pi ft}\,df$$

More precisely,

$$\lim_{F\to\infty} \int_{-F}^{F} X(f)e^{j2\pi ft}\,df = \frac{x(t^-) + x(t^+)}{2}$$

Note the lack of rigor or conditions, such as the Dirichlet conditions, for the convergence of the integral representing the inverse transform.

8.3   Examples of Fourier Transform Pairs

In this section we present three examples of transform pairs and observe some of their properties.

Example 8.1

Rectangular and triangular pulses

Find the Fourier transforms of

a. $x(t) = \begin{cases} V_0, & -\frac{\tau}{2} < t < \frac{\tau}{2} \\ 0, & \text{elsewhere} \end{cases}$

b. $x(t) = \begin{cases} V_1 \left(1 - 2\frac{|t|}{\tau}\right), & -\frac{\tau}{2} < t < \frac{\tau}{2} \\ 0, & \text{elsewhere} \end{cases}$

Solution

a. $$X(f) = \int_{-\tau/2}^{\tau/2} V_0 e^{-j2\pi ft} dt = \frac{V_0}{j2\pi f} \left(e^{j\pi f\tau} - e^{-j\pi f\tau}\right) = V_0 \frac{\sin(\pi f\tau)}{\pi f}$$

b. $$X(f) = V_1 \int_{-\tau/2}^{0} \left(1 + 2\frac{t}{\tau}\right) e^{-j2\pi ft} dt + V_1 \int_{0}^{\tau/2} \left(1 - 2\frac{t}{\tau}\right) e^{-j2\pi ft} dt$$

$$= \frac{2V_1}{\tau} \left[\frac{\sin(\pi f\tau/2)}{\pi f}\right]^2$$

The pairs are plotted in Figure 8.1(*a*) and (*b*), respectively. The signals are even functions: $x(t) = x(-t)$. The transforms are real functions of f. The transforms of these pulses will be discussed in more detail in section 8.18.

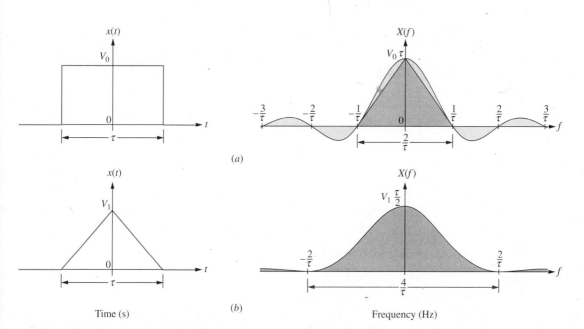

(a)

(b)

Time (s)   Frequency (Hz)

FIGURE 8.1 Two even time pulses (left column) and their Fourier transforms (right column). (*a*) Rectangular pulse. (*b*) Triangular pulse.

xample

8.2

Causal and anticausal exponential functions

Find the Fourier transforms of

$$a.\ x_1(t) = e^{-at}u(t), \quad a > 0$$

$$b.\ x_2(t) = e^{at}u(-t), \quad a > 0$$

Solution

Both functions are exponentials with a time constant $\tau = \frac{1}{a}$. They switch between zero and one at $t = 0$. Otherwise, they are absolute-integrable, well-behaved functions. Their transforms are

a. $X_1(f) = \displaystyle\int_0^\infty e^{-at} e^{-j2\pi ft} dt = \dfrac{1}{a + j2\pi f}, \quad -\infty < f < \infty$

b. $X_2(f) = \displaystyle\int_{-\infty}^0 e^{at} e^{-j2\pi ft} dt = \dfrac{1}{a - j2\pi f}, \quad -\infty < f < \infty$

The functions and their Fourier transforms are plotted in Figure 8.2(*a*) and (*b*). Note that $x_2(t) = x_1(-t)$. The transforms are related by $X_2(f) = X_1(-f)$.

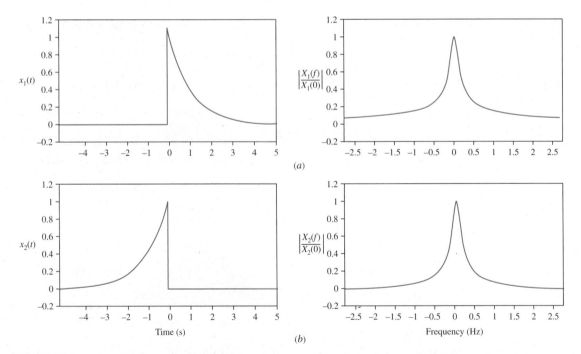

FIGURE 8.2 (*a*) and (*b*). One-sided exponential time functions (left column) and normalized magnitudes of their Fourier transforms (right column). (*a*) Causal function $x_1(t)$, (*b*) Anticausal function $x_2(t)$. The abscissa show times in seconds and frequencies in Herz. The time constants of all exponentials are 1 second. See Example 8.2.

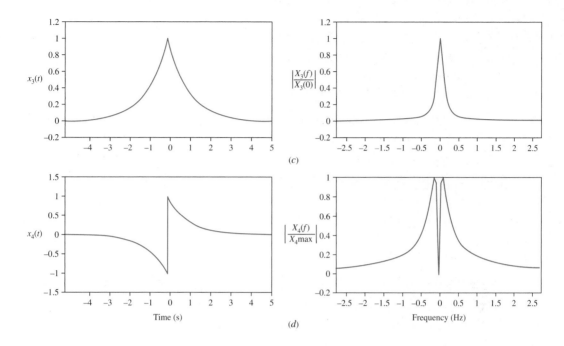

FIGURE 8.2 (*c*) and (*d*). Two-sided exponential time functions (left column) and normalized magnitudes of their Fourier transforms (right column). (*c*) Even function $x_3(t) = x_1(t) + x_2(t)$. Odd function $x_4(t) = x_1(t) + x_2(t)$. Note that the magnitude plot of the transform of the sum is narrower than the sum of the magnitudes of the transforms of its components. The abscissa show times in seconds and frequencies in hertz. The time constants of all exponentials are 1 second. For derivations of the transforms of the two-sided exponentials use linearity property; also see Examples 8.6 and 8.7.

8.4 Linearity Property

Linearity is an important property of the Fourier transform. It states that given

$$x(t) \iff X(f)$$

and

$$y(t) \iff Y(f)$$

then

$$ax(t) + by(t) \iff aX(f) + bY(f), \text{ for any } x, y, \text{ and constants } a \text{ and } b.$$

This property is derived from the definition of the Fourier transform.

Example
8.3

Find the Fourier transform of

$$x(t) = \begin{cases} 2, & -2 < t < 2 \\ -1, & 2 < |t| < 4 \\ 0, & \text{elsewhere} \end{cases}$$

Solution

Let $x(t) = x_1(t) - x_2(t)$, where $x_1(t) = \begin{cases} 3, & -2 < t < 2 \\ 0, & \text{elsewhere} \end{cases}$ and $x_2(t) = \begin{cases} 1, & -4 < t < 4 \\ 0, & \text{elsewhere} \end{cases}$

Using the linearity property and the result of Example 8.1 we find

$$X(f) = X_1(f) - X_2(f) = 3\frac{\sin(4\pi f)}{\pi f} - \frac{\sin(8\pi f)}{\pi f} = \frac{\sin(4\pi f)}{\pi f}[3 - 2\cos(4\pi f)]$$

8.5 Conjugate Symmetry

If $x(t)$ is real, then

$$x(t) = x^*(t) \Longleftrightarrow X(f) = X^*(-f)$$

In other words, if $x(t)$ is real then $\mathcal{RE}\{X(f)\}$ and $|X(f)|$ are even functions of f, while $\mathcal{IM}\{X(f)\}$ and $\angle X(f)$ are odd functions of f. Therefore, a real function $x(t)$ is completely specified from knowing $X(f)$ for $f > 0$. This knowledge could be in the form of the real and imaginary parts of $X(f)$ or its magnitude and phase.

8.6 Time Reversal

Let the Fourier transform of $x(t)$ be $X(f)$. Flip $x(t)$ around the vertical axis at $t = 0$ and you get a new function $x(-t)$. The Fourier transform of $x(-t)$ is $X(-f)$. Reversing a time function reverses its transform.

$$x(t) \Longleftrightarrow X(f)$$

$$x(-t) \Longleftrightarrow X(-f)$$

Proof

In the expression for the Fourier transform of $y(t) = x(-t)$ change the variable t to $-\tau$. The result is $X(-f)$.

$$Y(f) = \int_{-\infty}^{\infty} y(t)e^{-j2\pi ft}dt = \int_{-\infty}^{\infty} x(-t)e^{-j2\pi ft}dt = \int_{-\infty}^{\infty} x(\tau)e^{j2\pi f\tau}d\tau = X(-f)$$

Example

8.4

Fourier transform of even and odd parts of a rectangular pulse

Find the Fourier transform of $x(t) = u(t) - u(t - \tau/2)$. Then, use time-reversal property to find Fourier transforms of its even and odd parts.

Solution

a. $x(t) = \begin{cases} 1, & 0 < t < \frac{\tau}{2} \\ 0, & \text{elsewhere} \end{cases}$ $X(f) = \int_0^{\frac{\tau}{2}} e^{-j2\pi ft} dt = \frac{1}{j2\pi f}\left(1 - e^{-j\pi f\tau}\right)$

$$= \frac{\sin(\pi f\tau)}{2\pi f} - j\frac{\sin^2(\frac{\pi f\tau}{2})}{\pi f}$$

b. $x_e(t) = \frac{x(t) + x(-t)}{2},$ $X_e(f) = \frac{X(f) + X(-f)}{2} = \frac{\sin(\pi f\tau)}{2\pi f}\}$

c. $x_o(t) = \frac{x(t) - x(-t)}{2},$ $X_o(f) = \frac{X(f) - X(-f)}{2} = -j\frac{\sin^2(\frac{\pi f\tau}{2})}{\pi f}$

E*xample*

8.5

Exponential functions revisited

Consider the exponential functions of Example 8.2.

$$x_1(t) = e^{-at}u(t), \quad a > 0$$

$$X_1(f) = \frac{1}{a + j2\pi f} = \frac{a}{a^2 + 4\pi^2 f^2} - j\frac{2\pi f}{a^2 + 4\pi^2 f^2}$$

$$= \frac{1}{\sqrt{a^2 + 4\pi^2 f^2}} \angle - \tan^{-1}\left(\frac{2\pi f}{a}\right)$$

$$x_2(t) = e^{at}u(-t), \quad a > 0$$

$$X_2(f) = \frac{1}{a - j2\pi f} = \frac{a}{a^2 + 4\pi^2 f^2} + j\frac{2\pi f}{a^2 + 4\pi^2 f^2}$$

$$= \frac{1}{\sqrt{a^2 + 4\pi^2 f^2}} \angle \tan^{-1}\left(\frac{2\pi f}{a}\right)$$

Note that in both cases $\mathcal{RE}\{X(f)\}$ and $|X(f)|$ are even functions of f, while $\mathcal{IM}\{X(f)\}$ and $\angle X(f)$ are odd functions of f.

8.7 Waveform Symmetry

Even Function

A function $x(t)$ is even if $x(t) = x(-t)$. The Fourier transform of an even function is

$$X(f) = \int_{-\infty}^{\infty} x(t)e^{-j2\pi ft} dt = \int_0^{\infty} x(t)[e^{j2\pi ft} + e^{-j2\pi ft}] dt = 2\int_0^{\infty} x(t)\cos(2\pi ft) dt$$

If $x(t)$ is real and even, its transform $X(f)$ is a real and even function of f.

Example

8.6

Two-sided even exponential function

$$x(t) = e^{-a|t|}, \quad a > 0$$

$$X(f) = \int_{-\infty}^{0} e^{at} e^{-j2\pi ft} dt + \int_{0}^{\infty} e^{-at} e^{-j2\pi ft} dt$$

$$= \frac{1}{a - j2\pi f} + \frac{1}{a + j2\pi f} = \frac{2a}{a^2 + 4\pi^2 f^2}$$

The Fourier transform pair is plotted in Figure 8.2(c). The function $x(t)$ is related to $x_1(t)$ and $x_2(t)$ in Example 8.2, and their transforms $X_1(f)$ and $X_2(f)$, by

$$x(t) = x_1(t) + x_2(t)$$

$$X(f) = X_1(f) + X_2(f)$$

In addition, note that $X(f)$ is a real and even function of f: $X(f) = X(-f)$. This is to be expected as $x(t)$ is an even function.

Odd Function

A function $x(t)$ is said to be odd if $x(t) = -x(-t)$. The Fourier transform of an odd function is

$$X(f) = \int_{-\infty}^{\infty} x(t) e^{-j2\pi ft} dt = \int_{0}^{\infty} x(t) \left[-e^{j2\pi ft} + e^{-j2\pi ft} \right] dt$$

$$= -2j \int_{0}^{\infty} x(t) \sin(2\pi ft) dt$$

If $x(t)$ is real and odd, its transform $X(f)$ is purely imaginary and an odd function of f.

Example

8.7

Two-sided odd exponential function

$$x(t) = e^{-at} u(t) - e^{at} u(-t), \quad a > 0$$

$$X(f) = \int_{-\infty}^{0} -e^{at} e^{-j2\pi ft} dt + \int_{0}^{\infty} e^{-at} e^{-j2\pi ft} dt$$

$$= \frac{-1}{a - j2\pi f} + \frac{1}{a + j2\pi f} = \frac{-j4\pi f}{a^2 + 4\pi^2 f^2}$$

The Fourier transform pair is plotted in Figure 8.2(d). The function $x(t)$ is related to $x_1(t)$ and $x_2(t)$ in Example 8.2, and their transforms $X_1(f)$ and $X_2(f)$, by

$$x(t) = x_1(t) - x_2(t)$$

$$X(f) = X_1(f) - X_2(f)$$

In addition, note that $X(f)$ is an imaginary and odd function of f: $X(f) = -X(-f)$. This is to be expected as $x(t)$ is an odd function.

Summary of Symmetry Property

$$x(t) \text{ is real and even} \iff X(f) \text{ is real and even}$$

$$x(t) \text{ is real and odd} \iff X(f) \text{ is imaginary and odd}$$

8.8 Even and Odd Parts of Functions

Any function $x(t)$ may be represented by the sum of two components $x(t) = x_e(t) + x_o(t)$, where $x_e(t)$ is an even function [called the even part of $x(t)$] and $x_o(t)$ is an odd function [called the odd part of $x(t)$]. The even and odd parts of a function are uniquely determined from $x(t)$ by

$$x_e(t) = \frac{x(t) + x(-t)}{2}$$

$$x_o(t) = \frac{x(t) - x(-t)}{2}$$

Example 8.8

Derive the expressions for $x_e(t)$ and $x_o(t)$ from $x(t)$ and $x(-t)$, and relate them in the frequency domain.

Solution

Let $x(t) = x_e(t) + x_o(t)$. Then, $x(-t) = x_e(-t) + x_o(-t)$. But by definition, $x_e(-t) = x_e(t)$ and $x_o(-t) = -x_o(t)$. Therefore, $x(-t) = x_e(t) - x_o(t)$. By adding $x(t)$ and $x(-t)$ we get $x(t) + x(-t) = 2x_e(t)$. By subtracting $x(-t)$ from $x(t)$ we get $x(t) - x(-t) = 2x_o(t)$. These give the results that were stated before, that is

$$x_e(t) = \frac{x(t) + x(-t)}{2}$$

$$x_o(t) = \frac{x(t) - x(-t)}{2}$$

The even and odd parts of a real function are related to the real and imaginary parts of its transform.

$$x(t) \iff X(f) = \mathcal{RE}\{X(f)\} + j\mathcal{IM}\{X(f)\}$$

$$x(-t) \iff X(-f) = \mathcal{RE}\{X(f)\} - j\mathcal{IM}\{X(f)\}$$

$$x_e(t) = \frac{x(t) + x(-t)}{2} \iff \frac{X(f) + X(-f)}{2} = \mathcal{RE}\{X(f)\}$$

$$x_o(t) = \frac{x(t) - x(-t)}{2} \iff \frac{X(f) - X(-f)}{2} = j\mathcal{IM}\{X(f)\}$$

Example

8.9

Find the even and odd parts of the causal exponential function

$$x(t) = e^{-at}u(t), \quad a > 0$$

Solution

We first find $x(-t)$

$$x(-t) = e^{at}u(-t), \quad a > 0$$

from which

$$x_e(t) = \frac{x(t) + x(-t)}{2} = \frac{1}{2}\{e^{-at}u(t) + e^{at}u(-t)\} = \frac{1}{2}e^{-a|t|}$$

$$x_o(t) = \frac{x(t) - x(-t)}{2} = \frac{1}{2}\left[e^{-at}u(t) - e^{at}u(-t)\right]$$

Example

8.10

Find the Fourier transform of the even part of the causal exponential function $x(t) = e^{-at}u(t), \ a > 0$.

$$x_e(t) = \frac{x(t) + x(-t)}{2} = \frac{1}{2}e^{-a|t|}$$

Solution

From Example 8.6 we find the Fourier transform of $x_e(t)$ to be

$$X_e(f) = \frac{a}{a^2 + 4\pi^2 f^2}$$

The transforms are related by

$$X_e(f) = \frac{X(f) + X(-f)}{2} = \mathcal{RE}\{X(f)\}$$

Note that $X_e(f)$ is a real and even function of f: $X_e(f) = X_e(-f)$.

Example

8.11

Find the Fourier transform of the odd part of the causal exponential function $x(t) = e^{-at}u(t), \ a > 0$.

$$x_o(t) = \frac{1}{2}\left[e^{-at}u(t) - e^{at}u(-t)\right]$$

Solution

From Example 8.7 we find the Fourier transform of $x_o(t)$ to be

$$X_o(f) = \frac{-j2\pi f}{a^2 + 4\pi^2 f^2}$$

The transforms are related by

$$X_o(f) = \frac{X(f) - X(-f)}{2} = j\mathcal{IM}\{X(f)\}$$

Note that $X_o(f)$ is imaginary and an odd function of f: $X_o(f) = -X_o(-f)$.

8.9 Causal Functions

Time Domain

A function that is zero for $t < 0$ is called causal. A causal function is completely specified by its even or odd part. To find the casual function $x(t)$ in terms of $x_e(t)$ or $x_o(t)$, we note that:

1. For $t < 0$ we have $\qquad\qquad x(t) = x_e(t) + x_o(t) = 0$ which gives

    $$x_e(t) = -x_o(t), \quad t < 0$$

2. From summetry property we have $\qquad x_e(t) = x_e(-t)$ and

    $$x_o(t) = -x_o(-t) \quad \text{for all } t$$

3. Combining the above results we get $\qquad x_e(t) = x_o(t), \quad t > 0$

    $$x(t) = \begin{cases} 2x_e(t) = 2x_o(t), & t > 0 \\ 0, & t < 0 \end{cases}$$

Frequency Domain

In Example 8.8 it was observed that the even and odd parts of a (real) function are related to the real and imaginary parts of its transform by the following:

$$x(t) \iff X(f) = \mathcal{RE}\{X(f)\} + j\mathcal{IM}\{X(f)\}$$

$$x_e(t) \iff \mathcal{RE}\{X(f)\}$$

$$x_o(t) \iff j\mathcal{IM}\{X(f)\}$$

In addition, if the function is causal, the real and imaginary parts of the transform are not independent from each other either. Given either one, the other may be found through a trip to the time domain as shown in Example 8.12.

Example
8.12

Even and odd parts of a causal rectangular pulse in time and frequency

a. Find the even and odd parts of a causal rectangular pulse $x(t) = \begin{cases} 1, & 0 < t < \frac{\tau}{2} \\ 0, & \text{elsewhere} \end{cases}$

b. Find the Fourier transform of $x(t)$.

c. Find the Fourier transform of the even part of $x(t)$ and verify its equivalence with $\mathcal{RE}\{X(f)\}$.

d. Find the Fourier transform of the odd part of $x(t)$ and verify its equivalence with $j\mathcal{IM}\{X(f)\}$.

Solution

a. We first find

$$x(-t) = \begin{cases} 1, & -\frac{\tau}{2} < t < 0 \\ 0, & \text{elsewhere} \end{cases}$$

from which

$$x_e(t) = \frac{x(t) + x(-t)}{2} = \begin{cases} \frac{1}{2}, & -\frac{\tau}{2} < t < \frac{\tau}{2} \\ 0, & \text{elsewhere} \end{cases}$$

and

$$x_o(t) = \frac{x(t) - x(-t)}{2} = \begin{cases} -\frac{1}{2}, & -\frac{\tau}{2} < t < 0 \\ \frac{1}{2}, & 0 < t < \frac{\tau}{2} \\ 0, & \text{elsewhere} \end{cases}$$

b. $x(t) = \begin{cases} 1, & 0 < t < \frac{\tau}{2} \\ 0, & \text{elsewhere} \end{cases}$ $\qquad X(f) = \int_0^{\frac{\tau}{2}} e^{-j2\pi ft} dt = \frac{1}{j2\pi f}\left(1 - e^{-j\pi f\tau}\right)$

$$= \frac{\sin(\pi f\tau)}{2\pi f} - j\frac{\sin^2(\frac{\pi f\tau}{2})}{\pi f}$$

c. $x_e(t) = \begin{cases} \frac{1}{2}, & -\frac{\tau}{2} < t < \frac{\tau}{2} \\ 0, & \text{elsewhere} \end{cases}$ $\qquad X_e(f) = \frac{1}{2}\int_{\frac{-\tau}{2}}^{\frac{\tau}{2}} e^{-j2\pi ft} dt = \frac{\sin(\pi f\tau)}{2\pi f} = \mathcal{RE}\{X(f)\}$

d. $x_o(t) = \begin{cases} -\frac{1}{2}, & -\frac{\tau}{2} < t < 0 \\ \frac{1}{2}, & 0 < t < \frac{\tau}{2} \\ 0, & \text{elsewhere} \end{cases}$ $\qquad X_o(f) = -\frac{1}{2}\int_{\frac{-\tau}{2}}^{0} e^{-j2\pi ft} dt + \frac{1}{2}\int_0^{\frac{\tau}{2}} e^{-j2\pi ft} dt$

$$= -j\frac{\sin^2(\frac{\pi f\tau}{2})}{\pi f} = j\mathcal{IM}\{X(f)\}$$

Given $X_e(f)$ one can find its inverse $x_e(t)$, from which one can then find $x_o(t)$ and the transform $X_o(f)$. By a similar process one can find $X_e(f)$ from $X_o(f)$.

Example
8.13

Even and odd parts of a causal exponential function in time and frequency

Repeat Example 8.12 for the causal exponential function $x(t) = e^{-at}u(t)$, $\quad a > 0$.

Solution

a. $x(t) = e^{-at}u(t)$ $\qquad X(f) = \frac{1}{a + j2\pi f} = \frac{a}{a^2 + 4\pi^2 f^2} - j\frac{2\pi f}{a^2 + 4\pi^2 f^2}$

b. $x_e(t) = \frac{1}{2}e^{-a|t|}$ $\qquad X_e(f) = \frac{a}{a^2 + 4\pi^2 f^2} = \mathcal{RE}\{X(f)\}$

c. $x_o(t) = \frac{1}{2}\left[e^{-at}u(t) - e^{at}u(-t)\right]$ $\qquad X_o(f) = -j\frac{2\pi f}{a^2 + 4\pi^2 f^2} = j\mathcal{IM}\{X(f)\}$

As in Example 8.12, given $X_e(f) = \frac{a}{a^2+4\pi^2 f^2}$ one obtains $x_e(t) = \frac{1}{2}e^{-a|t|}$, from which one can then find

$$x_o(t) = \begin{cases} x_e(t) = \frac{1}{2}e^{-at} & t > 0 \\ -x_e(t) = -\frac{1}{2}e^{at} & t < 0 \end{cases} \quad \text{and } X_o(f) = -j\frac{2\pi f}{a^2 + 4\pi^2 f^2}$$

By a similar process one can find $X_e(f)$ from $X_o(f)$.

8.10 Time-Frequency Duality

The duality property of the Fourier transform states that

$$x(t) \Longleftrightarrow X(f)$$
$$X(t) \Longleftrightarrow x(-f)$$

To derive the duality property, start with the definition of the transform

$$X(f) = \int_{-\infty}^{\infty} x(\tau)e^{-j2\pi f\tau}d\tau$$

and change variable f to t to get

$$X(t) = \int_{-\infty}^{\infty} x(\tau)e^{-j2\pi t\tau}d\tau$$

Then change τ to $-f$ to obtain

$$X(t) = \int_{-\infty}^{\infty} x(-f)e^{j2\pi tf}df$$

Now compare the above expression with the definition of the inverse transform to verify that the function $x(-f)$ is the Fourier transform of $X(t)$. This property allows us to avoid evaluating complex integrals when deriving transform pairs. For examples see the following pairs in Appendix 8A: (b and c), (k and ℓ), (q and r).

Example

8.14

Duality and rectangular pulse

Use the duality property to find the transform of $y(t) = \sin(2\pi f_0 t)/(\pi t)$.

Solution

Start with
$$x(t) = \begin{cases} 1, & -\frac{\tau}{2} < t < \frac{\tau}{2} \\ 0, & \text{elsewhere} \end{cases} \Longrightarrow X(f) = \frac{\sin(\pi\tau f)}{\pi f}$$

Apply duality property: $X(t) = \frac{\sin(\pi\tau t)}{\pi t}$
$$\Longrightarrow x(-f) = \begin{cases} 1, & -\frac{\tau}{2} < f < \frac{\tau}{2} \\ 0, & \text{elsewhere} \end{cases}$$

Let $\tau = 2f_0$ $y(t) = \frac{\sin(2\pi f_0 t)}{\pi t}$
$$\Longrightarrow Y(f) = \begin{cases} 1, & -f_0 < f < f_0 \\ 0, & \text{elsewhere} \end{cases}$$

Example 8.15

Duality and exponential function

a. By direct application of the definition find the inverse transform of
$Y(f) = e^{-af} u(f)$.

b. Show that the result is in agreement with duality property.

Solution

a.
$$y(t) = \int_{-\infty}^{\infty} Y(f) e^{j2\pi ft} df = \int_{0}^{\infty} e^{-af} e^{j2\pi ft} df$$

$$= \int_{0}^{\infty} e^{(-a+j2\pi t)f} df = \frac{1}{a - j2\pi t}, \text{ all } t$$

b. Start with

$$x(t) = e^{at} u(-t) \implies X(f) = \frac{1}{a - j2\pi f}, \text{ all } f$$

Apply duality property:

$$X(t) = \frac{1}{a - j2\pi t}, \text{ all } t \impliedby x(-f) = e^{-af} u(f)$$

The time function obtained through application of the duality property is the same as that obtained by direct integration in part a. It is also noted that in this example $y(t)$ is a complex function of time.

8.11 Time Shift

A time delay, or time shift to the right, by t_0 seconds adds a negative value $-2\pi f t_0$ to the phase of the Fourier transform of a function.

$$x(t - t_0) \iff X(f) e^{-j2\pi f t_0}$$

The additional phase delay $2\pi f t_0$ is proportional to the frequency. Conversely, when going from the frequency domain to the time domain, a linear phase in the frequency domain translates into a constant time delay.

Example 8.16

Given

$$x(t) = \begin{cases} 1, & -\frac{\tau}{4} < t < \frac{\tau}{4} \\ 0, & \text{elsewhere} \end{cases} \implies X(f) = \frac{\sin(\pi f \tau/2)}{\pi f}$$

shift $x(t)$ to the right by $\tau/4$ and find its Fourier transform. Verify that the result is in agreement with the result obtained in Example 8.12(b).

Solution

$$y(t) = x(t - \tau/4) = \begin{cases} 1, & 0 < t < \frac{\tau}{2} \\ 0, & \text{elsewhere} \end{cases}$$

$$Y(f) = X(f)e^{-j\pi f\tau/2} = \frac{\sin(\pi f\tau/2)}{\pi f}e^{-j\pi f\tau/2}$$

$$= \frac{\sin(\pi f\tau)}{2\pi f} - j\frac{\sin^2(\frac{\pi f\tau}{2})}{\pi f} \quad \text{[same as in Example 8.12(b)]}$$

8.12 Frequency Shift

Multiplication of a time signal $x(t)$ by $e^{j2\pi f_0 t}$ shifts the transform by f_0 to the right:

$$x(t)e^{j2\pi f_0 t} \iff X(f - f_0)$$

The function $x(t)e^{j2\pi f_0 t}$ is not a real signal. However, by shifting $X(f)$ to the left and right and adding the results together, we obtain a real signal:

$X(f)$	$\iff$	$x(t)$
$X(f - f_0)$	$\iff$	$x(t)e^{j2\pi f_0 t}$
$X(f + f_0)$	$\iff$	$x(t)e^{-j2\pi f_0 t}$
$\dfrac{X(f - f_0) + X(f + f_0)}{2}$	$\iff$	$x(t)\left[\dfrac{e^{j2\pi f_0 t} + e^{-j2\pi f_0 t}}{2}\right] = x(t)\cos(2\pi f_0 t)$

The signal $x(t)\cos(2\pi f_0 t)$ is an amplitude-modulated (AM) signal. The sinusoid is the carrier of the information contained in $x(t)$. The last line in the above list states that amplitude modulation by $x(t)$ shifts $X(f)$ to the left and right, centering it at $\pm f_0$ where f_0 is the frequency of the sinusoidal carrier.

Example

8.17

Find the Fourier transform of $y(t) = x(t)\cos(2\pi f_c t)$, where $x(t) = \frac{\sin(2\pi f_0 t)}{\pi t}$, and obtain its frequency range if $f_0 = 3$ kHz and $f_c = 1$ MHz.

Solution

First use time-frequency duality, then frequency shift to obtain

$$x(t) = \frac{\sin(2\pi f_0 t)}{\pi t} \iff X(f) = \int_{-\infty}^{-\infty} \frac{\sin(2\pi f_0 t)}{\pi t}e^{-j2\pi ft}dt = \begin{cases} 1, & -f_0 < f < f_0 \\ 0, & \text{elsewhere} \end{cases}$$

$$y(t) = x(t)\cos(2\pi f_c t) \iff Y(f) = \frac{X(f - f_0) + X(f + f_0)}{2}$$

$$= \begin{cases} \frac{1}{2}, & f_c - f_0 < |f| < f_c + f_0 \\ 0, & \text{elsewhere} \end{cases}$$

8.13 Differentiation and Integration

Differentiation and integration operations on signals and their effects in the time domain were discussed in Chapter 1. Here we summarize their effects in the transform domain. By direct application of differentiation and integration to the inverse Fourier transform we get

$$\frac{dx}{dt} \iff j2\pi f X(f)$$

$$\int_{-\infty}^{t} x(\tau)d\tau \iff \frac{X(f)}{j2\pi f} + \frac{X(0)\delta(f)}{2}$$

The above spectral observation states that the high-frequency components of a signal are amplified by differentiation and attenuated by integration. Conversely, differentiation suppresses the DC value of a signal and integration accumulates it, making it more pronounced. These statements agree with direct observations in the time domain on the effects of differentiation and integration. For example, differentiation of a signal that contains high-frequency noise will degrade it because the noise is enhanced by the operation, and integration smoothes the noise out. Similarly, a small DC bias in the input of an op-amp integrator will build up in the output and will eventually saturate the op-amp.

8.14 Convolution Property

Convolution in the time domain is equivalent to multiplication in the frequency domain:

$$x(t) \star y(t) \iff X(f)Y(f)$$

Proof

Convolution of two functions $x(t)$ and $h(t)$ is defined by

$$y(t) = x(t) \star h(t) = \int_{-\infty}^{\infty} x(t - \tau)\, h(\tau)d\tau$$

The Fourier transform of $y(t)$ is

$$Y(f) = \int_{-\infty}^{\infty} y(t)e^{-j2\pi ft}dt = \int_{-\infty}^{\infty}\int_{-\infty}^{\infty} x(t-\tau)\, h(\tau)\, e^{-j2\pi ft}\, d\tau\, dt$$

To convert the double integral to two separate integrals with single variables we change the variable t to a new variable θ by $t = \tau + \theta$, $dt = d\theta$, and $t - \tau = \theta$. The terms containing the variables θ and τ in the double integral are then separated

$$Y(f) = \left[\int_{-\infty}^{\infty} x(\theta)\, e^{-j2\pi\theta f}\, d\theta\right] \times \left[\int_{-\infty}^{\infty} h(\tau)\, e^{-j2\pi f\tau}\, d\tau\right] = X(f)H(f)$$

The convolution property is of great importance because it establishes the frequency-domain relationship for LTI systems:

$$Y(f) = X(f)H(f)$$

$Y(f)$, $X(f)$, and $H(f)$ are Fourier transforms of the output, input, and impulse response, respectively.

8.15 Product Property

The product property is the dual of the convolution property. It states that multiplication in the time domain is equivalent to convolution in the frequency domain:

$$x(t)y(t) \quad \Longleftrightarrow \quad X(f) \star Y(f)$$

In other words, the Fourier transform of the product of two time functions is the same as the convolution of their Fourier transforms:

$$\int_{-\infty}^{\infty} x(t)y(t)e^{-j2\pi ft}dt = \int_{-\infty}^{\infty} X(f - \varphi)Y(\varphi)d\varphi$$

A special case exists when $y(t) = z^*(t + \tau)$, in which case,

$$\int_{-\infty}^{\infty} x(t)z^*(t + \tau)e^{-j2\pi ft}dt = \int_{-\infty}^{\infty} X(f - \varphi)Z^*(-\varphi)e^{j2\pi\varphi\tau}d\varphi$$

Changing the variable φ to $-\mu$, we obtain the generalized product property

$$\int_{-\infty}^{\infty} x(t)z^*(t + \tau)e^{-j2\pi ft}dt = \int_{-\infty}^{\infty} X(f + \mu)Z^*(\mu)e^{-j2\pi\mu\tau}d\mu$$

For example, the transform of amplitude-modulated signal $x(t)\cos(2\pi f_c t)$ is found by using product property.

$$x(t) \quad \Longrightarrow \quad X(f)$$

$$\cos(2\pi f_c t) \quad \Longrightarrow \quad \frac{\delta(f - f_c) + \delta(f + f_c)}{2} \quad \text{(See section 8.25.)}$$

$$x(t)\cos(2\pi f_c t) \quad \Longrightarrow \quad X(f) \star \left[\frac{\delta(f - f_c) + \delta(f + f_c)}{2}\right]$$

$$= \frac{1}{2}X(f - f_c) + \frac{1}{2}X(f + f_c)$$

8.16   Parseval's Theorem and Energy Spectral Density

The energy in a square-integrable signal is

$$\int_{-\infty}^{\infty} |x(t)|^2 dt$$

Parseval's theorem states that the expressions for energy in the time domain and the frequency domain are numerically identical. That is,

$$\int_{-\infty}^{\infty} |x(t)|^2 dt = \int_{-\infty}^{\infty} |X(f)|^2 df$$

Parseval's theorem may be derived from the generalized product property described in section 8.15 by letting $f = \tau = 0$ and $x(t) = z(t)$. Then, $x(t)z^*(t) = |x(t)|^2$ and

$$\int_{-\infty}^{\infty} |x(t)|^2 dt = \int_{-\infty}^{\infty} |X(f)|^2 df$$

For a square-integrable signal, $|X(f)|^2$ is called the energy spectral density. The energy at frequencies within a bandwidth of Δf centered at the frequency f is then $2|X(f)|^2 \Delta f$. The energy within the low-pass band of $f < f_0$ is obtained from the integral

$$\int_{-f_0}^{f_0} |X(f)|^2 df = 2 \int_{0}^{f_0} |X(f)|^2 df$$

Similarly, the energy within the band $f_1 < f < f_2$ is

$$2 \int_{f_1}^{f_2} |X(f)|^2 df$$

Example 8.18

Using Parseval's theorem compute the energy within the $0 \to 1$ kHz band of a 1 volt, $1 \mu s$ rectangular pulse and compare with its total energy.

Solution

$$X(f) = \frac{\sin(\pi f 10^{-6})}{\pi f}$$

$$X(f)\big|_{f=1\,kHz} = \frac{\sin(\pi 10^{-3})}{\pi 10^3} \approx 10^{-6} = X(0)$$

$$\text{Energy within the } 0 \to 1 \text{ kHz band} = 2 \int_{0}^{1,000} |X(f)|^2 df \approx 2 \times 10^{-9} \text{ J}$$

$$\text{Total energy in the pulse} = \int_{-\infty}^{\infty} x^2(t) dt = 10^{-6} \text{ J}$$

Energy within the 1-kHz band is $1/500$ of the total energy in the pulse, as one may estimate from Figure 8.1(a).

8.17 Summary of Fourier Transform Prop⌐

Some important properties of the Fourier transform are liste⌐

TABLE 8.1 Some Important Properties of the Fourier Tran⌐

	Property	Time Domain	⟺	Frequency ⌐
1	Definition	$x(t) = \int_{-\infty}^{\infty} X(f)e^{j2\pi ft}df$	⟺	$X(f) = \int_{-\infty}^{\infty} x(t)e^{-j2\pi ft}dt$
2	Linearity	$ax(t) + by(t)$	⟺	$aX(f) + bY(f)$
3	Zero time	$x(0)$	⟺	$\int_{-\infty}^{\infty} X(f)df$
4	Zero frequency	$\int_{-\infty}^{\infty} x(t)dt$	⟺	$X(0)$
5	Real	$x(t) = x^*(t)$	⟺	$X(f) = X^*(-f)$
6	Even	$x(t) = x(-t)$	⟺	$X(f) = X(-f)$
7	Odd	$x(t) = -x(-t)$	⟺	$X(f) = -X(-f)$
8	Duality	$X(t)$	⟺	$x(-f)$
9	Time shift	$x(t - t_0)$	⟺	$X(f)e^{-j2\pi ft_0}$
10	Frequency shift	$x(t)e^{j2\pi f_0 t}$	⟺	$X(f - f_0)$
11	Modulation	$x(t)\cos(2\pi f_0 t)$	⟺	$\dfrac{X(f - f_0) + X(f + f_0)}{2}$
12	Scale change	$x(at)$	⟺	$\frac{1}{|a|}X\left(\frac{f}{a}\right)$
13	Time reversal	$x(-t)$	⟺	$X(-f)$
14	Multiplication by t	$tx(t)$	⟺	$-\dfrac{1}{j2\pi}\dfrac{dX(f)}{df}$
15	Integration	$\int_{-\infty}^{t} x(\tau)d\tau$	⟺	$\dfrac{X(f)}{j2\pi f} + \dfrac{X(0)\delta(f)}{2}$
16	Differentiation	$\dfrac{dx}{dt}$	⟺	$j2\pi f X(f)$
17	Convolution	$\int_{-\infty}^{\infty} x(t - \tau)\, y(\tau)d\tau$	⟺	$X(f)Y(f)$
18	Multiplication	$x(t)y(t)$	⟺	$\int_{-\infty}^{\infty} X(f - \varphi)\, Y(\varphi)d\varphi$

8.18 Time-Limited Signals

A signal is called time limited if its energy is concentrated within a finite time interval (called its duration). In a strict sense, time-limited signals have zero value outside their duration, that is,

$$x(t) = \begin{cases} f(t), & -t_1 < t < t_2 \\ 0, & \text{elsewhere} \end{cases}$$

The transform of a strictly time-limited signal cannot be zero over a nonzero frequency interval. Theoretically, a time-limited signal contains all frequency components and occupies an infinite bandwidth, no matter how smoothly it grows or declines at the edges. The narrower the pulse is in the time domain, the wider its transform is in the frequency domain and vice versa. Therefore, a signal may only approximately be both time limited and band limited.

Time-limited signals are of much practical interest in communication and signal processing. For example, the digits in digital communication systems are encoded into such signals and then sent through the channel. Windows through which the data is selected and processed are another class of time-limited signals. Because of this, we discuss in detail the transforms of several such signals each containing a single pulse of duration τ. The time origin is chosen such that the signals are even or odd functions of time. We would like to see their transforms and how widespread their spectra are. This is signal design and of much interest in digital signal processing and digital communications.

Example 8.19

Rectangular pulse revisited

The Fourier transform of a rectangular pulse was found in Example 8.1(a).

$$x(t) = \begin{cases} 1, & -\frac{\tau}{2} < t < \frac{\tau}{2} \\ 0, & \text{elsewhere} \end{cases} \Longleftrightarrow X(f) = \frac{\sin(\pi f \tau)}{\pi f}$$

The Fourier transform pair is plotted in Figure 8.1(a). The nonzero segment of the time signal is a rectangular pulse of height 1 and duration τ. It has abrupt discontinuities at the two edges with an infinite rate of change. The transform $X(f)$ contains the following features.

1. $X(f)$ is a real and even function of f. It extends over an infinite frequency range. It switches between positive and negative values, creating an infinite number of positive and negative lobes. The zero-crossing between two neighboring lobes occurs at regular intervals $f = n/\tau$, where n is an integer in the range $-\infty < n < \infty, n \neq 0$.

2. The main lobe extends over the range $-1/\tau < f < 1/\tau$ which corresponds to the frequency band from 0 to f_0, where $f_0 = 1/\tau$. The major part of the energy in the signal is concentrated at low frequencies within the main lobe.

3. The value of the transform at $f = 0$ is $X(0) = \tau$ which is equal to the net area under $x(t)$. This compares with the periodic case, where X_0 is equal to the DC value of the signal.

4. From the equation for the inverse transform we deduce that

$$\int_{-\infty}^{\infty} X(f)df = x(0) = 1$$

The total area under $X(f)$ is equivalent to the area of the triangle encompassed by the main lobe. See Figure 8.1(a).

Triangular pulse revisited

Consider the isosceles triangle of height 1 and base τ.

$$x(t) = \begin{cases} 1 - 2\frac{|t|}{\tau}, & -\frac{\tau}{2} < t < \frac{\tau}{2} \\ 0, & \text{elsewhere} \end{cases}$$

The Fourier transform of the triangular pulse is

$$X(f) = \int_{-\tau/2}^{0} \left(1 + \frac{2t}{\tau}\right) e^{-j2\pi ft}\, dt + \int_{0}^{\tau/2} \left(1 - \frac{2t}{\tau}\right) e^{-j2\pi ft}\, dt = \frac{2}{\tau}\left[\frac{\sin(\pi f\tau/2)}{\pi f}\right]^2$$

The time function and its Fourier transform are plotted in Figure 8.1(b). The time function is continuous but its derivatives at the corners are discontinuous without being infinite. The transform $X(f)$ exhibits the following characteristics.

1. $X(f)$ is a real and even function of f extending over the infinite frequency range and containing an infinite number of lobes which are all non-negative. The zero tangents between two neighboring lobes occur at regular intervals $f = 2n/\tau$, where n is an integer $-\infty < n < \infty$.
2. The main lobe extends over the frequency range $-2/\tau < f < 2/\tau$.
3. The value of the transform at $f = 0$ is $X(0) = \tau/2$ which is equal to the total area under the triangular pulse.
4. Compared to the transform of the rectangular pulse of the same duration τ, the mathematical expression of $X(f)$ for the triangular pulse indicates that the main lobe is twice wider and lobes attenuate faster with frequency. See Figure 8.1(b).

Raised cosine pulse

The raised cosine pulse is defined by

$$x(t) = \begin{cases} 1 + \cos(\frac{2\pi t}{\tau}), & -\frac{\tau}{2} < t < \frac{\tau}{2} \\ 0, & \text{elsewhere} \end{cases}$$

The nonzero segment of the function is one cycle of a raised cosine. The pulse is the product of a uniform rectangular window of duration τ and a cosine function with period τ and unity DC value. At the edges both the function and its derivatives are continuous and zero. The transform is

$$X(f) = \int_{-\tau/2}^{\tau/2} \left[1 + \cos\left(\frac{2\pi t}{\tau}\right)\right] e^{-j2\pi ft}\, dt = \frac{\sin(\pi f\tau)}{\pi f}$$

$$+0.5\frac{\sin \pi\tau(f - f_0)}{\pi(f - f_0)} + 0.5\frac{\sin \pi\tau(f + f_0)}{\pi(f + f_0)}$$

where $f_0 = 1/\tau$ is the first zero-crossing of the transform of the rectangular window.

8.19 Windowing

A window is a finite-duration pulse through which we look at a data stream or function and its Fourier transform. The role of the window is to create and shape a finite segment of the sequence, be it data, a filter's impulse response, or its frequency response, and shape it in a way such that certain characteristics are met. Windows by definition have finite duration and, therefore, infinite bandwidth. Given the duration of the window, its shape has an important role in the function it fulfills.

The time functions for some familiar continuous-time windows and their Fourier transforms are listed in Table 8.2. For simplicity, the time origin is centered so that the windows are even functions. All windows are of width 2τ. The mathematical expressions are given for $-\tau < t < \tau$. Elsewhere, windows have zero value. Plots of the functions in the time and frequency domains are given in Figure 8.3. Note the effect of a window shape on its side-band frequencies. A comparison between a rectangular and Hanning window is shown in Figure 8.3(f).

TABLE 8.2 Continuous-Time Windows and Their Fourier Transforms

Window	Time Function	$\Longleftrightarrow$	Fourier Transform
Rectangular	1	$\Longleftrightarrow$	$\dfrac{\sin(2\pi f\tau)}{\pi f}$
Bartlett*	$1-\dfrac{|t|}{\tau}$	$\Longleftrightarrow$	$\dfrac{1}{\tau}\left[\dfrac{\sin(\pi f\tau)}{\pi f}\right]^2$
Hanning†	$0.5+0.5\cos\left(\dfrac{\pi t}{\tau}\right)$	$\Longleftrightarrow$	$0.5\dfrac{\sin(2\pi f\tau)}{\pi f}+0.25\dfrac{\sin 2\pi\tau\left(f-\frac{1}{2\tau}\right)}{\pi\left(f-\frac{1}{2\tau}\right)}+0.25\dfrac{\sin 2\pi\tau\left(f+\frac{1}{2\tau}\right)}{\pi\left(f+\frac{1}{2\tau}\right)}$
Hamming	$0.54+0.46\cos\left(\dfrac{\pi t}{\tau}\right)$	$\Longleftrightarrow$	$0.54\dfrac{\sin(2\pi f\tau)}{\pi f}+0.23\dfrac{\sin 2\pi\tau\left(f-\frac{1}{2\tau}\right)}{\pi\left(f-\frac{1}{2\tau}\right)}+0.23\dfrac{\sin 2\pi\tau\left(f+\frac{1}{2\tau}\right)}{\pi\left(f+\frac{1}{2\tau}\right)}$
Blackman	$0.42+0.5\cos\left(\dfrac{\pi t}{\tau}\right)$ $+0.08\cos\left(\dfrac{2\pi t}{\tau}\right)$	$\Longleftrightarrow$	$0.42\dfrac{\sin(2\pi f\tau)}{\pi f}+0.25\dfrac{\sin 2\pi\tau\left(f-\frac{1}{2\tau}\right)}{\pi\left(f-\frac{1}{2\tau}\right)}+0.25\dfrac{\sin 2\pi\tau\left(f+\frac{1}{2\tau}\right)}{\pi\left(f+\frac{1}{2\tau}\right)}$ $+0.04\dfrac{\sin 2\pi\tau\left(f-\frac{1}{\tau}\right)}{\pi\left(f-\frac{1}{\tau}\right)}+0.04\dfrac{\sin 2\pi\tau\left(f+\frac{1}{\tau}\right)}{\pi\left(f+\frac{1}{\tau}\right)}$

*A Bartlett window is triangular.

†A Hanning window is a raised cosine.

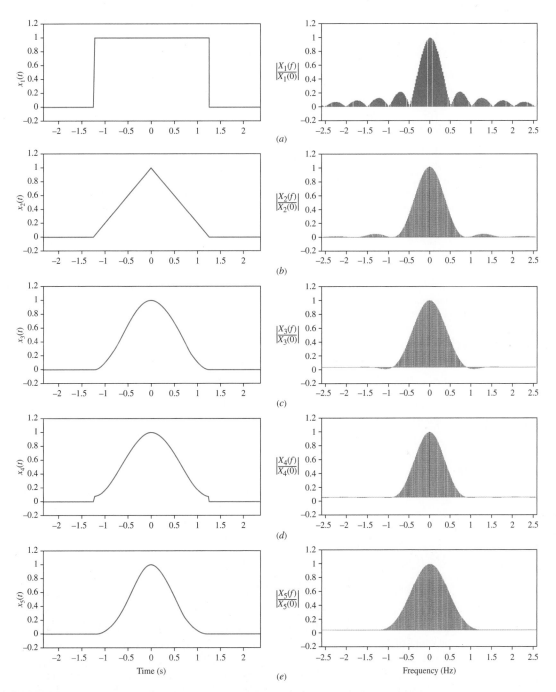

FIGURE 8.3 (*a*) to (*e*). Five finite-duration time windows (left column) and magnitudes of their Fourier transforms (right column). From the top: rectangular, Bartlett (triangular), Hanning (raised cosine), Hamming, and Blackman windows. Distinction between the transforms becomes more pronounced when a decibel scale is used.

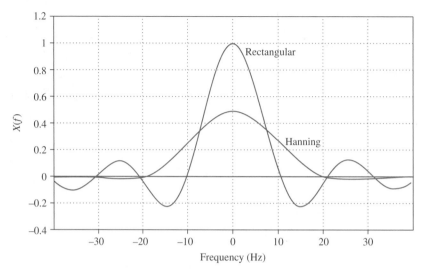

FIGURE 8.3 (*f*). Comparison of the Fourier transforms of rectangular and Hanning windows. Both windows are 10 msec wide with 1-volt maxima. The Hanning window has a smoother frequency characteristic with smaller side bands. For derivation of the Fourier transforms and a quantitative comparison see problem 7.

8.20 Band-Limited Signals

Low-Pass Signals

The Fourier transform of a strictly band-limited low-pass signal (also called a baseband signal) is zero for $|f| > f_0$. An example is the signal

$$x(t) = \frac{\sin(2\pi f_0 t)}{\pi t}$$

It has finite energy spread over an infinite duration. However, most of its energy is within the main lobe. Therefore, it is called a sinc pulse. The Fourier transform of $x(t)$ is

$$X(f) = \begin{cases} 1, & \text{for } -f_0 < f < f_0; \\ 0, & \text{elsewhere} \end{cases}$$

The frequencies are limited to 0 to f_0 and the signal is low pass.

Time-Limited, Band-Limited Signals

Strictly speaking, a signal can't be both time limited and band limited. However, in some applications, such as digital communication, we need time-limited signals with a small amount of energy outside a given frequency range. These may be called time-limited, band-limited signals. Examples of such low-pass signals are shown in Figure 8.3. The pulses shown in Figure 8.3 are strictly time limited, but not band limited, as their Fourier transforms extend to frequencies up to ∞. The magnitudes of the transforms diminish and can be ignored at high frequencies. A similar discussion also applies to bandpass signals as well.

Bandpass Signals

The frequency components of a bandpass signal are limited between f_1 and f_2 so that $X(f) = 0$ for $f_1 < |f| < f_2$. A bandpass signal can be converted to a low-pass signal modulating the amplitude of a carrier. The low-pass signal is in general a complex signal. It is real if the spectrum of the bandpass is symmetric around a center frequency as described in section 8.12. (For further details refer to Chapter 10 on time-domain sampling.)

Consider the infinite-duration toneburst and its Fourier transform

$$x(t) = \frac{\sin(2\pi f_0 t)}{\pi t} \cos(2\pi f_c t), \quad f_c > f_0$$

$$X(f) = \begin{cases} 1/2, & |f_c - f_0| < |f| < |f_c + f_0| \\ 0, & \text{elsewhere} \end{cases}$$

$x(t)$ is a bandpass signal limited to the band of $f_c - f_0$ to $f_c + f_0$ with a bandwidth of $2f_0$.

8.21 Paley-Wiener Theorem

The Paley-Wiener theorem[1] specifies necessary and sufficient conditions for a square-integrable function $|H(f)|$ to be the spectrum of a causal function. It states that if $h(t)$ is square-integrable and causal, then

$$\int_{-\infty}^{\infty} \frac{|\ln|H(f)||}{1 + f^2} df < \infty$$

The above condition is necessary and sufficient for an $|H(f)|$ to be the magnitude of the Fourier transform of a causal function (e.g., the impulse response of a realizable filter), and if it is satisfied, a phase function $\theta(f)$ may then be attached to $|H(f)|$ to obtain a causal $h(t)$. Conversely, if $|H(f)|$ is square-integrable but the above integral is not bounded, then a causal $h(t)$ may not be found regardless of the phase function associated with $H(f)$. Consequently, if $|H(f)| = 0$ for a nonzero frequency band, then the above integral becomes unbounded and the filter is not realizable.

The frequency response of the ideal low-pass filter is

$$H_a(f) = \begin{cases} 1, & |f| < f_0 \\ 0, & \text{elsewhere} \end{cases}$$

---

[1]See R.E.A.C. Paley and N. Wiener, *Fourier Transforms in the Complex Domain*, Vol. 19, American Mathematical Society (New York: Colloquium Publications, 1934), p. 16.

See Figure 8.4(a). The filter has a zero transition band. Show that the filter remains unrealizable even if the transition band is increased to be from f_0 to f_1 as shown by the example of the filter in Figure 8.4(b), (c), and (d).

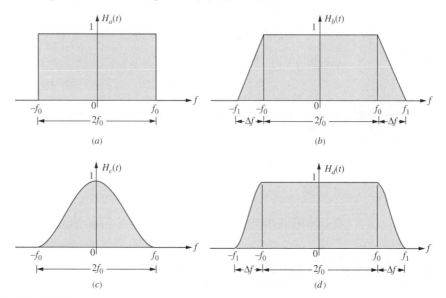

FIGURE 8.4 An ideal low-pass filter (**a**) remains unrealizable even if transition band is nonzero (**b,c**, and **d**).

Solution

The filter in (a) is unrealizable. This can be verified from the fact that its unit-impulse response is

$$h_a(t) = \frac{\sin(2\pi f_0 t)}{\pi t}$$

It extends from $-\infty$ to ∞ and would not become a causal function regardless of any amount of shift to the right. The filter in (b) may be constructed by subtracting the output of two filters with triangular frequency responses, both of which are also unrealizable by the same reason as (a). A smoother but finite-duration transition band [such as the graph in (c) and (d) in which $dX(f)/df$ is continuous] cannot make the filter causal either. The above conclusions are readily reached by applying the Paley-Wiener theorem according to which all of the above filters are unrealizable because $\ln|H(f)| = \infty$ over a nonzero range of frequency makes the integral given in that theorem unbounded.

8.22 Gibbs' Phenomenon

Consider a rectangular pulse $x(t)$ of width T and its transform $X(f)$:

$$x(t) = \begin{cases} 1, & -\frac{T}{2} < t < \frac{T}{2} \\ 0, & \text{elsewhere} \end{cases} \quad \Longleftrightarrow \quad X(f) = \frac{\sin(\pi f T)}{\pi f}$$

The pulse contains all frequencies $-\infty < f < \infty$, but most of its energy resides in the frequency band of the main lobe of the transform, $-1/T < f < 1/T$. Choose a finite segment of the transform, selected by a low-pass uniform window $H(f)$ in the frequency domain limited to the frequencies $-f_0 < f < f_0$:

$$H(f) = \begin{cases} 1, & -f_0 < f < f_0 \\ 0, & \text{elsewhere} \end{cases}$$

Call the segment $Y(f) = H(f)X(f)$. The signal $y(t)$ may be found from $Y(f)$ by direct application of the inverse transform. Since $Y(f)$ is band limited, the time function $y(t)$ is not time limited and it is not an exact replica of $x(t)$. This is expected. However, there is another observed effect, called Gibbs' phenomenon, as evidenced by the presence of horns at the edges of the pulse in $y(t)$ as shown in Figure 8.5(d). The Gibbs' phenomenon

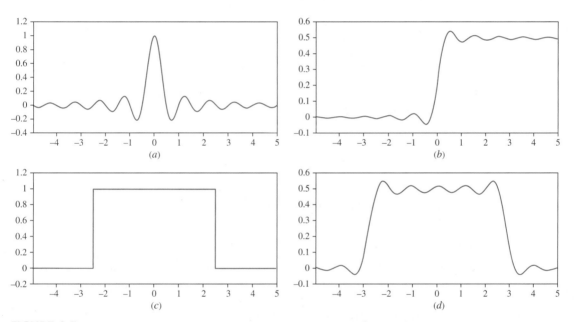

FIGURE 8.5 A sinc function $h(t)$ shown in (**a**) and its integral $g(t)$ in (**b**). Convolution of a rectangular pulse $x(t)$ shown in (**c**) and $h(t)$ produces the pulse $y(t)$ in (**d**) in which Gibbs' phenomenon is observed. The abscissa show times in seconds.

can be mathematically discerned if we find $y(t)$ from the convolution of $h(t)$ with $x(t)$, where $h(t)$ is the inverse Fourier transform of $H(f)$ (i.e., the impulse response of the low-pass filter). Note that $x(t) = u(t+T/2) - u(t-T/2)$ and $h(t) = \frac{\sin(2\pi f_0 t)}{\pi t}$. Therefore,

$$y(t) = h(t) \star x(t) = h(t) \star [u(t + T/2) - u(t - T/2)]$$

$$= h(t) \star u(t + T/2) - h(t) \star u(t - T/2)$$

$$= g(t + T/2) - g(t - T/2)$$

where $g(t) = h(t) \star u(t)$ is the unit-step response of the ideal low-pass filter shown in Figure 8.5(b).

$$g(t) = \int_{-\infty}^{t} h(\tau)d\tau = \int_{-\infty}^{t} \frac{\sin(2\pi f_0 \tau)}{\pi \tau} d\tau$$

$g(t)$ has undershoot and overshoot which cause ringing at transition instances in $y(t)$. As we increase the bandwidth of $H(f)$ by increasing f_0, the ripples become narrower, but their amplitudes do not diminish. They take the form of horns. In other words, $y(t)$ converges to $x(t)$ everywhere except at the transition neighborhood. The limit of $y(t)$ at $t = \pm T/2$ is 0.25 but the convergence is not uniform. The error cannot be reduced to a desired value by increasing f_0 and by considering a smaller transition neighborhood. Despite the persistence of the horns, the rms of error between $y(t)$ and $x(t)$ decreases as f_0 increases, and approaches zero when $f_0 \to \infty$.

8.23 Fourier Transform of Power Signals

Let $x(t)$ be a signal with finite average power so that

$$\lim_{T \to \infty} \frac{1}{2T} \int_{-T}^{T} |x(t)|^2 dt < \infty$$

Examples are (1) periodic signals, and (2) signals in communication and control, which are normally labeled as random signals and noise. The familiar version of the Fourier transform for these signals does not exist because their Fourier integrals do not converge. However, power signals may still be represented in the frequency domain. For the periodic signals this is done by employing singularity functions such as the impulse. For aperiodic power signals the subject may be approached in several ways. As an example, power signals are windowed in time and a finite segment is processed at a time, making it an energy signal with a well-defined Fourier transform. One case is processing of speech signals where, for example, a 200-msec. segment is transformed to the frequency domain. The Fourier integral, therefore, converges and a normalized Fourier transform for the segment of data under processing is defined by

$$X(f) = \frac{1}{2T} \int_{-T}^{T} x(t) e^{-j2\pi ft} dt$$

The normalization of the integral by the length factor allows comparison in the frequency domain, independent of the segment's length. In the case of a power signal with infinite length, if we can find an $X(f)$ by any means whose error, according to the following expression, can be reduced to any desired small amount by increasing T, then we may call that the Fourier transform of the signal $x(t)$.

$$\lim_{T \to \infty} \int_{-\infty}^{\infty} \left| X(f) - \frac{1}{2T} \int_{-T}^{T} x(t) e^{-j2\pi ft} dt \right|^2 df < \epsilon$$

8.24 Fourier Transform of Generalized Functions

In engineering, signals, waveforms, and functions are of interest for what they do. As long as two functions result in the same effect, they are appreciated as the same. Therefore, functions that differ by a finite value at a countably finite number of points are equivalent. An example is the pair $x_1(t)$ and $x_2(t)$ below:

$$x_1(t) = \sin(\omega t)$$

$$x_2(t) = \begin{cases} 1, & t = kT \\ \sin(\omega t), & \text{elsewhere} \end{cases}$$

where k is an integer and T is the period. By using generalized functions, such as the Dirac delta function $\delta(f)$, we can extend the Fourier transform to periodic signals and also to those signals for which the integral does not converge in the conventional sense.

The condition for existence of the generalized Fourier transform of a time function $x(t)$ is that $\int_{-\infty}^{\infty} x(t)\phi(t)dt$ be well defined, where $\phi(t)$ is any infinitely differentiable function that vanishes faster than any power of t as t increases. With this generalization, a function $x(t)$ and its Fourier transform $X(f)$ form a unique pair with symmetrical properties, including the guarantee that $x(t)$ can be obtained from $X(f)$.

Accordingly, a Fourier transform pair $x(t)$ and $X(f)$ is uniquely specified by either one of the following relationships:

$$X(f) = \int_{-\infty}^{\infty} x(t)e^{-j2\pi ft}dt$$

$$x(t) = \int_{-\infty}^{\infty} X(f)e^{j2\pi ft}df$$

Given the function in the time or frequency domain, the representation in the other domain may be found by searching for a function that satisfies any one of the above two equations. As expressed previously, this generalization allows us to specify Fourier transforms of functions for which the integral does not converge. For example, the Fourier transform of a periodic signal with period T is made of impulses in the frequency domain at the location of its harmonics, where the impulses are separated by $f_0 = \frac{1}{T}$ and T is the period. The magnitude of each impulse is the same as the coefficient of the Fourier series of the corresponding harmonic X_n

$$X(f) = \sum_{n=-\infty}^{\infty} X_n\delta\left(f - \frac{n}{T}\right), \quad X_n = \frac{1}{T}\int_{t_0}^{t_0+T} x(t)e^{-j2\pi nt/T}dt$$

8.25 Impulse Function and Operations

Impulse Function

The unit-impulse function (also called the Dirac delta, or δ, function) is defined in several ways. One is the limit of a narrow pulse with unit area, as pulse width is reduced to zero and height is increased to ∞. Another definition is given by its operation in the following integral:

$$\int_{-\infty}^{\infty} \phi(\tau)\delta(\tau - t)d\tau = \phi(t)$$

where $\phi(t)$ is an infinitely differentiable function that vanishes faster than any power of t as t increases. Accordingly, $\delta(t - t_0)$ is a unit-impulse function located at t_0.

Convolution with Impulse Function

Convolution of a function $h(t)$ with $\delta(t - t_0)$, a unit impulse at t_0, is

$$h(t) \star \delta(t - t_0) = \int_{-\infty}^{\infty} \delta(t - t_0 - \tau)h(\tau)d\tau$$

But by definition of $\delta(t)$,

$$\int_{-\infty}^{\infty} h(t)\delta(t - t_0)dt = h(t_0)$$

Therefore,

$$h(t) \star \delta(t - t_0) = h(t - t_0)$$

Convolution of $h(t)$ with a unit impulse located at t_0 shifts $h(t)$ by the amount t_0.

Fourier Transform of an Impulse Function

Consider a rectangular pulse $x(t)$ with unit area (width $= \tau$, height $= 1/\tau$), and its Fourier transform $X(f)$. As the pulse becomes narrower, the transform becomes wider. At the limit, the pulse becomes a unit impulse and its transform becomes equal to 1 for all f.

$$x(t) = \begin{cases} \frac{1}{\tau}, & -\frac{\tau}{2} < t < \frac{\tau}{2} \\ 0, & \text{elsewhere} \end{cases} \quad \Longleftrightarrow \quad X(f) = \frac{\sin(\pi f \tau)}{\pi f \tau}$$

$$\lim_{\tau \to 0} x(t) = \delta(t) \quad \Longleftrightarrow \quad \lim_{\tau \to 0} X(f) = 1$$

Fourier Transform of a DC Signal

Consider a rectangular pulse $x(t)$ (height 1, width τ) and its Fourier transform $X(f)$. As τ increases, the pulse becomes wider and the transform becomes narrower. As $\tau \to \infty$, the pulse becomes a DC signal and its transform becomes a unit impulse $\delta(f)$.

$$x(t) = \begin{cases} 1 & -\frac{\tau}{2} < t < \frac{\tau}{2} \\ 0, & \text{elsewhere} \end{cases} \quad \Longleftrightarrow \quad X(f) = \frac{\sin(\pi f \tau)}{\pi f}$$

$$\lim_{\tau \to \infty} x(t) = 1 \quad \Longleftrightarrow \quad \lim_{\tau \to \infty} X(f) = \delta(f)$$

Fourier Transform of Cosines and Sines

We use the time-shift property of the Fourier transform and the fact that the transform of $x(t) = 1$ is a unit impulse $\delta(f)$ to find the Fourier transform of $\cos(2\pi f_0 t)$ and $\sin(2\pi f_0 t)$.

$$1 \quad \Longleftrightarrow \quad \delta(f)$$

$$e^{j2\pi f_0 t} \quad \Longleftrightarrow \quad \delta(f - f_0)$$

$$e^{-j2\pi f_0 t} \quad \Longleftrightarrow \quad \delta(f + f_0)$$

$$\cos(2\pi f_0 t) = \frac{1}{2}[e^{j2\pi f_0 t} + e^{-j2\pi f_0 t}] \quad \Longleftrightarrow \quad \frac{1}{2}[\delta(f - f_0) + \delta(f + f_0)]$$

$$\sin(2\pi f_0 t) = \frac{1}{2j}[e^{j2\pi f_0 t} - e^{-j2\pi f_0 t}] \quad \Longleftrightarrow \quad \frac{1}{2j}[\delta(f - f_0) - \delta(f + f_0)]$$

Amplitude Modulation

Multiplication of a low-pass signal $x(t)$ by a sinusoid with frequency f_0 moves the frequency content of $x(t)$ upward by f_0 Hz. In terms of the Fourier transform,

$$x(t)\cos(2\pi f_0 t) \Longleftrightarrow \frac{X(f - f_0) + X(f + f_0)}{2}$$

This can be verified by using the convolution/multiplication property of Fourier transforms. Let $x(t)$ and $X(f)$ be a Fourier transform pair. Then,

$$\mathcal{FT}\{\cos(2\pi f_0 t)\} = \frac{\delta(f - f_0) + \delta(f + f_0)}{2}$$

and from the product property,

$$\mathcal{FT}\left\{x(t)\cos(2\pi f_0 t)\right\} = X(f) \star \left[\frac{\delta(f - f_0) + \delta(f + f_0)}{2}\right]$$

$$= \frac{1}{2}X(f - f_0) + \frac{1}{2}X(f + f_0)$$

The above result can also be found from a frequency translation as described below. Multiplication of a time signal $x(t)$ by $e^{j(2\pi f_0 t + \phi)}$ shifts the transform by f_0.

$$x(t)\left[e^{j(2\pi f_0 t + \phi)} + e^{-j(2\pi f_0 t + \phi)}\right] \Rightarrow e^{j\phi}X(f - f_0) + e^{-j\phi}X(f + f_0)$$

$$2x(t)\cos(2\pi f_0 t + \phi) \Rightarrow e^{j\phi}X(f - f_0) + e^{-j\phi}X(f + f_0)$$

8.26 Fourier Transform of Periodic Signals

With help from the generalized functions and impulses, we can introduce the Fourier transform of periodic signals. This will be done in three steps. First we will find the Fourier transform of a train of impulse. In the second step we generate a periodic signal and its Fourier transform by passing an impulse train through an LTI system. Finally, in the third step we derive the Fourier transform of the periodic signal.

Fourier Transform of a Periodic Impulse

Consider an infinite train of unit impulses with period T:

$$x(t) = \sum_{k=-\infty}^{\infty} \delta(t - kT)$$

For simplicity we have chosen the time origin such that one impulse occurs at $t = 0$. In this section we will show that the Fourier transform of $x(t)$ is a train of impulses of strength $1/T$ at $f = k/T = kf_0$ in the frequency domain. This will be done through two approaches. The first approach starts with the Fourier series expansion of a periodic impulse:

From Chapter 7
$$\sum_{k=-\infty}^{\infty} \delta(t - kT) = \frac{1}{T} \sum_{n=-\infty}^{\infty} e^{j2\pi nt/T}$$

But,
$$\mathcal{FT}\{1\} = \delta(f) \quad \text{and} \quad \mathcal{FT}\{e^{j2\pi nt/T}\} = \delta\left(f - \frac{n}{T}\right)$$

Therefore,
$$\mathcal{FT}\left\{\sum_{k=-\infty}^{\infty} \delta(t - kT)\right\} = \mathcal{FT}\left\{\frac{1}{T} \sum_{n=-\infty}^{\infty} e^{j2\pi nt/T}\right\}$$

$$= \frac{1}{T} \sum_{n=-\infty}^{\infty} \delta\left(f - \frac{n}{T}\right)$$

We observe that the Fourier transform of a periodic unit-impulse train with period T in the time domain is a periodic impulse train of strengths $1/T$ in the frequency domain with period $1/T$.

The second approach, less mechanistic and more instructive, uses the limit method. It starts with a finite segment of the time function containing $2M + 1$ impulses shifted by $\pm k/T$ $(k = 0, 1, 2, \ldots, M)$, to be called $x_M(t)$.

$$x_M(t) = \sum_{k=-M}^{M} \delta(t - kT)$$

Note that the Fourier transform of a shifted unit impulse is

$$\delta(t - kT) \Longleftrightarrow e^{-j2k\pi fT}$$

The Fourier transform of $x_M(t)$ is then

$$x_M(t) = \sum_{k=-M}^{M} \delta(t - kT)$$

$$X_M(f) = \sum_{k=-M}^{M} e^{-j2k\pi fT}$$

$$= e^{j2\pi f MT} + e^{j2\pi f(M-1)T} + \cdots + 1 + \cdots + e^{-j2\pi f(M-1)T} + e^{-j2\pi f MT}$$

$$= e^{j2\pi f MT} \left[1 + e^{-j2\pi fT} + \cdots + e^{-j2\pi f MT} + \cdots + e^{-j4\pi f(M-1)T} + e^{-j4\pi MfT} \right]$$

$$= e^{j2\pi f MT} \left[\frac{1 - e^{-j2\pi f(2M+1)T}}{1 - e^{-j2\pi fT}} \right] = \frac{\sin[(2M+1)\pi fT]}{\sin(\pi fT)}$$

$x_M(t)$ and its transform $X_M(f)$ are drawn for $T = 5$ seconds and $2M + 1 = 3$, 5 and 11 in Figure 8.6(a), (b), and (c), respectively. $X_M(f)$ is periodic with peaks of magnitude $(2M + 1)$ at $f = k/T$. Note that the location of the peaks is a function of T only (signal period) and is independent of M. Between the peaks there are zero-crossings at multiples of $1/[(2M + 1)T]$, that is, at

$$f = \pm \frac{k}{T(2M+1)}, \quad k = 0, 1, 2, \ldots$$

As M grows the location of peaks remain the same (i.e., at $f = k/T$) but their magnitudes $(2M+1)$ increases. The zero-crossings between the peaks also become more frequent. The lobes associated with the peaks become narrower and taller. However, the area associated with each lobe is $1/T$ as derived below.

$$\int_{-\frac{1}{2T}}^{\frac{1}{2T}} X_M(f) df = \int_{-\frac{1}{2T}}^{\frac{1}{2T}} \left[\sum_{k=-M}^{M} e^{-j2\pi kfT} \right] df = \sum_{k=-M}^{M} \int_{-\frac{1}{2T}}^{\frac{1}{2T}} e^{-j2\pi kfT} df = \int_{-\frac{1}{2T}}^{\frac{1}{2T}} df = \frac{1}{T}$$

If we let $M \to \infty$ we get $\qquad \lim_{M \to \infty} x_M(t) = x(t)$

$$\lim_{M \to \infty} X_M(f) = X(f) = \sum_{k=-\infty}^{\infty} \frac{1}{T} \delta\left(f - \frac{k}{T} \right)$$

The Fourier transform of a periodic train of unit impulses with period T in the time domain is a train of impulses of strength $1/T$ at multiples of $f_0 = 1/T$ in the f-domain. See Figure 8.7.

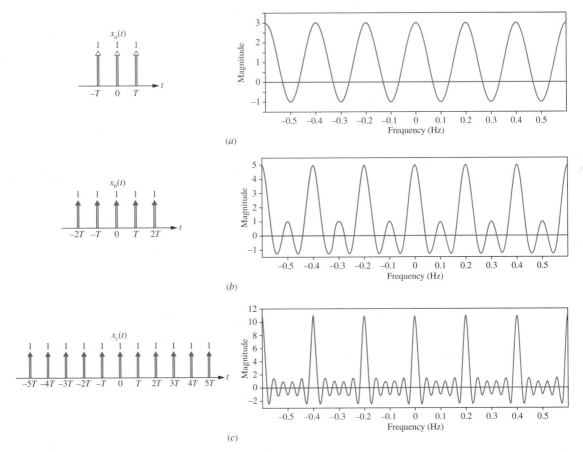

(a)

(b)

(c)

FIGURE 8.6 Impulse trains (in the left column) and their Fourier transforms (in the right column). A train of three impulses and its Fourier transform are shown in (a). The bottom two rows (b) and (c) show impulse trains containing 5 and 11 impulses, respectively, along with their Fourier transforms. The transforms are periodic and are made of pulses positioned at k/T, $T = 5$ sec. The pulses in the frequency domain become narrower as the number of impulses in the time domain increases.

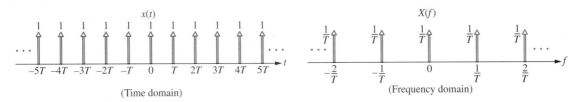

(Time domain)   (Frequency domain)

FIGURE 8.7 A periodic unit-impulse train with a period of T seconds (on the left) transforms into a periodic impulse train of strength $1/T$, spaced every $1/T$ Hz (on the right).

Example

8.24

Find the Fourier transform pairs given below.

a.
$$x(t) = \sum_{n=-\infty}^{\infty} \delta(t - 2n) \implies X(f) = \frac{1}{2} \sum_{k=-\infty}^{\infty} \delta\left(f - \frac{k}{2}\right)$$

b.
$$x(t) = \sum_{n=-\infty}^{\infty} (-1)^n \delta(t - n) \implies X(f) = \frac{1 - e^{-j2\pi f}}{2} \sum_{k=-\infty}^{\infty} \delta\left(f - \frac{k}{2}\right)$$

$$= \sum_{k=-\infty}^{\infty} \left(\frac{1 - e^{-j\pi k}}{2}\right) \delta\left(f - \frac{k}{2}\right)$$

$$= \sum_{k=\text{odd}}^{\infty} \delta\left(f - \frac{k}{2}\right)$$

$$= \sum_{k=-\infty}^{\infty} \delta\left(f - k - \frac{1}{2}\right)$$

Convolution of a Pulse with an Impulse Train

A periodic signal $y(t)$ with period T may be obtained by the convolution of a time-limited signal $h(t)$ with an infinite train $x(t)$ of unit impulses located at every T seconds:

$$y(t) = h(t) \star x(t) = h(t) \star \sum_{k=-\infty}^{\infty} \delta(t - k/T) = \sum_{k=-\infty}^{\infty} h(t - k/T)$$

where $h(t)$ is one cycle of $y(t)$.

$$h(t) = \begin{cases} y(t), & 0 \le t < T \\ 0, & \text{elsewhere} \end{cases}$$

Representing Periodic Signals by Transforms

We use the observation made on convolution of a pulse with an impulse train and the convolution property of the Fourier transform to find the Fourier transform of a periodic signal $y(t)$. The Fourier transform of $y(t)$ is found by multiplying the Fourier transforms of $h(t)$ and $x(t)$ with each other:

$$X(f) = \sum_{n=-\infty}^{\infty} \frac{1}{T} \delta\left(f - \frac{n}{T}\right)$$

$$Y(f) = H(f)X(f) = H(f) \sum_{n=-\infty}^{\infty} \frac{1}{T} \delta\left(f - \frac{n}{T}\right)$$

$$= \frac{1}{T} \sum_{n=-\infty}^{\infty} H(f)\big|_{f=\frac{n}{T}} \delta\left(f - \frac{n}{T}\right)$$

But

$$H(f) = \int_{-\infty}^{\infty} h(t)e^{-j2\pi ft}dt = \int_{0}^{T} y(t)e^{-j2\pi ft}dt$$

and

$$Y_n = \frac{1}{T}\int_{0}^{T} y(t)e^{-j2\pi nt/T}dt$$

Therefore

$$Y_n = \frac{1}{T}H(f)\Big|_{f=\frac{n}{T}}$$

$$Y(f) = \sum_{n=-\infty}^{\infty} Y_n\delta\left(f - \frac{n}{T}\right)$$

The Fourier transform of the periodic signal is obtained from the Fourier transform of a single cycle of the signal by sampling the latter at $f = 1/T$ intervals and multiplying the samples by a factor $1/T$, then representing them by impulses in the frequency domain positioned at the frequencies of its harmonics, that is, at $f = k/T \equiv kf_0$, k an integer. The strength of each impulse is the Fourier series coefficient of the periodic signal. See Figure 8.8.

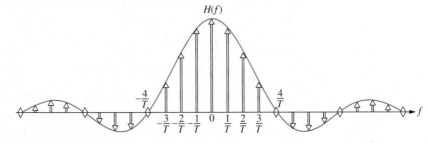

FIGURE 8.8 Repetition of a time-domain pulse produces sampling of its transform in the frequency domain. The samples are multiplied by a factor $1/T$. They are presented by impulses, if the Fourier transform formulation is used, or as the Fourier coefficients, if the Fourier series representation is used. This figure shows the Fourier transform, in the form of impulses, of a periodic rectangular pulse with a period T and a 25% duty cycle (pulse duration $= T/4$). The Fourier transform of the single rectangular pulse shapes the envelope of the impulses.

Example
8.25

a. Determine the Fourier transform of the single rectangular pulse (width $= \tau$, height $= 1$).

b. Repeat the pulse of part a every T seconds ($T > \tau$) to generate the periodic pulse train. Determine the exponential Fourier series coefficients of the periodic pulse train and its Fourier transform, and relate them to the Fourier transform of the single pulse.

Solution

Let $x(t) = \begin{cases} 1, & -\frac{\tau}{2} < t < \frac{\tau}{2} \\ 0, & \text{elsewhere,} \end{cases}$ and $y(t) = x(t) \star \sum_{k=-\infty}^{\infty} \delta(t - kT) = \sum_{k=-\infty}^{\infty} x(\cdot - kT)$

Then,

a. $X(f) = \int_{-\frac{\tau}{2}}^{\frac{\tau}{2}} e^{-j2\pi ft} dt = \dfrac{\sin(\pi f \tau)}{\pi f}$

b. $Y_n = \dfrac{1}{T} \int_{-\frac{\tau}{2}}^{\frac{\tau}{2}} e^{-\frac{j2\pi nt}{T}} dt = \dfrac{\sin(\frac{\pi n \tau}{T})}{\pi n}$

$$Y(f) = X(f) \times \left[\frac{1}{T} \sum_{n=-\infty}^{\infty} \delta\left(f - \frac{n}{T}\right) \right] = \sum_{n=-\infty}^{\infty} \frac{\sin(\frac{\pi n \tau}{T})}{\pi n} \delta\left(f - \frac{n}{T}\right)$$

$$= \sum_{n=-\infty}^{\infty} Y_n \delta\left(f - \frac{n}{T}\right)$$

8.27 Concluding Remarks

The Laplace transform, as we saw in Chapter 6 and will further examine in Chapter 9, provides a very powerful tool for the analysis and design of LTI systems. The Fourier transform may be considered a special case of the Laplace transform in which s is replaced by $j2\pi f$. In fact, many results obtained in this chapter could be derived using the Laplace transform. However, in one's mind, one may form the impression that the Fourier transform doesn't capture the frequency components of nonperiodic signals, as the Fourier series does for periodic signals. (Each coefficient in the Fourier series represents a measurable physical harmonic.) One may, therefore, question the need for special considerations accorded to the Fourier transform in frequency-domain analysis. The explanation for needing such a transform resides in, among other factors, the following: (1) Measurements and identification of systems use sinusoids and their results are formulated in terms of frequency spectra of signals and frequency responses of systems. (2) Numerical computations, simulations, and operations of digital devices make extensive use of Fourier transforms. This chapter introduced an analytic and theoretical basis for the Fourier transform for such applications and provided a foundation for its extension to the discrete-time domain.

Appendix 8A A Short Table of Fourier Transform Pairs

	Time Domain	$\Longleftrightarrow$	**Frequency Domain**
a)	$x(t) = \int_{-\infty}^{\infty} X(f)e^{j2\pi ft} df$	$\Longleftrightarrow$	$X(f) = \int_{-\infty}^{\infty} x(t)e^{-j2\pi ft} dt$
b)	1, all t	$\Longleftrightarrow$	$\delta(f)$
c)	$\delta(t)$	$\Longleftrightarrow$	1, all f

(Continued)

	Time Domain	$\Longleftrightarrow$	**Frequency Domain**
d)	$\cos(2\pi f_0 t)$	$\Longleftrightarrow$	$\dfrac{\delta(f - f_0) + \delta(f + f_0)}{2}$
e)	$\sin(2\pi f_0 t)$	$\Longleftrightarrow$	$\dfrac{\delta(f - f_0) - \delta(f + f_0)}{2j}$
f)	$u(t)$	$\Longleftrightarrow$	$\dfrac{\delta(f)}{2} + \dfrac{1}{j2\pi f}$
g)	$u(-t)$	$\Longleftrightarrow$	$\dfrac{\delta(f)}{2} - \dfrac{1}{j2\pi f}$
h)	$\text{sgn}(t) = u(t) - u(-t)$	$\Longleftrightarrow$	$\dfrac{1}{j\pi f}$
i)	$e^{-\alpha t} u(t)$	$\Longleftrightarrow$	$\dfrac{1}{\alpha + j2\pi f}, \ \alpha > 0$
j)	$t e^{-\alpha t} u(t)$	$\Longleftrightarrow$	$\dfrac{1}{(\alpha + j2\pi f)^2}$
k)	$e^{-\alpha|t|}$	$\Longleftrightarrow$	$\dfrac{2\alpha}{\alpha^2 + 4\pi^2 f^2}$
l)	$\dfrac{2\tau}{\tau^2 + 4\pi^2 t^2}$	$\Longleftrightarrow$	$e^{-\tau|f|}, \ \tau > 0$
m)	$e^{-\alpha t} u(t) - e^{\alpha t} u(-t)$	$\Longleftrightarrow$	$\dfrac{-j4\pi \alpha^2 f}{\alpha^2 + 4\pi^2 f^2}$
n)	$e^{-\pi t^2}$	$\Longleftrightarrow$	$e^{-\pi f^2}$
o)	$e^{-\alpha t} \cos(2\pi f_0 t) u(t)$	$\Longleftrightarrow$	$\dfrac{\alpha + j2\pi f}{(\alpha + j2\pi f)^2 + 4\pi^2 f_0^2}$
p)	$\displaystyle\sum_{k=-\infty}^{\infty} \delta(t - kT)$	$\Longleftrightarrow$	$\dfrac{1}{T} \displaystyle\sum_{k=-\infty}^{\infty} \delta(f - k/T)$

q)	$\begin{cases} 1, & -\tau/2 < t < \tau/2 \\ 0, & \text{elsewhere} \end{cases}$	$\Longleftrightarrow$	$\dfrac{\sin(\pi \tau f)}{\pi f}$

r)	$\dfrac{\sin(2\pi f_0 t)}{\pi t}$	$\Longleftrightarrow$	$\begin{cases} 1, & -f_0 < f < f_0 \\ 0, & \text{elsewhere} \end{cases}$

8.28 Problems

Notations

Energy signal	$\int_{-\infty}^{\infty} |x(t)|^2 dt < \infty.$
Fourier transform of $x(t)$	$X(f) = \int_{-\infty}^{\infty} x(t)e^{-j2\pi ft} dt.$
Inverse Fourier transform of $X(f)$:	$x(t) = \int_{-\infty}^{\infty} X(f)e^{j2\pi ft} df.$
Energy in the signal	$\int_{-\infty}^{\infty} |x(t)|^2 dt = \int_{-\infty}^{\infty} |X(f)|^2 df.$
Energy within a band from f_1 to f_2	$2\int_{f_1}^{f_2} |X(f)|^2 df.$
Frequency response of an LTI system	$H(f) = \int_{-\infty}^{\infty} h(t)e^{-j2\pi ft} dt,$

$h(t)$ is the system's unit-impulse response.

Useful Formulae

$$\int te^{\alpha t} dt = \frac{1}{\alpha^2}e^{\alpha t}(\alpha t - 1)$$

$$\int e^{\alpha t} \sin(\beta t)dt = \frac{e^{\alpha t}}{\alpha^2 + \beta^2}[\alpha \sin(\beta t) - \beta \cos(\beta t)]$$

$$\int e^{\alpha t} \cos(\beta t)dt = \frac{e^{\alpha t}}{\alpha^2 + \beta^2}[\alpha \cos(\beta t) + \beta \sin(\beta t)]$$

$$\int t \sin(\alpha t)dt = \frac{1}{\alpha^2} \sin(\alpha t) - \frac{t}{\alpha} \cos(\alpha t)$$

$$\int t \cos(\alpha t)dt = \frac{1}{\alpha^2} \cos(\alpha t) + \frac{t}{\alpha} \sin(\alpha t)$$

Solved Problems

1. a. Write the expression for the Fourier transform of a single even rectangular pulse of height V_0 and width τ such as that shown in Figure 8.1(a). The equation for the pulse is

$$x(t) = \begin{cases} V_0, & \text{for } -\frac{\tau}{2} < t < \frac{\tau}{2} \\ 0, & \text{elsewhere} \end{cases}$$

Observe that $X(f)$ is a real and even function of f, in agreement with expectations based on $x(t)$ being a real and even function of time.

b. Plot $X(f)$ as a function of f and specify values for the locations of the zero-crossings and $X(0)$.

c. Show that the total area under $X(f)$ is equal to the area of the triangle enclosed by the main lobe. From the above result deduce that

$$\int_{-\infty}^{\infty} \frac{\sin(\alpha t)}{t} dt = \pi$$

where α is a constant.

d. Obtain the Fourier transform of the rectangular pulse of Figure 8.1(a) from

$$X(f) = 2V_0 \int_0^{\frac{\tau}{2}} \cos(2\pi f t)\, dt$$

and show that the Fourier transform of a real and even function of time may be obtained from

$$X(f) = 2 \int_0^\infty x(t)\cos(2\pi f t)\, dt$$

Solution

a. $X(f) = V_0 \frac{\sin(\pi f \tau)}{\pi f}$. It is observed that $X(f)$ is the ratio of two real and odd functions of f, therefore, it is a real and even function of f.

b. See Figure 8.1(a). The zero-crossings of $X(f)$ are at $\sin(\pi f \tau) = 0$ or $f = \pm k/\tau$, where k is a nonzero integer. At $f = 0$ we obtain $X(0) = $ area under the time pulse $= V_0 \tau$.

c. Using the property $\int_{-\infty}^\infty X(f)df = x(0)$ we find $\int_{-\infty}^\infty X(f)df = V_0$. In other words the total (algebraic) area

under $X(f)$ of an even rectangular pulse is equal to the area of the shaded triangle enclosed by the main lobe in

Figure 8.1(a). From the above integral we find $\int_{-\infty}^\infty V_0 \frac{\sin(\pi f \tau)}{\pi f} df = V_0$ from which $\int_{-\infty}^\infty \frac{\sin(\alpha f)}{f} df = \pi$.

d.

$$X(f) = V_0 \int_{-\frac{\tau}{2}}^{\frac{\tau}{2}} e^{-j2\pi f t}\, dt = V_0 \int_{-\frac{\tau}{2}}^{\frac{\tau}{2}} \cos(2\pi f t)\, dt - jV_0 \int_{-\frac{\tau}{2}}^{\frac{\tau}{2}} \sin(2\pi f t)\, dt$$

$$= 2V_0 \int_0^{\frac{\tau}{2}} \cos(2\pi f t)\, dt = V_0 \left[\frac{\sin(2\pi f t)}{\pi f}\right]_0^{\frac{\tau}{2}} = V_0 \frac{\sin(\pi f \tau)}{\pi f}$$

The above formulation can be generalized for any even function $x(t) = x(-t)$:

$$X(f) = \int_{-\infty}^\infty x(t)e^{-j2\pi f t}\, dt = \int_{-\infty}^\infty x(t)\cos(2\pi f t)\, dt - j\int_{-\infty}^\infty x(t)\sin(2\pi f t)\, dt$$

$$= \int_{-\infty}^\infty x(t)\cos(2\pi f t)\, dt = 2\int_0^\infty x(t)\cos(2\pi f t)\, dt$$

2. Let $y(t) = x(t+\tau) + x(t) + x(t-\tau)$, where $x(t) = \begin{cases} V_0, & -\frac{\tau}{2} < t < \frac{\tau}{2} \\ 0, & \text{elsewhere} \end{cases}$ is the rectangular pulse of Figure 8.1(a).

Using the time shift and linearity properties of the Fourier transform, construct $Y(f)$ from $X(f)$. Show that $Y(f) = 3X(3f)$ and observe that $Y(f)$ is the same as the Fourier transform of a rectangular pulse of the same height as $x(t)$ but three times as wide.

Solution

$$y(t) = x(t+\tau) + x(t) + x(t-\tau)$$

$$Y(f) = X(f)e^{j2\pi f \tau} + X(f) + X(f)e^{-j2\pi f \tau} = X(f)[1 + 2\cos(2\pi f \tau)]$$

But $\qquad X(f) = V_0 \frac{\sin(\pi f \tau)}{\pi f}$ and $1 + 2\cos(2\pi f \tau) = 3\cos^2(\pi f \tau) - \sin^2(\pi f \tau)$

Therefore, $\quad Y(f) = V_0 \frac{3\cos^2(\pi f \tau)\sin(\pi f \tau) - \sin^3(\pi f \tau)}{\pi f} = V_0 \frac{\sin(3\pi f \tau)}{\pi f} = 3X(3f)$

3. **a.** Find the Fourier transforms of the causal rectangular pulse $x(t) = \begin{cases} 1 & 0 < t < \frac{\tau}{2} \\ 0 & \text{elsewhere} \end{cases}$.

b. Find the Fourier transforms of $x_e(t) = \dfrac{x(t) + x(-t)}{2}$ (the even part of x) and $x_o(t) = \dfrac{x(t) - x(-t)}{2}$ (the odd part of x) and verify their equivalence with the real and imaginary parts of $X(f)$, respectively.

Solution

a. $x(t) = \begin{cases} 1, & 0 < t < \frac{\tau}{2} \\ 0, & \text{elsewhere} \end{cases}$ $X(f) = \int_0^{\frac{\tau}{2}} e^{-j2\pi ft}\, dt = \dfrac{\sin\frac{\pi f \tau}{2}}{\pi f} e^{-j\frac{\pi f \tau}{2}} = \dfrac{\sin(\pi f \tau)}{2\pi f} - j\dfrac{\sin^2(\frac{\pi f \tau}{2})}{\pi f}$

b. $x_e(t) = \begin{cases} \frac{1}{2}, & -\frac{\tau}{2} < t < \frac{\tau}{2} \\ 0, & \text{elsewhere} \end{cases}$ $X_e(f) = \dfrac{\sin(\pi f \tau)}{2\pi f} = \mathcal{RE}\{X(f)\}$

$x_o(t) = \begin{cases} \frac{-1}{2}, & -\frac{\tau}{2} < t < 0 \\ \frac{1}{2}, & 0 < t < \frac{\tau}{2} \\ 0, & \text{elsewhere} \end{cases}$ $X_o(f) = \dfrac{\sin(\frac{\pi f \tau}{2})}{2\pi f}\left[e^{-j\frac{\pi f \tau}{2}} - e^{j\frac{\pi f \tau}{2}} \right] = -j\dfrac{\sin^2(\frac{\pi f \tau}{2})}{\pi f} = \mathcal{IM}\{X(f)\}$

4. Find the Fourier transform of the sawtooth pulse shown in Figure 8.9.

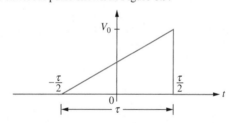

FIGURE 8.9

Solution

$$X(f) = \int_{-\frac{\tau}{2}}^{\frac{\tau}{2}} x(t) e^{-j2\pi ft}\, dt = \int_{-\frac{\tau}{2}}^{\frac{\tau}{2}} \left(\frac{V_0}{2} + \frac{V_0}{\tau} t \right) e^{-j2\pi ft}\, dt$$

$$= \frac{V_0}{2} \int_{-\frac{\tau}{2}}^{\frac{\tau}{2}} e^{-j2\pi ft}\, dt + \frac{V_0}{\tau} \int_{-\frac{\tau}{2}}^{\frac{\tau}{2}} t e^{-j2\pi ft}\, dt$$

$$= \frac{V_0}{2} \frac{\sin(\pi f \tau)}{\pi f} + \frac{V_0}{\tau} \int_{-\frac{\tau}{2}}^{\frac{\tau}{2}} t e^{\alpha t}\, dt, \quad \text{where} \quad \alpha = -j2\pi f$$

But $\displaystyle\int t e^{\alpha t}\, dt = \frac{1}{\alpha^2} e^{\alpha t}(\alpha t - 1)$

Therefore, $X(f) = \dfrac{V_0}{2} \dfrac{\sin(\pi f \tau)}{\pi f} + \dfrac{V_0}{\alpha \tau} \left[e^{\alpha t} \left(t - \dfrac{1}{\alpha} \right) \right]_{t=-\frac{\tau}{2}}^{\frac{\tau}{2}}$

$$= \frac{V_0}{2} \frac{\sin(\pi f \tau)}{\pi f} + j\frac{V_0}{2}\left[\frac{\cos(\pi f \tau)}{\pi f} - \frac{1}{\tau} \frac{\sin(\pi f \tau)}{(\pi f)^2} \right]$$

In summary $X(f) = \mathcal{RE}\{X(f)\} + j\mathcal{IM}\{X(f)\}$

$$\mathcal{RE}\{X(f)\} = \frac{V_0}{2}\frac{\sin(\pi f\tau)}{\pi f}, \qquad \mathcal{IM}\{X(f)\} = \frac{V_0}{2}\left[\frac{\cos(\pi f\tau)}{\pi f} - \frac{1}{\tau}\frac{\sin(\pi f\tau)}{(\pi f)^2}\right]$$

5. Find the Fourier transform of the time derivative of the $x(t)$ in problem 4.

Solution

Using $X(f)$ found in problem 4 and applying the differentiation property of the Fourier transform (shown in Table 8.1) we find

$$Y(f) = j2\pi f X(f) = j2\pi f\left\{\frac{V_0}{2}\frac{\sin(\pi f\tau)}{\pi f} + j\frac{V_0}{2}\left[\frac{\cos(\pi f\tau)}{\pi f} - \frac{1}{\tau}\frac{\sin(\pi f\tau)}{(\pi f)^2}\right]\right\}$$

$$= V_0\left[\frac{\sin(\pi f\tau)}{\pi f\tau} - \cos(\pi f\tau)\right] + jV_0\sin(\pi f\tau)$$

6. One may suppose that the time derivative of the sawtooth pulse of Figure 8.13 is the even rectangular pulse

$y_i(t) = \begin{cases} \frac{V_0}{\tau}, & -\frac{\tau}{2} < t < \frac{\tau}{2} \\ 0, & \text{elsewhere} \end{cases}$ and expect to obtain $Y(f) = Y_i(f) = \dfrac{V_0}{\tau}\dfrac{\sin(\pi f\tau)}{\pi f}$ (not in agreement with the

answer in problem 5). Show that the above supposition is incomplete and correct it.

Solution

The above supposition misses the negative impulse, with strength V_0 at $t = \tau/2$, in the derivative of $x(t)$. The sawtooth pulse $x(t)$ shown in Figure 8.9 and the correct expressions for its derivative $y(t)$ and $Y(f)$ are given below.

$$x(t) = \begin{cases} \frac{V_0}{2} + \frac{V_0}{\tau}t, & -\frac{\tau}{2} < t < \frac{\tau}{2} \\ 0, & \text{elsewhere} \end{cases}$$

$$y(t) = \frac{dx}{dt} = y_i(t) - V_0\delta\left(t - \frac{\tau}{2}\right) \quad \text{where } y_i(t) = \begin{cases} \frac{V_0}{\tau}, & -\frac{\tau}{2} < t < \frac{\tau}{2} \\ 0, & \text{elsewhere} \end{cases}$$

$$Y(f) = Y_i(f) - V_0 e^{-j\pi f\tau} = \frac{V_0}{\tau}\frac{\sin(\pi f\tau)}{\pi f} - V_0\cos(\pi f\tau) + jV_0\sin(\pi f\tau)$$

Inclusion of the downward impulse has completed the expression for $y(t)$ resulting in $Y(f)$ in agreement with the answer obtained in problem 5.

7. Find the Fourier transform of the raised cosine pulse (also called Hanning window) τ seconds wide:

$$x(t) = \begin{cases} 0.5(1 + \cos 2\pi f_0 t), & -\frac{\tau}{2} < t < \frac{\tau}{2} \\ 0, & \text{elsewhere} \end{cases}, \quad \text{where } f_0 = \frac{1}{\tau}$$

Solution

Define $p(t)$ to be the rectangular pulse

$$p(t) = \begin{cases} 1, & -\frac{\tau}{2} < t < \frac{\tau}{2} \\ 0, & \text{elsewhere} \end{cases} \quad \text{and } P(f) = \frac{\sin(\pi\tau f)}{\pi f}$$

Then

$$x(t) = 0.5p(t)(1 + \cos 2\pi f_0 t)$$

$$X(f) = 0.5\left[P(f) + 0.5P(f + f_0) + 0.5P(f - f_0)\right]$$

$$= \frac{\sin(\pi\tau f)}{2\pi f} + \frac{\sin\pi\tau(f + f_0)}{4\pi(f + f_0)} + \frac{\sin\pi\tau(f - f_0)}{4\pi(f - f_0)}$$

But $\qquad \sin \pi \tau (f \pm f_0) = \sin(\pi \tau f \pm \pi) = -\sin(\pi \tau f)$

Therefore, $\qquad X(f) = \dfrac{\sin(\pi \tau f)}{2\pi f}\left[1 - \dfrac{f}{2(f - f_0)} - \dfrac{f}{2(f + f_0)}\right] = \dfrac{\sin(\pi \tau f)}{2\pi f}\left[\dfrac{1}{1 - \left(\frac{f}{f_0}\right)^2}\right]$

8. Find the time function whose Fourier transform is a raised cosine pulse given below (a $2f_0$–Hz wide Hanning window in the frequency domain):

$$X(f) = \begin{cases} 0.5(1 + \cos \frac{\pi f}{f_0}), & |f| < f_0 \\ 0, & \text{elsewhere} \end{cases}$$

Solution

Consider the time function $p(t) = \dfrac{\sin(2\pi f_0 t)}{\pi t}$ and $P(f) = \begin{cases} 1, & -f_0 < f < f_0 \\ 0, & \text{elsewhere} \end{cases}$

Then $\qquad X(f) = 0.5 P(f) \left(1 + \cos \dfrac{\pi f}{f_0}\right)$

$$x(t) = 0.5\left[p(t) + 0.5p\left(t + \dfrac{\tau}{2}\right) + 0.5p\left(t - \dfrac{\tau}{2}\right)\right]$$

$$= \dfrac{\sin(2\pi f_0 t)}{2\pi t} + \dfrac{\sin 2\pi f_0(t + \frac{\tau}{2})}{4\pi(t + \frac{\tau}{2})} + \dfrac{\sin 2\pi f_0(t - \frac{\tau}{2})}{4\pi(t - \frac{\tau}{2})}, \quad \text{where } \tau = \dfrac{1}{f_0}$$

But $\qquad \sin 2\pi f_0\left(t \pm \dfrac{\tau}{2}\right) = \sin 2\pi f_0 t \pm \pi) = -\sin(2\pi f_0 t)$

Therefore, $\qquad x(t) = \dfrac{\sin(2\pi f_0 t)}{2\pi t}\left[1 - \dfrac{t}{2\left(t + \frac{\tau}{2}\right)} - \dfrac{t}{2\left(t - \frac{\tau}{2}\right)}\right] = \dfrac{\sin(2\pi f_0 t)}{2\pi f}\left[\dfrac{1}{1 - (2f_0 t)^2}\right]$

9. Fourier transform of a decaying sinusoid

The decaying sinusoid $e^{-\alpha t} \cos(2\pi f_0 t)u(t)$ is often used to model some signals and systems. Find its Fourier transform.

Solution

The Fourier transform may be obtained by using the frequency shift property.

$$e^{-\alpha t}u(t) \Rightarrow \dfrac{1}{\alpha + j2\pi f}$$

$$e^{-\alpha t}\cos(2\pi f_0 t)u(t) \Rightarrow \dfrac{\alpha + j2\pi f}{(\alpha + j2\pi f)^2 + 4\pi^2 f_0^2}$$

Note: The above function is absolutely integrable and so the Fourier transform can also be obtained from the Laplace transform by replacing s with $j2\pi f$.

Chapter Problems

10. Determine the Fourier transform of a real and even rectangular pulse with height V_0 and width $\tau/2$.

11. Determine the Fourier transform of a real and even rectangular pulse with height V_0 and width 2τ.

12. Determine and sketch the Fourier transform of the rectangular pulse shown in Figure 8.1(a) for V_0=10 V and the following values of τ:

  a. 100 msec
  b. 10 msec
  c. 1 msec
  d. 100 μsec
  e. 10 μsec
  f. 1 μsec

13. Repeat problem 12 for the following values of τ and V_0:

a. $\tau = 100$ msec	and	$V_0 = 10$ mV	d. $\tau = 100\ \mu$sec	and	$V_0 = 10$ V
b. $\tau = 10$ msec	and	$V_0 = 100$ mV	e. $\tau = 10\ \mu$sec	and	$V_0 = 100$ V
c. $\tau = 1$ msec	and	$V_0 = 1$ V	f. $\tau = 1\ \mu$sec	and	$V_0 = 1$ kV

14. In Figure 8.1(a), determine values for τ that result in zero-crossings of $X(f)$ at (a) $f = n$ kHz and (b) $f = 10n$ kHz, where $n = 1, 2, 3 \ldots$.

15. In Figure 8.1(a), determine V_0 so that $X(0) = 2 \times 10^{-3}$V $\times$ sec given (a) $\tau = 1$ msec and (b) $\tau = 100\ \mu$s.

16. Determine the Fourier transform of the single rectangular pulse

$$x(t) = \begin{cases} \frac{1}{\tau}, & \text{for } -\frac{\tau}{2} < t < \frac{\tau}{2} \\ 0, & \text{elsewhere} \end{cases}$$

and verify that $X(0) = 1$ regardless of τ.

17. Determine the Fourier transform of the single rectangular pulse of problem 16 for the following values of τ:

  a. 100 sec   b. 1 sec   c. 10 msec. Find $\lim_{\tau \to 0} X(f)$.

18. Use the duality property of the Fourier transform and the fact that the transform of a unit impulse is 1 to show that the Fourier transform of $x(t) = 1$ is $\delta(f)$.

19. Show that

$$\lim_{\tau \to 0} \left[\frac{\sin(\pi f \tau)}{\pi f \tau} \right] = 1 \quad \text{(Fourier transform of an impulse)}$$

20. Show that

$$\lim_{\tau \to \infty} \left[\frac{\sin(\pi f \tau)}{\pi f} \right] = \delta(f) \quad \text{(Fourier transform of a DC signal)}$$

21. Find the energy in the single rectangular pulse of Figure 8.1(a) for

  a. $V_0 = 1$ V and $\tau = 1$ msec
  b. $V_0 = 2$ V and $\tau = 250\ \mu$sec
  c. $V_0 = 5$ V and $\tau = 40\ \mu$sec

22. Find the energy in the frequency band from 0 to 10 Hz in a single rectangular pulse [as in Figure 8.1(a)] for

  a. $V_0 = 1$ V and $\tau = 1$ msec
  b. $V_0 = 2$ V and $\tau = 250\ \mu$sec
  c. $V_0 = 5$ V and $\tau = 40\ \mu$sec

23. Find the percentage of energy in the frequency band from 0 to 10 Hz in a single rectangular pulse [as in Figure 8.1(*a*)] for

a. $\tau = 1$ msec
b. $\tau = 250$ μsec
c. $\tau = 40$ μsec

24. Find the ratio of the amount of energy residing within the frequency range of the main lobe of the Fourier transform of a rectangular pulse to the total energy in the pulse.

25. Generalize problem 2 to show that

$$y(t) = \sum_{n=-k}^{k} x(t - n\tau) \implies Y(f) = \left[1 + 2\sum_{n=1}^{k} \cos(2\pi n f \tau)\right] X(f) = (2k+1)X[(2k+1)f]$$

26. Let $x(t)$ be the rectangular pulse of Figure 8.1(*a*) with $V_0 = 1$ V and $\tau = 100$ msec. Find and plot the Fourier transform of

$$y(t) = \sum_{n=-k}^{k} x(t - n)$$

for

a. $k = 1$
b. $k = 2$
c. $k = 10$

27. Repeat problem 26 for

$$y(t) = \sum_{n=-k}^{k} x(t - nT)$$

where $T = 500$ msec.

28. Let $x(t)$ be the rectangular pulse of Figure 8.1(*a*) with $V_0 = 1$ V and $\tau = 100$ msec. Find and plot the Fourier transform of

$$y(t) = \sum_{n=-k}^{k} (-1)^n x(t - n\tau)$$

for

a. $k = 1$
b. $k = 2$
c. $k = 10$

29. Let $x(t)$ be the rectangular pulse of Figure 8.1(*a*) with $V_0 = 1$ V and $\tau = 100$ msec. Find and plot the Fourier transform of

$$y(t) = \sum_{n=-k}^{k} (-1)^n x(t - n)$$

for

a. $k = 1$
b. $k = 2$
c. $k = 10$

30. Repeat problem 29 for

$$y(t) = \sum_{n=-k}^{k} (-1)^n x(t - nT)$$

with $T = 500$ msec.

31. Let $x(t)$ be the rectangular pulse of Figure 8.1(a) with $V_0 = 1$ V and $\tau = 2$ sec. A periodic waveform $y(t)$ is generated by repeating $x(t)$ every $T = 8$ seconds. Find and plot the Fourier transform of $y(t)$.

32. Repeat problem 31 for $\tau = 2$ and $T = 8$, both in msec.

33. Let $x(t)$ be the rectangular pulse of Figure 8.1(a) with $V_0 = 1$ V and $\tau = 2$ s. Find and plot the Fourier transform of

$$y(t) = \sum_{n=-\infty}^{k=\infty} (-1)^n x(t - nT)$$

with $T = 8$ s.

34. Repeat problem 33 for $\tau = 2$ and $T = 8$, both im msec.

35. A periodic waveform $y(t)$ is generated by repeating the rectangular pulse of Figure 8.1(a) every T seconds, where $T > \tau$.

$$y(t) = \sum_{n=-\infty}^{\infty} x(t - nT)$$

Find a mathematical expression for the Fourier transform of $y(t)$ in terms of V_0, τ, and T.

36. Repeat problem 35 for

$$y(t) = \sum_{n=-\infty}^{\infty} (-1)^n x(t - nT)$$

37. Find the peak and average power in the periodic rectangular pulse train with period $T = 500$ msec, for
   a. $V_0 = 1$ V and $\tau = 1$ msec
   b. $V_0 = 2$ V and $\tau = 250$ μsec
   c. $V_0 = 5$ V and $\tau = 40$ μsec

38. Find the average power in the frequency band from 0 to 10 Hz for the periodic rectangular pulse train with period $T = 500$ msec, for
   a. $V_0 = 1$ V and $\tau = 1$ msec
   b. $V_0 = 2$ V and $\tau = 250$ μsec
   c. $V_0 = 5$ V and $\tau = 40$ μsec

39. Find the percentage of average power in the frequency band from 0 to 10 Hz for a periodic rectangular pulse train with period $T = 500$ msec, for
   a. $\tau = 1$ msec
   b. $\tau = 250$ μsec
   c. $\tau = 40$ μsec

40. Find the Fourier transform of the step response of an ideal low-pass filter for which $H(f) = u(f + 1) - u(f - 1)$. *Hint:* Use the integration property of the Fourier transform.

41. Find the Fourier transform of the pulse $x(t)$ in Figure 8.10, where $\tau = 1.1$ msec, $V_0 = 1$ V, $\tau_1 = 100$ μsec, and $V_1 = 10$ V. Sketch $X(f)$ and label its important points.

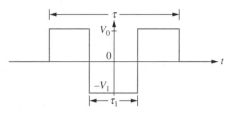

FIGURE 8.10

42. Find the Fourier transform of

$$y(t) = \int_{-\infty}^{t} x(t) dt$$

in which $x(t)$ is the pulse shown in Figure 8.10 (with $\tau = 1.1$ msec, $V_0 = 1$ V, $\tau_1 = 100$ μsec, and $V_1 = 10$ V). Sketch $Y(f)$ and label its important points.

43. The unit-impulse response of an LTI system is

$$h(t) = \begin{cases} 1, & -1 < t < 1 \\ -1, & -2 < t < -1 \text{ and } 1 < t < 2 \\ 0, & \text{elsewhere} \end{cases}$$

Find and plot the magnitude function $|H(f)|$. Show that the system is a kind of high-pass filter that approximates a differentiator within a frequencies range. Determine your criteria for the validity of the approximation and find the acceptable range of operating frequencies.

44. Find the Fourier transform of

$$x(t) = \begin{cases} V_0, & \text{for } -\frac{\tau}{2} < t < 0 \\ -V_0, & \text{for } 0 < t < \frac{\tau}{2} \\ 0, & \text{elsewhere} \end{cases}$$

See Figure 8.11(a). Verify that $X(f)$ is an imaginary and odd function of f, in agreement with expectations based on $x(t)$ being a real and odd function.

45. a. Verify that for the pulse of Figure 8.11(a), $X(0) = 0$ regardless of τ and V_0 values.
   b. Generalize the above observation to every $x(t)$, which is a real and odd function of time.

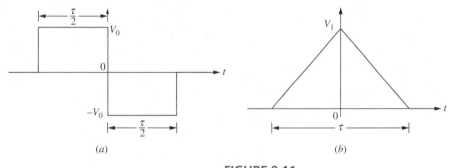

(a)   (b)

FIGURE 8.11

46. a. Show that the Fourier transform of the pulse of Figure 8.11(a) may be obtained from

$$X(f) = 2jV_0 \int_0^{\tau/2} \sin(2\pi ft)\, dt$$

b. Assume an $x(t)$ is a real and odd function of the real variable t. Show that its Fourier transform may be obtained from

$$X(f) = -2j \int_0^\infty x(t) \sin(2\pi ft)\, dt$$

47. Find the Fourier transform of an even triangular pulse of height V_1 and base τ as in Figure 8.11(b). Sketch the transform and label its important points.

48. a. Note that the convolution of a rectangular pulse with itself produces a triangular pulse. From the above observation and using the convolution property of the Fourier transform, derive the Fourier transform of the triangular pulse of height 1 and base τ and verify that it is in agreement with the result obtained by direct integration in problem 47.

b. Find the percentage of the energy residing within the frequency range of the main lobe of the transform of a triangular pulse. Compare with the percentage of energy residing within the frequency range of the main lobe of the transform of a rectangular pulse of the same width.

49. Find the Fourier transform of an even triangular pulse of height V_1 and base 2τ. Sketch the transform and label its important points.

50. Assume the triangle of Figure 8.11(b) is the integral of the pulse in Figure 8.11(a).

a. Specify V_1 [in Figure 8.11(b)] in terms of V_0 and τ [in Figure 8.11(a)].

b. Verify that the answers obtained in problems 44 and 47 are in compliance with the derivative property of the Fourier transform.

51. Find the Fourier transform of the sawtooth pulse shown in Figure 8.12(a).

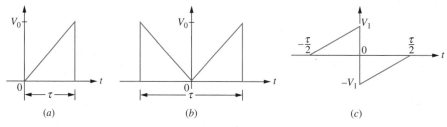

(a)   (b)   (c)

FIGURE 8.12

52. Find the Fourier transform of the even pulse shown in Figure 8.12(b).

53. Find the Fourier transform of the odd pulse shown in Figure 8.12(c).

54. Find the Fourier transform of the odd pulses shown in Figure 8.13(a).

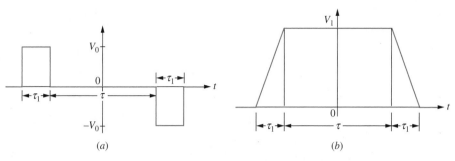

(a)   (b)

FIGURE 8.13

55. Find the Fourier transform of the even trapezoidal pulse shown in Figure 8.13(b).

56. Assume the trapezoidal pulse of Figure 8.13(b) is the integral of the pulses in Figure 8.13(a).

a. Specify V_1 [in Figure 8.13(b)] in terms of V_0 and τ_1 [in Figure 8.13(a)].

b. Verify that the answers obtained in problems 54 and 55 are in compliance with the derivative property of the Fourier transform.

57. Find the Fourier transform of a single cycle of a sinusoidal pulse

$$y(t) = \begin{cases} -V_0 \sin \frac{2\pi t}{\tau}, & \text{for } -\frac{\tau}{2} < t < \frac{\tau}{2} \\ 0, & \text{elsewhere} \end{cases}$$

as in Figure 8.14(a), where $\tau = 1$ msec. Sketch $Y(f)$ and label its important points.

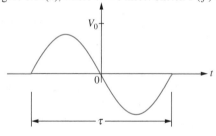
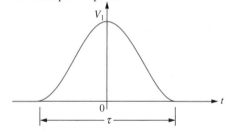

FIGURE 8.14

58. Find the Fourier transform of a single raised cosine pulse

$$y(t) = \begin{cases} \frac{V_1}{2} \left[1 + \cos \frac{2\pi t}{\tau} \right], & \text{for } -\frac{\tau}{2} < t < \frac{\tau}{2} \\ 0, & \text{elsewhere} \end{cases}$$

as in Figure 8.14(b), where $\tau = 1$ msec. Sketch $Y(f)$ and label its important points.

59. Assume the pulse of Figure 8.14(b) is the integral of the pulse in Figure 8.14(a).

a. Specify V_1 [in Figure 8.14(b)] in terms of V_0 and τ [in Figure 8.14(a)].

b. Verify that the answers obtained in problems 57 and 58 are in compliance with the derivative property of the Fourier transform.

60. Find the Fourier transform of the waveform made of a pair of half-cycle sinusoidal pulses as shown in Figure 8.15(*a*).

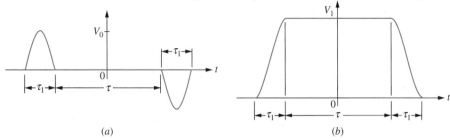

(*a*)   (*b*)

FIGURE 8.15

61. Find the Fourier transform of the trapezoid-like pulse shown in Figure 8.15(*b*). The leading and trailing edges are raised cosines.

62. Assume the trapezoidal pulse of Figure 8.15(*b*) is the integral of the pulses in Figure 8.15(*a*).
   a. Specify V_1 [in Figure 8.15(*b*)] in terms of V_0 and τ_1 [in Figure 8.15(*a*)].
   b. Verify that the answers obtained in problems 60 and 61 are in compliance with the derivative property of the Fourier transform.

63. a. Find the Fourier transform of a single pulse $x(t)$ shown in Figure 8.16 with $\tau = 5$ μs. Sketch $X(f)$ and label its important points.
   b. Find the Fourier series coefficients of a periodic waveform generated by repeating the pulse of part a every $T = 50$ μs. *Hint*: Use the superposition property of the Fourier transform.

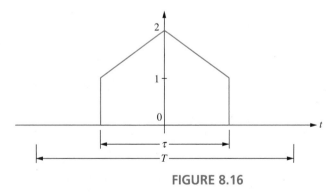

FIGURE 8.16

64. Repeat problem 63 for $\tau = 1$ msec and
   a. $T = 1$ msec
   b. $T = 10$ msec
   c. $T = 100$ msec

65. Repeat problem 63 for $T = 1$ msec and
   a. $\tau = 1$ μsec
   b. $\tau = 10$ μsec
   c. $\tau = 100$ μsec

66. Verify the Fourier transforms of the continuous-time windows of Table 8.2 in section 8.19 of the text.

67. Find and plot the Fourier transforms of the pulses described below. All pulses are of width 2τ. The mathematical expressions are given for $-\tau < t < \tau$. Elsewhere, the pulses have zero value.

a. A cosine pulse: $x(t) = \cos\left(\frac{\pi t}{2\tau}\right)$ b. A parabolic pulse: $x(t) = 1 - \left(\frac{t}{\tau}\right)^2$

c. An arc pulse: $x(t) = \sqrt{R^2 - t^2} - \sqrt{R^2 - \tau^2}$ d. A dome pulse: $x(t) = \left(\frac{t}{\tau}\right)^4 - 2\left(\frac{t}{\tau}\right)^2 + 1$

Compare the results with the Fourier transforms of the windows of section 8.19.

68. Find the Fourier transform of $x(t) = \frac{\sin(100\pi t)}{\pi t}$.

69. Find and plot the Fourier transform of the Lanczos window given by

$$w(t) = \begin{cases} \sin(\frac{\pi t}{\tau})/(\frac{\pi t}{\tau}), & -\tau < t < \tau \\ 0, & \text{elsewhere} \end{cases}$$

Compare with the Fourier transforms of the windows of section 8.19.

70. Find the Fourier transforms of the time functions given below, where $u(t)$ is the unit-step function and $\tau > 0$. If a transform does not exist, give a reason.

a. $x(t) = 1$, all t.   b. $x(t) = u(t)$

c. $x(t) = \frac{1}{2}\left[u(t) - u(-t)\right]$ d. $x(t) = tu(t)$

e. $x(t) = |t|$ f. $x(t) = \frac{1}{\sqrt{|t|}}$

g. $x(t) = e^{-\frac{t}{\tau}}u(t)$ h. $x(t) = te^{-\frac{t}{\tau}}u(t)$

i. $x(t) = e^{-|\frac{t}{\tau}|} = e^{-|f_o t|}$, where $f_o = \frac{1}{\tau}$ j. $x(t) = \frac{2\tau}{\tau^2 + 4\pi^2 t^2} = \frac{2f_o}{1 + 4\pi^2 f_o^2 t^2}$, where $f_o = \frac{1}{\tau}$

k. $x(t) = e^{-\frac{t}{\tau}}u(t) - e^{\frac{t}{\tau}}u(-t)$ l. $x(t) = e^{-\pi t^2/\tau^2}$

71. Find the Fourier transforms of the following functions:

a. $x(t) = \sin(2\pi f_0 t) + \cos(2\pi f_0 t)$ b. $x(t) = \sin(2\pi f_1 t) + \cos(2\pi f_2 t)$

c. $x(t) = \frac{2}{\pi}\left[\cos(\omega_0 t) - \frac{1}{3}\cos(3\omega_0 t) + \frac{1}{5}\cos(5\omega_0 t)\right]$ d. $x(t) = e^{-\frac{t}{\tau}}\cos(\omega_0 t)u(t)$

72. Find the Fourier transforms of

a. $x(t) = \frac{\sin(2\pi f_0 t)}{\pi t}$ b. $x(t) = 2\left[\frac{\sin(2\pi f_0 t)}{\pi t}\right]\cos(2\pi f_c t)$, where $f_c > f_0$

c. $x(t) = \left[\frac{\sin(2\pi f_0 t)}{\pi t}\right]^2$ d. $x(t) = 2\left[\frac{\sin(2\pi f_0 t)}{\pi t}\right]^2\cos(2\pi f_c t)$, where $f_c > f_0$

73. Find the Fourier transform of

$$x(t) = \frac{1}{2} + 2\sum_{n=1}^{N}\left[\frac{\sin(\frac{\pi n}{2})}{\pi n}\right]\cos(n\omega_0 t)$$

The above function is the sum of the first $N + 1$ terms of the trigonometric Fourier series expansion of a periodic square pulse with period $T = \frac{2\pi}{\omega_0}$.

74. Find the Fourier transform of

$$x(t) = \frac{\tau}{T} + 2\sum_{n=1}^{N}\left[\frac{\sin(\frac{\pi n \tau}{T})}{\pi n}\right]\cos\left(\frac{2\pi n t}{T}\right), \quad \text{where } T > \tau$$

The above function is the sum of the first $N + 1$ terms of the trigonometric Fourier series of a periodic rectangular pulse with period T and pulse width τ.

75. One period of a periodic signal is $x(t) = \sqrt{100 - t^2}$, $-10 \le t \le 10$. Find its Fourier transform.

76. Long-tone burst with a rectangular envelope.

   a. Find the Fourier transform of $x(t) = a(t)\cos 2\pi f_c t$, where $a(t)$ is a single rectangular pulse as shown in Figure 8.1(a) with $V_0 = 1$ V and $\tau = 10$ msec and $f_c = 5$ kHz. Sketch $X(f)$ and label important points.
   b. Is $x(t)$ band limited?
   c. Find the total energy in $x(t)$.
   d. Find the percentage of the energy contained within the bandwidth from 4.9 kHz to 5.1 kHz.

77. Short-tone burst with a triangular envelope. Repeat problem 76 for a tone burst with a triangular envelope such as that shown in Figure 8.1(b) with $V_1 = 1$ V and $\tau = 1$ msec.

78. Long-tone burst with a trapezoidal envelope. Repeat problem 76 for a tone burst with a trapezoidal envelope such as shown in Figure 8.13(b) with rise and fall times of 0.5 msec each, a total tone duration of 11 msec, and a maximum amplitude of $V_1 = 1$ V.

79. Short-tone burst with a raised cosine envelope. Repeat problem 76 for a tone burst having a raised cosine envelope such as that shown in Figure 8.14(b) with $V_1 = 1$ V and $\tau = 1$ msec.

80. Long-tone burst with raised cosine rise and fall times. Repeat problem 76 for a tone burst having rise and fall times of 0.5 msec each in the shape of a raised cosine, a total tone duration of 11 msec, and a maximum amplitude of $V_1 = 1$ V. See Figure 8.15(b).

81. Two-tone signal. Consider two-tone signals $x_1(t)$ and $x_2(t)$ defined below:

   a. $x_1(t) = s_1(t)\cos(2\pi f_1 t)$, where $s_1(t)$ is a 1-V, 1-msec rectangular pulse and $f_1 = 1,070$ Hz
   b. $x_2(t) = s_2(t)\cos(2\pi f_2 t)$, where $s_2(t)$ is a 1-V, 3-msec rectangular pulse and $f_2 = 1,270$ Hz.

   Find the Fourier transforms of $x_1(t)$, $x_2(t)$, and $y(t) = x_1(t) \star x_2(t)$ and plot them.

82. Repeat problem 81 for $f_1 = 2025$ Hz and $f_2 = 2,225$ Hz and compare with the results of that problem.

83. Find the Fourier transform of $x(t) = a(t)\cos(2\pi f_c t)$, where $a(t)$ is a periodic rectangular pulse (1 V, 1 msec wide, $T = 5$ msec) and $f_c = 100$ kHz. Sketch $X(f)$ and label the important points. Find the average power in $x(t)$ and the percentage of power contained within the bandwidth from 99 kHz to 101 kHz.

84. Repeat problem 83 for $x(t) = [a(t) + 1]\cos(2\pi f_c t)$.

85. Find and plot the Fourier transform of

$$h(t) = \frac{1}{2}\delta(t + T) + \delta(t) + \frac{1}{2}\delta(t - T)$$

Hint: Use the shift property of the Fourier transform.

86. Find and plot the Fourier transform of

$$h(t) = \delta(t + T) + \delta(t) + \delta(t - T)$$

87. Find and plot the Fourier transform of a finite set of $2N + 1$ unit impulses positioned every T seconds from $-NT$ to NT.

$$N(t) = \sum_{n=-N}^{N} \delta(t - nT)$$

Hint: Use the shift property of the Fourier transform and the finite sum identity

$$\sum_{n=0}^{2N} a^n = \frac{1 - a^{2N+1}}{1 - a}$$

88. Find and plot the Fourier transform of $s(t)$, an infinite train of alternating positive/negative unit impulses positioned every T seconds from $-\infty$ to ∞.

$$s(t) = \sum_{n=-\infty}^{\infty} (-1)^n \delta(t - nT)$$

89. a. Find the Fourier transform of a single rectangular pulse $x(t)$ as shown in Figure 8.1(*a*) with $V_0 = 1$ V and $\tau = 1$ msec.
   b. Pass $x(t)$ through an LTI system with frequency response

$$H(f) = \begin{cases} 1, & \text{for } |f| < 10 \text{ Hz} \\ 0, & \text{elsewhere} \end{cases}$$

By way of approximation, find and sketch $y(t)$, the output of the filter, in the form of a time function.

90. a. A signal is modeled by $x(t) = e^{-|t/\tau|}$, where $\tau = 1$ msec. The signal is passed through an ideal low-pass filter with unity gain and cutoff frequency $f_0 = 10$ Hz. Find the output $y(t)$ as a time function. You may use reasonable approximations.
   b. The total energy in $x(t)$ is $\int_{-\infty}^{\infty} |x(t)|^2 dt$. Find the ratio of total energy in $y(t)$ to total energy in $x(t)$.

91. a. Find the Fourier series coefficients of a periodic square pulse $x(t)$ with period $T = 1$ msec and a 50% duty cycle ($\tau = T/2$, zero DC level and 1 volt peak-to-peak value).
   b. Pass $x(t)$ through a low-pass LTI system with cutoff frequency $f_0 = 10$ Hz. Find the output of the system in the form of a time function. Find the average power of the output.

92. Repeat part b of problem 91 if $x(t)$ is a periodic triangular waveform with period $T = 1$ msec, a 50% duty cycle, and 1 volt base-to-peak value.

93. Assume $x(t)$ is a periodic square pulse with period $T = 1$ msec, zero DC level, and 1 volt peak-to-peak value. Pass $x(t)$ through an LTI system for which the Fourier transform of its unit-impulse response (called the frequency response) is shown in Figure 8.4(*b*) with a transitional cutoff frequency from $f_0 = 9$ Hz to $f_1 = 10$ Hz. Find the output of the filter in the form of a time function.

94. The frequency response of an LTI system is

$$H(f) = \begin{cases} 1, & \text{for } |f| < 1 \text{ kHz} \\ 0, & \text{elsewhere} \end{cases}$$

The input to the system is a periodic rectangular pulse train ($\tau = 2$ and $T = 8$ both in msec, and $V_0 = 1$ volt).
a. Find X_n and Y_n. b. Find total average power in $y(t)$.

95. Repeat problem 94 for

$$H(f) = \begin{cases} 1, & \text{for } 1 < |f| < 2 \text{ kHz} \\ 0, & \text{elsewhere} \end{cases}$$

96. A periodic function $s(t)$ assumes integer values only and can switch from one value to another only at $t = k\tau$ where k is an integer. One period of $s(t)$ (with $\tau = 1$ and $T = 10$ both in seconds) is represented by

$$s(t) = \{0, -1, -1, 2, 2, 2, 2, , -1, -1, 0\}$$
$$\uparrow$$

Simliarly, another periodic function $n(t)$ is represented by

$$n(t) = \{-1, 1, -1, -1, 1.5, 1.5, -1, -1, 1, -1\}$$
$$\uparrow$$

In the above representations the $\uparrow$ indicates the time origin.

a. Find Fourier series coefficients of $s(t)$, $n(t)$ and $x(t) = s(t) + n(t)$. *Hint:* Use the superposition property or a computer.

b. Find the average powers P_s, P_n, and P_x during one period.

c. Pass $s(t)$, $n(t)$, and $x(t)$ through an ideal bandpass filter with a 0.1-Hz passband around the center frequency of $0.1n$ Hz (where n is an integer). Repeat part b for the outputs of filter with $n = 1, 2, 3, 4$. Determine n, which maximizes the ratio of signal power to noise power at the output of the filter.

97. A full-wave rectifier is connected through a diode to a parallel RC filter (with $R = 10 \text{ k}\Omega$ and $C = 10 \text{ }\mu F$). See Figure 8.17(*a*). The input voltage to the diode-filter combination is a full-wave rectified sinusoid with an rms value of 15 V at 60 Hz. The threshold voltage for the diode is 0.7 V:

$$\begin{cases} i = 0, & \text{if } v < 0.7 \text{ V} \\ v = 0.7 \text{ V}, & \text{if } i > 0 \end{cases}$$

The diode's characteristic is shown in Figure 8.17(*b*). Find the percentage of power in the first three nonzero harmonics of the voltage at the output of the filter.

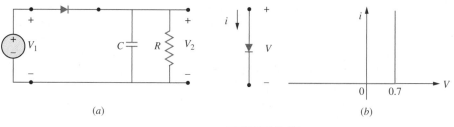

(*a*)   (*b*)

FIGURE 8.17

8.29   Project: Spectral Analysis Using a Digital Oscilloscope

Summary

In this experiment you will measure the spectra of several signals and compare them with theory. Measurement is done on a segment of the waveform observed and recorded through a window. The experiment may be performed by simulation or in real time. For simulation one may use a computation engine such as Matlab, Mathcad, Spice, or a signal processing software package. For real-time analysis one may use a general-purpose laboratory instrument such as an oscilloscope with spectral analysis capability, or a special-purpose instrument such as a spectrum analyzer or a DSP board. Using an oscilloscope is recommended in this project.

Important Reminder

Spectral analysis by digital tools (such as digital scope, spectrum analyzer, and digital computers) is done by finding the DFT/FFT of the sampled signal. A high sampling rate (such as the rate used by the scope in this experiment) makes the DFT/FFT a good measure of the Fourier transform of the continuous-time signal. In this experiment signals are sampled by the digital oscilloscope at sufficiently high rates for them to be treated as if they were continuous-time signals. In all plots, scales should be given in terms of the continuous-time frequency (Hz).

Preparation

Numerical Spectra of Single Pulses

Find the Fourier transforms $H(f)$ of single pulses specified in Table 8.3 and enter them in that table. Plot $H(f)$ in the last column with labels and scales.

TABLE 8.3 Spectra of Single Pulses $V_{\text{base-to-peak}} = 1$ V and Their Plots (To be completed by student)

$h(t)$	$H(f)$	Plot of $H(f)$
A single rectangular pulse, duration $= 500\ \mu s$		
A single rectangular pulse, duration $= 200\ \mu s$		
A single triangular pulse, duration $= 1$ ms		

Spectra of Periodic Pulses

Consider three periodic waveforms $x_1(t)$, $x_2(t)$, and $x_3(t)$ specified in Table 8.4. Note that each of these repetitive pulses is produced by the convolution of a single pulse (from Table 8.4) with a train of unit inpulses every T seconds.

$$x(t) = h(t) \star \sum_{-\infty}^{-\infty} \delta(t - nT)$$

$$X(f) = H(f) \times \frac{1}{T}\sum_{-\infty}^{-\infty} \delta(f - n/T) = \sum_{-\infty}^{-\infty} X_n \delta(f - f_0), \text{ where } X_n = \frac{1}{T}H(nf_0)$$

where $f_0 = 1/T$. Model each $x(t)$ by an even function of time and find the exponential form of its Fourier coefficients X_n. Show that X_n is a real and even function of n. Compute the first nine exponential Fourier coefficients X_n, $1 \le n \le 9$, of the periodic waveforms in Table 8.4 and enter the results in that table. For each waveform overlay X_n (periodic) on the graph of its corresponding $H(f)$ (single pulse). Show that because X_n is real we have

$$x(t) = X_0 + 2\sum_{n=1}^{\infty} X_n \cos\left(2\pi n \frac{t}{T}\right)$$

TABLE 8.4 Exponential Fourier Coefficients of the First Nine Harmonics of Periodic Waveforms (To be completed by student)

x(t)	X_1	X_2	X_3	X_4	X_5	X_6	X_7	X_8	X_9
Periodic square $V_{\text{base-to-peak}} = 1$ V duration = 500 μs period = 1 ms									
Periodic rectangular $V_{\text{base-to-peak}} = 1$ V duration = 200 μs period = 1 ms									
Periodic triangular $V_{\text{base-to-peak}} = 1$ V duration = 1 ms period = 1 ms									

Measurements and Analysis

In this project you will use periodic signals (sine, rectangular, and triangular). Feed the signal from the function generator to channel 1 of the scope and trigger the scope trace from that signal. Choose FFT under the Math menu. FFT first multiplies the data displayed on the scope by a window of the same size, takes its spectrum, and displays it on the scope using a dB scale. You will choose the range of displayed frequencies, vertical gain (dB), and window type. Three window types are offered (Hanning, Flat Top, and Rectangular). For each waveform you will choose appropriate parameters to display the FFT, as illustrated in the following sections. Note that you may use the scope in the single trigger mode and retain the data in its memory while you vary the *sweep speed* (in reality, the length of the displayed segment), adjust the dB gain of the FFT and its offset level, and choose a new range of displayed frequencies, or a new window type.

Measuring the Spectrum of a Sinusoidal Wave

Prepare the following setup.

Signal: sinusoidal, 1 kHz, $V_{\text{peak-to-peak}} = 5$ V, $V_{DC} = 0$ V

Sweep: single, 5 ms/div, data length = 50 ms

FFT: frequency span = 10 kHz, center frequency = 5 kHz, scale = 20 dB, offset = 0

Window: Hanning

Set the sweep speed of the oscilloscope to 5 ms/div. Use a single sweep to capture 50 ms of input to channel 1. The oscilloscope shows the spectrum in dB units, referenced to an rms value of 1 V. The readings are, therefore, referred to as dB volt (or dBV). A sinusoidal waveform with an rms value of 1 V results in 0 dBV (or simply dB). For example:

$$v(t) = \sqrt{2}\cos(\omega t) \implies V_{\text{rms}} = 1 \text{ V} \implies 0 \text{ dBV}$$
$$v(t) = 10\sqrt{2}\cos(\omega t) \implies V_{\text{rms}} = 10 \text{ V} \implies 20 \text{ dBV}$$
$$v(t) = V_0\cos(\omega t) \implies V_{\text{rms}} = V_0/\sqrt{2} \implies 20\log V_0 - 3 \text{ dBV}$$
$$v(t) = 2.5\cos(\omega t) \implies V_{\text{rms}} = 1.77 \text{ V} \implies 4.95 \text{ dBV}$$

Record the location of the peak of the spectrum and its dB value. Enter them along with theoretical values in Table 8.5. Compare with theory. Define the percentage relative difference ϵ by

$$\epsilon \text{ (in \%)} = \frac{\text{Measured value} - \text{Theory}}{\text{Theory}} \times 100$$

TABLE 8.5 The Measured Spectrum of a 1 kHz Sinusoid and Its Comparison with Theory (To be completed by student)

	Measurement	Theory	Relative Difference ϵ
X_1 in dBV			
f_1 in kHz			

Measuring the Spectrum of a Periodic Square Wave

Repeat the above for a square wave. For this purpose you need to change only the function generator output from sinusoidal to square. All other settings on the function generator and on the scope remain the same as before. The function generator sends out a 1-kHz, square wave with and $V_{\text{peak-to-peak}} = 5$ V, $V_{DC} = 0$ V. Record the location of the first nine peaks (in kHz) and their value (in dB) corresponding to the nine harmonics. Note the negligible values of the even harmonics. Enter your measurements along with theoretical values in Table 8.6. Compare measured values with theory. Note that the DC value of the square wave is zero and is not included in Table 8.6 because its value displayed by the spectrum analyzer is not reliable.

TABLE 8.6 The Measured Spectrum of a 1-kHz Square Wave and Its Comparison with Theory (To be completed by student)

	X_1, f_1	X_2, f_2	X_3, f_3	X_4, f_4	X_5, f_5	X_6, f_6	X_7, f_7	X_8, f_8	X_9, f_9
Measured									
Theory									
ϵ_X									
ϵ_f									

Measuring the Spectrum of a Periodic Triangular Wave

Repeat the above for a 1-kHz triangular wave. Again, you need to change only the function generator output to a triangular waveform. Record the values of the nine harmonics in the spectrum in dB and enter your measurements along with the previously measured values for the square waveform in Table 8.7. Note the negligible values of the even harmonics and the rapid decrease of the amplitude of odd harmonics (compared to the square wave). Relate these observations to the theoretical relationship between the spectra of the square and triangular waves.

TABLE 8.7 A Comparison of Harmonics in 1-kHz Triangular and Square Waves Measured by the Spectrum Analyzer (To be completed by student)

Waveform	X_1	X_2	X_3	X_4	X_5	X_6	X_7	X_8	X_9
Square wave									
Triangular wave									

Measuring the Spectrum of a Periodic Rectangular Wave

Capture the spectrum of a 1-kHz rectangular wave with $V_{\text{peak-to-peak}} = 5$ V at 20% duty cycle. Use a 10-ms Hanning window. Count the number of harmonics before the first zero-crossing in the envelope of the spectrum and measure their locations. Relate to theory. Enter the results in Table 8.8.

TABLE 8.8 The Measured Spectrum of a 1-kHz Rectangular Waveform at 20% Duty Cycle (To be completed by student)

Number of Harmonics Before the First Zero-Crossing	Estimated Location of the First Zero-Crossing, in kHz	Theoretical Location of the First Zero-Crossing, in kHz

Discussion and Conclusions

Average Power in Harmonics of a Periodic Square Wave

From the measured spectrum of the square wave determine the average power in the principal harmonic. What percentage of the total power in the square wave resides in the principal harmonic? What percentage is in the third harmonic?

Average Power in Harmonics of a Periodic Triangular Wave

Repeat the above for the triangular waveform of "Measuring the Spectrun of a Periodic Triangular Wave."

Approximate Bandwidth

Theoretically, the periodic square and triangular waves have infinite bandwidth because their spectra continue to $f = \infty$. From your measurements of the spectrum of a periodic square wave and of a periodic triangular wave, determine what perecentage of the power of a 1-kHz square wave resides in frequencies above 10 kHz; what percentage of the power of a 1-kHz triangular wave resides above 10 kHz?

Overall Conclusions

Summarize your overall conclusions drawn from this experiment, especially the relationship between the pulse width in the time domain and its bandwidth. Illustrate by examples from the measurement results.

9

System Function, the Frequency Response, and Analog Filters

Contents

Introduction and Summary

LTI systems may be analyzed in both the time and frequency domains. In the time domain, the output is found by solving the input-output differential equation or by convolution of the input with the unit-impulse response, $y(t) = x(t) \star h(t)$.[1] Earlier we introduced an object called the system function and showed it by $H(s)$. The formal definition of $H(s)$ was presented in Chapter 6 based on the Laplace transform operation. We used the Laplace transform to convert the convolution $y(t) = x(t) \star h(t)$ into a multiplication of the transforms $Y(s) = X(s)H(s)$. Likewise, the Laplace transform was used to convert the differential equation in the time domain to a linear equation in the $s-$domain. The operations involving the Laplace transform represent analysis in the frequency domain. Both approaches are closely related and provide the same result. This chapter discusses the analysis of LTI systems in the frequency domain. It is concerned with $H(s)$ and the parallels between the time and frequency domain solution methods. First, it describes the system function and looks at it from several points of view in order to learn about its overall place in system analysis. It reveals aspects of $H(s)$ that bind time and frequency analysis together. Specifically, it shows how the system function is obtained, some of its properties and capabilities, its relationship with other features that describe a system, and its applications. Second, it introduces the frequency response, methods for plotting it, poles and zeros of the system, and its vectorial interpretation. Second-order systems—dominant pair of poles, filters, and feedback system—are the third topic covered. Examples are chosen from simple electrical systems that require basic familiarity with linear circuits.

---

[1] The system may instead be characterized by its unit-step response. See Example 9.1.

9.1   What Is a System Function?

System Function *H(s)* is the Scale Factor for e^{st}

This is a direct consequence of linearity and time-invariance properties, which make e^{st} the eigenfunctions of LTI systems. (An eigenfunction of a system keeps its functional form when passing through the sytem.) We have seen (Chapter 3, section 3.9) that the response of an LTI system to an exponential input e^{st} is He^{st}, where the scale factor H is, in general, a function of s:

$$e^{st} \implies H(s)e^{st}$$

The fact that an exponential input e^{st} resulted in the response He^{st} was also obtained in the case of linear differential equation with constant coefficients. In either case $H(s)$ is called the *system function* and exists for almost all LTI systems[2] (specifically, for all LTI systems of interest in electrical engineering). When it exists, the system function totally characterizes the system.

H(s) for Lumped and Distributed Systems

The class of LTI systems made of lumped discrete linear elements (such as linear circuits) is specified by linear differential equations with constant coefficients. In this case, the system function becomes the ratio of two polynomials in s. Counter to these systems is the class of systems with distributed elements such as transmission lines, or delay systems. This chapter will mainly address cases in which the system function is a ratio of polynomials.

How to Find *H(s)*

When it exists, the system function may be obtained from

1.   An input-output pair.[3]
2.   The system's unit-impulse response.
3.   The input-output differential equation.
4.   The frequency response.

These methods are closely related to each other and produce the same result. They are described below.

H(s) Is the Ratio *Y(s)/X(s)*

This is often presented as a formal definition of system function. Here are several more examples.

---

[2]An example of an LTI system without a unit-impulse response or a system function is

$$y(t) = \frac{x(t) + x(-t)}{2}$$

[3]A sinusoidal input produces a sinusidal output with the same frequency but possibly a change in the magnitude and phase. This information given at a fixed frequency is not enough to describe the system, unless it is provided for the full frequency range of $0 \leq f \leq \infty$, from which the frequency response is formed.

The response of a system to a unit-step input is measured and modeled by $e^{-at}u(t)$. Assuming an LTI find its $H(s)$.

Solution

Laplace transforms will be used.

$$x(t) = u(t) \quad \Longrightarrow \quad X(s) = \frac{1}{s}$$

$$y(t) = e^{-at}u(t) \quad \Longrightarrow \quad Y(s) = \frac{1}{s+a}$$

$$H(s) = \frac{Y(s)}{X(s)} \quad \Longrightarrow \quad H(s) = \frac{s}{s+a}$$

A Brief Table of Laplace Transform Pairs

The following is a table of some Laplace transform pairs for use in this chapter. $x(t)$ is defined for $-\infty < t < \infty$.

	$x(t) = \mathcal{L}^{-1}[X(s)]$	$\Longleftrightarrow$	$X(s) = \int_{-\infty}^{\infty} x(t)e^{-st}dt$	Region of Convergence
1	$\delta(t)$	$\Longleftrightarrow$	1	all s
2	$u(t)$	$\Longleftrightarrow$	$\dfrac{1}{s}$	$\mathcal{RE}[s] > 0$
3	$u(-t)$	$\Longleftrightarrow$	$\dfrac{-1}{s}$	$\mathcal{RE}[s] < 0$
4	$e^{at}u(t)$	$\Longleftrightarrow$	$\dfrac{1}{s-a}$	$\mathcal{RE}[s] > a$
5	$e^{bt}u(-t)$	$\Longleftrightarrow$	$\dfrac{-1}{s-b}$	$\mathcal{RE}[s] < b$
6	$e^{-c|t|}$	$\Longleftrightarrow$	$\dfrac{-2c}{s^2-c^2}$	$-c < \mathcal{RE}[s] < c$
7	$\sin(\omega t)u(t)$	$\Longleftrightarrow$	$\dfrac{\omega}{s^2+\omega^2}$	$\mathcal{RE}[s] > 0$
8	$\cos(\omega t)u(t)$	$\Longleftrightarrow$	$\dfrac{s}{s^2+\omega^2}$	$\mathcal{RE}[s] > 0$
9	$\sin(\omega t)u(-t)$	$\Longleftrightarrow$	$\dfrac{-\omega}{s^2+\omega^2}$	$\mathcal{RE}[s] < 0$
10	$\cos(\omega t)u(-t)$	$\Longleftrightarrow$	$\dfrac{-s}{s^2+\omega^2}$	$\mathcal{RE}[s] < 0$
11	$e^{at}\sin(\omega t)u(t)$	$\Longleftrightarrow$	$\dfrac{\omega}{(s-a)^2+\omega^2}$	$\mathcal{RE}[s] > a$

$x(t) = \mathcal{L}^{-1}[X(s)]$	$\Longleftrightarrow$	$X(s) = \int_{-\infty}^{\infty} x(t)e^{-st}dt$	Region of Convergence
12	$e^{at}\cos(\omega t)u(t)$	$\Longleftrightarrow$	$\dfrac{s-a}{(s-a)^2 + \omega^2}$	$\mathcal{RE}[s] > a$
13	$e^{at}\cos(\omega t + \theta)u(t)$	$\Longleftrightarrow$	$\dfrac{(s-a)\cos\theta - \omega\sin\theta}{(s-a)^2 + \omega^2}$	$\mathcal{RE}[s] > a$
14	$e^{bt}\sin(\omega t)u(-t)$	$\Longleftrightarrow$	$-\dfrac{\omega}{(s-b)^2 + \omega^2}$	$\mathcal{RE}[s] < b$
15	$e^{bt}\cos(\omega t)u(-t)$	$\Longleftrightarrow$	$-\dfrac{s-b}{(s-b)^2 + \omega^2}$	$\mathcal{RE}[s] < b$
16	$e^{bt}\cos(\omega t + \theta)u(-t)$	$\Longleftrightarrow$	$-\dfrac{(s-b)\cos\theta - \omega\sin\theta}{(s-b)^2 + \omega^2}$	$\mathcal{RE}[s] < b$
17	$t^n e^{at}u(t)$	$\Longleftrightarrow$	$\dfrac{n!}{(s-a)^{n+1}}$	$\mathcal{RE}[s] > a$

H(s) is the Laplace Transform of h(t)

The system function is the Laplace transform of the unit-impulse response. This is a consequence of $H(s)$ being the ratio of the output transform to that of the input. The Laplace transform of a unit impulse is 1, which makes $H(s)$ the Laplace transform of $h(t)$.

Example **9.2**

Find $H(s)$ of the system with the unit-impulse response $h(t) = te^{-at}u(t)$.

Solution
From the table of Laplace transform pairs we find

$$H(s) = \frac{1}{(s+a)^2}, \quad \mathcal{RE}[s] > -a$$

Example **9.3**

Find $H(s)$ of the system with the unit-impulse response $h(t) = e^{-|t|}$.

Solution
From the table of Laplace transform pairs we find

$$H(s) = \frac{-2}{s^2 - 1}, \quad -1 < \mathcal{RE}[s] < 1$$

Example **9.4**

Find the system function of the LTI system with the unit-impulse response

a. $h_a(t) = e^{-|t|}\cos t$
b. $h_b(t) = e^{-|t|}\sin t$

Solution

From the table of Laplace transform pairs we have

a. $h_{a1}(t) = e^{-t}\cos(t)u(t)$ $\qquad \Longrightarrow H_{a1}(s) = \dfrac{s+1}{(s+1)^2+1},$ $\qquad\qquad \mathcal{RE}[s] > -1$

$h_{a2}(t) = e^{t}\cos(t)u(-t)$ $\qquad \Longrightarrow H_{a2}(s) = -\dfrac{s-1}{(s-1)^2+1},$ $\qquad\qquad \mathcal{RE}[s] < 1$

$h_a(t) = h_{a1}(t) + h_{a2}(t) = e^{-|t|}\cos(t) \Longrightarrow H_a(s) = H_{a1}(s) + H_{a2}(s) = -2\dfrac{s^2-2}{s^4+4}, \quad -1 < \mathcal{RE}[s] < 1$

b. $h_{b1}(t) = e^{-t}\sin(t)u(t)$ $\qquad \Longrightarrow H_{b1}(s) = \dfrac{1}{(s+1)^2+1},$ $\qquad\qquad \mathcal{RE}[s] > -1$

$h_{b2}(t) = e^{t}\sin(t)u(-t)$ $\qquad \Longrightarrow H_{b2}(s) = -\dfrac{1}{(s-1)^2+1},$ $\qquad\qquad \mathcal{RE}[s] < 1$

$h_b(t) = h_{b1}(t) + h_{b2}(t) = e^{-|t|}\sin(t) \Longrightarrow H_b(s) = H_{b1}(s) + H_{b2}(s) = \dfrac{-4s}{s^4+4}, \qquad -1 < \mathcal{RE}[s] < 1$

H(s) Is Found from the Differential Equation and Vice Versa

Consider the LTI differential equation

$$\frac{d^n y}{dt^n} + a_{n-1}\frac{d^{n-1}y}{dt^{n-1}} + \cdots + a_1\frac{dy}{dt} + a_0 y = e^{st}$$

We have seen that the solution to the above equation is

$$y(t) = \frac{1}{s^n + a_{n-1}s^{n-1} + \cdots + a_1 s + a_0}e^{st}$$

Extending the above observation, we noted that the response of the differential equation

$$\frac{d^n y}{dt^n} + a_{n-1}\frac{d^{n-1}y}{dt^{n-1}} + \cdots + a_1\frac{dy}{dt} + a_0 y = b_m\frac{d^m x}{dt^m} + b_{m-1}\frac{d^{m-1}x}{dt^{m-1}} + \cdots + b_1\frac{dx}{dt} + b_0 x$$

to an exponential input is an exponential

$$X_0 e^{st} \quad \Longrightarrow \quad H(s)X_0 e^{st}$$

where $H(s)$ is the system function

$$H(s) = \frac{b_m s^m + b_{m-1}s^{m-1} + \cdots + b_1 s + b_0}{s^n + a_{n-1}s^{n-1} + \cdots + a_1 s + a_0}$$

$H(s)$ is easily obtained from the differential equation by employing the notation s for a first-order derivative, s^2 for a second-order derivative, and so on.

$$H(s) = \frac{B(s)}{A(s)}$$

$$B(s) = b_m s^m + b_{m-1}s^{m-1} + \cdots + b_1 s + b_0$$

$$A(s) = s^n + a_{n-1}s^{n-1} + \cdots + a_1 s + a_0$$

Conversely we can construct the differential equation and its block diagram from a given $H(s)$, the ratio of two polynomials in s.

Example

9.5

Find the input-output differential equation of the LTI system with the system function

$$H(s) = \frac{Y(s)}{X(s)} = 1 + s + s^2$$

$$Y(s) = (1 + s + s^2)X(s) = X(s) + sX(s) + s^2 X(s)$$

$$y(t) = x(t) + \frac{dx(t)}{dt} + \frac{dx^2(t)}{dt^2}$$

Example

9.6

Find the input-output differential equation of a causal LTI system with the system function

$$H(s) = \frac{Y(s)}{X(s)} = \frac{s-2}{s^2 + 2s + 5}$$

Solution

$$(s^2 + 2s + 5)Y(s) = (s-2)X(s)$$

$$\frac{d^2 y(t)}{dt^2} + 2\frac{dy(t)}{dt} + 5y(t) = \frac{dx(t)}{dt} - 2x(t)$$

Poles and Zeros

In this book we are interested in LTI systems with $H(s) = B(s)/A(s)$, where $A(s)$ and $B(s)$ are polynomials in s. The roots of the numerator polynomial are called the *zeros* of the system. Let z_k be a zero of the system. At $s = z_k$ we have $H(z_k) = 0$ and the input $x(t) = e^{z_k t}$, for all t, will result in $y(t) = 0$.

Similarly, the roots of the denominator polynomial are called the *poles* of the system. Let p_k be a pole of the system. At $s = p_k$ we have $H(p_k) = \infty$ and the input $x(t) = e^{p_k t}$ will result in $y(t) = \infty$.

Example

9.7

Find the zeros of the system described by $H(s) = s^2 + s + 1$.

Solution
The zeros of the system are roots of

$$s^2 + s + 1 = 0 \implies z_{1,2} = -\frac{1}{2} \pm j\frac{\sqrt{3}}{2} = e^{\pm j 120°}$$

Example

9.8

Find the poles and zeros of the system described by

$$H(s) = \frac{s}{s^2 + 2s + 2}$$

Solution

The zeros of the system are roots of $s = 0$ $\implies z_1 = 0$

The poles of the system are roots of $s^2 + 2s + 2 = 0 \implies p_{1,2} = -1 \pm j$
$$= \sqrt{2}e^{\pm j135°}$$

Find the poles and zeros of the system described by

$$H(s) = \frac{s - 2}{s^2 + 2s + 5}$$

Solution

The zeros of the system are roots of $s - 2 = 0$ $\implies z_1 = 2$

The poles of the system are roots of $s^2 + 2s + 5 = 0 \implies p_{1,2} = -1 \pm j2$
$$= \sqrt{5}e^{\pm j116.6°}$$

Find the system function with two zeros at ± 1, a pair of poles at $re^{\pm j\theta}$, and $|H(s)|\big|_{s=jr} = 1$.

Solution

$$H(s) = k\frac{(s + 1)(s - 1)}{(s - re^{j\theta})(s - re^{-j\theta})} = k\frac{s^2 - 1}{s^2 - (2r\cos\theta)s + r^2}$$

$$|H(s)|\big|_{s=jr} = k\frac{1 + r^2}{2r^2\cos\theta} = 1 \implies k = \frac{2r^2\cos\theta}{1 + r^2}$$

Contribution of Poles and Zeros to the System Function

A zero at z_k contributes $(s - z_k)$ to the numerator of the system function. Similarly, a pole at p_k contributes $(s - p_k)$ to its denominator. If the coefficients of the system function are real, the poles and zeros are either real or complex conjugates. A pair of complex conjugate roots of the numerator or denominator at $re^{\pm j\theta}$ contribute

$$(s - re^{j\theta})(s - re^{-j\theta}) = s^2 - (2r\cos\theta)s + r^2$$

to the numerator or the denominator of the system function, respectively. The significance of poles and zeros locations becomes clear when we work with the frequency response.

9.2   The Time Response May Be Obtained from $H(s)$

The time response of an LTI system is obtained either through convolution of the input with the unit-impulse response or solution of the input-output differential equation, both of which may be obtained from $H(s)$. It is, therefore, expected that the system function contain all the information needed to find the time response to a given input. Clearly, with a known input $x(t)$ we can find its Laplace transform $X(s)$, multiply it by $H(s)$ to find $Y(s)$, from which $y(t)$ can be found. In this section, without going through the above steps, we will see how elements of the time response (natural frequencies, homogeneous and particular responses, boundary values, and the total response) may be obtained from $H(s)$. This can be done, often without resorting to formal tools and systematic methods [i.e., without finding $h(t)$ to do convolution, or solving the differential equation, or using the Laplace transform formulation].

We start with the observation that the system function provides the particular response to an input expressed by a linear combination of exponentials with constant weighting factors. (An example would be finding the AC steady-state response, where the input comprises two complex exponentials.) We then additionally note that the denominator of the system function contains the characteristic equation of the system whose roots are the natural frequencies providing the homogeneous response. In fact, the complete response to an exponential input may readily be written from the system function, if the initial conditions are known. This property makes $H(s)$ a powerful tool in the analysis of LTI systems. In the following three examples we find responses of the first-order LTI system $H(s) = 1/(s + 0.6421)$ to three different inputs.

Example

9.11

Given $H(s) = 1/(s + 0.6421)$, find the system's response to $x(t) = \cos t$, $-\infty < t < \infty$.

Solution
The response consists of the particular solution only, which in this case is the sinusoidal steady-state response. To find it we note that the angular frequency of the sinusoidal input is 1, which corresponds to $s = j$. The output is obtained from the input scaled by the system function evaluated at $s = j$.

$$H(s)\big|_{s=j} = H(j) = \frac{1}{0.6421 + j} = 0.8415e^{-j}$$

The input and the response may be written as

$$x(t) = \cos(t) = \mathcal{RE}\{e^{jt}\}$$

$$H(j) = \frac{1}{0.6421 + j} = 0.8415e^{-j}$$

$$y(t) = \mathcal{RE}\{H(j)e^{jt}\}$$

$$= \mathcal{RE}\{0.8415e^{-j} \times e^{jt}\} = 0.8415\cos(t - 1)$$

The system scales down the input amplitude by the factor 0.8415 and introduces a 1-second delay.

xample

9.12

Given $H(s) = 1/(s + 0.6421)$, find the system's response to $x(t) = \cos(t)u(t)$.

Solution

The particular solution was found in Example 9.11, $y_p(t) = 0.8415 \cos(t - 1)$. To obtain the total response, we need to add the homogeneous solution (if any) to it. The system has a single pole at $s = -0.6421$, which gives rise to the homogeneous solution $y_h(t) = Ce^{-0.6421t}$, where C is a constant. The total response is

$$y(t) = y_h(t) + y_p(t) = \left[Ce^{-0.6421t} + 0.8415 \cos(t - 1)\right] u(t)$$

Using the boundary condition $y(0^+) = C + 0.8415 \cos(-57.3°) = 0$, we find $C = -0.4547$. The total response, therefore, is

$$y(t) = \left[-0.4547e^{-0.6421t} + 0.8415 \cos(t - 1)\right] u(t)$$

xample

9.13

Given $H(s) = 1/(s + 0.6421)$ and the input $x(t) = \alpha\delta(t) + \cos(t)u(t)$, find the constant α such that for $t \geq 0$ the system's response contains the sinusoidal steady-state only.

Solution

The impulse of strength α evokes the additional response $\alpha e^{-0.6421t}u(t)$. Using the result of Example 9.12, we observe that $\alpha = 0.4547$ will result in neutralization of the homogeneous response.

Alternate Solution

In order to have no homogeneous response to the input $\cos(t)u(t)$ we need to create the appropriate initial condition $y(0^+) = 0.4547$ for it. This is achieved by the impulse $0.4547\delta(t)$, which arrives before the sinusoid and eliminates the need for the homogeneous response.

9.3   The Frequency Response $H(\omega)$

The response of an LTI system to a sinusoidal input is sinusoidal with the same frequency. Generally, the system changes the magnitude and phase of the sinusoidal input and the change is frequency dependent. The frequency response determines the changes in magnitude and phase of the sinusoidal input when it goes through the LTI system. The frequency response of an LTI system has two parts: a magnitude response and a phase response, both of which are functions of the frequency ω. Therefore,

$$\cos \omega t \implies |H(\omega)| \cos(\omega t + \theta)$$

The magnitude and phase of the frequency response can be combined as a complex function and shown by $H(\omega) = |H(\omega)|\angle\theta$.

From another point of view, the magnitude and phase of a sinusoidal function can be shown by a complex number called the complex amplitude or the phasor. The sinusoidal input $x(t)$ and output $y(t)$ are then represented by their complex amplitudes, or phasors, X and Y. The frequency response may be defined as the ratio of the output phasor to the input phasor.

$$H(\omega) \;=\; \frac{Y}{X} = |H(\omega)|e^{j\theta(\omega)}$$
$$\cos(\omega t) \Longrightarrow |H(\omega)|\cos(\omega t + \theta)$$

The frequency response of an LTI system may be measured experimentally from sinusoidal input-output pairs, without knowledge of the system's internal structure or its mathematical model. It may also be obtained from the system function, the unit-impulse response, or from the input-output differential equation.

$H(\omega)$ Characterizes the System

The frequency response isn't just useful in finding a system's response to a single sinusoid. It provides the system's response to any input that can be expressed as a sum of sinusoids (i.e., almost all signals of interest in engineering and sciences). In fact, from the frequency response one can obtain the system function, the input-output differential equation, and the response to any input in a causal system.

$H(\omega)$ May Be Obtained from $H(s)$[4]

The frequency response of an LTI system may be found from its system function $H(s)$ by setting $s = j\omega$

$$H(\omega) = H(s)\big|_{s=j\omega} = |H(\omega)|e^{j\theta(\omega)} = |H(\omega)|\angle\theta(\omega)$$

To verify the above statement we observe the following:

1. $H(s)$ is the scale factor for the exponential input:

$$e^{st} \Longrightarrow H(s)e^{st}$$

2. The real part of the response to an input is equal to the response of the system to the real part of that input:

$$x(t) \quad \Longrightarrow \quad y(t)$$
$$\mathcal{RE}\{x(t)\} \quad \Longrightarrow \quad \mathcal{RE}\{y(t)\}$$

3. A cosine signal is the real part of an exponential signal:

$$\cos(\omega t) = \mathcal{RE}\{e^{j\omega t}\}$$

---

[4] $H(s)$ is a real function of the complex number s. $H(\omega)$ is a complex function of a real number ω. From the substitution $s = j\omega$ one can rightly conclude that the frequency response is a real function of $j\omega$ and represent it by $H(j\omega)$, as done by some authors. For simplicity, we avoid this representation and show the frequency response by $H(\omega)$, but remember that it is a complex function of ω. It is defined on its own as a characteristic of the system which shows the change in magnitude and phase of a sinusoid signal passing through the system and not just a derivation from $H(s)$. Because of its close relationship with the system function, we use the symbol H doubly to represent both.

The above observations are summarized below:

$$e^{j\omega t} \implies H(\omega)e^{j\omega t}$$
$$\mathcal{RE}\{e^{j\omega t}\} \implies \mathcal{RE}\{H(\omega)e^{j\omega t}\}$$
$$\Downarrow \qquad\qquad \Downarrow$$
$$\Downarrow \qquad\qquad \Downarrow$$
$$\Downarrow \qquad\qquad \Downarrow$$
$$\cos(\omega t) \implies |H(\omega)|\cos(\omega t + \theta)$$

Moreover, as will be seen below, the square of the magnitude of the frequency response is found from

$$|H(\omega)|^2 = H(\omega)H^*(\omega) = H(s)H(-s)|_{s=j\omega}$$

where $H^*(\omega)$ is the complex conjugate of $H(\omega)$.

Finding $|H(\omega)|^2$ from $H(s)H(-s)$

The square magnitude of the frequency response is $|H(\omega)|^2 = H(\omega)H^*(\omega)$. But

$$H(\omega) = H(s)|_{s=j\omega}$$
$$H^*(\omega) = H(s)|_{s=-j\omega} = H(-s)|_{s=j\omega}$$
$$|H(\omega)|^2 = H(\omega)H^*(\omega) = H(s)H(-s)|_{s=j\omega}$$

This provides an easy way to obtain the magnitude of the frequency response from the system function.

Example 9.14

A zero (or a pole) at $s = a$ contributes the term $(s - a)$ to the numerator (or the denominator if a pole) of $H(s)$. Its contribution to the square of the magnitude of $H(\omega)$ is

$$|H(\omega)|^2 = (s - a)(-s - a)|_{s=j\omega} = (a^2 - s^2)|_{s=j\omega} = a^2 + \omega^2$$

Example 9.15

A pair of conjugate zeros (or poles) at $re^{\pm j\theta}$ contribute the following terms to $H(s)$.

$$(s - re^{j\theta})(s - re^{-j\theta}) = s^2 - (2r\cos\theta)s + r^2$$

Their contribution to the square of the magnitude of $H(\omega)$ is

$$|H(\omega)|^2 = [s^2 - (2r\cos\theta)s + r^2][s^2 + (2r\cos\theta)s + r^2]|_{s=j\omega}$$
$$= [s^4 - (2r^2\cos 2\theta)s^2 + r^4]|_{s=j\omega}$$
$$= \omega^4 + (2r^2\cos 2\theta)\omega^2 + r^4$$

Symmetry Properties of $H(\omega)$

Let $H(\omega) = |H|\angle\theta$, where $|H|$ and θ are its magnitude and phase angle, respectively. $|H|$ is an even function and θ is an odd function of ω. Knowledge of $|H|$ and θ for $\omega > 0$, therefore, provides full information about the system. Furthermore, if the system is causal, the magnitude and phase are related to each other.

*E*xample
9.16

Find and plot the magnitude and phase of

a. $H_1(s) = 1/(s + 1)$.

b. $H_2(s) = s/(s + 1)$. Verify that the magnitudes are even and phases are odd functions of ω.

Solution

a. $H_1(s) = \dfrac{1}{s + 1}$, $H_1(\omega) = \dfrac{1}{1 + j\omega}$, $|H_1| = \dfrac{1}{\sqrt{1 + \omega^2}}$, $\angle H_1 = -\tan^{-1}\omega$

b. $H_2(s) = \dfrac{s}{s + 1}$, $H_2(\omega) = \dfrac{j\omega}{1 + j\omega}$, $|H_2| = \sqrt{\dfrac{\omega^2}{1 + \omega^2}}$, $\angle H_2 = \begin{cases} \pi/2 - \tan^{-1}\omega & \omega > 0 \\ -\pi/2 - \tan^{-1}\omega & \omega < 0 \end{cases}$

Plots in Figure 9.1 exhibit even symmetry of magnitude and odd symmetry of phase angles. The symmetry properties may also be verified by observing that magnitude

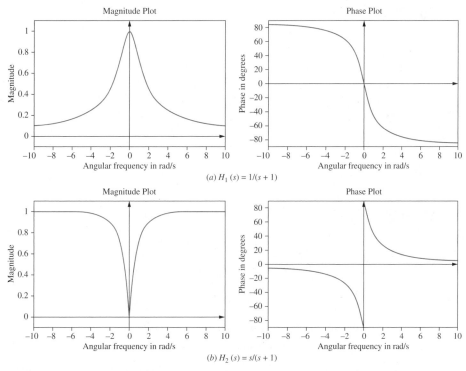

(a) $H_1(s) = 1/(s + 1)$

(b) $H_2(s) = s/(s + 1)$

FIGURE 9.1 Symmetry properties of the frequency response. Magnitude and phase plots of the frequency response of $H_1(s) = 1/(s + 1)$ are shown in the upper row for $-10 < \omega < 10$ and those of $H_2(s) = s/(s + 1)$ in the lower row. Magnitudes are aligned on the left side and phases on the right side to show their even and odd symmetry properties, respectively.

is a function of ω^2 (which is an even function of ω) and phase is a function of $\tan^{-1}(\omega)$ (which is an odd function of ω). One may also reach the same conclusion by substituting $-\omega$ for ω and observing no change in $|H|$ and sign reversal in $\angle H$.

Finding $H(s)$ from $|H(\omega)|^2$

Given an $|H(\omega)|$ we can apply a reverse process to find $H(s)$ of the associated causal stable system.

$$H(s)H(-s) = |H(\omega)|^2\Big|_{\omega=-js}$$

The resulting product $H(s)H(-s)$ has symmetrical poles in the LHP and RHP. By choosing the LHP poles we obtain $H(s)$ of the stable system whose magnitude frequency response is $|H(\omega)|$. [5]

*E*xample
9.17

Find the $H(s)$ of a causal stable system whose magnitude-squared frequency response is $|H(\omega)|^2 = 1/(1 + \omega^4)$.

Solution

$$H(s)H(-s) = \frac{1}{1 + \omega^4}\Big|_{\omega=-js} = \frac{1}{1 + s^4}$$

The right-half plane (RHP) roots of the equation $1 + s^4 = 0$ are at $s_{1,2} = e^{\pm j\pi/4}$
The left-half plane (LHP) roots of the equation $1 + s^4 = 0$ are at $s_{3,4} = e^{\pm j3\pi/4}$

The LHP roots will be chosen as the poles of the stable system. The system function is

$$H(s) = \frac{1}{(s - e^{j3\pi/4})(s - e^{-j3\pi/4})} = \frac{1}{s^2 - (2\cos\frac{3\pi}{4})s + 1} = \frac{1}{s^2 + \sqrt{2}s + 1}$$

This is called a second-order low-pass Butterworth filter.

9.4 Plotting $H(\omega)$

$H(\omega)$ is a complex function of ω. (And in this book it is a real function of $j\omega$ unless specified otherwise.) Graphically, $H(\omega)$ can be shown in several ways, some of which are

1. The magnitude and phase plots as functions of ω.
2. The magnitude and phase in the form of a polar plot.
3. The plots of the real and imaginary parts as functions of ω.

We will do the magnitude and phase plots. The frequency axis ω may be in linear or log scale. The magnitude $|H|$ may be shown in linear or log scale [in $20\log|H|$ units called

---

[5]To obtain a unique solution, we also need to know whether the system's zeros are in the LHP or RHP.

decibell (dB)]. The phase $\angle H$ is shown in degrees. The special case of the twin plots of $20 \log |H(\omega)|$ and $\angle H(\omega)$ versus $\log \omega$ is called the Bode plot, which we generally use unless something different is required.

Computers and calculators provide the preferred methods for plotting, and due to their convenience and speed they can also facilitate qualitative exploration of a system's performance under various conditions. But plotting by computer doesn't provide much insight into some interesting aspects of the system (e.g., approximations within a frequency range, the asymptotic behavior, the effect of proximity to a pole or zero, the 3-dB break points, and so on). A similar rational applies for the vectorial interpretation of the frequency response, which is the subject of the next section. The present section uses plotting $H(\omega)$ as a context for gaining such an insight. It starts with contribution from a single pole or zero and concludes with sketching a Bode plot by hand, which requires a better understanding of the role of the system's elements in forming the frequency response.

Bode Plot: What Is It?

The Bode plot of the frequency response $H(\omega)$ of an LTI system is the graph of $20 \log |H(\omega)|$ (magnitude in dB) and $\angle H(\omega)$ (phase angle) both plotted versus $\log \omega$. In some cases a reference level H_R [e.g., $|H(0)|$ or $|H_{max}|$] is used in drawing the plot. The magnitude of the Bode plot is then $20 \log |H(\omega)/H_R|$, where 0 dB indicates the reference level. Because of the nature of the logarithm, contributions from individual zeros add to and those of poles subtract from the plots. We start by examining the effect of a single zero or pole on the Bode plot.

Bode Plot: A Single Zero

Consider the system function $H(s) = 1 + s/\omega_0$. Note that $H(s)$ has a single zero at $s = -\omega_o$. The frequency response, its magnitude (in dB), and its phase are

$$H(\omega) = 1 + j(\omega/\omega_0)$$
$$20 \log |H(\omega)| = 720 \log |1 + j(\omega/\omega_0)| = 10 \log[1 + (\omega/\omega_0)^2]$$
$$\angle H(\omega) = \angle[1 + j(\omega/\omega_0)] = \tan^{-1}(\omega/\omega_0)$$

We want to plot $20 \log |H(\omega)|$ and $\angle H(\omega)$ vs. $\log \omega$. At very low frequencies where $\omega << \omega_0$, we may drop $j(\omega/\omega_0)$ against 1.[6] Then,

$$H(\omega) \approx 1, \quad 20 \log |H| \approx 0, \quad \angle H \approx 0$$

The low-frequency asymptote of the magnitude and phase are 0 dB and $0°$, respectively. At very high frequencies where $\omega >> \omega_0$, we may drop 1 against $j(\omega/\omega_0)$. Then,

$$H(\omega) \approx j(\omega/\omega_0), \quad 20 \log |H| \approx 20 \log(\omega/\omega_0) \text{ dB}, \quad \angle H \approx 90°$$

The high-frequency asymptotes of the magnitude and phase are $20 \log(\omega/\omega_0)$ dB and $90°$, respectively. The high-frequency asymptote of the magnitude plotted versus $\log \omega$

---

[6]For a better approximation at $\omega << \omega_0$, see Table 9.1.

is a line with a 20-dB/decade slope (6 dB/ octave) which passes through 0 dB at $\omega = \omega_0$. At $\omega = \omega_0$, called the break frequency, we have

$$H(\omega_0) = 1 + j = \sqrt{2}\angle 45°, \quad 20\log|H| = 3 \text{ dB}, \quad \angle H = 45°$$

Note that the break frequency is at the zero of $H(s)$. Also note that at $\omega = \omega_0/2$ and $\omega = 2\omega_0$, the magnitude is $20\log(5/4) = 1$ dB and $20\log\sqrt{5} = 7$ dB, respectively, each of which are 1 dB above the value of their respective asymptotes. The phases at these frequencies are $\tan^{-1}(1/2) = 26.5°$ and $\tan^{-1}\{2\} = 63.5°$, respectively. These results, summarized in Table 9.1, can be used to sketch the magnitude and phase plots of Figure 9.2.

TABLE 9.1 Magnitude and Phase of $H(\omega) = 1 + j(\omega/\omega_0)$

Frequency ω	0	$\omega \ll \omega_0$	$\omega_0/2$	ω_0	$2\omega_0$	$\omega \gg \omega_0$	∞
Magnitude, $20\log|H(\omega)|$ in dB	0	$\approx 4.343\left(\dfrac{\omega}{\omega_0}\right)^2$	1	3	7	$\approx 20\log\left(\dfrac{\omega}{\omega_0}\right)$	∞
Phase, $\angle H(\omega)$ in degrees	0	$\approx \dfrac{180}{\pi}\dfrac{\omega}{\omega_0}$	26.5°	45°	63.5°	$\approx \left(90° - \dfrac{180}{\pi}\dfrac{\omega_o}{\omega}\right)$	90°

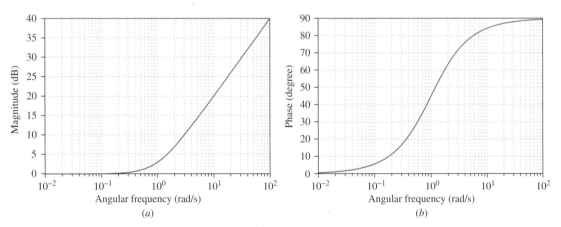

FIGURE 9.2 Bode plots of the magnitude **(a)** and phase **(b)** of $1 + j(\omega/\omega_0)$ versus ω/ω_0 for the range $0.01 < \omega/\omega_0 < 100$. The low-frequency asymptote of the magnitude plot is a flat 0-dB line. Its high-frequency asymptote is the line with a slope of 20 dB per decade (6 dB per octave). These two asymptotes intersect at ω_0 on the frequency axis, called the break frequency, where the magnitude is 3 dB and the phase is 45°. An examination of the plots within the range $0.5 < \omega/\omega_0 < 2$ shows that at both ends of this range the magnitude is approximately 1 dB above the asymptote and the phase is 26.6° away from it. More exactly,

at $\omega = \omega_0/2$ Magnitude = 0.969 dB,   phase = 26.56°
at $\omega = 2\omega_0$ Magnitude = 6.99 dB,   phase = 63.44°

Note: For $\omega \ll \omega_0$ we have approximated $20 \log |H(\omega)|$ by its Taylor series expansion.

$$\ln(1 + x) = x - \frac{x^2}{2} + \frac{x^3}{3} - \frac{x^4}{4} + \cdots + (-1)^{n-1}\frac{x^n}{n} + \cdots$$

$$\ln x = \ln(10) \times \log x = 2.3 \log x$$

$$10 \log \left[1 + \left(\frac{\omega}{\omega_0} \right)^2 \right] \approx \frac{10}{2.3} \left(\frac{\omega}{\omega_0} \right)^2 \approx 4.343 \left(\frac{\omega}{\omega_0} \right)^2 \quad \text{for } \omega \ll \omega_0$$

For example, let $\omega = 0.1$ and $\omega_0 = 1$. By the above approximation we have $10 \log[1 + (\omega/\omega_0)^2] \approx 0.04343$. The exact value is $10 \log(1 + 0.01) = 0.0432$.

Bode Plot: A Single Pole

Consider a system with a single pole at $s = -\omega_o$. The system function and frequency response are

$$H(s) = \frac{1}{1 + (s/\omega_0)}$$

$$H(\omega) = \frac{1}{1 + j(\omega/\omega_0)}$$

$$20 \log |H(\omega)| = -20 \log |1 + j(\omega/\omega_0)| = -10 \log[1 + (\omega/\omega_0)^2]$$

$$\angle H(\omega) = -\angle [1 + j(\omega/\omega_0)] = -\tan^{-1}(\omega/\omega_0)$$

The Bode plot of $1/[1 + j(\omega/\omega_0)]$ is similar to that of $1 + j(\omega/\omega_0)$ except for an opposite sign. Flip the magnitude and phase plots of the system with a zero (Figure 9.2) downward around the ω axis to see the contribution from a pole. The low-frequency asymptotes of the magnitude and phase are 0 dB and $0°$, respectively. The high-frequency asymptotes of the magnitude and phase are $-20 \log(\omega/\omega_0)$ dB and $-90°$, respectively. The high-frequency asymptote of the magnitude plotted versus $\log \omega$ is a line with a -20 dB/decade slope. At ω_0 the magnitude is -3 dB and the phase is $-45°$. At $\omega_0/2$ and $2\omega_0$ the magnitude is 1 dB below the value of the respective asymptotes. The phase at those two frequencies is $-26.5°$ and $-63.5°$, respectively. These observations are summarized in Table 9.2.

TABLE 9.2 Magnitude and Phase of $H(\omega) = \frac{1}{1+j(\omega/\omega_0)}$

Frequency ω	0	$\omega \ll \omega_0$	$\omega_0/2$	ω_0	$2\omega_0$	$\omega \gg \omega_0$	∞
Magnitude, $20 \log|H(\omega)|$ in dB	0	$\approx -4.343 \left(\frac{\omega}{\omega_0} \right)^2$	-1	-3	-7	$\approx -20 \log \left(\frac{\omega}{\omega_0} \right)$	$-\infty$
Phase, $\angle H(\omega)$ in degrees	0	$\approx -\frac{180}{\pi}\frac{\omega}{\omega_0}$	$-26.5°$	$-45°$	$-63.5°$	$\approx \left(\frac{180}{\pi}\frac{\omega_o}{\omega} - 90° \right)$	$-90°$

Summary of Asymptotic Behavior

The asymptotic behavior of the Bode plot of a system with a single zero or a pole is summarized below.

$$20 \log |H(\omega)| = \pm 10 \log \left[1 + \left(\frac{\omega}{\omega_0} \right)^2 \right] \approx \pm \begin{cases} 0 & \omega << \omega_0 \text{ (low-frequency asymptote)} \\ 3 \text{ dB} & \omega = \omega_0 \text{ (break frequency)} \\ 20 \log \left(\dfrac{\omega}{\omega_0} \right) & \omega >> \omega_0 \text{ (high-frequency asymptote)} \end{cases}$$

$$\angle H(\omega) = \pm \tan^{-1} \left(\frac{\omega}{\omega_0} \right) \approx \pm \begin{cases} 0 & \omega << \omega_0 \text{ (low-frequency asymptote)} \\ \pm \dfrac{\pi}{4} & \omega = \omega_0 \text{ (break frequency)} \\ \pm \dfrac{\pi}{2} & \omega >> \omega_0 \text{ (high-frequency asymptote)} \end{cases}$$

in which the upper signs refer to a zero and the lower to a pole.

Example 9.18

Find and sketch the frequency response of an LTI system with a pole in the left half of the s-plane at $s = -\sigma$, a zero at $s = 0$, and a high-frequency gain of 1.

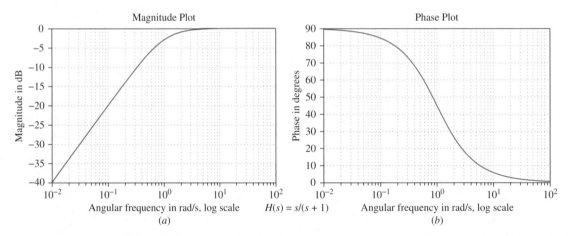

FIGURE 9.3 Magnitude *(a)* and phase *(b)* of $H(s) = s/(s+1)$ in log scales. The magnitude plot shows $20 \log |H(\omega)|$ (in dB) and the phase plot shows $\angle H(\omega)$ (in degrees). Both are plotted versus $\log \omega$. This is called a Bode plot. The system is high-pass with a high-frequency gain of 1 (corresponding to 0 dB). Its half-power frequency is at $\omega = 1$ rad/s where the magnitude plot shows -3 dB. Compare the Bode plot with the linear plot given in Figure 9.1*(b)*. Note the advantage of the Bode plot over the linear plot in its ability to cover a larger range of frequency and provide more detail at low frequencies. In addition, at the lower and higher limits of ω the magnitude plot [$20 \log |H(\omega)|$ versus $\log \omega$] becomes straight lines as seen in this figure. This property makes Bode plot an attractive tool for visualization of the frequency response and will be used in the rest of this chapter.

Solution

$$H(s) = \frac{s}{s + \sigma} \qquad H(\omega) = \frac{j\omega}{j\omega + \sigma}$$

$$|H(\omega)|^2 = \frac{\omega^2}{\omega^2 + \sigma^2} \qquad \angle H(\omega) = 90° - \tan^{-1}\left(\frac{\omega}{\sigma}\right)$$

The system is a high-pass filter with -3-dB frequency at $\omega = \sigma$. Its frequency response is plotted in Figure 9.3 (in log scale) for $\sigma = 1$. The linear-scale plots of the same system were shown in Figure 9.1(b).

Bode Plot: A Pair of Poles

Consider the second-order stable system function with two complex poles at p_1 and p_2, no finite zeros, and unity DC gain. The system function may be written in the form

$$H(s) = \frac{\omega_0^2}{s^2 + 2\zeta\omega_0 s + \omega_0^2}$$

where $\omega_0 = \sqrt{p_1 p_2}$ is called the undamped natural frequency and $\zeta = -(p_1 + p_2)/(2\omega_0)$ is called the damping ratio. Dividing the numerator and denominator by ω_0^2 and substituting for $s = j\omega$ we have

$$H(\omega) = \frac{1}{1 - \left(\frac{\omega}{\omega_0}\right)^2 + j2\zeta\left(\frac{\omega}{\omega_0}\right)}$$

At very low frequencies where $\omega << \omega_0$, $H(\omega) \approx 1$, which results in the low-frequency asymptote of $20\log|H| = 0$ dB and $\angle H = 0$. At very high frequencies where $\omega >> \omega_0$, $H(\omega) \approx -(\omega_0/\omega)^2$, which results in the high-frequency asymptote of $20\log|H| = -40\log(\omega/\omega_0)$ and $\angle H = -180°$. The high-frequency asymptote appears as a straight line with a slope of -40 dB/decade when it is plotted versus $\log(\omega/\omega_0)$. At the break frequency $\omega = \omega_0$, $H(\omega) = -j/(2\zeta)$, which results in $20\log|H| = -20\log(2\zeta)$ and $\angle H = -90°$.

In signal processing applications, the above system is specified by ω_0 and the quality factor $Q = 1/(2\zeta)$. The system function and frequency response may be written in terms of ω_0 and Q:

$$H(s) = \frac{\omega_0^2}{s^2 + \frac{\omega_0}{Q}s + \omega_0^2} \quad \text{and} \quad H(\omega) = \frac{1}{1 - \left(\frac{\omega}{\omega_0}\right)^2 + \frac{j}{Q}\left(\frac{\omega}{\omega_0}\right)}$$

The break frequency and asymptotes are the same as before. The magnitude at ω_0 is $|H(\omega_0)| = Q$, equal to the quality factor, or, expressed in dB units, $20 \log |H(\omega_0)| = 20 \log Q$ dB.

The above formulations apply to the second-order system for all values of ζ or Q (i.e., regardless of whether the poles are real or complex numbers). For the case of complex conjugate poles $p_{1,2} = -\sigma \pm j\omega_d$, $\omega_0 = \sqrt{\sigma^2 + \omega_d^2}$, and $\zeta = \sigma/\omega_0$. The relationship between a pair of complex conjugate poles, the natural frequency ω_0, the damping ratio ζ, the quality factor Q, and the time responses is summarized in Figure 9.4 (see also sections 9.6 and 9.7 to 9.21).

poles	$p_{1,2} = -\sigma \pm j\omega_d = \omega_0 e^{\pm j\theta}$
undamped natural frequency	$\omega_0 = \sqrt{\sigma^2 + \omega_d^2}$
system function	$H(s) = \dfrac{\omega_0^2}{s^2 + 2\sigma s + \omega_0^2}$
filtering and signal processing	$H(s) = \dfrac{\omega_0^2}{s^2 + \frac{\omega_0}{Q}s + \omega_0^2}$
quality factor	$Q = \dfrac{\omega_0}{2\sigma} = \dfrac{1}{2\zeta}$
damping state	$\begin{cases} \text{overdamped} & Q < 0.5, \ \zeta > 1 \\ \text{critically damped} & Q = 0.5, \ \zeta = 1 \\ \text{underdamped} & Q > 0.5, \ \zeta < 1 \end{cases}$
control applications	$H(s) = \dfrac{\omega_0^2}{s^2 + 2\zeta\omega_0 s + \omega_0^2}$
damping ratio	$\zeta = \dfrac{\sigma}{\omega_0}$
unit-step response	$g(t) = \left[1 - \dfrac{1}{\sqrt{1-\zeta^2}} e^{-\sigma t} \cos(\omega_d t - \phi)\right] u(t)$
phase angle of step response ϕ	$\sin\phi = \zeta$
percentage overshoot in $g(t)$	$\rho = 100 \times e^{-\frac{\pi\zeta}{\sqrt{1-\zeta^2}}}$
unit-impulse response	$h(t) = \dfrac{\omega_0}{\sqrt{1-\zeta^2}} e^{-\sigma t} \sin\omega_d t \ u(t)$

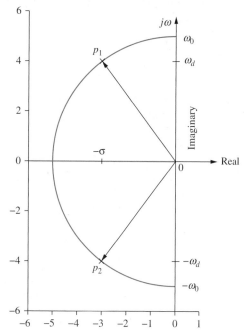

Pole Plot of $H(s) = 1/(s^2 + 6s + 25)$

FIGURE 9.4 Relationship between a pair of complex poles, $H(s)$, ζ, Q, and the time responses. This figure is plot of the poles of $H(s) = 1/(s^2 + 6s + 25)$.

Example

9.19

Plot the poles, and time and frequency responses of the following two second-order systems:

$$H_1(s) = \frac{1}{s^2 + 5s + 1} \quad \text{and} \quad H_2(s) = \frac{1}{s^2 + 0.1s + 1}$$

Discuss the effect of poles locations on systems' responses.

Solution

The poles, quality factor, damping ratio, and state of the systems are summarized in the table below. The poles, step responses, and magnitude and phase Bode plots of the second-order systems are shown in Figure 9.5.

System	Poles	Quality Factor	Damping Ratio	Damping State
$H_1(s) = \dfrac{1}{s^2 + 5s + 1}$	$-4.7913, \; -.2087$	$Q = 0.2$	$\zeta = 2.5$	overdamped
$H_2(s) = \dfrac{1}{s^2 + 0.1s + 1}$	$-0.05 \pm j0.9987 = e^{\pm j92.8°}$	$Q = 10$	$\zeta = 0.05$	underdamped

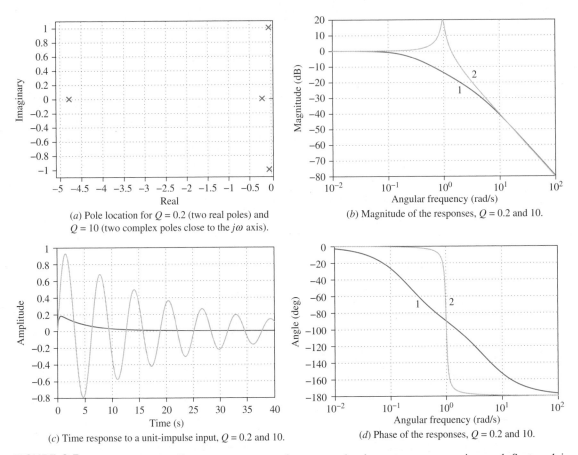

(a) Pole location for $Q = 0.2$ (two real poles) and $Q = 10$ (two complex poles close to the $j\omega$ axis).

(b) Magnitude of the responses, $Q = 0.2$ and 10.

(c) Time response to a unit-impulse input, $Q = 0.2$ and 10.

(d) Phase of the responses, $Q = 0.2$ and 10.

FIGURE 9.5 Pole plots and time/frequency responses of two second-order systems are superimposed. System 1 is overdamped and functions as a low-pass filter. System 2 has a high Q and can function as a bandpass filter.

In H_1 the poles are real and negative, Q is low, ζ is high, and the system is heavily overdamped. The unit-impulse response shows no oscillations and the magnitude plot of the frequency response remains below the low-frequency asymptote (DC level) at all frequencies. In contrast, $H_2(s)$ has two complex conjugate poles at an angle close to the $j\omega$ axis, Q is high, and ζ is low. The system is greatly underdamped (damping ratio $\zeta = 0.05$) and the system is highly oscillatory. This is evidenced by oscillations in its unit-impulse response and also a sharp and selective magnitude response around $\omega = 1$. The magnitude at $\omega = 1$ is equal to $20 \log Q = 20$ dB. The magnitude goes above the low-frequency asymptote.

Summary
A second-order system with two poles and no finite zeros is specified by its undamped natural frequency ω_0, quality factor Q, and DC gain K.

$$H(s) = K \frac{\omega_0^2}{s^2 + \frac{\omega_0}{Q}s + \omega_0^2}$$

The quality factor Q shapes the system's time and frequency responses. If Q is low (i.e., in an overdamped system with negative real poles) the system is low-pass. If Q is high (i.e., in an underdamped system with complex conjugate poles at an angle close to the $j\omega$ axis) the system has a sharp and selective magnitude frequency response around its undamped natural frequency, functioning as a resonator.

Bode Plot: Several Poles or Zeros

The system function of a realizable system made of lumped linear elements is in the form of the ratio of two polynomials with real coefficients. The poles and zeros of the system function are, therefore, either real or complex conjugate pairs. The system function may be written in the form of

$$H(s) = H_o \frac{(s - z_1)(s - z_2) \cdots (s - z_k) \cdots (s - z_M)}{(s - p_1)(s - p_2) \cdots (s - p_\ell) \cdots (s - p_N)}$$

where z_k and p_ℓ are the zeros and poles, of the system, respectively. The magnitude (in dB) and the phase are

$$20 \log |H(\omega)| = 20 \log |H_o| + \sum_{i=1}^{M} 20 \log |j\omega - z_k| - \sum_{i=1}^{N} 20 \log |j\omega - p_\ell|$$

$$\angle H(\omega) = \sum_{k=1}^{M} \angle(j\omega - z_k) - \sum_{\ell=1}^{N} \angle(j\omega - p_\ell)$$

Contributions to Bode plots from zeros add, while those of poles subtract. Consequently, the Bode plot of a system may be found by algebraically adding plots of its cascaded subsystems H_i.

$$H = H_1 \times H_2 \times H_3 \cdots \times H_i \cdots \times H_T$$

Then

$$20 \log |H| = \sum_{i=1}^{T} 20 \log |H_i| \quad \text{and} \quad \angle H = \sum_{i=1}^{T} \angle H_i$$

where T is the total number of subsystems. Theoretically, one may partition a system with M zeros and N poles into $T = M + N$ subsystems, each made of a single zero or pole. In practice, one may expand $H(s)$ into its first-order and second-order subsystems, in which case case $T \leq N$. Example 9.20 illustrates construction of the Bode plot of a third-order system from its subsystems.

Example

9.20

Plot the magnitude and phase of

$$H(s) = \frac{10^6}{s^3 + 200s^2 + 2 \times 10^4 s + 10^6} = \frac{10^6}{(s + 100)(s^2 + 100s + 10{,}000)}$$

Solution

The system function may be expressed as $H(s) = H_1(s) \times H_2(s)$, where

$$H_1(s) = \frac{10^2}{s + 100}, \qquad H_1(\omega) \text{ has a break point at } \omega_0 = 100.$$

$$H_2(s) = \frac{10^4}{s^2 + 100s + 10{,}000}, \qquad H_2(\omega) \text{ has a break point at } \omega_0 = 100 \text{ and } Q = 1.$$

Using the procedures described previously, the individual Bode plots for each segment of the system function may be constructed separately and the results added. See Table 9.3 and Figure 9.6.

TABLE 9.3 Adding Bode Plots of Two Subsystems H_1 and H_2 to Obtain the Bode Plot of the System $H = H_1 \times H_2$

System	DC Gain	ω_0	dB* at ω_0	High-Frequency Slope
H_1	0 dB	100 r/s	−3 dB	−20 dB/decade
H_2	0 dB	100 r/s	0 dB	−40 dB/decade
$H_1 \times H_2$	0 dB	100 r/s	−3 dB	−60 dB/decade

*Relative to the DC level.

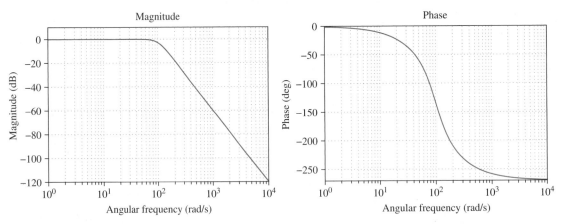

FIGURE 9.6 Bode plot of $H = \dfrac{10^6}{s^3 + 200s^2 + 10^4 s + 10^6} = \dfrac{10^2}{s + 100} \times \dfrac{10^4}{s^2 + 100s + 10^4}$ may be obtained from the plots of its cascaded subsystems. For details see Example 9.20.

Bode Plot: Graphing by Computer

Bode plots of a system (along with other characteristics such as step response, impulse response, pole-zero plots, etc.) may be conveniently obtained through the use of a computer software package. See Example 9.21 and problem 28.

Example
9.21

Given the system function

$$H(s) = \frac{s+3}{(s+15)(s+500)}$$

Find the magnitude and phase of its frequency response as functions of ω. Reduce them to polynomial forms. Write a computer program to plot the above magnitude and phase in the form of Bode plots.

Solution

$$H(s) = \frac{s+3}{s^2 + 515s + 7{,}500}, \qquad H(\omega) = \frac{j\omega + 3}{7{,}500 - \omega^2 + j515\omega}$$

$$|H(\omega)|^2 = \frac{\omega^2 + 9}{\omega^4 + 250{,}225\omega^2 + 5{,}625 \times 10^4},$$

$$\angle H(\omega) = \tan^{-1}\left(\frac{\omega}{3}\right) - \tan^{-1}\left(\frac{515\omega}{7{,}500 - \omega^2}\right)$$

$$H(s) = \frac{s+3}{(s+15)(s+500)}, \qquad H(\omega) = \frac{j\omega + 3}{(j\omega + 15)(j\omega + 500)}$$

$$20\log|H(\omega)| = 10\log(\omega^2 + 9) - 10\log(\omega^2 + 225) - 10\log(\omega^2 + 250{,}000)$$

$$\angle H(\omega) = \tan^{-1}\left(\frac{\omega}{3}\right) - \tan^{-1}\left(\frac{\omega}{15}\right) - \tan^{-1}\left(\frac{\omega}{500}\right)$$

Using the program given below one can plot $20\log|H(\omega)|$ and $\angle H(\omega)$ versus $\log\omega$.

```
clear
w=logspace( -1,4,1001);
s=i*w;
H=(s+3)./((s+15).*(s+500));
plot(log10(w),20*log10(abs(H)));
grid
plot(log10(w),angle(H));
grid
```

Other commands that can be more convenient or flexible for generating Bode plots are often used. The Bode plot for the current system function may also be generated by the following program.

```
num=[1 3];
den=[1 515 7500];
sys=tf(num,den); %tf constructs the transfer (i.e., sys-
tem) function
grid
bode(sys)
```

The following set of commands can provide more flexibility and is used for most Bode plots in this chapter.

```
w=logspace( -1,4,1001);
num=[1 3];
den=[1 515 7500];
[mag,angle]=Bode(num,den,w);
semilogx(w,20*log10(mag));
semilogx(w,angle);
```

Bode Plot: Sketching by Hand

Bode plots of a system may be constructed and sketched readily based on contributions from poles and zeros. For simplicity, we start with systems having real-valued poles and zeros that are apart from each other. To sketch the Bode plot you may follow the steps below.

Step 1. Specify the poles and zeros of the system. Determine the desired range of the frequency to be plotted.

Step 2. Use semilog paper. The log scale will be used for the horizontal axis (frequency ω in rad/s or f in Hz). The vertical axis will have a uniform scale (for magnitude in dB and phase in degrees or radians). Label the frequency axis from left to right (representing increasing frequency) to satisfy the desired range of the plot. Label the vertical axes from bottom to top (representing increasing values) to satisfy the desired range of the magnitude or angle. The magnitude and phase may be plotted on separate graphs or share the same set of axes.

Step 3. Mark the frequency axis at points corresponding to the value of the system's poles and zeros (or their absolute value if they are complex-valued). These will constitute break points in the asymptotic plots produced by zeros and poles. They will simply be called *zeros* and *poles*. Each will contribute to the plot as explained previously. Their contributions in dB and radians will add or subtract.

Step 4. Starting from a low frequency, draw the asymptotic lines for the magnitude plot. These are a series of lines with slopes of $\pm 20k$ dB per dacade, $k = 0, 1, 2, \ldots$. When encountering a break point corresponding to a *zero*, the slope of the asymptote will increase by 20 dB/decade (6 dB/octave) over the previous segment. A break point corresponding to a *pole* will pull down the asymptote by 20 dB/decade. The change will double if a double pole or zero (or a complex-valued pair) is encountered.

Step 5. Similarly, sketch the phase contribution of a pole or zero using $\pm 90°$ asymptotic lines. Then add them algebraically. When encountering a zero the phase will increase toward a $90°$ asymptote. A pole will pull down the asymptote by a $-90°$ asymptote. The change will double if a double pole or zero (or a complex-valued pair) is encountered.

Step 6. At a break point corresponding to a simple zero or pole the magnitude and phase plots deviate from the asymptotes by ± 3 dB and $\pm 45°$, respectively. At a break point corresponding to a pair of complex zeros or poles the magnitude deviation from the asymptote is $\pm 20 \log Q$ dB.

Step 7. The accuracy of the sketches can be improved by using the values of magnitude and phase at frequencies half and twice the break frequencies: 1 dB above the asymptote for a zero or 1 dB below it for a pole.

Contributions to Bode plots

The system function $H(s) = 50(s + 10)/[(s + 1)(s + 200)]$ has a zero at $s = -10$, two poles at -1 and -200, and a gain factor 50. Plot its Bode diagram and graphically show individual contributions from the zero and the poles to it.

Solution

The system function has a zero at $s = -10$, two poles at -1 and -200, and a gain factor 50. The magnitude plots are shown in Figure 9.7(a). The line labeled 1 shows contribution from $1/(s+1)$ (the first pole at $s = -1$). It has a DC gain of 0 dB, a break point at $\omega = 1$, and a downward asymptote with -20-dB slope. Line number 2 is due to $s + 10$ (the zero at $s = 10$). It has a DC gain of $20 \log 10 = 20$ dB, a break point at $\omega = 10$, and an upward asymptote with 20-dB slope. Contribution from $1/(s + 200)$ (the second pole) has a DC gain of $-20 \log 200 = -46$ dB, a break point at $\omega = 200$, and a -20 dB/decade asymptote. The line number 4 is the magnitude plot for $H(s)$. It is equal to the algebraic addition of lines 1, 2 and 3, plus $20 \log 50 = 34$ dB due to the gain factor. The DC gain of the system is $H(s)|_{s=0} = 2.5$, corresponding to 8 dB. This is equal to the algebraic sum of individual DC components from the poles, zero, and the gain factor, $34 - 0 + 20 - 46 = 8$ dB. In a similar way, contributions from the poles and the zero to the phase plot are shown in Figure 9.7(b), labeled as in the magnitude plot. Contribution from the first pole at $s = 1$ is shown by line number 1, which goes from nearly 0 at low frequencies to nearly $-90°$ at high frequencies (with $-45°$ dB at the break point $\omega = 1$). The zero at $s = 10$ contributes line number 2, which grows from nearly 0 at low frequencies to nearly $90°$ at high frequencies (with $45°$ dB at the break point $\omega = 10$). The effect of the second pole is similar to that of the first pole except for a shift of 200 rad/s along the frequency axis. The gain factor contributes zero to the phase. The total phase plot is shown by line number 4, equal to the algebraic summation of 1, 2, and 3. In summary the frequency response is obtained by algebraic addition of individual components. When the poles and the zero are far enough from each other, this can be done visually and result in a sketch by hand of the Bode plot, as shown in Example 9.23.

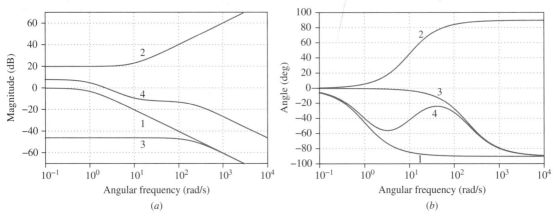

FIGURE 9.7 Contributions to the magnitude **(a)** and phase **(b)** of $H(s) = \frac{50(s+10)}{(s+1)(s+200)}$ (Example 9.22).

Example 9.23

Sketch the Bode plots for

$$H(s) = \frac{10^4(s+3)}{(s+50)(s+600)}$$

Solution

The system has a zero at $s = -3$ and two poles at $s = -50$ and -600. The break frequencies are at $\omega = 3$ rad/s (zero), and $\omega = 50, 600$ rad/s (poles). The desired range of frequency is taken to be 1 rad/s to 10 krad/s, requiring semilog paper with a minimum of 4 decades. See Figure 9.8. The poles and the zero are distant from each other, increasing the accuracy of the sketch. The magnitude is sketched in Figure 9.8(a). The low-frequency magnitude asymptote is 0 dB. The magnitude plot first encounters the zero at $\omega = 3$. The slope of the asymptote thus increases to 20 dB/decade (line number 1). At $\omega = 50$, it encounters a pole and the slope is reduced by 20 dB/decade, resulting in a flat line (number 2). At $\omega = 600$, it encounters the second pole and the slope is reduced by another 20 dB/decade, resulting in a line with a slope of -20 dB/decade, which forms the high-frequency asymptote (line number 3). These asymptotes are shown in Figure 9.8(a) by thin lines, marked 1, 2, and 3, respectively.

At the $\omega = 1.5, 3$, and 6 (related to the zero), the magnitude is 1, 3, and $6 + 1 = 7$ dB, respectively. At the $\omega = 25, 50$, and 100 (related to the first pole), the magnitude is below the asymptote by 1, 3, and 1 dB, respectively. Similarly, at $\omega = 300, 600$, and 1,200 (related to the second pole), the magnitude is below the asymptote by 1, 3, and 1 dB, respectively, as shown in Figure 9.8(a). The above values are added and the points are then connected by a smooth curve to form a sketch of the magnitude plot. The phase plot is obtained by sketching the contributions from the poles and zeros and adding them up. See Figure 9.8(b).

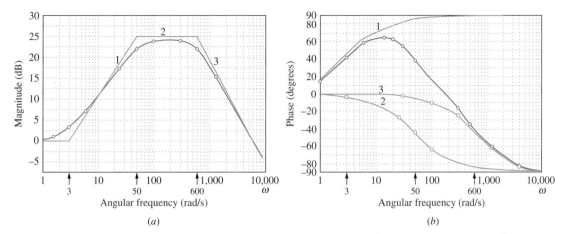

FIGURE 9.8 Sketching the magnitude **(a)** and phase **(b)** of $H(s) = \frac{10^4(s+3)}{(s+50)(s+600)}$ based on its components.

9.5   Vectorial Interpretation of $H(s)$ and $H(\omega)$

A rational system function with M zeros and N poles may be written in terms of its poles and zeros as follows:

$$H(s) = H_0 \frac{\Pi_k(s - z_k)}{\Pi_\ell(s - p_\ell)}$$

where z_k with $k = 1, 2, \cdots M$ are zeros of the system, p_ℓ with $\ell = 1, 2, \ldots, N$ are poles of the system, and H_0 is a gain factor. To simplify the discussion, assume H_0 is a positive number. A zero at z_k contributes $(s - z_k)$ to the numerator of the system function. In the s-plane this is a vector drawn from the point z_k (the zero) to a point s (where the system function is to be evaluated). Similarly, a pole at p_ℓ contributes $(s - p_\ell)$ to the denominator of the system function. In the s-plane this is a vector drawn from the pole at p_ℓ to point s. Let $\overline{B}_k = (s - z_k)$ designate the vector from z_k to s, and $\overline{A}_\ell = (s - p_\ell)$ designate the vector from p_ℓ to s. Then

$$H(s) = H_0 \frac{\overline{B}_1 \times \overline{B}_2 \cdots \times \overline{B}_k \cdots \times \overline{B}_M}{\overline{A}_1 \times \overline{A}_2 \cdots \times \overline{A}_\ell \cdots \times \overline{A}_N} = H_0 \frac{\Pi_k \overline{B}_k}{\Pi_\ell \overline{A}_\ell}$$

$$|H(s)| = H_0 \frac{|\overline{B}_1| \times |\overline{B}_2| \cdots \times |\overline{B}_k| \cdots \times |\overline{B}_M|}{|\overline{A}_1| \times |\overline{A}_2| \cdots \times |\overline{A}_\ell| \cdots \times |\overline{A}_N|} = H_0 \frac{\Pi_k |\overline{B}_k|}{\Pi_\ell |\overline{A}_\ell|}$$

$$\angle H(s) = \left[\angle \overline{B}_1 + \angle \overline{B}_2 \cdots + \angle \overline{B}_k \cdots + \angle \overline{B}_M \right] - \left[\angle \overline{A}_1 + \angle \overline{A}_2 \cdots + \angle \overline{A}_\ell \cdots + \angle \overline{A}_N \right]$$

$$= \sum_{k=1}^{M} \angle \overline{B}_k - \sum_{\ell=1}^{N} \angle \overline{A}_\ell$$

The above method provides a graphical technique for evaluating $H(s)$ at a desired point in the s-plane, and also a qualitative observation of its properties.

Determination of $H(\omega)$ Using Vectorial Interpretation

The frequency response is found by evaluating $H(s)$ on the $j\omega$ axis. Starting at the point $s = 0$ ($\omega = 0$) and moving upward toward $\omega = \infty$, we can sketch $H(\omega)$ using its vectorial interpretation. Because vectors from poles are in the denominator, the magnitude of the frequency response is increased when the operating frequency ω approaches the neighborhood of a pole. Similarly, because the zero vectors are in the numerator, the magnitude of the frequency response is decreased when the operating frequency approaches a zero's neighborhood. The qualitative insight obtained from the vectorial interpretation of $H(\omega)$ is especially useful in approximating its behavior (e.g., its peak and 3-dB bandwidth frequencies) near a pair of poles close to the $j\omega$ axis. Example 9.24 illustrates this property.

Example **9.24**

Consider a bandpass filter having a single zero at the origin and a pair of poles at $-0.1 \pm j10$. The filter is described by the system function

$$H(s) = K\frac{s}{s^2 + 0.2s + 100}$$

a. Determine the constant factor K so that the maximum gain of $|H(\omega)|$ is 5.

b. Set $K = 1$ and use the vectorial approach to determine or approximate important features of $H(\omega)$ (e.g., the maximum value of the magnitude and its frequency, the 3-dB lower and upper frequencies, and the 3-dB bandwidth).

c. Compare the approximate values obtained from the vectorial representation obtained in part b with their more exact values obtained by computation.

Solution

a. The frequency response and its magnitude-squared function are

$$H(\omega) = \frac{jK\omega}{-\omega^2 + j0.2\omega + 100}$$

$$|H(\omega)|^2 = \frac{K^2\omega^2}{(100 - \omega^2)^2 + (0.2\omega)^2}$$

$|H(\omega)|$ attains its maximum at $\omega = 10$, where $|H|_{Max} = 5K$, resulting in $K = 1$.

b. The system has a pair of poles at $-0.1 \pm j10$ (shown by A_1 and A_2 in Figure 9.9) and a zero at the origin (shown by B). The frequency response, evaluated for a frequency shown by point C on the $j\omega$ axis, is obtained from

$$H(\omega) = \frac{\overline{BC}}{\overline{A_1C} \times \overline{A_2C}}$$

Examples of vectorial components of $H(\omega)$ are shown in Figure 9.9(a) (for $\omega = 5$) and Figure 9.9(b) (for $\omega = 10$), while Figure 9.9(c) shows the vectors

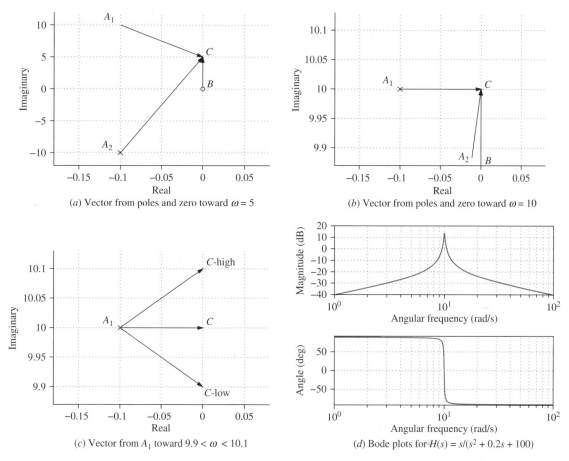

(a) Vector from poles and zero toward $\omega = 5$

(b) Vector from poles and zero toward $\omega = 10$

(c) Vector from A_1 toward $9.9 < \omega < 10.1$

(d) Bode plots for $H(s) = s/(s^2 + 0.2s + 100)$

FIGURE 9.9 This figure analyzes and plots the frequency response of the bandpass system $H(s) = s/(s^2 + 0.2s + 100)$ in relation to its vectorial interpretation. The system has two poles at $-0.1 \pm 10j$ (shown by A_1 and A_2) and a zero at the origin (shown by B). Point C represents the frequency at which $H(\omega)$ is evaluated. Then, $H(\omega) = \overline{BC}/[\overline{A_1C} \times \overline{A_2C}]$. **(a)** shows the vectors' configuration for $\omega = 5$. As one moves upward on the $j\omega$ axis, the vector $\overline{A_1C}$ decreases causing $|H(\omega)|$ to increase, until at $\omega = 10$ the vector $\overline{A_1C}$ reaches its minimum of 0.1 [as seen from **(b)**] and $|H(\omega)|$ attains its maximum of $|H_{Max}| = 5$. At $\omega = 10 \pm 0.1$, **(c)**, the length of $\overline{A_1C}$ is increased by a factor of $\sqrt{2}$, causing $|H(\omega)|$ to be reduced by that same factor, which brings it to 3 dB below its maximum. Therefore, the lower and upper 3-dB bandwidth frequencies are $\omega_\ell \approx 9.9$ and $\omega_h \approx 10.1$. The approximation is due to the fact that in the above process as ω moves closer to the pole at A_1, it gets farther away from the pole at A_2, and the vector $\overline{A_2C}$ grows slightly. The accuracy of the approximation increases the closer the pole's angle becomes to the $j\omega$ axis (or, equivalently, at higher-quality factors). At $Q > 20$, the approximation is considered acceptable. In the current case $Q = 50$, and 3-dB bandwidth of $\Delta\omega = 0.2$ is evenly divided between the lower and higher half-power frequencies. Computer-plotted magnitude and phase responses are shown in **(d)** for $1 < \omega < 100$ rad/s. Note the sharp magnitude peak at $\omega = 10$. See Example 9.24.

from A_1 to $\omega = 10$ and $\omega = 10 \pm 0.1$. To sketch $H(\omega)$ using vectorial interpretation we start with point C at the origin $\omega = 0$ and move upward toward $\omega = \infty$.

At $\omega = 0$ we have $\overline{BC} = 0$, which results in $H(\omega) = 0$. As one moves upward on the $j\omega$ axis, $\overline{BC}$ grows and $\overline{A_1C}$ becomes smaller, Figure 9.9(a), resulting in the growth of $|H(\omega)|$.[7] The vector A_1C is shortest at $\omega = 10$ with a length of 0.1. Because it is in the denominator of the $H(s)$ it produces the maximum of the magnitude response at that frequency. At $\omega = 10 \pm 0.1$ that vector becomes $0.1\sqrt{2}$ units long and produces the half-power points [see Figure 9.9(c)]. The 3-dB bandwidth is, therefore, $\Delta\omega = 0.2$ centered at $\omega = 10$, both in rad/s.

At $\omega = 10$, Figure 9.9(b), we have

$$\overline{BC} = 10\angle 90°, \quad \overline{A_1C} \approx 0.1\angle 0° \quad \text{and} \quad \overline{A_2C} \approx 20\angle 90°$$

$$H(10) = \frac{10\angle 90°}{0.1\angle 0° \times 20\angle 90°} = 5$$

Note that despite the two approximations $\overline{A_1C}$ and $\overline{A_2C}$ at $\omega = 10$, the maximum value $|H_{Max}(\omega)| = 5$ and its location at $\omega = 10$ are exact. In the neighborhood of the pole (near $\omega = 10$) the pole influences the frequency response the most. As said before, the 3-dB band may be obtained from the pole-zero plot by noting that at $\omega = 10 \pm 0.1$ the length of the vector $\overline{AC}$ from the nearby pole is increased by a factor $\sqrt{2}$ from its minimum value of 0.1, resulting in a 3-dB decrease in the magnitude. Thus,

$$\omega_\ell \approx 9.9, \quad \omega_h \approx 10.1, \quad \text{and} \quad \Delta\omega = 0.2, \quad \text{all in radians/s}$$

The exact values are found from a computer graph:

$$\omega_\ell = 9.9005, \quad \omega_h = 10.1005, \quad \text{and} \quad \Delta\omega = 0.2$$

The Bode plot of the system is given in Figure 9.9(d).
vector A_1C is shortest at $\omega = 10$

9.6 Second-Order Systems

In this section we consider a stable system with two poles in the left-half of the s-plane. Contribution from the poles to the system function is represented by

$$H(s) = \frac{b}{s^2 + 2as + b}$$

where a and b are positive real numbers. The system has two poles at $s_{1,2} = -a \pm \sqrt{a^2 - b}$, which could be real or complex. The poles are in the left-half of the s-plane and the system is stable. The position of the poles determines whether the system's

---

[7]In this process $\overline{A_2C}$ also grows but not enough to offset the reduction in $\overline{A_1C}$, except when we get close to $\omega = 10$.

unit-step response is oscillatory or not. Depending on the sign of $a^2 - b$, we recognize the following three cases.

Case 1. $\mathbf{a^2 > b}$. The system has two distinct negative real poles on the negative real axis at $s_{1,2} = -\omega_{1,2}$, where $\omega_{1,2} = a \pm \sqrt{a^2 - b}$. Note that $\omega_1\omega_2 = b$ and $\omega_1 + \omega_2 = 2a$. The system function is called *overdamped*.

Case 2. $\mathbf{a^2 = b}$. The system has a negative pole of order 2, that is, two identical poles on the negative real axis at $s = -\omega_0$, where $\omega_0 = a = \sqrt{b}$. The system function is called *critically damped*.

Case 3. $\mathbf{a^2 < b}$. The system has two complex conjugate poles with negative real parts, $s_{1,2} = -\sigma \pm j\omega_d$, where $\sigma = a$ and $\omega_d = \sqrt{b - a^2}$. Its step response is oscillatory and the system function is called *underdamped*.

Because of their simplifying features, second-order systems are widely used in modeling and analysis of systems. Consider the system having a pair of LHP poles and no zeros, with input $x(t)$, output $y(t)$, unit-impulse response $h(t)$, and unit-step response $g(t)$. The system may be represented in one of the following interrelated forms:

1. Pole locations ($p_{1,2}$)
2. Undamped natural frequency and damping ratio (ω_0, ζ) or
3. Damped natural frequency and step response overshoot (ω_d, ρ)
4. Undamped natural frequency and quality factor (ω_0, Q)
5. Frequency response $H(\omega)$

The first three of these representations are given below for the case of a complex conjugate pair of poles (see also Figure 9.4).

1. **By the location of the poles.** Let the poles be at $p_{1,2} = -\sigma \pm j\omega_d = \omega_0 e^{\pm j\theta}$. Then

$$H(s) = \frac{\omega_0^2}{(s+\sigma - j\omega_d)(s+\sigma+j\omega_d)} = \frac{\omega_0^2}{s^2+2\sigma s+\omega_0^2}, \quad \text{where } \omega_0^2 = \sigma^2 + \omega_d^2$$

2. **By the undamped natural frequency ω_0 and the damping ratio ζ.** The system function may be expressed as a function of the undamped natural frequency ω_0 and damping ratio ζ:

$$H(s) = \frac{b}{s^2 + 2as + b} = \frac{\omega_0^2}{s^2 + 2\zeta\omega_0 s + \omega_0^2}$$

where ζ and ω_0 are positive numbers. The system has a pair of complex conjugate poles at $p_{1,2} = -\sigma \pm j\omega_d$, where

$$\sigma = \zeta\omega_0 \quad \text{and} \quad \omega_d = \omega_0\sqrt{1 - \zeta^2}$$

Note that by substitution we can verify that the undamped natural frequency is

$$\sigma^2 + \omega_d^2 = \zeta^2\omega_0^2 + (1 - \zeta^2)\omega_0^2 = \omega_0^2$$

The poles are on a semicircle in the LHP with radius ω_0, at an angle $\phi = \sin^{-1}\zeta$ with the imaginary axis. See Figure 9.4. In the above formulation $\omega_0^2 = b$ is called the undamped natural frequency, ω_d is called the damped natural frequency,

$\zeta = a/\sqrt{b}$ is called the damping ratio, and the poles are at $-\omega_0(\zeta \pm \sqrt{\zeta^2 - 1})$. The value of ζ determines the damping state of the system function:

$$
\begin{cases}
\zeta > 1, & \text{overdamped} \\
\zeta = 1, & \text{critically damped} \\
\zeta < 1, & \text{underdamped}
\end{cases}
$$

The response of the system to a unit-step input is

$$
g(t) = \left[1 - \frac{1}{\sqrt{1-\zeta^2}} e^{-\sigma t} \cos(\omega_d t - \phi) \right] u(t)
$$

Note that ϕ, the phase angle of the step response, is the same as the angle of the poles with the imaginary axis. The percentage overshoot in the step response is

$$
\rho = 100 \times e^{-\frac{\pi\zeta}{\sqrt{1-\zeta^2}}}
$$

The unit-impulse response of the system is

$$
h(t) = \frac{d}{dt} g(t) = \frac{\omega_0}{\sqrt{1-\zeta^2}} e^{-\sigma t} \sin \omega_d t \, u(t)
$$

3. **By the undamped natural frequency ω_0 and the quality factor Q.** In signal processing applications and filter design the above system function is expressed in terms of the undamped natural frequency ω_0 and quality factor Q, both of which are real and positive numbers

$$
H(s) = \frac{b}{s^2 + 2as + b} = \frac{\omega_0^2}{s^2 + \frac{\omega_0}{Q}s + \omega_0^2} = \frac{1}{\left(\frac{s}{\omega_0}\right)^2 + \frac{1}{Q}\frac{s}{\omega_0} + 1}
$$

In the above formulation $\omega_0^2 = b$ is the undamped natural frequency and $Q = \sqrt{b}/(2a)$ is called the quality factor. The poles are at $s_{1,2} = -\frac{\omega_0}{2Q}(1 \pm \sqrt{1 - 4Q^2})$, which, depending on the value of Q, are either real or complex. It is seen that the quality factor of a system is related to its damping ratio by the equation $Q = 1/(2\zeta)$. This representation becomes helpful in the construction of Bode plots and filter design. The quality factor Q specifies the damping state of the system and shapes the system's responses in the time and frequency domains. The following table summarizes the role of Q in determining the location of poles, and whether the system is overdamped, critically damped, or underdamped (oscillatory).

Q	Poles	System's Step Response	Frequency Response
$Q < 0.5$	two distinct real poles	overdamped	low-pass
$Q = 0.5$	two identical real poles	critically damped	low-pass
$Q > 0.5$	two complex conjugate poles	underdamped	low-pass/bandpass

Bode Plot for a Second-Order System

In this section we plot the Bode diagram for the second-order system

$$H(s) = \frac{\omega_0^2}{s^2 + \frac{\omega_0}{Q}s + \omega_0^2}$$

and discuss the effect of the quality factor. As derived previously, the value of the magnitude at ω_0 is $|H(\omega_0)| = Q$, equal to the quality factor, or expressed in dB units, $20\log|H(\omega_0)| = 20\log Q$ dB. A higher Q produces a sharper peak at ω_0. An example is shown in Figure 9.10 for $Q = 0.2,\ 0.707,\ 5,\ 10,$ and 50. For $Q < 0.5$, the poles are real (the system is overdamped) and the magnitude doesn't exceed the asymptote. For $Q = 0.5$ the system is critically damped and has a double pole at $-\omega_0$ (the break frequency), resulting in a -6-dB magnitude at that frequency. For $Q > 0.5$ the system has a pair of complex conjugate poles at $\sigma \pm j\omega_d$ (the system is underdamped). The magnitude at ω_0 is equal to $20\log Q$ (when Q is expressed in dB). If $Q > 1$, the plot goes above the low-frequency asymptote. This phenomenon is present in other classes of second-order systems (such as bandpass, high-pass, bandstop) which have a set of zeros different from the above system function.

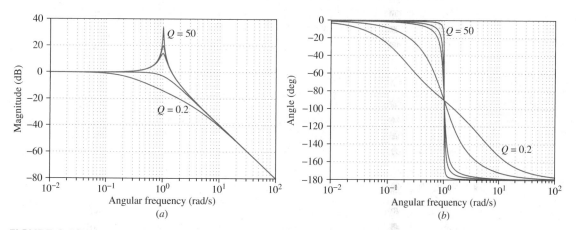

FIGURE 9.10 Magnitude *(a)* and phase plots *(b)* of the low-pass second-order system $H(s) = 1/(s^2 + \frac{1}{Q}s + 1)$ for $Q = 0.2,\ 0.707,\ 5,\ 10,$ and 50.

9.7 Dominant Poles

If the frequencies of signals that operate on the system are confined to the neighborhood of a pole p_k (or a zero z_k), the vector $s - p_\ell$ (or $s - z_k$) will have the biggest rate of change within that neighborhood and will influence and dominate the behavior of the system most. The case of dominant poles is especially of interest in system modeling because it allows us to reduce a complex system to a first-order or a second-order model. The simplified model will include only the dominant pole (and its conjugate if the pole is complex), with a gain adjusted to match the original system.

Single Dominant Pole

We first examine the case of a single dominant pole that leads to a system model of first order. Consider a stable system containing M zeros and $N + 1$ poles, including a negative pole at $-\sigma$ in the LHP. The system function is

$$H(s) = H_0 \frac{(s - z_1)(s - z_2) \cdots (s - z_k) \cdots (s - z_M)}{(s - p_1)(s - p_2) \cdots (s - p_\ell) \cdots (s - p_N)} \times \frac{1}{s + \sigma} = H_N(s) \times \frac{1}{s + \sigma}$$

The multiplying factor $H_N(s)$ is the contribution from all poles and zeros other than the pole at $-\sigma$. The above formulation, so far, uses no approximation. Now, let the operating frequency s be near the pole $-\sigma$ (and away from all other poles and zeros). Then $H_N(s) \approx H_N$ (in general, a complex number) may be considered almost a constant factor and $H(s)$ be simplified to

$$H(s) \approx \frac{H_N}{s + \sigma}$$

The new gain factor H_N embodies the total effect of all other poles and zeros on $H(s)$. Those poles and zeros being away from the operating complex frequency s will exert almost a fixed effect, making H_N nearly a constant. The complex system is then modeled by a first-order system.

Example
9.25

Modeling the 741 op-amp by its first-order pole

The 741 op-amp has a pole at $f = 5$ Hz and additional ones at higher frequencies, with the second pole being at 1 MHz. The DC gain of the op-amp (called the open-loop DC gain) is 2×10^5. At 5 Hz the gain is reduced by 3 dB. It attenuates at a rate of 20 dB per decade of frequency, such that at 1 MHz (the location of the second pole) the gain is one. The op-amp is not used at high frequencies because its gain becomes very low. Therefore, the first-order pole dominates the operation of the op-amp and it suffices in the analysis and design of the circuit. To verify this, we plot and compare the following frequency responses:

Model 1 (first-order model): $\qquad H_1(f) = \dfrac{200{,}000}{1 + j\frac{f}{5}}$

Model 2 (second-order model): $\quad H_2(f) = \dfrac{200{,}000}{(1 + j\frac{f}{5})(1 + j\frac{f}{10^6})}$

Figure 9.11 shows the Bode plots and step responses of both models, superimposed for comparison. The Bode plots are identical up to 1 MHz. The second pole at 1 MHz introduces an additional attenuation of 20 dB per decade in the magnitude and a phase angle which leads to an eventual phase of $-180°$ at $\omega = \infty$. The step responses are almost identical.

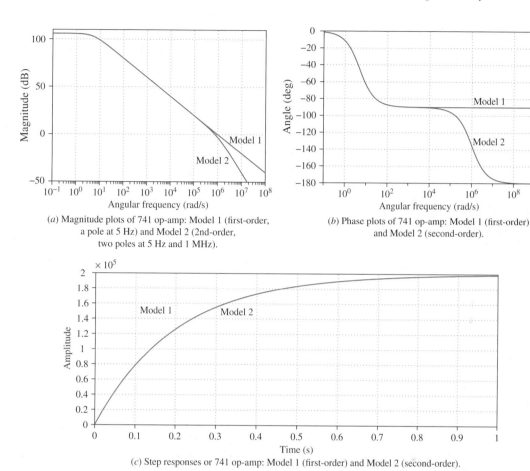

(a) Magnitude plots of 741 op-amp: Model 1 (first-order, a pole at 5 Hz) and Model 2 (2nd-order, two poles at 5 Hz and 1 MHz).

(b) Phase plots of 741 op-amp: Model 1 (first-order) and Model 2 (second-order).

(c) Step responses or 741 op-amp: Model 1 (first-order) and Model 2 (second-order).

FIGURE 9.11 (Example 9.25). Bode plots and step responses of the single-pole model (at 5 Hz, Model 1) and the two-poles model (at 5 Hz and 1 MHz, Model 2) of a 741 op-amp, superimposed for comparison. Adding the second pole (at 1 MHz) doesn't influence the system's performance up to 1 MHz as seen from the Bode plots. Beyond 1 MHz, the pole at 1 MHz introduces an additional attenuation of 20 dB per decade in the magnitude *(a)*, and adds to the negative phase angle leading to an eventual phase of $-180°$ at $\omega = \infty$ *(b)*. The step responses are almost identical *(c)*.

Pair of Complex Conjugate Dominant Poles

In a similar way, we derive the second-order model of a system with a complex conjugate pair of dominant poles. Consider a stable physical system containing M zeros and $N+2$ poles, including a pair of poles at $-\sigma \pm j\omega_d$ in the LHP. The system function is

$$H(s) = H_0 \frac{(s - z_1)(s - z_2) \cdots (s - z_k) \cdots (s - z_M)}{(s - p_1)(s - p_2) \cdots (s - p_\ell) \cdots (s - p_N)} \times \frac{1}{s^2 + 2\sigma s + \omega_0^2}$$

$$= H_N(s) \times \frac{1}{s^2 + 2\sigma s + \omega_0^2} \quad \text{where } \omega_0^2 = \sigma^2 + \omega_d^2$$

The gain factor $H_N(s)$ is the contribution from all poles and zeros other than the pair at $-\sigma \pm j\omega_d$. Now, let the operating frequency s be near the poles $-\sigma \pm j\omega_d$ (and away from all other poles and zeros). Then $H_N(s) \approx H_N$ may be considered almost a constant factor and $H(s)$ be simplified to

$$H(s) \approx \frac{H_N}{s^2 + 2\sigma s + \omega_0^2}$$

The new gain factor H_N embodies the total effect of all other poles and zeros on $H(s)$. Those poles and zeros being away from the complex operating frequency s will exert almost a fixed effect, making H_N nearly a constant (and, in general, a complex number).

Example 9.26

Modeling a third-order system by its dominant pair of poles

Consider the bandpass filter having a single zero at the origin, two poles at $-0.1 \pm j10$, and a third pole at -50. The filter is described by the system function

$$H(s) = K\frac{s}{(s^2 + 0.2s + 100)(s + 50)}$$

a. Find the constant factor K so that the maximum value of $|H(\omega)|$ is 5.

b. Let $K = 10\sqrt{26}$. Create its pole-zero plot in the s-plane. Plot its step and frequency responses. Determine the maximum gain, the frequency at which it occurs, and the 3-dB bandwidth. Construct a second-order model $\hat{H}(s)$ of the filter made of its dominant pair of poles (and the zero at the origin), with a maximum gain equal to that of the filter. Plot the step and frequency responses of the model. Determine the frequency at which the gain is maximum, and the 3-dB bandwidth. Visually inspect the reponses of the two systems, then comment on the validity of the second-order model.

Solution

a. The system has a pair of poles at $p_{1,2} = -0.1 \pm j10$, a remote pole at $p_3 = -50$ and a zero at $z_1 = 0$. From the vectorial composition of $H(\omega)$ we can conclude that $|H|_{Max} = 5$ will be at $\omega = 10$.

$$|H|_{Max} = K\left|\frac{j\omega}{(-\omega^2 + j0.2\omega + 100)(j\omega + 50)}\right|_{\omega=10} = \frac{K}{2\sqrt{26}} = 5$$

from which $K = 10\sqrt{26}$.

b. At frequencies near $p_{1,2} = -0.1 \pm j10$ the poles dominate the behavior of the system. The system function of the filter may be written as

$$H(s) = \frac{K}{s + 50} \times \frac{s}{s^2 + 0.2s + 100}$$

We will investigate the effect of the remote pole at -50 by using a vectorial interpretation of $H(s)$. Let C designate the location of s on the $j\omega$ axis at which

$H(\omega)$ is being evaluated. The four vectors contributing to the system's frequency response are

$$s = \overline{BC}, \quad s - p_1 = \overline{A_1 C}, \quad s - p_2 = \overline{A_2 C}, \quad s - p_3 = \overline{A_3 C}$$

The term $s + 50$ corresponds to the vector $\overline{A_3 C}$ from the pole at -50 to point C on the $j\omega$ axis. The vectorial representation of the system function on the $j\omega$ axis is

$$H(s) = \frac{K}{s + 50} \times \frac{s}{s^2 + 0.2s + 100}$$

$$H(\omega) = \frac{K}{A_3 C} \times \frac{s}{s^2 + 0.2s + 100}\bigg|_{s=j\omega}$$

Within the neighborhood of $\omega \approx 10 \pm 0.1$, the vector $\overline{A_3 C}$ varies slightly, causing a negligible change in $H(\omega)$. Evaluating $\overline{A_3 C}$ at $\omega = 10$ and substituting in the above equation we obtain the second-order model

$$\hat{H}(\omega) \approx \frac{10\sqrt{26}}{j10 + 50} \times \frac{s}{s^2 + 0.2s + 100}\bigg|_{s=j\omega}$$

$$\approx e^{-j0.0628\pi} \frac{s}{s^2 + 0.2s + 100}\bigg|_{s=j\omega}$$

The model's step and frequency responses are superimposed on the system's responses in Figure 9.12.

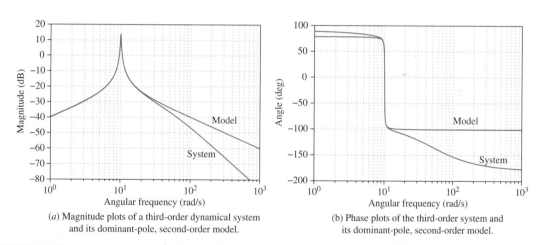

(a) Magnitude plots of a third-order dynamical system and its dominant-pole, second-order model.

(b) Phase plots of the third-order system and its dominant-pole, second-order model.

FIGURE 9.12 (For Example 9.26). Responses of a third-order system $H_1(s) = 10\sqrt{26}\, s/[(s + 50)(s^2 + 0.2s + 100)]$ and its dominant-poles model $H_2(s) = e^{-j11.3°}\, s/(s^2 + 0.2s + 100)$ superimposed. The system has three poles at $p_{1,2} = -0.1 \pm j10$ and $p_3 = -50$. The model with $p_{1,2} = -0.1 \pm j10$ approximates the frequency response of the system closely in the neighborhood of $\omega = 10$ rad/s. The third pole at $s = -50$ produces a break point at $\omega = 50$ and introduces additional attenuation of 20 dB per decade in the magnitude *(a)*, and adds to the negative phase angle leading to an eventual phase of $-180°$ at $\omega = \infty$ *(b)*. The step responses are almost identical *(c)*.

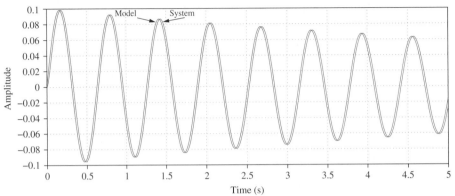

(c) Step responses of the third-order system and its dominant-pole, second-order model.

FIGURE 9.12 *(Continued)*

9.8 Stability and Oscillations

Stability

If bounded inputs to a linear system produce bounded outputs, the system is said to be BIBO-stable. In other words, the system's response to every bounded input would remain bounded. No bounded input can make the output diverge to an infinitely large value. The zero input response (and the natural response) of a BIBO system diminishes toward zero with time. A continuous-time LTI system is BIBO-stable if its poles (natural frequencies) are in the left half of the s-plane. A related property is that the unit-impulse response of a BIBO system is absolutely integrable.

$$\int_{-\infty}^{\infty} |h(t)| dt < M, \quad \text{where } M \text{ is any large number}$$

In the frequency domain, a pair of complex conjugate poles in the LHP causes the system to be underdamped (oscillatory). Poles at an angle close to the $j\omega$ axis make the system more underdamped (smaller damping ratio) and more oscillatory. But until the poles move out of the LHP the system remains stable. Poles on the $j\omega$ axis produce sustained oscillations in the absence of an input to the circuit (but initiated by noise), turning the system into an oscillator. Poles in the RHP make the system unstable. See project 2 on active bandpass filter at the end of this chapter.

Oscillators

A pair of complex conjugate poles at $-\sigma \pm j\omega_d$ contributes the term $Ae^{-\sigma t} \cos(\omega_d t + \theta)$ to the natural response of the system. With the poles in the LHP, the above term decays with the time constant $1/\sigma$. Move the poles toward the imaginary axis (e.g., by feedback) and oscillations last longer. The closer the poles get to the $j\omega$ axis, the longer it takes for the oscillations to die out. Place the poles on the $j\omega$ axis and oscillations sustain themselves. Example 9.27 illustrates the above concept.

Example

9.27

a. In the circuit of Figure 9.13(a) (called the Wien-bridge or *RC* oscillator) assume an ideal op-amp and find V_2/V_1. Short the input and find conditions for sustained oscillation [i.e., $v_2(t) \neq 0$ when $v_1(t) = 0$].

b. Redraw the circuit as Figure 9.13(b) with the noninverting op-amp replaced by its equivalent ideal amplifier with gain $K = 1 + R_2/R_1$. Find conditions for sustained oscillations in $v_2(t)$ (with $v_1(t) = 0$) and find its frequency.

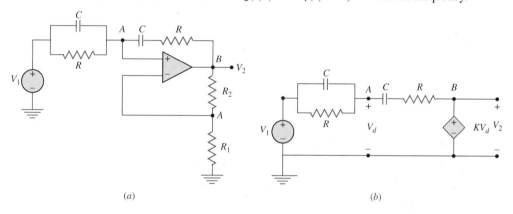

(a)   (b)

FIGURE 9.13 Wien-bridge (*RC*) oscillator using an op-amp **(a)** and its equivalent circuit model **(b)**.

Solution

a. Let the impedances of the parallel *RC* and the series *RC* be

$$Z_1 = \frac{R}{1 + RCs} \quad \text{and} \quad Z_2 = \frac{1 + RCs}{Cs},$$

respectively. To obtain V_2/V_1 in Figure 9.13(a), we find V_A by dividing V_2 between Z_1 and Z_2 and equating that to the division of V_2 between R_1 and R_2.

Divide V_2 between R_1 and R_2: $\qquad V_A = \dfrac{R_1}{R_1 + R_2} V_2 = \beta V_2, \quad \text{where } \beta = \dfrac{R_1}{R_1 + R_2}$

Divide V_2 between Z_1 and Z_2: $\qquad V_A = \dfrac{Z_1}{Z_1 + Z_2} V_2 + \dfrac{Z_2}{Z_1 + Z_2} V_1$

Equate the above expressions for V_A: $\qquad \beta V_2 = \dfrac{Z_1}{Z_1 + Z_2} V_2 + \dfrac{Z_2}{Z_1 + Z_2} V_1$

Resulting voltage transfer function: $\qquad \dfrac{V_2}{V_1} = \dfrac{(1 + RCs)^2}{\beta R^2 C^2 s^2 + (3\beta - 1)RCs + \beta}$

$$\frac{V_2}{V_1} = \frac{\frac{1}{\beta}\left(1 + \frac{s}{\omega_0}\right)^2}{\left(\frac{s}{\omega_0}\right)^2 + \frac{1}{Q}\left(\frac{s}{\omega_0}\right) + 1},$$

$$\omega_0 = \frac{1}{RC} \quad \text{and} \quad Q = \frac{\beta}{3\beta - 1}$$

The circuit is a second-order system with an undamped natural frequency $\omega_0 = 1/RC$ and a quality factor $Q = \beta/(3\beta - 1)$. Note that the gain of the noninverting op-amp is $1/\beta$. As $\beta \to 1/3$, the quality factor becomes larger and the poles move toward the $j\omega$ axis. At $\beta = 1/3$ the poles become purely imaginary at $s = \pm j\omega_0$ on the $j\omega$ axis. The system oscillates at ω_0 in the absence of an input.

b. In the circuit of Figure 9.13(b) set $V_1 = 0$ and use voltage division to find

$$V_A = \frac{Z_1}{Z_1 + Z_2} V_2$$

But

$$V_A = \frac{V_2}{K}$$

Therefore,

$$V_2[Z_2 + Z_1(1 - K)] = 0$$

Substitute for Z_1 and Z_2:

$$V_2 \frac{R^2 C^2 s^2 + RCS(3 - k) + 1}{(1 + RCs)Cs} = 0$$

With $K = 3$ the numerator of the coefficient of V_2 in the above equation becomes zero at $s = j/RC$, allowing oscillations with a frequency $\omega = 1/RC$ to be established in $v_2(t)$ and sustained despite zero input.

9.9 Analog Filters

What Is a Filter?

Linear dynamical systems are, in general, frequency selective. Their response to a sinusoidal excitation is a sinusoid with the same frequency. The amplitude and phase of the response depend on the frequency of excitation. This is in agreement with our experience from physical linear systems such as electrical, mechanical, thermal, or other systems with inertial or energy storage elements. The impedances of the elements of such systems vary with frequency and the system may filter some frequencies with various degrees of attenuation or amplification. From this point of view, every linear system is a filter. However, the name filter is applied to circuits, devices, and systems specifically designed to exhibit a prescribed frequency selectivity such as passing a certain range of frequencies (called passband) or blocking another range of frequencies (called stopband).

Other filter types of interest are those with prescribed frequency characteristics used as compensators (to improve the frequency characteristics of a system), equalizers (to provide amplitude and phase equalization), active tone controllers (to boost or dampen a range of frequencies), integrators or differentiators of signals (for control purposes), and so on. Some examples of these filters will be included in this chapter.

Filter Description

Filters are generally described by their system function $H(s)$ or frequency response $H(\omega)$. The magnitude of the frequency response produces the gain or attenuation in the amplitude of the signal and its phase produces delay. The gain is represented by $20\log|H(\omega)|$ in dB units. Attenuation is given by $-20\log|H(\omega)|$, also in dB units. A plot of magnitude and phase versus the frequency is provided by the Bode diagram.

Filter Classification: The Magnitude Factor

From some applications' point of view filters are classified based on their passband or stopband properties. We recognize five basic classes of filters: all-pass, low-pass, high-pass, bandpass, and bandstop. An all-pass filter passes all frequencies equally but with different phase angles. A low-pass filter passes frequency components below a certain range. A high-pass filter passes frequencies above a certain range. A bandpass filter passes frequencies within a band, and a bandstop filter blocks frequencies within a band. Ideally, this occurs in the passband $H(f) = 1$ and in the stopband $H(f) = 0$. Such filters are called *ideal filters* as their frequency responses assume one of two levels with a sharp vertical transition. They are not physically realizable.

The Phase Factor

In addition to changing the amplitude of a sinusoidal signal, a filter may also change its phase (which translates into a delay if the filter is realizable; see below). Although the magnitude change is often of primary consideration in filter applications, the phase change also needs to be taken into account in order to avoid or assess signal distortion. For example, a filter with a flat magnitude within a signal's bandwidth and a phase proportional to frequency (to be called linear phase) will not cause distortion in the signal but will rather delay it. If the phase is not linear, despite the zero magnitude change, the signal may experience distortion. The phase change may become the primary consideration and concern if the information carried by the signal is embedded in the phase. Without specification of the phase, the filters with an ideal magnitude response are called brick-wall filters, which is more descriptive.

Linear Phase

The frequency response $H(\omega) = |H|e^{j\theta}$ with $\theta = \tau\omega$, where τ is a constant for all ω, is called linear phase. It changes the sinusoid $A\cos(\omega t)$ into $A|H|\cos(\omega t + \theta) = A|H|\cos(\omega t + \tau\omega) = A|H|\cos\omega(t + \tau)$. The effect is a time-shift τ, which is independent of the frequency. In that case, all frequency components of the signal are shifted by the same amount. If the filter is realizable, the time shifts produced by a linear phase filter becomes a constant delay for all frequencies. In addition, if the magnitude is constant, the signal will be delayed without any change. In more general cases, where the delay created by the filter is frequency dependent, we express the phenomen through group delay and phase delay. That subject is introduced in section 9.26 of this chapter.

Distortion

A signal passing through a system generally changes its shape. The change may be caused by the nonlinearity of the filter, in which case additional frequency components will appear at the output. This is called *nonlinear distortion*. In the case of an LTI system, no additional frequency components may appear at the output. But, the change in the magnitude and/or phase of frequency components of the signal may cause distortion in the signal. This is called *linear distortion*. When the magnitude of the frequency response is not constant over the signal's bandwidth, various frequency components will experience different amplitude gains, leading to magnitude distortion. Phase distortion occurs when the phase of the frequency response is not proportional to the frequency.

Realization and Synthesis

A filter whose unit-impulse response is zero for $t < 0$ is called realizable. An example is a practical filter made of physical elements. The time-domain condition given above translates into the frequency-domain condition for realizability by the Paley-Wiener theorem, which is briefly discussed in section 8.21 of Chapter 8. Based on that theorem, the ideal filters are not realizable. But we can approximate the frequency response of an ideal filter with that of a practical filter with any desired level of tolerance and accuracy. Moreover, in practice we want the filter to be stable and buildable from lumped physical elements. These conditions require that we approximate the frequency response of the ideal filter by that of a system function, which is a ratio of two polynomials with real coefficients and poles in the LHP. Filter synthesis will not be addressed in this textbook.

Practical Filters

The system function of a practical filter, which is made of lumped linear elements, is a rational function (a ratio of two polynomials). The degree of the denominator polynomial is called the order of the filter. Because of the practical importance of first- and second-order filters and their widespread applications, we discuss these two classes of filters in detail and present examples of circuits to realize them. It is noted that first- and second-order filters are special cases of first- and second-order LTI systems. The properties and characteristics of these systems (such as impulse and step responses, natural frequencies, damping ratio, quality factor, overdamping, underdamping, and critical damping) apply to these filters as well and will be addressed.[8]

---

[8]For more information on active filter implementation see Sergio Franco, "Design with Operational Amplifiers and Analog Integrated Circuits," McGraw-Hill Series in Electrical Engineering, McGraw-Hill, Inc. 2001.

9.10   First-Order Low-Pass Filters

The system function and frequency response of a first-order, low-pass filter are

$$H(s) = \frac{1}{1 + \tau s} = \frac{1}{1 + \frac{s}{\omega_0}}$$

$$H(\omega) = \frac{1}{1 + j\tau\omega} = \frac{1}{1 + j\frac{\omega}{\omega_0}}$$

The filter has a pole at $s = -\omega_0$. Table 9.4 gives the magnitude and phase of the frequency response for five values of ω/ω_0. The half-power frequency (also called 3-dB attenuation frequency, corner frequency, break frequency, cutoff frequency, or break point) is equal to the magnitude of the pole, ω_o. The Bode diagram plots $20 \log |H(\omega)|$ on the vertical axis in uniform scale of dB versus $\log \omega$ on the horizontal axis. The low-frequency asymptote of the plot is a horizontal line at 0 dB. The high-frequency asymptote is a line with a slope of -20 dB per decade, corresponding to -6 dB per octave. These two asymptotes intersect at ω_0. At one octave below ω_0 (i.e., at $\omega_0/2$), the gain is -1 dB, which is one dB below the value of the asymptote. At one octave above ω_0 (i.e., at $2\omega_0$), the gain is -7 dB, which is one dB below the value of the asymptote. In summary, the actual gain at half or twice the cutoff frequency deviates from the value of the asymptote at that frequency by 1 dB. (For the low-pass filter it is below the asymptote and for the high-pass filter it is above it.) The phase angle varies from $0°$ at $f = 0$ to $-90°$ at $f = \infty$. The phase at ω_0 is $-45°$. These numbers provide enough data to sketch the Bode plot with good approximation.

TABLE 9.4 Measurements of the Magnitude and Phase of the Frequency Response of a First-Order Low-Pass Filter

$\frac{\omega}{\omega_0}$	$|H(\omega)|$	$20 \log |H(\omega)|$ in dB	$\theta°$
0	1	0	0
$\frac{1}{2}$	0.89	-1	-26.6
1	$\frac{\sqrt{2}}{2}$	-3	-45
2	0.45	-7	-63.4
∞	0	$-\infty$	-90

Example

9.28

First-order low-pass filters may be constructed from a series RC circuit. RC filters are common in low-pass filtering. Here we discuss an example in detail. In the circuit of Figure 9.14(a) input voltage $v_1(t)$ is connected to a series combination of an 800-Ω resistor and a 1-μF capacitor. The capacitor voltage $v_2(t)$ is connected to a voltage

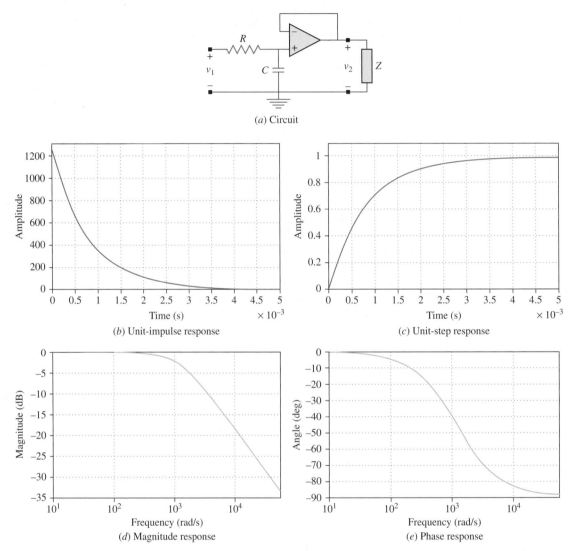

FIGURE 9.14 A first-order low-pass RC filter with a buffer. **(a)** The circuit, **(b)** unit-impulse response, **(c)** unit-step response, **(d)** magnitude plot of Bode diagram, and **(e)** phase plot of Bode diagram (dB and degrees vs. log ω, respectively).

follower (the noninverting input of an op-amp with the output of the op-amp short-circuited to its inverting input). The filter's output is picked up at the op-amp's output. Specify the filtering operation on $v_1(t)$. Find the system function $H(s) = V2/V1$, its poles and zeros, its responses to unit-impulse and unit-step inputs. its frequency response and the cutoff frequency f_0 in terms of R and C. Plot its Bode diagram and find its attenuation in dB at 10, 100, 200, 400, and 4,000 Hz.

Solution

Assuming that the op-amp is ideal, it functions as a voltage follower, preventing the impedance Z from loading the RC circuit. The ideal op-amp, therefore, has no effect on the frequency response of the RC circuit, and in this analysis we need not consider it. At low frequencies the capacitor has a high impedance. It functions as an open circuit. The current is small and the capacitor voltage is nearly the same as the input voltage. At high frequencies the capacitor behaves as a short circuit and its voltage is very small. The circuit is, therefore, a low-pass filter. The system function $H(s) = V2/V1$ is found by dividing the input voltage between the impedances of the capacitor and the resistor

$$H(s) = \frac{\frac{1}{Cs}}{R + \frac{1}{Cs}} = \frac{1}{1 + \tau s}, \quad \text{where } \tau = RC = 0.8 \text{ msec}$$

The filter has a pole at

$$s = -\frac{1}{\tau} = -\frac{10^6}{800} = -1{,}250 \text{ rad/s}$$

and no zero of finite value. The input-output differential equation and responses to unit-impulse and unit-step inputs are given below.

Differential equation $\qquad \dfrac{dv_2}{dt} + \dfrac{1}{RC}v_2 = \dfrac{1}{RC}v_1$

The unit-impulse response $\qquad h(t) = \dfrac{1}{RC}e^{-t/\tau}u(t) \qquad$ See Figure 9.14(b).

The unit-step response $\qquad g(t) = (1 - e^{-t/\tau})u(t) \qquad$ See Figure 9.14(c).

The frequency response is

$$H(\omega) = \frac{1}{1 + j\frac{\omega}{\omega_0}}, \quad \omega_0 = \frac{1}{\tau} = 1{,}250 \text{ rad/s}$$

The frequency response may also be written as

$$H(f) = \frac{1}{1 + j\frac{f}{f_0}}, \quad f_0 = \frac{\omega_0}{2\pi} \approx 199 \text{ Hz}$$

The Bode diagram is given in Figure 9.14(d) and (e). The half-power frequency with 3-dB attenuation is at $f_0 = \omega_0/2\pi = 199$ Hz. The DC attenuation is 0 dB. At 10 Hz, which is far from the half-power frequency, the attenuation is still nearly zero. At 100 Hz, which is half of the half-power frequency, the attenuation is 1 dB. At 400 Hz, which is twice the half-power frequency, the value of the asymptote is 6 dB (remember that the slope of the asymptote is -6 dB per octave) and the filter's attenuation is $1 + 6 = 7$ dB. At 4,000 Hz, the attenuation is almost the same as the value of the asymptote, which is $6 + 20 = 26$ dB.

Hardware Implementation

Analog filters may be implemented by passive or active circuits. (Exceptions are cases that require amplification to achieve a certain gain.) In this way, filters are classified as passive or active. Passive filters require no power source, are more robust and last longer, but are vulnerable to loading effects and may require inductors which may become bulky. Active filters can do without inductors, provide a better input impedance, and separate the load from the filter by employing an op-amp. However, they require a power source and are less robust. Examples of both types of implementations will be presented. The first-order low-pass system function of $H(s) = 1/(1 + \tau s)$ may be realized by the passive circuits of Figure 9.15(b) and (c), or by the active circuit of Figure 9.15(d).

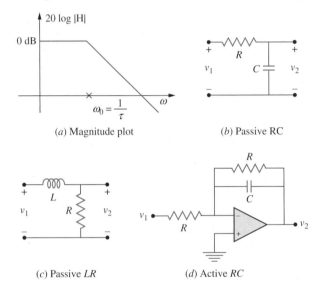

(a) Magnitude plot   (b) Passive RC

(c) Passive LR (d) Active RC

FIGURE 9.15 First-order low-pass filters.

9.11 Integrators

The system function of an integrator is $1/s$ and its magnitude Bode plot looks like that in Figure 9.16(a). Integration may be done by an RC (or RL) circuit and an op-amp. Here are two implementations.

A Perfect RC Integrator

A capacitor accumulates electric charge and develops a voltage proportional to it. The charge on a capacitor is proportional to the integral of the current that has passed through it.

$$v(t) = \frac{1}{C} \int_{-\infty}^{t} i(\theta)d\theta = v(0^+) + \frac{1}{C} \int_{0^+}^{t} i(\theta)d\theta$$

The capacitor may, therefore, function as an integrator of the input voltage if that voltage is proportionally converted to a current and passed through the capacitor.

The RC circuit of Figure 9.15(b) is a perfect integrator of the input current but not of the input voltage. The current in the circuit is proportional to the resistor voltage, which is the difference between the input and output voltages. Therefore, when converting the input voltage to the current, part of the input is withheld by the capacitor and the current in the circuit is not proportional to the input voltage. To avoid this effect, we employ an op-amp, which linearly converts the voltage signal to a current signal and sends it through the capacitor for integration. The circuit is shown in Figure 9.16(b). The inverting input of the op-amp is at ground potential and the current in the input resistor is proportional to the input voltage, that is, $i = v_1/R$. Because the inverting input of the op-amp draws no current, all of the above current passes through the capacitor and is absorbed by the op-amp output. The voltage at the op-amp's output is the same as the capacitor voltage and equals the integral of its current:

$$v_2(t) = -\frac{1}{C} \int_{-\infty}^{t} i(\theta)d\theta = -\frac{1}{RC} \int_{-\infty}^{t} v_1(\theta)d\theta$$

The circuit works as a perfect integrator for all frequencies. This behavior may also be deduced directly from the system function of the inverting op-amp circuits of Figure 9.16(b) and (c), or from their frequency responses.

$$H_b(s) = -\frac{Z_C}{Z_R} = -\frac{1}{RCs} \qquad\qquad H_c(s) = -\frac{Z_R}{Z_L} = -\frac{R}{Ls}$$

$$H_b(\omega) = \frac{-1}{jRC\omega} = j\frac{\omega_0}{\omega} = \frac{\omega_0}{\omega}\angle 90°, \qquad H_c(\omega) = \frac{-R}{jL\omega} = j\frac{\omega_0}{\omega} = \frac{\omega_0}{\omega}\angle 90°,$$

$$\text{where } \omega_0 = \frac{1}{RC}, \qquad\qquad\qquad \text{where } \omega_0 = \frac{L}{R}.$$

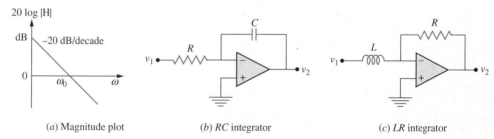

(a) Magnitude plot (b) RC integrator (c) LR integrator

FIGURE 9.16 Integrators.

A Perfect *LR* Integrator

Similarly, a perfect integrator may be constructed from an RL circuit and an op-amp as shown in Figure 9.16(c). The current in the inductor of Figure 9.16(c) is proportional to the integral of the input voltage.

$$i(t) = \frac{1}{L} \int_{-\infty}^{t} v_1(t)dt$$

If the op-amp is ideal, all of that current passes through the feedback resistor, which converts it to $v_2(t)$.

$$v_2(t) = -Ri(t) = -\frac{R}{L} \int_{-\infty}^{t} v_1(t)dt$$

9.12 First-Order High-Pass Filters

High-pass filters may be analyzed by an approach similar to that for low-pass filters. The transfer function of a first-order high-pass filter is

$$H(s) = \frac{\tau s}{1 + \tau s}$$

The Bode magnitude plot is shown in Figure 9.17(a). The 3-dB break frequency is $\omega_0 = 1/\tau$. Attenuation at $\omega_0/2$ and $2\omega_0$ is 1 dB above the attenuation value of the asymptote. Examples of passive RL and CR first-order high-pass filters are given in Figure 9.17(b) and (c). The active high-pass filter is given in Figure 9.17(d).

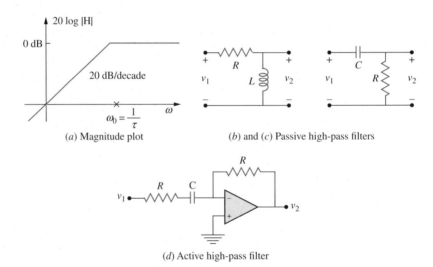

(a) Magnitude plot   (b) and (c) Passive high-pass filters

(d) Active high-pass filter

FIGURE 9.17 First-order high-pass filter.

Example
9.29

The high-pass filter transfer function $H(s) = s/(1{,}000 + s)$ is realized by the circuit of Figure 9.17(c) with $C = 1\ \mu\text{F}$ and $R = 1\ k\Omega$. Let $v_1(t) = \cos(2\pi f t)$. Find $v_2(t)$ for a) $f = 1$ Hz, b) $f = 10$ kHz.

a. $f = 1$ Hz, $H(f) = 0.00628\angle 89.6°$

$$v_2(t) = -0.00628\cos(2\pi f t + 89.6°) \approx 0.00628\sin(2\pi f t) = \frac{-1}{1{,}000}\frac{dv_1}{dt}$$

b. $f = 10$ kHz, $H(f) = 0.99987\angle 0.9°$

$$v_2(t) = -0.99987\cos(2\pi f t + 0.9°) \approx -\cos(2\pi f t) = -v_1(t)$$

The normalized Bode plot of the high-pass filter (i.e., magnitude and phase plotted versus ω/ω_0) is given in Figure 9.3. Scaling up the frequency axis in that plot by factor $\omega_0 = 1,000$ will provide the Bode diagram for the high-pass filter of Example 9.29. Within the frequency neighborhood of 1 Hz, the filter of Example 9.29 functions as a differentiator with 60-dB attenuation. Within the 10-kHz neighborhood it is an all-pass filter with unity gain and zero phase shift. At 10 kHz, it passes the signal without change in magnitude or phase. (The above conclusions may also be reached by inspecting the RC circuit.) At 1 Hz, almost all of the input voltage goes to the capacitor that dominates the impedance of the circuit. The capacitor current becomes proportional to the derivative of the input voltage. Consequently, the output voltage across the resistor becomes proportional to the derivative of the input voltage. At low frequencies, therefore, the circuit behaves as a differentiator. At high frequencies the capacitor behaves like a short-circuited path and all of the input voltage is transferred to the output with no change. The circuit works like an all-pass filter.

9.13 Differentiators

A high-pass filter exhibits the differentiation property at low frequencies. However, an ideal differentiator needs to operate perfectly over the whole frequency range. The system function of an ideal differentiator is, therefore, $H(s) = s$. The magnitude frequency response of an ideal differentiator is shown in Figure 9.18(a). The two circuits in Figure 9.18(b) and (c) perform differentiation. Their input-output relationships are

$$CR \text{ differentiator:} \quad v_2(t) = -RC\frac{dv_1}{dt}$$

$$RL \text{ differentiator:} \quad v_2(t) = -\frac{L}{R}\frac{dv_1}{dt}$$

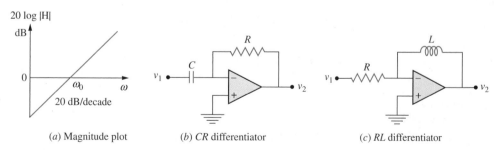

(*a*) Magnitude plot   (*b*) *CR* differentiator   (*c*) *RL* differentiator

FIGURE 9.18 Differentiators.

9.14 First-Order All-Pass Phase Shifters

The following filter has a constant gain H_0 irrespective of the frequency. Its phase, however, varies with frequency. It is called a phase shifter.

$$H(s) = H_0 \frac{s - \omega_0}{s + \omega_0}$$

$$H(\omega) = H_0 \frac{j\left(\frac{\omega}{\omega_0}\right) - 1}{j\left(\frac{\omega}{\omega_0}\right) + 1}$$

The first-order phase shifter is a stable system with a pole at $-\omega_0$ and a zero at ω_0. Figure 9.19(a) (on the left side) shows pole-zero plots for the case of $\omega_0 = 1$, 2, 5, and 10. The vectorial interpretation of $H(\omega)$ shown on the figure reveals the all-pass nature of the filter. Figure 9.19(b) (on the right side) shows phase versus frequency for the above four cases. In all cases the phase shift is 90° at ω_0. Within the neighborhood of ω_0, a frequency perturbation $\Delta\omega$ proportionally results in a phase perturbation almost proportional to $-\omega/\omega_0$. This translates into a time-shift perturbation of $1/\omega_0$.

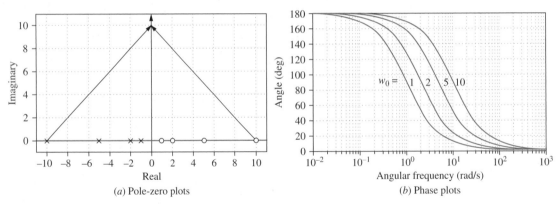

(a) Pole-zero plots

(b) Phase plots

FIGURE 9.19 Pole-zero and phase plots of first-order phase shifter $H(s) = (s - \omega_0)/s + \omega_0)$ for four pole locations: $\omega_0 = 1, 2, 5$, and 10. Within the neighborhood of ω_0 it functions as a delay element of $\tau = \pi/(2\omega_0)$.

The first-order phase shifter may be implemented by a passive circuit as the one in Figure 9.20(a) (with gain restriction), or by an active circuit as given in Figure 9.20(b).

a. Derive the input-output relationships in Figure 9.20(a) and (b) to show that:

Figure 9.20(a): $\dfrac{V_2}{V_1} = 0.5 \dfrac{1 - RCS}{1 + RCS}$ and Figure 9.20(b): $\dfrac{V_2}{V_1} = \dfrac{1 - RCs}{1 + RCs}$

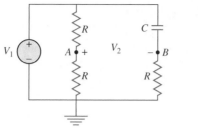
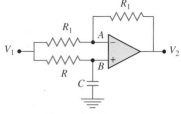

(a) Passive phase shifter   (b) Active phase shifter

FIGURE 9.20 Two circuits for the first-order phase shifter.

b. Consider the phase shifter $V_2/V_1 = (1{,}000 - s)/(1{,}000 + s)$. Let $v_1(t) = \cos 995t + \cos 1{,}000t + \cos 1005t$. Show that $v_2(t) \approx v_1(t - \tau)$ and find τ.

c. Repeat for $v_1(t) = \cos t \cos 1{,}000t$.

Solution

a. Figure 9.20(a): $V_A = \dfrac{V_1}{2}$, $V_B = \dfrac{RCs}{1 + RCs} V_1$, $V_2 = V_A - V_B$

$$= 0.5 \frac{1 - RCs}{1 + RCs} V_1$$

Figure 9.20(b): $V_A = \dfrac{V_1 + V_2}{2}$, $V_B = \dfrac{V_1}{1 + RCs}$, $V_A = V_B$, $V_2 = \dfrac{1 - RCs}{1 + RCs} V_1$

b. For the input $v_1(t) = \cos(\omega t)$, the output is

$v_2(t) = \cos(\omega t - \theta) = \cos \omega(t - \tau)$, where $\theta = 2 \tan^{-1}\left(\frac{\omega}{1{,}000}\right)$ and $\tau = \theta/\omega$.

These are calculated and the results are given below.

Amplitude	1	1	1
ω (in radians/sec)	995	1,000	1,005
θ (in degrees)	90.287°	90°	89.714°
θ (in radians)	1.575	1.570	1.565
$\tau = \theta/\omega$ (in μs)	1,583	1,570	1,558

$$v_2(t) = \cos 995(t - 0.001583) + \cos 1{,}000(t - 0.001570)$$
$$+ \cos 1005(t - 0.001558) \approx v_1(t - \tau), \quad \tau = 1.57 \text{ msec}$$

c.
$$v_1(t) = \cos t \cos 1{,}000t = \frac{\cos 999t + \cos 1{,}001t}{2}$$

$$v_2(t) = \frac{\cos 999(t - \tau) + \cos 1{,}001(t - \tau)}{2}$$

$$= \cos(t - \tau) \cos 1{,}000(t - \tau) = v_1(t - \tau)$$

9.15   Lead and Lag Compensators

Lead and lag compensators are first-order filters with a single pole and zero, chosen such that a prescribed phase lead and lag is produced in a sinusoidal input. In this way, when placed in a control loop, they reshape an overall system function to meet desired characteristics. Electrical lead and lag networks can be made of passive RLC elements, or employ operational amplifiers. In this section we briefy describe their system functions and frequency responses.

Lead Network

A lead network made of passive electrical elements is shown in Figure 9.21, along with its pole-zero and Bode plots.

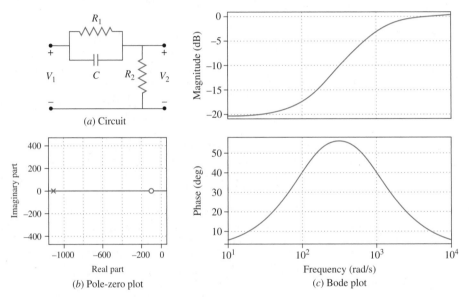

(a) Circuit

(b) Pole-zero plot

(c) Bode plot

FIGURE 9.21 Lead network, $H(s) = \alpha \dfrac{s + \omega_1}{\alpha s + \omega_1}$, $\alpha < 1$. This figure is for $\omega_1 = 100$ and $\alpha = 1/11$. See Example 9.31.

The network's input and output voltages are related by the following differential equation:

$$\frac{1}{\alpha} v_2 + \tau_1 \frac{dv_2}{dt} = v_1 + \tau_1 \frac{dv_1}{dt}$$

where

$$\tau_1 = R_1 C \quad \text{and} \quad \alpha = \frac{R_2}{R_1 + R_2}$$

From the above differential equation, as well as the system function, frequency response, and the Bode plots, the network's response to a sinusoidal input leads the input by a positive phase angle at all frequencies, hence the lead network. The system function and frequency response are

$$H(s) = \frac{V_2}{V_1} = \alpha \frac{1 + \tau_1 s}{1 + \alpha \tau_1 s} = \alpha \frac{s + \omega_1}{\alpha s + \omega_1}, \quad \text{where } \omega_1 = \frac{1}{\tau_1}$$

$$H(\omega) = \alpha \frac{1 + j(\omega/\omega_1)}{1 + j\alpha(\omega/\omega_1)}$$

The DC gain is $\alpha = R_2/(R_1 + R_2)$ and the high-frequency gain is unity (0 dB). Note that $\alpha < 1$ and, when expressed in dB, it is a negative number. We may also employ the low-frequency attenuation factor $1/\alpha$ and express it in dB by $A = -20 \log \alpha$ (dB). The system function has a zero at $z = -\omega_1$ and a pole at $p = -\omega_1/\alpha$. The pole of the system is located to the left of its zero, both being on the negative real axis in the s-plane (see Figure 9.21b). The magnitude is normally expressed in dB.

$$20 \log |H(\omega)| = 20 \log \alpha + 20 \log \sqrt{1 + (\omega/\omega_1)^2} - 20 \log \sqrt{1 + \alpha^2(\omega/\omega_1)^2} \quad \text{(dB)}$$

One may sketch the magnitude of the frequency response in dB and its phase in degrees on semilog paper [i.e., $20 \log |H(\omega)|$ and $\angle H(\omega)|$ vs. $\log \omega$] by the following qualitative argument. The low-frequency asymptote of the magnitude plot is the horizontal line at $20 \log \alpha$ dB. The high-frequency asymptote is the horizontal line at 0 dB. The magnitude plot is a monotonically increasing curve. Traversing from low frequencies toward high frequencies it first encounters ω_1 (the zero). The zero pushes the magnitude plot up with a slope of 20 dB per decade. As the frequency increases it encounters ω_1/α (the pole). The pole pulls the magnitude plot down with a slope of -20 dB per decade and, therefore, starts neutralizing the upward effect of the zero. The magnitude plot is eventually stablized at the 0 dB level. The zero and the pole are the break frequencies of the magnitude plot. If $\alpha << 1$, the pole and the zero are far enough from each other and constitute frequencies of 3-dB deviation from the asymptotes. At ω_1 (the zero) the magnitude is 3 dB above the low-frequency asymptote and the phase is 45°. At ω_1/α (the pole) the magnitude is 3 dB below 0 dB (the high-frequency asymptote) and the phase is $90° - 45° = 45°$. The maximum phase lead in the output occurs at $\omega_m = \omega_1/\sqrt{\alpha}$, which is the geometric mean of the two break frequencies. These are summarized in Table 9.5 in which $A = -20 \log \alpha$.

TABLE 9.5 Magnitude and Phase of the Frequency Responses of a Lead Network with a Large Pole-Zero Separation

Frequency Response	$\omega << \omega_1$ (low frequencies)	ω_1 (the zero)	$\omega_1/\sqrt{\alpha}$ (geometric mean)	ω_1/α (the pole)	$\omega >> \omega_1/\alpha$ (high frequencies)
$|H(\omega)|$	$\approx -A$ dB	$\approx -A + 3$ dB	$-\dfrac{A}{2}$ dB	≈ -3 dB	≈ 0 dB
$\angle H(\omega)$	≈ 0	$\approx 45°$	$90° - 2\tan^{-1}\sqrt{\alpha}$	$\approx 45°$	≈ 0

The phase of $H(\omega)$ varies from 0 (at low frequencies, $\omega = 0$) to a maximum lead of $90° - 2\tan^{-1}\sqrt{\alpha}$ at $\omega_m = \omega_1/\sqrt{\alpha}$ and returns back to zero at high frequencies. For derivation of the maximum phase see problems 10 and 46.

Example
9.31

Consider the lead network of Figure 9.21(*a*) with $R_1 = 10\ k\Omega$, $R_2 = 1\ k\Omega$, and $C = 1\ \mu F$. Let $v_1 = \cos \omega t$ be the input and $v_2 = V_2 \cos(\omega t + \theta)$ be the output.

a. Find V_2 and θ for ω at 1, 100, 331.6, 1,100, and 10^5, all in rad/s.

b. Find the system function. Show that the phase lead is maximum at $\omega_m = 100\sqrt{11} = 331.6$ rad/s and find its value. Find the DC and high-frequency gains (in dB), along with the break frequencies.

Solution

a. Let $V_1 = 1$ and V_2 be the complex amplitudes of the input and output voltages, respectively. Then by voltage division, we have

$$\frac{V_2}{V_1} \equiv H(s)\Big|_{s=j\omega} = \frac{R_2}{R_2 + Z(s)}\Big|_{s=j\omega}, \quad \text{where } Z(s) = \frac{1}{1 + R_1 Cs}$$

is the impedance of R_1 and C in parallel. After substituting for R_1, R_2, C, and V_1 in the above, we have

$$|V_2| = \alpha \frac{\omega^2 + \omega_1^2}{\alpha\omega^2 + \omega_1^2}$$

$$\theta = \tan^{-1}\left(\frac{\omega}{\omega_0}\right) - \tan^{-1}\left(\frac{\alpha\omega}{\omega_0}\right)$$

where

$$\alpha = \frac{R_2}{R_1 + R_2} = \frac{1}{11} \quad \text{and} \quad \omega_1 = \frac{1}{R_1 C} = 100 \text{ rad/s}$$

The magnitude and phase of the sinusoidal response at the given frequencies are shown in the table below.

ω (rad/s)	1	100	331.6	1,100	10^5
V_2 (volts)	0.091	0.128	0.3015	0.71	1
θ (degrees)	0.5°	39.8°	56.4°	39.8°	0.5°

b. The system function is

$$H(s) = \alpha \frac{s + \omega_1}{\alpha s + \omega_1}$$

The DC gain is $20 \log(1/11) = -48$ dB. The maximum phase occurs at $\omega = \omega_1/\sqrt{\alpha} = 100/\sqrt{11} = 331.66$ rad/s. The value of the maximum lead is $90° - 2\tan^{-1}\sqrt{\alpha} = 90° - 2\tan^{-1}\sqrt{(1/11)} = 56.44°$. The frequency response is plotted in the form of a Bode plot. The network has a zero at -100 and a pole at $-1,100$. See Figure 9.21(*b*) and (*c*). A vectorial interpretation of the system function confirms the above specifications.

Lag Network

A lag network is shown in Figure 9.22.

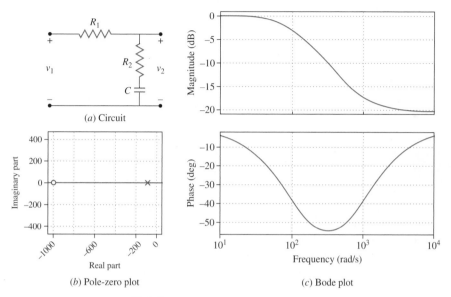

(a) Circuit

(b) Pole-zero plot

(c) Bode plot

FIGURE 9.22 Lag network, $H(s) = \dfrac{\alpha s + \omega_2}{s + \omega_2}$, $\alpha < 1$. This figure is for $\omega_1 = 90.9$ and $\alpha = 1/11$. See Example 9.32.

In many respects characteristics of the lag network are mirror images of those of the lead network. The input-output differential equation is

$$v_2 + \tau_2 \frac{dv_2}{dt} = v_1 + \alpha \tau_2 \frac{dv_1}{dt}$$

where

$$\tau_2 = (R_1 + R_2)C \quad \text{and} \quad \alpha = \frac{R_2}{R_1 + R_2}$$

The system function and frequency response are

$$H(s) = \frac{V_2}{V_1} = \frac{1 + \alpha \tau_2 s}{1 + \tau_2 s} = \frac{\alpha s + \omega_2}{s + \omega_2}, \quad \text{where} \quad \omega_2 = \frac{1}{\tau_2}$$

$$H(\omega) = \frac{1 + j\alpha(\omega/\omega_2)}{1 + j(\omega/\omega_2)}$$

The DC gain is unity (0 dB) and the high-frequency gain is $\alpha = R_2/(R_1 + R_2)$. The system has a pole at $p = -\omega_2$ and a zero at $z = -\omega_2/\alpha$ to the left of the pole, both being on the negative real axis in the s-plane, [see Figure 9.22(b)]. The magnitude plot of the lag network displays a horizontal flip of that of the lead network, and the phase is a vertical flip of that of the lead the network. The magnitude and phase of the frequency responses of a lag network with a large zero-pole separation are given below for several frequencies of interest. Here, as for the lead network $A = -20 \log \alpha$.

Frequency Response	$\omega \ll \omega_2$ (low frequencies)	ω_2 (the pole)	$\omega_2/\sqrt{\alpha}$ (geometric mean)	ω_2/α (the zero)	$\omega \gg \omega_2$ (high frequencies)
$\lvert H(\omega) \rvert$	≈ 0 dB	≈ -3 dB	$-\dfrac{A}{2}$ dB	$\approx -A + 3$ dB	$\approx -A$ dB
$\angle H(\omega)$	≈ 0	$\approx -45°$	$2\tan^{-1}\sqrt{\alpha} - 90°$	$\approx -45°$	≈ 0

The phase of $H(\omega)$ varies from 0 (at low frequencies, $\omega = 0$) to a minimum (i.e., a maximum phase lag) of $(2\tan^{-1}\sqrt{\alpha} - 90°)$ at $\omega_m = \omega_2/\sqrt{\alpha}$ and returns back to zero at high frequencies.

Example **9.32**

Consider the lag network of Figure 9.22(a) with $R_1 = 10\ k\Omega$, $R_2 = 1\ k\Omega$, and $C = 1\ \mu F$. Let $v_1 = \cos\omega t$ be the input and $v_2 = V_2\cos(\omega t - \theta)$ be the output. Find V_2 and θ for $\omega = 1, 90.9, 301.5, 1{,}000$, and 10^5, all in rad/s. Plot the frequency response in the form of Bode plot.

Solution

In this case we find $\alpha = 1/11$ and $\omega_2 = 90.9$ rad/s. The magnitude and phase of the response are given in the table below.

ω rad/s	1	90.9	301.5	1,000	10^5
V_2 volts	1	0.71	0.3015	0.128	0.091
θ degrees	0.5°	39.8°	56.4°	39.8°	0.5°

The frequency response is plotted in the form of a Bode plot in Figure 9.22(c). Parallels with the Bode diagram of Figure 9.21(c) are observed.

9.16 Summary of First-Order Filters

The transfer function $H(s)$ of a first-order stable filter has a single pole with a negative real value. Table 9.6 summarizes characteristics of seven first-order filter types: low-pass, integrator, high-pass, differentiator, all-pass, lead, and lag filters.

9.17 Second-Order Low-Pass Filters

A second-order low-pass filter has two poles and no zeros at finite frequencies. Its system function and frequency response are

$$H(s) = \frac{b}{s^2 + 2as + b} = \frac{1}{(\frac{s}{\omega_0})^2 + \frac{1}{Q}\frac{s}{\omega_0} + 1}, \quad \text{where } Q = \frac{\sqrt{b}}{2a} \text{ and } \omega_0 = \sqrt{b}$$

$$H(\omega) = \frac{b}{b - \omega^2 + j2a\omega} = \frac{1}{1 - (\frac{\omega}{\omega_0})^2 + \frac{j}{Q}\frac{\omega}{\omega_0}}$$

TABLE 9.6 Summary of First-Order Filters

	Type	$H(s)$	$H(\omega)$	Break Frequency	Pole-Zero	Asymptotic Bode Plot
a.	Low-pass	$\dfrac{1}{1+\tau s}$	$\dfrac{1}{1+j\frac{\omega}{\omega_0}}$	$\omega_0 = \dfrac{1}{\tau}$		
b.	Integrator	$\dfrac{1}{\tau s}$	$-j\dfrac{\omega_0}{\omega}$	$\omega_0 = \dfrac{1}{\tau}$		
c.	High-pass	$\dfrac{\tau s}{1+\tau s}$	$\dfrac{j\frac{\omega}{\omega_0}}{1+j\frac{\omega}{\omega_0}}$	$\omega_0 = \dfrac{1}{\tau}$		
d.	Differentiator	τs	$j\dfrac{\omega}{\omega_0}$	$\omega_0 = \dfrac{1}{\tau}$		
e.	All-pass	$\dfrac{1-\tau s}{1+\tau s}$	$\dfrac{1-j\frac{\omega}{\omega_0}}{1+j\frac{\omega}{\omega_0}}$	none		
f.	Lead	$\dfrac{1+b\tau s}{1+\tau s}$	$\dfrac{1+j\frac{\omega}{\omega_1}}{1+j\frac{\omega}{\omega_2}}$	$\omega_1 = \dfrac{1}{b\tau},\ \omega_2 = \dfrac{1}{\tau}$		
g.	Lag	$\dfrac{1+\tau s}{1+b\tau s}$	$\dfrac{1+j\frac{\omega}{\omega_1}}{1+j\frac{\omega}{\omega_2}}$	$\omega_1 = \dfrac{1}{\tau},\ \omega_2 = \dfrac{1}{b\tau}$		

The basic features of the above system function were briefly discussed previously.[9] The second-order systems described in section 9.6 are low-pass (except when the quality factor is high). Here we consider the frequency response behavior of such systems. As before, based on the quality factor Q, we recognize three cases.

$Q < 0.5$

The filter has two distinct poles on the negative real axis at $s_{1,2} = -\omega_{1,2} = -a \pm \sqrt{a^2 - b}$. See Figure 9.23($a$) (left). Note that $\omega_1 \omega_2 = b$ and $\omega_1 + \omega_2 = 2a$. This is an overdamped second-order system. Its frequency response may be written as

$$H(\omega) = \frac{1}{1 + j\left(\frac{\omega}{\omega_1}\right)} \times \frac{1}{1 + j\left(\frac{\omega}{\omega_2}\right)}$$

The filter is equivalent to the cascade of two first-order low-pass filters with 3-dB frequencies at ω_1 and ω_2, discussed previously. The magnitude Bode plot is shown on Figure 9.23(a) (right). The plot has three asymptotic lines, with slopes of 0, -20, and -40 dB/decade, for low, intermediate, and high frequencies, respectively.

$Q = 0.5$

The filter has a pole of order 2 on the negative real axis at $s = -\omega_0$, where $\omega_0 = a = \sqrt{b}$. See Figure 9.23(b) (left). This is a critically damped system. The frequency response may be written as

$$H(\omega) = \frac{1}{1 - \left(\frac{\omega}{\omega_0}\right)^2 + j2\left(\frac{\omega}{\omega_0}\right)} = \frac{1}{1 + j\left(\frac{\omega}{\omega_0}\right)} \times \frac{1}{1 + j\left(\frac{\omega}{\omega_0}\right)}$$

The filter is equivalent to the cascade of two identical first-order low-pass filters with 3-dB frequencies at ω_0. The Bode plot is shown on Figure 9.23(b) (right). The low-frequency asymptote is at 0 dB. The high-frequency asymptote is a line with a slope of -40 dB per decade. The break frequency is at ω_0. Attenuation at ω_0 is 6 dB.

$Q > 0.5$

The filter has two complex conjugate poles with negative real parts, $s_{1,2} = -\sigma \pm j\omega_d$, where $\sigma = a$ and $\omega_d = \sqrt{b - a^2}$. The frequency response may be written as

$$H(\omega) = \frac{1}{1 - \left(\frac{\omega}{\omega_0}\right)^2 + j\left(\frac{2a}{\omega_0}\right)\left(\frac{\omega}{\omega_0}\right)}$$

This is an underdamped system. Its poles are shown in Figure 9.23(c) (left) and its Bode plot is on Figure 9.23(c) (right).

---

[9]For the representation of a second-order system by its natural frequency, damping ratio, quality factor, and pole locations, along with the relationship of these entities to each other, see section 9.6 and Figure 9.4 in this chapter.

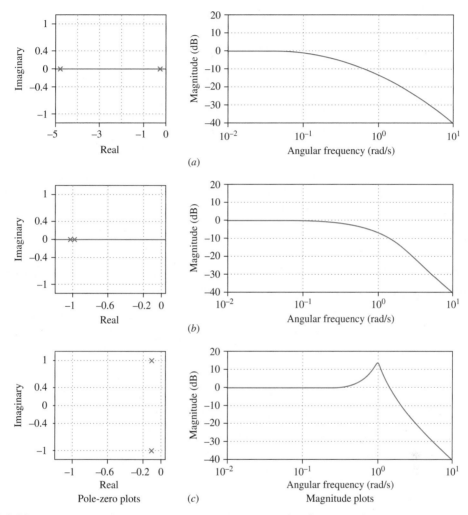

FIGURE 9.23 Summary of second-order low-pass filter $H(s) = 1/(s^2 + \frac{1}{Q}s + 1)$: *(a)* $Q = 0.2$, overdamped with two distinct negative real poles at -4.7913 and -0.2087. *(b)* $Q = 0.5$, critically damped with a repeated negative real pole at -1. *(c)* $Q = 5$, underdamped with a pair of complex conjugate poles at $(-0.1 \pm j0.995)$.

The above three situations may be analyzed in a unified manner using the filter's Q and ω_0 model. The gain of the frequency response in dB is

$$20 \log |H(\omega)| = -20 \log \left| 1 - \left(\frac{\omega}{\omega_0} \right)^2 + \frac{j}{Q} \frac{\omega}{\omega_0} \right|$$

For $\omega << \omega_0$, the terms ω/ω_0 and $(\omega/\omega_0)^2$ may be dropped, and the gain is

$$20 \log |H(\omega)| \approx -20 \log 1 = 0 \text{ dB}$$

The low-frequency asymptote is, therefore, the 0 dB horizontal line.

For $\omega \gg \omega_0$, the terms ω/ω_0 and 1 may be dropped and

$$20 \log |H(\omega)| \approx -40 \log \left(\frac{\omega}{\omega_0} \right)$$

The high-frequency asymptote is, therefore, a line with a slope of -40 dB per decade. At ω_0 the gain is

$$20 \log |H(\omega_0)| = -20 \log \frac{1}{Q} = 20 \log Q,$$

which is equal to the quality factor expressed in dB. For $Q = \sqrt{2}/2$, the gain is $20 \log(\sqrt{2}/2) = -3$ dB. For $Q > 1$, the gain in dB becomes positive. For high Q the gain peaks at ω_0 and the filter exhibits a behavior similar to resonance. See Example 9.33.

The circuit of Figure 9.24(a) is called a Sallen-Key low-pass filter. In this example we choose $m = n = 1$, making it an equal-component filter. The system function is found by applying basic circuit theory. It is

$$\frac{V_2}{V_1} = \frac{1}{\beta} \frac{1}{\left(\frac{s}{\omega_0} \right)^2 + \frac{1}{Q} \left(\frac{s}{\omega_0} \right) + 1},$$

where $\omega_0 = \dfrac{1}{RC}$, $\beta = \dfrac{R_1}{R_1 + R_2}$ is the feedback factor, and $Q = \dfrac{\beta}{3\beta - 1}$.

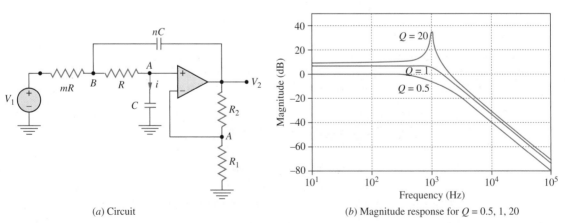

(a) Circuit (b) Magnitude response for $Q = 0.5, 1, 20$

FIGURE 9.24 An equal-component Sallen-Key low-pass filter and its magnitude frequency responses. (See Example 9.33.)

Let $R = 159\ \Omega$ and $C = 1\ \mu F$. Plot the magnitude response, $20\log|H(f)|$ versus $\log f$, for $Q = 0.5$, 1, and 20. Examine the magnitude response for $\beta = 1$, 0.63, 0.5, 0.4, 0.35, 0.339, 0.335 and discuss the performance of the filter for the above feedback factors. At what value of β does the circuit become a Butterworth filter, that is, poles at $\frac{\sqrt{2}}{2}\omega_0(-1 \pm j)$?

Solution

With $R = 159\ \Omega$ and $C = 1\ \mu F$ the natural frequency is 1 kHz. The feedback factor controls the location of the poles and, consequently, Q. See the results below.

β	1	0.63	0.5	0.4	0.35	0.339	0.335
Q	0.5	0.707	1	2	7	19.94	67

The superimposed magnitude plots are shown in Figure 9.24(b). The DC gain is $1/\beta$. At 1 kHz the value of the magnitude response is $20\log Q - 20\log \beta = 20\log(Q/\beta) = -20\log(3\beta - 1)$. Note that an increase in the feedback factor causes an increase in Q but decreases the DC gain. A higher Q produces a larger peak. For example, for $\beta = 0.335$ we have $Q = 67$ and the magnitude plot will read a peak of $46 \approx 20\log Q - 20\log \beta = 20\log(Q/\beta)$ dB. For a Butterworth filter $Q = \sqrt{2}/2$, which requires $\beta = \frac{1}{3-\sqrt{2}} = 0.63$. For $\beta = 1$ we have $Q = 0.5$ and the filter is a critically damped, second-order system with an attenuation of $-20\log(Q/\beta) = 6$ dB.

9.18 Second-Order High-Pass Filters

The system function and frequency response of a second-order high-pass filter may be written as

$$H(s) = \frac{s^2}{s^2 + 2as + b} = \frac{\left(\frac{s}{\omega_0}\right)^2}{\left(\frac{s}{\omega_0}\right)^2 + \frac{1}{Q}\left(\frac{s}{\omega_0}\right) + 1}, \quad \text{where } Q = \frac{\sqrt{b}}{2a} \text{ and } \omega_0 = \sqrt{b}$$

$$H(\omega) = \frac{-\omega^2}{b - \omega^2 + j2a\omega} = \frac{-\left(\frac{\omega}{\omega_0}\right)^2}{1 - \left(\frac{\omega}{\omega_0}\right)^2 + \frac{j}{Q}\frac{\omega}{\omega_0}}$$

The system has a double zero at $s = 0$ (DC) and two poles in the LHP. It works as the opposite of the low-pass filter. The frequency response has a zero magnitude at $\omega = 0$, which grows to 1 at $\omega = \infty$. The low-frequency magnitude asymptote is a line with a slope of 40 dB per decade. The high-frequency asymptote is the 0 dB horizontal line. Between the two limits (especially near ω_0) the frequency response is shaped by the location of the poles. As in any second-order system, we recognize the three cases of the filter being overdamped, critically damped, and underdamped. Here again, the frequency behavior of the filter may be analyzed in a unified way in terms of ω_0 and Q. Results are summarized in Table 9.7.

TABLE 9.7 Magnitude and Phase of the High-Pass Filter $H(s) = s^2/(s^2 + \omega_0 s/Q + \omega_0^2)$

Frequency ω	0	$\omega \ll \omega_0$	ω_0	$\omega \gg \omega_0$	∞
Magnitude, $20 \log |H(\omega)|$ in dB	$-\infty$	$\approx 40 \log \left(\dfrac{\omega}{\omega_0}\right)$	$20 \log Q$	0	0
Phase, $\angle H(\omega)$ in degrees	0	≈ -180	$90°$	0	0

*E*xample
9.34

The circuit of Figure 9.25(a) is an equal-component Sallen-Key high-pass filter. By applying circuit theory rules its system function is found to be

$$\frac{V_2}{V_1} = \frac{1}{\beta} \frac{\left(\frac{s}{\omega_0}\right)^2}{\left(\frac{s}{\omega_0}\right)^2 + \frac{1}{Q}\left(\frac{s}{\omega_0}\right) + 1},$$

where

$$\omega_0 = \frac{1}{RC}, \quad \beta = \frac{R_1}{R_1 + R_2}$$

is the feedback factor, and

$$Q = \frac{\beta}{3\beta - 1}.$$

Plot the magnitude response, $20 \log |H(f)|$ versus $\log f$, for $Q = 0.5$, 1, and 20. Examine the magnitude response for $\beta = 1$, 0.63, 0.5, 0.4, 0.35, 0.339, 0.335 and discuss the performance of the filter for the above feedback factors. At what value of β does the circuit become a second-order Butterworth filter?

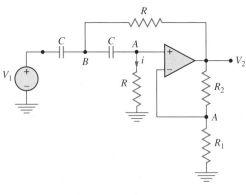

(a) Circuit

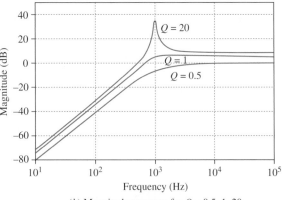

(b) Magnitude response for $Q = 0.5, 1, 20$

FIGURE 9.25 An equal-component Sallen-Key high-pass filter and its magnitude frequency responses. (See Example 9.34.)

Solution

With $R = 159\ \Omega$ and $C = 1\ \mu F$ the natural frequency is 1 kHz. The feedback factor controls the location of the poles and, consequently, Q. The superimposed magnitude plots are shown in Figure 9.25(b). The high-frequency gain is $1/\beta$. At 1 kHz the value of the magnitude response is $20 \log Q - 20 \log \beta = -20 \log (3\beta - 1)$. Note that an increase in the feedback factor causes an increase in Q but decreases the high frequency gain. A higher Q produces a larger peak. For example, at $\beta = 0.335$ we have $Q = 67$ and the magnitude plot will read $46 \approx 20 \log Q - 20 \log \beta = 20 \log (Q/\beta)$ dB. For a Butterworth filter $Q = \sqrt{2}/2$, which requires $\beta = \frac{1}{3-\sqrt{2}} = 0.63$.

9.19 Second-Order Bandpass Filters

The system function and frequency response of the basic second-order bandpass filter are

$$H(s) = \frac{2as}{s^2 + 2as + b} = \frac{\frac{1}{Q}\left(\frac{s}{\omega_0}\right)}{\left(\frac{s}{\omega_0}\right)^2 + \frac{1}{Q}\left(\frac{s}{\omega_0}\right) + 1}, \quad \text{where } Q = \frac{\sqrt{b}}{2a} \text{ and } \omega_0 = \sqrt{b}$$

$$H(\omega) = \frac{j2a\omega}{b - \omega^2 + j2a\omega} = \frac{\frac{j}{Q}\left(\frac{\omega}{\omega_0}\right)}{1 - \left(\frac{\omega}{\omega_0}\right)^2 + \frac{j}{Q}\left(\frac{\omega}{\omega_0}\right)}$$

The filter has a zero at $s = 0$ (DC) and two poles at $s_{1,2} = -\frac{\omega_0}{2Q}\left(1 \pm \sqrt{1 - 4Q^2}\right)$, which, depending on the value of Q, are either real or complex. The frequency response has zero magnitude at $\omega = 0$ and ∞. It attains its peak at ω_0, sometimes called the resonance frequency, where $|H(\omega_0)| = 1$. The half-power bandwidth $\Delta\omega$ is defined as the frequency range within which $|H(\omega)/H(\omega_0)| \geq \sqrt{2}/2$ (i.e., the gain is within 3 dB below its maximum, and thus called the 3-dB bandwidth). The lower and upper limits of the half-power frequency band are

$$\omega_{h,\ell} = \omega_0 \left[\sqrt{1 + \frac{1}{4Q^2}} \pm \frac{1}{2Q}\right] = \frac{\omega_0}{2Q}\left[\sqrt{4Q^2 + 1} \pm 1\right]$$

$$\Delta\omega = \omega_h - \omega_\ell = \frac{\omega_0}{Q}$$

In the present analysis we observe parallels with the cases of low-pass and high-pass filters, and, depending on the value of the quality factor Q (which controls the location of a filter's poles and thus its bandwidth), we recognize the three familiar states of overdamped ($Q < 0.5$), critically damped ($Q = 0.5$), and underdamped ($Q > 0.5$). The filter then becomes wideband (low Q) or narrowband (high Q). The sharpness of the peak is determined by the quality factor Q. In what follows we will discuss the shape of the Bode plot for three regions of Q values.

Q < 0.5

The system has two distinct negative real poles at $s_{1,2} = -\frac{\omega_0}{Q}(1 \pm \sqrt{1 - 4Q^2}) = -\omega_{1,2}$. Note that $\omega_1\omega_2 = \omega_0^2$ and $\omega_1 + \omega_2 = \frac{2\omega_0}{Q}$.

$$H(\omega) = \frac{1}{Q} \frac{j(\omega/\omega_0)}{[1 + j(\omega/\omega_1)][1 + j(\omega/\omega_2)]}$$

The slopes of the asymptotic lines in the magnitude Bode plot are 20, 0, and -20 dB/decade for low, intermediate, and high frequencies, respectively. The filter is a wideband bandpass filter. It is equivalent to the cascade of a first-order high-pass filter and a first-order low-pass filter with separate break points.

Q = 0.5

The filter has a double pole at $s = \omega_0$. The frequency response may be written as

$$H(\omega) = \frac{2j(\omega/\omega_0)}{1 - (\omega/\omega_0)^2 + j2(\omega/\omega_0)} = 2\frac{j(\omega/\omega_0)}{1 + j(\omega/\omega_0)} \times \frac{1}{1 + j(\omega/\omega_0)}$$

The asymptotic slopes of the plot are 20 dB/decade (low frequencies) and -20 dB/decade (high frequencies). The filter is bandpass, equivalent to the cascade of a first-order high-pass filter and a first-order low-pass sharing the same break frequency ω_0.

Q > 0.5

The filter has two complex conjugate poles with negative real parts $s_{1,2} = -\sigma \pm \omega_d$, where $\sigma = \frac{\omega_0}{(2Q)}$ and $\omega_d = \frac{\omega_0}{(2Q)}\sqrt{4Q^2 - 1}$. Note that $\sigma^2 + \omega_d^2 = \omega_0^2$. The asymptotic slopes of the Bode plot are 20 dB/decade (low frequency) and -20 dB/decade (high frequency).

High Q

For a bandpass system with high Q (e.g., $Q \geq 10$) the high and low 3-dB frequencies are approximately symmetrical on the upper and lower sides of the center frequency:

$$\omega_{h,\ell} \approx \omega_0 \pm \frac{\omega_0}{2Q}$$

A vectorial interpretation of $H(\omega)$ is especially illuminating in observing the above approximation.

The circuit of Figure 9.26(a) is an infinite gain, multiple feedback bandpass filter. In this example we choose equal-value capacitors $C_1 = C_2 = C$. Find the system function $H(s) = V_2/V_1$ and write it as a bandpass filter. Determine its quality factor and center frequency as functions of circuit elements, and specify them for $R_1 = 159\ \Omega$, $R_2 = 100R_1$, and $C = 0.1\ \mu F$. Plot the magnitude Bode plot $20\log|H(f)|$ versus $\log f$ and discuss the effect of increasing R_2.

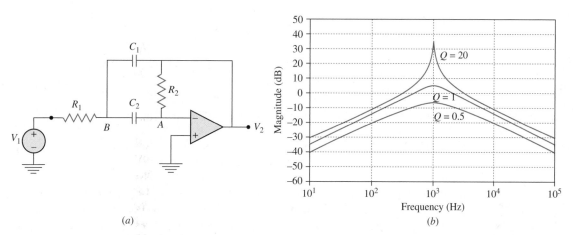

(a)   (b)

FIGURE 9.26 *(a)* An infinite gain, multiple feedback bandpass filter *(a)*. *(b)* Magnitude responese of a bandpass filter for $Q = 0.5$, 1, and 20.

Solution

The op-amp is operating in a noninverting configuration with a gain of $V_2/V_A = 1 + R_2/R_1 = 1/\beta$. From the above observation and from KCL at nodes A and B we obtain the following three equations:

KCL at node A: $(0 - V_B)Cs + \frac{0 - V_2}{R_2} = 0$

KCL at node B: $\frac{(V_B - V_1)}{R_1} + (V_B - 0)Cs + (V_B - V_2)Cs = 0$

After eliminating V_B between the above two equations we are left with

$$H(s) = \frac{V_2}{V_1} = -\frac{R_2Cs}{R_1R_2C^2s^2 + 2R_1Cs + 1}$$

$$= -G_0\frac{\frac{1}{Q}\left(\frac{s}{\omega_0}\right)}{\left(\frac{s}{\omega_0}\right)^2 + \frac{1}{Q}\left(\frac{s}{\omega_0}\right) + 1},$$

where $G_0 = \frac{1}{2}\frac{R_2}{R_1}$, $\omega_0 = \frac{1}{\sqrt{R_1R_2C}}$ and $Q = \frac{1}{2}\sqrt{\frac{R_2}{R_1}}$.

With $R_1 = 159\ \Omega$, $R_2 = 100R_1$, and $C = 0.1\ \mu F$ the peak frequency is 1 kHz, $G_0 = 50$, and $Q = 5$. The magnitude plot of a band pass filter for $Q = 0.5$, 1, and 20 is shown in Figure 9.26(b). The quality factor and the peak value of the magnitude plot both increase with increasing R_2. The filter becomes more selective as shown.

9.20 Second-Order Notch Filters

The system function and frequency response of a second-order notch filter may be written as

$$H(s) = \frac{s^2 + b}{s^2 + 2as + b} = \frac{(\frac{s}{\omega_0})^2 + 1}{(\frac{s}{\omega_0})^2 + \frac{1}{Q}(\frac{s}{\omega_0}) + 1}, \quad \text{where } Q = \frac{\sqrt{b}}{2a} \text{ and } \omega_0 = \sqrt{b}$$

$$H(\omega) = \frac{b - \omega^2}{b - \omega^2 + j2a\omega} = \frac{1 - (\frac{\omega}{\omega_0})^2}{1 - (\frac{\omega}{\omega_0})^2 + \frac{j}{Q}\frac{\omega}{\omega_0}}.$$

The filter has two zeros at $\pm j\omega_0$ (notch frequency) and two poles in the LHP. It works as the opposite of the bandpass filter. The frequency response has a zero magnitude at ω_0. The low- and high-frequency magnitude asymptotes are 0-dB horizontal lines. Between the two limits, especially near ω_0, the frequency response is shaped by the location of the poles. As in any second-order system, we recognize the three cases of the filter being overdamped, critically damped, and underdamped. The sharpness of the dip at the notch frequency is controlled by Q. Higher Qs produce narrower dips. As in the case of bandpass filters we can define a 3-dB band for the dip. In this case the notch band identifies frequencies around ω_0 within which the attenuation is greater than 3 dB. The notch filter described above is functionally equivalent to subtracting the output of a bandpass filter from the input signal traversing through a direct path as in Figure 9.27.

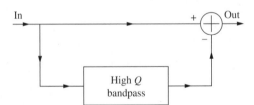

FIGURE 9.27 Functional block diagram and realization of a notch filter.

The circuit of Figure 9.28(a) is an equal-component twin-T notch filter. Show that the system function is

$$\frac{V_2}{V_1} = \frac{(\frac{s}{\omega_0})^2 + 1}{(\frac{s}{\omega_0})^2 + \frac{1}{Q}(\frac{s}{\omega_0}) + 1},$$

where

$$\omega_0 = \frac{1}{RC}, \quad Q = \frac{1}{4(1 - \beta)} \quad \text{and} \quad \beta = \frac{R_1}{R_1 + R_2}$$

is the feedback factor.

Let $R = 159 \, \Omega$ and $C = 1 \, \mu F$. Plot $20 \log |H(f)|$ versus $\log f$ for $\beta = 0.5$, 0.875, and 0.975. Discuss the performance of the filter for the above feedback factors.

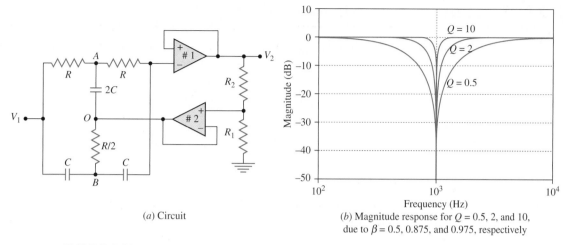

(a) Circuit

(b) Magnitude response for $Q = 0.5$, 2, and 10, due to $\beta = 0.5$, 0.875, and 0.975, respectively

FIGURE 9.28 An equal-component twin-T notch filter and its magnitude frequency responses.

Solution

Both op-amps are voltage followers. Op-amp 2 provides the voltage $V_O = V_2 \times R_1/(R_1 + R_2) = \beta V_2$ at its output. From the above observation and from KCL at nodes A, B, and the noninverting input of op-amp 1 we obtain the following four equations:

Output of op-amp 2: $\qquad\qquad\qquad\qquad V_O = \beta V_2$

KCL at node A: $\qquad\qquad\qquad \dfrac{V_A - V_1}{R} + \dfrac{V_A - V_2}{R} + (V_A - V_O)2Cs = 0$

KCL at node B: $\qquad\qquad\qquad (V_B - V_1)Cs + (V_B - V_2)Cs + \dfrac{V_B - V_O}{R/2} = 0$

KCL at the noninverting input of op-amp 1: $\qquad \dfrac{V_2 - V_A}{R} + (V_2 - V_B)Cs = 0$

After eliminating V_A, V_B, and V_O between the above four equations we are left with

$$\frac{V_2}{V_1} = \frac{R^2 C^2 s^2 + 1}{R^2 C^2 s^2 + 4(1 - \beta)RCs + 1}$$

$$= \frac{1 + \left(\frac{s}{\omega_0}\right)^2}{\left(\frac{s}{\omega_0}\right)^2 + \frac{1}{Q}\frac{s}{\omega_0} + 1}, \quad \text{where } \omega_0 = \frac{1}{RC}, \quad Q = \frac{1}{4(1 - \beta)}, \text{ and } \beta = \frac{R_1}{R_1 + R_2}$$

With $R = 159\ \Omega$ and $C = 1\ \mu F$ the notch frequency is 1 kHz. As in Examples 9.33 and 9.34, the feedback factor controls the location of the poles and, consequently, Q. The three feedback factors $\beta = 0.5$, 0.875, and 0.975, result in $Q = 0.5$, 2, and 10, respectively. The superimposed magnitude plots are shown in Figure 9.28(b). At 1 kHz the value of the magnitude response is always zero. Higher Qs produce narrower 3-dB bandwidths.

9.21 Second-Order All-Pass Filters

The second-order all-pass filter has a constant gain H_0 and a phase that varies with frequency. The system function and frequency response are

$$H(s) = \frac{(\frac{s}{\omega_0})^2 - \frac{1}{Q}(\frac{s}{\omega_0}) + 1}{(\frac{s}{\omega_0})^2 + \frac{1}{Q}(\frac{s}{\omega_0}) + 1}$$

$$H(\omega) = \frac{1 - (\frac{\omega}{\omega_0})^2 - \frac{j}{Q}(\frac{\omega}{\omega_0})}{1 - (\frac{\omega}{\omega_0})^2 + \frac{j}{Q}(\frac{\omega}{\omega_0})}$$

The operation of the above all-pass filter may be realized by passing the input signal through a bandpass filter (with a gain 2) and subtracting the output from the signal, as in Figure 9.29. The filter has a pair of poles in the LHP and a pair of zeros in the RHP mirror-imaging the poles with respect to the $j\omega$ axis. Pole-zero location and phase response depend on Q and ω_0.

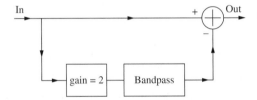

FIGURE 9.29 Functional block diagram and realization of a second-order all-pass filter. The system function of the block diagram is

$$H(s) = 1 - 2\frac{2as}{s^2 + 2as + b} = \frac{s^2 - 2as + b}{s^2 + 2as + b}$$

9.22 Contribution from a Zero

A zero at $s = -\omega_0$ contributes the normalized factor $(1 + s/\omega_0)$ to the system function and adds $10 \log[1 + (\omega/\omega_1)^2]$ dB to the magnitude response. At high frequencies this translates into $+20$ dB per decade. The effect of the zero's contribution at low or intermediate frequencies will depend on the location of the zero with respect to the poles and the Q factor. In this section, by way of an example, we illustrate how a zero reshapes the frequency response of a second-order system function.

Example **9.37**

Consider the second-order filter with a zero at $s = -\omega_1$ characterized by

$$H(s) = \frac{(\frac{s}{\omega_1}) + 1}{(\frac{s}{\omega_0})^2 + \frac{1}{Q}\frac{s}{\omega_0} + 1}$$

$$20 \log|H(\omega)| = 20 \log\left|1 + j\frac{\omega}{\omega_1}\right| - 20 \log\left|1 - \left(\frac{\omega}{\omega_0}\right)^2 + \frac{j}{Q}\frac{\omega}{\omega_0}\right|$$

in which the DC gain is one. The introduction of the zero increases the magnitude response by $10 \log[1 + (\omega/\omega_1)^2]$ dB. At low frequencies ($\omega << \omega_1$), the increase is negligible and the frequency response remains the same as that of the system without a zero. At $\omega = \omega_1$, the increase is 3 dB. At high frequencies ($\omega >> \omega_1$), the increase is $\approx 20 \log \omega$, which corresponds to 20 dB per decade, reducing the slope of the high-frequency asymptote by that amount. How influential the magnitude increase is in reshaping the frequency response at low and intermediate frequencies depends on the location of the zero with respect to the poles, and the Q factor. This is illustrated for zeros at $\omega_1 = 0.01$, 10, and ∞ (no zero) in Figure 9.30(a) ($Q = 0.1$) and Figure 9.30(b) ($Q = 20$). In all cases the undamped natural frequency of the system is 1 rad/s.

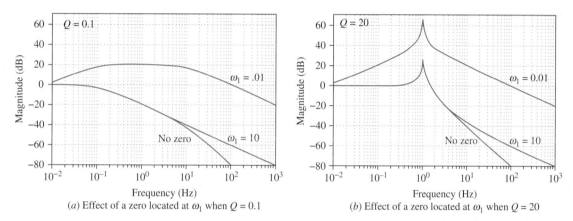

(a) Effect of a zero located at ω_1 when $Q = 0.1$ (b) Effect of a zero located at ω_1 when $Q = 20$

FIGURE 9.30 Adding a zero to a second-order system (DC gain $= 1$, $\omega_0 = 1$, $Q = 0.1$ and 20) reshapes its magnitude response. The zero is placed at $\omega_1 = 10$, 0.01, or ∞ (no zero). The slope of the high-frequency asymptote is reduced from -40 dB/ decade (for the system with no zero) to -20 dB/decade (for the system with a finite zero). A zero placed far from the origin doesn't influence the system's behavior at low and intermediate frequencies. When the zero is close to the origin the magnitude of the system's frequency response is increased by 20 dB over most of the frequency range.

9.23   Series and Parallel *RLC* Circuits

A circuit made of three elements R, L, C and an input source constitutes a second-order filter. The source-free circuit of the filter can have one of two configurations: *RLC* in series or *RLC* in parallel. The input signal may come from a voltage source or a current source. Depending on where in the circuit the input signal arrives and where the output signal is picked up, we can have a low-pass, high-pass, or bandpass filter. In this section we summarize the filtering effect in two basic configurations: (1) *RLC* in series with a voltage source and (2) *RLC* in parallel with a current source.

Consider an RLC circuit in series with a voltage source input as in Figure 9.31(a). Three element voltages v_R, v_L, and v_C constitute the low-pass, high-pass, and bandpass outputs, respectively. To see this, we construct the three system functions summarized

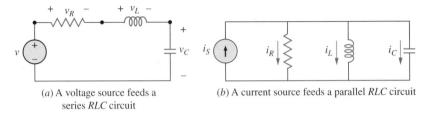

(a) A voltage source feeds a
series *RLC* circuit

(b) A current source feeds a parallel *RLC* circuit

FIGURE 9.31 Filtering by *RLC* circuits.

in Table 9.8(a). A similar situation arises when a current source (providing the input signal) is placed in parallel with an *RLC* circuit. See Figure 9.31(b). Currents in R, L, and C constitute the low-pass, high-pass and bandpass outputs, as summarized by the system functions in Table 9.8b.

TABLE 9.8a. Filtering in a Series *RLC* Circuit [Figure 9.31(a)]

Filter Type	$H(s)$		
Low-pass	$H_C(s) = \dfrac{V_c}{V_s}$	$= \dfrac{1}{LCs^2 + RCs + 1}$	$= \dfrac{1}{(\frac{s}{\omega_0})^2 + \frac{1}{Q}\frac{s}{\omega_0} + 1}$
High-pass	$H_L(s) = \dfrac{V_L}{V_s}$	$= \dfrac{LCs^2}{LCs^2 + RCs + 1}$	$= \dfrac{(\frac{s}{\omega_0})^2}{(\frac{s}{\omega_0})^2 + \frac{1}{Q}\frac{s}{\omega_0} + 1}$
Bandpass	$H_R(s) = \dfrac{V_R}{V_s}$	$= \dfrac{RCs}{LCs^2 + RCs + 1}$	$= \dfrac{\frac{1}{Q}\frac{s}{\omega_0}}{(\frac{s}{\omega_0})^2 + \frac{1}{Q}\frac{s}{\omega_0} + 1}$

$$\omega_0 = \frac{1}{\sqrt{LC}} \text{ and } Q = \frac{1}{R}\sqrt{\frac{L}{C}}$$

TABLE 9.8b. Filtering in a Parallel *RLC* Circuit [Figure 9.31(b)]

Filter Type	$H(s)$		
Low-pass	$H_L(s) = \dfrac{I_L}{I_s}$	$= \dfrac{1}{LCs^2 + \frac{L}{R}s + 1}$	$= \dfrac{1}{(\frac{s}{\omega_0})^2 + \frac{1}{Q}\frac{s}{\omega_0} + 1}$
High-pass	$H_C(s) = \dfrac{I_C}{I_s}$	$= \dfrac{LCs^2}{LCs^2 + \frac{L}{R}s + 1}$	$= \dfrac{(\frac{s}{\omega_0})^2}{(\frac{s}{\omega_0})^2 + \frac{1}{Q}\frac{s}{\omega_0} + 1}$
Bandpass	$H_R(s) = \dfrac{I_R}{I_s}$	$= \dfrac{\frac{L}{R}s}{LCs^2 + \frac{L}{R}s + 1}$	$= \dfrac{\frac{1}{Q}\frac{s}{\omega_0}}{(\frac{s}{\omega_0})^2 + \frac{1}{Q}\frac{s}{\omega_0} + 1}$

$$\omega_0 = \frac{1}{\sqrt{LC}} \text{ and } Q = R\sqrt{\frac{C}{L}}$$

Example
9.38

The frontal stage of a hypothetical AM radio receiver is modeled as a current source feeding a parallel *RLC* circuit in resonance and generating a voltage signal. The *RLC* circuit is used as a tuning device to select the signal from the desired station and attenuate signals from other stations. This example explores the minimum frequency separation between two adjacent stations if selection of the station is not enhanced by other tuning or filtering stages in the receiver.

An AM station, WGB1, broadcasts at the center frequency $f_1 = 1$ MHz with a 3-dB bandwidth of 10 kHz bandwidth. A second AM station, WGB2, broadcasts at the center frequency f_2 with the same power and bandwidth as WGB1. The parallel *RLC* circuit is used to tune to WGB1. It is desired that when the receiver is tuned to WGB1, the f_2 component of the voltage signal picked up by it be 60 dB below the f_1 component. Find the minimum frequency separation between the two stations.

Solution
From Table 9.8b the system function and frequency response of a parallel *RLC* circuit are found to be

$$H(s) = \frac{V}{I_s} = \frac{\frac{R}{Q}\frac{s}{\omega_0}}{\left(\frac{s}{\omega_0}\right)^2 + \frac{1}{Q}\frac{s}{\omega_0} + 1}$$

$$|H(f)|^2 = \frac{R^2}{1 + Q^2\left(\frac{f}{f_0} - \frac{f_0}{f}\right)^2}$$

$$20\log|H(f)| = 20\log R - 10\log\left[1 + Q^2\left(\frac{f}{f_0} - \frac{f_0}{f}\right)^2\right]$$

The quality factor of the *RLC* tuning curve of WGB1 is $Q = f_1/\Delta f = 100$. Its attenuation in dB at frequency f is

$$10\log\left[1 + Q^2\left(\frac{f}{f_1} - \frac{f_1}{f}\right)^2\right]$$

The minimum frequency separation between the two stations is found by setting the above attenuation to 60 dB, which results in $f_{2(low)} = 0.1 f_1 = 100$ kHz and $f_{2(high)} = 10.1 f_1 = 10.1$ MHz.

9.24 Summary of Second-Order Filters

The system function of a stable second-order filter is

$$H(s) = \frac{B(s)}{s^2 + 2as + b}$$

where, as before, a and b are positive real numbers, and $B(s)$ is a polynomial in s of an order not higher than 2. The pole-zero configuration of the filter determines its type and forms its frequency response. The polynomial $B(s)$ produces zeros and results in the filter becoming low-pass, high-pass, all-pass, bandpass, or bandstop. These are summarized in Table 9.9 using the ω_0 and Q model. Figure 9.32 shows superimposed magnitude and phase responses of four basic types of second-order filters for $Q = 0.7$, 2, and 20.

TABLE 9.9 Second-Order Filters

Filter Type	$H(s)$	$H(\omega)$
Low-pass	$\dfrac{1}{(\frac{s}{\omega_0})^2 + \frac{1}{Q}\frac{s}{\omega_0} + 1}$	$\dfrac{1}{1 - (\frac{\omega}{\omega_0})^2 + \frac{j}{Q}\frac{\omega}{\omega_0}}$
High-pass	$\dfrac{(\frac{s}{\omega_0})^2}{(\frac{s}{\omega_0})^2 + \frac{1}{Q}\frac{s}{\omega_0} + 1}$	$\dfrac{(\frac{\omega}{\omega_0})^2}{1 - (\frac{\omega}{\omega_0})^2 + \frac{j}{Q}\frac{\omega}{\omega_0}}$
All-pass	$\dfrac{(\frac{s}{\omega_0})^2 - \frac{1}{Q}\frac{s}{\omega_0} + 1}{(\frac{s}{\omega_0})^2 + \frac{1}{Q}\frac{s}{\omega_0} + 1}$	$\dfrac{1 - (\frac{\omega}{\omega_0})^2 - \frac{j}{Q}\frac{\omega}{\omega_0}}{1 - (\frac{\omega}{\omega_0})^2 + \frac{j}{Q}\frac{\omega}{\omega_0}}$
Bandpass	$\dfrac{\frac{s}{\omega_0}}{(\frac{s}{\omega_0})^2 + \frac{1}{Q}\frac{s}{\omega_0} + 1}$	$\dfrac{\frac{j\omega}{\omega_0}}{1 - (\frac{\omega}{\omega_0})^2 + \frac{j}{Q}\frac{\omega}{\omega_0}}$
Notch	$\dfrac{(\frac{s}{\omega_0})^2 + 1}{(\frac{s}{\omega_0})^2 + \frac{1}{Q}\frac{s}{\omega_0} + 1}$	$\dfrac{1 - (\frac{\omega}{\omega_0})^2}{1 - (\frac{\omega}{\omega_0})^2 + \frac{j}{Q}\frac{\omega}{\omega_0}}$

9.25 Group and Phase Delay

The phase shift experienced by a sinusoid in passing through an LTI system is frequency dependent. It is the phase of the frequency response of the system at that frequency. A phase shift may be expressed as a time shift; $\cos(\omega t \pm \theta) = \cos \omega(t \pm \tau)$, where ω is the angular frequency of the sinusoid in rad/s and $\tau = \theta/\omega$ in seconds. If the LTI system is physically realizable, the time shift becomes a delay. In a linear-phase system $\theta = \tau\omega$, which results in a constant delay at all frequencies. In a more general case, depending on the phase function of the frequency response of the system through which a signal passes, different frequency components experience different delays. Group delay t_g and phase delay t_p, defined below, can provide some measure of changes in the shape of signals.

$$t_g(f) = -\frac{1}{2\pi}\frac{d\theta(f)}{df} \quad \text{and} \quad t_p(f) = -\frac{1}{2\pi}\frac{\theta(f)}{f}$$

The following two examples illustrate simple cases of group delay.

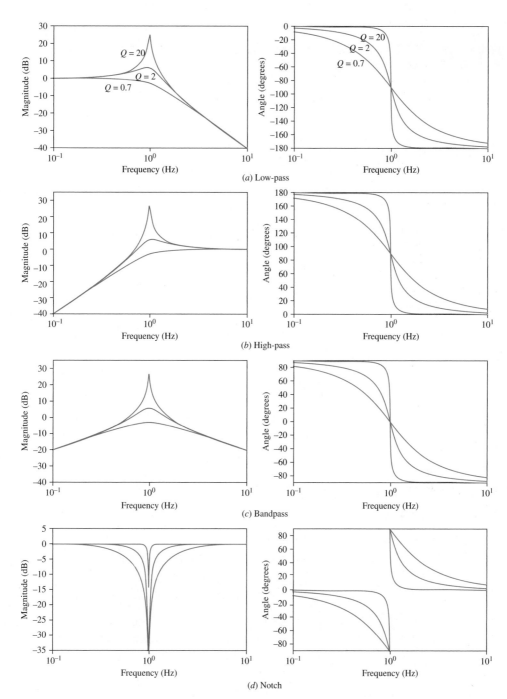

FIGURE 9.32 Superimposed frequency responses of four types of filters for $Q = 0.7$, 2, and 20. Magnitude plots are in the left and phase plots in the right.

Example

9.39

A linear, time-invariant, all-pass, unity-gain filter introduces a phase lag θ in a sinusoidal signal passing through it:

$$v_i(t) = V \cos \omega t \implies v_o(t) = V \cos(\omega t - \theta)$$

for all V and ω. Given $v_i(t) = (1 + 2 \cos \omega_0 t) \cos \omega_c t$, where $\omega_0 = 10^7$ and $\omega_c = 10^8$, find $v_o(t)$ for the following three filters:

1. Linear phase $\theta = 10^{-9} \omega$.
2. Constant phase $\theta = 10^{-1}$.
3. Linear phase with a constant bias $\theta = (10^{-9}\omega + 10^{-1})$. θ and ω are in radians and radians per second, respectively.

Solution

Assuming $\omega_l = \omega_c - \omega_0 = 0.9 \times 10^8$ and $\omega_h = \omega_c + \omega_0 = 1.1 \times 10^8$, the input signal may be expanded into its sinusoidal components

$$v_i(t) = \cos \omega_l t + \cos \omega_h t + \cos \omega_c t$$

The output is

$$v_o(t) = \cos(\omega_l t - \theta_l) + \cos(\omega_h t - \theta_h) + \cos(\omega_c t - \theta_c)$$

a. In the case of a linear phase filter, $\theta_l = 10^{-9}\omega_l$, $\theta_h = 10^{-9}\omega_h$, and $\theta_c = 10^{-9}\omega_c$. The output is

$$v_o(t) = \cos(\omega_l t - 10^{-9}\omega_l) + \cos(\omega_h t - 10^{-9}\omega_h) + \cos(\omega_c t - 10^{-9}\omega_c)$$

$$= \cos \omega_l (t - 10^{-9}) + \cos \omega_h (t - 10^{-9}) + \cos \omega_c (t - 10^{-9})$$

$$= v_i(t - 10^{-9})$$

Because of the linear phase property, the output is a delayed version of the input. The delay is equal to the slope of the phase lag versus the ω curve, which is equal to 10^{-9} seconds.

b. In the case of a filter with constant phase, $\theta_l = \theta_h = \theta_c = 0.1$.

$$v_o(t) = \cos(\omega_l t - 0.1) + \cos(\omega_h t - 0.1) + \cos(\omega_c t - 0.1)$$

$$= \cos \omega_l (t - 1.11 \times 10^{-9}) + \cos \omega_h (t - 0.91 \times 10^{-9}) + \cos \omega_c (t - 10^{-9})$$

Because of the constant phase, the time delay depends on the frequency. Higher frequencies experience smaller delays than lower frequencies, producing a distortion. The time delays are given below.

ω in rad/s	delay in sec
$\omega_\ell = 0.9 \times 10^8$	1.11×10^{-9}
$\omega_c = 10^8$	10^{-9}
$\omega_h = 1.1 \times 10^8$	0.91×10^{-9}

c. In the case of a linear phase with a constant bias we have

$$v_o(t) = \cos(\omega_l t - \theta_\ell) + \cos(\omega_h t - \theta_h) + \cos(\omega_c t - \theta_c)$$

where

$$\theta_l = 10^{-9}\omega_l + 0.1 = 0.19$$
$$\theta_h = 10^{-9}\omega_h + 0.1 = 0.21$$
$$\theta_c = 10^{-9}\omega_c + 0.1 = 0.2$$

Substituting for θ we get

$$\begin{aligned}
v_o(t) &= \cos(\omega_l t - 0.19) + \cos(\omega_h t - 0.21) + \cos(\omega_c t - 0.2) \\
&= \cos\omega_l(t - 2.11 \times 10^{-9}) + \cos\omega_h(t - 1.9 \times 10^{-9}) \\
&\quad + \cos\omega_c(t - 2 \times 10^{-9})
\end{aligned}$$

As in case b, the three frequency components in $v_o(t)$ undergo three different delays as given below.

ω in rad/s	delay in sec
$\omega_\ell = 0.9 \times 10^8$	2.11×10^{-9}
$\omega_c = 10^8$	2×10^{-9}
$\omega_h = 1.1 \times 10^8$	1.9×10^{-9}

By way of simplification, one may consider that the total signal is delayed by an average of 2×10^{-9} seconds. However, in this case, the nonuniformity in delays doesn't produce destructive distortion. By combining the low- and high-frequency components in $v_o(t)$ we find

$$v_o(t) = [1 + 2\cos\omega_0(t - 11 \times 10^{-9})]\cos\omega_c(t - 2 \times 10^{-9})$$

Note that the amplitude $1 + 2\cos\omega_0 t$ is delayed by 11 nanoseconds, becoming $1 + 2\cos\omega_0(t - 11 \times 10^{-9})$, but experiences no distortion. The carrier $\cos\omega_c t$ is delayed by 2 nanoseconds and becomes $\cos\omega_c(t - 2 \times 10^{-9})$. This phenomenon is generalized in the next example.

Example
9.40

A simple case of group delay

Consider a linear filter that has unity gain within the passband $\omega_c \pm \Delta\omega$ and introduces a phase lag θ in a sinusoidal signal passing through it. In other words,

$$v_i(t) = V\cos(\omega t) \quad \Longrightarrow \quad v_o(t) = V\cos(\omega t - \theta)$$

for all V and $(\omega_c - \Delta\omega) < \omega < (\omega_c + \Delta\omega)$. Within the above band the phase lag is $\theta = \theta_c + \tau_0(\omega - \omega_c)$. Let the input to the filter be an amplitude-modulated waveform

$$v_i(t) = [1 + 2A\cos(\omega_0 t)]\cos(\omega_c t)$$

where $\cos(\omega_c t)$ is the carrier and $1 + 2A\cos(\omega_0 t)$ is the modulating signal with $\omega_0 < \Delta\omega$. Find the output $v_o(t)$.

Solution

Expand the input signal into its sinusoidal components:

$$v_i(t) = A\cos(\omega_\ell t) + A\cos(\omega_h t) + \cos(\omega_c t)$$

where

$$\omega_\ell = \omega_c - \omega_0$$
$$\omega_h = \omega_c + \omega_0$$

The input signal, therefore, falls within the passband. The output is

$$v_o(t) = A\cos(\omega_\ell t - \theta_\ell) + A\cos(\omega_h t - \theta_h) + \cos(\omega_c t - \theta_c)$$

where

$$\theta_\ell = \theta_c + \tau_0(\omega_\ell - \omega_c) = \theta_c - \tau_0\omega_0$$
$$\theta_h = \theta_c + \tau_0(\omega_h - \omega_c) = \theta_c + \tau_0\omega_0$$

Combining the first two terms of $v_o(t)$ and noting that

$$\omega_c = \frac{\omega_h + \omega_\ell}{2}$$

$$\omega_0 = \frac{\omega_h - \omega_\ell}{2}$$

we get

$$v_o(t) = 2A\cos\left(\omega_0 t - \frac{\theta_h - \theta_\ell}{2}\right)\cos\left(\omega_c t - \frac{\theta_h + \theta_\ell}{2}\right) + \cos(\omega_c t - \theta_c)$$

Define $\tau_c = \theta_c/\omega_c$. Then

$$\theta_h + \theta_\ell = 2\theta_c = 2\tau_c\omega_c$$
$$\theta_h - \theta_\ell = 2\theta_0 = 2\tau_0\omega_0$$

Therefore,

$$v_o(t) = 2A\cos(\omega_0 t - \theta_0)\cos(\omega_c t - \theta_c) + \cos(\omega_c t - \theta_c)$$
$$= 2A\cos\omega_0(t - \tau_0)\cos\omega_c(t - \tau_c) + \cos(\omega_c t - \tau_c)$$
$$= [1 + 2A\cos\omega_0(t - \tau_0)]\cos\omega_c(t - \tau_c)$$

The modulating signal and the carrier are delayed by τ_0 and τ_c, respectively. In general, the modulating signal may contain many frequency components. As long as they remain within the band $2\Delta\omega$, all will be delayed by the same amount τ_0, which makes the modulating signal undistorted. The carrier $\cos\omega_c t$ is delayed by τ_c, a different amount. This, however, doesn't introduce any distortion in the modulating signal. As a group, the total signal is delayed by an average of τ_c.

9.26 Measurements, Identification, and Modeling

By measurement we can find the magnitude and phase of the frequency response and decide on break points if the system function is to be rational.

Example

9.41

Magnitude response of a filter is measured and plotted in Figure 9.33. Its salient features are summarized below. Model it by using a rational system function.

Frequency, rad/s	DC	$\omega_\ell = 3.75$	$\omega_{Max} = 5.612$	$\omega_h = 7$	High frequency
Magnitude (dB)	0	3.3	6.3	3.3	−40 dB/decade

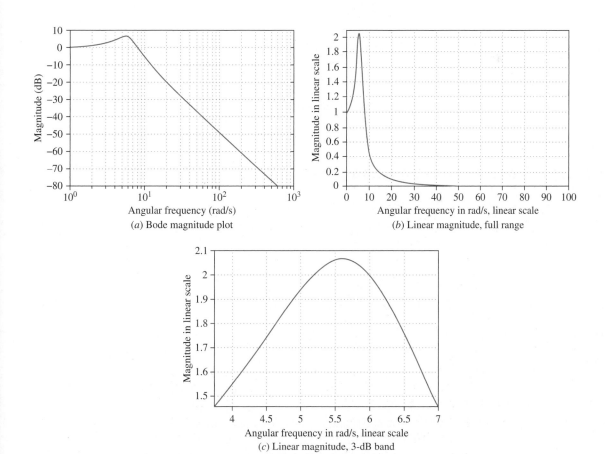

(a) Bode magnitude plot

(b) Linear magnitude, full range

(c) Linear magnitude, 3-dB band

FIGURE 9.33 Magnitude measurement of the frequency response of a filter to be modeled by a rational function.

Solution

The high-frequency asymptote of the magnitude response indicates that the filter has two more poles than zeros. We start with the simplest model, which is a second-order low-pass filter with no zeros in the finite s-plane shown by the normalized system function

$$H(s) = \frac{H_0}{\left(\frac{s}{\omega_0}\right)^2 + \frac{1}{Q}\frac{s}{\omega_0} + 1} \quad \text{and} \quad H(\omega) = \frac{H_0}{1 - \left(\frac{\omega}{\omega_0}\right)^2 + \frac{j}{Q}\frac{\omega}{\omega_0}}$$

Using the values given in the above table and applying the above equation as a first-order approximation, we estimate $Q = 2.066$ and $\omega_0 = 5.612$. This is not a high Q and the estimate is not satisfactory. We need to determine, in terms of Q and ω_0, the frequency that maximizes $|H(\omega)|^2$, or equivalently, minimizes its denominator. This is done by finding the roots of

$$\frac{d}{d\omega}\left[x^4 + \left(\frac{1}{Q^2} - 2\right)x^2 + 1\right] = 0, \quad \text{where} \quad x = \frac{\omega}{\omega_0}$$

from which we find $Q = 2.03$ and $\omega_0 = 5.986$. We choose $Q = 2$ and $\omega_0 = 6$ and the system function will then be

$$H(s) = \frac{36}{s^2 + 3s + 36}$$

The above model turns out to be the system function that generated plots of Figure 9.33. Further modeling exercise is available in Project 4 at the end of this chapter.

9.27 System Interconnection

Several systems may be connected and combined to form a bigger system. The system function of the combined system is found from a knowledge of its subsystems and their interconnection. Three basic connections are recognized:

A Series Connection

This exists when two subsystems are cascaded. The output of the first subsystem provides the input to the second subsystem. The second subsystem doesn't load the first subsystem. See Figure 9.34.

$$H(s) = H_1(s) \times H_2(s) \quad \Longleftrightarrow \quad h(t) = h_1(t) \star h_2(t)$$

The series connection, in general, retains the poles and zeros of its subsystems except in cases of pole-zero cancellation, examples of which will be given in the next section. The magnitude and phase of the frequency responses are multiplied together when they are expressed in linear scale. The magnitude responses are added together when expressed in dB (i.e., log scale such as in Bode plots).

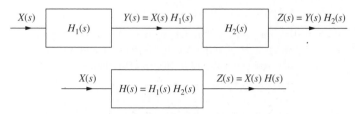

FIGURE 9.34 A series connection of two LTI systems H_1 and H_2 produces an equivalent LTI system with $H(s) = H_1(s) \times H_2(s)$. The unit-impulse response of the equivalent system is $h(t) = h_1(t) \star h_2(t)$.

A Parallel Connection

This exists when two subsystems receive the same input and their outputs are added together to provide the overall system's output. See Figure 9.35. This arrangement doesn't affect the performance of individual subsystems.

$$H(s) = H_1(s) + H_2(s) \iff h(t) = h_1(t) + h_2(t)$$

The parallel combination retains the poles of its subsystems except in possible pole-zero cancellation cases. The magnitudes and phases of the frequency responses, in general, neither multiply nor add (except in special cases).

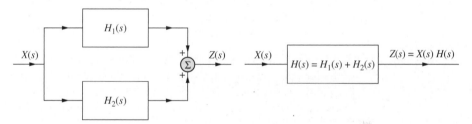

FIGURE 9.35 A parallel connection of two LTI systems H_1 and H_2 (*left*) produces an equivalent LTI system with $H(s) = H_1(s) + H_2(s)$ (*right*). The unit-impulse response of the equivalent system is $h(t) = h_1(t) + h_2(t)$.

A Feedback Connection

This is shown in Figure 9.36, and discussed in the next section.

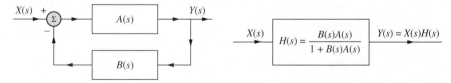

FIGURE 9.36 A feedback system (*left*) and its equivalent (*right*). $A(s)$ is the system function of the forward path (also called the open-loop transfer function) and $B(s)$ is the system function of the feedback path. The closed-loop system function is $H(s) = A(s)/[1 + B(s)A(s)]$.

A Note on Loading Effect

Loading may occur when subsystems are connected together. It is important to pay attention to possible loading effects from one subsystem to another. Ideally, connecting two subsystems should not modify their system functions and should not change their individual performances. In electric circuits this is achieved by placing a buffer (normally an op-amp) between the two subsystems. If this is not practical, the loading effect should be taken into consideration when specifying the system function of each subsystem (e.g., termination effect in transmission lines).

9.28 Feedback[10]

A feedback system is shown in Figure 9.36. The closed-loop system function is found by

$$Y(s) = [X(s) - Y(s)B(s)]A(s)$$

$$H(s) = \frac{Y(s)}{X(s)} = \frac{A(s)}{1 + B(s)A(s)}$$

$A(s)$ is called the forward path (or open-loop transfer function) and $B(s)$ is the system function of the feedback path. The product $A(s)B(s)$ is called the loop gain. For large loop gains we may appproximate $H(s)$ by

$$H(s) = \frac{A(s)}{1 + B(s)A(s)} \approx \frac{1}{B(s)} \quad \text{if} \quad |A(s)B(s)| >> 1$$

In such a case, the closed-loop system becomes less dependent on the system function of the forward path and more dependent on the system function of the feedback path, becoming approximately equal to its inverse. This fundamental property of negative feedback allows us to control pole-zero locations according to desired specifications. In this way we may:

1. Reduce sensitivity to change in $A(s)$ or its uncertainty.
2. Increase the bandwidth.
3. Improve the step response; for example, reduce its rise time, oscillations, and overshoot.
4. Control the system's performance.

Another model of a feedback control system is shown in Figure 9.37, in which the subsystem $C(s)$ is called the controller. The overall system function of Figure 9.37 is similarly found to be

$$H(s) = \frac{A(s)C(s)}{1 + A(s)C(s)}$$

---

[10]For several historical accounts on the origin and history of feedback control see the following: Otto Mayr, *The Origins of Feedback Control* (Cambridge: MIT. Press, 1970). Gordon S. Brown, in *Scientific American*, 1951. Stuart Bennet, "A Brief History of Automatic Control," *IEEE Control Systems*, June 1996, pp. 17–25.

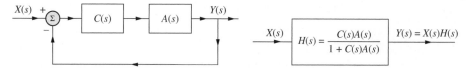

FIGURE 9.37 A feedback system with a controller (*left*) and its equivalent (*right*). The subsystem $C(s)$ is called the controller. The overall system function is $H(s) = C(s)A(s)/[1 + C(s)A(s)]$.

A third feedback model is shown in Figure 9.38 which contains a feedback path $B(s)$ representing sensors and a controller $C(s)$. The overall system function of Figure 9.38 is

$$H(s) = \frac{A(s)C(s)}{1 + A(s)B(s)C(s)}$$

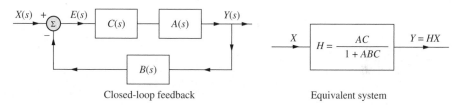

Closed-loop feedback   Equivalent system

FIGURE 9.38 A feedback control system with a sensor $B(s)$ and controller $C(s)$ (*left*). The overall system function is $H(s) = A(s)C(s)/[1 + A(s)B(s)C(s)]$ (*right*).

Feedback: Effect on Sensitivity

Negative feedback can reduce the sensitivity of the closed-loop system to variations in the forward path. We will now illustrate this effect. The measure of sensitivity S of a function $y = f(x)$ to variations in x is defined by

$$S_x^y = \frac{\text{Percentage change produced in } y}{\text{Percentage change in } x} = \frac{dy/y}{dx/x} \approx \frac{\Delta y/y}{\Delta x/x}$$

Now consider a feedback system with forward gain A and feedback factor β. The closed-loop gain of the system and its sensitivity to variations in A and β are

$$H \qquad = \frac{A}{1 + \beta A}$$

$$dH \qquad = \frac{dA - A^2 d\beta}{(1 + \beta A)^2}$$

$$\left.\frac{dH}{H}\right|_{\beta = \text{constant}} = \frac{1}{1 + \beta A}\left(\frac{dA}{A}\right) \Longrightarrow S_A^H = \frac{dH/H}{dA/A} = \frac{1}{1 + \beta A}$$

$$\left.\frac{dH}{H}\right|_{A = \text{constant}} = \frac{-\beta A}{1 + \beta A}\left(\frac{d\beta}{\beta}\right) \Longrightarrow S_\beta^H = \frac{dH/H}{d\beta/\beta} = \frac{-\beta A}{1 + \beta A}$$

xample

9.42

Let $A = 2,000$ and $\beta = 0.0495$ resulting in $H = 2,000/(1+0.0495 \times 2,000) = 20$. At that setting, the sensitivity of H to A is $S_A^H = 1/1 + 0.0495 \times 2,000) \approx 0.01$. That means a fractional change of $\epsilon\%$ in A results in a fractional change of $0.01\epsilon\%$ in H. For example, let A be increased from 2000 to 2001 (a fractional change of $100 \times (2,001 - 2,000)/2,000 = 0.05\%$). The result is an increase in H from 20 to 20.0001 (a fractional change of $100 \times (20.0001 - 20)/20 = 0.0005\%$).

Note that the sensitivity, as defined above, applies to small changes (dH and dA). For the effect of large changes (such as a doubling of A) one needs to devise an integration method if the above sensitivity index is used. Alternatively, one may use a new measure (as in Examples 9.46 and 9.47). Two practical applications, illustrated in Examples 9.43 and 9.44, measure performance of the feedback systems using the index $\mathcal{E} = 100 \times (H - H_d)/H_d$ which shows the percentage deviation of H from the desired value H_d.

xample

9.43

A power amplifier

A power amplifier has a nominal (and desired) gain of $A = 20$. But its actual gain varies between 10 and 30 (a $\pm 50\%$ variation). See Figure 9.39(a). To reduce the above variations, we place the power amplifier in a feedback loop containing a sensor (with a feedback factor B) and a proportional controller (preamplifier with a gain of C) as shown in Figure 9.39(b). Determine B and C so as to bring down the variation in the overall gain of the closed-loop system to less than 1% of the desired value.

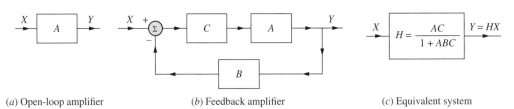

(a) Open-loop amplifier   (b) Feedback amplifier   (c) Equivalent system

FIGURE 9.39 A power amplifier in open-loop and with feedback. Negative feedback reduces the sensitivity of the system to variations in the power amplifier's gain. In this figure, $Y/X = (AC)/(1 + ABC)$. For example, let the open-loop gain of the power amplifier be $A = 20 \pm 50\%$. It is desired to have a power gain of $20 \pm 1\%$ using the above power amplifier. The system of (b), with a feedback coefficient of $B = 0.0495$ and a voltage preamplifier with a gain of $C = 100$, would result in an overall gain of $H = 2,000A/(20 + 99A)$. For $A = 10, \ 20, \ 30$ we will have $H = 19.802, \ 20.000, \ 20.067$, respectively, which is within $20 \pm 1\%$.

Solution

We start with the approximation that $H = (AC)/(1 + ABC) \approx 1/B = 20$ if $ABC \gg 1$. We choose a feedback factor $B = 0.0495$. The gain of the preamplifier C may then be computed from the expression for H or found by trial and error. Choosing $C = 100$ will result in the following satisfactory values for H.

A	H	$\mathcal{E}\%$
10	19.802	−0.99
20	20	0
30	20.067	+0.335

where $\mathcal{E} = 100 \times (H - 20)/20$ is the percentage deviation of H from the desired gain of 20.

Closed-loop operation of a noninverting amplifier

A noninverting op-amp circuit and its equivalent circuit are shown in Figure 9.40 a and b, respectively.

a. Find the closed-loop gain $H = V_2/V_1$ as a function of A, R_1, and R_2. Model it as a feedback loop and specify its open-loop gain and the feedback factor.

b. A closed-loop gain of 5 is desired regardless of A. Choose $R_1 = 1$ and $R_2 = 4$ both in $K\Omega$ and evaluate H for

$A = 5, \ 10, \ 100, \ 1{,}000, \ 10{,}000, \ 50{,}000, \ 100{,}000, \ \infty$. Define

$$\mathcal{E} = 100 \times \frac{H - 5}{5}$$

as the percentage deviation of H from the desired gain of 5. For each value of A compute $\mathcal{E}$ and enter the results in a table. Comment on the effect of feedback on the closed-loop gain and its sensitivity to variations in A.

Solution

a. From Figure 9.40(b): $V_d = V_1 - \beta V_2$ and $V_2 = AV_d$, resulting in $H = \frac{A}{1+\beta A}$, where $\beta = \frac{R_1}{R_1+R_2}$.

b. $\beta = .2$, $H = \frac{A}{1+0.2A}$, and $\mathcal{E} = \frac{-100}{1+0.2A}\%$. The table below shows values of H and $\mathcal{E}$ for given values of A.

A	5	10	100	1,000	10,000	50,000	100,000	∞
H	2.5	3.3333	4.7619	4.9751	4.9975	4.9995	4.9997	5.00
$\mathcal{E}\%$	−50%	−33.333%	−4.762%	-0.4975%	−0.04997%	−0.01%	−0.005%	0

As A increases, $|\mathcal{E}\%|$ decreases. At high values of A, the closed-loop gain changes very little even when A is increased considerably. For example a fivefold increase in A, from $A = 10{,}000$ to $A = 50{,}000$, increases H from 4.9975 (with a percentage relative error of $\mathcal{E} \approx -0.05\%$) to 4.9995 (with a percentage relative error of $\mathcal{E} = -0.01\%$). Feedback reduces sensitivity of the overall gain to variaitions in A by the factor $1 + \beta A$.

(*a*) Noninverting amplifier
with feedback

(*b*) Circuit model

(*c*) System model

FIGURE 9.40 (Example 9.44) The 741 op-amp is supplied with negative feedback to its inverting terminal through the R_1/R_2 voltage divider. The amplifier now may be modeled by a closed-loop system such as shown in Figure 9.36. The feedback ratio is $\beta = 0.1$ (i.e., the input to the op-amp is now $V_1 - \beta V_2$). The new closed-loop system function is $H(s) = a_1\omega_1/(s + \omega_1)$. The feedback moves the pole from 5 Hz to 1 MHz, thus reducing the time constant from 31.83 msec to 1.59 μsec. This increases the bandwidth 2×10^5 times at the cost of reducing the DC gain by that same factor.

Feedback: Effect on Frequency Response

Feedback can increase the bandwidth of a system.[11] To illustrate the effect of feedback on the frequency response, we examine the frequency response of a 741 op-amp, in both open-loop and closed-loop configurations. The 741 op-amp is modeled by a DC gain of 2×10^5 and a pole at $f = 5$ Hz (3-dB bandwidth). The product DC gain $\times$ bandwidth is $200{,}000 \times 5 = 10^6$. It is called the *gain-bandwidth product* (GBP) and comprises a specification of the op-amp. The GBP is the frequency at which the gain becomes one (i.e., where the op-amp's magnitude plot intersects the 0-dB level). Feedback pushes the pole of the closed-loop system away from the $j\omega$ axis, increasing the bandwidth and reducing the gain. In the closed-loop feedback system of Example 9.45, the GBP remains the same regardless of the feedback coefficient.

Example
9.45

Open-loop frequency response of a 741 op-amp

Typically, a 741 op-amp (Figure 9.41*a*) has a DC gain of 2×10^5, a first pole at $f = 5$ Hz and a second pole at 1 MHz. It is modeled[12] by the circuit of Figure 9.41*b* having the system function

$$A(s) = \frac{V_2}{V_1} = \frac{a_0}{1 + \left(\frac{s}{\omega_0}\right)},$$

where $a_0 = 2 \times 10^5$ and $\omega_0 = 10\pi$ rad/s (corresponding to 5 Hz)
The system model is shown in Figure 9.41(*c*). Find and sketch its Bode plot.

Solution
The frequency response, expressed as a function of frequency in Hz, is

$$A(f) = \frac{10^6}{5 + jf} = \frac{200{,}000}{1 + jf/5}$$

---

[11] For a brief historical background that led to development of feedback amplifiers see William Siebert, *Circuits, Signals and Systems* (Cambridge: MIT Press, 1986): p. 145.

[12] See section 9.7 on approximation of a system function by its dominant pole(s).

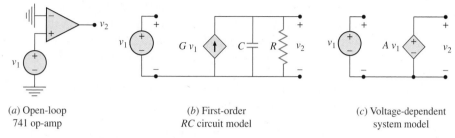

(*a*) Open-loop
741 op-amp

(*b*) First-order
RC circuit model

(*c*) Voltage-dependent
system model

FIGURE 9.41 The 741 op-amp in open-loop configuration. The amplifier *(a)* is approximated by a first-order model $A(s) = a_0\omega_0/(s + \omega_0)$ where $a_0 = 2 \times 10^5$ and $\omega_0 = 10\pi$ rad/s (corresponding to 5 Hz). The model may be shown by a dependent current source feeding a parallel *RC* *(b)*, or equivalently, by a dependent voltage source with a frequency dependent gain of $A(s)$ *(c)*.

The Bode plots are shown in Figure 9.42(*a*) (magnitude) and Figure 9.42(*b*) (phase). The curves are superimposed on the Bode plots of the closed-loop system of Figure 9.40(*a*) for comparison. The open-loop DC gain is $20 \log 200,000 = 106$ dB, which constitutes the low-frequency asymptote. The break frequency is at 5 Hz, where the magnitude is 3 dB below the DC gain; that is, at $106 - 3 = 103$ dB. The open-loop 3-dB bandwidth is 5 Hz. The GBP is $5 \times 200,000 = 10^6$.

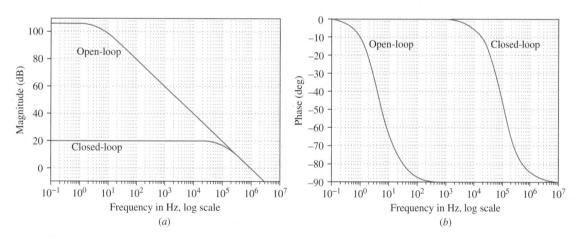

FIGURE 9.42 Frequency responses of 741 op-amp in open-loop and closed-loop configuration. Bode plots for magnitude *(a)* and phase *(b)* of a 741 op-amp in open-loop configuration $H(f) = 10^6/(5 + jf)$ versus f in Hz, DC gain = 106 dB, BW = 5 Hz, discussed in Example 9.45. Curves for the closed-loop configuration of the op-amp in the circuit of Figure 9.40(*a*) (DC gain = 20 dB, BW = 100 kHz) are given for comparison. Gain-bandwidth product (GBP) of the op-amp is 10^6.

The high-frequency asymptote of the magnitude Bode plot is a line with a -20 dB/decade slope passing through the point with the coordinates (106 dB, 5 Hz). At 2.5 Hz and 10 Hz, the magnitude is -1 and -7 dB, respectively, 1 dB below the value of the asymptote at these frequencies. The high-frequency asymptote reaches the 0-dB level (which corresponds to unity gain) at 1 MHz, which is the gain-bandwidth product of the 741 op-amp. It is noted that the product of gain and frequency along the segment of its magnitude plot with -20 dB slope (5 Hz to 1 MHz band) remains constant at 10^6.

The phase plot shown in Figure 9.42(b) goes from $0°$ to $-90°$. The inflexion point is at 5 Hz where the phase is $-45°$. At 2.5 Hz and 10 Hz, the phase is $-26.5°$ and $-63.5°$, respectively. Note that Figure 9.42 plots $20\log|H(f)|$ and $\angle H(f)$ versus $\log f$ (labeled in Hz). The labels on the frequency axis could be changed to $2\pi f$ without any other change in the graphs. The break frequency would then be 31.4 rad/s.

Closed-loop frequency response of a 741 op-amp in noninverting configuration

Find and plot the frequency response of the 741 op-amp circuit of Figure 9.40(a) in the form of a Bode plot and compare with the frequency response of the open-loop op-amp.

Solution

In linear circuits the op-amp is never used in the open-loop form. For one thing, the linear model of Figure 9.41 is valid only when the output remains between saturation levels $\pm V_0$. In the open-loop, because of the huge gain, the output immediately saturates, swinging between the high and low saturation levels. For another thing, the open-loop bandwidth is very small (≈ 5 Hz). Placing the op-amp in a feedback loop trades the high gain for other desired features such as a broader bandwidth (or equivalently a smaller time constant). In Figure 9.40(a) we already have seen a 741 op-amp in a simple closed-loop with a resistive feedback to its inverting input. Its circuit model is shown in Figure 9.40(b). The closed-loop system function may be derived from Figure 9.40(b) by using circuit laws. It may also be derived from the feedback model of Figure 9.36, with the feedback coefficient being $\beta = R_1/(R_1 + R_2)$. The resulting closed-loop system function of Figure 9.40(c) is

$$H(s) = \frac{A(s)}{1 + \beta A(s)} \approx \frac{a_1}{1 + \left(\frac{s}{\omega_1}\right)}, \quad \text{where } \beta = \frac{R_1}{R_1 + R_2}, \quad a_1 = \frac{1}{\beta} \quad \text{and } \omega_1 = a_0 \beta \omega_0$$

The frequency response of the circuit of Figure 9.40(a), expressed as a function of f (in Hz), is

$$H(f) = \frac{\frac{1}{\beta}}{1 + j\left(\frac{f}{a_0 \beta f_0}\right)}$$

where a_0 and ω_0 are the DC gain and the pole of the 741 op-amp, respectively, and $\beta = R_1/(R_1 + R_2)$ is the feedback factor. As an example, let $a_0 = 2 \times 10^5$, $\omega_0 = 10\pi$,

$R_1 = 1\ K\Omega$ and $R_2 = 9\ K\Omega$. Then $\beta = 0.1$ and the closed-loop system function becomes

$$H(s) = \frac{10}{1 + (\frac{s}{\omega_1})} \qquad \omega_1 = 200{,}000\pi$$

Feedback moves the pole to $-\omega_1 = -200{,}000\pi$ (corresponding to 100 kHz).

The magnitude and phase plots of the closed-loop frequency response for $\beta = 0.1$ are superimposed on Figure 9.42, for comparison with the open-loop frequency response. The 3-dB frequency of the closed-loop circuit is at $f_1 = a_0\beta f_0 = 2 \times 10^5 \times 0.1 \times 5 = 10^5$ Hz, which is the location of the new pole. Negative feedback has pushed away the pole from the proximity of the origin and consequently has increased the 3-dB bandwidth to 100 kHz. The price is a reduction of the passband gain from 106 dB to 20 dB. Note that the gain-bandwidth product (GBP = DC gain $\times$ bandwidth = $10 \times 10^5 = 10^6$) remains at 1 MHz, the same as in the open-loop configuration.

Summary
The gain-bandwidth product of the amplifiers in Examples 9.45 and 9.46 remained the same under open-loop and closed-loop conditions. The closed-loop system traded the gain for higher bandwidth. These effects can be summarized as shown in the following:

Open-loop system: $\qquad A(\omega) = \dfrac{a_0}{1 + j(\frac{\omega}{\omega_0})}, \quad GBP_o = a_0\omega_0$

Closed-loop system: $\qquad H(\omega) = \dfrac{A(\omega)}{1 + \beta A(\omega)} = \dfrac{a_c}{1 + j(\frac{\omega}{\omega_c})}$

$$a_c = \frac{a_0}{1 + \beta a_0}, \quad \omega_c = (1 + \beta a_0)\omega_0,$$

$$GBP_c = a_c\omega_c = a_0\omega_0 = GBP_o$$

Feedback: Effect on Time Response
By modifying the pole-zero configuration, negative feedback can shape a system's step response toward a desired form. The following example shows how feedback reduces the time constant and rise time.

Example **9.47**

Find, plot, examine, and compare the step responses of a 741 op-amp in open-loop (Figure 9.41a), and closed-loop (Figure 9.40a) configurations.

Open-loop
The 741 op-amp is modeled by the system function

$$A(s) = \frac{V_2}{V_1} = \frac{a_0}{1 + (\frac{s}{\omega_0})},$$

where $a_0 = 2 \times 10^5$ and $\omega_0 = 10\pi$ rad/s (corresponding to 5 Hz).

Its response to a 5 μV step input voltage is an exponential function

$$g(t)_{\text{openloop}} = (1 - e^{-t/\tau_0})u(t)$$

going from 0 to a final steady-state value of 1 volt with a time constant of $\tau_0 = 1/(10\pi) \approx 31.8$ msec. The 50% delay and the 10% to 90% rise time in the step response are measured from the plot (generated by Matlab). They are 22 msec, and 70 msec, respectively. These are called the open-loop specifications.

Closed-loop

The 741 is configured as in Figure 9.40(a) with a resistive feedback to its inverting input. The closed-loop system function was earlier found to be

$$H(s) = \frac{A(s)}{1 + \beta A(s)} \approx \frac{a_1}{1 + (\frac{s}{\omega_1})}, \quad \text{where } a_1 = \frac{1}{\beta} \text{ and } \omega_1 = a_0\beta\omega_0$$

As an example, let $R_1 = 1\ K\Omega$ and $R_2 = 9\ K\Omega$. Then $\beta = 0.1$ and the closed-loop specifications become

System function: $$H(s) = \frac{10}{1 + (\frac{s}{\omega_1})} \quad \omega_1 = 200{,}000\pi$$

Response to a 100-mV step: $$g(t)_{\text{closedloop}} = (1 - e^{-t/\tau_1})u(t),$$
$$\text{where } \tau_1 = 1/(200{,}000\pi) = 1.59\mu s$$

The 50% delay in the closed-loop step response is now 1.1 μsec. Similarly, the 10% to 90% rise time is 3.5 μsec. These constitute specifications of the closed-loop circuit and are in agreement with theoretical expectations from the feedback. The delay and rise time in the step response are reduced by a factor of 20,000. The feedback reduced the DC gain and increased the bandwidth by that factor.

Feedback: Effect on Pole-Zero Configuration

A control system is made of a sensor in the feedback path and a controller in the forward path, as shown in Figure 9.38. Using these elements one can modify the pole-zero configuration of the system. This will change the system's response in both time and frequency domains. [In the time domain, the aim may be to shape the step response toward a desired form; (e.g., reduce oscillations and rise time, and eliminate overshoot.)] The following example shows the use of a controller to eliminate a 60% overshoot in the step response of a system.

Example 9.48

Consider the system

$$A(s) = \frac{Y(s)}{X(s)} = \frac{177}{s^2 + 4.25s + 177}$$

a. Find and plot its poles and unit-step response. Verify that the step response has a 60% overshoot.

b. To reduce the rise time and the overshoot, it is recommended that we place the system in the feedback control loop of Figure 9.37 containing the PID controller

$$C(s) = k_p + \frac{k_i}{s} + k_d s$$

where k_p, k_i, and k_d are constants called controller parameters. Using an approach of your choice, determine the controller parameters such that the overshoot/undershoot of the unit-step response and its rise time are reduced, making it closer to a step. Suggestion: Start with $k_p = 4.1$, $k_i = 3.8$, and $k_d = 1.5$.[13]

c. Improve the controller by fine-tuning it.

Solution

a. Poles are found from

$$s^2 + 4.5s + 177 = 0, \quad p_{1,2} = -2.25 \pm j13.11 = \sqrt{177}e^{\pm j99.73°}$$

The poles are at an angle close to the $j\omega$ axis, giving rise to a 60% overshoot in the system's unit-step response. The poles and the unit-step response of the open-loop system are shown in Figure 9.43(a). The 60% overshoot is seen and measured in the figure.

b. We start with the suggested controller $C_1(s)$, resulting in the closed-loop system function $H_1(s)$.

$$C_1(s) = 4.1 + \frac{3.8}{s} + 1.5s = \frac{1.5s^2 + 4.1s + 3.8}{s}$$

$$H_1(s) = \frac{C_1(s)A(s)}{1 + C_1(s)A(s)} = \frac{265.5s^2 + 725.7s + 672.6}{s^3 + 269.75s^2 + 902.7s + 672.6}$$

The closed-loop system has three negative real poles at -266.37, -2.26, and -1.11, a pair of complex conjugate zeros at $-1.37 \pm j0.82$, and a unity DC gain. The pole-zero configuration and the step response of the closed-loop system are shown in Figure 9.43(b). The rise time in the step response is greatly reduced and the overshoot is totally eliminated. An undershoot of about 15% is present.

c. To reduce deviations of the step response from the step function, we fine-tune the controller by trial and error and choose the following controller which results in the closed-loop system function given below.

$$C_2(s) = 200 + \frac{200}{s} + 20s = 20\frac{s^2 + 10s + 10}{s}$$

$$H_2(s) = \frac{C_2(s)A(s)}{1 + C_2(s)A(s)} = 3540\frac{s^2 + 10s + 10}{s^3 + 3,544s^2 + 35,577s + 35,400}$$

---

[13]See some papers on PID controller design, for example, those by Ziegler, Nichols, and Phillips.

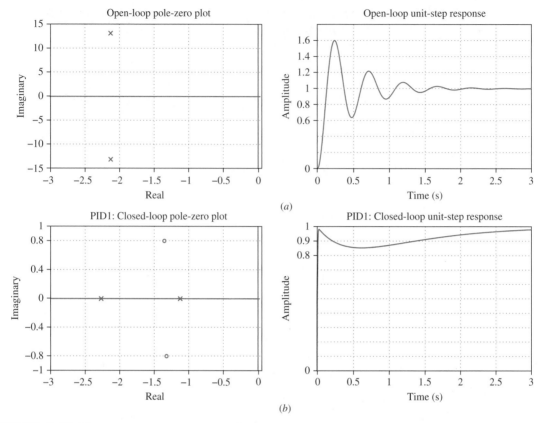

FIGURE 9.43 *(a)* The poles of $A(s) = 177/(s^2 + 4.25s + 177)$ at $-2.125 \pm j13.13$ produce a 60% overshoot in its step response.
(b) With PID1 controller $C_1(s) = 4.1 + 3.8/s + 1.5s$ (Figure 9.37) the closed-loop system function becomes

$$H_1(s) = \frac{C_1(s)A(s)}{1 + C_1(s)A(s)} = \frac{265.5s^2 + 725.7s + 672.6}{s^3 + 269.75s^2 + 902.7s + 672.6}$$

with poles at -266.37, -2.26, -1.11, and zeros at $-1.37 \pm j0.82$, with a reduced rise time and $\approx 15\%$ undershoot.

The closed-loop system has three negative real poles at $-3{,}534.2$, -8.9, -1.1, two negative real zeros at -8.873, -1.127, and a unity DC gain. The pole at $-3{,}534.2$ is far away from the origin and has very small effect on the system's step response (except at near transition time $t = 0$ where very high frequencies are developed). The other two poles at 8.9 and -1.1 are nearly cancelled by the system's zeros at -8.873 and -1.127. This makes the unit-step response very close to a unit step, within a deviation limited

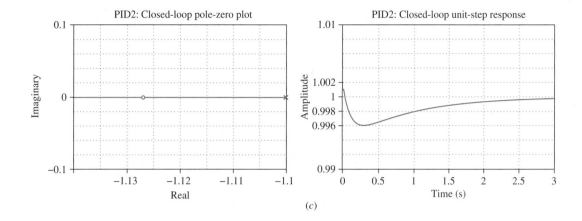

FIGURE 9.43 *(c)* Fine-tuning the controller to PID2 with $C_2(s) = 200 + \frac{200}{s} + 20s$ results in

$$H_2(s) = \frac{C_2(s)A(s)}{1 + C_2(s)A(s)} = 3{,}540 \frac{s^2 + 10s + 10}{s^3 + 3{,}544s^2 + 35{,}577s + 35{,}400}$$

with poles at -3534.2, -8.9, -1.1, and zeros at -8.873, -1.127, which nearly cancel two of the poles. With the third pole being far away, the step response becomes a step within a deviation of $\pm 0.3\%$. See Example 9.48.

to $\pm 0.3\%$. The pole-zero configuration and the step response of the closed-loop system with the above controller are shown in Figure 9.43(c).

9.29 Pole-Zero Cancellation

In combining two or more systems, a pole may be removed due to introduction of a zero. This is called pole-zero cancellation. A simple example is to place a system $(s - s_0)H_1(s)$ in series with another system $H_2(s)/(s - s_0)$. The first system has a zero and the second system has a pole at s_0. The system function of their cascaded combination is

$$H(s) = (s - s_0)H_1(s) \times \frac{H_2(s)}{s - s_0} = H_1(s) \times H_2(s)$$

The zero and the pole at s_0 have cancelled each other. This phenomenon may be shown both in the time and s-domains, as illustrated in Examples 9.49 and 9.50.

Example **9.49**

Pole-zero cancellation in an input impedance

a. Apply a unit-step voltage to the terminals of the circuit of Figure 9.44(a) and find the currents in the inductor, the capacitor, and the total current entering the circuit.

b.   Find conditions that the circuit elements must satisfy for the total current to be a step function.

c.   Show that under the conditions found in part b, the circuit performs as a resistor for all frequencies. Find the equivalent resistor.

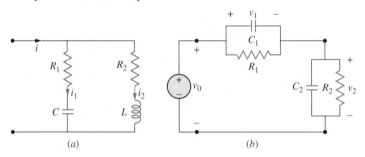

(a)   (b)

FIGURE 9.44 Pole-zero cancellation. **(a)** From its terminals the circuit appears as a resistor at all frequencies if $R_1 = R_2 = \sqrt{\frac{L}{C}} = R$. Then $Z = R$. See Example 9.49. **(b)** The circuit functions as a resistive voltage divider at all frequencies if $R_1 C_1 = R_2 C_2$. Then $v_2 = R_2 v_0 / (R_1 + R_2)$. See Example 9.50.

Solution

a.   The time constants of the capacitive and the inductive branches of the circuit are, $R_1 C$ and L/R_2, respectively. The responses of the circuit to an applied unit-step voltage are

$$i_1 = \frac{1}{R_1} e^{-\frac{t}{R_1 C}} u(t)$$

$$i_2 = \frac{1}{R_2} \left[1 - e^{-\frac{R_2}{L} t} \right] u(t)$$

$$i = i_1 + i_2 = \frac{1}{R_1} e^{-\frac{t}{R_1 C}} u(t) + \frac{1}{R_2} \left[1 - e^{-\frac{R_2}{L} t} \right] u(t)$$

b.   For the total current to be a unit-step function one needs

$$\begin{cases} R_1 = R_2 \\ R_1 C = L/R_2 \end{cases} \quad \Longrightarrow \quad R_1 = R_2 = \sqrt{\frac{L}{C}}$$

c.   Using the impedance (or admittance) concept in the s-domain we can write

Impedance of the series RC branch $\quad Z_1 = R_1 + \dfrac{1}{Cs} = \dfrac{R_1 Cs + 1}{Cs}$

Impedance of the series RL branch $\quad Z_2 = R_2 + Ls$

Admittance of the RC and RL branches in parallel: $\quad Y = \dfrac{1}{Z_1} + \dfrac{1}{Z_2} = \dfrac{Cs}{R_1 Cs + 1} + \dfrac{1}{R_2 + Ls}$

$$Y = \frac{1}{R_1} \times \frac{LCs^2 + (R_1 + R_2)Cs + 1}{LCs^2 + (R_2 C + L/R_1)s + R_2/R_1}$$

For Y to be a constant, the numerator should cancel the denominator. This happens when

$$\begin{cases} R_1 = R_2 \\ (R_1 + R_2)C = R_2C + L/R_1 \end{cases} \implies R_1 = R_2 = \sqrt{\frac{L}{C}}$$

In summary, the following conditions on element values make the two-terminal circuit equivalent to a resistor at all frequencies.

$$R_1 = R_2 = \sqrt{\frac{L}{C}} = RZ = R$$

Pole-zero cancellation in transfer functions

In the circuit of Figure 9.44(*b*) obtain the system functions V_1/V_0 and V_2/V_0, where v_0 is the input, and v_1 and v_2 are two outputs. Find condition(s) that the circuit elements must satisfy to make $v_1(t) = \alpha v_0(t)$ and $v_2(t) = \beta v_0(t)$ for any arbitrary input. Then find α and β. Show that under those conditions the circuit behaves as a resistive voltage divider for all frequencies.

Solution

We find V_1/V_0 and V_2/V_0 from voltage division in the complex frequency domain

$$Z_1 = \frac{R_1}{1 + R_1C_1s} \quad \text{and} \quad Z_2 = \frac{R_2}{1 + R_2C_2s}$$

$$\frac{V_1}{V_0} = \frac{Z_1}{Z_1 + Z_2} = \frac{R_1(1 + R_2C_2s)}{R_1(1 + R_2C_2s) + R_2(1 + R_1C_1s)} = \frac{R_1(1 + \tau_2 s)}{R_1(1 + \tau_2 s) + R_2(1 + \tau_1 s)}$$

$$\frac{V_2}{V_0} = \frac{Z_2}{Z_1 + Z_2} = \frac{R_2(1 + R_1C_1s)}{R_1(1 + R_2C_2s) + R_2(1 + R_1C_1s))} = \frac{R_2(1 + \tau_1 s)}{R_1(1 + \tau_2 s) + R_2(1 + \tau_1 s)}$$

where $\tau_1 = R_1C_1$ and $\tau_2 = R_2C_2$ are time constants of Z_1 and Z_2, respectively. For $v_1(t) = \alpha v_0(t)$ and $v_2(t) = \beta v_0(t)$, we need $\tau_1 = \tau_2$ and

$$R_1C_1 = R_2C_2$$

$$\frac{V_1}{V_0} = \alpha = \frac{R_1}{R_1 + R_2}$$

$$\frac{V_2}{V_0} = \beta = \frac{R_2}{R_1 + R_2}$$

Note that $\alpha + \beta = 1$ or, equivalently, $v_1(t) + v_2(t) = v_0(t)$.

9.30 Inverse Systems

Let a signal $x(t)$ pass through a system with the unit-impulse response $h(t)$, producing the output $y(t) = x(t) \star h(t)$. Passing $y(t)$ through a second system with the unit-impulse response $\hat{h}(t)$, such that $h(t) \star \hat{h}(t) = \delta(t)$, will reverse the effect of the first system resulting in the output $x(t)$:

$$y(t) \star \hat{h}(t) = [x(t) \star h(t)] \star \hat{h}(t) = x(t) \star [h(t) \star \hat{h}(t)] = x(t) \star \delta(t) = x(t)$$

See Figure 9.45. The two systems are said to be inverses of each other. In a sense, the inverse system performs a deconvolution of $x(t)$ and $h(t)$. If the two systems are causal, their system functions will be $H(s)$ and $1/H(s)$, respectively. Then

$$H(s) \times \frac{1}{H(s)} = 1$$

which correspondes to the overall unit-impulse response $\delta(t)$. In summary

$$
\begin{array}{ccc}
h(t) & \Longleftrightarrow & H(s) \\
\hat{h}(t) & \Longleftrightarrow & \frac{1}{H(s)} \\
h(t) \star \hat{h}(t) = \delta(t) & \Longleftrightarrow & H(s) \times \frac{1}{H(s)} = 1
\end{array}
$$

$\xrightarrow[\text{Input}]{X(s)}$	$H(s)$	$\xrightarrow{X(s)\,H(s)}$	$\dfrac{1}{H(s)}$	$\xrightarrow[\text{Output}]{X(s)}$
	System		Inverse system	

FIGURE 9.45 Two systems $h(t)$ and $\hat{h}(t)$ are inverses of each other if $h(t) \star \hat{h}(t) = \delta(t)$. One system reverses the effect of the other. For causal systems this translates into system functions $H(s)$ and $1/H(s)$, respectively.

Example

9.51

The series CR in section a of Figure 9.46 with $C = 1\ \mu$F and $R = 1\ \text{M}\Omega$ is a high-pass filter. Apply an input voltage to it. The filter's output is the voltage across the resistor. Connect an inverse filter in series with the above to reverse its effect and determine its element values.

Solution

The unit-impulse response and system functions of the high-pass and inverse filters are

$$h(t) = \delta(t) - e^{-t}u(t) \quad \Longrightarrow \quad H(s) = 1 - \frac{1}{s+1} = \frac{s}{s+1}$$

$$\Big\Downarrow$$

$$\hat{h}(t) = \delta(t) + u(t) \quad \Longleftarrow \quad \frac{1}{H(s)} = 1 + \frac{1}{s}$$

The inverse filter is realized in section c of Figure 9.46. It amplifies, with an appropriate gain, the low-frequency components of the signal which were attenuated by the high-pass filter. Note that in the time domain we confirm the same result as in the frequency domain.

$$h(t) \star \hat{h}(t) = \left[\delta(t) - e^{-t}u(t) \right] \star [\delta(t) + u(t)]$$

$$= \delta(t) \star [\delta(t) + u(t)] - e^{-t}u(t) \star [\delta(t) + u(t)]$$

$$= \delta(t) + u(t) - e^{-t}u(t) + e^{-t}u(t) \star u(t)$$

$$= \delta(t) + u(t) - e^{-t}u(t) + e^{-t}u(t) - u(t)$$

$$= \delta(t)$$

The cascade of the high-pass filter and its inverse is shown in Figure 9.46. The voltage follower between the high-pass RC filter and its inverse acts as a buffer.

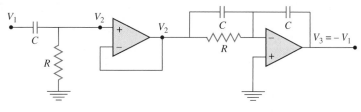

(a) High-pass filter   (b) Buffer   (c) Inverse of high-pass filter

FIGURE 9.46 A high-pass CR filter with the system function $H(s) = s/(s+\alpha)$ (shown in section a) is cascaded with its inverse with the system function $1/H(s) = 1 + \alpha/s$ (section c). The voltage follower (section b) provides a no-load buffer between the filter and its inverse.

*E*xample
9.52

A first-order active low-pass filter (with time constant $\tau = RC$) is cascaded with its inverse. See Figure 9.47. Show that the overall system function is 1.

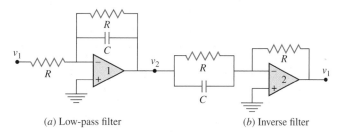

(a) Low-pass filter   (b) Inverse filter

FIGURE 9.47 An active low-pass filter with the system function $H_1(s) = -1/(s + \alpha)$ (shown in section a) is cascaded with its inverse $H_2(s) = 1/H_1(s) = -(s + \alpha)$ (section b). The overall system function is $H(s) = H_1(s) \times H_2(s) = 1$.

Solution

The op-amps are configured in inverting circuit configuration. Their system functions and the overall system functions are

$$H_1(s) = \frac{-1}{s+\alpha}, \quad \text{where } \alpha = \frac{1}{RC} = \frac{1}{\tau}$$

$$H_2(s) = -(s+\alpha)$$

$$H(s) = H_1(s) \times H_2(s) = \left[\frac{-1}{s+\alpha}\right] \times [-(s+\alpha)] = 1$$

See also discussions on deconvolution and inverse systems in the discrete-time systems.

9.31 Concluding Remarks

Performance of an LTI system may be analyzed in the time domain (using its unit-impulse response, unit-step response, convolution, and the differential equation). The analysis may also be done in the frequency domain (using the Laplace transform, system function, and the frequency response). This chapter highlighted the relationship between the above methods, emphasized the role of poles and zeros in building bridges between them, and integrated those analysis methods. In addition to quantitative methods, the chapter also provided some qualitative insights into the methods of analysis and design. It presents a capstone chapter for several topics of interest in system analysis.

9.32 Problems

Solved Problems

1. In the circuit of Figure 9.48(a), $R = 1\ k\Omega$, $L = 1$ H, and $C = 1{,}000\ \mu F$.

   a. Verify that the system function is

$$H(s) = \frac{V_2}{V_1} = \frac{s}{s^2 + s + 1{,}000}$$

   b. Plot the Bode diagram.

   c. Let $v_1(t)$ be a 5-Hz periodic square pulse with a base-to-peak value of 1 V. Using the Bode diagram find the average power in $v_2(t)$ within the band of $f < 30$ Hz.

Solution

Assume $x(t) = v_1(t)$ is the input to the system and $y(t) = v_2(t)$ the output.

   a.

$$H(s) = \frac{V_2}{V_1} = \frac{Z_{LC}}{R + Z_{LC}}$$

$$Z_{LC} = \frac{1}{Cs + 1/(Ls)} = \frac{Ls}{1 + LCs^2}$$

$$H(s) = \frac{Ls}{R(1 + LCs^2) + Ls} = \frac{1}{RC}\frac{s}{s^2 + \frac{1}{RC}s + \frac{1}{LC}}$$

$$RC = 1, \quad LC = 10^{-3}$$

$$H(s) = \frac{s}{s^2 + s + 1{,}000}$$

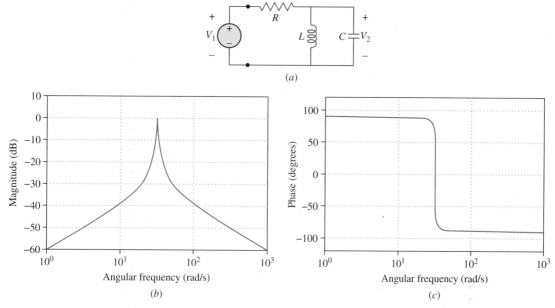

FIGURE 9.48 A narrowband RLC bandpass filter and its frequency characteristics. **(a)** The circuit. **(b)** Magnitude plot. **(c)** Phase plot. The high $Q = 31.6$ results in a narrowband $\Delta f = 0.159$ Hz centered at $f_0 = 5.033$ Hz. The filter is not exactly tuned to the frequency of a 5-Hz square wave. The narrow bandwidth attenuates the principal harmonic of the square wave input, hence reducing average power in $v_2(t)$.

b. $H(\omega) = \dfrac{j\omega}{-\omega^2 + j\omega + 1{,}000} = \dfrac{\frac{j}{Q}\left(\frac{\omega}{\omega_0}\right)}{-\left(\frac{\omega}{\omega_0}\right)^2 + \frac{j}{Q}\left(\frac{\omega}{\omega_0}\right) + 1}$,

where $Q = \omega_0$ and $\omega_0^2 = 1{,}000$. Hence, $\omega_0 = 31.62$ rad/s or $f_0 = \omega_0/(2\pi) = 5.033$ Hz.
The system has a quality factor of $Q = \sqrt{1{,}000} = 31.62$ and a bandwidth of $\Delta\omega = \frac{\omega_0}{Q} = 1$ rad/s or $\Delta f = 0.159$ Hz. The low-frequency asymptote (at $\omega \ll \omega_0$) is

$$20\log|H(\omega)| \approx 20\log\left(\frac{\omega}{\omega_0}\right) - 20\log Q = 20\log\left(\frac{\omega}{\omega_0}\right) - 30$$

The peak value (at $\omega = \omega_0$) is $20\log|H(\omega)| = 0$ dB. The high-frequency asymptote (at $\omega \gg \omega_0$) is

$$20\log|H(\omega)| \approx -20\log\left(\frac{\omega}{\omega_0}\right) - 20\log Q = -20\log\left(\frac{\omega}{\omega_0}\right) - 30$$

The low- and high-frequency asymptotes have slopes of ± 20 dB/decade, respectively. They intercept ω_0 at $20\log Q = 30$ dB below the peak. The Bode diagram is shown on Figure 9.48.
c. Harmonics of $x(t)$ are at $f_n = 5n$ Hz or $\omega_n = 10\pi n$ with

$$X_n = \begin{cases} \frac{1}{2} & n = 0 \\ \frac{\sin\left(\frac{\pi n}{2}\right)}{\pi n} & n \neq 0 \end{cases}.$$

Harmonics of the output $y(t)$ are $Y_n = X_n H(10\pi n)$. They are summarized below along with their power.

n	0	1	2	3	4	5	6
f (Hz)	0	5	10	15	20	25	30
X_n	$\dfrac{1}{2}$	$\dfrac{1}{\pi}$	0	$-\dfrac{1}{3\pi}$	0	$\dfrac{1}{5\pi}$	0
$\|H\|$ (dB)	$-\infty$	-0.69		-38.45		-43.56	
$\|H\|$ (linear)	0	0.9236		0.0119		0.0066	
$\|Y_n\| = \|X_n\| \times \|H\|$	0	0.294	0	0.0013	0	0.0004	0
$P_n = 2\|Y_n\|^2$	0	0.17286	0	0.0000034	0	0.00000032	0

$$P = \sum_{n=-5}^{5} P_n \approx P_1 = 0.17286 \text{ W}$$

A simplifying approximation of $f_0 \approx 5$ Hz would assume 0-dB attenuation at 5 Hz. It would then result in

$$P \approx \frac{2}{\pi^2} = 0.20264, \text{ and a percentage error of } \frac{20{,}264 - 17{,}286}{17{,}286} \times 100\% = 17.23\%$$

The resulting high percentage error is due to the narrow bandwidth of $\Delta f = 0.159$ at $f_0 = \omega_0/(2\pi) = 5.033$ Hz. The Matlab commands given below plot the Bode diagram shown in Figure 9.48.

```
num=[1 0];
den=[1 1 1000];
sys=tf(num,den)
figure(1)
grid
bode(sys)
title('Bode plot of H= s/(s^2+s+1000)')
```

2. In the circuit of Figure 9.48(a), $R = 1\ k\Omega$, $L = 1$ H, and $C = 250\ \mu$F.

   a. Verify that the system function is $H(s) = \dfrac{V_2}{V_1} = \dfrac{4s}{s^2 + 4s + 4{,}000}$

   b. Write a program to plot the Bode diagram.

   c. Let $v_1(t)$ be a 10-Hz periodic square pulse with a 1-V base-to-peak value. Using the Bode diagram find the average power in $v_2(t)$ within the band of $f < 60$ Hz. Compare the results with those of problem 1.

Solution

As in problem 1 assume $x(t) = v_1(t)$ is the input to the system and $y(t) = v_2(t)$ the output.

   a. $H(s) = \dfrac{V_2}{V_1} = \dfrac{1}{RC} \dfrac{s}{s^2 + \frac{1}{RC}s + \frac{1}{LC}}$, $RC = 0.25$, $LC = 0.25 \times 10^{-3}$. Hence, $H(s) = \dfrac{4s}{s^2 + 4s + 4{,}000}$.

   b. $H(\omega) = \dfrac{j4\omega}{-\omega^2 + j4\omega + 4{,}000} = \dfrac{\frac{j4}{\omega_0}\left(\frac{\omega}{\omega_0}\right)}{-\left(\frac{\omega}{\omega_0}\right)^2 + \frac{j4}{\omega_0}\left(\frac{\omega}{\omega_0}\right) + 1} = \dfrac{\frac{j}{Q}\left(\frac{\omega}{\omega_0}\right)}{-\left(\frac{\omega}{\omega_0}\right)^2 + \frac{j}{Q}\left(\frac{\omega}{\omega_0}\right) + 1}$,

   where $Q = \frac{\omega_0}{4}$ and $\omega_0^2 = 4{,}000$. Therefore, $\omega_0 = 63.25$ rad/s or $f_0 = \omega_0/(2\pi) = 10.066$ Hz. The system has a quality factor of $Q = \sqrt{4{,}000}/4 = 15.81$ and a bandwidth of $\Delta\omega = \frac{\omega_0}{Q} = 4$ rad/s or $\Delta f = 0.76$ Hz. The low-frequency asymptote (at $\omega \ll \omega_0$) is

$$20 \log |H(\omega)| \approx 20 \log \left(\frac{\omega}{\omega_0}\right) - 20 \log Q = 20 \log \left(\frac{\omega}{\omega_0}\right) - 23.98$$

The peak value (at $\omega = \omega_0$) is $20 \log |H(\omega)| = 0$ dB. The high-frequency asymptote (at $\omega \gg \omega_0$) is

$$20 \log |H(\omega)| \approx -20 \log \left(\frac{\omega}{\omega_0}\right) - 20 \log Q = -20 \log \left(\frac{\omega}{\omega_0}\right) - 23.98$$

The low- and high-frequency asymptotes have slopes of ± 20 dB/decade, respectively. They intercept ω_0 at $20 \log Q = 23.98$ dB below the peak. The following modification in the Matlab program of problem 1 plots the Bode diagram.

```
num=[4 0];
den=[1 4 4000];
```

The plots are similar to the diagram for $H(s) = s/(s^2 + s + 1,000)$ (shown in Figure 9.48) in that it shows a bandpass system with a relatively high Q of 15.81. But it differs from Figure 9.48 in that the location of the peak is shifted to $f_0 = 10.066$ Hz with a broader bandwidth of $\Delta f = 0.76$ Hz. The effect of the broader bandwidth will be seen shortly in part c.

c. Harmonics of $x(t)$ are at $f_n = 10n$ Hz or $\omega_n = 20\pi n$ with

$$X_n = \begin{cases} \frac{1}{2} & n = 0 \\ \frac{\sin(\frac{\pi n}{2})}{\pi n} & n \neq 0 \end{cases}$$

Harmonics of the output $y(t)$ are $Y_n = X_n H(20\pi n)$. They are summarized below along with their power.

n	0	1	2	3	4	5	6
f (Hz)	0	10	20	30	40	50	60
X_n	$\frac{1}{2}$	$\frac{1}{\pi}$	0	$-\frac{1}{3\pi}$	0	$\frac{1}{5\pi}$	0
$|H|$ (dB)	$-\infty$	-0.1831		-32.4299		-37.5432	
$|H|$ (linear)	0	0.9791		0.0239		0.0133	
$|Y_n| = |X_n| \times |H|$	0	0.3117	0.0000	-0.0025	-0.0000	0.0008	0.0000
$P_n = 2|Y_n|^2$	0	0.1943	0.0000	0.0000	0.0000	0.0000	0.0000

$$P = \sum_{n=-5}^{5} P_n \approx P_1 = 0.1943 \text{ W}$$

A simplifying approximation of $f_0 \approx 10$ Hz would assume 0-dB attenuation at 10 Hz. It would then result in

$$P \approx \frac{2}{\pi^2} = 0.20264, \text{ and a percentage error of } \frac{20,264 - 19,430}{19,430} \times 100\% = 4.2\%.$$

3. In the circuit of Figure 9.49,

a. find the system function $H(s) = \frac{V_2}{V_1}$

b. determine element values that make it a second-order Butterworth filter with a 3-dB frequency at 1 kHz.

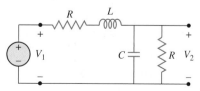

FIGURE 9.49 A passive second-order low-pass filter.

Solution

a. $Z_1 = R + Ls$

b. $\dfrac{2}{LC} = \omega_0^2 = (2\pi \times 10^3)^2$

$Z_2 = \dfrac{R}{1 + RCs}$

$\dfrac{R}{L} + \dfrac{1}{RC} = \sqrt{2}\omega_0 = 2\sqrt{2}\pi \times 10^3$

$\dfrac{V_2}{V_1} = \dfrac{Z_2}{Z_2 + Z_1}$

For a Butterworth filter $H(s) = k\dfrac{\omega_0^2}{s^2 + \sqrt{2}\omega_0 s + \omega_0^2}$

$\quad = \dfrac{R}{(R + Ls)(1 + RCs) + R}$

For example $R = 1k\Omega$

$\quad = \dfrac{1}{LC}\dfrac{1}{s^2 + \left(\dfrac{R}{L} + \dfrac{1}{RC}\right)s + \dfrac{2}{LC}}$

$L = 1/(\sqrt{2}\pi)\text{H} \approx 225\text{ mH}, C = 1/(\sqrt{2}\pi)\ \mu\text{F} \approx 225\text{ nF}$

Note: The filter is passive. Its DC gain is -6 dB.

4. **Graphing a system's responses by computer.** Qualitatively analyze the following second-order systems. Plot the pole locations, the frequency response, and the unit-impulse response. (This problem revisits Example 9.19 in this chapter.)

$$H_1(s) = \frac{1}{s^2 + 5s + 1} \quad \text{(with } \omega_0 = 1 \text{ and } Q = 0.2\text{)}$$

$$H_2(s) = \frac{1}{s^2 + 0.1s + 1} \quad \text{(with } \omega_0 = 1 \text{ and } Q = 10\text{)}$$

Solution

$H_1(s)$ has two negative real poles at -4.7913 and -0.2087. It is heavily overdamped (damping ratio $\zeta = 2.5$). The unit-impulse response shows no oscillations and the magnitude plot of the frequency response remains below the DC level at all frequencies. In contrast, $H_2(s)$ has two complex conjugate poles at $-0.05 \pm j0.9987 = e^{\pm j116.6°}$ (at an angle close to the $j\omega$ axis). The system is greatly underdamped (damping ratio $\zeta = 0.05$). This is evidenced by oscillations in its unit-impulse response and also a sharp and selective magnitude response around $\omega = 1$. Pole plots, Bode plots, and step responses are shown in Figure 9.15. The Matlab program is given below.

```
t=linspace(0,40,1001);
w=logspace( -2,3,801);

%Second-order with Q=0.2 (overdamped, functions as a lowpass)
num1=[1];
den1=[1 5 1];
[poles1,zeros1]=pzmap(num1,den1)
[mag1,angle1]=Bode(num1,den1,w);
y1 =impulse(num1,den1,t);

%Second-order with Q=10 (underdamped, functions as a bandpass)
num2=[1];
den2=[1  0.1 1];
[poles2,zeros2]=pzmap(num2,den2)
[mag2,angle2]=Bode(num2,den2,w);
y2 =impulse(num2,den2,t);
```

```
%Plots

figure(1)
plot(real(poles1),imag(poles1),'x',real(zeros1),imag(zeros1),'o',...
     real(poles2),imag(poles2),'x',real(zeros2),imag(zeros2),'o');
title('Pole-Zero Plot');
xlabel('Real');
ylabel('Imaginary');
axis([-5.1  0.05 -1.1  1.1]);
grid;

figure(2)
semilogx(w,20*log10(mag1),w,20*log10(mag2));
title('Magnitude of Bode Plot');
ylabel('Magnitude (dB)');
xlabel('Angular Frequency (rad/s)');
axis([0.01 100  -80 20]);
grid;

figure(3)
semilogx(w,angle1,w,angle2);
title('Phase of Bode Plot');
xlabel('Angular Frequency (rad/s)');
ylabel('Angle (deg)');
axis([0.01 100   -180  0]);
grid;

figure(4)
plot(t,y1,t,y2);
title('Impulse Response');
xlabel('Time (s)');
ylabel('Amplitude');
grid;
```

5. From a visual inspection of the Bode diagram plotted for the system

$$H(s) = 1,000\frac{s+3}{(s+15)(s+500)}$$

it may appear that the system function has a flat magnitude and linear phase within the neighborhood of $60 < \omega < 100$. If the assumption is correct, a signal with frequency components within that neighborhood should experience only a delay (and no distortion) when it passes through the system. Test the validity of the above assumption by passing the signal $x(t) = \cos(20\pi t) + \cos(25\pi t) + \cos(30\pi t)$ through the above system and examining the output.

Solution

The sinusoidal input-output relationship is

$$\cos(\omega t) \implies |H(\omega)| \cos(\omega t + \theta) = |H(\omega)| \cos \omega(t + \tau), \quad \text{where } \theta = \angle H \text{ and } \tau = \theta/\omega$$

Values of $|H|$, $\angle H$, and τ are computed for $f = 10$, 12.5, and 15 Hz and entered in the table below.

f in Hz	$|H|$	θ in rad	τ in msec
10 Hz	1.93	0.0616	0.98
12.5 Hz	1.94	−0.0053	−0.067
15 Hz	1.94	−0.0603	−0.64

$$y(t) \approx 1.94 \left[\cos 20\pi(t + 0.98 \text{ msec}) + \cos 25\pi(t - 0.067 \text{ msec}) + \cos 30\pi(t - 0.64 \text{ msec})\right]$$

The assumption is, therefore, not true because the three sinusoids experience different delays.

6. The circuit of Figure 9.24(a) is a Sallen-Key low-pass filter.

a. Show that

$$\frac{V_2}{V_1} = \frac{H_0}{s^2 + \frac{\omega_0}{Q}s + \omega_0^2}, \quad \text{where } H_0 = \frac{k}{mnR^2C^2}, \quad k = 1 + \frac{R_2}{R_1}, \quad \omega_0 = \frac{1}{mnRC}, \quad \text{and } Q = \frac{\sqrt{mn}}{1 + m + (1 - k)mn}$$

b. Compare with the equal-component circuit analyzed in Example 9.33.

Solution

a. From voltage division by the $R_1 R_2$ feedback path we have

$$V_A = \frac{V_2}{1 + \alpha} \quad \text{where } \alpha = \frac{R_2}{R_1} \quad (1)$$

From KVL along the path B-A-$ground$ we obtain

$$V_A = \frac{V_B}{1 + \tau s}, \quad \text{where } \tau = RC \quad (2)$$

By combining (1) and (2) we find

$$V_B = \frac{1 + \tau s}{1 + \alpha} V_2 \quad (3)$$

Now write KCL at node B and collect terms:

$$V_B \left[\frac{1}{m} + (1 + n\tau s) - \frac{1}{1 + \tau s}\right] - \frac{V_1}{m} - V_2 n\tau s = 0 \quad (4)$$

Eliminate V_B between (3) and (4) to obtain

$$\frac{V_2}{V_1} = \left(\frac{1 + \alpha}{mn\tau^2}\right) \frac{1}{s^2 + \frac{1 + m - \alpha mn}{mn\tau}s + \frac{1}{mn\tau^2}}$$

Compare with the low-pass system function

$$\frac{V_2}{V_1} = \frac{H_0}{s^2 + \frac{\omega_0}{Q}s + \omega_0^2}$$

to obtain

$$H_0 = \left(\frac{1 + \alpha}{mn\tau^2}\right), \quad \omega_0 = \frac{1}{\sqrt{mn}\tau} \quad \text{and} \quad Q = \frac{\sqrt{mn}}{1 + m + -\alpha mn}$$

b. Compared with the equal-component circuit of Example 9.33, the above circuit provides an additional degree of freedom to control G_0, ω_0, and Q.

7. The circuit of Figure 9.26(a) is an infinite gain, multiple-feedback bandpass filter. In Example 9.35 we considered the special case of equal-value capacitors. In this problem you will analyze the circuit by allowing for unequal capacitors. Let $R_2 = aR_1$ and $C_2 = bC_1$.

a. Show that

$$H(s) = -H_0 \frac{s}{s^2 + \frac{\omega_0}{Q}s + \omega_0^2}, \quad \text{where} \quad H_0 = \frac{1}{RC}, \quad \omega_0 = \frac{1}{RC\sqrt{ab}}, \quad \text{and} \quad Q = \frac{\sqrt{ab}}{1+b}$$

b. Compare with the case of $b = 1$ analyzed in Example 9.35. Find component values for a bandpass filter with $Q = 100$ at 60 Hz.

Solution

a. Node A is at zero voltage because the noninverting input of the op-amp is grounded. The current in R_2 is V_2/R_2 directed downward toward node A and going through C_2 because the op-amp doesn't draw any current. This results in

$$V_B = -\frac{V_2}{R_2 C_2 s}$$

Now write KCL at node B

$$\left(V_1 + \frac{V_2}{R_2 C_2 s}\right) \frac{1}{R} + \frac{V_2}{R_2} + \left(V_2 + \frac{V_2}{R_2 C_2 s}\right) C_1 s = 0$$

Let $R_1 = R$, $R_2 = aR$, $C_1 = C$, $C_2 = bC$, and collect terms to obtain

$$\frac{V_2}{V_1} = \frac{-RCs}{R^2 C^2 s^2 + \left(\frac{1+b}{ab}\right) RCs + \frac{1}{ab}}, \quad a \geq 0, \ b \geq 0$$

Rewriting it in the familiar form for a bandpass system function results in

$$H(s) = -H_0 \frac{s}{s^2 + \frac{\omega_0}{Q}s + \omega_0^2}, \quad \text{where} \quad H_0 = \frac{1}{RC}, \quad \omega_0 = \frac{1}{RC\sqrt{ab}}, \quad \text{and} \quad Q = \frac{\sqrt{ab}}{1+b}$$

b. It may appear that unequal capacitors provide us with some advantages in choosing component values. This is not always the case. An example is a bandpass filter with $Q = 100$ at 60 Hz. In order to minimize a we need to choose $b = 1$ which results in $a = 40{,}000$ and $RC = 13{,}263 \times 10^{-9}$. We can choose $C = 1\ \mu\text{F}$, $R_1 = 13.263\ \Omega$, and $R_2 = 530.5$ kΩ. But another choice is $C = 100$ nF, $R_1 = 132.63\ \Omega$, and $R_2 = 5.3$ MΩ.

8. Write and execute a Matlab program to plot the Bode diagrams for

$$H(s) = \frac{s^2 - 1}{s^2 + \frac{1}{Q}s + 1}, \quad Q = 0.01, \ 0.05, \ 0.2, \ 0.5, \ \sqrt{2}/2, \ 1, \ 2, \ 5, \ 10, \ 50$$

Explain the shape of the plots in terms of interconnection of two subsystems.

Solution
The following Matlab program superimposes the Bode plots for the given values of Q. The system function is $H = H_{hp} - H_{\ell p}$ where H_{hp} and $H_{\ell p}$ are high-pass and low-pass second-order filters, respectively, having the same quality factor and $\omega_0 = 1$ rad/s.

```
w=logspace(-2,3,801);
Q=[0.01 0.05 0.2 0.5  0.7 1 2 5 10 50];
num=[1 0 -1]; hold on
for i=1:10;
den=[1 1/Q(i) 1];
```

```
Bode(num,den,w);
end
title('Bode Plots for H(s)=(s^2-1)/(s^2+s/Q+1), Q=.01, .05, .2, .5, .7,
1, 2, 5, 10, 50');
grid; hold off
```

9. A two-dimensional ultrasonic-sound-emitting device is composed of four transmitters placed in a two-dimensional plane and operates at 30 kHz. Each transmitter receives a 30-kHz signal phase-shifted by θ_i, $i = 0, 1, 2, 3$. The first-order phase shifter circuit of Figure 9.20(b) is used to produce the shifts. Find the values of R and C corresponding to the following shifts: 0, 105°, 150°, 175°.

Solution

The system function and frequency response of the circuit of Figure 9.20(b) are

$$H(s) = \frac{1 - RCs}{1 + RCs}, \quad H(f) = e^{-j\theta}, \quad \theta = 2\tan^{-1}(2\pi f RC)$$

At $f = 30$ kHz $RC = \frac{1}{2\pi f} \tan\left(\frac{\theta}{2}\right) = 530.5 \times 10^{-8} \tan\left(\frac{\theta}{2}\right)$

$\theta_1 = 105°$, $RC = 530.5 \times 10^{-8} \tan 52.5° = 691 \times 10^{-8}$, $C = 10$ nF, $R = 691\ \Omega$

$\theta_1 = 150°$, $RC = 530.5 \times 10^{-8} \tan 75° = 1980 \times 10^{-8}$, $C = 10$ nF, $R = 1,980\ \Omega$

$\theta_1 = 175°$, $RC = 530.5 \times 10^{-8} \tan 87.5° = 12150 \times 10^{-8}$, $C = 10$ nF, $R = 12,150\ \Omega$

10. **Proof of phase lead and lag location and its value by analytical approach.**

a. Express the phase of the lead network of Figure 9.21(a) in one or more analytical forms.
b. Show that maximum value of the phase is $90° - 2\tan^{-1}\sqrt{\alpha}$ at $\omega_m = \omega_1/\sqrt{\alpha}$.

Solution

a. The system function is

$$H(s) = \alpha \frac{1 + \frac{s}{\omega_1}}{1 + \alpha \frac{s}{\omega_1}}$$

Without loss of generality (and for convenience), we work with the following normalized frequency response in which the variable s represents s/ω_0 for the filter of Figure 9.21(a).

$$H(s) = \alpha \frac{1 + s}{1 + \alpha s}$$

$$H(\omega) = \alpha \frac{1 + j\omega}{1 + j\alpha\omega} = |H(\omega)|e^{j\theta}$$

$$\theta = \theta_1 - \theta_2, \quad \text{where } \theta_1 = \tan^{-1}(\omega) \text{ and } \theta_2 = \tan^{-1}(\alpha\omega)$$

But

$$\tan\theta = \tan(\theta_1 - \theta_2) = \frac{\tan\theta_1 - \tan\theta_2}{1 + \tan\theta_1 \tan\theta_2}$$

Therefore,

$$\tan\theta = \frac{(1 - \alpha)\omega}{1 + \alpha\omega^2}$$

b. To obtain some insight into the variation of θ we need to look at its derivative with respect to ω. And to search for the possible occurrence of a maximum value for θ we need to find out if the derivative can be set to zero.

Sine $\tan\theta$ is a monotonically increasing function of θ we start with its derivative with respect to ω.

$$\frac{d}{d\omega}(\tan\theta) = \frac{d}{d\omega}\left[\frac{(1-\alpha)\omega}{1+\alpha\omega^2}\right] = \frac{1-\alpha\omega^2}{(1+\alpha\omega^2)^2}$$

But

$$\frac{d}{d\omega}(\tan\theta) = \frac{d}{d\theta}(\tan\theta)\frac{d\theta}{d\omega}$$

and

$$\frac{d}{d\theta}(\tan\theta) = \frac{1}{\cos^2\theta} \neq 0$$

Therefore, when

$$1-\alpha\omega^2 = 0, \quad\text{or}\quad \omega = \frac{1}{\sqrt{\alpha}},$$

then

$$\frac{d\theta}{d\omega} = 0.$$

At that point

$$\tan(\theta_M) = \frac{1}{2}\left[\frac{1}{\sqrt{\alpha}} - \sqrt{\alpha}\right]$$

Alternatively,

$$\theta = \tan^{-1}(\omega) - \tan^{-1}(\alpha\omega)$$

At the maximum

$$\theta_M = \tan^{-1}\left(\frac{1}{\sqrt{\alpha}}\right) - \tan^{-1}(\sqrt{\alpha})$$

But

$$\tan^{-1}\left(\frac{1}{\sqrt{\alpha}}\right) = \cot^{-1}(\sqrt{\alpha}) = 90° - \tan^{-1}(\sqrt{\alpha})$$

Therefore,

$$\theta_M = 90° - 2\tan^{-1}(\sqrt{\alpha})$$

Denormalization brings the frequency of θ_M back at $\omega_m = \omega_1/\sqrt{\alpha}$.

Yet Another Form

The maximum phase for the circuit of Figure 9.21(a) is equivalently specified by the following expression (see problem 46 in this chapter):

$$\sin(\theta_M) = \frac{1-\alpha}{1+\alpha}$$

To show the equivalence of the above expression with the previously derived ones, we note that:

$$\cos(\theta_M) = \frac{2\sqrt{\alpha}}{1+\alpha}, \quad \tan(\theta_M) = \frac{1-\alpha}{2\sqrt{\alpha}} = \frac{1}{2}\left[\frac{1}{\sqrt{\alpha}} - \sqrt{\alpha}\right]$$

11. The frequency response of a first-order low-pass filter is $H(f) = 1/(1 + j2\pi\tau f)$, where τ is the time constant of the filter. Find the group and phase delays of the filter at the following frequencies and evaluate them for $\tau = 1$ msec.

a. $f_a = \dfrac{0.1}{2\pi\tau}$

b. $f_b = \dfrac{1}{2\pi\tau}$

c. $f_c = \dfrac{10}{2\pi\tau}$

Solution

$$\theta = -\tan^{-1}(2\pi f\tau), \qquad \frac{d\theta}{df} = -\frac{2\pi\tau}{1 + (2\pi f\tau)^2}$$

$$t_p = -\frac{\theta}{2\pi f} = \frac{\tan^{-1}(2\pi f\tau)}{2\pi f}, \qquad t_g = -\frac{1}{2\pi}\frac{d\theta}{df} = \frac{\tau}{1 + (2\pi f\tau)^2}$$

a. $2\pi f_a\tau = 0.1$ $t_p = \dfrac{\tan^{-1}(0.1)}{0.1}\tau \approx \tau,$ $t_g = \dfrac{\tau}{1 + 0.01} \approx \tau$

b. $2\pi f_b\tau = 1$ $t_p = \dfrac{\pi}{4}\tau \approx 0.785\tau,$ $t_g = 0.5\tau$

c. $2\pi f_c\tau = 10$ $t_p = \dfrac{\tan^{-1}(10)}{10}\tau \approx 0.147\tau,$ $t_g = \dfrac{\tau}{1 + 100} \approx 0.01\tau$

Both delays decrease with frequency, but the group delay decreases faster than the phase delay. For a system with a 1-msec. time constant, the delays are

a. $f_a \approx 15.91$ Hz $t_p \approx 1$ msec $t_g \approx 1$ msec

b. $f_b \approx 159.1$ Hz $t_p = 785\ \mu s$ $t_g = 785\ \mu s$

c. $f_b \approx 1{,}591$ Hz $t_p \approx 147\ \mu s$ $t_g \approx 10\ \mu s$

12. Let a narrowband signal $x(t) = a(t)\cos(2\pi f_0 t)$ with frequencies in the neighborhood of f_0 pass through a filter with the known frequency response $H(f) = |H(f)|e^{j\theta(f)}$. Express $y(t)$ in terms of $x(t)$, t_g, and t_p.

Solution

The Fourier transform of $x(t)$ is

$$X(f) = \frac{A(f + f_0) + A(f - f_0)}{2}$$

Assume that within the signal's bandwidth the frequency response is smooth enough so that $|H(f)|$ may be considered a constant and $\theta(f)$ can be approximated by its first-order Taylor series expansion.

$$|H(f)| \approx |H(f_0)| = H_0, \quad \theta(f) = \theta_0 + \theta_0'(f - f_0) \text{ where } \theta_0 = \theta(f_0) \text{ and } \theta_0' = \frac{d\theta(f)}{df}\bigg|_{f=f_0}$$

The output of the filter is

$$Y(f) = X(f)H(f) \approx \frac{A(f + f_0) + A(f - f_0)}{2} H_0 e^{j\theta_0 + j\theta_0'(f - f_0)} = H_0\frac{A(f + f_0) + A(f - f_0)}{2} e^{j(\theta_0 - \theta_0' f_0)} e^{j\theta_0' f}$$

Substituting for $\theta_0' = 2\pi t_g$ and $\theta_0 = 2\pi f_0 t_p$ we obtain

$$Y(f) = H_0\frac{A(f + f_0) + A(f - f_0)}{2} e^{j2\pi f t_g} e^{j2\pi f_0(t_p - t_g)}, \quad y(t) = H_0 a(t - t_g)\cos\left[2\pi f_0(t - t_p)\right]$$

13. Feedback. The block diagram of a system is shown in Figure 9.50(a).

   a. Determine the system function $H_1 = Y/X$ and find the DC gain.

   b. A disturbance w arrives at the system as shown in Figure 9.50(b). Find the overall system function that includes the contributions from x and w. Note that $Y = H_1 X + H_2 W$. Then, with x and w assumed to be constants, find y as a function of w. Plot y versus w for $x = 1, 2, 3, 4$, and 5.

   c. In order to reduce the effect of w on y, a feedback path β and an amplifier k are added to the system as shown in Figure 9.50(c). Find the new system function. Suggest values for β and k to demonstrate the new system can reduce the effect of w. Then, plot the output y as a function of w.

   d. Choose β and k in Figure 9.50(c) so that the effect of w on y is reduced by a factor of 0.02 compared to Figure 9.50(b).

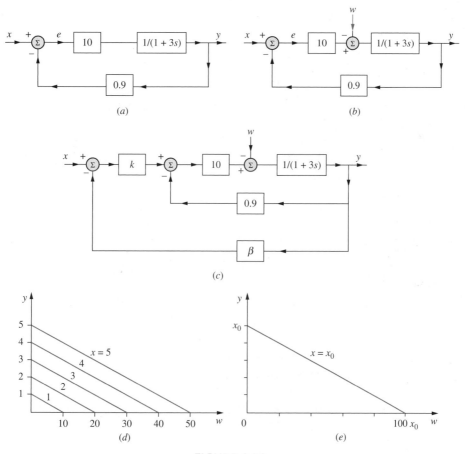

FIGURE 9.50

Solution

a. $H_1(s) = \dfrac{\frac{10}{1+3s}}{1 + \frac{9}{1+3s}} = \dfrac{10}{10 + 3s}$. DC gain $= H(0) = 1$.

b. With $w = 0$, the system is reduced back to Figure 9.50(a) for which we have found $H_1(s) = 10/10 + 3s$. With $x = 0$, the system will be reduced to a forward path of $1/(1 + 3s)$ and a negative feedback path of 9. Due to the negative sign on the arrival point of w, the system function relating y to w becomes

$$H_2(s) = \frac{Y(s)}{W(s)} = -\frac{\frac{1}{1+3s}}{1 + \frac{9}{1+3s}} = -\frac{1}{10 + 3s}$$

With both x and w present, the output is

$$Y(s) = H_1(s)X(s) + H_2(s)W(s) = \frac{10}{10 + 3s}X(s) - \frac{1}{10 + 3s}W(s)$$

At the DC steady state, $s = 0$ and, with both x and w being constants, we get $y = x - \frac{1}{10}w$ [see Figure 9.50(d)].

c. The system function in Figure 9.50(c) is easily obtained as

$$Y(s) = \frac{k}{1 + \beta k}X(s) - \frac{1}{10(1 + \beta k)}W(s)$$

In the DC steady state:

$$y = \frac{k}{1 + \beta k}x - \frac{1}{10(1 + \beta k)}w$$

To keep a unity gain from x to y [as in the system of Figure 9.50(a)], let $k/(1 + \beta k) = 1$, which results in $k = 1/(1 - \beta)$. As long as $\beta k > 1$, the effect of the disturbance in Figure 9.50(c) compared to Figure 9.50(b) is reduced. As an example, let $\beta k = 9$, $k = 10$, and $\beta = 0.9$. Then, at the DC steady state $y = x - 0.01w$. See Figure 9.50(e).

d. $\dfrac{\frac{1}{10(1+\beta k)}}{0.1} = \dfrac{1}{1 + \beta k} = 0.02$, $\beta k = 49$. With $\dfrac{k}{1 + \beta k} = 1$, we get $k = 1 + \beta k = 1 + 49 = 50$ and $\beta = \dfrac{49}{50} = 0.98$.

Chapter Problems

Note: In the following problems circuit elements have constant values, unless specified otherwise.

14. In the series RLC circuit of Figure 9.51(a) $L = 1$ mH, $C = 100$ nF, $v_s(t)$ is the input and element voltages $v_1(t)$, $v_2(t)$, and $v_3(t)$ (across R, L, and C, respectively) are the outputs.

a. Sketch by hand, in the form of Bode diagram, the magnitude and phase of three system functions

$$H_1(s) = \frac{V_1}{V_s}, \quad H_2(s) = \frac{V_2}{V_s}, \quad H_3(s) = \frac{V_3}{V_s}$$

for each of the following three cases

$$i)\ R = 2\ k\Omega, \quad ii)\ R = 200\ \Omega, \quad iii)\ R = 20\ \Omega$$

b. Repeat part a using a computer. Compare with the hand plots.

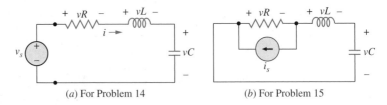

(a) For Problem 14   (b) For Problem 15

FIGURE 9.51 Two circuits whose system functions have the same set of poles.

15. (a) In the series RLC circuit connect a current source $i_s(t)$ in parallel with the resistor as shown in Figure 9.51(b) and consider it to be the input. Let an element's current or voltage be the output (six possible outputs). Apply element values given in problem 14 and use a computer to plot the Bode diagram for each system function representing an input-output pair. Examine the plots and relate your observations to the results of problem 14. (b) Place the input current source in parallel with the inductor and repeat part a. (c) Place the input voltage source in parallel with the capacitor and repeat part a.

16. In the parallel *RLC* circuit of Figure 9.52(a) $L = 1$ mH, $C = 100$ nF, $i_s(t)$ is the input and the element currents $i_1(t)$, $i_2(t)$, and $i_3(t)$ (in the resistor, inductor, and the capacitor, respectively) are the outputs.

a. Sketch by hand, in the form of Bode diagram, the magnitude and phase of three system functions

$$H_1(s) = \frac{I_1}{I_s}, \quad H_2(s) = \frac{I_2}{I_s}, \quad H_3(s) = \frac{I_3}{I_s}$$

for each of the following three cases

$$i) \ R = 5 \ \Omega, \quad ii) \ R = 50 \ \Omega, \quad iii) \ R = 500 \ \Omega$$

b. Repeat part a using a computer. Compare with the hand plots.

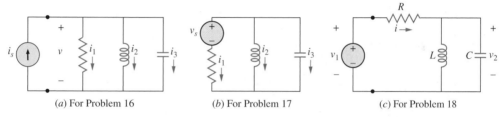

(a) For Problem 16   (b) For Problem 17   (c) For Problem 18

FIGURE 9.52 Three circuits whose system functions have the same set of poles.

17. (a) In the parallel *RLC* circuit connect a voltage source $v_s(t)$ in series with the resistor as shown in in of Figure 9.52(b) and consider it to be the input. Let an element's current or voltage be the output (six possible outputs). Apply element values given in problem 16 and use a computer to plot the Bode digram for each system function representing an input-output pair. Examine the plots and relate your observations to the results of problem 16. (b) Place the input voltage source in series with the inductor and repeat part (a). (c) Place the input voltage source in series with the capacitor and repeat part a.

18. Consider the circuit of Figure 9.52(c) with $R = 500 \ \Omega$, $L = 1$ mH, and $C = 10 \ \mu$F. (a) Plot, by hand, the magnitude and phase of the system function $H(s) = V_2/V_1$ in the form of Bode diagram. (b) The steady-state response of the circuit to $v_1(t) = \cos(\omega t)$ is $v_2(t) = A \cos(\omega t - \theta)$. Using the Bode plot estimate A and θ for $\omega = 10^3$, 5×10^3, 10^4, 2×10^4, and 10^5, all in rad/s and compare with actual values obtained from the system function.

19. The only poles of an LTI system are at $-1 \pm j$. The system has no zeros at finite values of s. Its steady-state response to a unit-step input is 5.

a. Find the system function $H(s)$ and plot its Bode diagram.
b. The response of the system to the input $\cos \sqrt{2}t$ is $A \cos(\sqrt{2}t - \theta)$. Find A and θ.
c. Move the poles to new locations $p_{1,2} = 2e^{\pm j\phi}$ while the DC steady-state response to unit step remains unchanged. The response of the new system to the input $\cos \omega t$ is $A \cos(\omega t - \theta)$, where A and θ are functions of ω. Choose ϕ so that A attains its maximum of 14 dB at $\omega = \sqrt{2}$. Then repeat part a.

20. a. A system $H_1(s)$ has a pair of poles at $-1.25 \pm j50$, a zero at the origin, and $|H_1(\omega)|_{Max} = 1$. Plot its magnitude and phase and find its -3 dB-band.

b. Another system $H_2(s)$ has two pairs of poles at $-1.25 \pm j49$ and $-1.25 \pm j51$, a double zero at the origin, and $|H_2(\omega)|_{Max} = 1$. Plot its magnitude and phase and find its -3-dB band.

c. Compare $H_1(\omega)$ and $H_2(\omega)$ in the neighborhood of $\omega = 50$.

21. a. A second-order bandpass system $H_1(s)$ has a zero at the origin, $Q = 20$, $\omega_0 = 50$, and $|H_1(\omega)|_{Max} = 1$. Determine its pole locations at $-\sigma \pm j\omega_d$, where $\omega_0^2 = \sigma^2 + \omega_d^2$. Plot its magnitude and phase and find its -3-dB band.

b. A fourth-order system $H_2(s)$ is constructed by cascading two second-order systems of the type in part a. but with $\omega_0 = 49$ and $\omega_0 = 51$, respectively. Plot its magnitude and phase and find its -3-dB band.

c. Compare $H_1(\omega)$ and $H_2(\omega)$ in the neighborhood of $\omega = 50$.

22. Sketch the magnitude and phase of the frequency response of the following systems in the form of Bode plots.

a. $H(s) = \dfrac{s+3}{(s+15)(s+500)}$

b. $H(s) = \dfrac{100(s+15)}{(s+3)(s+500)}$

c. $H(s) = \dfrac{10(s+15)}{(s+30)(s+500)}$

23. For each system given below sketch the magnitude and phase of $H(\omega)$ in the form of Bode plots and estimate their values at $\omega = 100, 200, 300,$ and $1,000$. Compare with calculated values.

a. $H(s) = \dfrac{s+1}{(s+15)(s+100)}$

b. $H(s) = \dfrac{2,000(s+1)}{(s+20)(s+100)}$

c. $H(s) = \dfrac{20(s+50)}{(s+10)(s+100)}$

24. The unit-step response of an LTI system is $(1 + e^{-2t})u(t)$. Find its system function and plot its Bode diagram.

25. Consider a first LTI system with the unit-step response $h_1(t) = (1 + e^{-2t})u(t)$ and a second LTI system with the unit-impulse response $h_2(t) = e^{-t}u(t)$.

a. Find their system functions $H_1(s)$ and $H_2(s)$, their poles and zeros, and plot their frequency responses.

b. Place the two systems in cascade form, then find the overall system function, its poles and zeros and frequency response. Relate them to those of systems $H_1(s)$ and $H_2(s)$.

c. Discuss possible differences between the $H_1 H_2$ and $H_2 H_1$ configurations.

26. Given the system function

$$H(s) = 5\frac{s+20}{s^2 + 2s + 100}$$

a. Find its poles and zeros.

b. Sketch the Bode plot of the frequency response $H(\omega)$.

c. Find and plot its unit-step response.

d. Examine the shape of the step response and relate its important characteristics to the frequency response.

27. Using a computer calculate the table entries in the solutions for problems 1 and 2.

28. Plot the magnitude and phase of the system $H(s) = 50(s+10)/(s+1)(s+200)$ using the following three Matlab programs. Discuss advantages or shortcomings of each program.

```
%C9P28a.m
w=linspace(0.1,1000,10001);
s=i*w; H=50*(s+10)./((s+1).*(s+200));
mag=abs(H); phase=((atan(w/10)-atan(w)-atan(w/200))*180/pi;
figure(1); plot(log10(w),20*log10(mag)); grid
figure(2); plot(log10(w),phase); grid


%C9P28b.m
w=logspace(-1,3,10001);
s=j*w; H=50*(s+10)./((s+1).*(s+200));
mag=abs(H); phase=(atan(w/10)-atan(w)-atan(w/200))*180/pi;
figure(1); semilogx(w,20*log10(mag)); axis([0.1 1000 -30 10]); grid;
figure(2); semilogx(w,phase); axis([0.1 1000 -90 0]); grid;


%C9P28c.m
w=linspace(0, 100, 10000);  num=[1 10]; den=[1  201 200];
[H,w]=freqs(num,den,10001); %H=freqs(num,den,w);
plot(w,50*abs(H)); grid; axis([0 100 0 2.1 ])
ylabel('Magnitude in Linear Scale')
xlabel('Angular Frequency in rad/sec, Linear Scale')
title('Magnitude of (s+10)/(s^2+201s+200) in Linear Scale')
```

29. Illustrate individual contributions from each pole and zero of the system function $H(s) = 50(s+10)/(s+1)(s+200)$ to the Bode diagram of its frequency response. For each one verify that the break-point frequency and asymptotic lines are in agreement with the analytic expression for the plot.

30. The signal $x(t)$ passes through the sytem $H(s)$, both are given below. Find and plot the output $y(t)$.

$$x(t) = \frac{\cos(2\pi f_c t)\sin(2\pi f_0 t)}{\pi t}, \quad f_c = 60 \text{ Hz}, \ f_0 = 10 \text{ Hz, and } H(s) = \frac{s+1}{(s+10)(s+5,000)}$$

31. A system has two poles at $-1 \pm j314$, a zero at -1, and a DC gain of 0 dB.

   a. Find $H(s)$.

   b. Plot the Bode plot of $H(\omega)$.

   c. Using $H(s)$, find the system's steady-state response to a 50-Hz sinusoid with 100-μV peak-to-peak amplitude.

   d. Evaluate $H(\omega)$ at 50 Hz using the graphical method of vectorial components and use it to find the answer of part c.

   e. Use a Bode plot to find the answer of part c. Compare results in c, d, and e.

32. Sketch the magnitude and phase of the frequency responses of the following systems

   a. $H(s) = \dfrac{2s}{s^2 + 2s + 2}$

   b. $H(s) = \dfrac{s - 10}{(s + 10)(s^2 + 2s + 101)}$

   c. $H(s) = \dfrac{10,001}{s^2 + 2s + 10,001}$

33. a. Sketch Bode plots of the magnitude (with 1-dB accuracy) and phase of the frequency responses of the following two systems:

(i) $H_1(s) = \dfrac{400(s+1)}{(s+4)(s^2+2s+101)}$

(ii) $H_2(s) = \dfrac{2(s^2+2s+101)}{(s+2)(s+101)}$

b. Estimate $|H_1(\omega)|$ at $\omega = 2$ and $\omega = 300$.

c. Estimate $|H_2(\omega)|$ at $\omega = 4$ and $\omega = 200$.

d. Compare the estimations with their corresponding calculated values.

Note: In the circuits of problems 34 to 41 the 741 op-amps are modeled by an open-loop DC gain of 2×10^5, a pole at $f = 5$ Hz, infinite input impedance, and zero output impedance.

34. The integrator of Figure 9.53(a) uses an input resistor $R_1 = 1\ k\Omega$. The feedback capacitor is $1\ \mu F \le C \le 10\ \mu F$ and can be adjusted in $1\ \mu F$ steps. For the two cases given below find $H(s) = \frac{V_2}{V_1}$, sketch the Bode plots, and compare results: a. op-amp is ideal, b. op-amp is a 741 model.

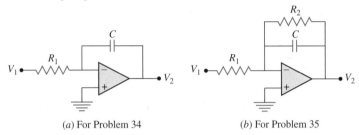

(a) For Problem 34   (b) For Problem 35

FIGURE 9.53 Integrators

35. The leaky integrator of Figure 9.53(b) uses an input resistor $R_1 = 1\ k\Omega$, and a feedback capacitor $1\ \mu F \le C \le 10\ \mu F$ (in $1\ \mu F$ steps) in parallel with $R_2 = 10\ k\Omega$. For the two cases given below find $H(s) = \frac{V_2}{V_1}$, sketch the Bode plot, and compare results: a. op-amp is ideal; b. op-amp is a 741 model.

36. The inverting amplifier of Figure 9.54(a) uses a 741 op-amp, an input resistor $R_1 = 1\ k\Omega$, and a feedback resistor R_2. Find $H(s) = V_2/V_1$ for $1\ k\Omega \le R_2 \le 10\ k\Omega$ and sketch the Bode plot of the frequency response using $1\ k\Omega$ steps for R_2.

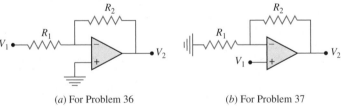

(a) For Problem 36   (b) For Problem 37

FIGURE 9.54 Amplifiers

37. A noninverting amplifier circuit uses a 741 op-amp, a grounding resistor $R_1 = 1\ k\Omega$, and a feedback resistor R_2. See Figure 9.54(b). Find $H(s) = V_2/V_1$ for $1\ k\Omega \le R_2 \le 10\ k\Omega$ and sketch the Bode plot using $1\ k\Omega$ steps for R_2.

38. a. A noninverting amplifier with a 64-dB DC gain uses a 741 op-amp and two resistors R_1 and R_2. See Figure 9.55(a). Plot its magnitude response and determine the 3-dB bandwidth.

b. The 64-dB DC gain may also be realized by cascading two stages with gains of 26 dB and 38 dB, respectively, as seen in Figure 9.55(b). Sketch the magnitude response plots for each stage and the cascaded system as a whole. Determine the 3-dB bandwidth of the cascaded system and compare with part a.

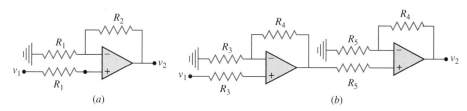

(a)   (b)

FIGURE 9.55 For Problem 38

39. A second-order system has a zero at the origin and two conjugate poles at $1\angle\theta$. Find $H(s)$ in terms of θ. Sketch its Bode plot for $90° < \theta < 180°$ using $10°$ steps. Discuss the correlation of these entities with those of section 9.20 and Figure 9.26.

40. Consider the circuit of Figure 9.56.

  a. Write KCL equations at nodes A, B, and C. Show that

$$H(s) = \frac{V_2}{V_1} = \frac{\tau_2 s}{\tau_1^2 \tau_2 s^3 + (\tau_1^2 + 3\tau_1\tau_2 + \alpha\tau_1\tau_2)s^2 + (3\tau_1 + (1 + 2\alpha)\tau_2)s + 1}$$

where $\tau_1 = R_1 C_1$, $\tau_2 = R_2 C_2$, and $\alpha = R_1/R_2$.

  b. Let $R_1 = 470\ \Omega$, $R_2 = 4{,}700\ \Omega$, $C_1 = 10$ nF, and $C_2 = 47$ nF, find $H(s)$ and its poles. Sketch its Bode magnitude and phase plots.

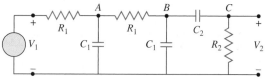

FIGURE 9.56 For Problem 40

41. In the circuit of Figure 9.57(a) show that $V_2/V_1 = 1/(s^2 + \sqrt{2}s + 1)$ using the following three approaches:

  a. Circuit approach, by applying KCL to nodes at the inverting inputs of the op-amps.

  b. System approach, by considering op-amps 1 and 2 as the forward path and op-amp 3 as the feedback. The system model is shown in Figure 9.57(b).

  c. Time-domain approach, by noting that the voltage at node B is $-v_2'$ and the circuit implements the differential equation $v_2'' = v_1 - \sqrt{2} - v_2' - v_2$.

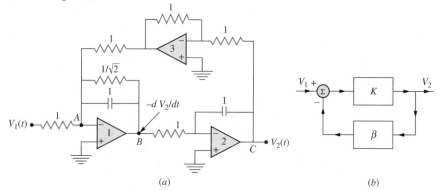

(a)   (b)

FIGURE 9.57 For Problem 41

42. Show that the circuits of Figure 9.58(a) and (b) have the same system function $V_2/V_1 = 1/(s^2 + 2s + 1)$.

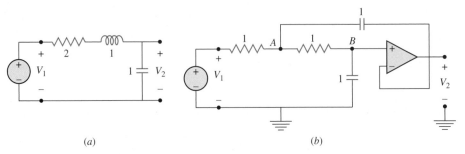

(a)   (b)

FIGURE 9.58 For Problem 42

43. A three-terminal circuit N with transfer function $\beta(s) = \dfrac{V_{out}}{V_{in}}$, as in Figure 9.59(a), is placed in the feedback path of a noninverting amplifier as in Figure 9.59(b). The op-amp is ideal except for having an open-loop gain of k.

a. Show that the system function of the overall system is

$$H(s) = \frac{V_2}{V_1} = \frac{k}{1 + \beta k} \approx \frac{1}{\beta} \quad \text{if } \beta k \gg 1$$

b. Let $k = 200,000$ and N be the bandpass circuit as in Figure 9.59(c). Show that $H(s)$ is a notch filter and find its notch frequency and the 3-dB bandwidth.

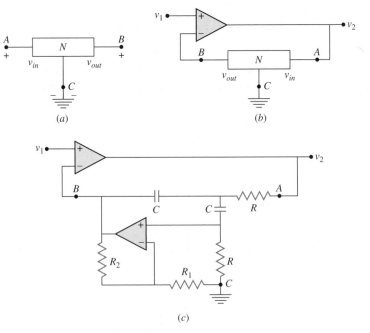

(a)   (b)

(c)

FIGURE 9.59 For Problem 43

44. The standard RIAA[14] cartridge preamplifier boosts low frequencies to compensate for their attenuation in the process of manufacturing records. It provides the following gains:

a. 17 dB (20 Hz to 50 Hz)
b. 0 dB (500 Hz to 2,120 Hz)
c. −13.7 dB at 10 kHz

Its frequency response may be considered to follow the asymptotic lines shown in Figure 9.60(a), with break points at 50 Hz (a pole, drop 3 dB), 500 Hz (a zero, add 3 dB), 2,120 Hz (a pole, drop 3 dB), and a roll-off of 20 dB per decade after that.

a. Show that the noninverting amplifier of Figure 9.60(b) can approximate the above gain provided the impedance of the feedback circuit is

$$z(s) = 10R\frac{1 + \tau_2 s}{(1 + \tau_1 s)(1 + \tau_3 s)}, \qquad \text{where } \tau_1 = 3 \text{ msec}, \tau_2 = 300 \text{ } \mu\text{sec}, \text{ and } \tau_3 = 75 \text{ } \mu\text{sec}.$$

b. Show that either of the two circuits in Figure 9.60(c) and (d) can perform the desired preamplification.

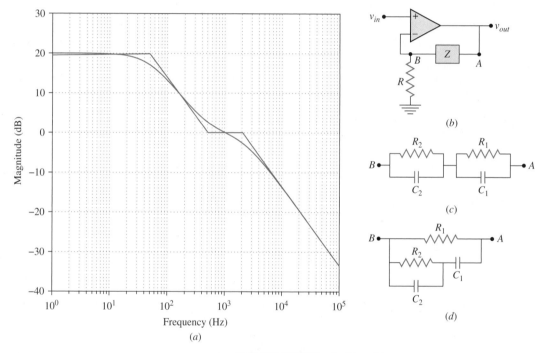

(a)

(b)

(c)

(d)

FIGURE 9.60 For Problem 44

45. Find the unit-step responses of the lead and lag networks of, respectively, Figure 9.21(a) and Figure 9.22(a) with $R_1 = 10 \text{ } k\Omega$, $R_2 = 1 \text{ } k\Omega$, and $C = 1 \text{ } \mu\text{F}$. Compare the responses and relate their salient features to their frequency responses.

46. Consider the lead compensator of Figure 9.21(a) with $H(s) = \alpha(s + \omega_1)/(\alpha s + \omega_1)$. Draw its frequency response $H(\omega)$ in polar form. Show that the plot is an upper half-circle with its center at $(1 + \alpha)/2$ on the real axis. From

---

[14]Recording Industry Association of America.

the origin draw a tangent line to the circle and show that the angle between that tangent and the real axis is θ_M, the maximum phase offered by the compensator, and that

$$\sin(\theta_M) = \frac{1-\alpha}{1+\alpha}$$

47. Repeat problem 46 for the lag compensator network of Figure 9.22(a).

48. A finite-duration integrator is made of an integrator, a delay element T, and a subtractor. Its unit-impulse response and system function are

$$h(t) = u(t) - u(t-T) \quad \text{and} \quad H(s) = \frac{1 - e^{-sT}}{s}$$

The system function has a pole and a zero at the origin which cancel each other. Let the integrator's duration be $T = 1$ and its input $x(t) = e^{-t}$.

a. Find the output $y(t)$.

b. Expand $H(s)$ by its Taylor series and approximate it by the first n terms in the expansion. Find the approximate $y(t)$ for $n = 1, 2, 3 \cdots 8$ and compare with the exact output of the finite integrator. Determine n if a relative error of 1% in y is acceptable.

49. The ear canal functions as a resonant cavity for sound waves originating from the outside of the external ear and impinging on the tympanic membrane (eardrum). Its typical contributions to pressure gain for a sound presented at $45°$ azimuth at various frequencies are listed in the table below, with a maximum gain of 20 dB at 3 kHz.[15]

Frequency, kHz	0.2	0.7	1	1.2	2	3	5	10
Gain, dB	2	8	6	8	14	20	15	2

Construct a system function to approximate the above magnitude response and design a circuit to implement it.

50. Plot the frequency response of the systems below. In each case determine if the system is low-pass, bandpass, bandstop, high-pass, or none of these. In the case of low-pass and high-pass systems, find the 3-dB frequencies. In the case of bandpass and bandstop systems, determine their 3-dB bands.

a. $i)$ $H_1(s) = \dfrac{s^2}{s^2 + 1.272s + 0.81}$ $ii)$ $H_2(s) = \dfrac{s^2}{s^2 + 0.157s + 0.81}$

b. $i)$ $H_1(s) = \dfrac{177}{s^2 + 4.25s + 177}$ $ii)$ $H_2(s) = \dfrac{265s^2 + 725.7s + 672.6}{s^3 + 269.75S^2 + 902.7s + 672.6}$

c. $i)$ $H_1(s) = \dfrac{1}{s^2 + 1.99s + 1}$ $ii)$ $H_2(s) = \dfrac{1}{s^2 + 0.2s + 1}$

d. $i)$ $H_1(s) = \dfrac{1}{s^2 + 5s + 1}$ $ii)$ $H_2(s) = \dfrac{1}{s^2 + 0.1s + 1}$

e. $i)$ $H_1(s) = \dfrac{s}{s^2 + 0.1s + 1}$ $ii)$ $H_2(s) = \dfrac{10^6}{(s + 100)(s^2 + 100s + 10{,}000)}$

---

[15] The numbers are adopted from W.D. Keidel, W.D. Neff, eds. *Handbook of Sensory Physiology* (Vol. 1). (New York: Springer-Verlag, 1974), p. 468.

51. A physical plant is modeled by the third-order system function

$$\text{Model 1:} \quad H(s) = \frac{8,850}{s^3 + 55s^2 + 380s + 8,850}$$

a. Find its poles and recognize the dominant pair with regard to frequencies on the $j\omega$ axis.
b. Construct a second-order model $\hat{H}(s)$ made of the dominant pair of poles (to be called Model 2).
c. Plot the unit-step frequency response of Model 2 and visually compare it with that of Model 1.
d. By a qualitative inspection determine where (in the time and frequency domains) differences between the two models are small and negligable, and where they are considerable and pronounced.
e. Qualitatively illustrate the effect of the third pole on the time and frequency responses by an example. Then devise a measure to quantify the effect.

52. Find and sketch the frequency response of an LTI system with a pair of poles in the left-half plane at $s_{1,2} = -\sigma \pm j\omega_d$, a zero at $s = 0$, and a maximum gain of H_M.

53. Find the frequency response of an LTI system with two poles at $0.9e^{\pm j135°}$, a double zero at the origin, and unity gain at very high frequencies. Sketch its magnitude-squared $|H(\omega)|^2$ plot as a function of ω and evaluate it at $\omega = 0$ (DC), $\omega = 0.9$, and $\omega = \infty$. Determine the -3-dB frequency (where $|H(\omega)|^2 = 0.5$).

54. Find the frequency response of an LTI system with two poles at $0.9e^{\pm j95°} = -0.0157 \pm j0.8965$, a double zero at the origin, and a multiplying factor of unity in the expression for $H(s)$. Sketch its magnitude-squared $|H(\omega)|^2$ plot as a function of ω and evaluate it at $\omega = 0$ (DC), $\omega = 0.9$, and $\omega = \infty$. Determine the -3-dB frequencies and its bandwidth.

55. A causal system is described by the system function

$$H(s) = \frac{s}{s^2 + 0.1s + 100}$$

Use the vectorial interpretation of $H(s)$ to evaluate $H(\omega)$. Show that the system is a bandpass filter. Find the 3-dB frequencies ω_ℓ and ω_h. Evaluate $H(\omega)$ at $\omega = 0$, 10, ω_ℓ, ω_h, and ∞. Sketch the magnitude and phase of $H(\omega)$.

56. Find and sketch the frequency response of an LTI system with a pair of poles in the left-half plane at $s_{1,2} = -\sigma \pm j\omega_d$, no finite zeros, and unity DC gain. Find $H(s)$ and the frequency response. Plot magnitude and phase responses (along with pole locations and the step response) for the following two sets of pole locations:

a. $s_{1,2} = -0.995 \pm j0.1$ (away from the $j\omega$ axis).   b. $s_{1,2} = -0.1 \pm j0.995$ (close to the $j\omega$ axis).

Explain the shape of each plot in terms of the vectorial interpretation of $H(s)$ and the effect of pole-zero location.

57. Two LTI systems are cascaded. System 1 is a second-order bandpass (with a pair of poles at $p_{1,2} = 0.9e^{\pm\theta}$, a zero at the origin, and a unity peak gain). System 2 is a unity-gain differentiator. For $\theta = 100°$, $135°$, and $170°$ obtain the following responses, plot them, and compare.

a. The impulse and step responses of system 1.
b. The impulse and step responses of the cascaded systems.
c. The Bode plots of systems 1, 2, and their cascade.

9.33 Project 1: *RC/CR* Passive Bandpass Filter

I. Summary
II. Theory
III. Prelab
IV. Measurements and Analysis
V. Discussion and Conclusions
      Appendix 9A: Hyperbolic Functions
      Appendix 9B: Bandpass System

I. Summary

In this project you will measure and model the time and frequency responses of a passive bandpass system and compare them with theory. The system is made of a low-pass *RC* stage cascaded by a high-pass *CR* stage. It has two poles on the negative real axis in the left-half plane and a zero at the origin. You will note that the circuit remains always overdamped, with no resonance. The quality factor of the filter remains low. You will explore, by analytic method and measurement, the effect of the pole location on the system's frequency and step responses. You will discover their relationship with the system's rise time, fall time, and bandwidth.

Equipment

One function generator
One oscilloscope
Resistors (15 kΩ, 1.5 kΩ, and 150 Ω, two of each)
Two 10-nF capacitors
Software tools such as Matlab

Assignments and Reporting

(for a three-hour work) Read and do items 8 to 15 and 19 to 23 of this project.

II. Theory

This part summarizes the time and frequency responses of the passive *RC/CR* bandpass filter of Figure 9.61. Most of the steps are already carried out and the results are shown for verification.

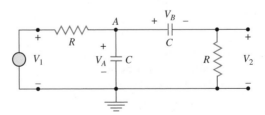

FIGURE 9.61 A passive bandpass *RC/CR* filter.

1. System Function and State Variables

Show that

$$H(s) = \frac{V_2}{V_1} = \frac{(\frac{s}{\omega_0})}{(\frac{s}{\omega_0})^2 + 3(\frac{s}{\omega_0}) + 1}$$

where $\omega_0 = 1/RC$. The system has two poles at $s_{1,2} = (-\alpha \pm \beta)\omega_0$, where $\alpha = 1.5$ and $\beta = \sqrt{1.25}$. Substituting the above values for α and β we get $s_1 = -0.382\omega_0$ and $s_2 = -2.618\omega_0$. The system has a zero at the origin. Show the location of the poles and zero on the s-plane.

Let the capacitor voltages be V_A and V_B as shown in Figure 9.61. These two voltages determine the state of the system at any moment. They are called state variables of the system. Show that their transfer functions are

$$H_A(s) = \frac{V_A}{V_1} = \frac{1 + \frac{s}{\omega_0}}{(\frac{s}{\omega_0})^2 + 3(\frac{s}{\omega_0}) + 1}$$

$$H_B(s) = \frac{V_B}{V_1} = \frac{1}{(\frac{s}{\omega_0})^2 + 3(\frac{s}{\omega_0}) + 1}$$

$H_A(s)$ and $H_B(s)$ have the same two poles as $H(s)$. The system's output may be written in terms of its state variables: $v_2 = v_a - v_b$. Note that v_a and v_b are both low-pass signals whose transfer functions were previously found in the experiment on second-order low-pass systems, which will not be repeated here. The present experiment will be limited to the study of v_2, a bandpass signal.

2. Frequency Response

Express the frequency response as a function of (ω/ω_0) and show that

$$H(\omega) = \frac{j(\omega/\omega_0)}{1 - (\omega/\omega_0)^2 + j3(\omega/\omega_0)}$$

$$|H(\omega)|^2 = \frac{(\omega/\omega_0)^2}{1 + 7(\omega/\omega_0)^2 + (\omega/\omega_0)^4}$$

$$\angle H(\omega) = \begin{cases} 90° - \tan^{-1}\frac{3(\omega/\omega_0)}{1-(\omega/\omega_0)^2} & 0 < \omega/\omega_o < 1 \\ 0° & \omega/\omega_o = 1 \\ -90° - \tan^{-1}\frac{3(\omega/\omega_0)}{1-(\omega/\omega_0)^2} & 1 < \omega/\omega_o < \infty \end{cases}$$

Plot the magnitude response $|H(\omega)|$ versus (ω/ω_0) for $0.1 \le (\omega/\omega_0) \le 10$. Use a log scale for the frequency axis and a linear scale for $|H|$, where $V_{pp} = 10$ V at the input.

3. Bode Plot

Express the frequency response in decibel (dB) and plot it in the form of a Bode plot. Show that

$$20\log|H(\omega)| = 20\log(\omega/\omega_0) - 10\log[1 + 7(\omega/\omega_0)^2 + (\omega/\omega_0)^4]$$

Express the frequency response in a Bode diagram. Specifically, show that

$$20\log|H(\omega)| = \begin{cases} 20\log(\omega/\omega_0) & \text{low-frequency asymptote, } \omega << \omega_1 \text{ (break point at } \omega_1 = 0.382\omega_0) \\ -12.5 \text{ dB} & \text{at } \omega = 0.303\omega_0, \text{ is 3 dB below } H_{Max} \\ -9.5 \text{ dB} & \text{at } \omega = \omega_0, \text{ is } H_{Max} \\ -12.5 \text{ dB} & \text{at } \omega = 3.303\omega_0, \text{ is 3 dB below } H_{Max} \\ -20\log(\omega/\omega_0) & \text{high-frequency asymptote, } \omega >> \omega_2 \text{ (break point at } \omega_2 = 2.618\omega_0) \end{cases}$$

Plot $20 \log |H(\omega)|$ (in dB) versus $\log(\omega/\omega_0)$ for $0.1 \le (\omega/\omega_0) \le 10$. Verify that the low-frequency asymptote is a line whose slope is 20 dB per decade (terminating at the first break frequency ω_1); that the high-frequency asymptote is also a line whose slope is -20 dB per decade (starting at the second break frequency ω_2); and that attenuation at ω_0 is $20 \log 3 = 9.5$ dB, where $|H(\omega)|$ is at its maximum value.[16]

4. Bandwidth

By convention, the 3-dB bandwidth (equivalently, the half-power bandwidth) is the frequency band within which the magnitude response is less than 3 dB below its maximum. Show that the maximum magnitude of the frequency response is at ω_0; that the half-power frequencies (3 dB below the maximum) are $\omega_{\ell,h} = (1.5 \pm \sqrt{3.25})\omega_0$, or $\omega_\ell = 0.303\omega_0$ and $\omega_h = 3.303\omega_0$; that the system's 3-dB bandwidth is $\Delta\omega = \omega_h - \omega_\ell = 3\omega_0$; and, its quality factor is $Q = \omega_0/\Delta\omega = 1/3$.

5. Unit-Impulse Response

Argue, without solving the system's input-output differential equation, that the unit-impulse response is $h(t) = C_1 e^{s_1 t} + C_2 e^{s_2 t}$ where s_1 and s_2 are system's poles, $s_{1,2} = (-\alpha \pm \beta)\omega_0$, $\alpha = 1.5$ and $\beta = \sqrt{1.25}$. Then by observing system's behavior during the transition from $t = 0^-$ to $t = 0^+$ conclude that $h(0^+) = \omega_0$ and $dh(t)/dt|_{t=0^+} = -3\omega_0^2$. From these boundary conditions[17] find

$$C_{1,2} = \frac{\omega_0}{2\beta}[\beta \pm (\alpha - 3)]$$

$$h(t) = \frac{\omega_0}{2\beta}e^{-\alpha\omega_0 t}\left[(\beta - \alpha)e^{\beta\omega_0 t} + (\beta + \alpha)e^{-\beta\omega_0 t}\right]u(t)$$

$$= \frac{\omega_0}{\beta}e^{-\alpha\omega_0 t}\left[\beta \cosh(\beta\omega_0 t) - \alpha \sinh(\beta\omega_0 t)\right]u(t)$$

Substituting for $\alpha = 1.5$ and $\beta = \sqrt{1.25}$ we get

$$h(t) = \frac{\omega_0}{\sqrt{5}}e^{-1.5\omega_0 t}\left[(\sqrt{1.25} - 1.5)e^{\sqrt{1.25}\omega_0 t} + (\sqrt{1.25} + 1.5)e^{-\sqrt{1.25}\omega_0 t}\right]u(t)$$

$$= \frac{2\omega_0}{\sqrt{5}}e^{-1.5\omega_0 t}\left[\sqrt{1.25} \cosh(\sqrt{1.25}\omega_0 t) - 1.5 \sinh(\sqrt{1.25}\omega_0 t)\right]u(t)$$

$$= \frac{2\omega_0}{\sqrt{5}}e^{-1.5\omega_0 t}\sinh(\sqrt{1.25}\omega_0 t + 0.9624)u(t)$$

$$= \omega_0\left(1.1708e^{-2.618\omega_0 t} - 0.1708e^{-0.382\omega_0 t}\right)u(t)$$

Sketch $h(t)$.

6. Unit-Step Response

Argue, without solving the system's input-output differential equation, that the unit-step response is $g(t) = D_1 e^{s_1 t} + D_2 e^{s_2 t}$. Then, by observing the system's behavior during the transition from $t = 0^-$ to $t = 0^+$, conclude that $g(0^+) = 0$ and $dg(t)/dt|_{t=0^+} = \omega_0$. From these boundary conditions find $D_{1,2} = \pm 1/(2\beta)$ and

$$g(t) = \frac{1}{2\beta}e^{-\alpha\omega_0 t}\left(e^{\beta\omega_0 t} - e^{-\beta\omega_0 t}\right)u(t) = \frac{1}{\beta}e^{-\alpha\omega_0 t}\sinh(\beta\omega_0 t)u(t)$$

---

[16]For a general formulation of a bandpass system see Appendix 9B at the end of this project.

[17]The unit-impulse response may also be obtained by taking the inverse Laplace transform of $H(s)$, yielding $C_{1,2} = \omega_0(\beta \mp \alpha)/(2\beta)$ which is the same as the result obtained from applying the initial conditions.

Substituting for $\alpha = 1.5$ and $\beta = \sqrt{1.25}$ we get

$$g(t) = \frac{1}{\sqrt{5}}e^{-1.5\omega_0 t}\left(e^{\sqrt{1.25}\omega_0 t} - e^{-\sqrt{1.25}\omega_0 t}\right)u(t)$$

$$= \frac{2}{\sqrt{5}}e^{-1.5\omega_0 t}\sinh(\sqrt{1.25}\omega_0 t)u(t)$$

$$= 0.4472(e^{-0.382\omega_0 t} - e^{-2.618\omega_0 t})u(t)$$

Plot or sketch the step response.
In summary,

$$\text{Unit-Impulse response:} \quad h(t) = \omega_0\left(1.1708e^{-2.618\omega_0 t} - 0.1708e^{-0.382\omega_0 t}\right)u(t)$$

$$\text{Unit-step response:} \quad g(t) = 0.4472(e^{-0.382\omega_0 t} - e^{-2.618\omega_0 t})u(t)$$

Note that $g(t) = \int_{-\infty}^{t} h(t)dt$, or equivalently, $h(t) = \frac{d}{dt}[g(t)]$.[18]

7. Pulse Response, Rise Time, Fall Time, and Droop

The rise time T_r is the time it takes the pulse response to go from 10% to 90% of its final value. Similarly, the fall time T_f is the time it takes the pulse response to be reduced from 90% to 10% of its height. Droop is the percentage amount of sag the response shows during the pulse.

III. Prelab

8. Planning the Experiment

Read Parts IV and V and plan the experiment.

9. Derivation of System Function and Frequency Response

Show that in the circuit of Figure 9.61,

$$H(s) = \frac{V_2}{V_1} = \frac{\left(\frac{s}{\omega_0}\right)}{\left(\frac{s}{\omega_0}\right)^2 + 3\left(\frac{s}{\omega_0}\right) + 1}$$

$$H(\omega) = \frac{j(\omega/\omega_0)}{1 - (\omega/\omega_0)^2 + j3(\omega/\omega_0)}$$

10. Finding $|H|_{Max}$

For $R = 15\ k\Omega$ and $C = 10$ nF show that $|H|$ obtains its maximum at $f_0 = 1,061$ Hz. Then find the maximum value, also in dB.

11. Computing Theoretical dB Values of $|V_2/V_1|$

Given $f_0 = 1,061$ Hz, compute theoretical values of $|H(f)| = |V_2/V_1|$ in dB for the frequencies listed in Table 9.1 and enter them in their appropriate places in that table.

IV. Measurements and Analysis

Construct the circuit of Figure 9.61 using $R = 15\ k\Omega$ and $C = 10$ nF. This will provide $\omega_0 = 1/RC = 10^5/15 = 6,666.7$ rad/s, corresponding to $f_0 = 1,061$ Hz, and a 3-dB bandwidth $\Delta f = 3f_0 = 3,183$ Hz.

---

[18]Note that $g(t)\big|_{bandpass} = \frac{1}{\omega_0}h(t)\big|_{lowpass}$, both of second order.

12. Recording the Frequency Response

Record the responses of the circuit to sinusoidal inputs ($V_{pp} = 10$ V, $V_{DC} = 0$) at $f = 100$ Hz to 10 kHz as identified in Table 9.10. Measure magnitude and phase of the responses and compute the attenuation in dB. Enter measured and computed values in Table 9.10. Plot the magnitude of the frequency response for the range $100\ Hz \leq f \leq 10$ kHz. Use log scale for the frequency axis and uniform scale for the magnitude axis. To obtain a plot of the magnitude response you may program the sinusoidal generator to sweep through the desired frequency range using a log scale, and display a single sweep of the output on the oscilloscope.

The numerical value of V_2/V_1 as a function of frequency is also obtainable from the screen. Accurately measure the 3-dB bandwidth. Compare the measured frequency response with theory (item 2). The comparison may be done by using the relative difference between theory and measured values of V_2/V_1, defined by:

$$\text{Relative difference (\%)} = \frac{(\text{measured} - \text{theory})}{\text{theory}} \times 100$$

The difference between theory and measured values of the frequency response may also be expressed in dB:

$$\text{dB difference} = |V_2/V_1|_{\text{dB measured}} - |V_2/V_1|_{\text{dB theory}}$$

The above measures are related to each other by

$$\text{dB difference} = 20 \log_{10}\left[(\text{relative difference}) + 1\right]$$

Enter results in Table 9.10. Draw the measured frequency response in the form of a Bode plot.

TABLE 9.10 Measurements of Magnitude and Phase of the Frequency Responses and Comparison with Theory

Frequency	100 Hz	200 Hz	330 Hz	1 kHz	3.3 kHz	5 kHz	10 kHz
$|V_2|$ (volts) measured							
$|V_2/V_1|$ measured							
$|V_2/V_1|$ (dB) measured							
$(\angle V_2 - \angle V_1)°$ measured							
$|V_2/V_1|$ theory							
$|V_2/V_1|$ (dB) theory							
relative difference (%)							
dB difference							

13. Recording an Impulse Response

Record the system's impulse response. For this purpose program the function generator to produce a train of very narrow pulses (e.g., $< 1\%$ duty cycle) at a very low frequency $f = 100$ Hz. Measure its time constants. Model the impulse response by a mathematical expression and compare with 5.

14. Recording a Step Response

Record the system's step response and model it by a mathematical expression and compare with 6.

15. Recording a Pulse Response

Record the response to a single rectangular pulse ($V_{pp} = 10$ V, DC level $= 0$, duration $= n\tau, n = 1, 3, 5$ where $\tau = RC$). Note that for $R = 15\ K\Omega$ and $C = 10$ nF we have $\tau = 150$ ns. A pulse may be created as a repetitive rectangular wave at a low-duty cycle (e.g., at 20%). In each case qualitatively verify that the pulse response is the superposition of responses to a step and a delayed negative step. Measure the rise and fall time of the response in μs, and its droop in %. Droop is defined in item 7 in this project. Enter your answers in the space below.

Answers:

n	T_r in μs	T_f in μs	Droop in %
1			
3			
5			

16. Recording Response to Repetitive Square Pulses

Record one cycle of the time response to a repetitive train of square pulses ($V_{pp} = 1$ V, $V_{DC} = 0$) at $f = 100$ Hz, 500 Hz, 1 kHz, 2 kHz, and 10 kHz. Observe and describe the trend in the response pattern as frequency increases.

17. Filtering White Noise

Pass a white noise through the system. Observe and qualitatively describe the filtering effect on the frequency components of the noise. Repeat for $R = 150$ kΩ and 1.5 MΩ.

18. Changing the Bandwidth

Change both resistors in the circuit of Figure 9.61 to $R = 150$ kΩ. Capture the new step and frequency responses on the scope. Measure the new bandwidth. Compare with theory.

V. Discussion and Conclusions

19. Poles/Zero Location and Step Response

From the pole-zero plot reason the validity of the shape of the step response.

20. Poles/Zero Location and Bandpass Property

From the pole-zero plot conclude that the system is a bandpass filter.

21. Bandwidth

The 3-dB bandwidth was defined in item 4 in this project. From the frequency response measurements determine the upper and lower 3-dB frequencies and the bandwidth. Compare with theory.

22. Transition Band

Define the transition band to be where -20 dB $< 20 \log |H(\omega)| < -1$ dB. Using the measured Bode plot, determine the upper and lower transition bands of the filter.

23. Overall Conclusions

Summarize your overall conclusions drawn from this experiment, especially the relationship between the system function, poles and zeros, frequency response, impulse response, step response, and pulse response.

APPENDIX 9A

Hyperbolic Functions

Define $\cosh(x)$ (pronounced *cosine hyperbolic of x*) and $\sinh(x)$ (*sine hyperbolic of x*) by the following equations

$$\begin{cases} \cosh(x) = \frac{e^x + e^{-x}}{2} \\ \sinh(x) = \frac{e^x - e^{-x}}{2} \end{cases} \implies \begin{cases} e^x = \cosh(x) + \sinh(x) \\ e^{-x} = \cosh(x) - \sinh(x) \end{cases}$$

The following properties are derived from the above definitions.

$$\begin{cases} \cosh(-x) = \cosh(x) \\ \sinh(-x) = -\sinh(x) \end{cases} \quad \text{and} \quad \begin{cases} \cosh(jx) = \cos(x) \\ \sinh(jx) = j\sin(x) \end{cases}$$

$$\cosh(x + y) = \cosh(x)\cosh(y) + \sinh(x)\sinh(y)$$
$$\sinh(x + y) = \cosh(x)\sinh(y) - \sinh(x)\cosh(y)$$

Example

Find θ in $x(t) = \sqrt{1.25}\cosh(\sqrt{1.25}\omega_0 t) - 1.5\sinh(\sqrt{1.25}\omega_0 t) = \sinh(\sqrt{1.25}\omega_0 t + \theta)$.

Let

$$\begin{cases} \sqrt{1.25} = \sinh(\theta) \text{ and } 1.5 = \cosh(\theta) \\ e^\theta = \cosh(\theta) + \sinh(\theta) = 1.5 + \sqrt{1.25} = 2.62 \end{cases} \implies \theta = Ln[2.62] = 0.9624$$

Then

$$x(t) = \sqrt{1.25}\cosh(\sqrt{1.25}\omega_0 t) - 1.5\sinh(\sqrt{1.25}\omega_0 t)$$
$$= \sinh(\theta)\cosh(\sqrt{1.25}\omega_0 t) - \cosh(\theta)\sinh(\sqrt{1.25}\omega_0 t)$$
$$= \sinh(\sqrt{1.25}\omega_0 t + \theta) = \sinh(\sqrt{1.25}\omega_0 t + 0.9624)$$

APPENDIX 9B

Bandpass System

The system function

$$H(s) = \frac{ks}{s^2 + bs + \omega_0^2}$$

where b is a positive number, represents a bandpass system. The magnitude frequency response $|H(\omega)|$ reaches its maximum at ω_0 (center frequency), where $|H|_{Max} = |H(\omega_0)| = |k|/b$. The 3-dB frequencies ω_ℓ and ω_h are defined to be the frequencies at which $|H(\omega)| = |H|_{Max}/\sqrt{2} = |k|/(\sqrt{2}b)$. The 3-dB bandwidth is $\Delta\omega = \omega_h - \omega_\ell = b$. The quality factor Q is defined by $Q = \omega_0/\Delta\omega = \omega_0/b$. The upper and lower 3-dB frequencies are

$$\omega_{\ell,h} = \frac{\mp b + \sqrt{b^2 + 4\omega_0^2}}{2}$$

$$= \omega_0 \left(\mp \frac{1}{2Q} + \sqrt{1 + \left(\frac{1}{2Q}\right)^2} \right)$$

$\omega_0 = \sqrt{\omega_\ell \omega_h}$ is the geometric mean of the 3-dB frequencies.

9.34 Project 2: Active Bandpass Filter

I. Summary

In this experiment you will measure and model the time and frequency responses of a narrowband second-order active filter and compare them with theory. By changing the gain of the amplifier, you will be able to move the poles of the system closer to or farther away from the $j\omega$ axis, thus controlling the bandwidth (or the quality factor) of the filter. By observing the time responses you will discover the relationship between the natural frequencies, step response, repetitive pulse response, resonance, and the frequency response.

Equipment

One function generator
One oscilloscope
Three 15-kΩ resistors
One 10-kΩ resistor
One potentiometer (or variable resistor box), 0–50 kΩ
Two 10-nF capacitors
One op-amp, LM148 or a similar one
One adjustable dual DC power supply for the op-amp, $\pm$12 V

II. Theory

Consider the circuit of Figure 9.62 with $R = 15$ kΩ, $C = 10$ nF, $R_1 = 10$ kΩ, and an adjustable R_2. Note that the circuit is constructed from a passive bandpass section (RC/CR) on the left, a noninverting amplifier with gain $k = 1 + \frac{R_2}{R_1}$ [equivalently, $R_2 = (k - 1)R_1$] and a resistive feedback path. In the prelab you will consider the circuit for $k = 1$, 1.1715, 2.5, 3, 3.5, 3.9, 4, 4.1. At $k = 1.1715$ the circuit is *critically damped*. At $k = 4$ the poles are on the $j\omega$ axis and the circuit becomes an oscillator.

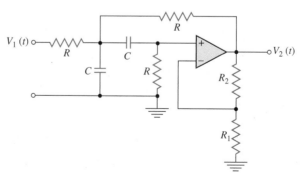

FIGURE 9.62 A second-order active bandpass filter.

Note: You may simulate the circuit (using Spice, MicroCap, etc.) or use a computation package (Matlab, Mathcad, Maple, etc.) to obtain the plots.

1. System Function

Find the system function $H(s) = V_2/V_1$ (e.g., by writing a KCL equation at node A). Show that

$$H(s) = V_2/V_1 = \frac{kRCs}{R^2C^2s^2 + (4-k)RCs + 2}$$

$$= \frac{k}{\sqrt{2}} \frac{\left(\frac{s}{\omega_0}\right)}{\left(\frac{s}{\omega_0}\right)^2 + \frac{1}{Q}\left(\frac{s}{\omega_0}\right) + 1}$$

where

$$k = 1 + \frac{R_2}{R_1}, \quad \omega_0 = \frac{\sqrt{2}}{RC}, \quad \text{and} \quad Q = \frac{\sqrt{2}}{4-k}$$

Note that when $k < 4$, the system is stable and the bandpass summary given in Appendix 9B of project 1 applies. In doing the prelab, you may use the results given in that appendix to compute indicators of the frequency response such as bandwidth and 3-dB frequencies.

2. Poles and Zero(s)

Find the system's poles and zeros. Determine the natural frequencies $s_{1,2} = -\alpha \pm j\omega_d$ and $\omega_o = \sqrt{\alpha^2 + \omega_d^2}$. Show that

$$\alpha = \frac{\omega_0}{2Q} \quad \text{and} \quad \omega_d = \frac{\omega_0}{2Q}\sqrt{4Q^2 - 1}$$

For $k = 1, 1.1715, 2.5, 3, 3.5, 3.9, 4, 4.1$ compute the locations of $s_{1,2}$ and plot them on the s-plane. Note how the poles migrate toward $j\omega$ axis as k is increased. For the k values listed above find $R_2, Q, s_{1,2} = \alpha \pm j\omega_d, \omega_0$ and enter your answers in a table and call it Table 9.11.

TABLE 9.11 Quality Factor and Pole Locations for Eight Values of Feedback Gain

k	R_2 (kΩ)	$Q = \sqrt{2}/(4-k)$	ω_0	$s_{1,2}$
1				
1.1715				
2.5				
3				
3.5				
3.9				
4				
4.1				

3. Unit-Step Response

Let $g(t)$ designate the unit-step response; that is, $v_2(t) = g(t)$, where $v_1(t) = u(t)$, a 1-V step. From item 2 above conclude that

$$g(t) = Ae^{-\alpha t}\cos(\omega_d t + \theta)$$

Sketch, by hand or by computer, the first 5 msec ($0 \leq t \leq 5$ msec) of the unit-step response for the case of $k = 3.9$.

4. Sketch of the Frequency Response from the Pole-Zero Plot

Using the pole-zero plot at $k = 3.9$, and based on a vectorial interpretation of $H(\omega)$, sketch, by hand, the magnitude and phase of the frequency response for $200\pi < \omega < 10^4 \times 2\pi$ (100 Hz to 10 kHz). You should obtain a bandpass frequency response. Use log scale for ω (or f) and uniform scale for $|H|$.

5. Frequency Response Indicators

Using $H(s)$ given in item 1, find the mathematical expression for the frequency response $H(\omega)$. Express the frequency response as a function of ω/ω_0 (or f/f_0). You should obtain a bandpass frequency response. For $k = 3.9$, plot it using a computer. You may use $f = \omega/2\pi$ for the frequency axis (i.e., to identify important points by f_0, f_ℓ, f_h, and, $\Delta f = f_h - f_\ell$). Use log scale for ω (or f) and uniform scale for $|H|$.

Show that for all $k < 4$ (stable system), $|H|_{Max} = k/(4-k)$ and it occurs at $\omega_0 = \sqrt{2}/RC$ (or $f_0 = \omega_0/2\pi$). Determine $\Delta\omega$. For high Q approximate half-power frequencies ω_ℓ and ω_h and enter your answers in Table 9.12. You may use the Appendix 9B at the end of project 1 in this chapter.

TABLE 9.12 Resonance Indicators for Eight Values of Feedback Gain

k	Q	$\|H\|_{Max} = k/(4k)$	ω_0	$\Delta\omega$	ω_ℓ	ω_h
1						
1.1715						
2.5						
3						
3.5						
3.9						
4						
4.1						

6. Stability

For $k \geq 4$ discuss location of the poles, shape of step and impulse responses, and show that the system becomes unstable.

7. Filter Summary at $k = 3.9$

Important response characteristics of the circuit of Figure 9.62 at $k = 3.9$ are summarized in the table below for comparison with measurement results.

k	R_2, kΩ	Q	$s_{1,2}$, neper	$\|H\|_{Max}$	f_0, Hz	Δf, Hz	f_ℓ, Hz	f_h, Hz
3.9	29	14.14	$-3{,}333.8 \pm j8{,}819$	39	1,500.53	106.1	1,447.5	1,553.6

III. Prelab

8. Planning the Experiment

Read Parts IV and V and plan the experiment.

9. Deriving the System Function

Derive the system function given in item 1.

10. Computing Q, ω_0, and Pole Locations
Complete Table 9.11.

11. Computing $|H|_{Max}$, $\Delta\omega$, ω_ℓ, and ω_h
Complete Table 9.12.

IV. Measurements and Analysis
Use the circuit of Figure 9.62 with element values $R = 15$ kΩ, $C = 10$ nF, $R_1 = 10$ kΩ, and an adjustable R_2. Measurement of response parameters is best done through the oscilloscope display. When required, capture responses from the oscilloscope and include their prints in your report.

12. Recording the Step Response and Natural Frequencies
Set $k = 3.9$ and record the first 5 msec of the step response ($0 \le t \le 5$ msec). You may use a square-pulse input signal (10 Hz, 200 mV). From the scope or the recorded step response specify the natural frequencies and compare with item 2. Model the step response by a mathematical expression and compare with item 3. Describe their differences and explain possible reasons behind them.

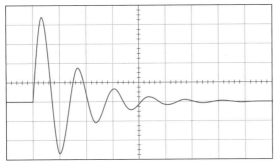

(*a*) Sweep = 500 μsec/div, Ch1 = 2 V/div. Ch2 = 1 V/div

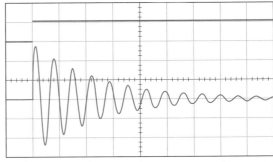

(*b*) Sweep = 1 msec/div, Ch1 = 200 mV/div. Ch2 = 200 mV/div

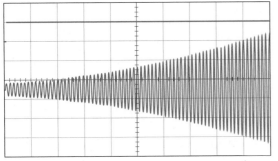

(*c*) Sweep = 5 msec/div, Ch1 = 200 mV/div. Ch = 2 V/div

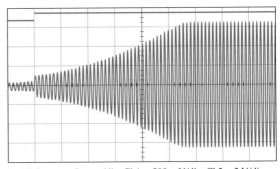

(*d*) Sweep = 5 msec/div, Ch1 = 500 mV/div. Ch2 = 2 V/div

FIGURE 9.63 Step responses of the bandpass filter. As the feedback gain increases the system's poles (a pair of complex conjugates in the LHP) move closer to the $j\omega$ axis. Oscillations become more pronounced but are still decaying (from *a* to *b*). In *(c)* the pair of complex poles have just moved to the RHP and oscillations grow rather than decay. In *(d)* the amplitude of oscillations has grown to a constant value imposed by op-amp saturation voltage. Step inputs are on channel 1 and responses are on channel 2. The frequency of oscillations is 1.5 kHz.

13. Effect of the Gain on the Step Response

Starting from low gain ($k = 1$), gradually increase k and qualitatively observe and record its effect (DC steady state, frequency of oscillations, damping ratio) on the unit-step response. Compare with Table 9.11. Examples of step response for four values of feedback are given in Figure 9.63.

14. Recording the Frequency Response

Set $k = 3.9$. Measure and record the magnitude of the frequency response for the range 100 Hz $\leq f \leq$ 10 kHz. You may use a sinusoidal input signal at a constant amplitude and measure the output. Plot $|H(\omega)|$. Use log scale for the frequency axis and uniform scale for the magnitude axis. Compare with the result in item 4. An example of the recording of the frequency response is shown in Figure 9.64.

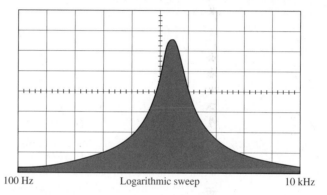

| 100 Hz | Logarithmic sweep | 10 kHz |

FIGURE 9.64 Sinusoidal output of the bandpass filter as a function of frequency obtained by a single log sweep. Time scale: 500 msec/division.

15. Effect of the Gain on the Frequency Response

Starting from low gain ($k = 1$), gradually increase k and qualitatively observe and record its effect (f_0, f_ℓ and f_h, Δf and Q) on the frequency response. Compare with Table 9.12.

16. Response to the Repetitive Square Pulse

At $k = 3.9$ measure and record the time response to a repetitive train of square pulses ($V_{pp} = 200$ mV, $V_{DC} = 0$ V). Sweep the frequency from $f = 100$ Hz to 10 kHz. Record responses to one cycle of the square pulse input at $f_0/10$, $f_0/5$, $f_0/3$, f_0, and $2f_0$ (5 records). Measure the gain at f_0 and compare it with the gain of the frequency response at f_0 (item 12).

17. Response to the Square Pulse Around f₀ Hz

At $k = 3.9$ observe and visually examine the time response to a repetitive train of square pulses ($V_{pp} = 200$ mV, $V_{DC} = 0$ V) at the frequencies $f_0 \pm 5$ Hz. Relate your observations to the natural frequency of the system. See how the oscillations merge to form a quasi-sinusoidal response.

18. Response to the Square Pulse Around f₀/3 Hz

At $k = 3.9$ set the square pulse input at $\approx f_0/3$ Hz and visually examine the response. Gradually and slowly change the frequency of the input (within the range of $f_0/3 \pm 50$ Hz). You should see an increase in the response amplitude at $f_0/3$. Explain this effect by observing that f_0, where the maximum of the magnitude of the frequency response occurs, is the third harmonic of the square pulse input.

V. Discussion and Conclusions

19. Gain, Pole-Zero Locations, Bandpass Property, and Stability

Discuss the effect of the gain on the bandpass characteristics of the filter (center frequency, bandwidth, quality factor). Discuss under what conditions the system becomes unstable and plot an example of the step response.

20. Overall Conclusions

Discuss the capabilities of the current active bandpass filter. Compare with the passive bandpass filter of Project 1.

9.35 Project 3: An Active Filter with Bandpass/Lowpass Outputs

  I. Summary
  II. Theory
  III. Prelab
  IV. Measurements and Analysis
  V. Discussion and Conclusions

I. Summary

In this experiment you will measure and model the time and frequency responses of a second-order system with one input and two outputs (low-pass and bandpass). See the circuit of Figure 9.65. By adjusting a single feedback resistor in the circuit you can move the poles of the system closer to or farther away from the $j\omega$ axis and control the bandwidth (or the quality factor) of the system. By analyzing time responses you will observe the system's poles and discover the relationship between the natural frequencies, step responses, repetitive pulse responses, resonance, and the frequency responses.

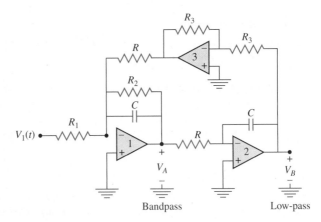

FIGURE 9.65 Active filter with bandpass/low-pass outputs.

Equipment

One function generator (Agilent 33120A or a similar one)
One oscilloscope (Agilent 54622A or a similar one)
Three 15-kΩ resistors
One 1.5-kΩ resistor

Two identical resistors (e.g., 10 kΩ) for unity gain inverting op-amp
One potentiometer (or variable resistor box), 0–50 kΩ
Two 10-nF capacitors
Three op-amps, LM148 or similar
One adjustable dual DC power supply for the op-amp, ± 12 V

Assignments and Reporting

Read and do items 7–10 and 13–19 in this project.

II. Theory

1. The System

The system is implemented by the circuit of Figure 9.65 which is made of three op-amps, all operating in the inverting configuration. Op-amp 1 is a summing leaky integrator and produces the bandpass output. It has two inputs: one through the system's input, the other through the feedback from the low-pass output. Op-amp 2 integrates the bandpass signal and produces the low-pass output. Op-amp 3 is a unity-gain inverting amplifier that feeds back the low-pass signal to op-amp 1 (the bandpass stage). The system has one input, $v_1(t)$, which arrives at the inverting terminal of op-amp 1. It has two outputs: $v_A(t)$ (a bandpass signal) and $v_B(t)$ (a low-pass signal).

The system has two poles (natural frequencies) at $s_{1,2} = -\alpha \pm j\omega_d$. They are observable throughout the system, including at the bandpass and low-pass outputs. The undamped natural frequency is $\omega_o = \sqrt{\alpha^2 + \omega_d^2}$. By adjusting the resistor R_2 (in the feedback path of op-amp 1) you can place the poles of the system at desired locations, move them closer to or farther away from the $j\omega$ axis, and control the bandwidth (or the quality factor) of the system. In the prelab section, we will use $Q = 5$, $Q = 10$, and 20. In the measurement section you will use $Q = 10$ (with the additional Q values of 5 and 20 optional).

2. System Functions

Consider the circuit of Figure 9.65 with $R = 15$ kΩ, $C = 10$ nF, $R_1 = 1.5$ kΩ, and $R_2 = QR$. Find the two system functions $H_A(s) = V_A/V_1$ and $H_B(s) = V_B/V_1$. Show that

$$\text{Bandpass system function: } H_A(s) = \frac{V_A}{V_1} = -\left(\frac{R}{R_1}\right) \frac{\frac{s}{\omega_0}}{\left(\frac{s}{\omega_0}\right)^2 + \frac{1}{Q}\left(\frac{s}{\omega_0}\right) + 1} = -\left(\frac{R_2}{R_1}\right) \frac{\frac{1}{Q}\left(\frac{s}{\omega_0}\right)}{\left(\frac{s}{\omega_0}\right)^2 + \frac{1}{Q}\left(\frac{s}{\omega_0}\right) + 1}$$

$$\text{Low-pass system function: } H_B(s) = \frac{V_B}{V_1} = -\frac{1}{RCs}H_A(s) = \left(\frac{R}{R_1}\right) \frac{1}{\left(\frac{s}{\omega_0}\right)^2 + \frac{1}{Q}\left(\frac{s}{\omega_0}\right) + 1}$$

where $\omega_0 = \frac{1}{RC}$, $Q = \frac{R_2}{R}$ and $|H_A|_{.Max} = \frac{R_2}{R_1}$, $|H_B|_{.Max} = \text{DC gain} = \frac{R}{R_1}$

Note: Items 3, 4, and 5 should be carried out for $Q = 5$, 10, and 20.

3. Poles and Zeros

Find the poles of the system. Note that the poles are the same in H_A and H_B. These are the system's natural frequencies which we show by $s_{1,2} = -\alpha \pm j\omega_d$. Show that

$$\alpha = \frac{\omega_0}{2Q}$$

$$\omega_d = \frac{\omega_0}{2Q}\sqrt{4Q^2 - 1}$$

$$\omega_o = \sqrt{\alpha^2 + \omega_d^2}$$

Compute $s_{1,2}$ for $Q = 5, \ 10, \ 20$ and show them on the s-plane. Note what happens to the natural frequencies when Q increases.

Find the zero(s) of H_A and H_B. Verify that they may be obtained directly from the circuit.

4. Unit-Step Responses

From item 2 conclude that the unit-step responses have the following form:

$$Ce^{-\alpha t} \cos(\omega_d t + \theta) + D$$

where D is the DC steady-state part of a response. Sketch, by hand or by computer, 10 msec of unit-step responses $(0 \le t \le 10 \text{ msec})$. Note the effect of increasing Q on the responses.

5. Sketch of H(ω)

Based on a vectorial interpretation of $H(\omega)$, sketch by hand the magnitude and phase of the frequency responses. Use log scale for ω and uniform scale for $|H|$.

6. Indicators of Frequency Responses

Using $H_A(s)$ and $H_B(s)$ given in part a, find the mathematical expression for the bandpass and low-pass frequency responses for $Q = 5, 10,$ and 20. Show that the maximum of $|H_A|$ occurs at $\omega_0 = 1/RC$. For high Q approximate half-power frequencies ω_ℓ and ω_h. Plot the frequency responses using a computer. Use log scale for ω and uniform scale for $|H|$.

III. Prelab

7. Planning the Experiment

Read Parts IV and V and plan the experiment.

8. Deriving System Functions

Derive the system functions given in item 2.

9. Computing Frequency Response Indicators

Compute indicators of the frequency responses for $Q = 10$.

IV. Measurements and Analysis

Use the circuit of Figure 9.65 with element values $R = 15 \text{ k}\Omega$, $C = 10 \text{ nF}$, $R_1 = 1.5 \text{ k}\Omega$, and $R_2 = QR$. Measurements will be done at $Q = 10$.

10. Obtaining Step Responses

At $Q = 10$ record the step responses (at A and B) for $0 \le t \le 10$ msec. You may use a square pulse input signal (10 Hz, 100 mV). Observe that the two step responses have the same natural frequencies. From the record measure their natural frequencies and compare with Part II. Model the step responses by mathematical expressions and compare with item 3.

11. Effect of Q on the Step Responses

Qualitatively observe and record the effect (DC steady state, ω_0, α, and ω_d) of increasing Q on the unit-step response. Compare with theory.

12. Measuring Frequency Responses

At $Q = 10$ measure and record the magnitude of the frequency responses $H_A(\omega)$ and $H_B(\omega)$ for the range $100\text{ Hz} \le f \le 10\text{ kHz}$. You may use a sinusoidal input signal at a constant amplitude and measure the output. Plot the magnitude and phase of $H(\omega)_A$ and $H_B(\omega)$. Use log scale for the frequency axis and uniform scale for the magnitude axis. Compare with the result in item 6.

13. Bandpass Output

At $Q = 10$ measure and record the magnitude of the frequency response $H_A(\omega)$. From frequency response measurements obtain the frequency at which $|H_A|$ is at its maximum and call it ω_0. Obtain half-power frequencies ω_ℓ and ω_h. Compute bandwidth $\Delta\omega = \omega_h - \omega_\ell$ and $Q = \omega/\Delta\omega$. Find the phase shift at ω_ℓ, ω_0, and ω_h. Compare with the results obtained in the prelab.

14. Lowpass Output

At $Q = 10$ measure and record the magnitude of the frequency response $H_B(\omega)$. Obtain its 3-dB (half-power) frequency and call it ω_c. Find the phase shift at ω_c. Compare with theory.

15. Effect of Q on Frequency Responses

Qualitatively observe and record the effect of increasing Q on the frequency responses (ω_0, ω_ℓ and ω_h, ω_c, $\Delta\omega$ and Q). Compare with prelab.

16. Response to Repetitive Square Pulse

At $Q = 10$ measure and record 10 msec of time response to a repetitive train of square pulses ($V_{pp} = 100\text{ mV}$, $V_{DC} = 0\text{ V}$). Sweep the frequency from $f = 100\text{ Hz}$ to 10 kHz and observe the trend in the response.

IV. Discussion and Conclusions
17. Conditions for Resonance

From the pole-zero plot determine the range of Q for which the system can resonate.

18. Interpretation of Pulse Response

Relate your observations in item 17 to the unit-step response, frequency response, and natural frequencies of the circuit.

19. Overall Conclusions

Summarize your overall conclusions drawn from this experiment, especially the relationship between the system function, poles and zeros, frequency response, step response, and pulse response.

9.36 Project 4: Examples of Modeling LTI Systems

Summary

This project explores modeling two LTI systems based on the magnitude of their frequency responses.

System 1

The magnitude frequency response of a system is measured and recorded in Table 9.13.

TABLE 9.13 Measurements Giving Salient Frequency Features of a Realizable LTI System

Frequency in kHz	DC	5 Hz	1 kHz	2 kHz	3 kHz	5 kHz	10 kHz	20 kHz	50 kHz
$|H(f)|$ in dB	-20	-19	-17	-13	-11	-7	-3	-1	-0.5
$\angle H(f)$ in degrees									

Plot the magnitude response in the form of a Bode plot. Approximate the plot by asymptotic lines and identify possible pole(s) and zero(s) of the system. From that approximation construct a causal stable rational system function $H(s)$. Realize the above $H(s)$ as a voltage transfer function in a circuit with a minimum number of passive elements and find their values.

System 2

The small-signal magnitude response of a single-stage transistor amplifier is measured and plotted in Figure 9.66. Its salient features are also summarized in Table 9.14. Using the above information, model the amplifier by a rational system function. Assume a realizable and stable system. Consider several possible phase functions. Discuss the order of the system and the consequences of zero placement in the RHP and/or LHP, considering several phase functions. Fine-tune the model by adjusting the zero(s) location to within 0.1 dB.

TABLE 9.14 Measurements Giving Salient Frequency Features of a Small-Signal Common-Emitter Amplifier

ω (rad/s)	$< 10^6$	3.2×10^7	10^9	4.5×10^9	10^{10}	1.1×10^{11}	10^{12}	10^{14}
Magnitude, dB	28 DC level	25 suggested 1st break point	-1.7	-20 suggested 2nd break point	-29	-65 suggested 3rd break point	-88	-128 very high frequency
Slope, dB/decade	0		-20		-40		-20	-20

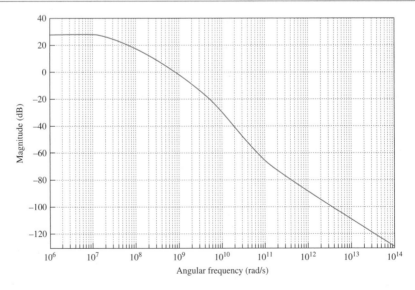

FIGURE 9.66 Small-signal magnitude frequency response of the amplifier obtained by measurement.

Chapter 10

Time-Domain Sampling and Reconstruction

Contents

Introduction and Summary

The frequency band occupied by a signal provides some information about the way the signal is expected to change from one moment to another. The value of the signal at any time may be estimated with various amounts of error from a finite or infinite number of observations. These observations are called samples. The chapter begins with the sampling theorem for low-pass signals, developed in the frequency domain. The frequency-domain presentation of the sampling process provides a simple and convenient method to discuss phenomena such as aliasing and leakage, sampling nearly low-pass signals, effect of the sampling rate, and nonideal reconstruction filters. These phenomena are discussed and illustrated by a multitude of examples. The chapter then extends the discussion to bandpass signals and their sampling. The first project at the end of the chapter suggests researching practical considerations and recommendations for professional standards in sampling neuroelectric signals such as EEG, EMG, and EKG. The second project experiments with and explores some real-time sampling and aliasing of signals.

10.1 Converting from Continuous Time to Discrete

Many continuous-time signals are represented by their samples taken at regular intervals. Traditional measurement and recording of atmospheric data (temperature, humidity, and pressure) at weather stations are done at specified intervals of time. The Dow Jones Industrial Average index (DJIA) is computed every minute and also obtained as the daily, monthly, and annual values. Similar observations apply to examples presented in Chapter 1 of this book, such as sunspot numbers, CO_2 content of the air, and vibrations of the earth. An intuitive commonsense rationale for recording the samples rather than continuously is that such signals on average do not change in time faster than at a given rate. Their frequency components are confined within a band. The signals are band limited. By sampling, a continuous-time function is converted to a sequence of numbers, which represent the value of the function at sampling instances. When sampling is done uniformly every T seconds, T is called sampling interval, and its inverse is called sampling rate, $F_s = 1/T$ in units of samples per seconds (also called Hz). The sequence of samples is called the discrete-time signal.

A general requirement in sampling a continuous-time signal is that it can be recovered from its samples. The sampling theorem[1] establishes the relationship between the bandwidth of the signal and the minimum sampling rate which can provide reconstruction of the original signal without distortion. For a low-pass signal with frequencies not higher than f_0 the acceptable sampling rate for signal recovery is greater than twice the highest frequency present in the signal, $F_s > 2f_o$. ($F_s = 2f_0$ is called Nyquist rate.) The theorem also provides method of reconstruction and states that the highest frequency in the reconstructed signal is less than or equal to F_s. Moreover, if the Nyquist rate is not satisfied, not only frequencies higher than $F_s/2$ are lost (cannot be

---

[1] The sampling theorem is originally credited to H. Nyquist and C. Shannon. See the following two papers: H. Nyquist, "Certain Topics in Telegraph Transmission Theory," *Transactions of the AIEE*, Feb. 1928, pp. 617–644. Reprinted in the *Proceedings of the IEEE*, 90, no. 2 (February 2002). C.E. Shannon, "Communication in the Presence of Noise," *Proceedings of the IRE*, 37, no. 1 (January 1949), pp. 10–21. Reprinted in the *Proceedings of the IEEE*, 86, no. 2 (February 1998).

recovered by reconstruction process), but also these frequencies fold back as the low end of the band (near DC) and produce a phenomenon called aliasing. Consequently, another requirement for sampling is that the continuous-time signal be low-pass filtered to limit it to a frequency band, no more than the band that contains the information of interest.

A familiar example in sampling is that of audio signals such as speech and music, which are initially recorded in the form of continuous-time electrical waveform obtained by a transducer such as a microphone. The information carried by normal speech (including speaker recognition) is confined to a band of less than 3 kHz. For that purpose the sampling rate of 8 kHz can be used. A sampling rate of over 40 kHz preserves all audio components that the human ear can hear. Another example in sampling is digital recording of seismograms. They use sampling rates from 0.01 Hz to 0.1, 1, 40, 80, and 100 Hz. Other examples of practical interest are digital recordings of neuroelectric signals. For example, a frequency band of 0.5 to 50 Hz is satisfactory for routine examination of electroencephalograms (EEG signals). The sampling rate, then, can be 100 Hz for that purpose, while a higher bandwidth (called full-band EEG) would use a higher sampling rate. Similar considerations apply to electrocardiograms (EKG, bandwidth $= 1$ Hz to 30 Hz) and electromyogram (EMG, bandwidth $= 4$ Hz to 2 kHz, also up to 20 kHz) and other electrical signals recorded from the central nervous system.

Notations

Sampling a continuous-time function $x(t)$ every T seconds generates a sequence of numbers whose values are $x(nT)$. The sequence is called a discrete-time function. In this book the sample sequence is shown by $x(n)$. It is understood that $x(n)$ represents a function of the discrete variable n.

Example **10.1**

Consider the continuous-time low-pass signal

$$x(t) = \frac{\sin(2\pi f_0 t)}{\pi t}, \quad f_0 = 1 \text{ kHz}, \quad -\infty < t < \infty$$

The signal is band limited to 1 kHz. Sample $x(t)$ at the rate of $F_s = 2(1 + \alpha) f_0$ samples per second. List the sequence of the first 13 samples (starting from $t = 0$) rounded to the nearest integer for the following values of

a. $\alpha = 1$

b. $\alpha = 3$

Solution
Let the sampling interval be $T = 1/F_s$. Substitute $t = nT$ in the expression for $x(t)$ to obtain

$$x(n) \equiv x(t)|_{t=nT} = \frac{\sin(2\pi f_0 nT)}{\pi nT} = 2 f_0 \frac{\sin\left(\frac{\pi n}{1+\alpha}\right)}{\frac{\pi n}{1+\alpha}}, \quad -\infty < n < \infty$$

a. $\alpha = 1$, $x(nT) = 2{,}000\dfrac{\sin\left(\frac{\pi n}{2}\right)}{\frac{\pi n}{2}}$

$x(n) = \{2{,}000,\, 1{,}273,\, 0,\, -424,\, 0,\, 254,\, 0,\, -182,\, 0,\, 141,\, 0,\, -115,\, 0,\, \cdots\}$
    ↑

b. $\alpha = 3$, $x(nT) = 2{,}000\dfrac{\sin\left(\frac{\pi n}{4}\right)}{\frac{\pi n}{4}}$

$x(n) = \{2{,}000,\, 1{,}800,\, 1{,}273,\, 600,\, 0,\, -360,\, -424,\, -257,\, 0,\, 200,\, 254,\, 164,\, 0,\, \cdots\}$
    ↑

The sample values are shown as an array. The up-arrow under the first sample indicates $n = 0$. Because $x(t)$ is an even function sample values for $n < 0$ are the same as those for $n > 0$. For recovery of $x(t)$ from $x(n)$ see section 10.3.

Sample the continuous-time low-pass signal

$$x(t) = \frac{\sin(2\pi f_0 t)}{\pi t}, \qquad f_0 = 1 \text{ kHz}, \quad -\infty < t < \infty$$

at the rate of $2f_0$ Hz. List the sequence of the first 10 samples (starting from $t = 0$) and argue that the original signal may not be reconstructed from its samples.

Solution

From Example 10.1 we have $x(nT) = 2{,}000\dfrac{\sin(\pi n)}{\pi n}$, $x(n) = \{1, 0, 0, 0, 0, 0, 0, 0, 0,$
                                                                                              ↑

$0 \cdots\} = \begin{cases} 1 & n = 0 \\ 0 & \text{elsewhere} \end{cases}$ Unless more exact information about the spectrum of $x(t)$ is known, it cannot be recovered from the above samples. A sampling rate greater than $2f_0$ is needed.

Example 10.3

Sampling a speech signal

Consider a nondeterministic signal such as speech. The value of the signal at a given time may not be exactly specified from its past samples or even from its complete past. Past history is used to derive some statistical averages. However, it is known that on average the rate of change of the signal does not exceed a known limit and the value of the signal at any moment may be estimated with some accuracy from samples taken prior to that moment. Speech is such a signal. An example of sampled speech is shown in Figure 1.7 of Chapter 1. The phrase *a cup of hot tea* is sampled at the rate of 22,050 samples per second. The abscissa in that figure shows the time in seconds. The trace in (*a*) is approximately 1.4 seconds long and shows the complete phrase. The trace in (*b*) is 240 msec long and shows the signal produced from the enunciation *hot.* Note the structure of the signal in (*b*): a sequence of similar but not identical wavelets that may be modeled by sinusoids with decreasing amplitudes. Also note that due to its low-pass property, the amplitude of speech signal is correlated with its values within its neighborhood and may be estimated from past samples. The correlation between neighboring samples is evident in trace (*c*), which exhibits 80 msec of the enunciation *hot.* Trace (*d*) shows the fine structure of the signal.

Summary

For sampling a continuous-time signal and its reconstruction, two questions need to be answered. The first question is, what minimum sampling rate guarantees the error-free reconstruction of the signal? As one may suspect, the minimum rate depends on how fast the signal may change with time. Fast-changing signals need faster sampling rates. This question is answered by the sampling theorem. The second question is, how does the reconstruction error depend on the number of samples used, and the weight given to each sample? This can be answered by the reconstruction method. In this chapter we will examine the sampling process and reconstruction method applied to the following three classes of signals:

1. Strictly band limited low-pass signals
2. Nearly low-pass signals (not strictly band limited)
3. Bandpass signals

10.2 Mathematical Representation of Sampling

Sampling a continuous-time signal $x(t)$ means registering the values of the function at certain countable instances. This converts $x(t)$ to the sequence of its values at the sampling times, generally shown as $x(n)$, a function of a discrete variable. Uniform sampling of $x(t)$ every T seconds (a sampling rate of $F_s = 1/T$ Hz) generates the sequence $x(nT)$. This sequence may be shown as a discrete-time signal $x(n)$. In this notation, it is understood that the discrete-time signal $x(n)$ is equal to the values of $x(t)$ at sampling times, or $x(n) \equiv x(nT)$.

Sampled Function in the Continuous Time

Samples may be represented as a function of time (a continuous variable) as in Figure 10.1. Such a representation helps illuminate conditions for reconstruction of $x(t)$ from its samples. To avoid confusion between the discrete-time function $x(n)$ and the continuous-time representations of the samples we will assign the name $y(t)$ to the continuous-time domain representation of the sampled function. Mathematically, uniform sampling may be modeled as multiplication of $x(t)$ by a sampling function $s(t)$, a train of very narrow pulses, ideally unit impulses, spaced every T seconds.

$$s(t) = \sum_{n=-\infty}^{\infty} \delta(t - nT)$$

The sampled function is

$$y(t) = x(t)s(t) = x(t) \sum_{n=-\infty}^{\infty} \delta(t - nT) = \sum_{n=-\infty}^{\infty} x(nT)\delta(t - nT)$$

$y(t)$ is a train of impulses spaced every T seconds. The strength of an impulse at $t = nT$ is equal to the magnitude of $x(t)$ at that point. We would like to find conditions and methods for recovering $x(t)$ from $y(t)$.

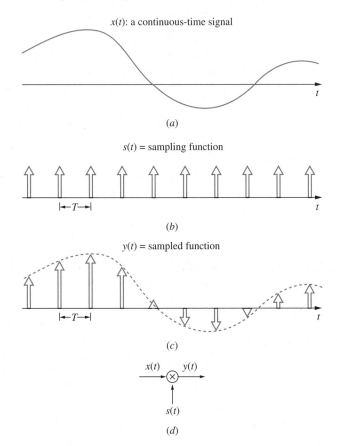

FIGURE 10.1 Representation of sampling in the time domain. Sampling a continuous-time signal, $x(t)$ shown in **(a)**, is mathematically equivalent to multiplying it with a train of unit impulses called the sampling function $s(t)$ shown in **(b)**. The sampled signal $y(t)$ is a train of impulses whose strength assume values of $x(t)$ at the sampling moments: $y(t) = x(t)s(t)$, shown in **(c)**. The operation is represented in **(d)**.

Fourier Transform of a Sampled Function

Multiplication of $x(t)$ and $s(t)$ in the time domain is equivalent to their convolution in the frequency domain. That is, $Y(f) = X(f) \star S(f)$, where $Y(f)$, $X(f)$, and $S(f)$ are Fourier transforms of $y(t)$, $x(t)$, and $s(t)$, respectively.

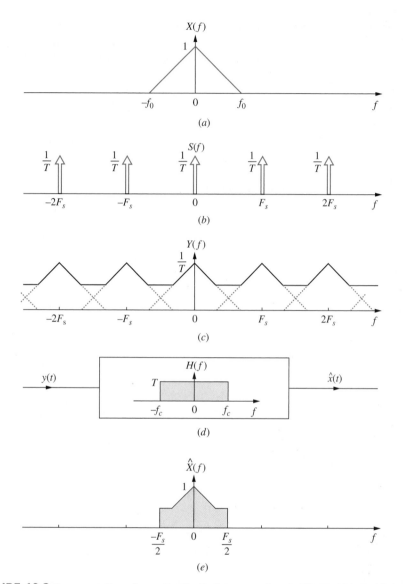

FIGURE 10.2 Representation of sampling in the frequency domain The Fourier transform of a low-pass continuous-time signal is shown by $X(f)$ in **(a)**. The sampling function $s(t)$ is made of a train of unit impulses with period T. Its Fourier transform is a train of impulses with period $F_s = 1/T$ in the frequency domain, shown by $S(f)$ in **(b)**. The Fourier transform $Y(f)$ of the sampled signal is obtained by convolving $X(f)$ with $S(f)$. Because of the periodicity of $S(f)$, $Y(f)$ is also periodic with period F_s. Under certain conditions $X(f)$ can be recovered by low-pass filtering $Y(f)$. In this figure $X(f)$ extends beyond $F_s/2$ and may not be recovered from $Y(f)$ without error. The case of error-free recovery is presented in Figure 10.4.

For illustration purposes, consider $X(f)$ shown in Figure 10.2(a). The Fourier transform of a train of unit impulses spaced every T seconds in the time domain is a train of impulses of strength $1/T$, spaced every $1/T$ Hz in the frequency domain (Figure 10.2b). Therefore,

$$s(t) = \sum_{n=-\infty}^{\infty} \delta(t - nT) \implies S(f) = \frac{1}{T} \sum_{n=-\infty}^{\infty} \delta\left(f - \frac{n}{T}\right)$$

and

$$Y(f) = X(f) \star S(f) = X(f) \star \left[\frac{1}{T} \sum_{n=-\infty}^{\infty} \delta\left(f - \frac{n}{T}\right)\right]$$

But

$$X(f) \star \delta\left(f - \frac{n}{T}\right) = X\left(f - \frac{n}{T}\right)$$

Therefore,

$$Y(f) = \frac{1}{T} \sum_{n=-\infty}^{\infty} X\left(f - \frac{n}{T}\right)$$

See Figure 10.2(c). It is seen that the Fourier transform of $y(t)$ is a periodic function obtained by adding the periodic repetitions, every $1/T$ Hz, of $(1/T)X(f)$. The number of samples per second is called the sampling rate and is shown by $F_s = 1/T$. F_s is the period of $Y(f)$.

Example
10.4

Consider a continuous-time triangular pulse signal $x(t)$ (1-volt height and 50-second base). The signal and its magnitude-normalized Fourier transform $X(f)$ are shown in Figure 10.3(a). The pulse is sampled every T seconds, corresponding to a rate of $F_s = 1/T$ Hz. The sampled signal $y(t)$ is shown in b, c, and d for three sampling rates ($F_s = 0.125$, 0.25, and 0.5 Hz, respectively), along with its Fourier transform. As expected $Y(f)$ is periodic with period F_s, that provides separation between neighboring lobes as shown in b, c, and d. As will be seen in the next section the above $x(t)$ may not be recovered error-free by sending $y(t)$ through a low-pass filter with cutoff frequency $f_c = F_s/2$, regardless of how high the sampling rate is.

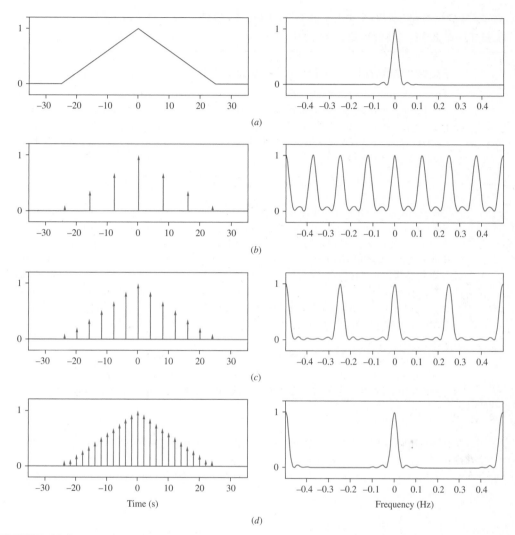

FIGURE 10.3 Sampling a nearly low-pass signal and its Fourier transform. A 50-second triangular pulse $x(t)$ and its magniutde-normalised Fourier transform are shown in *(a)*, left and right sides, respectively. The major energy of the pulse is concentrated in a main lobe within the low-pass band $|f| < 0.04$ Hz. The pulse is sampled at rates $F_s = 0.125,\ 0.25,$ and 0.5 Hz [shown in *(b)*, *(c)*, and *(d)*, respectively, left column]. The Fourier transforms of sampled signals are normalized in order to have the peak value 1 and are shown to the right of each signal. The transforms are F_s-periodic and exhibit repetitions of the main lobe in the transform domain. A higher sampling rate pushes the main lobes away from each other and results in lower reconstruction errors.

a. The continuous-time signal, a 50-second triangular pulse.
b. $T = 8$ seconds, $F_s = 0.125$ Hz, low separation between lobes results in the highest reconstruction error.
c. $T = 4$ seconds, $F_s = 0.25$ Hz, higher separation between lobes results in a lower reconstruction error.
d. $T = 2$ seconds, $F_s = 0.5$ Hz, highest separation between lobes results in the lowest reconstruction error.

10.3 Sampling and Reconstruction of Strictly Low-Pass Signals

Presentation in the Frequency Domain

The frequency components of a strictly band limited low-pass signal are limited to a maximum frequency f_0 so that $X(f) = 0$ for $|f| > f_0$. In the previous section we noted that $Y(f)$ is periodic with a period F_s Hz. By choosing a sampling rate $F_s > 2f_0$ the interference between each repetition of $X(f - \frac{n}{T})$ and its neighbors in creating $Y(f)$ is reduced to zero and

$$Y(f) = \frac{1}{T}X(f), \quad -\frac{F_s}{2} < f < \frac{F_s}{2}$$

An example is shown in Figure 10.4. In that figure, $X(f)$ is the Fourier transform of the low-pass signal (a), $S(f)$ is the Fourier transform of the sampling function (b), and $Y(f) = S(f) \star X(f)$ is the Fourier transform of the sampled signal (c). The original signal $x(t)$ can then be recovered by passing the sampled function $y(t)$ through an ideal low-pass filter with a gain of T and a cutoff at f_c, where $f_0 < f_c < (F_s - f_0)$. The filter extracts $X(f)$ from $Y(f)$ while eliminating its repetitions. The output of the filter is $x(t)$. Compare with Figure 10.2. For error-free reproduction, assuming an ideal low-pass filter, the sampling rate has to be greater than the Nyquist rate, $F_s > 2f_0$.

Presentation in the Time Domain

In the time domain, the output, $\hat{x}(t)$, of the reconstruction filter may be expressed as the convolution of the sampled function $y(t)$ with the filter's impulse response $h(t)$.

$$\hat{x}(t) = y(t) \star h(t)$$

The impulse response of an ideal low-pass filter with cutoff frequency at f_c and gain T is

$$h(t) = T\frac{\sin(2\pi f_c t)}{\pi t}$$

Therefore,

$$\hat{x}(t) = \sum_{-\infty}^{\infty} Tx(nT)\frac{\sin 2\pi f_c(t - nT)}{\pi(t - nT)}$$

Under the conditions $F_s > 2f_0$ and $f_0 < f_c < (F_s - f_0)$ we obtain $\hat{x}(t) = x(t)$ (see the frequency domain presentation in Figure 10.4). The above formula interpolates $x(t)$ from an infinite number of its samples, past and present. The filter is sometimes called the interpolation filter and the above equation is called interpolation formula. We will call it reconstruction or recovery filter (rather than the interpolation filter) in order to distinguish its function from the process of increasing number of samples by interpolation from existing samples.[2] The gain of the reconstruction filter is T and is

---

[2]Interpolation is briefly introduced in Chapter 16.

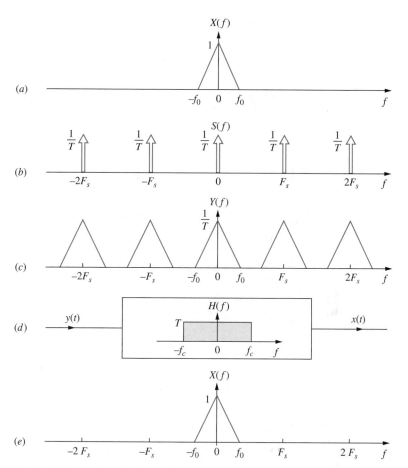

FIGURE 10.4 Frequency-domain representation of sampling and error-free reconstruction process. A strictly low-pass signal $[X(f) = 0, \ |f| > f_0]$ may be reconstructed (also said to be recovered) from its samples by passing them through an ideal low-pass filter with cutoff frequency at f_c $[H(f) = 0, \ |f| > f_c]$. Reconstruction is error-free if samples are taken at a rate higher than the Nyquist rate $(F_s \geq 2f_0)$, and if the cutoff frequency of the ideal filter is $f_0 \leq f_c \leq (F_s - f_0)$ as seen in this figure. For practical reasons, one often chooses $f_c = F_s/2$.

chosen to compensate for the gain $1/T$ introduced in $y(t)$ by sampling. The cutoff frequency of the reconstruction filter is generally chosen to be $f_c = F_s/2$. In that case

$$h(t) = \frac{\sin \pi F_s t}{\pi F_s t}, \quad x(t) = \sum_{-\infty}^{\infty} x(n) \frac{\sin \pi F_s (t - nT)}{\pi F_s (t - nT)}$$

The recovery formula states that $x(t)$ may be built from blocks of $x(n)h(t - nT)$. The building blocks are formed by shifting $h(t)$ (whose central lobe is T seconds wide) by nT units of time and multiplying it by sample values $x(n)$. Note that the above structure places the peak of each shifted $h(t)$ at a sampling instance with its zero-crossings at the neighboring sampling instances:

$$h(t) = \begin{cases} 1, & t = 0 \\ 0, & t = nT \neq 0 \end{cases}$$

Therefore, at $t = nT$ the value of the above sum is exactly equal to the value of the sample at that point. Figure 10.5 simulates the reconstruction process by such an ideal filter in the time domain.

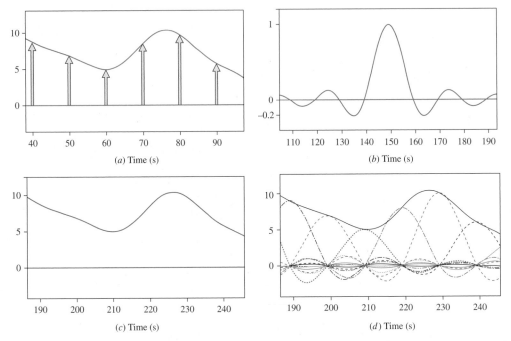

FIGURE 10.5 Time-domain representation of reconstruction process. This figure illustrates, in the time domain, how a signal is reconstructed from unit-impulse responses of the reconstructed filter, weighted by sample values and added together. All abscissa are time in seconds. *(a)* the signal, *(b)* filter's unit-impulse response, *(c)* the output of the filter, *(d)* reconstruction of the signal by the building blocks.

The signal shown in *(a)* is sampled every 10 seconds, $F_s = 0.1$ Hz. The sampled signal is made of impulses whose strengths are equal to the samples. The ideal reconstruction filter is low-pass with cutoff at $f_c = F_s/2 = 0.05$ Hz. The unit-impulse response of such a filter is $h(t) = 10\sin(0.1\pi t)/(\pi t)$. In the simulation, to make $h(t)$ realizable it is shifted by 150-second to the right as shown in *(b)*. Note that the base of the main lobe of $h(t)$ is 20 seconds wide, corresponding to a cutoff frequency at 0.05 Hz. The output of the filter is shown in *(c)*, to be compared with the signal shown in *(a)*. The output is the sum of contributions from individual samples, which are formed by shifting the unit-impulse response by nT units of time and multiplying it by sample values $x(n)$, which we will call the building blocks $x(n)h(t - 150 - nT)$. These are shown in *(d)*. Note that the zero-crossings of each building block occur exactly at the locations of all other samples. Therefore, at each sample time the output has one contribution, that of its own sample. The reconstruction is therefore exact.

10.4 Sample and Hold

In the previous section we used an ideal filter to reconstruct the continuous-time signal from its samples. The method produces the exact value of $x(t)$ from its samples. It, however, requires access to all samples during $-\infty < t < \infty$. This is because the unit-impulse response of the ideal low-pass filter used in that reconstruction process extends from $-\infty$ to ∞. That unit-impulse response, in time and frequency domains, is shown in Figure 10.6(*a*). Several other types of filter are used to reconstruct a continuous-time signal from the samples. The simplest one, often used in digital-to-analog converters,

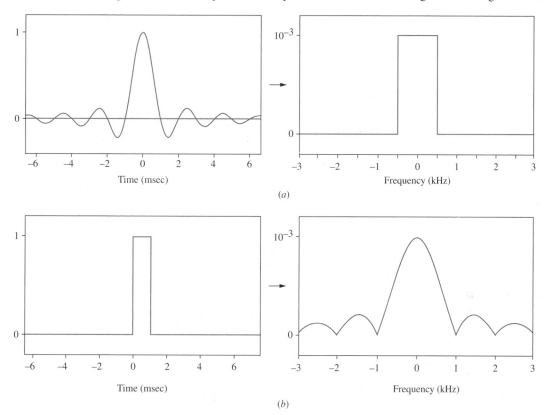

FIGURE 10.6 Reconstruction filters. The response characteristics of two types of reconstruction filters, intended for a sampling rate $F_s = 1$ kH, are shown. The filters' impulse responses are in the left-hand side and the normalized magnitude of their Fourier transforms are in the right-hand side. Time is in msec and frequency is in KHz. The upper row shows characteristics of an ideal low-pass filter with a cutoff frequency at $F_s/2 = 500$ Hz and the lower row shows those of a 1-msec S/H filter.

a. Ideal low-pass filter $h(t) = \dfrac{\sin(1000\pi t)}{1000\pi t}$ $H(f) = \begin{cases} 10^{-3}, & -500 < f < 500 \\ 0, & \text{elsewhere} \end{cases}$

b. S/H filter $h(t) = \begin{cases} 1, & 0 < t < 1 \text{ msec} \\ 0, & \text{elsewhere} \end{cases}$ $H(f) = \dfrac{\sin(\pi f/1{,}000)}{\pi f} e^{-j\pi f/1{,}000}$

is the method of sample and hold, shown by S/H. In the sample and hold method, instead of an ideal reconstruction low-pass filter we use the filter with an impulse response of

$$h(t) = \begin{cases} 1, & 0 < t < T \\ 0, & \text{elsewhere} \end{cases}$$

where T is the sampling interval. The frequency response of the sample and hold filter is

$$H(f) = \int_{-\infty}^{\infty} h(t)e^{-j2\pi ft}\,dt = \int_{0}^{T} e^{-j2\pi ft}\,dt = \frac{\sin(\pi fT)}{\pi f}e^{-j\pi fT} \quad \text{where } T = \frac{1}{F_s}$$

The unit-impulse response and the frequency response of the sample and hold filter are shown in Figure 10.6(*b*) and are to be compared with the ideal filter of Figure 10.6(*a*). The signal produced by the S/H process will acquire a stepwise structure, as seen in the lower trace of Figure 10.24(*a*), with sharp discontinuities that mainfest themselves as high frequencies. The reconstructed signal, therefore, will contain high frequencies that don't exist in the original signal. These high frequencies may be reduced by increasing the sampling rate (or by increasing the number of samples by a process called interpolation) and by adding a low-pass filter following the sample and hold.

Example **10.5**

Consider a sample and hold exercise where a 10-Hz sinusoidal signal, Figure 10.7(*a*), is sampled at the rate of F_s. Samples are then passed through a hold filter followed by a first-order low-pass filter [time constant = 1 msec, Figure 10.7(*b*)] for smoothing. Figure 10.7(*c*) shows the output of the hold filter (on the left-hand side) and that of the low-pass smoothing filter (on the right-hand side) for $F_s = 100$ Hz. Doubling the sampling rate to $F_s = 200$ Hz reduces the high frequencies in the output of the filters as shown in Figure 10.7(*d*). Increasing sampling rate by a factor 5, to $F_s = 500$ Hz, results in Figure 10.7(*e*), in which the reconstructed signal is closer to $x(t)$ than in Figure 10.7(*c*) and (*d*). The sampling and reconstructions in Figure 10.7 are summarized by the following.

Original signal	$x(t) = \cos(20\pi t)$	Figure 10.7(*a*)
Sampled signal	$\sum_{-\infty}^{\infty} \cos(20\pi nT)\delta(t - nT),$	$T = \dfrac{1}{F_s}$
S/H filter	$h_1(t) = u(t) - u(t - T)$	
Low-pass filter	$h_2(t) = e^{-1000t}u(t)$	Figure 10.7(*b*)
S/H and the low-pass filter	$h(t) = h_1(t) \star h_2(t)$	
Reconstructed signal	$\sum_{-\infty}^{\infty} \cos(20\pi nT)h(t - nT)$	Figure 10.7(*c*), $F_s = 100$ Hz
		Figure 10.7(*d*), $F_s = 200$ Hz
		Figure 10.7(*e*), $F_s = 500$ Hz

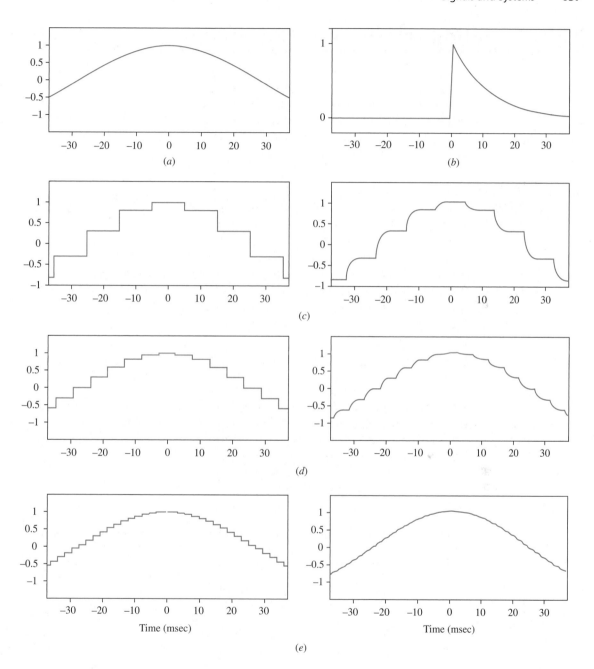

FIGURE 10.7 Sampling and reconstruction of a sinusoid by sample-and-hold followed by low-pass filtering, for three sampling rates. Increasing the number of samples reduces the high-frequency components generated in a sample and hold (S/H) operation, and provides looser requirements for a low-pass filter to follow it. For detail see Example 10.5.

10.5 Sampling Nearly Low-Pass Signals

Consider a signal $x(t)$ in which the major frequency components, but not all of them, are limited to below frequency f_0. The signal is not strictly low-pass as it contains some power in the frequency range beyond f_0, which diminishes as the frequency increases. Sample $x(t)$ uniformly every T sec to obtain $y(t)$. Can $x(t)$ be recovered error-free from its samples? Strictly speaking, as we have seen in the previous sections, the answer is negative, as illustrated by Example 10.6 below.

Example 10.6

Consider a two-sided even exponential signal with a time constant of τ sec, and its Fourier transform.

$$x(t) = e^{-|t|/\tau}, \quad X(f) = \frac{2\tau}{1 + 4\pi^2 \tau^2 f^2}$$

The signal is not strictly a band limited low-pass one. However, its Fourier transform diminishes as the frequency increases. Sample $x(t)$ uniformly every T sec, corresponding to a sampling rate of $F_s = 1/T$ Hz. The sampled function $y(t)$ and its Fourier transform, as derived in section 10.2, are

$$y(t) = \sum_{n=-\infty}^{\infty} e^{-\frac{T|n|}{\tau}} \delta(t - nT), \quad Y(f) = \frac{1}{T} \sum_{n=-\infty}^{\infty} X\left(f - \frac{n}{T}\right)$$

$Y(f)$ is the scaled sum of periodic repetitions of $X(f)$ every F_s Hz. To recover the original signal from its samples, $y(t)$ is passed through an ideal low-pass filter with gain T and cutoff frequency $f_c = F_s/2$ Hz. The output of the filter is named $z(t)$. The Fourier transform of the output of the filter is

$$Z(f) = \sum_{n=-\infty}^{\infty} X\left(f - \frac{n}{T}\right), \quad |f| < f_c$$

$$= 0, \quad \text{elsewhere}$$

Due to the interference by the tails of neighboring lobes in $Y(f)$ (which cannot be totally eliminated by the filter), $Z(f)$ cannot be a duplicate of $X(f)$ and $z(t)$ cannot be an error-free reconstruction of $x(t)$. For example, at $f_c = F_s/2$ we will have

$$Z(f_c) = 2\mathcal{RE}\{X(f_c) + X(3f_c) + X(5f_c) + \cdots\}$$

$$= 2\tau \left[\frac{1}{1 + \beta^2} + \frac{1}{1 + 9\beta^2} + \frac{1}{1 + 25\beta^2} + \cdots\right]$$

$$= 2\tau \sum_{n=1, n \text{ odd}}^{\infty} \frac{1}{1 + \beta^2 n^2}, \quad \text{where } \beta = \pi\tau F_s$$

At sufficiently high sampling rates, however, the spacing F_s in the frequency domain becomes large and the interference between the tails of repeated $X(f - n/T)$ becomes negligible, reducing the reconstruction error. A numerical example is shown in

Figure 10.8. A double-sided even exponential signal $x(t) = e^{-|t|/4}$, along with its normalized Fourier transform is shown in (*a*). The signal is sampled every 4 seconds, corresponding to a rate of $F_s = 0.25$ Hz in (*b*). The sampled signal, $y(t)$, is on the left-hand side, where impulse samples are shown by arrows. The Fourier transform of $y(t)$ is shown on the right. As expected, the Fourier transform of $y(t)$ is periodic in the frequency domain, with period 0.25 Hz. The sampled function in (*b*) is passed through an ideal low-pass filter with cutoff frequency at 0.125 Hz. The Fourier transform, $Z(f)$, of the filter's output is shown on the right in (*c*). The time-domain representation of the filter's output, $z(t)$, is shown on the left, to be compared with the original signal in (*a*). It is seen that recovery of the above low-pass signal (which is not strictly band limited) contains errors even when the reconstruction filter is ideal.

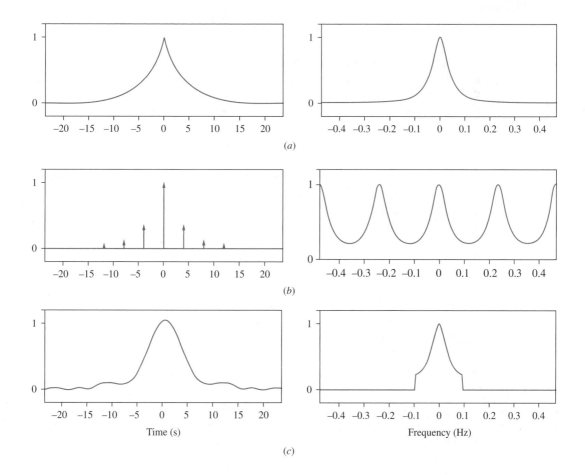

FIGURE 10.8 Error in signal recovery. *(a)* A double-sided even exponential function $x(t)$ and its Fourier transform. *(b)* Sampled function $y(t)$ and its Fourier transform. *(c)* Reconstructed signal $z(t)$ and its Fourier transform. The reconstruction has errors because $x(t)$ is not strictly band limited and low-pass. See Example 10.6.

Reducing Error in Signal Recovery by Increasing the Sampling Rate

A low-pass signal, which is not strictly band limited, may be recovered with less error if higher sampling rates are used. This is because higher sampling rates provide more spacing in the frequency domain between the repeated segments of the Fourier transform of the sampled signal, thus reducing overlap between neighboring segments. This is illustrated in the following example.

Revisiting Example 10.4

In Example 10.4 we sampled a 50-second continuous-time triangular pulse $x(t)$ at three different rates. Because the pulse is time limited, its transform is not band limited. Theoretically, the signal may not be recovered from its samples error-free, regardless of how high the sampling rate is. However, as the major share of the energy in the signal is found in frequencies within the central lobe (below 0.04 Hz) and higher sampling rates provide larger separations between the repetitions of the main lobe in the frequency domain, the result is lower error at the output of the reconstruction filter, which low-pass filters the central lobe in $Y(f)$ from its neighboring repetitions. This is illustrated in Figure 10.9. The time functions are on the left with their normalized Fourier transforms on the right. The $x(t)$ and its Fourier transform are shown in (a). The reconstruction error is considerable in (b) (sampled every 12 sec) but almost negligible in (c) (sampled every 2 sec, six times faster than in (b)).

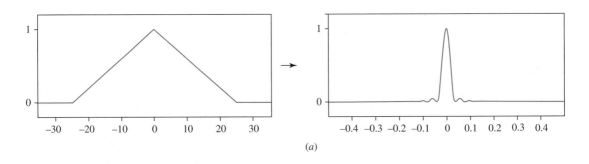

(a)

FIGURE 10.9 Signal recovery at higher sampling rate. This figure, to be continued on the next page, samples a 50-second triangular pulse, shown in **(a)**, at intervals $T = 12$, and 2-second [shown in **(b)** and **(c)**, respectively] and reconstructs it by passing the samples through an ideal low-pass filter with cutoff frequency at $F_s/2$. It visually demonstrates the effect of higher sampling rate in reducing recovery error. Time functions are shown in the left column with their Fourier transforms in the right column. The magnitudes of the transforms are all normalized to have maxima 1. The time scale is in seconds and the frequency scale is in Hz. The cutoff frequency of the low-pass reconstruction filter is $F_s/2$. The reconstruction error is considerable in **(b)** but almost negligible in **(c)**.

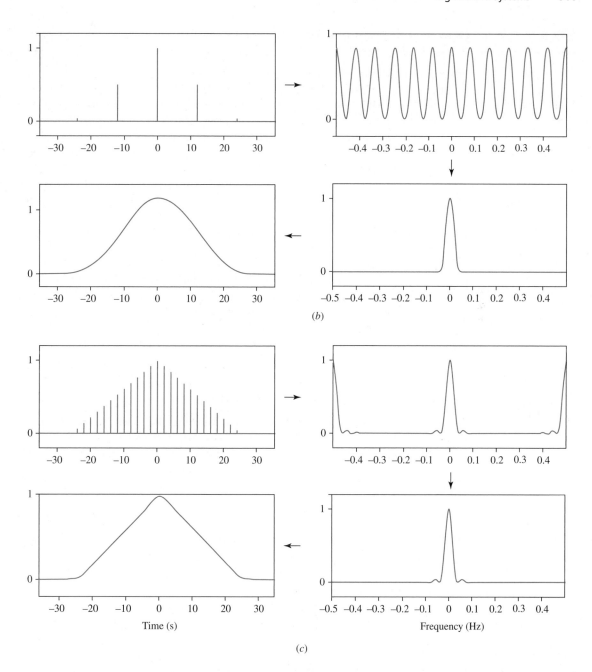

(b)

(c)

Time (s)   Frequency (Hz)

FIGURE 10.9 (*Continued*)

10.6   Aliasing and Leakage

From an examination of the Fourier transform of the output of the reconstruction filter we observe that the time function can become different from the original signal in two respects. One possible difference is at low frequencies where some of the high-frequency components of $x(t)$ are shifted to lower frequencies. This is called *aliasing*. This effect is due to the fact that the original signal is not band limited. The reconstruction filter, even if it is ideal, cannot eliminate aliasing unless the original signal is band limited and sampling rate is greater than or equal to the Nyquist rate.

The second possible difference between the reconstructed signal and the original signal may be found at high frequencies where because of windowing operations the spectra spread and the high frequencies leak back to the reconstructed signal. By making the low-pass reconstruction filter close to an ideal filter and processing long segments of data we can reduce the leakage.

In summary, for an error-free sampling and recovery operation, it is important that two conditions be satisfied. One condition is that the signal be low pass and the sampling rate be at least twice the highest frequency present in it (the Nyquist rate). If this condition is not met, aliasing occurs. To eliminate aliasing, we make the signal strictly low pass by passing it through a low-pass filter, called an anti-aliasing filter, before sampling. The second condition is that the reconstruction operation be ideal, otherwise leakage occurs.

Example **10.8**

Illustration in the frequency domain

A sinusoidal signal $x(t) = 2\cos(2\pi f_0 t)$ is sampled at the rate F_s Hz. To recover the signal, the samples are passed through an ideal low-pass filter with a gain $T = 1/F_s$ and cutoff frequency slightly greater than f_0. Find the output of the filter for

a. $F_s = 4f_0$
b. $F_s = 1.5f_0$
c. $F_s = 1.1f_o$

In each case, determine if aliasing has occurred or not.

Solution
Following the approach of sections 10.2 and 10.3 we first observe the Fourier transforms of the filter's input and output.

$$x(t) = 2\cos 2\pi f_0 t, \qquad X(f) = \delta(f + f_0) + \delta(f - f_0), \qquad \text{Two unit impulses at } \pm f_0,$$

$$s(t) = \sum_{n=-\infty}^{\infty} \delta(t - nT), \quad S(f) = \frac{1}{T}\sum_{k=-\infty}^{\infty} \delta(f - kF_s), \quad T = \frac{1}{F_s}, \qquad \text{Impulses at } kF_s, \ k \text{ integer}$$

$$y(t) = x(t)s(t), \qquad Y(f) = X(f) \star S(f) = \frac{1}{T}\sum_{k=-\infty}^{\infty} X(f - kF_s), \quad \text{Impulses at } \pm(kF_s \pm f_0)$$

Convolution of $X(f)$ with $S(f)$ generates spectral line at $(-2F_s \pm f_0)$, $(-F_s \pm f_0)$, $\pm f_0$, $(F_s \pm f_0)$, $(2F_s \pm f_0)$, ... The components that are at f_0 or below it pass through the filter and become parts of the reconstructed signal. The results are summarized in the following table.

	F_s	Frequency Components in $y(t)$	Output	Comments
a.	$4f_0$	$\pm f_0,\ \pm 3f_0,\ \pm 5f_0,\ \pm 7f_0,\ \cdots$	$2\cos 2\pi f_0 t$	Exact recovery, see Figure 10.10(a).
b.	$1.5f_0$	$\pm 0.5f_0,\ \pm f_0,\ \pm 2f_0,\ \pm 2.5f_0,\ \cdots$	$2\cos \pi f_0 t + 2\cos 2\pi f_0 t$	Aliasing, see Figure 10.10(b).
c.	$1.1f_0$	$\pm 0.1f_0,\ \pm f_0,\ \pm 1.2f_0,\ \pm 2.1f_0,\ \cdots$	$2\cos 0.2\pi f_0 t + 2\cos 2\pi f_0 t$	Aliasing.

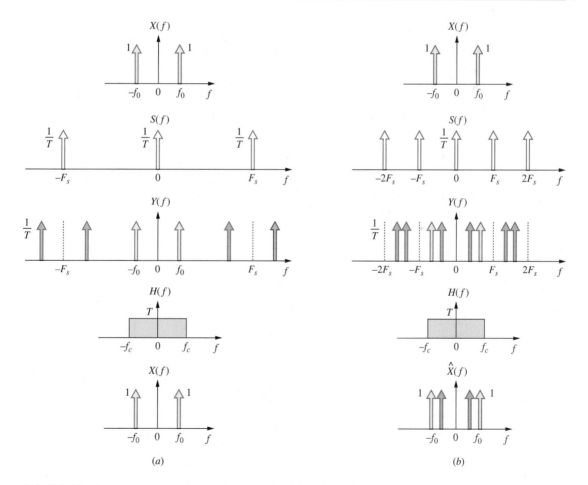

FIGURE 10.10 Sampling and reconstruction of $\cos(2\pi f_0 t)$. *(a)* The signal may be reconstructed from samples taken at the rate $F_s = 4f_0$. *(b)* At the rate $F_s = 1.5f_0$ aliasing adds a low-frequency component at $0.5f_0$. See Example 10.8.

In all cases the $\pm f_0$ components of $y(t)$ are due to convolution of $X(f)$ with the impulse in $S(f)$ at the origin. The convolution of $X(f)$ with the impulses at $\pm F_s$ generates two impulses at $\pm(F_s - f_0)$. To avoid aliasing, these should not pass through the reconstruction filter, that is $F_s - f_0 > f_0$ or $F_s > 2 f_0$. This condition is satisfied in part a, where $F_s = 4 f_0$ and $\pm(F_s - f_0) = 3 f_0$ (outside the filter's range). In part b, where $F_s = 1.5 f_0$, $\pm(F_s - f_0) = \pm 0.5 f_0$ passes through the filter and contributes $2 \cos \pi f_0 t$ to the reconstructed signal. Similarly, at the rate $F_s = 1.1 f_0$ (part c) we have $\pm(F_s - f_0) = \pm 0.1 f_0$ which appears as $2 \cos 0.2 \pi f_0 t$ in the output. In the above discussion $X(f)$ was made of a pair of impulses representing $x(t)$, $-\infty < t < \infty$. As for leakage, if the signal is windowed its spectrum spreads and some of the high frequencies may then leak through the filter, resulting in non-ideal reconstruction.

10.7   Frequency Downshifting

Low-sampling rates produce aliasing. In some cases this may be a desirable feature. The aliasing produced by a low-sampling rate (i.e., lower than the Nyquist rate) may be used to shift down the frequency band of a periodic signal. A *sequential repetitive sampling* performed at much lower than the Nyquist rate downshifts the spectrum of a high-frequency signal, allowing its observation and recording by instruments that have a low bandwidth. Example 10.9 below illustrates the concept of frequency downshifting. Frequency downshifting is further illustrated in Example 10.10 and problems 5, 8, and 9 at the end of this chapter.

Consider the day/night temperature profile at a given location. This may be measured in one of two ways. In one approach measurements are taken every hour on the hour, 24 measurements in 24 hours. See Table 10.1(a) and Figure 10.11(a). The temperature between the hours may then be estimated by a straight line interpolation of the two neighboring measurements, or by a second-order polynomial interpolation using three measurements, and so on.

In the second approach, 24 measurements may be taken starting at the 00 hour with a 25-hour sampling interval, and taking a total of 24 days. In the absence of a noticeable temperature trend during the 24-day measurement period, compressing the time scale by a factor of 24 will yield a profile similar to that of the first approach. See Table 10.1(b) and Figure 10.11(b).

TABLE 10.1a Sampling Temperature Every Hour

Day, Hour	Temperature in C°
1−00	20
1−01	19
1−02	18
1−03	18
1−04	19
1−05	19
1−06	20
1−07	20
1−08	21
1−09	22
1−10	24
1−11	26
1−12	28
1−13	30
1−14	32
1−15	32
1−16	31
1−17	30
1−18	29
1−19	27
1−20	25
1−21	23
1−22	21
1−23	20

TABLE 10.1b Sampling Temperature Every 25 Hours

Day, Hour	Temperature in C°
1−00	20
2−01	19
3−02	18
4−03	19
5−04	19
6−05	20
7−06	21
8−07	22
9−08	22
10−09	25
11−10	26
12−11	28
13−12	31
14−13	32
15−14	33
16−15	31
17−16	30
18−17	29
19−18	28
20−19	26
21−20	24
22−21	22
23−22	22
24−23	21

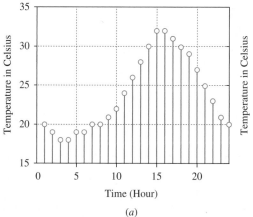

(a)

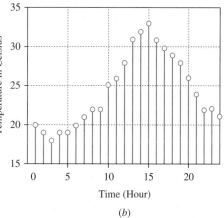

(b)

FIGURE 10.11 Sampling temperature. Temperature data given in Table 10.1 are plotted in this figure. *(a)* Temperature measurements at a location sampled every hour show the 24-hour trend. *(b)* Under stable day-to-day weather conditions the above trend may also be deduced from measurements taken at the much lower rate of one sample every 25 hours during a period of 24 days. The sampling process shown in *(b)* illustrates the concept of *frequency downshifting*.

Frequency downshifting may be examined by sampling a sinusoidal signal at a low rate and reconstructing it by low-pass-filtering the samples (e.g., by connecting the consecutive samples). Refer to Figure 10.12. Sample a sinusoidal signal (period $= T_0$) every $T = (1 + \alpha)T_0$ seconds. The time function and its sample values are

$$x(t) = \sin\left(\frac{2\pi t}{T_0}\right)$$

$$t = nT = n(1 + \alpha)T_0$$

$$x(n) = x(t)\big|_{t=nT} = \sin\left(\frac{2\pi nT}{T_0}\right) = \sin[2\pi(1 + \alpha)n] = \sin(2\pi\alpha n)$$

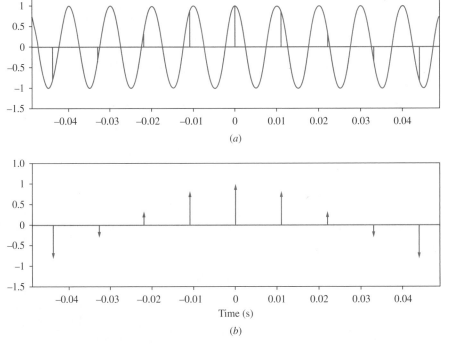

FIGURE 10.12 Frequency downshifting. **(a)** A 100-Hz sinusoidal signal with period $=$ 10 msec is sampled every 11 msec ($F_s = 90.91$ Hz). In this way the succeeding samples jump one period plus 1 msec (away from the previous samples). After 10 samples are taken, the cycle repeats, forming a periodic sequence with period 110 msec. **(b)** Sample values are identical with those taken from a sinusoidal signal with period $= 110$ msec. The signal to be reconstructed from the samples is a sinusoid with the frequency $f = 100 - 90.91 = 9.09$ Hz (period $= 110$ msec). In the above exercise, the frequency of the sinusoid is reduced by a factor $\alpha/(1 + \alpha)$, $\alpha = 0.1$. The following two examples provide further illustrations, in the time and frequency domains, of the effects of sampling a sinusoidal signal at a low rate.

In constructing a continuous-time signal from $x(n)$ we revert to $n = t/T$.

$$x(n) = \sin(2\pi\alpha n)$$

$$n = \frac{t}{T}$$

$$\hat{x}(t) = x(n)\big|_{n=t/T} = \sin\left(\frac{2\pi\alpha t}{T}\right)$$

$$= \sin\left[\frac{\alpha}{1+\alpha}\frac{2\pi t}{T_0}\right] = x\left[\frac{\alpha}{1+\alpha}t\right]$$

Example 10.11

A 900-Hz continuous-time sinusoidal signal $x(t) = \cos(1,800\pi t)$ is sampled by a periodic narrow rectangular pulse train (10 μs wide and 10 V base-to-peak) at the rate of $F_s = 1$ kHz, which is less than twice its frequency. Samples are then passed through an ideal low-pass filter with a DC gain of 10 and a cutoff frequency of 500 Hz. Show that the filter's output is $z(t) = \cos(200\pi t)$.

Solution

The sampling pulse can be modeled by an impulse whose strength is $10 \times 10^{-5} = 10^{-4}$. To sample $x(t)$ we, therefore, multiply it by the sampling signal $s(t)$ at the 1-kHz rate, producing the sampled function $y(t)$. In the time domain

$$x(t) = \cos(1,800\pi t)$$

$$s(t) = 10^{-4} \sum_{n=-\infty}^{\infty} \delta\left(t - \frac{n}{1,000}\right)$$

$$y(t) = x(t)s(t) = 10^{-4} \sum_{n=-\infty}^{\infty} \cos(1.8\pi n)\delta\left(t - \frac{n}{1,000}\right)$$

In the frequency domain

$$X(f) = \frac{1}{2}[\delta(f - 900) + \delta(f + 900)]$$

$$S(f) = 10^3 \times 10^{-4} \sum_{k=-\infty}^{\infty} \delta(f - 1,000k)$$

$$Y(f) = X(f) \star S(f) = X(f) \star 10^{-1} \sum_{k=-\infty}^{\infty} \delta(f - 1,000k)$$

$$= \frac{1}{20} \sum_{k=-\infty}^{\infty} [\delta(f - 1,000k - 900) - \delta(f - 1,000k + 900)]$$

The only component that passes through the low-pass filter with cutoff frequency at 500 Hz is the one at 100 Hz [to be called $z(t)$]. Other frequencies are all blocked. The output of the low-pass filter is

$$Z(f) = 10 \times \frac{1}{20}[\delta(f - 100) + \delta(f + 100)]$$

$$z(t) = \cos(200\pi t)$$

E*xample*

10.12

An amplitude-modulated bandpass signal is modeled by

$$x(t) = a(t)\cos(2000\pi t), \quad a(t) = \left[\frac{\sin(100\pi t)}{100\pi t}\right]^2$$

a.   Show that $X(f) = 0$ for $900 < f < 1{,}100$ Hz.

b.   The signal $x(t)$ is sampled at the rate of 1 kHz and then passed through an ideal low-pass filter with cutoff frequency at 500 Hz and 60 dB attenuation. Show that the filter's output is $a(t)$.

Solution

a.   The Fourier transform $A(f)$ of $a(t)$ is a triangle with the base $-100 < f < 100$ Hz and height $A(0) = 0.01$. See Figure 10.13(a). The low-pass signal $a(t)$ modulates the 1-kHz sinusoidal carrier and produces $x(t)$. The Fourier transform of the carrier is made of two impulses at $\pm 1{,}000$ Hz. See Figure 10.13(b). Modulation shifts $A(f)$ to the locations of those two impulses

$$X(f) = \frac{1}{2}[A(f - 1{,}000) + A(f + 1{,}000)]$$

limiting its frequency band to $(1{,}000 - 100) < f < (1{,}000 + 100)$, or 900 to 1,100 Hz. See Figure 10.13(c).

b.   By sampling $x(t)$ at the rate of 1 kHz, we obtain $y(t) = x(t)s(t)$, where

$$s(t) = \sum_{n=-\infty}^{\infty} \delta\left(t - \frac{n}{1{,}000}\right)$$

$$S(f) = 1{,}000 \sum_{k=-\infty}^{\infty} \delta(f - 1{,}000k)$$

$$Y(f) = X(f) \star S(f) = X(f) \star \left[1{,}000 \sum_{k=-\infty}^{\infty} \delta(f - 1{,}000k)\right]$$

$$= \frac{1{,}000}{2} \sum_{k=-\infty}^{\infty} [A(f - 1{,}000 - 1{,}000k) + A(f + 1{,}000 - 1{,}000k)]$$

The ideal low-pass filter with cutoff at 500 Hz blocks all components of $Y(f)$ except the triangle centered around origin generated from $k = \pm 1$. The filter's output is, therefore, $a(t)$.

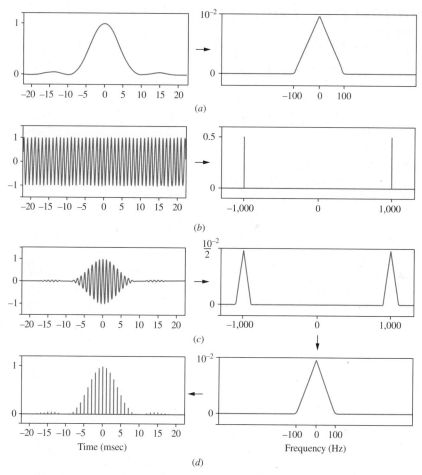

FIGURE 10.13 Downshifting an AM signal. This figure illustrates downshifting the continuous spectrum of an amplitude-modulated (AM) bandpass signal $x(t)$.

$$x(t) = a(t)\cos(2{,}000\pi t), \quad \text{where } a(t) = \left[\frac{\sin(100\pi t)}{100\pi t}\right]^2$$

Time functions are shown in the left-hand column with the time scale in seconds. Transforms are normalized and are shown in the right-hand column, across from their respective time functions, with the frequency scale in Hz.

a. The modulating waveform $a(t)$ is a low-pass signal with a triangular spectrum, limited to $|f| < 100$ Hz.

b. The carrier signal is $\cos 2{,}000\pi t$ with two spectral lines at ± 1 kHz.

c. The modulation operation shifts the spectrum of $a(t)$ to center around ± 1 kHz. The AM signal $x(t)$ is, therefore, bandpass limited to $1{,}000 \pm 100$ Hz.

d. To downshift the AM signal, $x(t)$ is sampled at the rate $F_s = 1$ kHz and the sampled signal $y(t)$ is passed through an ideal low-pass filter. The output of the filter has a triangular spectrum identical with $A(f)$. The filter's output is $a(t)$, a downshifted $x(t)$.

10.8 Summary of Sampling and Reconstruction Process

A block diagram of sampling a low-pass signal and its reconstruction is shown in Figure 10.14(a). First, the signal to be sampled is passed through a low-pass filter to cutoff (or reduce) undesired high-frequency components (e.g., noise or high-frequency components which carry no useful information). This filter helps set the sampling rate and prevents aliasing, thus is called an anti-aliasing filter. It could be a simple first-order RC (active or passive) filter, and is found in almost all sampling devices such as analog-to-digital (A/D) converters.

Next, the signal $x(t)$ is multiplied by a train of unit impulses every T seconds, called the sampling function and designated by $s(t)$.

$$s(t) = \sum_{n=-\infty}^{\infty} \delta(t - nT), \; S(f) = F_s \times \sum_{k=-\infty}^{\infty} \delta(f - kF_s)$$

where $F_s = 1/T$ Hz is the sampling rate. The resulting sampled signal, which is a continuous-time function, is designated by $y(t)$.

$$y(t) = x(t) \times s(t) = x(t) \times \sum_{n=-\infty}^{\infty} \delta(t - nT) = \sum_{n=-\infty}^{\infty} x(nT)\delta(t - nT)$$

$$Y(f) = X(f) \star S(f) = X(f) \star \left[F_s \times \sum_{k=-\infty}^{\infty} \delta(f - kF_s) \right] = F_s \times \sum_{k=-\infty}^{\infty} X(f - kF_s)$$

Impulse samples then pass through a low-pass filter with the unit-impulse response $h(t)$. The output of the filter is

$$z(t) = h(t) \star y(t) = h(t) \star \sum_{n=-\infty}^{\infty} x(nT)\delta(t - nT) = \sum_{n=-\infty}^{\infty} x(nT)h(t - nT)$$

$$Z(f) = F_s \times H(f) \times \sum_{k=-\infty}^{\infty} X(f - kF_s)$$

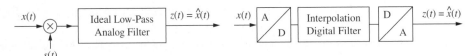

(a) Representation in the continuous-time domain.   (b) Representation in the digital domain employing A/D converter, digital filter, and D/A converter.

FIGURE 10.14 Block diagram summary of sampling and reconstruction process in continous- and discrete-time domains. Reconstruction is done by passing $y(t)$ through a filter with unit-impulse response $h(t)$. Filter's output is $z(t)$.

Depending on the sampling rate and the reconstruction filter, $z(t)$ may be equal or close to $x(t)$. We designate the rms of reconstruction error by $\mathcal{E}$ and define it by:

$$\text{For energy signals:} \quad \mathcal{E}^2 = \int_{-\infty}^{\infty} |x(t) - z(t)|^2 dt$$

$$\text{For power signals:} \quad \mathcal{E}^2 = \lim_{T \to \infty} \frac{1}{2T} \int_{-T}^{T} |x(t) - z(t)|^2 dt$$

Note that the independent variable throughout the block diagram of Figure 10.14(a) is continuous time. Each sample is represented by an impulse whose strength is equal to the value of the sample. All functions [$x(t)$, $s(t)$, $y(t)$, and $z(t)$] are functions of continuous time and have generalized Fourier transforms. The analysis uses continuous-time mathematics.

When processing the samples by digital devices such as computers, two additional specifications enter the process. First, each sample is represented by a number equal to its value. The sampled function is, therefore, represented by a sequence of numbers called the discrete-time signal $x(n)$. The independent variable is the discrete-time n, which is also an integer. Second, the magnitude of $x(n)$ is discretized and shown by a binary word of length N (corresponding to 2^N discrete levels). The resulting signal is called a digital signal and the devices that perform the operation and its converse are called analog-to-digital (A/D) and digital-to-analog (D/A) converters, respectively. The process is shown in Figure 10.14(b). The digital filter included in Figure 10.14(b) can perform decimation, interpolation, and filtering. It can simplify the analog reconstruction filter placed at the end of the process. The topic of A/D and D/A conversion will not be discussed in the present chapter.

10.9 Complex Low-Pass Signals

We have seen that the Fourier transform of a real valued function $v(t)$ holds the conjugate symmetry: $V(f) = V^*(-f)$. If $V(f) \neq V^*(-f)$, then $v(t)$ is a complex function of time and may be written as $v(t) = v_c(t) + jv_s(t)$. A class of signals of practical interest in communication and signal processing employ complex low-pass signal models to represent more general bandpass signals, as will be seen in section 10.10. In this section we illustrate the complex low-pass signal by way of an example.

Example 10.13

The Fourier transform of a complex low-pass signal $v(t)$ is

$$V(f) = \begin{cases} a, & -f_\ell < f < 0 \\ b, & 0 < f < f_h \\ 0, & \text{elsewhere} \end{cases}$$

The bandwidth is limited to $-f_\ell < f < f_h$ as shown in Figure 10.15(a).

a. Find $v(t)$ and show how it may be written as $v(t) = v_c(t) + jv_s(t)$.
b. Find $v_c(t)$ and $v_s(t)$ when $f_\ell = f_h = f_0$.

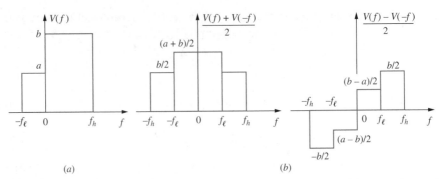

FIGURE 10.15 Fourier transform of a complex-valued low-pass signal *(a)*, and its even and odd components *(b)*.

Solution

a. We find $v(t)$ by taking the inverse Fourier transform of $V(f)$

$$
v(t) = a \int_{-f_\ell}^{0} e^{j2\pi ft}\,df + b \int_{0}^{f_h} e^{j2\pi ft}\,df
$$

$$
= \frac{1}{\pi t}\left[a e^{-j\pi f_\ell t}\sin(\pi f_\ell t) + b e^{j\pi f_h t}\sin(\pi f_h t)\right]
$$

$$
= \frac{1}{2\pi t}\left[a\sin(2\pi f_\ell t) + b\sin(2\pi f_h t)\right] + \frac{j}{\pi t}\left[-a\sin^2(\pi f_\ell t) + b\sin^2(\pi f_h t)\right]
$$

Alternative Approach

Split $V(f)$ into two components and take their inverse Fourier transforms as shown below.

$$
V_1(f) = \begin{cases} a, & -f_\ell < f < 0 \\ 0, & \text{elsewhere} \end{cases} \quad\Longrightarrow\quad v_1(t) = a\frac{\sin(\pi f_\ell t)}{\pi t}e^{-j\pi f_\ell t}
$$

$$
V_2(f) = \begin{cases} b, & 0 < f < f_h \\ 0, & \text{elsewhere} \end{cases} \quad\Longrightarrow\quad v_2(t) = b\frac{\sin(\pi f_h t)}{\pi t}e^{j\pi f_h t}
$$

$$
V(f) = V_1(f) + V_2(f) \quad\Longrightarrow\quad v(t) = v_1(t) + v_2(t)
$$

b. When $f_\ell = f_h = f_0$ we have

$$
v(t) = \frac{1}{\pi t}(a+b)\sin(\pi f_0 t)\cos(\pi f_0 t) - \frac{j}{\pi t}(a-b)\sin(\pi f_0 t)\sin(\pi f_0 t)
$$

$$
= (a+b)\frac{\sin(2\pi f_0 t)}{2\pi t} - j(a-b)\frac{\sin^2(\pi f_0 t)}{\pi t}
$$

from which

$$
v_c(t) = (b+a)\frac{\sin(2\pi f_0 t)}{2\pi t}
$$

$$
v_s(t) = (b-a)\frac{\sin^2(\pi f_0 t)}{\pi t}
$$

$$
v(t) = v_c(t) + jv_s(t)
$$

Note: In this example $v_c(t)$ and $v_s(t)$ are also the inverse transforms of the odd and even parts of $V(f)$, respectively.

10.10 Bandpass Signals and Their Sampling

Bandpass Signals

The spectrum of a bandpass signal $x(t)$ is limited to frequencies between f_l and f_h

$$X(f) = 0, \quad |f| < f_\ell \text{ and } |f| > f_h$$

In communication engineering, bandpass signals are also called *narrowband signals*. Examples are amplitude-modulated signals and the general case of single side-band signals.

Amplitude Modulation

Amplitude-modulated signals constitute a special case of narrowband signals. They are represented by $x(t) = a(t)\cos(2\pi f_c t)$, where $a(t)$ is a real-valued low-pass signal, $A(f) = A^*(-f)$, with frequencies limited to $0 \le f \le f_0$ and $f_c > f_0$. $X(f)$ is then band limited to $f_c - f_0 \le |f| \le f_c + f_0$.

General Case

In general, a bandpass signal $x(t)$ may be represented by two amplitude-modulated components sharing the same center frequency f_c but 90° out of phase with each other:

$$x(t) = v_c(t)\cos(2\pi f_c t) - v_s(t)\sin(2\pi f_c t)$$

This is called the *quadrature carrier* description of the bandpass signal. The signal $v(t) = v_c(t) + jv_s(t)$ is called the complex low-pass signal. $v_c(t)$ and $v_s(t)$ modulate the envelopes of the carriers (at f_c) and are low-pass signals with bandwidth $B/2$ Hz, where $B = f_h - f_\ell$. They are called the in-phase and the quadrature parts, respectively.

Alternately, a bandpass signal may be represented by an amplitude- and angle-modulated carrier at f_c

$$
\begin{aligned}
x(t) &= a(t)\cos[2\pi f_c t + \phi(t)] \\
&= a(t)\cos\phi(t)\cos(2\pi f_c t) - a(t)\sin\phi(t)\sin(2\pi f_c t) \\
&= v_c(t)\cos(2\pi f_c t) - v_s(t)\sin(2\pi f_c t)
\end{aligned}
$$
$$v_c(t) = a(t)\cos\phi(t)$$
$$v_s(t) = a(t)\sin\phi(t)$$

Finally, a bandpass signal may be represented as the real part of a complex exponential (carrier) at f_c whose amplitude is modulated by the complex baseband signal $v(t)$:

$$x(t) = \mathcal{RE}\{v(t)e^{j2\pi f_c t}\}$$

where $v(t) = v_c(t) + jv_s(t)$.

E*xample*

10.14

Let $v(t)$ be a real-valued low-pass signal with $V(f) = 0$, $|f| > f_0$. Show that the amplitude-modulated signal $x(t) = v(t) \cos(2\pi f_c t)$ is a bandpass signal. Specify the frequency band of $x(t)$ if $f_0 = 3$ and $f_c = 12$, both in kHz.

Solution

$V(f) = 0, \quad |f| > f_0$

$X(f) = \frac{1}{2} V(f)^* [\delta(f + f_c) + \delta(f - f_c)] = \frac{1}{2} [V(f - f_c) + V(f + f_c)].$

$X(f) = 0, |f_c - f_0| < |f| < |f_c + f_0|.$ The frequency band is from $f_c - f_0$ to $f_c + f_0$.

For $f_0 = 3$ kHz and $f_c = 12$ kHz the frequency band of $x(t)$ is from 9 to 15 kHz.

E*xample*

10.15

Consider the bandpass signal

$$x(t) = a_1 \cos(\omega_1 t) + a_2 \cos(\omega_2 t) + a_3 \cos(\omega_3 t)$$

where $\omega_1 < \omega_2 < \omega_3$. Express $x(t)$ in term of its quadrature components $x(t) = v_c \cos \omega_c t - v_s \sin \omega_c t$.

Solution

The Fourier transform of $x(t)$ is limited to $\omega_1 \leq \omega \leq \omega_3$. Let $\omega_c = (\omega_3 + \omega_1)/2$ and $\omega_0 = (\omega_3 - \omega_1)/2$. Then, $\omega_1 = \omega_c - \omega_0$ and $\omega_3 = \omega_c + \omega_0$. In addition, let $\omega_d = \omega_2 - \omega_c$. Then

$x(t) = a_1 \cos(\omega_1 t) + a_2 \cos(\omega_2 t) + a_3 \cos(\omega_3 t)$

$\quad = a_1 \cos(\omega_c - \omega_0)t + a_2 \cos(\omega_c + \omega_d)t + a_3 \cos(\omega_c + \omega_0)t$

$\quad = a_1(\cos \omega_c t \, \cos \omega_0 t + \sin \omega_c t \, \sin \omega_0 t) + a_2(\cos \omega_c t \, \cos \omega_d t - \sin \omega_c t \, \sin \omega_d t)$

$\quad \quad + a_3(\cos \omega_c t \, \cos \omega_0 t - \sin \omega_c t \, \sin \omega_0 t)$

$\quad = [(a_1 + a_3) \cos \omega_0 t + a_2 \cos \omega_d t] \cos \omega_c t$

$\quad \quad + [(a_1 - a_3) \sin \omega_0 t - a_2 \sin \omega_d t] \sin \omega_c t$

$\quad = v_c \cos \omega_c t - v_s \sin \omega_c t$

where

$$v_c = (a_3 + a_1) \cos \omega_0 t + a_2 \cos \omega_d t$$
$$v_s = (a_3 - a_1) \sin \omega_0 t + a_2 \sin \omega_d t$$

E*xample*

10.16

Finding the quadrature components of a bandpass signal

The Fourier transform of a bandpass signal is

$$X(f) = \begin{cases} \frac{|f| - f_\ell}{f_h - f_\ell}, & f_\ell < |f| < f_h \\ 0, & \text{elsewhere} \end{cases}$$

See Figure 10.16(a). Express $x(t)$ in the *quadrature carrier* form.

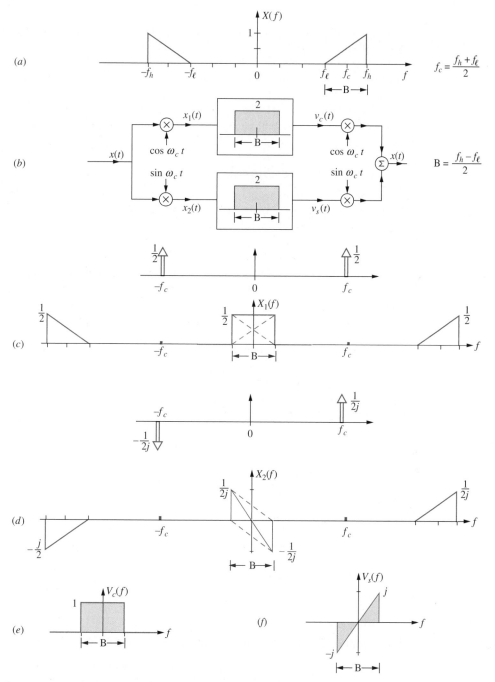

FIGURE 10.16 A single-side-band signal $X(f)$ [shown in **(a)**] is constructed from an in-phase component $V_c(f)$ [shown in **(e)**] and a quadrature-phase component $V_s(f)$ [shown in **(f)**]. See Example 10.16.

Solution

The signal is band limited to $f_\ell < f < f_h$. It is to be shown by $x(t) = v_c(t) \cos(2\pi f_c t) - v_s(t) \sin(2\pi f_c t)$. It is not required for f_c to be the center frequency of the bandwidth. Presently, we choose $f_c = (f_h + f_\ell)/2$ and implement the sequence of operations shown in Figure 10.16(b). The Fourier transforms of $x_1(t) = x(t) \cos(2\pi f_c t)$ and $x_2(t) = x(t) \sin(2\pi f_c t)$ are shown in Figure 10.16(c) and (d), respectively. The quadrature elements $v_c(t)$ and $v_s(t)$ are obtained at the outputs of low-pass filters with gain 2. Their Fourier transforms are shown in Figure 10.16(e) and (f), respectively.

Sampling Bandpass Signals

Obviously a bandpass signal $x(t) = v_c \cos \omega_c t - v_s \sin \omega_c t$ can be faithfully reconstructed from its samples taken at the rate of $2f_h$, where f_h is the highest frequency in $x(t)$. This is in general a high rate. In choosing it we are ignoring the bandpass property of the signal; that is, the fact that $X(f) = 0$ for $|f| < f_\ell$. In this section we summarize a method for sampling the bandpass signal $x(t) = v_c \cos \omega_c t - v_s \sin \omega_c t$ at the rate of $2B$, where B is the signal's bandwidth (or close to it) and faithfully reconstructing it. The method described in this section assumes uniform sampling.

To sample $x(t)$ at the low rate one may first downshift the in-phase and quadrature-phase components of $x(t)$ to obtain v_c and v_s. One then can sample each of the two signals at their Nyquist rate of B Hz or a total of $2B$ samples per second. See Figure 10.17. Alternately, one may sample $x(t)$ directly at rates ranging from a certain minimum (close to $2B$) to its Nyquist rate of $2f_h$. If $x(t)$ is sampled directly, not all rates within the above range are acceptable, as summarized below.

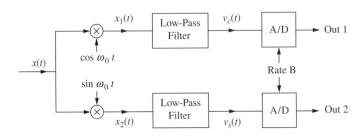

FIGURE 10.17 Sampling a bandpass signal by frequency downshifting.

Minimum Rate

To determine the minimum sampling rate we recognize the following two cases.

1. If f_h/B is an integer, then sample $x(t)$ at the rate of $2B$ Hz to obtain the discrete-time sequence $x(nT)$ where $T = 1/(2B)$. In addition it can be shown that the even-numbered samples of $x(t)$ produce samples of v_c and the odd-numbered samples produce samples of v_s. See the next item.

2. If f_h/B is not an integer, sample $x(t)$ at the rate of $(2f_h/N)$ Hz where N is the integer part of f_h/B. This is equivalent to increasing the bandwidth to B' so that $f_h/B' = N$ becomes an integer and then sampling at the new rate of $2B'$ Hz. The new f_h' and f_ℓ' are

$$f_h' = f_c + B' \text{ and } f_\ell' = f_c - B'$$

The new center frequency $f_c' = f_h - B'/2$ is used in the interpolation formula to find $x(t)$ from the samples.

Acceptable Rates

A bandpass signal with $f_l < |f| < f_h$ and $B = f_h - f_\ell$ may be sampled at a rate F_s and reconstructed by passing the samples through a bandpass filter, if

$$\frac{2f_h}{k} \leq F_s \leq \frac{2f_\ell}{(k-1)} \quad k = 1, 2, \dots, N$$

where N is the integer part of f_h/B. For $k = N$, the sampling rate becomes the minimum rate derived previously. For $k = 1$, the sampling rate becomes the Nyquist rate of $2f_h$.

In summary, it may be shown that the minimum sampling rate for the unique reconstruction of a bandpass signal with bandwidth B is $2B < F_s < 4B$.

Low-Pass Samples Are Obtained from Sampling Bandpass Signals

The bandpass signal $x(t)$ with frequencies from f_ℓ to f_h has the bandwidth $B = f_h - f_\ell$ and the center frequency $f_c = (f_\ell + f_h)/2$. The signal and its samples taken at the rate of $2B$ (sampling interval $T = 1/2B$) are listed below.

$$x(t) = v_c(t) \cos(2\pi f_c t) - v_s(t) \sin(2\pi f_c t)$$

$$x(nT) = v_c(nT) \cos(2\pi f_c nT) - v_s(nT) \sin(2\pi f_c nT)$$

Consider the case where the upper frequency of the signal is a multiple of its bandwidth; that is, $f_h/B = k$ is an integer.[3] Then

$$f_c = \frac{f_\ell + f_h}{2} = f_h - \frac{B}{2} = \left(k - \frac{1}{2}\right)B$$

Substituting for f_c we get

$$x(nT) = v_c(nT) \cos \frac{\pi n(2k-1)}{2} - v_s(nT) \sin \frac{\pi n(2k-1)}{2}$$

$$= \begin{cases} (-1)^{n/2} v_c(nT) & n \text{ even} \\ (-1)^{k+(n+3)/2} v_s(nT) & n \text{ odd} \end{cases}$$

The even-numbered samples of $x(t)$, taken at the rate of $2B$, provide samples of $v_c(t)$ and the odd-numbered samples of $x(t)$ provide samples of $v_s(t)$.

---

[3]If that is not the case, we increase the bandwidth B to a new value B' such that $f_h/B' = k$ becomes an integer.

10.11 Reconstruction of Bandpass Signals

The bandpass signal $x(t)$ described in section 10.10 may be recovered from its samples $x(nT)$. To derive the recovery equation, we first reconstruct the low-pass signals $v_c(t)$ and $v_s(t)$ from their samples, which were provided by $x(nT)$ in the previous section:

$$x(nT) = \begin{cases} (-1)^{n/2}v_c(nT), & n \text{ even} \\ (-1)^{k+(n+3)/2}v_s(nT), & n \text{ odd} \end{cases}$$

We then use the reconstructed $v_c(t)$ and $v_s(t)$ to find $x(t)$. The equation that gives us $x(t)$ in terms of its samples is

$$x(t) = v_c(t)\cos(2\pi f_c t) - v_s(t)\sin(2\pi f_c t)$$

$$= \sum_{n=-\infty}^{\infty} x(nT) \frac{\sin\frac{\pi(t-nT)}{2T}}{\frac{\pi(t-nT)}{2T}} \cos 2\pi f_c(t - nT)$$

where $T = 1/(2B)$. Derivation of the above equation is left as an exercise.

Example
10.17

Find the minimum sampling rate for a bandpass signal with $f_c = 900$ kHz and $B = 8$ kHz.

$$f_\ell = f_c - \frac{B}{2} = 900 - \frac{8}{2} = 896 \text{ kHz}$$

$$f_h = f_c + \frac{B}{2} = 900 + \frac{8}{2} = 904 \text{ kHz}$$

Since $f_h/B = 904/8 = 113$ is an integer, we sample $x(t)$ at the rate of $2B = 16$ kHz or 16,000 samples per second.

Example
10.18

Find the minimum sampling rate for a bandpass signal with $f_c = 910$ kHz and $B = 8$ kHz.

$$f_\ell = 910 - \frac{8}{2} = 906 \text{ kHz} \quad \text{and} \quad f_h = 910 + \frac{8}{2} = 914 \text{ kHz}$$

Solution
$f_h/B = 914/8 = 114.25$ is not an integer. We sample at the rate of $2f_h/114 = 16,035$ samples per second. The new bandwidth is $B' = f_h/114 = 8.0175$ kHz and as expected the sampling rate is $2B' = 16,035$ samples per second. The new center frequency is $f_c' = f_h - B'/2 = 909.99$ kHz.

Example

Find the minimum sampling rate for a bandpass signal with $f_\ell = 6$ kHz and $f_h = 8$ kHz. Construct the spectrum of a sampled signal. Reconstruct the original signal by bandpass filtering.

Solution

$B = f_h - f_\ell = 2$ kHz. $f_h/B = 8/2 = 4$, which is an integer. The signal may be sampled at the rate of $2f_h/4 = 4$ kHz. The spectrum of the sampled signal is shown in Figure 10.18. It is seen that bandpass filtering will reproduce the original signal. The center frequency of the recovery filter is 7 kHz.

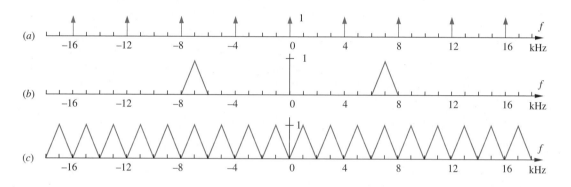

FIGURE 10.18 Sampling and reconstructing a bandpass signal. A bandpass signal (6 kHz < f < 8 kHz) may be recovered from its samples taken at the minimum rate of 4 kHz.
(a) $S(f)$ is the Fourier transform of $s(t)$, a periodic train of impulses in the frequency domain spaced every $F_s = 4$ kHz.
(b) $X(f)$ is the Fourier transform of the continuous-time bandpass signal $x(t)$ (6 kHz < f < 8 kHz).
(c) $Y(f) = X(f) \star S(f)$ is the Fourier transform of the sampled signal $y(t) = x(t)s(t)$.
From the frequency-domain representations shown above, it is observed that $x(t)$ may be recovered by passing $y(t)$ through a bandpass filter (6 kHz < f < 8 kHz).

Example
10.20

a. Sample the bandpass signal of Example 10.19 at the rate of $F_s = 5$ kHz. Examine the spectrum of the sampled signal and show that the original continuous-time signal may not be faithfully reconstructed by bandpass filtering of samples taken at this rate.

b. Repeat for $F_s = 6$ kHz and verify that the rate is acceptable.

c. Repeat for $F_s = 7$ kHz and verify that the rate is not acceptable except when the signal is a real-valued amplitude-modulated (AM) signal.

Solution

All frequencies shown are in kHz. The constraint on F_s is

$$\frac{16}{k} \le F_s \le \frac{12}{(k-1)}, \quad k = 1, 2, 3, 4$$

$$k = 1, \quad 16 \le F_s$$
$$k = 2, \quad 8 \le F_s \le 12$$
$$k = 3, \quad 5.33 \le F_s \le 6$$
$$k = 4, \quad 4 \le F_s \le 4$$

a. The spectrum of the sampled signal is shown in Figure 10.19(a). $F_s = 5$ is not acceptable.

b. $F_s = 6$ falls within the limits given above and is acceptable. See Figure 10.19(b) for the spectrum of samples.

c. $F_s = 7$ does not fall within the limits given above and the rate is not acceptable. See Figure 10.19(c). The above rate is, however, acceptable in the case of an AM signal. See Figure 10.19(d).

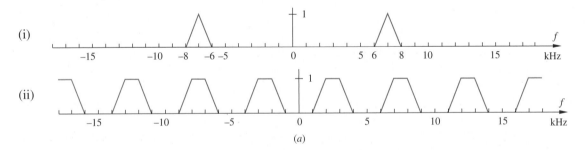

(a)

(*a*) A bandpass signal (6 kHz $< f <$ 8 kHz) may not always be recovered from samples taken at the rate of 5 kHz, although this rate is higher than the minimum. The rate does not satisfy the condition for sampling (and recovery of) a bandpass signal as described in the text. In this figure (and also in Figure 10.19*b*, *c*, and *d*) traces shown in (*i*) and (*ii*) are $X(f)$ and $Y(f)$, respectively, as described in the legend of Figure 10.18. From the frequency domain representations shown above, it is observed that passing $y(t)$ through a bandpass filter (6 kHz $< f <$ 8 kHz) will not produce $x(t)$.

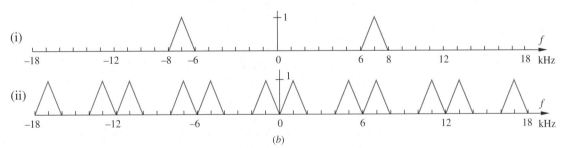

(b)

(**b**) The 6-kHz rate is acceptable for sampling (and recovery of) a bandpass signal (6 kHz $< f <$ 8 kHz).

FIGURE 10.19 Sampling a bandpass signal (6 kHz $< f <$ 8 kHz).

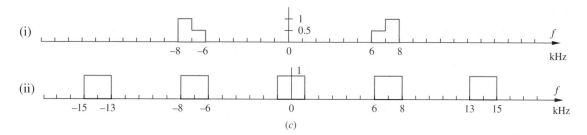

(i)

(ii)

(c)

(c) The 7-kHz rate is not always acceptable for sampling a bandpass signal with 6 kHz $< f <$ 8 kHz [except when the signal is amplitude modulated, see (*d*)].

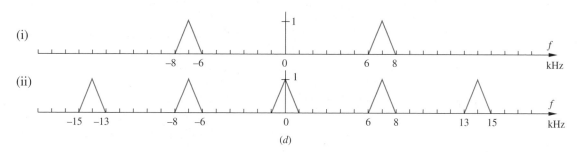

(i)

(ii)

(d)

(d) The 7-kHz rate is acceptable for sampling an AM bandpass signal with 6 kHz $< f <$ 8kHz.

FIGURE 10.19 *(Continued)*

Appendix 10A Additional Notes on Sampling

Nonuniform and Random Sampling

A band-limited signal may be recovered from its samples taken nonuniformly, provided the average rate satisfies the Nyquist sampling criteria.

Samples may also be taken randomly. In some cases this becomes necessary in order to eliminate biasing statistical averages of the signal when derived from its samples. As an example consider a sinusoidal signal riding on a DC level, $x(t) = A\cos(2\pi f_0 t + \theta) + B$. The DC value B may be estimated from the average of N samples of $x(t)$.

$$B \approx \frac{1}{N}\sum_{i=1}^{N} x_i(t)$$

For the above to be an unbiased estimate, samples should be taken randomly. A similar condition applies when $x(t)$ is a stochastic process.

Random sampling may also be applied to reduce the sampling rate. An example is found in *random repetitive sampling*, where high sampling rates (e.g., required for signals with high bandwidth) are avoided by taking samples randomly, but repeatedly. The signal may then be reconstructed from the samples if a time reference (such as level triggering a digital scope by the signal to be sampled) can be applied to all samples.

Sampling Stochastic Signals

We have developed a sampling and reconstruction process for deterministic signals. But, by definition, the value of a deterministic signal at any moment is predictable once the signal is determined. So why do we need to sample it further? The real application and usefulness of a sampling process comes in play when signals are not predictable. Such signals are called stochastic. Unlike deterministic signals, the value of a stochastic signal is not predictable except statistically and in its average sense. In this section we briefly summarize the extension of sampling to certain classes of stochastic signals which are classified as stationary processes.

Statistical averages of a stationary process $x(t)$ are provided by its autocorrelation $\gamma_{x,x}(t)$ which is a deterministic function, or equivalently by its power density spectrum $\Gamma_{x,x}(f)$ which is the Fourier transform of the autocorrelation. If the power density spectrum of a stationary process is zero for frequencies above f_0 the process is band limited (which we have designated as low pass). Such a process may be represented by

$$\hat{x}(t) = \sum_{n=-\infty}^{\infty} x(nT) \frac{\sin 2\pi f_0(t - nT)}{2\pi f_0(t - nT)}$$

where $T = 1/(2 f_0)$ is the sampling interval and f_0 is the highest frequency present in the autocorrelation function. The samples $x(nT)$ are random variables whose joint probability density is described by the statistical averages of the process. The equivalence shown by the above equation is in the sense that the expected value of the mean square error between the two sides of the equation is zero.

$$E\{|x(t) - \hat{x}(t)|^2\} = 0$$

10.12   Problems

Solved Problems

Note: For a block diagram of the sampling process, filtering, reconstruction, and notations, see Figure 10.14.

1. A 1-kHz sinusoidal signal is sampled at $t_1 = 0$ and $t_2 = 250$ μs. The sample values are $x_1 = 0$ and $x_2 = -1$, respectively. Find the signal's amplitude and phase.

Solution

$$x(t) = A \cos(2000\pi t + \theta)$$

at t_1 $A \cos(\theta) = 0$

at t_2 $A \cos(\pi/2 + \theta) = -A \sin(\theta) = -1$

The answer: $A = 1$, $\theta = \frac{\pi}{2}$

2. A continuous-time sinusoidal signal $x(t) = \cos(2\pi f_0 t)$ with unknown f_0 is sampled at the rate of 1,000 samples per second resulting in the discrete-time signal $x(n) = \cos(4\pi n/5)$. Find f_0.

Solution

$$x(n) = x(t)\Big|_{t=n\Delta t} = \cos(2\pi f_0 t)\Big|_{t=10^{-3}n} = \cos(2\pi f_0 10^{-3} n) = \cos(4\pi n/5), \quad f_0 = 400 \text{ Hz}$$

But $\quad \cos(4\pi n/5) = \cos(2k\pi n + 4\pi n/5)$.

Therefore, $\quad f_0 = (1{,}000k + 400)$ Hz, $\ k = 0, 1, 2, 3 \cdots$

3. A continuous-time periodic signal $x(t)$ with unknown period is sampled at the rate of 1,000 samples per second resulting in the discrete-time signal $x(n) = A\cos(4\pi n/5)$. Can you conclude with certainty that $x(t)$ is a single sinusoid?

Solution

The answer is no. Sampling the sum of two sinusoids with frequencies 400 and 1,400 Hz at the 1-kHz rate produces $x(n) = A\cos(4\pi n/5)$.

4. Consider a sinusoidal signal $x(t) = A\sin(2\pi f_0 t + \theta)$ with a known frequency. Show that the signal may be completely specified from two *independent* samples x_1 and x_2 taken at t_1 and t_2, respectively, during a period.

Solution

Given x_1 and x_2, the amplitude, A, and phase, θ, can be determined from the following equations:[4]

$$x_1 = A\sin(2\pi f_0 t_1 + \theta)$$
$$x_2 = A\sin(2\pi f_0 t_2 + \theta)$$

Several examples on sampling a sinusoidal signal at $f_0 = 100$ MHz are discussed below.

a. $x_1 = 2.5$ at $2\pi f_0 t_1 = 0°$ (at the time origin) and $x_2 = 2.5$ at $2\pi f_0 t_2 = 120°$. Then

$$A\sin\theta = 2.5$$
$$A\sin(120° + \theta) = 2.5$$

from which we find $A = 5$, $\theta = 30°$, and $x(t) = 5\sin(2\pi f_0 t + 30°)$.

b. $x_1 = 1$ at $2\pi f_0 t_1 = 30°$ and $x_2 = 2$ at $2\pi f_0 t_2 = 45°$. Then

$$A\sin(30° + \theta) = 1$$

$$A\sin(45° + \theta) = 2$$

from which we find $x \approx 4.12\sin(2\pi f_0 t - 16°)$.

c. $x_1 = 2.5$, $x_2 = 4.9$, $f_0 = 100$ MHz, and $t_2 - t_1 = 2$ ns. Let $2\pi f_0 t_1 + \theta = \alpha$. At 100 MHz, the 2 *ns* time lapse between the two samples corresponds to $72°$. Then

$$\text{at } t_1: \quad A\sin(2\pi f_0 t_1 + \theta) = A\sin\alpha = 2.5$$

$$\text{at } t_2: \quad A\sin(2\pi f_0 t_2 + \theta) = A\sin[2\pi f_0 t_1 + \theta + 2\pi f_0(t_2 - t_1)] = A\sin[\alpha + 2\pi f_0(t_2 - t_1)]$$
$$= A\sin(\alpha + 72°) = 4.9$$

from which we find $x(t) = 5\sin(2\pi f_0 t + 30°)$, where the time origin is set at the moment of the first sample. If the time origin is preset at another moment, then the phase angle may not be determined unless t_1 is also specified.

---

[4]Exceptions are when $x_1 = x_2 = 0$ or $t_2 - t_1$ is a half-period, in which case the two samples don't provide independent information.

d. $x_1 = 2.5$, $x_2 = -2.5$, $f_0 = 100$ MHz, and $t_2 - t_1 = 5$ ns. At 100 MHz, the 5 *ns* time lapse between the two samples corresponds to 180°. Then

$$A \sin \alpha = 2.5$$
$$A \sin(\alpha + 180°) = -2.5$$

The two samples don't produce two independent equations and $x(t)$ doesn't have a unique answer.

5. Sequential repetitive sampling, revisited. A continuous-time signal $x(t) = \cos(2\pi f_0 t)$ with known frequency is sampled uniformly every $(1 + \alpha)T$ seconds, where $T = 1/f_0$ and α is a proper number $\alpha < 0.5$.

a. Show that the sampling rate is $F_s = f_0/(1 + \alpha)$ and sample values are given by $x(n) = \cos(2\pi \alpha n)$.

b. Impulse samples are passed through an ideal low-pass analog filter with unity gain and a cutoff frequency f_c, where

$$\left(\frac{\alpha}{1+\alpha}\right) f_0 < f_c < \left(\frac{1-\alpha}{1+\alpha}\right) f_0$$

Find the output of the filter and show that it is a sinusoid at a frequency $(\frac{\alpha}{1+\alpha}) f_0$.

c. Determine α so that the above sampling and reconstruction downshifts the frequency by a factor 10.

Solution

Continuous time: $x(t) = \cos(2\pi f_0 t)$, $X(f) = \dfrac{1}{2}[\delta(f + f_0) + \delta(f - f_0)]$

a. Sampling rate: $F_s = \dfrac{1}{(1+\alpha)T} = \dfrac{f_0}{(1+\alpha)}$

Discrete time: $x(n) = \cos(2\pi f_0 t)\Big|_{t=n/F_s} = \cos\left(2\pi f_0 \dfrac{n}{F_s}\right) = \cos[2\pi n(1+\alpha)] = \cos(2\pi n\alpha)$

b. Sampling function: $s(t) = \displaystyle\sum_{n=-\infty}^{\infty} \delta\left(t - \dfrac{n}{F_s}\right)$, $S(f) = F_s \displaystyle\sum_{k=-\infty}^{\infty} \delta(f - kF_s)$

Sampled function: $y(t) = x(t) \times s(t)$, $Y(f) = X(f)^* S(f)$

$$= \dfrac{1}{2} F_s \left[\sum_{k=-\infty}^{\infty} \delta(f - kF_s + f_0) + \sum_{k=-\infty}^{\infty} \delta(f - kF_s - f_0)\right]$$

The Fourier transform of $y(t)$ is made of impulses of strength $F_s/2$ located at

$$\pm \dfrac{\alpha}{1+\alpha} f_0, \ \pm \dfrac{1-\alpha}{1+\alpha} f_0, \ \pm f_0, \ \pm \dfrac{1+2\alpha}{1+\alpha} f_0, \ \pm \dfrac{2-\alpha}{1+\alpha} f_0, \ldots, \pm \dfrac{k+1+\alpha}{1+\alpha} f_0, \ldots, k = 0, 1, 2, \ldots$$

The output of the low-pass filter is

$$Z(f) = \dfrac{1}{2} F_s \left[\delta\left(f + \dfrac{\alpha}{1+\alpha} f_0\right) + \delta\left(f - \dfrac{\alpha}{1+\alpha} f_0\right)\right]$$

$$z(t) = F_s \cos\left(2\pi \dfrac{\alpha}{1+\alpha} f_0 t\right) = F_s x\left(\dfrac{\alpha}{1+\alpha} t\right)$$

c. $\dfrac{\alpha}{1+\alpha} = 0.1$, $\alpha = \dfrac{1}{9}$

6. The signal $x(t) = \cos 500\pi t + \cos 200\pi t$ is sampled by multiplying it with a train of unit impulses at the rate of F_s impulses per second where $F_s > 500$. Impulse samples are then passed through a first-order analog low-pass *RC* filter with time constant $\tau = 2$ msec. Since the analog filter is nonideal, its output $z(t)$ contains components at frequencies higher than 250 Hz, which cause distortion. Define a crosstalk index, η, to be the ratio of the power in $f > 250$ Hz to the power in $f < 250$ Hz. Find the above index for

a. $F_s = 1{,}000$
b. $F_s = 2{,}000$

Interpret the differences in results of a and b.

Solution

The unit impulse and the frequency responses of the RC filter are:

$$h(t) = e^{-500t}u(t), \quad H(\omega) = \frac{1}{500 + j\omega}, \quad H(f) = \frac{500^{-1}}{1 + j2\pi \left(\frac{f}{500}\right)}, \quad |H(f)|^2 = \frac{500^{-2}}{1 + 4\pi^2 \left(\frac{f}{500}\right)^2}$$

The Fourier transform of the impulse-sampled signal is made of impulses with equal strength at the frequencies below

a. $F_s = 1{,}000$, $f = 100, \ 250, \ 750, \ 900, \ 1{,}100, \ 1{,}250$ Hz, $\cdots$
b. $F_s = 2{,}000$, $f = 100, \ 250, \ 1{,}750, \ 1{,}900, \ 2{,}100, \ 2{,}250$ Hz, $\cdots$

The RC filter makes the powers proportional to $|H(f)|^2$. Therefore,

a. $F_s = 1{,}000$ Hz, $\eta \approx \dfrac{\text{power in the 750 and 900 Hz}}{\text{power in the 100 and 250 Hz}}$

$$= \frac{1/(1 + 4\pi^2 \times 1.5^2) + 1/(1 + 4\pi^2 \times 1.8^2)}{1/(1 + 4\pi^2 \times 0.2^2) + 1/(1 + 4\pi^2 \times 0.5^2)} \approx 4\% \approx -14 \text{ dB}$$

b. $F_s = 2{,}000$ Hz, $\eta \approx \dfrac{\text{power in the 1,750 and 1,900 Hz}}{\text{power in the 100 and 250 Hz}}$

$$= \frac{1/(1 + 4\pi^2 \times 3.5^2) + 1/(1 + 4\pi^2 \times 3.8^2)}{1/(1 + 4\pi^2 \times 0.2^2) + 1/(1 + 4\pi^2 \times 0.5^2)} \approx 0.8\% \approx -21 \text{ dB}$$

Doubling the sampling rate reduces the crosstalk index by a factor 5 (≈ -7 dB). It should be noted that due to its nonuniform gain over the frequency range, the RC filter introduces additional distortion in the recovery of $x(t)$.

7. Bandpass sampling. The spectrum of a bandpass signal is shown in Figure 10.20(a) with $f_\ell = 7$ kHz and $f_h = 9$ kHz.

a. Find the minimum sampling rate. Show the spectrum of the sampled signal and its reconstruction by bandpass filtering.
b. Find the range of acceptable sampling rates.

Solution

We proceed as in examples in section 10.11.

a. $B = f_h - f_l = 2$ kHz and $f_h/B = 9/2 = 4.5$, the integer part of which is $N = 4$. The signal is sampled at the rate of $2f_h/N = 18/4 = 4.5$ kHz. See Figure 10.20. The spectrum of the sampled signal is shown in Figure 10.20(c). It is seen that bandpass filtering ($f_c = 8$ kHz, $B = 2$ kHz) will reproduce the original signal.
b. The sampling rate should satisfy the following relationship:

$$\frac{18}{k} \le F_s \le \frac{14}{(k-1)}, \quad k = 1, 2, 3, 4, \qquad \begin{cases} k = 1, & 18 \le F_s \\ k = 2, & 9 \le F_s \le 14 \\ k = 3, & 6 \le F_s \le 7 \\ k = 4, & 4.5 \le F_s \le 4.66 \end{cases}$$

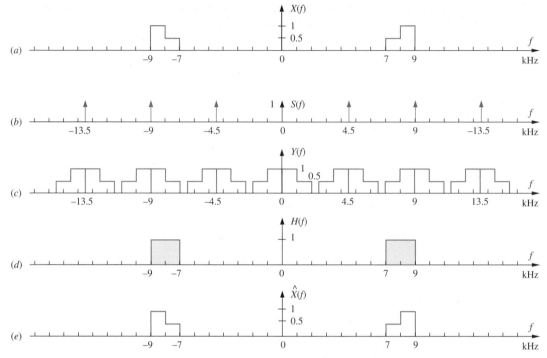

FIGURE 10.20 The minimum sampling rate for a bandpass signal and the characteristics of the reconstruction filter can be derived from the spectrum of samples. This figure illustrates sampling and recovery of a bandpass signal $x(t)$ whose Fourier transform $X(f)$ (shown in **a**) is limited to 7 kHz $< f <$ 9 kHz. When sampled at the rate of 4.5 kHz, its Fourier transform is repeated at 4.5-kHz intervals. The Fourier transforms of the sampling function and the impulse samples are shown in **(b)** and **(c)**, respectively. To recover $x(t)$, samples are sent through a bandpass filter $H(f) = 1$, limited to 7 kHz $< f <$ 9 kHz, shown in **(d)**. The Fourier transform of the filter's output is shown in **(e)**.

8. Sample a 1-Hz continuous-time sinusoidal signal $x(t) = \sin 2\pi t$ every T seconds. Reconstruct the signal from its samples, take the Fourier transform of the reconstructed signal and display its magnitude for the following sampling intervals:

a. $T = 0.1$
b. $T = 0.65$
c. $T = 0.85$
d. $T = 0.95$

Discuss the relationship between the position of spectral lines of the reconstructed signal with the sampling rate.

Solution
The sampled signal and the Fourier transform of the reconstructed signal, that is, the part of the transform of the sampled function that is filtered through a $\pm F_s/2$ low-pass reconstructed filter are shown in Figure 10.21.

a. $T = 0.1$ s. The sampling rate is $F_s = 10$ Hz, which is more than twice the frequency of the signal. As expected, the Fourier transform shows two spectral lines at ± 1 Hz, Figure 10.21(a).

b. $T = 0.65$ s. The sampling rate is $F_s = 1.5385$ Hz, less than the Nyquist rate. The spectral lines of the reconstructed signal appear at $|F_s - 1| = 0.5385$ Hz, Figure 10.21(b).

c. $T = 0.85$ s, $F_s = 1.1765$ Hz, and the reconstructed spectral lines are at $|F_s - 1| = 0.1765$ Hz, Figure 10.21(c).

d. $T = 0.95$ s, $F_s = 1.0526$ Hz, and we observe two spectral lines at $|F_s - 1| = 0.0526$ Hz, Figure 10.21(d).

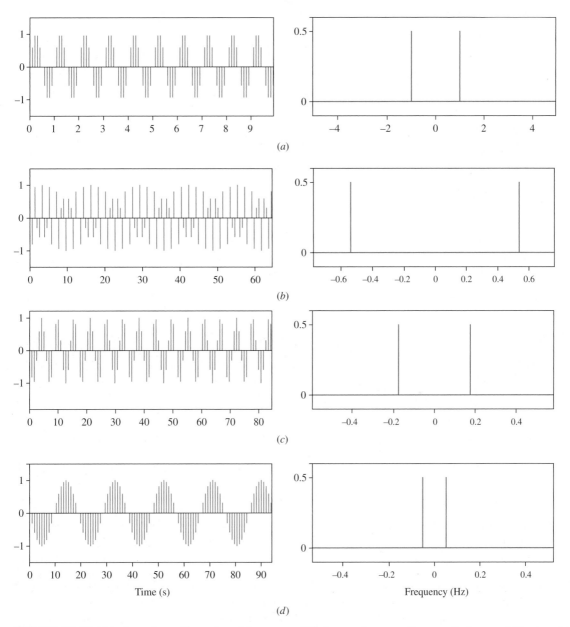

FIGURE 10.21 This figure is an illustration of frequency shift due to a low sampling rate. It shows a 1-Hz sinusoid sampled at four different rates (left column) along with the resulting transforms (right column). Relevant information is summarized below.

The above observations are verified by convolution of the Fourier transforms of $x(t)$ and the sampling impulse train at the F_s rates.

	Sampling Interval, T	Sampling Rate, $F_s = 1/T$	Reconstructed Frequency
(a)	100 msec	10 Hz	1 Hz
(b)	650 msec	1.5385 Hz	$1.5385 - 1 = 0.5385$ Hz
(c)	850 msec	1.1765 Hz	$1.1765 - 1 = 0.1765$ Hz
(d)	950 msec	1.0526 Hz	$1.0526 - 1 = 0.0526$ Hz

Note that in case a, where the sampling rate is greater than 2 Hz, the frequency of the reconstructed signal is the same as that of the original continuous-time signal at 1 Hz. But in cases b, c, and d, where the sampling rate is less than 2 Hz, aliasing occurs and spectral lines appear at $|F_s - 1|$.

9. **Frequency downshifting.** Take a continuous-time sinusoidal signal at the frequency $f = 1\ Hz$.

a. Sample it every 0.01 second (equivalently, at the rate of 100 samples per second). Record 101 samples and plot the sampled data versus time.

b. Repeat for a sampling interval of 1.01 seconds. Interpret the plots in light of the sampling rates.

Solution

The Matlab program to implement the above procedure is shown below.

```
t=linspace(0,1,101); x=sin(2*pi*t);
subplot(2,1,1); plot(t,x); grid
title('Plot of a 1 Hz sinusoid sampled every 10 msec., sampling rate=100 Hz.');
%
t=linspace(0,100,100); x=sin(2*pi*t);
subplot(2,1,2); plot(t,x); grid
title('Plot of a 1 Hz sinusoid sampled every 1.01 sec., sampling rate=0.99 Hz.');
```

a. $F_s = 100$ Hz. The command $t = linspace(0, 1, 101)$ produces 101 sampling instances in 1 second (one sample every 10 msec), including one sample at $t = 0$ and one sample at $t = 1$, for a total of 100 intervals. The sampling interval is, therefore, 10 msec, corresponding to a sampling rate of 100 Hz. The resulting plot shows a 1-Hz sinusoid.

b. $F_s = 0.99$ Hz. The command $t = linspace(0, 100, 100)$ produces 100 sampling instances in 100 seconds, including one sample at $t = 0$ and one sample at $t = 100$, for a total of 99 intervals. The sampling interval is, therefore, $100/99 = 1.01$ seconds, corresponding to a sampling rate of 0.99 Hz. The resulting plot shows a 0.01 Hz sinusoid.

The figure generated by the above program illustrates the sampling of a sinusoidal signal (with frequency f_0 Hz) at the rate F_s and reconstructing it by interpolation; for example, through the *plot* command in Matlab. It shows, that when $F_s < 2f_0$, the reconstructed frequency is shifted to $|F_s - f_0|$, causing aliasing. The 1-Hz sinusoid $x = \sin(2\pi t)$ is sampled at the rates of $F_s = 100$ Hz in (a) and $F_s = 0.99$ Hz in (b). The frequency of the sinusoidal plot in (b) is $|0.99 - 1| = 0.01$ Hz. In summary, the plot in (a) shows one cycle of a 1-Hz sinusoid. The plot in (b) is similar to (a) except for the time scale, which indicates a sinusoid at 0.01 Hz. Time has slowed down by factor 100.

Chapter Problems

10. **Sampling a sinusoid with known frequency.** The frequency of a sinusoidal signal $x(t)$ is known to be 1 kHz. The signal is sampled at two instances t_1 and t_2 resulting in sample values x_1 and x_2, respectively. For the following cases, express the signal in the form $x(t) = A \cos(2\pi f_0 t - \theta)$ where $A > 0$.

a. $\begin{cases} t_1 = 0 & x_1 = 1 \\ t_2 = 250 \ \mu s & x_2 = 1.732 \end{cases}$
b. $\begin{cases} t_1 = 0 & x_1 = 0 \\ t_2 = 583 \ \mu s & x_2 = 1 \end{cases}$
c. $\begin{cases} t_1 = 0 & x_1 = 0 \\ t_2 = 1{,}083 \ \mu s & x_2 = 0.5 \end{cases}$

d. $\begin{cases} t_1 = 250 \ \mu s & x_1 = 1 \\ t_2 = 333 \ \mu s & x_2 = 0 \end{cases}$
e. $\begin{cases} t_1 = 100 \ \mu s & x_1 = 3 \\ t_2 = 600 \ \mu s & x_2 = -3 \end{cases}$
f. $\begin{cases} t_1 = 50 \ \mu s & x_1 = 2 \\ t_2 = 1{,}300 \ \mu s & x_2 = -4 \end{cases}$

11. Extend problem 10 to formulate a general solution for possible determination of magnitude and phase of a sinusoidal signal with known frequency from two samples taken at t_1 and t_2.

12. The frequency of a sinusoidal signal is known to be 1 kHz. The signal is sampled at two moments 250 microseconds apart. Can you find signal's amplitude and/or its phase from the above two measurements? Show how to do it or why it can't be done.

13. **Attenuation and path delay.** A microwave ranging system transmits a sinusoidal signal $s(t) = \cos(2\pi f_0 t)$ with a known frequency. The received signal is $x(t) = As(t - \tau)$ where A is the path attenuation and τ is the path delay. The goal is to determine A and τ from a finite number of samples taken from $x(t)$. Discuss possible strategies for determination of A and τ in the following two measurement scenarios:

a. Two measurements of x_1 and x_2 taken at t_1 and t_2

b. One measurement of x_1 at t_1, and the time between consecutive zero-crossings of $s(t)$ and $x(t)$

14. **Modeling a periodic signal.** A periodic signal $x(t)$ of unknown period is to be modeled as $\hat{x}(t) = A \sin(2\pi f_0 t)$ by choosing a zero-crossing as $t = 0$. The goal is to find A and f_0 from a finite set of measurements. For the following three cases determine if the above goal may be achieved, and if so find the model.

a. $\begin{cases} t_1 = 500 \ \mu s & x_1 = 2.5 \\ x_{Max} = 5 \end{cases}$

b. $\begin{cases} t_1 = 166.7 \ \mu s & x_1 = 0.5 \\ t_2 = 250 \ \mu s & x_2 = 1 \end{cases}$

c. $\begin{cases} t_1 = 16.67 \ \mu s & x_1 = 0.5 \\ t_2 = 500 \ \mu s & x_2 = 1 \end{cases}$

15. A periodic signal $x(t)$ with unknown period is to be modeled as $\hat{x}(t) = A \sin(2\pi f_0 t)$ by choosing a zero-crossing as $t = 0$. Show that A and f_0 may not always be uniquely determined from a finite number of samples taken from $x(t)$.

16. A periodic signal $x(t)$ with an unknown period is to be modeled as $\hat{x}(t) = A \sin(2\pi f_0 t - \theta)$. The goal is to find A, f_0, and θ from a finite set of measurements. For the following three cases determine if the above goal may be achieved, and if so find the model.

a. $\begin{cases} t_1 = 0 & x_1 = 0 \text{ and } \frac{dx}{dt}\big|_{t=0} = 5{,}000\pi \text{ V/s.} \\ t_2 = 250 \ \mu s & x_2 = -1.4142 \end{cases}$

b. $\begin{cases} t_1 = 0 & x_1 = 0 \\ t_2 = 583 \ \mu s & x_2 = 1 \end{cases}$

c. $\begin{cases} t_1 = 0 & x_1 = 0 \\ t_2 = 1{,}083 \ \mu s & x_2 = 0.5 \end{cases}$

17. Modeling an exponential signal. A signal decays exponentially from an unknown initial value to a zero final value. The decay rate and the initial value at $t = 0$ are not known. Two samples taken 1 msec apart result in $x_1 = 3$ and $x_2 = 1$.

a. Find the time constant of the signal.

b. Can you find the value of the signal at $t = 0$? If the answer is *yes*, find it. If the answer is *no*, specify additional information needed.

18. Impulse function. The unit-impulse function $\delta(t)$ is defined by the *distribution*

$$\phi(t_0) = \int_{-\infty}^{\infty} \delta(t - t_0)\phi(t)dt \ \text{ for all } t_0$$

where $\phi(t)$ is any well-behaved (also called good) function. The above equation is called sifting property of $\delta(t)$.

a. Use the sifting property of $\delta(t)$ to prove or disprove the following equations.

$$\delta(at) = \frac{\delta(t)}{|a|}, \ \ a \neq 0$$

$$\delta(at + b) = \frac{\delta(t - b/a)}{|a|}, \ \ a \neq 0$$

$$\delta[(t - t_1)(t - t_2)] = \frac{\delta(t - t_1) + \delta(t - t_2)}{|t_1 - t_2|} \ \ t_1 \neq t_2$$

b. Prove or disprove that for an integer n

$$\delta\left(\sin\frac{2\pi t}{T}\right) = \frac{1}{2\pi} \sum_{n=-\infty}^{\infty} \delta\left(t - \frac{nT}{2}\right)$$

$$\delta\left(\cos\frac{2\pi t}{T}\right) = \frac{1}{2\pi} \sum_{n=-\infty}^{\infty} \delta\left(t - \frac{(2n + 1)T}{4}\right)$$

19. Impulse sampling. Prove or disprove that

$$x(t) \times \sum_{n=-\infty}^{\infty} \delta(t - nT) = \frac{1}{T}x(t) + \frac{2}{T} \sum_{n=1}^{\infty} x(t) \cos\left(\frac{2\pi nt}{T}\right)$$

20. Sample the signal $x(t) = \cos 20\pi t$ at the rate of F_s. The resulting impulse samples are called $y(t)$. For the following rates find and plot the time-domain and frequency-domain representation of $y(t)$:

a. $F_s = 30$ Hz

b. $F_s = 40$ Hz

c. $F_s = 50$ Hz

21. The signal $x(t) = \cos(1{,}800\pi t - \pi/6)$ is sampled uniformly at the rate of 1 kHz and passed through an ideal low-pass filter with a DC gain of 0.001 and a cutoff frequency of 500 Hz. Find the filter's output.

22. a. The Fourier transform of a continuous-time signal is

$$X(f) = \begin{cases} 1 & |f| < 1 \\ 0, & \text{elsewhere} \end{cases}$$

Find $x(t)$ and plot it. Find the minimum sampling rate from which $x(t)$ may be recovered. Determine sample values $x(nT)$ for $-5 \le n \le 5$.

   b. Repeat for

$$X(f) = \begin{cases} \frac{1}{2} + |f|, & 0 < |f| < \frac{1}{2} \\ 0, & \text{elsewhere} \end{cases}$$

23. **Ideal interpolation filter.** Sample the signal $x(t) = \cos(2\pi f_0 t)$ at the rate F_s and pass the impulse samples $y(t)$ through a filter. For the following two sets of f_0 and F_s determine the frequency response of the filter that would reconstruct $x(t)$ from the samples with no error:

   a. $f_0 = 1.1$ kHz and $F_s = 1$ kHz
   b. $f_0 = 1$ kHz and $F_s = 1.1$ kHz

24. Sample the signal $x(t) = \cos 2\pi t$ at the rate of F_s. Reconstruct the signal by sending the impulse samples $y(t)$ through an ideal low-pass filter with the cutoff frequency at $F_s/2$ and the gain= $1/F_s$. The output of the filter is $z(t)$. For the following sampling rates find and plot the time-domain and frequency-domain representations of $y(t)$ and $z(t)$:

   a. $F_s = 3$ Hz
   b. $F_s = 4$ Hz
   c. $F_s = 5$ Hz

   Compare with problem 20.

25. **First-order interpolation filter.** Repeat problem 24 after replacing the ideal low-pass filter with a realizable first-order low-pass filter having 3-dB attenuation at 1 Hz and a DC gain = $\sqrt{2}$.

26. **Second-order interpolation filter.** Repeat problem 24 after replacing the ideal low-pass filter with a realizable second-order low-pass Butterworth filter having 3-dB attenuation at 1 Hz and a DC gain = $\sqrt{2}$. The system function for the filter is $H(s) = \sqrt{2}/(s^2 + \sqrt{2}s + 1)$.

27. **nth-order interpolation filter.** Sample $x(t) = \cos 2\pi t$ at the rate of 3 Hz and send the impulse samples through a realizable nth-order Butterworth low-pass filter having 3-dB attenuation at 1 Hz and a DC gain = $\sqrt{2}$. Filter's output is $z(t)$. Find the magnitude Fourier transform of $z(t)$ for $n = 2, 3, 4, 5$. The magnitude frequency response of the nth-order Butterworth filter is $|H(f)|^2 = 2/(1 + f^{2n})$.

28. **High-attenuation interpolation filter.** Sample $x(t) = \cos 2\pi t$ at the rate of F_s and reconstruct it by sending the impulse samples through a low-pass filter having a DC gain = $\sqrt{2}$, 3-dB attenuation at 1 Hz, and 140-dB attenuation per decade beyond 1 Hz. For the following sampling rates find the magnitude Fourier transforms of the reconstructed signal $z(t)$:

   a. $F_s = 3$ Hz
   b. $F_s = 4$ Hz
   c. $F_s = 5$ Hz

29. **Distortion.** Sample $x(t) = \cos 2\pi t$ at the rate of F_s and send the impulse samples through a realizable 1st-order low-pass filter having 3-dB attenuation at 1 Hz and a DC gain = $\sqrt{2}$. Filter's output is $z(t)$. Because the filter is not ideal, $z(t)$ contains distortions in the form of additional frequency components. Define the distortion index in $z(t)$ to be the ratio of the sum of power in frequencies higher than 1 Hz to the total power in $z(t)$. Determine the minimum sampling rate such that the distortion index in $z(t)$ remains below 60 dB.

30. Sample $x(t) = \cos 2\pi t$ at the rate of 3 Hz and send the impulse samples through a realizable nth-order Butterworth low-pass filter having 3-dB attenuation at 1 Hz. Determine minimum filter order such that distortion index at the output of the filter (defined in problem 29) remains below 60 dB. The attenuation by the nth-order Butterworth filter is $10 \log(1 + f^{2n})$ (dB).

31. A continuous-time analog signal $x(t) = \cos 500\pi t + \cos 200\pi t$ is sampled at the rate of F_s samples per second. The impulse samples are then passed through a low-pass analog filter with unity DC gain and cutoff frequency $f_c = 260 \, Hz$. Find the output of the filter and show its relationship to $x(t)$ for

  a. $F_s = 600$
  b. $F_s = 400$

32. Frequency multiplication. Sample $x(t) = \cos 3\pi t$ at the rate of 4 Hz and send impulse samples through an ideal low-pass filter with the cutoff frequency at f_c and gain $= 1/4$. Find and plot the Fourier transform of the reconstructed signal $z(t)$ and its time expression for $f_c = 2, \, 3, \, 4, \, 5, \, 6$ Hz.

33. Aliasing. Sample $x(t) = \cos(1.2\pi t)$ at the rate of F_s and send impulse samples through an ideal low-pass filter with the cutoff frequency at $F_s/2$ and gain $= 1/F_s$. Find and plot the time-domain and frequency-domain expressions for the sampled signal $y(t)$ and the reconstructed signal $z(t)$ for

  a. $F_s = 1$ Hz
  b. $F_s = 3$ Hz

34. Sample $x(t) = \cos(2\pi f_0 t)$ at the rate of 4 Hz and send impulse samples through an ideal low-pass filter with the cutoff frequency at 2 Hz and gain $= 0.25$. Find the Fourier transform of the reconstructed signal $z(t)$ and its time expression for $f_0 = 1, \, 3, \, 4, \, 5$ Hz.

35. A finite segment $(0 \leq t \leq T)$ of a continuous-time signal $x(t)$ is sampled at the precise rate of 1,000 samples per second resulting in a sequence of numbers $x(n)$, $n = 0, \ldots 1,023$. Using the finite set of the samples we want to model $x(t)$ within the interval $0 < t < T$ by a sinusoid $\hat{x}(t) = A \cos(2\pi f_0 t)$ such that the error defined by the following ϵ is minimized.

$$\epsilon = \int_0^T |x(t) - \hat{x}(t)|^2 dt$$

  a. Show how A and f_0 may be found and if the answer is unique.
  b. Can you find the smallest f_0 with 100% certainty? In each case if the answer is no, describe why. If the answer is yes, find it.

36. Sampling a low-pass signal. A signal $x(t) = \frac{\sin(1,000\pi t)}{\pi t}$ is sampled at the rate of F_s and sent through a unity-gain ideal low-pass filter with the cutoff frequency at $F_s/2$. Find and plot the Fourier transform of the reconstructed signal $z(t)$ at filter's output if

  a. $F_s = 20$ kHz
  b. $F_s = 2$ kHz
  c. $F_s = 1$ kHz
  d. $F_s = 800$ Hz

37. A signal $x(t) = \frac{\sin(1,000\pi t)}{\pi t}$ is sampled at the rate of F_s and sent through a realizable first-order low-pass filter having 3-dB attenuation at 500 Hz and a DC gain $= \sqrt{2}$. Find and plot the Fourier transform of the reconstructed signal $z(t)$ if

  a. $F_s = 20$ kHz
  b. $F_s = 2$ kHz

c. $F_s = 1$ kHz

d. $F_s = 800$ Hz

38. Reconstruction error. A signal $x(t) = \frac{\sin(1,000\pi t)}{\pi t}$ is sampled at the rate of F_s kHz and sent through a realizable first-order low-pass filter with a unity DC gain and 3-dB attenuation at 500 Hz. Find an expression for reconstruction error (as defined in section 10.8 of this chapter) for

a. $F_s = 20$ kHz

b. $F_s = 2$ kHz

Hint: Use Parseval's theorem to compute the error.

39. Frequency downshifting. This problem extends problem 5 to sampling intervals being several times greater than the period of the continuous-time signal. Sample $x(t) = \cos(2\pi f_0 t)$ uniformly every $\Delta = (M + \alpha)T$ seconds where M is a fixed positive integer, $\alpha < 0.5$, and $T = 1/f_0$. Sampling instances are, therefore, at $t = (M + \alpha)nT$.

a. Show that the sequence of samples $x(n)$ is periodic and determine its period.

b. Determine the spectrum of impulse samples.

c. Let $f_0 = 100$ GHz. Determine α and M so that the sampling and reconstruction downshifts the frequency of $x(t)$ to 100 MHz.

40. Extend problem 5 to the case of a band limited periodic continuous-time signal $x(t)$ with period T. Let the highest frequency component of the signal be f_h and let the lowest frequency be f_ℓ. The aim is to sample the signal at a much lower rate than the Nyquist rate and still obtain a representation in the form of $x(t/g)$ where g is a frequency downshift factor. Examine use of the sampling interval $\Delta = (1 + \alpha)T$ which is higher than the period of the signal by the amount αT where $0 < \alpha < 1$. Based on f_ℓ and f_h discuss conditions on α, f_c, and the resulting frequency downshift factor g.

41. Sample and hold. A signal $x(t)$ is sampled every T seconds and is kept at that level until the next sample. The operation is called sample and hold. Find the Fourier transform of the stepwise signal obtained by the sample and hold operation.

42. Sample the signal $x(t) = \sin 200\pi t$ at intervals T and hold for the duration τ. The continuous signal generated by sample and hold operation is called $z(t)$. For $T = \tau = 1$ msec:

a. Find and plot the spectrum of $z(t)$.

b. Find the frequency response of a filter that would recover $x(t)$ from $z(t)$ (if that can be done).

43. Repeat problem 42 for the following cases and compare results.

a. $T = \tau = 0.1$ msec

b. $T = 1$ and $\tau = 0.5$ msec

c. $T = 0.1$ and $\tau = 0.05$ msec

44. Rectangular pulse. A continuous-time analog signal $x(t)$ consists of a single 1-volt rectangular pulse lasting slightly more than 3 msec. The signal is sampled at the rate of F_s samples per second by multiplying it with a train of impulses. The samples are then passed through a second-order analog low-pass Butterworth filter with unity DC gain and half-power frequency at 500 Hz. Find the output of the filter and the total energy difference compared to the original analog pulse for

a. $F_s = 1,000$ Hz

b. $F_s = 10,000$ Hz

Compare results obtained in a and b and discuss the source of differences.

45. Digital filtering. A continuous-time pulse

$$x(t) = \begin{cases} 1, & -5.1 \text{ msec} \leq t \leq 5.1 \text{ msec} \\ 0, & \text{elsewhere} \end{cases}$$

is sampled at the rate of 200 samples per second with one sample at $t = 0$.

a. Show that the resulting continuous-time signal is represented by $\delta(t + T) + \delta(t) + \delta(t - T)$ where $T = 5$ msec and that the above information may also be represented by $x(n) = \{1, 1, 1\}$.

b. Impulse samples are passed through an LTI system with the unit-sample response $h(t) = \delta(t) + 2\delta(t - T) + \delta(t - 2T)$. Find the output of the LTI system.

c. The LTI system is followed by a hold filter with $h(t) = u(t) - u(t - T)$. Plot the output $y(t)$ of the hold filter.

d. Show that the above analysis and results may be represented by discrete-time notations as follows:

$$x(n) = \{1, \underset{\uparrow}{1}, 1\}$$

$$h(n) = \{1, \underset{\uparrow}{2}, 1\}$$

$$y(n) = \{1, 3, \underset{\uparrow}{4}, 3, 1\}$$

In solving the succeeding problems you may use the above discrete-time representations.

46. A continuous-time analog signal $x(t) = \cos 500\pi t + \cos 200\pi t$ is sampled at the rate of F_s samples per second. The resulting discrete signal $x(n)$ is passed through a digital filter with

$$h(n) = \{1, 2, \underset{\uparrow}{3}, 2, 1\}$$

Find the output $y(n)$ and the reconstructed $y(t)$ for

a. $F_s = 1,000$ Hz
b. $F_s = 2,000$ Hz

Compare $y(n)$ obtained in a and b, and discuss sources of their possible differences.

47. An AM signal is given by $x(t) = a(t) \cos(2\pi f_c t)$ where a(t) is a low-pass signal band limited to $[0, B]$ Hz and $f_c > 2B$. Show that the minimum sampling rate from which the signal can be recovered is $2B$. Show how $x(t)$ may be recovered from its samples taken at the minimum rate. Interpret the results in both the time and frequency domains.

48. AM signal. A signal is modeled by $x(t) = a(t) \cos(2\pi f_c t)$ where $a(t)$ is a real-valued low-pass signal limited to $[0, 3]$ kHz and $f_c = 1$ MHz. Devise a scheme for sampling the signal at the 6 kHz rate and its reconstruction, which would result in $x(t)$.

49. AM signal. A signal is modeled by $x(t) = [A + \mu a(t)] \cos(2\pi f_c t)$ where A and μ are constant, $a(t)$ is a real-valued low-pass signal limited to $[0, 3]$ kHz and $f_c \geq 500$ kHz. Multiply $x(t)$ by a sinusoidal signal at $(f_c + 455)$ kHz and pass it through an ideal bandpass filter (center frequency at 455 kHz, bandwidth $= 6$ kHz). Devise a sampling scheme that results in $x(t)$ independent of f_c.

50. Bandpass signal. The amplitude-modulated signal

$$x(t) = a(t) \cos(2,000\pi t), \quad a(t) = \left[\frac{\sin 200\pi t}{\pi t} \right]^2$$

is passed through an ideal low-pass filter with unity gain and cutoff at 1 kHz. Express the filter's output as

$y(t) = v_c \cos(2\pi f_c t) - v_s \sin(2\pi f_c t)$. Find f_c, $v_c(t)$ and $v_s(t)$. Sample $y(t)$ at the minimum required rate and relate the samples to $v_c(t)$ and $v_s(t)$. Find and plot the sequence of samples $y(n)$ for $-15 \le n \le 15$.

51. Bandpass signal. The Fourier transform of a bandpass signal $y(t)$ is

$$Y(f) = \begin{cases} \frac{3}{2} - |f|, & \frac{1}{2} < |f| < 1 \\ 0, & \text{elsewhere} \end{cases}$$

a. Express $y(t)$ in the form of $y(t) = v_c \cos(2\pi f_c t) - v_s \sin(2\pi f_c t)$. Find f_c, $v_c(t)$ and $v_s(t)$.
b. Sample $y(t)$ at the minimum required rate and relate the samples to $v_c(t)$ and $v_s(t)$. Plot $y(n)$ for $-15 \le n \le 15$.
c. Given that $y(t)$ is the output of an ideal low-pass filter with unity gain and the cutoff frequency at $f = 1$ Hz, with input $2x(t) \cos 2\pi t$, find $x(t)$.

52. Bandpass signal. The Fourier transform of a bandpass signal is given by $X(f)$ in Figure 10.16(a).

a. Express it as $x(t) = v_c(t) \cos(2\pi f_0 t) - v_s(t) \sin(2\pi f_0 t)$ where $f_0 = (f_h + f_\ell)/2$ and find $v_c(t)$ and $v_s(t)$.
b. Specify the minimum and acceptable sampling rates for $f_\ell = 9$ and $f_h = 12$, both in kHz.

53. Bandpass signal. Assume the spectrum of the bandpass signal of Figure 10.16(a) is limited to $f_\ell = 9$ and $f_h = 12$, both in kHz, and that the signal is described by $x(t) = v_c(t) \cos(2\pi f_0 t) - v_s(t) \sin(2\pi f_0 t)$ with $f_0 = f_\ell$. Specify the minimum and acceptable sampling rates for the bandpass signal and discuss its relation with the rates obtained in problem 53.

54. A signal is modeled by

$$x(t) = \sum_{n=1}^{11} 2n \cos[2,000(n+99)\pi t]$$

Find and plot its Fourier transform $X(f)$ and determine sampling schemes at low rate from which $x(t)$ may be reconstructed.

55. The Fourier transform of a periodic continuous-time signal is

$$X(f) = \sum_{k=1}^{11} \{k\delta[f - 1,000(k+99)] + (11-k)\delta[f + 1,000(k+99)]\}$$

Plot $X(f)$. Express $x(t)$ in the time domain and determine sampling rates from which $x(t)$ may be reconstructed.

56. Complex-valued low-pass signals. Given the Fourier transforms express the time functions as $x(t) = v_c(t) + jv_s(t)$ for the following cases.

a. $X(f) = \delta(f - 1,000)$
b. $X(f) = \delta(f + 1,000)$
c. $X(f) = \delta(f - 2,000) + \delta(f + 1,000)$
d. $X(f) = \delta(f - 1,000) + \delta(f + 2,000)$
e. $X(f) = \delta(f - 1,000) + 2\delta(f + 2,000)$
f. $X(f) = 2\delta(f - 1,000) + \delta(f + 1,000)$

57. A bandpass signal is given by $x(t) = \mathcal{RE}\{a(t)e^{j2\pi f_c t}\}$ where $a(t)$ is a complex-valued low-pass signal whose Fourier transform is

$$A(f) = \begin{cases} \beta + \alpha f, & |f| < f_0 \\ 0, & |f| > f_0 \end{cases}$$

in which α and β are constants.

a. Express the low-pass signal as $a(t) = v_c(t) + jv_s(t)$.

b. For the following sets of parameter values determine acceptable sampling rates and sampling schemes. Show how $x(t)$ is recovered.

	f_c	f_0	α	β
Set 1:	1 MHz	3 kHz	1	2
Set 2:	1 MHz	3 kHz	0	1
Set 3:	1 MHz	6 kHz	1	2
Set 4:	1 MHz	6 kHz	0	2
Set 5:	10 MHz	3 kHz	1	2
Set 6:	10 MHz	3 kHz	0	1
Set 7:	10 MHz	6 kHz	1	2
Set 8:	10 MHz	6 kHz	0	2

58. A bandpass signal is given by $x(t) = \mathcal{RE}\{a(t)e^{j2\pi f_c t}\}$ where $a(t)$ is a complex-valued low-pass signal whose Fourier transform is

$$A(f) = \begin{cases} 1, & -f_\ell < f < f_h \\ 0, & \text{elsewhere} \end{cases}$$

in which f_ℓ and f_h are positive constants.

a. Express the low-pass signal as $a(t) = v_c(t) + jv_s(t)$.

b. For the following sets of parameter values determine acceptable sampling rates and sampling schemes. Show how $x(t)$ is recovered.

	f_c	f_ℓ	f_h
Set 1:	1 MHz	1 kHz	2 kHz
Set 2:	1 MHz	0 kHz	2 kHz
Set 3:	10 MHz	1 kHz	2 kHz
Set 4:	10 MHz	0 kHz	2 kHz

59. A continuous-time function $h(t)$ is known to be low pass so that its Fourier transform is zero for $|f| > f_0$.

a. Sampling $h(t)$ at the Nyquist rate of $2f_0$ produces $h(n) = d(n)$. Find $h(t)$.

b. Sampling $h(t)$ at twice the Nyquist rate (i.e., at $4f_0$ samples per second) yields

$$h(n) = \frac{\sin(\frac{\pi n}{2})}{\frac{\pi n}{2}}$$

Find $h(t)$. Show that the additional samples obtained in part b are not needed for recovering $h(t)$.

10.13   Project 1: Sampling Neuroelectric Signals

In this project you will investigate bandwidths and sampling rate requirements for neuroelectric signals. See the sections on EEG, EMG, EKG, and other neuroelectric signals in Chapter 1. For each signal type you start with a review of standards currently used in conventional clinical applications. You then seek information on possible advantages in using broader frequency bands (at the low and high frequencies) in a new system. The sources for your investigation are: (1) technical data and specifications of the commercially available systems and devices; (2) recommendations by the relevant professional societies; and (3) research papers about salient features of these signals and their relation to clinical applications. The outcome of the project is the set of your recommendations to be considered in the design of a commercial system for digital analysis of EEG, EMG, and EKG signals in clinical and research applications.

10.14   Project 2: Time-Domain Sampling

Summary

In this project you will first sample the rotary motion of a fan that may be modeled as a single sinusoidal signal. Then, using software and the computer, you will generate sampled versions of several continuous-time functions that model signals of interest in signal processing and find their Fourier representations. You will start with a high-sampling rate and examine the effect of reducing it below the Nyquist rate. In this way you will investigate aliasing and frequency downshifting. Finally, you will use a digital sampling device such as a digital oscilloscope, sound card, DFT spectrum analyzer, or a DSP board to sample analog signals in real time and examine the effect of the sampling on the resulting signal.

Equipment and Tools

> Variable-speed electric fan, circular disks, and auto transformer
> Adjustable strobe light
> Software package for signal processing
> Hardware platform (PC, sound card or DSP board)
> Function generator and spectrum analyzer (used in "Real-Time Sampling, Reconstruction, and Aliasing")

Introduction

Mathematical software packages as well as display and DSP packages can plot mathematical functions; for example, $y = f(t)$. In actuality, what is plotted is the discrete function $y(n) = f(n\Delta t)$. For example, you may need to specify the initial and final value of t, and the number of sample points N within that range to be used in the plot. Alternatively, you may need to specify the sampling interval (time resolution). Use the plot command to investigate the effect of sampling a high-frequency signal at a low rate. The effect may be called frequency downshifting or aliasing. Plots of a continuous-time sinusoidal signal made throughout this chapter as well as in problems show the above effect.

Prelab Exercises

Exercise 1. Sampling a Sinusoid at a Low Rate

Sample a 10-kHz sinusoid every $\Delta t = 100/99$ seconds ($F_s = 0.99$ Hz) and plot it. The following Matlab program illustrates the above exercise.

```
t=linspace(0,100,100);
y=sin(20000*pi*t);
plot(t,y); grid;
ylabel('Magnitude'); xlabel('Time in Seconds');
title('Plot of 10 kHz sinusoid sampled at the rate of 0.99 Hz.');
```

Analyze the above program, run it, and examine the output plot. Determine the frequency downshift and relate it to the sampling rate.

Exercise 2. Sampling a Sinusoid at the Zero-Crossing Time

Examine the following Matlab file. Run it and interpret the output plot.

```
t=linspace(0,100,101);
y=sin(2*pi*t);
plot(t,y); grid;
ylabel('Magnitude'); xlabel('Time in Seconds');
```

Measurements and Analysis

Sampling the Rotary Motion of a Fan

In this section you use a strobe light to explore the effect of the sampling rate on the perceived rotary motion of a fan. A circular disk marked by single or multiple spokes is mounted on the axis of the fan to observe the rotation and measure its speed. The schematic block diagram of the experimental setup is shown in Figure 10.22. Do not look directly at the strobe when it is flashing.

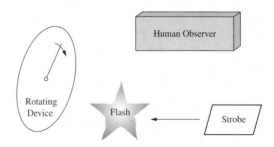

FIGURE 10.22 Sampling a rotary motion by strobe light.

Single-Spoke Disk. Mount the single-spoke circular disk on the axis of the fan. Set the auto transformer output that feeds the fan to 120 volts and run the fan at the high-speed setting. Wait until the fan achieves its final speed. Sample the rotary motion by strobe. With the fan running at the constant speed, vary the rate of the strobe and explore your perception of the disk's speed and direction of its rotation. Measure the fan's RPM by adjusting the strobe's rate so that the disk appears stationary. Describe your method. At what rate(s) do you see a single stationary spoke? At what rates do you see two stationary spokes 180 degrees apart? At what rate do you have four spokes 90 degrees apart? Describe, quantitatively, the perceived motion of the fan and its direction and speed in terms of the strobe rate.

Multiple-Spoke Disk. Repeat for the multiple-spoke disk (marked by three spokes 120 degrees apart) with the fan at the high-speed setting and the autotransformer output at 120 volts.

a. **Monition.** Beware of possible slippage of the cardboard disk on the shaft. The fan may not slow down immediately or appreciably when you switch the setting (high, medium, low) because it is not under load from the air blades.

b. **Reminiscense.** Remember the backward-appearing motion of wagon wheels and airplane blades in the movies or in reality.

Simulating the Fan/Strobe Experiment

Using a software package such as Matlab, simulate the fan/strobe experiment of section 1 at the highest rotation speed.

Sampling a Rectangular Pulse

Using a computer software package, sample a 1,200-msec rectangular pulse at three rates: $F_s = 20$, 10, and five samples per second. Produce Figure 10.23 with two columns and three rows showing sampled signals (on the left-hand column) and their Fourier transforms (on the right-hand side). The Fourier transform of a continuous-time rectangular pulse is a *sinc* function with infinite bandwidth. The Fourier transforms of the sampled pulses are periodic functions of frequency with a period F_s. Show one cycle of each is shown on the right-hand side of Figure 10.23.

+--+
| |
| This figure has been masked so as to be recreated by the student. |
| |
+--+

FIGURE 10.23

Figure 10.23. Sampling a 1,200-msec rectangular pulse at three sampling rates and their transforms.

   a. $F_s = 20$ samples per second
   b. $F_s = 10$ samples per second
   c. $F_s = 5$ samples per second

Examine the time and frequency plots in Figure 10.23 and satisfy yourself that they meet theoretical expectations. Then, generate a rectangular pulse (pulse width $\tau = 500$ msec, pulse period $T = 8$ seconds). Sample the signal at the following rates. Find and plot the Fourier transform of each sampled signal.

   W1: $F_s = 256$ samples/sec
   W3: $F_s = 64$ samples/sec
   W5: $F_s = 16$ samples/sec
   W7: $F_s = 8$ samples/sec

Do the following and answer these questions:

   1. In each case read (from the plots) the highest frequency content of the transform.
   2. Double the data length to $T = 16$ seconds while keeping the pulse width at $\tau = 500$ msec. What is the effect on the transform? Does the highest frequency change? Does the frequency resolution change?
   3. Verify the relationship between T, Δf, f_0, and F_s.

Aliasing

Write a computer program to illustrate, in the time and frequency domains, observations of Example 10.7 in this chapter. Run the program on the computer and summarize the result.

Aliasing

Write a computer program to sample a 1-kHz square wave every Δt, and take its transform. Do that for

   a. $\Delta t = 50 \ \mu S$
   b. $\Delta t = 2$ msec

Interpret the frequency representation of the sampled signals.

Real-Time Sampling, Reconstruction, and Aliasing

The concepts of time-domain sampling, signal reconstruction, and frequency aliasing may be experienced and illustrated in real time by using a hardware device that has analog-to-digital and digital-to-analog converters. Examples of such devices are computer sound cards, digital signal processing (DSP) boards, or data acquisition cards. Use a sound card or a DSP board and a software program to sample signals in real time and filter them by digital filters. As a starting point, sample a 12-kHz sinusoidal signal at the 5-kHz rate and run it through a first-order low-pass filter. Change the frequency of the sine wave without changing the sampling rate or the cutoff frequency of the filter. Observe the analog output and relate it to the sampling rate.

An example is shown in Figure 10.24, where the footprint of the reconstruction low-pass filter appears in the step-wise output sinusoid. Figures 10.24, 10.25, and 10.26 illustrate aliasing and frequency downshifting. These figures were produced by using Texas Instruments' TMS320C6713 DSK board.

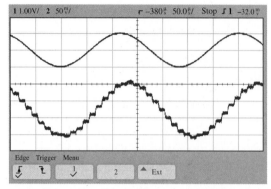
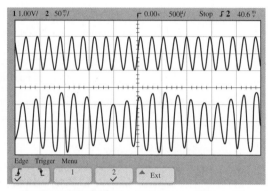

(*a*) Ch1 1 V/div. Ch2 50 mV/div. Sweep 50 μsec/div.   (*b*) Ch1 1 V/div. Ch2 50 mV/div. Sweep 500 μsec/div.

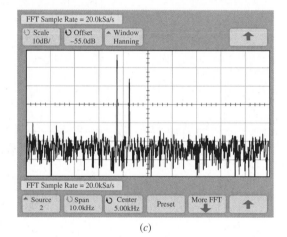

(*c*)

FIGURE 10.24 Sampling and reconstruction of a 4.25-kHz sinusoidal signal by the TMS320C6713 DSK board at the rate of 8,000 samples per seconds (8 kHz). Time sweeps are shown in *(a)* (fast one) and *(b)* (slow one). In both cases, upper traces show the sampled signal (the input to the board) and lower traces show the reconstructed signal (the output of the board). Note the footprint of the low-pass reconstruction filter (sample and hold) in the form of steps in *(a)*, to be compared with Figure 10.7 in this chapter. Also note the beat frequency effect seen in *(b)*, which is the time-domain manifestation of the consequence of an insufficient sampling rate. The effect can be seen more constructively through the spectrum of the reconstructed signal, shown in *(c)*. The spectral lines shown in *(c)* are at 4.25 kHz and 3.75 kHz. Note the frequency difference between the input and the output (aliasing), which is the frequency-domain manifestation of the consequence of insufficient sampling rate. As seen, the reconstructed signal in fact contains two components: one at 4.25 kHz (the original frequency) and a second at 3.75 kHz (the difference between 4.25 kHz and 8 kHz). Note that in this case, the downshifted frequency component is stronger than the component at the original frequency.

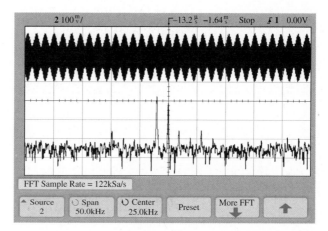

FIGURE 10.25 Another illustration of aliasing due to an insufficient sampling rate. A 25-kHz sinusoid is sampled at the rate of 48 kHz and reconstructed by passing the samples through a low-pass filter. The spectrum of the reconstructed signal, on the lower trace, shows one spectral line at 25 kHz (the original frequency) and another at 23 kHz (the difference between 25 kHz and 48 kHz). The frequency axis is at 5 kHz/div, with the center frequency at 25 kHz. The time trace (on the top) exhibits the beat effect.

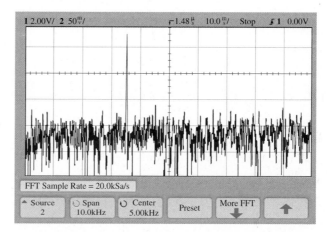

FIGURE 10.26 Aliasing leads to frequency downshifting. When a 4.5-kHz sinusoid is sampled at the 8 kHz rate and reconstructed by the DSK board, the output is a 3.5-kHz sinusoid. The original component at 4.5 kHz is totally eliminated by the low-pass reconstruction filter. The output spectrum shows a single line at 3.5 kHz.

Discussion and Conclusions

Classify and summarize your observations from this project. Relate your observations and results to theory.

Chapter **11**

Discrete-Time Signals

Contents

Introduction and Summary

A discrete signal is a sequence of numbers $\{x_n\}$, where the index n is an integer. In this book a discrete signal is considered a function of the integer variable n, $-\infty < n < \infty$, and shown by $x(n)$. This representation is applied to actual discrete signals and also to the mathematical functions which model them. Many discrete signals are produced by sampling a continuous-time signal at regular intervals. They are, appropriately, called discrete-time signals. Some discrete signals are produced by sampling other variables, such as space, or may be inherently discrete. However, for historical reasons, the term *discrete time* is used for all *discrete signals*. In this book *discrete signals* and *discrete-time signals* mean the same thing and the terms will be used interchangeably.

A discrete-time signal may be represented by a sequence of impulses in the continuous-time domain and, in theory, be subject to the tools and techniques developed for analysis in that domain. A simpler and more efficient approach, however, uses tools specially developed for the discrete-time domain. Such tools stand on their own and need not be derived from continuous-time domain. It may even appear advantageous to start with the analysis of signals and systems in the discrete-time domain for several reasons. First, discrete-time operations on signals are conceptually easier to understand. Second, computer software and computational tools for discrete-time signals are easier to use and more readily available. Third, digital hardware often constitute the platforms to implement a design. Fourth, in many cases discrete-time systems provide more design flexibility. This book provides parallel paths for the two domains, bridged by Chapter 10 on time-domain sampling. This chapter introduces discrete-time signals not as a special case of continuous-time functions but on an equal footing with them. It introduces basic notations and some operations on discrete-time signals, which are then used within the rest of the book. In addition, some examples and problems illustrate elementary forms of the filtering operation.

11.1 Domain and Range of Discrete Signals

The domain of discrete-time signals is the set of real integers $-\infty < n < \infty$. In some cases, the actual signal has a finite duration or we are only interested in a finite segment of it. The finite-duration signal is then modeled by a mathematical function that (1) may assign to it a zero value outside of its duration, (2) may repeat the signal in the form of a periodic function, or (3) may keep the signal as is, a finite-sequence signal, and ignore modeling it over the infinite domain.

The range of discrete-time signals is the continuous space of real or complex numbers $-\infty < x < \infty$. In this book x is, in general, an analog real number, unless specified otherwise. The range of actual signals is limited, but the model is not.

11.2 Actual Signals and Their Mathematical Models

In this chapter, and throughout most of the book, we deal with mathematical models of actual signals. Models may be simple elementary functions, some of which are summarized in the next section, or probabilistic models and stochastic processes. How accurately do models represent the signals and how critical is the accuracy of the model? The answer lies in the circumstances under which the signal and its model are used. In fact, a simplified model of a complicated signal may be preferred to a more accurate one if it does its intended job.

A temperature profile

Temperature measurements taken at regular 6-hour intervals at a weather station for a duration of 12 months are regarded as a discrete sequence. The temperature profile may be modeled by a periodic function that accounts for day-night and seasonal variations. The model is often satisfactory, even though the temperature profile itself is not deterministic but includes a random component. See Example 10.9 in Chapter 10 for more details.

Sunspots

The monthly recordings of the number of sunspots is an example of an actual discrete signal of finite length. Sunspot numbers were recorded starting in the year 1700. See Figure 1.1 in Chapter 1. The signal may be described by mathematical models that exhibit some periodicity. But because of their random nature, the models cannot accurately predict future numbers, but only probabilistically and in terms of statistical averages.[1]

Sampled speech signal

A speech signal is a continuous-time voltage recorded by a microphone. Its discrete-time version is a sequence of samples taken at a specified rate. An example was given in Chapter 1.

---

[1]For example, see the paper by A. J. Izenman, J. R. Wolf, and J. A. Wolfer, *An Historical Note on the Zurich Sunspot Relative Numbers,* J. R. Statist. Soc., Part A, vol. 146, 1983, pp. 311–318. The data is often called Wolfer numbers. For the data, models, and other references on sunspots, you may search Google under *wolfer+zurich+sunspots.*

The above three examples illustrate actual discrete-time signals that exhibit some randomness. They are modeled as stochastic processes (not to be discussed here). Once a segment of such a signal is observed or recorded, the random property vanishes and the signal becomes a known time series.

11.3 Some Elementary Functions

The unit-sample, unit-step, unit-ramp, sinusoidal, exponential, and sinc functions are often used as the elementary building blocks for modeling signals.

Unit Sample

The unit-sample function $d(n)^2$ is defined by

$$d(n) = \begin{cases} 1, & n = 0 \\ 0, & n \neq 0 \end{cases}$$

Unit Step

The unit-step function $u(n)$ is defined by

$$u(n) = \begin{cases} 1, & n \geq 0 \\ 0, & n < 0 \end{cases}$$

Unit Ramp

The unit-ramp function $r(n)$ is defined by

$$r(n) = \begin{cases} n, & n \geq 0 \\ 0, & n < 0 \end{cases}$$

The unit-sample, unit-step, and unit-ramp functions are plotted in Figure 11.1(*a*), (*b*), and (*c*), respectively.

Note: The unit-sample, unit-step, and unit-ramp functions are related to each other. One way to express their relationship is by $d(n) = u(n) - u(n-1)$ and $r(n) = nu(n)$. Other ways, which employ difference and summing operators, will be shown in Chapter 12.

Sinusoid

The sinusoidal function is defined by $x(n) = X_0 \cos(\omega n + \theta)$. It is periodic if $\omega = 2\pi/N$, N being an integer, in which case N is the period. See Figure 11.2(*a*). A sinusoid $X_0 \cos(\omega n + \theta)$ is completely specified by three parameters X_0, θ, and ω. X_0 and θ are combined as the complex amplitude $\mathbf{X} = X_0 e^{j\theta} = X_0 \angle \theta$, also called the phasor.

---

[2] The unit-sample function is sometimes shown by $\delta(n)$ and called the unit-impulse function (as in the continuous-time domain). Such usage may cause confusion with a sample of infinite value. In this book we generally represent the unit-sample function by $d(n)$.

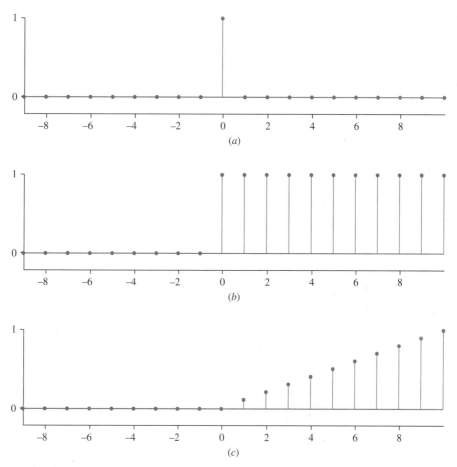

FIGURE 11.1 *(a)* The unit-sample function $d(n)$. *(b)* The unit-step function $u(n)$. *(c)* The unit-ramp function $r(n)$.

Exponential

The exponential function is $x(n) = X_0 e^{\sigma n}$, where X_0 and σ are constant real numbers. It grows exponentially with increasing n if $\sigma > 0$, and decays if $\sigma < 0$. For $\sigma = 0$, we have a DC-level function [$x(n) = X_0$ for all n]. See Figure 11.2(b). Other examples of exponential functions are found through the rest of this chapter; for example, see Figures. 11.7 and 11.11.

Exponentially Growing or Decaying Sinusoid

A sinusoidal function whose amplitude varies exponentially is given by

$$x(n) = X_0 e^{\sigma n} \cos(\omega n + \theta)$$

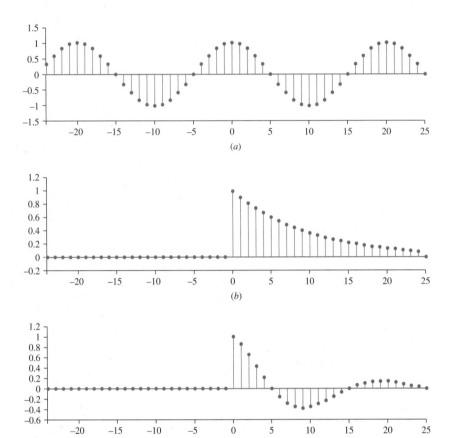

FIGURE 11.2 *(a)* The period of the sinusoidal function $\cos(\frac{\pi n}{10})$ is 20. *(b)* The one-sided decaying exponential function shown is $x(n) = e^{-n/10}u(n)$ and has a time constant equal to 10. *(c)* Multiplication of functions in *(a)* and *(b)* produces the causal function $x(n) = e^{-n/10}\cos(\pi n/10)u(n)$ shown. Its envelope has a time constant equal to 10. The function may be expressed as $x(n) = \mathcal{RE}\{e^{0.1(-1+j\pi)n}\}u(n)$.

If $\sigma > 0$, the amplitude will grow exponentially. If $\sigma < 0$, the amplitude will decay exponentially. See Figure 11.2(c). For $\sigma = 0$, the amplitude of the sinusoid is constant.

Complex Exponential

The complex exponential function is

$$x(n) = X_0 e^{\sigma n} e^{j(\omega n + \theta)}$$

where X_0, σ, ω, and θ are constant real numbers. The first part of the above expression, $X_0 e^{\sigma n}$, represents an exponential amplitude. The second part of the expression, $e^{j(\omega n + \theta)}$, represents the periodic behavior of the function. The exponential function may also be written as

$$x(n) = X_0 e^{j\theta} e^{(\sigma + j\omega)n}$$

Its real part

$$\mathcal{RE}\,[x(n)] = X_0 e^{\sigma n} \cos(\omega n + \theta)$$

is a sinusoid with exponential growth (if $\sigma > 0$) or decay (if $\sigma < 0$). For $\sigma = 0$, the complex exponential function becomes

$$x(n) = \mathbf{X} e^{j\omega n}$$

Its real and imaginary parts then become sinusoids with constant amplitudes.

The complex exponential function and its real part are completely specified by the four parameters X_0, θ, σ, and ω. These parameters are combined in pairs as the complex amplitude $\mathbf{X} = X_0 e^{j\theta}$, also called the phasor $\mathbf{X} = X_0 \angle \theta$, and the complex frequency $s = \sigma + j\omega$.

Exponential and sinusoidal functions are important building blocks in the analysis of signals and linear systems and are encountered frequently.

Sinc Function

The discrete-time unit-sinc function is defined by $\frac{\sin n}{n}$. It is an even function with a maximum magnitude of $x(0) = 1$. The plot of the sinc function exhibits positive and negative lobes of equal width but diminishing height as n increases, with the central lobe being the biggest one. The central lobe (with nonzero values) extends over $-3 \le n \le 3$. The regular zero-crossings produce a sort of periodicity in the function. However, the sinc function is not periodic.

A more general form of the discrete-time sinc function is

$$x(n) = \frac{\sin(\omega n)}{\omega n}, \quad -\infty < n < \infty$$

where ω represents the angular frequency with which the function switches between positive and negative values when creating lobes. In summary, a discrete sinc function may be specified completely by (1) the number of samples in its main lobe and (2) a scale factor, such as, $x(0)$. Two sinc functions are plotted in Figure 11.3.

*E*xample **11.4**

A sinc function is given by

$$x(n) = \frac{\sin(\omega n)}{\omega n}, \quad -\infty < n < \infty$$

a. Let $\omega = 1$. Show that the central lobe contains seven nonzero samples $(-3 \le n \le 3)$.

b. Let $\omega = 2\pi/M$, where M is an integer. Show that the numerator is periodic (period $= M$) but $x(n)$ is not.

c. Let $\omega = 2\pi/M$, where M is an even integer. Show that the main lobe contains $M - 1$ nonzero samples, and the zero-crossings occur at $n = \pm kM/2$.

d. Let $\omega = 2\pi/M$, where M is an odd integer. Show that the central lobe contains M nonzero samples.

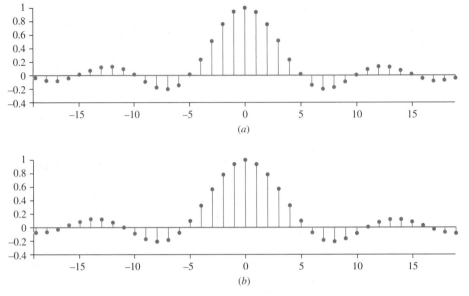

FIGURE 11.3 A discrete sinc function is specified completely by the number of samples in the main lobe N and its value at $n = 0$. **(a)** The sinc function with $N = 9$ and $x(0) = 1$ contains positive and negative lobes diminishing with $1/n$. Regular zero-crossings occur at $n = \pm5$, $\pm10, \pm15, \dots \pm 5k$, where k is a nonzero integer. Mathematically, the function is $x(n) = \sin(\pi n/5)/(\pi n/5)$. **(b)** A sinc function with $N = 11$ and $x(0) = 1$ is expressed by $x(n) = \sin(2\pi n/11)/(2\pi n/11)$. Note that the location of the first zero-crossing in the envelope is at $5 < n < 6$.

Solution

a.   For $\sin n = 0$ we need $n = \pm\pi$. Because n needs to be an integer, the switch in the sign occurs between $n = 3$ and 4 (similarly, between $n = -3$ and -4). The central lobe contains seven samples.

b.   Assuming $\omega = 2\pi/M$, the numerator of the sinc function becomes

$$\sin(\omega n)|_{\omega=2\pi/M} = \sin(2\pi n/M) = \sin(2\pi(n + M)/M)$$

showing a period of M. But $\frac{\sin(\omega n)}{\omega n}$ decreases with n, making it nonperiodic.

c.   Zero-crossings occur at $2\pi n/M = \pm k\pi$ or $n = \pm kM/2$, where k is an integer. With M being even, $n = \pm kM/2$ becomes an integer indicating zero-crossings at $n = \pm kM/2$. The central lobe with nonzero samples includes $-M/2 \le n \le M/2$, for a total of $M - 1$ samples.

d.   Asssuming M is an odd number, $\pm M/2$ is not an integer and the switch in the sign of $x(n)$ occurs between $n = (M - 1)/2$ and $n = (M + 1)/2$. The central lobe will then contain $2\frac{M-1}{2} + 1 = M$ samples.

11.4 Summary of Elementary Functions

Table 11.1 gives a summary of eight elementary functions often used as models for actual signals.

TABLE 11.1 Several Elementary Functions

Signal	Mathematical Description
Unit Sample	$d(n) = \begin{cases} 1, & n = 0 \\ 0, & n \neq 0 \end{cases}$
Unit Step	$u(n) = \begin{cases} 1, & n \geq 0 \\ 0, & n < 0 \end{cases}$
Unit Ramp	$r(n) = \begin{cases} n, & n \geq 0 \\ 0, & n < 0 \end{cases}$
Sinusoid	$x(n) = X_0 \cos\left(\dfrac{2\pi n}{N} + \theta\right)$, N is an integer and the period
Exponential	$x(n) = X_0 e^{\sigma n}$, $\begin{cases} \sigma < 0 \text{ (decaying)} \\ \sigma = 0 \text{ (constant)} \\ \sigma > 0 \text{ (growing)} \end{cases}$
Decaying Sinusoid	$X_0 e^{\sigma n} \cos(\omega n + \theta)$, $\sigma < 0$
Complex Exponential	$x(n) = X_0 e^{\sigma n} e^{j(\omega n + \theta)}$
	$e^{j\omega n} = \cos(\omega n) + j \sin(\omega n)$
Sinc Function	$x(n) = \dfrac{\sin(\omega n)}{\omega n}$

11.5 Periodicity and Randomness

Discrete signals are classified as

Periodic

Almost-periodic

Nonperiodic

Random

The first three groups are subclasses of deterministic signals.

Periodic Signals

A signal $x(n)$ is periodic with period N if $x(n) = x(n + N)$ for all n. The signal is then referred to as N-periodic. See section 11.6 for examples.

Sum of Two Periodic Signals

The sum of two discrete-time periodic signals is periodic. The period is the least common multiple of the two periods. This is in variance with the case of continuous-time signals where the sum of two periodic signals may be nonperiodic.

$\sin(\pi n/5)$ is 10-periodic and $\sin(2\pi n/15)$ is 15-periodic. Their sum

$$x(n) = \sin\left(\frac{\pi n}{5}\right) + \sin\left(\frac{2\pi n}{15}\right)$$

is 30-periodic.

$\sin(\pi n/5)$ is 10-periodic and $\sin(2\pi n/11)$ is 11-periodic. Their sum

$$x(n) = \sin\left(\frac{2\pi n}{10}\right) + \sin\left(\frac{2\pi n}{11}\right)$$

is 110-periodic. Similarly, the signal

$$x(n) = \sin\left(\frac{2\pi n}{100}\right) + \sin\left(\frac{2\pi n}{101}\right)$$

is periodic with period $N = 10,100$.

Almost-Periodic Signals

A discrete-time signal may not be periodic, as no integer N may be found that would satisfy the periodicity equation in its exact form, $x(n) = x(n + N)$ for all n. However, the signal may exhibit certain periodicity, such as in its envelope. We call such classes of signals almost-periodic. Examples are discrete-time signals obtained by sampling a continuous-time periodic signal at a rate such that the ratio of the sampling interval to the period of the continuous-time signal is not a rational number.

Example **11.7**

Consider $x(n) = \sin(\frac{n}{M})$, where M is an integer. For $x(n)$ to be periodic with integer period N we need $x(n) = x(n + N)$ or

$$\sin\left(\frac{n + N}{M}\right) = \sin\left(\frac{n}{M}\right), \quad \Rightarrow \quad \frac{N}{M} = 2\pi, \ \text{ or } \ N = 2\pi M$$

This is not possible because both N and M are integers. However, as M becomes large, the percentage difference between $2\pi M$ and the integer nearest to it becomes small and $\sin(\frac{n}{M})$ looks more periodic. Examples of this phenomenon are found in section 11.7.

Nonperiodic Signals

Most discrete-time signals are nonperiodic. Examples are the sinc function or a finite pulse. They may have finite or infinite energy.

Random Signals

A discrete-time signal obtained by sampling a random source (such as microphone sound or hourly readings of weather temperature) is a random signal. Random signals are modeled by their statistical averages and stochastic processes. An example is $x(n) = a \sin(bn)$, where a and b are random variables.

Gaussian Random Signals

The random sequence $x(n) = a$, where a is a Gaussian random variable and the samples are independent of each other, is called white Gaussian noise. An example is shown in Figure 11.4(a). Filtering white Gaussian noise creates a correlation between the samples. Figure 11.4(b) shows the sequence of Figure 11.4(a) passed through a low-pass filter with cutoff at $\pi/10$ radians.

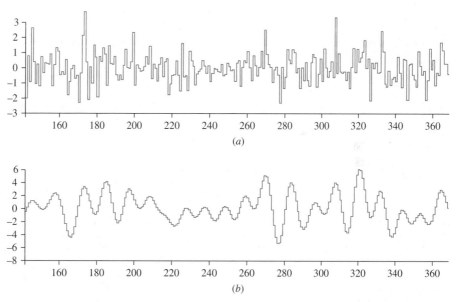

FIGURE 11.4 The sequence in *(a)* is a section of a white Gaussian process. Each sample is a Gaussian random variable (zero mean and unity variance) and the samples are independent of each other. The sequence is called white Gaussian noise. Passing a white Gaussian noise through a band-limited filter creates a correlation between the samples. The sequence shown in *(b)* is produced by passing *(a)* through a low-pass digital filter with cutoff at $\pi/10$ radians.

11.6 Examples of Periodic Signals

In Examples 11.8, 11.9, and 11.10, one period of an N-periodic signal is specified.

Example **11.8**

$x(n)$ is a periodic rectangular pulse train with period N and pulse width M. See Figure 11.5(a).

$$x(n) = \begin{cases} 1, & \text{for } 0 \le n < M \\ 0, & \text{for } M \le n < N \end{cases} \quad \text{(defined for 1 period)}$$

Example **11.9**

$x(n)$ is a periodic exponential pulse.

$$x(n) = \begin{cases} 0.5^n, & \text{for } 0 \le n < M \\ 0, & \text{for } M \le n < N \end{cases} \quad \text{(defined for 1 period)}$$

Example **11.10**

The repeated tone burst $x(n)$ is N-periodic. See Figure 11.5(b).

$$x(n) = \begin{cases} \sin(\omega_0 n), & \text{for } 0 \le n < M \\ 0, & \text{for } M \le n < N \end{cases} \quad \text{(defined for 1 period)}$$

Example **11.11**

The signal

$$x(n) = \frac{\sin(2\pi n/5)}{\sin(\pi n/20)}$$

is 40-periodic, $x(n) = x(n + 40)$. See Figure 11.5(c). The signal seems like a sinc function that repeats itself every 40 samples. However, it is not a periodic sinc function.

Example **11.12**

The signal $x(n)$ given in Figure 11.5(d) is N-periodic; that is, $x(n) = x(n + N)$. The signal is the sum of identical sinc functions repeated at intervals of N samples each. The mathematical expression of the signal is

$$x(n) = \sum_{k=-\infty}^{\infty} \frac{\sin \frac{2\pi(n-kN)}{M}}{\pi(n - kN)}, \quad -\infty < n < \infty$$

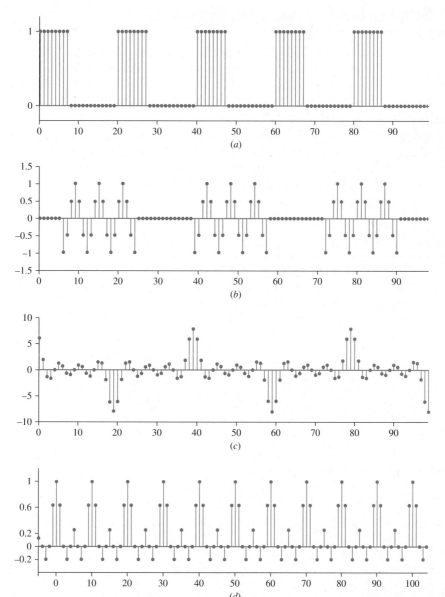

FIGURE 11.5 *(a)* A periodic pulse train, period = 20, pulsewidth = 8, duty cycle = 40%. *(b)* The repeated tone burst shown in this figure is 33-periodic. The period of the tone is 6. *(c)* The signal is 40-periodic. It resembles a sinc function that is repeated every 40 samples, alternating between positive and negative values. However, the signal is not a periodic sinc function. It is generated from

$$x(n) = \frac{\sin(2\pi n/5)}{\sin(\pi n/20)}$$

The numerator $\sin(2\pi n/5)$ is 5-periodic and the denominator $\sin(\pi n/20)$ is 40-periodic. *(d)* The signal is 10-periodic. It is generated from unit-sinc functions (with 3 samples in the main lobe) shifted 10 samples and added together.

11.7 Sources of Discrete Signals

Some discrete signals are produced by sampling a continuous-time signal $x(t)$ every Δt seconds, corresponding to the rate of $F_s = 1/\Delta t$ samples per second (also called Hz). The result is $x(n\Delta t)$, which we designate as the discrete-time signal $x(n)$. The above class of discrete-time signals also includes signals that are functions of space.

Example
11.13

Continuous-time functions $x(t)$ shown in Table 11.2 are sampled at the rate of 100 Hz (every 10 msec, $t = n\Delta t = n/100$, with a sample at $t = 0$) to produce the discrete-time functions $x(n)$. In each case find $x(n)$, then specify whether the function is periodic and, if it is, find the period (in either domain).

Solution
See Table 11.2.

TABLE 11.2 Generating Discrete-Time Signals (For Example 11.13)

	Continuous Time			Sampling	Discrete Time		
$x(t)$	Periodic?	Period (sec)	Frequency (Hz)	Rate in Hz $\Longrightarrow$	$x(n)$	Periodic?	Period (N)
$e^{-10t}u(t)$	no	—	—	100	$e^{-n/10}u(n)$	no	—
$\dfrac{\sin t}{t}$	no	—	—	100	$\dfrac{100}{n}\sin\left(\dfrac{n}{100}\right)$	no	—
$\dfrac{\sin(100\pi t)}{100\pi t}$	no	—	—	100	$\dfrac{\sin(\pi n)}{\pi n}$	no	—
$\sin(500\pi t)$	yes	0.004	250	100	0	—	—
$\cos(500\pi t)$	yes	0.004	250	100	$(-1)^n$	yes	2
$\sin(100\pi t)$	yes	0.02	50	100	0	—	—
$\cos(100\pi t)$	yes	0.02	50	100	$(-1)^n$	yes	2
$\sin(10\pi t)$	yes	0.2	5	100	$\sin\left(\dfrac{\pi n}{10}\right)$	yes	20
$\cos(10\pi t)$	yes	0.2	5	100	$\cos\left(\dfrac{\pi n}{10}\right)$	yes	20
$\sin(\pi t)$	yes	2	0.5	100	$\sin\left(\dfrac{\pi n}{100}\right)$	yes	200
$\sin\left(\dfrac{\pi t}{10}\right)$	yes	20	0.05	100	$\sin\left(\dfrac{\pi n}{1000}\right)$	yes	2,000
$\sin(3t)$	yes	$2\pi/3$	≈ 0.48	100	$\sin(0.03n)$	almost	≈ 209
$\sin\left(\dfrac{t}{3}\right)$	yes	6π	≈ 0.05	100	$\sin\left(\dfrac{n}{300}\right)$	almost	$\approx 1,885$
$\sin(314t)$	yes	$2\pi/314$	≈ 50	100	$\sin(3.14n)$	almost	≈ 2
$\sin(3.14t)$	yes	$2\pi/3.14$	≈ 0.5	100	$\sin(0.0314n)$	almost	≈ 200

Example
11.14

Consider the continuous-time sinusoidal signal $x(t) = \sin\left(\frac{2pt}{T}\right)$. The signal is sampled at the rate of F_s samples per second, where $N = F_s T$ is an integer. Examine the effect of the following values of p on $x(n)$.

1. $p = 3$
2. $p = 3.14$
3. $p = 3.1416$
4. $p = \pi$

Solution

The discrete-time signal is obtained by setting $t = n/F_s$:

$$x(n) = \sin\left(\frac{2pn}{TF_s}\right) = \sin\left(\frac{2pn}{N}\right)$$

$x(n)$ is periodic with period N for $p = \pi$ only. Otherwise, it is almost-periodic. As p approaches π, $x(n)$ appears *closer* to being periodic.

11.8 Representation of Discrete Signals

Discrete signals may be represented in one of several ways, such as in analytic closed form or an array made of a string of numbers, as illustrated in the following examples.

Analytic Representation

A combination of elementary mathematical functions (such as the unit-sample, unit-step, unit-ramp, sinusoidal, exponential, and similar functions) may be used to represent discrete-time signals. (See section 11.3 for details on such elementary functions.) A discrete-time signal may also be represented in a segmented form. For example,

$$x(n) = u(n) - 2u(n-3) + u(n-6) = \begin{cases} 1, & 0 \le n < 3 \\ -1, & 3 \le n < 6 \\ 0, & \text{elsewhere} \end{cases}$$

Array Representation

A discrete-time signal may be represented by an array in the form of a string of numbers

$$x(n) = \{\ldots, x_{-5}, x_{-4}, x_{-3}, x_{-2}, x_{-1}, \underset{\uparrow}{x_0}, x_1, x_2, x_3, x_4, x_5, \ldots\}$$

where n is the domain of the signal and x_k is the value of its kth element, $x_k = x(n)|_{n=k}$. The underset up-arrow $\uparrow$ indicates the location of the origin $n = 0$. Dots at the left and right ends of a sequence represent continuation of the trend set by adjacent samples. Here are two examples.

$$x(n) = \{\ldots, 1/16, 1/8, 1/4, 1/2, \underset{\uparrow}{1}, 1/2, 1/4, 1/8, 1/16, \ldots\} = 0.5^{|n|}$$

$$x(n) = \{\ldots, 1/81, 1/27, 1/9, 1/3, \underset{\uparrow}{1}, 1/2, 1/4, 1/8, 1/16, \ldots\}$$

$$= 3^n u(-n) + 0.5^n u(n) - d(n)$$

For a one-sided signal, the side that is made of zeros may be truncated for simplicity as shown in the following examples.

$$u(n+1) = \{1, \underset{\uparrow}{1}, 1, 1, 1, \ldots\}$$

$$u(-n+2) = \{\ldots, 1, 1, 1, \underset{\uparrow}{1}, 1, 1\}$$

$$0.5^n u(n) = \{\underset{\uparrow}{1}, 0.5, 0.25, 0.125, \ldots\}$$

When the signal has a finite length N, the string representation may be further reduced to the following:

$$x(n) = \{x_{-k}, x_{(-k+1)}, \ldots, x_{-1}, \underset{\uparrow}{x_0}, x_1, \ldots, x_{(\ell-1)}, x_\ell\}, \quad N = k + \ell + 1$$

The above string contains the values of $x(n)$ for $-k \leq n \leq \ell$. Its values outside the above interval may be zero, a repetition of the given string, or of no interest. The state of the signal outside the specified interval may be clarified, if needed, either explicitly by an additional statement or by the context of the problem. For example,

1. $x(n) = \{8, 4, 2, \underset{\uparrow}{1}, 0.5, 0.25, 0.125\} = 0.5^n[u(n-4) - u(n-3)]$

2. $x(n) = \{\underset{\uparrow}{1}, 2, 3, 2, 1\}$ (and we don't care about the outside interval)

3. $x(n) = \{\underset{\uparrow}{0}, 1, 2, 1, 0, -1, -2, -1\}$ and $x(n) = x(n+8)$ (periodic)

Note that only periodic case 3 above requires an explicit statement. Otherwise, the finite sequence by itself carries all the information of interest.

Example 11.15

The signal $x(n) = u(n) - u(n-3)$, where $u(n)$ is the unit-step function starting at the origin can be stated as

$$x(n) = \begin{cases} 1, & 0 \leq n < 3 \\ 0, & \text{elsewhere} \end{cases}$$

The domain of x is $-\infty < n < \infty$. $x(n)$ can also be represented by $x(n) = \{\ldots, 0, \underset{\uparrow}{1}, 1, 1, 0, \ldots\}$ or by $x(n) = \{\underset{\uparrow}{1}, 1, 1\}$ and zero elsewhere. The up-arrow identifies the origin, $n = 0$.

E xample
11.16

The signal $h(n) = (0.5)^n u(n)$ may also be represented by $h(n) = \{\ldots, 0, 1, 0.5, 0.25,$ $0.125, \ldots\}$. The domain of h is $-\infty < n < \infty$. The signal is causal ($h(n) = 0$ for $n < 0$) and is reduced by a factor of 0.5 for each positive n increment. The signal may also be shown by the array $h(n) = \{1, .5, .25, .125, \ldots\}$.

Circular Display

A finite sequence of length N (or a periodic signal) may also be represented by a circular display of N numbers around a circle at $2\pi/N$ degree intervals. Figure 11.6 is the circular display of the sequence in Example 11.17. The circular display is especially helpful in visualizing circular shifts and convolutions. See Chapter 13.

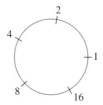

FIGURE 11.6 Circular display of $s(n) = \{1, 2, 4, 8, 16\}$, a finite sequence of length 5. The first element in the sequence corresponds to $n = 0$ and is placed at a zero angle on the circle. The display increments n samples in the counterclockwise direction.

E xample
11.17

The signal $s(n)$ shown by the array $s(n) = \{1, 2, 4, 8, 16\}$ is a finite-power signal of length 5. The status of the signal outside the interval is left open. The signal is displayed in Figure 11.6. The first element in the sequence corresponds to $n = 0$ and is placed at a zero angle on the circle. The display increments the location of $s(n)$ in the counterclockwise direction by $2\pi/5 = 72°$.

11.9 Digital Signals

A discrete-time signal that can assume only discrete values is called a digital signal. In signal processing we use digital signals, and digital signal processing (DSP) has become synonymous with the processing of discrete-time signals. In practice, DSP is normally performed by digital devices on binary numbers. Therefore, the bulk of digital signals is represented by binary numbers.

E xample
11.18

Consider a discrete-time signal whose amplitude varies between 0 and 1 volt, $0 \leq x(n) < 1$. The signal amplitude is digitized into four levels 0, 1, 2, and 3, and represented as a two-bit binary number $x(n) = y_1 y_0$, where y_0 and y_1 are the two bits representing the digitized signal. See Table 11.3.

TABLE 11.3 Digitized Signal (For Example 11.18)

Signal Amplitude	Level Assignment	Binary Representation
$0 \leq x(n) < 0.25$	0	00
$0.25 \leq x(n) < 0.5$	1	01
$0.5 \leq x(n) < 0.75$	2	10
$0.75 \leq x(n) < 1$	3	11

Here is an example.

$$x(n)=\{0.81, 0.63, 0.32, 0.17, 0.92, 0.48, 0.54\}_{10} \overset{\text{digitized}}{\Longrightarrow} \{11, 10, 01, 00, 11, 01, 10\}_2$$

A 3-bit *word* is needed to represent a signal that is digitized into eight levels. A digital signal containing 256 levels needs an 8-bit *word* representation. An N-bit *word* represents a signal with 2^N discrete levels.

11.10 Energy and Power Signals

The total energy in a signal is defined by

$$\text{Total energy} = \sum_{n=-\infty}^{\infty} x(n)x^*(n) = \sum_{n=-\infty}^{\infty} |x(n)|^2$$

where $x^*(n)$ is the complex conjugate of $x(n)$. If the total energy is finite, the signal is called an *energy signal*. Energy signals are not necessarily of finite length. Examples are sinc functions and signals whose amplitudes are modulated by exponential decay.

The average power over a segment of length N, from n_0 to $(n_0 + N - 1)$, is defined by

$$\text{Average power} = \frac{1}{N} \sum_{n=n_0}^{n_0+N-1} |x(n)|^2$$

If a signal has infinite energy, it might have finite power, in which case it is called a power signal. Periodic signals are power signals. Random signals with infinite duration (and infinite energy) are modeled as power signals. Nonperiodic signals can be power signals, too.

11.11 Time Reversal

In the analysis of discrete-time signals and digital signal processing we often encounter time reversal, which converts $x(n)$ to $x(-n)$, where n is the argument of the function. When a signal $x(n)$ is represented by a mathematical formula, change the sign of n to $-n$ and will you have the formula for $x(-n)$. Time reversal flips the plot of $x(n)$ around the origin, $n = 0$, to produce the plot of $x(-n)$. A graphical visualization is often helpful especially when time reversal and shift are combined. The following four examples illustrate time reversal.

Example

11.19

Given $x(n) = 2^n u(-n) + 0.75^n u(n)$, see Figure 11.7(a). Find and plot $x(-n)$, and compare with the plot of $x(n)$.

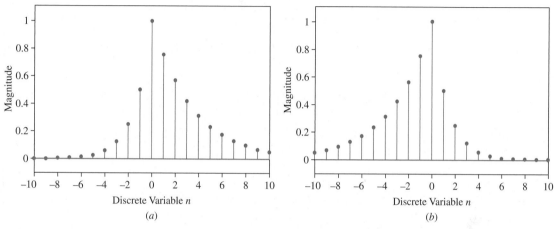

(a)   *(b)*

FIGURE 11.7 *(a)* The two-sided exponential function $x(n) = 2^n u(-n) + 0.75^n u(n)$ is shown. *(b)* Plot of the time-reversed function $x(-n) = 2^{-n} u(n) + 0.75^{-n} u(-n)$.

Solution

Change n to $-n$ to find $x(-n) = 2^{-n} u(n) + 0.75^{-n} u(-n)$ This is plotted in Figure 11.7(b) Note that time reversal flips the plot of the function around the vertical at $n = 0$.

Example

11.20

Five examples of time reversal are given in Table 11.4.

TABLE 11.4 Examples of Time Reversal

$x(n) = 2^{-n} u(n)$	$x(-n) = 2^n u(-n)$
$x(n) = \begin{cases} n, & n \geq 3 \\ 0, & \text{elsewhere} \end{cases}$	$x(-n) = \begin{cases} -n, & n \leq -3 \\ 0, & \text{elsewhere} \end{cases}$
$x(n) = \begin{cases} 0.5^n, & n < -2 \\ 0, & \text{elsewhere} \end{cases}$	$x(-n) = \begin{cases} 0.5^{-n} & n > 2 \\ 0, & \text{elsewhere} \end{cases}$
$x(n) = \begin{cases} \cos\left(\frac{n}{2}\right), & 2 \leq n \leq 5 \\ 0, & \text{elsewhere} \end{cases}$	$x(-n) = \begin{cases} \cos\left(\frac{n}{2}\right) & -5 \leq n \leq -2 \\ 0, & \text{elsewhere} \end{cases}$
$x(n) = \begin{cases} \cos\left(\frac{\pi n}{8}\right), & 0 \leq n \leq 4 \\ -\cos\left(\frac{\pi n}{8}\right), & -4 \leq n < 0 \\ 0, & \text{elsewhere} \end{cases}$	$x(-n) = \begin{cases} \cos\left(\frac{\pi n}{8}\right), & -4 \leq n \leq 0 \\ -\cos\left(\frac{\pi n}{8}\right), & 0 < n \leq 4 \\ 0, & \text{elsewhere} \end{cases}$

Example 11.21

Time reversal has no effect on the functions shown in Table 11.5. $x(n) = x(-n)$. The functions are called even.

TABLE 11.5 Time Reversing an Even Function Doesn't Change It

$x(n) = 2^{-|n|}$	$x(-n) = 2^{-|n|}$
$x(n) = 2^n u(-n) + 2^{-n} u(n)$	$x(-n) = 2^{-n} u(n) + 2^n u(-n)$
$x(n) = \begin{cases} -n, & n \le -3 \\ n, & n \ge 3 \\ 0, & \text{elsewhere} \end{cases}$	$x(-n) = \begin{cases} n, & n \ge 3 \\ -n, & n \le -3 \\ 0, & \text{elsewhere} \end{cases}$
$x(n) = \begin{cases} \cos\left(\frac{\pi n}{8}\right), & -4 \le n \le 4 \\ 0, & \text{elsewhere} \end{cases}$	$x(-n) = \begin{cases} \cos\left(\frac{\pi n}{8}\right) & -4 \le n \le 4 \\ 0, & \text{elsewhere} \end{cases}$
$x(n) = \begin{cases} \sin\left(\frac{\pi n}{8}\right), & 0 < n \le 4 \\ -\sin\left(\frac{\pi n}{8}\right), & -4 \le n < 0 \\ 0, & \text{elsewhere} \end{cases}$	$x(-n) = \begin{cases} -\sin\left(\frac{\pi n}{8}\right), & -4 \le n < 0 \\ \sin\left(\frac{\pi n}{8}\right), & 0 < n \le 4 \\ 0, & \text{elsewhere} \end{cases}$

Example 11.22

Time reversal produces a sign change in the functions shown in Table 11.6. $x(n) = -x(-n)$. The functions are called odd.

TABLE 11.6 Time Reversing an Odd Function Changes Its Sign Only

$x(n) = -2^n u(-n) + 2^{-n} u(n)$	$x(-n) = -2^{-n} u(n) + 2^n u(-n)$
$x(n) = \begin{cases} n, & n \le -3 \\ n, & n \ge 3 \\ 0, & \text{elsewhere} \end{cases}$	$x(-n) = \begin{cases} -n, & n \ge 3 \\ -n, & n \le -3 \\ 0, & \text{elsewhere} \end{cases}$
$x(n) = \begin{cases} 1, & 0 < n \le 3 \\ -1, & -3 \le n < 0 \\ 0, & \text{elsewhere} \end{cases}$	$x(-n) = \begin{cases} 1, & -3 \le n < 0 \\ -1, & 0 < n \le 3 \\ 0, & \text{elsewhere} \end{cases}$
$x(n) = \begin{cases} \sin\left(\frac{\pi n}{8}\right), & -4 \le n \le 4 \\ 0, & \text{elsewhere} \end{cases}$	$x(-n) = \begin{cases} -\sin\left(\frac{\pi n}{8}\right) & -4 \le n \le 4 \\ 0, & \text{elsewhere} \end{cases}$
$x(n) = \begin{cases} \cos\left(\frac{\pi n}{8}\right), & 0 < n \le 4 \\ -\cos\left(\frac{\pi n}{8}\right), & -4 \le n < 0 \\ 0, & \text{elsewhere} \end{cases}$	$x(-n) = \begin{cases} \cos\left(\frac{\pi n}{8}\right), & -4 \le n < 0 \\ -\cos\left(\frac{\pi n}{8}\right), & 0 < n \le 4 \\ 0, & \text{elsewhere} \end{cases}$

Summary

In the mathematical expression for $x(n)$, change n to $-n$ and you have a mathematical expression for $x(-n)$. Flip the graph of $x(n)$ horizontally around $n = 0$ and you have the graph for $x(-n)$.

11.12 Time Shift

Substituting n by $n - k$ in the mathematical expression for $x(n)$ produces a shift of k units in it. When k is a positive number the shift delays the signal by k units and slides the plot of $x(n)$ to the right by k units. A negative k advances it and slides the plot to the left. For the case of sinusoidal functions, a time shift translates into a phase shift.

Example **11.23**

Three examples of a time shift are given in Table 11.7.

TABLE 11.7 Examples of Time Shift

$x(n) = 2^{-\lvert n \rvert}$	$x(n-2) = 2^{-\lvert n-2 \rvert}$
$x(n) = 2^{-n}u(n)$	$x(n+3) = \dfrac{2^{-n}}{8}u(n+3)$
$x(n) = \begin{cases} \cos\left(\frac{\pi n}{8}\right), & -4 \le n \le 4 \\ 0, & \text{elsewhere} \end{cases}$	$x(n-4) = \begin{cases} \sin\left(\frac{\pi n}{8}\right) & 0 \le n \le 8 \\ 0, & \text{elsewhere} \end{cases}$

A shift of four time units in $\cos(\pi n/8)$ (third part of Table 11.7) translated into a $90°$ phase shift and has converted it to $\sin(\pi n/8)$.

Example **11.24**

Time shift in periodic signals

After N shifts a periodic signal with period N reverts to its original form. An example is given for $x(n) = \cos(\pi n/2)$ in Table 11.8.

TABLE 11.8 Time Shifts in the Sinusoid $\cos(\pi n/2)$

Function		Representation
$x(n)$	$= \cos\left(\dfrac{\pi n}{2}\right)$	$= \{\ldots, -1, 0, \overbrace{1}^{}, 0, -1, 0, 1, 0, \ldots\}$ (one period)
$x(n-1)$	$= \cos\left(\dfrac{\pi n}{2} - \dfrac{\pi}{2}\right)$	$= \{\ldots, 0, -1, \overbrace{0}^{}, 1, 0, -1, 0, 1, \ldots\}$ (one period)
$x(n-2)$	$= \cos\left(\dfrac{\pi n}{2} - \pi\right)$	$= \{\ldots, 1, \;\; 0, \overbrace{-1}^{}, 0, 1, 0, -1, 0, \ldots\}$ (one period)
$x(n-3)$	$= \cos\left(\dfrac{\pi n}{2} - \dfrac{3\pi}{2}\right)$	$= \{\ldots, 0, \;\; 1, \overbrace{0}^{}, -1, 0, 1, 0, -1, \ldots\}$ (one period)
$x(n-4)$	$= \cos\left(\dfrac{\pi n}{2} - 2\pi\right)$	$= \{\ldots, -1, 0, \overbrace{1}^{}, 0, -1, 0, 1, 0, \ldots\}$ (one period)

It is easily seen that one only needs to attend to a finite segment of the signal seen through a one-period-long window (e.g., from $n = 0$ to 3 as shown in Table 11.8). See the close parallels with circular shift discussed in section 11.15.

11.13 Combination of Time Reversal and Shift

Some mathematical operations may require shifting a signal and reversing it. The order according to which the operations are applied become important. In this section we begin with examples and then provide guidelines.

Example
11.25

Several time shifts and time reversals of the unit-step function are given in Table 11.9. Plots of the functions are found in Figure 11.8(*a*) to (*f*). Note that a delay shifts the plot to the right and an advance shifts it to the left. Also note that the time-reversal operation is pivoted around $n = 0$.

TABLE 11.9 Time Shifts and Time Reversals of the Unit-Step Function $u(n)$

Function	Description	Representation	Figure
$u(n)$	Unit step	$\{\cdots 0, 0, 0, 0, 0, 0, 1, 1, 1, 1, 1, 1, 1\cdots\} = \begin{cases} 1, & n \geq 0 \\ 0, & n < 0 \end{cases}$	Figure 11.8(*a*)
$u(-n)$	Time reversal	$\{\cdots 1, 1, 1, 1, 1, 1, 1, 0, 0, 0, 0, 0, 0\cdots\} = \begin{cases} 1, & n \leq 0 \\ 0, & n > 0 \end{cases}$	Figure 11.8(*b*)
$u(n-3)$	Delay (3 units)	$\{\cdots 0, 0, 0, 0, 0, 0, 0, 0, 0, 0, 1, 1, 1, 1\cdots\} = \begin{cases} 1, & n \geq 3 \\ 0, & n < 3 \end{cases}$	Figure 11.8(*c*)
$u(-n-3)$	Time reversal	$\{\cdots 1, 1, 1, 1, 0, 0, 0, 0, 0, 0, 0, 0, 0\cdots\} = \begin{cases} 1, & n \leq -3 \\ 0, & n > -3 \end{cases}$	Figure 11.8(*d*)
$u(n+4)$	Advance (4 units)	$\{\cdots 0, 0, 1, 1, 1, 1, 1, 1, 1, 1, 1, 1, 1\cdots\} = \begin{cases} 1, & n \geq -4 \\ 0, & n < -4 \end{cases}$	Figure 11.8(*e*)
$u(-n+4)$	Time reversal	$\{\cdots 1, 1, 1, 1, 1, 1, 1, 1, 1, 1, 1, 0, 0\cdots\} = \begin{cases} 1, & n \leq 4 \\ 0, & n > 4 \end{cases}$	Figure 11.8(*f*)

The validity of the array representations in Table 11.9 and the plots of Figure 11.8 is checked by applying the definition of the unit-step function.

Exchanging the Order of Time Reversal and Shift

Note that changing the order of *time shift* and *time reversal* would lead to the same result only if at the same time *delay* and *advance* are also interchanged. As an example, $u(-n-3)$ may be obtained by a delay/reversal sequence; first introduce a three-unit delay in $u(n)$ (by changing n to $n-3$) to produce $u(n-3)$, then reverse $u(n-3)$ (by changing n to $-n$) to obtain $u(-n-3)$. Conversely, $u(-n-3)$ may be obtained by a reversal/advance

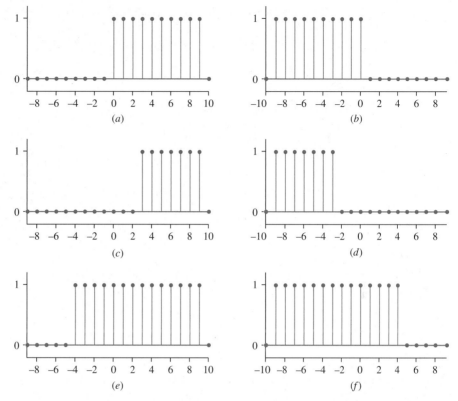

FIGURE 11.8 *(a)* $u(n)$, *(b)* $u(-n)$, *(c)* $u(n-3)$, *(d)* $u(-n-3)$, *(e)* $u(n+4)$, *(f)* $u(-n+4)$

sequence; first reverse $u(n)$ (by changing n to $-n$) to obtain $u(-n)$, then introduce a three-unit advance (by changing n to $n+3$) in it to get $u(-(n+3)) = u(-n-3)$.

Example **11.26**

Shifts and reversals of a right-sided exponential function are given below.

a. One-sided exponential function

$$x(n) = 2^{-n}u(n) = \begin{cases} 2^{-n}, & n \geq 0 \\ 0, & n < 0 \end{cases} = \{\cdots 0, 0, 0, \underset{\uparrow}{1}, 1/2, 1/4, 1/8, 1/16, 1/32, \ldots\}$$

b. Its shift to the right by three units to produce a delay

$$x(n-3) = 2^{-(n-3)}u(n-3) = \begin{cases} 8 \times 2^{-n}, & n \geq 3 \\ 0, & n < 3 \end{cases}$$
$$= \{\cdots 0, 0, \underset{\uparrow}{0}, 0, 0, 1, 1/2, 1/4, 1/8, 1/16, 1/32, \ldots\}$$

c.   Its shift to the left by four units to produce an advance

$$x(n+4) = 2^{-(n+4)}u(n+4) = \begin{cases} \frac{2^{-n}}{16}, & n \geq -4 \\ 0, & n < -4 \end{cases}$$

$$= \{\cdots 0, 0, \underset{\uparrow}{1}, 1/2, 1/4, 1/8, 1/16, 1/32, \ldots\}$$

d.   Reversal of $x(n)$ around the point $n = 0$ to produce

$$x(-n) = 2^n u(-n) = \begin{cases} 2^n, & n \leq 0 \\ 0, & n > 0 \end{cases}$$

$$= \{\cdots 1/32, 1/16, 1/8, 1/4, 1/2, \underset{\uparrow}{1}, 0, 0, 0 \cdots\}$$

e.   Reversal of $x(n-3)$ around the point $n = 0$ to produce

$$x(-n-3) = 2^{(n+3)}u(-n-3) = \begin{cases} 8 \times 2^n, & n \leq -3 \\ 0, & n > -3 \end{cases}$$

$$= \{\cdots 1/32, 1/16, 1/8, 1/4, 1/2, 1, 0, \underset{\uparrow}{0}, 0, 0 \cdots\}$$

f.   Reversal of $x(n+4)$ around the point $n = 0$ to produce

$$x(-n+4) = 2^{(n-4)}u(-n+4) = \begin{cases} \frac{2^n}{16}, & n \leq 4 \\ 0, & n > 4 \end{cases}$$

$$= \{\cdots 1/32, 1/16, 1/8, 1/4, 1/2, 1, 0, 0, 0, 0 \cdots\}$$
$$\phantom{= \{\cdots 1/32, 1/16, 1/8, 1/4, 1/2, 1, 0,}\underset{\uparrow}{}$$

11.14 Time Scaling and Transformation

Time Scaling

Changing the variable n in the expression for a discrete-time function to an, where a is a constant, generates a new function $x(an)$. The operation is called time scaling. For example, in the function $x(n) = 2^{-n}u(n)$ replace n by $2n$ and you will get the new time-scaled function $x(2n) = 2^{-2n}u(2n) = 4^{-n}u(n)$. Time scaling compresses the time axis (if $a > 1$) or expands it (if $a < 1$).

11.27

Given $x(n) = \{\ldots, 0, 0, \underset{\uparrow}{1}, 1, 1, 1, 1, 1, 1, 0, 0, \ldots\}$, find $y(n) = x(2n)$.

Solution

$$\vdots$$
$$y(0) = x(0) = 1$$
$$y(1) = x(2) = 1$$
$$y(2) = x(4) = 1$$
$$y(3) = x(6) = 1$$
$$y(4) = x(8) = 0$$
$$\vdots$$
$$y(n) = \{\ldots, 0, 0, \underset{\uparrow}{1}, 1, 1, 1, 0, 0, \ldots\}$$

Alternative Solution

Describe $x(n)$ by the following expression and change n to $2n$.

$$x(n) = \begin{cases} 1, & 0 \le n \le 6 \\ 0, & \text{elsewhere} \end{cases}$$

$$x(2n) = \begin{cases} 1, & 0 \le 2n \le 6 \\ 0, & \text{elsewhere} \end{cases} = \begin{cases} 1, & 0 \le n \le 3 \\ 0, & \text{elsewhere} \end{cases}$$

Yet Another Solution

$$x(n) = u(n) - u(n - 7)$$
$$x(2n) = u(2n) - u(2n - 7) = u(n) - u(n - 4)$$

Time Transformation

This operation converts the discrete variable n to another discrete variable $m = an + b$, where a and b are constant numbers. The operation, therefore, combines the three operations of reversal, shift, and scaling through a single formula.

*Example* **11.28**

Given

$$x(n) = \begin{cases} n, & 0 \le n \le 6 \\ 0, & \text{elsewhere} \end{cases}$$

find $y(n) = x(-2n + 3)$.

Solution

$$y(n) = x(-2n + 3) = \begin{cases} -2n + 3, & 0 \le -2n + 3 \le 6 \\ 0, & \text{elsewhere} \end{cases}$$

$$= \begin{cases} -2n + 3, & -1 \le n \le 1 \\ 0, & \text{elsewhere} \end{cases} = \begin{cases} 5, & n = -1 \\ 3, & n = 0 \\ 1, & n = 1 \\ 0, & \text{elsewhere} \end{cases}$$

Alternative Solution

Let $x(n) = n[u(n) - u(n - 7)]$. Then,

$$x(-2n + 3) = (-2n + 3)[u(-2n + 3) - u(-2n + 3 - 7)]$$

$$= (-2n + 3)[u(-n + 1) - u(-n - 2)] = \begin{cases} 5, & n = -1 \\ 3, & n = 0 \\ 1, & n = 1 \\ 0, & \text{elsewhere} \end{cases}$$

Verification

$$x(n) = \{\ldots, 0, 0, \underset{\uparrow}{0}, 1, 2, 3, 4, 5, 6, 0, 0, \ldots\}$$

$$y(n) = x(-2n + 3)$$

$$\vdots$$

$$y(-2) = x(4 + 3) = 0$$
$$y(-1) = x(2 + 3) = 5$$
$$y(0) = x(3) = 3$$
$$y(1) = x(-2 + 3) = 1$$
$$y(2) = x(-4 + 3) = 0$$

$$\vdots$$

$$y(n) = \{\ldots, 0, 5, \underset{\uparrow}{3}, 1, 0, 0, \ldots\}$$

11.15 Circular Shift

The circular shift operation is applied to finite-duration sequences. The idea behind it is simple. The sequence is shifted to the right by the specified number of positions. Those members of the sequence that "fall off" the tail end as a result of the shift are tacked back on to the start of the sequence. As such, they "wrap around." A similar situation applies to a left shift. This is best illustrated by an example.

Example
11.29

Consider a finite-duration sequence such as $x(n) = \{\underset{\uparrow}{1}, 2, 4, 8, 16\}$ shown by the circular display in Figure 11.9(a). A circular shift of the above sequence by k units to the right produces a new sequence as shown in the table below for $k = 1, \ldots, 5$.

$$k = 1, \quad x(n-1)|_{\text{circular}} = \{16, \underset{\uparrow}{1}, 2, 4, 8\} \quad \text{Figure 11.9(b)}$$

$$k = 2, \quad x(n-2)|_{\text{circular}} = \{\underset{\uparrow}{8}, 16, 1, 2, 4\} \quad \text{Figure 11.9(c)}$$

$$k = 3, \quad x(n-3)|_{\text{circular}} = \{\underset{\uparrow}{4}, 8, 16, 1, 2\} \quad \text{Figure 11.9(d)}$$

$$k = 4, \quad x(n-4)|_{\text{circular}} = \{\underset{\uparrow}{2}, 4, 8, 16, 1\} \quad \text{Figure 11.9(e)}$$

$$k = 5, \quad x(n-5)|_{\text{circular}} = \{\underset{\uparrow}{1}, 2, 4, 8, 16\} \quad \text{Figure 11.9(f)}$$

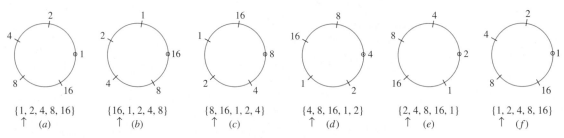

FIGURE 11.9 Circular shift.

It is seen that a finite sequence with N samples reverts to its original form after N circular shifts, as was seen for linear shift in a periodic signal with a period N (Example 11.24). The circular shifts of $s(n) = \{1, 2, 4, 8, 16\}$ by k units to the right are displayed in Figure 11.9 for $k = 1, \ldots 5$. The fifth shift reverts the signal to its original form. Circular shift will be discussed in more detail in Chapter 17 along with application.

11.16   Even and Odd Functions

A function $x(n)$ is even if $x(n) = x(-n)$ for all n. See Figure 11.10(a). Similarly, a function $x(n)$ is odd if $x(n) = -x(-n)$ for all n. See Figure 11.10(b). Any function $x(n)$ may be represented by the sum of two components as $x(n) = x_e(n) + x_o(n)$, where $x_e(n)$ is an even function, called the even part of $x(n)$, and $x_o(n)$ is an odd function, called the odd part of $x(n)$. The functions in Figure 11.10(a) and (b) are the even and odd parts of the function shown in Figure 11.10(c). The even and odd parts of a function are uniquely determined from $x(n)$ by the following:

$$x_e(n) = \frac{x(n) + x(-n)}{2}$$

$$x_o(n) = \frac{x(n) - x(-n)}{2}$$

Note: The value of an odd function at the origin is zero, $x_o(0) = 0$.

The expressions given above for $x_e(n)$ and $x_o(n)$ can be derived by the following. Let $x(n) = x_e(n) + x_o(n)$. Then, $x(-n) = x_e(-n) + x_o(-n)$. But by definition, $x_e(-n) = x_e(n)$ and $x_o(-n) = -x_o(n)$. Therefore, $x(-n) = x_e(n) - x_o(n)$. By adding $x(n)$ and $x(-n)$ we get $x(n) + x(-n) = 2x_e(n)$. By subtracting $x(-n)$ from $x(n)$ we get $x(n) - x(-n) = 2x_o(n)$. These give the results stated previously:

$$x_e(n) = \frac{x(n) + x(-n)}{2}$$

$$x_o(n) = \frac{x(n) - x(-n)}{2}$$

Example **11.30**

Find the even and odd parts of $x(n) = \{2, 2, 2\}$ $[x(n) = 0$ outside the bracket].

Solution
We first find $x(-n) = \{2, 2, 2\}$. Then,

$$x_e(n) = \frac{x(n) + x(-n)}{2} = \{1, 1, 2, 1, 1\}$$

$$x_o(n) = \frac{x(n) - x(-n)}{2} = \{-1, -1, 0, 1, 1\}$$

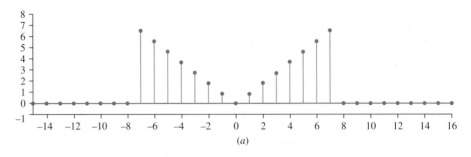

(a)

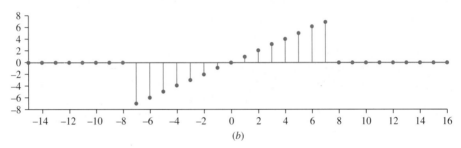

(b)

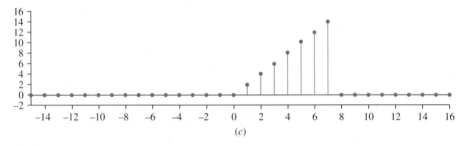

(c)

FIGURE 11.10 *(a)* An even function. *(b)* An odd function. *(c)* The function whose even and odd parts are shown in *(a)* and *(b)*.

Find the even and odd parts of $h(n) = \{1, 2, \underset{\uparrow}{3}, 2, 1\}$ [$h(n) = 0$ outside the bracket].

Solution

$h(n)$ is an even function because $h(-n) = h(n) = \{1, 2, \underset{\uparrow}{3}, 2, 1\}$. Therefore, $h_o(n) = 0$. This result may also be derived from the equations

$$h_e(n) = \frac{h(n) + h(-n)}{2} = \{1, 2, \underset{\uparrow}{3}, 2, 1\}$$

$$h_o(n) = \frac{h(n) - h(-n)}{2} = 0, \quad \text{all } n$$

Example
11.32

Find the even and odd parts of $x(n) = \{1, \underset{\uparrow}{3}, 3, 2\}$.

Solution

We first find $x(-n) = \{2, 3, \underset{\uparrow}{3}, 1\}$. Then,

$$x_e(n) = \frac{x(n) + x(-n)}{2} = \{1, 2, \underset{\uparrow}{3}, 2, 1\}$$

$$x_o(n) = \frac{x(n) - x(-n)}{2} = \{-1, -1, \underset{\uparrow}{0}, 1, 1\}$$

Example
11.33

Find the even and odd parts of

$$h(n) = \begin{cases} 0.5^n, & \text{for } n \geq 0 \\ 0, & \text{for } n < 0 \end{cases}$$

Solution

We first find $h(-n)$

$$h(-n) = \begin{cases} 0, & \text{for } n > 0 \\ 0.5^{-n}, & \text{for } n \leq 0 \end{cases}$$

The even part is

$$h_e(n) = \frac{h(n) + h(-n)}{2} = \begin{cases} \frac{1}{2}(0.5)^n, & \text{for } n > 0 \\ 1, & \text{for } n = 0 \\ \frac{1}{2}(0.5)^{-n}, & \text{for } n < 0 \end{cases}$$

and the odd part is

$$h_o(n) = \frac{h(n) - h(-n)}{2} = \begin{cases} \frac{1}{2}(0.5)^n, & \text{for } n > 0 \\ 0, & \text{for } n = 0 \\ -\frac{1}{2}(0.5)^{-n}, & \text{for } n < 0 \end{cases}$$

Note that $h_e(n) + h_o(n) = (0.5)^n u(n)$.

Even and Odd Parts of a Causal Function

Signals that are zero for $n < 0$ are called *causal* (and those which are zero for $n > 0$ are *anticausal*). It can be verified that a causal discrete-time function $h(n)$ is related to its even or odd parts by the following equations:

$$h(n) = 2h_e(n)u(n) - h_e(0)d(n)$$

$$h(n) = 2h_o(n)u(n) + h(0)d(n)$$

where $u(n)$ and $d(n)$ are the unit-step and unit-sample functions, respectively. From the above equations we conclude that a causal function $h(n)$ may be obtained from its even part $h_e(n)$ or from its odd part $h_o(n)$ and $h(0)$. In other words, any one of the three functions given below can specify the other two.

1. $h(n)$
2. $h_e(n)$
3. $h_o(n)$ and $h(0)$

The causal $x(n) = e^{-n/3}u(n) = (0.7163)^n u(n)$ and the anticausal $x(-n)=e^{n/3}u(-n)$ $=(0.7163)^{-n}u(-n)$ exponential functions are shown in Figure 11.11 along with the even and odd parts of $x(n)$.

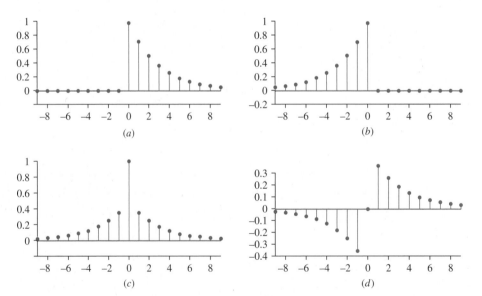

FIGURE 11.11 **(a)** The causal exponential function $x(n) = e^{-n/3}u(n) = (0.7163)^n u(n)$. **(b)** The anticausal exponential function $x(-n) = e^{n/3}u(-n) = (0.7163)^{-n}u(-n)$. **(c)** The even part of $x(n)$ is $[x(n) + x(-n)]/2 = 0.5e^{-|n|/3}$. **(d)** The odd part of $x(n)$ is

$$[x(n) - x(-n)]/2 \triangleq 0.5e^{-|n|/3} \times \text{sgn}(n) - 0.5d(n)$$

where sgn(n) (pronounced as *signum* and standing for *sign of*) is defined by

$$\text{sgn}(n) = \begin{cases} 1, & \text{for } n \geq 0 \\ -1, & \text{for } n < 0 \end{cases}$$

11.17 Windows

The domain of mathematical models of real signals is $-\infty < n < \infty$. However, in practice we process finite-length segments of a signal at a time. Mathematically, this corresponds to multiplying the original signal by a window, which is nonzero within an interval and zero outside that interval. A window is, therefore, a finite-duration pulse through which we look at a data stream or function. The role of the window is to create and shape a finite segment of the sequence in a way that certain characteristics are met. Given the duration of the window, its shape has an important role in its function. Five examples of even windows of length $2M + 1$ are listed in Table 11.10.

TABLE 11.10 Five Commonly Used Discrete-Time Windows

Window Name	$-M \le n \le M$ (Even Function)	$0 \le n \le N - 1$ (Causal Function)
Rectangular (uniform)	1	1
Triangular (Bartlett)	$1 - \dfrac{\lvert n \rvert}{M}$	$1 - \dfrac{2\left\lvert n - \frac{N-1}{2}\right\rvert}{N-1}$
Hanning	$0.5 + 0.5\cos\left(\dfrac{\pi n}{M}\right)$	$0.5 - 0.5\cos\left(\dfrac{2\pi n}{N-1}\right)$
Hamming	$0.54 + 0.46\cos\left(\dfrac{\pi n}{M}\right)$	$0.54 - 0.46\cos\left(\dfrac{2\pi n}{N-1}\right)$
Blackman	$0.42 + 0.5\cos\left(\dfrac{\pi n}{M}\right) + 0.08\cos\left(\dfrac{2\pi n}{M}\right)$	$0.42 - 0.5\cos\left(\dfrac{2\pi n}{N-1}\right) + 0.08\cos\left(\dfrac{4\pi n}{N-1}\right)$

Example
11.35

a. Shift the even Hanning, Hamming, and Blackman windows of Table 11.10 by M units to the right to make them N-point causal functions with $N = 2M + 1$. Find mathematical expressions for the shifted functions in terms of N and show that they are in agreement with those given for the causal functions in that table.

b. Write a computer program to evaluate the window functions in Table 11.10.

c. Obtain numerical values of the even functions for $M = 5$ and the causal functions for $N = 11$. Briefly summarize their time-domain characteristics.

d. Evaluate the causal functions of Table 11.10 for $N = 10$. Compare with the values obtained in part c for $N = 11$ and observe their differences.

Solution

a. Replace n by $n - M$. This generates N-point long causal functions, where $N = 2M + 1$ or $M = (N - 1)/2$.

$$\cos\left(\frac{\pi n}{M}\right), \quad -M \le n \le M \Longrightarrow \cos\left(\frac{\pi(n - M)}{M}\right) = -\cos\left(\frac{\pi n}{M}\right)$$

$$= -\cos\left(\frac{2\pi n}{N - 1}\right), \quad 0 \le n \le N - 1$$

$$\cos\left(\frac{2\pi n}{M}\right), \quad -M \le n \le M \Longrightarrow \cos\left(\frac{2\pi(n - M)}{M}\right)$$

$$= \cos\left(\frac{2\pi n}{M}\right) = \cos\left(\frac{4\pi n}{N - 1}\right), \quad 0 \le n \le N - 1$$

b. See below.

```
% Even window functions (2M+1 points from -M to M)
M=5; n=0:M;
tri1=1-n/M
han1=0.5+0.5*cos(pi*n/M)
ham1=0.54+0.46*cos(pi*n/M)
blk1=0.42+0.5*cos(pi*n/M)+0.08*cos(2*pi*n/M)
%
% Causal window functions (N points  from 0 to N-1)
N=11; n=0:N-1;
tri2=1+2*(n-(N-1)/2)/(N-1)
han2=0.5-0.5*cos(2*pi*n/(N-1))
ham2=0.54-0.46*cos(2*pi*n/(N-1))
blk2=0.42-0.5*cos(2*pi*n/(N-1))+0.08*cos(4*pi*n/(N-1))
```

c. The 11-point windows are symmetrical. The values for half of each window are given below.

Triangular:	1	0.8	0.6	0.4	0.2	0
Hanning:	1	0.9045	0.6545	0.3455	0.0955	0
Hamming:	1	0.9121	0.6821	0.3979	0.1679	0.08
Blackman:	1	0.8492	0.5098	0.2008	0.0402	0

The triangular, Hanning, and Blackman windows all taper off to a zero value at their edge. The Hamming window does not. The frequency characteristics of these windows will be discussed in Chapter 16.

d. Having an even number of samples, the windows now become half-symmetric as observed below.

Triangular:	0	0.2222	0.4444	0.6667	0.8889	0.8889	0.6667	0.4444	0.2222	0
Hanning:	0	0.1170	0.4132	0.7500	0.9698	0.9698	0.7500	0.4132	0.1170	0
Hamming:	0.08	0.1876	0.4601	0.7700	0.9723	0.9723	0.7700	0.4601	0.1876	0.08
Blackman:	0	0.0509	0.2580	0.6300	0.9511	0.9511	0.6300	0.2580	0.0509	0

11.18 Signal Processing and Filtering

Signals are processed for various purposes such as filtering, detection, prediction, range measurement, image and speech recognition, medical and dignostic purposes, decision making, and the like. These may require two types of operations: (1) algebraic operations such as addition, subtraction, mutiplication, division; and (2) logic operations such as AND, OR, NAND, NOR, and so on. In this book we concentrate on the first category. Since most processing is done by digital tools such as digital computers, we normally call it digital signal processing (DSP). DSP tools and techniques allow us to handle complex DSP problems from design to simulation, testing, and real-time processing. The basic mathematical tools and techniques for working with discrete signals are covered in the remainder of this book. The following examples illustrate some bare-bones signal processing operations dealing with filtering and smoothing.

*E**xample*
11.36

Consider the discrete-time sinusoidal signal $x(n) = \cos(\alpha n)$, where α is a constant. Show that

a. $\dfrac{x(n+k) + x(n-k)}{2} = \cos(\alpha k) \cos(\alpha n) = \cos(\alpha k) x(n) = Hx(n)$

b. $\dfrac{x(n) + x(n-2k)}{2} = \cos(\alpha k) \cos(\alpha n - \alpha k) = \cos(\alpha k) x(n-k) = Hx(n-k)$

where $H = \cos(\alpha k)$ is a constant.

Solution
To show a,

$$x(n+k) = \cos[\alpha(n+k)] = \cos(\alpha n) \cos(\alpha k) - \sin(\alpha n) \sin(\alpha k)$$
$$x(n-k) = \cos[\alpha(n-k)] = \cos(\alpha n) \cos(\alpha k) + \sin(\alpha n) \sin(\alpha k)$$
$$x(n+k) + x(n-k) = 2\cos(\alpha k) \cos(\alpha n) = 2\cos(\alpha k) x(n) = 2Hx(n)$$

To show b, in a let $n + k = \ell$. Then, $n = \ell - k$ and

$$x(\ell) + x(\ell - 2k) = 2\cos(\alpha k) \cos(\alpha \ell - \alpha k) = 2Hx(\ell - k)$$

An appropriate choice of delay such that $\alpha k = \pi/2$ (or odd multiples of $\pi/2$) will result in $H = 0$ and allows us to filter out frequencies at α, as illustrated in Examples 11.37 and 11.38.

*E**xample*
11.37

The information-carrying signal to be picked up by a sensor is a sinusoid at an unknown frequency $F < 600$ Hz. The sensor also picks up a 60-Hz disturbance. The sensor's output is, therefore, $x(t) = A \cos(2\pi F t) + B \cos(120\pi t)$. The first term constitutes the signal of interest and the second term is the undesired disturbance. Devise a simple DSP operation that *notches out* the sinusoidal interference and keeps the signal.

Solution

We first sample $x(t)$ at the rate of 1,200 samples per second, twice the maximum frequency of the information-carrying sinusoid to preserve the information in it (see Chapter 10). By setting $t = n\Delta t = n/1,200$ in $x(t)$ we obtain a discrete-time data sequence $x(n) = A\cos(\omega n) + B\cos(\alpha n)$, where $\omega = 2\pi F/1200$ and $\alpha = \pi/10$ are the angular frequencies in the discrete domain. To filter out the $\cos(\alpha n)$, we apply the shift-and-add operation used in Example 11.36.

$$\frac{x(n) + x(n - 2k)}{2} = A\frac{\cos(\omega n) + \cos(\omega n - 2\omega k)}{2} + B\frac{\cos(\alpha n) + \cos(\alpha n - 2\alpha k)}{2}$$

$$= A\cos(\omega k)\cos\omega(n - k) + B\cos(\alpha k)\cos\alpha(n - k)$$

The operation delays both the signal and disturbance by k units and changes their magnitudes by the constant values $\cos(\omega k)$ and $\cos(\alpha k)$, respectively. By choosing $k = 5$ we will have $\alpha k = 5\pi/10 = \pi/2$, which results in $\cos(\alpha k) = 0$. This blocks the 60 Hz interference (i.e., notches it out). Then, the output of the digital filter becomes $y(n) = A\cos(\omega k)\cos\omega(n - 5)$. The five-unit delay in the discrete signal corresponsds to a delay of $\tau = 5/1,200$ seconds or 416.6 μs in the continuous-time domain and can be accommodated. But, more importantly, the filter attenuates the signal by the factor $\cos(5\omega)$, corresponding to $\cos(\pi F/120)$ in the continuous-time domain. Unless F is known the strength of the signal, therefore, cannot be determined. Moreover, the frequency-dependent amplitude change produced by the above filtering operation blocks the signal at $F = 120$ and 240 Hz.

Example **11.38**

Consider a signal $x(t)$ made of the sum of two sinusoids at frequencies F_1 and F_2.

$$x(t) = A_1\cos(2\pi F_1 t) + A_2\cos(2\pi F_2 t)$$

To block the F_2 component and retain as much of the F_1 component as possible, we first sample $x(t)$ at the rate of F_s using an analog-to-digital converter (A/D) to obtain $x(n)$. We then use a shift-and-add operation (shown by the block diagram of Figure 11.12) followed by a digital-to-analog converter (D/A). Find the amount of shift necessary to block the component at F_2. Then determine the filter's effect on the component at F_1.

Solution

The discrete-time data is

$$x(n) = A_1\cos(\omega_1 n) + A_2\cos(\omega_2 n), \quad \text{where } \omega_1 = 2\pi\left(\frac{F_1}{F_s}\right) \text{ and } \omega_2 = 2\pi\left(\frac{F_2}{F_s}\right)$$

Following the procedure used in Examples 11.36 and 11.37

$$x(n) + x(n - 2k) = 2A_1\cos(\omega_1 k)\cos[\omega_1(n - k)] + 2A_2\cos(\omega_2 k)\cos[\omega_2(n - k)]$$

To block the ω_2 component we need $\cos(\omega_2 k) \cos[\omega_2(n - k)] = 0$ for all n, which is achieved if

$$\cos(\omega_2 k) = 0, \quad \omega_2 k = \frac{\pi}{2}, \quad k = \frac{\pi}{2\omega_2} = \frac{1}{4}\frac{F_s}{F_2}$$

The component at ω_1 will be multiplied by a factor

$$2\cos(\omega_1 k)\Big|_{k=\frac{\pi}{2\omega_2}} = 2\cos\left(\frac{\pi}{2}\frac{\omega_1}{\omega_2}\right) = 2\cos\left(\frac{\pi}{2}\frac{F_1}{F_2}\right)$$

In summary, assuming unity gains for the A/D and D/A converters, the continuous-time signal at the output of the system of Figure 11.12 will be

$$y(t) = 2A_1 \cos\left(\frac{\pi}{2}\frac{F_1}{F_2}\right) \cos\left[2\pi F_1(t - \tau)\right] \quad \text{where } \tau = \frac{1}{4F_2}$$

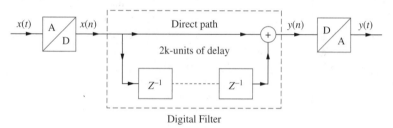

FIGURE 11.12 Block diagram of a digital signal processing system made of an A/D converter, a simple digital filter, and a D/A converter. The filter has a direct and an indirect path of k-unit delay (a unit delay is shown by the symbol z^{-1}). Input and output signals are in the continuous-time domain. Filtering is done in the discrete-time domain. See Example 11.38.

Summary of Examples 11.36–38

The simple digital filters discussed in Examples 11.36 to 11.38 are summarized below.

Example 11.36:

$$x(n) = \cos(\alpha n).$$

$$\frac{x(n) + x(n - 2k)}{2} = Hx(n - k), \qquad \text{where } H = \cos(\alpha k)$$

Example 11.37:

$$x(t) = A\cos(2\pi Ft) + B\cos(120\pi t).$$

$$x(n) = A\cos(\omega n) + B\cos(\alpha n), \qquad \text{where } \omega = 2\pi\frac{F}{1,200} \text{ and } \alpha = \frac{\pi}{10}$$

$$\frac{x(n) + x(n - 10)}{2} = HA\cos\omega(n - 5), \quad \text{where } H = \cos(5\omega)$$

Example 11.38:

$$x(t) = A\cos(2\pi F_1 t) + B\cos(2\pi F_2 t).$$

$$x(n) = A\cos(\omega_1 n) + B\cos(\omega_2 n), \qquad \text{where } \omega_1 = 2\pi\frac{F_1}{F_s}, \ \omega_2 = 2\pi\frac{F_2}{F_s}.$$

$$\frac{x(n) + x(n - 2k)}{2} = HA\cos\omega_1(n - k), \qquad \text{where } H = \cos\left(\frac{\pi}{2}\frac{F_1}{F_2}\right) \text{ and } k = \frac{1}{4}\frac{F_s}{F_2}$$

$$y(t) = 2A_1\cos\left(\frac{\pi}{2}\frac{F_1}{F_2}\right)\cos[2\pi F_1(t - \tau)], \text{ where } \tau = \frac{1}{4F_2}$$

11.19 Problems

Solved Problems

1. Write a Matlab program to generate $u(n - N)$, $N1 \leq n \leq N2$, for $N = 2$, $N1 = -5$, and $N2 = 10$. Plot it as stem and save the plot as an encapsulated postscript file.

Solution

```
N1=-5; N2=10; N=2; n=N1:N2;
x=[zeros(1,-N1+N),ones(1,N2-N+1)];
stem(n,x)
axis([N1  N2  -.2 1.2]); grid
print -dpsc Ch11_p1a.eps
```

2. Write a Matlab program to generate $nu(n - N)$, $N1 \leq n \leq N2$, for $N = -2$, $N1 = -5$, and $N2 = 10$. Then plot the result.

Solution

```
N1=-5; N2=10; N=-2; n=N1:N2;
x=n.*[zeros(1,-N1+N),ones(1,N2-N+1)];
stem(n,x)
axis([N1  N2+1  N-1 N2+1]); grid
```

3. Write a Matlab program to generate and plot $\cos(\omega_0 n)$, $-20 \leq n \leq 50$, for

a. $\omega_0 = 0.125$
b. $\omega_0 = 0.25\pi$

Solution

```
N1=-20; N2=50; n=N1:N2;
xa=cos(n/8);     figure;  stem(n,xa);  grid;  axis([N1 N2+1 -1.2 1.2]);
xb=cos(pi*n/4); figure;  stem(n,xb);  grid;  axis([N1 N2+1 -1.2 1.2]);
```

4. Write a Matlab program to generate $2^{-n}u(n - N)$, $N1 \leq n \leq N2$, for $N = 2$, $N1 = -5$, and $N2 = 10$. Then plot the result.

Solution

```
N1=-5; N2=10; N=2; n=N1:N2;
x=2.^(-n).*[zeros(1,-N1+N),ones(1,N2-N+1)];
stem(n,x); grid; axis([N1  N2  -.05 .3])
```

5. Write a Matlab program to generate and plot $e^{-an} \cos(\omega n + \theta)$, $N1 \le n \le N2$, for $a = 0.05$, $\omega = 1/4$, $\theta = \pi/3$, $N1 = -10$, and $N2 = 50$.

Solution

```
N1=-10; N2=50; n=N1:N2; a=.05; w=1/4; theta=pi/3;
x=exp(-a*n).*cos(w*n+theta);  stem(n,x); grid; axis([N1 N2 -1 1.5])
```

6. Write a Matlab program to generate and plot $V_0 e^{-n/\tau} \cos(\omega n + \theta)u(n)$, $N1 \le n \le N2$, for $V_0 = 1$, $\tau = 20$, $\omega = 0.25$, $\theta = 0$, $N1 = -10$, and $N2 = 50$.

Solution

```
clear; V0=1; N1=-10; N2=50; N=-N1+N2; tau=20; w=1/4; theta=0; n=N1:N2
for i=1:N+1
    if n(i)>=0;
        x(i)=1;
    else
        x(i)=0;
    end
end
v=V0*exp(-n/tau).*cos(w*n+theta); v1=v.*x;
stem(n,v1); axis([N1  N2 -1  V0+.5])
```

7. Write a Matlab program to generate and plot $V_0 e^{-n/\tau} \cos(\omega n + \theta)u(an + b)$, $N1 \le n \le N2$, for $\tau = 20$, $\omega = 0.25$, $\theta = 0$, $a = 1$, $b = 0$, $N1 = -10$, and $N2 = 50$.

Solution

```
clear; V0=1; N1=-10; N2=50; N=-N1+N2; tau=20; w=1/4; theta=0; n=N1:N2;
a=1; b=0;
for i=1:N+1
    if a*n(i)+b >=0;
        x(i)=1;
    else
        x(i)=0;
    end
end
v=V0*exp(-n/tau).*cos(w*n+theta); v1=v.*x;
stem(n,v1); grid; axis([N1  N2 -1  V0+.5])
```

8. Write a Matlab program to generate and plot $y(n) = x(2n - 10)$, where $x(n) = e^{-n/20} \cos(0.25n + \pi/3)u(n)$, $-10 \le n \le 50$.

Solution

```
V0=1; N1=-10; N2=50; N=-N1+N2; tau=20; w=1/4; theta=pi/3; n=N1:N2
a=2; b=-10;
for i=1:N+1
    if a*n(i)+b >=0;
        x(i)=1;
    else
        x(i)=0;
    end
end
```

```
end
v=V0*exp(-(a*n+b)/tau).*cos(w*(a*n+b)+theta); v1=v.*x;
stem(n,v1); axis([N1   N2 -1   V0+.5])
```

9. Write a Matlab program to generate and plot $y(n) = x(-3n + 10)$, where $x(n) = e^{-n/20} \cos(0.25n + \pi/3)u(n)$, $-10 \le n \le 50$.

Solution

In the solution given for problem 8, replace the second line with

```
a=-3;  b=10
```

10. Write a Matlab program to generate and plot the three discrete waveforms whose envelopes are shown in Figure 11.13. Each waveform contains two cycles of a periodic function with a period of $N = 100$.

   a. A rectangular waveform with a base-to-peak value of 0 to 1 and a 40% duty cycle.

   b. A sawtooth waveform with a peak-to-peak value of -1 to 1, spending 40% of the time in the state with positive slope.

   c. An exponentially varying waveform with a peak-to-peak value of -1 to 1, spending 40% of the time as 1.05^{-n} and 60% of the time as $-(1.1)^{-n}$.

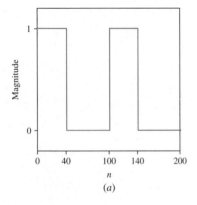
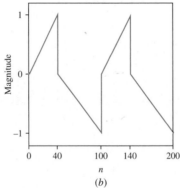
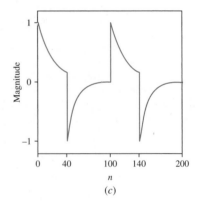

FIGURE 11.13

Solution

   a. The following program generates and plots functions of Figure 11.13 sampled at the 1 Hz rate. Change the command "plot" to "stem" to see the discrete-time display.

```
k2=2;      % Number of cycles
N=100;     % Period
d=0.4;     % Duty cycle
N1=N*d;        n1=0:N1-1;
N2=N*(1-d);  n2=0:N2-1;
n=0:k*N-1;
% a) Rectangular waveform:
x1=ones(1,N1); x2=zeros(1,N2); x=[x1, x2];
for j=1:k;
    for i=1:N;
```

```
        y(i+(j-1)*N)=x(i);
    end
  end
  plot(n,y);
```

b. Repeat with

```
x1=n1.*ones(1,N1)/N1; x2=-n2.*ones(1,N2)/N2; x=[x1,x2];
```

c. Repeat with

```
x1=1.05.^(-n1); x2=-1.1.^(-n2); x=[x1,x2];
```

Chapter Problems

11. Sketch the following discrete-time signals as functions of n and express them in the following forms: (1) array and (2) weighted sum of unit-sample functions.

a. $x(n) = \begin{cases} 2, & -2 \le n \le 4 \\ 0, & \text{elsewhere} \end{cases}$

b. $x(n) = \begin{cases} 1, & |n| \le 3 \\ 0, & \text{elsewhere} \end{cases}$

c. $x(n) = \begin{cases} n, & |n| \le 3 \\ 0, & \text{elsewhere} \end{cases}$

d. $x(n) = \begin{cases} 2-n, & |n| \le 3 \\ 0, & \text{elsewhere} \end{cases}$

e. $x(n) = \begin{cases} 2, & 0 \le n \le 99 \\ 0, & \text{elsewhere} \end{cases}$

f. $x(n) = \begin{cases} 0, & n \le 0 \\ n, & 1 \le n \le 2 \\ 3, & n \ge 3 \end{cases}$

g. $x(n) = \begin{cases} 0, & n < 0 \\ 2, & 0 \le n \le 5 \\ 3, & n > 5 \end{cases}$

h. $x(n) = \begin{cases} 0, & n < 0 \\ n, & 0 \le n \le 5 \\ 10 - n, & 6 \le n \le 9 \\ 0, & n \ge 10 \end{cases}$

12. Express the signals of problem 11 as sums of unit steps and unit ramps.

13. The signal $x(n) = \{1, 1, 1, -1, -1, 1, -1\}$ has seven samples at $0 \le n \le 6$ and is zero elsewhere.

a. Sketch $x(n)$.

b. Show that $x(n)$ may also be described by any of the following three forms.
   (i) $x(n) = d(n) + d(n-1) + d(n-2) - d(n-3) - d(n-4) + d(n-5) - d(n-6)$.
   (ii) $x(n) = u(n) - 2u(n-3) + 2u(n-5) - 2u(n-6) + u(n-7)$.

   (iii) $x(n) = \begin{cases} 0, & n < 0 \\ 1, & 0 \le n \le 2 \\ -1, & 3 \le n \le 4 \\ 1, & n = 5 \\ -1, & n = 6 \\ 0, & n > 6 \end{cases}$

c. Which of the above four representations is easiest for you to work with? Explain.

Problems 14–27

The discrete signals in problems 14 through 27 are represented by sequences of numbers, along with four possible mathematical expressions of which only one (or none) is correct. Mark the correct answer.

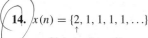

14. $x(n) = \{2, 1, 1, 1, 1, \ldots\}$
$\uparrow$

   a. $d(n) + u(n-1)$
   b. $d(n) + u(n)$
   c. $2u(n-1)$
   d. $d(n-1)$
   e. None of the above

15. $x(n) = \{1, 1, 1, 1, 1, \ldots\}$
$\uparrow$

   a. $u(n+1)$
   b. $u(n-1)$
   c. $d(n)$
   d. $d(n-1)$
   e. None of the above

16. $x(n) = \{1, 1, 2, 2, 2, \ldots\}$
$\uparrow$

   a. $d(n)$
   b. $2d(n) + u(n-1)$
   c. $u(n) + u(n-2)$
   d. $u(n) + u(n-3)$
   e. None of the above

17. $x(n) = \{\ldots, 1, 1, 1, 1, 1\}$
$\uparrow$

   a. $u(-n)$
   b. $u(n-1)$
   c. $u(-n+1)$
   d. $u(-n-1)$
   e. None of the above

18. $x(n) = \{1, 1\}$
$\uparrow$

   a. $d(n) + d(n-1)$
   b. $u(n-1) - u(n)$
   c. $u(n+1) - u(n-1)$
   d. $d(n+1) + d(n-1)$
   e. None of the above

19. $x(n) = \{\ldots, 8, 4, 2, 1, 2, 4, 8, \ldots\}$
$\uparrow$

   a. $2^n u(n) + 2^{-n} u(-n) - d(n)$
   b. $2^{|n|} + d(n)$

   c. 2^n
   d. 0.5^n
   e. None of the above

20. $x(n) = \{1, 1, 2, 4, 8, \ldots\}$
$\uparrow$

   a. $0.5^{-n} u(n) + d(n)$
   b. $0.5^{-n} u(n+1)$
   c. $2^n u(n+1)$
   d. $0.5^{(n+1)} u(n+1)$
   e. None of the above

21. $x(n) = \{\ldots, 1/8, 1/4, 1/2, 1\}$
$\uparrow$

   a. $2^{-n} u(n)$
   b. $0.5^{-n} u(n)$
   c. $2^n u(-n)$
   d. $0.5^n u(-n)$
   e. None of the above

22. $x(n) = \{2, 1, 1/2, 1/4, 1/8, \ldots\}$
$\uparrow$

   a. $0.5^n u(n) + d(n)$
   b. $2 \times (0.5)^n u(n+1)$
   c. $0.5^n u(n+1)$
   d. $0.5^{(n+1)} u(n+1)$
   e. None of the above

23. $x(n) = \{1, 1/2, 1/4, 1/8, 1/16\}$
$\uparrow$

   a. $0.5^n u(n) + d(n)$
   b. $2 \times (0.5)^n u(n+1)$
   c. $0.5^n u(n+1)$
   d. $0.5^{(n+1)} u(n+1)$
   e. None of the above

24. $x(n) = \{1, -1/2, 1/4, -1/8, 1/16, -1/32, \ldots\}$
$\uparrow$

   a. $0.5^n u(n) + d(n)$
   b. $2 \times 0.5^n u(n+1)$
   c. $0.5^n u(n+1)$
   d. $0.5^{(n+1)} u(n+1)$
   e. None of the above

25. $x(n) = \{\ldots, 1/8, 1/4, 1/2, 1, 1/2, 1/4, 1/8, \ldots\}$

   a. $0.5^{-n}u(n) + 0.5^{n}u(-n)$

   b. $2^{-n}u(n) + 2^{n}u(-n)$

   c. $0.5^{n}u(n) + 0.5^{-n}u(-n)$

   d. $2^{n}u(n) + 2^{-n}u(-n)$

   e. None of the above

26. $x(n) = \{\ldots, 1/27, 1/9, 1/3, 1, 1/2, 1/4, 1/8, \ldots\}$

   a. $3^{n}u(n) + (-3)^{n}u(-n)$

   b. $3^{-n}u(n) + 3^{n}u(-n)$

   c. $(1/3)^{n}u(n) + (1/3)^{-n}u(-n)$

   d. $(1/3)^{n}u(n) + (1/3)^{-n}u(-n)$

   e. None of the above

27. $x(n) = \{1/27, 1/9, 1/3, 1\}$

   a. $3^{n}[u(n + 4) - u(n)]$

   b. $3^{-n}[u(n) + u(-n + 4)]$

   c. $(1/3)^{n}u(n) + (1/3)^{-n}u(-n + 4)$

   d. $(1/3)^{n}u(n + 4) + (1/3)^{-n}u(-n)$

   e. None of the above

28. Evaluate, sketch, and compare the following discrete-time ramp signals within the range $-6 \leq n \leq 6$.

   a. $r(n) - r(n - 3)$, where $r(n)$ is a unit ramp.

   b. $r(n)[u(n) - u(n - 4)]$

   c. $[r(n) - r(n - 3)][u(n) - u(n - 4)]$

   d. $r(n)[u(n) - u(n - 3)]$

29. Evaluate and sketch the following discrete-time signals within the range $-6 \leq n \leq 6$.

   a. $\cos n$ b. $\cos(3n)$ c. $\cos(\pi n/2)$ d. $\cos(\pi n/3)$

   e. $\cos n + \sin n$ f. $\cos n + \cos(3n)$ g. $\cos(\pi n/2) + \cos(\pi n/3)$ h. $\cos n + \cos(\pi n)$

30. Evaluate and sketch the following discrete-time signals within the range $-6 \leq n \leq 6$.

   a. $3^{-n}u(n - 3)$ b. $(-0.3)^{n}u(n + 3)$ c. $0.4^{n}\sin(\pi n/6)u(n - 3)$ d. $0.4^{n}\cos(\pi n/6)u(n + 3)$

31. Determine the function that models each discrete signal in Table 11.11A. Pick the number referring to the correct formula from Table 11.11B.

TABLE 11.11A Signal	
Letter	Signal
a.	$\{1, 1, 1, 1, 1, \ldots\}$
b.	$\{\ldots, 1, 1, 1, 1, 1\}$
c.	$\{1, 1\}$
d.	$\{\ldots, 1, 1, 1, 1, 1, 1, 1, \ldots\}$
e.	$\{1, 1, 1, 1, 1\}$
f.	$\{1, 1, 1, 1, 1\}$
g.	$\{1, 2, 2, 2, 2, \ldots\}$

TABLE 11.11B Function	
Number	Function $x(n)$
1	$u(n) + u(n - 1)$
2	$u(n + 3) - u(n - 2)$
3	$u(n + 1) - u(n - 4)$
4	$u(-n + 1)$
5	1
6	$u(n - 1)$
7	$d(n) + d(n - 1)$
8	$u(n - 1) - u(n)$
9	$u(n + 1) - u(n - 1)$
10	None of the above

32. Determine the function that models each discrete signal in Table 11.12A. Pick the number refering to the correct formula from Table 11.12B.

TABLE 11.12A Signal

Letter	Signal
a.	$\{\ldots, 1/16, 1/8, 1/4, 1/2, 1\}$
b.	$\{\ldots, 1/16, 1/8, 1/4, 1/2, 1, 1/2, 1/4, 1/8, 1/16\cdots\}$
c.	$\{1/8, 1/4, 1/2, 1, 1/2, 1/4, 1/8\}$
d.	$\{1/8, 1/4, 1/2, 1, 2, 4, 8\}$
e.	$\{1, -1/2, 1/4, -1/8, 1/16, -1/32\cdots\}$
f.	$\{2, 4, 8, 16, \ldots\}$
g.	$\{\ldots, 1/16, 1/8, 1/4, 1/2, 2, 1/2, 1/4, 1/8, 1/16, \ldots\}$

TABLE 11.12B Function

Number	Function $x(n)$
1	$2^{-|n|}$
2	$0.5^{-n}u(n)$
3	$2^n u(-n)$
4	$0.5^n[u(n+3) - u(n-4)]$
5	$2^n[u(n+3) - u(n-4)]$
6	$(-0.5)^n u(n)$
7	$2^{(n+1)}u(n)$
8	$2^{-n}u(n) + 2^n u(-n)$
9	$2^n u(n+3)$
10	None of the above

33. Evaluate and sketch the following discrete-time signals within the range $-6 \le n \le 6$.

a. $2^{-n}u(n-1)$ b. $(-0.5)^n u(n+1)$ c. $\cos(\pi n/2)u(n)$ d. $\cos(\pi n/4)u(n)$
e. $0.5^n \sin(\pi n/3)u(-n)$ f. $0.5^n \cos(\pi n/3)u(n)$ g. $2^n \cos(\pi n/3)u(-n)$ h. $2^{|n|}\cos(\pi n/3)$

34. Evaluate and sketch the following discrete-time sinc signals within the range $-6 \le n \le 6$.

a. $\dfrac{\sin n}{n}$ b. $\dfrac{\sin(n/6)}{n/6}$ c. $\dfrac{\sin(2n/5)}{n}$ d. $\dfrac{\sin(\pi n/2)}{\pi n/2}$ e. $\dfrac{\sin(2\pi n/5)}{2\pi n/5}$

35. Evaluate and sketch the following sinc signals within the range $-6 \le n \le 6$.

a. $\dfrac{\sin(n-3)}{(n-3)}$ b. $\displaystyle\sum_{k=-\infty}^{\infty} \dfrac{\sin(n-3k)}{(n-3k)}$ c. $\displaystyle\sum_{k=-\infty}^{\infty} \dfrac{\sin(n-4k)}{(n-4k)}$

d. $\dfrac{\sin[\pi(n-3)/2]}{(n-3)}$ e. $\displaystyle\sum_{k=-\infty}^{\infty} \dfrac{\sin[\pi(n-3k)/2]}{(n-3k)}$ f. $\displaystyle\sum_{k=-\infty}^{\infty} \dfrac{\sin[2\pi(n-4k)/5]}{(n-4k)}$

36. Evaluate and sketch the following signals within the range $-6 \le n \le 6$.

a. $\dfrac{\sin(\pi n)}{\sin\left(\frac{\pi n}{8}\right)}$ b. $\dfrac{\sin n}{\sin\left(\frac{\pi n}{8}\right)}$ c. $\dfrac{\sin n}{\sin\left(\frac{n}{8}\right)}$

37. For each of the four sinc functions given below find the locations of the samples at which (or closest to them) the zero-crossings occur. Obtain the number of nonzero samples within each of the 10 lobes of the function for $n > 0$, including the main lobe. Briefly summarize your observations and conclusion on the pattern of the alternating lobes.

a. $\dfrac{\sin n}{n}$ b. $\dfrac{\sin \pi n}{n}$ c. $\dfrac{\sin \pi n/3}{n}$ d. $\dfrac{\sin 2\pi n/5}{n}$

38. Consider the discrete-time signal $\cos(\omega_0 n)$, where ω_0 is a constant. Evaluate and sketch it for $0 \le n \le 12$. Assume

a. $\omega_0 = 1$
b. $\omega_0 = \pi/3$

39. Show that the discrete-time signal $\cos(\omega_0 n)$ is periodic only if $\omega_0 = 2\pi/N$, where N is an integer, in which case N is the period of the signal.

40. Show that the discrete-time signal $x(n) = e^{j2\pi kn/N}$, where k and N are integers, is periodic and find its fundamental period.

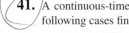

41. A continuous-time function $x(t)$ is sampled every 1 msec resulting in the discrete-time function $x(n)$. For the following cases find $x(n)$ and determine if it is periodic:

   a. $x(t) = 2^{-1000t} u(t)$
   b. $x(t) = e^{-100t} u(t)$
   c. $x(t) = \cos(5\pi t + \pi/4)$
   d. $x(t) = \sin(\sqrt{2}\pi t)$

42. The following continuous-time functions are sampled every 1 msec. Find their discrete-time counterparts $x_1(n)$ and $x_2(n)$. Compare results and explain the reason behind the possible similarities.

   a. $x_1(t) = \cos(200\pi t)$
   b. $x_2(t) = \cos(2{,}000\pi t/9)$

43. An analog periodic signal $x(t)$ with period P is sampled every T seconds. Find the conditions on P and T that produce a periodic $x(n)$.

Suggestion and Hint

Use the Fourier transform of the sampled signal in the continous-time domain (see Chapter 10). Sampling is done by multiplying $x(t)$ by a periodic sampling function $s(t)$ made of a train of unit impulses:

$$s(t) = \sum_{n=-\infty}^{\infty} \delta(t - nT)$$

The sampled function

$$y(t) = x(t)s(t) = \sum_{n=-\infty}^{\infty} x(nT)\delta(t - nT)$$

is a train of impulses each, with strength $x(nT)$ and spaced apart every T seconds. Examine $Y(f)$, the Fourier transform of $y(t)$, and test conditions for its periodicity.

44. a. Sample the continuous-time functions in Table 11.2 of this chapter at the rate of $F_s = 10$ Hz and compare the resulting discrete-time functions with those found for $F_s = 100$ (Example 11.13).
   b. Repeat for $F_s = 1$.

45. Consider the discrete-time signal $x(n) = \{1, 2, 3, 4, 5, 4, 3, 2, 1\}$.

   a. Sketch $x(n)$.
   b. Find and sketch $y(n) = x(n) - x(n - 2)$.

46. For the periodic pulse given below, find and sketch $x(n) - x(n - 1)$. Interpret your results.

$$x(n) = \begin{cases} 2, & 0 \leq n \leq 1 \\ -1, & 2 \leq n \leq 5 \end{cases} \quad \text{defined for one period and } x(n) = x(n+6)$$

47. Consider the sinusoidal signal $x(n) = \cos(2\pi n/N)$ and let $y(n) = x(n) + x(n-1) + x(n-2)$. Evaluate and sketch $x(n)$ and $y(n)$ for $0 \leq n \leq 6$. Assume

   a. $N = 2$
   b. $N = 6$

48. Consider the discrete-time periodic pulse $x(n)$ given in problem 47. Let $y(n) = x(n) + y(n-1)$ and $y(0) = 0$. Find and sketch $y(n)$. **Hint:** First note that

$$y(n) = \sum_{k=-\infty}^{n} x(k)$$

49. Given $x(n) = n[u(n) - u(n-4)]$, sketch and label the following functions, in which k is a positive integer.

a. $x(n \pm 3)$ b. $x(n \pm k)$ c. $x(-n \pm 3)$ d. $x(-n \pm k)$

50. Consider the discrete-time function $x(n) = u(n) - 2u(n-1) + d(n-3)$. Sketch and label

a. $x[-(n-2)]$ b. $x(-n-2)$ c. $x(n)d(n-1)$

51. Given $x(n) = \{1, 2, 3, 4\}$, find and sketch

a. $x(n \pm 1)$ b. $x(-n)$ c. $x(-n \pm 1)$ d. $x(-n \pm 2)$

52. Given $x(n) = 0.5^n u(n)$, find and sketch

a. $x(n \pm 1)$ b. $x(-n)$ c. $x(-n \pm 1)$ d. $x(-n \pm 2)$

53. Given $x(n) = 0.5^n[u(n) - u(n-4)]$, find and sketch

a. $x(n \pm 1)$ b. $x(-n)$ c. $x(-n \pm 1)$ d. $x(-n \pm 2)$

54. Given $x(n) = 0.5^n u(n)$, sketch and label the following functions

a. $x(n-3)$ b. $x(3-n)$ c. $x(n+3)$ d. $x(-n-3)$

55. Given $x(n) = 0.5^n u(n) + 3^n u(-n)$, sketch and label the following functions

a. $x(n-1)$ b. $x(1-n)$ c. $x(n+1)$ d. $x(-n-1)$

56. Given $x(n) = 0.5^{-n^2}$, find and sketch the following functions for $-4 \le n \le 4$.

a. $x(n-1)$ b. $x(-n)$ c. $x(-n+1)$ d. $x(-n-1)$

57. Given $x(n) = 0.5^{-|n|}$, find and sketch the result of the following two operations:

a. First delay $x(n)$ by 2 units, then time reverse.
b. Time reverse $x(n)$ first, then advance by 2 units.

58. Repeat problem 57 for the following functions:

a. $x(n) = 0.5^{-n} u(n) + 0.5^n u(-n)$
b. $x(n) = 0.5^{-n} u(n) - 0.5^n u(-n)$

59. Using the graphical method, demonstrate that the following two operations on a discrete-time function $x(n)$ are equivalent, both resulting in $x(-n+k)$.

a. First delay $x(n)$ by k units, then time reverse.
b. Time reverse $x(n)$ first, then advance by k units.

60. Show that $d(n) = d(-n)$. Therefore, $d(n-k) = d(k-n)$, where k is an integer constant.

61. Show that $d(kn) = d(n)$, where k is an integer constant.

62. Show that $d(kn + a) = d(n + a/k)$, where k and a are integer constants.

63. Show that $x(n)d(n - k) = x(k)d(n - k)$, where k is an integer constant.

64. Given $x(n) = u(n)$ and $h(n) = 0.5^n[u(n) - u(n - 3)]$,

   a. Find, sketch, and label the following set of functions $x(n)h(k - n)$, $k = -2, -1, 0, 1, 2, 3, 4, 5, 6$.
   b. Find, sketch, and label

$$y(k) = \sum_{n=-\infty}^{\infty} x(n)h(k - n), \quad k = -2, -1, 0, 1, 2, 3, 4, 5, 6$$

65. Let $p_k(n) = d(n)d(n - k)$, where k is an integer (shifting the unit sample to the left or right).

   a. Find and sketch $p_k(n)$ for $k = 0$.
   b. Repeat for $k \neq 0$.
   c. Using the collection of $p_k(n)$ sketches found in a and b, define a new function

$$\rho(k) = \sum_{n=-\infty}^{\infty} p_k(n), \quad -\infty < k < \infty$$

   with k as the independent variable. Evaluate and sketch $\rho(k)$ versus k.

66. Consider the single rectangular pulse $x(n) = \{\ldots, 0, 0, 1, 1, 1, 1, 1, 1, 0, 0, \ldots\}$.

   a. Define a new discrete function $p_k(n) = x(n)x(n - k)$, where n is the independent variable and k is an integer, $-\infty < k < \infty$. Find and sketch the collection of $p_k(n)$ for $-\infty < k < \infty$.
   b. Define a new function

$$\rho(k) = \sum_{n=-\infty}^{\infty} p_k(n)$$

   with k as the independent variable. Evaluate and sketch $\rho(k)$ versus k.

67. Repeat problem 66 for

   a. $x(n) = \{\ldots, 0, 0, 1, 1, 1, 1, 0, 0, \ldots\}$
   b. $x(n) = \{\ldots, 0, 0, 1, 1, 1, -1, -1, 1, -1, 0, 0, \ldots\}$

68. Repeat problem 66 for $x(n) = u(n)$ and show that $\rho(k) = \infty$ for all k.

69. Find the even and odd parts of $h(n) = 0.5^n u(n - 1)$ and compare with those of $h(n) = 0.5^n u(n)$ found in Example 11.33.

70. Let $x(n) = e^{j\omega_0 n}$, k be an integer and $A = \cos(\omega_0 k)$. Show that

$$\frac{x(n + k) + x(n - k)}{2} = Ax(n)$$

$$\frac{x(n) + x(n - 2k)}{2} = Ax(n - k)$$

71. Let

$$x(n) = \sum_{m=-\infty}^{\infty} X_m e^{jm\omega_0 n}$$

where the X_m coefficients are constant and m is an integer. Find $y(n) = \frac{x(n)+x(n-2k)}{2}$ and determine conditions for $y(n) = x(n-k)$.

72. Consider the discrete-time signal $x(n) = \cos(\pi n/8)$. Let

$$y(n) = 0.25x(n) + 0.225[x(n-1) + x(n+1)] + 0.159[x(n-2) + x(n+2)] + 0.075[x(n-3) + x(n+3)]$$

Find a closed form expression for $y(n)$ in terms of $x(n)$. Evaluate and sketch $y(n)$.

73. Repeat problem 72 for (a) $x(n) = \cos(\pi n/4)$, (b) $x(n) = \cos(3\pi n/4)$.

11.20 Project: An Introduction to Discrete-Time Signal Processing

Introduction and Summary

In section 11.18 a simple digital filter was used to notch out a sinusoidal disturbance, while attempting to keep the signal of interest. The notching operation changed the amplitude of the signal (and blocked all frequencies that were harmonics of the disturbance). In theory, if, as in Examples 11.36 and 11.37, the signals of interest were pure sinusoids at known frequencies, the frequency-dependent amplitude change could be compensated for by a simple amplification. In practice, however, signals of interest are not made of simple sinusoids. Signal structure changes continually and unpredictably with time.[3] As a result, the frequency-dependent attenuation produces distortion in the signal and makes $y(n)$ an unfaithful representation of $s(n)$. To reduce the distortion in $s(n)$ while blocking the interference, we may employ a sharper notch filter with a narrow bandwidth.

In this project you will explore filtering operations that separate signals from noise or distrurbance. In the continuous-time domain, the data is given by $x(t) = s(t) + w(t)$. $s(t)$ is the signal and $w(t)$ is the noise or disturbance. After sampling, it becomes $x(n) = s(n) + w(n)$. The project includes two procedures. In the first procedure, $s(t)$ contains frequencies from 0 to f_c and $w(t)$ is a sinusoid (to be called the disturbance) with known frequency $0 < f_0 < f_c$. The aim is to block (notch out) the disturbance by a digital notch filter with little distortion of the signal. Ideally, the filter's output should be $Ks(n-k)$, where K and k are the constant gain and delay, respectively. The second procedure is concerned with extracting a single sinusoid from the data, using a narrowband digital filter tuned to the frequency of the sinusoid. For example, the signal $s(t)$ is a single sinusoid embedded in a background noise $w(t)$ and it is desired to extract the signal from noise. Another example, which falls within the second procedure, is to separate a certain frequency component of a signal or noise.

Procedure 1. Notching the Disturbance

A continuous-time signal $s(t)$ is recorded together with a 60-Hz sinusoidal interference, resulting in $x(t) = s(t) + \cos(120\pi t)$ (to be called the data). The bandwidth of $s(t)$ is 0 to 600 Hz. It is desired to notch out the sinusoidal interference from the data without doing excessive damage to $s(t)$.

---

[3]Signals, in general, are stochastic and unpredictable.

A Simple Notch Filter

Sample $x(t)$ at the rate of 1,200 samples per second, twice the bandwidth of $s(t)$, and obtain a discrete-time data sequence $x(n) = s(n) + \cos(\alpha n)$, where $\alpha = \pi/10$. To cancel $\cos(\alpha n)$, we first apply a simple shift-and-add notch filter as in Example 11.36.

$$\frac{x(n) + x(n - 2k)}{2} = \frac{s(n) + s(n - 2k)}{2} + \frac{\cos(\alpha n) + \cos(\alpha n - 2\alpha k)}{2}$$

$$= \frac{s(n) + s(n - 2k)}{2} + \cos(\alpha k)\cos(\alpha n - \alpha k)$$

$$= \frac{s(n) + s(n - 2k)}{2} + A\cos\alpha(n - k)$$

Let $A = \cos(\alpha k)$. By choosing $k = 5$ we will have $\alpha k = \pi/2$, which results in $A = 0$, thus blocking (or notching out) the 60-Hz interference. The above filter is not satisfactory because its gain, which affects the frequency components of $s(n)$, depends on frequency. The result is that even though the filter cancels the disturbance, its output $\frac{s(n)+s(n-2k)}{2}$ is not a reproduction of the signal but rather a distortion of it.

Ideal Low-Pass Filter

To reduce signal distortion we use a filter with a narrow notch. For this purpose we introduce the ideal low-pass filter and use it within the rest of the project. An ideal low-pass filter with a cutoff angular frequency ω_0 has a constant gain at frequencies below the cutoff frequency and a zero gain at frequencies above it. The filter is specified by the discrete-time function $h(n)$ (called the unit-sample response) given below.

$$h(n) = \frac{\sin(\omega_0 n)}{\pi n}, \quad -\infty < n < \infty$$

The filter performs the following operation on a signal $x(n)$ and produces an output $y(n)$:

$$y(n) = \sum_{k=\infty}^{\infty} x(k)h(n - k)$$

In practice, a finite segment of $h(n)$, seen through a window, may be used. It is shifted to the right to make it causal and realizable.

An Improved Notch Filter

Consider a discrete-time signal $x(n) = s(n) + \cos(7\pi n/16)$, where $s(n)$ is the signal of interest and $\cos(7\pi n/16)$ is the disturbance. The goal is to block or attenuate the disturbance by filtering. For that purpose, use a notch filter with the unit-sample response

$$h(n) = [h_1(n) + d(n) - h_2(n)]\, w(n)$$

where h_1 and h_2 are unit sample responses of ideal low-pass filters (see part a) with cutoff frequencies at $\omega_1 = \pi/4$ and $\omega_2 = 5\pi/8$, respectively. $w(n)$ is the M-tap window specified by

$$w(n) = \begin{cases} 0.54 + 0.46\cos(\frac{2\pi n}{M}), & -N \leq n \leq N, \quad M = 2N + 1 \\ 0, & \text{elsewhere} \end{cases}$$

The output of the filter is the sum of the weighted samples $x(n-k)$, $-N \leq k \leq N$, according to the following:

$$y(n) = \sum_{k=-N}^{N} h(k)x(n-k)$$

$$= h_0 x(n)$$
$$+ h_1[x(n+1) + x(n-1)]$$
$$+ h_2[x(n+2) + x(n-2)]$$
$$+ h_3[x(n+3) + x(n-3)]$$
$$+ h_4[x(n+4) + x(n-4)]$$
$$+ h_5[x(n+5) + x(n-5)]$$
$$\cdots$$
$$+ h_N[x(n+N) + x(n-N)]$$

1. Show that the filter is capable of blocking the sinusoidal disturbances at $\omega = 7\pi/16$.
2. Examine the effect of the filter on the sinusoidal disturbance. For this purpose let $x(n) = \cos(7\pi n/16)$ (i.e., no signal). Apply the filter and record the output. Devise a measure to evaluate the peformance of the filter in achieving the goal. Explore the effect of filter length M on the output. Start with $M = 11, 17$, and 101. Compare with the simple notch filter of part a.
3. Examine the effect of the filter on the signal. For this purpose let $x(n) = \cos(\omega n)$ (i.e., no disturbance). Show that $y(n) = A\cos(\omega n + \theta)$. Find and sketch A and θ as functions of $0 < \omega < \pi$. Compare with the plot for the notch filter of part a. Conclude, by appropriate reasoning, that the filter discussed in this part is improved in that it procduces less distortion in the signal compared with the simple notch filter of part a.

Suggestion

Use a computer. To evaluate the output you may use the Matlab command *conv*, which stands for convolution.

Procedure 2. Extracting the Principal Harmonic of a Signal

Let $x(n)$ be the periodic pulse signal with period $N = 5$.

$$x(n) = \{1, 1, 1, 0, 0\} \text{ and } x(n) = x(n+5)$$

The goal is to extract its principal harmonic by filtering. Use a bandpass filter with the unit-sample response

$$h(n) = [h_2(n) - h_1(n)] w(n)$$

where h_1 and h_2 are unit-sample responses of ideal low-pass filters with cutoff frequencies at $\omega_1 = 3\pi/10$ and $\omega_2 = 5\pi/10$, respectively. $w(n)$ is the M-tap window of procedure 1. The output of the filter is the sum of weighted samples $x(n-k)$, $-N \leq k \leq N$, as in Procedure 1.

Show that the filter is capable of blocking all frequencies except the sinusoidal at $\omega = 2\pi/5$. Devise a measure to evaluate the peformance of the filter in achieving the goal. Explore the effect of filter length M on the output. Start with $M = 11, 17$, and 101. The mathematical explanation for the bandpass filtering property of the above $h(n)$ will be covered in Chapter 16. The following Matlab file may be used:

```
% Procedure 1.
clear; N=8; M=2*N+1; n=-N:N; hamming=0.54+.46*cos(2*pi*n/M);
omega1=pi/4;;   h1=sin(omega1*n)./(pi*n);     h1(N+1)=omega1/pi;
omega2=5*pi/8; h2=-sin(omega2*n)./(pi*n);    h2(N+1)=1-omega2/pi;
h=(h2+h1).*hamming;
```

```
q=5;  m=-q*N:q*N;  x=cos(7*pi*m/16);  y=conv(x,h);
%
% Procedure 2.
omega1=3*pi/10;;   h1=sin(omega1*n)./(pi*n);      h1(N+1)=omega1/pi;
omega2=5*pi/10;    h2=sin(omega2*n)./(pi*n);      h2(N+1)=omega2/pi;
h=(h2-h1).*hamming;
p=[1 1 1 0 0 ];  x=[p p p p p p p];  y=conv(x,h);
```

Chapter 12

Linear Time-Invariant Discrete-Time Systems

Contents

Introduction and Summary

From a theoretical perspective, a discrete-time system may be considered to be a special case of the continuous-time system which samples the continuous-time input, obtains the output, and then keeps it until the next input sample. In such an approach, time would still play the role of the independent variable and the analysis tools (such as the transforms) would be derived from those in the continuous-time domain. Such an approach, however, can easily be bypassed. Discrete-time systems may be analyzed, synthesized, and designed without falling back on a background from the continuous-time domain. The necessary analysis tools (such as the transform, system function, frequency response, filtering operations, signal processing, and feedback) may be developed on their own. While from the operational point of view some degree of separation between the two domains may be useful in reducing possible confusions in applying the tools and methods, a certain degree of connection does seem to enhance the intuitive understanding of the basic concepts. The discussion in Chapter 3 introduced systems in general (and LTI systems in particular) using such a connected approach. This chapter presents the definitions and elements of discrete-time systems in stand-alone format in order to better serve the operational aspect and analysis methods that are developed in future chapters.

After summarizing the linearity and time-invariance properties, the chapter introduces the unit-sample response as an important tool for finding the output of the system to any input, in the form of convolution sum. The convolution sum is discussed in detail in the next chapter. Other system specifications such as responses to unit-step inputs, power signals, sinusoids, and the description by difference equations are then briefly introduced. Discrete LTI operators and system classifications as FIR and IIR, recursive and nonrecursive, are also introduced. The overview is intended to provide an introduction at this early stage of the discussion. These topics will be encountered again and in more detail in later chapters.

12.1 Linear Time-Invariant (LTI) Discrete-Time Systems

A discrete-time system is made of an input space, an output space, and a mapping rule. Input, output, and internal states are functions of a discrete variable that assumes only integer values. The system changes its output at a regular interval only and not continuously. Many discrete systems are models of continuous-time systems that switch their state at the time of a clock pulse. For such systems, the term *discrete time* seems appropriate. However, this term is also applied to physical phenomena whose variables are not time but other parameters such as space. Also, some physical phenomena are, by their very nature, discrete. Nevertheless, in this book we use the terms *discrete* and *discrete time* interchangeably.

The linearity and time-invariance properties of LTI systems have already been discussed in detail in Chapter 3. Here we summarize them for the discrete-time domain.

Linearity

Let two arbitrary inputs x_1 and x_2 produce y_1 and y_2, respectively. If the input $x = ax_1 + bx_2$ produces $y = ay_1 + by_2$ for all constants a and b, the system is called linear.

$$x_1(n) \quad\quad \Longrightarrow \quad y_1(n)$$
$$x_2(n) \quad\quad \Longrightarrow \quad y_2(n)$$
$$ax_1(n) + bx_2(n) \quad \Longrightarrow \quad ay_1(n) + by_2(n)$$

Time Invariance

Let an arbitrary input $x(n)$ produce the output $y(n)$. If the input $x(n-k)$ produces the output $y(n-k)$ for all k, the system is called time invariant.

$$x(n) \implies y(n)$$
$$x(n-k) \implies y(n-k)$$

Linear Time-Invariant (LTI) Discrete-Time Systems

A system that is both linear and time invariant is called LTI and satisfies the following property for all x_1, x_2, a, b, k.

$$x_1(n) \implies y_1(n)$$
$$x_2(n) \implies y_2(n)$$
$$ax_1(n-k) + bx_2(n-k) \implies ay_1(n-k) + by_2(n-k)$$

Delay operator

The delay operator is defined by the equation $y(n) = x(n-N)$, where N is an integer constant. The operation shifts the signal N units to the right if $N > 0$. For $N < 0$ the signal is shifted to the left by N units and is said to be advanced. The operation is linear and time invariant as verified by the following.

$$x_1(n) \implies y_1(n) = x_1(n-N)$$
$$x_2(n) \implies y_2(n) = x_2(n-N)$$
$$ax_1(n-k) + bx_2(n-k) \implies ax_1(n-N-k) + bx_2(n-N-k)$$
$$= ay_1(n-k) + by_2(n-k),$$
$$\text{for all } x_1, x_2, a, b, \text{ and } k.$$

A difference operator

The backward-difference operator is defined by the equation

$$y(n) = x(n) - x(n-1)$$

Verification of its linearity and time invariance is similar to that in Example 12.1. The backward-difference operator is the counterpart of differentiation operation in the continuous-time domain. As an example, application of the backward-difference operator on the unit-step function results in the unit-sample function.

$$d(n) = u(n) - u(n-1)$$

As another example the backward difference operating on a unit-ramp function produces a delayed unit-step function,

$$u(n-1) = nu(n) - (n-1)u(n-1)$$

The backward-difference operator is also one approximation to differentiation operation in the continuous-time domain as shown in section 12.5 of this chapter, along with several other discrete-time LTI operators and their block diagram representations.

A moving average

An example of a discrete-time system is taking the average of samples of a discrete-time signal within a window of fixed size, assigning the average as the output, shifting the window, and repeating the operation. One type of moving average may be described as the following input-output relationship:

$$y(n) = \frac{1}{N} \sum_{k=M}^{M+N-1} x(n-k)$$

in which $x(n)$ is the input, $y(n)$ is the output, M is the location of trailing edge (beginning) of the window, and N is its width. For example, the weekly average price of a certain item computed from

$$y(n) = \frac{1}{7} \sum_{k=0}^{6} x(n-k)$$

gives the average price for today and the past six days. In the above system samples are given equal weight. They influence the averaging result process uniformly. The averaging window is called a *uniform window* and the operation may be labeled a *simple moving average*. In another type of averaging, a weighting function may give samples different weights. For example, the most recent samples may be considered more relevant and given an exponential weighting function. The system is then described by the following relationship:

$$y(n) = \frac{1}{N} \sum_{k=M}^{M+N-1} x(n-k)e^{-(k-M)}$$

The above formulation, which still represents an LTI system, is also called convolution of the input signal with system's unit-sample response.

A credit account

Pat has established a credit account with a bank that requires five monthly payments of 0.25 dollars each for every dollar of purchase. Statements are issued at the end of each month and the payments for charges during that month begin in the following month. Let $x(n)$ be the amount charged to the acount in the nth month and $y(n)$ be the amount due in that month. Express $y(n)$ in terms of $x(n)$.

Solution

$y(n) = 0.25\,[x(n-1) + x(n-2) + x(n-3) + x(n-4) + x(n-5)]$. For a numerical illustration of this system see Example 4.5 in Chapter 4.

xample

12.5

Testing a system for time invariance

A unit sample arriving at $n = k$ to a linear system evokes the response $0.5^n u(n - k)$. Is the system time invariant?

Solution

The system is not time invariant, as shown by the following counterexample. Let $h_0(n)$ be the response to $d(n)$ and $h_1(n)$ be the response to $d(n - 1)$.

$$d(n) \quad \Longrightarrow \quad h_0(n) = 0.5^n u(n)$$

$$d(n - 1) \quad \Longrightarrow \quad h_1(n) = 0.5^n u(n - 1) = \tfrac{1}{2}(0.5)^{n-1} u(n - 1)$$

Note that $h_1(n) \neq h_0(n - 1)$. A one-unit shift in the input has produced a one-unit shift in the output, plus a gain factor of $\frac{1}{2}$.

Alternative Solution to Example 12.5

Using the array display notation we have:

$$d(n) \quad \Longrightarrow \quad \{0, \; \underset{\uparrow}{1}, \; 0.5, \; 0.25, \; 0.125, \; \cdots\}$$

$$d(n - 1) \quad \Longrightarrow \quad \{0, \; \underset{\uparrow}{0}, \; 0.5, \; 0.25, \; 0.125, \; \cdots\}$$

It is observed that a unit shift in the input does not produce a unit shift in the system's response. The system is, therefore, not time invariant.

xample

12.6

Testing a system for time invariance

The response of a linear system to a unit sample arriving at $n = k$ is $h(n, k) = 0.5^{(n-k)} u(n - k)$. Is the system time invariant?

Solution

The system is time invariant. This may be deduced directly from the observation $h(n, k) = h(n - k)$. To see how, let $k_2 = k_1 + N$ and note that

$$d(n - k_1) \quad \quad \quad \quad \Longrightarrow \quad h(n - k_1)$$

$$d(n - k_2) = d(n - k_1 - N) \quad \Longrightarrow \quad h(n - k_2) = h(n - k_1 - N)$$

A shift of N units in the input has produced a shift of N units in the output, and nothing else.

12.2 The Unit-Sample Response

The unit-sample response is a most powerful description tool for linear systems. It completely specifies a linear system, whether time invariant or not. It is briefly considered here, through two examples for both time-variant and time-invariant linear systems. The unit-sample response of LTI systems will be encountered throughout the rest of the book.

The Unit-Sample Response of Linear Systems, h(n, k)

A unit-sample function occurring at time k is shown by $d(n - k)$. Let such a function be the input to a linear system. Let the system's output at time n to $d(n - k)$ be called $h(n, k)$. Since any arbitrary input may be expressed as a weighted sum of unit samples, the resulting output may be computed from the weighted sum of $h(n, k)$. It is, therefore, clear that if one knows the unit-sample responses of a linear system, one can compute its output to any input.

Example **12.7**

A unit sample arriving at $n = k$ to a linear system evokes the response $0.5^n u(n - k)$.

a. Find its response $y(n)$ to the input $x(n) = d(n) + 4d(n - 1) - 2d(n - 2)$.

b. Given the above input find a closed-form mathematical expression for $y(n)$ for $n \geq 2$ and evaluate it at $n = 2$.

Solution

Using linearity property we find

a. $y(n) = 0.5^n u(n) + 4 \times 0.5^n u(n - 1) - 2 \times 0.5^n u(n - 2)$

$$= 0.5^n \left[u(n) + 4u(n - 1) - 2u(n - 2) \right]$$

$$= d(n) + 2.5d(n - 1) + 3(0.5)^n u(n - 2)$$

b. $y(n) = 0.5^n (1 + 4 - 2) = 3(0.5)^n, \quad n \geq 2$

$$y(2) = 3 \times 0.5^2 = 0.75$$

Alternative Solution to Example 12.7

Using the array display notation we have

$d(n)$	$\Longrightarrow$	$\{0, \; 1, \; 0.5, \; 0.25, \; 0.125, \; \cdots\}$
$4d(n - 1)$	$\Longrightarrow$	$4 \times \{0, \; 0, \; 0.5, \; 0.25, \; 0.125, \; \cdots\}$
$-2d(n - 2)$	$\Longrightarrow$	$-2 \times \{0, \; 0, \; 0, \; 0.25, \; 0.125, \; \cdots\}$
$d(n) + 4d(n - 1) - 2d(n - 2) \; \Longrightarrow$		$\{0, \; 1, \; 2.5, \; 0.75, \; 0.375, \; \cdots\}$

The last array showing the total response may be written as $d(n) + 2.5d(n - 1) + 3(0.5)^n u(n - 2)$.

The Unit-Sample Response of LTI Systems, h(n)

In the case of LTI systems $h(n, k)$ depends on $(n - k)$. The system can then be specified completely by $h(n)$, its response to $d(n)$. Based on linearity and time-invariance properties the output of an LTI system may be computed from the weighted sum of shifted

$h(n)$. Time invariance greatly facilitates this task. In this book we mainly deal with LTI systems, representing the unit-sample response by $h(n)$.

Example 12.8

Consider an LTI system having the unit-sample response $h(n) = \{\underset{\uparrow}{1}, -1\}$. Find $y(n)$ for:

a. $x(n) = \{\underset{\uparrow}{1}, 1\}$

b. $x(n) = \{\underset{\uparrow}{1}, -1\}$

c. $x(n) = \{\underset{\uparrow}{1}, -1, -1\}$.

Solution

a. $x(n) = d(n) + d(n-1)$

$y(n) = h(n) + h(n-1) = \{\underset{\uparrow}{1}, -1\} + \{\underset{\uparrow}{0}, 1, -1\} = \{\underset{\uparrow}{1}, 0, -1\}$

b. $x(n) = d(n) - d(n-1)$

$y(n) = h(n) - h(n-1) = \{\underset{\uparrow}{1}, -1\} - \{\underset{\uparrow}{0}, 1, -1\} = \{\underset{\uparrow}{1}, -2, 1\}$

c. $x(n) = d(n) - d(n-1) - d(n-2)$

$y(n) = h(n) - h(n-1) - h(n-2) = \{\underset{\uparrow}{1}, -1\} - \{\underset{\uparrow}{0}, 1, -1\} - \{\underset{\uparrow}{0}, 0, 1, -1\}$

$= \{\underset{\uparrow}{1}, -2, 0, 1\}$

Example 12.9

The response of an LTI system to a unit sample arriving at $n = k$ is $h(n, k) = 0.5^{(n-k)}u(n-k)$.

a. Find its response $y(n)$ to the input $x(n) = d(n) + 4d(n-1) - 2d(n-2)$.

b. Given the above input, for $n \geq 2$ express $y(n)$ by a closed-form mathematical formula, evaluate it at $n = 2$, and compare with the result obtained in Example 12.7.

Solution

a. $y(n) = 0.5^n u(n) + 4 \times 0.5^{n-1} u(n-1) - 2 \times 0.5^{n-2} u(n-2)$

$= 0.5^n [u(n) + 8u(n-1) - 8u(n-2)]$

$= d(n) + 4.5d(n-1) + (0.5)^n u(n-2)$

b. $y(n) = 0.5^n, \quad n \geq 2$

$y(2) = 0.25$

Alternative Solution to Example 12.9

As in Example 12.7, we use the array display notation to obtain:

$d(n)$	$\Longrightarrow$	$\{0,\ \underset{\uparrow}{1},\ 0.5,\ 0.25,\ 0.125,\ \cdots,\ \cdots,\ \cdots\}$
$4d(n-1)$	$\Longrightarrow$	$4 \times \{0,\ \underset{\uparrow}{0},\ 1,\ 0.5,\ 0.25,\ 0.125,\ \cdots,\ \cdots\}$
$-2d(n-2)$	$\Longrightarrow$	$-2 \times \{0,\ \underset{\uparrow}{0},\ 0,\ 1,\ 0.5,\ 0.25,\ 0.125,\cdots\}$

$d(n)+4d(n-1)-2d(n-2)$	$\Longrightarrow$	$\{0,\ \underset{\uparrow}{1},\ 4.5,\ 0.25,\ 0.125,\ \cdots,\ \cdots,\ \cdots\}$

Note that $y(2)$ is smaller than its counterpart in Example 12.7, as one may expect from the unit-sample responses.

Obtaining the Unit-Sample Response of an LTI System from a Given Input-Output Pair

We have noted that an LTI system is completely specified by its unit-sample response, $h(n)$.[1] Computation of the output from $h(n)$ and $x(n)$ is called convolution and will be discussed in Chapter 13. Other methods include input-output difference equation and representation of system structure by block diagram.

Equivalently, knowing a single input-output pair of an LTI system enables us to predict the output to any input.[2] This may be called *reverse engineering* as shown in the following example.

Example **12.10**

By using reverse calculation find the unit-sample response of LTI systems for which input-output pairs are given as

a.　An FIR system:　$x(n) = \{\underset{\uparrow}{1}, -1\} \implies y(n) = \{\underset{\uparrow}{1}, 0, -1\}$

b.　An IIR system:　$x(n) = \{\underset{\uparrow}{1}, -1\} \implies y(n) = d(n)$

---

a.
$$\begin{aligned}
y_0 &= h_0 x_0 & h_0 \times 1 &= 1 & &\longrightarrow & h_0 &= 1 \\
y_1 &= h_0 x_1 + h_1 x_0 & -1 + h_1 &= 0 & &\longrightarrow & h_1 &= 1 \\
y_2 &= h_0 x_2 + h_1 x_1 + h_2 x_0 & 0 - 1 + h_2 &= -1 & &\longrightarrow & h_2 &= 0 \\
y_3 &= h_0 x_3 + h_1 x_2 + h_2 x_1 + h_3 x_0 & 0+0+0+h_3 &= 0 & &\longrightarrow & h_3 &= 0
\end{aligned}$$
$\cdots$ $\qquad\qquad\qquad\qquad\qquad\qquad\qquad\qquad\qquad\qquad\qquad$ $\ldots\ldots$

the unit-sample response is $\qquad\qquad\qquad\qquad\qquad\qquad\qquad$ $h(n) = \{\underset{\uparrow}{1}, 1\}$

---

b.
$$\begin{aligned}
y_0 &= h_0 x_0 & h_0 \times 1 &= 1 & &\longrightarrow & h_0 &= 1 \\
y_1 &= h_0 x_1 + h_1 x_0 & -1 + h_1 &= 0 & &\longrightarrow & h_1 &= 1 \\
y_2 &= h_0 x_2 + h_1 x_1 + h_2 x_0 & 0 - 1 + h_2 &= 0 & &\longrightarrow & h_2 &= 1 \\
y_3 &= h_0 x_3 + h_1 x_2 + h_2 x_1 + h_3 x_0 & 0+0-1+h_3 &= 0 & &\longrightarrow & h_3 &= 1
\end{aligned}$$
$\cdots$ $\qquad\qquad\qquad\qquad\qquad\qquad\qquad\qquad\qquad\qquad\qquad$ $\ldots\ldots$

the unit-sample response is $\qquad\qquad\qquad\qquad\qquad$ $h(n) = \{\underset{\uparrow}{1}, 1, 1\cdots\} = u(n)$

---

[1] There are exceptions, but they are of no interest to our present discussion.

[2] Except when the input is a power series.

Note that the output of an IIR system may be a finite sequence as shown in case *b* of the present example.

12.3 Response of LTI Discrete-Time Systems to Power Signals and Sinusoids

Using linearity and time invariance we can show that the input $x(n) = z^n$, where z is a constant, evokes the output $y(n) = Hz^n$, where the scale factor H is a function of z.[3] To show this, and to find the scale factor $H(z)$, start with the unit-sample response $h(n)$ (the first entry in Table 12.1) and go through steps 2 to 8.

TABLE 12.1 Using LTI Properties to Derive System's Response to Power Signals

Step	Operation	Property	Input	$\Longrightarrow$	Output
1	Unit-impulse response		$d(n)$	$\Longrightarrow$	$h(n)$
2	Shift k units	Time invariance	$d(n - k)$	$\Longrightarrow$	$h(n - k)$
3	Multiplication by z^k	Proportionality	$z^k d(n - k)$	$\Longrightarrow$	$z^k h(n - k)$
4	Add over k	Superposition	$\displaystyle\sum_{k=-\infty}^{\infty} z^k d(n - k)$	$\Longrightarrow$	$\displaystyle\sum_{k=-\infty}^{\infty} z^k h(n - k)$
5	Let $k = n - m$	Change of variable	$\displaystyle\sum_{m=-\infty}^{\infty} z^{(n-m)} d(m)$	$\Longrightarrow$	$\displaystyle\sum_{m=-\infty}^{\infty} z^{(n-m)} h(m)$
6	Factor out z^k		$\displaystyle z^n \sum_{m=-\infty}^{\infty} z^{-m} d(m)$	$\Longrightarrow$	$\displaystyle z^n \sum_{m=-\infty}^{\infty} z^{-m} h(m)$
7	Sifting by $d(m)$		z^n	$\Longrightarrow$	$\displaystyle z^n \sum_{m=-\infty}^{\infty} z^{-m} h(m)$
8	$\displaystyle H(z) = \sum_{m=-\infty}^{\infty} z^{-m} h(m)$ Define $H(z)$		z^n	$\Longrightarrow$	$H(z) z^n$

Note that the above result is derived directly from LTI system properties and not from the z-transform. The latter is discussed in Chapter 15, where it is shown that $H(z)$ not only specifies the system's output to z^n, but to any other input as well.

Example
12.11

The unit-sample response of an LTI system is $h(n) = \{1, 2, 1\}$. Find its response to $x(n) = 0.5^n$.

---

[3] This is true for almost all LTI systems of interest in engineering and signal processing. Exceptions are those systems that have no unit-sample response.

Solution

From Table 12.1 the output is

$$y(n) = 0.5^n \sum_{m=0}^{2} 0.5^{-m} h(m) = 0.5^n \left[1 + 2 \times 0.5^{-1} + 0.5^{-2} \right] = 9 \times 0.5^n$$

The response is also obtained by superposition, using the linearity and time-invariance property of the system.

$$y(n) = x(n) + 2x(n-1) + x(n-2) = 0.5^n + 2 \times 0.5^{n-1} + 0.5^{n-2} = 9 \times 0.5^n$$

Response to Sinusoids

Let $z = e^{j\omega}$. The response to $\cos \omega n$ is found from the following:

Input	$\Longrightarrow$	Output
z^n	$\Longrightarrow$	$H(z)z^n$
$e^{j\omega n}$	$\Longrightarrow$	$H(\omega)e^{j\omega n}$
$\cos(\omega n) = \mathcal{RE}\left[e^{j\omega n} \right]$	$\Longrightarrow$	$\mathcal{RE}\left[H(\omega)e^{j\omega n} \right] = |H(\omega)| \cos(\omega n + \theta)$

In the above

$$H(\omega) = H(z)\Big|_{z=e^{j\omega}} = |H(\omega)|\angle\theta$$

The response to a sinusoid is a sinusoid with the same frequency but possibly different magnitude and phase. It should be noted that the scale factors $H(z)$ and $H(\omega)$ are two different functions. It is for convenience and simplicity that the same notation H is used to represent the scale factor when we move from z to ω.

Example
12.12

The backward-difference operator was introduced in Example 12.2. Find its response to the input $\cos(\omega n)$.

Solution

The backward-difference operator and its sinusoidal response are given below.

Input	$\Longrightarrow$	Output
$x(n)$	$\Longrightarrow$	$x(n) - x(n-1)$
$e^{j\omega n}$	$\Longrightarrow$	$e^{j\omega n} - e^{j\omega(n-1)} = H(\omega)e^{j\omega n}$, where $H(\omega) = (1 - e^{-j\omega}) = 2\sin\left(\frac{\omega}{2}\right) e^{j\left(\frac{\pi}{2} - \frac{\omega}{2}\right)}$
$\cos(\omega n) = \mathcal{RE}\left[e^{j\omega n} \right]$	$\Longrightarrow$	$\mathcal{RE}\left[H(\omega)e^{j\omega n} \right] = |H(\omega)| \cos(\omega n + \theta) = -2\sin\left(\frac{\omega}{2}\right) \sin\left(n\omega - \frac{\omega}{2}\right)$

Note that for small ω the response is $\approx -\omega \sin \omega n$, which is the derivative of the input with respect to n. The above is an analysis of the backward-difference operator in the frequency domain. A time-domain analysis, producing the same result, will be given in section 12.5.

12.4 Some Properties and Classifications of Discrete-Time LTI Systems

Two properties of an LTI system, which are of primary interest, are its causality and stability. These are briefly introduced below.

Causality

In a causal system the output at any time depends on the past and present values of the input only. The future values of the input to a causal system do not affect its output. For LTI systems this condition translates to $h(n) = 0$, $t < 0$, which is a necessary and sufficient condition for causality. The causality defined by the above criteria is a technical definition only and should not be considered synonomous with physical reality. Only when the independent variable n represents samples of real time, is any physical system a causal system and vice versa. (Because of this connection a causal system is then also called physically realizable.) A system that is not causal may still be physically realizable. Examples are systems in which the independent variable represents factors other than time (e.g., physical space or a location in a data sequence).

Stability

Stability implies that the value of the output doesn't grow to infinity if the input is less than a fixed finite valued and less. A system is called BIBO-stable (standing for *Bounded-Input Bounded-Output*) if every bounded input results in a bounded output. The BIBO stability condition for a discrete-time system is that the unit-sample response be absolutely summable.

$$\sum_{n=-\infty}^{\infty} |h(n)| < B, \quad \text{where } B \text{ is any large number with a fixed finite value.}$$

Input-Output Integration Property

Integration of the input of an LTI system leads to integration of its output. For discrete-time systems integration becomes summation

$$x(n) \implies y(n)$$

$$\sum_{-\infty}^{n} x(k) \implies \sum_{-\infty}^{n} y(k)$$

As an example, the unit-step response $g(n)$ may be obtained from the unit-sample $h(n)$ response by

$$d(n) \implies h(n)$$

$$u(n) = \sum_{-\infty}^{n} d(k) \implies g(n) = \sum_{-\infty}^{n} h(k)$$

The above property often simplifies analysis and solution of LTI systems.

FIR and IIR Systems

An LTI system whose unit-sample response is a sequence with finite length is called a Finite Impulse Response (FIR) system. In contrast, a system whose unit-sample response is a sequence with infinite length is called an Infinite Impulse Response (IIR) system. The distinction between these two types is seen throughout the analysis, design, and implementation of discrete-time systems, especially in digital signal processing.

Example **12.13**

The first three LTI systems given below are FIR and the last three are IIR systems. The systems represented by two-sided $h(n)$ are not causal.

$$\text{FIR:} \quad h(n) = \begin{cases} u(n) - u(n-2) = \{\underset{\uparrow}{1}, 1, 1\} \\[2mm] (-0.5)^n \, [u(n) - u(n-4)] = \{\underset{\uparrow}{1}, -1/2, 1/4, -1/8\} \\[2mm] (0.5)^{|n|} \, [u(n+3) - u(n-4)] = \{1/8, 1/4, 1/2, \underset{\uparrow}{1}, 1/2, 1/4, 1/8\} \end{cases}$$

$$\text{IIR:} \quad h(n) = \begin{cases} 2u(n) - u(n-2) = \{\underset{\uparrow}{2}, 2, 1, 1, 1, 1, \cdots\} \\[2mm] (-0.5)^n u(n) = \{\underset{\uparrow}{1}, -1/2, 1/4, -1/8, 1/16, -1/32, \cdots\} \\[2mm] (0.5)^{|n|} = \{\cdots, 1/32, 1/16, 1/8, 1/4, 1/2, \underset{\uparrow}{1}, 1/2, \\[1mm] \qquad\qquad 1/4, 1/8, 1/16, 1/32, \cdots\} \end{cases}$$

Recursive and Nonrecursive Systems

Let a system be described by the difference equation

$$y(n) = \sum_{k=0}^{M} b_k x(n-k) - \sum_{k=1}^{N} a_k y(n-k)$$

where dependence of $y(n)$ on its past values is expressed explicity through the terms $a_k y(n-k)$. The system and the equation describing it, are called recursive. If coefficients a_k, $k = 1, 2, \ldots$ are all zero then

$$y(n) = \sum_{k=0}^{M} b_k x(n-k)$$

and the system is called nonrecursive. The distinction between these two types is also reflected in the block diagram representation described in section 12.6.

12.5 Discrete LTI Operators and Difference Equations

Operators are elementary systems that transform one discrete signal to another. The following operators are some of the basic building blocks of discrete systems.

1. The unit delay operator $D[\cdot]$ shifts the signal one unit to the right. It is also shown by the symbol z^{-1}.

$$\text{Unit delay operator: } D[x(n)] = x(n-1)$$

2. The unit advance operator $A[\cdot]$ shifts the signal one unit to the left. It is also shown by the symbol z.

$$\text{Unit advance operator: } A[x(n)] = x(n+1)$$

3. The gain operator $G[\cdot]$ multiplies all elements of the discrete signal by a factor G (positive or negative).

$$\text{Gain: } G[x(n)] = G \cdot x(n)$$

4. The sum operator adds two discrete signals term by term.

The graphical symbols for the unit delay, unit advance, gain, and sum operator are shown in Figure 12.1.

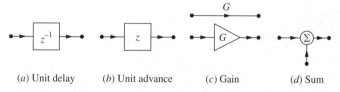

(*a*) Unit delay   (*b*) Unit advance   (*c*) Gain   (*d*) Sum

FIGURE 12.1 Symbols for the *(a)* unit delay , *(b)* unit advance, *(c)* gain, and *(d)* sum operator.

5. The difference operators described below evaluate increase (or decrease) in the discrete-time signals.

The forward-difference operator $\Delta[\cdot]$ evaluates the increase in $x(n+1)$ over $x(n)$.

$$\Delta x(n) = x(n+1) - x(n)$$

The backward-difference operator $\nabla[\cdot]$ evaluates the increase in $x(n)$ over $x(n-1)$.

$$\nabla x(n) = x(n) - x(n-1)$$

Difference operators may be applied to a signal repeatedly. The second-order forward-difference operator Δ^2 is

$$\Delta^2 x(n) = \Delta[\Delta x(n)] = \Delta[x(n+1) - x(n)] = \Delta x(n+1) - \Delta x(n)$$

$$= [x(n+2) - x(n+1)] - [x(n+1) - x(n)]$$

$$= x(n+2) - 2x(n+1) + x(n)$$

The second-order backward-difference operator ∇^2 is

$$\nabla^2 x(n) = \nabla[\nabla x(n)] = \nabla[x(n) - x(n-1)] = \nabla x(n) - \nabla x(n-1)$$

$$= [x(n) - x(n-1)] - [x(n-1) - x(n-2)]$$

$$= x(n) - 2x(n-1) + x(n-2)$$

As seen, these operators are constructed from a combination of delay and advance operators. Their graphical representation in the form of block diagrams, discussed in section 12.6, are shown in Figure 12.2.

Show that the forward- and backward-difference operators may be expressed in terms of advance or delay operators in the following way:

Solution

$$\Delta = A - 1, \quad \text{and} \quad \nabla = 1 - D$$

$$\Delta x(n) = x(n+1) - x(n) = A[x(n)] - x(n) = [A-1]x(n)$$

$$\nabla x(n) = x(n) - x(n-1) = x(n) - D[x(n)] = [1-D]x(n)$$

Find the response of the backward-difference operator ∇ to $x(n) = \cos(\omega n)$.

Solution

$$\nabla x(n) = \cos(\omega n) - \cos[\omega(n-1)] = (1 - \cos \omega)\cos(\omega n) - \sin \omega \sin(\omega n)$$

$$= 2\sin\left(\frac{\omega}{2}\right)\left[\sin\left(\frac{\omega}{2}\right)\cos(\omega n) - \cos\left(\frac{\omega}{2}\right)\sin(\omega n)\right]$$

$$= -2\sin\left(\frac{\omega}{2}\right)\sin\left(\omega n - \frac{\omega}{2}\right). \quad \text{See also Example 12.12.}$$

LTI Systems and Difference Equations

The linear constant-coefficients difference equation, expressed in any of the following three forms,

$$y(n) + a_1 y(n-1) \cdots + a_N y(n-N) = b_0 x(n) + b_1 x(n-1) \cdots + b_M x(n-M)$$

$$y(n) = \sum_{k=0}^{M} b_k x(n-k) - \sum_{k=1}^{N} a_k y(n-k)$$

$$\sum_{k=0}^{N} a_k y(n-k) = \sum_{k=0}^{M} b_k x(n-k)$$

describes a linear time-invariant discrete-time system with $x(n)$ being the input, $y(n)$ the output, and a_k and b_k constants (with $a_0 = 1$). Most LTI discrete-time systems are represented by linear difference equations with constant coefficients such as the above. For brevity the equation may be written as

$$D^N[y(n)] = D^M[x(n)]$$

$$D^N[y(n)] = y(n) + a_1 y(n-1) \cdots + a_N y(n-N)$$

$$D^M[x(n)] = b_0 x(n) + b_1 x(n-1) \cdots + b_M x(n-M)$$

D^N and D^M are linear difference operators made of shifts and gains operating on $y(n)$ and $x(n)$, respectively. Using the symbol z^{-1} to represent unit delay, we represent these operators by

$$D^N = 1 + a_1 z^{-1} + \cdots + a_N z^{-N}$$

$$D^M = b_0 + b_1 z^{-1} + \cdots + b_M z^{-M}$$

The difference equation is then written as

$$[1 + a_1 z^{-1} + \cdots + a_N z^{-N}]y(n) = [b_0 + b_1 z^{-1} + \cdots + b_M z^{-M}]x(n)$$

In the above equations, N terms of y and M terms of x are included in the N-th order equation. Contributions from $x(n)$ may be lumped together as $q(n)$, with the quotation written as

$$y(n) + a_1 y(n-1) \cdots + a_N y(n-N) = q(n)$$

$$q(n) = b_0 x(n) + b_1 x(n-1) \cdots + b_M x(n-M)$$

You can easily see that the difference equation is made of the basic discrete operators described previously. Solution methods are provided in Chapter 14.

12.6 Block Diagram Representation

LTI discrete systems whose input-output relationships are given by difference equations may be represented by an interconnection of unit delays, gains, and adders, called block diagram. Each connection node of the diagram is associated with an internal variable. The gains are written on the paths that connect the nodes. A unit delay is shown by a block of z^{-1}. The input and output are indicated on the diagram. Conventionally, the input enters from the left side and the output is picked up on the right side of the block diagram. Block diagrams for difference operators are presented in Figure 12.2. Flow of data in nonrecursive diagrams is in the forward direction only, from left to right as illustrated by Example 12.18 (Figure 12.3a). Block diagram of recursive systems contain both feedforward and feedback paths as illustrated by Example 12.19 (Figure 12.3b).

xample
12.16

Figure 12.2(a) to 12.2(d) show the block diagrams for the following operators:

a. Forward-difference operator $\Delta x(n) = x(n+1) - x(n)$
b. Backward-difference operator $\nabla x(n) = x(n) - x(n-1)$
c. Second-order forward-difference operator $\Delta^2 x(n) = x(n+2) - 2x(n+1) + x(n)$
d. Second-order backward-difference operator $\nabla^2 x(n) = x(n) - 2x(n-1) + x(n-2)$

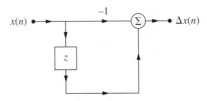   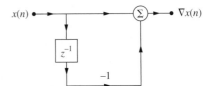

(a) The forward-difference operator
$\Delta x(n) = x(n+1) - x(n)$.

(b) The backward-difference operator
$\nabla x(n) = x(n) - x(n-1)$.

FIGURE 12.2 *(a)*, *(b)*. The forward- *(a)* and backward- *(b)* difference operators.

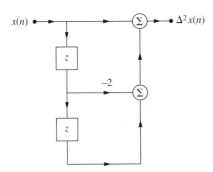   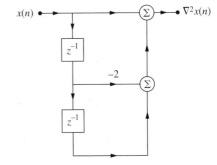

(c) The second-order forward-difference operator
$\Delta^2 x(n) = x(n+2) - (n+1) + x(n)$.

(d) The second-order backward-difference operator
$\nabla^2 x(n) = x(n) - 2x(n-1) + x(n-2)$.

FIGURE 12.2 *(c)*, *(d)*. The second-order forward- *(c)* and backward- *(d)* difference operators.

xample
12.17

A discrete-time system is described by the nonrecursive equation $y(n) = kx(n)$. The block diagram of such a system was shown in Figure 12.1(c). The system provides amplification (or attenuation) in $x(n)$. Note that an error in k proportionally causes the same amount of error in the output. For example 1% error in k produces 1% error in y.

Example 12.18

Figure 12.3(a) shows the block diagram for the nonrecursive difference equation

$$y(n) = \frac{x(n) + x(n-1)}{2}$$

The output at each moment is the average of the current and previous inputs. The input reaches the output through two paths, both in the forward direction. This system provides a smoothing operation on $x(n)$.

Example 12.19

An LTI system is described by the recursive difference equation

$$y(n) = k[x(n) - \beta y(n-1)]$$

In this system the output sample is scaled by a factor β and feedback to the input. The output is then formed by amplifying the difference signal $x(n) - \beta y(n-1)$. The system is shown by the block diagram of Figure 12.3(b) in which the input affects the output through two paths, one direct and one through feedback, making the system recursive. Establishing the feedback path reduces sensitivity of the output to variations in k. Compare with the sensitivity in the amplifier of Example 12.17.

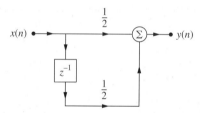

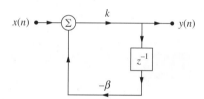

(a) Example 12.18. The LTI system $y(n) = \dfrac{x(n) + x(n-1)}{2}$ shown by the block diagram is nonrecursive. The input reaches the output through two paths both in the forward direction. The value of the output at each moment is the average of the current input and its previous value. This system provides smoothing on $x(n)$.

(b) Example 12.19. Block diagram of the LTI system $y(n) = k\{x(n) - \beta y(n-1)\}$ contains a direct path and one with feedback making the system recursive. Establishing the feedback path reduces sensitivity of the input-output relationship to variations in k.

FIGURE 12.3 Block diagram representations of **(a)** a nonrecursive and **(b)** a recursive LTI system.

Example 12.20

Figure 12.4 shows block diagrams made of unit delays, gains, and adders of the following systems.

a. $y(n) = b_0 x(n) + b_1 x(n-1)$

b. $y(n) + a_1 y(n-1) = b_0 x(n)$

c. $y(n) + a_1 y(n-1) = b_0 x(n) + b_1 x(n-1)$

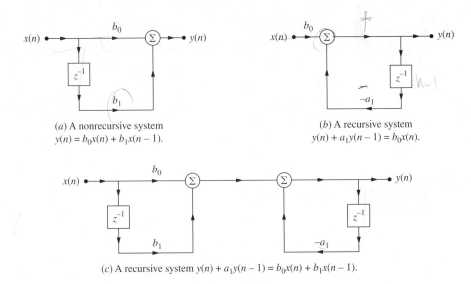

(a) A nonrecursive system
$y(n) = b_0 x(n) + b_1 x(n - 1)$.

(b) A recursive system
$y(n) + a_1 y(n - 1) = b_0 x(n)$.

(c) A recursive system $y(n) + a_1 y(n - 1) = b_0 x(n) + b_1 x(n - 1)$.

FIGURE 12.4 Block diagram representations of three LTI systems.

Block diagram representation is not unique. A system may be represented by one of several different block diagrams some of which use fewer elements. See Example 12.21 below. However, a block diagram specifies the system uniquely.

Example **12.21**

Verify that the input-output relationship of the system shown in Figure 12.5 is

$$y(n) + a_1 y(n - 1) = b_0 x(n) + b_1 x(n - 1) \quad (1)$$

Therefore, the two block diagrams of Figure 12.4(c) and Figure 12.5 represent the same system.

Solution

Let $v(n)$ designate the internal variable at the output of the first adder on the diagram of Figure 12.5. From the feedback loop on the left side of the block diagram we get:

$$v(n) = x(n) - a_1 v(n - 1)$$

or

$$x(n) = v(n) + a_1 v(n - 1) \quad (2)$$

From the feedforward loop on the right side of the block diagram we get:

$$y(n) = b_0 v(n) + b_1 v(n - 1) \quad (3)$$

Substituting $x(n)$ and $y(n)$ from (2) and (3) in the two sides of the input-output equation (1) we get

$$b_0 x(n) + b_1 x(n - 1) = b_0[v(n) + a_1 v(n - 1)] + b_1[v(n - 1) + a_1 v(n - 2)]$$
$$= b_0 v(n) + (a_1 b_0 + b_1) v(n - 1) + a_1 b_1 v(n - 2) \quad (4)$$

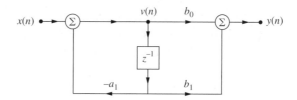

FIGURE 12.5 The input-output relationship of the system shown by the block diagram is $y(n) + a_1 y(n-1) = b_0 x(n) + b_1 x(n-1)$. This is the same difference equation given in Example 12.20(c). Therefore, the two block diagrams of Figure 12.4(c) and Figure 12.5 represent the same system.

$$y(n) + a_1 y(n-1) = [b_0 v(n) + b_1 v(n-1)] + a_1[b_0 v(n-1) + b_1 v(n-2)]$$

$$= b_0 v(n) + (a_1 b_0 + b_1)v(n-1) + a_1 b_1 v(n-2) \qquad (5)$$

The two sides are equal and the input-output equation is verified.

Show that the following two input-output equations are equivalent.

$$y(n) = x(n) + x(n-1)$$

$$y(n) - y(n-1) = x(n) - x(n-2)$$

Substituting for $y(n)$ from the first equation into the second equation we verify that

$$y(n) - y(n-1) = [x(n) + x(n-1)] - [x(n-1) + x(n-2)] = x(n) - x(n-2)$$

12.7 Analysis and Solution Methods

One goal of systems analysis is to develop and apply methods for computing and predicting its response to inputs of interest. In the previous sections we have seen examples of discrete-time systems, mostly described by their input-output relationship

$$f[y(n), y(n-1), \ldots, x(n), x(n-1), \ldots] = 0$$

Obviously, from the above equation we may evaluate $y(n)$ from past values of y and x by numerical methods. The numerical method may be applied regardless of the system being linear or time invariant. Numerical methods may seem attractive, especially if computers are being used, but rarely provide a broader insight. For LTI systems several analyical methods are used to solve for the response, some of which were illustrated through examples in this chapter. These methods are listed below.

1. Time-domain solution by *convolution*. This method is based on direct application of linearity property and uses the unit-sample response. The unit-sample response

of LTI systems will be encountered throughout the rest of the book. Discrete convolution will be discussed in Chapter 13.

2. Time-domain solution by *difference equation*. Difference equations and their solution methods are discussed in Chapter 14.

3. Frequency-domain solution by the *z-transform*.

4. Frequency-domain solution by the *Fourier transform*.

Description of signals and systems in the frequency domain will be discussed in Chapters 15–17 while, as a capstone, Chapter 18 provides an integration of multiple methods.

12.8 Problems

Solved Problems

1. The response of a system to a unit sample arriving at time $n = k$ is $d(n - k)$ for all integers k.

$$d(n - k) \implies d(n - k), \quad -\infty < k < \infty$$

It is, however, not known how the system would respond to an input made of more than one unit sample. Can it be concluded that its response to a unit step is also a unit step?

Solution
The answer is negative. Only if the system is linear will its unit-step response be a unit step.

2. The response of a system to a sample of size x arriving at time $n = k$ is known to be $xh(n - k)$ for all k and x, where k is an integer, $-\infty < k < \infty$, and x is a number that can assume any value so that $-\infty < x < \infty$.

$$xd(n - k) \implies xh(n - k)$$

Based on the above, can it be deduced that (1) the system is time invariant? (2) the system is linear?

Solution
The answers to both questions are negative. The shift property is valid in the case of an impulse input. This does not mean that it holds true for all inputs. Similarly, the proportionality property doesn't imply the superposition property in the case of two inputs.

3. In problem 2 assume a linear system. Can it be deduced that the system is time invariant?

Solution
To test for time invariance, we shift an input $x(n)$ and examine the output $y(n)$. We first express the input as a sum of weighted samples.

$$x(n) = \sum_{m=-\infty}^{\infty} x_m d(n - m) \qquad \implies \quad y(n) = \sum_{m=-\infty}^{\infty} x_m h(n - m)$$

$$x_2(n) = x(n - k) = \sum_{m=-\infty}^{\infty} x_m d(n - m - k) \implies y_2(n) = \sum_{m=-\infty}^{\infty} x_m h(n - m - k) = y(n - k) \text{ all } k$$

The system is time invariant because a shift of k units in the input produces a shift of the same amount in the output and nothing else.

4. In an LTI system, given $h(n) = \{1, -1, 1\}$, find the output for: a) $x(n) = \{1, 1\}$ and b) $x(n) = \{1, -1\}$.

Solution

a. $x(n) = d(n) + d(n-1)$

   $y(n) = h(n) + h(n-1) = \{1, -1, 1\} + \{0, 1, -1, 1\} = \{1, 0, 0, 1\}$

b. $x(n) = d(n) - d(n-1)$

   $y(n) = h(n) - h(n-1) = \{1, -1, 1\} - \{0, 1, -1, 1\} = \{1, -2, 2, -1\}$

5. The unit-step response of an LTI system is $(2 - 2^{-n})\, u(n)$. Note that it takes a very long time (theoretically, $n = \infty$) for the response to reach its steady-state value 2. It is desired to reduce the transition time to one sample. For this purpose we first convert the unit step to the following input:

$$x(n) = \begin{cases} a, & n = 0 \\ 1, & n \geq 1 \\ 0, & n < 0 \end{cases}$$

and then apply it to the system. Using the linearity and time-invariance properties, determine the parameter a such that $y(n) = 2$ for $n \geq 1$.

Solution

Designate the unit-step response by $g(n)u(n)$, where $g(n) = 2 - 2^{-n}$. The unit-sample response is $h(n) = g(n)u(n) - g(n-1)u(n-1)$.

The input is $\qquad\qquad\qquad\qquad x(n) = ad(n) + u(n-1)$.

The output, given the above input, is $\qquad y(n) = ah(n) + g(n-1)u(n-1)$

$$= ag(n)u(n) - ag(n-1)u(n-1) + g(n-1)u(n-1)$$

$$= \begin{cases} 0, & n = 0 \\ ag(n) + (1-a)g(n-1), & n \geq 1 \end{cases}$$

Substitute $g(n) = 2 - 2^{-n}$ into the above to find $y(n) = \begin{cases} 0, & n = 0 \\ 2 + (a-2)2^{-n}, & n \geq 1 \end{cases}$

For $n \geq 1$, we want $y(n) = 2$, which requires $a = 2$. The first sample $ad(n)$ sets up the appropriate initial conditions for the elimination of the transient part of the response.

6. Find the input-output relationship of the discrete-time system given in Figure 12.6.

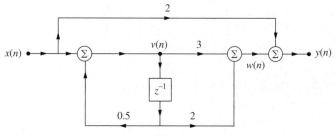

FIGURE 12.6 The input-output relationship of the system shown is $y(n) - 0.5y(n-1) = 5x(n) + x(n-1)$.

First Approach

Let $w(n)$ designate the internal variable at the output of the second adder. Then,

from the block diagram, $y(n) = 2x(n) + w(n)$ (6)

from Example 12.21, $w(n) = 0.5w(n-1) + 3x(n) + 2x(n-1)$ (7)

To find the relationship between $x(n)$ and $y(n)$, we need to eliminate $w(n)$ and $w(n-1)$.

from (6) and using time invariance, $w(n-1) = y(n-1) - 2x(n-1)$ (8)

from (7) in combination with (6), $w(n-1) = 2w(n) - 6x(n) - 4x(n-1)$

$$= 2y(n) - 10x(n) - 4x(n-1)$$ (9)

By setting (8) = (9), we get

$$y(n) - 0.5y(n-1) = 5x(n) + x(n-1)$$

Second Approach

Let $v(n)$ designate the internal variable between the first and second adders on the left side of Figure 12.6. Then,

from the forward loop on the left, $v(n) = x(n) + 0.5v(n-1)$ (10)

from the feedback loop on the right, $y(n) = 3v(n) + 2x(n) + 2v(n-1)$ (11)

To eliminate $v(n)$ and $v(n-1)$ from (10) and (11) we apply a method similar to the first approach and obtain the same result:

$$y(n) - 0.5y(n-1) = 5x(n) + x(n-1)$$

The validity of the above relationship between x and y may be verified in a way similar to that used in Example 12.21.

Third Approach. Employing the z-Operator

The z operator introduced in sections 12.5 and 12.6 may be used as a tool for a systematic (and more convenient) way of obtaining the input-output relationship. The formal approach is called the z-transform and is discussed in Chapter 15. Here we briefly demonstrate the use of z^{-1} in solving the present problem. We show the unit delay by the operator z^{-1} and apply it to equations (6) and (7), transforming them into (6a) and (7a), respectively.

from the block diagram, $Y = 2X + W$ (6a)

from the problem statement, $W = 0.5z^{-1}W + 3X + 2z^{-1}X$ (7a)

By eliminating W from (6a) and (7a) and translating z^{-1} back to a unit delay in the n-domain we have

$$(1 - 0.5z^{-1})Y = (5 + z^{-1})X$$
$$y(n) - 0.5y(n-1) = 5x(n) + x(n-1)$$

In the second approach, applying the z^{-1} operator transforms equations (10) and (11) into (10a) and (11a), respectively.

from the forward loop on the left, $V = X + 0.5z^{-1}V$ (10a)

from the feedback loop on the right, $Y = 3V + 2X + 2z^{-1}V$ (11a)

By eliminating V from (10a) and (11a) and translating z^{-1} back to a unit delay in the n-domain we get the same result. In summary,

input-output relationship in the z-domain: $(1 - 0.5z^{-1})Y = (5 + z^{-1})X$ (12)

input-output relationship in the n-domain: $y(n) - 0.5y(n-1) = 5x(n) + x(n-1)$ (13)

This concludes three approaches to the solution of this problem.

Chapter Problems

7. Determine if the systems described by the following input-output relationships are linear or time invariant. If LTI, find their unit-sample responses.

a. $y(n) = x(n-2)$ b. $y(n) = x(n) + x(-n)$

c. $y(n) = x(n) + u(n)$ d. $y(n) = x(n) + x(-n+1)$

e. $y(n) = x(n) + 2x(n-1)$ f. $y(n) = x(n) - x(n-1)$

g. $y(n) = x(n)x(n-1) + x(n-2)$ h. $y(n) = 2x(n) - nx(n+1)$

i. $y(n) = 2x(n) + n$ j. $y(n) = |x(n)|$

k. $y(n) = ny(n-1) + 2x(n)$ l. $y(n) = y(n-1) + nx(n)$

m. $y(n) = y(n-1) + 2x(n)$ n. $y(n) = x(n) + x(n-1)y(n-1)$

8. Repeat problem 7 for the systems described by the following input-output relationships.

a. $y(n) = \sum_{k=-1}^{100}(k-1)x(n-k)$ b. $y(n) = 2^n\sum_{k=-2}^{3}x(n+k)$

c. $y(n) = (n-1)\sum_{k=0}^{5}x(n-k)$ d. $y(n) = \sum_{k=3}^{5}0.9^k x(n+k)$

e. $y(n) = \sum_{k=-\infty}^{0}x(k)$ f. $y(n) = \sum_{k=-\infty}^{n}x(k)$

g. $y(n) = \sum_{k=0}^{n}x(k)$ h. $y(n) = \sum_{k=n}^{n+4}x(k)$

i. $y(n) = \sum_{k=n-4}^{n+4}x(k)$ j. $y(n) = \sum_{k=n-2}^{n}x(k)$

k. $y(n) = \sum_{k=1}^{n+4}x(k)$ l. $y(n) = \sum_{k=-\infty}^{\infty}x(k)$

9. Let $x(n)$ be the input and $y(n)$ the output of discrete-time systems described by the following difference equations. Determine if the systems are linear or time invariant. If linear, find their unit-sample responses.

a. $y(n) = y(n-1) + 2x(n)$ b. $y(n) = y(n-1) + x(n) + u(n)$

c. $y(n) = x(-n)$ d. $y(n) = x(n)y(n) - nx(n-1)$

e. $y(n) = x(n)d(n)$ f. $y(n) = 3^n x(n) + x(n-1)$

g. $y(n) = x(n-1)d(n)$ h. $y(n) = -3^n y(n) + x(n)$

i. $y(n) = y(n-1) + x(-n)$ j. $y(n) = -y(n-1) + 3^n x(n) + x(n-1)$

k. $y(n) = nx(n)$ l. $y(n) = 3^n y(n-1) + 3^n x(n) + x(n-1)$

m. $y(n) = y(n-1) + nx(n)$ n. $y(n) = -(0.99)^n y(n) + x(n)$

10. Find the input-output difference equation and the unit-step response of an LTI system whose unit-sample response is:

a. $h(n) = \{3, 2, 1\}$
         ↑

b. $h(n) = \{1, 1, 1\}$
       ↑

11. The input-output difference equation of a causal LTI system and its input are given below.

$$y(n) = 1.3y(n-1) - 0.4y(n-2) + x(n) - 2x(n-1)$$

$$x(n) = \{1, 5, -1, 3, 1, 2, -1, 3, 5, 5, 4, -5, \cdots\}$$
         ↑

Find the first five nonzero samples of $y(n)$ by recursive numerical method. Note that $y(n) = 0$, $n < 0$.

12. A causal LTI system is specified by the difference equation

$$y(n) = 3x(n) - 2x(n-1) + x(n-2)$$

a. Sketch the system's block diagram using gains and unit delays.
b. Find $h(n)$.
c. For $x(n) = \{1, 3, -1, 5, -2, 3, 1, -4\}$ find $y(n)$, $n \leq 10$.
       ↑

13. The unit-sample response of an LTI system is $h(n) = (0.5)^n u(n)$. Find its response to:

a. $x(n) = 2[d(n) - d(n-11)]$
b. $x(n) = 2[u(n) - u(n-11)]$

14. A causal system is represented by the difference equation $y(n) + 0.3y(n-1) - 0.4y(n-2) = x(n)$.

a. Find the first five terms in $h(n)$.
b. Given $x(n) = 9u(n)$, find the first five terms in $y(n)$.

15. A causal system is represented by the difference equation $y(n) - 0.2y(n-1) = x(n)$, where $x(n) = \cos(\omega n)$, $-\infty < n < \infty$. Show that $y(n) = H\cos(\omega n + \theta)$ and find H and θ for:

a. $\omega = \pi$
b. $\omega = \pi/2$

16. The input to an LTI system with a unit-sample response $h(n) = 3d(n) + 3d(n-1)$ is $x(n) = \{1, 1, 1, 1\}$. Sketch
                                                                  ↑
and label completely the system output $y(n)$.

17. Consider the following discrete-time signal:

$$x(n) = \begin{cases} 0 & n < 0 \\ 2 & 0 \leq n \leq 4 \\ 1 & n > 4 \end{cases}$$

a. Express $x(n)$ as an algebraic sum of weighted discrete-time step functions $u(n)$.
b. $x(n)$ is the input to an LTI system with unit-sample response $h(n) = u(n)$. Find its output.

18. Find the first five terms in $y(n)$, the output of a causal system given by $y(n) - 0.5y(n-1) = d(n)$.

19. Find the first five terms in $y(n)$, the output of a causal system given by $y(n) - 0.5y(n-1) = u(n)$.

20. The unit-step response of an LTI system is $g(n) = (2 - 0.5^n)\, u(n)$. Find and sketch its response to $x(n) = d(n) + d(n - 1)$.

21. Find the first five terms in $y(n)$, the output of a causal system given by the following difference equations and compare results.

   a. $y(n) + 2y(n - 1) = u(n)$
   b. $y(n) - 2y(n - 1) = u(n)$
   c. $y(n) + 0.5y(n - 1) = u(n)$
   d. $y(n) - 0.5y(n - 1) = u(n)$

22. Find the first five terms in $y(n)$, the output of a causal system given by

   a. $y(n) + 2y(n - 1) = u(n) - u(n - 5)$
   b. $y(n) - 2y(n - 1) = u(n) - u(n - 5)$
   c. $y(n) + 0.5y(n - 1) = u(n) - u(n - 5)$
   d. $y(n) - 0.5y(n - 1) = u(n) - u(n - 5)$

23. For each LTI system specified by the difference equations given below determine if the system is FIR or IIR. Then identify its block diagram from Figure 12.7, and place appropriate gains on the paths. If the block diagram of a system is not found in Figure 12.7, draw it.

   (i) $y(n) = ax(n) + bx(n - 1)$
   (ii) $y(n) = ax(n) + bx(n - 1) + cx(n - 2)$
   (iii) $y(n) + cy(n - 1) = ax(n) + bx(n - 1)$
   (iv) $y(n) + cy(n - 1) = ax(n) + x(n - 2)$
   (v) $y(n) + cy(n - 1) = ax(n) + bx(n - 1) + dx(n - 2)$
   (vi) $y(n) + cy(n - 1) = ax(n) + bx(n - 1) + x(n - 2)$
   (vii) $y(n) + cy(n - 1) + y(n - 2) = ax(n) + bx(n - 1)$
   (viii) $y(n) + cy(n - 2) = ax(n) + bx(n - 1) + cx(n - 2)$
   (ix) $y(n) + cy(n - 1) + y(n - 2) = ax(n) + bx(n - 1)$
   (x) $y(n) - cy(n - 1) - dy(n - 2) = ax(n) + bx(n - 1)$
   (xi) $y(n) + cy(n - 1) + dy(n - 3) = ax(n - 1) + bx(n - 2)$
   (xii) $y(n) + cy(n - 1) + dy(n - 3) = ax(n) + bx(n - 2)$

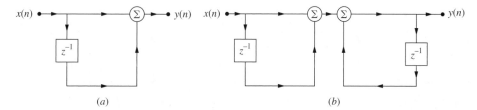

FIGURE 12.7 Block diagrams to be used in problem 23.

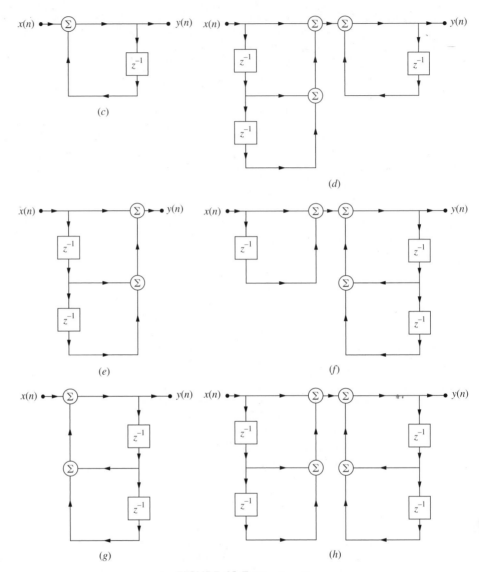

FIGURE 12.7 (*Continued*)

12.9 Project: Deadbeat Control

Introduction and Summary

The discrete-time system introduced at the beginning of this project doesn't go to rest (i.e., the internal variables of the system and its output don't become zero) even when the input has ceased to exist for a long time. Procedure 1 explores the performance of the system and shows that the system is marginally stable (i.e., the output blows up for a certain group of inputs even when the input is bounded). To make the system stable and have it return to a zero state in a finite number of steps, a feedback path is suggested. Procedure 2 explores the effect of the feedback on performance of the system. It shows that the system becomes a FIR system. (FIR systems are inherently stable.) At the end, the project ponders the usability of the approach in making a broader class of systems stable.

Procedure 1

A single-input, single-output discrete-time system is described by the block diagram of Figure 12.8(a), where $x(n)$ is the input, $y(n)$ is the output, and $x_i(n)$, $i = 1, 2, 3$, are the internal so-called state variables of the system which represent its state. The system's output is a linear combination of its state variables, $y(n) = ax_1(n) + bx_2(n) + cx_2(n)$. All connections within the block diagram have a unity gain. In this procedure you compute the output for several inputs and initial conditions. For the purpose of numerical calculations, let $a = b = c = 1$. (It is recommended that you write a computer program for doing the work.)

a. Obtain $y(n)$ for the following inputs: (i) $x(n) = d(n)$

(ii) $x(n) = u(n) - u(n - k)$, $k = 2, 5, 10, 20$

b. Repeat for sinusoidal inputs: (iii) $x(n) = \cos(n\pi)u(n)$

(iv) $x(n) = \cos\left(\frac{n\pi}{2}\right)u(n)$

(v) $x(n) = \cos\left(\frac{2\pi n}{3}\right)u(n)$

c. Repeat for step and ramp inputs: (vi) $x(n) = u(n)$

(vii) $x(n) = nu(n)$

d. Repeat for zero input: (viii) $x(n) = 0$, $n \geq 0$, $x_1(0) \neq 0$, $x_2(0) \neq 0$, $x_3(0) \neq 0$

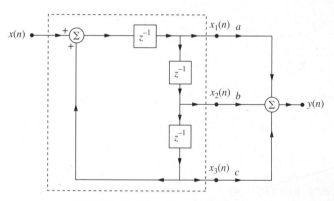

FIGURE 12.8(a) The original system.

Analysis and Conclusion for Procedure 1

Document and summarize your observations. Is the system stable? Is the system FIR or IIR? Under a zero-input condition, how long does it take to go to a zero state (part d above)?

Procedure 2

In order to reduce the transition time in going to a zero state, it is suggested that we establish a negative feedback path with unity gain from $x_3(n)$ to the input, as seen in Figure 12.8(b). Show that given such feedback, the system will be reduced to the FIR block diagram of Figure 12.8(c). Then, repeat procedure 1. Show that the system would require only three steps to go to a zero state.

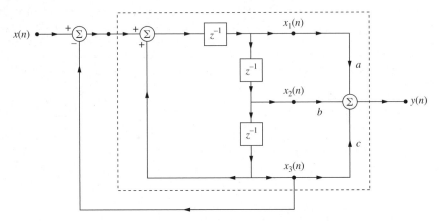

FIGURE 12.8(b) The system with negative feedback.

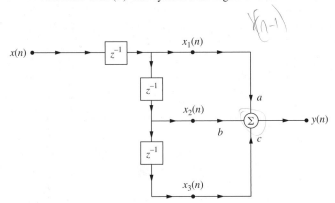

FIGURE 12.8(c) An equivalent system for Figure 12.8(b).

Analysis and Conclusion

Is the feedback system stable? Under a zero-input condition, how long does it take to go to a zero state (input d in procedure 1)? The feedback approach in this project converts the IIR system of procedure 1 to an FIR system, which is inherently stable. Discuss how the present approach (that of establishing a feedback path) could provide a general approach to a broader class of unstable systems.[4]

---

[4]For further discussion, see the article by H. Seraji, "Deadbeat Control of Discrete-Time Systems Using Output Feedback," *International Journal of Control*, 21, No. 2 (1975), pp. 213–23.

Addendum

The following Matlab code generates input-output plots used in procedure 1.

```
clear; k=100; n=1:k; x=0*n;
for i=5:5;
    x(i)=1;
  end
y=0*n;
for i=4:k;
    y(i)= y(i-3)+x(i-3);
end
figure(1); stem(n,x); axis([0 30 -1 2]); grid;
figure(2); stem(n,y); axis([0 30 -1 2]); grid;
%
for i=5:15;
    x(i)=1;
end
y=0*n;
for i=4:k;
    y(i)= y(i-3)+x(i-3);
end
figure(3); stem(n,x); axis([0 30 -1 2]); grid;
figure(4); stem(n,y); axis([0 30 -1 5]); grid;
%
x=cos(pi*n);
for i=1:5;
    x(i)=0;
  end
y=0*n;
for i=4:k;
    y(i)= y(i-3)+x(i-3);
end
figure(5); stem(n,x); axis([0 30 -2 2]); grid;
figure(6); stem(n,y); axis([0 30 -2 2]); grid;
%
x=cos(pi*n/2);
for i=1:5;
    x(i)=0;
  end
y=0*n;
for i=4:k;
    y(i)= y(i-3)+x(i-3);
end
figure(7); stem(n,x); axis([0 30 -2 2]); grid;
figure(8); stem(n,y); axis([0 30 -2 2]); grid;
%
x=cos(2*pi*n/3);
for i=1:5;
```

```
   x(i)=0;
 end
y=0*n;
for i=4:k;
   y(i)= y(i-3)+x(i-3);
end
figure(9);  stem(n,x); axis([0 30 -1 2]);  grid;
figure(10); stem(n,y); axis([0 30 -6 10]); grid;
%
x=0*n;
for i=6:k;
   x(i)=1;
 end
y=0*n;
for i=4:k;
   y(i)= y(i-3)+x(i-3);
end
figure(11); stem(n,x); axis([0 30 -1 2]);  grid;
figure(12); stem(n,y); axis([0 30 -1 10]); grid;
%
x=n-5;
for i=1:5;
   x(i)=0;
 end
y=0*n;
for i=4:k;
   y(i)= y(i-3)+x(i-3);
end
figure(13); stem(n,x); axis([0 30 -5 30]);  grid;
figure(14); stem(n,y); axis([0 30 -10 100]); grid;
```

Chapter 13

13

Discrete Convolution

Contents

Introduction and Summary

Convolution is a mathematical implementation of the superposition property. Given the unit-sample response of a linear system, convolution enables us to obtain the response to an arbitrary input. An arbitrary input can be expressed as a sum of weighted and delayed unit samples. If a system is linear and time invariant, the response can also be expressed by the sum of the unit-sample responses, weighted and delayed accordingly. The summation operation that expresses the output is called convolution. Evaluation of the convolution is done either in the time domain or the transform domain. This chapter introduces time-domain methods. Transform methods will be discussed in later chapters on the z-transform and Fourier transform.

The chapter begins with the definition of discrete convolution in LTI systems, and then shows how the response to an arbitrary input is found by convolving it with the system's unit-sample response. Computation of the convolution sum is illustrated by the numerical method (when the input signal is a time series) or the analytic method (when the input signal is specified in analytic form). The chapter then presents the distributive, associative, and commutative properties of convolution. The convolution of two finite-length sequences is evaluated as the set of coefficients in the product of two polynomials, providing a bridge to the transform domain.

An example of convolution for a linear time-varying system is also included in the chapter. The chapter project illustrates, by way of an example, recovery of a signal through the concepts of deconvolution and inverse systems.

13.1   Linear Convolution and LTI Systems

Definition of Linear Convolution

The convolution of two discrete-time signals $x(n)$ and $h(n)$ is defined by

$$y(n) = \sum_{k=-\infty}^{\infty} x(k)h(n-k)$$

and is shown by $y(n) = x(n) \star h(n)$. The discrete convolution defined by the above sum is called linear convolution, to be distinguished from circular convolution, which is performed circularly on two finite sequences.[1] The discrete convolution is a linear operation. It parallels convolution of continuous-time signals and shares its properties.

Application in Linear Systems Analysis

Let $h(n)$ be the response of a discrete-time LTI system to a unit-sample input. The response of the system to the input $x(n)$ may be evaluated by convolution, that is, $y(n) = x(n) \star h(n)$. To show this, express $x(n)$ as the sum of unit samples that are shifted k units and weighted by $x(k)$. Because of time invariance, a shift in the input creates a similar shift in the output. Therefore, the response of the system to $d(n-k)$ is $h(n-k)$. Similarly, because of linearity, the response to $x(k)d(n-k)$ is $x(k)h(n-k)$. By superposition, the response to $x(n)$ is, therefore,

$$y(n) = \sum_{k=-\infty}^{\infty} x(k)h(n-k)$$

These steps are summarized in Table 13.1.

---

[1] Circular convolution is used in conjunction with the discrete fourier transform (DFT) and will be discussed in Chapter 17, along with its applications.

TABLE 13.1 Superposition of Sample Responses

Input	$\Longrightarrow$	Output
$d(n)$	$\Longrightarrow$	$h(n)$
$d(n-k)$	$\Longrightarrow$	$h(n-k)$
$x(k)d(n-k)$	$\Longrightarrow$	$x(k)h(n-k)$

$$x(n) = \sum_{-\infty}^{\infty} x(k)d(n-k) \Longrightarrow y(n) = \sum_{-\infty}^{\infty} x(k)h(n-k)$$

Special Cases

If the system is causal $h(n-k) = 0$ for $k > n$, then the upper limit of the sum will be n.

$$y(n) = \sum_{k=-\infty}^{n} x(k)h(n-k)$$

Similarly, if the input is causal $x(k) = 0$ for $k < 0$, then the lower limit of the sum will be 0.

$$y(n) = \sum_{k=0}^{\infty} x(k)h(n-k)$$

The response of a causal system to a causal input is, therefore,

$$y(n) = \sum_{k=0}^{n} x(k)h(n-k)$$

13.2 Properties of Convolution

The following three important properties may be directly derived from the definition of convolution. These properties apply equivalently to LTI systems, as illustrated in Figure 13.1.

Convolution is commutative: $x(n) \star h(n) = h(n) \star x(n)$. The order of $x(n)$ and $h(n)$ in the convolution sum may be reversed and the result remains the same. To verify this property, start with the convolution sum and then change the summing variable k to a new variable p so that $k = n - p$ and $n - k = p$. Substituting for k and $n - k$, we have

$$y(n) = \sum_{k=-\infty}^{\infty} x(k)h(n-k) = \sum_{p=-\infty}^{\infty} x(n-p)h(p) = \sum_{k=-\infty}^{\infty} h(k)x(n-k)$$

Therefore, $x(n) \star h(n) = h(n) \star x(n)$. In an LTI system, the unit-sample response and the input may exchange places and produce the same output, as shown in Figure 13.1(a).

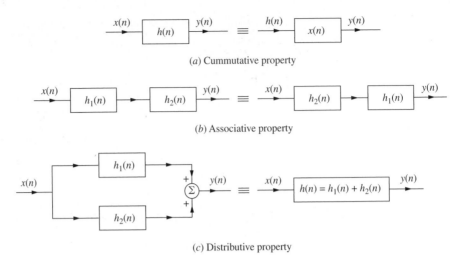

(a) Cummutative property

(b) Associative property

(c) Distributive property

FIGURE 13.1 Three properties of convolution.

Convolution is associative: $[x(n) \star h_1(n)] \star h_2(n) = x(n) \star [h_1(n) \star h_2(n)]$. Two cascaded LTI systems are equivalent to an LTI system with $h(n) = h_1(n) \star h_2(n) = h_2(n) \star h_1(n)$. Consequently, the two systems can be interchanged, as shown in Figure 13.1(b).

Convolution is distributive: $x(n) \star [h_1(n) + h_2(n)] = x(n) \star h_1(n) + x(n) \star h_2(n)$. Consequently, two LTI systems that are in parallel are equivalent to one LTI system with $h(n) = h_1(n) + h_2(n)$, as shown in Figure 13.1(c).

13.3 Solution by Numerical Method

In this section we present the numerical method for computing the convolution. This method may also be extended to linear time-variant systems as shown in section 13.7. For simplicity, we assume that $x(n)$ and $h(n)$ are finite sequences of lengths $N + 1$ and $M + 1$, respectively:

$$x(n) = \{x_0, x_1, x_2, \cdots, x_N\}$$

$$h(n) = \{h_0, h_1, h_2, \cdots, h_M\}$$

To find $y(n) = x(n) \star h(n)$, we first write $x(n)$ as

$$x(n) = \sum_{k=0}^{N} x(k)d(n - k)$$

We then find the system's response to each sample as shown in Table 13.2.

TABLE 13.2 Superposition of Responses to Individual Input Samples

Input	$\Longrightarrow$	Output
$x_0 d(n)$	$\Longrightarrow$	$x_0 h(n)$
$x_1 d(n-1)$	$\Longrightarrow$	$x_1 h(n-1)$
$x_2 d(n-2)$	$\Longrightarrow$	$x_2 h(n-2)$
$x_3 d(n-3)$	$\Longrightarrow$	$x_3 h(n-3)$
$\vdots$	$\vdots$	$\vdots$
$x(N) d(n-N)$	$\Longrightarrow$	$x_N h(n-N)$

$$x(n) = \sum_{0}^{N} x_k d(n-k) \quad \Longrightarrow \quad y(n) = \sum_{0}^{N} x_k h(n-k)$$

In order to illustrate how $y(n)$ is obtained by adding the elements in the nth column, Table 13.3 expands the summation operation of Table 13.2. In summary,

$$y(n) = x_0 h_n + x_1 h_{n-1} + \cdots + x_{n-1} h_1 + x_n h_0$$

TABLE 13.3 Implementation of $y(n) = x(n) \star h(n) = \{x_0, x_1, x_2, x_3, \cdots\} \star \{h_0, h_1, h_2, h_3, \cdots\}$

	h_0	h_1	h_2	h_3	h_4	h_5	h_6	$\cdots$	h_M		
x_0	$x_0 h_0$	$x_0 h_1$	$x_0 h_2$	$x_0 h_3$	$x_0 h_4$	$x_0 h_5$	$x_0 h_6$	$\cdots$	$x_0 h_M$		
x_1		$x_1 h_0$	$x_1 h_1$	$x_1 h_2$	$x_1 h_3$	$x_1 h_4$	$x_1 h_5$	$\cdots$	$x_1 h_{(M-1)}$	$\cdots$	
x_2			$x_2 h_0$	$x_2 h_1$	$x_2 h_2$	$x_2 h_3$	$x_2 h_4$	$\cdots$	$x_2 h_{(M-2)}$	$\cdots$	$\cdots$
x_3				$x_3 h_0$	$x_3 h_1$	$x_3 h_2$	$x_3 h_3$	$\cdots$	$x_3 h_{(M-3)}$	$\cdots$	$\cdots$
x_4					$x_4 h_0$	$x_4 h_1$	$x_4 h_2$	$\cdots$	$x_4 h_{(M-4)}$	$\cdots$	$\cdots$
x_5						$x_5 h_0$	$x_5 h_1$	$\cdots$	$x_5 h_{(M-5)}$	$\cdots$	$\cdots$
x_6							$x_6 h_0$	$\cdots$	$x_6 h_{(M-6)}$	$\cdots$	$\cdots$
$\vdots$								$\cdots$	$\cdots$	$\cdots$	$\cdots$
x_N								$\cdots$		$\cdots$	$x_N h_M$
	y_0	y_1	y_2	y_3	y_4	y_5	y_6	$\cdots$	y_M	$\cdots$	y_{M+N}

Sample Calculation

$$y_6 = x_0 h_6 + x_1 h_5 + x_2 h_4 + x_3 h_3 + x_4 h_2 + x_5 h_1 + x_6 h_0$$

Example
13.1

Find the linear convolution of $y(n) = x(n) \star h(n)$ given

$$x(n) = \begin{cases} \frac{2}{\sqrt{3}} \sin(\frac{\pi n}{3}), & 0 \leq n \leq 5 \\ \\ 0, & \text{elsewhere} \end{cases} \quad \text{and} \quad h(n) = \{\underset{\uparrow}{1}, 2, 3, 2, 1\}$$

Solution

The input is the finite sequence $x(n) = \{\underset{\uparrow}{0}, 1, 1, 0, -1, -1\}$. Table 13.4 implements $x(n) \star h(n)$ where rows represent the output of the system to individual input samples.

TABLE 13.4 $y(n) = x(n) \star h(n) = \{\underset{\uparrow}{0}, 1, 1, 0, -1, -1, 0\} \star \{\underset{\uparrow}{1}, 2, 3, 2, 1\}$

	h_0	h_1	h_2	h_3	h_4	h_5	h_6	h_7	h_8	h_9	h_{10}
	1	2	3	2	1	0	0	0	0	0	0
$x_0 = 0$	0	0	0	0	0	0					
$x_1 = 1$	0	1	2	3	2	1	0				
$x_2 = 1$		0	1	2	3	2	1	0			
$x_3 = 0$			0	0	0	0	0	0	0		
$x_4 = -1$				0	-1	-2	-3	-2	-1	0	
$x_5 = -1$					0	-1	-2	-3	-2	-1	0
$x_6 = 0$						0	0	0	0	0	0
	$\Downarrow$	$\Downarrow$	$\Downarrow$	$\Downarrow$	$\Downarrow$	$\Downarrow$	$\Downarrow$	$\Downarrow$	$\Downarrow$	$\Downarrow$	$\Downarrow$
	0	1	3	5	4	0	-4	-5	-3	-1	0
	y_0	y_1	y_2	y_3	y_4	y_5	y_6	y_7	y_8	y_9	y_{10}

Sample Calculation

$$y_5 = x_0 h_5 + x_1 h_4 + x_2 h_3 + x_3 h_2 + x_4 h_1 + x_5 h_0 = 0 + 1 + 2 + 0 - 2 - 1 + 0 = 0$$

13.4 Product Property

In section 13.3 we found the elements of the output sequence produced by the convolution of two finite sequences $x(n)$ and $h(n)$. These are listed in Table 13.5.

It is seen that y_k is the sum of all products $x_i \times h_j$ for which $i + j = k$. Here is an example:

$$y_5 = x_0 h_5 + x_1 h_4 + x_2 h_3 + x_3 h_2 + x_4 h_1 + x_5 h_0$$

TABLE 13.5 Results of Convolving Two Finite Sequences $x(n)$ and $h(n)$

$$y_0 = x_0 h_0$$
$$y_1 = x_0 h_1 + x_1 h_0$$
$$y_2 = x_0 h_2 + x_1 h_1 + x_2 h_0$$
$$y_3 = x_0 h_3 + x_1 h_2 + x_2 h_1 + x_3 h_0$$
$$y_4 = x_0 h_4 + x_1 h_3 + x_2 h_2 + x_3 h_1 + x_4 h_0$$
$$y_5 = x_0 h_5 + x_1 h_4 + x_2 h_3 + x_3 h_2 + x_4 h_1 + x_5 h_0$$
$$\vdots$$
$$y_N = x_0 h_N + x_1 h_{(N-1)} + x_2 h_{(N-2)} + \cdots + x_N h_0$$
$$\vdots$$
$$y_{(N+M)} = x_0 h_{(N+M)} + x_1 h_{(N+M-1)} + x_2 h_{(N+M-2)} + \cdots + x_{(N+M)} h_0$$

Momentary Diversion

Now consider two finite polynomials $X(z^{-1})$ and $H(z^{-1})$ of orders N and M, respectively.

$$X(z^{-1}) = x_0 + x_1 z^{-1} + x_2 z^{-2} + x_3 z^{-3} + \cdots + x_N z^{-N}$$
$$H(z^{-1}) = h_0 + h_1 z^{-1} + h_2 z^{-2} + h_3 z^{-3} + \cdots + h_M z^{-M}$$

The product $Y(z^{-1})$ is a polynomial of order $(M+N)$

$$Y(z^{-1}) = X(z^{-1})H(z^{-1}) = \sum_{n=0}^{N+M} y(n) z^{-n}$$

Coefficient $y(k)$ is the sum of all products $x_i \times h_j$ for which $i + j = k$. In other words, the coefficient of z^{-k} in the product $Y(z^{-1}) = X(z^{-1})H(z^{-1})$ are obtained from the convolution of $x(n) \star h(n)$ and vice versa. This is illustrated in Table 13.6. This property may be used to analytically evaluate the convolution.

TABLE 13.6 Product of $X(z^{-1})$ and $H(z^{-1})$

	h_0	$h_1 z^{-1}$	$h_2 z^{-2}$	$h_3 z{-}3$	$\cdots$	$h_M z^{-M}$
x_0	$x_0 h_0$	$x_0 h_1 z^{-1}$	$x_0 h_2 z^{-2}$	$x_0 h_3 z^{-3}$	$\cdots$	$x_0 h_M z^{-M}$
$x_1 z^{-1}$	$x_1 h_0 z^{-1}$	$x_1 h_1 z^{-2}$	$x_1 h_2 z^{-3}$	$x_1 h_3 z^{-4}$	$\cdots$	$x_1 h_M z^{-(M+1)}$
$x_2 z^{-2}$	$x_2 h_0 z^{-2}$	$x_2 h_1 z^{-3}$	$x_2 h_2 z^{-4}$	$x_2 h_3 z^{-5}$	$\cdots$	$x_2 h_M z^{-(M+2)}$
$x_3 z^{-3}$	$x_3 h_0 z^{-3}$	$x_3 h_1 z^{-4}$	$x_3 h_2 z^{-5}$	$x_3 h_3 z^{-6}$	$\cdots$	$x_3 h_M z^{-(M+3)}$
$x_4 z^{-4}$	$x_4 h_0 z^{-4}$	$x_4 h_1 z^{-5}$	$x_4 h_2 z^{-6}$	$x_4 h_3 z^{-7}$	$\cdots$	$x_4 h_M z^{-(M+4)}$
$\vdots$	$\cdots$	$\cdots$	$\cdots$	$\cdots$	$\cdots$ $\cdots$	
$x_N z^{-N}$	$x_N h_0 z^{-N}$	$x_N h_1 z^{-(N+1)}$	$x_N h_2 z^{-(N+2)}$	$x_N h_3 z^{-(N+3)}$	$\cdots$	$x_N h_M z^{-(M+N)}$

As an example, collecting the third-order terms we get

$$(x_0 h_3 + x_1 h_2 + x_2 h_1 + x_3 h_0) z^{-3} = y_3 z^{-3}$$
$$(x_0 h_3 + x_1 h_2 + x_2 h_1 + x_3 h_0) = y_3$$

Example

13.2

Find $y(n) = \{\underset{\uparrow}{1}, 2, 3, 3, 2, 1\} \star \{\underset{\uparrow}{1}, -1, 1, -1, 1, -1\}$.

Solution

$$Y(z^{-1}) = H(z^{-1})X(z^{-1})$$

$$= \left(1 + 2z^{-1} + 3z^{-2} + 3z^{-3} + 2z^{-4} + z^{-5}\right)\left(1 - z^{-1} + z^{-2} - z^{-3} + z^{-4} - z^{-5}\right)$$

$$= 1 + z^{-1} + 2z^{-2} + z^{-3} + z^{-4} - z^{-6} - z^{-7} - 2z^{-8} - z^{-9} - z^{-10}$$

$$y(n) = \{\underset{\uparrow}{1}, 1, 2, 1, 1, 0, -1, -1, -2, -1, -1\}$$

Example

13.3

Find $y(n) = x(n) \star h(n)$ when $x(n) = a^n u(n)$ and $h(n) = b^n u(n)$ by applying the product property in the z-domain.

Solution

$$X(z^{-1}) = \sum_{n=0}^{\infty} x(n)z^{-n} = \sum_{n=0}^{\infty} a^n z^{-n} = \frac{1}{1 - az^{-1}}$$

$$H(z^{-1}) = \sum_{n=0}^{\infty} h(n)z^{-n} = \sum_{n=0}^{\infty} b^n z^{-n} = \frac{1}{1 - bz^{-1}}$$

$$Y(z^{-1}) = X(z^{-1}) \times H(z^{-1}) = \frac{1}{(1 - az^{-1})(1 - bz^{-1})}$$

We can break $Y(z^{-1})$ into two partial fractions:

$$Y(z^{-1}) = \frac{A}{1 - az^{-1}} + \frac{B}{1 - bz^{-1}}$$

$$A = \frac{a}{a - b} \quad \text{and} \quad B = \frac{-b}{a - b}$$

We expand the two fractional components of $Y(z)$ into their infinite series, from which their time functions may be deduced.

$$Y_1(z^{-1}) = \frac{A}{1 - az^{-1}} = A\left(1 + az^{-1} + a^2 z^{-2} + \cdots\right) = A\sum_{n=0}^{\infty}(az^{-1})^n$$

$$\implies y_1(n) = \frac{a}{a - b}a^n u(n) = \frac{a^{n+1}}{a - b}u(n)$$

$$Y_2(z^{-1}) = \frac{B}{1 - bz^{-1}} = B\left(1 + bz^{-1} + b^2 z^{-2} + \cdots\right) = B\sum_{n=0}^{\infty}(bz^{-1})^n$$

$$\implies y_2(n) = -\frac{b}{a - b}b^n u(n) = -\frac{b^{n+1}}{a - b}u(n)$$

By adding the two components we have (see Chapter 15 for the formal derivations)

$$y(n) = y_1(n) + y_2(n) = \frac{1}{(a-b)} \left[a^{(n+1)} - b^{(n+1)} \right] u(n)$$

13.5   Solution by Analytical Method

When $x(n)$ and $h(n)$ are expressed analytically, the convolution sum may be obtained analytically and in a closed form, as shown in Examples 13.4 to 13.7.

*E*xample

13.4

Convolution with the unit-sample function

Find (a) $x(n) \star d(n)$ and (b) $x(n) \star d(n-N)$, where $d(n) = \begin{cases} 1, & n = 0 \\ 0, & \text{elsewhere} \end{cases} \equiv \{\underset{\uparrow}{1}\}$

is the unit-sample function and N is a constant integer.

Solution

a. $x(n) \star d(n) = \displaystyle\sum_{k=-\infty}^{\infty} x(k)d(n-k) = x(n)$

b. $x(n) \star d(n-N) = \displaystyle\sum_{k=-\infty}^{\infty} x(k)d(n-k-N) = x(n-N)$

Convolution of a discrete-time sequence with a unit sample positioned at N shifts the sequence to the position of the unit sample.

*E*xample

13.5

Find $y(n) = x(n) \star h(n)$ when $x(n) = \sin\left(\dfrac{\pi n}{2}\right)$ and $h(n) = \{1, 2, 1\}$.
${}_{\uparrow}$

Soution

Using the result of Example 13.4 we find

$$y(n) = x(n) + 2x(n-1) + x(n-2) = \sin\left(\frac{\pi n}{2}\right) + \sin\left(\frac{\pi(n-1)}{2}\right) + \sin\left(\frac{\pi(n-2)}{2}\right)$$

$$= -2\cos\left(\frac{\pi n}{2}\right)$$

*E*xample

13.6

Find $y(n) = x(n) \star h(n)$ when $x(n) = a^n u(n)$ and $h(n) = b^n u(n)$.

Solution

$$y(n) = \sum_{k=-\infty}^{\infty} x(k)h(n-k) = \sum_{k=-\infty}^{\infty} a^k u(k) b^{n-k} u(n-k)$$

But $u(k) = 0$ when $k < 0$ and $u(n - k) = 0$ when $k > n$. Therefore, because x and h are both causal functions, the convolution summation will span the range 0 to n:

$$y(n) = \sum_{k=0}^{n} a^k b^{(n-k)} = b^n \sum_{k=0}^{n} a^k b^{-k}$$

$$= b^n \sum_{k=0}^{n} \left(\frac{a}{b}\right)^k = b^n \frac{1 - \left(\frac{a}{b}\right)^{n+1}}{1 - \left(\frac{a}{b}\right)}, \quad n \geq 0$$

$$y(n) = \frac{1}{(a-b)} \left[a^{(n+1)} - b^{(n+1)}\right] u(n)$$

Example 13.7

Find $y(n) = x(n) \star h(n)$ where $x(n) = a^n u(-n)$ and $h(n) = b^n u(n)$.

Solution

The convolution sum is

$$y(n) = \sum_{k=-\infty}^{\infty} h(k)x(n - k) = \sum_{k=-\infty}^{\infty} b^k u(k) a^{n-k} u(-n + k)$$

But $u(k) = 0$ when $k < 0$. Therefore,

$$y(n) = \sum_{k=0}^{\infty} b^k a^{n-k} u(-n + k)$$

Also $u(-n + k) = 0$ when $k < n$. Therefore, the limits and result of the summation become:

For $n < 0$ $y(n) = \sum_{k=0}^{\infty} b^k a^{(n-k)} = \frac{a}{a-b} a^n$

For $n \geq 0$ $y(n) = \sum_{k=n}^{\infty} b^k a^{(n-k)} = \frac{a}{a-b} b^n$

$$y(n) = \frac{1}{a-b} \left[a^n u(-n) + b^n u(n) - d(n)\right]$$

The above results are valid only if $a > b$. Otherwise, the convolution sum becomes unbounded.

Comment

Limits of the summation can also be obtained by the graphical method.

Example 13.8

Find $y(n) = x(n) \star h(n)$, where $x(n) = e^{-\alpha n} u(n)$ and $h(n) = e^{-\beta n} u(n)$.

Solution

From Example 13.6 we have

$$a^n u(n) \star b^n u(n) = \left[\frac{a}{a-b} a^n + \frac{b}{b-a} b^n\right] u(n)$$

Let $a = e^{-\alpha}$ and $b = e^{-\beta}$. Then

$$e^{-\alpha n}u(n) \star e^{-\beta n}u(n) = \left[\frac{1}{1 - e^{(\alpha - \beta)}} e^{-\alpha n} + \frac{1}{1 - e^{(\beta - \alpha)}} e^{-\beta n} \right] u(n)$$

Example

13.9

It is desired to find $y(n) = x(n) \star h(n)$, where $x(n) = e^{-\alpha n}u(n)$ and $h(n) = e^{-\beta n}\cos(\omega n)u(n)$. Note that $h(n) = \mathcal{RE}\{e^{-(\beta + j\omega)n}u(n)\}$.

a. Show that $y(n) = \mathcal{RE}\{e^{-\alpha n}u(n) \star e^{-z_0 n}u(n)\}$, where $z_0 = \beta + j\omega$.

b. Using the above property find $y(n) = e^{-n}u(n) \star e^{-n}\cos(n)u(n)$.

Solution

a.
$$e^{-\alpha n}u(n) \star e^{-z_0 n}u(n) = e^{-\alpha n}u(n) \star e^{-\beta n}[\cos(\omega n) - j\sin(\omega n)]u(n)$$
$$= e^{-\alpha n}u(n) \star e^{-\beta n}\cos(\omega n)u(n)$$
$$- j e^{-\alpha n}u(n) \star e^{-\beta n}\sin(\omega n)u(n)$$

Therefore, $e^{-\alpha n}u(n) \star e^{-\beta n}\cos(\omega n)u(n) = \mathcal{RE}\{e^{-\alpha n}u(n) \star e^{-z_0 n}u(n)\}$

b. Using the results of Example 13.8 we find

$$C(n) = e^{-n}u(n) \star e^{-(1+j)n}u(n) = \left(\frac{1}{1 - e^{-j}} e^{-n} + \frac{1}{1 - e^{j}} e^{-(1+j)n} \right) u(n)$$
$$= \left[(0.5 - 0.9152j)e^{-n} + (0.5 + 0.9152j)e^{-(1+j)n} \right] u(n)$$
$$y(n) = e^{-n}u(n) \star e^{-n}\cos(n)u(n) = \mathcal{RE}\{C(n)\}$$
$$= 0.5e^{-n}(1 + \cos n + 1.83 \sin n)u(n)$$
$$= [0.5e^{-n} + 1.043e^{-n}\sin(n + 28.6°)]u(n)$$

Example

13.10

The unit-sample response of an LTI system is $h(n) = 0.5^n u(n)$. Find its response to $x(n) = 3^{-|n|}$.

Solution

The input can be expressed as $x(n) = 3^n u(-n) + 3^{-n}u(n) - d(n)$. The elements of the response corresponding to the above input components are, respectively,

$$y_1(n) = 3^n u(-n) \star 0.5^n u(n) = 1.2 \left[3^n u(-n) + 0.5^n u(n) - d(n) \right]$$
$$y_2(n) = 3^{-n}u(n) \star 0.5^n u(n) = \left(3 \times 0.5^n - 2 \times 3^{-n} \right) u(n)$$
$$y_3(n) = d(n) \star 0.5^n u(n) = 0.5^n u(n)$$
$$y(n) = y_1(n) + y_2(n) - y_3(n) = 1.2 \times 3^n u(-n) + \left[3.2 \times 0.5^n - 2 \times 3^{-n} \right] u(n) - 1.2d(n)$$
$$= \begin{cases} 1.2 \times 3^n & n \le 0 \\ 3.2 \times 0.5^n - 2 \times 3^{-n} & n \ge 0 \end{cases}$$

13.6 **Graphical Convolution**

The convolution

$$y(n) = \sum_{k=-\infty}^{\infty} x(k)h(n-k)$$

may be performed graphically by the *flip-shift-multiply-add* method described below.

Step 1 : Flip $h(k)$ around $k = 0$ to obtain $h(-k)$.

Step 2 : Shift $h(-k)$ by n units to obtain $h(n-k)$. A positive n shifts $h(-k)$ to the right. A negative n shifts $h(-k)$ to the left. Start with $n = -\infty$. With n kept fixed, perform tasks in steps 3 and 4 for k ranging from its lowest to highest value.

Step 3 : Multiply $h(n-k)$ by $x(k)$ term by term. Obtain the product $x(k)h(n-k)$ for all k.

Step 4 : Add $x(k)h(n-k)$ for all k. Call the sum $y(n)$ for the given n.

Step 5 : Go back to step 2, increment n by one unit and repeat the cycle until $y(n)$ is found for all n.

The graphical method can also determine the limits when the convolution is evaluated analytically.

Find the convolution of

$$y(n) = \sum_{k=-\infty}^{\infty} h(k)x(n-k), \quad \text{where } x(n) = u(n) - u(n-8), \quad \text{and } h(n) = e^{-\frac{n}{3}}u(n)$$

by the graphical method and compare with the result obtained by the analytical method.

Solution

The graphical method was implemented on a computer, resulting in the displays of Figure 13.2.

The steps in Figure 13.2 are: (*a*) Display of $h(k)$. (*b*) Display of $x(k)$. (*c*) Flip and shift $x(k)$ by n units to obtain $x(-n+k)$, here $n = 4$ units to the right. (*d*) Multiply $h(k) \times x(n-k)$. (*e*) Add $y(n) = \sum h(k)x(n-k)$. For example, the sum of the sample values for $n = 4$, shown in (*d*) is $y(4) = 2.8614$. The values of $y(n)$ obtained by the computer and rounded to the nearest second decimal are

$$y(n) = \{\cdots, 0, \underset{\uparrow}{1}, \ 1.72, \ 2.23, \ 2.6, \ 2.86, \ 3.05, \ 3.19, \ 3.28, \ 2.35, \ 1.69, \ 1.17, \ 0.8,$$

$$0.54, \ 0.35, \ 0.22, \ 0.12, \ 0.05, \ 0, \cdots\}$$

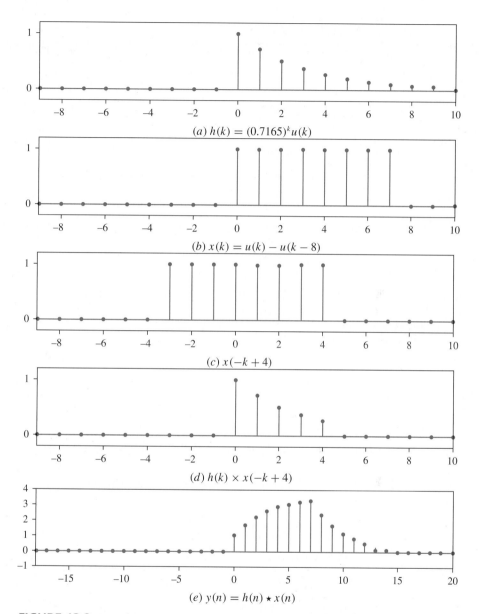

FIGURE 13.2 The graphical visualization of the convolution in Example 13.11 showing the steps to *flip-shift-multiply-add*. In this example, $x(n) = u(n) - u(n-8)$, $h(n) = e^{-\frac{n}{3}}u(n)$ and $y(n) = x(n) \star h(n)$.

Analytic Solution

The convolution result shown in (e) may be expressed analytically by

$$y(n) = y_1(n) - y_1(n - 8)$$

$$y_1(n) = u(n) \star e^{-\frac{n}{3}} u(n) = \sum_{k=0}^{n} e^{-\frac{n-k}{3}} = \left(3.5277 - 2.5277e^{-\frac{n}{3}}\right) u(n)$$

$$y(n) = \begin{cases} 0, & n < 0 \\ 3.5277 - 2.5277e^{-\frac{n}{3}}, & 0 \le n \le 7 \\ 33.8511e^{-\frac{n}{3}}, & n \ge 8 \end{cases}$$

The values obtained by the graphical method are in agreement with the analytic method.

Find $y(n) = h(n) \star x(n)$ where

$$h(n) = \{\underset{\uparrow}{1}, 1, 1, -1, -1, 1, -1\}$$

$$x(n) = \{\underset{\uparrow}{0}, 1, -1, 1, 1, -1, 1, -1, -1, 1, -1, -1, 1, 1, 1, -1, 1, 1, -1, -1, 1, 1, -1,$$

$$1, -1, -1, 1, 1, 1, 1, -1, -1, 1\}$$

Solution

Table 13.7 displays the first 16 elements of $x(k)$. It reverses $h(k)$, shifts it by n units, and registers $h(n - k)$ under $x(k)$. For each n, $y(n)$, obtained by multiplying $x(k)$ with $h(n - k)$ and summing the elements in the product, is shown by the right-side column of Table 13.7. The result is

$$y(n) = \{\underset{\uparrow}{0}, 1, 0, 1, 0, 1, 2, -5, 1, -1, -3, 3, -3, 1, 7, -1, -1, 1, 1,$$

$$-1, -5, 3, 3, -1, -3, 1, -1, -1, 7, 1, -1, -1, -3, 0, 3, -2, -1, 2, -1\}$$

Sample Calculation

$$y_7 = \sum_k x(k)h(7 - k) = x_1h_6 + x_2h_5 + x_3h_4 + x_4h_3 + x_5h_2 + x_6h_1 + x_7h_0$$

$$= (1)(-1) + (-1)(1) + (1)(-1) + (1)(-1) + (-1)(1) + (1)(1) + (-1)(1)$$

$$= -1 - 1 - 1 - 1 - 1 + 1 - 1 = -5$$

TABLE 13.7 Implementation of the Convolution of Example 13.12

$x(k) \Rightarrow$	↓ 0 1 −1 1 1 −1 1 −1 −1 1 −1 −1 1 1 1 −1	
n $h(n-k)$		$y(n)$
↓ ⇓		⇓
0 $h(-k)$	· 1 1 1	0 ←
1 $h(1-k)$	·−1 1 1 1	1
2 $h(2-k)$	·−1 −1 1 1 1	0
3 $h(3-k)$	· 1 −1 −1 1 1 1	1
4 $h(4-k)$	−1 1 −1 −1 1 1 1	0
5 $h(5-k)$	−1 1 −1 −1 1 1 1	1
6 $h(6-k)$	−1 1 −1 −1 1 1 1	2
7 $h(7-k)$	−1 1 −1 −1 1 1 1	−5
8 $h(8-k)$	−1 1 −1 −1 1 1 1	1
9 $h(9-k)$	−1 1 −1 −1 1 1 1	−1
10 $h(10-k)$	−1 1 −1 −1 1 1 1	−3
11 $h(11-k)$	−1 1 −1 −1 1 1 1	3
12 $h(12-k)$	−1 1 −1 −1 1 1 1	−3
13 $h(13-k)$	−1 1 −1 −1 1 1 1	1
14 $h(14-k)$	−1 1 −1 −1 1 1 1	7
15 $h(15-k)$	−1 1 −1 −1 1 1 1	−1
⋮ ⋮		

Example 13.13

Find $y(n) = x(n) \star h(n)$, where $x(n) = e^{-|\frac{n}{3}|}$ and $h(n) = u(n) - u(n-5) = \{\underset{\uparrow}{1}, 1, 1, 1, 1\}$, by the graphical method and compare with the result obtained by the analytical method. The graphical method was implemented on a computer, resulting in the displays of Figure 13.3, where $a = e^{-1/3} = 1.3956$.

Solution

The convolution sum is $y(n) = \displaystyle\sum_{k=-\infty}^{\infty} x(k)h(n-k)$. It adds up the product $x(k)h(n-k)$ over the dummy variable k. To find the limits of the summation and evaluate its value first plot $x(k)$ and $h(k)$ as a function of k as done in Figure 13.3(a) and (b). Then flip $h(k)$ around the vertical axis to get $h(-k)$, shift it by n units (right or left) to get

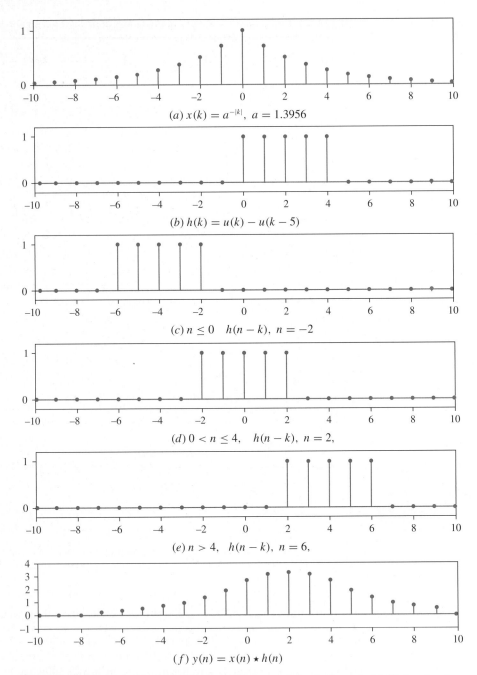

(a) $x(k) = a^{-|k|}$, $a = 1.3956$

(b) $h(k) = u(k) - u(k-5)$

(c) $n \leq 0$ $h(n-k)$, $n = -2$

(d) $0 < n \leq 4$, $h(n-k)$, $n = 2$,

(e) $n > 4$, $h(n-k)$, $n = 6$,

(f) $y(n) = x(n) \star h(n)$

FIGURE 13.3 The graphical visualization of the convolution in Example 13.13 showing the steps to *flip-shift-multiply-add*. In this example, $x(n) = e^{-|\frac{n}{3}|}$, $h(n) = u(n) - u(n-5)$ and $y(n) = x(n) \star h(n)$.

$h(n-k)$, multiply by $x(k)$ to get $x(k)h(n-k)$, and add the elements to get the value of $y(n)$ for the given shift. See Figure 13.3. In this example

$$y(n) = \sum_{k=n-4}^{n} x(k)$$

is a moving average of $x(n)$ under a five-samples wide uniform window. We recognize the following three regions:

(i) $n \leq 0$ (shown in Figure 13.3c for $n = -2$), $y(n) = \sum_{k=n-4}^{n} a^k = 2.8614a^n$, $n \leq 0$

(ii) $0 < n \leq 4$ (shown in Figure 13.3d for $n = 2$),   derivation is left as a problem

(iii) $n > 4$ (shown in Figure 1.3e for $n = 6$), $y(n) = \sum_{k=n-4}^{n} a^{-k} = 10.8553a^{-n}$, n

Note: $y(0) = y(4) = 2.8614$, $y_{Max} = y(2) = 1 + 2(a^{-1} + a^{-2}) = 3.46$, and $y(n+2)$ is an even function. The graphical method was implemented on a computer, resulting in the displays of Figure 13.3. Readings off the computer output are in agreement with the results obtained by analytical solution.

13.7 Convolution in Linear Time-Varying Systems

A discrete-time system may be linear but change with time. Its unit-sample response then depends on the arrival time of the unit sample. An example is a system represented by a linear difference equation whose coefficients are n-dependent. To illustrate the use of convolution for linear time-varying systems, let the response of the system at time n to a unit-sample input arriving at time k be designated by $h(n, k)$. Using the linearity property, the output $y(n)$ may be computed as shown below.

Input	$\Longrightarrow$	Output
$d(n - k)$	$\Longrightarrow$	$h(n, k)$
$x(k)d(n - k)$	$\Longrightarrow$	$x(k)h(n, k)$
$x(n) = \sum_{k=-\infty}^{\infty} x(k)d(n - k)$	$\Longrightarrow$ $y(n) = \sum_{k=-\infty}^{\infty} x(k)h(n, k)$	

Convolution for linear time-variant systems is, therefore,

$$y(n) = \sum_{k=-\infty}^{\infty} x(k)h(n, k)$$

Example

13.14

In a linear time-varying system the response at time n to a unit-sample input arriving at time k is given by

$$h(n, k) = \begin{cases} ka^n u(n - k), & k \geq 0 \\ 0, & k < 0 \end{cases}$$

Find $g(n, k)$, the response at time n to a unit step arriving at time $k > 0$.

Solution

A unit step arriving at time k is shown by $u(n - k)$. It may be expressed as a train of unit samples starting from time $n = k$.

$$u(n - k) = \sum_{p=k}^{\infty} d(n - p)$$

This is the sum of unit samples arriving at time p from $p = k$ to $p = \infty$. Each sample produces a response $h(n, p)$. Since the system is linear, the response to $u(n - k)$ is found by superposition of the system response to the samples:

$$g(n, k) = \sum_{p=k}^{\infty} h(n, p)$$

Having assumed a causal system, $h(n, p) = 0$ for $n < p$,

$$g(n, k) = \sum_{p=k}^{n} h(n, p)$$

To evaluate the sum, change the variable p to $(n - q)$ so that $q = (n - p)$ with q being the new summation index. The lower and upper limits of the above summation are found from

$$
\begin{aligned}
\text{Variable:} \quad & p \Rightarrow (n - q) \\
\text{Lower limit:} \quad & k \Rightarrow (n - k) \\
\text{Upper limit:} \quad & n \Rightarrow 0
\end{aligned}
$$

The summation becomes

$$g(n, k) = \sum_{q=n-k}^{0} h[n, (n - q)]$$

Reversing the order of summation we get

$$g(n, k) = \sum_{q=0}^{n-k} h[n, (n - q)]$$

Substituting for $h(n, k) = ka^n$ we get

$$g(n, k) = \sum_{q=0}^{n-k} (n - q)a^n = a^n \sum_{q=0}^{n-k} (n - q)$$

This is a finite sum of $(n - k + 1)$ elements that starts from n and decrements in steps of one until it becomes equal to k. To evaluate the sum note that

i) The value of the first term is n.

ii) The value of the last term is k.

iii) There are $(n - k + 1)$ terms in the sum.

iv) The sum of the first and last terms is $n + k$.

v) The sum of the second and penultimate term is $(n - 1) + [n - (n - k - 1)] = n + k$, etc.

vi) Therefore,

$$\sum_{q=0}^{n-k}(n - q) = \frac{(n - k + 1)(n + k)}{2}$$

vii) Finally, the response of the system at time n to a unit step arriving at time $k > 0$ is

$$g(n, k) = \frac{(n - k + 1)(n + k)}{2}a^n u(n - k)u(k)$$

Alternative Solution

The time-varying convolution may also be evaluated from the array shown in Table 13.8. The array has two dimensions indexed by k (for the rows, $0 \leq k < \infty$) and n (for the columns, $0 \leq n < \infty$). Each array entry represents $h(n, k)$, the system's response at time n to a unit sample arriving at time k.

TABLE 13.8 Implementation of Convolution for a Time-Varying System

		Response at time n to a unit sample arriving at time k								
		0	**1**	**2**	**3**	**4**	**5**	**6**	$\cdots$	**M**	$\cdots$
$d(n)$	$\Rightarrow$	0	0	0	0	0	0	0	$\cdots$	0	$\cdots$
$d(n - 1)$	$\Rightarrow$		a	a^2	a^3	a^4	a^5	a^6	$\cdots$	a^M	$\cdots$
$d(n - 2)$	$\Rightarrow$			$2a^2$	$2a^3$	$2a^4$	$2a^5$	$2a^6$	$\cdots$	$2a^M$	$\cdots$
$d(n - 3)$	$\Rightarrow$				$3a^3$	$3a^4$	$3a^5$	$3a^6$	$\cdots$	$3a^M$	$\cdots$
$d(n - 4)$	$\Rightarrow$					$4a^4$	$4a^5$	$4a^6$	$\cdots$	$4a^M$	$\cdots$
$d(n - 5)$	$\Rightarrow$						$5a^5$	$5a^6$	$\cdots$	$5a^M$	$\cdots$
$d(n - 6)$	$\Rightarrow$							$6a^6$	$\cdots$	$6a^M$	$\cdots$
$\vdots$									$\cdots$	$\cdots$	$\cdots$
$d(n - N)$	$\Rightarrow$									Na^M	$\cdots$

The output at n_0 to a step function arriving at k_0 is the sum of the elements of the vertical strip with $n = n_0$ and $k \geq k_0$. This strip is made of the elements of the n_0-th column for which $k \geq k_0$. As an example, the response at time $n = 5$ to a step that has arrived at $k = 2$ is the sum of the third through sixth elements, counting downward, in the $n = 5$ column and is evaluated to be

$$g(n, k)\big|_{n=5,k=2} = 2a^5 + 3a^5 + 4a^5 + 5a^5$$

$$= a^5(2 + 3 + 4 + 5) = 14a^5$$

As another example, the response at time $n = 6$ to a step that has arrived at $k = 4$ is

$$g(6, 4) = 4a^6 + 5a^6 + 6a^6 = 15a^6$$

The closed form of the response is

$$g(n, k) = ka^n + (k + 1)a^n + (k + 2)a^n + \cdots + na^n$$

$$= a^n \sum_{\ell=k}^{n} \ell = \frac{(n - k + 1)(n + k)}{2} a^n, \ n \geq k \geq 0$$

Example 13.15

Find $y(n)$, the response of the linear time-varying system of Example 13.14 to the pulse

$$x(n) = \begin{cases} 1, & 0 \leq n < N \\ 0, & \text{elsewhere} \end{cases}$$

Solution

The pulse is N samples wide. It may be expressed as the sum of two steps $x(n) = u(n) - u(n - N)$. From Example 13.14 the response at time n to a unit step arriving at time k is

$$g(n, k) = \frac{(n - k + 1)(n + k)}{2} a^n u(n - k)$$

Because of linearity

$$y(n) = g(n, 0) - g(n, N)$$

$$= \frac{(n + 1)n}{2} a^n u(n) - \frac{(n - N + 1)(n + N)}{2} a^n u(n - N)$$

and so

$$y(n) = \begin{cases} 0, & n < 0 \\ \dfrac{n(n + 1)}{2} a^n & 0 \leq n < N \\ \dfrac{N(N - 1)}{2} a^n & n \geq N \end{cases}$$

13.8 Deconvolution

The convolution operation allows us to find the output of an LTI system given its unit-sample response and the input. Can the reverse be done? In other words, can we find an input that would produce a desired output? Can we *recover* the input by observing the output? The answer is often *yes*, especially in the case of FIR systems, and is accomplished through deconvolution. Here we present the idea by a simple example. Formal methods will be devised later on in Chapter 15. See also the project at the end of this chapter.

Example

13.16

An input-output pair in a discrete-time LTI system is

$$\{\underset{\uparrow}{1}, 0, 1\} \implies \{\underset{\uparrow}{3}, 2, 4, 2, 1\}$$

a.   Find the unit-sample response $h(n)$.
b.   Find the input that produces $\{\underset{\uparrow}{3}, 5, 3, 1\}$ at the output.

Solution

a.   Using the information previously organized in Table 13.5, we find that

$$h_0 = y_0/x_0 = 3/1 = 3$$
$$h_1 = (y_1 - x_1 h_0)/x_0 = (2 - 0 \times 3)/1 = 2$$
$$h_2 = (y_2 - x_1 h_1 - x_2 h_0)/x_0 = (4 - 0 \times 2 - 1 \times 3)/1 = 1$$
$$h_3 = (y_3 - x_1 h_2 - x_2 h_1 - x_3 h_0)/x_0 = (2 - 0 \times 1 - 1 \times 2)/1 = 0$$
$$h_4 = (y_4 - x_1 h_3 - x_2 h_2 - x_3 h_1 - x_4 h_0)/x_0 = (1 - 0 \times 1 - 1 \times 1)/1 = 0$$
$$h_5 = (y_5 - x_1 h_4 - x_2 h_3 - x_3 h_2 - x_4 h_1 - x_5 h_0)/x_0 = 0$$

The result is $h(n) = \{\underset{\uparrow}{3}, 2, 1\}$.

b.   Again, from Table 13.5 we find

$$x_0 = y_0/h_0 = 3/3 = 1$$
$$x_1 = (y_1 - x_0 h_1)/h_0 = (5 - 1 \times 2)/3 = 1$$
$$x_2 = (y_2 - x_0 h_2 - x_1 h_1)/h_0 = (3 - 1 \times 1 - 1 \times 2)/3 = 0$$
$$x_3 = (y_3 - x_0 h_3 - x_1 h_2 - x_2 h_1)/h_0 = (1 - 1 \times 0 - 1 \times 1)/3 = 0$$
$$x_4 = (y_4 - x_0 h_4 - x_1 h_3 - x_2 h_2 - x_3 h_1)/h_0 = 0$$
$$x_5 = (y_5 - x_0 h_5 - x_1 h_4 - x_2 h_3 - x_3 h_2 - x_4 h_1)/h_0 = 0$$

The result is $x(n) = \{\underset{\uparrow}{1}, 1\}$.

xample

13.17

Use the product property, in the z-domain, of convolution to solve Example 13.16.

Solution

a.
$$X(z) = 1 + z^{-2}$$
$$Y(z) = 3 + 2z^{-1} + 4z^{-2} + 2z^{-3} + z^{-4}$$
$$H(z) = Y(z)/X(z) = \frac{3 + 2z^{-1} + 4z^{-2} + 2z^{-3} + z^{-4}}{1 + z^{-2}}$$
$$= 3 + 2z^{-1} + z^{-2} \quad \text{(by long division)}$$
$$h(n) = \{\underset{\uparrow}{3}, 2, 1\}.$$

b.
$$X(z) = \frac{3 + 5z^{-1} + 3z^{-2} + z^{-3}}{3 + 2z^{-1} + z^{-2}} = 1 + z^{-1} \quad \text{(by long division)}$$
$$x(n) = \{\underset{\uparrow}{1}, 1\}.$$

13.9 Inverse Systems

It is sometimes desired to cancel the effect of a particular system. An example is the equalizer inserted in an audio system to compensate for the weak frequency response of a recording head. Another example is the averaging and smearing effects produced by a sensor that has limited resolution. A third example is the degradation (blaring) of an image due to the motion of the object. Still another example is the dynamics of a transducer, such as a robotic arm, which converts a command signal into a desired trajectory. In such cases one would like to eliminate the distortions and imperfections in the system by neutralizing its dynamics (such as those of a microphone, a camera, or a robotic arm). This is done by introducing an inverse effect.

Mathematically, given an LTI system with unit-sample response $h(n)$, we search for another LTI system with unit-sample response $\overline{h}(n)$ such that the overall unit-sample response of their cascade arrangement becomes $h(n) \star \overline{h}(n) = d(n)$. $\overline{h}(n)$ is then called the inverse of $h(n)$. The concept of an inverse system is closely related to deconvolution. We illustrate this concept by an example. More discussion will be presented later. See also the project at the end of this chapter.

xample

13.18

The unit-sample response of a discrete-time LTI system is $h(n) = (0.5)^n(n)$. Find the unit-sample response of its inverse system.

Solution

This problem may be formulated as in Example 13.16. That is, given the system with an input-output pair

$$d(n) \implies (0.5)^n u(n)$$

find the input that produces $d(n)$ at the system's output. Using the approach of Example 13.5 we find that $\{1, -0.5\}$ is such an input. If we construct an LTI system with $\overline{h}(n) = \{1, -0.5\}$ (to be called the inverse system) and place it in cascade fashion in front of the original system (as shown in Figure 13.4), the overall unit-sample response will be $d(n)$. This may also be verified directly by evaluating

$$h(n) \star \overline{h}(n) = (0.5)^n u(n) \star \{1, -0.5\} = (0.5)^n u(n) - 0.5(0.5)^{n-1}u(n-1)$$

$$= (0.5)^n[u(n) - u(n-1)] = d(n)$$

We will revisit inverse systems and deconvolution in the z-domain in Chapter 15.

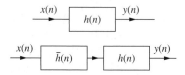

FIGURE 13.4 The two systems $h(n) = (0.5)^n u(n)$ and $\overline{h}(n) = \{1, -0.5\}$ are inverses of each other. Cascading them produces a system in which $y(n) = x(n)$.

Example 13.19

Draw the block diagram of the cascade of the two systems of Example 13.18, $\overline{h}(n) = \{1, -0.5\}$ followed by $h(n) = (0.5)^n u(n)$, and show that the unit-sample response of the overall system is $d(n)$.

Solution

The first system, with the unit-sample response $\overline{h}(n) = \{1, -0.5\}$, is described by the difference equation $w(n) = x(n) - 0.5x(n-1)$, where $x(n)$ is the input and $w(n)$ is the output. Its block diagram has a forward path. The second system, with the unit-sample response $h(n) = (0.5)^n u(n)$ and input coming from the output of the first system, is described by $y(n) = w(n) + 0.5y(n-1)$. Its block diagram has a feedback path. Eliminating $w(n)$ between the two difference equations, we obtain

$$y(n) - 0.5y(n-1) = x(n) - 0.5x(n-1) \implies y(n) = x(n)$$

The above derivations are summarized in the table below. See also Figure 13.5(a).

System	Unit-Sample Response	Input	Output	Difference Equation
1	$\overline{h}(n) = \{1, -0.5\}$	$x(n)$	$w(n)$	$w(n) = x(n) - 0.5x(n-1)$
2	$h(n) = (0.5)^n u(n)$	$w(n)$	$y(n)$	$y(n) = w(n) + 0.5y(n-1)$
$1 \to 2$	$\overline{h}(n) \star h(n) = d(n)$	$x(n)$	$y(n)$	$y(n) = x(n)$

Example

13.20

Draw the block diagram of the cascade of the two systems of Example 13.18, $h(n) = (0.5)^n u(n)$ followed by $\bar{h}(n) = \{\underset{\uparrow}{1}, -0.5\}$, and show that the unit-sample response of the overall system is $d(n)$. Then, conclude that the block diagram of Figure 13.5(a) may be reduced to Figure 13.5(c).

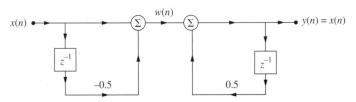

(a) Cascading $\bar{h}(n)$ with $h(n)$.

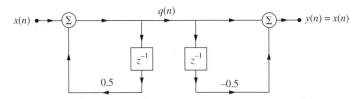

(b) Cascading $h(n)$ with $\bar{h}(n)$ is equivalent to the system in (a).

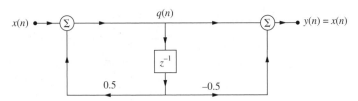

(c) Combining the unit delays in (b). The system is, therefore, equivalent to (a).

$x(n) \bullet$ ———————————————— $\bullet\, y(n) = x(n)$

(d) Equivalent system for (a), (b), and (c).

FIGURE 13.5 Cascading two inverse systems with unit-sample responses $h(n)$ and $\bar{h}(n)$.

$$h(n) = 0.5^n u(n)$$

$$\bar{h}(n) = \{\underset{\uparrow}{1}, -0.5\}$$

$$h(n) \star \bar{h}(n) = \bar{h}(n) \star h(n) = 0.5^n u(n) - 0.5 \times 0.5^{(n-1)} u(n-1) = d(n)$$

The three systems **(a)**, **(b)**, and **(c)** are equivalent to **(d)**.

Solution

By an approach similar to that of Example 13.19 we write the difference equations and construct the cascade block diagram of Figure 13.5(b).

$q(n) = x(n) + 0.5q(n-1)$ from the feedback loop on the left side of Figure 13.5(b)

$y(n) = q(n) - 0.5q(n-1)$ from the feedforward loop on the right side of

Figure 13.5(b)

Eliminating $q(n)$ between the two difference equations we obtain $y(n) = x(n)$. By combining the unit delays in Figure 13.5(b) we obtain Figure 13.5(c), which makes it equivalent to Figure 13.5(a). The above derivations are summarized below.

System	Unit-Sample Response	Input	Output	Difference Equation
1	$h(n) = (0.5)^n u(n)$	$x(n)$	$q(n)$	$q(n) = x(n) + 0.5q(n-1)$
2	$\overline{h}(n) = \{1, -0.5\}$	$q(n)$	$y(n)$	$y(n) = q(n) - 0.5q(n-1)$
$1 \to 2$	$h(n) \star \overline{h}(n) = d(n)$	$x(n)$	$y(n)$	$y(n) = x(n)$

13.10 Problems

Solved Problems

1. Given two finite-duration sequences $x(n) = \{1, 2, 3, 2, 1\}$ and $h(n) = \{1, 0, 1\}$.

a. Compute the convolution $y(n) = x(n) \star h(n)$ by superposition in the time domain.
b. Find $y(n)$ by the product property in the z-domain.

Solution

a. By superposition $y(n) = x(n) + x(n-2)$ and

$x(n)$	1 2 3 2 1 0 0
$x(n-2)$	0 0 1 2 3 2 1
$y(n)$	1 2 4 4 4 2 1

b. In the z-domain:

$$H(z) = 1 + z^{-2}$$

$$X(z) = 1 + 2z^{-1} + 3z^{-2} + 2z^{-3} + z^{-4}$$

$$Y(z) = H(z)X(z) = \left(1 + z^{-2}\right)\left(1 + 2z^{-1} + 3z^{-2} + 2z^{-3} + z^{-4}\right)$$

$$= 1 + 2z^{-1} + 4z^{-2} + 4z^{-3} + 4z^{-4} + 2z^{-5} + z^{-6}$$

$$y(n) = \{1, 2, 4, 4, 4, 2, 1\}$$

2. Convolution of two cyclic causal functions. Find $y(n) = x(n) \star h(n)$ where $x(n)$ and $h(n)$ are causal functions given below.

$$x(n) = \{1, 1, 1, 1, 0, 0, 1, 1, 1, 1, 0, 0 \cdots\} = \begin{cases} x(n+6) & n \geq 0 \\ 0 & n < 0 \end{cases}$$

$$h(n) = \{1, -1, 1, -1, 1, -1, \cdots\} = \begin{cases} h(n+2) & n \geq 0 \\ 0 & n < 0 \end{cases}$$

Solution

$$y(n) = \sum_{k=0}^{\infty} [x(n-k) - x(n-k-1)] = \{1, 0, 1, 0, 0, 0, 1, 0, 1, 0, 0, 0, 1, 0, 1, 0, 0, 0, 1, 0, 1, 0, 0, 0 \cdots\}$$

3. Integration. Find $y(n) = \sin\left(\frac{\pi n}{4}\right) u(n) \star u(n)$.

Solution

$x(n)$ is 8-periodic with zero mean: $x(n) = \{0, 0.707, 1, 0.707, 0, -0.707, -1, -0.707\}$

$$y(n) = \sum_{k=-\infty}^{n} x(n) \text{ is 8-periodic:} \qquad y(n) = \{0, 0.707, 1.707, 2.414, 2.414, 1.707, 0.707, 0\}$$

4. Initial conditions. The input to a discrete-time LTI system with the unit-sample response $h(n) = a^n u(n), \quad a < 1$, is $x(n) = Ad(n+1) + u(n)$. Determine the factor A so that $y(n)$ is a constant for $n \geq 0$.

Solution

$$y(n) = x(n) \star h(n) = Ah(n+1) + \sum_{k=0}^{n} a^k = Aa^{n+1} u(n+1) + \frac{1 - a^{n+1}}{1 - a} u(n)$$

For $A = \dfrac{1}{1-a}$ we have $y(n) = \dfrac{1}{1-a}, \quad n \geq -1$

5. The unit-sample response of a causal system is $h_1(n) = [5(0.5)^n - 4(0.4)^n]u(n)$. Find $h_2(n)$ such that $h(n) = h_1(n) \star h_2(n) = d(n)$.

Solution

Apply the z operator to $h_1(n)$ and $h(n) = h_1(n) \star h_2(n) = d(n)$, and use the product property.

$$H_1(z^{-1}) = \sum_{n=0}^{\infty} h_1(n)z^{-n} = \frac{5}{1 - 0.5z^{-1}} - \frac{4}{1 - 0.4z^{-1}} = \frac{1}{1 - 0.9z^{-1} + 0.2z^{-2}}$$

$$H(z^{-1}) = \sum_{n=0}^{\infty} h(n)z^{-n} = \sum_{n=0}^{\infty} [h_1(n) \star h_2(n)]z^{-n} = H_1(z^{-1})H_2(z^{-1}) = 1$$

Therefore, $H_2(z^{-1}) = \dfrac{1}{H_1(z^{-1})}] = 1 - 0.9z^{-1} + 0.2z^{-2}$

$$h_2(n) = d(n) - 0.9d(n-1) + 0.2d(n-1)$$

Verification: $h_1(n) \star h_2(n) = h_1(n) - 0.9h_1(n-1) + 0.2h_1(n-1) = \begin{cases} 1 & n = 0 \\ 0 & \text{elsewhere} \end{cases}$

Chapter Problems

6. In an LTI system $x(n)$ is the input and $h(n)$ is the unit-sample response. Find and sketch the output $y(n)$ for the following cases:

Input		Unit-Sample Response
$x(n) = \{1, 1, 1, 1, 1\}$	and	$h(n) = \{1, 1, 1, 1, 1\}$
$x(n) = \{1, 1, 1\}$	and	$h(n) = \{1, 1, 1\}$
$x(n) = \{1, 2, 3, 2, 1\}$	and	$h(n) = \{1, 2, 3, 2, 1\}$
$x(n) = \{1, 1, 1, 1, 1\}$	and	$h(n) = \{1, -1\}$

7. Repeat problem 6 for the following cases:

Input		Unit-Sample Response
$x(n) = \cos\left(\frac{\pi n}{6}\right) u(n)$	and	$h(n) = \{1, -1\}$
$x(n) = \cos\left(\frac{\pi n}{6}\right)$	and	$h(n) = \{1, -1\}$
$x(n) = \cos\left(\frac{\pi n}{2}\right)$	and	$h(n) = \{0.5, 0.5\}$
$x(n) = \cos\left(\frac{n}{2}\right)$	and	$h(n) = \{0.5, 0, -0.5\}$

8. Repeat problem 6 for the following cases:

Input		Unit-Sample Response
$x(n) = \{1, 2, 1\}$	and	$h(n) = \{1, -0.5, 1\}$
$x(n) = \{1, 1, 1\}$	and	$h(n) = \{1, 0, 1\}$
$x(n) = \{1, 1, 1\}$	and	$h(n) = \{1, 0, 1\}$
$x(n) = \{1, 2, 3, 2, 1\}$	and	$h(n) = \{1, 0, 1\}$

9. Repeat problem 6 for the following cases:

Input		Unit-Sample Response
$x(n) = \cos\left(\frac{\pi n}{6}\right) u(n)$	and	$h(n) = \{1, -1, 1\}$
$x(n) = \cos\left(\frac{\pi n}{6}\right)$	and	$h(n) = \{1, -1, 1\}$
$x(n) = \cos\left(\frac{\pi n}{2}\right)$	and	$h(n) = \{1, 2, 1\}$
$x(n) = 1 + \cos\left(\frac{n}{2}\right)$	and	$h(n) = \{1, 2, 1\}$

10. The unit-sample response of an LTI system is $h(n) = \{1, 2, 3, 2, 1\}$. Find $y(n) = x(n) \star h(n)$ for the following inputs.

a. $x(n) = \{\underset{\uparrow}{1}, 1, 1, 1, 1, 1, 1\}$

b. $x(n) = \{\underset{\uparrow}{1}, 0, 1, 0, 1, 0, 1, 0, 1\}$

c. $x(n) = \{\underset{\uparrow}{1}, -1, 1, -1, 1, -1, 1\}$

d. $x(n) = \{\underset{\uparrow}{1}, 0, -1, 0, -1, 0, -1, 0, 1\}$

e. $x(n) = \{\underset{\uparrow}{1}, 0, 0, 1, 0, 0, 1, 0, 0, 1, 0, 0, 1\}$

f. $x(n) = \{\underset{\uparrow}{1}, 0, 0, -1, 0, 0, 1, 0, 0, -1, 0, 0, 1\}$

g. $x(n) = \{\underset{\uparrow}{1}, 0, 0, 0, 1, 0, 0, 0, 1, 0, 0, 0, 1, 0, 0, 0, 1\}$

h. $x(n) = \{\underset{\uparrow}{1}, 0, 0, 0, 0, 1, 0, 0, 0, 0, 1, 0, 0, 0, 0, 1, 0, 0, 0, 0, 1\}$

i. $x(n) = \{\underset{\uparrow}{1}, 0, 0, 0, 0, 0, 1, 0, 0, 0, 0, 0, 1, 0, 0, 0, 0, 0, 1, 0, 0, 0, 0, 0, 1\}$

11. In an LTI system $x(n)$ is the input and $h(n)$ is the unit-sample response. Find and sketch the output $y(n)$ for the following cases:

Input		Unit-Sample Response
$x(n) = 0.3^n u(n)$	and	$h(n) = 0.4^n u(n)$
$x(n) = 0.5^n u(n)$	and	$h(n) = 0.6^n u(n)$
$x(n) = 0.5^{|n|}$	and	$h(n) = 0.6^n u(n)$

12. Repeat problem 11 for the following cases:

Input		Unit-Sample Response
$x(n) = 2[u(n) - u(n - 11)]$	and	$h(n) = 0.5^n u(n)$
$x(n) = 2[u(n) - u(n - 11)]$	and	$h(n) = (-0.5)^n u(n)$
$x(n) = u(n) - u(n - 5)$	and	$h(n) = 0.5^{|n|} u(n)$
$x(n) = u(n) - u(n - 5)$	and	$h(n) = (-0.5)^{|n|} u(n)$

13. Using the convolution properties show that

a. $h(n) \star d(n) = h(n)$

b. $h(n) \star d(n - k) = h(n - k)$

c. $h(n) \star [d(n - k) + d(n - \ell)] = h(n - k) + h(n - \ell)$

d. $h(n) \star \sum_{\ell=k_1}^{k_2} d(n - \ell) = \sum_{\ell=k_1}^{k_2} h(n - \ell)$

where n, k, k_1, k_2, and ℓ are all integers.

14. First show that

   a. $d(n - k) \star h(n) = h(n - k)$

   b. $[x(k)d(n - k)] \star h(n) = x(k)h(n - k)$

Then, using the above results, derive the convolution sum

$$x(n) \star h(n) = \sum_{k=-\infty}^{\infty} x(k)h(n - k)$$

15. A continuous-time analog signal $x(t) = \cos 500\pi t + \cos 200\pi t$ is sampled at the rate of 1,000 samples per second.

   a. Find $x(n)$.

   b. The resulting discrete signal $x(n)$ is passed through a filter with unit-sample response $h(n) = \{1, 2, 3, 2, 1\}$.
     Find the output $y(n)$ and its analog reconstruction $y(t)$.

16. A discrete-time signal $x(n) = 1 + \cos(\omega_0 n)$ passes through a linear filter with unit-sample response $h(n) = \{1, \frac{1}{2}, 1\}$.
Find the output of the filter. For what value of ω_0 will the output be zero?

17. An input $x(n) = \{1, 1, 1, 1\}$ passes through the following two filters.

$$h_1(n) = \begin{cases} 0.81^n, & 0 \le n \le 2 \\ 0, & \text{otherwise} \end{cases} \quad \text{and} \quad h_2(n) = \begin{cases} 0.9^n, & 0 \le n \le 5 \\ 0, & \text{otherwise} \end{cases}$$

Find the outputs. What can you deduce about $y_2(n)$, given $y_1(n)$?

18. Two groups of LTI systems are specified in the following table by their unit-sample responses. By pairing members of the two groups, determine which pair can be inverse systems.

Group 1	Group 2
$h_a(n) = u(n)$	$h_1(n) = d(n)$
$h_b(n) = 0.5u(n) + 0.5u(n - 2)$	$h_2(n) = d(n) - d(n - 1)$
$h_c(n) = [1 - \frac{(-1)^n}{3}]u(n) - \frac{1}{3}d(n)$	$h_3(n) = 0.5d(n) - 0.5d(n - 2)$

19. Given $h(n) = \{1, 2, 3, 2, 1\}$, find $y(n) = x(n) \star h(n)$ for each $x(n)$ given below.

   a. Four unit samples spaced by seven zero samples,

$$x(n) = d(n) + d(n - 8) + d(n - 16) + d(n - 24) = \sum_{k=0}^{3} d(n - 8k).$$

   b. Four unit samples spaced by N zero samples,

$$x(n) = d(n) + d(n - N) + d(n - 2N) + d(n - 3N) = \sum_{k=0}^{3} d(n - kN).$$

c. $(\ell + 1)$ unit samples spaced by N zero samples,

$$x(n) = d(n) + d(n - N) + d(n - 2N) + \cdots + d(n - \ell N) = \sum_{k=0}^{\ell} d(n - kN)$$

In each case compute average and total powers in the output.

20. In an LTI system $h(n) = \{-1, 0, 2, 3, 4, 3, 2, 0, -1\}$ and the input is a periodic rectangular pulse train with period N and pulse width M. Find and plot $y(n) = x(n) \star h(n)$ for the following cases.

a. $N = 12$, $M = 1$, $x(n) = \begin{cases} 1, & n = 0 \\ 0, & 1 \leq n \leq 11 \end{cases}$, defined for 1 period

b. $N = 8$, $M = 1$, $x(n) = \begin{cases} 1, & n = 0 \\ 0, & 1 \leq n \leq 7 \end{cases}$, defined for 1 period

c. $N = 6$, $M = 1$, $x(n) = \begin{cases} 1, & n = 0 \\ 0, & 1 \leq n \leq 5 \end{cases}$, defined for 1 period

d. $N = 12$, $M = 2$, $x(n) = \begin{cases} 1, & n = 0, 1 \\ 0, & 2 \leq n \leq 11 \end{cases}$, defined for 1 period

e. $N = 8$, $M = 2$, $x(n) = \begin{cases} 1, & n = 0, 1 \\ 0, & 2 \leq n \leq 7 \end{cases}$, defined for 1 period

f. $N = 6$, $M = 2$, $x(n) = \begin{cases} 1, & n = 0, 1 \\ 0, & 2 \leq n \leq 5 \end{cases}$, defined for 1 period

13.11 Project: Deconvolution and Inverse Systems

Introduction and Summary

This project illustrates simple cases of signal recovery through the concept of deconvolution and inverse systems. It is designed around the discussion in sections 13.8 and 13.9. It contains three procedures, the first two of which are theoretical (with possible applications). The third procedure simulates experiments in parallel with the theory. Effects of deviations from theoretical requirements are explored qualitatively by visual displays in simulation. Such deviations can also be quantitatively measured.

Procedure 1. A Thought Experiment in Deconvolution

Consider a hypothetical earth observation satellite that uses a digital camera with a pixel resolution of 100×100 m. This means that the output at any moment of each charged coupled device (CCD element or pixel) is a voltage proportional to the average of light intensity within the 100×100 meters square field of view (to be called the receptive field in this procedure). If the camera were stationary with respect to the earth, the above resolution would correspond to a spatial sampling rate of every 100 meters, and the averaging property of the CCD element would result in a permanent loss of information below the specified resolution. In this procedure you will explore to see if by sweeping the camera over the field of view and taking overlapping pictures such that the consecutive receptive fields of each pixel overlap, one may increase the resolution of the picture by deconvolution.

For simplicity, let us consider a camera containing a row of CCD elements with a spatial resolution of X_0 meters. The camera sweeps its footprint at a speed of ℓ meters per second in a direction perpendicular to its CCD row and takes consecutive pictures every T seconds such that the distance X_0 is swept by M pictures. The output of the CCD row is called $y(n)$. It is the average of light intensity in M neighboring subpixels. In summary, let the light intensity within a subpixel be $x(n)$ and a CCD element at each instance see M subpixels. Then, $y(n) = x(n) \star h(n)$. We want to recover $x(n)$ from $y(n)$.

Let $h(n) = \{1, 1, 1\}$. Given $y(n)$ and $h(n)$, we may find $x(n)$, the light intensity within a subpixel. Following the deconvolution process of section 13.8, obtain the input $x(n)$ which produces the output $y(n) = x(n) \star h(n)$ given below:

a. $y(n) = \{1, 1, 1, 1, 2, 2, 2, 1, 2, 2, 2, 2, 2, 3, 2, 1, 1, 1, 2, 1, 1\}$
    $\quad\quad\uparrow$

b. $y(n) = \{1, 1, -1, 1, 2, 6, 4, 3, 0, -3, -3, -1, 4, 8, 8, 8, 3, 0, -3, -2\}$
    $\quad\quad\uparrow$

Procedure 2. Inverse System, Signal Recovery, Deblurring

A digital camera takes a picture of a one-dimensional object described by $x(n) = \{x_0, x_1, x_2, x_3, x_4, \cdots, x_N\}$ where x_k is the sample average light intensity in the kth element. If the object were stationary, each sample in $x(n)$ would map into a single CCD element in the digital image. However, the object is moving to the right while the camera aperture is open. As a result, M samples in $x(n)$ pass in front of a CCD element which senses and registers their total average, causing the picture to become blurred. The operation may be modeled by a digital filter with unit-sample response $h(n)$ which convolves with the input $x(n)$ to produce $y(n) = x(n) \star h(n)$ on the CCD array.

a. Assuming uniform motion and $M = 3$, show that $h(n) = \{1, 1, 1\}$. Then find camera outputs $y_1(n)$ and $y_2(n)$
    $\quad\quad\quad\quad\quad\quad\quad\quad\quad\quad\quad\quad\quad\quad\quad\quad\quad\quad\quad\quad\uparrow$
    corresponding to (i) $x_1(n) = \{1, 1, 1, 1, 1\}$ and (ii) $x_2(n) = \{1, 2, 1\}$, respectively.
    $\quad\quad\quad\quad\quad\quad\quad\quad\quad\uparrow\quad\quad\quad\quad\quad\quad\quad\quad\quad\quad\uparrow$

b. Find the unit-sample response $g(n)$ of an inverse system that would cancel the effect of the above $h(n)$. The inverse operation should satisfy the following equation $x(n) = y(n) \star g(n)$. Proceed as follows:

  (i)   As a starting point, show that $g(n) = \{1, -1, 0, 1, -1, 0, 1, -1, 0, 1, -1, 0 \cdots\}$ inverts the output $y_1(n) = \{1, 2, 3, 3, 3, 2, 1\}$ to the input $x_1(n) = \{1, 1, 1, 1, 1\}$.
        $\quad\uparrow\quad\quad\quad\quad\quad\quad\quad\quad\quad\quad\quad\quad\quad\quad\uparrow$

  (ii) As a second test, apply the above $g(n)$ to $y_2(n)$ to verify its inverse property.

  (iii) Construct a new set of input-output pairs and test their inversion using the above $g(n)$.

  (iv) Derive the above $g(n)$ by reverse computation of the convolution operation. However, this is a tedious process. One may develop easier methods of finding inverse systems by using the z-transform (Chapter 15).

Procedure 3. Simulation and Display by Computer

In this procedure you will simulate the passage of discrete-time signals through a digital filter. You will then examine the inverse of filtering operation in order to recover the signal from the filter's output. The unit-sample response of the filter is $h(n) = a^n u(n)$, $a < 1$. The inverse filter is $g(n) = d(n) - ad(n-1)$. The following program convolves $x(n)$ with a truncated $h(n)$ to find $y(n)$. It then convolves $y(n)$ with $g(n)$ to obtain the output $z(n)$, to be compared with $x(n)$. For a quick and qualitative comparison of x and z, visual displays of x, y, and z can be provided through the Matlab command *image*, as shown in the programs below.

a. Run the following program, analyze it, and examine the error between $x(n)$ and $z(n)$.

```
s=[1 2]; e=[0 0]; x=[e s e]; % Input
a=.8; k=2; n=1:k; h=a.^n;    % Filter
y=conv(h,x);
for i=2:length(y);
    z(i)=y(i)-a*y(i-1);
end
```

```
mx=max(x); my=max(y); mz=max(z);
p=zeros(length(z)-length(x));
pad=p(1,:);
xpad=[x pad];
xyz=[xpad/mx; y/my; z/mz];
image(50*xyz)
```

b. In the above program, substitute [e s s e] for x and repeat part a. Then repeat with $k = 5$.
c. Repeat for the set of parameters in the table below and enter your brief comments in the last column.

Filter Length	Input	Comments on the Output
$k = 5$	$x = $[e s s s s e]	
$k = 10$	$x = $[e s s s s e]	
$k = 10$	$x = $[e s s s s s s e]	
$k = 15$	$x = $[e s s s s s s e]	
$k = 20$	$x = $[e s s s s s s s s e]	

d. Replace the first line in the program given in (a) by the following:

```
s=[1 2 3 4 5 ]; e=[0 0 0 0 ]; x=[e s s s s s e];
```

Each image created by the program is composed of three horizontal segments made of vertical color bands. They represent three discrete signals. Each signal is a one-dimensional finite sequence $x(n)$. Signal's value is shown by color. Horizontal axis is the discrete variable n. The upper segment in each figure visualizes the input to the FIR filter specified by the program. The filter is a 20-tap FIR filter with the unit-sample response $h(n) = 0.8^n$, $n = 0, 1 \cdots, 19$. Filter's output is visualized on the middle segment. The signal is recovered from filter's output by deconvolution and is visualized on the lower segment. From a qualitative comparison of the upper segment with the lower segment in each example one may conclude a satisfactory signal recovery operation or not. Run the program and comment on the inverse output as an estimate of the initial input. Experiment with different lengths for the signal and filter. Experiment with different values for the filter's time constant. Create two figures to be called Figure A and B (not shown here). The input $x(n)$ for Figure A is

```
s=[1 2]; e=[0 0]; x=[e e s s s s s s s s s s s s s s s e e];
```

and for Figure B,

```
s=[1 2 3 4 5 ]; e=[0 0 0 0 ]; x=[e s s s s s e];
```

The figures you generate with the above parameters demonstrate possible signal recovery from filter's output by deconvolution. Comment on their performance.

e. The following program creates two-dimensional signals. Analyze and run it. Pass the signals through one-dimensional filters such as those introduced in a, followed by the appropriate inverse filters, and compare the output with the original signals.

```
m=50; n=1:m;
for i=1:2;
  for j=1:m;
     x(i,j)=n(j);
     end
  end
  for i=3:4;
  for j=1:m;
     x(i,j)=m-n(j);
```

```
            end
     end
%
for i=1:2;
  for j=1:m;
      y(i,j)=(m+n(j))/2;
      end
  end
  for i=3:4;
  for j=1:m;
      y(i,j)=m-n(j)/2;
      end
  end
figure(1); image(x)
figure(2); image(y)
xh=[x x x x]; yh=[y y y y];
figure(3); image(xh)
figure(4); image(yh)
figure(5); image([xh; xh; xh])
figure(6); image([yh; yh; yh])
```

Conclusions

Summarize your observations from each procedure. Combine your conclusions for the design of a satisfactory inverse system.

14

LTI Difference Equations

Contents

Introduction and Summary

The input-output description of a discrete-time LTI system in the time domain is almost always given by a linear difference equation with constant coefficients (to be called the LTI difference equation). The analytic method for those equations follows the same approach used for continuous-time LTI differential equations. In that approach we recognized the solution to be the sum of the particular solution (also called the forced response, reflecting the effect of the input) and the homogeneous solution (also called the natural response, made of the system's natural frequencies and reflecting its initial conditions). The role of the homogeneous solution was interpreted as accommodating the initial conditions. Similar parallels are observed in the solution of difference equations. After the solution methods are discussed, the chapter considers inputs of common interest such as the power, exponential, sinusoidal, unit-sample, and unit-step functions. It also introduces, informally and independent of the frequency domain, the concept of a system function as a scale factor that functions as the common thread relating the above responses.

14.1 What Is a Difference Equation?

A difference equation is a relationship between present and past values of two discrete-time signals $x(n)$ and $y(n)$. Most linear time-invariant discrete-time systems are described by a linear difference equation with constant coefficients

$$y(n) + a_1 y(n-1) \cdots + a_N y(n-N) = b_0 x(n) + b_1 x(n-1) \cdots + b_M x(n-M)$$

The equation is also written in the following two forms:

$$\sum_{k=0}^{N} a_k y(n-k) = \sum_{k=0}^{M} b_k x(n-k)$$

$$y(n) = \sum_{k=0}^{M} b_k x(n-k) - \sum_{k=1}^{N} a_k y(n-k)$$

where $x(n)$ is the input, $y(n)$ is the output, and a_k and b_k are constants (with $a_0 = 1$). The above difference equation is of order N and is recursive. It may be realized by an interconnection of unit delays, gains, and adders. If coefficients $a_k, k = 1, 2, \ldots$, are all zero, then

$$y(n) = \sum_{k=0}^{M} b_k x(n-k)$$

and the equation is nonrecursive. For further details and examples see sections 12.4 and 12.6 in Chapter 12.

Solution Methods

To find $y(n)$, one needs to know all the past values of $x(n)$, or know $x(n)$ starting at a given point plus the boundary conditions of $y(n)$. In either case, the difference equation may be solved through one of the following ways:

1. Numerical method.
2. Analytical method in the n-domain.

3. Convolution method.

4. z-transform method.

5. Fourier transform method.

Here we will discuss solutions by the numerical and n-domain analytical methods. (Convolution, z-transform, and Fourier transform methods are discussed under the broader topic of LTI systems, and apply to the difference equation as well.)

14.2 Numerical Solution

In this method, $y(n)$ is calculated from the weighted sums of its own past and the input's present and past values:

$$y(n) = -\sum_{k=1}^{N} a_k y(n-k) + \sum_{k=0}^{M} b_k x(n-k)$$

Calculation is repeated iteratively for the desired range of n. The method is computationally slow and the solution is not in closed form. However, it is applicable to nonlinear and time-varying difference equations and when input data is given in the form of an array of numbers.

Example **14.1**

Response to unit sample

Using the numerical method find the unit-sample response, $h(n)$, of the difference equation $y(n) + 2y(n-1) = d(n)$.

Solution

The response at time n is $y(n) = -2y(n-1) + d(n)$, which is the sum of $-2y(n-1)$ and $d(n)$. We form the set of arrays shown below to calculate the above sum for each n, incrementing n and iterating the calculation.

n	$d(n)$	$-2y(n-1)$	$y(n) = d(n) - 2y(n-1)$
-2	0	0	0
-1	0	0	0
0	1	0	1
1	0	-2	-2
2	0	4	4
3	0	-8	-8
4	0	16	16
5	0	-32	-32
...	...	...	...

[The closed-form solution of the above difference equation found by the analytic method in Example 14.3 is $h(n) = (-2)^n u(n)$.]

Response to unit step

Using the numerical method find $y(n)$ given

$$y(n) - 2y(n-1) - y(n-2) = 3x(n) + 5x(n-1) \text{ and } x(n) = u(n)$$

Solution

The solution is called the unit-step response of the system represented by the above difference equation. It may be written as

$$y(n) = 2y(n-1) + y(n-2) + 3u(n) + 5u(n-1)$$

We form the set of arrays shown below to calculate the above sum for each n, incrementing n and iterating the calculation.

n	$3u(n)$	$+5u(n-1)$	$+y(n-2)$	$+2y(n-1)$	$= y(n)$
−2	0	0	0	0	0
−1	0	0	0	0	0
0	3	0	0	0	3
1	3	5	0	6	14
2	3	5	3	28	39
3	3	5	14	78	100
4	3	5	39	200	247
5	3	5	100	494	602
...	...	...	...	...	...

The closed-form solution of the above difference equation is found by the analytic method in Example 14.8. The complete solution is

$$y(n) = \left[7.39(2.414)^n - 0.39(-0.414)^n - 4\right]u(n)$$

Note: The numerical method could be used, as easily, to solve nonlinear or time-varying equations. See section 14.16 for an example.

14.3   Analytic Solution in the *n*-Domain

Consider the N-th order linear difference equation with constant coefficients

$$y(n) + a_1 y(n-1) + \cdots + a_N y(n-N) = f(n), \quad n \geq 0$$

where $f(n)$ represents the total contribution from $x(n)$ and its derivatives and is called the forcing function. The objective is to find $y(n)$ for $n \geq 0$ so that (i) $y(n)$ satisfies the difference equation and (ii) $y(n)$ and its past $n - 1$ values are equal to a set of prescribed initial conditions. The total solution is written as the sum of two parts:

$$y(n) = y_h(n) + y_p(n)$$

where $y_h(n)$ is the homogeneous solution and $y_p(n)$ is the particular solution. The rationale behind the above solution is similar to what was discussed in the case of linear constant-coefficient differential equations (see Chapter 5). The particular solution satisfies the difference equation, but by itself does not necessarily meet the initial boundary conditions. The homogeneous solution is the solution of the equation with no input. It complements the particular solution in meeting the boundary conditions.

14.4 The Homogeneous Solution

The difference equation with a zero input

$$\sum_{k=0}^{N} a_k y(n-k) = 0$$

is called the homogeneous equation. One solution for the above equation is r^n, where r is found by substitution:

$$\sum_{k=0}^{N-1} a_k y(n-k) = \sum_{k=0}^{N} a_k r^{(n-k)} = r^n \sum_{k=0}^{N} a_k r^{-k} = 0$$

For r^n to be a solution of the homogeneous equation, r should be a solution of the equation

$$\sum_{k=0}^{N} a_k r^{-k} = 0$$

which is called the characteristic equation. The characteristic equation has N roots r_i, $i = 1, 2, \ldots, N$. Each term r_i^n can satisfy the homogeneous equation by itself. However, for the initial boundary conditions to be met, one needs all possible solutions in the form of

$$y_h(n) = \sum_{i=1}^{N} C_i r_i^n$$

Coefficients C_i must be set to certain values so that the total solution satisfies the boundary conditions.

Example 14.3

Find the unit-sample response, $h(n)$, of the difference equation $y(n) + 2y(n-1) = d(n)$.

Solution

Assuming a causal system, we note that $y(n) = 0, n < 0$. The characteristic equation is $r + 2 = 0$ with a root at $r = -2$. The homogeneous solution is $y_h(n) = C(-2)^n$.

The input for $n > 0$ is zero and, therefore, the particular solution for $n > 0$ is zero. The total solution is $y(n) = y_h(n) = C(-2)^n, \quad n > 0$. To determine C, we need $y(1)$, which is found from the difference equation.

$$y(n) = d(n) - 2y(n - 1)$$

$$y(0) = 1 - 2y(-1) = 1$$

$$y(1) = 0 - 2y(0) = -2$$

Substituting the above value of $y(1)$ in the total solution at $n = 1$, we find $C = 1$. Combining the above results we find the unit-sample response of the system to be $h(n) = (-2)^n u(n)$.

Observation

By using a less rigorous argument we find $y(n) = y_h(n) = C(-2)^n, \quad n \geq 0$, with the initial condition $y(0) = x(0) - 2y(-1) = 1$, which yields $C = 1$ and $h(n) = (-2)^n u(n)$.

14.5 The Particular Solution

The particular solution $y_p(n)$ is a function that satisfies the difference equation without consideration for boundary value conditions. The main requirement is that

$$y(n) + a_1 y(n - 1) \cdots + a_N y(n - N) = b_0 x(n) + b_1 x(n - 1) \cdots + b_M x(n - M)$$

The form of the particular solution depends on the forcing function. Some commonly encountered forcing functions and their corresponding particular solutions are given in Table 14.1. The unspecified constants k_i are found by substitution in the equation.

TABLE 14.1 A Short Table of Forcing Functions and Their Particular Solutions

Forcing Function $x(n)$ $\Longrightarrow$	Form of Particular Solution $y_p(n)$
a	k
a^n [see note below]	ka^n
$\sin(\omega n)$ or $\cos(\omega n)$	$k_1 \sin(\omega n) + k_2 \cos(\omega n)$
n^m	$k_0 + k_1 n + \cdots + k_m n^m$
$n^m a^n$	$a^n (k_0 + k_1 n + \cdots + k_m n^m)$
$a^n \sin(\omega n)$ or $a^n \cos(\omega n)$	$a^n [k_1 \sin(\omega n) + k_2 \cos(\omega n)]$
$n \sin(\omega n)$ or $n \cos(\omega n)$	$k_1 \sin(\omega n) + k_2 \cos(\omega n) + k_3 n \sin(\omega n) + k_4 n \cos(\omega n)$

Note: If a is a root of the characteristic equation, repeated m times, the particular solution is $kn^m a^n$ (see section 14.7).

14.6 The Total Solution

The total solution is

$$y(n) = y_p(n) + y_h(n) = y_p(n) + \sum_{i=1}^{N} C_i r_i^n$$

The parameters C_i are to be chosen so that the above solution meets the boundary conditions. For example, when $x(n)$ is given for $n \geq 0$, the conditions on $y(n)$ for $n < 0$ are $y(-N)$, $y(-N+1)$, $y(-N+2)$, ..., $y(-1)$. From these conditions, and by applying the difference equation, we derive values of $y(n)$ for $n \geq 0$; that is, $y(0)$, $y(1)$, $y(2)$, ..., $y(N-1)$. These values can then be inserted in the total solution to find the constants brought into it by the homogeneous part.

Example
14.4

Find the solution, to be called the unit-step response $g(n)$, of the difference equation given below.

$$y(n) + 2y(n-1) = u(n)$$

Solution
The homogeneous solution of this equation was found in Example 14.3 to be $y_h(n) = C(-2)^n$. The particular solution for $n \geq 0$ is a constant found by substitution $y_p(n) = 1/3$. The total solution is

$$y(n) = y_p(n) + y_h(n) = [1/3 + C(-2)^n]u(n)$$

with $y(0) = 1$ (found from the difference equation at $n = 0$). Applying the initial condition we get

$$y(0) = 1/3 + C = 1, \text{ from which } C = 2/3 \text{ and so } g(n) = \frac{1}{3}\left[1 + 2(-2)^n\right]u(n)$$

Example
14.5

Given $y(n) + 2y(n-1) = 1$, $n \geq 0$, and $y(-1) = 1/3$, find $y(n)$, $n \geq 0$.

Solution
Using the results of Example 14.4, we have $y(n) = 1/3 + C(-2)^n$, $n \geq 0$, with $y(0) = 1/3$ (found from the difference equation at $n = 0$). Applying the initial condition we get

$$y(0) = 1/3 + C = 1/3, \text{ from which } C = 0 \text{ and } y(n) = \frac{1}{3}, n \geq 0$$

Note that the homogeneous solution is zero.

14.7 Special Cases, Repeated Roots

The previous sections contain the basic approach sufficient to solve difference equations in most typical cases. However, in special cases the method is extended or tailored to the problem at hand. Some special cases are as follows:

1. The characteristic equation has one or more repeated roots.
2. The input is z^n, exponential, or sinusoidal.
3. The input contains power terms of roots of the characteristic equation.

Special treatment of the above cases often simplifies the analysis as discussed below.

Repeated Roots

A root of an equation that is repeated p times is called a multiple root of the p-th order. Contribution to the homogeneous solution from a multiple root r_k of order p is

$$(C_k + C_{k+1}n + C_{k+2}n^2 + \cdots + C_{k+p-1}n^{p-1})r_k^n$$

Example 14.6

Find $y(n)$ given by the difference equation

$$y(n) - y(n-1) + 0.25y(n-2) = u(n)$$

Solution

The characteristic equation is $r^2 - r + 0.25 = 0$. It has a repeated root at $r_{1,2} = 0.5$. The homogeneous solution is $y_h(n) = (C_1 + C_2n)0.5^n$. The particular solution is $y_p(n) = 4$. The total solution is $y(n) = \{(C_1 + C_2n)0.5^n + 4\}u(n)$. To find C_1 and C_2 we need $y(0)$ and $y(1)$. From the difference equation and knowing $y(-2) = y(-1) = 0$, we find $y(0) = 1$ and $y(1) = 1 + 1 = 2$. Now, from the total solution we have

$$\begin{cases} y(0) = C_1 + 4 \\ y(1) = 0.5(C_1 + C_2) + 4 \end{cases}$$

which, after substituting for $y(0)$ and $y(1)$, translate into the following two equations:

$$\begin{cases} C_1 + 4 = 1 \\ 0.5(C_1 + C_2) + 4 = 2 \end{cases}$$

From the above we find $C_1 = -3$ and $C_2 = -1$. The complete solution is, therefore,

$$y(n) = \left[(C_1 + C_2n)0.5^n + 4\right]u(n)$$

$$= \left[4 - (3+n)0.5^n\right]u(n)$$

14.8 Properties of LTI Difference Equations

A difference equation with constant real coefficients represents the input-output relationship of a linear time-invariant system. Some important properties of such an equation are summarized in Table 14.2.

TABLE 14.2 Properties of LTI Difference Equations with Real Coefficients

	Property	Input, $x(n)$	$\Rightarrow$	Output, $y(n)$
1	Linearity	$ax_1(n) + bx_2(n)$	$\Rightarrow$	$ay_1(n) + by_2(n)$
2	Delay	$x(n - n_0)$	$\Rightarrow$	$y(n - n_0)$
3	Real	$\mathcal{RE}[x(n)]$	$\Rightarrow$	$\mathcal{RE}[y(n)]$
4	Imaginary	$\mathcal{IM}[x(n)]$	$\Rightarrow$	$\mathcal{IM}[y(n)]$

The above properties can be proved by substitution in the equation. Property 1 is due to the lack of cross terms. Property 2 is due to the coefficients of the equation being constant. Properties 3 and 4 are due to the coefficients being real-valued numbers.

14.9 Response to z^n

The particular solution of the difference equation

$$y(n) + a_1 y(n - 1) + \cdots + a_N y(n - N) = z^n$$

is $y_p(n) = H \times z^n$. In this situation, z is a number independent of n. See section 14.5. The scale factor H is found by substituting $H \times z^n$ into the equation.

$$H \times \left[1 + a_1 z^{-1} + \cdots + a_N z^{n-N}\right] z^n = z^n$$

$$H = \frac{1}{1 + a_1 z^{-1} + \cdots + a_N z^{n-N}}$$

Regardless of the time interval during which the difference equation is valid, the particular response is proportional to the input. If the interval is infinite (steady-state condition), the total response is made of the particular response only and the total solution is proportional to the input. The proportionality factor depends on z and is shown by H. This is of special interest in that z^n plays the role of an Eigenfunction of the difference equation.

14.10 Response to the Complex Exponentials and Sinusoids

Complex Exponential Inputs

The solution of the difference equation

$$y(n) + a_1 y(n - 1) \cdots + a_N y(n - N) = e^{j\omega n}$$

is $He^{j\omega n}$. The scale factor H is found by substitution

$$H = \frac{1}{1 + a_1 e^{-j\omega} + \cdots + a_N e^{-j\omega N}}$$

The response is proportional to the input. The proportionality factor is a complex number H, which depends on ω. Note that

$$H(\omega) = H(z)\big|_{z=e^{j\omega}}$$

where $H(z)$ was found in section 14.9. In polar form,

$$H(\omega) = |H(\omega)|e^{j\theta}$$

where $|H(\omega)|$ is the magnitude and $\theta = \angle H(\omega)$ is the phase.

Sinusoidal Inputs

The response to a sinusoidal input may be derived from the response to a complex exponential input using the real property.

Input	$\Rightarrow$	Response
$e^{j\omega n}$	$\Rightarrow$	$H(\omega)e^{j\omega n} = |H(\omega)|e^{j(\omega n + \theta)}$
$\cos \omega n$	$\Rightarrow$	$|H(\omega)| \cos(\omega n + \theta)$

The difference equation scales the magnitude of the sinusoidal input by the factor $|H(\omega)|$ and adds $\theta = \angle H(\omega)$ to its phase angle. This is the steady-state response to a sinusoidal input and is an important feature of LTI systems. Note that $H(\omega)$ was found in section 14.10.

14.11 Unit-Step Response, $g(n)$

The response of the difference equation to a unit-step input is called the unit-step response and shown by $g(n)$.

$$y(n) + a_1 y(n-1) \cdots + a_N y(n-N) = u(n)$$

$$g(n) = \left[k_0 + \sum_{k=1}^{N} C_k r_k^n \right] u(n)$$

$$k_0 = \frac{1}{1 + a_1 + \cdots + a_N}$$

The solution contains N unknown coefficients C_k, $k = 1, 2, \ldots, N$. These are found from the N equations formed from the initial conditions $g(0)$, $g(1)$, $g(2)$, $\ldots$, $g(N-1)$.

$$g(0) = k_0 + \sum_{k=1}^{N} C_k$$

$$g(1) = k_0 + \sum_{k=1}^{N} C_k r_k$$

$$g(2) = k_0 + \sum_{k=1}^{N} C_k r_k^2$$

$$g(3) = k_0 + \sum_{k=1}^{N} C_k r_k^3$$

$$\vdots$$

$$g(N-1) = k_0 + \sum_{k=1}^{N} C_k r_k^{N-1}$$

Values for $g(0), g(1), g(2), \ldots, g(N-1)$ are derived from

$$y(n) = -\sum_{k=1}^{N} a_k y(n-k) + 1, \quad n \geq 0$$

using $y(-1) = y(-2) = y(-3) = \cdots = y(-N+1) = 0$ as shown below.

$$g(0) = 1$$

$$g(1) = 1 - a_1 y(0)$$

$$g(2) = 1 - a_1 y(0) - a_2 y(1)$$

$$g(3) = 1 - a_1 y(0) - a_2 y(1) - a_3 y(2)$$

$$\vdots$$

$$g(N-1) = 1 - \sum_{k=1}^{N-1} a_k y(k-1)$$

14.12 Unit-Sample Response, *h(n)*

The solution of the equation

$$y(n) + a_1 y(n-1) \cdots + a_N y(n-N) = d(n)$$

is called the unit-sample response. It is given by

$$h(n) = \sum_{k=1}^{N} C_k r_k^n u(n)$$

The solution contains N unknown coefficients $C_k, k = 1, 2, \ldots, N$. These are found from the N equations formed from the initial conditions $h(0), h(1), h(2), \ldots, h(N-1)$. These equations are

$$h(0) = 1 + \sum_{k=1}^{N} C_k$$

$$h(1) = \sum_{k=1}^{N} C_k r_k$$

$$h(2) = \sum_{k=1}^{N} C_k r_k^2$$

$$h(3) = \sum_{k=1}^{N} C_k r_k^3$$

$$\vdots$$

$$h(N-1) = \sum_{k=1}^{N} C_k r_k^{N-1}$$

Values for $h(0), \ h(1), \ h(2), \ldots, h(N-1)$ are derived from

$$y(n) = -\sum_{k=1}^{N} a_k y(n-k) + d(n)$$

using $y(-1) = y(-2) = y(-3) \cdots = y(-N+1) = 0$. Results are shown below.

$$h(0) = 1$$

$$h(1) = -a_1 y(0)$$

$$h(2) = -a_1 y(0) - a_2 y(1)$$

$$h(3) = -a_1 y(0) - a_2 y(1) - a_3 y(2)$$

$$\vdots$$

$$h(N-1) = -\sum_{k=1}^{N-1} a_k y(k-1)$$

14.13 Relation Between $h(n)$ and $g(n)$

Based on the linearity and time-invariance properties, the unit-sample response $h(n)$ may be derived from the unit-step response and vice versa as summarized in Table 14.3.

TABLE 14.3 Relationship Between the Unit-Sample and Unit-Step Responses

From $h(n)$ to $g(n)$		From $g(n)$ to $h(n)$	
Input $\implies$ **Response**		**Input** $\implies$ **Response**	
$d(n)$ $\implies$ $h(n)$		$u(n)$ $\implies$ $g(n)$	
$u(n) = \displaystyle\sum_{k=0}^{\infty} d(n-k)$ $\implies$ $g(n) = \displaystyle\sum_{k=0}^{\infty} h(n-k)$		$d(n) = u(n) - u(n-1)$ $\implies$ $h(n) = g(n) - g(n-1)$	

*E*xample

14.7

In Examples 14.3 and 14.4 we found responses of $y(n) + 2y(n-1) = x(n)$ to unit-sample and unit-step inputs.

$$d(n) \quad \implies \quad h(n) = (-2)^n u(n)$$

$$u(n) \quad \implies \quad g(n) = \tfrac{1}{3}\left[1 + 2(-2)^n\right]u(n)$$

Verify that $g(n) - g(n-1) = h(n)$.

Solution

$$g(n) = \frac{1}{3}\left[1 + 2(-2)^n\right]u(n) = d(n) + \frac{1}{3}\left[1 + 2(-2)^n\right]u(n-1)$$

$$g(n-1) = \frac{1}{3}\left[1 + 2(-2)^{n-1}\right]u(n-1)$$

$$g(n) - g(n-1) = d(n) + \frac{2}{3}\left[(-2)^n - (-2)^{n-1}\right]u(n-1) = d(n) + (-2)^n u(n-1)$$

$$= (-2)^n u(n) = h(n)$$

14.14 Use of Superposition

In sections 14.11 and 14.12 we found the solution of the difference equation when the right side of the equation was a unit step or unit sample. Now let us consider an LTI system (to be called the total system) described by the difference equation

$$y(n) + a_1 y(n-1) \cdots + a_N y(n-N) = b_0 x(n) + b_1 x(n-1) \cdots + b_M x(n-M)$$

To find the response of the above system to $x(n)$ we first solve for the following subsystem:

$$w(n) + a_1 w(n-1) \cdots + a_N w(n-N) = x(n)$$

We then use the linearity and time-invariance properties of the system to find its response in terms of $w(n)$.

$$y(n) = b_0 w(n) + b_1 w(n-1) \cdots + b_M w(n-M)$$

As an example, let the unit-step response of the subsystem be $g(n)$ (as found in section 14.11). The response of the total system to a unit-step input $x(n) = u(n)$ is then

$$y(n) = b_0 g(n) + b_1 g(n-1) \cdots + b_M g(n-M)$$

A similar approach may be employed to find the unit-sample response of the total system as a linear combination of the response of the subsystem to a unit-sample input.

a. Find the solution to the equation $y(n) - 2y(n-1) - y(n-2) = u(n)$ and call it $g(n)$.

b. Now consider the difference equation $y(n) - 2y(n-1) - y(n-2) = 3x(n) + 5x(n-1)$, $x(n) = u(n)$. The solution to this equation (derived by direct method in solved problem 4 at the end of the chapter) is

$$y(n) = [7.3891(2.4142)^n - 0.3891(-0.4142)^n - 4]u(n)$$

Verify that the above solution is equal to $3g(n) + 5g(n-1)$, where $g(n)$ is the unit-step response of part a.

Solution

a. The step response of the equation $y(n) - 2y(n-1) - y(n-2) = u(n)$ is

$$g(n) = \left[-\frac{1}{2} + C_1 \alpha^n + C_2 \beta^n \right] u(n) \text{ where } \alpha = 1 + \sqrt{2} \text{ and } \beta = 1 - \sqrt{2}.$$

C_1 and C_2 are found by applying the initial conditions $y(0)$ and $y(1)$, which are found from the difference equation:

$$y(0) = 2y(-1) + y(-2) + 1 = 1$$
$$y(1) = 2y(0) + y(-1) + 1 = 3$$

Applying the initial conditions we get

$$C_{1,2} = \frac{3 \pm 2\sqrt{2}}{4}$$

b. For $n \geq 1$ we have

$$3g(n) + 5g(n-1) = -\frac{3}{2} + 3C_1 \alpha^n + 3C_2 \beta^n$$
$$-\frac{5}{2} + 5C_1 \alpha^{(n-1)} + 5C_2 \beta^{(n-1)}$$
$$= -4 + \frac{36 + 25\sqrt{2}}{4}(1 + \sqrt{2})^{(n-1)}$$
$$+\frac{36 - 25\sqrt{2}}{4}(1 - \sqrt{2})^{(n-1)}$$
$$= -4 + 7.3891(2.412)^n - 0.3891(-0.4142)^n, \ n \geq 1$$

The above expression also provides a correct value for $3g(0) + 5g(-1)$. Therefore,

$$3g(n) + 5g(n-1) = [-4 + 7.3891(2.412)^n - 0.3891(-0.4142)^n]u(n)$$

This is the same solution obtained directly in solved problem 4 at the end of the chapter.

14.15 Zero-Input and Zero-State Responses

The superposition property (a consequence of linearity and time invariance) may be applied to combine the contributions to the response from the initial state (with zero input) and the input (with zero initial state). These are called *zero-input* and *zero-state* responses. Therefore, the difference equation

$$y(n) + a_1 y(n-1) \cdots + a_N y(n-N) = b_0 x(n) + b_1 x(n-1) \cdots$$
$$+ b_M x(n-M), \quad n \ge 0$$

with nonzero initial conditions $y(0), y(1), \ldots y(N-1)$ can be divided into two equations

(Zero-input:) $y(n) + a_1 y(n-1) \cdots + a_N y(n-N) = 0, \; y(0), \; y(1), \ldots y(N-1)$

(Zero-state:) $y(n) + a_1 y(n-1) \cdots + a_N y(n-N) = [b_0 x(n) + b_1 x(n-1) \cdots$
$+ b_M x(n-M)]u(n)$

Superposition of the solutions of the above two equations produces the complete solution to the original equation.

Given $y(n) + 2y(n-1) = 1$, $n \ge 0$, and $y(-1) = 1/3$, find the zero-input and zero-state responses for $n \ge 0$ and show that their superposition results in the total response obtained in Example 14.5.

Solution
Zero-input response. The zero-input response is found from

$$y(n) + 2y(n-1) = 0, \; n \ge 0, \; \text{and } y(-1) = 1/3$$

The solution is $y(n) = C(-2)^n, n \ge 0$, with $y(0) = -2/3$ (found from the difference equation at $n = 0$). Applying the initial condition, we get $C = -2/3$ and

$$y_1(n) = -(2/3)(-2)^n, \; n \ge 0 \quad \text{(zero-input response)}$$

Zero-state response. The zero-state response is found from

$$y(n) + 2y(n-1) = 1, \ n \geq 0, \ \text{and} \ y(-1) = 0$$

The solution is the unit-step response already found in Example 14.4.

$$y_2(n) = \frac{1}{3}\left[1 + 2(-2)^n\right], \ n \geq 0 \ \text{(zero-state response)}$$

The total solution is

$$y(n) = y_1(n) + y_2(n) = -(2/3)(-2)^n + 1/3 + (2/3)(-2)^n = 1/3, \ \ n \geq 0$$

The total solution is the same as the answer found in Example 14.5.

14.16 A Nonlinear Time-Varying Difference Equation

When an exact and closed-form solution is not needed, the response of a difference equation can be obtained by numerical methods. The method was illustrated for LTI difference equations in section 14.2 and the components of the solution were associated with the particular and homogeneous response. The method can be applied to nonlinear and time-varying equations, more conveniently as analytical solutions for many such equations are complex or don't exist. The response obtained by step-by-step computation may not explicitly reflect the various factors that influence the shape of the response, but computation can be carried out for a desired length of the response and its accuracy. The following is an example.

Find $y(n)$ given the equation

$$y(n) - 0.1ny(n-1) + y^2(n-2) = x(n) \ n \geq 0, \ y(-1) = 1, \ y(-2) = 1$$

where $x(n)$ is given by

$$x(n) = \{\underset{\uparrow}{1}, \ 2, \ -1.2, \ 0, \ 3, \ -3, \ 5, \ 0, \ -2, \ldots\}.$$

Solution
This is a nonlinear and time-varying equation. The nonlinearity is due to $y^2(n-2)$. The term $0.1ny(n-1)$ makes the equation time varying. The response at time n is computed from the sum

$$y(n) = x(n) - y^2(n-2) + 0.1ny(n-1) \ n \geq 0, \ y(-1) = 1, \ y(-2) = 1$$

We form the following set of arrays and calculate $y(n)$ iteratively.

n	$y(n-2)$	$y(n-1)$	$x(n)$	$-y^2(n-2)$	$+0.1ny(n-1)$	$= y(n)$	(comments)
-3	—	—	—	—	—	—	(not given)
-2	—	—	—	—	—	1	(given)
-1	—	1	—	—	—	1	(given)
0	1	1	1	-1	0	0	(computed)
1	1	0	2	-1	0	1	"
2	0	1	-1.2	0	0.2	-1	"
3	1	-1	0	-1	-0.3	-1.3	"
4	-1	-1.3	3	-1	-0.52	1.48	"
5	-1.3	1.48	-3	-1.69	0.74	-3.95	"
6	1.48	-3.95	5	-2.19	-2.37	0.44	"
7	-3.95	0.44	0	-15.6	0.31	-15.29	"
8	0.44	-15.29	-2	-0.19	-12.23	-14.42	"
$\cdots$	$\cdots$	$\cdots$	$\cdots$	$\cdots$	$\cdots$	$\cdots$	$\cdots$

14.17 Problems

Solved Problems

1. Find the homogeneous solution of the difference equation $y(n) + 2y(n-1) = x(n)$.

Solution

The characteristic equation is $r + 2 = 0$, or $r = -2$. The homogeneous solution is $y_h(n) = C(-2)^n$. The constant C is determined from the forcing function and $y(n)$ specified at a given time.

2. Fibonacci numbers $y(n)$ are described by the following equation:

$$y(n) = y(n-1) + y(n-2), \quad n \ge 0, \quad y(-1) = 1, \quad y(-2) = 0$$

Find a closed-form expression for $y(n)$, $n \ge 0$.

Solution

The characteristic equation is $r^2 - r - 1 = 0$, or $r_{1,2} = 0.5(1 \pm \sqrt{5})$. The solution is made of the homogeneous part only which is $y(n) = A(r_1)^n + B(r_2)^n$, $n \ge 0$. The constants A and B are found from the initial conditions:

$$y(0) = y(-1) + y(-2) = 1 \text{ and } y(1) = y(0) + y(-1) = 2$$

By applying the initial conditions we find A, B, and $y(n)$.

$$\begin{cases} y(0) = A + B = 1 \\ y(1) = 0.5(1 + \sqrt{5})A + 0.5(1 - \sqrt{5})B = 2 \end{cases} \implies \begin{cases} A = 0.5 + 0.3\sqrt{5} \\ B = 0.5 - 0.3\sqrt{5} \end{cases}$$

$$y(n) = \left(\frac{5 + 3\sqrt{5}}{10}\right)\left(\frac{1 + \sqrt{5}}{2}\right)^n + \left(\frac{5 - 3\sqrt{5}}{10}\right)\left(\frac{1 - \sqrt{5}}{2}\right)^n, \quad n \ge 0$$

3. Find the complete solution of the difference equation

$$y(n) + 1.5y(n-1) + 0.5y(n-2) = (0.3)^n, \ n \ge 0, \ y(-1) = y(-2) = 0.$$

Solution
The characteristic equation is $r^2 + 1.5r + 0.5 = 0$. The roots are $r_{1,2} = -1$ and -0.5. The homogeneous solution is

$$y_h(n) = C_1(-1)^n + C_2(-0.5)^n$$

The total solution is $y(n) = y_h(n) + y_p(n)$, where $y_p(n) = k(0.3)^n, \ n \ge 0$. To determine k we put $y_p(n)$ into the equation. Therefore,

$$k(0.3)^n + 1.5k(0.3)^{n-1} + 0.5k(0.3)^{n-2} = (0.3)^n$$

$$k(0.3)^2 + 1.5k(0.3) + 0.5k = (0.3)^2$$

or $k = 9/104$. The total solution is

$$y(n) = C_1(-1)^n + C_2(-0.5)^n + \frac{9}{104}(0.3)^n, \ n \ge 0$$

To determine C_1 and C_2, we need $y(0)$ and $y(1)$ which are found from applying the difference equation at $n = 0$ and $n = 1$ along with the given values for $y(-2) = y(-1) = 0$.

$$y(n) = (0.3)^n - 1.5y(n-1) - 0.5y(n-2), \ n \ge 0$$

$$y(0) = 1 - 1.5y(-1) - 0.5y(-2) = 1$$

$$y(1) = 0.3 - 1.5y(0) - 0.5y(-1) = -1.2$$

Substituting the above values for $y(0)$ and $y(1)$ in the total solution we find

$$y(0) = C_1 + C_2 + \frac{9}{104} = 1$$

$$y(1) = -C_1 - 0.5C_2 + \frac{9}{104}(0.3) = -1.2$$

from which we obtain

$$C_1 = \frac{480}{312} \approx 1.5385 \ \text{and} \ C_2 = -\frac{195}{312} = -0.625$$

The total solution is, therefore,

$$y(n) = \frac{1}{312}\left[480 \times (-1)^n - 195 \times (-0.5)^n + 27 \times (0.3)^n\right], \ n \ge 0$$

4. Find $y(n)$ given by $y(n) - 2y(n-1) - y(n-2) = 3x(n) + 5x(n-1), \ x(n) = u(n)$.

Solution
The homogeneous solution is $y_h(n) = C_1 r_1^n + C_2 r_2^n$, where $r_{1,2} = 1 \pm \sqrt{2}$ are the roots of the characteristic equation $r^2 - 2r - 1 = 0$.

The particular solution is $y_p(n) = k_0, \ n \ge 1$, where $k_0 = -4$ (by substitution). The total solution is then

$$y(n) = C_1(1 + \sqrt{2})^n + C_2(1 - \sqrt{2})^n - 4, \ n \ge 0$$

where C_1 and C_2 are found by applying the initial conditions $y(1)$ and $y(2)$. Assuming a causal system, we have $y(-1) = y(-2) = 0$. The values of $y(1)$ and $y(2)$ are then found from the difference equation

$$y(0) = 2y(-1) + y(-2) + 3x(0) + 5x(-1) = 0 + 0 + 3 + 0 = 3$$

$$y(1) = 2y(0) + y(-1) + 3x(1) + 5x(0) = 6 + 0 + 3 + 5 = 14$$

$$y(2) = 2y(1) + y(0) + 3x(2) + 5x(1) = 28 + 3 + 3 + 5 = 39$$

Applying the initial conditions we get

$$y(1) = C_1(1 + \sqrt{2}) + C_2(1 - \sqrt{2}) - 4 = 14$$
$$y(2) = C_1(1 + \sqrt{2})^2 + C_2(1 - \sqrt{2})^2 - 4 = 39$$

From the above we find $C_1 = 7.3891$ and $C_2 = -0.3891$. The complete solution is

$$y(n) = 7.3891(2.4142)^n - 0.3891(-0.4142)^n - 4, \quad n \geq 1$$

Note that the above expression also provides a correct value for $y(0)$. Therefore,

$$y(n) = [7.3891(2.4142)^n - 0.3891(-0.4142)^n - 4]u(n)$$

(Revisit this equation in Examples 14.2 and 14.8, and problem 34.)

5. The Chebyshev polynomial $C_n(x)$ is an nth-order polynomial of the continuous variable x. It is the solution to the following difference equation

$$c_{n+1}(x) = 2xc_n(x) - c_{n-1}(x), \quad n \geq 1, \quad c_0 = 1, \quad \text{and } c_1 = x$$

Find a closed-form expression for $C_n(x)$.
Hint: You may rewrite the equation in the familiar form

$$c(n + 1) = 2xc(n) - c(n - 1), \quad n \geq 1, \quad c(0) = 1, \quad \text{and } c(1) = x$$

Solution
The characteristic equation is $r^2 - 2xr + 1 = 0$, or $r_{1,2} = x \pm \sqrt{x^2 - 1}$. The solution is made of the homogeneous part only, which is $c(n) = A(r_1)^n + B(r_2)^n, \quad n \geq 1$. The constants A and B are found from the initial conditions. The initial conditions to be used for finding A and B are $c(0) = 1$ and $c(1) = x$. By substitution we get

$$c(0) = A + B = 1$$

$$c(1) = A\left(x + \sqrt{x^2 - 1}\right) + B\left(x - \sqrt{x^2 - 1}\right) = x$$

From the above we find $A = B = \frac{1}{2}$. The closed-form expression for the Chebyshev polynomials of order n is

$$c_n(x) = \frac{1}{2}\left[\left(x + \sqrt{x^2 - 1}\right)^n + \left(x - \sqrt{x^2 - 1}\right)^n\right], \quad n \geq 0$$

The first six Chebyshev polynomials computed from the above solution are the following:

n	$C_n(x)$
0	1
1	x
2	$2x^2 - 1$
3	$4x^3 - 3x$
4	$8x^4 - 8x^2 + 1$
5	$16x^5 - 20x^3 + 5x$

6. Find the complete solution of the difference equation $y(n) + 2y(n - 1) = x(n)$ for the following forcing functions:
a. $x(n) = 2u(n)$
b. $x(n) = 3nu(n)$
c. $x(n) = n^2u(n)$

d. $x(n) = (3n + 2)u(n)$

e. $x(n) = (3n + 1)u(n)$

f. $x(n) = (n^2 + 1)u(n)$

Solution

The total solution is $y(n) = y_h(n) + y_p(n)$, where $y_h(n) = C(-2)^n$ was found in problem 1. $y_p(n)$ is a polynomial in n whose coefficients are found by inserting it in the difference equation. The constant C in the homogeneous solution is then found from the boundary value $y(0)$. From the difference equation, $y(0) = x(0) - 2y(-1)$. But, $y(-1) = 0$. Therefore, the boundary condition to be used for finding C is $y(0) = x(0)$.

a. $y_p(n) = k_0$. By plugging $y_p(n) = k_0$ into the equation we obtain $k_0 + 2k_0 = 2$, or $k_0 = 2/3$. The total solution is

$$y(n) = \left[C(-2)^n + \frac{2}{3} \right] u(n)$$

The initial condition to be applied is $y(0) = 2 - 2y(-1) = 2$. The constant C is found from $y(0) = C + 2/3 = 2$, or $C = 4/3$. Therefore, the complete solution is

$$y(n) = \frac{2}{3} \left[1 + 2(-2)^n \right] u(n)$$

b. $y_p(n) = k_1 n + k_0$. After plugging $y_p(n)$ into the equation, we get $k_1 = 1$ and $k_0 = 2/3$. The total solution is

$$y(n) = \left[C(-2)^n + n + \frac{2}{3} \right] u(n)$$

with $y(0) = 0$. The constant C is found from $y(0) = C + \frac{2}{3} = 0$, or $C = -\frac{2}{3}$. The complete solution is

$$y(n) = \frac{2}{3} \left[1 + \frac{3}{2} n - (-2)^n \right] u(n)$$

c. $y_p(n) = k_2 n^2 + k_1 n + k_0$. After plugging $y_p(n)$ into the equation, we get $k_2 = 1/3$, $k_1 = 4/9$, and $k_0 = 2/27$. The total solution is

$$y(n) = \left[C(-2)^n + \frac{1}{3} n^2 + \frac{4}{9} n + \frac{2}{27} \right] u(n)$$

with the boundary value $y(0) = 0$. The constant C is found from $y(0) = C + 2/27 = 0$, or $C = -2/27$. The complete solution is

$$y(n) = \frac{1}{3} \left[n^2 + \frac{4}{3} n + \frac{2}{9} - \frac{2}{9}(-2)^n \right] u(n)$$

d. $y_p(n) = k_1 n + k_0$. By plugging this into the equation, we obtain $k_1 = 1$ and $k_0 = 4/3$. The total solution is

$$y(n) = \left[C(-2)^n + n + \frac{4}{3} \right] u(n)$$

and $y(0) = 2$. The constant C is found from $y(0) = C + 4/3 = 2$, or $C = 2/3$. The complete solution is

$$y(n) = \frac{2}{3} \left[(-2)^n + \frac{3}{2} n + 2 \right] u(n)$$

e. $y_p(n) = k_1 n + k_0$. By plugging this into the equation, we obtain $k_1 = 1$ and $k_0 = 1$. The total solution is

$$y(n) = \left[C(-2)^n + n + 1 \right] u(n)$$

with the boundary value $y(0) = 1$. The constant C is found from $y(0) = C + 1 = 1$, or $C = 0$. The complete solution is

$$y(n) = (n + 1)u(n)$$

The complete solution is made up of the particular part only. The particular solution meets the boundary conditions by itself and there is no need for a homogeneous part.

f. $y_p(n) = k_2 n^2 + k_1 n + k_0$. By plugging this into the equation, we obtain $k_2 = 1/3$, $k_1 = 4/9$, and $k_0 = 11/27$. The total solution is

$$y(n) = \left[C(-2)^n + \frac{1}{3}n^2 + \frac{4}{9}n + \frac{11}{27} \right] u(n)$$

with the boundary value $y(0) = 1$. The constant C is found from $y(0) = C + 11/27 = 1$, or $C = 16/27$. The complete solution is

$$y(n) = \frac{1}{3} \left[n^2 + \frac{4}{3}n + \frac{11}{9} + \frac{16}{9}(-2)^n \right] u(n)$$

The solution to parts d, e, and f may also be found from the solution to parts a, b, and c using the linearity property of the difference equation. The answers are summarized in the table below.

$x(n)$		$\Rightarrow$	$y(n)$	
$x_1(n)$	$= 2u(n)$	$\Rightarrow$	$y_1(n)$	$= \dfrac{2}{3}[1 + 2(-2)^n]u(n)$
$x_2(n)$	$= 3nu(n)$	$\Rightarrow$	$y_2(n)$	$= \dfrac{2}{3}\left[1 + \dfrac{3}{2}n - (-2)^n\right]u(n)$
$x_3(n)$	$= n^2 u(n)$	$\Rightarrow$	$y_3(n)$	$= \dfrac{1}{3}\left[n^2 + \dfrac{4}{3}n + \dfrac{2}{9} - \dfrac{2}{9}(-2)^n\right]u(n)$
$x_4(n)$	$= x_1(n) + x_2(n)$	$\Rightarrow$	$y_4(n)$	$= y_1(n) + y_2(n)$
	$= (2 + 3n)u(n)$			$= \dfrac{2}{3}\{1 + 2(-2)^n\}u(n) + \dfrac{2}{3}\left[1 + \dfrac{3}{2}n - (-2)^n\right]u(n)$
				$= \dfrac{2}{3}\left[(-2)^n + \dfrac{3}{2}n + 2\right]u(n)$
$x_5(n)$	$= 0.5x_1(n) + x_2(n)$	$\Rightarrow$	$y_5(n)$	$= 0.5y_1(n) + y_2(n)$
	$= (1 + 3n)u(n)$			$= \dfrac{1}{3}\{1 + 2(-2)^n\}u(n) + \dfrac{2}{3}\left[1 + \dfrac{3}{2}n - (-2)^n\right]u(n)$
				$= (n + 1)u(n)$
$x_6(n)$	$= 0.5x_1(n) + x_3(n)$	$\Rightarrow$	$y_6(n)$	$= 0.5y_1(n) + y_3(n)$
	$= (1 + n^2)u(n)$			$= \dfrac{1}{3}\{1 + 2(-2)^n\}u(n) + \dfrac{1}{3}\left[n^2 + \dfrac{4}{3}n + \dfrac{2}{9} - \dfrac{2}{9}(-2)^n\right]u(n)$
				$= \dfrac{1}{3}\left[n^2 + \dfrac{4}{3}n + \dfrac{11}{9} - \dfrac{16}{9}(-2)^n\right]u(n)$

7. Find the complete solution of the difference equation

$$y(n) + 1.5y(n - 1) + 0.5y(n - 2) = x(n) \quad \text{for}$$

a. $x(n) = 5$, $n > 0$, $y(0) = y(-1) = 1$.
b. $x(n) = 5$, $n > 0$. What values of $y(0)$ and $y(-1)$ result in $y_h(n) = 0$?
c. $x(n) = (-0.5)^n$, $n > 0$, $y(0) = y(-1) = 0$.

Solution

The characteristic equation is $r^2 + 1.5r + 0.5 = 0$. The roots are $r_{1,2} = -1, -0.5$. The homogeneous solution is

$$y_h(n) = C_1(-1)^n + C_2(-0.5)^n$$

The total solution is $y(n) = y_h(n) + y_p(n)$, where $y_p(n)$ for each case is as derived below.

a. The particular solution is $y_p(n) = k_0$, where k_0 is found by substitution

$$k_0 + 1.5k_0 + 0.5k_0 = 5, \quad k_0 = \frac{5}{3}$$

$$y(n) = C_1(-1)^n + C_2(-0.5)^n + \frac{5}{3}, \quad n > 0$$

To find C_1 and C_2 we need $y(1)$ and $y(2)$. These are found from

$$y(1) = 5 - 1.5y(0) - 0.5y(-1) = -C_1 - .5C_2 + \frac{5}{3} = 3$$

$$y(2) = 5 - 1.5y(1) - 0.5y(0) = C_1 + 0.25C_2 + \frac{5}{3} = 0$$

from which we obtain

$$C_1 = -2 \quad \text{and} \quad C_2 = \frac{4}{3}$$

The complete solution is, therefore,

$$y(n) = -2(-1)^n + \frac{4}{3}(-0.5)^n + \frac{5}{3}, \quad n > 0$$

b. The particular solution is $y_p(n) = \frac{5}{3}$. For $n > 0$, it is desired that $y(n) = y_p(n) = 5/3$. Values of $y(0)$ and $y(-1)$ that produce such a solution are found from

$$y(1) = -1.5y(0) - 0.5y(-1) + 5 = \frac{5}{3}$$

$$y(2) = -1.5y(1) - 0.5y(0) + 5 = \frac{5}{3}$$

Hence, $y(0) = y(-1) = 5/3$. We now check the complete solution for these conditions. The total solution is

$$y(n) = C_1(-1)^n + C_2(-0.5)^n + \frac{5}{3}, \quad n > 0$$

C_1 and C_2 are found from $y(1)$ and $y(2)$.

$$y(1) = -C_1 - .5C_2 + \frac{5}{3} = \frac{5}{3} \quad \text{and} \quad y(2) = C_1 + 0.25C_2 + \frac{5}{3} = \frac{5}{3}$$

from which we obtain $C_1 = C_2 = 0$. The complete solution is, therefore, $y(n) = 5/3$.

c. $y_p(n) = kn(-0.5)^n$. By plugging $y_p(n)$ into the equation, we get

$$kn(-0.5)^n + 1.5k(n-1)(-0.5)^{n-1} + 0.5k(n-2)(-0.5)^{n-2} = (-0.5)^n$$

or $k = -1$. The total solution is

$$y(n) = y_h(n) + y_p(n) = C_1(-1)^n + C_2(-0.5)^n - n(-0.5)^n, \quad n > 0$$

The boundary conditions are

$$y(1) = -1.5y(0) - 0.5y(-1) - 0.5 = -0.5$$
$$y(2) = -1.5y(1) - 0.5y(0) + 0.25 = 1$$

from which we obtain

$$y(1) = -C_1 - 0.5C_2 + 0.5 = -0.5$$
$$y(2) = C_1 + 0.25C_2 - 0.5 = 1$$

resulting in $C_1 = 2$ and $C_2 = -2$. The complete solution is then

$$y(n) = 2(-1)^n - (2+n)(-.5)^n, \quad n > 0$$

In summary:

a. $y(n) = -2(-1)^n + \frac{4}{3}(-0.5)^n + \frac{5}{3}, \quad n > 0$
b. $y(n) = 5/3, \quad n > 0$
c. $y(n) = 2(-1)^n - (2+n)(-0.5)^n, \quad n > 0$

8. Find the solution of the equation $y(n) - y(n-1) + .5y(n-2) = 0, \quad n \geq 0, \ y(-1) = y(-2) = 1$.

Solution

The characteristic equation is $r^2 - r + 0.5 = 0$. The roots are

$$r_{1,2} = 0.5(1 \pm j) = \frac{\sqrt{2}}{2}e^{\pm j\frac{\pi}{4}}$$

$y(n) = y_h(n) = C_1 r_1^n + C_2 r_2^n$. Since r_1 and r_2 are complex conjugates of each. C_1 and C_2 must also be complex conjugates then: $C_{1,2} = Ce^{\pm j\theta}$, where C and θ are real constants. The response may be written as

$$y(n) = C_1 r_1^n + C_2 r_2^n = Ce^{j\theta}\left(\frac{\sqrt{2}}{2}\right)^n e^{j\frac{\pi}{4}n} + Ce^{-j\theta}\left(\frac{\sqrt{2}}{2}\right)^n e^{-j\frac{\pi}{4}n}$$

$$= C\left(\frac{\sqrt{2}}{2}\right)^n \left\{ e^{j(\frac{\pi}{4}n+\theta)} + e^{-j(\frac{\pi}{4}n+\theta)} \right\}$$

$$= 2C\left(\frac{\sqrt{2}}{2}\right)^n \cos(\frac{\pi}{4}n + \theta)$$

The boundary conditions are

$$y(0) = y(-1) - 0.5y(-2) = 1 - 0.5 = 0.5 \text{ and } y(1) = y(0) - 0.5y(-1) = 0.5 - 0.5 = 0$$

The constants C and θ are found from the boundary conditions

$$y(0) = 2C\cos\theta = 0.5 \text{ and } y(1) = C\sqrt{2}\cos(\frac{\pi}{4} + \theta) = 0$$

$C = \frac{\sqrt{2}}{4}$ and $\theta = \frac{\pi}{4}$. The complete solution is, therefore,

$$y(n) = \left(\frac{\sqrt{2}}{2}\right)^{n+1} \cos\left[\frac{\pi}{4}(n+1)\right], \quad n \geq 0$$

9. An LTI system is represented by $y(n) + y(n-1) - y(n-2) = x(n)$. Determine the initial conditions $y(-1)$ and $y(-2)$ so that for $x(n) = 1, n \geq 0$, the response becomes $y(n) = k_0$. Then find k_0.

Solution

It is desired that the response be made of the particular solution only; that is, $y(n) = y_p(n) = k_0$. By plugging $y_p(n)$ into the equation, we get $y(n) = 1$. We need to find $y(-1)$ and $y(-2)$ that produce such a condition. From an examination of the difference equation for $n \geq 0$ we observe that $y(n) = -y(n-1) + y(n-2) + x(n)$ But, $y(n) = 1$ and $x(n) = 1$. Therefore, $y(n-2) = y(n-1)$.

By applying the above at $n = 1$ and $n = 0$, we find, respectively, $y(-1) = y(0) = 1$ and $y(-2) = y(-1) = 1$. These are the conditions required of $y(n)$ before the arrival of the input $x(n) = 1$ at $n = 0$.

To check the validity of the above conditions, we find the total solution and determine the constants in the homogeneous solution. The homogeneous solution is $y_h(n) = C_1 r_1^n + C_2 r_2^n$, where r_1 and r_2 are the roots of the characteristic equation $r^2 + r - 1 = 0$:

$$r_{1,2} = \frac{-1 \pm \sqrt{5}}{2}$$

The particular solution was found to be $y_p(n) = 1$ and the total response is then $y(n) = C_1(r_1)^n + C_2(r_2)^n + 1$, $n \geq 0$. The boundary conditions $y(0)$ and $y(1)$ are derived by applying the difference equation $y(n) = -y(n-1) + y(n-2) + x(n)$ at $n = 0$ and 1, resulting in $y(0) = y(1) = 1$. Applying these conditions to the total response we find C_1 and C_2 as shown below:

$$y(0) = C_1 + C_2 + 1 = 1 \text{ and } y(1) = C_1 r_1 + C_2 r_2 + 1 = 1$$
$$C_1 + C_2 = 0 \text{ and } C_1 r_1 + C_2 r_2 = 0$$
$$C_1 = 0 \text{ and } C_2 = 0$$

10. The response of a fifth-order linear time-invariant discrete system to a DC input is represented by the difference equation

$$y(n) + a_1 y(n-1) + a_2 y(n-2) + a_3 y(n-3) + a_4 y(n-4) + a_5 y(n-5) = 1, \quad n \geq 0$$

Determine values of $y(-1)$, $y(-2)$, $y(-3)$, $y(-4)$, and $y(-5)$ that produce $y(n) = k_0, n \geq 0$; that is, make the homogeneous part of the solution zero.

Solution

The particular solution is

$$y_p(n) = k_0 = \left[\sum_{i=1}^{5} a_i \right]^{-1}, \quad (\text{with } a_0 = 1)$$

To find $y(n), n = -1, -2, -3, -4, -5$, we need five independent equations. From the difference equation we note that

$$a_1 y(n-1) + a_2 y(n-2) + a_3 y(n-3) + a_4 y(n-4) + a_5 y(n-5) = 1 - y(n) = 1 - k_0, \quad n \geq 0$$

By applying the above at $n = 0, 1, 2, 3$, and 4, we get

$n = 0,$	$a_1 y(-1)$	$+a_2 y(-2)$	$+a_3 y(-3)$	$+a_4 y(-4)$	$+a_5 y(-5)$ = $1 - k_0$
$n = 1,$	$a_1 y(0)$	$+a_2 y(-1)$	$+a_3 y(-2)$	$+a_4 y(-3)$	$+a_5 y(-4)$ = $1 - k_0$
$n = 2,$	$a_1 y(1)$	$+a_2 y(0)$	$+a_3 y(-1)$	$+a_4 y(-2)$	$+a_5 y(-3)$ = $1 - k_0$
$n = 3,$	$a_1 y(2)$	$+a_2 y(1)$	$+a_3 y(0)$	$+a_4 y(-1)$	$+a_5 y(-2)$ = $1 - k_0$
$n = 4,$	$a_1 y(3)$	$+a_2 y(2)$	$+a_3 y(1)$	$+a_4 y(0)$	$+a_5 y(-1)$ = $1 - k_0$

Substituting for $y(0) = y(1) = y(2) = y(3) = y(4) = k_0$, we get

$$a_1 y(-1) + a_2 y(-2) + a_3 y(-3) + a_4 y(-4) + a_5 y(-5) = 1 - k_0$$
$$a_2 y(-1) + a_3 y(-2) + a_4 y(-3) + a_5 y(-4) = 1 - k_0(1 + a_1)$$

$$a_3 y(-1) + a_4 y(-2) + a_5 y(-3) = 1 - k_0(1 + a_1 + a_2)$$

$$a_4 y(-1) + a_5 y(-2) = 1 - k_0(1 + a_1 + a_2 + a_3)$$

$$a_5 y(-1) = 1 - k_0(1 + a_1 + a_2 + a_3 + a_4)$$

The answers are, therefore,

$$y(-1) = \frac{1}{a_5}[1 - k_0(1 + a_1 + a_2 + a_3 + a_4)]$$

$$y(-2) = \frac{1}{a_5}[1 - k_0(1 + a_1 + a_2 + a_3) - a_4 y(-1)]$$

$$y(-3) = \frac{1}{a_5}[1 - k_0(1 + a_1 + a_2) - a_3 y(-1) - a_4 y(-2)]$$

$$y(-4) = \frac{1}{a_5}[1 - k_0(1 + a_1) - a_2 y(-1) - a_3 y(-2) - a_4 y(-3)]$$

$$y(-5) = \frac{1}{a_5}[1 - k_0 - a_1 y(-1) - a_2 y(-2) - a_3 y(-3) - a_4 y(-4)]$$

Chapter Problems

In the following problems all systems are causal unless specified otherwise.

11. Find the homogeneous part of the solution of the following equations:

a. $y(n) \pm 0.5y(n-1) = x(n)$
b. $y(n) \pm 2y(n-1) = x(n)$
c. $y(n) - 0.9y(n-1) + 0.2y(n-2) = x(n)$
d. $y(n) - y(n-1) + y(n-2) = x(n)$

12. An LTI system is described by $y(n) - 0.5y(n-1) = x(n)$. Given $x(n) = 0, \ n \geq 0$ find $y(n)$ for $n \geq 0$ if it is known that

a. $y(-1) = 2$
b. $y(-1) = 1$

13. Given the input $x(n) = 0, \ n \geq 0$ to an LTI system it is desired that the output becomes $y(n) = 0.5^n$ for $n \geq 0$.

a. Find a difference equation that describes the system.
b. Determine $y(-1)$ which produces the desired output.

14. Find the solution of the following equations given $y(-1) = 1$.

a. $y(n) + 0.3y(n-1) = 0, \ n \geq 0$
b. $y(n) + 2y(n-1) = 0, \ n \geq 0$

15. Find the solution of the following equations given $y(-1) = 1$.

a. $y(n) - 0.3y(n-1) = 0, \ n \geq 0$
b. $y(n) - 2y(n-1) = 0, \ n \geq 0$

16. Find the solution of the following equations given $y(-2) = y(-1) = 1$.

a. $y(n) - 0.9y(n-1) + 0.2y(n-2) = 0, \ n \geq 0$
b. $y(n) - y(n-1) + y(n-2) = 0, \ n \geq 0$
c. $y(n) - y(n-1) + .5y(n-2) = 0, \ n \geq 0$

17. Find the unit-sample response of the LTI systems given by

    a. $h(n) + 2h(n - 1) = d(n)$
    b. $h(n) + 0.3h(n - 1) - 0.4h(n - 2) = d(n)$
    c. $h(n) + 1.3h(n - 1) + 0.4h(n - 2) = d(n)$
    d. $h(n) - h(n - 1) + 0.89h(n - 2) = d(n)$
    e. $h(n) - 0.2h(n - 1) + 0.17h(n - 2) = d(n)$

18. Find $y(n)$ given by $y(n) - 1.5y(n - 1) + 0.5y(n - 2) = 1$, $n > 0$ and $y(0) = y(-1) = 0$.

19. Find $y(n)$ given $y(n) + 0.3y(n - 1) - 0.4y(n - 2) = 9u(n)$.

20. The unit-sample response of an LTI system is

$$h(n) = \frac{10}{13} \left[(0.5)^{n+1} - (-0.8)^{n+1} \right] u(n)$$

Find its response to $x(n) = 9u(n)$ by convolution. Compare with the answer found for problem 19.

21. The unit-step response, $g(n)$, of an LTI system is related to its unit-sample response, $h(n)$, by $g(n) - g(n-1) = h(n)$. Find the unit-step response of an LTI system given its unit-sample response

$$h(n) = [(0.5)^{n+1} - (-0.8)^{n+1}]u(n)$$

Compare with the result of problem 20.

22. Find $y(n)$ given

    a. $y(n) + 2y(n - 1) = 2$, $n > 0$, $y(0) = 0$
    b. $y(n) + 10y(n - 1) = 10n$, $n \geq 0$, $y(0) = 0$
    c. $y(n) + 1.5y(n - 1) + 0.5y(n - 2) = 6u(n)$

23. Find the solution of the equation $y(n) + ay(n - 1) = x(n)$, $y(0) = 1$, $a = -0.1$ for

    a. $x(n) = 0$, $n \geq 0$
    b. $x(n) = 1$, $n \geq 0$
    c. $x(n) = n$, $n \geq 0$

24. Solve problem 23 with $a = -0.2$.

25. Solve problem 23 with $a = -0.9$.

26. Solve problem 23 with $a = 0.9$.

27. Find the solution of the equation $y(n) - 0.81y(n - 2) = x(n)$, $y(0) = y(1) = 1$, for

    a. $x(n) = 0$, $n \geq 0$
    b. $x(n) = 1$, $n \geq 0$
    c. $x(n) = n$, $n \geq 0$

28. Find $y(n)$ given by $y(n) - y(n - 1) + 0.09y(n - 2) = x(n)$, $y(0) = y(1) = 1$, for

    a. $x(n) = 0$, $n \geq 0$
    b. $x(n) = 1$, $n \geq 0$
    c. $x(n) = n$, $n \geq 0$

29. Find the solution of the equation $y(n) + 2y(n - 1) = x(n)$ for

    a. $x(n) = 0.5^n u(n)$
    b. $x(n) = [3n + 2 + (0.5)^n]u(n)$
    c. $x(n) = (-2)^n u(n)$

30. Find the steady-state part of the solution of the equation $y(n) + 2y(n - 1) = \cos(\omega n)u(n)$ for

    a. $\omega = \pi/10$
    b. $\omega = \pi/4$
    c. $\omega = \pi/2$
    d. $\omega = 3\pi/4$
    e. $\omega = 9\pi/10$

31. Find the steady-state part of the solution of the equation $y(n) + 1.5y(n - 1) + 0.5y(n - 2) = \cos(\omega n)u(n)$ for

    a. $\omega = \pi/10$
    b. $\omega = \pi/4$
    c. $\omega = \pi/2$
    d. $\omega = 3\pi/4$
    e. $\omega = 9\pi/10$

32. Find the steady-state part of $y(n)$ given $y(n) - 2y(n - 1) - y(n - 2) = 3x(n) + 5x(n - 1)$ for $x(n) = \cos(\omega n)u(n)$. Write the answer as a function of ω.

33. An LTI system is represented by the equation $y(n) + y(n - 1) - y(n - 2) = x(n)$. Determine initial conditions $y(-1)$ and $y(-2)$ so that for $x(n) = \cos n$, $n \geq 0$, the response becomes $y(n) = H\cos(n + \theta)$ where H and θ are constants.

34. Find $y(n)$ given

$$y(n) - 2y(n - 1) - y(n - 2) = 3x(n) + 5x(n - 1), \quad y(-1) = y(-2) = 0,$$

with $x(n) = 1$, $n \geq 0$, and $x(-1) = 0$

35. Find $y(n)$ given by $y(n) - 0.25y(n - 2) = x(n)$ for

    a. $x(n) = 0.5^n u(n)$
    b. $x(n) = (-0.5)^n u(n)$

36. Find $y(n)$ given $y(n) - 0.5y(n - 1) = 0.5^n$, $n \geq 0$, and $y(-1) = 1$.

37. The response of an LTI system to a 1-volt DC input is given by the 2nd-order equation

$$y(n) + a_1 y(n - 1) + a_2 y(n - 2) = 1, \quad n \geq 0$$

where a_1 and a_2 are real constants. Determine values of $y(-1)$, $y(-2)$ which produce $y(n) = k_0$, $n \geq 0$; that is, those values that make the homogeneous part of the solution zero. Then find k_0.

38. The response of an LTI system to a DC input is given by the Nth-order equation

$$y(n) + a_1 y(n - 1) + a_2 y(n - 2) + \cdots + a_i y(n - i) + \cdots + a_N y(n - N) = 1, \quad n \geq 0$$

Determine initial conditions $y(-1)$, $y(-2) \cdots y(-i) \cdots y(-N)$, which produce $y(n) = k_0$ at $n \geq 0$. Then find k_0.

39. a. Find the unit-step response, $g(n)$, of the causal system described by the difference equation $y(n) + 1.3y(n - 1) + 0.4y(n - 2) = u(n)$ using the following two methods:
      (i) Direct solution of the difference equation.
      (ii) Find the unit-sample response and then convolve it with $u(n)$.
    b. Check your results by verifying that $h(n) = g(n) - g(n - 1)$.

40. Find the solution of the equation $y(n) - 0.2y(n - 1) = x(n)$ for

    a. $x(n) = 8d(n)$

b. $x(n) = 8u(n)$

c. $x(n) = 0$, $n \geq 0$ and $y(-1) = 10$

d. $x(n) = 8$, $n \geq 0$ and $y(-1) = 10$

Then

(i) Verify that the response in part b is equal to $\sum_{k=0}^{n} y_a(k)$, where $y_a(n)$ is the response in part a. Explain why this should be the case.

(ii) Verify that the response in part d is equal to the sum of the responses in parts b (called zero-state) and c (called zero-input). Give reasons why this should be the case.

41. Find the solution of the equation $y(n) - y(n-1) + y(n-2) = x(n)$ for

a. $x(n) = d(n)$

b. $x(n) = u(n)$

c. $x(n) = 0$, $n \geq 0$ and $y(-2) = y(-1) = 1$

d. $x(n) = 1$, $n \geq 0$ and $y(-2) = y(-1) = 1$

Then

(i) Verify that the response in part b is equal to $\sum_{k=0}^{n} y_a(k)$, where $y_a(n)$ is the response in part a. Explain why this should be the case.

(ii) Verify that the response in part d is equal to the sum of the responses in parts b (zero-state) and c (zero-input). Give reason why this should be the case.

42. Find the solution of the equation $y(n) - y(n-1) + 0.5y(n-2) = \cos(\omega n)$ for

a. $\omega = \pi/10$

b. $\omega = \pi/4$

c. $\omega = 9\pi/10$

In each case find $y(-2)$ and $y(-1)$.

43. Find the solution of the equation $y(n) - y(n-1) + 0.5y(n-2) = \cos(\omega n)u(n)$ for

a. $\omega = \pi/10$

b. $\omega = \pi/4$

c. $\omega = 9\pi/10$

44. Find the solution of the equation $y(n) - y(n-1) + 0.5y(n-2) = 0$, $n \geq 0$, given

a. $y(-2)$ and $y(-1)$ found in problem 42, part a

b. $y(-2)$ and $y(-1)$ found in problem 42, part b

c. $y(-2)$ and $y(-1)$ found in problem 42, part c

45. Find the solution of the equation $y(n) - y(n-1) + 0.5y(n-2) = \cos(\omega n)$, $n \geq 0$ for

a. $\omega = \pi/10$ with $y(-2)$ and $y(-1)$ found in problem 42 part a

b. $\omega = \pi/4$ with $y(-2)$ and $y(-1)$ found in problem 42 part b

c. $\omega = 9\pi/10$ with $y(-2)$ and $y(-1)$ found in problem 42 part c

In each case verify that the response is the sum of the zero-state and zero-input responses found in problems 43 and 44, respectively.

46. Find the solution of the equation $y(n) - 0.6y(n-1) + 0.36y(n-2) = x(n)$ for

a. $x(n) = \cos(\omega n)$ with   (i) $\omega = \pi/12$,   (ii) $\omega = \pi/6$,   (iii) $\omega = 11\pi/12$

b. $x(n) = \cos(\omega n)u(n)$ with   (i) $\omega = \pi/12$,   (ii) $\omega = \pi/6$,   (iii) $\omega = 11\pi/12$

47. Repeat problem 46 for $y(n) - 0.6y(n-1) + 0.36y(n-2) = x(n) - x(n-1) + x(n-2)$

48. Repeat problem 46 for $y(n) - 2y(n-1) + 4y(n-2) = x(n)$

14.18   Project: Low-Pass Filtering by Difference Equation

Summary

This project first constructs a difference equation that describes a low-pass digital filter. The equation is then applied to a sound file. The sequence of sound samples constitute the input to the equation. The project requires you to write a Matlab program for numerically computing the output sequence. The filtering effect of the equation is tested by comparing the input and output spectra.

Procedure 1. Developing the Difference Equation

A second-order low-pass digital Butterworth filter can be constructed by transforming an analog counterpart (either manually using paper and pencil or using a computer command). In this project it is desired to low-pass-filter a sound file sampled at the rate of $F_s = 22,050$ Hz. The cutoff frequency of the filter is 1 kHz. For simplicity, a second-order Butterworth filter is considered. Use the Matlab command *butter* to find the system function $H(z)$ of such a filter. From the system function construct the difference equation.

Procedure 2. Applying the Difference Equation

Use a sound file (e.g., *cht5.wav* which has been used on other occasions in this book), to be called $x(n)$, as the input to the equation. The output $y(n)$ is obtained from the linear combination of its past values and the input, as prescribed by the difference equation. Write a Matlab program to iteratively compute the output and save it as a sound file.

Procedure 3. Changing the Cutoff Frequency and Filter Order

Repeat procedures 1–2 for a 500-Hz cutoff frequency employing a fifth-order filter or higher.

Conclusions

Compare the input-output spectra, the sounds of the input and output, and summarize your observations.

15

The *z*-Transform and Its Applications

Contents

Introduction and Summary

The *z*-transform of a discrete-time function is equivalent to the Laplace transform in the continuous-time domain. It is a convenient and powerful tool for the solution of difference equations and analysis of LTI systems. It converts the discrete convolution into a product operation and, thus, simplifies the analysis of discrete-time LTI systems, especially when applied to several interacting subsystems. The main purpose of this chapter is to introduce the method of the *z*-transform and its inverse, and to demonstrate its use. It starts with the bilateral transform, its inverse, and properties. The chapter then demonstrates applications such as evaluating a convolution, finding the unit-sample response, and solving the difference equation. The power of the *z*-transform is best manifested at the system level, for example, in relation to the system function, poles and zeros, stability, and the frequency response. A few examples of system-level applications are included in this chapter. More comprehensive and broader uses of the *z*-transform in LTI system analysis are found in Chapter 18. The unilateral *z*-transform is presented to illustrate application of the transform to system with initial conditions. Evaluating the inverse *z*-transform by the residue method, and the relationship between the *s*- and *z*-planes are given at the end. The chapter is appended by tables that summarize some frequently encountered *z*-transform pairs, properties, and theorems. The chapter ends with two projects on the design of digital notch filters in the *z*-domain.

15.1 Definition of the *z*-Transform

The two-sided (or bilateral) *z*-transform of a discrete sequence $x(n)$ is defined by

$$X(z) = \sum_{n=-\infty}^{\infty} x(n)z^{-n}$$

where z is a complex variable.

Rectangular pulse

Find the *z*-transform of $x(n) = u(n) - u(n-6)$.

Solution

$$X(z) = \sum_{n=0}^{5} z^{-n} = 1 + z^{-1} + z^{-2} + z^{-3} + z^{-4} + z^{-5}, \quad z \neq 0$$

$X(z)$ exists on the entire *z*-plane except at the origin.

Sum of a Geometric Series

The sum of an infinite geometric series

$$S = \sum_{n=0}^{\infty} a^n$$

is a finite number if $|a| < 1$. To find the sum, we note that

$$S = 1 + a + a^2 + a^3 + \cdots = 1 + a(1 + a + a^2 + a^3 + \cdots) = 1 + aS, \quad S = \frac{1}{1-a}, \quad |a| < 1$$

Similarly, the sum of a geometric series of finite length N is

$$S = \sum_{n=0}^{N-1} a^n = 1 + a + a^2 + a^3 + \cdots + a^{N-1} = \frac{1 - a^N}{1 - a}$$

These identities are useful in evaluating z-transforms (and discrete-time Fourier transforms) of discrete signals.

15.2 Region of Convergence

The z-transform does not exist if the sum becomes infinite. For the sum to converge, the variable z must be constrained to a region in the complex plane called the region of convergence (ROC). In Examples 15.2, 15.3, and 15.4 we examine the z-transforms (and their ROC) for a causal, an anticausal, and a double-sided sequence, respectively.

Causal exponential sequence

Find the z-transform of $x(n) = a^n u(n)$ and its ROC.

Solution

$$X(z) = \sum_{n=0}^{\infty} a^n z^{-n} = 1 + az^{-1} + a^2 z^{-2} + a^3 z^{-3} + \cdots$$

This is a geometric progression. For the sum to converge to a finite value, the magnitude of the elements of the series must diminish as n increases. This happens when $|az^{-1}| < 1$ or $|z| > |a|$. Then,

$$X(z) = \sum_{n=0}^{\infty} (az^{-1})^n = \frac{1}{1 - az^{-1}}, \quad |z| > |a|$$

The region of convergence of $X(z)$ in the z-plane is outside the circle with radius $|a|$. See Figure 15.1. Note that the series converges even for $|a| > 1$ (when a^n grows with n). The function of the ROC is to attenuate $a^n z^{-n}$, making it diminish rather than grow, thus causing the series to converge.

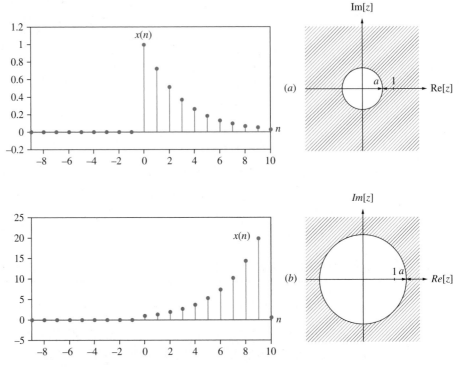

FIGURE 15.1 (For Example 15.2) The ROC of a causal function containing is outside a circle in the *z*-plane. This figure shows two causal exponential functions $a^n u(n)$ and their ROC being $|z| > a$. **(a)** A decaying function $x(n) = e^{-n/3} u(n) = (0.7165)^n u(n)$ with the ROC being $|z| > 0.7165$. **(b)** A growing function $x(n) = [e^{n/3} - 1]u(n) = [(1.3956)^n - 1]u(n)$ with the ROC being $|z| > 1.3956$. In both cases the time function is a causal one. It results in the ROC being outside the circle. For the growing exponential the radius of the circle becomes larger in order for the $a^n z^{(-n)}$ term to attenuate below a magnitude of 1, making the *z*-transform series converge.

*E*xample
15.3

Anticausal exponential sequence

Find the *z*-transform of $x(n) = b^n u(-n)$.

Solution

$$X(z) = \sum_{n=-\infty}^{0} b^n z^{-n} = \sum_{n=0}^{\infty} (b^{-1} z)^n = \frac{1}{1 - b^{-1} z}, \quad |z| < |b|$$

The region of convergence of $X(z)$ is inside the circle with radius $|b|$. See Figure 15.2.

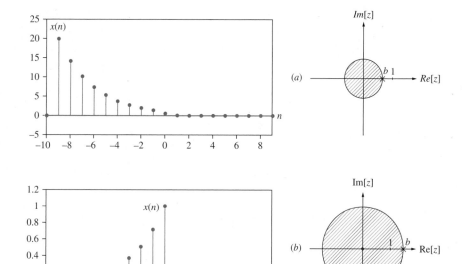

FIGURE 15.2 (For Example 15.3) The ROC of an anticausal function is inside a circle in the z plane. This figure shows two anticausal exponential functions containing $b^n u(-n)$ and their ROC being $|z| < b$. For $b = 0.7165$ [decaying with n shown in **(a)**] the ROC is $|z| < 0.7165$ and for $b = 1.3956$ [growing with n, shown in **(b)**] the circle becomes larger, $|z| < 1.3956$.

Two-sided exponential sequence

Find the z-transform of

$$x(n) = \begin{cases} a^n, & n > 1 \\ b^n, & n \leq 0 \end{cases}$$

Solution

$$X(z) = \sum_{n=-\infty}^{0} b^n z^{-n} + \sum_{n=1}^{\infty} a^n z^{-n}$$

$$= \sum_{n=-\infty}^{0} b^n z^{-n} + \sum_{n=0}^{\infty} a^n z^{-n} - 1$$

$$= -\frac{bz^{-1}}{1 - bz^{-1}} + \frac{1}{1 - az^{-1}} - 1$$

$$= \frac{(a-b)z^{-1}}{(1 - az^{-1})(1 - bz^{-1})}, \quad |a| < |z| < |b|$$

The region of convergence of $X(z)$ is inside the band between two circles with radii $|a|$ and $|b|$ and with $|b| > |a|$. See Figure 15.3. Note that the z-transform of $x(n)$ doesn't exist if $|b| \leq |a|$.

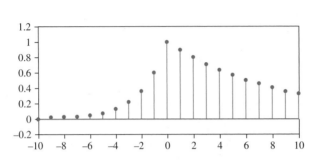
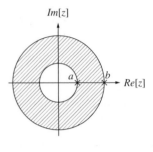

FIGURE 15.3 (For Example 15.4) The ROC of $a^n u(n - 1) + b^n u(-n)$ is $a < |z| < b$.

From Example 15.4 we conclude that the z-transform of $x(n) = a^n$, $-\infty < n < \infty$ doesn't exist.

Summary

The ROC is a connected region in the z-plane. It is either outside a circle (for causal sequences), inside a circle (for anticausal sequences), or limited to an annular region between two circles (for two-sided signals). By definition, the ROC cannot include a pole of $X(z)$. The inner bound of the ROC of causal sequences is the circle whose radius is the magnitude of the outermost pole (see Figure 15.1). Similarly, the outer bound of the ROC of an anticausal sequence is the circle whose radius is the magnitude of the innermost pole (see Figure 15.2). When the ROC is annular as in Figure 15.3, the outer poles characterize the sequence for $n < 0$ and the inner poles characterize it for $n > 0$.

15.3 More Examples

Truncated exponential sequence

Find the z-transform of $x(n) = a^n[u(n) - u(n - N)]$.

Solution

$$X(z) = \sum_{n=0}^{N-1} a^n z^{-n} = 1 + az^{-1} + a^2 z^{-2} + a^3 z^{-3} + \cdots + a^{(N-1)} z^{-(N-1)}$$

$$= \frac{1 - a^N z^{-N}}{1 - az^{-1}}, \quad z \neq 0$$

This is a finite sum expressed in the form of the ratio of two polynomials. The numerator polynomial has a zero at $z = a$ which cancels the pole in the denominator. Therefore, the ROC of $X(z)$ is all of the z-plane except $z = 0$.

Example 15.6

Two-sided even exponential sequence

Find the z-transform of $x(n) = \beta^{-|n|}$ and show that the result is in agreement with Example 15.4.

Solution

$$X(z) = \sum_{n=-\infty}^{\infty} \beta^{-|n|} z^{-n}$$

$$= \sum_{n=-\infty}^{0} \beta^n z^{-n} + \sum_{n=0}^{\infty} \beta^{-n} z^{-n} - 1$$

$$= \frac{-\beta z^{-1}}{1 - \beta z^{-1}} + \frac{1}{1 - \beta^{-1} z^{-1}} - 1$$

$$= \frac{(\beta^{-1} - \beta) z^{-1}}{(1 - \beta z^{-1})(1 - \beta^{-1} z^{-1})}, \qquad \left|\frac{1}{\beta}\right| < |z| < |\beta|$$

The region of convergence of $X(z)$ is inside the band between two circles with radii $|1/\beta|$ and $|\beta|$ and with $|\beta| > 1$. Note that the z-transform of $x(n) = \beta^{-|n|}$ doesn't exist if $|\beta| < 1$. These results can be derived from Example 15.4 by letting $b = \beta$ and $a = 1/\beta$.

Example 15.7

Causal sinusoidal sequence

Find the z-transform of $x(n) = \cos \omega n \, u(n)$.

Solution

$$x(n) = \frac{e^{j\omega n} + e^{-j\omega n}}{2} u(n)$$

$$X(z) = \frac{1}{2} \left[\frac{1}{1 - e^{j\omega} z^{-1}} + \frac{1}{1 - e^{-j\omega} z^{-1}} \right]$$

$$= \frac{1}{2} \frac{2 - (e^{j\omega} + e^{-j\omega}) z^{-1}}{1 - (e^{j\omega} + e^{-j\omega}) z^{-1} + z^{-2}}$$

$$= \frac{1 - (\cos \omega) z^{-1}}{1 - 2(\cos \omega) z^{-1} + z^{-2}}$$

The region of convergence is $|z| > 1$.

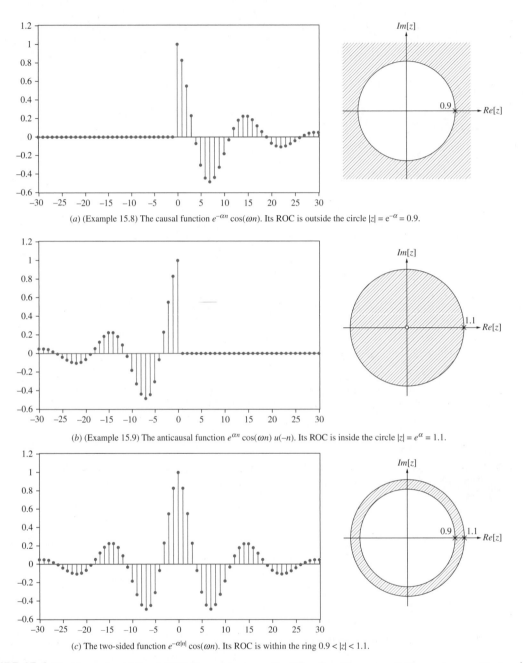

(a) (Example 15.8) The causal function $e^{-\alpha n}\cos(\omega n)$. Its ROC is outside the circle $|z| = e^{-\alpha} = 0.9$.

(b) (Example 15.9) The anticausal function $e^{\alpha n}\cos(\omega n)\,u(-n)$. Its ROC is inside the circle $|z| = e^{\alpha} = 1.1$.

(c) The two-sided function $e^{-\alpha|n|}\cos(\omega n)$. Its ROC is within the ring $0.9 < |z| < 1.1$.

FIGURE 15.4 (For Examples 15.8 and 15.9) Sinusoids at $\omega = 2\pi/15$ rad/s with exponentially varying amplitudes. *(a)* The causal function with exponentially decaying amplitude $e^{-0.1n}$. Its ROC is outside the circle at $|z| = e^{-0.1} = 0.9$. *(b)* The anticausal function with exponentially growing amplitude $e^{0.1n}$. Its ROC is inside the circle at $|z| = e^{0.1} = 1.1$. *(c)* The two-sided function. Its ROC is within the ring $0.9 < |z| < 1.1$. The ROC of a causal sinusoid with constant amplitude is $|z| > 1$, that of an anti-causal is $|z| < 1$. The function $x(n) = \cos(\omega n)$ doesn't have a *z*-transform.

xample

15.8

Causal exponential sinusoid

Find the z-transform of $x_1(n) = e^{-\alpha n} \cos(\omega n) \, u(n)$, Figure 15.4(a).

Solution

$$x_1(n) = e^{-\alpha n} \left[\frac{e^{j\omega n} + e^{-j\omega n}}{2} \right] u(n) = \left[\frac{e^{(-\alpha + j\omega)n} + e^{(-\alpha - j\omega)n}}{2} \right] u(n)$$

$$X_1(z) = \frac{1}{2} \left[\frac{1}{1 - e^{(-\alpha + j\omega)} z^{-1}} + \frac{1}{1 - e^{(-\alpha - j\omega)} z^{-1}} \right] = \frac{1 - e^{-\alpha} (\cos \omega) z^{-1}}{1 - 2 e^{-\alpha} (\cos \omega) z^{-1} + e^{-2\alpha} z^{-2}}$$

The region of convergence is $|z| > e^{-\alpha}$.

xample

15.9

Anticausal exponential sinusoid

Find the z-transform of $x_2(n) = e^{\alpha n} \cos(\omega n) \, u(-n)$, Figure 15.4(b).

Solution

$$x_2(n) = e^{\alpha n} \left[\frac{e^{j\omega n} + e^{-j\omega n}}{2} \right] u(-n) = \left[\frac{e^{(\alpha + j\omega)n} + e^{(\alpha - j\omega)n}}{2} \right] u(-n)$$

$$X_2(z) = \frac{1}{2} \left[\frac{1}{1 - e^{-(\alpha + j\omega)} z} + \frac{1}{1 - e^{-(\alpha - j\omega)} z} \right] = \frac{1 - e^{-\alpha} (\cos \omega) z}{1 - 2 e^{-\alpha} (\cos \omega) z + e^{-2\alpha} z^2}$$

The region of convergence is $|z| < e^{\alpha}$.

Note: The z-transform of $x(n) = e^{-\alpha |n|} \cos(\omega n)$ may be derived from results of Examples 15.8 and 15.9.

$$x(n) = x_1(n) + x_2(n) - d(n) \implies X(z) = \sum_{n=-\infty}^{\infty} x(n) z^{-n}$$

$$= \sum_{n=-\infty}^{0} x_1(n) z^{-n} + \sum_{n=0}^{\infty} x_2(n) z^{-n} - 1$$

The region of convergence is the intersection of the ROC of $X_1(z)$ and $X_2(z)$, $e^{-\alpha} < |z| < e^{\alpha}$.

ROC Summary

The region of convergence of the z-transform of a discrete-time function is contiguous and contains no poles. See Figure 15.5.

1. The ROC of a causal function is outside the circle bordering the outermost pole.
2. The ROC of an anticausal function is inside the circle bordering the innermost pole.
3. The ROC of a two-sided function is an annular ring containing no poles.

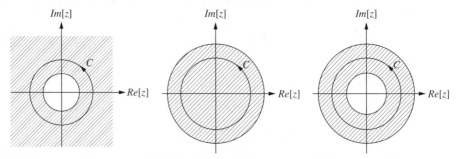

(*a*) ROC of a causal function   (*b*) ROC of an anticausal function   (*c*) ROC of a two-sided function

FIGURE 15.5 The three possible ROCs of a z-transform and the contour integrations for finding the inverse. See section 15.10.

15.4 Properties of the *z*-Transform

The z-transform is a member of a class of linear transformations and has the properties of that class. Here we briefly summarize several properties that are often used to facilitate taking the transform of new functions. The properties may be readily derived from the definition of the z-transform. The properties also allow the use of a table of transform pairs as a practical way to find the inverse transforms. [According to the uniqeness theorem, the relation between $x(n)$ and $X(z)$, with the ROC included, is one-to-one.] Appendix 15A summarizes properties and theorems of the z-transform. A table of transform pairs is given in Appendix 15B.

Linearity

The linearity is the basic and most important property of the z-transform. It is described below

$$
\begin{array}{lll}
x_1(n) & \Longleftrightarrow \quad X_1(z) & R_1 \\
x_2(n) & \Longleftrightarrow \quad X_2(z) & R_2 \\
ax_1(n) + bx_2(n) & \Longleftrightarrow \quad aX_1(z) + bX_2(z) & R_1 \cap R_2
\end{array}
$$

where $x_1(n)$ and $x_2(n)$ are any two finctions, a and b are any constants, and $R_1 \cap R_2$ is the intersection of R_1 and R_2. This property is derived directly from the definition of the z-transform as shown below. Let $y(n) = ax_1(n) + bx_2(n)$:

$$
Y(z) = \sum_{n=-\infty}^{\infty} \left[ax_1(n) + bx_2(n) \right] z^{-n} = a \sum_{n=-\infty}^{\infty} x_1(n)z^{-n} + b \sum_{n=-\infty}^{\infty} x_2(n)z^{-n}
$$

$$
= aX_1(z) + bX_2(z)
$$

In the above derivation no restrictions was placed on $x_1(n)$, $x_2(n)$, a, and b.

Example
15.10
Two-sided odd exponential function

Use linearity property to find the z-transform of $x(n) = a^n u(n) - a^{-n} u(-n)$.

Solution

$x(n)$ is the difference between the two time functions $a^n u(n)$ and $a^{-n} u(-n)$. Using the results of Examples 15.2 and 15.3, along with the linearity property, we get (see Figure 15.6)

$$a^n u(n) \quad\Longleftrightarrow\quad \frac{1}{1 - az^{-1}}, \quad |z| > |a|$$

$$a^{-n} u(-n) \quad\Longleftrightarrow\quad \frac{1}{1 - az}, \quad |z| < \frac{1}{|a|}$$

$$a^n u(n) - a^{-n} u(-n) \quad\Longleftrightarrow\quad \frac{1}{1 - az^{-1}} - \frac{1}{1 - az}$$

$$= \frac{(z - z^{-1})}{(z + z^{-1}) - (a + a^{-1})}, \quad |a| < |z| < \frac{1}{|a|}$$

For this to exist, we must have $|a| < 1$.

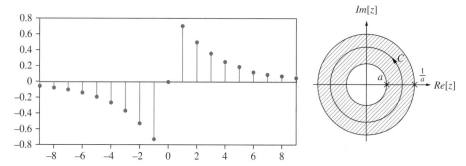

FIGURE 15.6 (For Example 15.10) Using the linearity property, the z-transform of $x(n) = a^n u(n) - a^{-n} u(-n)$ is found to be $X(z) = \frac{(z-z^{-1})}{(z+z^{-1})-(a+a^{-1})}$, $|a| < |z| < \frac{1}{|a|}$. In this figure $a = 0.716$.

Time Shift

Shifting $x(n)$ by k units multiplies $X(z)$ by z^{-k}.

$$x(n - k) \quad\Longleftrightarrow\quad X(z)z^{-k}$$

Conversely, multiplication of $X(z)$ by z^{-k} translates into k units of shift in the n-domain. The above results apply to both positive values of k (corresponding to a delay of k units) and negative values of k (corresponding to an advance of k units). To prove the shift

property let $y(n) = x(n - k)$. Then

$$Y(z) = \sum_{n=-\infty}^{\infty} x(n - k)z^{-n}. \text{ Let } n = m + k, \text{ and, } Y(z) = \sum_{m=-\infty}^{\infty} x(m)z^{-(m+k)}$$

$$= z^{-k} \sum_{m=-\infty}^{\infty} x(m)z^{-m} = z^{-k} X(z)$$

Shifting an anticausal function

Consider the following two anticausal functions and their z-transforms. (See the table of z-transform pairs in Appendix 15B.)

Item i in Appendix 15B: $x_1(n) = a^n u(-n)$ $\Longrightarrow X_1(z) = \dfrac{1}{1 - a^{-1}z}, \ |z| < |a|$

Item k in Appendix 15B: $x_2(n) = -a^n u(-n - 1) \Longrightarrow X_2(z) = \dfrac{1}{1 - az^{-1}}, \ |z| < |a|$

Note that $x_2(n) = -a^{-1}x_1(n+1)$. Verify that by applying the linearity and time-shift properties to $X_1(z)$ one obtains $X_2(z)$.

Solution
A unit shift to the left in the n-domain multiplies the transform by z. From this property and scaling by $-a^{-1}$ we expect $X_2(z) = -a^{-1}zX_1(z)$. To see if this is the case, we rewrite the given $X_2(z)$ in the following form:

$$X_2(z) = \frac{1}{1 - az^{-1}} = \frac{-a^{-1}z}{-a^{-1}z + 1} = -a^{-1}zX_1(z), \quad |z| < |a|$$

which is in agreement with our expectation for $X_2(z)$ obtained from the application of properties to $X_1(z)$.

Time Reversal
Reversing n converts $x(n)$ to $x(-n)$. The effect on the transform is

$$x(n) \quad \Longleftrightarrow \quad X(z)$$
$$x(-n) \quad \Longleftrightarrow \quad X(1/z)$$

If the ROC of $X(z)$ is R_x, the ROC of $X(1/z)$ is $1/R_x$. To prove the time-reversal property let $y(n) = x(-n)$. Then

$$Y(z) = \sum_{n=-\infty}^{\infty} x(-n)z^{-n}. \text{ Let } n = -m, \text{ and, } Y(z) = \sum_{m=\infty}^{-\infty} x(m)z^{m}$$

$$= \sum_{m=-\infty}^{\infty} x(m)z^{m} = X(1/z)$$

E *xample*

15.12

Time-reversing a causal function

Consider the following two functions and their z-transforms. (See table of z-transform pairs in Appendix 15B.)

Item h in Appendix 15B: $x_1(n) = a^n u(n) \implies X_1(z) = \dfrac{1}{1 - az^{-1}}, \; |z| > |a|$

Item i in Appendix 15B: $x_2(n) = a^n u(-n) \implies X_2(z) = \dfrac{1}{1 - a^{-1}z}, \; |z| < |a|$

Note that $x_2(n)$ can be obtained from $x_1(n)$ by a time reversal followed by changing a to a^{-1}. Verify that $X_2(z)$ can be obtained by applying the above transformations to $X_1(z)$.

Solution

Time reversal changes z to z^{-1}. Apply it to $x_1(n)$, then change a to a^{-1} and the following table of transform pairs are obtained:

$$a^n u(n) \quad \implies \quad \frac{1}{1 - az^{-1}}, \quad |z| > |a|$$

$$a^{-n} u(-n) \quad \implies \quad \frac{1}{1 - az}, \quad |z| < \left|\frac{1}{a}\right|$$

$$a^n u(-n) \quad \implies \quad \frac{1}{1 - a^{-1}z}, \quad |z| < |a|$$

The last line in the above is in agreement with the given transform pair for $x_2(n)$.

Convolution

Convolution of two discrete-time functions is equivalent to multiplication of their z-transforms.

$$x(n) \star y(n) \quad \Longleftrightarrow \quad X(z)Y(z)$$

To show this, note that:

$$y(n) = x(n) \star h(n) = \sum_{k=-\infty}^{\infty} x(n-k)\, h(k)$$

$$Y(z) = \sum_{n=-\infty}^{\infty} y(n)z^{-n} = \sum_{n=-\infty}^{\infty} \sum_{k=-\infty}^{\infty} x(n-k)\, h(k)\, z^{-n}$$

To convert the double sum to two separate sums with single variables, we replace the variable n by $(k + p)$, or equivalently $(n - k) = p$. The terms containing the variables p and k in the double sum are then separated

$$Y(z) = \left[\sum_{p=-\infty}^{\infty} x(p)\, z^{-p} \right] \left[\sum_{k=-\infty}^{\infty} h(k)\, z^{-k} \right] = X(z)H(z)$$

The convolution property establishes the input-output relationship $Y(z) = X(z)H(z)$, where $Y(z)$, $X(z)$, and $H(z)$ are z-transforms of the output, input, and unit-sample response of an LTI system, respectively.

Example
15.13

The unit-sample response of an LTI system may be found from the unit-step response by the difference equation $h(n) = g(n) - g(n-1)$. This is a consequence of linearity as shown below.

$$u(n) \qquad\qquad \Longrightarrow \qquad g(n)$$

$$d(n) = u(n) - u(n-1) \quad \Longrightarrow \quad h(n) = g(n) - g(n-1)$$

$g(n)$ and $h(n)$ are also related by the convolution $g(n) = u(n) \star h(n)$. Using the z-transform show that the two relationships are equivalent.

Solution
The difference equation that relates $g(n)$ to $h(n)$ can be derived from the convolution as shown below.

$$g(n) = u(n) \star h(n) \qquad\qquad \Longrightarrow \qquad G(z) = \frac{H(z)}{1 - z^{-1}}$$

$$H(z) = G(z)(1 - z^{-1}) = G(z) - z^{-1}G(z) \quad \Longrightarrow \quad h(n) = g(n) - g(n-1)$$

Example
15.14

Find the unit-sample response $h(n)$ of an LTI system having an input-output pair:

$$x(n) = [\underset{\uparrow}{1}, 2, 3, 4, 5, 6, 7, 8, 9, 10, 11, 12]$$

$$y(n) = [\underset{\uparrow}{0}, 1, 3, 6, 10, 15, 20, 25, 30, 35, 40, 45, 50, 42, 33, 23, 12]$$

Solution

$$X(z) = 1 + 2z^{-1} + 3z^{-2} + 4z^{-3} + \cdots + 11z^{-10} + 12z^{-11}$$

$$Y(z) = z^{-1} + 3z^{-2} + 6z^{-3} + 10z^{-4} + \cdots + 23z^{-15} + 12z^{-16}$$

$$H(z) = \frac{Y(z)}{X(z)} = \frac{z^{-1} + 3z^{-2} + 6z^{-3} + \cdots + 23z^{-15} + 12z^{-16}}{1 + 2z^{-1} + 3z^{-2} + 4z^{-3} + 10z^{-4} + \cdots + 11z^{-10} + 12z^{-11}}$$

By long division:

$$H(z) = z^{-1} + z^{-2} + z^{-3} + z^{-4} + z^{-5}$$

$$h(n) = [\underset{\uparrow}{0}, 1, 1, 1, 1, 1]$$

Multiplication by n

Multiplying the time function by n results in the following transform pairs:

$$nx(n) \iff -z\frac{dX(z)}{dz}$$

The ROC remains the same. To show the above, start with the derivative of $X(z)$:

$$\frac{dX(z)}{dz} = \frac{d}{dz}\sum_{-\infty}^{\infty} x(n)z^{-n} = -\sum_{-\infty}^{\infty} nx(n)z^{(-n-1)} = -z^{-1}\sum_{-\infty}^{\infty} nx(n)z^{-n}$$

and observe that

$$\sum_{-\infty}^{\infty} nx(n)z^{-n} = -z\frac{dX(z)}{dz}$$

Example 15.15

Given the transform pair

$$x(n) = a^n u(n) \implies X(z) = \frac{1}{1-az^{-1}}, \quad |z| > |a|$$

find the z-transform of $y(n) = na^n u(n)$ using the rule for multiplication by n.

Solution

$$X(z) = \frac{1}{1-az^{-1}}, \quad |z| > |a|$$

$$Y(z) = -z\frac{dX(z)}{dz} = -z\frac{d}{dz}\left[\frac{1}{1-az^{-1}}\right] = \frac{az^{-1}}{(1-az^{-1})^2}, \quad |z| > |a|$$

Multiplication by an Exponential a^n

Multiplying the time function by a^n results in the following transform pairs:

$$a^n x(n) \iff X(a^{-1}z)$$

The ROC becomes $|a|R_x$. The rule is derived through the following steps:

$$y(n) = a^n x(n) \quad Y(z) = \sum_{-\infty}^{\infty} a^n x(n)z^{-n} = \sum_{-\infty}^{\infty} x(n)\left(\frac{z}{a}\right)^{-n} = X\left(\frac{z}{a}\right)$$

Example 15.16

Revisiting Example 15.12, let

$$x_1(n) = a^n u(n) \implies X_1(z) = \frac{1}{1-az^{-1}}, \quad |z| > |a|$$

$$x_2(n) = a^n u(-n) \implies X_2(z) = \frac{1}{1-a^{-1}z}, \quad |z| < |a|$$

Note that $x_2(n)$ can be obtained from $x_1(n)$ through a time reversal followed by a^{2n} multiplication. Verify that by applying the rules to $X_1(z)$ one obtains $X_2(z)$.

Solution

Apply time reversal and multiplication by a^n to obtain the following table of transform pairs.

$$x(n) = a^n u(n) \quad \Longrightarrow \quad \frac{1}{1 - az^{-1}}, \qquad |z| > |a|$$

Time reversal:

$$x(-n) = a^{-n} u(-n) \quad \Longrightarrow \quad \frac{1}{1 - az}, \qquad |z| < \left|\frac{1}{a}\right|$$

Multiplication by a^{2n}:

$$y(n) = a^{2n} x(-n) = a^n u(-n) \quad \Longrightarrow \quad Y(z) = \frac{1}{1 - a\left(\frac{z}{a^2}\right)} = \frac{1}{1 - a^{-1}z}, \qquad |z| < |a|$$

The last line in the above equations is in agreement with the given transform pair for $a^n u(-n)$.

15.5 Inverse *z*-Transform

A sequence $x(n)$ may be obtained from its *z*-transform $X(z)$ using the following contour integral:

$$x(n) = \frac{1}{2\pi j} \oint_C X(z) z^{n-1} dz$$

(See section 15.10.) The integral is taken in the counterclockwise direction along a closed path in the ROC of the *z*-plane, enclosing the origin (e.g., a circle centered at $z = 0$). The contour integral may be evaluated using Cauchy's residue theorem, hence also called the residue method. The residue method will not be used as an operational tool for finding the inverse transform. It, however, enlightens the underlying principles used in other methods, such as when the ROC is an annular ring. Section 15.10 elaborates on the residue method and can be consulted when a deeper understanding of the subject is desired.

Another approach to finding the inverse of $X(z)$, if it is a rational function of z^{-1}, is to expand it into a power series of z^{-1}. The coefficients of the series constitute the time function. Expansion of a rational function may be done in one of several ways (e.g., by long division).

Example 15.17

Find the inverse *z*-transform of the following $X(z)$ by expanding it in terms of z^{-1}.

$$X(z) = \frac{3}{(1 - z^{-1})(1 + 2z^{-1})}, \qquad |z| > 2$$

Solution

$$X(z) = \frac{3}{(1 - z^{-1})(1 + 2z^{-1})} = \frac{3z^2}{z^2 + z - 2}$$

Using long division we obtain:

$$
\begin{array}{r}
3 - 3z^{-1} + 9z^{-2} - 15z^{-3} + 33z^{-4} - 63z^{-5} \cdots \\
\end{array}
$$

$$
z^2 + z - 2 \enclose{longdiv}{3z^2}
$$

$$
\begin{array}{r}
3z^2 + 3z - 6 \\
\hline
-3z + 6 \\
-3z - 3 + 6z^{-1} \\
\hline
9 - 6z^{-1} \\
9 + 9z^{-1} - 18z^{-2} \\
\hline
-15z^{-1} + 18z^{-2} \\
-15z^{-1} - 15z^{-2} + 30z^{-3} \\
\hline
33z^{-2} - 30z^{-3} \\
33z^{-2} + 33z^{-3} - 66z^{-4} \\
\hline
-63z^{-3} + 66z^{-4}
\end{array}
$$

$$ \cdots $$

$$ x(n) = \{\underset{\uparrow}{3}, -3, 9, -15, 33, -63, \ldots\} $$

The power series expansion method doesn't produce the time function in a closed form and is not an efficient method, except in special cases where an approximation is acceptable. The more powerful and popular method is by partial fraction expansion to be discussed in the next section and used throughout the rest of the chapter.

15.6 Partial Fraction Expansion Method

According to the uniqeness theorem, the relationship between $x(n)$ and $X(z)$ (the ROC included) is one-to-one: For each $x(n)$ we have only one $X(z)$, and, each $X(z)$ corresponds to only one $x(n)$. Therefore, the inverse of an $X(z)$ may be found through tables of transform pairs (Appendix 15B) in conjunction with a few basic properties and rules (Appendix 15A). A practical way of finding the inverse is to expand $X(z)$ into simple forms for which the time function may be deduced by inspection or from tables of transform pairs. Since the z-transform is a linear operation, the inverse of $X(z)$ is the sum of the inverses of its fractions. The method is similar to finding the inverse Laplace transform by partial fraction expansion. In this section we assume $X(z)$ is a rational function. The denominator is a polynomial in terms of z^{-1} and roots are either real or complex. The roots can also be repeated. [The roots of the denominator polynomial are also called the poles of $X(z)$.] We start with simple real-valued poles.

Remark

It is advantageous to write $X(z)$ and its expansions in terms of z^{-1}, rather than z, because of the ease of recognizing the correspondence between an exponential function and its transform:

$$a^n u(n) \implies \frac{1}{1 - az^{-1}}, \quad |z| > |a|.$$

Simple Real-Valued Roots

Each pole produces an exponential time function. The function is causal (right-sided) if the pole is encircled by the ROC. Otherwise, the function is anticausal (left-sided).

Find the inverse z-transform of $X(z)$ given below for three ROCs:

$$X(z) = \frac{3}{(1 - z^{-1})(1 + 2z^{-1})} \quad \begin{cases} \text{a)} & |z| > 2 \\ \text{b)} & 1 < |z| < 2 \\ \text{c)} & |z| < 1 \end{cases}$$

Solution
We expand $X(z)$ into two fractions and determine the inverses according to the ROCs.

$$X(z) = \frac{A}{1 - z^{-1}} + \frac{B}{1 + 2z^{-1}}$$

$$A = X(z)(1 - z^{-1})\big|_{z=1} = \frac{3}{1 + 2z^{-1}}\bigg|_{z=1} = 1$$

$$B = X(z)(1 + 2z^{-1})\big|_{z=-2} = \frac{3}{1 - z^{-1}}\bigg|_{z=-2} = 2$$

$$X(z) = \frac{1}{1 - z^{-1}} + \frac{2}{1 + 2z^{-1}}$$

a. $x_a(n) = u(n) + 2(-2)^n u(n)$,   item h in Appendix 15B

b. $x_b(n) = u(n) + (-2)^{n+1} u(-n - 1)$,   item k

c. $x_c(n) = -u(-n - 1) + (-2)^{n+1} u(-n - 1)$,   item k

Remark
When the degree of the numerator in z^{-1} is equal or higher than that of the denominator we need to first divide the denominator into the numerator, making the numerator of lesser degree than the denominator, then expand it. Example 15.19 illustrates the point.

Find the inverse z-transform of

$$X(z) = \frac{1 - 2z^{-1} + 4z^{-2}}{(1 - z^{-1})(1 + 2z^{-1})}, \quad |z| > 2$$

Solution
We first reduce the degree of the numerator by extracting the constant -2 from it.

$$X(z) = -2 + \frac{3}{(1 - z^{-1})(1 + 2z^{-1})}, \quad |z| > 2$$

Then we expand the remaining fraction into its fractions as was done in Example 15.18.

$$X(z) = -2 + \frac{A}{1 - z^{-1}} + \frac{B}{1 + 2z^{-1}}, \quad |z| > 2$$

$$= -2 + \frac{1}{1 - z^{-1}} + \frac{2}{1 + 2z^{-1}}, \quad |z| > 2$$

$$x(n) = -2d(n) + u(n) + 2(-2)^n u(n)$$

Remark

Standard tables of z-transform use polynomials of z^{-1}. Partial fraction expansion of an $X(z)$ which is expressed by polynomials of z will not produce fractions directly available from the tables. In such a case, we expand $X(z)/z$ instead, which can produce fractions in the familiar forms found in tables. This is illustrated in Example 15.20.

Example **15.20**

Express the $X(z)$ in Example 15.18 in terms of positive powers of z and find its inverse by partial fraction expansion.

Solution

Expressing $X(z)$ in terms of positive powers of z and then expanding it into fractions (not recommended, even though it produces the correct inverse) we get:

$$X(z) = \frac{3z^2}{(z-1)(z+2)} = 3 - \frac{3z - 6}{z^2 + z - 2} = 3 + \frac{1}{z - 1} - \frac{4}{z + 2}, \quad |z| > 2$$

The inverse functions for the second and the third terms in the above expansion are not readily available in tabular form. Therefore, we expand $X(z)/z$ and take the inverses:

$$\frac{X(z)}{z} = \frac{3z}{(z-1)(z+2)} = \frac{A}{z - 1} + \frac{B}{z + 2}$$

$$A = \left.\frac{3z}{z + 2}\right|_{z=1} = 1, \quad \text{and} \quad B = \left.\frac{3z}{z - 1}\right|_{z=-2} = 2$$

$$X(z) = \frac{z}{z - 1} + \frac{2z}{z + 2} = \frac{1}{1 - z^{-1}} + \frac{2}{1 + 2z^{-1}}, \quad |z| > 2$$

$$x(n) = u(n) + 2(-2)^n u(n)$$

Complex Roots

The z-transform of the exponential function a^n is valid regardless of a being real or complex. Therefore, an $X(z)$ with complex roots may be expanded into its partial fractions with each root treated individually. It is, however, preferable to combine each pair of complex conjugate roots, as found in tables of transform pairs (Appendix 15B). This is illustrated by the following two examples.

xample

15.21

Find the inverse of

$$X(z) = \frac{1}{1 + a^2 z^{-2}}, \quad |z| > |a|$$

Solution

By matching $X(z)$ with item p in the table of Appendix 15B we easily find $x(n) = a^n \cos(\pi n/2)u(n)$.

Remark

In Example 15.21 we could alternatively expand $X(z)$ in terms of the two poles and then find the inverses. This method gives a correct answer but is not recommended, except as an exercise as done below.

$$X(z) = \frac{1}{1 + a^2 z^{-2}} = \frac{1}{(1 - jaz^{-1})(1 + jaz^{-1})} = \frac{A}{1 - jaz^{-1}} + \frac{B}{1 + jaz^{-1}}, |z| > |a|$$

$$A = X(z)(1 - jaz^{-1})\big|_{z=ja} = \frac{1}{2}$$

$$B = X(z)(1 + jaz^{-1})\big|_{z=-ja} = \frac{1}{2}$$

$$x(n) = \frac{1}{2}\Big[(ja)^n + (-ja)^n\Big]u(n) = \frac{a^n}{2}\left(e^{j\pi n/2} + e^{-j\pi n/2}\right)u(n) = a^n \cos\left(\frac{\pi n}{2}\right)u(n)$$

xample

15.22

Find the inverse of

$$X(z) = \frac{2 - 2z^{-1} + 3z^{-2}}{1 - z^{-1} + 2z^{-2} + 4z^{-3}}, \quad |z| > 2$$

Solution

We first find the roots of the denominator.

$$1 - z^{-1} + 2z^{-2} + 4z^{-3} = 0, \quad z_1 = -1, \quad z_{2,3} = 1 \pm j\sqrt{3}$$

The denominator can be expanded as shown below.

$$X(z) = \frac{2 - 2z^{-1} + 3z^{-2}}{(1 + z^{-1})(1 - 2z^{-1} + 4z^{-2})}, \quad |z| > 2$$

By an inspection of the denominator we anticipate using the following two transform pairs from Appendix 15B:

Item h in Appendix 15B

$$a^n u(n) \quad \Longleftrightarrow \quad \frac{1}{1 - az^{-1}}, \quad |z| > a$$

Item p in Appendix 15B

$$a^n \cos(\omega n)u(n) \quad \Longleftrightarrow \quad \frac{1 - a(\cos\omega)z^{-1}}{1 - 2a(\cos\omega)z^{-1} + a^2z^{-2}} \qquad |z| > a$$

Therefore, we expand $X(z)$ in terms of the single pole z_1 and the combination of the complex conjugate pair $z_{2,3}$ as shown below.

$$X(z) = \frac{A}{1 + z^{-1}} + \frac{Bz^{-1} + C}{1 - 2z^{-1} + 4z^{-2}}, \quad |z| > 2$$

$$A = X(z)(1 + z^{-1})\big|_{z=-1} = \frac{2 - 2z^{-1} + 3z^{-2}}{1 - 2z^{-1} + 4z^{-2}}\bigg|_{z=-1} = 1$$

$$(Bz^{-1} + C)\big|_{z=1+j\sqrt{3}} = X(z)(1 - 2z^{-1} + 4z^{-2})\big|_{z=1+j\sqrt{3}}$$

$$= \frac{2 - 2z^{-1} + 3z^{-2}}{1 + z^{-1}}\bigg|_{z=1+j\sqrt{3}} = \frac{3 + j\sqrt{3}}{4}$$

By direct substitution:

$$(Bz^{-1} + C)\big|_{z=1+j\sqrt{3}} = \frac{1}{4} - j\frac{\sqrt{3}}{4}$$

Therefore, $B = -1$ and $C = 1$

$$X(z) = \frac{1}{1 + z^{-1}} + \frac{1 - z^{-1}}{1 - 2z^{-1} + 4z^{-2}}, \quad |z| > 2$$

$$x(n) = \left[(-1)^n + 2^n \cos\left(\frac{\pi n}{3}\right)\right]u(n)$$

Repeated Roots

A root that is repeated n times is called an nth order pole and contributes up to n fractions to the expansion. This is illustrated by the following example for $n = 2$.

Example
15.23

Find the inverse of

$$X(z) = \frac{1 - 0.6z^{-1} + 0.06z^{-2}}{1 - 1.7z^{-1} + 0.96z^{-2} - 0.18z^{-3}}, \quad |z| > 0.6$$

Solution

We write $X(z)$ as the ratio of two polynomials in z

$$X(z) = z\frac{z^2 - 0.6z + 0.06}{z^3 - 1.7z^2 + 0.96z - 0.18}, \quad |z| > 0.6$$

The roots of the denominator are at $z_1 = 0.5$, $z_{2,3} = 0.6$. The root at $z = 0.6$ is repeated. We now expand $X(z)/z$ into partial fractions.

$$\frac{X(z)}{z} = \frac{z^2 - 0.6z + 0.06}{(z - 0.5)(z - 0.6)^2} = \frac{A}{z - 0.5} + \frac{B}{(z - 0.6)^2} + \frac{C}{z - 0.6}$$

$$A = \frac{X(z)}{z}(z - 0.5)\Big|_{z=0.5} = \frac{z^2 - 0.6z + 0.06}{(z - 0.6)^2}\Big|_{z=0.5} = 1$$

$$B = \frac{X(z)}{z}(z - 0.6)^2\Big|_{z=0.6} = \frac{z^2 - 0.6z + 0.06}{z - 0.5}\Big|_{z=0.6} = 0.6$$

$$C = \frac{d}{dz}\left[\frac{X(z)}{z}(z - 0.6)^2\right]\Big|_{z=0.6} = \frac{d}{dz}\left[\frac{z^2 - 0.6z + 0.06}{z - 0.5}\right]\Big|_{z=0.6} = 0$$

$$\frac{X(z)}{z} = \frac{1}{z - 0.5} + \frac{0.6}{(z - 0.6)^2}$$

$$X(z) = \frac{1}{1 - 0.5z^{-1}} + \frac{0.6z^{-1}}{(1 - 0.6z^{-1})^2}$$

$$x(n) = \left[(0.5)^n + n(0.6)^n\right]u(n)$$

(For more details on partial fraction expansion, including the case of multiple-order poles, see related sections in the chapter on the Laplace transform.)

15.7 Application to Difference Equations

The z-transform may be applied to solve linear difference equations with constant coefficients. We start with the following class of difference equations

$$y(n) + a_1 y(n - 1) \cdots + a_N y(n - N) = b_0 x(n) + b_1 x(n - 1) \cdots + b_M x(n - M),$$
$$-\infty < n < \infty$$

where $x(n)$ is known for all n.[1] Taking the bilateral z-transform of both sides and making use of the shift property, we find

$$Y(z) + a_1 z^{-1} Y(z) \cdots + a_N z^{-N} Y(z) = b_0 X(z) + b_1 z^{-1} X(z) \cdots + b_M z^{-M} X(z)$$

$$\left(1 + a_1 z^{-1} \cdots + a_N z^{-N}\right) Y(z) = \left(b_0 + b_1 z^{-1} \cdots + b_M z^{-M}\right) X(z)$$

$$Y(z) = \frac{b_0 + b_1 z^{-1} \cdots + b_M z^{-M}}{1 + a_1 z^{-1} \cdots + a_N z^{-N}} X(z)$$

Let

$$H(z) = \frac{b_0 + b_1 z^{-1} \cdots + b_M z^{-M}}{1 + a_1 z^{-1} \cdots + a_N z^{-N}}$$

---

[1]One-sided z-transforms will be used to solve difference equations with initial conditions. See section 15.9.

Then, $Y(z) = H(z)X(z)$, which could be reverted to the time domain and $y(n)$, as demonstrated by the following three examples.

Example 15.24

Use the z-transform to find the solution to the difference equation

$$y(n) - 0.2y(n - 1) = x(n), \quad \text{where } x(n) = 0.5^n u(n)$$

Solution

Taking z-transforms of both sides of the difference equation, we get

$$Y(z) - 0.2z^{-1}Y(z) = X(z), \, Y(z) = H(z)X(z), \, \text{where } H(z) = \frac{1}{1 - 0.2z^{-1}}, \, |z| > 0.2$$

The z-transforms of $x(n)$ and $y(n)$ are

$$X(z) = \sum_{n=0}^{\infty} 0.5^n z^{-n} = \frac{1}{1 - 0.5z^{-1}}, \quad |z| > 0.5$$

$$Y(z) = H(z)X(z) = \frac{1}{(1 - 0.2z^{-1})(1 - 0.5z^{-1})}, \quad |z| > 0.5$$

Expanding $Y(z)$ into its partial fractions and taking their inverses results in

$$Y(z) = \frac{1}{(1 - 0.2z^{-1})(1 - 0.5z^{-1})} = \frac{A}{1 - 0.2z^{-1}} + \frac{B}{1 - 0.5z^{-1}}$$

$$A = Y(z)(1 - 0.2z^{-1}) \Big|_{z=0.2} = \frac{1}{1 - 0.5z^{-1}} \Big|_{z=0.2} = -\frac{2}{3}$$

$$B = Y(z)(1 - 0.5z^{-1}) \Big|_{z=0.5} = \frac{1}{1 - 0.2z^{-1}} \Big|_{z=0.5} = \frac{5}{3}$$

$$y(n) = \left[\frac{5}{3}(0.5^n) - \frac{2}{3}(0.2^n) \right] u(n)$$

Example 15.25

Find $y(n)$ in Example 15.24 if

$$x(n) = \begin{cases} 0.5^{-n}, & n < 0 \\ 0, & n \geq 0 \end{cases}$$

Solution

$$X(z) = \sum_{-\infty}^{-1} 0.5^{-n} z^{-n} = \frac{-1}{1 - 2z^{-1}}, \quad |z| < 2$$

$$Y(z) = H(z)X(z) = \frac{-1}{(1 - 2z^{-1})(1 - 0.2z^{-1})}, \quad 0.2 < |z| < 2$$

Expanding $Y(z)$ into its partial fractions and taking their inverses, we have

$$Y(z) = \frac{1}{9} \frac{1}{(1 - 0.2z^{-1})} - \frac{10}{9} \frac{1}{(1 - 2z^{-1})}, \quad 0.2 < |z| < 2$$

$$y(n) = \begin{cases} \left(\frac{1}{9}\right) 0.2^n, & n \geq 0 \\ -\left(\frac{10}{9}\right) 0.5^{-n}, & n < 0 \end{cases}$$

Example 15.26

Find $y(n)$ in Example 15.24 if $x(n) = 0.5^{|n|}$, $-\infty < n < \infty$.

Solution

$$X(z) = \sum_{-\infty}^{-1} 0.5^{-n} z^{-n} + \sum_{0}^{\infty} 0.5^n z^{-n} = \frac{-1}{1 - 2z^{-1}} + \frac{1}{1 - 0.5z^{-1}}, \quad 0.5 < |z| < 2$$

$$Y(z) = H(z)X(z) = \left[\frac{-1}{1 - 2z^{-1}} + \frac{1}{1 - 0.5z^{-1}} \right] \left(\frac{1}{1 - 0.2z^{-1}} \right)$$

$$= \frac{-1.5z^{-1}}{(1 - 2z^{-1})(1 - 0.5z^{-1})(1 - 0.2z^{-1})}, \quad 0.5 < |z| < 2$$

Expanding $Y(z)$ into its partial fractions and taking their inverses, we have

$$Y(z) = -\frac{10}{9} \frac{1}{(1 - 2z^{-1})} + \frac{5}{3} \frac{1}{(1 - 0.5z^{-1})} - \frac{5}{9} \frac{1}{(1 - 0.2z^{-1})}, \quad 0.5 < |z| < 2$$

$$y(n) = \begin{cases} \left(\frac{5}{3}\right) 0.5^n - \left(\frac{5}{9}\right) 0.2^n, & n \geq 0 \\ -\left(\frac{10}{9}\right) 0.5^{-n}, & n < 0 \end{cases}$$

Due to the linearity property, the response of the difference equation in Example 15.26 is the sum of its responses in Examples 15.24 and 15.25.

15.8 Application to the Analysis of LTI Systems

The z-domain is appropriately named the frequency domain for analysis of LTI systems, with $H(z)$ a sufficient instrument to describe the system. We have encountered $H(z)$ previously as the scale factor to power signals, exponential signals, and sinusoids (Chapter 13). In this chapter we applied the z-transform to the input-output difference equation and found $Y(z) = H(z)X(z)$. It is easily seen that $H(z)$ is the z-transform of the unit-sample response. [Let $x(n) = d(n)$, $X(z) = 1$ and get $Y(z) = H(z)$.] It is, therefore, not surprising that in the frequency domain $H(z)$ plays the role of $h(n)$ in the time domain. Theoretically, $H(z)$ and $h(n)$ are on a par in describing the system. In practice, $H(z)$ offers a much more powerful analysis tool as it provides explicit and immediate information about the system and the output. For example, take the case of a sinusoidal input. Remember that z is a complex variable and can be represented in the complex plane by a point $z = \rho e^{j\omega}$. For a sinusoidal signal with constant amplitude, z becomes restricted to the unit circle and $H(z)$ is reduced to the frequency response. Its magnitude and phase provide the magnitude and phase change in a sinusoidal input. As another example of the power of $H(z)$, consider its poles. They are identical to the natural frequencies of the system and provide information about the system's stability. As a third example, we note that $H(z)$ provides a theoretical model for an LTI system whose unit-sample response has been measured experimentally. Furthermore, $H(z)$ is used to develop structures for LTI system (including systems' interconnections and feedback). Its relationships with the Fourier transform is yet another factor that places the z-transform at the center of frequency-domain analysis. These and many more related topics are addressed in Chapter 18, to which the reader is referred. In this section, by way of Example 15.27, we familiarize the reader with some of the features and capabilities of $H(z)$.

Example 15.27

The unit-sample response of a causal discrete-time LTI system is measured and modeled by $h(n) = d(n) + u(n)$.

a. Find $H(z)$.

b. Find the system's input-output difference equation.

c. Find the unit-step response using the z-transform.

d. Using the z-transform, find the response of the system to $x(n) = \cos(\omega n)$ and evaluate it for (i) $\omega = \pi$ and (ii) $\omega = \pi/2$.

e. Is the system BIBO stable?

Solution

a. $H(z) = 1 + \dfrac{1}{1 - z^{-1}} = \dfrac{2 - z^{-1}}{1 - z^{-1}} = \dfrac{Y(z)}{X(z)}$

b. $(1 - z^{-1})Y(z) = (2 - z^{-1})X(z)$

$y(n) - y(n - 1) = 2x(n) - x(n - 1)$

c. $Y(z) = H(z)X(z) = \dfrac{2 - z^{-1}}{1 - z^{-1}} \times \dfrac{1}{1 - z^{-1}} = \dfrac{2 - z^{-1}}{(1 - z^{-1})^2}$

$Y(z) = \dfrac{2 - z^{-1}}{(1 - z^{-1})^2} = \dfrac{A}{(1 - z^{-1})^2} + \dfrac{B}{1 - z^{-1}}$

$A = Y(z)(1 - z^{-1})^2\big|_{z=1} = 2 - z^{-1}\big|_{z=1} = 1$

$\dfrac{d}{dz^{-1}}\left[B(1 - z^{-1})\right]\Big|_{z=1} = \dfrac{d}{dz^{-1}}\left[Y(z)(1 - z^{-1})^2\right]\Big|_{z=1}$

$= \dfrac{d(2 - z^{-1})}{dz^{-1}}\Big|_{z=1} = -1, \quad B = 1$

$Y(z) = \dfrac{1}{1 - z^{-1}} + \dfrac{1}{(1 - z^{-1})^2}$

$y(n) = u(n) + (n + 1)u(n + 1) = 2u(n) + nu(n)$

An alternative form of the partial fraction expansion for $Y(z)$ is

$$Y(z) = \frac{2}{1 - z^{-1}} + \frac{z^{-1}}{(1 - z^{-1})^2}$$

which directly results in the time function $y(n) = 2u(n) + nu(n)$.

d. $x(n) = \cos(\omega n) = \mathcal{RE}\{e^{j\omega n}\}, \quad y(n) = \mathcal{RE}\{He^{j\omega n}\}$

$$H = \frac{2 - z^{-1}}{1 - z^{-1}}\bigg|_{z=e^{j\omega}} = \frac{2 - e^{-j\omega}}{1 - e^{-j\omega}} = \frac{3}{2} - \frac{j}{2}\cot\left(\frac{\omega}{2}\right)$$

$$y(n) = \mathcal{RE}\left\{\left[\frac{3}{2} - \frac{j}{2}\cot\left(\frac{\omega}{2}\right)\right][\cos(\omega n) + j\sin(\omega n)]\right\}$$

$$= \frac{3}{2}\cos(\omega n) + \frac{1}{2}\cot\left(\frac{\omega}{2}\right)\sin(\omega n)$$

$$\omega = \pi, \quad x(n) = \cos(\pi n), \quad y(n) = \frac{3}{2}\cos(\pi n)$$

$$\omega = \frac{\pi}{2}, \quad x(n) = \cos\left(\frac{\pi n}{2}\right), \quad y(n) = \frac{3}{2}\cos\left(\frac{\pi n}{2}\right) + \frac{1}{2}\sin\left(\frac{\pi n}{2}\right)$$

e.   The system is not BIBO-stable. The bounded input $x(n) = u(n)$ gives rise to an output that grows with n without any bound. An LTI discrete-time system is BIBO-stable if its poles (natural frequencies) are inside the unit circle in the complex plane. The system in this problem has a single pole that is on the unit circle, making it not BIBO. At the same time its response to a sinusoid doesn't grow out of bounds, making it marginally stable. The relevant property in the time domain is that for a BIBO system,

$$\sum_n |h(n)| < M, \quad \text{where M is any (fixed) large number,}$$

which is not the case here.

15.9 One-Sided *z*-Transform

In some cases the input to a system is not known for $-\infty < n < \infty$ but only for $n \geq 0$. To find the output of the system, in addition to the input for $n \geq 0$, we need the state of the system before the arrival of the input, for example, at $n = -1, -2, \ldots$. In such a case (where the two-sided z-transform is not applicable), we use the one-sided z-transform defined by

$$X(z) = \sum_{n=0}^{\infty} x(n)z^{-n}$$

It is noted that the one-sided z-transform is not limited to causal functions. It considers only the values of the function for $n \geq 0$. Therefore, two different functions with the same right side have the same one-sided z-transform. An important difference between the two-sided and the one-sided z-transforms is the shift property to the right. In the case of the one-sided z-transform, a unit delay in time multiplies $X(z)$ by z^{-1} but adds the additionally shifted term $x(-1)$ to the transform. Similarly, right-shifting $x(n)$ by two units multiplies $X(z)$ by z^{-2} and adds the terms $x(-1)z^{-1} + x(-2)$ to the transform and so on.

$$x(n-1) \iff X(z)z^{-1} + x(-1)$$
$$x(n-2) \iff X(z)z^{-2} + x(-1)z^{-1} + x(-2)$$
$$\cdots \qquad \cdots \qquad \cdots$$
$$x(n-k) \iff X(z)z^{-k} + x(-1)z^{-(k-1)} + \cdots + x(-k)$$

This property is essential when solving difference equations by the one-sided z-transform method, as it automatically takes into account the effect of initial conditions.

Example
15.28

Use the z-transform to find the solution to the difference equation

$$y(n) - 0.2y(n-1) = 0.5^n, \quad n \geq 0, \quad \text{given} \ \ y(-1) = \frac{5}{9}$$

Solution

Taking the one-sided z-transforms of both sides of the equation we have

$$Y(z) - 0.2\left[z^{-1}Y(z) + y(-1)\right] = X(z)$$

$$Y(z) = H(z)X(z) + 0.2y(-1)H(z), \quad \text{where}$$

$$H(z) = \frac{1}{1 - 0.2z^{-1}}, \quad |z| > 0.2$$

$$X(z) = \frac{1}{1 - 0.5z^{-1}}, \quad |z| > 0.5$$

$$Y(z) = \frac{1}{(1 - 0.2z^{-1})(1 - 0.5z^{-1})} + \frac{1}{9}\frac{1}{(1 - 0.2z^{-1})}, \quad |z| > 0.5$$

The above expression for $Y(z)$ shows contributions from the input and the initial condition. The term $H(z)X(z)$ is the contribution from the input alone (zero-state response). The term $0.2y(-1)H(z)$ is the contribution from the initial condition (zero-input response). Expanding $Y(z)$ into its partial fractions and taking their inverses, we get

$$Y(z) = \frac{5}{3}\frac{1}{1 - 0.5z^{-1}} - \frac{5}{9}\frac{1}{1 - 0.2z^{-1}}, \quad |z| > 0.5$$

$$y(n) = \frac{5}{3}\left(0.5^n\right) - \frac{5}{9}\left(0.2^n\right), \quad n \geq 0$$

In this example the input for $n \geq 0$ is the same as in Example 15.24. As expected, the difference in the outputs is due to the nonzero initial condition.

15.10 Evaluating the Inverse *z*-Transform by the Residue Method

A sequence $x(n)$ may be obtained from its z-transform $X(z)$ using the following contour integral:

$$x(n) = \frac{1}{2\pi j} \oint_C X(z) z^{n-1} dz$$

The integral is taken in the counterclockwise direction along a closed path in the ROC, enclosing the origin (e.g., a circle centered at $z = 0$).

Proof

In the following integral substitute for $X(z)$. Then change the order of integration and summation.

$$I \equiv \oint_C X(z) z^{n-1} dz = \oint_C \left[\sum_{m=-\infty}^{\infty} x(m) z^{-m} \right] z^{n-1} dz = \sum_{m=-\infty}^{\infty} x(m) \oint_C z^{n-m-1} dz$$

Without loss of generality we let the integral contour to be a circle with radius ρ. On this contour we will have

$$z = \rho e^{j\theta}, \ dz = j\rho e^{j\theta} d\theta$$

$$I = \sum_{m=-\infty}^{\infty} j x(m) \rho^{n-m} \int_{\theta=0}^{2\pi} e^{j(n-m)\theta} d\theta$$

But

$$\int_{\theta=0}^{2\pi} e^{j(n-m)\theta} d\theta = \begin{cases} 2\pi, & m = n \\ 0, & m \neq n \end{cases}$$

Therefore

$$I = j2\pi x(n) \text{ and } x(n) = \frac{1}{2\pi j} \oint_C X(z) z^{n-1} dz$$

Evaluating the Integral by Residue Method

The contour integral given above may be evaluated using Cauchy's residue theorem, hence also called the residue method. According to the Cauchy residue theorem, the integral of a complex function $F(z)$ along a closed path in its region of convergence in the counterclockwise direction is equal to $2\pi j$ times the sum of the residues of $F(z)$ at the poles inside the area enclosed by the path.

$$\oint_C F(z) dz = 2\pi j \sum [\text{ residues of } F(z) \text{ at its poles inside C }]$$

If $F(z)$ has a simple pole at z_0,

$$\text{residue at } z_0 = F(z)(z - z_0)\big|_{z=z_0}$$

If $F(z)$ has a repeated pole of kth order at z_0 its residue at that pole is

$$\text{residue at } z_0 = \frac{1}{(k-1)!} \frac{d^{k-1} \left[F(z)(z-z_0)^k \right]}{dz^{k-1}} \Bigg|_{z=z_0}$$

In applying the integral theorem to $X(z)z^{n-1}$, we need to evaluate the integral for all values $-\infty < n < \infty$. We recognize the two cases $n > 0$ and $n < 0$ separately and examine the poles of $X(z)z^{n-1}$ inside the contour. For $n > 0$, the poles of $X(z)z^{n-1}$ are the same as the poles of $X(z)$. For $n < 0$, the function $X(z)z^{n-1}$ has multiple poles at the origin which have to be considered. The concluding result is that

$$x(n) = \frac{1}{2\pi j} \oint_C X(z)z^{n-1} dz$$

$$x(n) = \sum [\text{ residues of } X(z)z^{n-1} \text{ at poles inside C}] \ \text{ for } \ n > 0$$

$$x(n) = -\sum [\text{ residues of } X(z)z^{n-1} \text{ at poles outside C}] \ \text{ for } \ n < 0$$

See Figure 15.5. The value of the sequence at $n = 0$ is found by examining $X(z)$ at the origin. For a causal function,

$$X(z) = \sum_{n=0}^{\infty} x(n)z^{-n} = x(0) + x(1)z^{-1} + x(2)z^{-2} + x(3)z^{-3} + \cdots$$

from which $x(0) = X(z)\big|_{z^{-1}=0}$. For an anticausal function,

$$X(z) = \sum_{n=-\infty}^{0} x(n)z^{-n} = x(0) + x(-1)z + x(-2)z^2 + x(-3)z^3 + \cdots$$

from which $x(0) = X(z)\big|_{z=0}$.

In the following examples we apply the residue method to find the inverse of several z-transforms and compare them with the results obtained in Examples 15.2, 15.3, and 15.4.

Example 15.29

Causal function

Find the inverse z-transform of

$$X(z) = \frac{1}{1 - az^{-1}}, \quad |z| > |a|$$

Solution

$$x(n) = \frac{1}{2\pi j} \oint_C X(z)z^{n-1} dz = \frac{1}{2\pi j} \oint_C \frac{z^{n-1}}{1 - az^{-1}} dz$$

The contour of integration is in the ROC, that is, $|z| > |a|$. $X(z)$ has a single pole at $z = a$, which is inside the contour. The residue of $X(z)z^{n-1}$ at that pole is

$$\text{residue (at } z = a) = \frac{z^{n-1}(z-a)}{1-az^{-1}}\bigg|_{z=a} = \frac{(z-a)z^n}{z-a}\bigg|_{z=a} = a^n$$

Therefore, $x(n) = a^n$, $n > 0$. At $n = 0$

$$x(0) = \frac{1}{1-az^{-1}}\bigg|_{z^{-1}=0} = 1$$

$X(z)$ has no poles outside the contour resulting in $x(n) = 0$ for $n < 0$. In summary, $x(n) = a^n u(n)$. This is in agreement with the result of Example 15.2 (Figure 15.1).

Causal function

Find the inverse *z*-transform of

$$X(z) = \frac{az^{-1}}{1-az^{-1}}, \quad |z| > |a|$$

Solution
As in Example 15.29, the contour of integration is within the region $|z| > |a|$. $X(z)$ has a single pole at $z = a$, which is enclosed by the contour, and the residue of $X(z)z^{n-1}$ at that pole is

$$\text{residue (at } z = a) = \frac{az^{-1}z^{n-1}(z-a)}{1-az^{-1}}\bigg|_{z=a} = \frac{a(z-a)z^{n-1}}{z-a}\bigg|_{z=a} = a^n$$

Therefore, $x(n) = a^n$, $n > 0$. At $n = 0$

$$x(0) = \frac{az^{-1}}{1-az^{-1}}\bigg|_{z^{-1}=0} = 0$$

$X(z)$ has no poles outside the contour resulting in $x(n) = 0$ for $n < 0$. In summary, $x(n) = a^n u(n - 1)$. See Figure 15.7.

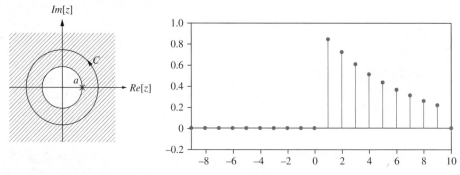

FIGURE 15.7 (For Example 15.30) The inverse of $X(z) = \frac{az^{-1}}{1-az^{-1}}$, $|z| > |a|$, is the causal function $x(n) = a^n u(n - 1)$.

Example 15.31

Anticausal function

Find the inverse z-transform of $X(z) = \dfrac{1}{1 - b^{-1}z}$, $|z| < |b|$

Solution

$$x(n) = \frac{1}{2\pi j} \oint_C \frac{z^{n-1}}{1 - b^{-1}z} dz$$

The contour of integration is in $|z| < |b|$. $X(z)$ has no poles inside the contour resulting in $x(n) = 0$ for $n > 0$. It has a single pole at $z = b$, which is outside the contour. The residue of $X(z)z^{n-1}$ at $z = b$ is

$$\text{residue (at } z = b) = \left. \frac{z^{n-1}(z-b)}{1 - b^{-1}z} \right|_{z=b} = \left. \frac{-b(z-b)z^{n-1}}{z - b} \right|_{z=b} = -b^n$$

Therefore, $x(n) = b^n$ for $n < 0$. We also note that for this anticausal function $x(0) = X(0) = 1$. In summary, $x(n) = b^n u(-n)$. This is in agreement with the result of Example 15.3.

Example 15.32

Anticausal function

Find the inverse z-transform of

$$X(z) = \frac{-1}{1 - bz^{-1}}, \quad |z| < |b|$$

Solution

The contour of integration is in $|z| < |b|$. $X(z)$ has no poles inside the contour resulting in $x(n) = 0$, $n > 0$. It has a single pole at $z = b$, which is outside the contour. The residue of $z^{n-1}X(z)$ at $z = b$ is

$$\text{residue (at } z = b) = \left. \frac{-z^{n-1}(z-b)}{1 - bz^{-1}} \right|_{z=b} = \left. \frac{-(z-b)z^n}{z - b} \right|_{z=b} = -b^n$$

Therefore, $x(n) = b^n$, $n < 0$. We also note that for this anticausal function $x(0) = X(0) = 0$. In summary, $x(n) = b^n u(-n - 1)$. See Figure 15.8.

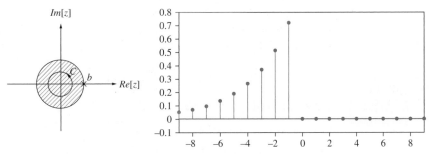

FIGURE 15.8 (For Example 15.32) The inverse of $X(z) = \frac{-1}{1 - bz^{-1}}$, $|z| < |b|$, is the anticausal function $x(n) = b^n u(-n - 1)$.

15.33

Two-sided function

Find the inverse z-transform of

$$X(z) = \frac{(a-b)z^{-1}}{(1-az^{-1})(1-bz^{-1})}, \quad |a| < |z| < |b|$$

Solution

For the ROC to exist, one must have $|b| > |a|$. The contour of integration is in the annular region $|a| < |z| < |b|$. $X(z)$ has a pole inside the contour at $z = a$ and the residue of $X(z)z^{n-1}$ at that pole is

$$\text{residue (at } z = a) = (z-a)z^{n-1} \left[\frac{(a-b)z^{-1}}{(1-az^{-1})(1-bz^{-1})} \right]_{z=a} = a^n$$

Therefore, $x(n) = a^n$, $n > 0$. To find $x(n)$ for $n < 0$ we note that $X(z)$ has a pole outside the contour at $z = b$ and the residue of $X(z)z^{n-1}$ at that pole is

$$\text{residue (at } z = b) = (z-b)z^{n-1} \left[\frac{(a-b)z^{-1}}{(1-az^{-1})(1-bz^{-1})} \right]_{z=b} = -b^n$$

Therefore, $x(n) = b^n$, $n < 0$. To find $x(0)$ we expand $X(z)$ into its causal and anticausal fractions:

$$X(z) = \frac{1}{1-az^{-1}} - \frac{1}{1-bz^{-1}}$$

The first term here is the z-transform of the causal part of $x(n)$ and the second term is that of its anticausal part. By setting $z = \infty$ in the causal part and $z = 0$ in the anticausal part of $X(z)$ we find their contributions to $x(0)$ to be

$$\frac{1}{1-az^{-1}} \bigg|_{z=\infty} = 1, \quad \text{and} \quad \frac{1}{1-bz^{-1}} \bigg|_{z=0} = 0$$

Consequently, $x(0) = 1 + 0 = 1$ and

$$x(n) = \begin{cases} a^n, & n \geq 1 \\ b^n, & n < 0 \end{cases}$$

which may be rewritten as $x(n) = a^n u(n-1) + b^n u(-n)$. This is in agreement with the result of Example 15.4.

15.34

Two-sided function

Find the inverse z-transform of

$$X(z) = \frac{az^{-1}}{1-az^{-1}} - \frac{1}{1-bz^{-1}}, \quad |a| < |z| < |b|. \quad \text{(See Figure 15.9)}.$$

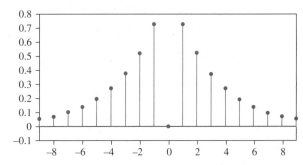

FIGURE 15.9 (For Example 15.34) The inverse of $X(z) = \frac{az^{-1}}{1-az^{-1}} - \frac{1}{1-bz^{-1}}$, $|a| < |z| < |b|$, is the two-sided function $x(n) = a^n u(n-1) + b^n u(-n-1) = a^n u(n) + b^n u(-n) - 2d(n)$. In Figure 15.9 $a = \frac{1}{b} = 0.716$.

Solution

$X(z)$ has a pole at $z = a$ (inside the integration contour) and a pole at $z = b$ (outside the integration contour).

$$\text{Residue (at } z = a) = (z - a)z^{n-1}\left[\frac{az^{-1}}{1-az^{-1}} - \frac{1}{1-bz^{-1}}\right]_{z=a} = a^n$$

Therefore, $x(n) = a^n$, $n > 0$. Also,

$$\text{residue (at } z = b) = (z - b)z^{n-1}\left[\frac{az^{-1}}{1-az^{-1}} - \frac{1}{1-bz^{-1}}\right]_{z=b} = -b^n$$

Therefore, $x(n) = b^n$, $n < 0$. To find $x(0)$ we examine the causal and anticausal parts of the function.

$$\text{From the causal part:} \quad \Rightarrow \quad \frac{az^{-1}}{1-az^{-1}}\Bigg|_{z=\infty} = 0$$

$$\text{From the anticausal part:} \quad \Rightarrow \quad \frac{1}{1-bz^{-1}}\Bigg|_{z=0} = 0$$

In either case the value of the function at $n = 0$ is found to be zero. In summary,

$$x(n) = \begin{cases} a^n, & n > 0 \\ b^n, & n < 0 \\ 0, & n = 0 \end{cases}$$

15.11 Relationship Between the *s*- and *z*-Planes

In this section we summarize the relationship between the Laplace transform of a continuous-time function and the *z*-transform of its sampled sequence. We will use Ω and F to represent continuous-time frequency.

The Laplace Transform

The two-sided Laplace transform of a continuous-time function $x(t)$ is defined by

$$X(s) = \int_{-\infty}^{\infty} x(t)e^{-st}\,dt$$

where s is a complex number normally written as $s = \sigma + j\Omega$. For $X(s)$ to exist, s may be limited to a region in the s-plane. An example of a Laplace transform pair is

$$x(t) = e^{at}u(t), \quad X(s) = \frac{1}{s-a}, \quad \sigma > a$$

The z-Transform

A continuous-time signal $x(t)$ may be uniformly sampled every T seconds to produce the discrete-time signal $x(n) \equiv x(nT)$. The z-transform $X(z)$ of the discrete-time signal may be evaluated directly. As an example,

$$x(n) = e^{aTn}u(n), \quad X(z) = \sum_{n=-\infty}^{\infty} x(n)z^{-n} = \sum_{n=0}^{\infty} e^{aTn}z^{-n} = \frac{1}{1 - e^{aT}z^{-1}}, \quad |z| > e^{aT}$$

Converting the Laplace Transform $X(s)$ to the z-Transform $X(z)$

The z-transform of a causal discrete-time signal may also be obtained from the Laplace transform of the continuous-time signal by

$$X(z) = \sum \left\{ \text{residues of } \frac{X(s)}{1 - e^{sT}z^{-1}} \text{ at all poles of } X(s) \right\}$$

For the anticausal signals one may use the time reversal property of the z-transform to find $X(z)$ (Appendix 15A).

Continuous-Time Representation of a Sampled Signal

The sampled signal may also be represented by a continuous-time function, $\hat{x}(t)$, made of a train of impulses spaced every T seconds and weighted by $x(nT)$:

$$\hat{x}(t) = x(nT)\delta(t - nT)$$

Then, $\hat{X}(s)$ may be obtained as

$$\hat{X}(s) = X(z)\big|_{z=e^{sT}}$$

The Laplace transform of $\hat{x}(t)$ is a function of $e^{-sT} = e^{-\sigma T}e^{-j\Omega T}$. It is seen that $\hat{X}(s)$ is a periodic function of Ω. In other words, $\hat{X}(s)$ repeats itself over horizontal strips of width 2π in the s-plane as shown in Figure 15.10. For further details, see Chapter 10 and 16.

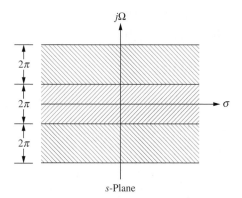

FIGURE 15.10 2π-wide horizontal strips in the s-plane representing periodicity of the Fourier transform of a sampled function in the continuous-time domain.

Mapping Regions of Convergences

Let

$$\begin{cases} z = \rho e^{j\theta} & \text{(in the } z\text{-plane)} \\ s = \sigma + j\Omega & \text{(in the } s\text{-plane)} \end{cases}$$

The s-to-z transformation produces the following equations

$$\begin{cases} z = e^{sT} \\ \rho e^{j\theta} = e^{\sigma T} e^{j\Omega T} \\ \rho = e^{\sigma T} \\ \theta = \Omega T \end{cases}$$

The left-half of the s-plane (LHP) maps onto the region inside the unit circle in the z-plane, because for $\sigma < 0$ we have $e^{\sigma T} < 1$. Similarly, the right-half s-plane (RHP) maps onto the region outside the unit circle in the z-plane. The $j\Omega$ axis in the s-plane maps onto the unit circle in the z-plane with the origin $s = 0$ going to $z = 1$. The LHP maps inside and the RHP outside the unit circle. The general mapping from the s-plane onto the z-plane is shown in Figure 15.11.

Given $X(s) = \dfrac{s+1}{(s+3)(s+4)}$, $\mathcal{RE}[s] > -3$, find the z-transform of the discrete signal.

Solution

$$X(z) = \sum \left\{ \text{residues of } \frac{s+1}{(s+3)(s+4)(1 - e^{sT}z^{-1})} \text{ at } s = -3, -4 \right\}$$

$$= \frac{s+1}{(s+4)(1 - e^{sT}z^{-1})} \bigg|_{s=-3} + \frac{s+1}{(s+3)(1 - e^{sT}z^{-1})} \bigg|_{s=-4}$$

$$= \frac{-2}{1 - e^{-3T}z^{-1}} + \frac{3}{1 - e^{-4T}z^{-1}}, \quad |z| > e^{-3T}$$

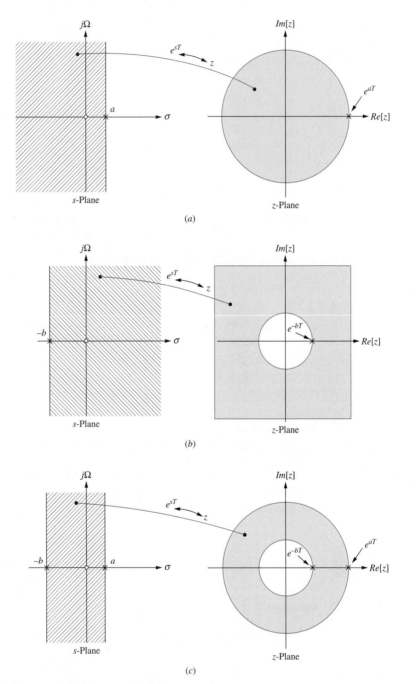

FIGURE 15.11 General mappings from the *s*-plane onto the *z*-plane: *(a)* anticausal function, *(b)* causal function, *(c)* two-sided function.

Example 15.36

a. Find the Laplace transform of $x(t) = \left[-2e^{-3t} + 3e^{-4t}\right]u(t)$.

b. Sample $x(t)$ at the rate of 10 samples per second and obtain its z-transform directly from $x(n)$.

c. Derive $X(z)$ from $X(s)$ and verify that it is the same result as in b.

Solution

a. The Laplace transform:

$$X(s) = \int_0^\infty [-2e^{-3t} + 3e^{-4t}]e^{-st}\,dt = \frac{s+1}{(s+3)(s+4)}, \quad \mathcal{RE}[s] > -3$$

b. The z-transform:

$$x(n) = x(t)\big|_{t=0.1n} = \left[-2e^{-0.3n} + 3e^{-0.4n}\right]u(n)$$

$$= \left[-2(0.7408)^n + 3(0.6703)^n\right]u(n)$$

$$X(z) = \frac{-2}{1 - 0.7408z^{-1}} + \frac{3}{1 - 0.6703z^{-1}}, \quad |z| > 0.7408$$

c. The z-transform from the Laplace transform:

$$X(z) = \sum \left\{\text{residues of } \frac{s+1}{(s+3)(s+4)(1 - e^{sT}z^{-1})} \text{ at } s = -3, -4\right\}$$

$$= \frac{s+1}{(s+4)(1 - e^{sT}z^{-1})}\bigg|_{s=-3,\ T=0.1} + \frac{s+1}{(s+3)(1 - e^{sT}z^{-1})}\bigg|_{s=-4,\ T=0.1}$$

$$= \frac{-2}{1 - 0.7408z^{-1}} + \frac{3}{1 - 0.6703z^{-1}}, \quad |z| > 0.7408$$

The expressions for the z-transform obtained in b and c are identical.

Example 15.37

Consider the bilateral signal

$$x(t) = e^{-a|t|}, \quad a > 0, \quad -\infty < t < \infty$$

a. Find the Laplace transform of $x(t)$ and determine its region of convergence.

b. Sample $x(t)$ every T seconds to find the discrete-time function $x(n) \equiv x(nT)$. Derive $X(z)$ from $X(s)$ and also by the direct method. Specify its ROC.

Solution

a. $X(s) = \int_{-\infty}^{\infty} e^{-a|t|} e^{st} \, dt = \dfrac{-1}{s-a} + \dfrac{1}{s+a} = \dfrac{-2a}{s^2 - a^2}$ $-a < \sigma < a$

b. $X^+(z) = \sum \left\{ \text{residue of } \dfrac{1}{(s+a)(1 - e^{sT} z^{-1})} \text{ at } s = -a \right\}$

$= \dfrac{1}{1 - e^{-aT} z^{-1}}, \quad |z| > e^{-aT}$

$X^-(z) = X^+\left(\dfrac{1}{z}\right) = \dfrac{1}{1 - e^{-aT} z}, \quad |z| < e^{aT}$

(using time-reversal property of the z-transform, Appendix 15A)

$X(z) = X^+(z) + X^-(z) - 1 = \dfrac{1}{1 - e^{-aT} z^{-1}} + \dfrac{1}{1 - e^{-aT} z} - 1$

$= \dfrac{(e^{-aT} - e^{aT}) z^{-1}}{(1 - e^{-aT} z^{-1})(1 - e^{aT} z^{-1})}, \quad e^{-aT} < |z| < e^{aT}$

By direct method:

$x(n) = e^{-aTn} u(n) + e^{aTn} u(-n) - d(n)$

$X(z) = \dfrac{1}{1 - e^{-aT} z^{-1}} + \dfrac{1}{1 - e^{-aT} z} - 1$

$= \dfrac{(e^{-aT} - e^{aT}) z^{-1}}{(1 - e^{-aT} z^{-1})(1 - e^{aT} z^{-1})}, \quad e^{-aT} < |z| < e^{aT}$

Appendix 15A Table of *z*-Transform Properties and Theorems

	Property	*n*-domain	$\Longleftrightarrow$	*z*-domain	ROC
1	Definition	$x(n) = \dfrac{1}{2\pi j} \oint X(z)z^{n-1}dz$	$\Longleftrightarrow$	$X(z) = \displaystyle\sum_{-\infty}^{\infty} x(n)z^{-n}$	R_x
2	Linearity	$ax(n) + by(n)$	$\Longleftrightarrow$	$aX(z) + bY(z)$	$R_x \cap R_y$
3	Time shift	$x(n - k)$	$\Longleftrightarrow$	$X(z)z^{-k}$	R_x
4	Time reversal	$x(-n)$	$\Longleftrightarrow$	$X(1/z)$	$1/R_x$
5	Multiplication by n	$nx(n)$	$\Longleftrightarrow$	$-z\dfrac{dX(z)}{dz}$	R_x
6	Conjugate	$x^*(n)$	$\Longleftrightarrow$	$X^*(z^*)$	R_x
7	Real	$\mathcal{RE}[x(n)]$	$\Longleftrightarrow$	$\dfrac{1}{2}[X(z) + X^*(z^*)]$	R_x
8	Imaginary	$\mathcal{IM}[x(n)]$	$\Longleftrightarrow$	$\dfrac{1}{2j}[X(z) - X^*(z^*)]$	R_x
9	Multiplication by a^n	$a^n x(n)$	$\Longleftrightarrow$	$X(a^{-1}z)$	$|a|R_x$
10	Convolution	$x(n) \star y(n)$	$\Longleftrightarrow$	$X(z)Y(z)$	$R_x \cap R_y$
11	Multiplication	$x(n)y(n)$	$\Longleftrightarrow$	$X(z) \star Y(z)$	
				$= \dfrac{1}{2\pi j} \oint_c X(v)\, Y(z/v)v^{-1}dv$	$R_x \cap R_y$

Theorems

Zero n, for causal functions:	$x(0) = \lim_{z\to\infty} X(z)$
Zero n, for anticausal functions:	$x(0) = \lim_{z\to 0} X(z)$
Zero z:	$X(0) = \displaystyle\sum_{-\infty}^{\infty} x(n)$
Parseval's theorem:	$\displaystyle\sum_{-\infty}^{\infty} |x(n)|^2 = \dfrac{1}{2\pi j} \oint_c X(z)X^*(1/z^*)z^{-1}dz$
	$\displaystyle\sum_{-\infty}^{\infty} x(n)y^*(n) = \dfrac{1}{2\pi j} \oint_c X(z)Y^*(1/z^*)z^{-1}dz$

Appendix 15B Table of *z*-Transform Pairs

$x(n) = \frac{1}{2\pi j} \oint_C X(z) z^{n-1} dz \Longleftrightarrow$		$X(z) = \sum_{-\infty}^{\infty} x(n) z^{-n}$	ROC
a.	$d(n)$	1	all z
b.	$d(n - n_0)$	z^{-n_0}	$z \neq 0 \ (n_0 > 0)$
			$z \neq \infty \ (n_0 < 0)$
c.	$\begin{cases} 1, & 0 \leq n \leq N - 1 \\ 0, & \text{elsewhere} \end{cases}$	$\dfrac{1 - z^{-N}}{1 - z^{-1}}$	$z \neq 0$
d.	$\begin{cases} 1, & -M \leq n \leq M \\ 0, & \text{elsewhere} \end{cases}$	$\dfrac{z^M - z^{-M}}{1 - z^{-1}}$	$z \neq 0$
e.	$\begin{cases} a^n, & 0 \leq n \leq N - 1 \\ 0, & \text{elsewhere} \end{cases}$	$\dfrac{1 - a^N z^{-N}}{1 - az^{-1}}$	$z \neq 0$
f.	$u(n)$	$\dfrac{1}{1 - z^{-1}}$	$|z| > 1$
g.	$u(-n)$	$\dfrac{-z^{-1}}{1 - z^{-1}} = \dfrac{1}{1 - z}$	$|z| < 1$
h.	$a^n u(n)$	$\dfrac{1}{1 - az^{-1}}$	$|z| > |a|$
i.	$a^n u(-n)$	$\dfrac{1}{1 - a^{-1}z} = \dfrac{-az^{-1}}{1 - az^{-1}}$	$|z| < |a|$
j.	$-a^n u(n - 1)$	$\dfrac{1}{1 - a^{-1}z} = \dfrac{-az^{-1}}{1 - az^{-1}}$	$|z| > |a|$
k.	$-a^n u(-n - 1)$	$\dfrac{1}{1 - az^{-1}}$	$|z| < |a|$
l.	$a^{-|n|} u(n), \ |a| > 1$	$\dfrac{1}{1 - a^{-1}z} + \dfrac{1}{1 - a^{-1}z^{-1}} - 1$	$\left|\dfrac{1}{a}\right| < |z| < |a|$
m.	$na^n u(n)$	$\dfrac{az^{-1}}{(1 - az^{-1})^2}$	$|z| > |a|$
n.	$\cos(\omega n) u(n)$	$\dfrac{1 - (\cos \omega) z^{-1}}{1 - 2(\cos \omega) z^{-1} + z^{-2}}$	$|z| > 1$
o.	$\sin(\omega n) u(n)$	$\dfrac{(\sin \omega) z^{-1}}{1 - 2(\cos \omega) z^{-1} + z^{-2}}$	$|z| > 1$
p.	$a^n \cos(\omega n) u(n)$	$\dfrac{1 - a(\cos \omega) z^{-1}}{1 - 2a(\cos \omega) z^{-1} + a^2 z^{-2}}$	$|z| > |a|$
q.	$a^n \sin(\omega n) u(n)$	$\dfrac{a(\sin \omega) z^{-1}}{1 - 2a(\cos \omega) z^{-1} + a^2 z^{-2}}$	$|z| > |a|$
r.	$a^n \cos(\omega n + \theta) u(n)$	$\dfrac{\cos \theta - a \cos(\omega - \theta) z^{-1}}{1 - 2a(\cos \omega) z^{-1} + a^2 z^{-2}}$	$|z| > |a|$
s.	$a^n \sin(\omega n + \theta) u(n)$	$\dfrac{\sin \theta + a \sin(\omega - \theta) z^{-1}}{1 - 2a(\cos \omega) z^{-1} + a^2 z^{-2}}$	$|z| > |a|$

15.12 Problems

Solved Problems

1. a. Find the z-transform of $x_1(n) = 2^n u(n)$ and indicate its region of convergence.
    b. Find the z-transform of $x_2(n) = 2^{-n} u(-n)$ and indicate its region of convergence.
    c. Define $x(n) = x_1(n) + x_1(n) = 2^{|n|}$. Does $x(n)$ have a z-transform? If the answer is positive, find $X(z)$ and if negative, state the reason why not.
    d. Let $x(n) = 2^{|n|}$ pass through an LTI system with the unit-sample response $h(n) = d(n) + d(n-1) + d(n-2)$. Find the output and specify its value for $-5 \le n \le 7$.

Solution

From the table of transform pairs of Appendix 15B we obtain

a. $X_1(z) = \dfrac{1}{1 - 2z^{-1}}$, $|z| > 2$.

b. $X_2(z) = \dfrac{1}{1 - 2^{-1}z} = \dfrac{2}{2 - z} = \dfrac{-2z^{-1}}{1 - 2z^{-1}}$, $|z| < 2$.

c. $x(n)$ doesn't have a z-transform because the intersection of $|z| > 2$ and $|z| < 2$ is null.

d. $y(n) = x(n) + x(n-1) + x(n-2) = 2^{|n|} + 2^{|n-1|} + 2^{|n-2|}$.

n	-5	-4	-3	-2	-1	0	1	2	3	4	5	6	7
$x(n)$	32	16	8	4	2	1	2	4	8	16	32	64	128
$x(n-1)$	64	32	16	8	4	2	1	2	4	8	16	32	64
$x(n-2)$	128	64	32	16	8	4	2	1	2	4	8	16	32
$y(n)$	224	112	56	28	14	7	5	7	14	28	56	112	224

Note: The lack of the z-transform of the input doesn't lead to a lack of a well-defined output.

2. Using a lookup table, find the inverse of the following bilateral z-transforms

a. $X_1(z) = \dfrac{z}{1 - 3z}$, ROC: $|z| > \dfrac{1}{3}$

b. $X_2(z) = \dfrac{z}{1 - 3z}$, ROC: $|z| < \dfrac{1}{3}$

Verify your answers by taking their transforms through the definition.

Solution

The ROCs for $x_1(n)$ and $x_2(n)$ are outside and inside, respectively, the circle in the z-plane with radius $1/3$, indicating a right-sided $x_1(n)$ and a left-sided $x_2(n)$. Rewrite the transforms as shown below and find their inverses using the lookup table of Appendix 15B. [Also, consider the shift property in the case of $X_2(z)$.]

a. $X_1(z) = \dfrac{z}{1 - 3z} = \dfrac{-3^{-1}}{1 - 3^{-1}z^{-1}}$, $|z| > \dfrac{1}{3}$, right-sided time function $x_1(n) = -3^{-(n+1)}u(n)$

b. $X_2(z) = \dfrac{z}{1 - 3z}$, $|z| < \dfrac{1}{3}$, left-sided time function $x_2(n) = 3^{-(n+1)}u(-n-1)$

To validate the above answers, we take their z-transforms and observe that they agree with what was given originally.

a. $X_1(z) = -\sum_{n=0}^{\infty} 3^{-(n+1)} z^{-n} = -\sum_{n=0}^{\infty} \frac{1}{3}(3z)^{-n} = -\frac{1}{3}\frac{1}{1-(3z)^{-1}} = \frac{z}{1-3z}$

b. $X_2(z) = \sum_{n=-\infty}^{-1} 3^{-(n+1)} z^{-n} = \frac{1}{3}\sum_{m=1}^{\infty}(3z)^m = \frac{1}{3}\left(\frac{1}{1-3z}-1\right) = \frac{z}{1-3z}$

3. Find all possible $h(n)$ functions, which may correspond to

$$H(z) = \frac{-3z}{z^2 - 3z + 2}$$

Solution

For convenience and ease of using the lookup table, we write $H(z)$ as a function of z^{-1} and expand it into its partial fractions.

$$H(z) = \frac{-3z^{-1}}{1 - 3z^{-1} + 2z^{-2}} = \frac{-3z^{-1}}{(1 - z^{-1})(1 - 2z^{-1})} = \frac{3}{1 - z^{-1}} - \frac{3}{1 - 2z^{-1}}$$

Then we recognize three possible regions of convergence in the z-plane, which yield the following three time functions.

$|z| > 2,$ ROC is outside the circle. The function is right-sided: $h_1(n) = 3(1 - 2^n) u(n)$

$1 < |z| < 2,$ ROC is an annular region. The function is two-sided: $h_2(n) = 3u(n) + 3(2^n)u(-n - 1)$

$|z| < 1,$ ROC is inside the circle. The function is left-sided: $h_3(n) = -3u(-n) + 3(2^n)u(-n).$

4. The z-transform of the impulse response of a causal LTI system is

$$H(z) = \frac{1}{2}\frac{z^{-1}}{z^{-2} - 4.5z^{-1} + 5}$$

a. Identify the region of convergence.
b. Find $h(n)$.
c. Is the system stable? Discuss it in the z-domain and n-domain.
d. Find an input $x(n)$ that would produce the output $y(n) = u(-n) + (0.5)^n u(n)$.

Solution

a. and b.

$$H(z) = \frac{1}{2}\frac{z^{-1}}{z^{-2} - 4.5z^{-1} + 5} = \frac{0.1z^{-1}}{(1 - 0.5z^{-1})(1 - 0.4z^{-1})} = \frac{1}{1 - 0.5z^{-1}} - \frac{1}{1 - 0.4z^{-1}}, \quad z > 0.5$$

$$h(n) = [0.5^n - 0.4^n]u(n)$$

c. The system is stable because the poles of the system function are inside the unit circle and, as a result, $h(n)$ is made of exponentially decaying exponentials.

d. We first find $Y(z)$ and its ROC.

$u(-n) \implies \dfrac{1}{1 - z}$ $\qquad\qquad |z| < 1$

$0.5^n u(n) \implies \dfrac{1}{1 - 0.5z^{-1}}$ $\qquad\qquad |z| > 0.5$

$y(n) = u(-n) + 0.5^n u(n) \implies Y(z) = \dfrac{1}{1 - z} + \dfrac{1}{1 - 0.5z^{-1}} = \dfrac{2 - 0.5z^{-1} - z}{(1 - z)(1 - 0.5z^{-1})}$ $\qquad 0.5 < |z| < 1$

$$X(z) = \frac{Y(z)}{H(z)} = \frac{2 - 0.5z^{-1} - z}{(1-z)(1-0.5z^{-1})} \times \frac{(1-0.5z^{-1})(1-0.4z^{-1})}{0.1z^{-1}} = \frac{(2-0.5z^{-1}-z)(1-0.4z^{-1})}{(1-z) \times 0.1z^{-1}}$$

$$= 10z - 14 + 2z^{-1} - \frac{3}{z-1}, \quad |z| < 1$$

$$x(n) = 10d(n+1) - 14d(n) + 2d(n-1) + 3u(-n)$$

Should We Work with z or z^{-1}?

In the z-domain analysis of discrete-time signals and systems, one may employ the variable z or z^{-1}, whichever makes the analysis easier. Below, we repeat problem 4 by employing polynomials of z.

$$H(z) = \frac{z}{10(z^2 - 0.9z + 0.2)} = \frac{z}{10(z-0.5)(z-0.4)} = \frac{z}{z-0.5} - \frac{z}{z-0.4}, \quad z > 0.5$$

$$h(n) = [0.5^n - 0.4^n]u(n)$$

$$Y(z) = \frac{-1}{z-1} + \frac{z}{z-0.5} = \frac{z^2 - 2z + 0.5}{(z-1)(z-0.5)}$$

$$X(z) = \frac{Y(z)}{H(z)} = \frac{10(z^2 - 2z + 0.5)(z - 0.4)}{(z-1)z} = \frac{10z^3 - 24z^2 + 13z - 2}{z^2 - z}$$

$$= 10z - 14 + 2z^{-1} - \frac{3}{z-1}, \quad |z| < 1$$

$$x(n) = 10d(n+1) - 14d(n) + 2d(n-1) + 3u(-n)$$

5. The unit-sample response of a causal discrete-time LTI system is measured and modeled by

$$h(n) = 0.4^n \cos\left(\frac{\pi n}{4}\right) u(n)$$

    a. Find $H(z)$.
    b. Find the system's input-output difference equation.
    c. Find the unit-step response by solving the difference equation.
    d. Find the unit-step response by the z-transform method.

Solution

    a. From the lookup table,

$$a^n \cos(\omega n)u(n) \quad \Longrightarrow \quad \frac{1 - a\cos\omega z^{-1}}{1 - 2a\cos\omega z^{-1} + a^2 z^{-2}}$$

$$h(n) = 0.4^n \cos\left(\frac{\pi n}{4}\right) u(n) \quad \Longrightarrow \quad H(z) = \frac{1 - 0.2\sqrt{2}z^{-1}}{1 - 0.4\sqrt{2}z^{-1} + 0.16z^{-2}} = \frac{Y(z)}{X(z)}$$

    b. From the unit-sample response,

$$Y(z)\left(1 - 0.4\sqrt{2}z^{-1} + 0.16z^{-2}\right) = X(z)\left(1 - 0.2\sqrt{2}z^{-1}\right)$$

$$y(n) - 0.4\sqrt{2}y(n-1) + 0.16y(n-2) = x(n) - 0.2\sqrt{2}x(n-1)$$

    c. The characteristic equation is $r^2 - 0.4\sqrt{2}r + 0.16 = 0$. Its roots are $r_{1,2} = 0.4e^{\pm j\pi/4}$. The particular solution to a unit-step input is

$$H(z)|_{z=1} = \frac{1 - 0.2\sqrt{2}}{1 - 0.4\sqrt{2} + 0.16} = 1.207$$

The overall response to a unit-step input is

$$g(n) = 1.207 + C(0.4)^n \cos\left(\frac{\pi n}{4} + \theta\right)$$

The constants C and θ are found by substituting the initial conditions into the solution. The initial conditions are

$$g(0) = 1 \text{ and } g(1) = 1 - 0.2\sqrt{2} + 0.4\sqrt{2} = 1 + 0.2\sqrt{2}.$$

Therefore,

$$\begin{cases} g(0) = 1 = 1.207 + C\cos(\theta) = 1 \\ g(1) = 1 + 0.2\sqrt{2} = 1.207 + 0.4C\cos(\frac{\pi}{4} + \theta) \end{cases} \implies \begin{cases} C\cos(\theta) = -0.207 \\ C\cos(\frac{\pi}{4} + \theta) = 0.1896 \end{cases} \implies \begin{cases} \theta = 66.4° \\ C = -0.52 \end{cases}$$

$$g(n) = \left[1.207 - 0.52(0.4)^n \cos\left(\frac{\pi n}{4} + 66.4°\right)\right] u(n)$$

d. First, find the z-transform of the unit-step response and expand it into partial fractions that could be matched with some entries in the lookup table.

$$g(n) = h(n) \star u(n)$$

$$G(z) = H(z) \times \frac{1}{1 - z^{-1}} = \frac{1 - 0.2\sqrt{2}z^{-1}}{(1 - z^{-1})(1 - 0.4\sqrt{2}z^{-1} + 0.16z^{-2})}$$

$$= \frac{1.207}{1 - z^{-1}} + \frac{-0.207 + 0.193z^{-1}}{1 - 0.4\sqrt{2}z^{-1} + 0.16z^{-2}}$$

From the table,

$$a^n \cos(\omega n + \theta)u(n) \qquad \implies \qquad \frac{\cos\theta - a\cos(\omega - \theta)z^{-1}}{1 - 2a\cos\omega z^{-1} + a^2 z^{-1}}$$

$$-0.52(0.4)^n \cos\left(\frac{\pi n}{4} + 66.4°\right) \qquad \Longleftarrow \qquad \frac{-0.207 + 0.193z^{-1}}{1 - 0.4\sqrt{2}z^{-1} + 0.16z^{-2}}$$

The unit-step response in the z and n domain is, therefore,

$$G(z) = \frac{1.207}{1 - z^{-1}} + \frac{-0.207 + 0.193z^{-1}}{1 - 0.4\sqrt{2}z^{-1} + 0.16z^{-2}}$$

$$g(n) = \left[1.207 - 0.52(0.4)^n \cos\left(\frac{\pi n}{4} + 66.4°\right)\right] u(n)$$

Chapter Problems

6. Find $X(z)$ and the region of convergence for each $x(n)$ given in Table 15.1. You may quote $X(z)$ from Table 15.2 and the ROC from Table 15.3.

TABLE 15.1 Functions		
x(n)	**X(z)**	**ROC**
$a^n u(n)$		
$a^{-n} u(n)$		
$a^n u(-n)$		
$a^{-n} u(-n)$		

TABLE 15.2 X(z)	
	X(z)
1	$-az^{-1}/(1 - az^{-1})$
2	$-a^{-1}z^{-1}/(1 - a^{-1}z^{-1})$
3	$1/(1 - az^{-1})$
4	$1/(1 - a^{-1}z^{-1})$

TABLE 15.3 ROC	
	ROC
A	$|z| > |a|$
B	$|z| > 1/|a|$
C	$|z| < |a|$
D	$|z| < 1/|a|$

7. a. Find the z-transform of $x_1(n) = 6^n u(n)$ and indicate its region of convergence on the z-plane.

b. Find the z-transform of $x_2(n) = 6^{-n} u(-n)$ and indicate its region of convergence on the z-plane.

8. Find $X(z)$ and the region of convergence for each $x(n)$ function of Table 15.4. You may quote $X(z)$ from Table 15.5 and the ROC from Table 15.6.

TABLE 15.4 Functions

$x(n)$	$X(z)$ ROC
$(0.5)^n u(n)$	
$(0.5)^n u(-n)$	
$(0.5)^{-n} u(n)$	
$(0.5)^{-n} u(-n)$	
$(-0.5)^n u(n)$	
$(-0.5)^n u(-n)$	
$(-0.5)^{-n} u(n)$	
$(-0.5)^{-n} u(-n)$	
$(2)^n u(n)$	
$(2)^n u(-n)$	
$(2)^{-n} u(n)$	
$(2)^{-n} u(-n)$	
$(-2)^n u(n)$	
$(-2)^n u(-n)$	
$(-2)^{-n} u(n)$	
$(-2)^{-n} u(-n)$	

TABLE 15.5 $X(z)$

	$X(z)$
1	$2z^{-1}/(1 + 2z^{-1})$
2	$1/(1 + 0.5z^{-1})$
3	$-0.5z^{-1}/(1 - 0.5z^{-1})$
4	$1/(1 - 2z^{-1})$
5	$0.5z^{-1}/(1 + 0.5z^{-1})$
6	$-2z^{-1}/(1 - 2z^{-1})$
7	$1/(1 + 2z^{-1})$
8	$1/(1 - 0.5z^{-1})$
9	None of the above

TABLE 15.6 ROC

	ROC
A	$|z| < 0.5$
B	$|z| > 0.5$
C	$|z| < 2$
D	$|z| > 2$

9. Find $H(z)$ for

a. $h(n) = 0.3^n u(n)$

b. $h(n) = 0.3^{-n} u(-n)$

10. Find the z-transform of $x(n) = 2^{-3|n|}$ and specify its region of convergence.

11. Consider the following discrete-time signal

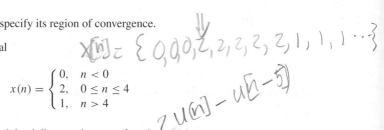

$$x(n) = \begin{cases} 0, & n < 0 \\ 2, & 0 \le n \le 4 \\ 1, & n > 4 \end{cases}$$

a. Express $x(n)$ as an algebraic sum of weighted discrete-time step functions $u(n)$.

b. Find $X(z)$ as a function of z^{-1}. Simplify your answer in the form of a ratio of two polynomials.

12. A discrete-time function $x(n)$ is given by:

$$x(n) = \begin{cases} 0, & n < 0 \\ 1, & 0 \le n \le 4 \\ 2, & n > 4 \end{cases}$$

a. Express $x(n)$ as a sum of step functions.

b. Find $X(z)$ in the form of the ratio of two polynomials and specify the region of convergence.

13. The unit-sample response $h(n)$ of a digital filter is

$$h(n) = \begin{cases} 1, & 0 \le n < N \\ 0, & \text{elsewhere} \end{cases}$$

Find a closed-form expresion for $H(z)$ and determine its ROC.

14. Given $X(z) = -1/(3 - z^{-1})$, $|z| < \frac{1}{3}$, find its inverse z-transform $x(n)$ and plot it for $-5 \le n \le 5$.

15. Given

$$X(z) = \frac{1 + 2z^{-1} + z^{-2}}{(1 - 0.5z^{-1})(1 - z^{-1})}$$

find $x(n)$ for the regions of convergence in Table 15.7. You may quote $x(n)$ from Table 15.8.

TABLE 15.7 ROC

ROC
a $0.5 < |z| < 1$
b $|z| > 1$
c $|z| < 0.5$

TABLE 15.8 Time Functions

$x(n)$
1    $2d(n) - 9(0.5)^n u(n) + 8u(n)$
2    $2d(n) - 9(0.2)^n u(n) + 8u(n)$
3    $2d(n) + 9(0.5)^n u(-n-1) - 8u(-n-1)$
4    $2d(n) - 9(0.5)^n u(n) - 8u(-n-1)$
5    $2d(n) - 9(0.5)^n u(n) + 8u(-n)$

16. Find a causal signal $x(n)$ whose z-transform is

$$X(z) = \frac{1}{1 - 0.25z^{-2}}$$

17. Find $h(n)$, the inverse z-transform of $H(z)$, for

a. $H(z) = \dfrac{1}{1 - 0.25z^{-1}}$, $|z| < 0.25$

b. $H(z) = \dfrac{1}{1 - 0.25z^{-1}}$, $|z| > 0.25$

18. Determine, in closed form, the impulse response $h(n)$ of a causal filter characterized by

$$H(z) = \frac{1}{1 - 1.5z^{-1} + 0.5z^{-2}}$$

Draw the block diagram that implements the above filter.

19. Given

$$H(z) = \frac{1}{1 - 0.5z^{-1}} + \frac{1}{1 - 0.4z^{-1}}$$

find $h(n)$ for the three ROC given below.

a. $|z| > 0.5$
b. $|z| < 0.4$
c. $0.4 < |z| < 0.5$

20. The system function of a causal system is given by

$$H(z) = \frac{z^{-1}}{1 - 0.5z^{-1}}$$

a. Find the unit-sample response of the system, $h(n)$.
b. Find the unit-step response of the system and its steady state value.
c. Find the unit-ramp response of the system and its steady state $y_{ss}(n)$.
d. Find the response of the system to a left-sided unit step $x(n) = u(-n)$.

21. Find the overall unit-sample response of the negative feedback system in which the unit-sample response of the forward path is $h_1(n) = 0.3^n u(n)$ and that of the feedback path is $h_2(n) = d(n-1)$.

22. Find the unit-sample response of the negative-feedback system made of $h_1(n) = 0.4^n u(n)$ and $h_2(n) = d(n-1)$ when

a. $h_1(n)$ is in the forward path and $h_2(n)$ is in the feedback path
b. $h_2(n)$ is in the forward path and $h_1(n)$ is in the feedback path.

23. The transfer function of a causal system is given by $H(z) = z^{-1}/(1 + z^{-1})$.

a. Find the response of the system to a unit step $u(n)$.
b. Find the steady-state response of the above system to a sinusoidal input $\sin(\omega_0 n)$.

24. The unit-sample response of an LTI system is $h(n) = 0.5^n u(n)$.

a. Find the transfer function of the system.
b. Examine the system's stability using $H(z)$.
c. Given an input $x(n) = 0.3^n u(n)$, find $Y(z)$ and $y(n)$.

25. a. Find the z-transform of $x(n) = 3^{-|n|}$, $-\infty < n < \infty$.
b. The above $x(n)$ passes through an LTI system with the unit-sample response $h(n) = 0.5^n u(n)$. Find $y(n)$ for all n.

26. The unit-sample response of an LTI system is $h(n) = 2^{-n} u(n)$.

a. Find the system function $H(z)$.
b. For $x(n) = u(n)$ find $Y(z)$, $y(n)$, and $y_{ss}(n)$.

27. The autocorrelation function, $c(n)$, of a time function $h(n)$ is defined by

$$c(n) = \sum_{k=-\infty}^{\infty} h(k)h(k+n)$$

a. Let $C(z)$ be the z-transform of $c(n)$. Show that

$$C(z) = H(z)H(z^{-1})$$

b. Can you use the above to find $c(n)$ for $h(n) = 2^n u(n)$?
c. Use the above to find $c(n)$ for $h(n) = \sqrt{3}(0.5)^n u(n)$.
d. Can you find two different time functions that produce the same $c(n)$? If the answer is positive provide an example, and if it is negative discuss why not.

28. The system function of a continuous-time LTI system is:

$$H(s) = \frac{s+1}{s^2 + 2s + 2}$$

The unit-impulse response and unit-step response of the system are shown by $h(t)$ and $g(t)$, respectively. It is desired to transform the above system function to a discrete-time system. The unit-sample and unit-step responses of the discrete system will be shown by $h(n)$ and $g(n)$, respectively.

a. Sample the unit-impulse response to produce a discrete-time system for which $h_1(n) \equiv h(nT)$. Call its system function $H_1(z)$. The method is called impulse-invariance transformation.

b. Sample the unit-step response to produce a discrete-system for which $g_2(n) \equiv g(nT)$. Call its system function $H_2(z)$. The method is called step-invariance transformation.

c. Determine the unit-step response of the first system $g_1(n)$ and the unit-sample response of the second system $h_2(n)$. Remember the relation between unit-sample and unit-step responses.

$$g(n) = \sum_{k=-\infty}^{n} h(k), \quad \text{and} \quad g(t) = \int_{-\infty}^{t} h(\tau)d\tau$$

Which of the following statements are true? Explain.

(i) $h_1(n) = h_2(n)$

(ii) $g_1(n) = g_2(n)$

(iii) None of the above

29. The unit-impulse response of a continuous-time system is $h(t) = e^t u(t)$. Find its Laplace transform and the region of convergence. $h(t)$ is sampled at the rate of 5 samples per sec producing the discrete-time unit-sample response $h(n)$. Find the z-transform $H(z)$ and the region of convergence. Determine important features in mapping from s- to z-domain.

30. The Laplace transform of a time function is given by

$$X(s) = \frac{2s + 7}{s^2 + 7s + 12}$$

along its region of convergence $\mathcal{R}$. The time function is sampled at the rate of 100 samples per sec, producing the discrete-time function $x(n)$. Find the z-transform $X(z)$ and its region of convergence for when

a. $\mathcal{R}$ is $\sigma < -4$

b. $\mathcal{R}$ is $-4 < \sigma < -3$

c. $\mathcal{R}$ is $\sigma > -3$

15.13 Project 1: FIR Filter Design by Zero Placement

Summary

In this project you will design an FIR filter by placing a pair of zeros of the system function on the unit circle at the discrete frequency corresponding to a 1-kHz continuous-time sinusoid, and explore the advantages and limitations of this design method. The zeros block 1-kHz signals. The function of the filter is to extract the signal from the data by removing additive sinusoidal disturbances at 1 kHz. The filter's operation will be tested off-line on synthetic data and on sound files sampled at 8 kHz using Matlab. The real-time operation of the filter will then be evaluated by running it on a DSP platform such as the Texas Instruments' Starter Kit, and its frequency response will be compared with theory.

Equipment and Tools

Software package for signal processing
Hardware platform (PC, sound card, or DSP board)
Microphone

Assignments and Reporting.   Read and do the Prelab, Filtering, and Discussion and Conclusions.

Theory

Contributions from the Zeros to the Frequency Response.   The frequency of a continuous-time sinusoidal waveform is mapped onto the unit circle of the z-plane. To block a certain frequency, the filter should have a zero on the unit circle at that frequency. Zeros of filters with real coefficients are complex conjugates of each other and a pair of zeros at $e^{\pm j\omega_0}$ will contribute the multiplier term

$$(1 - e^{j\omega_0}z^{-1})(1 - e^{-j\omega_0}z^{-1}) = 1 - 2\cos\omega_0 z^{-1} + z^{-2}$$

to $H(z)$. If a filter has no poles, $H(z)$ is a polynomial and the filter is an FIR filter. An FIR notch filter with real coefficients and zeros on the unit circle has linear phase.

Prelab

Generating a Gaussian Signal.   Using Matlab generate and save a 1,024-point-long signal made of a zero-mean Gaussian random process with a variance of 1 V and uniform spectral density over the bandwidth 0 to 4 kHz. The sampling rate should be 8 kHz.

Design of an FIR Notch Filter.   Design a 3-tap notch filter to meet the following analog specifications:

> Filter: 3-tap FIR
> Sampling rate $= 8$ kHz
> Notch frequency $= 1$ kHz
> High frequency gain $= 1$

Place a pair of conjugate zeros at $z_{1,2} = e^{\pm j\omega_0}$, where ω_0 is the discrete notch frequency corresponding to 1 kHz. Find $H(z)$ and $h(n)$. Plot its frequency response using the uniform and dB scales. Measure the 3-dB bandwidth. Compute the percentage reduction in the power of the synthetic noise when it passes through the filter. (The Matlab file given below may be used.)

Partial Solution.   Specifications of the digital filter are $\omega_0 = 2\pi \times 1,000/8,000 = \pi/4$ and

$$H(z) = k(1 - 2\cos(\pi/4)z^{-1} + z^{-2}) = k(1 - \sqrt{2}z^{-1} + z^{-2}), \quad k = 1/(2 + \sqrt{2})$$

The above $H(z)$ has a pair of zeros at $e^{\pm j\pi/4}$ only and its high-frequency gain (at $z = -1$) is unity. The magnitude and phase of the frequency response of the filter are computed and plotted in Figure 15.12 by the Matlab file given below. It is noted that the 3-dB bandwidth of the filter is very large, making it in fact more like a high-pass filter.

```
w=linspace(0,pi,100);
z=exp(i*w); k=1/(2+sqrt(2));
H=k*(1-sqrt(2)*z.^(-1)+z.^(-2));
subplot(2,1,1);
plot(w/pi,abs(H)); grid
title('FIR notch filter, gk10d.m');
ylabel('Magnitude, linear scale');
xlabel('Angular frequency (normalized by Pi)');
subplot(2,1,2);
plot(w/pi,180/pi*angle(H)); grid
ylabel('Phase, degrees');
xlabel('Angular frequency (normalized by Pi)');
```

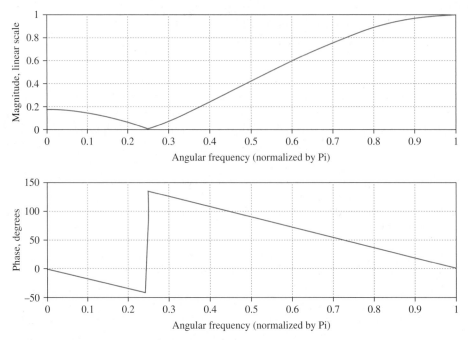

FIGURE 15.12 Magnitude and phase responses of the FIR notch filter with a pair of zeros at $e^{\pm j\pi/4}$.

Filtering

Filtering Synthetic Data.   Using Matlab, generate a 1,024-point-long sequence of data made of a 1-kHz sinusoidal disturbance (with rms = 0.1 V) added to a zero-mean Gaussian signal with a variance of 1 V and uniform spectral density over the bandwidth 0 to 4 kHz (as you did above). The sampling rate should be 8 kHz. Pass the data through the filter. Find the percentage rms difference between the output of the filter and its input.

Filtering Prerecorded Speech.   Repeat the section above using the prerecorded speech file.

Measuring the Frequency Response.   Load the filter coefficients into the Starter Kit board and measure the magnitude and phase of its frequency response from 0 to 4 kHz. Seven measurements taken at important points within the above range would suffice.

Filtering Live Speech.   Load the filter into the Starter Kit board and qualitatively explore its real-time performance on live speech (to which sinusidal disturbance may be added). For this purpose, add a live speech signal coming out of the microphone to a periodic disturbance obtained from the function generator. You may use an op-amp circuit. Normalize the speech signal to 1 V rms and add the 1-kHz sinusoid at 0.1 V rms. Repeat for higher and lower signal-to-noise ratios.

Discussion and Conclusions

Effect of Sampling Rate.   Discuss the effect of the sampling rate on the performance of the 3-tap FIR notch filter. For example, find the 3-dB bandwidth of the filter if the sampling rate is reduced to 4 kHz.

Overall Conclusions.   Qualitatively describe the deficiencies of the 3-tap FIR notch filter in this experiment and propose two other designs that would remedy them.

15.14 Project 2: IIR Filter Design by Pole/Zero Placement

Summary

In this project you will design a digital IIR notch filter that, when run under an 8-kHz sampling rate, will block 1-kHz sinusoids and have a bandwidth of 50 Hz. Blocking the sinusoid is accomplished by placing a pair of zeros at the notch frequency. The narrow bandwidth is achieved by placing a pair of poles inside the unit circle. The filter's operation will be tested off-line on sound files sampled at 8 kHz, using Matlab. Real-time operation of the filter will then be evaluated by running it on a DSP platform such as the Texas Instruments' Starter Kit, and its frequency response will be compared with the theory.

Equipment and Tools

> Software package for signal processing
> Hardware platform (PC, sound card, or DSP board)
> Microphone

Assignments and Reporting.   Read and do the following.

Theory

Contributions from Zeros and Poles to the Frequency Response.   The frequency of a continuous-time sinusoidal waveform is mapped onto the unit circle of the z-plane. To block a certain frequency, the filter should have a zero on the unit circle at that frequency. Zeros of filters with real coefficients are complex conjugates of each other and a pair of zeros at $e^{\pm j\omega_0}$ will contribute the following multiplier term to the numerator of $H(z)$:

$$(1 - e^{j\omega_0}z^{-1})(1 - e^{-j\omega_0}z^{-1}) = 1 - 2\cos\omega_0 z^{-1} + z^{-2}$$

The zeros at $e^{\pm j\omega_0}$ not only remove signals at the frequency ω_0, but also greatly attenuate the neighboring frequencies. This is because the bandwidth of the notch due to the zero is very wide. A pair of nearby poles at $\rho e^{\pm j\omega_0}$ will contribute

$$(1 - \rho e^{j\omega_0}z^{-1})(1 - \rho e^{-j\omega_0}z^{-1}) = 1 - 2\rho\cos\omega_0 z^{-1} + \rho^2 z^{-2}$$

to the denominator of the system function, narrowing the notch bandwidth. Recall the vectorial interpretation of the frequency response: the closer the pole is to the unit circle, the narrower the bandwidth will become. The poles, however, should be inside the unit circle because of stability conditions. The system function is, therefore,

$$H(z) = k\frac{1 - 2\cos\omega_0 z^{-1} + z^{-2}}{1 - 2\rho\cos\omega_0 z^{-1} + \rho^2 z^{-2}}$$

The factor k is chosen according to the desired gain at a given frequency (e.g., DC or high frequency).

Prelab

Design of an IIR Notch Filter.   Design an IIR notch filter to meet the following analog specifications:

> Notch frequency = 1 kHz
> 3-dB attenuation bandwidth = 50 Hz
> DC (or high frequency) gain = 1
> Sampling rate = 8 kHz

First compute the frequency-domain notch frequency ω_0 and the filter's bandwidth $\Delta\omega_0$. To block the 1-kHz sinusoidal disturbance, place two zeros at $z_{1,2} = e^{\pm j\omega_0}$, where ω_0 corresponds to 1 kHz. As seen in Project 1, the resulting FIR is not satisfactory. The notch bandwidth of the filter may be reduced by placing two conjugate poles at $\rho e^{\pm j\omega_0}$ near the zeros. The poles are, therefore, at the same angles as the zeros but inside the unit circle. The example given below provides a starting point for pole placement.

Partial Solution. Specifications of the digital filter are

$$\omega_0 = 2\pi \times 1{,}000/8{,}000 = \pi/4 \quad \text{and} \quad \Delta\omega_0 = 2\pi \times 50/8{,}000 = \pi/80$$

To start, consider an IIR filter with a pair of zeros on the unit circle at $e^{\pm j\pi/4}$ and a pair of poles inside the unit circle at $\rho e^{\pm j\pi/4}$.

$$H(z) = k\frac{1 - \sqrt{2}z^{-1} + z^{-2}}{1 - \rho\sqrt{2}z^{-1} + \rho^2 z^{-2}}$$

There exist systematic methods for determining ρ (e.g., transformation from an analog notch filter). Here, however, we find the appropriate ρ by trial and error, starting with $\rho = 0.96757$. The Matlab file for obtaining and ploting the magnitude response is shown below. The resulting magnitude plot is shown in Figure 15.13.

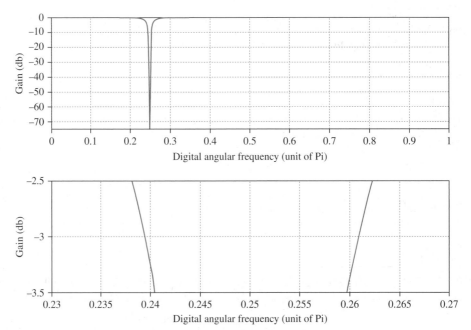

FIGURE 15.13 The magnitude response of the notch filter with a pair of zeros at $e^{\pm j\pi/4}$ and a pair of poles at $0.96757e^{\pm j\pi/4}$ is shown in the upper part. The notch frequency is at $\pi/4$. The magnification (lower section) shows the 3-dB bandwidth $\Delta\omega = 0.023 \times \pi = 0.125$ (desired to be $\pi/80 = 0.0125\pi \approx 0.04$ rad).

```
r=0.96757;
k=(1-sqrt(2)*r+r^2)/(2-sqrt(2));
num=[1 -sqrt(2) 1];
den=[1 -sqrt(2)*r r^2];
w=0:0.002*pi:pi;
H=k*freqz(num,den,w);
gain=20*log10(abs(H));
figure(1)
subplot(2,1,1)
```

```
plot(w/pi,gain); grid;
axis([0 1  -75 0])
title('magnitude plot');
xlabel('digital angular frequency (unit of pi)');
ylabel ('gain (db)');
subplot(2,1,2)
plot(w/pi,gain);grid;
axis([0.23 0.27  -3.5 -2.5])
%title('magnified plot')
xlabel('digital angular frequency (unit of pi)');
ylabel ('gain (db)');
```

It is noted that the 3-dB bandwidth of the above filter is greater than the desired specification. The poles need to be moved closer to the unit circle. Try $\rho = 0.98$, 0.982593, and 0.99. Determine your choice for the ρ value to be used in the next part (Filtering). Plot its frequency response using the dB scale. Measure the 3-dB bandwidth.

Filtering

Filtering Synthetic Data.   Using a software tool (e.g., Matlab), generate a 1,024-point-long sequence of data made of a 1-kHz sinusoidal disturbance (with rms = 0.1 V) added to a zero-mean Gaussian signal with a variance of 1 V and uniform spectral density over the bandwidth 0 to 4 kHz. The sampling rate should be 8 kHz. Pass the data through the filter. Find the percentage rms difference between the output of the filter and its input.

Filtering Prerecorded Speech.   Repeat the section above using the prerecorded speech file.

Measuring the Frequency Response.   Load the filter coefficients into the computer (hardware or software) and measure the magnitude and phase of its frequency response from 0 to 4 kHz. Seven measurements taken at important points within the above range would suffice.

Filtering Live Speech.   Load the filter into the computer and qualitatively explore its real-time performance on live speech (to which sinusidal disturbance may be added). For this purpose, add a live speech signal coming out of the microphone to a periodic disturbance obtained from the function generator. You may use an op-amp circuit. Normalize the speech signal to 1 V rms and add the 1-kHz sinusoid at 0.1 V rms. Repeat for higher and lower signal-to-noise ratios.

Discussion and Conclusions

Effect of Sampling Rate.   Discuss the effect of the sampling rate on the performance of the IIR notch filter. For example, find the 3-dB bandwidth of the filter if the sampling rate is reduced to 4 kHz.

Overall Conclusions.   Qualitatively describe possible deficiencies of the IIR notch filter in this experiment and propose other designs that would remedy them.

Discrete-Time Fourier Transform

Contents

Introduction and Summary

The discrete-time Fourier transform (DTFT) is defined for discrete signals. It is the tool for the analysis and design of linear discrete-time systems in the frequency domain. For example, the convolution between two signals in the discrete-time domain is equivalent to the multiplication of their DTFTs in the frequency-domain. To take another example, the frequency response of a digital filter is the DTFT of its unit-sample response. Yet a third example is that the DTFT, in its sampled form,[1] becomes the main tool for the analysis, design, detection, and coding of digitized audio and video signals.

In Chapter 8 we discussed Fourier transform (FT) of continuous-time signals, and in this chapter we present DTFT of discrete-time signals. As one may expect, the DTFT of a discrete-time signal is closely related to the Fourier transform (FT) the continuous-time signal from which the discrete signal was generated by sampling. Under certain conditions, the FT of a continuous-time signal can be approximated from the DTFT of its discrete counterpart. Conversely, the DTFT of the discrete signal may be obtained from the FT of the continuous-time signal. The relationship between the DTFT and FT, when applicable, is of great importance as it allows frequency analysis of real-world signals, and will be illustrated later in this chapter. Note, however, that some signals are inherently discrete, as for example, a sequence of numbers. The DTFT is, therefore, defined on its own merit. This chapter starts with the mathematical definition of the DTFT and its inverse, the inverse discrete-time Fourier transform (IDTFT), for energy signals, along with relevant examples and properties. By allowing for Dirac delta functions in the frequency domain, the DTFT is then extended to power signals. Some basic steps in rate conversion (such as zero-insertion, interpolation, and decimation) and their effects on a signal's DTFT are also discussed. Two projects are then suggested at the end of the chapter.

The transforms may be expressed as functions of angular frequency ω (in rad/s) or f (in Hz). These variables are not the same for the continuous-and discrete-time domains. Nevertheless, for simplicity, throughout this book we almost always will use lowercase letters for both time domains, as the context is clear and won't lend itself to confusion. However, to avoid confusion, when both domains are being discussed simultaneously, uppercase letters (F and Ω) are used for continuous-time signals and lowercase ones (f and ω) for discrete-time signals. Accordingly, when dealing with FT and DTFT simultaneously, we will use the following notations. The Fourier transform of a continuous-time signal $x(t)$, $-\infty < t < \infty$, will be shown by $X(F)$, where frequency F is a continuous variable, $-\infty < F < \infty$. The DTFT transforms a discrete-time signal $x(n)$, $-\infty < n < \infty$, where n is an integer, into its frequency-domain representation $X(\omega)$, $-\infty < \omega < \infty$, where ω is a continuous variable. $X(\omega)$ is, however, 2π-periodic [$X(f)$ is periodic with a period of 1 Hz]. Because of its periodicity, we examine only one period of $X(\omega)$; for example, for $-\pi < \omega < \pi$ [or $-0.5 < f < 0.5$ for $X(f)$]. From an analysis and computational point of view, the infinite-length discrete-time domain $-\infty < n < \infty$ is transformed into the finite-length continuous-frequency domain $-\pi < \omega < \pi$. Hence, one can say that sampling in the time domain makes representation in the frequency domain periodic.

When measuring the spectrum of a signal by a computer, the signal is sampled and a finite number of them are used. This is performed through an operation called discrete Fourier transform (DFT, to be distinguished from DTFT).[2] The DFT (to be

---

[1] See Chapter 17 on the DFT.

[2] See R. B. Blackman and J. W. Tukey, *The Measurment of Power Spectra* (New York: Dover Publications, 1958).

discussed in Chapter 17) transforms a finite sequence $x(n)$, $n = 0, 1, 2, \cdots, N - 1$, of length N into another finite sequence $X(k)$, $k = 0, 1, 2, \cdots, N - 1$, of the same length. The elements of the sequence $X(k)$ are samples of $X(\omega)$ [where $X(\omega)$ is the DTFT of $x(n)$, $\infty < n < \infty$] taken every $2\pi/N$ radians [i.e., N samples in one period of $X(\omega)$]. The DFT, therefore, performs sampling in the frequency domain. By sampling the DTFT, the DFT simplifies spectral analysis of finite data sequences. Moreover, evaluation of the DFT involves multiplications and additions only, which can be done efficiently both in real-time operations and in simulations. Real-time DFT is performed efficiently by DSP processors, with architectures specifically designed for such operations. The computationally efficient algorithm for evaluating the DFT and its inverse is the Fast Fourier Transform (FFT and IFFT), which is also used in off-line signal processing applications.

16.1 Definitions

DTFT

The DTFT of a discrete-time signal $x(n)$ is defined by

$$X(\omega) = \sum_{n=-\infty}^{\infty} x(n)e^{-j\omega n} \qquad (1)$$

where n is an integer and ω is a continuous real variable called the angular frequency.[3] $X(\omega)$ is a complex function and can be represented by its magnitude and phase

$$X(\omega) = |X(\omega)|e^{j\theta(\omega)}, \quad \text{where } \theta(\omega) = \angle X(\omega)$$

The magnitude and phase of $X(\omega)$ are real functions of the real variable ω, which ranges from $-\infty$ to ∞. However, the DTFT, $X(\omega)$, is periodic with period 2π because

$$e^{-j(\omega+2\pi)n} = e^{-j\omega n}e^{-j2\pi n} = e^{-j\omega n}$$

and, therefore, $X(\omega)$ is specifed for $-\pi < \omega < \pi$.[4]

IDTFT

The inverse of the DTFT may be found from

$$x(n) = \frac{1}{2\pi} \int_{-\pi}^{\pi} X(\omega)e^{j\omega n}d\omega \qquad (2)$$

---

[3]DTFT of $x(n)$ may also be expressed in terms of the variable $f = \omega/2\pi$. The DTFT pair would then be

$$X(f) = \sum_{n=-\infty}^{\infty} x(n)e^{-j2\pi f n} \text{ and } x(n) = \int_{-0.5}^{0.5} X(f)e^{j2\pi f n}df$$

In some applications, such as digital filter design, it is customary to represent the frequency by ω (rad/s). On the other hand, by using f (Hz) as the variable of the DTFT we are able to get rid of the factor 2π in many equations and identities.

[4]When the DTFT is expressed in terms of f, the period is $1\ Hz$ and $X(f)$ may be specified for $-0.5 < f < 0.5$.

To verify the above, substitute for $X(\omega)$ from (1) in the integral of (2) and switch the order of summation/integration to find

$$\frac{1}{2\pi} \int_{-\pi}^{\pi} X(\omega) e^{j\omega n} d\omega = \frac{1}{2\pi} \int_{-\pi}^{\pi} \left[\sum_{m=-\infty}^{\infty} x(m) e^{-j\omega m} \right] e^{j\omega n} d\omega$$

$$= \frac{1}{2\pi} \sum_{m=-\infty}^{\infty} \left[x(m) \int_{-\pi}^{\pi} e^{-j\omega(m-n)} d\omega \right]$$

$$= \frac{1}{2\pi} \sum_{m=-\infty}^{\infty} 2\pi x(m) d(m-n) = x(n)$$

Note that $e^{jn\omega}$ and $e^{jm\omega}$ are orthogonal to each other.

The pair $x(n)$ and $X(\omega)$ defined by (1) and (2) are unique. To find the inverse of $X(\omega)$, we find an $x(n)$ so that when used in (1) the result is the given $X(\omega)$.

Convergence of $X(\omega)$

For $X(\omega)$ to exist, the series in (1) should converge. If $x(n)$ is absolutely summable, that is,

$$\sum_{n=-\infty}^{\infty} |x(n)| \leq \infty$$

the series in (1) converges uniformly and the DTFT exists. For a square-summable signal (also called an energy signal) we have

$$\sum_{n=-\infty}^{\infty} |x(n)|^2 < \infty$$

and the series in (1) converges and the DTFT exists. The convergence is, however, not necessarily uniform. The case of periodic signals $x(n)$, where the series in (1) does not converge in the conventional sense but turns into impulses, will be discussed in sections 16.8 and 16.9.

Closed-Form Expressions

It is often desirable to obtain the DTFT function in closed form. The following equations may be used to find the sum of some finite and infinite series encountered in the evaluation of the DTFT.

The sum of an infinite geometric series converges if the magnitude of progression factor a is less than one:

$$\sum_{n=0}^{\infty} a^n = \frac{1}{1-a}, \quad \text{if } |a| < 1$$

The above equation may be obtained by a Taylor series expansion of $1/(1-a)$ around $a \approx 0$. Likewise, the sum of the first N terms of a geometric series may be found from

$$\sum_{n=0}^{N-1} a^n = \frac{1-a^N}{1-a}$$

For $a = e^{-j\omega}$,

$$\sum_{n=0}^{N-1} e^{-j\omega n} = \frac{1-e^{-jN\omega}}{1-e^{-j\omega}} = \frac{\sin(\frac{N\omega}{2})}{\sin\left(\frac{\omega}{2}\right)} e^{-j(\frac{N-1}{2})\omega}$$

Similarly,

$$\sum_{n=-M}^{M} e^{-j\omega n} = 1 + 2\sum_{n=1}^{M} \cos(\omega n) = \frac{\sin(M+\frac{1}{2})\omega}{\sin(\frac{\omega}{2})}$$

Note that while $|e^{-j\omega n}| = 1$ is a constant, the magnitude of the above sums vary with ω.

16.2 Examples of DTFT

The following examples that derive DTFTs of several elementary functions also illustrate a few of the properties. Because of periodicity plots of magnitude and phase of $X(\omega)$ are given for $-\pi < \omega < \pi$. More DTFT pairs are found in Appendix 16A.

Example
16.1

a. Find $X(\omega)$ for $x(n) = d(n)$.

b. Find $X(\omega)$ and plot its magnitude given

$$x(n) = \begin{cases} \frac{1}{2}, & n = 1, -1 \\ 0, & \text{elsewhere} \end{cases}$$

Solution

a. $X(\omega) = \displaystyle\sum_{n=-\infty}^{\infty} d(n)e^{-j\omega n} = 1e^{-j\omega n}\big|_{n=0} = 1$

b. $X(\omega) = \displaystyle\sum_{n=-\infty}^{\infty} x(n)e^{-j\omega n} = \frac{1}{2}e^{j\omega} + \frac{1}{2}e^{-j\omega} = \cos\omega$

See Figure 16.1(a). Within the period $-\pi < \omega < \pi$ the magnitude and phase of $X(\omega)$ are

$$|X| = |\cos\omega|, \text{ and } \theta = \begin{cases} 0, & |\omega| < \frac{\pi}{2} \\ \pi, & |\omega| > \frac{\pi}{2} \end{cases}$$

xample

16.2

Find $X(\omega)$ and plot its magnitude given

$$x(n) = \begin{cases} 1, & n = 0 \\ \frac{1}{2}, & n = 1, -1 \\ 0, & \text{elsewhere} \end{cases}$$

Solution

$$X(\omega) = \sum_{n=-\infty}^{\infty} x(n)e^{-j\omega n} = 1 + \frac{1}{2}e^{j\omega} + \frac{1}{2}e^{-j\omega} = 1 + \cos\omega$$

$$|X(\omega)| = 1 + \cos\omega, \text{ and } \theta(\omega) = 0 \text{ See Figure 16.1}(b).$$

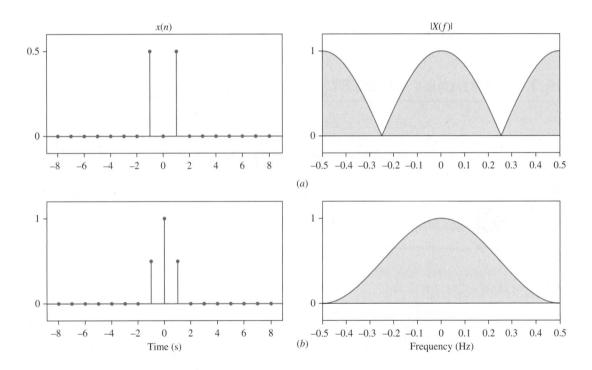

FIGURE 16.1 Two discrete-time functions (on the left) and one period of the normalized magnitude of their DTFT plotted for $-0.5 < f < 0.5$ (on the right, corresponding to $-\pi < \omega < \pi$). **(a)** $x(n) = \{0.5, 0, 0.5\}$. **(b)** $x(n) = \{0.5, 1, 0.5\}$.

xample

16.3

a. Find the DTFT of $x(n) = a^n u(n)$, $|a| < 1$.

b. Find the DTFT of $h(n) = 2^{-n}u(n)$ and plot its magnitude.

Solution

a. $X(\omega) = \displaystyle\sum_{n=0}^{\infty} a^n e^{-j\omega n} = \sum_{n=0}^{\infty} (ae^{-j\omega})^n = \dfrac{1}{1-ae^{-j\omega}} = \dfrac{1}{1-a\cos\omega + ja\sin\omega}$

The magnitude and phase of $X(\omega)$ are

$$|X|^2 = \dfrac{1}{1+a^2 - 2a\cos\omega} \quad \text{and} \quad \theta = -\tan^{-1}\left(\dfrac{a\sin\omega}{1-a\cos\omega}\right)$$

b. $X(\omega)$ is found by setting $a = 0.5$ in part a.

$$X(\omega) = \dfrac{1}{1 - \frac{1}{2}\cos\omega + j\frac{1}{2}\sin\omega}$$

The magnitude and phase of $X(\omega)$ are given below. See Figure 16.2(a).

$$|X|^2 = \dfrac{1}{5/4 - \cos\omega} \quad \text{and} \quad \theta = -\tan^{-1}\left(\dfrac{\sin\omega}{2 - \cos\omega}\right)$$

Example 16.4

Find the DTFT of $h(n) = (-0.5)^n u(n)$ and plot its magnitude.

Solution

$$H(\omega) = \sum_{n=0}^{\infty} (-0.5)^n e^{-j\omega n} = \dfrac{1}{1 + \frac{1}{2}e^{-j\omega}} = \dfrac{1}{1 + \frac{1}{2}\cos\omega - j\frac{1}{2}\sin\omega}$$

The magnitude and phase of $H(\omega)$ are given below. See Figure 16.2(b).

$$|H|^2 = \dfrac{1}{5/4 + \cos\omega} \quad \text{and} \quad \theta = \tan^{-1}\left(\dfrac{\sin\omega}{2 + \cos\omega}\right)$$

Example 16.5

Find the DTFT of $x(n) = (0.8)^n \cos\left(\frac{\pi n}{2}\right) u(n)$ and plot its magnitude.

Solution

The DTFT may be obtained by application of the definition.

$$x(n) = (0.8)^n \cos\left(\dfrac{\pi n}{2}\right) u(n) \Longrightarrow X(\omega) = \sum_{n=0}^{\infty} (0.8)^n \cos\left(\dfrac{\pi n}{2}\right) e^{-j\omega n}$$

$$= \sum_{n=0}^{\infty} \dfrac{1}{2}(0.8)^n \left[e^{j\frac{\pi n}{2}} + e^{-j\frac{\pi n}{2}}\right] e^{-j\omega n}$$

$$= \sum_{n=0}^{\infty} \dfrac{1}{2}(0.8)^n \left[e^{-j(\omega - \frac{\pi}{2})n} + e^{-j(\omega + \frac{\pi}{2})n}\right]$$

$$= \dfrac{1}{2}\sum_{n=0}^{\infty} (j0.8e^{-j\omega})^n + \dfrac{1}{2}\sum_{n=0}^{\infty} (-j0.8e^{-j\omega})^n$$

$$= \frac{1}{2}\left[\frac{1}{1 - j0.8e^{-j\omega}} + \frac{1}{1 + j0.8e^{-j\omega}}\right]$$

$$= \frac{1}{1 + 0.64e^{-j2\omega}} = \frac{1}{(1 + 0.64\cos 2\omega) - j0.64\sin 2\omega}$$

The magnitude and phase of $X(\omega)$ are given below. See Figure 16.2(c).

$$|H|^2 = \frac{0.7813}{1.1012 + \cos 2\omega} \quad \text{and} \quad \theta = \tan^{-1}\left(\frac{\sin 2\omega}{1.5625 + \cos 2\omega}\right)$$

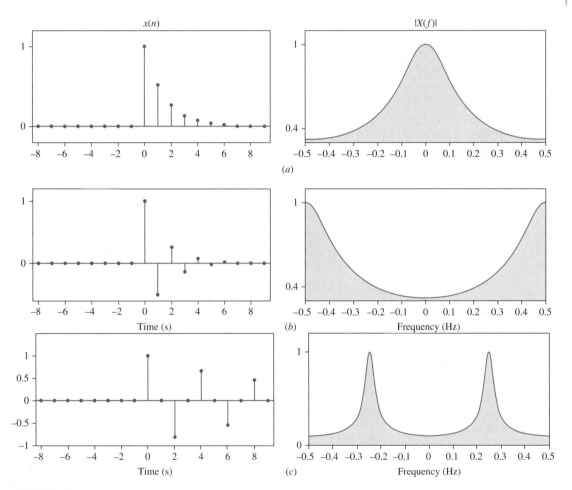

FIGURE 16.2 Three discrete-time causal functions (on the left) and one period of the normalized magnitude of their DTFT plotted for $-0.5 < f < 0.5$ (on the right, corresponding to $-\pi < \omega < \pi$).
(a) $x(n) = (0.5)^n u(n)$. The magnitude plot shows the low-pass property of the function.
(b) $x(n) = (-0.5)^n u(n)$. The magnitude plot shows the high-pass property of the function.
(c) $x(n) = (0.8)^n \cos\left(\frac{\pi n}{2}\right) u(n)$. The magnitude plot shows the bandpass property of the function.

16.3 The DTFT and the *z*-Transform

The discrete-time Fourier transform (DTFT) of a signal is closely related to its z-transform by the following:

$$X(\omega) = \sum_{n=-\infty}^{\infty} x(n)e^{-j\omega n}$$

$$X(z) = \sum_{n=-\infty}^{\infty} x(n)z^{-n}, \quad \text{ROC includes the unit circle.}$$

$$X(\omega) = X(z)|_{z=e^{j\omega}}$$

Existence of DTFT is contingent upon convergence of the sum. That condition requires the ROC of $X(z)$ to include the unit circle in the z-plane. In such a case, the DTFT can be derived from $X(z)$ simply by substituting z by $e^{-j\omega}$. In Example 16.6 to follow the DTFT of the decaying sinusoidal signal of Example 16.5 is derived from the z-transform. The use of z-transform is recommended as an easier method. Example 16.7 discusses finding DTFT of a two-sided function from the z-transform and condition for its existence.

Obtain the DTFT of $x(n) = (0.8)^n \cos\left(\frac{\pi n}{2}\right) u(n)$ (Example 16.5) from its z-transform.

Solution

The DTFT of $x(n)$ is found from its z-transform by setting $z = e^{j\omega}$ as shown below. From tables of the z-transform we have

$$x(n) = a^n \cos(\omega_0 n)u(n) \qquad \Longrightarrow \qquad X(z) = \frac{1 - a(\cos\omega_0)z^{-1}}{1 - 2a(\cos\omega_0)z^{-1} + a^2 z^{-2}}, \quad |z| > a$$

$$x(n) = (0.8)^n \cos\left(\frac{\pi n}{2}\right)u(n) \qquad \Longrightarrow \qquad X(z) = \frac{1}{1 + 0.64 z^{-2}}, \quad |z| > 0.8$$

$$X(\omega) = X(z)|_{z=e^{j\omega}} = \frac{1}{1 + 0.64 e^{-2j\omega}}$$

The result is the same that was found in Example 16.5. The signal $x(n)$ and the normalized magnitude $|X(\omega)|/|X_{Max}|$ have been plotted in Figure 16.2(c). The poles of $X(z)$ are at $z = \pm j0.8$. The maximum magnitude is at $f = \pm 0.25$ Hz (corresponding to $z = \pm j$ which is the closest point on the unit circle to the pole).

E*xample*

16.7

Use the z-transform to find the DTFT of $x(n) = a^n \cos\left(\frac{\pi n}{2}\right) u(n) + b^{-n} \cos\left(\frac{\pi n}{2}\right) u(-n)$ and discuss the conditions when it doesn't exist.

Solution

Let $x(n) = x_1(n) + x_2(n)$, where $x_1(n)$ and $x_2(n)$ are the left- and right-sided components of $x(n)$. Find the z-transform, $X_1(z)$ and $X_2(z)$, of each component and determine if the ROCs overlap. If $X(z)$ exists and the ROC includes the unit-circle in the z-plane, then substitute for $z = e^{j\omega}$ to obtain the DTFT. The above steps are shown below.

a. $x_1(n) = a^n \cos\left(\dfrac{\pi n}{2}\right) u(n)$ $\implies$ $X_1(z) = \dfrac{1}{1 + a^2 z^{-2}}, \quad |z| > |a|$

b. $x_2(n) = b^{-n} \cos\left(\dfrac{\pi n}{2}\right) u(-n)$ $\implies$ $X_2(z) = \dfrac{1}{1 + b^2 z^2}, \quad |z| < \left|\dfrac{1}{b}\right|$

c. $x(n) = x_1(n) + x_2(n)$ $\implies$ $X(z) = X_1(z) + X_2(z)$

$$= \frac{b^2 + 2z^{-2} + a^2 z^{-4}}{b^2 + (1 + a^2 b^2)z^{-2} + a^2 z^{-4}}, \quad |a| < |z| < \left|\frac{1}{b}\right|$$

$$X(\omega) = X(z)|_{z = e^{j\omega}}$$

$$= \frac{b^2 + 2e^{-j2\omega} + a^2 e^{-j4\omega}}{b^2 + (1 + a^2 b^2)e^{-j2\omega} + a^2 e^{-j4\omega}}, \quad |a| < 1 < \left|\frac{1}{b}\right|$$

16.4 Examples of IDTFT

E*xample*

16.8

Find the inverse DTFT of $X(\omega) = \begin{cases} 1, & -\omega_0 \le \omega \le \omega_0 \\ 0, & \text{elsewhere} \end{cases}$

Solution

$$x(n) = \frac{1}{2\pi} \int_{-\pi}^{\pi} X(\omega) e^{j\omega n} d\omega = \frac{1}{2\pi} \int_{-\omega_0}^{\omega_0} e^{j\omega n} d\omega$$

$$= \begin{cases} \frac{\omega_0}{\pi}, & n = 0 \\ \frac{\sin(\omega_0 n)}{\pi n}, & n \ne 0 \end{cases} \quad \text{See Figure 16.3.}$$

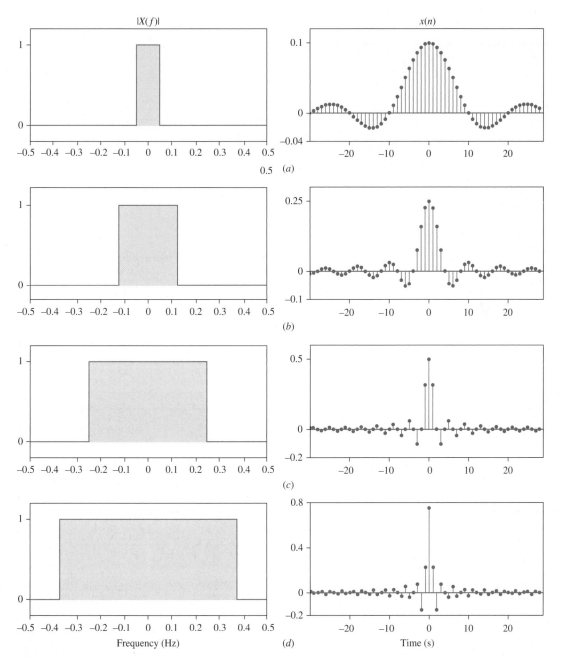

FIGURE 16.3 Low-pass DTFT pairs, $X(f)$ on the left and $x(n)$ on the right. Rows *(a)* to *(d)* show $X(f)$ and $x(n)$ for cutoff frequency $f_0 = 1/20, \ 1/8, \ 1/4,$ and $3/8$ Hz, respectively. See Example 16.8.

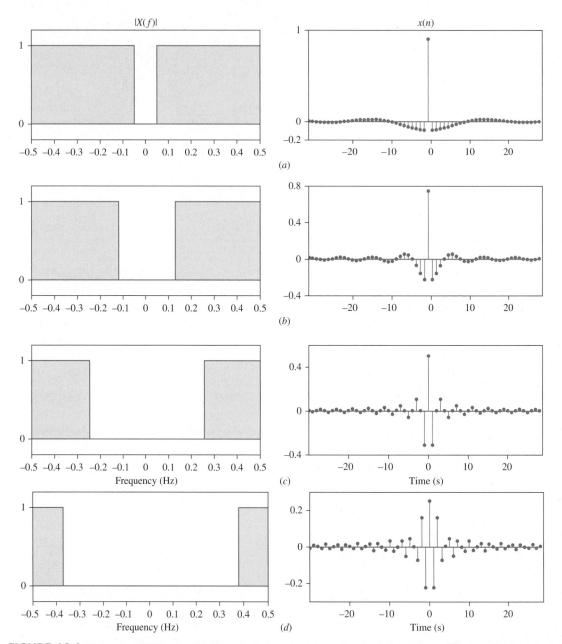

FIGURE 16.4 High-pass DTFT pairs $X(f)$, on the left, and $x(n)$, on the right. Rows **(a)** to **(d)** show $X(f)$ and $x(n)$ for cutoff frequencies $f_0 = 1/20,\ 1/8,\ 1/4$, and $3/8$ Hz, respectively. See Example 16.9.

Find the inverse DTFT of

$$X(\omega) = \begin{cases} 0, & \text{for } -\omega_0 \le \omega \le \omega_0 \\ 1, & \text{elsewhere} \end{cases}$$

Solution

$$x(n) = \frac{1}{2\pi} \int_{-\pi}^{\pi} X(\omega)e^{j\omega n} d\omega = \frac{1}{2\pi} \left[\int_{-\pi}^{-\omega_0} e^{j\omega n} d\omega + \int_{\omega_0}^{\pi} e^{j\omega n} d\omega \right]$$

$$= \begin{cases} 1 - \frac{\omega_0}{\pi}, & n = 0 \\ -\frac{\sin(\omega_0 n)}{\pi n}, & n \ne 0 \end{cases} \quad \text{See Figure 16.4.}$$

16.5 Rectangular Pulse

Rectangular pulses are encountered frequently in signal processing. In Examples 16.8 and 16.9 we worked with rectangular pulses representing ideal filters and found their corrsponding time-domain representations. In the time domain a rectangular pulse is used for window averaging, or simply to select a finite length of a function. In this section we examine frequency representation of a single time-domain rectangular pulse in more detail. First, in Example 16.10 we discuss the DTFT of an even pulse. Then, in Example 16.11 we consider a causal pulse.

Example **16.10**

Find and analyze the DTFT of the even rectangular pulse of width $N = 2M + 1$ which starts at $n = -M$ and ends at $n = M$, $x(n) = u(n + M) - u(n - M - 1)$, and summarize its salient features.

Solution

$$X(\omega) = \sum_{n=-M}^{M} e^{-j\omega n}$$

$$= e^{j\omega M} + e^{j\omega(M-1)} \cdots + e^{j\omega} + 1 + e^{-j\omega} \cdots + e^{-j\omega(M-1)} + e^{-j\omega M}$$

$$= 1 + 2\sum_{n=1}^{M} \cos(\omega n) = X_r(\omega)$$

Note that $X(\omega)$ is a real and even function [to be called $X_r(\omega)$], which can have positive and negative values. The phase of $X(\omega)$ is, therefore, either zero or $\pm\pi$. $X(\omega)$ of the even rectangular pulse may also be given by the following closed-form expression (see solved problem 4 for derivation):

$$X(\omega) = \sum_{n=-M}^{M} e^{-j\omega n} = \frac{\sin(M + \frac{1}{2})\omega}{\sin(\frac{\omega}{2})} = \frac{\sin(\frac{N\omega}{2})}{\sin(\frac{\omega}{2})} = X_r(\omega), \quad \text{where } N = 2M + 1 \text{ is odd}$$

Expressed as a function of f, the DTFT of the even square pulse of length $N = 2M+1$ is

$$X(f) = \frac{\sin(\pi N f)}{\sin(\pi f)}$$

Figure 16.5 Plots an even pulse of size $N = 9$ and its DTFT as a function of f for $-0.5 < f < 0.5$ Hz. The DTFT is a real and even function with $X(0) = N$. It contains a main lobe centered at $f = 0$ (and repeated at $f = \pm k$) with height N, base $2/N$, and unit area. The first zero-crossings occur at $\pm 1/N$. As pulse-width N is increased, the main lobe of $X(f)$ becomes narrower and taller, with the area under it remaining 1. At the limit $N \to \infty$ (DC signal), the main lobe becomes an impulse $\delta(f)$ (and repeats at $f = \pm k$).

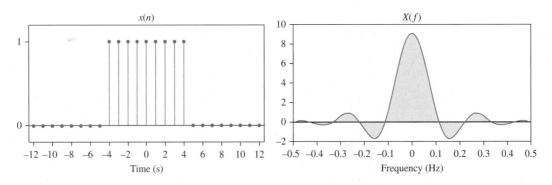

FIGURE 16.5 A rectangular pulse N unit samples long (here, $N = 9$) and its DTFT plotted as a function of f.

xample

Find the DTFT of the causal rectangular pulse $x(n) = u(n) - u(n - N)$, and summarizes its salient features.

Solution

$$X(\omega) = \sum_{n=0}^{N-1} e^{-j\omega n} = \frac{1 - e^{-j\omega N}}{1 - e^{-j\omega}} = e^{-j\omega \frac{(N-1)}{2}} \frac{\sin\left(\frac{N\omega}{2}\right)}{\sin\left(\frac{\omega}{2}\right)} = X_r e^{j\theta}$$

$$\text{where} \quad X_r = \frac{\sin\left(\frac{N\omega}{2}\right)}{\sin\left(\frac{\omega}{2}\right)}, \quad \text{and} \quad \theta = -\frac{(N-1)}{2}\omega$$

The DTFT of a causal square pulse of length N is expressed as the product of a real-valued function $X_r(\omega)$ and a phase-shift component $e^{j\theta}$. The real-valued function $X_r(\omega)$ is the same as found in Example 16.10. Shifting the pulse of Example 16.10 by $(N-1)/2$ units makes the pulse causal. This contributes the phase shift component $e^{j\theta}$ to the DTFT.

Summary

1. The DTFT of the even rectangular pulse made of $N = 2M + 1$ consecutive unit samples is

$$x(n) = u(n + M) - u(n - M - 1) \iff$$

$$X(\omega) = 1 + 2\sum_{n=1}^{M} \cos(\omega n) = \frac{\sin(\frac{N\omega}{2})}{\sin(\frac{\omega}{2})} = X_r(\omega), \quad X(f) = \frac{\sin(\pi N f)}{\sin(\pi f)}$$

It is a real and even function of ω with $X(0) = N$. Because its period is 2π and it is an even function of ω, we attend to it during the interval $0 \leq \omega \leq \pi$.

2. The DTFT of the causal rectangular pulse made of N consecutive unit samples is

$$x(n) = u(n) - u(n - N) \iff X(\omega) = X_r(\omega)e^{-j\omega\frac{(N-1)}{2}}$$

The phase component is caused by a right-shift transforming the even pulse into a causal one. (See section 16.6 on DTFT theorems and properties.)

3. The denominator of the above $X_r(\omega)$ is periodic with a period of 4π. The numerator is also periodic but with a shorter period of $4\pi/N$. $X(\omega)$, therefore, switches between positive and negative values with zero-crossings at $\omega = \pm 2k\pi/N$, $k = 1, 2, 3, 4, \cdots, M$, creating positive and negative lobes (each $2\pi/N$ wide) that repeat at $\omega = \pm 2k\pi/N$ intervals but diminish as $\omega \Rightarrow \pi$ (or $f \Rightarrow 0.5$). See Figure 16.5.

4. The main lobe of $X_r(\omega)$ is centered at $\omega = 0$ with $X(0) = N$. The first zero crossing occurs at $\omega = \pm 2\pi/N$. The base of the lobe is, therefore, $4\pi/N$ units wide. Consider the triangle under the main lobe [with apex at $X(0)$ and the two corners of the base at $\pm 2\pi/N$]. The area under the triangle is $\frac{N}{2} \times \frac{4\pi}{N} = 2\pi$. As pulse width N is increased, the main lobe of $X(\omega)$ becomes narrower and taller, with the area under it remaining 2π. As $N \to \infty$, the main lobe becomes an impulse $2\pi\delta(\omega)$.[5]

5. Observe parallels with the Fourier transform (FT) of a single even rectangular analog pulse of unit height in the continuous-time domain. The FT of such a pulse is $X(F)$ an even function of frequency with a its value at $F = 0$ equal to the area of the time-domain pulse. It has alternating positive and negative lobes which extend over the frequency range $-\infty \leq F \leq \infty$, diminishing toward zero with the increase in frequency. As pulse width increases, the FT becomes a unit impulse in the F-domain.

16.6 DTFT Theorems and Properties

The discrete-time Fourier transform is a member of a family of linear transforms, which includes the Fourier transform, the Laplace transform, and the z-transform. The DTFT,

---

[5]In the f-domain, the limit of $X(f)$ as $N \to \infty$ is $\delta(f)$.

therefore, shares many of their properties. Some of these theorems and properties are described in this section and tabulated in Appendixes 16B and 16C, respectively.

Linearity

The DTFT of $ax(n) + by(n)$ is $aX(\omega) + bY(\omega)$ for any $x(n)$ and $y(n)$ and any constants a and b.

$$x(n) \Longleftrightarrow X(\omega)$$
$$y(n) \Longleftrightarrow Y(\omega)$$
$$ax(n) + by(n) \Longleftrightarrow aX(\omega) + bY(\omega)$$

Real Signals

If $x(t)$ is real, then $X(\omega) = X^*(-\omega)$. This is the conjugate symmetry property for real signals and is shown by

$$x(n) = x^*(n) \Longleftrightarrow X(\omega) = X^*(-\omega)$$

Waveform Symmetry

The DTFT of a real and even signal is a real and even function. Likewise, the DTFT of a real and odd signal is a purely imaginary and odd function.

If $x(n)$ is real and even: $x(n) = x(-n) \Longleftrightarrow$ then $X(\omega)$ is real and even:
$$X(\omega) = X(-\omega)$$
If $x(n)$ is real and odd: $x(n) = -x(-n) \Longleftrightarrow$ then $X(\omega)$ is imaginary and odd:
$$X(\omega) = -X(-\omega)$$

Time and Frequency Reversal

Reversing time flips the signal around $n = 0$ and converts $x(n)$ to $x(-n)$. The DTFTs of $x(n)$ and $x(-n)$ are related by

$$x(n) \Longleftrightarrow X(\omega)$$
$$x(-n) \Longleftrightarrow X(-\omega)$$

The effect is symmetrical with respect to time and frequency. A reversal in the frequency domain results in a reversal in the time domain.

Time Shift

A delay of k units in $x(n)$ adds $(-\omega k)$ to the phase of its DTFT.

$$x(n) \Longleftrightarrow X(\omega)$$
$$x(n - k) \Longleftrightarrow X(\omega)e^{-j\omega k}$$

Frequency Translation

Multiplication of $x(n)$ by $e^{j\omega_0 n}$ shifts the transform by ω_0.

$$x(n) \qquad \Longleftrightarrow \qquad X(\omega)$$
$$x(n)e^{j\omega_0 n} \qquad \Longleftrightarrow \qquad X(\omega - \omega_0)$$

The product $x(n)e^{j\omega_0 n}$ is not a real signal. However, by shifting $X(\omega)$ to the left and to the right by the same amount ω_0 and adding the results together, one obtains the real signal $2x(n)\cos(\omega_0 n)$ in which $x(n)$ modulates the amplitude of $\cos(\omega_0 n)$.

$$x(n)e^{j\omega_0 n} \qquad \Longleftrightarrow \qquad X(\omega - \omega_0)$$
$$x(n)e^{-j\omega_0 n} \qquad \Longleftrightarrow \qquad X(\omega + \omega_0)$$
$$x(n)\cos(\omega_0 n) \qquad \Longleftrightarrow \qquad \tfrac{1}{2}\{X(\omega - \omega_0) + X(\omega + \omega_0)\}$$

Even and Odd Parts of a Signal

The even and odd parts of a real-valued signal are related to the $\mathcal{RE}$ and $\mathcal{IM}$ parts of its DTFT.

$$x(n) \qquad\qquad \Longleftrightarrow \qquad X(\omega) = \mathcal{RE}\{X(\omega)\} + j\mathcal{IM}\{X(\omega)\}$$
$$x(-n) \qquad\qquad \Longleftrightarrow \qquad X^*(\omega) = \mathcal{RE}\{X(\omega)\} - j\mathcal{IM}\{X(\omega)\}$$
$$x_e(n) = \frac{x(n) + x(-n)}{2} \qquad \Longleftrightarrow \qquad \frac{X(\omega) + X^*(\omega)}{2} = \mathcal{RE}\{X(\omega)\}$$
$$x_o(n) = \frac{x(n) - x(-n)}{2} \qquad \Longleftrightarrow \qquad \frac{X(\omega) - X^*(\omega)}{2} = \mathcal{IM}\{X(\omega)\}$$

From the symmetry property and the above it follows that for real $x(n)$

$$\mathcal{RE}\{X(\omega)\} = x(0) + 2\sum_{n=1}^{\infty} x_e(n)\cos(\omega n)$$

$$\mathcal{IM}\{X(\omega)\} = -2\sum_{n=1}^{\infty} x_o(n)\sin(\omega n)$$

$$x_e(n) = \frac{1}{\pi}\int_0^{\pi} \mathcal{RE}\{X(\omega)\}\cos(\omega n)d\omega$$

$$x_o(n) = -\frac{1}{\pi}\int_0^{\pi} \mathcal{IM}\{X(\omega)\}\sin(\omega n)d\omega$$

Causal Functions

A function that is zero for $n < 0$ is called causal. A causal discrete-time function is completely specified by its even part $x_e(n)$, or by its odd part $x_o(n)$ and $x_e(0)$.

$$x(n) = \begin{cases} 0, & n < 0 \\ x_e(0), & n = 0 \\ 2x_e(n) = 2x_o(n), & n > 0. \end{cases}$$

Proof

Let $x(n)$ be a causal function. Then,

For $n < 0$, $x_e(n) + x_o(n) = 0$,   therefore, $x_e(n) = -x_o(n), n < 0$.
For $n > 0$, $x_e(n) = x_o(n)$,   therefore, $x(n) = 2x_e(n) = 2x_o(n), \ n > 0$.
For $n = 0$, $x_e(0) = x(0)$,

From the above results and the symmetry property of the DTFT, we observe the following special relationships between a real-valued causal discrete-time signal $x(n)$ and its DTFT $X(\omega)$.

$$X(\omega) = \sum_{n=0}^{\infty} x(n)e^{j\omega n}, \quad \text{for } -\infty < \omega < \infty <$$

$$x(n) = \begin{cases} \frac{2}{\pi} \int_0^{\pi} \mathcal{RE}\{X(\omega)\} \cos(\omega n) d\omega = -\frac{2}{\pi} \int_0^{\pi} \mathcal{IM}\{X(\omega)\} \sin(\omega n) d\omega, & n > 0 \\ \frac{1}{\pi} \int_0^{\pi} \mathcal{RE}\{X(\omega)\} d\omega, & n = 0 \end{cases}$$

Convolution

Convolution of two discrete-time signals is equivalent to multiplication of their DTFTs in the frequency domain. This is called the convolution property of the DTFT.

$$x(n) \star y(n) \Longleftrightarrow X(\omega)Y(\omega)$$

This property may be proven by applying DTFT to the convolution sum. Convolution of two functions $x(n)$ and $h(n)$ is defined by

$$y(n) = x(n) \star h(n) = \sum_{k=-\infty}^{\infty} x(n-k)\, h(k)$$

The DTFT of $y(n)$ is

$$Y(\omega) = \sum_{n=-\infty}^{\infty} y(n)e^{-j\omega n} = \sum_{n=-\infty}^{\infty}\sum_{k=-\infty}^{\infty} x(n-k)\, h(k)\, e^{-j\omega n}$$

The double sum may be converted to the product of two separate sums by changing the variable n to a new variable m by $n = k + m$ so that $n - k = m$. The terms containing m and k in the double sum are then separated

$$Y(\omega) = \left[\sum_{-\infty}^{\infty} x(m)\, e^{-j\omega m} \right] \times \left[\sum_{-\infty}^{\infty} h(k)\, e^{-j\omega k} \right] = X(\omega)H(\omega)$$

This property is valuable in the analysis of LTI systems because it establishes the frequency-domain relationship between output, input, and the unit-sample response.

Product of Two Signals

Multiplication of two discrete-time signals is equivalent to the convolution of their DTFTs in the frequency domain. This property, which is called the product property of the DTFT, is the dual of the convolution property.

$$x(n)y(n) \iff \tfrac{1}{2\pi} X(\omega) \star Y(\omega) = X(f) \star Y(f)$$

In other words,

$$\sum_{-\infty}^{\infty} x(n)y(n)e^{-j\omega n} = \frac{1}{2\pi} \int_{-\pi}^{\pi} X(\omega - \phi)Y(\phi)d\phi \quad \text{or equivalently,}$$

$$\sum_{-\infty}^{\infty} x(n)y(n)e^{-j2\pi fn} = \int_{-0.5}^{0.5} X(f - \phi)Y(\phi)d\phi$$

This property is used in the modulation and correlation analysis of signals and LTI systems.

*E*xample
16.12

Apply the product property to find the DTFT of $0.5^n \cos(\pi n)u(n)$ and observe that the answer is the same as that found in Example 16.4 for $(-0.5)^n u(n)$.

Solution

$$0.5^n u(n) \implies \frac{1}{1 - 0.5e^{-j\omega}}$$

$$\cos(\pi n) \implies \pi [\delta(\omega + \pi) + \delta(\omega - \pi)]$$

$$0.5^n \cos(\pi n)u(n) \implies \frac{1}{2\pi} \left[\frac{1}{1 - 0.5e^{-j\omega}} \right] \star \pi[\delta(\omega + \pi) + \delta(\omega - \pi)]$$

But
$$\delta(\omega + \pi) \star \frac{1}{1 - 0.5e^{-j\omega}} \implies \frac{1}{1 - 0.5e^{-j(\omega + \pi)}} = \frac{1}{1 + 0.5e^{-j\omega}}$$

and
$$\delta(\omega - \pi) \star \frac{1}{1 - 0.5e^{-j\omega}} \implies \frac{1}{1 - 0.5e^{-j(\omega - \pi)}} = \frac{1}{1 + 0.5e^{-j\omega}}$$

Therefore, $0.5^n \cos(\pi n)u(n) \implies \dfrac{1}{1 + 0.5e^{-j\omega}}$

Conjugate Symmetry

A complex function $x(n)$ is *conjugate symmetric* if $x(n) = x^*(-n)$. When $x(n)$ is real and conjugate symmetric, then $x(n) = x(-n)$ and the function is *even*. It is seen that the even property is a special case of the conjugate symmetric property.

Similarly, a complex function $x(n)$ is *conjugate antisymmetric* if $x(n) = -x^*(-n)$. When $x(n)$ is real and conjugate antisymmetric, then $x(n) = -x(-n)$ and the function is *odd*. It is seen that the odd property is a special case of the conjugate antisymmetric property.

A complex function $x(n)$ may be written as the sum of a conjugate symmetric part [shown by $x_e(n)$] and an antisymmetric part [shown by $x_o(n)$].

$$x(n) = x_e(n) + x_o(n)$$

$$x_e(n) = \frac{x(n) + x^*(-n)}{2} \quad \text{and} \quad x_o(n) = \frac{x(n) - x^*(-n)}{2}$$

When $x(n)$ is real,

$$x_e(n) = \frac{x(n) + x(-n)}{2} \quad \text{and} \quad x_o(n) = \frac{x(n) - x(-n)}{2}$$

and $x_e(n)$ and $x_o(n)$ become the even and odd parts of $x(n)$, respectively. The conjugate symmetry property of the DTFT of complex and real signals is listed in Appendix 16B.

16.7 Parseval's Theorem and Energy Spectral Density

The quantity

$$\sum_{n=-\infty}^{\infty} |x(n)|^2$$

is called the energy in $x(n)$. If the energy is finite, the signal is called square summable. For square summable signals it may be shown that

$$\sum_{n=-\infty}^{\infty} |x(n)|^2 = \frac{1}{2\pi} \int_{-\pi}^{\pi} |X(\omega)|^2 d\omega = \int_{-0.5}^{0.5} |X(f)|^2 df$$

This is called Parseval's theorem for discrete signals. Hence, $|X(f)|^2$ is called the energy spectral density. When $x(n)$ is a real function, $X(f) = X^*(-f)$ and $|X(f)|^2$ is an even function of f. The total energy in $x(n)$ may be found from

$$\sum_{n=-\infty}^{\infty} |x(n)|^2 = 2 \int_{0}^{0.5} |X(f)|^2 df$$

Compare this with the one-sided spectral density of continuous-time signals.

Example

16.13

Find the total energy in the causal exponential signal $x(n) = a^n u(n)$, $|a| < 1$. Then obtain an expression for the energy in the frequency band 0 to 0.01 Hz and find its value if $a = 0.5$.

Solution

The causal signal is given in Figure 16.6(a) for $a = 0.5$.

$$E_{\text{total}} = \sum_{n=0}^{\infty} a^{2n} = \frac{1}{1-a^2}, \quad E_{(0 \to 0.01)} = 2 \int_{0}^{0.01} |X(f)|^2 df$$

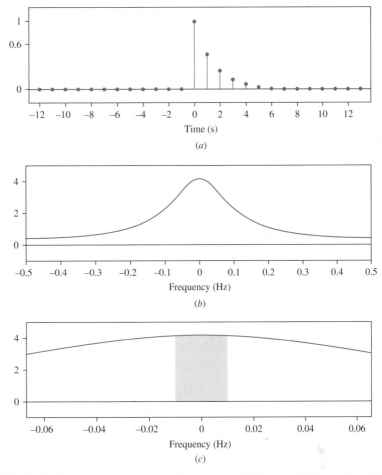

FIGURE 16.6 *(a)* The causal exponential signal $x(n) = (0.5)^n u(n)$. *(b)* One period of $|X(f)|^2$. *(c)* The shading shows a desired area under $|X(f)|^2$, which is numerically equal to the energy within the band -0.01 Hz $< f < 0.01$ Hz. See Example 16.13.

Substituting for $|X(f)|^2$ [from Example 16.3(a)] we have

$$E_{(0 \to 0.01)} = 2 \int_0^{0.01} \frac{1}{1 + a^2 - 2a \cos(2\pi f)} df$$

With $a = 0.5$, the values of the integrand at the lower and upper limits of the integral are, respectively,

$$\text{At } f = 0, \quad |X(f)|^2 = \frac{1}{1.25 - \cos 0^o} = 4$$

$$\text{At } f = 0.01, \quad |X(f)|^2 = \frac{1}{1.25 - \cos 3.6^o} = 3.968$$

We can obtain the energy in the frequency band 0 to 0.01 Hz by approximating the area of the shaded region in Figure 16.6(c). The shaded region is approximated as a rectangle with area

$$E = 2 \times 0.01 \times \frac{4 + 3.968}{2} \approx 0.08$$

This represents the total signal energy in the frequency band $-0.01 < f < 0.01$ Hz.

16.8 DTFT of Power Signals: The Limit Approach

Some signals may not be square summable, but their average power is finite.

$$\lim_{M \to \infty} \frac{1}{2M + 1} \sum_{n=-M}^{M} |x(n)|^2 < \infty$$

These are called *power signals*. The sum representing their DTFT doesn't converge in the familiar sense. However, by employing singularity functions [such as $\delta(\omega)$] we can generalize the DTFT definition to include power signals. The following three cases illustrate the concept and provide examples.

DC Signal

To find the DTFT of a DC signal we start with a rectangular pulse. In section 16.5 we found the DTFT of an even rectangular pulse.

$$x(n) = \begin{cases} 1, & -M \leq n \leq M \\ 0, & \text{elsewhere} \end{cases}$$

$$X(\omega) = \sum_{n=-M}^{M} e^{-j\omega n} = \frac{\sin(M + \frac{1}{2})\omega}{\sin(\frac{\omega}{2})}$$

$X(\omega)$ contains major lobes centered at $\omega = \pm 2k\pi$, interlaced with minor lobes and zero crossings at $\omega = \pm 2k\pi/N$. It was demonstrated that as the pulse in the time domain becomes wider, the major lobes centered at $\omega = \pm 2k\pi$ in the ω-domain become narrower and taller, and the minor lobes become smaller (with more frequent zero-crossings). As $M \to \infty, x(n) \to 1$ (a DC signal) and $X(\omega)$ becomes a train of impulses of strength 2π at $\omega = \pm 2k\pi$; that is, a collection of $2\pi\delta(\omega - 2k\pi), -\infty < k < \infty$. See Figure 16.7($a$).

$$\lim_{M \to \infty} x(n) = 1$$

$$\lim_{M \to \infty} X(\omega) = \lim_{M \to \infty} \sum_{n=-M}^{M} e^{-j\omega n} = \lim_{M \to \infty} \frac{\sin(M + \frac{1}{2})\omega}{\sin(\frac{\omega}{2})} = \sum_{k=-\infty}^{\infty} 2\pi\delta(\omega - 2k\pi)$$

$$\lim_{M \to \infty} X(f) = \sum_{k=-\infty}^{\infty} \delta(f - k)$$

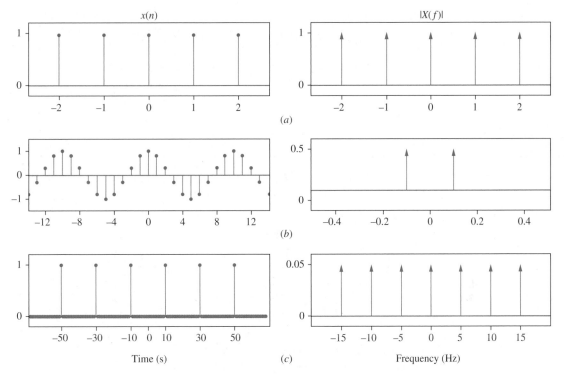

FIGURE 16.7 Three periodic discrete-time signals (left column) and their DTFT (right column). The DTFTs are made of impulses in the frequency domain.

(a) The DTFT of the DC signal $x(n) = 1$, all n, is a train of unit impulses separated by 1 Hz.

$$x(n) = \sum_{k=-\infty}^{\infty} d(n - k) \iff X(f) = \sum_{k=-\infty}^{\infty} \delta(f - k)$$

(b) The DTFT of a discrete-time sinusoidal signal with period N is a periodic pair of impulses positioned at the multiple frequencies of the signal $f = \pm(f_0 + k)$, where $f_0 = 1/N$. In this figure $N = 10$ and $f_0 = 0.1$ Hz. The strength of each impulse in the ω-domain is π (for $\cos \omega_0 n$) or $\pm j\pi$ (for $\sin \omega_0 n$). In the f-domain, the strengths of the impulses are $1/2$ [for $\cos(2\pi n/N)$] or $\pm j/2$ [for $\sin(2\pi n/N)$]. In summary, one cycle of the DTFTs is

$$\cos(\omega_0 n) = \cos(2\pi n/N) \implies \pi\delta(\omega - \omega_0) + \pi\delta(\omega + \omega_0) \quad \text{or} \quad \tfrac{1}{2}[\delta(f - f_0) + \delta(f + f_0)]$$

$$\sin(\omega_0 n) = \sin(2\pi n/N) \implies -j\pi\delta(\omega - \omega_0) + j\pi\delta(\omega + \omega_0) \quad \text{or} \quad \tfrac{1}{2j}[\delta(f - f_0) - \delta(f + f_0)]$$

(c) The DTFT of an infinite train of unit samples with period N and one sample per period in the n-domain (left side) is a train of impulses of strength $2\pi/N$ at $\omega = 2k\pi/N$ in the ω-domain or, equivalently, a train of impulses of strength $1/N$ at $f = k/N$ in the f−domain (right side).

$$\sum_{k=-\infty}^{\infty} d(n - kN) \iff \frac{2\pi}{N} \sum_{k=-\infty}^{\infty} \delta\left(\omega - \frac{2\pi k}{N}\right) = \frac{1}{N} \sum_{k=-\infty}^{\infty} \delta\left(f - \frac{k}{N}\right)$$

This figure is generated with $N = 20$.

Sinusoid

Consider a discrete-time sinusoidal signal with period N

$$x(n) = \cos(\omega_0 n) = \frac{e^{j\omega_0 n} + e^{-j\omega_0 n}}{2}, \quad \text{where } \omega_0 = \frac{2\pi}{N}$$

The DTFT of the sinusoidal signal is

$$X(\omega) = \lim_{M \to \infty} \sum_{n=-M}^{M} \frac{1}{2} e^{-j\omega n} \left(e^{j\omega_0 n} + e^{-j\omega_0 n} \right)$$

$$= \lim_{M \to \infty} \sum_{n=-M}^{M} \frac{1}{2} \left(e^{-j(\omega-\omega_0)n} + e^{-j(\omega+\omega_0)n} \right)$$

$$= \pi \sum_{k=-\infty}^{\infty} [\delta(\omega - \omega_0 - 2k\pi) + \delta(\omega + \omega_0 + 2k\pi)]$$

$$X(f) = \frac{1}{2} \sum_{k=-\infty}^{\infty} [\delta(f - f_0 - k) + \delta(f + f_0 + k)]$$

See Figure 16.7(b). The DTFT of a *cosine* is made of impulses of strength π at $\omega = \pm(\omega_0 + 2k\pi)$. For a sine, the strength of the impulses are $-j\pi$ (at $\omega = \omega_0$) and $j\pi$ (at $\omega = -\omega_0$). One cycle of the DTFTs is given below:

$$\cos(\omega_0 n), \quad -\infty < n < \infty \quad \Longleftrightarrow \quad \pi\delta(\omega + \omega_0) + \pi\delta(\omega - \omega_0)$$

$$\sin(\omega_0 n), \quad -\infty < n < \infty \quad \Longleftrightarrow \quad -j\pi\delta(\omega + \omega_0) + j\pi\delta(\omega - \omega_0)$$

Note the special case of $\cos(\pi n)$.

Train of Unit Samples

Consider an infinite train of unit samples, $x(n)$, with period N and one sample per period, Figure 16.7(c) (left side).

$$x(n) = \sum_{k=-\infty}^{\infty} d(n - kN)$$

For simplicity we have chosen the time origin such that one sample occurs at $n = 0$. We will show that the DTFT of $x(n)$ is a train of impulses, of strength 2π, at multiples of π/N in the ω-domain.

To derive the DTFT of $x(n)$, first consider a segment of $x(n)$ containing $2M + 1$ periods ($2M + 1$ nonzero samples), to be called $x_M(n)$.

$$x_M(n) = \sum_{k=-M}^{M} d(n - kN)$$

$x_M(n)$ is an energy signal and has a DTFT in the conventional sense. It contains $2M + 1$

unit samples at $n = \pm kN$, $k = 0, 1, 2, \cdots, M$. Its DTFT is

$$X_M(\omega) = \sum_{n=-\infty}^{\infty} x_M(n)e^{-j\omega n} = \sum_{n=-MN}^{MN} \left[\sum_{k=-M}^{M} d(n-kN)e^{-j\omega n} \right] = \sum_{n=-M}^{M} e^{-j\omega n N}$$

$$= e^{j\omega MN} + e^{j\omega(M-1)N} \cdots + 1 \cdots + e^{-j\omega(M-1)N} + e^{-j\omega MN}$$

$$= e^{j\omega MN} \left[1 + e^{-j\omega N} + e^{-2j\omega N} \cdots + e^{-j\omega 2MN} \right]$$

$$= e^{j\omega MN} \left[\frac{1 - e^{-j(2M+1)\omega N}}{1 - e^{-j\omega N}} \right] = \frac{\sin\left[\frac{\omega N(2M+1)}{2} \right]}{\sin\left(\frac{\omega N}{2} \right)}$$

It is periodic and contains major lobes with peaks of magnitude $(2M+1)$ centered at $\omega = 2k\pi/N$. Note that the location of the peaks is a function of N only (signal period) and independent of M. Between the major lobes there are minor lobes with zero-crossings at

$$\omega = \pm \frac{2k\pi}{N(2M+1)}, \quad k = 0, 1, 2, \cdots$$

As M grows, the locations of the peaks remain the same, that is, at $\omega = 2k\pi/N$, but their magnitudes $(2M+1)$ grow. The zero-crossings between the peaks also become more frequent. The major lobes associated with the peaks become narrower and taller. However, the area associated with each lobe is $\frac{2\pi}{N}$ as derived below.

$$\int_{-\frac{\pi}{N}}^{\frac{\pi}{N}} X_M(\omega)d\omega = \int_{-\frac{\pi}{N}}^{\frac{\pi}{N}} \left[\sum_{k=-M}^{M} e^{-j\omega k N} \right] d\omega = \sum_{k=-M}^{M} \int_{-\frac{\pi}{N}}^{\frac{\pi}{N}} e^{-j\omega k N} d\omega = \int_{-\frac{\pi}{N}}^{-\frac{\pi}{N}} d\omega = \frac{2\pi}{N}$$

If we let $M \to \infty$, we get

$$\lim_{M \to \infty} x_M(n) = x(n)$$

$$\lim_{M \to \infty} X_M(\omega) = X(\omega) = \frac{2\pi}{N} \sum_{k=-\infty}^{\infty} \delta\left(\omega - \frac{2k\pi}{N} \right)$$

Therefore, the DTFT of a train of unit samples at multiples of N in the n-domain is a train of impulses of strength $2\pi/N$ at $\omega = 2k\pi/N$ in the ω-domain or, equivalently, a train of impulses of strength $1/N$ at $f = k/N$ in the f-domain.

$$\sum_{k=-\infty}^{\infty} d(n-kN) \iff \frac{2\pi}{N} \sum_{k=-\infty}^{\infty} \delta\left(\omega - \frac{2\pi k}{N} \right) = \frac{1}{N} \sum_{k=-\infty}^{\infty} \delta\left(f - \frac{k}{N} \right)$$

See Figure 16.7(c).

Example
16.14

Two examples of DTFT pairs are given below.

a. $x(n) = \begin{cases} 1, & n \text{ even} \\ 0, & n \text{ odd} \end{cases} \quad \Longleftrightarrow \quad X(\omega) = \sum_{k=-\infty}^{\infty} \pi \delta(\omega - k\pi)$

b. $x(n) = (-1)^n \quad \Longleftrightarrow \quad X(\omega) = \sum_{k=-\infty}^{\infty} \pi \delta(\omega - k\pi)\left(1 - e^{-j\omega}\right)$

$$= \sum_{k=-\infty}^{\infty} 2\pi \delta[\omega - (2k+1)\pi]$$

16.9 DTFT of Periodic Signals: The Convolution Approach

A periodic signal $y(n)$ with period N is obtained by convolution of a time-limited signal $h(n)$ with an infinite train $x(n)$ of unit samples located at every multiple of N samples.

$$y(n) = h(n) \star x(n) = h(n) \star \sum_{k=-\infty}^{\infty} d(n - kN) = \sum_{k=-\infty}^{\infty} h(n - kN)$$

where $h(n)$ is one cycle of $y(n)$

$$h(n) = \begin{cases} y(n), & 0 \leq n < N \\ 0, & \text{elsewhere} \end{cases}$$

The DTFT of $y(n)$ is found by multiplying the DTFTs of $h(n)$ and $x(n)$

$$Y(\omega) = X(\omega)H(\omega)$$

In section 16.8 the DTFT of the train of unit samples was found to be

$$x(n) = \sum_{k=-\infty}^{\infty} d(n - k) \quad \Longrightarrow \quad X(\omega) = \frac{2\pi}{N} \sum_{k=-\infty}^{\infty} \delta\left(\omega - \frac{2k\pi}{N}\right)$$

$$\Longrightarrow \quad X(f) = \frac{1}{N} \sum_{k=-\infty}^{\infty} \delta\left(f - \frac{k}{N}\right)$$

Therefore,

$$Y(\omega) = \frac{2\pi}{N} H(\omega) \sum_{k=-\infty}^{\infty} \delta\left(\omega - \frac{2k\pi}{N}\right) = \frac{2\pi}{N} \sum_{k=-\infty}^{\infty} H\left(\frac{2k\pi}{N}\right) \delta\left(\omega - \frac{2k\pi}{N}\right)$$

$$Y(f) = \frac{1}{N} \sum_{k=-\infty}^{\infty} H\left(\frac{k}{N}\right) \delta\left(f - \frac{k}{N}\right)$$

The DTFT of the periodic signal is obtained by sampling the DTFT of a single cycle at $f = k/N$ and representing the samples by impulse functions.

DTFT of a Periodic Rectangular Pulse

Let $x(n)$ be a periodic rectangular pulse with period N. Each pulse has W samples. For simplicity, assume $x(n)$ is an even function of n. Here we find its DTFT using f to represent the frequency.

Let $h(n)$ represent the single pulse centered at $n = 0$. Then,

$$h(n) = \sum_{k=-M}^{M} d(n-k) \implies H(f) = \frac{\sin(W\pi f)}{\sin(\pi f)}, \quad \text{where } W = 2M + 1$$

$$x(n) = \sum_{k=-\infty}^{\infty} h(n - kN) \implies X(f) = \frac{1}{N} \sum_{k=-\infty}^{\infty} H\left(\frac{k}{N}\right) \delta(f - k/N)$$

$$= \frac{1}{N} \sum_{k=-\infty}^{\infty} \frac{\sin(W\pi k/N)}{\sin(\pi k/N)} \delta(f - k/N)$$

Alternate Approach

$$X(f) = \sum_{n=-M}^{M} \left[\frac{e^{-j2\pi nf}}{N} \sum_{k=-\infty}^{\infty} \delta(f - k/N) \right]$$

$$= \frac{1}{N} \sum_{k=-\infty}^{\infty} \delta(f - k/N) \left[\sum_{n=-M}^{M} e^{-j2\pi nf} \right]$$

$$= \frac{1}{N} \sum_{k=-\infty}^{\infty} \frac{\sin(W\pi k/N)}{\sin(\pi k/N)} \delta(f - k/N)$$

As expected, the DTFT of the periodic pulse train with period N is obtained by sampling the DTFT of a single cycle (at $\omega = 2k\pi/N$ or $f = k/N$) and representing the samples by impulse functions, Figure 16.8.

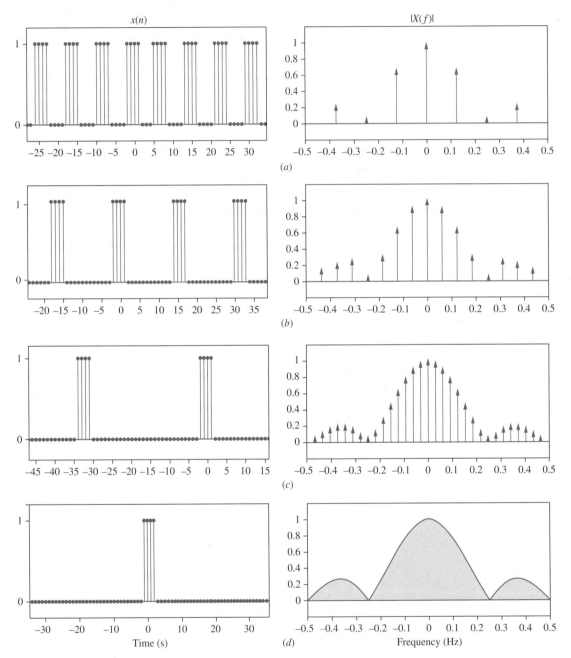

FIGURE 16.8 The DTFT of a periodic signal with period N (a, b, c) is derived by sampling the DTFT of a single cycle (d) at $f = kf_0$, where $f_0 = 1/N$, and representing the samples by impulse functions. In this figure: (a) $N = 8$, $f_0 = 0.125$ Hz, (b) $N = 16$, $f_0 = 0.0625$ Hz, and (c) $N = 32$, $f_0 = 31.25$ mHz.

16.10 Zero-Insertion

Imagine stretching the time axis of a discrete-time signal $x(n)$ in a way that it inserts $N - 1$ zeros ($N \geq 1$) at the newly generated spaces between neighboring samples, producing a new signal $y(n)$. An example of the effect of zero-insertion on $x(n) = (0.6)^n u(n)$ and its DTFT for $N = 2$ is shown in Figure 16.9. Another example is shown in Figure 16.10 for the *sinc-type* signal[6]

$$x(n) = \left[\frac{\sin(\frac{\pi n}{2})}{(\frac{\pi n}{2})} \right]^2$$

for $N = 5$. It resembles stretching the time axis by a factor of 5.

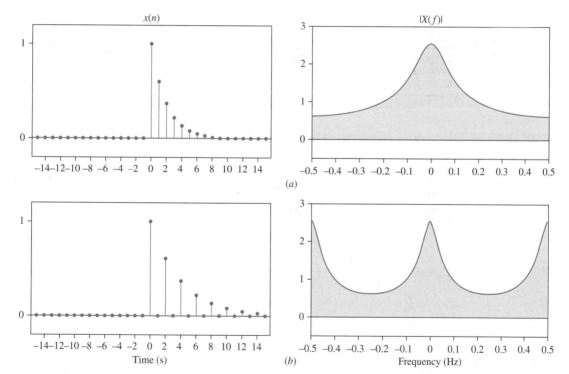

FIGURE 16.9 Inserting a zero between successive samples of $x(n) = (0.6)^n u(n)$ and its effect on the DTFT. Zero-insertion compresses the frequency axis by a factor of 2.

---

[6]For the definition of *sinc* signals see Chapter 10.

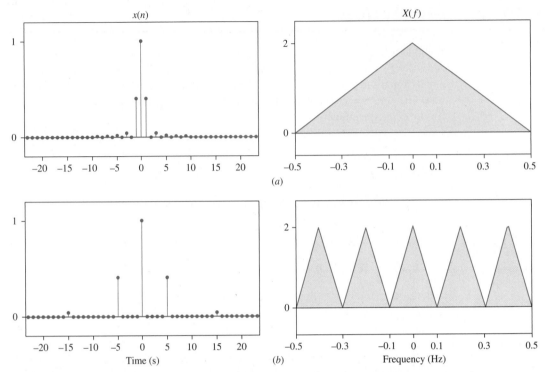

FIGURE 16.10 Inserting four zeros between successive samples of $x(n) = \sin^2(\pi n/2)/(\pi n/2)^2$ and its effect on the DTFT. It compresses the frequency axis by a factor of 5.

A zero-insertion process that is referenced at $n = 0$ may be summarized by:

$$\vdots$$

$$y(-N-2) = 0$$
$$y(-N-1) = 0$$
$$y(-N) \quad = x(-1)$$
$$y(-N+1) = 0$$
$$y(-N+2) = 0$$

$$\vdots$$

$$y(-2) \quad = 0$$
$$y(-1) \quad = 0$$
$$y(0) \quad = x(0)$$
$$y(1) \quad = 0$$
$$y(2) \quad = 0$$

$$\vdots$$

$$y(N-2) \quad = 0$$
$$y(N-1) \quad = 0$$
$$y(N) \quad = x(1)$$

$$y(N+1) = 0$$
$$y(N+2) = 0$$
$$\vdots$$

It can be shown that inserting $(N-1)$ zeros between samples in $x(n)$ that produces a new function $y(t)$ results in the following effect in the frequency domain:

$$x(n) \iff X(\omega)$$

$$y(n) \iff Y(\omega) = X(N\omega)$$

The proof is shown below.

$$X(\omega) = \sum_{n=-\infty}^{\infty} x(n)e^{-j\omega n}$$

$$Y(\omega) = \sum_{n=-\infty}^{\infty} y(n)e^{-j\omega n} = \sum_{k=-\infty}^{\infty} y(kN)e^{-j\omega kN}$$

$$= \sum_{k=-\infty}^{\infty} x(k)e^{-j\omega kN} = \sum_{k=-\infty}^{\infty} x(k)e^{-j(N\omega)k} = X(N\omega)$$

Inserting $N-1$ zeros between neighboring samples in the unit-sample response of a digital filter shrinks (compresses) the frequency axis by a factor N and replicates the frequency response. Knowing N, one can always unshrink $Y(\omega)$ and retrieve $X(\omega)$ from it. No information is lost by inserting zeros. This is hardly surprising. By removing zeros from $y(n)$ one can always retrieve the original signal $x(n)$.

Example 16.15

Take $x(n) = (0.5)^n u(n)$ and insert single zeros between its samples to create a new signal $y(n)$. Find $Y(\omega)$ and relate it to $X(\omega)$.

Solution

$$X(\omega) = \sum_{n=0}^{\infty} (0.5)^n e^{-j\omega n} = \frac{1}{1 - 0.5e^{-j\omega}}$$

$$Y(\omega) = \sum_{n=0}^{\infty} (0.5)^n e^{-j2\omega n} = \frac{1}{1 - 0.5e^{-j2\omega}} = X(2\omega)$$

Example 16.16

Now consider $h(n) = (-0.5)^n u(n)$ and note that $h(n) = y(n) - 0.5y(n-1)$, where $y(n)$ is the signal generated by inserting single zeros between samples of $x(n) = (0.25)^n u(n)$. Derive $H(\omega)$ from $X(\omega)$. Compare with the results obtained directly.

Solution

$$x(n) = (0.25)^n u(n) \qquad \Longrightarrow \quad X(\omega) = \frac{1}{1 - 0.25e^{-j\omega}}$$

$$y(n) \qquad\qquad \Longrightarrow \quad Y(\omega) = X(2\omega) = \frac{1}{1 - 0.25e^{-j2\omega}}$$

$$h(n) = y(n) - 0.5y(n-1) \quad \Longrightarrow \quad H(\omega) = Y(\omega) - 0.5e^{-j\omega}Y(\omega)$$

$$= (1 - 0.5e^{-j\omega})Y(\omega)$$

$$H(\omega) = \frac{1 - 0.5e^{-j\omega}}{1 - 0.25e^{-j2\omega}} = \frac{1}{1 + 0.5e^{-j\omega}}$$

The above result may also be derived directly from the definition of the DTFT

$$H(\omega) = \sum_{n=0}^{\infty} (-0.5)^n e^{-j\omega n} = \sum_{k=0}^{\infty} (0.5)^{2k} e^{-j2k\omega} - \sum_{k=0}^{\infty} (0.5)^{2k+1} e^{-j(2k+1)\omega}$$

$$= \sum_{k=0}^{\infty} \left(0.25e^{-j2\omega}\right)^k - 0.5e^{-j\omega} \sum_{k=0}^{\infty} \left(0.25e^{-j2\omega}\right)^k = \frac{1 - 0.5e^{-j\omega}}{1 - 0.25e^{-j2\omega}}$$

$$= \frac{1}{1 + 0.5e^{-j\omega}}$$

The result is in agreement with $H(\omega)$ found in Example 16.4.

16.11 Decimation

Decimation implies discarding some samples. For example, keeping every other sample and dropping the sample that is in between is called decimation by a factor of 2. To decimate by a factor of 10 (at which factor the expression is clearly appropriate), we would keep every tenth sample. Decimation by a factor D of an existing signal $x(n)$ generates a new signal $y(n) = x(Dn)$, where D is an integer greater than one. The operation keeps a sample and drops the next $D - 1$ samples in the original signal. It reduces the number of samples available for reconstruction of the original continuous-time signal. Mathematically it is equivalent to reducing the sampling rate by a factor of D. It is used in multirate signal processing where two or more signals sampled at different rates are being mixed with one another. It is also used in antialiasing filtering and frequency downshifting.

Decimation of a discrete-time signal may or may not cause aliasing. If a signal is sampled at a sufficiently higher-than-minimum rate (for example, at a rate greater than or equal to D-times the Nyquist rate), its decimation by a factor D doesn't cause aliasing. On the other hand, if a signal is sampled at the minimum (Nyquist) rate, its decimation would mean the loss of samples required for its reconstruction and, hence, aliasing occurs.

Another application of decimation is to reduce the complexity of the analog anti-aliasing filter. For this purpose, a simple low-pass analog filter (such as a first-order filter) is used in conjunction with a sampling rate much higher than necessary. The samples are then passed through a digital decimation/antialiasing filter.

As an example, consider the continuous-time signal

$$x(t) = \cos(1{,}000\pi t) + \cos(2{,}000\pi t) + \cos(3{,}000\pi t) + \cos(4{,}000\pi t)$$

The spectrum of $x(t)$ has four spectral lines at 500, 1,000, 1,500, and 2,000 Hz. Sample $x(t)$ at the 10-kHz rate and produce the following sequence

$$x(n) = x(t)|_{t=10^{-4}n} = \cos(\pi n/10) + \cos(2\pi n/10) + \cos(3\pi n/10) + \cos(4\pi n/10)$$

Then decimate $x(n)$ by a factor of 21 to obtain a new sequence of samples. Decimation reduces the sampling rate to $10{,}000/21 = 476.2$ and produces the sequence $y(n) = x(21n)$. Reconstruct a new time function $y(t)$ from the decimated samples. Take the Fourier transform of $y(t)$ and examine the relationship between $Y(f)$ and $X(f)$. Decimation downshifts the spectral lines to 23.8, 47.6, 71.4, and 95.23 Hz, respectively. [You may save $x(n)$ and $y(n)$ in the form of *.wav* files and play them back through a sound card to examine the effect of frequency downshifting. For further exploration see Project 2 at the end of this chapter.]

Examples of Decimation by a Factor D

Let a new signal $y(n)$ be generated from a given signal $x(n)$ according to the following rule:

$$y(n) = x(Dn)$$

where D is an integer greater than one. This is a special type of compression of the time axis of the discrete-time signal $x(n)$ and is called decimation by a factor D. It is equivalent to keeping a sample and dropping the next $D - 1$ samples in the original signal. A decimation process that is referenced at $n = 0$ may be summarized by

$\vdots$

keep $x(-D)$ $\Longrightarrow$ call it $y(-1)$
drop $x(-D+1)$
drop $x(-D+2)$

$\vdots$

drop $x(-1)$
keep $x(0)$ $\Longrightarrow$ call it $y(0)$
drop $x(1)$
drop $x(2)$

$\vdots$

drop $x(D-1)$
keep $x(D)$ $\Longrightarrow$ call it $y(1)$

drop $x(D+1)$
drop $x(D+2)$

$\vdots$

drop $x(2D-1)$
keep $x(2D)$ $\Longrightarrow$ call it $y(2)$
drop $x(2D+1)$
drop $x(2D+2)$

$\vdots$

drop $x(3D-1)$
keep $x(3D)$ $\Longrightarrow$ call it $y(3)$

$\vdots$

Decimating a signal resembles shrinking its time scale. An example of decimation by factors 2 and 5 for $x(n) = (0.6)^n u(n)$ and its effect in the frequency domain (with aliasing) are shown in Figure 16.11. Figure 16.12 shows decimation of $x(n) = \sin^2(\pi n/10)/(\pi n/10)^2$ by factors 2 and 4, which extend the DTFT of the newly generated signal to near the full range $-0.5 < f < 0.5$ (in Hz) with no aliasing. See Examples 16.17 and 16.18.

Loosely speaking, decimation expands the range of frequency components away from the fixed points $\pm k\pi$ on the ω axis. Because of this, aliasing may occur, leading to signal distortion and loss of information, in which case one cannot retrieve $X(\omega)$ from $Y(\omega)$.

Decimating $x(n)$ by a factor D will not lead to aliasing if $x(n)$ is band limited enough; that is, if $|X(\omega)| = 0, \quad \omega \geq \pi/D$. For such a class of signals it can be shown that decimation results in the following effect in the frequency domain:

$$x(n) \qquad \Longleftrightarrow \quad X(\omega)$$
$$y(n) = x(Dn) \quad \Longleftrightarrow \quad Y(\omega) = \tfrac{1}{D}X(\omega/D)$$

The proof is shown below.

$$X(\omega) = \sum_{n=-\infty}^{\infty} x(n)e^{-j\omega n}$$

$$Y(\omega) = \sum_{n=-\infty}^{\infty} y(n)e^{-j\omega n} = \sum_{n=-\infty}^{\infty} x(Dn)e^{-j\omega n}$$

$$= \frac{1}{D}X\left(\frac{\omega}{D}\right)$$

Therefore, $x(n)$ may be retrieved from $y(n) = x(Dn)$. These observations are not surprising as elaborated in the following two cases.

Case 1

Sampled at the minimum (Nyquist) rate, the spectrum of the resulting $x(n)$ occupies the full range of $-\pi < \omega < \pi$. Decimation of $x(n)$ by a factor D expands the frequency range of the signal in the discrete domain in both directions, leftward and rightward simultaneously, around the fixed points $\pm k\pi$ on the ω-axis. Because the original signal occupies the full frequency range, this causes aliasing. The spectrum $X(\omega)$ may not be derived from $Y(\omega)$. Decimation of $x(n)$ means loss of samples required for reconstruction of the original signal. It results in aliasing in the frequency domain. By removing samples from $x(n)$ one throws away valuable information that cannot be retrieved. As an example, consider a continuous-time low-pass signal sampled at a rate 1.5 times the minimum (Nyquist) rate. The result is $x(n)$ whose DTFT $X(f)$ occupies the discrete frequency range $-0.333 < f < 0.333$ Hz. Decimation of $x(n)$ by a factor 2 expands the frequency range of the signal in the discrete domain by factor 2, produces overlap with neighboring segments, and causes aliasing.

Case 2

Sampled at a sufficiently higher-than-minimum rate (for example, at D-times the Nyquist rate, to be called oversampling by a factor D), the spectrum of the resulting $x(n)$ occupies the range of $-\pi/D < \omega < \pi/D$. Decimation of $x(n)$ by a factor D expands the ω-axis but doesn't cause aliasing because of oversampling. The spectrum $X(\omega)$ may be derived from $Y(\omega)$. Decimation of $x(n)$ doesn't result in the loss of samples required for reconstruction of the original signal. By removing samples from $x(n)$ one throws away redundant information that still may be found in the remaing samples. As an example, consider a continuous-time low-pass signal oversampled by a factor 5 (5 times the Nyquist rate). The DTFT of the resulting $x(n)$ occupies the discrete frequency range $-0.1 < f < 0.1$ Hz. Decimation of $x(n)$ by a factor 2 expands the plot of $X(f)$ to $-0.2 < f < 0.2$ Hz but doesn't cause aliasing. $X(f)$ may still be derived from the DTFT of the decimated signal.

Example 16.17

The time constant of an exponentially decaying continuous-time function is $200\ \mu$sec. It is sampled every $100\ \mu$ sec.

a. Find $x(n)$ and $X(\omega)$.

b. Decimate $x(n)$ by a factor 2 and call the new function $y(n)$.

c. Find $Y(\omega)$ and relate it to $X(\omega)$.

Can $x(t)$ be reconstructed from $y(n)$?

Solution

$$x(n) = x(t)\Big|_{t=10^{-4}n} = e^{-0.5n}u(n) \approx (0.6)^n u(n),$$

$$X(\omega) = \sum_{n=0}^{\infty}(0.6)^n e^{-j\omega n} = \frac{1}{1 - 0.6e^{-j\omega}}$$

$$y(n) = x(2n) = (0.36)^n u(n),\ Y(\omega) = \sum_{n=0}^{\infty}(0.36)^n e^{-j\omega n} = \frac{1}{1 - 0.36e^{-j\omega}} \neq X\left(\frac{\omega}{2}\right)$$

See Figure 16.11. Note that from a knowledge of $Y(\omega)$ [equivalently, $y(n)$] one may not find $X(\omega)$ exactly. Therefore, $x(t)$ cannot be reconstructed from $y(n)$.

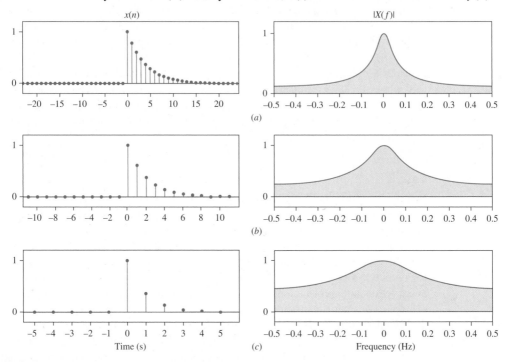

FIGURE 16.11 Decimation of $x(n) = (0.6)^n u(n)$ (left column) by factors of 2 and 4 and its effect on the DTFT (right column). Because $X(\omega)$ covers the full frequency range, decimation of $x(n)$ produces aliasing. **(a)** shows the original signal, while **(b)** and **(c)** give the decimated versions.

Example

16.18

Repeat Example 16.17 for the low-pass signal

$$x(t) = \frac{\sin^2(1{,}000\pi t)}{(1{,}000\pi t)^2}$$

and show that in this case $x(t)$ may be reconstructed from the decimated discrete-time function.

Solution

$$x(n) = x(t)\Big|_{t=10^{-4}n} = \left[\frac{\sin\left(\frac{\pi n}{10}\right)}{\frac{\pi n}{10}}\right]^2 \quad \text{See Figure 16.12(}a\text{) for } X(\omega).$$

$$y(n) = x(2n) = \left[\frac{\sin\left(\frac{\pi n}{5}\right)}{\frac{\pi n}{5}}\right]^2 \quad \text{[Except at } n = 0 \text{ where } y(0) = x(0)\text{]},$$

$$Y(\omega) = \frac{1}{2}X\left(\frac{\omega}{2}\right). \quad \text{See Figure 16.12(}b\text{).}$$

One may obtain $X(\omega)$ from $Y(\omega)$ [equivalently, $x(n)$ from $y(n)$]. Therefore, an exact $x(t)$ can be reconstructed from $y(n)$.

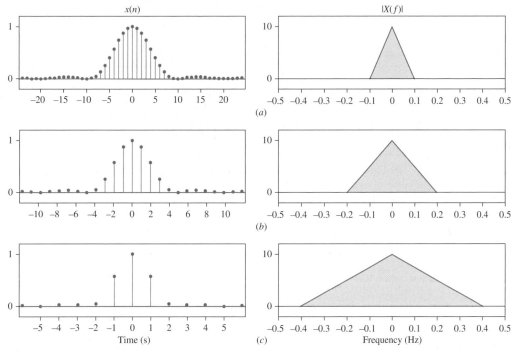

FIGURE 16.12 Decimation of $x(n) = \sin^2(\pi n/10)/(\pi n/10)^2$ (left column) by factors of 2 and 4 and its effect on the DTFT (right column). Because $X(\omega)$ covers one-fifth of the full frequency range, decimation by a factor of 2 or 4 extends its frequency with no aliasing. *(a)* shows the original signal, while *(b)* and *(c)* give the decimated versions.

Example
16.19

Consider the low-pass signal

$$x(n) = \frac{\sin(\omega_0 n)}{\pi n} \iff X(\omega) = \begin{cases} 1, & |\omega| \le \omega_0 \\ 0, & \text{elsewhere} \end{cases}$$

Assuming $\omega_0 \le \pi/4$, decimate $x(n)$ by a factor 2 to find $y(n)$ and its DTFT.

Solution

$$x(n) = \frac{\sin(\frac{\pi n}{4})}{\pi n}$$

$$y(n) = x(2n) = \frac{\sin(\frac{\pi n}{2})}{2\pi n}$$

$$Y(\omega) = \begin{cases} \frac{1}{2}, & |\omega| \le \pi/2 \\ 0, & \text{elsewhere} \end{cases}$$

Note that $Y(\omega) = \frac{1}{2} X\left(\frac{\omega}{2}\right)$. For the special case of $\omega_0 = \pi/2$ we get

$$y(n) = \frac{1}{2} d(n) \iff Y(\omega) = \frac{1}{2}$$

16.12 Interpolation

Interpolation may be viewed as the converse of decimation. It involves adding more samples. For example, adding an additional sample between two existing samples is called interpolation by a factor of 2. To interpolate by a factor 10 we would add 9 new samples for every existing one. Interpolation by factor I of an existing signal $x(n)$ generates a new signal $y(n)$ such that $x(n) = y(nI)$, where I is an integer greater than one. The operation keeps a sample and adds $I - 1$ new samples to the original signal. The value of a newly generated sample is determined by an interpolation rule such as one based on a linear or spline combination of the value of previously existing neighboring samples.

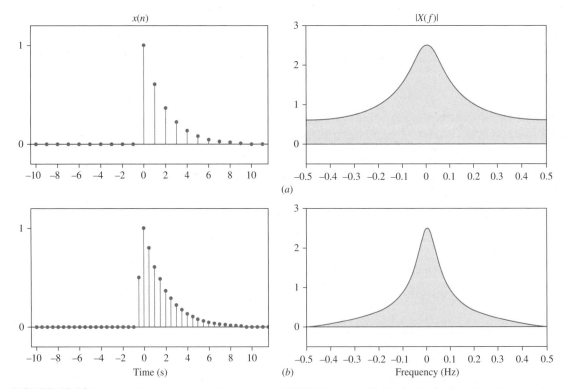

FIGURE 16.13 Interpolation by a factor of 2 narrows the DTFT of $x(n) = (0.6)^n u(n)$ by 2. Note the reduction of high frequencies in the DTFT. *(a)* shows the original signal, while *(b)* gives the interpolated version.

Interpolation increases the number of samples but is not exactly equivalent to having the sampling done at a higher rate. The new samples created by interpolation don't add any information to the signal because they are derived from the samples already present in the discrete-time signal, and not from the original signal in the continuous-time domain. In that sense distortions caused by an initially inadequate sampling rate can't be remedied by interpolation. Nonetheless, in addition to satisfying several other needs (e.g., matching a signal's rate to that of a system), increasing number of samples can simplify the structure of reconstruction filters as was shown in Example 10.5 of Chapter 10.

Examples of Interpolation by a Factor I

Visualize stretching the time axis of a discrete-time signal $x(n)$ in a way that $(I - 1)$, $I \geq 1$, new samples between successive samples are generated, producing a new signal $y(n)$. Unlike the case of zero-insertion, the value of a newly generated sample is determined by an interpolation rule, such as one based on a linear or spline combination of the values of previously existing neighboring samples. An example of interpolation of $x(n) = (0.6)^n u(n)$ and its effect on $X(\omega)$ for $I = 2$ is shown in Figure 16.13. Another example is shown in Figure 16.14 for the *sinc-type* signal $x(n) = 3 \sin(\pi n/3)/(\pi n)$ and $I = 5$. It resembles compressing the frequency axis by a factor 5, while keeping the

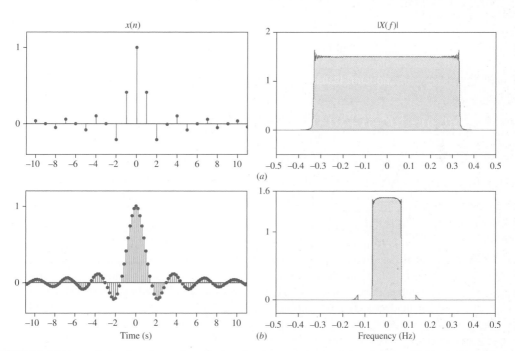

FIGURE 16.14 Interpolation by a factor 5 applied on the *sinc-type* signal $x(n) = 3 \sin(\pi n/3)/(\pi n)$. The effect on the DTFT is a compression of the frequency axis by a factor of 5 while the points at multiples of 0.5 Hz are kept fixed. *(a)* shows the original signal, while *(b)* gives the interpolated version.

points at multiples of 0.5 Hz fixed. Because of this, repetitions of $X(\omega)$ don't appear in the newly generated DTFT (unlike zero-insertion).

Several methods are used to generate the new samples from the neighborhood of the original samples. Examples are: *linear, spline,* or by increasing the sampling rate. In general, interpolation and decimation may be viewed as converse operations on discrete signals.

16.13 How Rate Conversion Reshapes DTFT

Decimation and interpolation of discrete-time signals (discussed in sections 16.11 and 16.12) are special cases of rate conversion. They become more meaningful when they are considered within the framework of sampling and reconstruction of the continuous-time signal from which the discrete-time signal originates. It is, therefore, constructive to refer to Figures 10.2 and 10.4 in Chapter 10 (modified here as Figure 16.15). In this section we represent the continuous-time frequency by F (in Hz).

Consider a continuous-time signal $x(t)$ with the highest frequency at F_0 [Figure 16.15(a)].[7] Let the Fourier transform of $x(t)$ be shown by $X(F)$, $X(F) = 0$ for $|F| > F_0$. Sample $x(t)$ at the rate of F_s samples per second by multiplying it with the train of unit impulses every $T = 1/F_s$ seconds. The time domain representation of the sampled signal is

$$\overline{x}(t) = x(t) \sum_{n=-\infty}^{\infty} \delta(t - nT) = \sum_{n=-\infty}^{\infty} x(nT)\delta(t - nT)$$

The Fourier transform of the sampled signal is

$$\overline{X}(F) = \frac{1}{T} \sum_{n=-\infty}^{\infty} X(F - nF_s)$$

Figure 16.15(b) shows $\overline{X}(F)$ for $F_s = 6F_0$ (i.e., oversampled by a factor of 3).

Decimating $x(n)$ by a factor D corresponds to reducing the sampling rate to F_s/D and may cause aliasing. This is illustrated in Figures 16.15(c) ($D = 2$, no aliasing) and 16.15(d) ($D = 4$, aliasing occurs).

Interpolation by a factor I corresponds to having increased the sampling rate to $F_s \times I$. If applied to the signal of Figure 16.15(d), it pulls the triangular lobes farther apart (see Figure 16.16). Also note that while interpolation doesn't produce aliasing, distortion, or an irreversible change in the signal, it doesn't add any new information to it either. It can't reverse an aliasing previously produced by a high decimation factor or an inadequately low sampling rate.

---

[7] Sampling the signal at the rate $2F_0$ eliminates frequencies higher than F_0 and causes aliasing if the continuous-time signal contains frequencies greater than F_0.

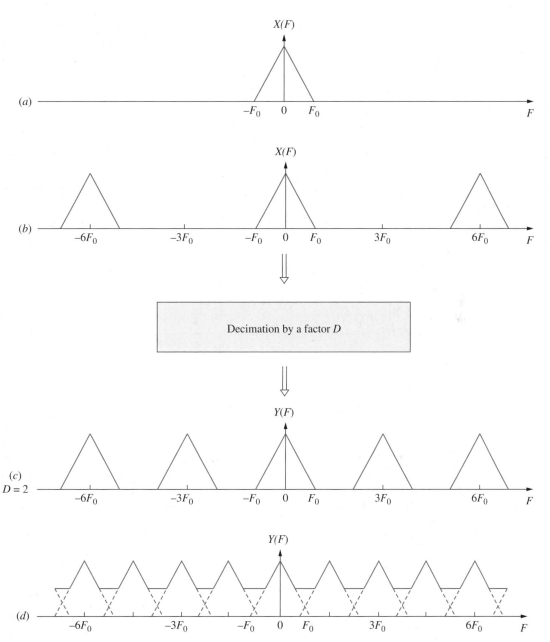

FIGURE 16.15 *(a)* Fourier transform of $x(t)$, a continuous-time low-pass signal with the highest frequency at F_0. The Nyquist rate for this signal is $2F_0$. *(b)* Sampling $x(t)$ uniformly by a train of unit impulses at the rate F_s (every $T = 1/F_s$ sec) produces $x(n)$. The sampled time function is $\bar{x}(t) = \sum x(nT)\delta(t - nT)$. Its Fourier transform is shown for $F_s = 6F_0$ (i.e., oversampled by a factor of 3). Because of the high sampling rate, no aliasing occurs. *(c)* Decimating $x(n)$ by the factor 2 corresponds to reducing the sampling rate to $F_s = 3F_0$, causing no aliasing. *(d)* Decimating $x(n)$ by the factor 4 corresponds to reducing the sampling rate to $F_s = 1.5F_0$, causing aliasing.

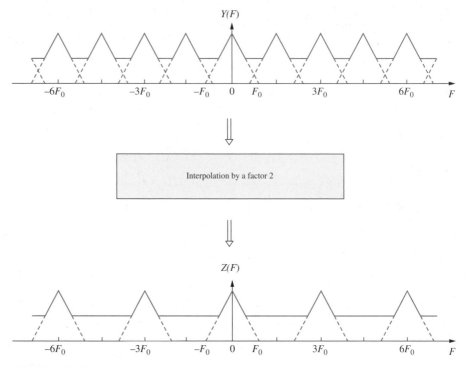

FIGURE 16.16 Interpolation appears to pull the triangular lobes farther apart but cannot remove aliasing. An interpolation by the factor 2 applied to the signal in *(a)* results in *(b)*.

Note that in the above discussion the frequency axis represents the actual frequency in the continuous-time domain. Decimation and interpolation don't change the frequency scale (assuming that the digital-to-analog conversion from the discrete to the continuous domains remains at a fixed rate). In Figures 16.15 and 16.16 it is F_s, which manipulates the frequency components of the sampled signal. Many signal processing computer packages contain decimation and interpolation tools, mostly labeling the frequency axis in the continuous-time domain.

Appendix 16A A Short Table of DTFT Pairs

DTFT Pairs	$-\pi \leq \omega \leq \pi$	$-0.5 \leq f \leq 0.5$
$x(n) = \dfrac{1}{2\pi}\displaystyle\int_{-\pi}^{\pi} X(\omega)e^{j\omega n}\,d\omega$ $\Longleftrightarrow$ $\;\;=\displaystyle\int_{-0.5}^{0.5} X(f)e^{j2\pi fn}\,df$	$X(\omega) = \displaystyle\sum_{-\infty}^{\infty} x(n)e^{-j\omega n}$	$X(f) = \displaystyle\sum_{-\infty}^{\infty} x(n)e^{-j2\pi fn}$
$1,\ \text{all } n$ $\Longleftrightarrow$	$2\pi\,\delta(\omega)$	$\delta(f)$
$d(n)$ $\Longleftrightarrow$	$1,\ \text{all } \omega$	$1,\ \text{all } f$
$d(n - n_0)$ $\Longleftrightarrow$	$e^{-jn_0\omega}$	$e^{-j2\pi n_0 f}$
$\begin{cases} 1, & 0 \leq n < N \\ 0, & \text{elsewhere} \end{cases}$ $\Longleftrightarrow$	$\dfrac{\sin\frac{N\omega}{2}}{\sin\frac{\omega}{2}}e^{-j(\frac{N-1}{2})\omega}$	$\dfrac{\sin(N\pi f)}{\sin(\pi f)}e^{-j(N-1)\pi f}$
$\begin{cases} 1, & -M \leq n \leq M \\ 0, & \text{elsewhere} \end{cases}$ $\Longleftrightarrow$	$\dfrac{\sin(M + \frac{1}{2})\omega}{\sin(\omega/2)}$	$\dfrac{\sin(2M + 1)\pi f}{\sin(\pi f)}$
$\dfrac{\sin(\omega_0 n)}{\pi n}$ $\Longleftrightarrow$	$\begin{cases} 1, & -\omega_0 \leq \omega \leq \omega_0 \\ 0, & \text{elsewhere} \end{cases}$	$\begin{cases} 1, & -f_0 \leq f \leq f_0,\ \ f_0 = \omega_0/(2\pi) \\ 0, & \text{elsewhere} \end{cases}$
$u(n)$ $\Longleftrightarrow$	$\dfrac{1}{1 - e^{-j\omega}} + \pi\delta(\omega)$	$\dfrac{1}{1 - e^{-j2\pi f}} + \dfrac{1}{2}\delta(f)$
$u(-n)$ $\Longleftrightarrow$	$\dfrac{1}{1 - e^{j\omega}} + \pi\delta(\omega)$	$\dfrac{1}{1 - e^{j2\pi f}} + \dfrac{1}{2}\delta(f)$
$u(n) - u(-n)$ $\Longleftrightarrow$	$-\dfrac{j\sin\omega}{1 - \cos\omega}$	$-\dfrac{j\sin(2\pi f)}{1 - \cos(2\pi f)}$
$u(n) - u(-n) + d(n)$ $\Longleftrightarrow$	$\dfrac{1 - \cos\omega - j\sin\omega}{1 - \cos\omega}$	$\dfrac{1 - \cos(2\pi f) - j\sin(2\pi f)}{1 - \cos(2\pi f)}$
$u(n) - u(-n) - d(n)$ $\Longleftrightarrow$	$\dfrac{\cos\omega - j\sin\omega - 1}{1 - \cos\omega}$	$\dfrac{\cos(2\pi f) - j\sin(2\pi f) - 1}{1 - \cos(2\pi f)}$
$\cos(\omega_0 n)$ $\Longleftrightarrow$	$\pi\,[\delta(\omega + \omega_0) + \delta(\omega - \omega_0)]$	$\dfrac{1}{2}[\delta(f + f_0) + \delta(f - f_0)],\ \ f_0 = \omega_0/(2\pi)$
$\sin(\omega_0 n)$ $\Longleftrightarrow$	$-j\pi\,[\delta(\omega + \omega_0) - \delta(\omega - \omega_0)]$	$\dfrac{1}{2j}[\delta(f + f_0) - \delta(f - f_0)],\ \ f_0 = \omega_0/(2\pi)$
$a^n u(n),\ \ |a| < 1$ $\Longleftrightarrow$	$\dfrac{1}{1 - ae^{-j\omega}}$	$\dfrac{1}{1 - ae^{-j2\pi f}}$
$a^{-n} u(-n),\ \ |a| < 1$ $\Longleftrightarrow$	$\dfrac{1}{1 - ae^{j\omega}}$	$\dfrac{1}{1 - ae^{j2\pi f}}$
$a^{|n|},\ \ |a| < 1$ $\Longleftrightarrow$	$\dfrac{1 - a^2}{1 + a^2 - 2a\cos\omega}$	$\dfrac{1 - a^2}{1 + a^2 - 2a\cos(2\pi f)}$
$na^n u(n),\ \ |a| < 1$ $\Longleftrightarrow$	$\dfrac{ae^{-j\omega}}{(1 - ae^{-j\omega})^2}$	$\dfrac{ae^{-j2\pi f}}{(1 - ae^{-j2\pi f})^2}$

Appendix 16B Symmetry Properties of the DTFT

Property	Time	$\Leftrightarrow$	Frequency	
$x(n)$ Is a Complex Function				
Basic pair	$x(n)$	$\Leftrightarrow$	$X(\omega)$	
Time reversal	$x(-n)$	$\Leftrightarrow$	$X(-\omega)$	
Conjugation	$x^*(n)$	$\Leftrightarrow$	$X^*(-\omega)$	
Real part of $x(n)$	$\mathcal{RE}\{x(n)\}$	$\Leftrightarrow$	$X_e(\omega)$	
Imaginary part of $x(n)$	$\mathcal{IM}\{x(n)\}$	$\Leftrightarrow$	$-jX_o(\omega)$	
Conjugate symm. part of $x(n)$	$x_e(n)$	$\Leftrightarrow$	$\mathcal{RE}\{X(\omega)\}$	
Conj. ant. symm. part of $x(n)$	$x_o(n)$	$\Leftrightarrow$	$j\mathcal{IM}\{X(\omega)\}$	
$x(n)$ is a Real Function				
Real function	$x(n) = x^*(n)$	$\Leftrightarrow$	$X(\omega) = X^*(-\omega)$	(Hermitian symmetry)
Even function	$x(n) = x(-n)$	$\Leftrightarrow$	$X(\omega)$ is real	
Odd function	$x(n) = -x(-n)$	$\Leftrightarrow$	$X(\omega)$ is imaginary	
Even part of $x(n)$	$x_e(n) = \dfrac{x(n) + x(-n)}{2}$	$\Leftrightarrow$	$\mathcal{RE}\{X(\omega)\}$	(real and even)
Odd part of $x(n)$	$x_o(n) = \dfrac{x(n) - x(-n)}{2}$	$\Leftrightarrow$	$j\mathcal{IM}\{X(\omega)\}$	(real and odd)

Appendix 16C Summary of DTFT Theorems

Theorem	Time	⇔ Frequency, ω	⇔ Frequency, f
Definition	$x(n) = \dfrac{1}{2\pi}\displaystyle\int_{-\pi}^{\pi} X(\omega)e^{j\omega n}\,d\omega$ ⇔ $X(\omega) = \displaystyle\sum_{-\infty}^{\infty} x(n)e^{-j\omega n}$		⇔ $X(f) = \displaystyle\sum_{-\infty}^{\infty} x(n)e^{-j2\pi fn}$
	$x(n) = \displaystyle\int_{-\pi}^{\pi} X(f)e^{j2\pi fn}\,df$		
Periodicity	$x(n)$	⇔ $X(\omega) = X(\omega + 2\pi)$	⇔ $X(f) = X(f + 1)$
Linearity	$ax(n) + by(n)$	⇔ $aX(\omega) + bY(\omega)$	⇔ $aX(f) + bY(f)$
Time shift	$x(n - n_0)$	⇔ $e^{-jn_0\omega}X(\omega)$	⇔ $e^{-j2\pi n_0 f}X(f)$
Frequency shift	$e^{j\omega_0 n}x(n)$	⇔ $X(\omega - \omega_0)$	⇔ $X(f - f_0), \quad f_0 = \omega_0/(2\pi)$
Modulation	$x(n)\cos\omega_0 n$	⇔ $\dfrac{X(\omega - \omega_0) + X(\omega + \omega_0)}{2}$	⇔ $\dfrac{X(f - f_0) + X(f + f_0)}{2}$
Time reversal	$x(-n)$	⇔ $X(-\omega)$	⇔ $X(-f)$
n-Multiplication	$nx(n)$	⇔ $j\dfrac{dX(\omega)}{d\omega}$	⇔ $\dfrac{j}{2\pi}\dfrac{dX(f)}{df}$
Decimation	$x(an)$	⇔ $\dfrac{1}{a}X(\omega/a)$	⇔ $\dfrac{1}{a}X(f/a)$
Conjugation	$x^*(n)$	⇔ $X^*(-\omega)$	⇔ $X^*(-f)$
Multiplication	$x(n)y(n)$	⇔ $\dfrac{1}{2\pi}X(\omega) * Y(\omega)$	⇔ $X(f) * Y(f)$
		$= \dfrac{1}{2\pi}\displaystyle\int_{-\pi}^{\pi} X(\phi)Y(\omega - \phi)\,d\phi$	$= \displaystyle\int_{-0.5}^{0.5} X(\phi)Y(f - \phi)\,d\phi$

Zero time
$$x(0) = \frac{1}{2\pi}\int_{-\pi/2}^{\pi/2} X(\omega)\,d\omega = \int_{-0.5}^{0.5} X(f)\,df$$

Zero frequency
$$X(0) = \sum_{-\infty}^{\infty} x(n)$$

Parseval's theorem
$$\sum_{-\infty}^{\infty} |x(n)|^2 = \frac{1}{2\pi}\int_{-\pi}^{\pi} |X(\omega)|^2\,d\omega = \int_{-0.5}^{0.5} |X(f)|^2\,df$$

$$\sum_{-\infty}^{\infty} x(n)y^*(n) = \frac{1}{2\pi}\int_{-\pi}^{\pi} X(\omega)Y^*(\omega)\,d\omega = \int_{-0.5}^{0.5} X(f)Y^*(f)\,df$$

16.14 Problems

Solved Problems

1. Find the DTFT of $x(n) = d(n+1) + d(n) + d(n-1)$ and plot its magnitude for $-\pi < \omega < \pi$.

$$x(n) = \begin{cases} 1, & n = -1, 0, 1 \\ 0, & \text{elsewhere} \end{cases}$$

Solution

$$X(\omega) = 1 + e^{j\omega} + e^{-j\omega} = 1 + 2\cos\omega$$

Within the period $-\pi < \omega < \pi$, the magnitude of $X(\omega)$ is given below. See Figure 16.17(a).

$$|X| = \begin{cases} 2\cos\omega + 1, & |\omega| < \frac{2\pi}{3} \\ 2|\cos\omega| - 1, & \text{elsewhere} \end{cases} \quad \text{and } \theta = \begin{cases} 0, & |\omega| < \frac{2\pi}{3} \\ \pi, & \text{elsewhere} \end{cases}$$

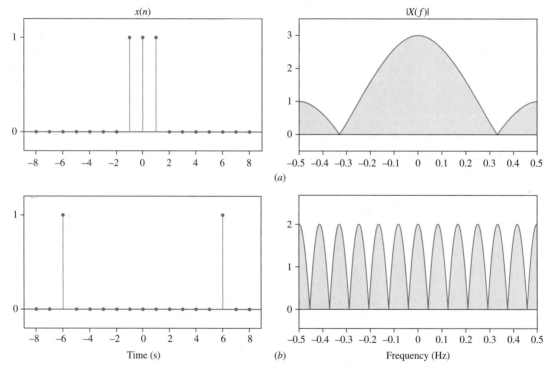

FIGURE 16.17 Two discrete-time signals (on the left) and their DTFT magnitudes (on the right). **(a)** $x(n) = \{1, 1, 1\}$. **(b)** $x(n) = d(n+6) + d(n-6)$.

2. Given $h(n) = d(n+k) + d(n-k)$, find $H(\omega)$ by applying the definition of the DTFT. Plot its magnitude for $k = 6$ and $-\pi < \omega < \pi$. Determine its period and zero-crossings.

Solution

$H(\omega) = e^{j\omega k} + e^{-j\omega k} = 2\cos(k\omega)$. One period of $H(\omega)$ is $2\pi/k$. Zero-crossings are at $\omega = \pm m\pi/(2k)$, where m is an odd integer. For $k = 6$, see Figure 16.17(b). The zero-crossings are at $\omega = \pm m\pi/12$, with m odd.

3. The unit-sample response of a discrete-time LTI system is $h(n) = d(n) + d(n - 8)$. A sinusoidal signal $x(n) = \cos(2\pi n/N)$ is applied to the input. Find the output

  a. by the frequency-domain method (DTFT)

  b. by the time-domain method (convolution)

For what values of N does the filter block the input (i.e., the output is zero)?

Solution

a. DTFT:

$$H(\omega) = 1 + e^{-j8\omega} = 2\cos(4\omega)e^{-j4\omega}$$

$$x(n) = \frac{1}{2}\left[e^{j\omega_0 n} + e^{-j\omega_0 n}\right], \quad \text{where } \omega_0 = \frac{2\pi}{N}$$

$$X(\omega) = \pi\left[\delta(\omega + \omega_0) + \delta(\omega - \omega_0)\right]$$

$$Y(\omega) = X(\omega)H(\omega) = 2\pi\cos(4\omega_0)\left[e^{j4\omega_0}\delta(\omega + \omega_0) + e^{-j4\omega_0}\delta(\omega - \omega_0)\right]$$

$$y(n) = 2\cos(4\omega_0)\cos\left[\omega_0(n - 4)\right]$$

b. Convolution: $y(n) = x(n) + x(n - 8) = \cos(\omega_0 n) + \cos\left[\omega_0(n - 8)\right]$

However, $\cos a + \cos b = 2\cos\left(\dfrac{a+b}{2}\right)\cos\left(\dfrac{a-b}{2}\right)$

Therefore, $y(n) = 2\cos(4\omega_0)\cos\left[\omega_0(n - 4)\right]$.

The system multiplies the signal by the gain factor $2\cos(4\omega_0)$ and introduces a phase (lag) of $-4\omega_0$. The gain factor is zero when $\omega_0 = \pm m\pi/8$, with m an odd integer. In the present case with $\omega_0 = 2\pi/N$, the blockage occurs when $N = 16$.

4. Single rectangular pulse. Apply the definition to derive a closed-form expression for the DTFT of a single rectangular pulse which is N unit samples long.

Solution

  a. **Causal pulse.** Assume the pulse starts at $n = 0$.

$$x(n) = \begin{cases} 1, & 0 \le n < N \\ 0, & \text{otherwise} \end{cases}$$

The DTFT of $x(n)$ is

$$X(\omega) = \sum_{n=0}^{N-1} e^{-j\omega n} = \frac{1 - e^{-jN\omega}}{1 - e^{-j\omega}} = \frac{\sin(\frac{N\omega}{2})}{\sin(\frac{\omega}{2})}e^{-j(\frac{N-1}{2})\omega}$$

Note that $X(\omega)$ may be written as

$$X(\omega) = X_r(\omega)e^{-j(\frac{N-1}{2})\omega}, \quad X_r(\omega) = \frac{\sin(\frac{N\omega}{2})}{\sin(\frac{\omega}{2})}$$

where $X_r(\omega)$ is a real function. The magnitude and phase of $X(\omega)$ are

$$|X| = |X_r| = \left|\frac{\sin(\frac{N\omega}{2})}{\sin(\frac{\omega}{2})}\right|, \quad \theta = -\left(\frac{N - 1}{2}\right)\omega \pm k\pi$$

The phase is a piecewise linear function of ω.

b. **Even pulse.** Shift the causal pulse of width $N = 2M + 1$ to the left by M units to obtain the even pulse. The shift adds $M\omega = (\frac{N-1}{2})\omega$ to the phase of its DTFT, resulting in

$$X(\omega) = \left[\frac{\sin(\frac{N\omega}{2})}{\sin(\frac{\omega}{2})} e^{-j(\frac{N-1}{2})\omega} \right] \times e^{j(\frac{N-1}{2})\omega} = \frac{\sin(\frac{N\omega}{2})}{\sin(\frac{\omega}{2})}$$

Figures 16.18(a) and (b) display $x(n)$ for $N = 5$ and 31, respectively, and normalized $|X(f)|$ for $-0.5 < f < 0.5$ Hz.

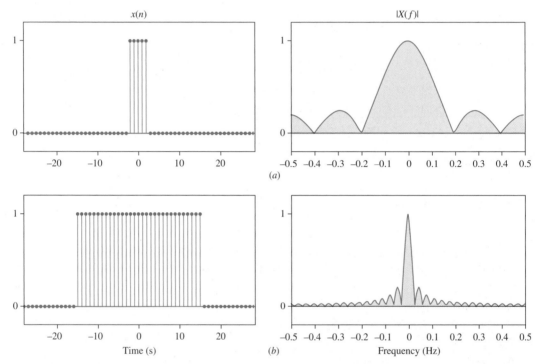

FIGURE 16.18 *(a)*, *(b)*. Two rectangular pulses of width N and the normalized magnitudes of their DTFTs, $|X(f)|$, for $-0.5 < f < 0.5$. *(a)* $N = 5$. *(b)* $N = 31$.

5. A triangular pulse. Find the DTFT of the even isosceles triangular pulse of height 1 and base $N = 2M + 1$ that contains $(N - 2) = (2M - 1)$ nonzero unit samples, the two extreme corners of the base having zero values.

Solution

The mathematical expression for the pulse is

$$x(n) = \begin{cases} 1 + \frac{n}{M}, & -M \le n \le 0 \\ 1 - \frac{n}{M}, & 0 \le n \le M \\ 0, & \text{elsewhere} \end{cases}$$

The DTFT of $x(n)$ is

$$X(\omega) = \sum_{n=-M}^{M} x(n)e^{-j\omega n} = 1 + 2 \sum_{n=1}^{M} (1 - \frac{n}{M}) \cos(\omega n)$$

Note that $X(\omega) = X_r(\omega) \ge 0$ and $\theta(\omega) = 0$. For $M = 2$, we get $X(\omega) = 1 + \cos \omega$, as was derived in Example 16.2 directly. Figure 16.19 displays $X_r(\omega)$ for $M = 15$.

Alternate Approach

$X(\omega)$ may also be found in closed form by noting that the triangular pulse may be generated by convolving an even rectangular pulse of width M with itself. The DTFT of the triangular pulse is, therefore, the square of the DTFT of the rectangular pulse. $X(\omega)$ is a real and even function of ω with period 2π. Within one period it contains $2N$ lobes, which are all nonnegative, but become tangent to the ω-axis at regular intervals $\omega = \pm 2\pi k/N$. See Figure 16.19

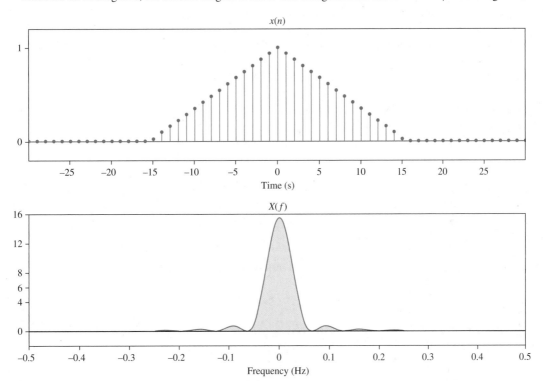

FIGURE 16.19 An even isosceles triangular pulse of height 1 and base $N = 2M + 1$ is expressed by

$$x(n) = \begin{cases} 1 + \frac{n}{M}, & -M \le n \le 0 \\ 1 - \frac{n}{M}, & 0 \le n \le M \\ 0, & \text{elsewhere} \end{cases}$$

The pulse contains $(N - 2) = (2M - 1)$ nonzero unit samples, the two extreme corners of the base having zero values. The DTFT of $x(n)$ is

$$X(\omega) = 1 + 2 \sum_{n=1}^{M} (1 - \frac{n}{M}) \cos(\omega n)$$

Note that $X(\omega) = X_r(\omega) \ge 0$ and $\theta(\omega) = 0$. The DTFT is nonnegative and its first zero occurs at $f = \pm 1/(2N)$ (compare with the DTFT of a rectangular pulse). The DTFT of the triangular pulse is also equal to the square of the DTFT of a rectangular pulse of length M. Expressed as a function of f, it can, therefore, be written as

$$X(f) = \frac{\sin^2(\pi M f)}{\sin^2(\pi f)}$$

This figure displays $X(f)$, $-0.5 < f < 0.5$, for $M = 15$.

Chapter Problems

6. Find the DTFT of the following single rectangular pulses.

a. $x(n) = d(n) + d(n-1) \equiv \{\cdots 0, 1, 1, 0, \cdots\}$

b. $x(n) = d(n) + d(n-1) + d(n-2) \equiv \{\cdots 0, 1, 1, 1, 0, \cdots\}$

c. $x(n) = d(n) + d(n-1) + d(n-2) + d(n-3) \equiv \{\cdots 0, 1, 1, 1, 1, 0, \cdots\}$

7. Find the DTFT of the following single rectangular pulses.

a. $x(n) = \{\cdots 0, 0.5, 1.5, 2, 1.5, 0.5, 0, \cdots\}$

b. $x(n) = \{\cdots 0, 0.5, 1, 1.5, 2, 1.5, 1, 0.5, 0, \cdots\}$

8. Find the DTFT of the following single rectangular pulses.

a. $x(n) = \{\cdots 0, 0.5, 1.5, 2, 2, 1.5, 0.5, 0, \cdots\}$

b. $x(n) = \{\cdots 0, 0.5, 1, 1.5, 2, 2, 1.5, 1, 0.5, 0, \cdots\}$

9. Find the DTFT of $x(n) = \begin{cases} 1, & 0 \le n < N \\ 0, & \text{elsewhere} \end{cases}$ and plot $|X(\omega)|$ for

a. $N = 16$

b. $N = 17$

c. $N = 30$

10. Find the DTFT of $x(n) = \begin{cases} a^{-n}, & 0 \le n < N \\ 0, & \text{elsewhere} \end{cases}$, $a = 1.396$, and plot $|X(\omega)|$ for

a. $N = 3$

b. $N = 4$

c. $N = 30$

11. Find the DTFT of $x(n) = \begin{cases} e^{-\frac{n}{3}}, & 0 \le n < N \\ 0, & \text{elsewhere} \end{cases}$ and plot $|X(\omega)|$ for

a. $N = 3$

b. $N = 4$

c. $N = 30$

12. Find the DTFT of $x(n) = \alpha^n u(n)$ and plot its magnitude and phase for

a. $\alpha = 0.95$

b. $\alpha = 0.8$

c. $\alpha = 0.5$

13. Find the DTFT of $x(n) = e^{-\frac{n}{3}} u(n)$ and plot its magnitude and phase.

14. Find the DTFT of $x(n) = (\alpha)^n \cos(\omega_0 n) u(n)$ and plot its magnitude and phase for the following parameters:

a. $\alpha = 0.5$ and $\omega_0 = \dfrac{\pi}{4}$ b. $\alpha = 0.5$ and $\omega_0 = \dfrac{\pi}{2}$

c. $\alpha = 0.5$ and $\omega_0 = \dfrac{3\pi}{4}$ d. $\alpha = 0.5$ and $\omega_0 = \pi$

e. $\alpha = 0.9$ and $\omega_0 = \dfrac{\pi}{4}$ f. $\alpha = 0.9$ and $\omega_0 = \dfrac{\pi}{2}$

g. $\alpha = 0.9$ and $\omega_0 = \dfrac{3\pi}{4}$ h. $\alpha = 0.9$ and $\omega_0 = \pi$

i. $\alpha = 0.95$ and $\omega_0 = \dfrac{\pi}{4}$ j. $\alpha = 0.95$ and $\omega_0 = \dfrac{\pi}{2}$

k. $\alpha = 0.95$ and $\omega_0 = \dfrac{3\pi}{4}$ l. $\alpha = 0.95$ and $\omega_0 = \pi$

m. $\alpha = 0.99$ and $\omega_0 = \dfrac{\pi}{4}$ n. $\alpha = 0.99$ and $\omega_0 = \dfrac{\pi}{2}$

o. $\alpha = 0.99$ and $\omega_0 = \dfrac{3\pi}{4}$ p. $\alpha = 0.99$ and $\omega_0 = \pi$

15. Discrete-time windows. A window is a finite-duration pulse. Some familiar windows of interest are the following: rectangular, triangular (or Bartlett), Blackman, Hamming, and Hanning. Their time functions are listed in the table below. Find their DTFT and plot their magnitude using dB scale.

Mathematical expressions for discrete-time windows. All windows are N samples wide and specified for $0 \le n < N$. Elsewhere, the windows have zero values.

Window Type	Mathematical Expression in the Time Domain, $0 \le n < N$
Rectangular	$w(n) = 1$
Triangular	$w(n) = \begin{cases} \frac{2n}{N-1}, & 0 \le n < \frac{N-1}{2} \\ 2 - \frac{2n}{N-1}, & \frac{N-1}{2} \le n < N \end{cases}$
Hanning	$w(n) = 0.5 - 0.5\cos\left(\dfrac{2\pi n}{N-1}\right)$
Hamming	$w(n) = 0.54 - 0.46\cos\left(\dfrac{2\pi n}{N-1}\right)$
Blackman	$w(n) = 0.42 - 0.5\cos\left(\dfrac{2\pi n}{N-1}\right) + 0.08\cos\left(\dfrac{4\pi n}{N-1}\right)$

16. Trapezoid window. In Figure 16.20 let $\tau_1 = 100$ and $\tau = 800$, both in msec. Sample the functions shown in (a) and (b) at the rate of 100 samples per second. Find their DTFT and plot their magnitude. Note that the window in Figure 16.20(b) is the integral of the function in (a). Can you relate the two DTFTs? What is their relationship?

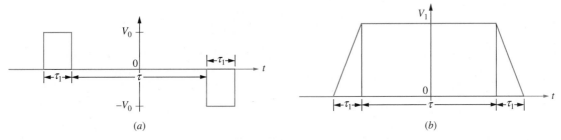

(a)   (b)

FIGURE 16.20

17. Sample $x(n)$ and $y(n)$ in Figure 16.21 at the rate of 20 samples per second, with one sample at $t = 0$. Obtain their DTFT and find a relationship between them.

(a)

$$x(t) = \begin{cases} \sin 2\pi t, & 0 \le t < 0.5 \\ 0, & 0.5 \le t < 1.5 \\ \sin 2\pi t, & 1.5 \le t < 2 \\ 0, & \text{elsewhere} \end{cases}$$

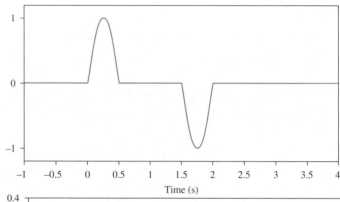

(b)

$$y(t) = \int_{-\infty}^{t} x(t)dt$$

$$= \begin{cases} \frac{1}{2\pi}(1 - \cos 2\pi t), & 0 \le t < 0.5 \\ \frac{1}{\pi}, & 0.5 \le t < 1.5 \\ \frac{1}{2\pi}(1 - \cos 2\pi t), & 1.5 \le t < 2 \\ 0, & \text{elsewhere} \end{cases}$$

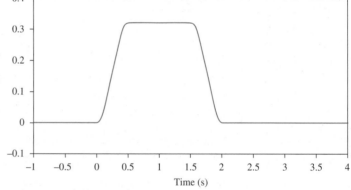

FIGURE 16.21

18. Construct the DTFT of $x(n) = 1 + 2\cos(\pi n)$ from its components to show

$$X(f) = \sum_{k=-1}^{1} \delta(f - k/2)$$

19. Find the DTFT of $\cos(2\pi n/N)$ for
   a. $N = 2$
   b. $N = 4$

20. Verify the DTFT pairs given below.

   a. $\sum_{k} d(n - k) \implies \sum_{k} \delta(f - k)$

b. $\displaystyle\sum_k d(n - kN) \implies \frac{1}{N} \sum_k \delta(f - k/N)$

c. $\cos(\pi n) \implies \displaystyle\frac{1}{2} \sum_{k \text{ odd}} \delta(f - k/2)$

21. Waveforms with even symmetry. Three periodic waveforms having even symmetry are shown in Figure 16.22. Sample them at the rate of 5 samples per second with a sample taken at $t = 0$ and find the DTFT of the discrete-time functions.

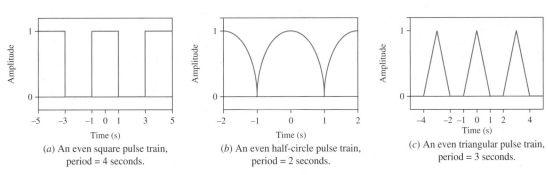

(a) An even square pulse train, period = 4 seconds.

(b) An even half-circle pulse train, period = 2 seconds.

(c) An even triangular pulse train, period = 3 seconds.

FIGURE 16.22 Three waveforms with even symmetry: $x(t) = x(-t)$.

22. Waveforms with odd symmetry. Three periodic waveforms having odd symmetry are shown in Figure 16.23. Sample them at the rate of 5 samples per second with a sample taken at $t = 0$ and find the DTFT of the discrete-time functions.

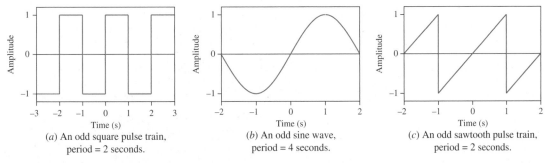

(a) An odd square pulse train, period = 2 seconds.

(b) An odd sine wave, period = 4 seconds.

(c) An odd sawtooth pulse train, period = 2 seconds.

FIGURE 16.23 Three waveforms with odd symmetry: $x(t) = -x(-t)$.

23. Half-wave symmetry. A function $x(t)$ is said to have half-wave symmetry if $x(t) = -x(t + T/2)$, where T is the period. The periodical waveforms shown in Figure 16.24 have half-wave symmetry. Sample them at the rate of 5 samples per second with a sample taken at $t = 0$ and find the DTFT of the discrete-time functions. Find if the DTFT $X_n = 0$ for n even, and the DTFT contains only odd harmonics.

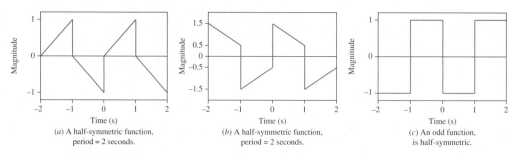

(a) A half-symmetric function,   (b) A half-symmetric function,   (c) An odd function,
period = 2 seconds.   period = 2 seconds.   is half-symmetric.

FIGURE 16.24 Three waveforms with half-wave symmetry: $x(t) = -x(t + T/2)$.

24. Five periodic functions $x_i(t), i = 1, \cdots, 5$ shown in Figure 16.25(a) to (e), respectively, are specified by $T = 10$ msec, $\tau = 1$ msec, and $V_0 = 5$ V.

a. Sample them at the rate of 5 kHz, with one sample at $t = 0$, and find their DTFTs.

b. Define $\zeta(t)$ (pronounced *zeta of t*) to be a single 1-msec rectangular pulse with a height of 1.

$$\zeta(t) = \begin{cases} 1, & 0 \le t < 1 \text{ msec} \\ 0, & \text{elsewhere} \end{cases}$$

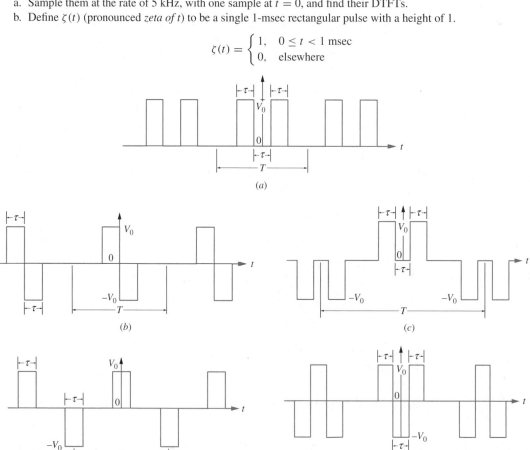

(a)

(b)

(c)

(d)

(e)

FIGURE 16.25

Similarly, sample $\zeta(t)$ with one sample at $t = 0$, call it $\zeta(n)$ and find its DTFT. Express each sampled function in Figure 16.25 in terms of $\zeta(n)$. Then investigate the relationships between their DTFT and DTFT of $\zeta(n)$.

25. In the periodic waveforms of Figure 16.25 let $V_0 = 1$ and $\tau = 100$ msec. Sample them at the rate of 1 kHz with a sample taken at $t = 0$ and find the DTFT of the discrete-time functions for

   a. $T = 10\tau$
   b. $T = 20\tau$

26. Three periodic functions $x_1(t)$, $x_2(t)$, and $x_3(t)$ shown in Figure 16.26(a), (b), and (c), respectively, are specified by $T = 10$ msec, $\tau = 1$ msec, and $V_0 = 5$ V.

   a. Sample the functions at the rate of 5 kHz with one sample at $t = 0$. Call the sampled functions $x_1(n)$, $x_2(n)$, and $x_3(n)$ and find their DTFTs.

   b. Define $\xi(t)$ (pronounced *cai of t*) to be a single right-angle triangular pulse with a height of 1 and a base of τ:

$$\xi(t) = \begin{cases} t/\tau, & 0 \leq t < \tau \\ 0, & \text{elsewhere} \end{cases}$$

Similarly, sample $\xi(t)$ with one sample at $t = 0$, call it $\xi(n)$ and find its DTFT. Express each sampled function in Figure 16.26 in terms of $\xi(n)$. Then investigate the relationships between the DTFTs of the sampled functions and DTFT of $\xi(n)$.

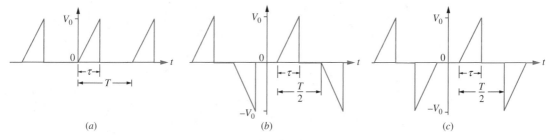

FIGURE 16.26

27. a. Sample the half-wave rectifed sinusoid with a base-to-peak voltage of 0 to V_o volts and frequency f (Figure 16.27a) at the rate of $20f$ Hz. Choose the time origin so that the function is even with one sample at $t = 0$. Find the DTFT of the sampled function.

   b. Repeat for the full-wave rectified waveform [Figure 16.27(b)].

   c. Repeat again for the waveform of Figure 16.27(c).

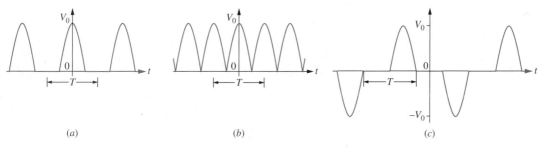

FIGURE 16.27

28. In the periodic waveforms of Figure 16.28 let $T = 1$ sec. Sample them at the rate of 25 samples per second with a sample taken at $t = 0$ and find the DTFT of the discrete-time functions for

    a. $\tau = 400$ msec
    b. $\tau = 500$ msec
    c. $\tau = 600$ msec

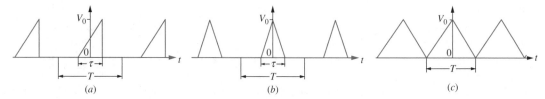

FIGURE 16.28

29. The periodical waveforms of Figure 16.29 have $T = 2$ and $\tau = 0.5$ seconds. Sample them at the rate of 10 samples per second with a sample taken at $t = 0$ and find the DTFT of the discrete-time functions.

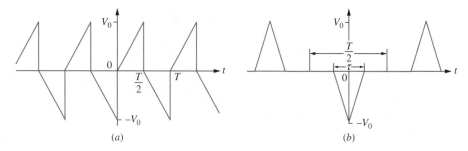

FIGURE 16.29

30. Find $h(n)$ of an FIR filter with $H(\omega) = 0$ at $\omega = \pm 105^0, \pm 135^0$, and $\pm 165^0$. The FIR systems were introduced in Chapter 11.

31. Find $h(n)$ of an FIR filter with $H(\omega) = 0$ at $\omega = \pm 30^0, \pm 60^0$, and $\pm 90^0$.

32. Find $H(\omega) = H_r(\omega)e^{j\Theta(\omega)}$ and plot $|H(\omega)|$ for the 21-term FIR defined by

$$h(n) = \{4, 0, -5, 0, 6, 0, -11, 0, 32, 0, -50, 0, 32, 0, -11, 0, 6, 0, -5, 0, 4\}$$
$$\uparrow$$

33. The system function of a discrete LTI filter $H(z) = \frac{Y(z)}{X(z)}$ has two poles at $p_{1,2} = \rho e^{\pm j\omega_0}$ and no zeros.

    a. Write the expression for $H(z)$ as a function of z^{-1}.
    b. Find $H(\omega) = H_r(\omega)e^{j\theta(\omega)}$ and plot $H_r(\omega)$ and $\theta(\omega)$ for $\omega_0 = \frac{\pi}{4}$ and $\rho = 0.6, 0.8, 0.9$, and 0.95. Specify magnitude and phase of $H(\omega)$ at $\omega = \frac{\pi}{4}$. For each ρ, determine the maximum of the magnitude.

34. The $H(z)$ of a discrete LTI filter has two poles at $p_{1,2} = .95e^{\pm j\frac{\pi}{4}}$ and two zeros at $z = 1, -1$.

    a. Write the expression for $H(z)$ as a function of z^{-1}.
    b. Find and plot $H(\omega) = H_r(\omega)e^{j\theta(\omega)}$.
    c. Discuss the differences between this filter and that of problem 33.

35. Let $h(n) = b^{-n}u(n)$, where $b = 1.369$.

    a. Find $H(\omega)$.

    b. Define

$$h_1(n) = \begin{cases} b^{-n}, & \text{for } 0 \le n < M \\ 0, & \text{elsewhere} \end{cases}$$

    Sketch $h_1(n)$ for $M = 2$ and $M = 30$.

    c. Find $H_1(\omega)$ as a function of M.

    d. Compare $|H_1(\omega)|$ and $|H(\omega)|$ for low M ($M = 2$) and high M ($M = 30$).

    e. Define $h_2(n) = b^{-n}$ for $0 \le n < M$ and $h_2(n + M + 1) = h_2(n)$ for all n. Plot $h_2(n)$ for $M = 2$ and $M = 30$.

    f. Compare $|H_2(\omega)|$ and $|H(\omega)|$ for low M ($M = 2$) and high M ($M = 30$).

36. Given the two filters:
$$\begin{cases} \text{Filter 1:} & h_1(n) = \{0, 1, 1, 1\} \\ & \quad\quad\quad\quad\quad\uparrow \\ \text{Filter 2:} & h_2(n) = \{0, 1, 0, 1, 0, 1\} \\ & \quad\quad\quad\quad\quad\uparrow \end{cases}$$

    a. Find $H_1(z^{-1})$ and $H_2(z^{-1})$.

    b. Find and plot their frequency responses $H_1(\omega)$ and $H_2(\omega)$ in the form of $H(\omega) = H_r e^{j\theta}$. In each case evaluate $H_r(\omega)$ and θ for $\omega = 0$, $\pi/4$, $\pi/2$, $3\pi/4$, and π.

    c. What is the relationship between $H_1(\omega)$ and $H_2(\omega)$?

    d. The above filters operate with A/D and D/A at a 48-kHz sampling rate. The analog input is $x(t) = \cos(2\pi f t)$. Find the analog outputs $y_1(t)$ and $y_2(t)$ for (i) $f = 6$ kHz, (ii) $f = 12$ kHz.

37. Four FIR filters are specified by their unit-sample responses given below.

$$\begin{cases} \text{Filter 1:} & h_1(n) = \{1, -1\} \\ & \quad\quad\quad\quad\uparrow \\ \text{Filter 2:} & h_2(n) = \{1, 1\} \\ & \quad\quad\quad\quad\uparrow \\ \text{Filter 3:} & h_3(n) = \{1, -1, 1\} \\ & \quad\quad\quad\quad\uparrow \\ \text{Filter 4:} & h_4(n) = \{1, 1, 1\} \\ & \quad\quad\quad\quad\uparrow \end{cases}$$

Find the DTFT of each filter and plot its magnitude and phase. Comment on the type of each filter.

38. The continuous-time signal $x(t) = 1 + \cos(10\pi t) + \cos(20\pi t)$ is sampled at the rate of 40 Hz and passed through a filter characterized by the system function $H(z)$. The filter's output, $y(n)$, is converted back to a continuous-time signal at the same sampling rate of 40 Hz, to become $y(t)$. Find $y(n)$ and $y(t)$ for the following systems:

    a. $H_1(z) = \dfrac{1 - z^{-1}}{1 - 0.5z^{-1}}$

    b. $H_2(z) = 1 + z^{-1} + z^{-2}$

39. The unit-sample response of a digital filter is $h(n) = \{1, 2, 1\}$.
                                                                     ↑

    a. Find $H(\omega) = H_r(\omega)e^{-j\theta(\omega)}$. Sketch $H_r(\omega)$ and $\theta(\omega)$, $-\pi < \omega < \pi$.

    b. A continuous-time rectangular pulse

$$x(t) = \begin{cases} 1 & \text{for } -5.1 \text{ msec} < t < 5.1 \text{ msec} \\ 0 & \text{elsewhere} \end{cases}$$

    is sampled uniformly such that one sample occurs at $t = 0$. The resulting discrete signal $x(n)$ is passed through the above filter. The output $y(n)$ is then converted to an analog signal. The same sampling rate R samples per second (Hz) is used in A/D and D/A conversion. Find and sketch $y(n)$ and $y(t)$ for (i) $R = 200$ Hz and (ii) $R = 2{,}000$ Hz. Observe and discuss their similarities to $x(n)$ and $x(t)$.

c. Repeat part b for a discrete-time sinusoidal input generated by sampling $\cos(1{,}000\pi t)$ at the rate of (i) $R = 200$ Hz and (ii) $R = 2{,}000$ Hz.

40. The unit-sample response of an FIR filter is $h(n) = \{1, 0, -2, 0, 3, 4, 3, 0, -2, 0, 1\}$.
   $\uparrow$

a. Find $H(\omega)$ and determine the filter's type (low-pass, bandpass, band-stop, or high-pass).
b. Is the filter linear phase? Why or why not?
c. Determine the time delay produced in a sinusoidal signal sampled at the rate of 8 kHz and passed through the filter.

41. Two FIR filters are specified by the unit-sample responses $h_1(n)$ and $h_2(n)$, where

$$h_1(n) = \begin{cases} (0.64)^n & 0 \le n < 3 \\ 0 & \text{otherwise} \end{cases} \quad \text{and} \quad h_2(n) = \begin{cases} (0.8)^n & 0 \le n < 5 \\ 0 & \text{otherwise} \end{cases}$$

Discuss similarities and differences between $H_1(\omega)$ and $H_2(\omega)$. What can you deduce from $H_1(\omega)$ about $H_2(\omega)$?

42. An FIR filter is specified by

$$h(n) = \begin{cases} (0.9)^n, & \text{for } 0 \le n < N; \\ 0, & \text{elsewhere}. \end{cases}$$

Find $H(\omega)$. Plot $|H(\omega)|^2$ for $N = 5$ and $N = 10$. Compare results. What happens to the frequency response when N goes to infinity?

43. In a perfect integrator $y(t) = \int x(t)dt$. Three digital integrators approximate the above by

a. $y(n) = x(n) + y(n-1)$
b. $y(n) = 0.5x(n) + 0.5x(n-1) + y(n-1)$
c. $y(n) = 0.333x(n) + 1.333x(n-1) + 0.333x(n-2) + y(n-2)$

In each case find $H(w)$. Plot its magnitude and phase. Compare with the frequency response of a perfect integrator.

44. In a perfect differentiator $y(t) = \frac{dx(t)}{dt}$. Two digital differentiators approximate the above by

a. $y(n) = x(n) - x(n-1)$
b. $y(n) = 0.5x(n) - 0.5x(n-2)$

In each case find $H(w)$. Plot its magnitude and phase. Compare with the frequency response of a perfect differentiator.

45. The impulse response of an analog filter is $h(t) = e^{-1{,}000t}u(t)$. Sample it at the rate of $F_s = 10$ kHz.

a. Find the resulting $h(n)$.
b. Find $H(z)$ and determine its poles and zeros.
c. Find $|H(\omega)|$ and the 3-dB attenuation digital frequency.
d. Let the above digital filter be connected to A/D and D/A operating at the same rate of $F_s = 10$ kHz to filter analog signals. Find the corresponding 3-dB attenuation analog frequency and compare it with the 3-dB attenuation frequency of the original analog filter [i.e., the filter with $h(t) = e^{-1{,}000t}u(t)$.]

46. A continuous-time analog signal $x(t) = \cos 500\pi t + \cos 200\pi t$ is sampled at the rate of 1,000 samples per second.

a. Find $x(n)$.
b. The resulting discrete signal $x(n)$ is passed through a filter with unit-sample response $h(n) = \{1, 2, 3, 2, 1\}$.
   $\uparrow$
   Find the output $y(n)$ and its analog reconstruction $y(t)$.

16.15 Project 1: Windows

Introduction and Summary

1. Objectives

In this project you will explore the basic features of some popular windows used in digital signal processing and filtering. You will then specify some quantitative characteristics of the windows and use them in a comparative analysis of the windows' effects in digital signal processing. To review the windows and their transforms you may refer to the project in Chapter 7 titled "Computer Explorations in Fourier Analysis" or the project in Chapter 8 called "Spectral Analysis Using a Digital Oscilloscope."

2. Simulation Tools

You may use Matlab, Scilab, or other packages for simulation purposes. The software packages contain commands that create familiar discrete-time windows of desired lengths. For a higher resolution in the frequency domain, you need to pad the windows with additional zeros, thus increasing data length T and reducing $\Delta f = 1/T$.

3. Reporting and Deliverables

During this project you will generate Figures A–E. Your report should contain the results of items 6-14.

Theory

4. What Is a Window?

A window is a finite-duration pulse through which we look at a time function or its Fourier transform. The role of a window is to create and shape a finite segment of the sequence (be it data, a filter's impulse response, or its frequency response) and do so in a way that certain characteristics are met. By definition, time windows have finite duration and, therefore, infinite bandwidth. Given the duration of the window, its shape has an important role in its function.

From the theory of the Fourier integral it is known that the transform of a time-limited signal cannot be zero over a nonzero frequency interval. Theoretically, a time-limited signal contains all frequency components and occupies an infinite bandwidth. The narrower the pulse is in the time domain, the wider its transform is in the frequency domain and viceversa. Therefore, a signal may only approximately be both time-limited and bandlimited.

5. Time-Frequency Relationship

Consider a time function $x(t)$ and a time window $w(t)$ [with Fourier transforms $X(f)$ and $W(f)$, respectively]. Windowing $x(t)$ by $w(t)$ produces a finite duration time function $x(t) \times w(t)$. The Fourier transform of the windowed function is $X(f) \star W(f)$, where $\star$ represents the convolution operation.

$$x(t) \times w(t) \Longleftrightarrow X(f) \star W(f)$$

A similar result is obtained when the Fourier transform of a time function is windowed by a $W(f)$ with finite bandwidth.

$$X(f) \times W(f) \Longleftrightarrow x(t) \star w(t)$$

6. Continuous-Time Windows

The time functions for some familiar continuous-time windows and their Fourier transforms are listed in Table 16.1. For simplicity, the time origin is centered so that the windows are even functions. All windows are τ msec wide. The mathematical expressions are given for $-\tau/2 < t < \tau/2$. Elsewhere, windows have zero values. Plot five time windows and their transforms and label them Figure A.

TABLE 16.1 Continuous-Time Windows and Their Fourier Transforms

Window	Time Function, $-\tau/2 < t < \tau/2$	$\Longleftrightarrow$	Fourier Transform
Rectangular	1	$\Longleftrightarrow$	$\dfrac{\sin(\pi f \tau)}{\pi f}$
Bartlett	$1 - 2\dfrac{\lvert t \rvert}{\tau}$	$\Longleftrightarrow$	$\dfrac{2}{\tau}\left[\dfrac{\sin(\pi f \tau/2)}{\pi f \tau/2}\right]^2$
Hanning	$0.5 + 0.5\cos\left(\dfrac{2\pi t}{\tau}\right)$	$\Longleftrightarrow$	$0.5\dfrac{\sin(\pi f \tau)}{\pi f} + 0.25\dfrac{\sin \pi \tau (f - \frac{1}{\tau})}{\pi(f - \frac{1}{\tau})}$ $+0.25\dfrac{\sin \pi \tau (f + \frac{1}{\tau})}{\pi(f + \frac{1}{\tau})}$
Hamming	$0.54 + 0.46\cos\left(\dfrac{2\pi t}{\tau}\right)$	$\Longleftrightarrow$	$0.54\dfrac{\sin(\pi f \tau)}{\pi f} + 0.23\dfrac{\sin \pi \tau (f - \frac{1}{\tau})}{\pi(f - \frac{1}{\tau})}$ $+0.23\dfrac{\sin \pi \tau (f + \frac{1}{\tau})}{\pi(f + \frac{1}{\tau})}$
Blackman	$0.42 + 0.5\cos\left(\dfrac{2\pi t}{\tau}\right) + 0.08\cos\left(\dfrac{4\pi t}{\tau}\right)$	$\Longleftrightarrow$	$0.42\dfrac{\sin(\pi f \tau)}{\pi f} + 0.25\dfrac{\sin \pi \tau (f - \frac{1}{\tau})}{\pi(f - \frac{1}{\tau})}$ $+0.25\dfrac{\sin \pi \tau (f + \frac{1}{\tau})}{\pi(f + \frac{1}{\tau})} + 0.04\dfrac{\sin \pi \tau (f - \frac{2}{\tau})}{\pi(f - \frac{2}{\tau})}$ $+0.04\dfrac{\sin \pi \tau (f + \frac{2}{\tau})}{\pi(f + \frac{2}{\tau})}$

7. Frequency Response of the Hanning Window

In this section we derive the Fourier transform of a continuous-time Hanning window of duration τ. The window is defined by the raised cosine pulse given by

$$x(t) = \begin{cases} 1 + \cos(\frac{2\pi t}{\tau}), & -\frac{\tau}{2} < t < \frac{\tau}{2} \\ 0, & \text{elsewhere} \end{cases}$$

The pulse is the product of a uniform rectangular window of duration τ and a cosine function with period τ and unity DC value. At the edges, the window function and its derivatives are continuous and zero. The transform is

$$X(f) = \int_{-\tau/2}^{\tau/2}\left[1 + \cos\left(\frac{2\pi t}{\tau}\right)\right]e^{-j2\pi f t}dt = \frac{\sin(\pi f \tau)}{\pi f} + 0.5\frac{\sin \pi \tau (f - f_0)}{\pi(f - f_0)} + 0.5\frac{\sin \pi \tau (f + f_0)}{\pi(f + f_0)}$$

where $f_0 = 1/\tau$ is the first zero-crossing of the transform of the rectangular window. Note that the Fourier transform of the above window is not band limited. Plot three Hanning windows (5, 2, and 1 msec wide) and their Fourier transforms and label them Figure B. Observe that, as expected, narrower windows have wider bandwidths.

8. Discrete-Time Windows

The domain of discrete-time signals is $-\infty < n < \infty$, where n is an integer. In practice, however, we process finite-length segments of a signal at any given time. Mathematically, this corresponds to multiplying the original signal by a discrete-time window of finite length. Five examples of discrete-time windows are listed in Table 16.2 (left column). In practice, windows are causal functions as specified in the right-hand column. The central column specifies the noncausal symmetric version of the windows with $2M + 1$ samples (the time origin is chosen at the center to make the windows even functions of n). Plot three discrete-time windows (rectangular, triangular, and Hanning) nine samples wide, along with their transforms.

TABLE 16.2 Discrete-Time Windows

Window Name	Even Function, $-M \le n \le M$	Causal Function, $0 \le n \le N-1$
Rectangular	1	1
Bartlett	$1 - \dfrac{\|n\|}{M}$	$1 - \dfrac{2\left\|n - \dfrac{N-1}{2}\right\|}{N-1}$
Hanning	$0.5 + 0.5\cos\left(\dfrac{\pi n}{M}\right)$	$0.5 - 0.5\cos\left(\dfrac{2\pi n}{N-1}\right)$
Hamming	$0.54 + 0.46\cos\left(\dfrac{\pi n}{M}\right)$	$0.54 - 0.46\cos\left(\dfrac{2\pi n}{N-1}\right)$
Blackman	$0.42 + 0.5\cos\left(\dfrac{\pi n}{M}\right) + 0.08\cos\left(\dfrac{2\pi n}{M}\right)$	$0.42 - 0.5\cos\left(\dfrac{2\pi n}{N-1}\right) + 0.08\cos\left(\dfrac{4\pi n}{N-1}\right)$

Measurements and Analysis

9. Operation of Time-Domain Windows

    a. Produce a worksheet to show a rectangular window nine samples wide (in the left-hand column) along with the magnitude of its transform (in the right-hand column). Label it Figure C, part a.

    b. Create a triangular window of the same size as in a and repeat part a. Label it Figure C, part b.

    c. Create a Hanning window of the same width and repeat part a. Label it Figure C, part c.

    d. Qualitatively compare the side-lobes in the above three windows.

    e. Develop a quantitative index (or indices) for measuring the effect of a window in the frequency domain, to be used throughout this project.

Suggestion. One such measure may be developed by defining passband and stop-band attenuations, then finding the passband, stop-band, and transition band frequencies for each window. Another measure could be the relative amount of power (or energy) in each band.

10. Gibbs' Phenomenon.

    a. Use the Fourier series coefficients of a periodic rectangular pulse to obtain the sum of its first N nonzero harmonics. This corresponds to a windowing operation in the frequency domain. Note that as the size of the window increases, the sum of harmonics approach the rectangular form, except at the transition instances, where it exhibits horns. This is called Gibbs' phenomenon. Can you get rid of the horns by adding more harmonics? By changing the size of the window, produce several examples (label them Figure D) and observe the persistence of the Gibbs' phenomenon.

b. Investigate the effect of applying the Hanning window in reducing Gibbs' phenomenon (label them Figure E). Determine the cost of reducing the horns by increasing the transition band defined in item 9 "Operation of Time Domain Windows" of this project.

c. Repeat part b by applying a Hamming window.

11. Mathematical Interpretation of Gibbs' Phenomenon

Consider the following three time-functions: (1) an infinite-duration sinc function $x(t)$, (2) a finite-width rectangular window $w(t)$ which is 2τ seconds wide, and (3) their convolution $y(t) = x(t) \star w(t)$.

$$x(t) = \frac{\sin(2\pi f_0 t)}{\pi t}, \quad -\infty < t < \infty$$

$$w(t) = u(t + \tau) - u(t - \tau) = \begin{cases} 1, & -\tau < t < \tau \\ 0, & \text{elsewhere} \end{cases}$$

$$y(t) = x(t) \star w(t)$$

$$= x(t) \star \left[u(t + \tau) - u(t - \tau) \right]$$

$$= x(t) \star u(t + \tau) - x(t) \star u(t - \tau)$$

$$= \int_{-\infty}^{t} x(\theta + \tau) d\theta - \int_{-\infty}^{t} x(\theta - \tau) d\theta$$

The above derivation is illustrated in Figure 16.30. The horns seen in $y(t)$ are interpreted as manifestations of Gibbs' phenomenon (due to multiplication of the low-pass $X(f)$ by $W(f)$ of the rectangular window). In this section we look into the root cause of the horns that appear in $y(t)$.

a. Examine the above equations, then note that the convolution of $x(t)$ with a unit-step function located at $\pm\tau$ amounts to its integral shifted by $\pm\tau$. This is because (1) a unit-step is the integral of a unit-impulse and (2) convolution of $x(t)$ with a unit-impulse results in shifting $x(t)$ to the location of the unit-impulse.

b. Note that integration of the sinc function unavoidably produces ripples that are more pronounced around the transition instances (in this discussion, at $\pm\tau$). The ripples become narrower for a narrower sinc function $x(t)$ but never disappear, especially in the neighborhood where they remain pronounced. Therefore, increasing the width of $w(t)$ smoothens the convolution pulse except at the transition times $\pm\tau$ where horns persist.

c. Now look at the situation in the frequency domain. Examine the Fourier transforms $X(f)$, $W(f)$, $Y(f)$, and their relationship.

$$X(f) = u(t + f_0) - u(f - f_0) = \begin{cases} 1, & -f_0 < f < f_0 \\ 0, & \text{elsewhere} \end{cases}$$

$$W(f) = \frac{\sin(2\pi f \tau)}{\pi f}, \quad -\infty < f < \infty$$

$$Y(f) = X(f) \times W(f) = \begin{cases} \frac{\sin(2\pi f \tau)}{\pi f}, & -f_0 < f < f_0 \\ 0, & \text{elsewhere} \end{cases}$$

$Y(f)$ is produced through windowing $W(f)$ by the rectangular window $X(f)$. Because of Gibbs' phenomenon, its inverse $y(t)$ is expected to contain horns, as in fact it does.

d. In summary, the horns are developed by the integration of $x(t)$. They are interpreted as consequences of windowing by the rectangular window $X(f)$ in the frequency domain. The result is Gibbs' phenomenon in the time domain.

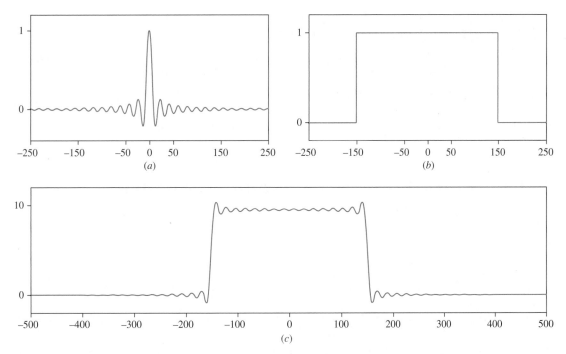

FIGURE 16.30 Manifestation of Gibbs' phenomenon due to convolution of a sinc function with a rectangular window. **(a)** A discrete-time sinc function $x(n) = \frac{sin(2\pi n/19)}{2\pi n/19}$. **(b)** An even rectangular window $w(n) = u(n + 151) - u(n - 151)$ that is 301 samples long. **(c)** $y(n) = x(n) \star w(n)$ corresponding to multiplication $X(f) \times W(f)$ creates horns. For interpretation see the text.

12. Operation of Frequency-Domain Windows

a. Create $h_d(n) = sin(\pi n/4)/(\pi n)$, $-\infty < n < \infty$. This is the unit-sample response of an ideal (and desired) low-pass filter with the DTFT specified by

$$H_d(\omega) = \begin{cases} 1, & -\frac{\pi}{4} < \omega < \frac{\pi}{4} \\ 0, & \text{elsewhere} \end{cases}$$

b. Take a finite segment of $h_d(n)$ of length $2M + 1$

$$h(n) = \begin{cases} \frac{sin(\pi n/4)}{\pi n}, & -M \le n \le M \\ 0, & \text{elsewhere} \end{cases}$$

and obtain its DTFT. Do you see Gibbs' phenomenon in the frequency domain due to the finite length of $h(n)$? Report your observations.

c. Construct $h(n)$ for $M = 15$ and convolve it with the speech file $cht5.wav$. What does it do to the speech file? What is the cutoff frequency?

Discussion and Conclusions

13. Summary of Windows' Effects

Summarize the pay-off obtained by applying a window and the cost associated with it. Use the indices you defined in item 9 "Operation of Time-Domain Windows" (e.g., based on the shape of the frequency response of a window and its effect on increasing the transition band).

14. Conclusions

In 50 words or less summarize the need, cost, and criteria for the choice of window type.

16.16   Project 2: Decimation and Frequency Downshifting

1. Introduction and Summary

This project illustrates the decimation effect in the frequency domain. It can be reproduced using a computer program following the steps described below. Reconstruction of the signal from the samples is assumed to be done through an ideal low-pass filter with a cutoff frequency equal to half the sampling rate. The project contains three steps. The three figures to be produced at the completion of the project illustrate how decimation of a sampled signal may cause frequency downshifting. You will sample sinusoidal signals and tones, compute their Fourier transforms, and display their magnitudes. You will then decimate the signals and observe the effect on their transforms.

2. Step a. Generating Tones

Start with four continuous-time sinusoids $x(t) = \cos(1{,}000\pi k t)$, $k = 1, 2, 3, 4$. Sample them at the sampling rate of $F_s = 10{,}000$ samples per second, or 10 kHz (i.e., set $t = n/10{,}000$), to generate four sequences $x_k(n)$. Then compute their Fourier transforms.

$x(t)$	$x(n)$	$\Longrightarrow$	$X(f)$
$x_1(t) = \cos(1{,}000\pi t)$,	$\cos(\pi n/10)$	$\Longrightarrow$	$\frac{1}{2}[\delta(f - 500) + \delta(f + 500)]$
$x_2(t) = \cos(2{,}000\pi t)$,	$\cos(2\pi n/10)$	$\Longrightarrow$	$\frac{1}{2}[\delta(f - 1{,}000) + \delta(f + 1{,}000)]$
$x_3(t) = \cos(3{,}000\pi t)$,	$\cos(3\pi n/10)$	$\Longrightarrow$	$\frac{1}{2}[\delta(f - 1{,}500) + \delta(f + 1{,}500)]$
$x_4(t) = \cos(4{,}000\pi t)$,	$\cos(4\pi n/10)$	$\Longrightarrow$	$\frac{1}{2}[\delta(f - 2{,}000) + \delta(f + 2{,}000)]$

When a computer is used, computation is done on finite sequences of length N and the Fourier transforms are approximated by discrete Fourier transforms. As expected, the spectral lines should occur at 500, 1,000, 1,500, and 2,000 Hz. Produce Figure A with two columns and four rows showing the sampled signals (on the left-hand column) and their Fourier transforms (on the right-hand side). The Fourier transforms of the sampled signals are periodic functions of frequency with a period F_s. Show one cycle of each on the right-hand side of Figure A.

This figure has been masked so as to be recreated by the student.

FIGURE A Sinusoids sampled at a high rate. In Figure A four continuous-time sinusoids $x(t) = \cos(1,000\pi kt)$, $k = 1, 2, 3, 4$, corresponding to 500, 1,000, 1,500, and 2,000 Hz, are sampled at the rate of $F_s = 10$ kHz, generating four sequences $x_k(n)$, shown in the left column. The transforms of the sampled signals are shown in the right column. The time scale is in seconds and the frequency scale is in Hz. Transforms are computed from finite sequences (10,000 samples) and show spectral lines at 500, 1,000, 1,500, and 2,000 Hz, as expected. Also observe that the spectral lines are singular with no sidebands because in all four cases the data sequence is windowed by a rectangular window that contains an integer number of full cycles of the sinusoids.

3. Step b. Concatenation of Tones and Decimation by a Factor of 3

Construct a new time series $x_5(n)$ by concatenating $x_1(n)$, $x_2(n)$, $x_3(n)$, and $x_4(n)$, and then take its Fourier transform to obtain $X_5(f)$. Note that $x_5(n)$ contains $4N$ samples but the sampling rate is still 10 kHz. Its Fourier transform contains spectral lines at 500, 1000, 1500, and 2000 Hz with the same magnitudes as found in step a. Produce Figure B with two columns and two rows, showing $x_5(n)$ and $X_5(f)$ in the upper row. (The lower row will show the decimated signal and its transform, as described below.)

Starting from its beginning, decimate $x_5(n)$ by a factor of 3 to obtain a new sequence $x_6(n)$. This effectively reduces the sampling rate to 10 kHz/3 = 3,333 Hz. Compute the Fourier transform of $x_6(n)$ and identify its frequency components (locations and amplitudes). Pay attention to frequency components which don't exist in $x_5(n)$. Note that spectral lines at 500, 1,000, 1,500 Hz remain where they were, but the fourth line (originally at 2,000 Hz) is now shifted down to $3,333 - 2,000 = 1,333$ Hz.

You may also listen to the effect of decimation by saving $x_5(n)$ and $x_6(n)$ and playing them back through the sound card. The files may be saved as sound files in the *.wav* format. The audio effect shows shifting of the 2-kHz tone to 1,333 Hz.

This figure has been masked so as to be recreated by the student.

FIGURE B Decimation by a factor of 3. Construct Figure B in the following form:

1. The upper row shows the time series $x_5(n)$ [on the left-hand side, generated by concatenating $x_1(n)$, $x_2(n)$, $x_3(n)$, and $x_4(n)$] and its transform (on the right-hand side). The transform shows spectral lines at 500, 1,000, 1,500, and 2,000 Hz, with the same magnitudes as found in Figure A. Note that the spectral lines are not singular anymore, but have developed sidebands due to a windowing effect.

2. The lower row shows a new sequence $x_6(n)$ and its transform, obtained from decimating $x_5(n)$ by a factor of 3. This decimation effectively reduces the sampling rate to 10 kHz/3 = 3,333 Hz. The new sampling rate still remains greater than twice the frequencies 500, 1,000, 1,500 Hz, leaving the location of their spectral lines unchanged. However, the fourth line (originally at 2,000 Hz) is now shifted down to $3,333 - 2,000 = 1,333$ Hz.

4. Step c. Summing the Tones and Decimation by a Factor of 7

Construct a new time series $x_7(n) = x_1(n) + x_2(n) + x_3(n) + x_4(n)$. Note that $x_7(n)$ contains N samples at the 10-kHz rate. Take its Fourier transform to obtain $X_7(f)$ and relate it (magnitude and locations of its peaks) to $X_k(f)$, k = 1, 2, 3, 4.

Decimate $x_7(n)$ by a factor of 7 to obtain a new sequence $x_8(n)$. This effectively reduces the sampling rate to $\frac{10}{7}$ kHz = 1,428.5 Hz. Compute the Fourier transform of $x_8(n)$ and identify its frequency components (magnitude and locations of its peaks). Note that the spectral lines are now at 71.5, 428.5, 500, and 571.5 Hz. Again, for audio examination, you may save $x_7(n)$ and $x_8(n)$ in the form of a *.wav* file and play them back through the sound card.

> This figure has been masked so as to be recreated by the student.

FIGURE C Decimation by a factor of 7. Construct Figure C in the following form:

1. The upper row shows a new time series $x_7(n) = x_1(n) + x_2(n) + x_3(n) + x_4(n)$ (left-hand side), and its transform (right-hand side). Again, the transform shows spectral lines at 500, 1,000, 1,500, and 2,000 Hz.
2. The lower row shows a new sequence $x_8(n)$ (in the left-hand side) obtained from decimating $x_7(n)$ by the factor 7. This decimation effectively reduces the sampling rate to 10 kHz/7 = 1,428.5 Hz. The transform of $x_8(n)$ is shown in the right-hand side. Note that the spectral lines are now centered at 71.5, 428.5, 500, and 571.5 Hz, as summarized in the table below.

Signal	Original Frequency f_0	Effective Sampling Rate	Reconstructed Frequency
x_1	500Hz	1,428.5Hz	500Hz
x_2	1,000Hz	1,428.5Hz	$\|1,428.5 - 1,000\| = 428.5$Hz
x_3	1,500Hz	1,428.5Hz	$\|1,428.5 - 1,500\| = 71.5$Hz
x_4	2,000Hz	1,428.5Hz	$\|1,428.5 - 2,000\| = 571.5$Hz

Note that when $2f_0 < 1428.5$, the frequency of the reconstructed signal is the same as that of the original continuous-time signal (f_0). But when $2f_0 > 1,428.5$, aliasing occurs and spectral lines appear at $\|1,428.5 - f_0\|$.

5. Conclusions

For each of the above decimations examine the magnitudes and locations of the spectral lines. Observe and measure aliasing. Calculate expected aliasing in terms of the decimation factor and convince yourself that they meet theoretical expectations. Note that reconstruction filters eliminate frequency components which are above half of the sampling rate.

Chapter 17

Discrete Fourier Transform

Contents

Introduction and Summary

The discrete Fourier transform (DFT) operates on a finite set of numbers or a finite segment of a discrete signal. Physically, the DFT of a discrete-time function $x(n)$ may be viewed as its frequency-domain representation and used as a tool to approximate the DTFT, FT, or the spectrum of the continuous-time function $x(t)$ [from which $x(n)$ is produced]. Mathematically, the DFT may be formulated by extending the classical Fourier transform of continuous-time signals to a finite-length discrete-time sequence through sampling and windowing. Such an approach may appear logical and attractive, but it draws in nonessential discussions and details that are marginal to the subject.

Alternatively, the DFT can be introduced as an operation specified by a transformation matrix that converts a sequence $x(n)$ to another sequence $X(k)$ of the same length. Parallels with the classical Fourier transform and physical interpretations can be introduced after sufficient familiarity with the subject has been acquired. The present chapter approaches the DFT through the latter avenue. It starts with the definitions of the DFT and its inverse, followed by examples. One goal is to bring to the student's attention the importance and advantages of the DFT (and its implementation by the fast Fourier transform [FFT]) in practical signal analysis, specifically in speeding up the processing. The DFT's important application is in performing circular convolution by FFT, from which linear convolution is obtained. To that effect, the chapter discusses the circular convolution of two time-domain signals and its implementation through the DFT. Mathematical properties of the DFT and its relation with the DTFT are then described. The principles behind the FFT and its efficient method of performing DFT are briefly discussed. Finally, the four methods of Fourier analysis—Fourier series (FS), Fourier transform (FT), discrete-time Fourier transform (DTFT), and discrete Fourier transform (DFT)—are brought together at the end of this chapter.

17.1 Definitions

The DFT transforms a finite-length sequence $x(n)$, $0 \le n \le (N-1)$, into another sequence $X(k)$, $0 \le k \le (N-1)$, of the same length, where

$$X(k) = \sum_{n=0}^{N-1} x(n)e^{-j2\pi nk/N}, \quad k = 0, 1, 2, \ldots N - 1 \tag{17.1}$$

Both sequences are N points long. $X(k)$ is called the N-point DFT of $x(n)$ and $x(n)$ is called the inverse discrete Fourier transform (IDFT) of $X(k)$. The latter may be recovered from $X(k)$ by

$$x(n) = \frac{1}{N} \sum_{k=0}^{N-1} X(k)e^{j2\pi kn/N}, \quad n = 0, 1, 2, \ldots N - 1 \tag{17.2}$$

Customarily, $x(n)$ represents a function in the time domain and $X(k)$ is referred to as its representation in the frequency domain. The sequences $x(n)$ and $X(k)$ always begin with the zeroth term. They may be real-valued or complex numbers. In this chapter we will focus mostly on a real-valued $x(n)$.

17.2 Examples of the DFT

The examples in this section demonstrate the summation method of finding the DFT. They also demonstrate some properties that are summarized at the end of the section. The reader may want to consult these properties while doing the examples.

E*xample*

17.1

Find the DFT of the following sequences:

a. $x(n) = \{\underset{\uparrow}{1}, 1, 0\}$

b. $x(n) = \{\underset{\uparrow}{1}, 1, 0, 0\}$

These are called the 3-point and 4-point DFT, respectively, of the sequence $\{\underset{\uparrow}{1}, 1\}$.

Solution

a. $x(n) = \{\underset{\uparrow}{1}, 1, 0\}$

$$X(k) = \sum_{n=0}^{1} e^{-j2\pi nk/3} = 1 + e^{-j2\pi k/3} = e^{-j\pi k/3}\left(e^{j\pi k/3} + e^{-j\pi k/3}\right)$$

$$= 2\cos\left(\frac{\pi k}{3}\right)e^{-j\pi k/3}, \quad k = 0, 1, 2$$

$$= \{\underset{\uparrow}{2}, e^{-j\pi/3}, e^{j\pi/3}\} = \left\{\underset{\uparrow}{2}, \frac{1 - \sqrt{3}j}{2}, \frac{1 + \sqrt{3}j}{2}\right\}$$

Note that $X(0) = 2$ and $X(1) = X^*(2)$.

b. $x(n) = \{\underset{\uparrow}{1}, 1, 0, 0\}$

$$X(k) = \sum_{n=0}^{1} e^{-j2\pi nk/4} = 1 + e^{-j\pi k/2} = e^{-j\pi k/4}\left(e^{j\pi k/4} + e^{-j\pi k/4}\right)$$

$$= 2\cos\left(\frac{\pi k}{4}\right)e^{-j\pi k/4}, \quad k = 0, 1, 2, 3$$

$$= \{\underset{\uparrow}{2}, \sqrt{2}e^{-j\pi/4}, 0, \sqrt{2}e^{j\pi/4}\} = \{\underset{\uparrow}{2}, (1 - j), 0, (1 + j)\}$$

Note that $X(0) = 2$ and $X(1) = X^*(3)$.

E*xample*

17.2

Find the 3-point DFT of

a. $x(n) = \{\underset{\uparrow}{1}, 0, 1\}$

b. $x(n) = \{\underset{\uparrow}{1}, 0, -1\}$

Solution

a. $x(n) = \{\underset{\uparrow}{1}, 0, 1\}$

$$X(k) = \sum_{n=0}^{2} x(n)e^{-j2\pi nk/3} = 1 + e^{-j4\pi k/3} = 2\cos\left(\frac{2\pi k}{3}\right)e^{-j2\pi k/3}, \quad k = 0, 1, 2$$

$$= \{\underset{\uparrow}{2}, -e^{-j2\pi/3}, -e^{j2\pi/3}\} = \left\{\underset{\uparrow}{2}, \frac{1 + \sqrt{3}j}{2}, \frac{1 - \sqrt{3}j}{2}\right\}$$

Note that $X(0) = 2$ and $X(1) = X^*(2)$.

b. $x(n) = \{1, 0, -1\}$
$\quad\quad\quad\quad\quad\uparrow$

$$X(k) = \sum_{n=0}^{2} x(n)e^{-j2\pi nk/3} = 1 - e^{-j4\pi k/3} = 2j \sin\left(\frac{2\pi k}{3}\right) e^{-j2\pi k/3}, \quad k = 0, 1, 2$$

$$= \{0, j\sqrt{3}e^{-j2\pi/3}, -j\sqrt{3}e^{j2\pi/3}\} = \left\{0, \frac{3 - j\sqrt{3}}{2}, \frac{(3 + j\sqrt{3})}{2}\right\}$$

Note that $X(0) = 0$ and $X(1) = X^*(2)$.

Example 17.3

Even and odd sequences

Find the 3-point DFT of

a. $x(n) = \{0, 1, 1\}$
$\quad\quad\quad\quad\quad\uparrow$
b. $x(n) = \{0, 1, -1\}$
$\quad\quad\quad\quad\quad\uparrow$

Solution

a. $x(n) = \{0, 1, 1\}$ (an even sequence)
$\quad\quad\quad\quad\quad\uparrow$

$$X(k) = \sum_{n=0}^{2} x(n)e^{-j2\pi nk/3} = e^{-j2\pi k/3} + e^{-j4\pi k/3}$$

$$= 2\cos\left(\frac{\pi k}{3}\right) e^{-j\pi k}, \quad k = 0, 1, 2$$

$$= \left\{2, -1, -1\right\}$$
$\quad\;\;\uparrow$

Note that $X(0) = 2$ and $X(1) = X(2)$.

b. $x(n) = \{0, 1, -1\}$ (an odd sequence)
$\quad\quad\quad\quad\quad\uparrow$

$$X(k) = \sum_{n=0}^{2} x(n)e^{-j2\pi nk/3} = e^{-j2\pi k/3} - e^{-j4\pi k/3}$$

$$= 2j \sin\left(\frac{\pi k}{3}\right) e^{-j\pi k}, \quad k = 0, 1, 2$$

$$= \left\{0, -j\sqrt{3}, j\sqrt{3}\right\}$$
$\quad\;\;\uparrow$

Note that $X(0) = 0$ and $X(1) = -X(2)$.

For the definitions of even and odd sequences and their DFT properties see section 17.6

xample

17.4

Zero padding

This example illustrates the effect of padding a sequence with additional zeros on the right side to make it N elements long. Find the 8-point and 16-point DFTs of $x(n)$ in Example 17.1, plot their magnitudes, and comment on the effect of increasing the DFT length.

Solution

a. The 8-point DFT of the sequence is obtained as follows. See Figure 17.1(a).

$$x(n) = \{\underset{\uparrow}{1}, 1, 0, 0, 0, 0, 0, 0\}$$

$$X(k) = \sum_{n=0}^{1} e^{-j2\pi nk/8} = 1 + e^{-j\pi k/4} = e^{-j\pi k/8}\left(e^{j\pi k/8} + e^{-j\pi k/8}\right)$$

$$= 2\cos\left(\frac{\pi k}{8}\right)e^{-j\pi k/8}, \quad k = 0, 1, 2, \ldots, 7$$

$$= \{\underset{\uparrow}{2}, (1.707 - j0.707), (1 - j), (0.293 - j0.707), 0, (0.293 + j0.707), (1 + j), (1.707 + j0.707)\}$$

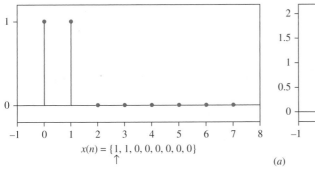

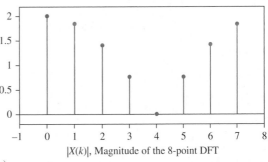

$x(n) = \{\underset{\uparrow}{1}, 1, 0, 0, 0, 0, 0, 0\}$

$|X(k)|$, Magnitude of the 8-point DFT

(a)

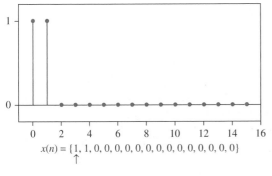

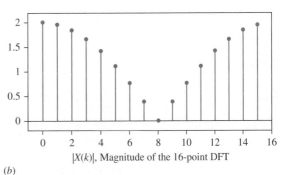

$x(n) = \{\underset{\uparrow}{1}, 1, 0, 0, 0, 0, 0, 0, 0, 0, 0, 0, 0, 0, 0, 0\}$

$|X(k)|$, Magnitude of the 16-point DFT

(b)

FIGURE 17.1 $x(n)$ (in the left column) and its N-point DFT (in the right column). **(a)** $N = 8$. **(b)** $N = 16$.

b. The 16-point DFT of the sequence is obtained as follows. See Figure 17.1(b).

$$x(n) = \{\underset{\uparrow}{1}, 1, 0, 0, 0, 0, 0, 0, 0, 0, 0, 0, 0, 0, 0, 0\}$$

$$X(k) = \sum_{n=0}^{1} e^{-j2\pi nk/16} = 1 + e^{-j\pi k/8} = e^{-j\pi k/16}\left(e^{j\pi k/16} + e^{-j\pi k/16}\right)$$

$$= 2\cos\left(\frac{\pi k}{16}\right)e^{-j\pi k/16}, \quad k = 0, 1, 2, \dots, 15$$

Note that in both cases, $X(0) = 2$. Also note the conjugate property $X(n) = X^*(N - n)$ [because $x(n)$ is real]. Zero padding has increased resolution in the frequency domain without affecting the envelope.

xample

17.5

Increasing the number of samples

a. Find the N-point DFT of a rectangular pulse that is 3 samples wide
b. Find the N-point DFT of a rectangular pulse that is M samples wide ($N \geq M$).
c. Plot its magnitude for $M = 3$, 4 and $N = 32$ and observe the effect of increasing the number of samples in the pulse.

Solution

a. $X(k) = 1 + e^{-j2\pi k/N} + e^{-j4\pi k/N} = e^{-j2\pi k/N}\left(e^{j2\pi k/N} + 1 + e^{-j2\pi k/N}\right)$

$$= \left(1 + 2\cos\frac{2\pi k}{N}\right)e^{-j\frac{2\pi k}{N}}, \quad k = 0, 1, 2, \dots, N-1 \text{ (frequently-used simplification method)}$$

b. $X(k) = \sum_{n=0}^{M-1} e^{-j2\pi nk/N} = \dfrac{1 - e^{-j2\pi Mk/N}}{1 - e^{-j2\pi k/N}}$

$$= \frac{\sin(\pi Mk/N)}{\sin(\pi k/N)}e^{-j\frac{\pi(M-1)k}{N}}, \quad k = 0, 1, \dots N-1$$

c. See Figure 17.2.

Example 17.5 illustrates the effect of the number of samples in the pulse on the DFT. We will discuss this effect later in the chapter in light of the sampling rate. Here it suffices to mention that a higher sampling rate allows higher frequencies to appear in the DFT. The center represents higher frequencies (see section 17.7). Matlab code for computation and plotting of the DFT of a rectangular pulse is provided in problem 10 in order to explore the effect of varying its parameters.

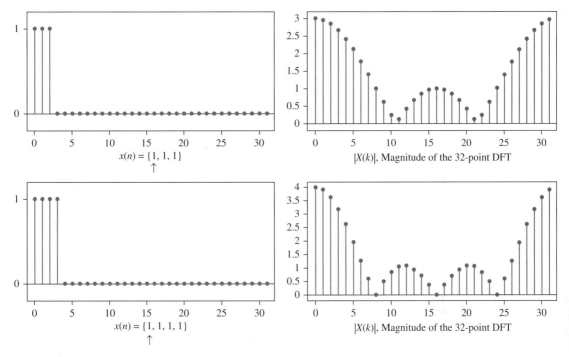

FIGURE 17.2 Rectangular pulses 3 and 4 samples wide (left) and plots of the magnitude of their 32-point DFT's (right).

Example 17.6

Exponential pulse

Find the N-point DFT of $h(n) = a^n$, $0 \le n \le (N-1)$.

Solution

$$X(k) = \sum_{n=0}^{N-1} a^n e^{-j2\pi nk/N} = \frac{1-a^N}{1-ae^{-j2\pi k/N}}, \quad k = 0, 1, \ldots, N-1$$

(The pulse may be expressed as an exponential function $e^{-n/\theta}$.) In problem 10 you will use a Matlab program to compute and plot the DFT of an exponential pulse. You will explore the effect of varying the pulse parameters.

Example 17.7

Sinusoids

Find the 6-point DFT of

a. $\cos(2\pi n/3)$

b. $\cos(\pi n/3)$

Solution

a.
$$\sum_{n=0}^{5} \cos\left(\frac{2\pi n}{3}\right) e^{-j2\pi kn/6}$$

$$= \underbrace{1}_{n=0} + \underbrace{\cos(2\pi/3)e^{-j\pi k/3}}_{n=1} + \underbrace{\cos(4\pi/3)e^{-j2\pi k/3}}_{n=2} + \underbrace{e^{-j\pi k}}_{n=3}$$
$$+ \underbrace{\cos(8\pi/3)e^{-j4\pi k/3}}_{n=4} + \underbrace{\cos(10\pi/3)e^{-j5\pi k/3}}_{n=5}$$

$$= \underbrace{1}_{n=0} + \underbrace{\cos(2\pi/3)e^{-j\pi k/3} + \cos(10\pi/3)e^{-j5\pi k/3}}_{n=1 \text{ and } 5}$$
$$+ \underbrace{\cos(4\pi/3)e^{-j2\pi k/3} + \cos(8\pi/3)e^{-j4\pi k/3}}_{n=2 \text{ and } 4} + \underbrace{e^{-j\pi k}}_{n=3}$$

$$= 1 - \cos(\pi k/3) - \cos(2\pi k/3) + \cos(\pi k), \quad k = 0, 1, \ldots, 5$$
$$= \{\underset{\uparrow}{0}, \ 0, \ 3, \ 0, \ 3, \ 0\}$$

b.
$$\sum_{n=0}^{5} \cos\left(\frac{\pi n}{3}\right) e^{-j2\pi kn/6}$$

$$= \underbrace{1}_{n=0} + \underbrace{\cos(\pi/3)e^{-j\pi k/3}}_{n=1} + \underbrace{\cos(2\pi/3)e^{-j2\pi k/3}}_{n=2} - \underbrace{e^{-j\pi k}}_{n=3}$$
$$+ \underbrace{\cos(4\pi/3)e^{-j4\pi k/3}}_{n=4} + \underbrace{\cos(5\pi/3)e^{-j5\pi k/3}}_{n=5}$$

$$= \underbrace{1}_{n=0} + \underbrace{\cos(\pi/3)e^{-j\pi k/3} + \cos(5\pi/3)e^{-j5\pi k/3}}_{n=1 \text{ and } 5}$$
$$+ \underbrace{\cos(2\pi/3)e^{-j2\pi k/3} + \cos(4\pi/3)e^{-j4\pi k/3}}_{n=2 \text{ and } 4} - \underbrace{e^{-j\pi k}}_{n=3}$$

$$= 1 + \cos(\pi k/3) - \cos(2\pi k/3) - \cos(\pi k), \quad k = 0, 1, \ldots, 5$$
$$= \{\underset{\uparrow}{0}, \ 3, \ 0, \ 0, \ 0, \ 3\}$$

17.3 Examples of the IDFT

Example

17.8

Find the 6-point IDFT of

a. $X(k) = \{\underset{\uparrow}{0}, \ 0, \ 3, \ 0, \ 3, \ 0\}$

b. $X(k) = \{\underset{\uparrow}{0}, \ 3, \ 0, \ 0, \ 0, \ 3\}$

Solution

a.
$$x(n) = \frac{1}{6} \sum_{k=0}^{5} X(k)e^{j2\pi kn/6}$$

$$= \underbrace{\frac{1}{2}e^{j2\pi n/3}}_{k=2} + \underbrace{\frac{1}{2}e^{j4\pi n/3}}_{k=4} = \cos(2\pi n/3), \quad n = 0, 1, \ldots, 5$$

b.
$$x(n) = \frac{1}{6} \sum_{k=0}^{5} X(k)e^{j2\pi kn/6}$$

$$= \underbrace{\frac{1}{2}e^{j\pi n/3}}_{k=1} + \underbrace{\frac{1}{2}e^{j5\pi n/3}}_{k=5} = \cos(\pi n/3), \quad n = 0, 1, \ldots, 5$$

Example

17.9

Determine coefficients $h(n)$ of a low-pass linear-phase FIR filter of length 16 for which samples of the frequency response at regular intervals of $2\pi k/16$ are

$$|H(k)| = \begin{cases} 1, & k = 0, 1, 2 \\ 0.5, & k = 3 \\ 0, & k = 4, 5, 6, 7 \end{cases}$$

Note that for this problem you need to complete the above sequence in a way such that its inverse DFT produces a real $h(n)$. It requires conjugate symmetry to satisfy the real-valued condition of the inverse function and a phase angle of $e^{-j\pi k}$ (corresponding to an 8-unit shift in time). For the above two considerations we start with the following:

$$|H(k)| = \{\underset{\uparrow}{1}, 1, 1, .5, 0, 0, 0, 0, 0, 0, 0, 0, 0, .5, 1, 1\}$$
$$H(k) = \{\underset{\uparrow}{1}, -1, 1, -.5, 0, 0, 0, 0, 0, 0, 0, 0, 0, -.5, 1, -1\}$$

Solution

$$h(n) = \frac{1}{16} \sum_{k=0}^{15} X(k)e^{j2\pi kn/16}$$

$$= \frac{1}{16} \left[\underbrace{1}_{k=0} - \underbrace{e^{j\pi n/8}}_{k=1} + \underbrace{e^{j\pi n/4}}_{k=2} - \underbrace{0.5e^{j3\pi n/8}}_{k=3} - \underbrace{0.5e^{j13\pi n/8}}_{k=13} + \underbrace{e^{j7\pi n/4}}_{k=14} - \underbrace{e^{j15\pi n/8}}_{k=15} \right]$$

$$= \frac{1}{16} \left[\underbrace{1}_{k=0} - \underbrace{e^{j\pi n/8} - e^{j15\pi n/8}}_{k=1 \text{ and } 15} + \underbrace{e^{j\pi n/4} + e^{j7\pi n/4}}_{k=2 \text{ and } 14} - \underbrace{0.5e^{j3\pi n/8} - 0.5e^{j13\pi n/8}}_{k=3 \text{ and } 13} \right]$$

$$= \frac{1}{16} \left[1 - 2\cos\left(\frac{\pi n}{8}\right) + 2\cos\left(\frac{\pi n}{4}\right) - \cos\left(\frac{3\pi n}{8}\right) \right], \quad n = 0, 1, \ldots, 15$$

$$= \{0,\ 0.011486,\ 0.018306,\ -0.015981,\ -0.0625,\ -0.035795,\ 0.106694,$$
$$0.290291,\ 0.375,\ 0.290291,\ 0.106694,\ -0.035795,\ -0.0625,$$
$$-0.015981,\ 0.018306,\ 0.011486\}$$

For further elaboration on this example see problem 11.

Summary of Some Observations on DFT Pairs

1. In all cases, $X(0)$ is the sum of the samples in the sequence (corresponding to the DC level).

2. When the time sequence $x(n)$ is real valued (as in the above examples), $X(k) = X^*(N - k)$ (conjugate property). In such a case half of the $X(k)$ sequence is enough to specify the entire DFT.

3. When $x(n)$ is real and even, $X(k) = X(N - k)$. When it is real and odd, $X(k) = -X(N - k)$ (symmetry property). The symmetry property of a sequence is defined with reference to $n = 0$.

4. Zero padding in the time domain increases the number of samples and brings a higher frequency resolution to the DFT. However, it doesn't change its envelope or the frequency range.

5. A higher sampling rate increases the number of samples and results in a higher frequency range for the DFT.

17.4   Time Reversal and Circular Shift

Time reversal and shift operations are encountered often in the analysis of LTI systems. For finite-length sequences, the domain is restricted to $n = 0, \ldots, N - 1$ and analysis in the frequency domain is performed by DFT. In such cases, time reversal and shift are defined with reference to the circular display as shown in this section. Alternatively, they may be interpreted as time reversal and shift of signals with infinite domain (as illustrated in Examples 17.10 and 17.11).

Time Reversal

Time reversal of a discrete signal $x(n)$ is done by changing n to $-n$. This flips the display of the sequence around the origin. The present chapter is concerned with finite-duration signals constrained to $n = 0, \ldots N - 1$. The signals are shown by a circular display with the positive direction for n being the counterclockwise direction. Time reversal flips the circular display around $n = 0$ (the point at the zeroth angle). The time-reversed sequence is found from the original sequence by $x(N - n)$.

Example **17.10**   Given $x(n) = \{1, 2, 3, 4, 5, 6, 7, 8\}$, find $x(-n)$ and its circular display. (Circular display of a finite-length sequence was introduced in section 11.8 of Chapter 11.)

Solution

$x(-n) = \{1, 8, 7, 6, 5, 4, 3, 2\}$. See the circular display in Figure 17.3. Note that $x(-n) = x(8 - n)$.

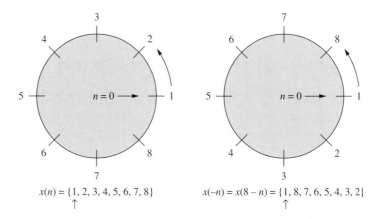

$x(n) = \{1, 2, 3, 4, 5, 6, 7, 8\}$

$x(-n) = x(8 - n) = \{1, 8, 7, 6, 5, 4, 3, 2\}$

FIGURE 17.3 The sequence $x(n) = \{1, 2, 3, 4, 5, 6, 7, 8\}$ (left) and its time-reversal $x(-n) = \{1, 8, 7, 6, 5, 4, 3, 2\}$ (right).

Alternative Visualization of Time Reversal

Assume an N-periodic signal specified in one period by $x(n)$ and displayed linearly. Time-reverse the periodic signal around the origin and take the segment that is seen through the finite-duration window $n = 0, \ldots N - 1$. The lines shown below use this alternative approach to obtain the time reversal of the sequence of Example 17.10. The result is $x(-n) = x(N - n)$, the same as obtained by using the circular display.

Assumed 8-periodic signal: $\{\cdots 1, 2, 3, 4, 5, 6, 7, 8, 1, 2, 3, 4, 5, 6, 7, 8, 1, 2, 3, 4, 5, 6, 7, 8, 1 \cdots\}$

Its time reversal: $\{\cdots 1, 8, 7, 6, 5, 4, 3, 2, 1, 8, 7, 6, 5, 4, 3, 2, 1, 8, 7, 6, 5, 4, 3, 2, 1 \cdots\}$

$x(-n) = x(8 - n)$: $\{1, 8, 7, 6, 5, 4, 3, 2\}$

Circular Shift

An N-point sequence is circularly (or cyclically) shifted m units to the right if all items in the sequence are shifted to the right by m units and the rightmost overflowing m elements are fed back into the newly created leftmost m empty spaces. Table 17.1 illustrates circular shifts to the right and left.

Similarly, $x(n)$ is circularly shifted m units to the left if all items in the sequence are shifted to the left by m units and the leftmost overflowing m elements are fed back into the newly created rightmost m empty spaces.

TABLE 17.1(a) Circular Shifts to the Right for the Sequence $x(n)$

$$x(n) = \{x_0, \quad x_1, \quad x_2, \quad x_3, \quad \cdots \quad x_{(N-2)}, \quad x_{(N-1)}\}$$
$$\uparrow$$
$$x(n-1) = \{x_{(N-1)}, \ x_0, \quad x_1, \quad x_2, \quad \cdots \quad x_{(N-3)}, \quad x_{(N-2)}\}$$
$$\uparrow$$
$$x(n-2) = \{x_{(N-2)}, \ x_{(N-1)}, \quad x_0, \quad x_1, \quad \cdots \quad x_{(N-4)}, \quad x_{(N-3)}\}$$
$$\uparrow$$
$$x(n-3) = \{x_{(N-3)}, \ x_{(N-2)}, \quad x_{(N-1)}, \quad x_0, \quad \cdots \quad x_{(N-5)}, \quad x_{(N-4)}\}$$
$$\uparrow$$
$$x(n-m) = \{x_{(N-m)}, \ x_{(N-m+1)}, \ x_{(N-m+2)}, \ x_{(N-m+3)} \quad \cdots \quad x_{(N-m-2)}, \ x_{(N-m-1)}\}$$
$$\uparrow$$

TABLE 17.1(b) Circular Shifts to the Left for the Sequence $x(n)$

$$x(n) = \{x_0, \quad x_1, \quad x_2, \quad \cdots \quad x_{(N-3)}, \quad x_{(N-2)}, \quad x_{(N-1)}\}$$
$$\uparrow$$
$$x(n+1) = \{x_1, \quad x_2, \quad x_3, \quad \cdots \quad x_{(N-2)}, \quad x_{(N-1)}, \quad x_0\}$$
$$\uparrow$$
$$x(n+2) = \{x_2, \quad x_3, \quad x_4, \quad \cdots \quad x_{(N-1)}, \quad x_0, \quad x_1\}$$
$$\uparrow$$
$$x(n+3) = \{x_3, \quad x_4, \quad x_5, \quad \cdots \quad x_0, \quad x_1, \quad x_2\}$$
$$\uparrow$$
$$x(n+m) = \{x_m, \quad x_{(m+1)}, \quad x_{(m+2)}, \quad \cdots \quad x_{(m-3)}, \quad x_{(m-2)}, \quad x_{(m-1)}\}$$
$$\uparrow$$

TABLE 17.1(c) Shifting $x(n)$ by m Units to the Right and Left

$x(n) =$	x_0 ↑	x_1	x_2	$\cdots$	x_m	$\cdots$	$x_{(N-2)}$	$x_{(N-1)}$
$x(n-m) =$	$x_{(N-m)}$ ↑	$x_{(N-m+1)}$	$\cdots$	$x_{(N-1)}$	x_0	x_1	$\cdots$	$x_{(N-m-1)}$
$x(n+m) =$	x_m ↑	$x_{(m+1)}$	$\cdots$	$x_{(N-m-2)}$	$x_{(N-m-1)}$	$x_{(N-m)}$	$\cdots$	$x_{(m-1)}$

In summary, a shift to the right by m units results in $x(n-m)$. An opposite shift to the left produces $x(n+m)$. After N shifts, a sequence of length N becomes the same as it was initially: $x(n+N) = x(n) = x(n-N)$.

Given $x(n) = \{\underset{\uparrow}{1}, 2, 3, 4, 5, 6, 7, 8\}$

a. Circularly shift it 3 units to the right to obtain $x(n-3)$

b. Circularly shift it 3 units to the left to obtain $x(n+3)$

Solution
See Figure 17.4.

a. $x(n - 3) = \{6, 7, 8, 1, 2, 3, 4, 5\}$
       ↑

b. $x(n + 3) = \{4, 5, 6, 7, 8, 1, 2, 3\}$
       ↑

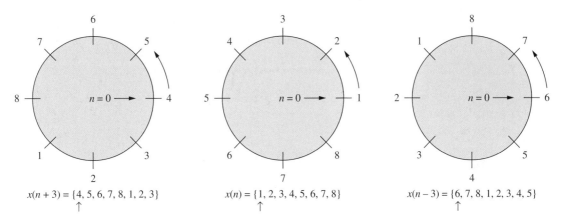

$x(n + 3) = \{4, 5, 6, 7, 8, 1, 2, 3\}$ $x(n) = \{1, 2, 3, 4, 5, 6, 7, 8\}$ $x(n - 3) = \{6, 7, 8, 1, 2, 3, 4, 5\}$
        ↑   ↑   ↑

FIGURE 17.4 Circularly shifting $x(n) = \{1, 2, 3, 4, 5, 6, 7, 8\}$ (center) to the left by 3 units produces $x(n + 3) =$
                                        ↑
$\{4, 5, 6, 7, 8, 1, 2, 3\}$ (left). A 3-unit circular shift to the right produces $x(n - 3) = \{6, 7, 8, 1, 2, 3, 4, 5\}$ (right).
↑   ↑

Alternative Visualization of Circular Shift

Given a finite-duration sequence $x(n)$, $n = 0, \ldots N - 1$, assume an N-periodic signal specified in one period by $x(n)$ and displayed linearly. Shift the periodic signal by $\pm m$ units. The segment that is seen through the finite-duration window $n = 0, \ldots N - 1$ is the circularly shifted $x(n \pm m)$. The lines shown below use this alternative approach to find circular shifts of the sequence of Example 17.11. The result is the same as that obtained by using the circular display.

Assume 8-periodic signal: $\{\cdots 1, 2, 3, 4, 5, 6, 7, 8, \underset{\uparrow}{1}, 2, 3, 4, 5, 6, 7, 8, 1, 2, 3, 4, 5, 6, 7, 8, 1 \cdots\}$

3-unit shift to the right: $\{\cdots 6, 7, 8, 1, 2, 3, 4, 5, \underset{\uparrow}{6}, 7, 8, 1, 2, 3, 4, 5, 6, 7, 8, 1, 2, 3, 4, 5, 6 \cdots\}$

$x(n - 3)$: $\{\underset{\uparrow}{6}, 7, 8, 1, 2, 3, 4, 5\}$

3-unit shift to the left: $\{\cdots 4, 5, 6, 7, 8, 1, 2, 3, \underset{\uparrow}{4}, 5, 6, 7, 8, 1, 2, 3, 4, 5, 6, 7, 8, 1, 2, 3, 4 \cdots\}$

$x(n + 3)$: $\{\underset{\uparrow}{4}, 5, 6, 7, 8, 1, 2, 3\}$

xample

17.12

Find one complete cycle of $x(n) = \{1, 1, 1, 1, 0, 0, 0, 0\}$ shifted to the right and in steps of one.

$$x(n) = \{\underset{\uparrow}{1}, 1, 1, 1, 0, 0, 0, 0\}$$

$$x(n-1) = \{\underset{\uparrow}{0}, 1, 1, 1, 1, 0, 0, 0\}$$

$$x(n-2) = \{\underset{\uparrow}{0}, 0, 1, 1, 1, 1, 0, 0\}$$

$$x(n-3) = \{\underset{\uparrow}{0}, 0, 0, 1, 1, 1, 1, 0\}$$

$$x(n-4) = \{\underset{\uparrow}{0}, 0, 0, 0, 1, 1, 1, 1\}$$

$$x(n-5) = \{\underset{\uparrow}{1}, 0, 0, 0, 0, 1, 1, 1\}$$

$$x(n-6) = \{\underset{\uparrow}{1}, 1, 0, 0, 0, 0, 1, 1\}$$

$$x(n-7) = \{\underset{\uparrow}{1}, 1, 1, 0, 0, 0, 0, 1\}$$

$$x(n-8) = \{\underset{\uparrow}{1}, 1, 1, 1, 0, 0, 0, 0\}$$

After eight unitary shifts to the right, $x(n-8) = x(n)$ and the sequence becomes the same as it was initially.

xample

17.13

Find one complete cycle of $x(n) = \{x_0, x_1, x_2, x_3, x_4, x_5, x_6, x_7\}$ shifted to the left, in steps of one, so that $x(n+8) = x(n)$.

$$x(n) = \{x_0, x_1, x_2, x_3, x_4, x_5, x_6, x_7\}$$

$$x(n+1) = \{x_1, x_2, x_3, x_4, x_5, x_6, x_7, x_0\}$$

$$x(n+2) = \{x_2, x_3, x_4, x_5, x_6, x_7, x_0, x_1\}$$

$$x(n+3) = \{x_3, x_4, x_5, x_6, x_7, x_0, x_1, x_2\}$$

$$x(n+4) = \{x_4, x_5, x_6, x_7, x_0, x_1, x_2, x_3\}$$

$$x(n+5) = \{x_5, x_6, x_7, x_0, x_1, x_2, x_3, x_4\}$$

$$x(n+6) = \{x_6, x_7, x_0, x_1, x_2, x_3, x_4, x_5\}$$

$$x(n+7) = \{x_7, x_0, x_1, x_2, x_3, x_4, x_5, x_6\}$$

$$x(n+8) = \{x_0, x_1, x_2, x_3, x_4, x_5, x_6, x_7\}$$

After eight unitary shifts to the left, $x(n+8) = x(n)$, and the sequence becomes the same as it was initially.

17.5 Circular Convolution

The circular convolution of two sequences $x(n)$ and $h(n)$ of finite length N is defined by

$$y(n) = x(n) \otimes h(n) = \sum_{k=0}^{N-1} x(k)h(n-k)$$

where $h(n-k)$ is found by time reversal and circular shift. Circular convolution is shown by the symbol $\otimes$. The evaluation of the circular convolution requires steps similar to linear convolution. These steps are the following:

1. Time reversal of $h(k)$ to produce $h(-k)$
2. A circular shift to produce $h(n-k)$
3. Multiplication of $x(k)$ by $h(n-k)$
4. Summation of the N terms

These steps may be visualized and carried out more conveniently by a circular display of the two signals as illustrated in the following example.

E*xample*
17.14

Find $y(n) = x(n) \otimes h(n)$, where $x(n) = \{1, 1, 0, 0\}$ and $h(n) = \{1, 1, -1, -1\}$.

Solution
Using the circular display as a tool, we compute $y(n)$:

$y(0) = 1 - 1 + 0 + 0 = 0$ [See Figure 17.5(a).]
$y(1) = 1 + 1 + 0 + 0 = 2$ [See Figure 17.5(b).]
$y(2) = -1 + 1 + 0 + 0 = 0$ [See Figure 17.5(c).]
$y(3) = -1 - 1 + 0 + 0 = -2$ [See Figure 17.5(d).]

The output sequence is $y(n) = \{0, 2, 0, -2\}$.

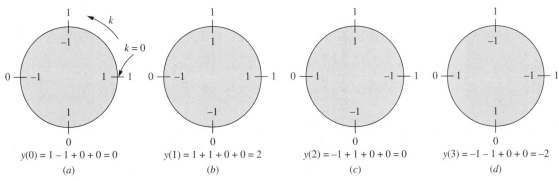

$y(0) = 1 - 1 + 0 + 0 = 0$

(a)

$y(1) = 1 + 1 + 0 + 0 = 2$

(b)

$y(2) = -1 + 1 + 0 + 0 = 0$

(c)

$y(3) = -1 - 1 + 0 + 0 = -2$

(d)

FIGURE 17.5 This figure displays steps in circular convolution $y(n) = x(n) \otimes h(n) = \sum_{k=0}^{N-1} x(k)h(n-k)$, where $x(n) = \{1, 1, 0, 0\}$ and $h(n) = \{1, 1, -1, -1\}$. $x(k)$ is displayed on the outside and $h(n-k)$ is displayed inside the circle for $n = 0, 1, 2$, and 3. They are multiplied term by term and added to produce $y(n)$.

Zero Padding

When one sequence is longer than the other, the shorter sequence should be padded with enough zeros on the right side to make it the same size as the longer sequence.

Find $y(n) = x(n) \otimes h(n)$, where

$$x(n) = \{\underset{\uparrow}{1}, 1, 0, 0, 0, 0\} \quad \text{and} \quad h(n) = \{\underset{\uparrow}{1}, 1, -1, -1\}.$$

Solution

$x(n)$ has six samples and $h(n)$ has four. $x(n) \otimes h(n)$ is a 6-point circular convolution. We first pad $h(n)$ with two zeros on the right to make it 6 points long, then perform circular convolution (following steps similar to Figure 17.5) to find

$$y(n) = \{\underset{\uparrow}{1}, 1, 0, 0, 0, 0\} \otimes \{\underset{\uparrow}{1}, 1, -1, -1, 0, 0\} = \{\underset{\uparrow}{1}, 2, 0, -2, -1, 0\}.$$

DFT and Circular Convolution

The circular convolution of two time-domain signals is equivalent to multiplication of their DFTs in the frequency domain.

$$x(n) \otimes y(n) \quad \Longleftrightarrow \quad X(k)Y(k)$$

This may be proved by taking the DFT of the circular convolution sum. Let

$$y(n) = x(n) \otimes h(n) = \sum_{k=0}^{N-1} x(k)h(n-k)$$

Taking the DFT of $y(n)$ results in

$$Y(k) = \sum_{n=0}^{N-1} y(n)e^{-j2\pi nk/N} = \sum_{n=0}^{N-1}\sum_{m=0}^{N-1} x(m)h(n-m)e^{-j2\pi nk/N}$$

The double sum may be converted to a product of two separate sums by changing the variable n to a new variable p such that $n = m + p$. The terms containing the variables p and m in the double sum are then separated and so

$$Y(k) = \sum_{m=0}^{N-1} x(m)e^{-j2\pi mk/N} \times \sum_{p=0}^{N-1} h(p)e^{-j2\pi pk/N} = X(k)H(k)$$

This property is of great importance because DFT operations, when done through the fast Fourier transform (FFT) algorithm, are completed much faster than convolutions. (See Example 17.20.)

Obtaining Linear Convolution from Circular Convolution

The circular convolution of two finite sequences may be performed by multiplying their DFTs in the frequency domain and then taking the IDFT of the result. This becomes very efficient and fast if the DFTs are obtained through FFT algorithm. However, the output of a discrete-time LTI system is not the circular convolution of the input sequence with the sequence of the unit-sample response, but rather the *linear* convolution of the two. When both sequences are finite and approximately the same length, we can pad both of them with enough zeros on both sides and perform circular convolution (through DFT and IDFT). With enough zero padding, the result of circular convolution will be the same as linear convolution.

Example **17.16**

a. Using the linear convolution operation find $y(n) = x(n) \star h(n)$, where

$$x(n) = \{0, 0.1678, 0.6821, 1, 1, 0.6821, 0.1678\} \text{ and}$$
$$\uparrow$$
$$h(n) = \{0, -0.2122, 0, 0.6366, 1, 0.6366, 0, -0.2122\}$$
$$\uparrow$$

b. Minimally pad $x(n)$ and $h(n)$ to find $y(n)$ from $x(n) \otimes h(n)$ using the DFT.

Solution

a. $x(n)$ has 7 samples and $h(n)$ has eight. Their linear convolution will have 14 samples.

$$y(n) = x(n) \star h(n)$$
$$= \{0, 0, -0.03562, -0.1448, -0.1053, 0.3899,$$
$$\uparrow$$

$$1.2809, 2.0353, 2.0353, 1.2809, 0.3899, -0.1053, -0.1448, -0.0356\}$$

b. The circular convolution should have at least 14 samples. To use the DFT, we pad $x(n)$ with 7 zeros and $h(n)$ with 6 zeros [to get $\psi(n)$ and $\zeta(n)$, respectively]. Using DFT/IDFT we find the same result as the linear convolution.

$$\psi(n) \otimes \zeta(n) = \{0, 0, -0.03562, -0.1448, -0.1053, 0.3899, 1.2809, 2.0353,$$
$$\uparrow$$

$$2.0353, 1.2809, 0.3899, -0.1053, -0.1448, -0.0356\}$$

Padding sequences with large numbers of zeros reduces the efficacy of convolution through DFT and is not recommended. Instead, when one sequence is short and the other is long (or infinite), we obtain the output of an LTI system $y(n) = x(n) \star h(n)$ from the circular convolution of short segments. Two such methods are described in section 17.9.

17.6 Properties of the DFT

The discrete Fourier transform is a member of the family of linear transforms that includes the Fourier transform, Laplace transform, z-transform, and discrete-time Fourier transform. It, therefore, shares many of their properties. Some of these properties are described below and summarized in Table 17.2.

Linearity Property

The N-point DFT of $ax(n) + by(n)$ is $aX(k) + bY(k)$ for any $x(n)$ and $y(n)$ of the same length N and any constants a and b.

$$
\begin{aligned}
x(n) &\iff X(k) \\
y(n) &\iff Y(k) \\
ax(n) + by(n) &\iff aX(k) + bY(k)
\end{aligned}
$$

Periodicity

The DFT and IDFT operations are concerned with finite-length sequences. Therefore, $x(n)$ and $X(k)$ are not periodic functions per se. However, because the summations in (1) and (2) contain $e^{-j2\pi nk/N}$ and $e^{j2\pi kn/N}$, which are periodic with period N, one can talk of a cyclic property for $x(n)$ and $X(k)$ and consider the following periodic functions that extend beyond the finite length of the sequence:

$$
x(n) = x(n + N), \quad \text{for all } n, \quad \text{and} \quad X(k) = X(k + N), \quad \text{for all } k.
$$

The circular shift of the finite sequence would then become a linear shift of the newly considered periodic function. A linear display of the periodic function may be generated by rolling the wheel of the circular display of the function along a linear axis (picture the impressions left by a cylindrical seal rolling on a flat piece of paper). The circular display of $x(n)$ or $X(k)$ introduced in the previous section is nonetheless helpful in visualizing the concepts of cyclicity, circular shift, and circular convolution.

Time Reversal

Reversing time flips the signal around $n = 0$ and produces $x(-n) = x(N - n)$ (due to its periodicity). The effect in the frequency domain is

$$
x(-n) = x(N - n) \iff X(-k) = X(N - k)
$$

To prove the above relationship, start with

$$
\mathcal{DFT}\{x(-n)\} = \sum_{n=0}^{N-1} x(N - n) e^{-j2\pi nk/N}
$$

and let $N - n = m$. Then,

$$
\mathcal{DFT}\{x(-n)\} = \sum_{m=1}^{N} x(m) e^{-j2\pi(N-m)k/N} = \sum_{m=1}^{N} x(m) e^{-j2\pi m(-k)/N} = X(-k) = X(N - k)
$$

Time-Domain Conjugate Property

The DFT of the complex conjugate of a signal is related to the DFT of the signal by

$$
x^*(n) \iff X^*(-k) = X^*(N - k)
$$

The above may be derived by applying the DFT definition

$$\mathcal{DFT}\{x^*(n)\} = \sum_{n=0}^{N-1} x^*(n)e^{-j2\pi nk/N} = \sum_{n=0}^{N-1} x^*(n)e^{j2\pi n(-k)/N}$$

$$= \left[\sum_{n=0}^{N-1} x(n)e^{-j2\pi n(-k)/N}\right]^* = X^*(-k)$$

DFT of a Real-Valued Signal

From the time-domain conjugate property we deduce that when $x(n)$ is real then $X(k) = X^*(-k)$.

$$x(n) = x^*(n) \iff X(k) = X^*(N-k)$$

In such a case, the real part of $X(k)$ (also its magnitude) is an even function and its imaginary part (also its phase) is an odd function. The signal can be completely specified by $X(k)$, $k = 0, 1, \ldots (N/2 + 1)$ (N even) or $(N+1)/2$ (N odd).

Circular Shift

Circular shift of m units multiplies $X(k)$ by $e^{-j2\pi mk/N}$ and, therefore, subtracts $2\pi mk/N$ from its phase.

$$x(n-m) \iff X(k)e^{-j2\pi mk/N}$$

Frequency Shift

Multiplication of $x(n)$ by $e^{j2\pi np/N}$ shifts its DFT by p units to the right.

$$x(n)e^{j2\pi np/N} \iff X(k-p)$$

Modulation

The n-domain signal corresponding to a shift to the left or the right in the frequency domain is not a real signal. However, by shifting $X(k)$ to the left and the right and adding the results, we obtain a real signal modulated by a carrier.

$$x(n)e^{j2\pi np/N} \iff X(k-p)$$
$$x(n)e^{-j2\pi np/N} \iff X(k+p)$$
$$x(n)\cos\left(\frac{2\pi np}{N}\right) \iff \frac{X(k-p) + X(k+p)}{2}$$

Waveform Symmetry

The symmetry properties of a function introduce features in its DFT, which simplify calculations. Two such symmetry properties are listed below.

Even Sequence

If a sequence is symmetric about point 0 on the circular display, then it is called an *even sequence*, in which case $x(N - n) = x(n)$.

Odd Sequence

If a sequence is antisymmetric about point 0 on the circular display, then it is called an *odd sequence*, in which case $x(N - n) = -x(n)$.

The DFT of a real and even sequence is a real and even sequence. Likewise, the DFT of a real and odd sequence is a purely imaginary and odd sequence.

$$\text{Real and even sequence: } x(N - n) = x(n) \quad \Longleftrightarrow \quad X(k) = X(-k)$$

$$\text{Real and odd sequence: } x(N - n) = -x(n) \quad \Longleftrightarrow \quad X(k) = -X(-k)$$

Circular Convolution and Product Property

The DFT of the circular convolution of two sequences was shown (section 17.5) to be equal to the product of the DFTs:

$$x(n) \otimes y(n) \quad \Longleftrightarrow \quad X(k)Y(k)$$

Similarly, the DFT of the product of two time sequences can be obtained from the circular convolution of their DFTs:

$$x(n)y(n) \quad \Longleftrightarrow \quad X(k) \otimes Y(k) = \frac{1}{N} \sum_{m=0}^{N-1} X(m)Y(k - m)$$

Parseval's Theorem

The energy in a discrete-time signal $x(n)$, $n = 0, 1, \ldots, N - 1$, is defined by

$$E_x = \sum_{n=0}^{N-1} |x(n)|^2$$

Parseval's theorem expresses E_x in terms of the DFT coefficients $X(k)$. It states that

$$\sum_{n=0}^{N-1} |x(n)|^2 = \frac{1}{N} \sum_{k=0}^{N-1} |X(k)|^2$$

The above summations are referred to as the energy in the time and frequency domains, respectively. $|X(k)|^2$ may be called the energy density spectrum (spectrum of energy density, energy spectral density). The time signal can be a complex-valued function of n and, therefore, in general $|x(n)|^2 = x(n)x^*(n)$. Parseval's theorem is one of the set of relations between the time- and frequency-domain representations of two finite-length sequences.

$$\sum_{n=0}^{N-1} x(n)y(n) = \frac{1}{N} \sum_{k=0}^{N-1} X(k)Y(-k) \qquad \text{and} \qquad \sum_{n=0}^{N-1} x(n)y^*(n) = \frac{1}{N} \sum_{k=0}^{N-1} X(k)Y^*(k)$$

To derive Parseval's relations, express the DFT of the product $x(n)y(n)$ as the convolution of their DFTs.

$$\text{DFT}\{x(n)y(n)\} = X(k) \otimes Y(k)$$

$$\Downarrow \qquad\qquad \Downarrow$$

$$\sum_{n=0}^{N-1} x(n)y(n)e^{-j2\pi kn/N} = \frac{1}{N}\sum_{m=0}^{N-1} X(m)Y(k-m)$$

Then set $k = 0$ to find

$$\Downarrow \qquad\qquad \Downarrow$$

$$\sum_{n=0}^{N-1} x(n)y(n) = \frac{1}{N}\sum_{m=0}^{N-1} X(m)Y(-m)$$

For $y(n) = x^*(n)$ we have $Y(m) = X^*(-m)$, $Y(-m) = X^*(m)$, and

$$\sum_{n=0}^{N-1} x(n)x^*(n) = \frac{1}{N}\sum_{m=0}^{N-1} X(m)X^*(m)$$

$$\sum_{n=0}^{N-1} |x(n)|^2 = \frac{1}{N}\sum_{m=0}^{N-1} |X(m)|^2$$

Example 17.17

Verify Parseval's theorem for the sequences in Example 17.1.

Solution

a. $\quad x(n) = \{\underset{\uparrow}{1}, 1, 0\} \qquad\qquad \Longrightarrow \sum_{n=0}^{2} |x(n)|^2 = 2$

$$X(k) = \left\{\underset{\uparrow}{2}, \frac{1-\sqrt{3}j}{2}, \frac{1+\sqrt{3}j}{2}\right\} \Longrightarrow \frac{1}{3}\sum_{k=0}^{2} |X(k)|^2 = \frac{1}{3}[4+1+1] = 2$$

Parseval's theorem: $\displaystyle\sum_{n=0}^{2} |x(n)|^2 = \frac{1}{3}\sum_{k=0}^{2} |X(k)|^2 = 2$

b. $\quad x(n) = \{\underset{\uparrow}{1}, 1, 0, 0\} \qquad\qquad \Longrightarrow \sum_{n=0}^{3} |x(n)|^2 = 2$

$$X(k) = \{\underset{\uparrow}{2}, (1-j), 0, (1+j)\} \Longrightarrow \frac{1}{4}\sum_{k=0}^{3} |X(k)|^2 = \frac{1}{4}[4+2+0+2] = 2$$

Parseval's theorem: $\displaystyle\sum_{n=0}^{3} |x(n)|^2 = \frac{1}{4}\sum_{k=0}^{3} |X(k)|^2 = 2$

TABLE 17.2 Summary of DFT Properties

Definition	$x(n) = \dfrac{1}{N}\sum\limits_{k=0}^{N-1} X(k)e^{j2\pi kn/N}$	$\Longleftrightarrow$	$X(k) = \sum\limits_{n=0}^{N-1} x(n)e^{-j2\pi nk/N}$

Linearity	$ax(n) + by(n)$	$\Longleftrightarrow$	$aX(k) + bY(k)$
Conjugation	$x^*(n)$	$\Longleftrightarrow$	$X^*(-k) = X^*(N-k)$
Real	$x(n) = x^*(n)$	$\Longleftrightarrow$	$X(k) = X^*(-k)$
Real and even	$x(N-n) = x(n)$	$\Longleftrightarrow$	$X(k) = X(-k)$
Real and odd	$x(N-n) = -x(n)$	$\Longleftrightarrow$	$X(k) = -X(-k)$
Time reversal	$x(-n) = x(N-n)$	$\Longleftrightarrow$	$X(-k) = X(N-k)$
Circular shift	$x(n-m)$	$\Longleftrightarrow$	$X(k)e^{-j2\pi mk/N}$
Frequency shift	$x(n)e^{j2\pi np/N}$	$\Longleftrightarrow$	$X(k-p)$
Modulation	$x(n)\cos\left(\dfrac{2\pi np}{N}\right)$	$\Longleftrightarrow$	$\dfrac{X(k-p) + X(k+p)}{2}$
Circular convolution	$x(n) \otimes y(n) = \sum\limits_{m=0}^{N-1} x(m)y(n-m)$	$\Longleftrightarrow$	$X(k)Y(k)$
Product property	$x(n)y(n)$	$\Longleftrightarrow$	$X(k) \otimes Y(k) = \dfrac{1}{N}\sum\limits_{m=0}^{N-1} X(m)Y(k-m)$
Parseval's Theorem	$\sum\limits_{n=0}^{N-1} x(n)y(n) = \dfrac{1}{N}\sum\limits_{k=0}^{N-1} X(k)Y(-k)$ $\sum\limits_{n=0}^{N-1} |x(n)|^2 = \dfrac{1}{N}\sum\limits_{k=0}^{N-1} |X(k)|^2$	and	$\sum\limits_{n=0}^{N-1} x(n)y^*(n) = \dfrac{1}{N}\sum\limits_{k=0}^{N-1} X(k)Y^*(k)$

17.7 Relation Between the DFT and DTFT

The DFT is the sampled version of the DTFT. The N-point DFT of a sequence $x(n)$, $n = 0, 1, 2, \ldots N-1$, generates samples of the DTFT of $x(n)$, zero-padded to extend it to $-\infty < n < \infty$. The samples are taken every $1/N$ Hz in the f-domain over the span of 1 Hz. The DFT is, therefore, analogous to a sampling instrument in the frequency domain.

*E***xample**

17.18

A finite-duration sequence $x(n)$, $n = 0, 1, 2, \ldots, N-1$, [and, consequently, its N-point DFT sequence $X(k)$] is given.

a. Zero-pad $x(n)$ in both directions in order to extend it to $-\infty < n < \infty$ and call the result $y(n)$. Relate $Y(\omega)$, the DTFT of $y(n)$, to $X(k)$.

b. Repeatedly append $x(n)$ to itself to obtain the periodic signal $z(n)$ shown below.

$$z(n) = x(n), \quad n = 0, 1, 2, \ldots, N-1$$
$$z(n+N) = z(n), \quad -\infty < n < \infty$$

Relate $Z(\omega)$, the DTFT of $z(n)$, to the DFT of $x(n)$.

Solution

a.
$$X(k) = \sum_{n=0}^{N-1} x(n)e^{-j2\pi nk/N}, \quad k = 0, 1, 2, \ldots N-1$$

$$Y(\omega) = \sum_{n=-\infty}^{\infty} y(n)e^{-j\omega n} = \sum_{n=0}^{N-1} x(n)e^{-j\omega n}$$

$$X(k) = Y(\omega)\big|_{\omega=\frac{2\pi k}{N}}, \quad k = 0, 1, 2, \ldots N-1$$

Remember that $Y(\omega)$ is 2π−periodic and, therefore, there is no need to sample it outside the DFT window. The $X(k)$ elements are samples of $Y(\omega)$ taken at $\frac{2\pi k}{N}$, $k = 0, 1, 2, \ldots N-1$ [during the first period of $Y(\omega)$]. In addition, $Y(\omega)$ doesn't contain any information beyond sample values available as $X(k)$.[1] The zero padding doesn't add any new information either.

b.
$$Z(\omega) = Y(\omega) \star \sum_{k=-\infty}^{\infty} \frac{2\pi}{N}\delta\left(\omega - \frac{2k\pi}{N}\right) = \sum_{k=-\infty}^{\infty} \frac{2\pi}{N}Y\left(\frac{2k\pi}{N}\right)\delta\left(\omega - \frac{2k\pi}{N}\right)$$

$$= \sum_{k=-\infty}^{\infty} \frac{2\pi}{N}X(k)\delta\left(\omega - \frac{2k\pi}{N}\right)$$

$Z(\omega)$, the DTFT of $z(n)$, is an N-periodic sequence of impulses of strength $\frac{2\pi}{N}X(k)$ located at $\omega = 2k\pi/N$.

Example 17.19

a. Find $Y(\omega)$, the DTFT of $y(n) = \{\ldots, 0, 0, 0, \underset{\uparrow}{1}, 1, 1, 1, 1, 0, 0, 0, \ldots\}$ and plot its magnitude as a function of $f = \omega/(2\pi)$ for $-0.25 < f < 1.25$.

b. Find $X(k)$, the 20-point DFT of $x(n) = \{\underset{\uparrow}{1}, 1, 1, 1, 1, \}$. Plot its magnitude for $0 \le k \le 19$ and note its relation with the frequency f, which ranges from 0 to 1.

c. Compare plots in a and b and verify that $X(k) = Y(\omega)|_{\omega=\pi k/10}$.

Solution

a. $Y(\omega) = \sum_{n=0}^{4} e^{-j\omega n} = \frac{1 - e^{-j5\omega}}{1 - e^{-j\omega}} = \frac{\sin\left(\frac{5\omega}{2}\right)}{\sin\left(\frac{\omega}{2}\right)}e^{-j2\omega}$. See Figure 17.6(a).

---

[1]The fact that $X(k)$ completely specifies $Y(\omega)$ is a consequence of the sampling theorem applied to $Y(\omega)$. See the sampling theorem in C. E. Shannon, "Communication in the Presence of Noise," *Proceedings of the IRE*, vol. 37, no. 1, pp. 10–21, January 1949. Reprinted in the *Proceedings of the IEEE*, vol. 86, no. 2, February 1998.

b.
$$X(k) = \sum_{n=0}^{4} e^{-j2\pi nk/20} = \frac{1 - e^{-j\pi k/2}}{1 - e^{-j\pi k/10}} = \frac{\sin\left(\frac{\pi k}{4}\right)}{\sin\left(\frac{\pi k}{20}\right)} e^{-j\left(\frac{\pi k}{5}\right)}$$

$$= Y(\omega)\Big|_{\omega = \pi k/10}, \quad k = 0, \dots, 19. \quad \text{See Figure 17.6(b).}$$

c.

k	ω	f	$X(k) = Y(\omega)$
0	0	0	5
1	0.1π	0.05	$4.5201\angle - 36°$
2	0.2π	0.1	$3.2360\angle - 72°$
...			
m	$m\,\pi/10$	$0.05m$	See parts a and b above.
...			
18	1.8π	0.9	$3.2360\angle 72°$
19	1.9π	0.95	$4.5201\angle 36°$
			Note that $X(k) = X^*(20 - k)$

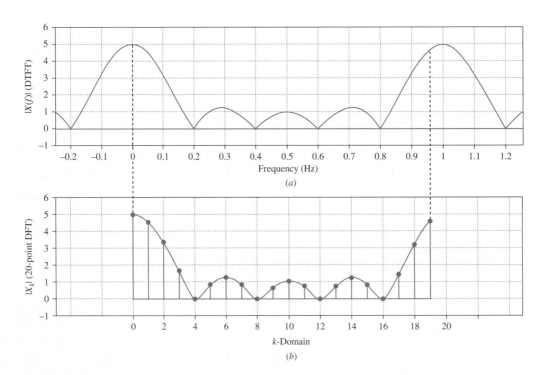

FIGURE 17.6 *(a)* Magnitude plot of the DTFT of the time function $y(n) = \{\dots, 0, 0, 0, 1, 1, 1, 1, 1, 0, 0, 0, \dots\}$.
(b) Magnitude plot of the 20-point DFT of the sequence $x(n) = \{1, 1, 1, 1, 1\}$ (padded with 15 zeros) is superimposed on $|Y(f)|$. The DTFT magnitude is the envelope of the magnitude of the DFT sequence. The elements of $X(k)$ are samples of $Y(f)$ taken every $1/20$ Hz (that is, at $f = k/20, k = 0, \dots 19$). The transforms were obtained and plotted by computer.

17.8   Fast Fourier Transform (FFT)

Direct computation of the sum in (1) for a single value of k takes N multiplications and $N - 1$ additions. Computation of the entire N-point DFT for $k = 0, \ldots, N - 1$ would take N^2 multiplications and $N(N - 1)$ additions. The above numbers can be reduced by making use of the periodic property of the DFT. An approach that dramatically reduces the number of multiplications and additions needed to compute the DFT is called the fast Fourier transform (FFT). For example, the FFT reduces the total number of multiplications and additions by factors of 683 and 341, respectively, for a 4,096-point data sequence. The FFT motif is a repeated DFT operation on small sequences (down to 2-point DFT) and combination of the results to produce the N-point DFT of the original sequence. The following introduces an FFT concept which uses decimation-in-time and assumes N is a power of 2.

Start with the definition of DFT:

$$X(k) = \sum_{n=0}^{N-1} x(n) e^{-j2\pi nk/N} = \sum_{n=0}^{N-1} x(n) W_N^{nk}, \text{ where } W_N = e^{-j2\pi/N} \text{ and } W_N^{nk} \equiv (W_N)^{nk}$$

Split $x(n)$ into its even-numbered and odd-numbered points and call them $a(n)$ and $b(n)$, respectively:

$$a(n) = x(2n) \qquad \text{(the even-numbered points)}, \quad n = 0, 1, \ldots, \frac{N}{2} - 1$$

$$b(n) = x(2n + 1) \quad \text{(the odd-numbered points)}, \quad n = 0, 1, \ldots, \frac{N}{2} - 1$$

For example, split $x(n) = \{\underset{\uparrow}{1}, 2, 3, 4\}$ into $a(n) = \{\underset{\uparrow}{1}, 3\}$ and $b(n) = \{\underset{\uparrow}{2}, 4\}$, see Figure 17.7($a$). Splitting $x(n)$ divides the DFT summation into two parts with contributions from the even-numbered and odd-numbered points, respectively.

$$X(k) = \sum_{n=0}^{N/2-1} a(n) W_N^{2nk} + \sum_{n=0}^{N/2-1} b(n) W_N^{(2n+1)k}, \quad k = 0, 1, 2, \ldots N - 1$$

Since $W_N^2 = W_{N/2}$ the above equation becomes

$$X(k) = \sum_{n=0}^{N/2-1} a(n) W_{N/2}^{nk} + W_N^k \sum_{n=0}^{N/2-1} b(n) W_{N/2}^{nk}, \quad k = 0, 1, 2, \ldots N - 1$$

The first summation on the right side of the above equation is $N/2$-point DFT of $a(n)$, to be designated as $A(k)$. Similarly, the second summation is $N/2$-point DFT of $b(n)$, to be designated as $B(k)$. Then,

$$X(k) = A(k) + W_N^k B(k), \quad k = 0, 1, 2, \ldots N - 1$$

Because

$$A(k) = A(k + N/2), \quad B(k) = B(k + N/2), \quad \text{and } W_N^{k+N/2} = -W_N^k$$

the equation for $X(k)$ can be separated into

$$X(k) = A(k) + W_N^k B(k), \quad k = 0, 1, 2, \ldots N/2 - 1$$

$$X(k + N/2) = A(k) - W_N^k B(k), \quad k = 0, 1, 2, \ldots N/2 - 1$$

The N-point DFT of $x(n)$ can be obtained by combining the $N/2$-point DFTs of $a(n)$ and $b(n)$. The butterfly flowgraph for the combination is shown in Figure 17.7(b).

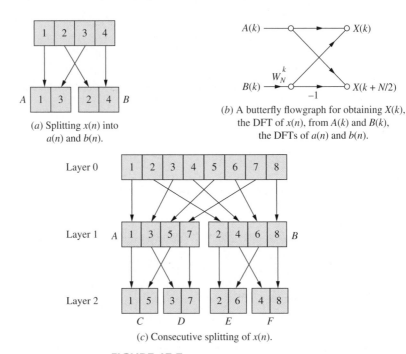

(a) Splitting $x(n)$ into $a(n)$ and $b(n)$.

(b) A butterfly flowgraph for obtaining $X(k)$, the DFT of $x(n)$, from $A(k)$ and $B(k)$, the DFTs of $a(n)$ and $b(n)$.

(c) Consecutive splitting of $x(n)$.

FIGURE 17.7 Layering in FFT algorithm.

Further splitting of $a(n)$ and $b(n)$ into even- and odd-numbered sequences, and applying the butterfly flowgraph to their DFT would save more computation time as smaller sequences need less computation time for their DFT. For example, layer 1 in Figure 17.7(c) shows splitting the sequence $x(n) = \{1, 2, 3, 4, 5, 6, 7, 8\}$ into the odd-numbered sequence $a(n) = \{1, 3, 5, 7\}$ and the even-numbered sequence $b(n) = \{2, 4, 6, 8\}$. Layer 2 repeats the process and produces $c(n) = \{1, 5\}$, $d(n) = \{3, 7\}$, and, $e(n) = \{2, 6\}$, $f(n) = \{4, 8\}$, respectively, each only two elements long. In this way $X(k)$ can be found by applying butterfly flowgraph to smaller sequences, taking less time to compute. See problem 13 in this chapter.

Computational Savings

Direct computation of N-point DFT requires N^2 multiplications. If obtained by the flowgraph of Figure 17.7 the total number of multiplications will be reduced to

$2 \times (N/2)^2 + N/2 = N^2/2 + N/2$, which is a reduction by a factor 2 for large N. From another angle, having obtained the $N/2$-point DFTs of $a(n)$ and $b(n)$ we would need only N more additions and $N/2$ more multiplications to obtain the N-point DFT of $x(n)$. Further reduction can be achieved by applying the above procedure to $a(n)$ and $b(n)$ and starting with four $N/4$-point DFTs. By repeatedly dividing the time sequence to smaller size one can start with 2-point DFT (called radix 2), achieving a maximum computational saving. Table 17.3 shows the number of multiplications and additions required for the DFT and FFT operations on sequences of data ranging in length from 4 to 4,096 points.

TABLE 17.3 Comparing the DFT and FFT Operations[2]

Data Length N	Multiplications		Additions	
	DFT	FFT	DFT	FFT
4	16	4	12	8
8	64	12	56	24
16	256	32	240	64
32	1,024	80	992	160
64	4,096	192	4,032	384
128	16,384	448	16,256	896
256	65,536	1,024	65,280	2,048
512	262,144	2,304	261,632	4,608
1,024	1,048,576	5,102	1,047,552	10,240
2,048	4,194,304	11,264	4,192,256	22,528
4,096	16,777,216	24,576	16,773,120	49,152

*E*xample
17.20

Measure, as accurately as you can, and compare computation times needed to perform the DFT and FFT on a finite sequence.

Solution

Results from an experiment carried out on a personal computer running at 400-MHz speed under Windows 7 are shown below.

```
w1=gnorm(8192,1,0,1)%Generates 8192 random numbers, N(0,1).
w2=fft(w1)          %Takes FFT of w1 (about a split second).
w3=dft(w1)          %Takes DFT of w2 (about 13 seconds).
```

[2]M. T. Jong, *Methods of Discrete Signals and System Analysis* (New York: McGraw-Hill Book Company, 1982).

17.9 Linear Convolution from Circular

Two methods for obtaining linear convolution by using circular convolution are described below.

Method 1

One recipe for obtaining the linear convolution of $x(n)$ with $h(n)$ by way of their circular convolution is the following (also called "overlap and save"):

1. Choose a segment of $x(n)$ of finite length L. (For the FFT operation, L needs to be a power of two. However, the present method does not require that.)

2. Let $h(n)$ be N samples long. Pad it with $L - N$ zeros to make it L samples long. Call the padded sequence $g(n)$.

$$g(n) = h(n), \quad n = 0, 1, 2, \ldots, N - 1$$

$$g(n) = 0, \quad n = N, N + 1, \ldots, L - 1$$

3. Partition $x(n)$ into segments of length L such that the first $N - 1$ samples in a segment overlap the last $N - 1$ samples in the previous segment. Segments are then indexed by the superscript i to form the collection $x^i(n)$:

$$x^i(n), \quad n = 0, 1, 2, \ldots, (L - 1), \quad i = 0, 1, 2, \ldots, \infty$$

4. Apply circular convolution to obtain $y^i(n) = h(n) \otimes x^i(n)$. This may be done by calculation of the convolution sum in the n-domain or by multiplication of DFTs and the taking of the inverse transform of the results. The resulting output segment $y^i(n)$ has L samples. Discard the first $N - 1$ samples as they do not belong to $x(n){\star}h(n)$. Save the last $(L - N + 1)$ samples.

$$y^i(n) = \{\underbrace{y^i(0), y^i(1), y^i(2), \ldots, y^i(N - 2)}_{\text{discard}}, \underbrace{y^i(N - 1), y^i(N), \ldots, y^i(L - 1)}_{\text{save}}\}$$

The last $(L - N + 1)$ samples in $y^i(n)$ are placed serially in a sequence to form $y(n)$ (linear convolution). This is the contribution of the i-th segment in $x(n)$ to $y(n)$. Arrange these contribution in series to form $y(n)$.

Example
17.21

Find $y(n) = x(n) \star h(n)$ by (a) direct calculation of the linear convolution sum and (b) circular convolution applying the "overlap and save" method. The input and unit-sample functions are

$$h(n) = \{a_0, a_1, a_2\} \text{ and } x(n) = \{b_0, b_1, b_2, b_3, b_4, b_5, b_6, b_7, b_8, b_9, \cdots\}$$

Solution

a. The result of the linear convolution is

$$y(n) = \sum_{k=0}^{n} h(k)x(n-k)$$

$$y(0) = a_0 b_0$$

$$y(1) = a_0 b_1 + a_1 b_0$$

$$y(2) = a_0 b_2 + a_1 b_1 + a_2 b_0$$

$$y(3) = a_0 b_3 + a_1 b_2 + a_2 b_1$$

$$\vdots$$

$$y(i) = a_0 b_i + a_1 b_{i-1} + a_2 b_{i-2}$$

b. To obtain the above results by circular convolution, we first partition $x(n)$ into segments of length $L \geq 3$. In this example we want to perform circular convolution directly and not through FFT. Therefore, L does not have to be a power of 2. Let $L = 3$. Because $h(n)$ is also 3 samples long, no padding is necessary. We now partition $x(n)$ into segments each of length 3 with a 2-point overlap between two consecutive sequences.

Segment	Samples
0	$x^0(n) = \{0,\ 0,\ b_0\}$
1	$x^1(n) = \{0,\ b_0,\ b_1\}$
2	$x^2(n) = \{b_0,\ b_1,\ b_2\}$
3	$x^3(n) = \{b_1,\ b_2,\ b_3\}$
$\vdots$	$\vdots$
i	$x^i(n) = \{b_{i-2},\ b_{i-1},\ b_i\}$
$\vdots$	$\vdots$

We perform circular convolution of

$$h(n) \otimes x^i(n) = \sum_{k=0}^{N-1} h(k)x^i(n-k)$$

where $x^i(n-k)$ is found by time reversal and circular shift, starting with the first segment $x^0(n)$.

From segment 0 we get

$$\left. \begin{aligned} y^0(0) &= a_1 b_0 \\ y^0(1) &= a_2 b_0 \end{aligned} \right\} \text{throw away}$$

$$y^0(2) = a_0 b_0 \quad \} \text{ keep as } y(0).$$

From segment 1 we get

$$\left. \begin{aligned} y^1(0) &= a_1b_1 + a_2b_0 \\ y^1(1) &= a_0b_0 + a_2b_1 \end{aligned} \right\} \text{ throw away}$$

$$y^1(2) = a_0b_1 + a_1b_0 \quad \} \text{ keep as } y(1).$$

$\vdots$

From segment i we get

$$\left. \begin{aligned} y^i(0) &= \cdots\cdots\cdots \\ y^i(1) &= \cdots\cdots\cdots \end{aligned} \right\} \text{ throw away}$$

$$y^i(2) = a_0b_i + a_1b_{i-1} + a_2b_{i-2} \quad \} \text{keep as } y(i).$$

$\vdots$

The result is $y(n) = \{y^0(2), \ y^1(2), \ y^2(2), \ y^3(2), \ y^4(2), \ \ldots, \ y^i(2), \ \cdots\}$, where

$$y(0) = y^0(2) = a_0b_0$$
$$y(1) = y^1(2) = a_0b_1 + a_1b_0$$
$$y(2) = y^2(2) = a_0b_2 + a_1b_1 + a_2b_0$$
$$y(3) = y^3(2) = a_0b_3 + a_1b_2 + a_2b_1$$

$\vdots$

$$y(i) = y^i(2) = a_0b_i + a_1b_{i-1} + a_2b_{i-2}$$

$\vdots$

This is the same result obtained through linear convolution in part a.

Note: To make L a power of 2 (e.g., 4, 8, 16, 32, 64, ...), pad the shorter sequence $h(n)$ with the required number of zeros.

Method 2

Another recipe for obtaining linear convolution of $x(n)$ with $h(n)$ by way of their circular convolution is the following (called "overlap and add"):

1. Construct the $x^i(n)$ segments by taking $(L - N + 1)$ samples for each segment from the long sequence $x(n)$ and padding them with $(N - 1)$ zeros on the right side. You will then get a sequence of L samples, the last $N - 1$ of which are zeros.
2. Pad $h(n)$ with $(L - N)$ zeros on the right side to make it L samples long, the last $(L - N)$ of which are zeros.
3. Apply circular convolution to the two sequences.
4. Place the first $(L - N + 1)$ samples in the output.
5. The remaining $(N - 1)$ samples overlap the $(N - 1)$ samples of the subsequent segment and are added to it.

Example

17.22

Apply the "overlap and add" method and circular convolution to find the linear convolution $y(n) = x(n) \star h(n)$, where

$$h(n) = \{a_0, \ a_1, \ a_2\}$$
$$\text{and} \quad x(n) = \{b_0, \ b_1, \ b_2, \ b_3, \ b_4, \ b_5, \ b_6, \ b_7, \ b_8, \ b_9, \ \cdots\}$$

Solution

We have $N = 3$, $L = 3$, $L - N + 1 = 1$. To form the input segments take one sample from the long sequence and pad it with two zeros. Do not pad $h(n)$.

Segment	Samples
0	$x^0(n) = \{b_0, \ 0, \ 0\}$
1	$x^1(n) = \{b_1, \ 0, \ 0\}$
2	$x^2(n) = \{b_2, \ 0, \ 0\}$
3	$x^3(n) = \{b_3, \ 0, \ 0\}$
$\vdots$	$\vdots$
i	$x^i(n) = \{b_i, \ 0, \ 0\}$
$\cdots$	$\cdots$

Now perform the circular convolution $y^i(n) = h(n) \otimes x^i(n)$ starting with the first segment $x^0(n)$.

Segment 0 is $x^0(n) = \{b_0, 0, 0\}$, from which we get:

$y^0(0) = a_0 b_0$ (keep)
$y^0(1) = a_1 b_0$ (add to the next)
$y^0(2) = a_2 b_0$ (add to the element after *that*)

Segment 1 is $x^1(n) = \{b_1, 0, 0\}$, from which we get:

$y^1(0) = a_0 b_1$ (keep)
$y^1(1) = a_1 b_1$ (add to the next)
$y^1(2) = a_2 b_1$ (add to the element after *that*)

$\vdots$

Segment i is $x^i(n) = \{b_i, 0, 0\}$, from which we get:

$y^i(0) = a_0 b_i$ (keep)
$y^i(1) = a_1 b_i$ (add to the next)
$y^i(2) = a_2 b_i$ (add to the element after *that*)

$\vdots$

The result is $y(n) = \{y(0),\ y(1),\ y(2),\ y(3),\ y(4),\ \ldots,\ y(i),\ \cdots\}$, where

$$y(0) = y^0(0) = a_0 b_0$$
$$y(1) = y^0(1) + y^1(0) = a_1 b_0 + a_0 b_1$$
$$y(2) = y^0(2) + y^1(1) + y^2(0) = a_2 b_0 + a_1 b_1 + a_0 b_2$$
$$y(3) = y^1(2) + y^2(1) + y^3(0) = a_2 b_1 + a_1 b_2 + a_0 b_3$$
$$\vdots$$
$$y(i) = y^{i-2}(2) + y^{i-1}(1) + y^i(0) = a_2 b_{i-2} + a_1 b_{i-1} + a_0 b_i$$

This is the same result obtained through the method 1 and the direct linear convolution operation.

17.10   DFT in Matrix Form

The DFT relationship

$$X(k) = \sum_{n=0}^{N-1} x(n) e^{-j2\pi nk/N}, \quad k = 0, 1, 2, \ldots N - 1$$

may also be given in matrix form. Let $\mathbf{x}$ and $\mathbf{X}$ be vectors, both of size N, representing the signal in the time (n) and frequency (k) domains, respectively. Their relationship in matrix form is given by

$$\mathbf{X} = \mathbf{W_N} \times \mathbf{x}$$

The transformation matrix $\mathbf{W_N}$ is an $N \times N$ matrix whose elements are defined as follows:

$$w_{n,k} = e^{-j2\pi nk/N} = w^{nk}, \quad \text{where} \quad w = e^{-j2\pi/N}$$

$$\mathbf{W_N} = \begin{bmatrix} 1 & 1 & 1 & \ldots & 1 \\ 1 & w & w^2 & \ldots & w^{(N-1)} \\ \vdots & & & & \\ 1 & w^i & w^{2i} & \ldots & w^{(N-1)i} \\ \vdots & & & & \\ 1 & w^{N-1} & w^{2(N-1)} & \ldots & w^{(N-1)(N-1)} \end{bmatrix}$$

Note that w^{nk} is periodic with period N. The transformation matrix, therefore, contains a repetition within its structure, which can be used for an efficient computation algorithm. The following examples illustrate the transformation matrix and its application.

Example **17.23**

Construct $\mathbf{W_4}$ and use it to find the 4-point DFT of $x(n) = \{1, 1, 0, 0\}$.
$\uparrow$

Solution

The elements of the 4×4 DFT matrix are

$$w^{nk} = e^{-j\pi nk/2} = (-j)^{nk}$$

$$w^0 = w^4 = w^8 = 1$$

$$w^1 = w^5 = w^9 = -j$$

$$w^2 = w^6 = -1$$

$$w^3 = w^7 = j$$

See Figure 17.8(a). Note the periodicity. The transformation matrix is

$$\mathbf{W_4} = \begin{bmatrix} 1 & 1 & 1 & 1 \\ 1 & w & w^2 & w^3 \\ 1 & w^2 & w^4 & w^6 \\ 1 & w^3 & w^6 & w^9 \end{bmatrix} = \begin{bmatrix} 1 & 1 & 1 & 1 \\ 1 & -j & -1 & j \\ 1 & -1 & 1 & -1 \\ 1 & j & -1 & -j \end{bmatrix}$$

The DFT vector is

$$\mathbf{X(k)} = \begin{bmatrix} X(0) \\ X(1) \\ X(2) \\ X(3) \end{bmatrix} = \begin{bmatrix} 1 & 1 & 1 & 1 \\ 1 & -j & -1 & j \\ 1 & -1 & 1 & -1 \\ 1 & j & -1 & -j \end{bmatrix} \times \begin{bmatrix} 1 \\ 1 \\ 0 \\ 0 \end{bmatrix} = \begin{bmatrix} 2 \\ (1-j) \\ 0 \\ (1+j) \end{bmatrix}$$

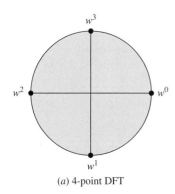

(a) 4-point DFT

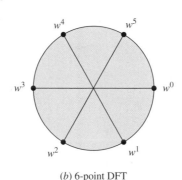

(b) 6-point DFT

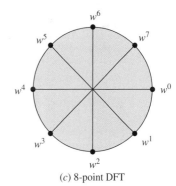

(c) 8-point DFT

FIGURE 17.8 Elements of DFT transformation.

Example **17.24**

Construct $\mathbf{W_6}$ and use it to find the DFT of the 6-point window

$$x(n) = \{0.31, 0.77, 1, 1, 0.77, 0.31\}$$
$\uparrow$

Solution

Let $w = e^{-j\pi/3} = (1 - j\sqrt{3})/2$. See Figure 17.8(b). The elements of the transformation matrix are

$$w^0 = w^6 = w^{12} = w^{18} = w^{24} = w^{30} = 1$$

$$w^1 = w^7 = w^{13} = w^{19} = w^{25} = w^{31} = e^{-j\pi/3} = \frac{1 - j\sqrt{3}}{2}$$

$$w^2 = w^8 = w^{14} = w^{20} = w^{26} = w^{32} = e^{-j2\pi/3} = \frac{-1 - j\sqrt{3}}{2}$$

$$w^3 = w^9 = w^{15} = w^{21} = w^{27} = w^{33} = e^{-j\pi} = -1$$

$$w^4 = w^{10} = w^{16} = w^{22} = w^{28} = w^{34} = e^{-j4\pi/3} = \frac{-1 + j\sqrt{3}}{2}$$

$$w^5 = w^{11} = w^{17} = w^{23} = w^{29} = w^{35} = e^{-j5\pi/3} = \frac{1 + j\sqrt{3}}{2}$$

The DFT vector is

$$\mathbf{X(k)} = \begin{bmatrix} 1 & 1 & 1 & 1 & 1 & 1 \\ 1 & w & w^2 & -1 & w^4 & w^5 \\ 1 & w^2 & w^4 & 1 & w^2 & w^4 \\ 1 & -1 & 1 & -1 & 1 & -1 \\ 1 & w^4 & w^2 & 1 & w^4 & w^2 \\ 1 & w^5 & w^4 & -1 & w^2 & w \end{bmatrix} \times \begin{bmatrix} 0.31 \\ 0.77 \\ 1 \\ 1 \\ 0.77 \\ 0.31 \end{bmatrix} = \begin{bmatrix} 4.16 \\ 1.195\angle -150° \\ 0.23\angle -120° \\ 0 \\ 0.23\angle 120° \\ 1.195\angle 150° \end{bmatrix}$$

*E*xample
17.25

Construct $\mathbf{W_8}$ and use it to find the DFT of the 8-point window

$$x(n) = \{0.146, 0.5, 0.853, 1, 1, 0.853, 0.5, 0.146\}$$
$$\uparrow$$

Solution

Let $w = e^{-j\pi/4} = \sqrt{2}(1 - j)/2$. See Figure 17.8(c). The elements of the transformation matrix are

$$w^0 = w^8 = w^{16} = w^{24} = w^{32} = w^{40} = w^{48} = 1$$

$$w^1 = w^9 = w^{17} = w^{25} = w^{33} = w^{41} = w^{49} = e^{-j\pi/4} = \frac{\sqrt{2}}{2}(1 - j)$$

$$w^2 = w^{10} = w^{18} = w^{26} = w^{34} = w^{42} = e^{-j\pi/2} = -j$$

$$w^3 = w^{11} = w^{19} = w^{27} = w^{35} = w^{43} = e^{-j3\pi/4} = -\frac{\sqrt{2}}{2}(1 + j)$$

$$w^4 = w^{12} = w^{20} = w^{28} = w^{36} = w^{44} = e^{-j\pi} = -1$$

$$w^5 = w^{13} = w^{21} = w^{29} = w^{37} = w^{45} = e^{-j5\pi/4} = \frac{\sqrt{2}}{2}(-1 + j)$$

$$w^6 = w^{14} = w^{22} = w^{30} = w^{38} = w^{46} = e^{-j3\pi/2} = j$$

$$w^7 = w^{15} = w^{23} = w^{31} = w^{39} = w^{47} = e^{-j7\pi/4} = \frac{\sqrt{2}}{2}(1 + j)$$

The DFT vector is

$$
\mathbf{X(k)} =
\begin{bmatrix}
1 & 1 & 1 & 1 & 1 & 1 & 1 & 1 \\
1 & w & -j & w^3 & -1 & w^5 & j & w^7 \\
1 & -j & -1 & j & 1 & -j & -1 & j \\
1 & w^3 & j & w & -1 & w^7 & -j & w^5 \\
1 & -1 & 1 & -1 & 1 & -1 & 1 & -1 \\
1 & w^5 & -j & w^7 & -1 & w & j & w^3 \\
1 & j & -1 & -j & 1 & j & -1 & -j \\
1 & w^7 & j & w^5 & -1 & w^3 & -j & w
\end{bmatrix}
\times
\begin{bmatrix}
0.146 \\
0.5 \\
0.853 \\
1 \\
1 \\
0.853 \\
0.5 \\
0.146
\end{bmatrix}
=
\begin{bmatrix}
5 \\
1.85\angle -157.5° \\
0.293\angle -135° \\
0 \\
0 \\
0 \\
0.293\angle 135° \\
1.85\angle 157.5°
\end{bmatrix}
$$

17.11 Conclusions: FS, FT, DTFT, and DFT

We have discussed Fourier analysis in four forms for four types of signals. These are listed in Table 17.4. The table also illustrates relationships between the signal and its various transforms.

TABLE 17.4 Signals in the Time and Frequency Domains, and Their Relationships

Time-Domain Signals	Transform
Continuous time, periodic signals	Fourier series (FS)
Continuous-time signals	Fourier transform (FT)
Discrete-time signals	Discrete-time Fourier transform (DTFT)
Finite sequences	Discrete Fourier transform (DFT)

Fourier Series (FS) $\Longleftrightarrow$ **Fourier Transform (FT)**

Signal

Discrete Fourier Transform (DFT) $\Longleftrightarrow$ **Discrete-Time Fourier Transform (DTFT)**

The transforms may be expressed as functions of angular frequency ω (in rad/s) or f (in Hz). These variables are not the same for the continuous- and discrete-time domains. To avoid confusion, especially when both domains are being discussed simultaneously, uppercase letters (F and Ω) are often used for continuous-time signals and lowercase letters (f and ω) are allocated to discrete-time signals. Throughout our discussions we almost always have used lowercase letters for both time domains, as the context was clear and didn't lend itself to confusion.

The exponential Fourier series expansion transforms periodic continuous-time signals $x(t)$, $-\infty < t < \infty$, into Fourier series with coefficients X_k, where k is an integer, $-\infty < k < \infty$. In the new domain, integers k represent harmonics or the discrete-frequency domain.

The Fourier transform converts nonperiodic, continuous-time signals $x(t)$, $-\infty < t < \infty$, into $X(F)$, where frequency F is a continuous variable, $-\infty < F < \infty$. Under Dirichlet conditions and for energy signals, the transform is a bounded function and exists. By admitting the singularity function $\delta(F)$ into the picture, the Fourier transform is generalized to include power and periodic signals. Under this generalization, the Fourier transform of a periodic signal consisting of a train of impulses in the F-domain placed at $F = kF_0$, $-\infty < k < \infty$, where k is an integer, $F_0 = 1/T$ is the principal frequency of the periodic function, and T is its period. The strength of the impulse at $F = kF_0$ is X_k, the exponential Fourier series coefficient at that harmonic. The frequency domain is continuous, but the spectrum is made of discrete lines at the harmonics. In summary, sampling in the frequency domain makes the time function periodic.

The DTFT transforms a discrete-time signal $x(n)$, $-\infty < n < \infty$, where n is an integer, into its frequency domain representation $X(\omega)$, $-\infty < \omega < \infty$, where ω is a continuous variable. $X(\omega)$ is, however, 2π-periodic [$X(f)$ is periodic with a period of 1 Hz]. Because of its periodicity, we only examine one period of $X(\omega)$, for example, for $-\pi < \omega < \pi$ [or $-0.5 < f < 0.5$ for $X(f)$]. From an analysis and computational point of view, the infinite-length discrete-time domain $-\infty < n < \infty$ is transformed into the finite-length continuous-frequency domain $-\pi < \omega < \pi$. Hence, one can say that sampling in the time domain makes representation in the frequency domain periodic.

The DFT transforms a finite sequence $x(n)$, $n = 0, 1, 2, \ldots, N - 1$, of length N into another finite sequence $X(k)$, $k = 0, 1, 2, \ldots, N - 1$, of the same length. The elements of the sequence $X(k)$ are samples of $X(\omega)$ [where $X(\omega)$ is the DTFT of $x(n)$, $\infty < n < \infty$] taken every $2\pi/N$ radians [i.e., N samples in one period of $X(\omega)$]. The DFT, therefore, performs sampling in the frequency domain. By sampling the DTFT, the DFT simplifies spectral analysis of finite data sequences. Moreover, evaluation of the DFT involves multiplications and additions only, which can be done efficiently both in real-time operations and in simulations. Real-time DFT is performed efficiently by DSP processors, with architectures specifically designed for such operations. The computationally efficient algorithm for evaluating the DFT and its inverse is the fast Fourier transform (FFT and IFFT), which is also used in off-line signal processing applications.

17.12 Problems

Solved Problems

1. Find the following DFTs:

   a. 7-point DFT of $x_1(n) = \{\underset{\uparrow}{1}, 1, 1, 0, 0, 0, 0\}$

   b. 8-point DFT of $x_2(n) = \{\underset{\uparrow}{1}, 1, 1, 0, 0, 0, 0, 0\}$

   c. 9-point DFT of $x_3(n) = \{\underset{\uparrow}{1}, 1, 1, 0, 0, 0, 0, 0, 0\}$

   d. N-point DFT of $x_4(n) = \{\underset{\uparrow}{1}, 1, 1\}$

Solution

a. $X_1(k) = \displaystyle\sum_{n=0}^{2} e^{-j2\pi nk/7} = 1 + e^{-j2\pi k/7} + e^{-j4\pi k/7} = \left(1 + 2\cos\frac{2\pi k}{7}\right) e^{-j2\pi k/7}, \qquad k = 0, 1, \ldots, 6$

b. $X_2(k) = \displaystyle\sum_{n=0}^{2} e^{-j2\pi nk/8} = 1 + e^{-j\pi k/4} + e^{-j\pi k/2} = \left(1 + 2\cos\frac{\pi k}{4}\right) e^{-j\pi k/4}, \qquad k = 0, 1, \ldots, 7$

c. $X_3(k) = \displaystyle\sum_{n=0}^{2} e^{-j2\pi nk/9} = 1 + e^{-j2\pi k/9} + e^{-j4\pi k/9} = \left(1 + 2\cos\frac{2\pi k}{9}\right) e^{-j2\pi k/9}, \qquad k = 0, 1, \ldots, 8$

d. $X_4(k) = \displaystyle\sum_{n=0}^{2} e^{-j2\pi nk/N} = 1 + e^{-j2\pi k/N} + e^{-j4\pi k/N} = \left(1 + 2\cos\frac{2\pi k}{N}\right) e^{-j2\pi k/N}, \qquad k = 0, 1, \ldots, N-1$

2. Find the N-point DFT of $x(n) = \{\underset{\uparrow}{1}, 1, 1, 1\}$.

Solution

$$X(k) = \sum_{n=0}^{3} e^{-j2\pi nk/N} = 1 + e^{-j2\pi k/N} + e^{-j4\pi k/N} + e^{-j6\pi k/N}$$

$$= 2\left(\cos\frac{\pi k}{N} + \cos\frac{3\pi k}{N}\right) e^{-j3\pi k/N} = 4\cos\left(\frac{\pi k}{N}\right)\cos\left(\frac{2\pi k}{N}\right) e^{-j3\pi k/N}, \ \ k = 0, 1, 2, \ldots, N-1$$

3. Find DFTs of the following triangular windows with even symmetry.

   a. 7-point DFT of $x_1(n) = \left\{\underset{\uparrow}{0}, \ 0.5, \ 1.5, \ 2, \ 1.5, \ 0.5, \ 0\right\}$

   b. 9-point DFT of $x_2(n) = \left\{\underset{\uparrow}{0}, \ 0.5, \ 1, \ 1.5, \ 2, \ 1.5, \ 1, \ 0.5, \ 0\right\}$

Solution

a. $X_1(k) = \displaystyle\sum_{n=0}^{6} x(n) e^{-j2\pi nk/7}$

$$= 0.5 e^{-j2\pi k/7} + 1.5 e^{-j4\pi k/7} + 2 e^{-j6\pi k/7} + 1.5 e^{-j8\pi k/7} + 0.5 e^{-j10\pi k/7}$$

$$= e^{-j6\pi k/7}(0.5 e^{j4\pi k/7} + 1.5 e^{j2\pi k/7} + 2 + 1.5 e^{-j2\pi k/7} + 0.5 e^{-j4\pi k/7})$$

$$= \left(2 + 3\cos\frac{2\pi k}{7} + \cos\frac{4\pi k}{7}\right) e^{-j6\pi k/7}$$

b. $X_2(k) = \sum_{n=0}^{8} x(n)e^{-j2\pi nk/9}$

$= 0.5e^{-j2\pi k/9} + e^{-j4\pi k/9} + 1.5e^{-j6\pi k/9} + 2e^{-j8\pi k/9} + 1.5e^{-j10\pi k/9} + e^{-j12\pi k/9} + 0.5e^{-j14\pi k/9}$

$= e^{-j8\pi k/9} \left(0.5e^{j6\pi k/9} + e^{j4\pi k/9} + 1.5e^{j2\pi k/9} + 2 + 1.5e^{-j2\pi k/9} + e^{-j4\pi k/9} + 0.5e^{-j6\pi k/9} \right)$

$= \left(2 + 3\cos\frac{2\pi k}{9} + 2\cos\frac{4\pi k}{9} + \cos\frac{6\pi k}{9} \right) e^{-j8\pi k/9}$

4. Find DFTs of the following windows with even symmetry.

    a. 8-point DFT of $x_1(n) = \left\{ 0,\ 0.5,\ 1.5,\ 2,\ 2,\ 1.5,\ 0.5,\ 0 \right\}$

    b. 10-point DFT of $x_2(n) = \left\{ 0,\ 0.5,\ 1,\ 1.5,\ 2,\ 2,\ 1.5,\ 1,\ 0.5,\ 0 \right\}$

Solution

    a. $X_1(k) = \sum_{n=0}^{7} x(n)e^{-j2\pi nk/8}$

$= 0.5e^{-j\pi k/4} + 1.5e^{-j2\pi k/4} + 2e^{-j3\pi k/4} + 2e^{-j4\pi k/4} + 1.5e^{-j5\pi k/4} + 0.5e^{-j6\pi k/4}$

$= e^{-j7\pi k/8} \left(0.5e^{j5\pi k/8} + 1.5e^{j3\pi k/8} + 2e^{j\pi k/8} + 2e^{-j\pi k/8} + 1.5e^{-j3\pi k/8} + 0.5e^{-j5\pi k/8} \right)$

$= \left(4\cos\frac{\pi k}{8} + 3\cos\frac{3\pi k}{8} + \cos\frac{5\pi k}{8} \right) e^{-j7\pi k/8}$

    b. $X_2(k) = \sum_{n=0}^{9} x(n)e^{-j2\pi nk/10}$

$= 0.5e^{-j\pi k/5} + e^{-j2\pi k/5} + 1.5e^{-j3\pi k/5} + 2e^{-j4\pi k/5} + 2e^{-j5\pi k/5} + 1.5e^{-j6\pi k/5} + e^{-j7\pi k/5} + 0.5e^{-j8\pi k/5}$

$= e^{-j9\pi k/10} \left(0.5e^{j7\pi k/10} + e^{j5\pi k/10} + 1.5e^{j3\pi k/10} + 2e^{j\pi k/10} + 2e^{-j\pi k/10} + 1.5e^{-j3\pi k/10} + e^{-j5\pi k/10} + 0.5e^{-j7\pi k/10} \right)$

$= \left(4\cos\frac{\pi k}{10} + 3\cos\frac{3\pi k}{10} + 2\cos\frac{5\pi k}{10} + \cos\frac{7\pi k}{10} \right) e^{-j9\pi k/10}$

5. Find the 128-point DFT of

    a. $x_1(n) = \begin{cases} 1,\ 0 \le n \le 15 \\ 0,\ 16 \le n \le 127 \end{cases}$

    b. $x_2(n) = \begin{cases} 1,\ \ 0 \le n \le 16 \\ 0,\ \ 17 \le n \le 127 \end{cases}$

Solution

    a. $X_1(k) = \sum_{n=0}^{15} e^{-j2\pi nk/128} = \dfrac{1 - e^{-j\pi k/4}}{1 - e^{-j\pi k/64}} = \dfrac{\sin(\pi k/8)}{\sin(\pi k/128)} e^{-j15\pi k/128}$

    b. $X_2(k) = \sum_{n=0}^{16} e^{-j2\pi nk/128} = \dfrac{1 - e^{-j\pi 17k/64}}{1 - e^{-j\pi k/64}} = \dfrac{\sin(17\pi k/128)}{\sin(\pi k/128)} e^{-j\pi k/8}$

6. Find the N-point DFT of the following $x(n)$ for $N = 60,\ 120,$ and 240.

$$x(n) = \begin{cases} 1, & 0 \le n \le 29 \\ 0, & 30 \le n \le N-1 \end{cases}$$

Solution

$$X(k) = \sum_{n=0}^{29} e^{-j2\pi nk/N} = \frac{1 - e^{-j60\pi k/N}}{1 - e^{-j2\pi k/N}} = \frac{\sin(30\pi k/N)}{\sin(\pi k/N)} e^{-j29\pi k/N}$$

7. Find the 3-point DFT of $h(n) = a^{-n},\ 0 \le n \le 2$, for $a = 1.369$.

Solution

$$H(k) = \sum_{n=0}^{2} 1.369^{-n} e^{-j2\pi kn/3}, \quad k = 0, 1, 2$$

$$H(0) = \sum_{n=0}^{2} 1.369^{-n} = 1 + 0.73 + 0.533 = 2.263$$

$$H(1) = \sum_{n=0}^{2} 1.369^{-n} e^{-j2\pi n/3} = 1 + 0.73 e^{-j2\pi/3} + 0.533 e^{-j4\pi/3} = 0.369 - j0.17 = 0.406\angle-24.7°$$

$$H(2) = \sum_{n=0}^{2} 1.369^{-n} e^{-j4\pi n/3} = 1 + 0.73 e^{-j4\pi/3} + 0.533 e^{-j8\pi/3} = 0.369 + j0.17 = 0.406\angle24.7°$$

Note that $H(2) = H^*(1)$.

8. Find the N-point DFT of $h(n) = e^{-\frac{n}{5}},\ 0 \le n \le N$, for

a. $N = 3$
b. $N = 30$

Solution

a. For $N = 3$,

$$H(k) = \sum_{n=0}^{2} e^{-\frac{n}{5}} e^{-j2\pi nk/3}$$

$$H(0) = \sum_{n=0}^{2} e^{-\frac{n}{5}} = 1 + 0.8187 + 0.6703 = 2.489$$

$$H(1) = \sum_{n=0}^{2} e^{-\frac{n}{5}} e^{-j2\pi n/3} = 1 + 0.8187 e^{-j2\pi/3} + 0.6703 e^{-j4\pi/3} = 0.2555 - j0.1285 = 0.286\angle-26.7°$$

$$H(2) = \sum_{n=0}^{2} e^{-\frac{n}{5}} e^{-j4\pi n/3} = 1 + 0.8187 e^{-j4\pi/3} + 0.6703 e^{-j8\pi/3} = 0.2555 + j0.1285 = 0.286\angle26.7°$$

Note that $H(1) = H^*(2)$.

b. For $0 \leq n \leq 29$ we use the approach taken in Example 17.6. Here, $h(n) = e^{-\frac{n}{5}} = a^{-n}$ with $a = e^{\frac{1}{5}} = 1.2214$.
Hence,

$$H(k) = \sum_{n=0}^{29} a^{-n} e^{-j2\pi nk/30} = \frac{1 - a^{-30}}{1 - a^{-1} e^{-j2\pi k/30}} = \frac{0.9975}{1 - (0.8187)e^{-j2\pi k/30}}$$

9. The following Matlab code computes the 3-point DFT of $h(n) = \{0, \underset{\uparrow}{1}, 1\}$ and displays $h(n)$ and $|H(k)|$,
$k = 0, 1, 2$.

```
h=[0,1,1]; H=fft(h)
N=length(h); n=0:N-1; M=max(H);
figure(1); stem(n,h); axis([-1 N -.2 1.2]); grid
figure(2); stem(n,abs(H)); axis([-1 N -.2 M+.2]); grid
```

10. The Matlab codes in this problem compute the N-point DFTs of rectangular and exponential functions that are
M-samples long. The functions and their transforms are then displayed in the time and frequency domains. Run the
programs and explore the results for various values of M and N.

Rectangular Pulse

```
M=6; N=100;
h=[ones(1,M) zeros(1, N-M)];
H1=fft(h); H2=fftshift(H1); n=0:N-1;
figure(1); stem(n,h);        axis([-1 N -.2 1.2]);  grid
figure(2); stem(n,abs(H1)); axis([-1 N -.2 M+.2]); grid
figure(3); stem(n,abs(H2)); axis([-1 N -.2 M+.2]); grid
```

Alternatively, use the following commands to generate h and $H1$. Then mask figure (1).

```
h=ones(1,M); H1=fft(h,N)
```

Exponential Function

```
M=16; N=16; m=0:M-1; n=0:N-1; a=2; %a=1+j;
h=[a.^(-m/10) zeros(1,N-M)]; H=fft(h);
figure(1); stem(n,abs(h)); axis([0 N -.2 1.2]); grid
figure(2); stem(n,abs(H)); axis([0 N -.2 abs(max(H))+.2]); grid
```

11. The following Matlab code obtains the IDFT of the 16-point $H(k)$ in Example 17.9 by three methods. It then pads
the $h(n)$ by 30 zeros and takes its 46-point DFT. See Figure 17.9.

```
clear
H=[1 -1 1 -0.5 zeros(1,9) -0.5 1 -1];
% Method 1. Inverse of H by ifft command.
h=ifft(H);
% Method 2. Inverse of H by summation.
h1=zeros(1,16);
for n=1:16
  hh=0;
```

```
  for k=1:16
    hh=hh+H(k)*exp(i*pi*(k-1)*(n-1)/8);
  end
  h1(n)=hh/16;
end
% Method 3. Inverse of H by analytic expression.
n=0:15; h2=(1-2*cos(pi*n/8)+2*cos(pi*n/4)-cos(3*pi*n/8))/16;
h3=sin(pi*(n-8)/2)./(pi*(n-8)); h3(9)=1/2;
% Computing differences
e1=h1-h; e2=h2-h; e3=h3-h;
% Padding h for a better frequency resolution.
hp=[h zeros(1,30)]; k=0:45; HP=fft(hp);
% Plots
figure(1); stem(n,real(h));   axis([0 16 -.2 .5]);   grid
figure(2); stem(n,H);         axis([0 16 -1.2 1.2]); grid
figure(3); stem(k,real(HP));  axis([0 46 -1.2 1.2]); grid
figure(4); stem(k,abs(HP));   axis([0 46 -.2 1.2]);  grid
figure(5); stem(n,h3);        axis([0 16 -.2 1.2]);  grid
```

a. Run the program and compare the results obtained by the three methods.

b. Change H in the second line of the program to

```
H=[1, 1, 1, 0.5, zeros(1,9), 0.5, 1, 1]
```

and run the program. Describe the results and explain the reason for choosing the initial vector for H.

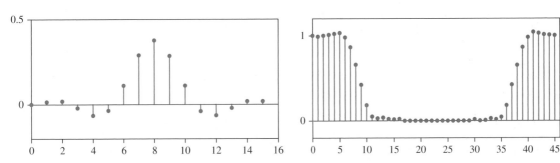

FIGURE 17.9 A 16-tap low-pass filter $h(n)$ (left) and the magnitude of $H(k)$ the 46-point DFT of padded $h(n)$ (right). Padding with zeros is done to increase the frequency resolution. See Example 17.9 and problem 11.

12. Parseval's theorem. Given $x(n) = \{0, \underset{\uparrow}{1}, 1\}$ and $y(n) = \{0, \underset{\uparrow}{j}, -j\}$, verfiy that

$$\sum_{n=0}^{2} x(n)y(n) = \frac{1}{3}\sum_{k=0}^{2} X(k)Y(-k) \qquad \text{and} \qquad \sum_{n=0}^{2} x(n)y^*(n) = \frac{1}{3}\sum_{k=0}^{2} X(k)Y^*(k)$$

Solution

From Example 17.3 we have $X(k) = 2\cos\left(\frac{\pi k}{3}\right)e^{-j\pi k}$ and $Y(k) = -2\sin\left(\frac{\pi k}{3}\right)e^{-j\pi k}$, $k = 0, 1, 2$.

$$x(n)y(n) = \{0, j, -j\} \qquad \Longrightarrow \qquad \sum_{n=0}^{2} x(n)y(n) = 0$$

$$x(n)y^*(n) = \{0, -j, j\} \qquad \Longrightarrow \qquad \sum_{n=0}^{2} x(n)y^*(n) = 0$$

$$X(k)Y(-k) = 2\sin\left(\frac{2\pi k}{3}\right) \qquad \Longrightarrow \qquad \frac{1}{3}\sum_{k=0}^{2} X(k)Y(-k) = \frac{1}{3}(0 + \sqrt{3} - \sqrt{3}) = 0$$

$$X(k)Y^*(k) = -2\sin\left(\frac{2\pi k}{3}\right) \qquad \Longrightarrow \qquad \frac{1}{3}\sum_{k=0}^{2} X(k)Y(-k) = \frac{1}{3}(0 - \sqrt{3} + \sqrt{3}) = 0$$

13. FFT algorithm. Consider the finite sequence $x(n) = \{1, 1, 0, 0\}$.

    a. Find $X(k)$, the 4-point DFT of $x(n)$.

    b. Show how an FFT algorithm would evaluate the DFT.

    c. Apply the algorithm to $x(n)$ and obtain the same results as in part a.

Solution

    a. $X(k) = \sum_{n=0}^{3} x(n)e^{-j\pi nk/2} = 1 + e^{-j\pi k/2}$, $k = 0, 1, 2, 3$, $X(k) = \{2, (1 - j), 0, (1 + j)\}$

    b. Split $x(n)$ of length N into two sequences $a(n) = x(2n)$ and $b(n) = x(2n + 1)$, each of length $N/2$. Let

$$A(k) = \mathcal{DFT}[a(n)]$$
$$B(k) = \mathcal{DFT}[b(n)]$$

Then

$$X(k) = A(k) + e^{-j2\pi k/N}B(k) \quad k = 0, 1, \ldots, \frac{N}{2} - 1$$

$$X(N/2 + k) = A(k) - e^{-j2\pi k/N}B(k), \quad k = 0, 1, \ldots, \frac{N}{2} - 1$$

In this problem $N = 4$, $X(k) = A(k) + e^{-j\pi k/2}B(k)$, $k = 0, 1$ and $X(2 + k) = A(k) - e^{-j\pi k/2}B(k)$, $k = 0, 1$

$$
\begin{array}{ll}
a(n) = \{1, 0\} & A(k) = \mathcal{DFT}\{1, 0\} = \{1, 1\} \\
b(n) = \{1, 0\} & B(k) = \mathcal{DFT}\{1, 0\} = \{1, 1\} \\
k = 0, & X_0 = 1 + e^0 = 2 \\
k = 1, & X_1 = 1 + e^{-j\pi/2} = 1 - j \\
k = 2, & X_2 = 1 - e^{-j\pi} = 0 \\
k = 3, & X_3 = 1 - e^{-j3\pi/2} = 1 + j
\end{array}
$$

Chapter Problems

14. Find the 4-point DFTs of the sequences given below.

    a. $x_1(n) = \{1, 0, 0, 0\}$ b. $x_2(n) = \{1, 1, 0, 0\}$

    c. $x_3(n) = \{1, 1, 1, 0\}$ d. $x_4(n) = \{1, 1, 1, 1\}$

15. Find the 5-point DFTs of the sequences given below.

  a. $x_1(n) = \{1, 1, 0, 0, 0\}$ b. $x_2(n) = \{1, 1, 1, 0, 0\}$

  c. $x_3(n) = \{1, 1, 1, 1, 0\}$ d. $x_4(n) = \{1, 1, 1, 1, 1\}$

16. Find the 9-point DFTs of the sequences given below.

  a. $x_1(n) = \{1, 2, 1, 0, 0, 0, 0, 0, 0\}$ b. $x_2(n) = \{1, 2, 3, 2, 1, 0, 0, 0, 0\}$

  c. $x_3(n) = \{1, 2, 3, 4, 3, 2, 1, 0, 0\}$ d. $x_3(n) = \{1, 2, 3, 4, 5, 4, 3, 2, 1\}$

17. From the definition find the 5-point DFT of $x_e(n)$, $x_o(n)$ and $x(n)$ given below. In each case observe that (i) $x(n) = x_e(n) + x_o(n)$, (ii) $X_e(k)$ is real-valued and $X_o(k)$ is imaginary. Find the relation between $X(k)$, $X_e(k)$, $X_o(k)$.

  a. $x_e(n) = \{1, 5, -2, -2, 5\}$ $x_o(n) = \{0, 5, -2, 2, -5\}$ $x(n) = \{1, 10, -4, 0, 0\}$

  b. $x_e(n) = \{2, 7, 7, 7, 7\}$ $x_o(n) = \{0, -3, -1, 1, 3\}$ $x(n) = \{2, 4, 6, 8, 10\}$

  c. $x_e(n) = \{2, 7, 7, 7, 7\}$ $x_o(n) = \{0, 3, 1, -1, -3\}$ $x(n) = \{2, 10, 8, 6, 4\}$

18. For each $x(n)$ given below do the following:

  a. Fnd $X(k) = \mathcal{DFT}\{x(n)\}$.
  b. Take the inverse of its real part and examine it to see if the inverse is equal to $x_e(n) = \frac{x(n)+x(-n)}{2}$.
  c. Take the inverse of its imaginary part and examine it to see if the inverse is equal to $x_o(n) = \frac{x(n)-x(-n)}{2}$.

  a. $x(n) = \{1, 1, 1, 1\}$

  b. $x(n) = \{2, 3, 4, 3, 2\}$

  c. $x(n) = \{1, 2, 3, 4, 3, 2, 1\}$

19. An LTI system is specified by the unit-sample response $h(n)$ and an input $x(n)$.

  a. Find the DFTs of $h(n) = \{1, -1, 1, 0\}$ and $x(n) = \{1, 2, 2, 1\}$.
  b. Let $Y(k) = X(k) \times H(k)$. Find $y(n)$ the IDFT of $Y(k)$.
  c. Note that $y(n)$ is the circular convolution of $x(n)$ and $h(n)$. Discuss and show why it is not the output of the LTI system.
  d. Show that padding with a single zero and taking 5-point DFT/IDFT of the functions involved doesn't provide the output. Determine the minimum padding for which the result of the circular convolution becomes equal to linear convolution.

20. a. Find $X_a(k)$ and $X_b(k)$, the DFTs of $x_a(n) = \{1, 1, 0, 0\}$ and $x_b(n) = \{1, 0, 0, 1\}$.

  b. Find the relationship between the magnitudes of $X_a(k)$ and $X_b(k)$ and discuss the reason behind it.

21. a. Find DTFT of $x(n) = \{\cdots 0, 0, 1, 2, 1, 0, 0\cdots\}$. Call it $X(\omega)$.

  b. Find 3-point DFT of $\{1, 2, 1\}$. Call it $X(k) = \{X_0, X_1, X_2\}$.

  c. Explain and verify the relationship between $x(\omega)$ and $X(k)$.

22. The unit-sample response of a filter is $h(n) = \{\cdots 0, 0, 1, 1, 1, 1, 0, 0\cdots\}$.

  a. Find the system's frequency response $H(\omega) = \mathcal{DTFT}\{h(n)\} = H_r(\omega)e^{j\Theta(\omega)}$.
  b. Sketch $H_r(\omega)$ and $\Theta(\omega)$ for $-\pi < \omega < \pi$.

c. Let $H(k)$ be the 4-point DFT of the finite sequence $\{1, 1, 1, 1\}$. It is expected that $H(k) = H(\omega)\big|_{\omega=\pi k} = [4, 0, 0, 0]$. Obtain $H(k)$ by applying the DFT definition and verify that it is in agreement with samples of $H(\omega)$.

d. Let the input to the filter be $x(n) = \cos(\frac{\pi n}{3})$. Find the system's steady-state response in the form of $y(n) = A\cos(\omega_0 n + \Theta)$.

23. Find the N-point DFT of $\{1, 1\}$ for $N = 6,\ 8,\ 10,$ and 64. Plot their magnitude and discuss the effect of increasing N.

24. Obtain a closed-form expression for the N-point DFT of

$$x(n) = \begin{cases} a^n, & 0 \le n \le (M-1) \\ 0, & M \le n \le (N-1) \end{cases}$$

Plot the magnitude of the DFT for the following parameter values:

a. $N = M = 2, 3, 4, 10, 100$
b. $N = 100$ and $M = 2, 3, 4, 10$

25. Obtain a closed-form expression for the DFT of

$$x(n) = \begin{cases} \cos(2\pi n/3), & 0 \le n \le 3 \\ 0, & 4 \le n \le (N-1) \end{cases}$$

Plot $x(n)$ and the magnitude of the DFT for $N = 4,\ 6,\ 8$. Discuss the reason behind possible differences between the DFTs in this problem and Example 17.7a.

26. Obtain a closed-form expression for the DFT of

$$x(n) = \begin{cases} \cos(2\pi n/3), & 0 \le n \le (M-1) \\ 0, & M \le n \le (N-1) \end{cases}$$

Plot $x(n)$ and the magnitude of the DFT for the following parameter values:

a. $N = M = 2, 5, 8, 11$
b. $N = 101$ and $M = 2, 5, 8, 11$.

27. Low-pass filter. Obtain a closed-form expression for the DFT of

$$h(n) = \begin{cases} \sin\left[\frac{\pi}{2}(n-3)\right]/\pi(n-3), & 0 \le n \le 6 \\ 0, & 7 \le n \le (N-1) \end{cases}$$

Plot $x(n)$ and the magnitude of the DFT for the following:

a. $N = 7$
b. $N = 120$

Hint: Examine the following Matlab code and use it to plot the DFT magnitudes.

```
M=7; m=1:M; omega=pi/2;
p=sin(omega*(m -(M+1)/2))./(pi*(m-(M+1)/2)); p((M+1)/2)=omega/pi;
N=120; n=0:N-1;
h=[p zeros(1,N-M)];
H=fft(h);
figure(1); stem(n,h); axis([-1 M+1 -.2 1.2]); grid
figure(2); stem(n,abs(H)); axis([-1 N -.2 max(abs(H))+.2]); grid
```

28. Replace the first two lines in the Matlab code of problem 27 by the following two lines and run it.

```
M=6; m=1:M; omega=pi/2;
p=sin(omega*(m -M/2))./(pi*(m-M/2)); p(M/2)=omega/pi;
```

Comment on the plots.

29. Use the Matlab code given in problem 27 to plot the magnitude of the DFT of $h(n)$ in that problem for the following:

a. $M = 11$ and $\omega_0 = \dfrac{\pi}{4}$ b. $M = 11$ and $\omega_0 = \dfrac{3\pi}{4}$

c. $M = 21$ and $\omega_0 = \dfrac{\pi}{4}$ d. $M = 21$ and $\omega_0 = \dfrac{3\pi}{4}$

e. $M = 31$ and $\omega_0 = \dfrac{\pi}{4}$ f. $M = 31$ and $\omega_0 = \dfrac{3\pi}{4}$

30. Bandpass filter. Obtain the DFT of

$$h(n) = \begin{cases} \dfrac{\sin[\omega_2(n-3)]}{\pi(n-3)} - \dfrac{\sin[\omega_1(n-3)]}{\pi(n-3)}, & 0 \le n \le 6 \\ 0, & 7 \le n \le (N-1) \end{cases}$$

where $\omega_1 = 5\pi/12$ and $\omega_2 = 7\pi/12$. Plot $x(n)$ and the magnitude of its DFT for

a. $N = 7$
b. $N = 120$

Hint: Examine the following Matlab code and use it to plot the DFT magnitudes.

```
M=7; m=1:M; omega0=pi/2; domega=pi/6;
omega1=omega0-domega/2; omega2=omega0+domega/2;
p1=sin(omega1*(m -(M+1)/2))./(pi*(m-(M+1)/2)); p1((M+1)/2)=omega1/pi;
p2=sin(omega2*(m -(M+1)/2))./(pi*(m-(M+1)/2)); p2((M+1)/2)=omega2/pi;
p=p2-p1;
N=120; n=0:N-1;
h=[p zeros(1,N-M)];
H=fft(h);
figure(1); stem(n,h); axis([-1 M+1 -.2 .2]); grid
figure(2); stem(n,abs(H)); axis([-1 N -.2 max(abs(H))+.2]); grid
```

31. a. Replace the first line in the Matlab code of problem 30 by the following line and run it. Comment on the changes you see in the plots.

```
M=21; m=1:M; omega0=pi/2; domega=pi/6;
```

b. Repeat for $M = 49$.

32. a. Replace the first line in the Matlab code of problem 30 by the following line and run it. Comment on the changes you see in the plots.

```
M=21; m=1:M; omega0=pi/4; domega=pi/6;
```

b. Repeat for $M = 49$.

33. a. Replace the first line in the Matlab code of problem 30 by the following line and run it. Comment on the changes you see in the plots.

```
M=21; m=1:M; omega0=3pi/4; domega=pi/6;
```

b. Repeat for $M = 49$.

34. This problem designs a linear-phase bandpass FIR filter with nine taps using rectangular window method. The desired frequency response within the range $-\pi < \omega < \pi$ is

$$H_d(\omega) = \begin{cases} 1 & \pi/5 \leq |\omega| \leq \pi/3 \\ 0 & \text{elsewhere} \end{cases}$$

For simplicity we represent the unit-sample response in the noncausal form shown below.

$$h(n) = \{h_4, \ h_3, \ h_2, \ h_1, \ \underset{\uparrow}{h_0}, \ h_1, \ h_2, \ h_3, \ h_4\}$$

Follow steps described below.

a. Find $h_a(n)$ of a linear-phase 9-tap low-pass FIR filter with cutoff frequency at $\omega_a = \pi/5$.
b. Find $h_b(n)$ of a linear-phase 9-tap low-pass FIR filter with cutoff frequency at $\omega_b = \pi/3$.
c. Using $h_a(n)$ and $h_b(n)$ find $h(n)$.
d. Obtain and plot DFT of $h(n)$. Compare with desired frequency response.
e. Increase the number of taps and observe how it brings the DFT of $h(n)$ closer to the desired frequency response except at the cutoff frequency. Explain the reason and suggest a solution.

35. Using the approach of problems 28 and 31 write a Matlab code to design M-tap high-pass digital filters with cutoff frequency at ω_0. Find $h(n)$ and plot $|H(k)|$ for following sets of parameters:

a. $M = 11$ and $\omega_0 = \dfrac{\pi}{4}$ b. $M = 11$ and $\omega_0 = \dfrac{\pi}{2}$ c. $M = 11$ and $\omega_0 = \dfrac{3\pi}{4}$

d. $M = 21$ and $\omega_0 = \dfrac{\pi}{4}$ e. $M = 21$ and $\omega_0 = \dfrac{\pi}{2}$ f. $M = 21$ and $\omega_0 = \dfrac{3\pi}{4}$

g. $M = 31$ and $\omega_0 = \dfrac{\pi}{4}$ h. $M = 31$ and $\omega_0 = \dfrac{\pi}{2}$ i. $M = 31$ and $\omega_0 = \dfrac{3\pi}{4}$

36. Obtain the N-point DFTs of the following three functions defined for $n = 0, 1, \ldots, N - 1$.

a. $x_1(n) = \cos\left(\dfrac{2\pi n}{N}\right)$

b. $x_2(n) = \sin\left(\dfrac{2\pi n}{N}\right)$

c. $x_3(n) = e^{j2\pi n/N}$

37. **Windows.** Obtain the N-point DFTs of the following windows: rectangular, Bartlett, Hanning, Hamming, and Blackman. All windows are defined for $0 \leq n \leq N - 1$. Window functions are found in section 11.17 of Chapter 11.

38. Obtain a closed-form expression for the N-point DFT of

$$h(n) = \rho^n \cos(\omega_0 n), \ 0 \leq n \leq (N - 1)$$

Plot $x(n)$ and the magnitude of the DFT for the following parameter values:

a. $\rho = 0.9, \ \omega_0 = \pi/4$ b. $\rho = 0.95, \ \omega_0 = \pi/4$
c. $\rho = 0.9, \ \omega_0 = \pi/2$ d. $\rho = 0.95, \ \omega_0 = \pi/2$
e. $\rho = 0.9, \ \omega_0 = 3\pi/4$ f. $\rho = 0.95, \ \omega_0 = 3\pi/4$

39. Circular convolution. Obtain the circular convolution of $h(n) = \{\underset{\uparrow}{1}, 1, 1, 1\}$ with itself.

40. Circular convolution. Obtain the circular convolution between $h(n) = \{\underset{\uparrow}{1}, 1, 1, 1, 1\}$ and $x(n) = \{\underset{\uparrow}{1}, -1, 1, -1, 1\}$.

41. Circular convolution. Obtain the circular convolution between $h(n) = 0.8^n$, $n = 0, 1, \ldots 7$ and $x(n) = \{\underset{\uparrow}{1}, -1\}$ padded by 6 zeros.

42. Linear convolution from circular. Apply the "overlap and save" method and circular convolution to find the linear convolution $y(n) = x(n) \star h(n)$, where

$$h(n) = \{a_0, a_1, a_2, a_3\}$$

and $\quad x(n) = \{b_0, b_1, b_2, b_3, b_4, b_5, b_6, b_7, b_8, b_9, \ldots\}$

43. Linear convolution from circular. Apply the "overlap and add" method and circular convolution to find the linear convolution $y(n) = x(n) \star h(n)$, where

$$h(n) = \{a_0, a_1, a_2, a_3\}$$

and $\quad x(n) = \{b_0, b_1, b_2, b_3, b_4, b_5, b_6, b_7, b_8, b_9, \ldots\}$

44. FFT algorithm. Repeat problem 13 for the finite-length sequence $x_n = \{\underset{\uparrow}{1}, 2, 3, 4\}$.

17.13 Project: DFT Spectral Analysis

Summary

In this project you will use a special-purpose instrument called a spectrum analyzer to obtain and measure the spectrum of signals in real time. Measurement is done on a segment of the waveform observed and recorded through a window. Three types of windows are used: uniform (rectangular), Hanning, and flat-top. You will examine and investigate the effect of the window on the spectrum and develop insights into their choice.

Equipment

    1 function generator
    1 oscilloscope
    1 spectrum analyzer

Assignments and Reporting   This project has 26 items. Read and do items 12–16, 22(c)–24.

Theory

You need to be familiar with the theory of Fourier analysis for periodic and aperiodic signals summarized below.

1. Fourier Transform

The Fourier transform of a continuous-time function $x(t)$ is defined by

$$X(f) = \int_{-\infty}^{\infty} x(t)e^{-j2\pi ft}\, dt \iff x(t) = \int_{-\infty}^{\infty} X(f)e^{j2\pi ft} df$$

In practice, the following two factors affect the computation of $X(f)$:

  i. The integral is evaluated by a digital device. The continuous-time signal $x(t)$ is sampled and converted to a discrete-time signal $x(n)$ and the integral is replaced by a sum.

ii. The summation cannot be extended to future values of $x(n)$ because the future hasn't arrived. It cannot include all the past values of $x(n)$ because of limitations on acceptable delay, memory, and computation time. The summation is, therefore, done over a finite segment of the data.

In summary, the infinite Fourier integral is replaced by a finite sum. A similar situation arises when taking the inverse transform; $X(f)$ is sampled and a finite number of samples are used. These operations are performed by the discrete Fourier transform (DFT) and the inverse discrete Fourier transform (IDFT).

2. Windows

As we have stated, the domain of the mathematical models of a real signal is $-\infty < t < \infty$. However, in practice we process finite-length segments of a signal at any given time. Mathematically, this corresponds to multiplying the original signal by a window that is nonzero within an interval and zero outside that interval. A window is, therefore, a finite-duration pulse through which we look at a data stream or function. It may be a time window or a frequency window. The role of a time window is to create and shape a finite segment of the data or a filter's impulse response in a way that certain characteristics are met. The narrower the window is in the time domain, the wider its transform is in the frequency domain and vice versa. Given the duration of a window, its shape has an important role in its function. The time functions for some familiar continuous-time windows and their Fourier transforms are listed in Table 17.5 below. For simplicity, the time origin in the continuous-time windows is chosen such that the windows are even functions of time.

TABLE 17.5 Mathematical Expressions for Continuous-Time Windows and Their Fourier Transforms. All windows are τ seconds wide and are specified for $-\tau/2 < t < \tau/2$. Elsewhere, the windows have zero values.

Window	Time Function	$\Longleftrightarrow$	Fourier Transform
Rectangular	1	$\Longleftrightarrow$	$\dfrac{\sin(\pi f \tau)}{\pi f}$
Bartlett	$1 - 2\dfrac{\lvert t \rvert}{\tau}$	$\Longleftrightarrow$	$\dfrac{2}{\tau}\left[\dfrac{\sin(\pi f \tau/2)}{\pi f}\right]^2$
Hanning	$0.5 + 0.5\cos\left(\dfrac{2\pi t}{\tau}\right)$	$\Longleftrightarrow$	$0.5\dfrac{\sin(\pi f \tau)}{\pi f} + 0.25\dfrac{\sin \pi \tau(f - \frac{1}{\tau})}{\pi(f - \frac{1}{\tau})} + 0.25\dfrac{\sin \pi \tau(f + \frac{1}{\tau})}{\pi(f + \frac{1}{\tau})}$
Blackman	$0.42 + 0.5\cos\left(\dfrac{2\pi t}{\tau}\right) + 0.08\cos\left(\dfrac{4\pi t}{\tau}\right)$	$\Longleftrightarrow$	$0.42\dfrac{\sin(\pi f \tau)}{\pi f} + 0.25\dfrac{\sin \pi \tau(f - \frac{1}{\tau})}{\pi(f - \frac{1}{\tau})} + 0.25\dfrac{\sin \pi \tau(f + \frac{1}{\tau})}{\pi(f + \frac{1}{\tau})}$
			$+0.04\dfrac{\sin \pi \tau(f - \frac{2}{\tau})}{\pi(f - \frac{2}{\tau})} + 0.04\dfrac{\sin \pi \tau(f + \frac{2}{\tau})}{\pi(f + \frac{2}{\tau})}$

3. Fourier Transform of a Windowed Signal

Let $y(t) = x(t) \times w(t)$ be a finite segment of a signal $x(t)$ seen through a window $w(t)$. The Fourier transform of $y(t)$ is obtained from the convolution of the Fourier transforms of x and w. $Y(f) = X(f) \star W(f)$. These are summarized in Table 17.6 along with an example of a tone burst signal.

Mathematically, windowing modulates the amplitude of a sinusoid, shifting its spectrum to the location of the frequency of the sinusoid.

TABLE 17.6 Windowed Signals in Time and Frequency

Function	Time Domain		Frequency Domain
Signal	$x(t)$	$\Longrightarrow$	$X(f)$
Window	$w(t)$	$\Longrightarrow$	$W(f)$
Windowed signal	$y(t) = x(t)w(t)$	$\Longrightarrow$	$Y(f) = X(f) \star W(f)$
Tone	$x(t) = 2\cos(2\pi f_0 t)$	$\Longrightarrow$	$X(f) = \delta(f + f_0) + \delta(f - f_0)$
Tone burst	$y(t) = 2w(t)\cos(2\pi f_0 t)$	$\Longrightarrow$	$Y(f) = W(f + f_0) + W(f - f_0)$
Rectangular window, τ sec	$w(t) = \begin{cases} 1, & -\tau/2 < t < \tau/2 \\ 0, & \text{elsewhere} \end{cases}$	$\Longrightarrow$	$W(f) = \dfrac{\sin(\pi f \tau)}{\pi f}$
Rectangular tone burst	$y(t) = \begin{cases} 2\cos(2\pi f_0 t), & -\tau/2 < t < \tau/2 \\ 0, & \text{elsewhere} \end{cases}$	$\Longrightarrow$	$Y(f) = \dfrac{\sin[\pi\tau(f + f_0)]}{\pi(f + f_0)}$ $+ \dfrac{\sin[\pi\tau(f - f_0)]}{\pi(f - f_0)}$

4. Comparative Analysis of Windows

The spectrum analyzer allows you to choose one of three windows: Hanning, flat-top, and rectangular. The Hanning window produces a sharper peak at the harmonics and is used for best accuracy in measuring the frequency of a peak. The flat-top window is a special feature of the spectral analyzer. It produces, as the name implies, a broader peak and is preferred for best accuracy in measuring the value of the peak. All of these windows are called low-pass windows because they have a low-pass spectrum. The spectra may be roughly specified by their 3-dB bandwidths, where $20\log|W(f)/W(0)| = -3$ dB, and by the first frequency zero-crossing, where $W(f) = 0$. These specifications depend on the window's size and shape. Table 17.7 summarizes these specifications for rectangular, Bartlett, Hanning, Blackman, and flat-top windows. A flat-top window is specific to the spectrum analyzer. Note that from here on a window's width in the continuous-time domain is shown by T, which produces a frequency resolution $\Delta f = 1/T$.

TABLE 17.7 Comparing the salient features of five windows (height = 1, size = T seconds): G_0 in dB = window's DC gain, ΔG in dB = drop in gain at the second lobe with reference to the peak gain at $f = 0$, f_0 = the first zero-crossing, BW = 3-dB bandwidth.

Window Type	G_0	ΔG	f_0	3-db BW
Rectangular	0 dB	-13.35 dB	$\dfrac{1}{T}$	$\dfrac{0.44}{T}$
Bartlett	-6 dB	-26.7 dB	$\dfrac{2}{T}$	$\dfrac{0.44}{T}$
Hanning	-6 dB	-31.37 dB	$\dfrac{2}{T}$	$\dfrac{0.72}{T}$
Blackman	-7.54 dB	-58.5 dB	$\dfrac{3}{T}$	$\dfrac{0.44}{T}$
Flat-top		-90 dB	$\dfrac{5}{T}$	$\dfrac{1.8}{T}$

The magnitude frequency responses of four discrete-time windows are shown in Figure 17.10. Abscissa is discrete frequency in Hz. Ordinate is magnitude in dB. The Matlab file for generating Figure 17.10 is given in Attachment A at the end of this project.

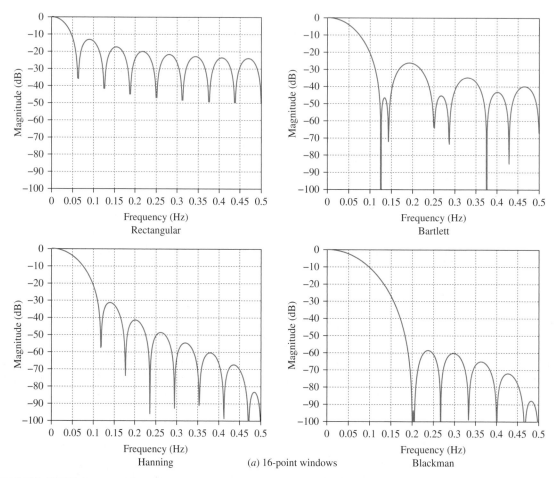

FIGURE 17.10 Spectra of four discrete-time windows: rectangular, Bartlett, Hanning, and Blackman. Abscissa's unit is $F_s/2$ Hz. The ordinate unit is dB.

5. Obtaining a Fourier Transform by Digital Hardware (DFT)

The spectrum of a continuous-time signal may be obtained by using a digital device such as a general-purpose digital computer, special-purpose digital signal processing hardware (DSP chips and boards), or by a stand-alone digital instrument such as the spectrum analyzer used in this experiment. The spectrum analyzer is a digital instrument that computes and displays the DFT of a signal in real time. This involves the following three steps:

(a) Sampling. The incoming continuous-time signal $x(t)$ is sampled at the rate of F_s samples per second (or equivalently with a sampling interval $\Delta t = 1/F_s$ seconds). This results in a discrete-time sequence $x(n)$. Any frequency content of $x(t)$ that is greater than $F_s/2$ Hz is not recoverable from $x(n)$. The sampling rate, therefore, should be twice the highest frequency present in $x(t)$. If this condition is not met, the higher frequencies are reflected back, causing aliasing.

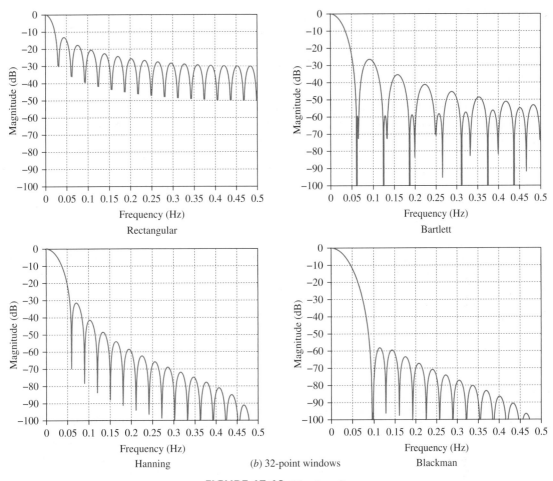

FIGURE 17.10 (*Continued*)

(*b*) *Windowing.*   A finite segment of length N of the discrete signal $x(n)$ is chosen and labeled x_n, $n = 0, 1, 2 \ldots, (N -$ 1). This operation is equivalent to windowing by a rectangular window of size N. Other windows, such as a Hanning window, multiply the sequence x_n by their corresponding weighting function. The result is saved in the memory of the digital device for computation. An N-point discrete signal corresponds to a segment of length $T = N \Delta t$ seconds of the signal in the continuous-time domain.

(*c*) *Computation.*   The following sum is evaluated and labeled as X_k:

$$X_k = \sum_{n=0}^{N-1} x(n)e^{-j2\pi kn/N}, \quad k = 0, 1, 2 \cdots (N - 1)$$

X_k is called the discrete Fourier transform (DFT) of $x(n)$. $k = 0$ corresponds to the DC value of the spectrum and $k = N - 1$ corresponds to the highest frequency $F_s/2$. The frequency resolution is, therefore, $\Delta f = F_s/N$.

6. Fast Fourier Transform

When window size N is a power of 2 (such as $N = 128, 256, 512, 1,024, \ldots$) a fast algorithm is employed to evaluate the DFT, in which case the operation is called the fast Fourier transform (FFT). Note that if x_n is a sequence of real numbers (as real-time functions are), X_k will be complex with a magnitude and phase. The spectrum analyzer computes both the magnitude and phase of the spectrum.

7. Time and Frequency Resolution

By sampling every Δt seconds (or equivalently at the rate of $F_s = 1/\Delta t$ samples per seconds), the spectrum analyzer converts a continuous-time signal to a discrete-time signal. The time resolution is $\Delta t = 1/F_s$. An N-point discrete signal corresponds to a segment of length $T = N\Delta t = N/F_s$ of the signal in the continuous-time domain. On the other hand, as shown in item 5(c) of this project, the frequency resolution is $\Delta f = F_s/N$. In summary

$$\Delta f = \frac{1}{T}, \quad \Delta t = \frac{1}{F_s}$$

The highest frequency in the computed spectrum is $F_s/2$. To increase the frequency span of the spectrum we need to use a higher sampling rate.

The lowest frequency, past the DC component, in the computed spectrum is $\Delta f = 1/T$, where T is the size of the window (length of data used for computation of the spectrum). For a better frequency resolution (lower Δf), a wider window is needed; that is, we need to take in a longer segment of the signal.

8. Periodicity in Time and Frequency

Mathematically, sampling in time creates periodicity in the frequency domain. Similarly, the DFT creates a discrete frequency domain, corresponding to a periodicity in the time domain. It is, therefore, sometimes said that sampling in one domain creates periodicity in the other.

9. The Spectrum Analyzer

The spectrum analyzer samples the signal at the rate of $F_s = 4.096 \times$ frequency span, twice the minimum rate (called the Nyquist rate), and windows it. The time window has 1,024 points. It then computes both the magnitude and phase of X_k, where $k = 0, 1, 2, \ldots (N-1)/2$, corresponding to frequencies $f = 0, \Delta f, 2\Delta f, 3\Delta f, \ldots F_s/2$. $F_s/2$ is the frequency span chosen on the front panel and is the highest frequency in the sampled signal. The screen display shows the magnitude and phase of the one-sided spectrum of the periodic extension of the windowed signal. The display is, therefore, made of a set of discrete points. The display labels the frequency axis in Hz. The displayed magnitude value at each frequency represents the amplitude of the component of the signal at that frequency. The numerical values of the display at each point can be estimated from the screen. Their more accurate values are also readable through a marker that can give their value in linear scale (volts, rms) or dBV (2 dB or 10-dB scales). In summary, the spectrum analyzer shows the magnitudes V_k and phases θ_k of the periodic signal $v(t)$

$$v(t) = V_0 + \sum_{k=1}^{(N-1)/2} V_k \cos\left(\frac{2\pi kt}{T} + \theta_k\right)$$

in which $V_k = 2|X_k|$, $\theta_k = \angle X_k$, and one period of $v(t)$ is captured by a sweep of length T.

10. The DFT of Windowed Signals

The spectrum analyzer computes the discrete Fourier transform (DFT) of the discrete-time window. DFT is a discrete representation in the frequency domain. It is computed directly as explained in the "Comparative Analysis of Windows" section above, but is also obtained by sampling the Fourier transform of the continuous-time signal at $\Delta f = 1/T$ Hz, where T is the window size. Consequently, the spectrum of a sinusoid at frequency f_0 obtained by the spectrum analyzer under a window size T is made of lines that are sample values of the continuous spectra. See Figure 17.11 in Attachment B at the end of this project. In the case of the uniform window all samples of the spectrum are zero except for a single line at f_0. Under a Hanning window, the DFT of the sinusoid displays three spectral lines (at $f_0 \pm \Delta f$). The Blackman window will have five lines at $f_0 \pm k \Delta f$, $k = 1, 2$ Hz. The flat-top window has nine lines at $f_0 \pm k \Delta f$, $k = 1, , 2, 3, 4$ Hz. These are verified by measurements shown in Table 17.10 in Attachment B.

11. Real-Time Signal Processing

Strictly speaking, a real-time signal processing system would operate on the input sample immediately and have it contribute to the processing before the next sample arrives to the processor. An example is a window averager that, after the arrival of a sample, computes the average of the past M-samples, places the result in the output, and waits for the next sample to arrive. In conjunction with an analog-to-digital and a digital-to-analog converter, the above system performs similar to the real-time operation of an analog system.

Digital signal processing operations that are based on an analysis in the frequency domain first store an array of consecutive N incoming samples in a buffer, multiply the array by a window of the same size, and then take the DFT of the windowed signal for further processing. As time progresses, both the windowing of the incoming signal (with some overlap between neighboring windows) and the DFT operation are repeated. For example, in a typical Linear Predictive Coding (LPC) scheme of speech, signal windows are 80 msec long and consecutive windows may have 20-msec overlaps. In a sense and strictly speaking, such a frequency domain processing is not done in real time. However, it is called real-time processing. We, therefore, consider the operation of the spectrum anayzer to be in real time.

Prelab

12. The rms and dBV Values of a Sinusoid

Let $v(t) = V_0 \cos(2\pi f t + \theta)$. The rms and dBV (decibel volt) values of $v(t)$ are defined by

$$V_{\text{rms}} = \frac{1}{T} \int_{t_0}^{t_0+T} v^2(t)dt \quad \text{and} \quad V_{\text{dBV}} = 20 \log V_{\text{rms}}$$

where T is a multiple of the period of the sinusoid. From the above definition, for the sinusoidal waveform we find that $V_{\text{rms}} = V_0/\sqrt{2}$. Find the rms and dB values corresponding to the amplitudes given in Table 17.8. Complete the table.

TABLE 17.8 The rms and dB Values of a Sinusoid Computed from Its Amplitude V_0

V_0 (V) $\Longrightarrow$	1	2	5	10	20
V_{rms} $\Longrightarrow$					
V_{dBV} $\Longrightarrow$					

13. Spectrum of a Windowed Sinusoid

Multiply the sinusoidal signal ($V_{\text{rms}} = 1$ V, f $= 1$ kHz) by a window (peak $= 1$, T $= 20$ msec) to obtain $x(n) = w(t) \times \cos(2000\pi t)$. Find the spectrum of $x(t)$ for three windows: (1) Uniform, (2) Hanning, (3) Blackman. Plot the three $x(t)$ and their spectra.

Measurements and Analysis

14. Setting Up the Spectrum Analyzer

Feed the signal from the function generator to channel A of the spectrum analyzer. Start with the following settings on the spectrum analyzer.

Input:	Channel A, AC coupling
Trigger:	Repetitive, free run
Window:	Uniform
Average:	Off
Frequency span	0–25 kHz
Sensitivity (and Vernier):	30 dBV (Vernier on Cal.)
Display:	Channel A, Magnitude
Scale:	10 dB/DIV
Amplitude reference level:	Normal
Marker:	On

The above numerical settings appear on the spectrum analyzer display. In addition, the display shows the location of the marker on the frequency axis (in Hz). You are able to move the marker along the frequency axis using the knob. With the frequency span set at 25 kHz, the marker's frequency increment setting will be 100 Hz and is shown on the screen. At this setting answer the following questions:

1. What is the frequency resolution Δf?
2. How many discrete points are displayed on the frequency (horizontal) axis?
3. What is the sampling rate (number of samples per second)?
4. What is the sampling interval Δt?

15. Measuring the Spectrum of a Sinusoid

Set the function generator to 1 kHz, 2 $V_{\text{peak-to-peak}}$ sinusoidal signal with $V_{\text{DC}} = 0$. The mathematical expression for the above signal is $v(t) = \cos(2\pi f_0 t + \alpha)$, where $f_0 = 1{,}000$ Hz and α depends on the time reference (trigger moment). By pressing the Time button and holding it down observe the time display of the sinusoidal signal and answer the following questions.

1. How many cycles of the sinusoid do you see on the screen?
2. How many "msec" of the signal is displayed?
3. How many points on the horizontal axis are displayed on the screen?

16. Magnitude of the Spectrum

The spectrum analyzer display is set on channel A, magnitude only, at 10 dB per division. A single bar appears at 1 kHz, representing the magnitude of the sinusoidal input in dBV. The dBV value (which can also be picked up by the marker and displayed on the screen) is with reference to an rms value of 1 V. A sinusoidal waveform with an rms value of 1 V results in 0 dBV (or simply 0 dB). Note that the magnitude display is stable and doesn't change from one trigger to another. Note also that the peak of the spectrum is expected to be located at 1 kHz and its value is expected to be -3 dB (registered by the marker on the screen).

17. Doubling the Amplitude of the Sinusoid

Double the amplitude of the sinusoidal input and observe the change in the dBV of the peak of its spectrum. Compare with what you expect from theory.

18. Effect of Window Type

Change the window from uniform to Hanning, and then to flat-top. Note the change in the shape of the spectrum. The Hanning window broadens the spectral line. The flat-top window makes it even broader. Along with each window, an additional number called BW in Hz (standing for 3-dB bandwidth) appears at the lower-right corner of the screen. For each window find the following quantities and record them in Table 17.9:

1. Peak value, in dBV.
2. Peak location, in kHz.
3. Displayed BW, in Hz.

The two additional columns in Table 17.9 record the measured 3-dB bandwidth and the uncertainty in measured frequency. These are described below.

19. Measuring the 3-dB Bandwidth

The frequency resolution at the current setting (25-kHz frequency span) is 100 Hz. Therefore, it may not be possible to measure the 3-dB bandwith of the spectrum by simply moving the marker to a location 3 dB below its peak value (at −6 dB, see "Magnitude of the Spectrum," above). Instead, keep the marker fixed at 1,000 Hz and slowly reduce (or increase) the frequency of the signal generator away from the 1-kHz point, and watch the dBV read by the marker reduced. The frequencies $f_{1,2}$, at which the marker reads −6 dBV (at 3 dB below its peak value), are the low and high 3-dB attenuation frequencies. The 3-dB bandwidth is, therefore, $BW = f_2 - f_1$. This value is expected to be close to the displayed BW. In this manner measure the 3-dB bandwidth of each window and enter results in Table 17.9. For an example see Attachment B.

20. Uncertainty in Frequency

The frequency resolution in the current setting (25-kHz frequency span) is 100 Hz. The peak of the spectrum at 1 kHz does not necessarily indicate the actual frequency of the signal. The 100-Hz resolution creates an uncertainty in frequency measurement, or error, ϵ_f. To measure the frequency uncertainty, keep the marker at 1-kHz and slowly reduce (or increase) the frequency of the signal generator away from the 1-kHz point, and watch the dBV read by the marker remain unchanged. The limit frequencies $f_{a,b}$ at which the marker still remains at the original value of −3 dBV, are the low and high end points of the range of frequencies which are taken to be 1 kHz, and $\epsilon_f = |f_a - f_b|$ is defined as the frequency uncertainty (absolute value) in Hz. It is expected that the uncertainty in frequency measurement be less than or equal to the displayed frequency resolution, which at the current setting is 100 Hz. Measure the uncertainty in frequency for each window and enter results in Table 17.9.

TABLE 17.9 Salient Features of the Spectrum of a 1-kHz Sinusoidal Wave Under Three Windows, Displayed and Measured Bandwidths, and Uncertainty in Measured Frequency

Window	Peak Value, dBV	Peak Location, kHz	Displayed BW, Hz	Measured BW, Hz	ϵ_f, Hz
Uniform					
Hanning					
Flat-top					

21. Increasing Frequency Resolution

Change the frequency span setting of the spectrum analyzer to 0–2.5 kHz. This will reduce Δf to 10 Hz. All other settings remain as in the "Prelab." Repeat the procedures described in this section (items 13–16).

22. Measuring the Phase of a Sinusoid

Phase is a relative quantity. It depends on the time reference. The phase computed and displayed by the spectrum analyzer depends on the starting point of the captured trace (trigger moment). Because of the free-run triggering, the phase display does not remain stable and changes from one sweep to another. To be able to read a stable phase, we can change the repetitive trigger from free run to level, where the captured data is the same from one sweep to the next. Alternatively, we can use a single trigger (free run or level) and capture one set of data only. Examine the following three cases.

Repetitive Triggering at Peak Level.   Set the spectum analyzer display on channel A, phase only. Other settings remain as in the "Prelab." To capture a stable phase, change the trigger from free run to level. If the signal starts at the exact moment of its positive peak, the captured segment of the data is nearly a cosine function and the phase is zero. For this purpose, set the trigger level at the highest level of the signal such that just above that level it will not trigger. To visually examine the time waveform captured by the spectrum analyzer, momentarily press down the Time button on its front panel. Notice that phase is constant. Record the phase at the frequency where the spectrum is at its maximum.

Repetitive Triggering at Near-Zero Level.   Trigger at a level near zero with positive slope to capture a sine function and visually examine the captured time function. This gives rise to a stable 90° phase. Measure the phase at the frequency where the spectrum is at its maximum.

A Single Trigger Under Free Run.   Change the free-run trigger from repetitive to single. The spectrum analyzer captures the 1,024 samples (window size = 10 msec; explain the reason) in a single sweep, and then computes and displays the spectrum. Because the trigger is set on free run, the starting point of the windowed sinusoid is random and the phase is also a random number between ±180°. See the "Theory" and "Prelab" sections. Record the phase at the location of the magnitude peak, and repeat 10 times.

Discussion and Conclusions

23. Conclusions of Prelab and Measurements and Analysis

Based on the work done in the "Prelab" and "Measurements and Analysis" sections, compare the salient features of each window along with their advantages and disadvantages. Develop criteria for choosing an appropriate window for various spectral measurement and signal processing situations.

24. Overall Conclusions

Summarize your overall conclusions drawn from this experiment.

25. Attachment A

The Matlab file for generating Figure 17.10 is listed below.

```
%Spectra of four windows: Boxcar, Bartlett, Hanning, Blackman
n = 32;
w1=boxcar(n); w2=bartlett(n); w3=hanning(n); w4=blackman(n);
[w1,f]=freqz(w1/sum(w1),1,512,2);
[w2,f]=freqz(w2/sum(w2),1,512,2);
[w3,f]=freqz(w3/sum(w3),1,512,2);
[w4,f]=freqz(w4/sum(w4),1,512,2);
%The following figure overlays the 4 spectra
figure
plot(f,20*log10(abs([w1,w2,w3,w4])));
%The following figure creates 4 separate plots.
figure
```

```
subplot(2,2,1);
plot(f,20*log10(abs(w1))), title('32-points rectangular window');
subplot(2,2,2);
plot(f,20*log10(abs(w2))),title('32-point bartlett window');
subplot(2,2,3);
plot(f,20*log10(abs(w3))),title('32-point hanning window');
subplot(2,2,4);
plot(f,20*log10(abs(w4))),title('32-point blackman window');
```

26. Attachment B. Spectral Profiles of Windows

The DFT of a discrete-time window may be obtained by sampling the spectrum of its corresponding continuous-time window at a frequency spacing $\Delta f = 1/T$, where T is the window size. Figure 17.11 shows the spectral profiles (DFT) of four discrete-time windows superimposed on the spectral profiles of their corresponding continuous-time windows. The left column shows four windows and the right column shows their spectra. From the top: uniform (rectangular), Hanning, Blackman, and Bartlett (triangular). All windows are 10 msec wide. Under the DFT operation, the 10-msec data produces a sampling interval $\Delta f = 1/(10\ msec) = 100$ Hz in the frequency domain, also shown on the spectra. Note that in the case of the uniform window the 100-Hz sampling interval results in a single spectral line at the center frequency ($f = 0$ in this figure). Table 17.10 gives measured spectral profiles of the windows used by the spectrum analyzer.

TABLE 17.10 Measured Spectral Profiles, in dBV, of the Windows Used by the Spectrum Analyzer Operating on a Sinusoid ($V_{peak-to-peak} = 2$ V, $f = 1$ kHz). The star ⋆ indicates off-scale values.

Frequency (Hz) ⇒	500	600	700	800	900	1,000	1,100	1,200	1,300	1,400	1,500
Uniform window					⋆	−3.1	⋆				
Hanning window				⋆	−9.1	−3.1	−9.1	⋆			
Flat-top window	⋆	−37.2	−17.0	−7.1	−3.5	−3	−3.5	−7.1	−17	−37.2	⋆

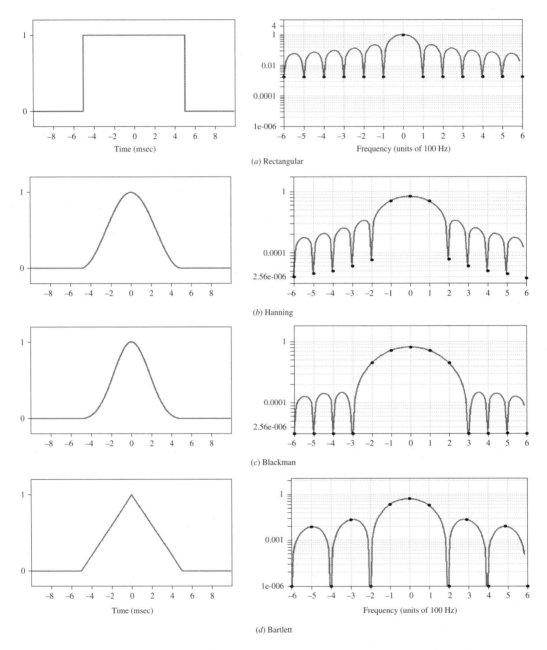

(a) Rectangular

(b) Hanning

(c) Blackman

(d) Bartlett

FIGURE 17.11 Spectral profiles (DFT) of four discrete-time windows superimposed on the spectral profiles (DTFT) of their corresponding continuous-time windows. Sampling DTFT at intervals Δf results in DFT. In this figure the windows are 10 msec wide, resulting in a sampling interval $\Delta f = 1/(10\ msec) = 100$ Hz in the frequency domain. Note that in the case of the uniform window the 100-Hz sampling interval leaves only a single spectral line intact. The windows are *(a)* rectangular, *(b)* Hanning, *(c)* Blackman, *(d)* Bartlett. The numbers on the ordinates in the right-hand column are normalized magnitudes of the Fourier transforms of corresponding windows on the left.

Chapter 18

System Function, the Frequency Response, and Digital Filters

Contents

Introduction and Summary

LTI systems may be analyzed in both time and frequency domains. In the time domain, the output is found by convolution of the input with the unit-sample response, $y(n) = x(n) \star h(n)$, or by solving the input-output difference equation. In the frequency domain, one uses the z-transform to convert the convolution $y(n) = x(n) \star h(n)$ into a multiplication of transforms $Y(z) = X(z)H(z)$. Both approaches are closely related and provide the same result. In this chapter we expand on the properties and capabilities of $H(z)$ and the parallels between the time and frequency-domain solutions. The chapter starts with a new look at the system function from a different, but familiar, perspective. It discusses the effects of poles and zeros in forming the system function, influencing its characteristics, and shaping the system's time response. The chapter then presents the frequency response, its relation to the system function, and the DTFT of the unit-sample response. The frequency response is the characteristic of an LTI system most closely encountered in practical and experimental applications. The chapter attends to this characteristic and its relationship to the continuous-time domain. In that respect, the discussion forms a basis for digital filtering and design which follows the discussion on the frequency response and is illustrated by several solved problems at the end of the chapter. As another application of the system function, system stability and inverse systems are briefy discussed. Three projects at the end of the chapter provide application examples.

18.1   The System Function $H(z)$

The System Function $H(z)$ Is the Scale Factor for z^n

We have seen (Chapter 12) that the response of an LTI system to the input z_0^n is $H z_0^n$, where z_0 is a constant and the scale factor H is, in general, a function of z_0:

$$z_0^n \implies H(z_0) z_0^n$$

$H(z)$ is called the *system function*. It may be found from the unit-sample response, from an input-output pair, or from the input-output difference equation. These will be described below.

$H(z)$ Is the z-Transform of $h(n)$

The system function is the z-transform of the unit-sample response. (See Chapter 15.)

The unit-sample response of an LTI system is

$$h(n) = \begin{cases} 1, & 0 \le n < M \\ 0, & \text{elsewhere} \end{cases}$$

Find its system function.

Solution

$$H(z) = \sum_{n=0}^{M-1} z^{-n} = \frac{1 - z^{-M}}{1 - z^{-1}}$$

xample

18.2

Find the system function of the LTI system with the unit-sample response $h(n) = a^n \cos(\omega n)u(n)$.

Solution

From the table of z-transform pairs in Chapter 15 (a portion of which is reproduced below), we find

$$H(z) = \frac{1 - a(\cos\omega)z^{-1}}{1 - 2a(\cos\omega)z^{-1} + a^2z^{-2}}$$

A Minitable of *z*-Transform Pairs

The following minitable of z-transform pairs will be used in this chapter.

$x(n)$	$\Longleftrightarrow$	$X(z)$	ROC
$u(n)$	$\Longleftrightarrow$	$\dfrac{1}{1 - z^{-1}}$	$\|z\| > 1$
$a^n u(n)$	$\Longleftrightarrow$	$\dfrac{1}{1 - az^{-1}}$	$\|z\| > \|a\|$
$a^n \cos(\omega n)u(n)$	$\Longleftrightarrow$	$\dfrac{1 - a(\cos\omega)z^{-1}}{1 - 2a(\cos\omega)z^{-1} + a^2z^{-2}}$	$\|z\| > \|a\|$
$a^n \sin(\omega n)u(n)$	$\Longleftrightarrow$	$\dfrac{a(\sin\omega)z^{-1}}{1 - 2a(\cos\omega)z^{-1} + a^2z^{-2}}$	$\|z\| > \|a\|$
$a^n \cos(\omega n + \theta)u(n)$	$\Longleftrightarrow$	$\dfrac{\cos\theta - a\cos(\omega - \theta)z^{-1}}{1 - 2a(\cos\omega)z^{-1} + a^2z^{-2}}$	$\|z\| > \|a\|$

$H(z)$ Is the Ratio $Y(z)/X(z)$

The system function is also the ratio of the z-transform of the output to the z-transform of the input. See Chapter 15.

xample

18.3

Find the system function of the LTI system with the given input-output pair.

$$x(n) = \{\underset{\uparrow}{1}, 1, 1\} \Longrightarrow y(n) = \{\underset{\uparrow}{1}, 3, 4, 3, 1\}$$

Solution

In the z-domain the input-output pair is

$$X(z) = 1 + z^{-1} + z^{-2} \Longrightarrow Y(z) = 1 + 3z^{-1} + 4z^{-2} + 3z^{-3} + z^{-4}$$

By long division we get

$$H(z) = \frac{Y(z)}{X(z)}$$
$$= \frac{1 + 3z^{-1} + 4z^{-2} + 3z^{-3} + z^{-4}}{1 + z^{-1} + z^{-2}}$$
$$= 1 + 2z^{-1} + z^{-2}$$

xample

18.4

Find the system function of the LTI system with the unit-step response

$$y(n) = \left[1.1908 - 0.67864 \times 0.5^n \cos(\pi n/4 + 73.67°)\right] u(n)$$

Solution

Using the minitable of z-transform pairs given above we find the z-transforms of the unit-step input, the unit-step response, and their ratio to be

$$X(z) = \frac{1}{1 - z^{-1}}$$

$$Y(z) = \frac{1.1908}{1 - z^{-1}} - 0.67864 \frac{\cos(73.67°) - 0.5\cos(\pi/4 - 73.67°)z^{-1}}{1 - \cos(\pi/4)z^{-1} + 0.25z^{-2}}$$

$$= \frac{1.1908}{1 - z^{-1}} - \frac{0.1908 - 0.2977z^{-1}}{1 - 0.7071z^{-1} + 0.25z^{-2}}$$

$$= \frac{1 - 0.3535z^{-1}}{(1 - z^{-1})(1 - 0.7071z^{-1} + 0.25z^{-2})}$$

$$H(z) = \frac{Y(z)}{X(z)} = \frac{1 - 0.3535z^{-1}}{1 - 0.7071z^{-1} + 0.25z^{-2}}$$

$H(z)$ Is Found from the Difference Equation and Vice Versa

We have seen (Chapter 14) that the solution of the LTI difference equation

$$y(n) + a_1 y(n - 1) \cdots + a_N y(n - N) = z_0^n$$

is $Y z_0^n$, where

$$Y = \frac{1}{1 + a_1 z_0^{-1} \cdots + a_N z_0^{-N}}$$

In the above, the variable n represents the discrete time and constitutes the variable of the difference equation, z_0 is a constant number and z_0^n is expressing powers of z_0 in terms of the variable n. Extending the above observation, we conclude that the response of the difference equation

$$y(n) + a_1 y(n - 1) \cdots + a_N y(n - N) = b_0 x(n) + b_1 x(n - 1) \cdots + b_N x(n - M)$$

to $x(n) = X_0 z^n$ is $y(n) = Y_0 z^n$. By substituting the above input-output pair in the equation we find $Y_0 = H(z)X_0$, where $H(z)$ is the *system function*. For systems described by an LTI difference equation such as the equation shown above, $H(z)$ is the ratio of two polynomials in z:

$$H(z) = \frac{b_0 + b_1 z^{-1} + b_2 z^{-2} \cdots + b_M z^{-M}}{1 + a_1 z^{-1} + a_2 z^{-2} \cdots + a_N z^{-N}}$$

$H(z)$ is easily obtained from the difference equation by employing the notation z^{-1} for a unit delay. The difference equation is then written as

$$D^N[y(n)] = D^M[x(n)]$$

where D^N and D^M are linear delay operators acting on $y(n)$ and $x(n)$, respectively. They are constructed from sums of z^{-k}, where z^{-k} represents a delay of k units.

$$D^M = b_0 + b_1 z^{-1} + b_2 z^{-2} + \cdots + b_M z^{-M}$$
$$D^N = 1 + a_1 z^{-1} + a_2 z^{-2} + \cdots + a_N z^{-N}$$

In light of the above, we can conversely construct the difference equation and its block diagram from a given $H(z)$, which is the ratio of two polynomials in z^{-1}.

Find the input-output difference equation of the LTI system with the system function

$$H(z) = \frac{Y(z)}{X(z)} = 1 + z^{-1} + z^{-2}$$

Solution

$$Y(z) = (1 + z^{-1} + z^{-2})X(z)$$
$$= X(z) + z^{-1}X(z) + z^{-2}X(z)$$
$$y(n) = x(n) + x(n-1) + x(n-2)$$

Find the input-output difference equation of an LTI system with the system function

$$H(z) = \frac{Y(z)}{X(z)} = \frac{1 - 0.3535z^{-1}}{1 - 0.7071z^{-1} + 0.25z^{-2}}$$

Solution

$$(1 - 0.7071z^{-1} + 0.25z^{-2})Y(z) = (1 - 0.3535z^{-1})X(z)$$
$$y(n) - 0.7071y(n-1) + 0.25y(n-2) = x(n) - 0.3535x(n-1)$$

The Time Response May Be Found from $H(z)$

The response of an LTI system is obtained either through convolution of the input with the unit-sample response or solution of the input-output difference equation, both of which may be obtained from $H(z)$. It is, therefore, expected that the system function contains all the information needed to find the response to a given input. In this section we will see how each component of the time response (natural frequencies, homogeneous and particular responses, boundary values, and total time response) may be derived from $H(z)$, often without resorting to the formal tools and systematic methods [i.e., without finding $h(n)$ to do convolution, or solving the difference equation, or using the z-transform formulation].

We start with the observation that the system function provides the particular response to an input expressed by a linear combination of power series (with constant weighting factors). An example is the AC steady-state response. We then additionally note that the denominator of the system function contains the characteristic equation of the system, which provides the natural frequencies and determines the homogeneous response. In fact, the complete response may readily be written from the system function if the input is given as a power series and the initial conditions are known. This property makes $H(z)$ a powerful tool in discrete-time LTI system analysis. In the following three examples we find responses of the first-order LTI system $H(z) = (1 + z^{-1})/(1 - z^{-1})$ to various inputs using the system function.

Example 18.7

Given $H(z) = (1+z^{-1})/(1-z^{-1})$, find the system's response to $x(n) = \cos(\pi n/2)$, $-\infty < n < \infty$.

Solution

The response consists of the particular solution only, which, in this case, is the sinusoidal steady-state response. To find it, we note that the frequency of the sinusoidal input is $\omega = \pi/2$, which corresponds to $z = j$ (because $z = e^{j\omega} = e^{j\pi/2} = j$). The value of $H(z)$ at $z = j$ is

$$H(z)|_{z=j} = \frac{1+z^{-1}}{1-z^{-1}}\bigg|_{z=j} = \frac{1-j}{1+j} = -j$$

The input and the response may be written as

$$x(n) = \cos(\pi n/2)$$
$$= \mathcal{RE}\{X_0 e^{j\omega n}\}, \quad X_0 = 1, \quad \omega = \pi/2$$
$$= \mathcal{RE}\{X_0 z^n\}, \quad z = e^{j\omega} = e^{j\pi/2} = j$$
$$y(n) = \mathcal{RE}\{H(z)X_0 z^n\}\big|_{z=j} = \mathcal{RE}\{j \times 1 \times e^{j\pi n/2}\} = \cos[\pi(n-1)/2] = \sin(\pi n/2)$$

Example 18.8

Given $H(z) = (1+z^{-1})/(1-z^{-1})$, find the system's response to $x(n) = \cos(\pi n/2)u(n)$.

Solution

The particular solution was found in Example 18.7 to be $y_p(n) = \sin(\pi n/2)$. We need to add to it the homogeneous solution (if any). The system has a single pole at $p = 1$, which gives rise to the homogeneous solution $y_h(n) = C(1)^n = C$. The total solution is $y(n) = y_h(n) + y_p(n) = C + \sin(\pi n/2)$. To find the constant C we need boundary value $y(0)$, which is obtained from the difference equation:

$$y(n) = y(n-1) + x(n) + x(n-1)$$
$$y(0) = y(-1) + x(0) + x(-1) = 0 + 1 + 0 = 1$$
$$y(n) = C + \sin(\pi n/2)$$

$$y(0) = 1 = C$$

$$y(n) = 1 + \sin(\pi n/2)$$

Therefore, $y(n) = [1 + \sin(\pi n/2)]u(n)$.

Example
18.9

Given $H(z) = (1+z^{-1})/(1-z^{-1})$ and the input $x(n) = \alpha d(n+1) + \cos(\pi n/2)u(n)$, find the constant α such that for $n \geq 0$ the system's response contains only the sinusoidal steady state.

Solution

In order to have no homogeneous solution to the input $\cos(\pi n/2)u(n)$, we need to create the appropriate initial condition(s) which, following the solution of Example 18.7, would be $y(0) = 0$. But, from the difference equation, we have

$$x(-1) = \alpha, \ x(0) = 1, \ y(-1) = \alpha$$

$$y(0) = y(-1) + x(0) + x(-1) = \alpha + 1 + \alpha = 2\alpha + 1$$

By setting $2\alpha + 1 = 0$ we get $\alpha = -0.5$. In other words, the input sample $\alpha d(n+1)$ arriving before the sinusoid is going to produce a zero initial condition at $n = 0$ if $\alpha = -0.5$.

18.2 Poles and Zeros

In this book we are interested in LTI systems with $H(z) = A(z)/B(z)$, where $A(z)$ and $B(z)$ are polynomials in z. The roots of the numerator polynomial are called the *zeros* of the system. Let z_0 be a zero of the system. At z_0 we have $H(z_0) = 0$ and the input $x(n) = z_0^n$ will result in $y(n) = 0$ for all n.

Similarly, the roots of the denominator polynomial are called the *poles* of the system. Let p_0 be a pole of the system. At p_0 we have $H(p_0) = \infty$ and the input $x(n) = p_0^n$ will result in $y(n) = \infty$ for all n.

Example
18.10

Find the poles and zeros of the system described by $H(z) = 1 + z^{-1} + z^{-2}$.

Solution

The system function is $H(z) = (z^2 + z + 1)/z^2$. The zeros and poles of the system are the roots of the numerator and denominator, respectively.

$$\text{zeros:} \ \ z^2 + z + 1 = 0 \ \implies \ z_{1,2} = -\tfrac{1}{2} \pm j\tfrac{\sqrt{3}}{2} = e^{\pm j120°}$$

$$\text{poles:} \ \ z^2 = 0 \qquad\qquad \implies \ p_{1,2} = 0$$

Example 18.11

Find the poles and zeros of the system described by

$$H(z) = \frac{1 - 0.3535z^{-1}}{1 - (\sqrt{2}/2)z^{-1} + 0.25z^{-2}} = \frac{z(z - 0.3535)}{z^2 - (\sqrt{2}/2)z + 0.25}$$

Solution

zeros of the system are roots of $z(z - 0.3535) = 0$ $\implies z_1 = 0.3535, \quad z_2 = 0$

poles of the system are roots of $z^2 - (\sqrt{2}/2)z + 0.25 = 0$ $\implies p_{1,2} = \frac{\sqrt{2}}{4} \pm j\frac{\sqrt{2}}{4} = \frac{1}{2}e^{\pm j45°}$

Example 18.12

An LTI system has one pair of poles at $p_{1,2} = \rho e^{\pm j\theta}$, two zeros at $z_{1,2} = \pm 1$, and $H(z)\big|_{z=e^{j\theta}} = 1$. Find the system function.

Solution

$$H(z) = k\frac{(1 - z^{-1})(1 + z^{-1})}{(1 - \rho e^{j\theta}z^{-1})(1 - \rho e^{-j\theta}z^{-1})}$$

$$= k\frac{1 - z^{-2}}{1 - 2\rho(\cos\theta)z^{-1} + \rho^2 z^{-2}}$$

To find k we note

$$H(e^{j\theta}) = k\frac{(1 - e^{-j\theta})(1 + e^{-j\theta})}{(1 - \rho e^{j\theta}e^{-j\theta})(1 - \rho e^{-j\theta}e^{-j\theta})}$$

$$= k\frac{1 - e^{-j2\theta}}{(1 - \rho)(1 - \rho e^{-j2\theta})} = 1$$

from which

$$k = (1 - \rho)\frac{1 - \rho e^{-j2\theta}}{1 - e^{-j2\theta}}$$

Contribution of Poles and Zeros to the System Function

We are interested in systems described by the ratio of two polynomials in z:

$$H(z) = \frac{b_0 + b_1 z^{-1} + b_2 z^{-2} \cdots + b_M z^{-M}}{1 + a_1 z^{-1} + a_2 z^{-2} \cdots + a_N z^{-N}}$$

For such systems a zero at z_0 contributes $(z - z_0) = z(1 - z_0 z^{-1})$ to the numerator of the system function. This may be easily verified by setting $z = z_0$ in the expression for $H(z)$. Similarly, a pole at p_0 contributes $(z - p_0) = z(1 - p_0 z^{-1})$ to the denominator. It is often desirable to express the system function in terms of z^{-1} rather than z. A system with M zeros at z_k, $k = 1, 2, \cdots M$, and N poles at p_k, $k = 1, 2, \cdots N$, is then described by

$$H(z) = H_0 z^{(M-N)} \frac{(1 - z_1 z^{-1})(1 - z_2 z^{-1}) \cdots (1 - z_M z^{-1})}{(1 - p_1 z^{-1})(1 - p_2 z^{-1}) \cdots (1 - p_N z^{-1})}$$

When $N > M$, the system has $N - M$ poles at the origin.

When $N < M$, the system has $M - N$ zeros at the origin.

When $N = M$, the system has no poles or zeros at the origin.

In the above expression, H_0 is a constant gain. The significance of poles and zeros at the origin, or their diminished role, will be seen in section 18.4.

Poles of FIR Systems Are at the Origin

By definition, the unit-sample response of an FIR system has a finite number of samples. The z-transform of $h(n)$, therefore, is a polynomial in z^{-1}. The poles are all at $z = 0$. If $H(z)$ of an FIR system is presented as the ratio of two polynomials in z^{-1} with some apparent nonzero poles, they will be canceled by the zeros of the numerator. See Example 18.13.

The unit-sample response of an LTI system is $h(n) = 2^n[u(n) - u(n - 4)]$. Find the system function and its zeros. Determine if it has any nonzero poles.

Solution

$$H(z) = \frac{1}{1 - 2z^{-1}} - \frac{2^4 z^{-4}}{1 - 2z^{-1}} = \frac{1 - 2^4 z^{-4}}{1 - 2z^{-1}}$$

Roots of the numerator are obtained from $1 - 2^4 z^{-4} = 0$. They are $z = 2$, $j2$, -2, and $-j2$. The denominator has a single root at $z = 2$, which is canceled by a zero, leaving the system with three zeros at $j2$, -2, and $-j2$.

Remark
The zeros of the above system may also be found directly from

$$h(n) = \{\underset{\uparrow}{1}, 2, 4, 8\}$$

$$H(z) = 1 + (2z)^{-1} + (2z)^{-2} + (2z)^{-3} = 0$$

To find the roots of the above equation, we employ the auxiliary variable $x = (2z)^{-1}$ and obtain $x^3 + x^2 + x + 1 = 0$, the roots of which are $x = -1$ and $\pm j$. The zeros of the system are, therefore, at $z = -2$ and $\pm j2$. The system has no poles.

Example

18.14

The unit-sample response of an LTI system is $h(n) = (0.9)^n[u(n) - u(n - 10)]$. Find the system function and its zeros. Determine if it has any poles.

Solution

$$H(z) = \frac{1}{1 - 0.9z^{-1}} - \frac{(0.9)^{10}z^{-10}}{1 - 0.9z^{-1}} = \frac{1 - (0.9)^{10}z^{-10}}{1 - 0.9z^{-1}}$$

Roots of the numerator are obtained from the equation

$$1 - 0.9^{10}z^{-10} = 0 \quad \text{or} \quad z^{10} = (0.9)^{10}$$

which gives $z = 0.9e^{\pm j2k\pi/10}$, $k = 0, 1, 2, \cdots, 5$. These are uniformly distributed on the circle with radius 0.9, starting at $z = 0.9$ and spaced every 36°. The zero at $z = 0.9$ (corresponding to $k = 0$) is canceled by the pole at that location, leaving the system with 9 zeros at $0.9e^{\pm j36°}$, $0.9e^{\pm j72°}$, $0.9e^{\pm j108°}$, $0.9e^{\pm j144°}$, and -0.9.

Explanation

The roots of the numerator polynomial are obtained from

$$1 - (0.9)^{10}z^{-10} = 0$$

$$1 - (0.9z^{-1})^{10} = 0$$

$$(0.9z^{-1})^{10} = 1 = e^{j2k\pi}$$

$$0.9z^{-1} = e^{j2k\pi/10}$$

$$z = 0.9e^{-j2k\pi/10}, \quad k = 0, 1, 2, \cdots, 9$$

System Stability and Pole Location

If bounded inputs to a linear system produce bounded outputs, the system is said to be BIBO-stable. The natural (zero-input) response of a BIBO system diminishes with time. An LTI discrete-time system is BIBO-stable if its poles (natural frequencies) are inside the unit circle in the complex plane. A related property is that in a BIBO system,

$$\sum_n |h(n)| < M, \quad \text{where } M \text{ is any number (as large as one wishes it to be).}$$

Inverse Causal Systems $H(z)$ and $1/H(z)$

Let a signal $x(n)$ pass through a system with the unit-sample response $h(n)$, producing the output $y(n) = x(n) \star h(n)$. Passing $y(n)$ through a second system with the unit-sample response $\overline{h}(n)$ such that $h(n) \star \overline{h}(n) = d(n)$, will reverse the effect of the first system resulting in the output $x(n)$:

$$y(n) \star \overline{h}(n) = [x(n) \star h(n)] \star \overline{h}(n) = x(n) \star [h(n) \star \overline{h}(n)] = x(n) \star d(n) = x(n)$$

See Figure 18.1. The two systems are said to be inverses of each other. In a sense, the inverse system performs a deconvolution of $x(n)$ with $h(n)$, which was analyzed in Chapter 13, sections 13.8 and 13.9.

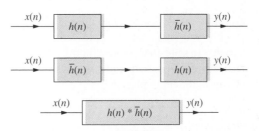

FIGURE 18.1 Two systems $h(n)$ and $\overline{h}(n)$ are inverses of each other if $h(n) \star \overline{h}(n) = d(n)$. In that case, $y(n) = x(n)$. One system reverses the effect of the other. For causal systems this translates into system functions $H(z)$ and $1/H(z)$.

If the two systems are causal, the system functions of $h(n)$ and $\overline{h}(n)$ will be $H(z)$ and $1/H(z)$, respectively.

$H(z) \times \dfrac{1}{H(z)} = 1$, which corresponds to the z-transform of unit-sample response $d(n)$.

Example 18.15

In Chapter 13, section 13.9 we considered the LTI system $h(n) = 0.5^n u(n)$. By applying deconvolution we found its inverse to be $\overline{h}(n) = \{1, -0.5\}$. In this example we apply the system function to obtain the same result.

$$h(n) = 0.5^n u(n) \implies H(z) = \frac{1}{1-0.5z^{-1}}$$
$$\Downarrow$$
$$\Downarrow$$
$$\overline{h}(n) = \{\underset{\uparrow}{1}, -0.5\} \impliedby 1/H(z) = 1 - 0.5z^{-1}$$

Note that
$$h(n) \star \overline{h}(n) = 0.5^n u(n) \star \{\underset{\uparrow}{1}, -0.5\} = 0.5^n u(n) - 0.5^n u(n-1) = \begin{cases} 1, & n = 0 \\ 0, & \text{elsewhere} \end{cases}$$

18.3 The Frequency Response $H(\omega)$

We have seen (Chapter 14) that the response of an LTI system, which is described by a linear differential equation with real constant coefficients, to a sinusoidal input is a sinusoid. The frequency response determines the relative changes in magnitude and phase of a sinusoid when it goes through an LTI system. It is expressed by its magnitude and phase, both of which are functions of the frequency ω. The magnitude and phase of the frequency response are combined as a complex function and shown by $H(\omega) = |H(\omega)|\angle\theta$. Therefore,

$$\cos\omega n \implies |H(\omega)|\cos(\omega n + \theta)$$

We often combine the magnitude and phase of a sinusoidal function into a complex number and call it the complex amplitude or phasor. The sinusoidal input $x(n)$ and output $y(n)$ are then represented by their complex amplitudes, or phasors, X and Y, respectively. The frequency response then may be defined as the ratio of the output phasor to the input phasor.

$$H(\omega) = \frac{Y}{X} = |H(\omega)|e^{j\theta(\omega)}$$

The frequency response of an LTI system may be measured experimentally without knowledge of its internal structure or its mathematical model. It may also be obtained from the system function, the unit-sample response, or the input-output difference equation, as will be seen in this chapter.

Frequency Response of a Uniform Window

The FIR system with an $h(n)$ made of a uniform window of M samples of size $1/M$ each is frequently encountered in the analysis of discrete-time systems. The system takes an average of M input samples within the window and assigns it as the output, then slides the window one sample and repeats the operation. The operation is called moving window averaging. In this section we review and summarize its properties in the z- and ω-domains. For simplicity, we set aside the division by M and consider the LTI system with the uniform window containing M unit samples to be the unit-sample response. For brevity, we call the system a uniform or rectangular window. It is frequently applied to signals when processing them in finite segments. The unit-sample response, output, system function, and frequency response of the uniform window of size M are listed below.

$$h(n) = \sum_{k=0}^{M-1} d(n-k)$$

$$y(n) = \sum_{k=0}^{M-1} x(n-k)$$

$$H(z) = \sum_{n=0}^{M-1} z^{-n} = 1 + z^{-1} + z^{-2} + z^{-3} + \cdots + z^{-(M-1)} = \frac{1 - z^{-M}}{1 - z^{-1}}$$

$$H(\omega) = \sum_{n=0}^{M-1} e^{-jn\omega} = 1 + e^{-j\omega} + e^{-j2\omega} + e^{-j3\omega} + \cdots + e^{-j(M-1)\omega} = \frac{1 - e^{-jM\omega}}{1 - e^{-j\omega}}$$

$$= \frac{\sin(M\omega/2)}{\sin(\omega/2)} e^{-j\omega\frac{(M-1)}{2}} = H_r(\omega)e^{j\theta\omega}, \quad \text{where } H_r(\omega) \text{ is a real function of } \omega.$$

The uniform window has no poles. (Poles at origin are not considered.) Its zeros are equally spaced around the circumference of the unit circle starting at $z = 1$. The $H_r\omega$

component of the frequency response is a real function of ω, made of alternating positive and negative lobes, which diminish in magnitude as ω increases. The zero-crossings occur at

$$\omega = \frac{2k\pi}{M}$$

As expected, $H(0) = M$. Plots of $H(\omega)$ for windows consisting of 2, 3, and 20 unit samples are shown in Figure 18.2(a), (b), and (c), respectively. In Examples 18.16 and 18.17 we derive $H(z)$ and $H(\omega)$ for narrow windows made of three and four unit samples, respectively. An example of a wide window (20 samples wide) is given in Example 18.18.

Find and plot the frequency response of the uniform window $h(n) = \{\underset{\uparrow}{1}, 1, 1\}$.

Solution

$$H(z) = 1 + z^{-1} + z^{-2}$$

$$H(\omega) = 1 + e^{-j\omega} + e^{-j2\omega} = e^{-j\omega}\left[e^{j\omega} + 1 + e^{-j\omega}\right] = (1 + 2\cos\omega)e^{-j\omega}$$

See Figure 18.2(b). Note that there are zero-crossings at $\omega = \pm 2\pi/3$. In the above expressions, $H(z)$ and $H(\omega)$ are written as sums because a small window size results in a small number of terms. They may equally be written in fractional form:

$$H(z) = \frac{1 - z^{-3}}{1 - z^{-1}} \quad \text{and} \quad H(\omega) = \frac{\sin(3\omega/2)}{\sin(\omega/2)}e^{-j\omega}$$

For large windows the fractional form is recommended. The fractional form also explicitly exhibits the zeros of the system.

Find the frequency response of the uniform window $h(n) = \{\underset{\uparrow}{1}, 1, 1, 1\}$.

Solution

$$H(z) = 1 + z^{-1} + z^{-2} + z^{-3}$$

$$H(\omega) = 1 + e^{-j\omega} + e^{-j2\omega} + e^{-j3\omega}$$

$$= e^{-j3\omega/2}\left[e^{3j\omega/2} + e^{j\omega/2} + e^{-j\omega/2} + e^{-3j\omega/2}\right]$$

$$= 2\left[\cos(\omega/2) + \cos(3\omega/2)\right]e^{-j3\omega/2}$$

The zero-crossings occur at $\omega = \pm\pi/2, \ \pm\pi$. In fractional form,

$$H(z) = \frac{1 - z^{-4}}{1 - z^{-1}} \quad \text{and} \quad H(\omega) = \frac{\sin(2\omega)}{\sin(\omega/2)}e^{-j3\omega/2}$$

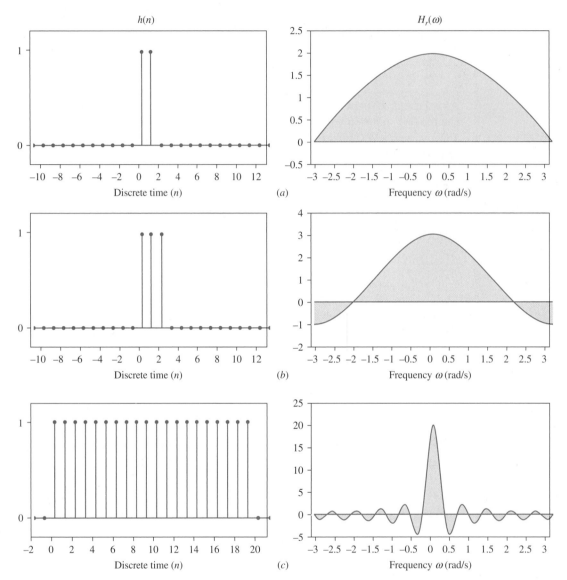

FIGURE 18.2 The unit-sample response $h(n)$ and the frequency response $H(\omega)$ of a uniform window that starts at $n = 0$ and consists of M unit samples are

$$h(n) = \begin{cases} 1, & 0 \le n < M \\ 0, & \text{elsewhere,} \end{cases} \qquad H(\omega) = \frac{1 - e^{-jM\omega}}{1 - e^{-j\omega}} = \frac{\sin(M\omega/2)}{\sin(\omega/2)} e^{-j\omega\frac{(M-1)}{2}} = H_r(\omega)e^{j\theta(\omega)}$$

$H_r(\omega)$ is a real function of ω. It can assume positive or negative values with alternating positive and negative lobes that diminish as ω increases. Its zero-crossings are at $\omega = 2k\pi/M$. This figure shows three $h(n)$ (left column) and their $H_r(\omega)$ (right column) for a 2-sample window *(a)*, a 3-sample window *(b)*, and a 20-sample window *(c)*. The frequency axis is $-\pi < \omega < \pi$ in rad/s.

Example

18.18

Find the frequency response of the uniform window containing 20 unit samples, $h(n) = u(n) - u(n - 20)$.

Solution

$$H(z) = \frac{1 - z^{-20}}{1 - z^{-1}} \quad \text{and} \quad H(\omega) = \frac{\sin(10\omega)}{\sin(\omega/2)} e^{-j10\omega}$$

The zero-crossings of $H(\omega)$ occur at $\omega = \pm k\pi/10$, k being a nonzero integer.

Observation

Zero-crossings of the frequency response of a uniform window of size M occur at $\pm 2k\pi/M$, $k \neq 0$. The main lobe of the frequency response is $4\pi/M$ rad/s wide. A wide window has a narrower lobe and vice versa. See Figure 18.2.

$H(\omega)$ May Be Obtained from $H\{z\}$

So far, we have developed $H(z)$ and $H(\omega)$ independently from each other. Note that we have used the same notation, H, for both the frequency response and the system function. This is no accident or oversight. As one can see from their definitions, the frequency response and the system function of an LTI system are related to each other. The frequency response $H(\omega)$ of a discrete-time system is a special case of $H(z)$ and, therefore, may be obtained from it by setting $z = e^{j\omega}$.

$$H(\omega) = H(z)\big|_{z=e^{j\omega}}$$

To derive the above result, observe that

1. $H(z)$ is the complex scale factor for z^n inputs.
2. $\cos \omega n = \mathcal{RE}\{e^{j\omega n}\}$
3. The real part of the response to an input is the response to the real part of that input: $\mathcal{RE}\{x(n)\} \Rightarrow \mathcal{RE}\{y(n)\}$.

These observations are summarized below:

$$z^n \Longrightarrow H(z)z^n$$
$$e^{j\omega n} \Longrightarrow H(\omega)e^{j\omega n}, \quad \text{where} \ H(\omega) = H(z)|_{z=e^{j\omega}}$$
$$\cos(\omega n) = \mathcal{RE}\{e^{j\omega n}\} \Longrightarrow \mathcal{RE}\{H(\omega)e^{j\omega n}\} = |H(\omega)|\cos(\omega n + \theta)$$

$H(z)$ is, therefore, a convenient representation of $H(\omega)$. They both represent the system in the frequency domain.

18.19

Find the frequency response of an FIR system with the system function $H(z) = 1 + z^{-1} + z^{-2} + z^{-3}$. Find the response of the system to the input $x(n) = \cos(\omega n)$. At what frequency is the above response zero?

Solution

$$H(\omega) = 1 + e^{-j\omega} + e^{-j2\omega} + e^{-j3\omega}$$

$$= e^{-j3\omega/2} \left(e^{j3\omega/2} + e^{j\omega/2} + e^{-j\omega/2} + e^{-j3\omega/2} \right)$$

$$= 2e^{-j3\omega/2} \left[\cos(3\omega/2) + \cos(\omega/2) \right]$$

$$= 4e^{-j3\omega/2} \cos(\omega) \cos(\omega/2) = H_r(\omega) e^{j\theta(\omega)}$$

$$H_r(\omega) = 4\cos(\omega)\cos(\omega/2) \quad \text{and} \quad \theta(\omega) = -3\omega/2$$

The output is

$$y(n) = H_r(\omega)\cos(\omega n + \theta) = 4\cos(\omega)\cos(\omega/2)\cos(\omega n - 3\omega/2)$$

The system blocks sinusoids at $\omega = \pi/2$ and π, where $H_r(\omega) = 0$.

18.20

Find the frequency response of an IIR system with two poles at $0.9e^{\pm j45°}$ and a double zero at the origin. See Figure 18.3(a). Find the square of its magnitude, $|H(\omega)|^2$, at $\omega = 0$ (DC), $\omega = \pi/4$ (closest to the poles), and $\omega = \pi$ (farthest away from the poles). Determine the 3-dB bandwidth. Sketch $|H(\omega)|^2$ versus ω.

Solution

Let the poles be at $\rho e^{\pm j\theta}$. First find $H(z)$, then set $z = e^{j\omega}$ to find $H(\omega)$. Assuming a unity gain factor, we have

$$H(z) = \frac{1}{(1 - \rho e^{j\theta} z^{-1})(1 - \rho e^{-j\theta} z^{-1})} = \frac{1}{1 - 2\rho\cos\theta z^{-1} + \rho^2 z^{-2}}$$

$$H(\omega) = \frac{1}{1 - 2\rho\cos\theta(\cos\omega - j\sin\omega) + \rho^2(\cos 2\omega - j\sin 2\omega)}$$

$$= \frac{1}{(1 - 2\rho\cos\theta\cos\omega + \rho^2\cos 2\omega) + j(2\rho\cos\theta\sin\omega - \rho^2\sin 2\omega)}$$

$$|H(\omega)|^2 = \frac{1}{(1 - 2\rho\cos\theta\cos\omega + \rho^2\cos 2\omega)^2 + (2\rho\cos\theta\sin\omega - \rho^2\sin 2\omega)^2}$$

$$= \frac{1}{(1 + 4\rho^2\cos^2\theta + \rho^4) - 4\rho(1 + \rho^2)\cos\theta\cos\omega + 2\rho^2\cos(2\omega)}$$

With $\rho = 0.9$ and $\theta = 45°$:

$$|H(\omega)|^2 = \frac{1}{3.2761 - 4.6075 \cos \omega + 1.62 \cos 2\omega}$$

$$|H(0)|^2 = \frac{1}{3.2761 - 4.6075 + 1.62} = 3.465$$

$$|H(\pi/4)|^2 = \frac{1}{3.2761 - 4.6075(\sqrt{2}/2) + 0} = 55.2318$$

$$|H(\pi)|^2 = \frac{1}{3.2761 + 4.6075 + 1.62} = 0.1052$$

The maximum of $|H|$ occurs at $\omega = 45°$. The 3-dB bandwidth is $\Delta\omega = \omega_h - \omega_\ell$, where ω_h and ω_ℓ are the upper and lower half-power frequencies, respectively.

$$|H(\omega_h)|^2 = |H(\omega_\ell)|^2 = \frac{1}{2}|H_{Max}|^2 = 55.2318/2 = 27.6159$$

ω_h and ω_ℓ are then the roots of the equation

$$\frac{1}{3.2761 - 4.6075 \cos \omega + 1.62 \cos(2\omega)} = 27.6159$$

$$\cos \omega - 0.3516 \cos(2\omega) = 0.7032$$

Because the poles are close to the unit circle, the behavior of $H(\omega)$ near a pole (i.e., in the vicinity of $\omega = 45°$) is governed mainly by that pole. In that neighborhood, variations in $H(\omega)$ can be approximated by variations in its distance from the pole, $(1 - p_1 z^{-1})$, as seen in Figure 18.3(b). For details of this approximate analysis see section 18.4 in this chapter, "Vectorial Interpretation of $H(z)$ and $H(\omega)$." Therefore, an approximate solution of the above equation will give us $\omega_{\ell,\,h} \approx \pi/4 \pm 0.1$ rad/s $= 45° \pm 5.73° = 39.27°$ and $50.73°$, where $0.1 = (1 - 0.9)$ is the shortest distance between a pole and the unit circle. The 3-dB bandwidth is $\Delta\omega \approx 0.2$ rad/s $= 50.73° - 39.27° = 11.46°$. More exact values are:

$$\omega_l = 38.2162°, \; \omega_h = 50.4865°, \; \Delta\omega = \omega_h - \omega_l = 12.2703°$$

$H(\omega)$ Is 2π-Periodic

The frequency response is a function of $e^{j\omega}$ as shown by

$$H(\omega) = H(z)\big|_{z=e^{j\omega}}$$

But $e^{j\omega} = e^{j(\omega+2\pi)}$, and, therefore, $H(\omega) = H(\omega+2\pi)$. Consequently, $H(\omega)$ is periodic with period 2π. In our analysis we consider $H(\omega)$ for $-\pi < \omega < \pi$ (which covers one period).

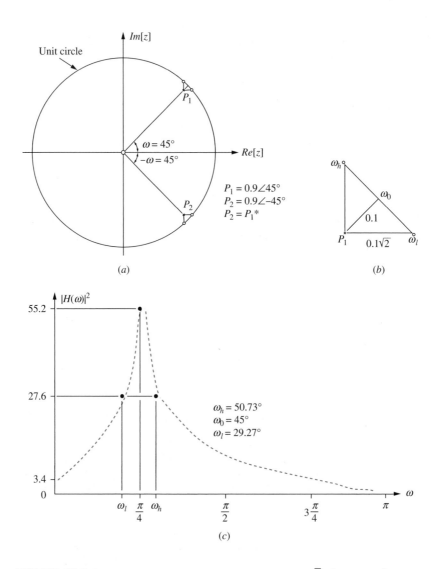

FIGURE 18.3 The system described by $H(z) = 1/(1 - 0.9\sqrt{2}z^{-1} + 0.81z^{-2})$ has two poles at $0.9e^{\pm j45°}$ and a double zero at the origin, as seen in **(a)**. $|H(\omega)|^2$ is plotted versus ω in **(c)**. The maximum of $|H|$ occurs at $\omega = 45°$, nearest the pole. The 3-dB bandwidth is approximately $\Delta\omega = 2(1 - 0.9) = 0.2$ rad/s $= 11.46°$, where $(1 - 0.9)$ is the shortest distance between each pole and the unit circle. Because the poles are close to the unit circle, the 3-dB bandwidth is almost evenly divided between the two sidebands, resulting in the lower and higher half-power frequencies $\omega_\ell = 45 - 11.46/2 = 39.27°$ and $\omega_h = 45 + 11.46/2 = 50.73°$. See **(b)**. The double zero at the origin has no effect on $|H(\omega)|^2$. For details see Example 18.20.

Example 18.21

Examine the periodicity of the frequency response $H(\omega) = 4e^{-j3\omega/2}\cos(\omega)\cos(\omega/2)$ obtained in Example 18.19.

Solution

$$
H(\omega + 2\pi) = 4e^{-j3(\omega+2\pi)/2}\cos(\omega + 2\pi)\cos[(\omega + 2\pi)/2]
$$

$$
= 4e^{-j3\omega/2}e^{-j3\pi}\cos(\omega + 2\pi)\cos(\omega/2 + \pi)
$$

$$
= -4e^{-j3\omega/2}\cos(\omega + 2\pi)\cos(\omega/2 + \pi)
$$

$$
= 4e^{-j3\omega/2}\cos(\omega)\cos(\omega/2) = H(\omega)
$$

Finding $|H(\omega)|^2$ from $H(z)H(z^{-1})$

The square of the magnitude of the frequency response is $|H(\omega)|^2 = H(\omega)H^*(\omega)$, where $H^*(\omega)$ is the complex conjugate of $H(\omega)$. However,

$$
H(\omega) = H(z)|_{z=e^{j\omega}}
$$

$$
H^*(\omega) = H(z)|_{z=e^{-j\omega}} = H(z^{-1})|_{z=e^{j\omega}}
$$

$$
|H(\omega)|^2 = H(\omega)H^*(\omega) = H(z)H(z^{-1})|_{z=e^{j\omega}}
$$

The above equation provides an easy way of obtaining the magnitude of the frequency response from the system function.

Example 18.22

A zero (or pole) at $z = a$ contributes $(1 - az^{-1})$ to the numerator (or denominator, if a pole) of $H(z)$. The contribution to the square of the magnitude of $H(\omega)$ is

$$
|H|^2 = (1 - az^{-1})(1 - az)|_{z=e^{j\omega}}
$$

$$
= 1 + a^2 - a(z + z^{-1})|_{z=e^{j\omega}}
$$

$$
= 1 + a^2 - 2a\cos\omega
$$

Example 18.23

A pair of conjugate zeros (or poles) at $\rho e^{\pm j\theta}$ contribute the following to $H(z)$:
$$
(1 - \rho e^{j\theta}z^{-1})(1 - \rho e^{-j\theta}z^{-1}) = 1 - 2\rho\cos\theta z^{-1} + \rho^2 z^{-2}
$$

Their contribution to $|H(\omega)|^2$ is

$$(1 - 2\rho \cos\theta z^{-1} + \rho^2 z^{-2})(1 - 2\rho \cos\theta z + \rho^2 z^2)|_{z=e^{j\omega}} =$$

$$1 + 4\rho^2 \cos^2\theta + \rho^4 - 2\rho(1 + \rho^2)\cos\theta(z + z^{-1}) + \rho^2(z^2 + z^{-2})|_{z=e^{j\omega}} =$$

$$(1 + 4\rho^2 \cos^2\theta + \rho^4) - 4\rho(1 + \rho^2)\cos\theta\cos\omega + 2\rho^2\cos(2\omega)$$

This is in agreement with the result derived directly in Example 18.20, where we can see the contributions of the conjugate zeros and poles.

Trajectory of ω in the z-Plane

The frequency response is obtained by $H(\omega) = H(z)|_{z=e^{j\omega}}$. In this substitution, the locus of $z = e^{j\omega}$ is the unit circle in the z-plane because $|z| = |e^{j\omega}| = 1$. The point representing $z = 1$ corresponds to $\omega = 0$ and $z = -1$ corresponds to $\omega = \pm\pi$. Conversely, a point at $\omega = 0$ on the unit circle is represented by $z = 1$ and $\omega = \pm\pi$ is represented by $z = -1$[1]. See Figure 18.4. As ω sweeps from 0 to π, its trajectory in the z-plane will be the upper-half of the unit circle, from $z = 1$ to $z = -1$, traversed in the counterclockwise direction. Likewise, as ω goes from 0 to $-\pi$, its trajectory will be the lower-half of the unit circle, from $z = 1$ to $z = -1$, traversed in the clockwise direction.

Some examples of ω (in rad/s) and their corresponding z-values on the unit circle are listed below.

ω	$\Longleftrightarrow$	z	
$\pm\pi/6$	$\Longleftrightarrow$	$e^{\pm j\pi/6}$	$= \sqrt{3}/2 \pm j1/2$
$\pm\pi/4$	$\Longleftrightarrow$	$e^{\pm j\pi/4}$	$= \sqrt{2}/2(1 \pm j)$
$\pm\pi/2$	$\Longleftrightarrow$	$e^{\pm j\pi/2}$	$= \pm j$
$\pm 2\pi/3$	$\Longleftrightarrow$	$e^{\pm j2\pi/3}$	$= -1/2 \pm j\sqrt{3}/2$
$\pm 3\pi/4$	$\Longleftrightarrow$	$e^{\pm j3\pi/4}$	$= \sqrt{2}/2(-1 \pm j)$
$\pm 5\pi/4$	$\Longleftrightarrow$	$e^{\pm j5\pi/4}$	$= \sqrt{2}/2(-1 \pm j)$
$\pm 9\pi/4$	$\Longleftrightarrow$	$e^{\pm j9\pi/4}$	$= \sqrt{2}/2(1 \pm j)$

---

[1]This property also illustrates the periodicity of the frequency response.

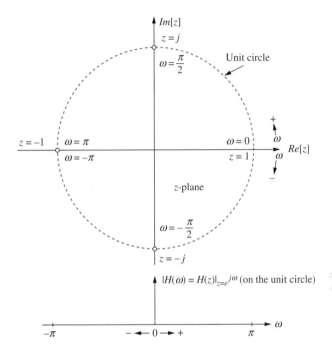

FIGURE 18.4 Visualization of z and ω in the z-plane under the transformation $z = e^{j\omega}$. The point $z = 1$ corresponds to $\omega = 0$ and $z = -1$ corresponds to $\omega = \pm\pi$. Conversely, the point at $\omega = 0$ on the unit circle is represented by $z = 1$ and $\omega = \pm\pi$ is represented by $z = -1$. As ω sweeps from 0 toward π, its trajectory in the z-plane will be the upper-half of the unit circle, from $z = 1$ to $z = -1$, traversed in the counterclockwise direction. Likewise, as ω goes from 0 to $-\pi$, its trajectory will be the lower-half of the unit circle, from $z = 1$ to $z = -1$, traversed in the clockwise direction.

18.4 Vectorial Interpretation of $H(z)$ and $H(\omega)$

A rational system function with M zeros and N poles may be written in terms of its poles and zeros:

$$H(z) = H_0 \frac{\Pi_k(z - z_k)}{\Pi_\ell(z - p_\ell)}$$

where z_k, $k = 1, 2, \cdots M$, are zeros of the system, p_ℓ, $\ell = 1, 2, \cdots N$, are poles of the system, and H_0 is a gain factor. To simplify the discussion, assume H_0 is a positive number. A zero at z_k contributes $(z - z_k)$ to the numerator of the system function. In the z-plane this is a vector drawn from the point z_k (the zero) to a point z (where the system function is to be evaluated). Similarly, a pole at p_ℓ contributes $(z - p_\ell)$ to the denominator of the system function. In the z-plane this is a vector drawn from the pole at p_ℓ to point z. See Figure 18.5. Let $\overline{B}_k = (z - z_k)$ designate the vector from z_k to z and $\overline{A}_\ell = (z - p_\ell)$ designate the vector from p_ℓ to z. Then,

$$H(z) = H_0 \frac{\overline{B}_1 \times \overline{B}_2 \times \overline{B}_3 \cdots \times \overline{B}_M}{\overline{A}_1 \times \overline{A}_2 \times \overline{A}_3 \cdots \times \overline{A}_N} = H_0 \frac{\Pi_k \overline{B}_k}{\Pi_\ell \overline{A}_\ell}$$

$$|H(z)| = H_0 \frac{|\overline{B}_1| \times |\overline{B}_2| \times |\overline{B}_3| \cdots \times |\overline{B}_M|}{|\overline{A}_1| \times |\overline{A}_2| \times |\overline{A}_3| \cdots \times |\overline{A}_N|} = H_0 \frac{\Pi_k |\overline{B}_k|}{\Pi_\ell |\overline{A}_\ell|}$$

$$\angle H(z) = \left[\angle \overline{B}_1 + \angle \overline{B}_2 + \angle \overline{B}_3 \cdots + \angle \overline{B}_M \right] - \left[\angle \overline{A}_1 + \angle \overline{A}_2 + \angle \overline{A}_3 \cdots + \angle \overline{A}_N \right]$$

$$= \sum_{k=1}^{M} \angle \overline{B}_k - \sum_{\ell=1}^{N} \angle \overline{A}_\ell$$

The above interpretation provides a graphical technique to qualitatively derive some properties of $H(z)$ or to evaluate its exact value at desired points in the z-plane. In the next section we apply the above interpretation to the points on the unit circle where the frequency response is found.

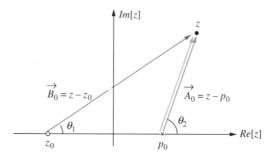

FIGURE 18.5 Vectorial interpretation of $H(z)$, with a zero at z_0 and a pole at p_0, using the pole- zero plot. The zero contributes $(z - z_0)$, the vector $\overline{B}_0$ drawn from the zero to z, to the numerator of the system function. Similarly, the pole contributes $(z - p_0)$, the vector $\overline{A}_0$ drawn from the pole to z, to the denominator.

$$\overline{B}_0 = (z - z_0) \text{ and } \overline{A}_0 = (z - p_0)$$

$$H(z) = H_0 \frac{z - z_0}{z - p_0} = H_0 \frac{1 - z_0 z^{-1}}{1 - p_0 z^{-1}} = H_0 \frac{\overline{B}_0}{\overline{A}_0}$$

$$|H(z)| = H_0 \frac{|\overline{B}_0|}{|\overline{A}_0|} \text{ and } \angle H(z) = \theta_1 - \theta_2, \text{ assuming } H_0 \text{ is a positive number.}$$

This interpretation provides a graphical technique for evaluation of $H(z)$ and its qualitative analysis.

Determination of $H(\omega)$ from the Pole-Zero Plot

The frequency response is found by evaluating $H(z)$ on the unit circle. Starting at the point $z = 1$ ($\omega = 0$) and moving counterclockwise on the unit circle toward the point $z = -1$ ($\omega = \pi$), we can obtain a complete picture of $H(\omega)$ from the pole-zero plot. Because vectors from poles are in the denominator, the magnitude of the frequency response is increased when the point on the unit circle representing that frequency approaches the neighborhood of a pole. Similarly, because the zero vectors are in the numerator, the magnitude of the frequency response is decreased when its representative point on the unit circle approaches a zero or its neighborhood. The following example illustrates the utility of the vectorial interpretation and its graphical technique.

Example **18.25**

A discrete-time causal system is described by the system function

$$H(z) = \frac{1 + z^{-1}}{1 - 0.9\sqrt{2}z^{-1} + 0.81z^{-2}}$$

Use the vectorial interpretation of $H(z)$ to evaluate $H(\omega)$. Show that the system is a bandpass filter. Find the 3-dB frequencies ω_ℓ and ω_h. Evaluate $H(\omega)$ at $\omega = 0$, ω_ℓ, $\pi/4$, ω_h, $\pi/2$, $3\pi/4$, and π. Sketch the magnitude and phase of $H(\omega)$.

Solution
The system function is

$$H(z) = \frac{1 + z^{-1}}{1 - 0.9\sqrt{2}z^{-1} + 0.81z^{-2}}$$

$$= \frac{1 + z^{-1}}{(1 - 0.9e^{j\pi/4}z^{-1})(1 - 0.9e^{-j\pi/4}z^{-1})}$$

$$= \frac{z(z + 1)}{(z - 0.9e^{j\pi/4})(z - 0.9e^{-j\pi/4})}$$

It has a pair of poles at $p_{1,2} = 0.9e^{\pm j\pi/4}$ [shown by points A_1 and A_2 on Figure 18.6(a)] and two zeros at $z = 0$ and $z = -1$ (shown by points B_0 and B_1, respectively). Let C designate the location of z on the unit circle at which $H(\omega)$ is evaluated. Then, at $z = e^{j\omega}$ we have

$$z = \overline{B_0 C}, \quad z + 1 = \overline{B_1 C}, \quad z - 0.9e^{j\pi/4} = \overline{A_1 C}, \quad z - 0.9e^{-j\pi/4} = \overline{A_2 C}$$

$$H(\omega) = H(z)\big|_{z=e^{j\omega}} = \frac{\overline{B_0 C} \times \overline{B_1 C}}{\overline{A_1 C} \times \overline{A_2 C}}$$

$$|H(\omega)| = |\overline{B_1 C}|/(|\overline{A_1 C}| \times \overline{A_2 C}|)$$

$$\angle H(\omega) = [\omega + \angle \overline{B_1 C}] - [\angle \overline{A_1 C} + \angle \overline{A_2 C}]$$

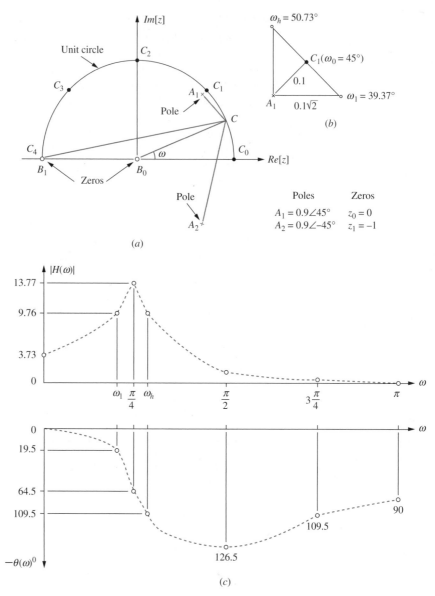

FIGURE 18.6 Evaluation of the frequency response of a system having a pair of poles at $p = 0.9e^{\pm j\pi/4}$ and two zeros at $z = 0$ and $z = -1$, using the vectorial interpretation: **(a)** Pole-zero plot; **(b)** finding the 3-dB frequency; **(c)** sketch of the frequency response. See Example 18.25.

At $\omega = \pi/4$ the magnitude of the vector $\overline{A_1 C} = 0.1\angle 45°$ is at its minimum, producing $|H(\omega)|_{Max}$ as $|H(\pi/4)| = 13.74\angle -64.5°$. In the vicinity of $\omega = \pi/4$ the vector $\overline{A_1 C}$ is the most influential of the four vectors in shaping $H(\omega)$. The other three vectors ($\overline{B_0 C}$, $\overline{B_1 C}$, and $\overline{A_2 C}$) don't change appreciably, and their contributions to $H(\omega)$ remain almost at a constant value.[2] A 3-dB frequency is where $\overline{A_1 C} = \sqrt{2} \times 0.1$. See Figure 18.6($b$). There exist two such points, ω_ℓ and ω_h, located at $\omega = \pi/4 \pm 0.1$ rad/s. More generally, we may conclude that in the neighborhood of $\omega = \pi/4$ (which includes the 3-dB frequencies ω_ℓ and ω_h) we will have

$$H(\omega) = \frac{1.38\angle -19.5°}{\overline{A_1 C}} \quad \text{and} \quad H(\pi/4) = \frac{1.38\angle -19.5°}{0.1\angle 45°} = 13.8\angle -64.5$$

$$|\overline{A_1 C_{\ell,h}}| \approx 0.1 \times \sqrt{2}$$

$$\omega_{\ell,h} \approx \pi/4 \pm 0.1 \text{ rad} = 39.3°, \text{ and } 50.7°$$

$$|H(\omega_{\ell,h})| \approx 9.76$$

$$\angle H(\omega_{\ell,h}) \approx -64.5 \pm 45° = -19.5° \text{ and } -109.5°$$

From the graphical representation of Figure 18.6(a) and (b) we have the following approximations:

Location of z	ω	$H(\omega)$	$|H(\omega)|$	$\angle H(\omega)$	Comments
C_0	$0°$	$\overline{B_0 C_0} \times \overline{B_1 C_0}/(\overline{A_1 C_0} \times \overline{A_2 C_0})$	3.73	$0°$	
C_ℓ	$39.3°$	$\overline{B_0 C_\ell} \times \overline{B_1 C_\ell}/(\overline{A_1 C_\ell} \times \overline{A_2 C_\ell})$	9.76	$-19.5°$	Lower 3-dB frequency, $\|H\| = \dfrac{\|H_{Max}\|}{\sqrt{2}}$
C_1	$45°$	$\overline{B_0 C_1} \times \overline{B_1 C_1}/(\overline{A_1 C_1} \times \overline{A_2 C_1})$	13.77	$-64.5°$	Near a pole, $\|H_{Max}\| = 13.77$
C_h	$50.7°$	$\overline{B_0 C_h} \times \overline{B_1 C_h}/(\overline{A_1 C_h} \times \overline{A_2 C_h})$	9.76	$-109.5°$	Upper 3-dB frequency, $\|H\| = \dfrac{\|H_{Max}\|}{\sqrt{2}}$
C_2	$90°$	$\overline{B_0 C_2} \times \overline{B_1 C_2}/(\overline{A_1 C_2} \times \overline{A_2 C_2})$	1.1	-126.5	
C_3	$135°$	$\overline{B_0 C_3} \times \overline{B_1 C_3}/(\overline{A_1 C_3} \times \overline{A_2 C_3})$	0.3	$-109.5°$	
C_4	$180°$	$\overline{B_0 C_4} \times \overline{B_1 C_4}/(\overline{A_1 C_4} \times \overline{A_2 C_4})$	0	$-90°$	Zero of the system

The frequency response is sketched in Figure 18.6(c).

18.5 Transforming Continuous Time to Discrete Time

In many applications, including digital signal processing, digital control, or digital communication, one may require to convert an existing continuous-time system to a discrete one, or design a new system in the discrete-time domain based on a design in the

---

[2] At $\omega = \pi/4$ we can find $(\overline{B_0 C} \times \overline{B_1 C})/\overline{A_2 C} \approx 1.38\angle -19.5°$. However, to find the 3-dB frequency we don't need to know this value.

continuous-time domain. In this migration one needs to determine transformation of both the signals and the systems. The sampling theorem provides us with the guidelines to convert a continuous-time signal to discrete time. The correspondence between the frequencies in these domains have been described and exemplified. In this section, using three examples, we first refresh the subject of mapping the frequency from continuous time to discrete time, as a result of uniform time-domain sampling.

With regard to systems we need to detemine how to convert or construct a discrete-time system that exhibits characteristics identical to or similar to its continuous-time counterpart. LTI systems are described by $H(s)$ (in the continuous-time domain) and $H(z)$ (in the discrete-time domain). Mathematically expressed, we need to determine mapping from the s-domain to the z-domain. The transformation should answer questions such as the following: What becomes of the continuous-time differential and integral operators? Do the discrete-time difference operators approximate time-derivative in the continuous-time domain satisfactorily? Does the $j\Omega$ axis in the s-plane (i.e., the frequency in the continuous-time) maps onto the unit circle in the z-plane (the frequency in the discrete time), and if so, is the mapping linear and uniform? How do the parameters of the transformation (e.g., sampling rate) affect the pole-zero locations and stability of the system? Does the left-half plane in the s-domain map onto the inside of the unit circle in the z-plane? In Chapter 15 (section 15.11) the relationship between the Laplace transform of a continuous-time function and the z-transform of its sampled sequence was discussed. This section first discusses the frequency mapping from Ω to ω (using Ω and F for continuous time, and ω and f for the discrete, respectively). It then briefly introduces impulse invariant and bilinear transformations to provide a window through which some light is shed on the subject. These two transformations will be used later in this chapter in relation to design of digital filters.

Mapping the Frequency from Continuous Time to Discrete Time

The frequency response of discrete-time systems is 2π-periodic in ω. Therefore, it is enough to consider it for one period, normally $-\pi < \omega < \pi$. What does this range mean in terms of the actual frequency in the continuous-time domain? The relation between the frequencies in the continuous-time (shown by Ω) and discrete-time (shown by ω) domains depends on the sampling scheme used and the transformation from continuous-time to discrete-time signals and systems. Here it suffices to say that under uniform sampling (e.g., at the rate of F_s samples per second, or Hz) the transformation from Ω (the continuous-time signal) to ω (the resulting discrete-time signal) is linear, $\Omega = F_s\omega$, such that $\omega = 0$ in the discrete-time analysis corresponds to a DC signal in the continuous-time domain and $\omega = 2\pi$ corresponds to a continuous-time sinusoidal signal at $\Omega = 2\pi F_s$ rad/s. Similarly, using F (the continuous-time signal) and f (the resulting discrete-time signal) we will have $F = F_s f$, such that $f = 0$ in the discrete-time analysis corresponds to a DC signal in the continuous-time domain and $f = 1$ corresponds to a continuous-time sinusoidal signal at $F = F_s$ (all in Hz).

The answer to the question posed at the beginning of this section may now be summarized as follows. The range $-\pi < \omega < \pi$ rad/s maps onto $-\pi F_s < \Omega < \pi F_s$ (equivalently, $-0.5 < f < 0.5$ Hz maps onto $-F_s/2 < F < F_s/2$). Sampling a continuous-time signal at the rate of F_s downshifts the frequency components above $F_s/2$. In this book the ratio F/F_s will be called the normalized frequency.

Example 18.26

The recording of a continuous-time voltage $x(t)$ contains a signal of interest $s(t)$ along with a 60-Hz disturbance $x(t) = s(t) + \alpha \cos(120\pi t + \phi)$. $x(t)$ is sampled at the rate of F_s samples per second and the samples are sent through a three-sample-wide window. Determine F_s so that the 60-Hz disturbance is eliminated in the output of the window.

Solution

The unit-sample response, system function, and frequency response of the discrete-time window are given below.

$$h(n) = \{\underset{\uparrow}{1}, 1, 1\}$$

$$H(z) = 1 + z^{-1} + z^{-2}, \quad H(\omega) = 1 + e^{-j\omega} + e^{-j2\omega} = (1 + 2\cos\omega)e^{-j\omega}$$

The frequency response has only 1 zero at $\omega = 2\pi/3$. The 60-Hz component ($\Omega = 120\pi$) will be eliminated if its ω in the discrete-time domain falls on the location of the zero of $H(\omega)$. The relationship between Ω (in continuous time) and ω (in discrete time) is $\Omega = F_s\omega$. Therefore, one needs $120\pi = F_s(2\pi/3)$, which yields $F_s = 180$ Hz.

Example 18.27

Extend Example 18.26 to the case where, when recording $x(t)$, a 60-Hz disturbance and its two harmonics are added to the desired signal $s(t)$.

$$x(t) = s(t) + \sum_{k=1}^{3} \alpha_k \cos(120k\pi t + \phi_k)$$

$x(t)$ is sampled at the rate of F_s and the samples are sent through a discrete-time system with the unit-sample response $h(n)$. Show that by choosing $F_s = 420$ Hz and $h(n) = \{1, 1, 1, \underset{\uparrow}{1}, 1, 1, 1\}$ one blocks the disturbances at 60, 120, and 180 Hz from reaching the output. Discuss parallels with Example 18.26.

Solution

For simplicity, we examine a noncausal window. This will not affect our solution of this example as we are interested in the magnitude of the frequency response only. The frequency response $H(\omega)$ of the seven-point uniform window is found below.

$$H(z) = z^3 + z^2 + z + 1 + z^{-1} + z^{-2} + z^{-3} = \frac{z^3 - z^{-4}}{1 - z^{-1}}$$

$$H(\omega) = e^{j3\omega} + e^{j2\omega} + e^{j\omega} + 1 + e^{-j\omega} + e^{-j2\omega} + e^{-j3\omega} = \frac{\sin(7\omega/2)}{\sin(\omega/2)}$$

From the above representation of $H(\omega)$ by the ratio of sinusoids, it is easily seen that the frequency response has 3 zeros at $\omega = 2\pi/7,\ 4\pi/7$ and $6\pi/7$.[3] But, $\Omega = F_s \times \omega$ and at $F_s = 420$ Hz the zeros of $H(\omega)$ translate to $\Omega = 120\pi,\ 240\pi$, and 360π, which correspond to 60, 120, and 180 Hz. The given filter blocks disturbances at the above harmonics.

Example **18.28**

Now consider another case of signal plus disturbance, similar to Example 18.26, where the recorded data is $x(t) = w(t) + \alpha \cos(\Omega t + \phi)$ and $\Omega = 120\pi$. However, in contrast to Example 18.26, here the signal of interest is the 60-Hz component and $w(t)$ is the disturbance to be reduced by filtering. The aim is to filter $x(t)$ such that $w(t)$ is attenuated and $\cos(\Omega t + \phi)$ is enhanced. Here again $x(t)$ is sampled at the rate of F_s samples per second. The samples are then sent through a discrete-time system with a pair of poles at $0.9e^{\pm j\theta}$ and 2 zeros at ± 1. Find F_s for $\theta = \pi/6,\ \pi/4,\ \pi/3$, and $\pi/2$. Discuss the choice of θ.

Solution

The discrete-time signal, system function, and frequency response are

$$x(n) = w(n) + \alpha \cos(\omega n + \phi), \quad \text{where } \omega = \frac{\Omega}{F_s}$$

$$H(z) = \frac{(1 - z^{-1})(1 + z^{-1})}{(1 - 0.9e^{j\theta}z^{-1})(1 - 0.9e^{-j\theta}z^{-1})} = \frac{1 - z^{-2}}{1 - 1.8(\cos\theta)z^{-1} + 0.81z^{-2}}$$

$$H(\omega) = \frac{1 - e^{-j2\omega}}{1 - 1.8(\cos\theta)e^{-j\omega} + 0.81e^{-j2\omega}}$$

The magnitude of the frequency response is at a maximum when $\omega = \theta$ (see Examples 18.20 and 18.25), which translates to $\Omega = F_s\theta$, from which $F_s = \Omega/\theta$. But $\Omega = 120\pi$. Therefore, $F_s = 120\pi/\theta$. For $\theta = \pi/6,\ \pi/4,\ \pi/3$, and $\pi/2$ we will have $F_s = 720,\ 480,\ 360$, and 240 Hz, respectively. The corresponding values of the frequency response are

θ	$\pi/6$	$\pi/4$	$\pi/3$	$\pi/2$
$|H|$	10.4828	10.5118	10.5215	10.5263
$\angle H$	5.2087°	3.0128°	1.7405°	0.0000°

The filter with $\theta = \pi/2$ introduces zero phase shift and will be chosen.

Sampling the Unit-Impulse Response: Impulse-Invariance Transformation

A continuous-time LTI system may be converted to a discrete-time system by sampling its time responses. The sampled time-response will then describe the discrete system.

---

[3]The alternate representation is $H(\omega) = 1 + 2\cos\omega + 2\cos(2\omega) + 2\cos(3\omega)$.

For example, sampling the unit-impulse response produces an impulse-invariant transformation. Here is an example.

The unit-impulse response $h(t) = e^{-t}u(t)$ of a continuous-time system is sampled every T sec. Find the system function $H(z)$ of the resulting discrete-time system and determine the location of its pole for $T = 3,\ 2,\ 1,\ 0.5,\ 0.1,\ 0.01,$ and 0.001 sec.

Solution

$$h(n) = h(t)\big|_{t=nT} = e^{-nT}u(n) = a^n u(n),\quad \text{where } a = e^{-T}.$$

$$\text{Note that } a < 1 \text{ for } T > 0.$$

$$H(z) = \frac{1}{1 - az^{-1}}, \quad \text{where } |z| < a \text{ is inside the circle.}$$

The discrete-time system has a pole at $z = a < 1$. The location of the pole depends on the sampling rate as shown below. For all sampling rates, the pole remains inside the unit circle indicating that $h(n)$ diminishes as n increases, making the discrete system stable.

$T = 1$ msec	pole at 0.999
$T = 10$ msec	pole at 0.99
$T = 100$ msec	pole at 0.905
$T = 500$ msec	pole at 0.607
$T = 1$ sec	pole at 0.368
$T = 2$ sec	pole at 0.135
$T = 3$ sec	pole at 0.05

Remark

$H(z)$ may also be found directly from $H(s)$ as described in Chapter 15, section 15.11.

Approximating Derivatives by Difference Operators

An LTI differential equation may be converted to a difference equation by sampling the time and then approximating the derivatives by difference operators. The pole-zero locations of the resulting discrete system will depend on the sampling scheme and its rate. Here are two examples.

Approximation by the forward-difference operator

The differential equation

$$\frac{dy(t)}{dt} + y(t) = x(t)$$

may be converted to a difference equation by sampling $x(t)$ and $y(t)$ every T seconds (i.e., by letting $t = nT$) and approximating the derivative by the following expression representing the forward-difference operator:

$$\frac{dy(t)}{dt} \approx \frac{y(t+T) - y(t)}{T}$$

The differential equation then becomes

$$\frac{y(t+T) - y(t)}{T} + y(t) = x(t)$$

$$y(t+T) - (1-T)y(t) = Tx(t)$$

With $t = nT$, we obtain the difference equation and system function

$$y(n+1) - (1-T)y(n) = Tx(n)$$

$$zY(z) - (1-T)Y(z) = TX(z)$$

$$H(z) = \frac{Tz^{-1}}{1 - \alpha z^{-1}}, \quad \text{where } \alpha = 1 - T$$

The discrete-time system has a pole, here at $z = 1 - T$. The location of the pole depends on the sampling rate as given below. However, at low sampling rates the pole may move outside the unit circle, indicating that $h(n)$ grows as n increases, making the discrete system unstable.

$T = 1$ msec	pole at 0.999
$T = 10$ msec	pole at 0.99
$T = 100$ msec	pole at 0.9
$T = 500$ msec	pole at 0.5
$T = 1$ sec	pole at 0
$T = 2$ sec	pole at -1
$T = 3$ sec	pole at -2

Remark
The forward-difference approximation of the derivative in this example is satisfactory at high sampling rates, but unacceptable at low rates.

Approximation by the backward-difference operator

The continuous-time differential equation

$$\frac{dy(t)}{dt} + y(t) = x(t)$$

may also be converted to a difference equation by sampling $x(t)$ and $y(t)$ every T seconds and approximating the derivative by the backward-difference operator:

$$\frac{dy}{dt} \approx \frac{y(n) - y(n-1)}{T}$$

Following the approach of Example 18.30 we obtain the following system function.

$$H(z) = \frac{Tz}{\beta z - 1}, \quad \text{where } \beta = 1 + T$$

The resulting discrete-time system has a pole inside the unit circle and is stable at all sampling rates.

Approximating Integral: The Case of The Bilinear Transformation

The following mapping from s to z

$$s = \frac{2}{T} \frac{1 - z^{-1}}{1 + z^{-1}}$$

(where T is the sampling interval) is called a bilinear tranformation. It is used as a main tool in the design of digital filters and control systems. An example of IIR filter design by bilinear transformation is given in section 18.10. The following example illustrates how the bilinear transformation is obtained by approximating an integral by the area of a trapezoid.

An example of bilinear transformation

Consider the differential equation

Time domain:	$\dfrac{dy(t)}{dt} + ay(t) = bx(t)$	(18.1a)
Frequency domain:	$sY(s) + aY(s) = bX(s)$	(18.1b)
System function:	$H(s) = \dfrac{Y(s)}{X(s)} = \dfrac{b}{s + a}$	(18.1c)

In Examples 18.30 and 18.31 we approximated derivatives of $y(t)$ by difference operators. In this example we approximate an integral element by a trapezoid. Start with

$$y(t) = \int_{t_0}^{t} y'(\tau)d\tau + y(t_0) \tag{18.2a}$$

$$y(nT) = \int_{(n-1)T}^{nT} y'(\tau)d\tau + y(nT - T) \tag{18.2b}$$

Approximate the value of the integral in equation (18.2b) by the area of a trapezoid $ABCD$. See Figure 18.7.

$$\int_{(n-1)T}^{nT} y'(\tau)d\tau \approx \left[y'(nT) + y'(nT - T) \right]\frac{T}{2} \tag{18.3}$$

Plug (18.3) in (18.2b) to find

$$y(nT) = \frac{y\prime(nT) + y\prime(nT - T)}{2}T + y(nT - T) \tag{18.4}$$

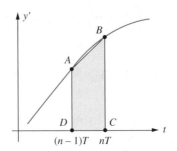

FIGURE 18.7 Approximating an integral element by the area of the trapezoid.

But from the differential equation we have

$$y\prime(nT) = bx(nT) - ay(nT) \tag{18.5a}$$

$$y\prime(nT - T) = bx(nT - T) - ay(nT - T) \tag{18.5b}$$

Now plug (18.5) in (18.4) and note that $x(nT) \equiv x(n)$, $y(nT) \equiv y(n)$.

$$y(n) = \frac{T}{2}\Big[bx(n) - ay(n) + bx(n - 1) - ay(n - 1)\Big] + y(n - 1) \tag{18.6}$$

Collect the y terms on the left side and x terms on the right.

Time domain $\qquad \Big[\frac{aT}{2} + 1\Big]y(n) + \Big[\frac{aT}{2} - 1\Big]y(n - 1) = b\frac{T}{2}[x(n) + x(n - 1)]$

$$\tag{18.7a}$$

Frequency domain $\quad \Big[\frac{aT}{2} + 1\Big]Y(z) + \Big[\frac{aT}{2} - 1\Big]Y(z)z^{-1} = b\frac{T}{2}(1 + z^{-1})X(z)$

$$\tag{18.7b}$$

System function $\quad H(z) = \dfrac{Y(z)}{X(z)} = \dfrac{b}{\dfrac{2}{T}\left(\dfrac{1 - z^{-1}}{1 + z^{-1}}\right) + a} \tag{18.7c}$

Compare equation (18.7c) with (18.1c) to derive the following s to z transformation

$$s = \frac{2}{T}\frac{1 - z^{-1}}{1 + z^{-1}} = \frac{2}{T}\frac{z - 1}{z + 1}$$

Bilnear transformation is commonly used in digital filter design.

18.6 Digital Filters

The name *digital filter* applies to a discrete-time system that performs one or more (possibly interrelated) operations such as:

Frequency selection

Smoothing

Separating a signal from disturbances

Reducing noise

Analyzing time-series data

Producing a certain phase shift or delay

Detecting presence of a known signal by template matching

Detecting random signals embedded in noise

The operation of a filter may also be related to some optimality or adaptation in such tasks as:

Prediction

Modeling and identification

Statistical analysis and estimation

This chapter provides an introduction to digital filters as frequency-selective systems with the input shown by $x(n)$ and the output by $y(n)$. The frequency response is used as the main tool for analysis and synthesis.

Filter Types

The filters discussed in this chapter are LTI discrete-time systems classifed based on the frequency response, $H(\omega)$.[4] As in the analog case, we consider five types of filters: (1) low-pass, (2) high-pass, (3) bandpass, (4) bandstop, and (5) all-pass. Ideally

$$H(\omega) = \begin{cases} 1, & \text{within the passband} \\ 0, & \text{within the stopband} \end{cases}$$

In addition to the above, we will also consider some other filter types such as differentiators. Note that the frequency response of the digital filter is periodic in ω with a period of 2π. Therefore, we consider only one period of $H(\omega)$, normally for $-\pi \leq \omega \leq \pi$. As in the analog case, ideal digital filters are not realizable because of their brickwall frequency response shape. They are approximated and realizable by practical filters.

Filter Description

Any one of the following functions can completely describe a digital filter.

1. Input-output difference equation
2. System function $H(z)$
3. Frequency response $H(\omega)$
4. Unit-sample response $h(n)$
5. Pole-zero locations plus a gain factor

---

[4]To be mathematically precise, the frequency response of a discrete-time LTI system should be written as $H(e^{j\omega})$ or $H(e^{j2\pi f})$. However, whether choosing ω or f as the frequency variable, throughout this textbook we show the frequency response function by $H(\omega)$ or $H(f)$.

Of the above, the frequency response $H(\omega)$ explicitly exhibits the characteristics normally considered to be of interest in a filter. The unit-sample response $h(n)$ serves as the end result in the design of a filter. The difference equation, system function $H(z)$, and the pole-zero map serve as analysis and design tools. A zero of $H(z)$ at z_0, which is close to the unit circle, will pull down the magnitude response for frequencies on the unit circle that are close to z_0. A pole pulls up the magnitude of the frequency response in its neighborhood. To block a certain frequency, the filter should have a zero on the unit circle at that frequency, and to enhance it, the filter should have a pole inside the unit circle near that frequency. Examples of simple low-pass and bandpass filters will be given in the next two sections. Examples of simple high-pass filter are found at the end of this chapter in the solved problems. In addition to the projects at the end of this chapter, several other chapters include projects on filtering. These are: an introduction to discrete-time signal processing in Chapter 11, low-pass filtering by difference equation in Chapter 14, and FIR and IIR notch filter design in Chapter 15.

Filter's Pole-Zero Locations and the Frequency Response

From the previous discussions about system function of an LTI system we have drawn qualitative observations concerning the effect of the locations of poles and zeros on the frequency response. In this section we discuss quantitatively the role poles and zeros play in shaping the filter's frequency response, suggesting some insight into the function and structure of filters and also tools for their design.

The system function $H(z)$ provides a filter's z-domain input-output relationship $Y(z) = H(z)X(z)$. Alternatively, it provides the scale factor, a more physically tangible measure in the time domain, for the input signal $x(n) = z^n$ to pass through the filter, resulting in the output $y(n) = H(z)z^n$. The complex number $z = \rho e^{j\omega}$, called the complex frequency, is represented by a point in the z-plane. For a discrete-time sinusoidal signal $x(n) = \mathcal{RE}\{e^{j\omega n}\}$ we have $z = e^{j\omega}$ and the signal is represented by a point on a circle with the radius $\rho = 1$, called the unit circle.[5] Hence, the unit circle plays a significant role in the analysis and design of LTI digital filters. Based on the above explanation we can explore the role played by poles and zeros on a filter's frequency response.

Effects of Zeros

Let z_0 be a zero of the system function; that is, $H(z_0) = 0$. The presence of the zero at z_0 prevents passage of the signal $x(n) = z_0^n$ through the filter. To block a certain frequency, the filter should have a zero on the unit circle at that frequency. The zero at $z_0 = e^{\pm j\omega_0}$ not only notches out the signal at the frequency ω_0, but also attenuates signals at the neighboring frequencies. [See the vectorial interpretation of $H(z)$.] The bandwidth of the notch will be wide unless it is compensated by nearby poles.

---

[5] Stated differently, the frequency of a continuous-time sinusoidal waveform is mapped onto the unit circle of the z-plane.

Effects of Poles

A pole pulls up the magnitude of the frequency response in its neighborhood. Poles are never placed on the unit circle or outside of it because of stability considerations. A pole placed nearby a zero reduces the attenuative effect of the zero. Therefore, one or more poles placed near a notching zero (the zero being on the unit circle but the poles inside it) narrow the notch bandwidth.

Contributions to the Frequency Response

Zeros (and poles) are the roots of the numerator (and denominator) of the system function. For a filter with real-valued coefficients, zeros (and poles) are either real-valued numbers or complex conjugate pairs. A real-valued zero at ρ multiplies the system function by $1 - \rho z^{-1}$ and a pole divides by it. A pair of zeros at $\rho e^{\pm j\omega_0}$ contributes the following multiplier term to the numerator (or to the denominator in case of poles) of the system function.

$$(1 - \rho e^{j\omega_0} z^{-1})(1 - \rho e^{-j\omega_0} z^{-1}) = 1 - 2\rho \cos \omega_0 z^{-1} + \rho^2 z^{-2}$$

A pair of zeros on the unit circle at $e^{\pm j\omega_0}$ multiplies the system function by $1 - 2\cos \omega_0 z^{-1} + z^{-2}$.

A Dominant Pair of Poles

A pair of poles at $\rho e^{\pm j\omega_0}$ close to the unit circle (ρ being nearly one) will dominate the frequency response in its neighborhood such that

$$H(\omega) \approx H_0 \frac{1}{1 - 2\rho \cos \omega_0 z^{-1} + \rho^2 z^{-2}}$$

which is similar to a second-order system (see parallels with the discussion on modeling an analog system with a pair of dominant poles, Chapter 9). The multiplier H_0 represents the effect of other poles and zeros, and will be almost a constant as long as the other poles and zeros remain far away. The dominant pole then pulls up the magnitude of the frequency response. The closer the pole is to the unit circle, the sharper the frequency response becomes in the neighborhood.

18.7 Simple Filters

Simple Low-Pass Filters

A simple low-pass filter can be constructed from a zero at $z = -1$ (which corresponds to $\omega = \pi$ and thus blocks high frequencies), a pole near $z = 1$ (which corresponds to DC, thus boosts up low frequencies), or their combinations. Three such examples are discussed below. Their graphical characteristics (i.e., the frequency response, unit-sample response, and pole-zero plot) visualize the basic features of low-pass filters. The examples also illustrate how the passband and stopband-edge frequencies are determined from the given attenuations A_p and A_s, respectively.

Example 18.33

A low-pass filter with a single zero

A realizable digital filter is described by $y(n) = x(n) + x(n-1)$.

a. Find the filter's $H(z)$, $H(\omega)$, $h(n)$, pole-zero locations, magnitude response, and the DC gain in dB.

b. Find the passband and stopband edge frequencies for $A_p = 3$ and $A_s = 20$, both in dB.

c. Repeat for $A_p = 1$ and $A_s = 40$.

Solution

a. $H(z) = 1 + z^{-1}$

A single zero at $z = -1$. (The pole at $z = 0$ has no effect on the frequency response and is ignored.)

$$H(\omega) = 1 + e^{-j\omega} = 2\cos\left(\frac{\omega}{2}\right) e^{-j\omega/2}$$

$$h(n) = d(n) + d(n-1)$$

$$20 \log |H(\omega)| = 20 \log \left[2\cos\left(\frac{\omega}{2}\right)\right] \approx 6 + 20 \log \cos\left(\frac{\omega}{2}\right) \text{ dB}$$

DC gain ≈ 6 dB

b. $A_p = 3$ dB, $6 + 20 \log \cos\left(\frac{\omega_p}{2}\right) = 6 - 3$, $\cos\left(\frac{\omega_p}{2}\right) = \frac{\sqrt{2}}{2}$, $\omega_p = 90°$

$A_s = 20$ dB, $6 + 20 \log \cos\left(\frac{\omega_s}{2}\right) = 6 - 20$, $\cos\left(\frac{\omega_s}{2}\right) = 0.1$, $\omega_s = 168.52°$

c. $A_p = 1$ dB, $6 + 20 \log \cos\left(\frac{\omega_p}{2}\right) = 6 - 1$, $\cos\left(\frac{\omega_p}{2}\right) = 0.8912$, $\omega_p = 53.94°$

$A_s = 40$ dB, $6 + 20 \log \cos\left(\frac{\omega_s}{2}\right) = 6 - 40$, $\cos\left(\frac{\omega_s}{2}\right) = 0.01$, $\omega_s = 178.85°$

Example 18.34

A low-pass filter with a single pole

Repeat Example 18.33 for the realizable digital filter described by $y(n) - 0.9y(n-1) = x(n)$.

Solution

a. $H(z) = \dfrac{1}{1 - 0.9z^{-1}}$

A single pole at $z = 0.9$. (The zero at $z = 0$ has no effect on the frequency response and is ignored.)

$$H(\omega) = \frac{1}{1 - 0.9e^{-j\omega}} = \frac{1}{(1 - 0.9\cos\omega) + j0.9\sin\omega}$$

$$= \frac{1}{\sqrt{1.81 - 1.8\cos\omega}} e^{-j\theta}, \quad \text{where } \theta = \tan^{-1}\left[\frac{0.9\sin\omega}{1 - 0.9\cos\omega}\right]$$

$$h(n) = (0.9)^n u(n)$$

$$20\log|H(\omega)| = -10\log(1.81 - 1.8\cos\omega) \text{ dB}$$

$$\text{DC gain} = -10\log 0.01 = 20 \text{ dB}$$

b. $A_p = 3$ dB, $-10\log(1.81 - 1.8\cos\omega_p) = 20 - 3 = 17$, $\cos\omega_p = 0.99447, \omega_p = 6.03°$

$A_s = 20$ dB, $-10\log(1.81 - 1.8\cos\omega_s) = 20 - 20 = 0$, $\cos\omega_s = 0.45, \omega_s = 63.25°$

c. $A_p = 1$ dB, $-10\log(1.81 - 1.8\cos\omega_p) = 20 - 1 = 19$, $\cos\omega_p = 0.99856,$ $\omega_p = 3.07°$

$A_s = 40$ dB, $-10\log(1.81 - 1.8\cos\omega_s) = 20 - 40 = -20$,

No answer; see below.

With regard to the no-answer in part c, having only a single pole at $z = 0.9$, the maximum attenuation provided by the filter to occur at $z = -1$ (corresponding to $\omega = \pi$), at which point $H(z) = \frac{1}{1.9} = 0.5263$ (or -5.57 dB), which is still greater than the -20 dB required by the $A_s = -40$-dB criterium in part c. In other words, the maximum high-frequency attenuation is 25.57 dB (below the DC gain) and doesn't reach the $A_s = 40$-dB level.

E*xample*

18.35   **A low-pass filter with a zero and a pole**

Repeat Example 18.33 for the realizable digital filter described by $y(n) - 0.9y$ $(n - 1) = x(n) + x(n - 1)$.

Solution

a. $H(z) = \dfrac{1 + z^{-1}}{1 - 0.9z^{-1}}$

A zero at $z = -1$ and a pole at $z = 0.9$.

$$H(\omega) = \frac{1 + e^{-j\omega}}{1 - 0.9e^{-j\omega}} = \frac{2\cos\left(\frac{\omega}{2}\right)}{\sqrt{1.81 - 1.8\cos\omega}}e^{-j\theta},$$

where $\theta = \dfrac{\omega}{2} + \tan^{-1}\left[\dfrac{0.9\sin\omega}{1 - 0.9\cos\omega}\right].$

$$h(n) = [d(n) + d(n-1)] * (0.9)^n u(n) = (0.9)^n u(n) + (0.9)^{n-1} u(n-1)$$

$$= d(n) + 1.9 \times (0.9)^{n-1} u(n-1)$$

$$20\log|H(\omega)| = 10\log\left[\frac{4\cos^2\left(\frac{\omega}{2}\right)}{1.81 - 1.8\cos\omega}\right]$$

$$= 6 + 20\log\cos\left(\frac{\omega}{2}\right) - 10\log(1.81 - 1.8\cos\omega)\ \text{dB}$$

DC gain $= 6 - 10\log 0.01 = 26$ dB

b. $20\log|H(\omega)| = 10\log\left[\dfrac{2(1 + \cos\omega)}{1.81 - 1.81\cos\omega}\right]$

$A_p = 3$ dB, $10\log\left[\dfrac{2(1 + \cos\omega_p)}{1.81 - 1.8\cos\omega_p}\right] = 26 - 3 = 23,$ $\cos\omega_p = 0.99445,\ \omega_p = 6.04°$

$A_s = 20$ dB, $10\log\left[\dfrac{2(1 + \cos\omega_s)}{1.81 - 1.8\cos\omega_s}\right] = 26 - 20 = 6,$ $\cos\omega_s = 0.56794,\ \omega_s = 55.39°$

c. $A_p = 1$ dB, $10\log\left[\dfrac{2(1 + \cos\omega_p)}{1.81 - 1.8\cos\omega_p}\right] = 26 - 1 = 25,$ $\cos\omega_p = .99853,\ \omega_p = 3.1°$

$A_s = 40$ dB, $10\log\left[\dfrac{2(1 + \cos\omega_s)}{1.81 - 1.8\cos\omega_s}\right] = 26 - 40 = -14,$ $\cos\omega_s = -0.9306,\ \omega_s = 158.53°$

Adding a zero at $z = -1$ has pegged the magnitude response to zero at $\omega = \pi$, and pulled it down at other frequencies, making it possible to achieve -40-dB attenuation within the range of $0 < \omega < \pi$.

Simple Bandpass Filters

A simple bandpass filter can be constructed by placing a pair of complex conjugate poles inside and near the unit circle [see vectorial interpretation of $H(z)$ in this chapter].

Adding two zeros at $z = \pm 1$ (corresponding to $\omega = 0$ and π) will completely block the DC and high frequencies. Such a simple bandpass filter will be of second-order and have a narrowband. The bandwidth may be increased by employing several poles within the desired passband, making it a higher-order filter. An example of a simple bandpass filter is given below.

Example 18.36

A bandpass filter

A digital filter has a pair of complex conjugate poles at $z = \pm j\rho$, where $\rho < 1$, 2 zeros at $z = \pm 1$, and $H(z)|_{z=\infty} = 1$.

a. Find $H(z)$ and $H(\omega) = |H(\omega)|\angle H(\omega)$. Specify the filter's important values [maximum value of $|H(\omega)|$, 3-dB frequencies (ω_ℓ and ω_h], and the bandwidth $\Delta\omega = \omega_h - \omega_\ell$ in terms of ρ, the pole's distance from the origin].

b. Repeat a for $\rho = 0.8, 0.9, 0.95$, and 0.99. Determine the limiting values for ω_ℓ, ω_h, and $\Delta\omega$ as the poles approach the unit circle.

c. Plot $|H(\omega)|$ and $\angle H(\omega)$ versus ω for $\rho = 0.8, 0.95$, and 0.99.

Solution

a.
$$H(z) = \frac{(1 - z^{-1})(1 + z^{-1})}{(1 - j\rho z^{-1})(1 + j\rho z^{-1})} = \frac{1 - z^{-2}}{1 + \rho^2 z^{-2}}$$

$$H(\omega) = \frac{1 - e^{-j2\omega}}{1 + \rho^2 e^{-j2\omega}} = \frac{(j2\sin\omega)e^{-j\omega}}{1 + \rho^2 e^{-j2\omega}} = \frac{j2\sin\omega}{e^{j\omega} + \rho^2 e^{-j\omega}}$$

$$= \frac{j2\sin\omega}{(1 + \rho^2)\cos\omega + j(1 - \rho^2)\sin\omega} = \frac{2|\sin\omega|}{\sqrt{(1 - \rho^2)^2 + 4\rho^2\cos^2\omega}} e^{j\theta}$$

where $|H(\omega)| = \dfrac{2|\sin\omega|}{\sqrt{(1 - \rho^2)^2 + 4\rho^2\cos^2\omega}}$, and

$$\theta = \begin{cases} -\tan^{-1}\left[\frac{1-\rho^2}{1+\rho^2}\tan\omega\right] + \frac{\pi}{2}, & 0 < \omega < \pi \\ -\tan^{-1}\left[\frac{1-\rho^2}{1+\rho^2}\tan\omega\right] - \frac{\pi}{2}, & -\pi < \omega < 0 \end{cases}$$

$$|H_{Max}| = H(z)\big|_{z=j} = \frac{2}{1 - \rho^2}$$

$$|H(\omega_\ell)| = |H(\omega_h)| = \frac{|H_{Max}|}{\sqrt{2}} = \frac{\sqrt{2}}{1 - \rho^2}$$

Find $\omega_{\ell,h}$ from $|H(\omega)|^2 = \dfrac{4\sin^2\omega}{(1 - \rho^2)^2 + 4\rho^2\cos^2\omega} = \dfrac{2}{(1 - \rho^2)^2}$

The answer is $\cos(\omega_{\ell,h}) = \pm\dfrac{(1 - \rho^2)}{\sqrt{2(1 + \rho^4)}}$

b. $\rho = 0.8,$ $H(\omega)\big|_{Max} = 5.56,$ $\omega_\ell = 77.62°,$ $\omega_h = 102.38°,$ and $\Delta\omega = 24.76° = 0.4321$ rad/s

$\rho = 0.9,$ $H(\omega)\big|_{Max} = 10.52,$ $\omega_\ell = 84.00°,$ $\omega_h = 95.99°,$ and $\Delta\omega = 11.99° = 0.2092$ rad/s

$\rho = 0.95,$ $H(\omega)\big|_{Max} = 20.51,$ $\omega_\ell = 87.07°,$ $\omega_h = 92.93°,$ and $\Delta\omega = 5.87° = 0.1024$ rad/s

$\rho = 0.99,$ $H(\omega)\big|_{Max} = 100.50,$ $\omega_\ell = 89.42°,$ $\omega_h = 90.58°,$ and $\Delta\omega = 1.16° = 0.0202$ rad/s

As the poles approach the unit circle, $\rho \to 1$ and

$$\cos(\omega_{\ell,h}) = \pm\frac{(1-\rho)(1+\rho)}{\sqrt{2(1+\rho^4)}} \to \pm(1-\rho), \quad \Delta\omega \to 2(1-\rho)$$

See Figure 18.8.

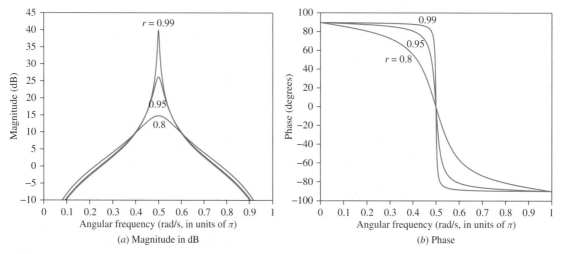

FIGURE 18.8 (For Example 18.36) Frequency response for the bandpass filter $H(z) = (1 - z^{-2})/(1 - \rho^2 z^{-2})$. Plots are for $\rho = 0.8, \; 0.95, \;$ and 0.99. Magnitudes are shown in **(a)** (dB units) and phase is in **(c)**. The magnitude peak (always at $\omega = 90°$) grows and becomes narrower as the poles approach the unit circle.

18.8 Filter Design

In filter design, one often begins with a desired frequency response $H_d(\omega)$ or some specifications related to it (e.g., the filter's passband and stopband frequencies and their attenuations, A_p and A_s, respectively). The conditions to be satisfied will be called design criteria. The aim is to find the $h(n)$ of a realizable filter (FIR or IIR), which meets some design criteria. The $h(n)$ and $H(\omega)$ are DTFT pairs as shown below.

$$H(\omega) = \sum_{n=0}^{\infty} h(n)e^{-j\omega n} \quad \Longleftrightarrow \quad h(n) = \frac{1}{2\pi}\int_{-\pi}^{\pi} H(\omega)e^{j\omega}d\omega$$

The acceptability of the design outcome may be given in terms of satisfying the desired specification (A_p and A_s) or a measure of error (such as the rms of error between the magnitude responses). The following sections briefly present some of the most commonly used methods in filter design. A method applicable to both FIR and IIR design is pole-zero placement. This method will be illustrated in the next section. Other methods specific to FIR or IIR cases are also used. Design methods to be discussed in the remainder of this chapter provide approximations to the desired filters. The following methods are not meant to provide an exhaustive picture of design methods and processes. They are to illustrate, by way of examples, some basic aspects of the design process. In FIR design we find the finite-length $h(n)$ such that the filter approximates the desired one and complies with constraints. The following two methods will be discussed.

$$\text{FIR design} \begin{cases} \text{a. Windowing method (also called truncating Fourier method)} \\ \text{b. Frequency sampling method} \end{cases}$$

In IIR design we find the system function $H(z)$ by transforming the $H(s)$ of an analog filter. The following two transformations will be discussed.

$$\text{IIR design} \begin{cases} \text{a. Impulse-invariance transformation from analog to digital} \\ \text{b. Bilinear transformation} \end{cases}$$

Ideal Low-Pass Filter

In filter design by windowing we begin with the unit-sample response of an ideal filter. A low-pass type is normally chosen. The frequency response of an ideal low-pass filter with the cutoff at ω_0 is

$$H_{\text{LP}}(\omega) = \begin{cases} 1, & -\omega_0 < \omega < \omega_0 \\ 0, & \text{elsewhere} \end{cases}$$

The unit-sample response of the filter is obtained by taking the inverse DTFT of $H(\omega)$.

$$h(n) = \frac{1}{2\pi} \int_{-\pi}^{\pi} H(\omega)e^{j\omega n}d\omega = \frac{1}{2\pi} \int_{-\omega_0}^{\omega_0} e^{j\omega n}d\omega = \begin{cases} \frac{\sin(\omega_0 n)}{\pi n}, & n \neq 0 \\ \frac{\omega_0}{\pi}, & n = 0 \end{cases}$$

The above $h(n)$ produces zero phase. Shift $h(n)$ to the right by k units of time to produce $\overline{h}(n) = h(n - k)$ and its transform $\overline{H}(\omega)$. The pair are

$$\overline{h}(n) = h(n-k) = \begin{cases} \frac{\sin[\omega_0(n-k)]}{\pi(n-k)} & n \neq k \\ \frac{\omega_0}{\pi} & n = k \end{cases} \implies \overline{H}(\omega) = \begin{cases} e^{-jk\omega} & -\omega_0 < \omega < \omega_0 \\ 0, & \text{elsewhere} \end{cases}$$

The shifted unit sample response is still symmetric (around k) and has a linear phase. It still is, however, infinitely long and no amount of shift to the right would make it causal. The ideal filter is, therefore, unrealizable. The design methods to be described in this chapter approximate an unrealizable ideal filter by a realizable one in which $h(n) = 0$ for $n < 0$.

An ideal high-pass filter can be constructed by $H_{HP}(\omega) = 1 - H_{LP}(\omega)$, which, in the time domain, translates into $h_{HP}(n) = d(n) - h_{LP}(n)$. Similarly, ideal bandpass and bandstop filters can be constructed from ideal low-pass filters.

18.9   Filter Design by Pole-Zero Placement

A zero of $H(z)$ at z_0 that is close to the unit circle will pull down the magnitude response for frequencies close to z_0, and a pole will pull up the magnitude response in its neighborhood. To block a certain frequency, the filter should have a zero on the unit circle at that frequency, and to enhance it, the filter should have a pole inside the unit circle near that frequency. These properties can be used to design a filter. Poles and zeros will be placed at such locations that the frequency response satisfies the design criteria. A good example is the design of notch filters. Examples of FIR and IIR notch filters are the subject of projects in Chapter 15 (also see problems 16 and 17 in this chapter). These examples show how adding poles can improve the filter's performance in meeting the design objectives and controlling the bandwidth. Example 18.36 discussed a narrow-band band-pass filter with a pair of poles. The following example illustrates a wideband bandpass filter design by the method of pole-zero placement.

The objective is to design a wideband bandpass filter with a passband of $\pi/4 < \omega < 3\pi/4$, by placing poles in the vicinity of the passband and zeros within or nearby the stopband.

Solution

a.   We place 10 zeros on the unit circle at $e^{\pm j\omega_k}$, where $\omega_k = \pm(1 + 2k)\pi/40$, $k = 0, 1, 2, 3$, and 4. The zeros on the right-half side of the circle produce a stopband from $\omega = 0$ to $\pi/4$. The mirror images, with respect to the j axis (the axis of $\omega = \pi/2$), of those zeros produce a stopband from $\omega = 3\pi/4$ to π. To provide the passband we place 10 poles at $0.9e^{\pm j\omega_k}$, where $\omega_k = \pi/2 \pm (1 + 2k)$ $\pi/40$, , $k = 0, 1, 2, 3$, and 4. These poles, in conjunction with their complex conjugates, provide the passband from $\omega = \pi/4$ to $3\pi/4$. The pole-zero constellation and the resulting magnitude response are shown in Figure 18.9(a) (upper row). Because the zeros are located on the unit circle, the magnitude plot has 10 zeros (shown by an infinite dB attenuation).

b.   We move the zeros of part a inside the unit circle to a distance of 0.7 from the origin and maintain the same angles as before. We also add 10 more zeros which are their mirror images with respect to the unit circle (at a distance of $1/0.7$ from the origin). The poles are kept inside the circle at their previous locations. There are no zeros on the unit circle. See Figure 18.9(b) (lower row). The Matlab program used to design the above filter can also be used to explore the effect on the filter's performance, of changing the number of poles or zeros and their distances from the origin.

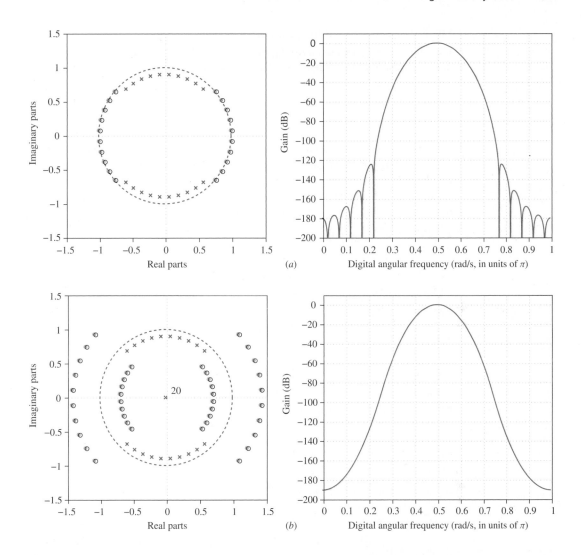

FIGURE 18.9 Two bandpass filters designed by the pole-zero placement method. See Example 18.37.

18.10 FIR Filter Design

The unit-sample response of an FIR filter has a finite length and its system function has no poles (poles at the origin are not considered). FIR filters may be designed based on one of three methods: (1) placing zeros of the system function at the desired locations in the z-plane (and no poles), (2) windowing a desired unit-sample response, and (3) sampling a desired frequency response. Examples of FIR filter design by zero-placement have already been given. In this section we introduce the two other methods.

FIR Filter Design by Windowing

In FIR design by windowing one takes the inverse DTFT of a desired frequency response (e.g., an ideal low-pass filter) and multiplies it by a finite-duration window. Truncating the unit-sample response of an ideal filter to make it of finite length is mathematically equivalent to multiplying it with a rectangular window of finite length M, the simplest of windows. It could be easily shown that for a given window length, a rectangular window minimizes the rms of error between the ideal and practical frequency responses. But the rectangular window has drawbacks and, therefore, other types are considered (with trade-offs). The rectangular window's major drawback is the ringing at the transition frequencies due to Gibbs' phenomenon, which cannot be overcome by increasing the window length. Therefore, in FIR design by windowing, given desired specifications (e.g., A_p, ω_p and A_s, ω_s) two questions are considered:

1. How long should the window be?
2. What type of window should be used?

Consideration of the above two factors requires a measure of error and a set of specifications to judge the performance of the actual filter compared with the desired frequency response.

Theory

Let $\bar{h}(n)$ and $\bar{H}(\omega)$ be the unit-sample and frequency responses of the desired filter, respectively. Let $h(n) = \bar{h}(n) \times w(n)$ be a finite segment of $\bar{h}(n)$ seen and weighted through a window $w(n)$ of length M. The frequency response of the FIR filter is obtained from the convolution of the frequency response of the window with that of the desired filter.

$$H(\omega) = \frac{1}{2\pi}\bar{H}(\omega) \star W(\omega)$$

As an example, consider an ideal low-pass filter $\bar{h}(n)$ with cutoff frequency at ω_0 and k units of delay. An M-tap low-pass FIR filter is obtained by multiplying $\bar{h}(n)$ with a rectangular window of length $M, k = (M-1)/2$. The following table summarizes the unit-sample and frequency responses of the ideal filter, the rectangular window, and the resulting FIR filter.

An M-Tap Low-Pass FIR Filter Under a Rectangular Window

Ideal low-pass delayed filter	$\bar{h}(n) = \frac{\sin[\omega_0(n-k)]}{\pi(n-k)}$ delay: $k = (M-1)/2$	$\bar{H}(\omega) = \begin{cases} e^{-jk\omega} & -\omega_0 < \omega < \omega_0 \\ 0, & \text{elsewhere} \end{cases}$
M-tap rectangular window	$w(n) = \begin{cases} 1, & 0 \le n \le M-1 \\ 0, & \text{elsewhere} \end{cases}$	$W(\omega) = \frac{\sin(\frac{\omega M}{2})}{\sin(\frac{\omega}{2})}e^{-jk\omega}$
M-tap low-pass FIR filter	$h(n) = \begin{cases} \frac{\sin[\omega_0(n-k)]}{\pi(n-k)}, & 0 \le n \le M-1 \\ 0, & \text{elsewhere} \end{cases}$	$H(\omega) = \frac{1}{2\pi}\bar{H}(\omega) \star W(\omega)$

Figure 18.10 shows, in the time and frequency domains, the ideal filter, the rectangular window, and the resulting FIR filter for $\omega = \pi/5$ and $M = 31$. Note the Gibbs' effect at the transition band of the FIR filter.

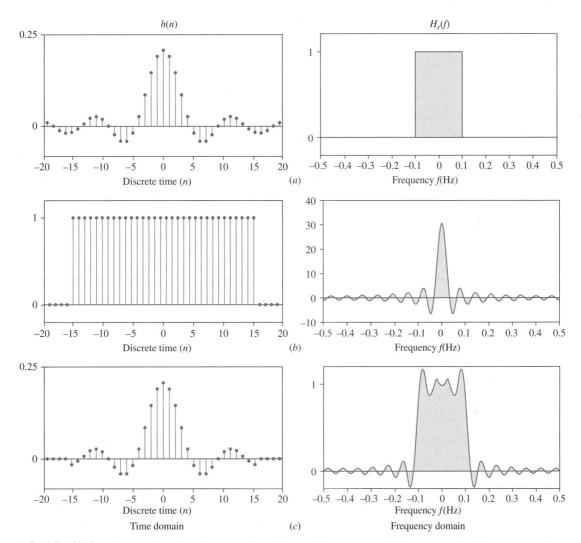

FIGURE 18.10 A discrete-time ideal low-pass filter with cutoff frequency at 0.1 Hz (top row); a 31-tap rectangular window (middle row); the resulting 31-tap FIR filter (bottom row) based on the rectangular window. Time-domain characteristics are shown in the left column. The frequency-domain characteristics (in the right column) are shown by $H(\omega) = H_r(\omega)e^{j\theta}$. Because of no time shift and even symmetry in $h(n)$, we have $\theta = 0$ and $H(\omega) = H_r(\omega)$.

In summary, the unit-sample response of the FIR filter is found by multiplying the unit-sample response of the corresponding ideal filter with a window of finite length M.

Several windows of interest in digital signal processing have been introduced already. Each window type provides a trade-off between attenuation in the stopband, ripples in the passband, and the length of the transition bandwidth. For example, a rectangular window has a sharp transition but contains ringing (Gibbs' effect). As $M \to \infty$, the transition band narrows and the frequency response of the FIR filter approaches that of the ideal filter except at the neighborhood of the cutoff frequency ω_0. The ringings persist regardless of how wide the window is. To remedy this, one may use a window type such as Hanning, Hamming, Blackman, or Kaiser, which eliminate the ringing but widen the transition band. Example 18.38 and Figure 18.11 illustrate the contrast between the rectangular and Hamming windows.

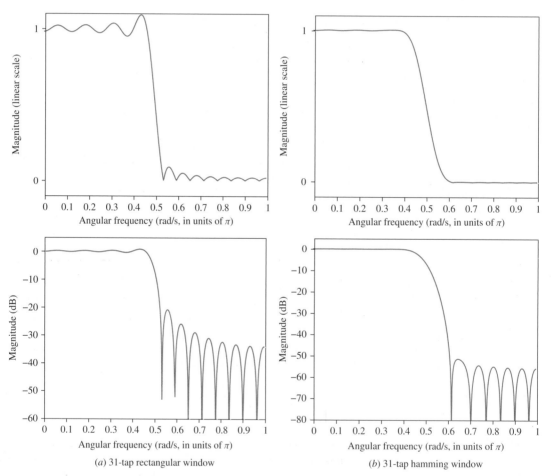

(a) 31-tap rectangular window

(b) 31-tap hamming window

FIGURE 18.11 (For Example 18.38) Magnitude responses of two low-pass FIR filters designed through windowing the unit-sample response of an ideal filter. Filter 1 (left column) uses a rectangular window and filter 2 uses the Hamming window. Both are 31 samples long. The top row shows the magnitudes in linear scale and the bottom in dB. Gibbs' phenomenon, which manifested itself in filter 1 (rectangular window), disappears in filter 2 (Hamming window). The transition band is enlarged when using a Hamming window.

Example
18.38

Two low-pass FIR filters were designed by windowing the unit-sample response of an ideal filter with cutoff frequency at $\omega_0 = \frac{\pi}{2}$. Filter 1 (left column in Figure 18.11) uses the rectangular window and filter 2 uses the Hamming window, both 31 samples long. Magnitude responses of the two filters are shown in Figure 18.11 (linear scale on the top row, and dB scale on the bottom). The ringing in the rectangular window shows the contrast between the two windows. Gibbs' phenomenon, which manifests itself in filter 1 (rectangular window), disappears in filter 2 (Hamming window). The transition band is enlarged under the Hamming window.

FIR Filter Design by the Frequency Sampling Method

The frequency response, being the DTFT of the unit-sample response, is a function of continuous variable ω and is periodic with a period 2π. For an FIR filter of length M, it is found from the expression

$$H(\omega) = \sum_{n=0}^{M-1} h(n)e^{-j\omega n}$$

where $h(n)$ is the unit-sample response of the filter. Consider samples of $H(\omega)$ taken uniformly every $\frac{2\pi}{M}$ rad/s at the following equally spaced frequencies:

$$\omega_k = \frac{2\pi k}{M}, \quad \begin{cases} k = 0, 1, \cdots, \frac{M-1}{2}, & M \text{ odd} \\ k = 0, 1, \cdots, \frac{M}{2}, & M \text{ even} \end{cases}$$

Call these H_k. This provides M samples within the range of $\omega = 0$ to 2π. Then,

$$H_k \equiv H(\omega_k) = \sum_{n=0}^{M-1} h(n)e^{-j\omega_k n} = \sum_{n=0}^{M-1} h(n)e^{-j\frac{2\pi kn}{M}}, \quad k = 0, 1, \cdots, M-1$$

The sampling produces M samples within the range of $\omega = 0$ to 2π, with the first one starting at $\omega = 0$ and the last one at $2\pi - 2\pi/M$. By knowing M sample values H_k we have M independent equations that can be solved for $h(n)$, $n = 1, 2, \cdots, M$. In fact, the equation relating H_k to $h(n)$ is the M-point DFT operation on $h(n)$. Taking the IDFT of the sequence of samples will give us $h(n)$.

$$h(n) = \frac{1}{M} \sum_{k=0}^{M-1} H_k e^{j\frac{2\pi kn}{M}}, \quad n = 0, 1, \cdots, M-1$$

Needless to say, if the unit-sample response of the filter is to be real valued, the set of H_k should possess properties expected from the DFT of a real-valued sequence. Also, needless to say that the frequency response of the design outcome is

$$H_s(\omega) = \sum_{n=0}^{M-1} h(n)e^{-j\omega n}$$

which is a function of the continuous variable ω. The values of $H_s(\omega)$ at $\omega = \frac{2\pi k}{M}$ are equal to the given sample values. At other frequencies, however, $H_s(\omega)$ differs from the values of $H(\omega)$.

18.11 IIR Filter Design

The unit-sample response of an IIR filter has an infinite length and its system function contains poles (and possibly zeros). An IIR filter may be designed by choosing the location of its poles and zeros such that the resulting frequency response acquires the desired characteristics. This method was illustrated in section 18.9 and has been used in some examples previously. An IIR filter is also designed by transforming an analog counterpart to the discrete-time domain. Two such methods will be presented in the following.

IIR Filter Design by Impulse-Invariant Transformation

In the impulse-invariant transformation from s to z, the continuous-time angular frequency Ω (asssociated with the prototype filter) and the discrete-time angular frequency ω (associated with the design objective) are related through the following linear relationship:

$$\Omega = \omega \times F_s, \quad \text{where } F_s \text{ is the sampling rate}$$

Consequently, the prototype analog filter is the same as a filter made of the digital filter operating along with the A/D and D/A converters working at the sampling rate of F_s. In the impulse-invariant transformation method, the unit-sample response of the digital filter is found by sampling the unit-impulse response of the analog filter every T second (a sampling rate of $F_s = 1/T$ Hz). To distinguish between the two domains, a subscript is added (c for continuous time and d for discrete time) in the time and frequency domains. Therefore, $h_d(n) = h_c(nT)$, where $h_d(n)$ is the unit-sample response of the digital filter and $h_c(t)$ is that of the analog filter. The system function of the digital filter $H_d(z)$ is then found by taking the z-transform of $h_d(n)$. $H_d(z)$ can also be found directly from $H_c(s)$. For this purpose, start with the example of exponential function

Continuous time: $h_c(t) = e^{at}u(t)$

$$H_c(s) = \frac{1}{s-a}, \quad \sigma > a$$

Discrete time: $h_d(n) = e^{aTn}u(n)$

$$H_d(z) = \sum_{n=-\infty}^{\infty} h_d(n)z^{-n} = \sum_{n=0}^{\infty} e^{aTn}z^{-n} = \frac{1}{1-e^{aT}z^{-1}}, \quad |z| > e^{aT}$$

An $H(s)$ of higher order can be expanded into its fractions and then transformed as above.

Example **18.39**

Find $h_d(n)$ and $H_d(z)$ of a first-order low-pass digital IIR filter, which is to have a 0-dB DC gain and a 3-db attenuation frequency of 4.8 kHz when interfaced with an ensemble of A/D and D/A converters operating at a 48-kHz sampling rate.

Solution

The 3-dB attenuation angular frequency of the digital filter is $\omega_0 = \Omega_0/F_s = (2\pi \times 4.8)/48 = 0.2\pi$. The performance of the ensemble is equivalent to the performance of an analog filter with

$$H(s) = \frac{\Omega_0}{s + \Omega_0}, \quad \text{where } \Omega_0 = 2\pi \times 4{,}800 = 9{,}600\pi$$

Consequently, the prototype analog filter has the same system function as the equivalent filter.

$$H_c(s) = \frac{\Omega_0}{s + \Omega_0}, \quad h_c(t) = \Omega_0 e^{-\Omega_0 t} u(t)$$

By sampling $h_c(t)$ every $T = \frac{1}{48}$ msec we find

$$h_c(nT) = \Omega_0 e^{-\omega_0 n} u(n), \quad \text{where } \omega_0 = \frac{\Omega_0}{48{,}000}$$

If the overall gain is adjusted through AD/DA segments, then the unit-sample response of the digital filter is

$$h_d(n) = e^{-\omega_0 n} u(n), \quad \text{with } H_d(z) = \frac{1}{1 - z_0 z^{-1}}, \quad \text{where } z_0 = e^{-\omega_0}$$

Note that a direct transformation of $H_c(s)$ will result in the above $H_d(z)$.

IIR Filter Design by Bilinear Transformation

The bilinear tranformation from s to z is given by

$$s = \frac{2}{T} \frac{1 - z^{-1}}{1 + z^{-1}}$$

In the case of real frequencies, $s = j\Omega$ and $z = e^{j\omega}$. The effect of the bilinear transformation on real frequencies is, therefore, obtained by substituting the above values for s and z in the transformation equation.

$$j\Omega = \frac{2}{T} \frac{1 - e^{-j\omega}}{1 + e^{-j\omega}} = j\frac{2}{T} \tan\left(\frac{\omega}{2}\right)$$

As seen below, the bilinear tranformation converts the continuous-time angular frequency Ω to the discrete-time angular frequency ω by the following nonlinear relationships:

$$\Omega = \frac{2}{T} \tan\left(\frac{\omega}{2}\right) \quad \text{(to be called frequency warping)}$$

$$\omega = 2 \tan^{-1}\left(\frac{\Omega T}{2}\right), \quad \text{where } T \text{ is the sampling interval.}$$

Consequently, the $H(s)$ that is be transformed needs to account for the aforementioned nonlinear effect by setting different specifications (e.g., a new 3-dB analog frequency Ω_0 for a first-order filter) while keeping the same functional form as the desired equivalent analog filter. This is illustrated in Example 18.40.

Example
18.40

Bilinear transformation of a first-order filter

In the present case, the 3-dB attenuation angular frequency of the digital filter is $\omega_d = 0.2\pi$. The warped 3-dB frequency is

$$\Omega_0 = \frac{2}{T} \tan\left(\frac{\omega_d}{2}\right) = \frac{2}{T} \tan(0.1\pi)$$

At this stage we refrain from substituting for T. Soon it will be noted that use of frequency warping and the bilinear transformation causes T to be canceled. The $H(s)$ to be transformed is, therefore,

$$H(s) = \frac{\Omega_0}{s + \Omega_0}$$

We next transform the above $H(s)$ to $H(z)$ by the following steps:

$$s = \frac{2}{T} \frac{1 - z^{-1}}{1 + z^{-1}}$$

$$H(z) = H(s)\Big|_{s = \frac{2}{T}\frac{1-z^{-1}}{1+z^{-1}}} = \frac{\Omega_0}{\frac{2}{T}\frac{1-z^{-1}}{1+z^{-1}} + \Omega_0}$$

$$= \frac{\frac{2}{T}\tan(0.1\pi)}{\frac{2}{T}\frac{1-z^{-1}}{1+z^{-1}} + \frac{2}{T}\tan(0.1\pi)}$$

$$= \frac{(1 + z^{-1})\tan(0.1\pi)}{(1 - z^{-1}) + (1 + z^{-1})\tan(0.1\pi)}$$

$$= 0.2452 \frac{1 + z^{-1}}{1 - 0.5095 z^{-1}}$$

We can see that in contrast with the impulse-invariant transformation, frequency specifications of the digital filter are independent of the sampling rate under which the filter operates.

18.12 Filter Structures

A filter processes the input and obtains the output through three types of operations: delay, multiplication, and addition, as evidenced from the difference equation, system function, or convolution sum. These operations are done by three elements that are either physical or conceptual, or implemented through hardware or software tools. The elements are represented in Figure 18.12.

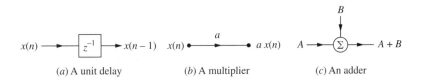

(a) A unit delay   (b) A multiplier   (c) An adder

FIGURE 18.12 Processing elements in a discrete-time system.

Implementation of a filter (as a hardware or by a software program) requires an interconnected set, or network, of the above elements, which according to the filter's description produces the output $y(n)$ to an incoming $x(n)$. Such a network is called filter's structure. Implementation may be achieved through different structures. Each structure is associated with the memory requirements, computational complexity, and accuracy limitations (i.e., the effect of finite-word length). Such considerations play a central role in choosing a structure that is best suited for a situation. An analysis of resources or design processes that would determine the best structure for a filter are not addressed in what follows. The aim is limited to presenting several commonly used structures for FIR and IIR filters, and briefly pointing out some obvious choice criteria.

FIR Filter Structure

The output of a causal FIR filter with a unit-sample response of length M may be obtained by convolving the input and the unit-sample response

$$y(n) = \sum_{k=0}^{N} h(k)x(n-k)$$

where $N = M - 1$ is the order of the filter, or equally by applying the system function

$$H(z) = \sum_{k=0}^{N} h(k)z^{-k}$$

In this section we present two types of structures for implementing FIR filters. These are

1. Direct-form structures (Figures 18.13, 18.14, and 18.15).
2. Cascade structures (Figure 18.16).

Direct-Form Structure for FIR Filters

This form directly implements a filter's equation. The multipliers in the structure are coefficients of the system function. The implementation is also called a tapped delay line or a transversal filter. Figure 18.13 is a direct-form structure for a filter

$$H(z) = h(0) + h(1)z^{-1} + h(2)z^{-2} + \cdots + h(N-1)z^{-(N-1)} + h(N)z^{-N}$$

where $N = M - 1$ is the order of the filter. They employ $M - 1$ delay elements, M multipliers, and $M - 1$ two-input adders.

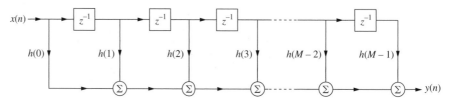

FIGURE 18.13 Realization of FIR filter of length M by a direct-form structure.

Example
18.41

The two structures shown in Figure 18.14(a) and (b) implement the FIR filter with $h(n) = \{\overset{\uparrow}{3}, 5, 3\}$ and the two structures shown in Figure 18.14(c) and (d) implement $h(n) = \{\overset{\uparrow}{3}, 5, 5, 3\}$.

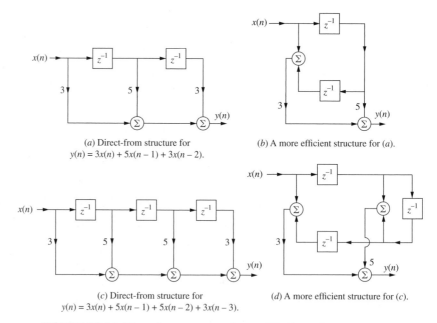

(a) Direct-from structure for
$y(n) = 3x(n) + 5x(n-1) + 3x(n-2)$.

(b) A more efficient structure for (a).

(c) Direct-from structure for
$y(n) = 3x(n) + 5x(n-1) + 5x(n-2) + 3x(n-3)$.

(d) A more efficient structure for (c).

FIGURE 18.14 Direct-form structures for the FIR filters of Example 18.41.

Cascade Structure for FIR Filters

The system function of an FIR filter may be factored as the product of smaller subsystems such as first-order and second-order sections.

$$H(z) = H_0(1 + \alpha_1 z^{-1} + \beta_1 z^{-2})(1 + \alpha_2 z^{-1} + \beta_2 z^{-2}) \cdots (1 + \alpha_K z^{-1} + \beta_K z^{-2})$$

Each subsystem can be realized by a direct-form structure and placed in cascade of each other. See Figure 18.15.

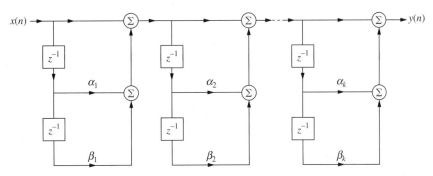

FIGURE 18.15 Cascade structure for FIR filter.

IIR Filter Structure

An IIR filter of order N is described by the difference equation

$$y(n) = -\sum_{k=1}^{N} a_k y(n-k) + \sum_{k=0}^{M-1} b_k x(n-k)$$

or by the system function

$$H(z) = \sum_{k=0}^{M-1} b_k z^{-k} \left/ \left(1 + \sum_{k=1}^{N} a_k z^{-k}\right)\right.$$

Two types of structures will be presented here to implement IIR filters. These are

1. Direct-form structures (Figures 18.16 and 18.17).
2. Cascade structures (Figure 18.18).

Direct-Form Structure for IIR Filters

The difference equation describing an IIR filter is

$$y(n) + \sum_{k=1}^{N} a_k y(n-k) = \sum_{k=0}^{M-1} b_k x(n-k) \equiv w(n)$$

where each side of the equation is also represented by a new variable $w(n)$. From the right-hand side we obtain the FIR equation

$$w(n) = \sum_{k=0}^{M-1} b_k x(n-k)$$

which can be implemented by the forward structure shown on the left side of Figure 18.16. From the left-hand side we obtain the IIR equation

$$y(n) + \sum_{k=1}^{N} a_k y(n-k) = w(n)$$

which can be implemented by the feedback structure shown on the right side of Figure 18.16. The two subsystems of Figure 18.16(a) are transposed and combined to form the structures of Figure 18.16(b) and (c).

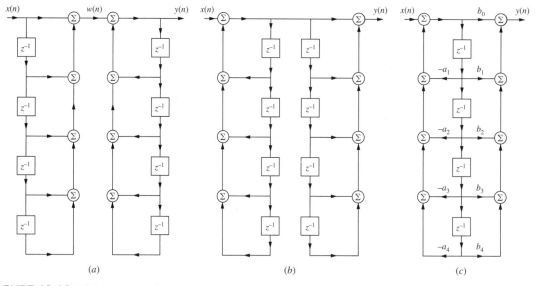

(a)   (b)   (c)

FIGURE 18.16 *(a)* Realization of an IIR filter by a set of feed-forward loops (left side) and feedback loops (right side) produces the direct-form I structure shown here. Transposing the two segments in *(a)* produces the structure shown in *(b)*. The two sets of unit delays in the forward and feedback segments of the structure in *(b)* may be combined, resulting in a smaller number of delay units as well as adders *(c)*.

*xample
18.42

An IIR filter is described by

$$y(n) + a_1 y(n-1) = b_0 x(n) + b_1 x(n-1)$$

Construct the block diagram of the right side of the equation as a feed-forward loop and that of the left side as a feedback loop. Cascade them as shown in Figure 18.17(a) and you have direct-form I structure of the IIR filter. Transpose the feed-forward and feedback equivalent block diagrams of the above filter.

Solution

Let the output of the first adder in Figure 18.17(a) be $w(n)$. The IIR filter consists of the following LTI systems in series.

Feed-forward system: $w(n) = b_0 x(n) + b_1 x(n-1)$, $\qquad H_1(z) = \dfrac{W(z)}{X(z)} = b_0 + b_1 z^{-1}$

Feedback system: $\quad y(n) = -a_1 y(n-1) + w(n)$, $\qquad H_2(z) = \dfrac{Y(z)}{W(z)} = \dfrac{1}{1 + a_1 z^{-1}}$

The IIR filter: $\quad y(n) = -a_1 y(n-1) + b_0 x(n) + b_1 x(n-1), \ H(z) = \dfrac{Y(z)}{X(z)} = \dfrac{b_0 + b_1 z^{-1}}{1 + a_1 z^{-1}}$

$$H(z) = H_1(z) H_2(z) = H_2(z) H_1(z)$$

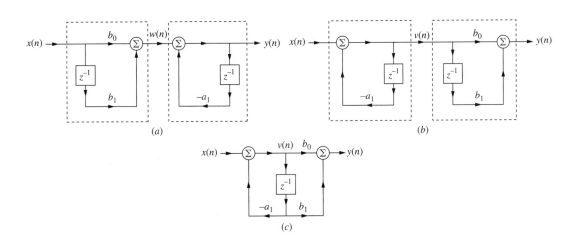

FIGURE 18.17 (a), (b), (c). Direct-form I structure for realizations of the IIR filter $y(n) + a_1 y(n-1) = b_0 x(n) + b_1 x(n-1)$. See Example 18.42.

The order of H_1 and H_2 may be exchanged resulting in Figure 18.17(b). This result may also be derived directly through convolution. Let $h_1(n)$ and $h_2(n)$ be the unit-sample responses of the feed-forward and the feedback segments, respectively. Then,

Feed-forward system: $w(n) = x(n) \star h_1(n)$

Feedback system: $y(n) = w(n) \star h_2(n)$

The IIR filter: $y(n) = [x(n) \star h_1(n)] \star h_2(n) = [x(n) \star h_2(n)] \star h_1(n)$

Due to the associative property of the convolution the order of the feed-forward and the feedback segments may be exchanged producing the block diagram of Figure 18.17(b). The two unit delays in Figure 18.17(b) perform a redundant function to Figure 18.1(b). They may be replaced by a single unit delay resulting in Figure 18.17(c). See also Example 12.21 in Chapter 12.

Cascade Structure for IIR Filters

An IIR system function can be written as a cascade of subsystems of lower order. This is done by factoring its numerator and denominator into polynomials of first or second order. As in the case of FIR, each subsystem can be implemented by a direct-form structure. Figure 18.18 shows implementation of

$$H(z) = \left(\frac{b_{01} + b_{11}z^{-1}}{1 + a_{11}z^{-1}} \right) \left(\frac{b_{02} + b_{12}z^{-1} + b_{22}z^{-2}}{1 + a_{12}z^{-1} + a_{22}z^{-2}} \right) \left(\frac{b_{03} + b_{13}z^{-1} + b_{23}z^{-2}}{1 + a_{13}z^{-1} + a_{23}z^{-2}} \right)$$

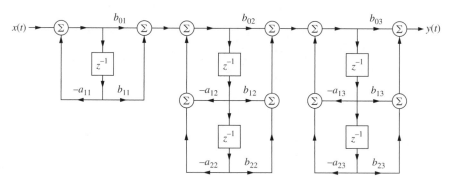

FIGURE 18.18 Cascade structure of a fifth-order IIR filter.

18.13   Problems

Solved Problems

Note: It is highly recommended that in doing the problems in this chapter, analytic solutions be supplemented with the use of computer code (such as Matlab) to facilitate further analysis, exploration, and design.

1. A causal LTI system is described by the input-output difference equation

$$y(n) - 0.3y(n-1) - 0.4y(n-2) = x(n)$$

a. Draw its block diagram as a direct-form structure.
b. Use the block diagram to determine the values of its unit-sample response $h(n)$ for $n = 0, 1, 2, 3$, and 4.
c. Determine its system function, pole(s), and zero(s).
d. Find a closed-form expression for $h(n)$ using the z-transform.
e. Do part d by solving the difference equation in the time domain.

Solution

a. In Figure 18.16(c), let $b_0 = 1$, $b_k = 0$, $k \geq 1$, $a_1 = -0.3$, $a_2 = -0.4$, and $a_k = 0$, $k \geq 3$.

b. $h(n) = \begin{cases} n < 0, \ x(n) = 0, & h(n) = 0 \\ n = 0, \ x(0) = 1, & h(0) = x(0) = 1 \\ n = 1, \ x(1) = 0, & h(1) = 0.3h(0) = 0.3000 \\ n = 2, \ x(2) = 0, & h(2) = 0.3h(1) + 0.4h(0) = 0.4900 \\ n = 3, \ x(3) = 0, & h(3) = 0.3h(2) + 0.4h(1) = 0.2670 \\ n = 4, \ x(4) = 0, & h(4) = 0.3h(3) + 0.4h(2) = 0.2761 \end{cases}$

c. $Y(z) - 0.3z^{-1}Y(z) - 0.4z^{-2}Y(z) = X(z)$, $H(z) = \dfrac{Y(z)}{X(z)} = \dfrac{1}{1 - 0.3z^{-1} - 0.4z^{-2}}$

Poles: $p^2 - 0.3p - 0.4 = 0$, $p_1 = 0.8$, $p_2 = -0.5$
Zeros: $z_0 = 0$

d. $H(z) = \dfrac{1}{1 - 0.3z^{-1} - 0.4z^{-2}} = \dfrac{A}{1 - 0.8z^{-1}} + \dfrac{B}{1 + 0.5z^{-1}}$

$A = H(z)(1 - 0.8z^{-1})\Big|_{z=0.8} = \dfrac{1}{1 + 0.5z^{-1}}\Big|_{z=0.8} = \dfrac{8}{13} = 0.6154$

$B = H(z)(1 + 0.5z^{-1})\Big|_{z=-0.5} = \dfrac{1}{1 - 0.8z^{-1}}\Big|_{z=-0.5} = \dfrac{5}{13} = 0.3846$

$h(n) = \left[0.6154(0.8)^n + 0.3846(-0.5)^n\right]u(n)$

e. $h(n) - 0.3h(n-1) - 0.4h(n-2) = d(n)$

$h(0) = 0.3h(-1) + 0.4h(-2) + 1 = 1$

$h(1) = 0.3h(0) + 0.4h(-1) = 0.3$

$h(n) = A(0.8)^n + B(-0.5)^n$, $n \geq 0$, $h(0) = 1$, $h(1) = 0.3$

A and B: $\begin{cases} n = 0, & A + B = 1 \\ n = 1, & 0.8A - 0.5B = 0.3 \end{cases} \implies A = \dfrac{8}{13}, \ B = \dfrac{5}{13}$

$y(n) = 0.6154(0.8)^n + 0.3846(-0.5)^n$, $n \geq 0$

2. Consider the causal system $y(n) - 0.3y(n-1) - 0.4y(n-2) = x(n)$ described in problem 1.

a. Given $y(-2) = 19/16$, $y(-1) = 7/4$, and $x(n) = 0$, $n \geq 0$, use the block diagram to compute $y(n)$ for $n = 0, 1, 2, 3$, and 4.

b. Given the conditions in (part a) find a closed-form expression for $y(n)$, $n \geq 0$, using the z-transform.

c. Do part b by solving the difference equation in the time domain.

Solution

a. $y(n) = \begin{cases} n = -2 & y(-2) = \frac{19}{16} \\ n = -1 & y(-1) = \frac{7}{4} \\ n = 0 & y(0) = 0.3y(-1) + 0.4y(-2) = 0.3 \times \frac{7}{4} + 0.4 \times \frac{19}{16} = 1 \\ n = 1 & y(1) = 0.3y(0) + 0.4y(-1) = 0.3 + 0.4 \times \frac{7}{4} = 1 \\ n = 2 & y(2) = 0.3y(1) + 0.4y(0) = 0.3 + 0.4 = 0.7 \\ n = 3 & y(3) = 0.3y(2) + 0.4y(1) = 0.21 + 0.4 = 0.61 \\ n = 4 & y(4) = 0.3y(3) + 0.4y(2) = 0.183 + 0.28 = 0.463 \end{cases}$

b. $y(n) - 0.3y(n-1) - 0.4y(n-2) = 0$, $n \geq 0$

$Y(z) - 0.3[z^{-1}Y(z) + y(-1)] - 0.4[z^{-2}Y(z) + y(-1)z^{-1} + y(-2)] = 0$

$Y(z) - 0.3 \left[z^{-1}Y(z) + \frac{7}{4} \right] - 0.4 \left[z^{-2}Y(z) + \frac{7}{4}z^{-1} + \frac{19}{16} \right] = 0$

$Y(z) = \frac{1 + 0.7z^{-1}}{1 - 0.3z^{-1} - 0.4z^{-2}} = \frac{1 + 0.7z^{-1}}{(1 - 0.8z^{-1})(1 + 0.5z^{-1})} = \frac{A}{1 - 0.8z^{-1}} + \frac{B}{1 + 0.5z^{-1}}$

$A = Y(z)(1 - 0.8z^{-1}) \Big|_{z=0.8} = \frac{1 + 0.7z^{-1}}{1 + 0.5z^{-1}} \Big|_{z=0.8} = \frac{15}{13} = 1.1538$

$B = Y(z)(1 + 0.5z^{-1}) \Big|_{z=-0.5} = \frac{1 + 0.7z^{-1}}{1 - 0.8z^{-1}} \Big|_{z=-0.5} = -\frac{2}{13} = -0.1538$

$y(n) = 1.1538(0.8)^n - 0.1538(-0.5)^n$, $n \geq 0$

c. $r^2 - 0.3r - 0.4 = 0$, $r = 0.8$, -0.5

$y(n) = A(0.8)^n + B(-0.5)^n$, $n \geq 0$, $y(0) = y(1) = 1$

Finding A and B: $\begin{cases} n = 0, & A + B = 1 \\ n = 1, & 0.8A - 0.5B = 1 \end{cases} \implies A = \frac{15}{13} = 1.1538, \ B = -\frac{2}{13} = -0.1538$

$y(n) = 1.1538(0.8)^n - 0.1538(-0.5)^n$, $n \geq 0$

3. Consider again the causal system of problem 1, $y(n) - 0.3y(n-1) - 0.4y(n-2) = x(n)$. Let it be known that (i) $y(n) = 0$ for $n < -1$, (ii) $y(-1) = 1$, and (iii) $x(n) = 0$, for $n \geq 0$.

a. Find $y(0)$ and $y(1)$.

b. Determine $y(n)$.

c. Argue that $x(n) = d(n+1)$ and, therefore, we expect $y(n) = h(n+1)$.

d. Verify that $y(n) = h(n+1)$, where $h(n)$ is the unit-sample response obtained in problem 1.

Solution

a. $y(0) = 0.3y(-1) + 0.4y(-2) = 0.3$

$y(1) = 0.3y(0) + 0.4y(-1) = 0.49$

b. $y(n) = A(0.8)^n + B(-0.5)^n$, $n \geq 0$, $y(0) = 0.3$, $y(1) = 0.49$

A and B: $\begin{cases} n = 0, & A + B = 0.3 \\ n = 1, & 0.8A - 0.5B = 0.49 \end{cases}$ $\implies A = \dfrac{64}{130} = 0.4923$, $B = -\dfrac{25}{130} = -0.1923$

$y(n) = \begin{cases} 0, & n < -1 \\ 1, & n = -1 \\ 0.4923(0.8)^n - 0.1923(-0.5)^n, & n \geq 0 \end{cases}$

c. From the system's causality and the given information that $y(n) = 0$ for $n < 1$, we conclude that $x(n) = 0$, $n < 1$. It is also given that $x(n) = 0$, $n \geq 1$. Therefore, $x(n) = X_0 d(n+1)$ and $y(n) = X_0 h(n+1)$. To determine X_0 we note that $y(-1) = 1$ and $h(0) = 1$. Therfore, $X_0 = 1$.

d. $y(n) = d(n+1) + [0.4923(0.8)^n - 0.1923(-0.5)^n]u(n)$

$h(n) = \left[0.6154(0.8)^n + 0.3846(-0.5)^n\right]u(n)$ from problem 1.

$h(n+1) = (0.6154 + 0.3846)d(n+1) + \left[0.8 \times 0.6154(0.8)^n - 0.5 \times 0.3846(-0.5)^n\right]u(n+1)$

$= d(n+1) + \left[0.4923(0.8)^n - 0.1923(-0.5)^n\right]u(n) = y(n)$

4. A discrete-time system is described by the system function $H(z) = 1 - z^{-1}$. Use the vectorial interpretation of $H(z)$ to evaluate $H(\omega)$ at $\omega = 0$, $\pi/4$, $\pi/2$, $3\pi/4$, and π. Sketch the magnitude and phase of $H(\omega)$.

Solution
The system function is $H(z) = 1 - z^{-1} = (z-1)/z$. It has a pole at $z = 0$ (shown by point A on Figure 18.19a) and a zero at $z = 1$ (shown by point B). Let C designate the location of z on the unit circle at which $H(\omega)$ is evaluated. Then, $H(\omega) = \overline{BC}/\overline{AC}$ and from the graphical representation of Figure 18.19(a) we have the follwing:

Location of z	ω	$H(\omega) = \overline{BC}/\overline{AC}$	$|H(\omega)| = |\overline{BC}|$	$\angle H(\omega) = \angle\overline{BC} - \omega$	Comments
C_0	0	$\overline{BC_0}/\overline{AC_0}$	0	$90°$	Zero of the system
C_1	$\pi/4$	$\overline{BC_1}/\overline{AC_1}$	0.8	$67.5°$	
C_2	$\pi/2$	$\overline{BC_2}/\overline{AC_2}$	$\sqrt{2}$	$45°$	3-dB frequency, $|H| = \frac{|H_{Max}|}{\sqrt{2}}$
C_3	$3\pi/4$	$\overline{BC_3}/\overline{AC_3}$	1.84	$22.5°$	
C_4	π	$\overline{BC_4}/\overline{AC_4}$	2	$0°$	$|H_{Max}| = 2$

The frequency response is plotted in Figure 18.19(b). The 3-dB (half-power) frequency is at $\omega = \pi/2$.

Remark
The system of problem 4 is a difference operator, $h(n) = d(n) - d(n-1)$. In the frequency domain, it can exhibit high-pass filtering, a property shared by continuous-time differentiators.

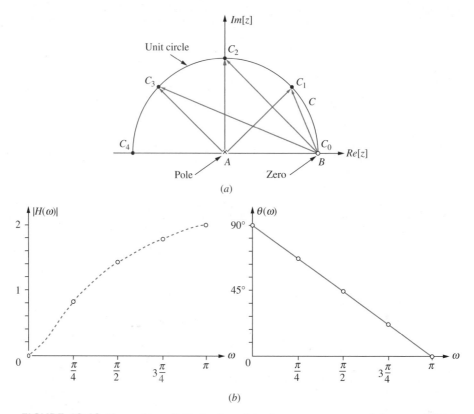

FIGURE 18.19 (For problem 4) Evaluation of the frequency response of the system $H(z) = 1 - z^{-1}$ at $\omega = 0$, $\pi/4$, $\pi/2$, $3\pi/4$, and π using the vectorial interpretation: **(a)** Pole-zero plot, **(b)** sketch of $H(\omega)$.

5. A discrete-time causal system is described by the system function

$$H(z) = \frac{1 + z^{-1}}{1 - 1.8z^{-1} + 0.81z^{-2}}$$

Use the vectorial interpretation of $H(z)$ to evaluate $H(\omega)$. Show that the system is a low-pass filter. Find the 3-dB frequency ω_α. Evaluate $H(\omega)$ at $\omega = 0$, ω_α, $\pi/4$, $\pi/2$, $3\pi/4$, and π. Sketch the magnitude and phase of $H(\omega)$.

Solution
The system function is

$$H(z) = \frac{1 + z^{-1}}{1 - 1.8z^{-1} + 0.81z^{-2}} = \frac{1 + z^{-1}}{(1 - 0.9z^{-1})^2} = \frac{z(z + 1)}{(z - 0.9)^2}$$

It has repeated poles at $p = 0.9$ (shown by point A on Figure 18.20a) and two zeros at $z = 0$ and $z = -1$ (shown by points B_0 and B_1, respectively). Let C designate the location of z on the unit circle at which $H(\omega)$ is evaluated.

Then, at $z = e^{j\omega}$ we have

$$z = \overline{B_0 C}, \quad z + 1 = \overline{B_1 C}, \quad z - 0.9 = \overline{AC}$$

$$H(\omega) = H(z)\Big|_{z=e^{j\omega}} = \frac{\overline{B_0 C} \times \overline{B_1 C}}{\overline{AC}^2}$$

$$|H(\omega)| = \frac{|\overline{B_1 C}|}{|\overline{AC}|^2}$$

$$\angle H(\omega) = \omega + \angle \overline{B_1 C} - 2\angle \overline{AC}$$

It is clear from the pole-zero plot that $|H(\omega)|$ is at its maximum when $\omega = 0$, where $H(0) = 200$. Nearby, in the vicinity of $\omega = 0$, the zero vectors $\overline{B_0 C}$ and $\overline{B_1 C}$ don't change appreciably, and their contribution to $H(\omega)$ remains almost at a constant value: $\overline{B_0 C} \times \overline{B_1 C} \approx 2$. Therefore, for small ω [which includes the 3-dB frequency ω_α, see Figure 18.20(b)] we will have

$$H(\omega) = \frac{2}{\overline{AC}^2}, \quad \text{and} \quad H(0) = \frac{2}{0.1^2} = 200$$

$$H(\omega_\alpha) = \frac{2}{\overline{AC_\alpha}^2} = \frac{200}{\sqrt{2}}$$

$$|\overline{AC_\alpha}| \approx 0.11892$$

$$|\overline{C_0 C_\alpha}| \approx \omega_\alpha = 3.9°$$

$$|H(\omega_\alpha)| \approx 141.42$$

$$\angle H(\omega_\alpha) \approx -2\tan^{-1}[\omega_\alpha/(1 - 0.9)] = -65.5°$$

From the graphical representation of Figure 18.20(a) we complete the table below [see Figure 18.20(c) and (d)].

Location of z	ω	$H(\omega)$	$|H(\omega)|$	$\angle H(\omega)$	Comments
C_0	0	$\overline{B_0 C_0} \times \overline{B_1 C_0}/\overline{AC_0}^2$	200	0°	$|H_{Max}| = 200$
C_α	3.9°	$\overline{B_0 C_\alpha} \times \overline{B_1 C_\alpha}/\overline{AC_\alpha}^2$	141.42	−63.8°	$|H| = \frac{|H_{Max}|}{\sqrt{2}}$, 3-dB frequency
C_1	45°	$\overline{B_0 C_1} \times \overline{B_1 C_1}/\overline{AC_1}^2$	3.44	−143°	
C_2	90°	$\overline{B_0 C_2} \times \overline{B_1 C_2}/\overline{AC_2}^2$	0.78	−129°	
C_3	135°	$\overline{B_0 C_3} \times \overline{B_1 C_3}/\overline{AC_3}^2$	0.25	−110°	
C_4	180°	$\overline{B_0 C_4} \times \overline{B_1 C_4}/\overline{AC_4}^2$	0	−90°	Zero of the system

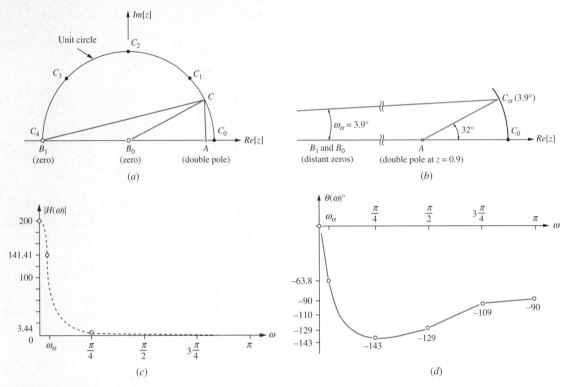

FIGURE 18.20 (For problem 5) Evaluation of the frequency response of a system having a repeated pole at $p = 0.9$ and two zeros at $z = 0$ and $z = -1$, by the vectorial approach: **(a)** Pole-zero plot; **(b)** finding the 3-dB frequency; **(c)** and **(d)** sketch of the frequency response.

Comparison with Theory

The square of the magnitude of the frequency response of the system of problem 5 is

$$|H(\omega)|^2 = \frac{2(1 + \cos \omega)}{(1.81 - 1.8 \cos \omega)^2}$$

The 3-dB frequency may be found, approximately, from

$$1.81 - 1.8 \cos \omega_\alpha = \sqrt{2} \times 10^{-2} \implies \omega_\alpha = 3.9°$$

The frequency response is shown in Figure 18.20(c) and (d).

6. The system of Example 18.25 has a zero at the origin. Move it to $z = 1$ to obtain a new system described by the system function

$$H(z) = \frac{1 - z^{-2}}{1 - 0.9\sqrt{2}z^{-1} + 0.81z^{-2}}$$

Use the vectorial interpretation of $H(z)$ to evaluate $H(\omega)$. Reason that $|H(\omega)|$ is at its maximum when $\omega = \pi/4$.

Find the 3-dB frequencies ω_ℓ and ω_h. Evaluate $H(\omega)$ at $\omega = 0$, ω_ℓ, $\pi/4$, ω_h, $\pi/2$, $3\pi/4$, and π. Compare with the frequency response obtained in Example 18.25.

Solution

The system function is

$$H(z) = \frac{(z-1)(z+1)}{(z-0.9e^{j\pi/4})(z-0.9e^{-j\pi/4})}$$

As in Example 18.25, this system has a pair of poles at $p_{a,2} = 0.9e^{\pm j\pi/4}$ but its zeros are at $z = \pm 1$. By using the pole-zero plot and an analysis similar to Example 18.25 we find

$$H(\omega) = \frac{\overline{B_1 C} \times \overline{C_0 C}}{\overline{A_1 C} \times \overline{A_2 C}} \quad \text{See Figure 18.6}$$

$$|H(\omega)| = \frac{|\overline{B_1 C}| \times |\overline{C_0 C}|}{|\overline{A_1 C}| \times |\overline{A_2 C}|}$$

$$\angle H(\omega) = [\angle \overline{B_1 C} + \angle \overline{C_0 C}] - \left[\angle \overline{A_1 C} + \angle \overline{A_2 C}\right]$$

$$H(\pi/4) = 10.5\angle 3°$$

$$\omega_{\ell,h} \approx \pi/4 \pm 0.1 \text{ rad} = 39.3°, \text{ and } 50.7°$$

$$|H(\omega_{\ell,h})| \approx 7.4$$

$$\angle H(\omega_{\ell,h}) \approx 3° \pm 45° = 48° \text{ and } -42°$$

From a vectorial representation similar to that of Figure 18.6 we have the following approximations:

ω	$|H(\omega)|$	$\angle H(\omega)$	Comments
$0°$	0	$90°$	Zero of the system
$\omega_\ell \approx 39.3°$	7.4	$48°$	Lower 3-dB frequency, $|H| = \frac{|H_{Max}|}{\sqrt{2}}$
$45°$	10.9	$3°$	$|H_{Max}| = 10.9$
$\omega_h \approx 50.7°$	7.4	$-42°$	Upper 3-dB frequency, $|H| = \frac{|H_{Max}|}{\sqrt{2}}$
$90°$	1.55	$-81.5°$	
$135°$	0.55	$-87°$	
$180°$	0	$-90°$	Zero of the system

More exact values are

ω		ω_ℓ	ω_0	ω_h			
ω	0	$39.61°$	$45°$	$51.64°$	$90°$	$135°$	$180°$
$|H|$	0	7.43	10.51	7.43	1.55	1.55	0
$\angle|H|°$	0	45.12	3.01	45.1	-81.5	-87	-90

Comparison with Example 18.25

The peak of the frequency response is still at $\omega = \pi/4$ (as was the case in Example 18.25). The zero at $\omega = 0$, however, has pulled down the frequency response in its neighborhood, reducing $|H|_{Max}$ from 13.77 (Example 18.25) to 10.9 (problem 6).

The following two problems illustrate parallels between the continuous-time and discrete-time operation of a finite integrator, and possible errors associated with it.

7. **A continuous-time finite-duration integrator.** Consider the continuous-time LTI system with the unit-impulse response consisting of an even square pulse lasting 2τ.

$$h_a(t) = \begin{cases} 1, & \text{for } -\tau \le t \le \tau \\ 0, & \text{elsewhere} \end{cases}$$

where the subscript a stands for *analog*. The system may be called a noncausal finite-duration analog integrator. Given $\tau = 1$ msec, find its frequency response $H_a(\Omega)$[6] and determine its value at $F = 0, 100, 400, 500, 600, 900$, and 1,000 Hz. Find its response to the input

$$x_a(t) = 1 + \cos(200\pi t) + \cos(800\pi t) + \cos(1,000\pi t) + \cos(1,200\pi t) + \cos(1,800\pi t) + \cos(2,000\pi t) \text{ V.}$$

Solution

$$H_a(\Omega) = \int_{-\infty}^{\infty} h_a(t)e^{-j\Omega t}\,dt = \int_{-\tau}^{\tau} e^{-j\Omega t}\,dt = 2\frac{\sin(\Omega\tau)}{\Omega}\bigg|_{\tau=10^{-3}} = 2\frac{\sin(\Omega/1,000)}{\Omega}$$

See Figure 18.21(a). The input components at $F = 0, 100, 400, 500, 600, 900$, and 1,000 Hz correspond to angular frequencies $\Omega = 2\pi F = 0, 200\pi, 800\pi, 1,000\pi, 1,200\pi, 1,800\pi$, and $2,000\pi$, respectively, and result in the following values for the frequency response and the output.

F (Hz)	$\Omega = 2\pi F$	Input (V)	$H_a = 2\sin(\Omega/1,000)/\Omega$		Output (μV)
0	0	1	0.002	$\longrightarrow$	2,000
100	200π	$\cos(200\pi t)$	$2\sin(0.2\pi)/(200\pi) = 1,871 \times 10^{-6}$	$\longrightarrow$	$1,871\cos(200\pi t)$
400	800π	$\cos(800\pi t)$	$2\sin(0.8\pi)/(800\pi) = 468 \times 10^{-6}$	$\longrightarrow$	$468\cos(800\pi t)$
500	$1,000\pi$	$\cos(1,000\pi t)$	$2\sin(\pi)/(1,000\pi) = 0$	$\longrightarrow$	0
600	$1,200\pi$	$\cos(1,200\pi t)$	$2\sin(1.2\pi)/(1,200\pi) = -312 \times 10^{-6}$	$\longrightarrow$	$-312\cos(1,200\pi t)$
900	$1,800\pi$	$\cos(1,800\pi t)$	$2\sin(1.8\pi)/(1,800\pi) = -208 \times 10^{-6}$	$\longrightarrow$	$-208\cos(1,800\pi t)$
1,000	$2,000\pi$	$\cos(2,000\pi t)$	$2\sin(2\pi)/(2,000\pi) = 0$	$\longrightarrow$	0

The system's output is $y_a(t) = 2,000 + 1,871\cos(200\pi t) + 468\cos(800\pi t) - 312\cos(1,200\pi t) - 208\cos(1,800\pi t)\mu$V. The components at 500 Hz and 1 kHz [located at the zeros of $H(\Omega)$] are eliminated.

---

[6]In order to distinguish the continuous-time frequency from the discrete-time frequency, we use uppercase letters (F and $\Omega = 2\pi F$) for the continuous-time domain.

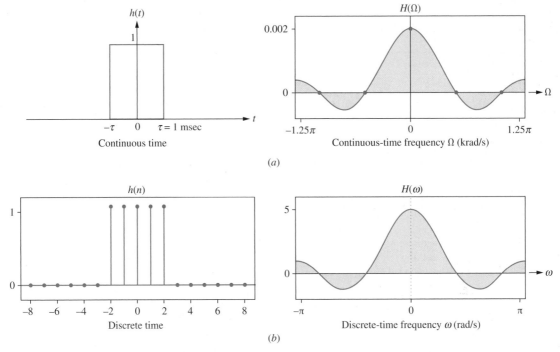

(a)

(b)

FIGURE 18.21 (a) (Problem 7) A noncausal, finite-duration, continuous-time integrator. Its unit-impulse response (shown on the left) consists of an even square pulse lasting slightly more than $2\tau = 2$ msec. $h_a(t) = u(t + \tau)u(t - \tau)$. Its frequency response (shown on the right) is $H_a(\Omega) = 2\sin(\Omega/1{,}000)/\Omega$, $-\infty < \omega < \infty$. Frequency axis is Ω in rad/s and displays the segment $-1{,}250\pi$ to $1{,}250\pi$. **(b)** (Problem 8). A noncausal, finite-duration, discrete-time integrator. Its unit-sample response, which consists of an even square pulse with 5 samples, is $h(n) = \{1, 1, 1, 1, 1\}$. Its frequency response is $H(\omega) = 1 + 2\cos\omega + 2\cos(2\omega) = [\sin(5\omega/2)/\sin(\omega/2)]$, $-\infty < \omega < \infty$. $H(\omega)$ is, however, 2π-periodic. Frequency axis is ω in rad/s, showing one period of $H(\omega)$ from $-\pi$ to π. Compare with the continuous-time integrator of **(a)**, nothing that $\omega = 2{,}000\,\omega$

8. A Discrete-time finite-duration integrator. The unit-impulse response of the continuous-time finite-duration integrator in problem 7 (lasting slightly more than 2 msec) and its input are sampled at the rate of $F_s = 2{,}000$ samples per sec (every 0.5 msec, $t = n/2{,}000$, with a sample at $t = 0$). The unit-sample response of the resulting discrete-time system is $h(n) = \{1, 1, 1, 1, 1\}$. Likewise, the resulting discrete-time input is

$$x(n) = 1 + \cos(0.1\pi n) + \cos(0.4\pi n) + \cos(0.5\pi n) + \cos(0.6\pi n) + \cos(0.9\pi n) + \cos(\pi n)$$

Find the frequency response $H(\omega)$ of the discrete-time system and its response to $x(n)$.

Solution

$$H(z) = z^{-2} + z^{-1} + 1 + z^{1} + z^{2}$$

$$H(\omega) = e^{-j2\omega} + e^{-j\omega} + 1 + e^{j\omega} + e^{j2\omega}$$

$$= 1 + 2\cos\omega + 2\cos(2\omega)$$

See Figure 18.21(b). The input components at $\omega = 0$, 0.1π, 0.4π, 0.5π, 0.6π, 0.9π, and π (corresponding to discrete frequencies $f = 0$, 0.05, 0.2, 0.25, 0.3, 0.45, and 0.5 Hz)[7] result in the following values for the frequency response and the output.

f (Hz)	$\omega = 2\pi f$	Input (V)	$H = 1 + 2\cos\omega + 2\cos(2\omega)$		Output (V)
0	0	1	5	$\longrightarrow$	5
0.05	0.1π	$\cos(0.1\pi n)$	$1 + 2\cos(0.1\pi) + 2\cos(0.2\pi) = 4.52$	$\longrightarrow$	$4.52\cos(0.1\pi n)$
0.2	0.4π	$\cos(0.4\pi n)$	$1 + 2\cos(0.4\pi) + 2\cos(0.8\pi) = 0$	$\longrightarrow$	0
0.25	0.5π	$\cos(0.5\pi n)$	$1 + 2\cos(0.5\pi) + 2\cos(\pi) = -1$	$\longrightarrow$	$-\cos(0.5\pi n)$
0.3	0.6π	$\cos(0.6\pi n)$	$1 + 2\cos(0.6\pi) + 2\cos(1.2\pi) = -1.236$	$\longrightarrow$	$-1.236\cos(0.6\pi n)$
0.45	0.9π	$\cos(0.9\pi n)$	$1 + 2\cos(0.9\pi) + 2\cos(1.8\pi) = 0.716$	$\longrightarrow$	$0.716\cos(0.9\pi n)$
0.5	π	$\cos(\pi n)$	$1 + 2\cos(\pi) + 2\cos(2\pi) = 1$	$\longrightarrow$	$\cos(\pi n)$

The output of the discrete-time system is $y(n)$. After going through a digital-to-analog converter operating at 2 kHz, $y(n)$ will be converted to an analog (continuous-time) signal $y_d(t)$. The lines below list $y(n)$ and $y_d(t)$. For comparison, $y_a(t)$ the analog output of the continuous-time system of problem 7, $y_a(t)$ is also listed.

$$y(n) = 5 + 4.52\cos(0.1\pi n) - \cos(0.5\pi n) - 1.2366\cos(0.6\pi n) + 0.716\cos(0.9\pi n) + \cos(\pi n)\text{V}$$

$$y_d(t) = 5 + 4.52\cos(200\pi t) - \cos(1000\pi t) - 1.2366\cos(1200\pi t) + 0.716\cos(1800\pi t) + \cos(2000\pi t)\ \text{V}$$

$$y_a(t) = 2000 + 1871\cos(200\pi t) + 468\cos(800\pi t) - 312\cos(1200\pi t) - 208\cos(1800\pi t)\ \mu\text{V}$$

The effect of aliasing is seen at higher frequencies and is especially pronounced at 500 Hz and 1 kHz.

9. Consider the discrete-time system

$$H(z) = 0.5\frac{1 + z^{-1}}{1 - z^{-1}}$$

A continuous-time signal $x(t) = \cos(20\pi t) + \cos(60\pi t)$ (with components at 10 Hz and 30 Hz) is sampled at the rate of F_s samples per second and the resulting discrete-time signal $x(n)$ is passed through $H(z)$. Find the output of the system for the following three sampling rates: (1) $F_s = 300$ Hz, (2) 600 Hz, and (3) 1,200 Hz. Note in all cases that the sampling rate is greater than twice the highest frequency in the signal, satisfying the Nyquist criteria. (In problem 10 you will compare these outputs with the output of a continuous-time integrator.)

Solution

The frequency response of the discrete-time system is

$$H(\omega) = 0.5\frac{1 + z^{-1}}{1 - z^{-1}}\bigg|_{z=e^{j\omega}} = -j0.5\cot(\omega/2) = 0.5\cot(\omega/2)\angle-90°$$

The discrete inputs at the three sampling rates are

a. $F_s = 300$ Hz $\quad\longrightarrow\quad x_a(n) = \cos(\pi n/15) + \cos(\pi n/5)$

b. $F_s = 600$ Hz $\quad\longrightarrow\quad x_b(n) = \cos(\pi n/30) + \cos(\pi n/10)$

c. $F_s = 1,200$ Hz $\quad\longrightarrow\quad x_c(n) = \cos(\pi n/60) + \cos(\pi n/20)$

---

[7]Note that $F = 2,000 \times f$.

The output components corresponding to each input component are found and given in the table below.

Input	ω	$H(\omega) = 0.5\cot(\omega/2)\angle{-90°}$	$\longrightarrow$	Output
$\cos(\pi n/60)$	$\pi/60$	$19.095\angle{-90°}$	$\longrightarrow$	$19.095\sin(\pi n/60)$
$\cos(\pi n/30)$	$\pi/30$	$9.541\angle{-90°}$	$\longrightarrow$	$9.541\sin(\pi n/30)$
$\cos(\pi n/20)$	$\pi/20$	$6.353\angle{-90°}$	$\longrightarrow$	$6.353\sin(\pi n/20)$
$\cos(\pi n/15)$	$\pi/15$	$4.757\angle{-90°}$	$\longrightarrow$	$4.757\sin(\pi n/15)$
$\cos(\pi n/10)$	$\pi/10$	$3.157\angle{-90°}$	$\longrightarrow$	$3.157\sin(\pi n/10)$
$\cos(\pi n/5)$	$\pi/5$	$1.539\angle{-90°}$	$\longrightarrow$	$1.539\sin(\pi n/5)$

Sample calculation:

$$x(n) = \cos(\pi n/60), \quad \omega = \pi/60, \quad H(\omega) = 0.5\cot(\pi/120)\angle{-90°} = 19.095\angle{-90°}$$

$$y(n) = |H|\cos(\pi n/60 + \angle H) = 19.095\cos(\pi n/60 - 90°) = 19.095\sin(\pi n/60)$$

The system's output and the ratio of the low- to high-frequency amplitudes are listed below for each rate.

a. $F_s = 300$ Hz $\longrightarrow$ $x(n) = \cos(\pi n/15) + \cos(\pi n/5)$

$\qquad\qquad\qquad\quad y(n) = 4.757\sin(\pi n/15) + 1.539\sin(\pi n/5)$

$\qquad\qquad\qquad\quad R_a =$ low-frequency to high-frequency ratio $= 4.757/1.539 = 3.091$

b. $F_s = 600$ Hz $\longrightarrow$ $x(n) = \cos(\pi n/30) + \cos(\pi n/10)$

$\qquad\qquad\qquad\quad y(n) = 9.541\sin(\pi n/30) + 3.157\sin(\pi n/10)$

$\qquad\qquad\qquad\quad R_b =$ low-frequency to high-frequency ratio $= 9.541/3.157 = 3.022$

c. $F_s = 1,200$ Hz $\longrightarrow$ $x(n) = \cos(\pi n/60) + \cos(\pi n/20)$

$\qquad\qquad\qquad\quad y(n) = 19.095\sin(\pi n/60) + 6.353\sin(\pi n/20)$

$\qquad\qquad\qquad\quad R_c =$ low-frequency to high-frequency ratio $= 19.095/6.353 = 3.006$

Note in all cases that the sampling rate is greater than twice the highest frequency in the signal. The differences observed in a to c are due to $H(z)$ being an approximation of a continuous-time integral. [At high sampling rates, $\Omega/F_s = \omega$ becomes small and the above system-frequency response becomes $H(\omega) \approx (1/\omega)\angle{-90°}$.]

10. The system of problem 9

$$H(z) = 0.5\frac{1 + z^{-1}}{1 - z^{-1}}$$

is a discrete-time integrator that approximates (with a scale factor) the continuous-time integrator

$$y(t) = \int_{-\infty}^{t} x(t)dt$$

In problem 9 the continuous-time signal $x(t) = \cos(20\pi t) + \cos(60\pi t)$ was sampled by analog-to-digital (A/D) converters operating under three sampling rates and these produced three discrete-time signals $x_a(n)$, $x_b(n)$, and $x_c(n)$. The discrete-time signals were then passed through the discrete-time integrator. Let the discrete-time outputs $y_a(n)$, $y_b(n)$, and $y_c(n)$ be converted back to analog signals $y_a(t)$, $y_b(t)$, and $y_c(t)$ by digital-to-analog (D/A) converters which in each case use the original sampling rate and produce analog signals proportional to the discrete signal. Compare $y_a(t)$, $y_b(t)$, and $y_c(t)$ with the output of the continuous-time integrator.

Solution

a. A/D and D/A at $F_s = 300$ Hz: $y_a(t) = -4.757 \sin(20\pi t) - 1.539 \sin(60\pi t)$ $R_a = 3.091$

b. A/D and D/A at $F_s = 600$ Hz: $y_b(t) = -9.541 \sin(20\pi t) - 3.159 \sin(60\pi t)$ $R_b = 3.020$

c. A/D and D/A at $F_s = 1{,}200$ Hz: $y_c(t) = -19.095 \sin(20\pi t) - 6.353 \sin(60\pi t)$ $R_c = 3.006$

d. Analog Integration: $y(t) = (1/20\pi) \sin(20\pi t) + (1/60\pi) \sin(60\pi t)$ $R = 3$

11. Filter description. Any one of the following functions can completely describe a digital filter.

a. The input-output difference equation
b. The system function $H(z)$
c. The frequency response $H(\omega)$
d. The unit-sample response $h(n)$
e. The pole-zero locations plus a gain factor.

Express the above functions in mathematical terms and relate them to each other.

Solution

a. Input-output equation: $y(n) + a_1 y(n-1) + \cdots + a_N y(n-N) = b_0 x(n) + b_1 x(n-1) + \cdots + b_M x(n-M)$

$$y(n) = -\sum_{k=1}^{N} a_k y(n-k) + \sum_{k=0}^{M} b_k x(n-k)$$

b. Unit-sample response: $x(n) = d(n)$, $h(n) = \{\cdots, h_{-k}, \cdots, h_{-2}, h_{-1}, \underset{\uparrow}{h_0}, h_1, h_2, \cdots, h_k, \cdots\} \equiv \sum_{-\infty}^{\infty} h_k d(n-k)$

$$h(n) = -\sum_{k=1}^{N} a_k h(n-k) + \sum_{k=0}^{M} b_k d(n-k)$$

c. System function: $H(z) = \dfrac{B(z^{-1})}{A(z^{-1})} = \dfrac{b_0 + b_1 z^{-1} + \cdots + b_M z^{-M}}{1 + a_1 z^{-1} + \cdots + a_N z^{-N}} = \sum_{k=-\infty}^{\infty} h_k z^{-k}$

d. Frequency response: $H(\omega) = \sum_{k=-\infty}^{\infty} h_k e^{-jk\omega} = H(z)\big|_{z=e^{j\omega}}$

e. Pole-zero locations: poles [roots of $A(z) = 0$], zeros [roots of $B(z) = 0$].

12. Four types of FIR filters designed by windowing

a. Design a 31-sample-long linear-phase low-pass FIR filter with a cutoff frequency at $\omega_0 = \pi/4$. Plot its magnitude frequency response and determine its passband and stopband frequencies (ω_p and ω_s, respectively) given $A_p = 1$ dB and $A_s = 30$ dB. Plot zeros and the unit-sample response of the filter and observe that they exhibit characteristics of a linear-phase filter.

The following Matlab codes determine the unit-sample response, the frequency response, and the plot filter characteristics.

```
w=0:0.002*pi:pi;
den=[1];
m=15;
```

```
M=2*m+1;
k=0:1:M-1;
zerostemline=0*k;
n=-m:m;
window=hamming(M)';
% window=0.54+0.46*cos(pi*n/m);% Constructs a Hamming window.
%
% a) Linear phase lowpass filter, Hamming window.
w0=pi/4;
num=window.*sin(w0*n)./(pi*n);
num(m+1)= w0/pi;
H=freqz(num,den,w);
gain=20*log10(abs(H));
%
% Magnitude plot in linear scale.
figure
plot(w/pi,abs(H),'linewidth',2);
axis([0 1 -0.1 1.1]);
grid
title('Magnitude response of a linear phase lowpass FIR filter (Hamming win-
dow, M=31)');
ylabel('Magnitude (linear scale)');
xlabel('Angular frequency (rad/s, in units of Pi)');
%
% Magnitude plot in dB scale.
figure
plot(w/pi,gain,'linewidth',2);
axis([0 1 -80 5]);
grid
title('Magnitude response of a linear phase lowpass FIR filter (Hamming win-
dow, M=31)');
ylabel('Magnitude (dB)');
xlabel('Angular frequency (rad/s, in units of Pi)');
%
% Phase plot
figure
plot(w/pi,180/pi*angle(H),'linewidth',2);
grid
title('Magnitude response of a linear phase lowpass FIR filter (Hamming win-
dow, M=31)');
ylabel('Phase (degrees)');
xlabel('Angular frequency (rad/s, in units of Pi)');
%
% PZ plot
figure
zplane(num,den);
axis([-1.75 1.75 -1.75 1.75]);
title('Pole-zero plot of a linear phase lowpass FIR filter (Hamming win-
dow, M=31)');
```

```
xlabel('Real parts');
ylabel('Imaginary parts');
%
% Unit-sample response plot
figure
hold on
stem(num,'fill','b');
plot(k,zerostemline,'b');
axis([0 M -0.05 1.1*num(m+1)]);
title('Unit sample response of a linear phase lowpass FIR filter (Ham-
ming window, M=31)');
xlabel('n');
ylabel('h(n)');
hold off
```

b. Repeat part a for a high-pass filter with a cutoff frequency at $\omega_0 = 3\pi/4$.

By substituting the following Matlab codes for the corresponding lines in the program of part a, one obtains the unit-sample response and the frequency response of the high-pass filter.

```
% b) Linear phase highpass filter, Hamming window.
w0=3*pi/4;
num=-window.*sin(w0*n)./(pi*n);
num(m+1)=1-w0/pi;
H=freqz(num,den,w);
```

c. Repeat part a for a bandpass filter with a center frequency at $\omega_0 = \pi/2$ and a bandwidth $\Delta\omega = \pi/2$.

By substituting the following Matlab codes for the corresponding lines in the program of part a, one obtains the unit-sample response and the frequency response of the bandpass filter.

```
% c) Linear phase bandpass filter.
w1=pi/4;
num1=window.*sin(w1*n)./(pi*n);
num1(m+1)= w1/pi;
w2=3*pi/4;
num2=window.*sin(w2*n)./(pi*n);
num2(m+1)= w2/pi;
num=num2-num1;
H=freqz(num,den,w);
```

d. Repeat part a for a bandstop filter with a center frequency at $\omega_0 = \pi/2$ and a bandwidth $\Delta\omega = \pi/2$.

By substituting the following Matlab codes for the corresponding lines in the program of part a, one obtains the unit-sample response and the frequency response of the bandstop filter.

```
% c) Linear phase bandstop filter.
w1=pi/4;
num1=window.*sin(w1*n)./(pi*n);
num1(m+1)= w1/pi;
w2=3*pi/4;
num2=window.*sin(w2*n)./(pi*n);
num2(m+1)= w2/pi;
num=num1-num2;
```

```
num(m+1)=1+num1(m+1)-num2(m+1);
H=freqz(num,den,w);
```

13. Given the unit-sample response of a filter $h(n) = \{1, 2, 1\}$ find its output to the following inputs:

  a. A single sinusoidal signal $x(n) = \cos(\omega n)$.

  b. A rectangular periodic pulse signal with the period $N = 8$ and a duty cycle 25%. A cycle of the signal is shown by

$$x(n) = \begin{cases} \{1, 1, 0, 0, 0, 0, 0, 0\} & 0 \le n \le 7 \\ \\ x(n+8) & \text{elsewhere} \end{cases}$$

Solution

  a. The frequency response of the filter is

$$H(\omega) = 1 + 2e^{-j\omega} + e^{-j2\omega} = 2(1 + \cos\omega)e^{-j\omega}$$

The phase function, $-\omega$, is a linear function of ω which translates into a unit delay in the sinusoid, regardless of its frequency. The output is $y(n) = 2(1 + \cos\omega)\cos[\omega(n-1)]$.

  b. The output is a periodic signal with the period $N = 8$. One period of the signal is $\{1, 3, 3, 1, 0, 0, 0, 0\}$.

14. **Filtering in time and frequency domains.** An 8-point time-series $x(n)$ (generated by a Gaussian random noise generator with zero mean and unity variance given below) passes through the three-tap FIR filter specified by the unit-sample response $h(n) = \{1, -1.5, 1\}$. Predict filter's output by (a) linear convolution and (b) circular convolution (i.e., DFT and IDFT), and compare the two predictions.

Solution

In the time domain the output is predicted by applying linear convolution

$$y_1(n) = \sum_{k=0}^{9} x(k)h(n-k) = x(n) - 1.5x(n-1) + x(n-2)$$

$y_1(n)$ is 10 points long and constitutes the true output. In the frequency domain the output is found by first padding $h(n)$ by 5 zeros to make it 8 points long. Then by taking its 8-point DFT, multiplying the DFT by the 8-point DFT of $x(n)$, and taking the IDFT to obtain $y_2(n)$. $y_2(n)$ is 8 points long and deviates from the output at the edges, too. These are shown below.

  (i) Input $x(n) = \{-0.854258, 0.622873, 0.934505, -1.500489, -0.513985, 1.357851, -0.592673,$

   $0.546175\}$

  (ii) Filter $h(n) = \{1, -1.5, 1\}$

  (iii) Output by convolution, $y_1(n) = \{-0.284753, 0.634753, -0.284688, -0.759791, 0.890418, 0.209447,$

   $-1.047811, 0.931012, -0.470645, 0.182058\}$

  (iv) Output by DFT-IDFT, $y_2(n) = \{-0.755398, 0.816812, -0.284688, -0.759791, 0.890418, 0.209447,$

   $-1.047811, 0.931012\}$

  (v) Difference, $y_1(n) - y_2(n) = \{-0.470645, 0.182059, 0, 0, 0, 0, 0, 0, -0.470645, 0.182058\}$

15. Example 18.36 in this chapter illustrated a simple bandpass filter with its center frequency at $\pi/2$. Example 18.37 illustrated design of a wideband bandpass filter by pole-zero placement. It placed 10 pairs of poles inside the unit circle all at the radius 0.9 on both sides of the center frequency $\pm\omega = \pi/2$ equally distanced from each other. The angular distance between two neighboring poles is $9°$. Similarly, it placed 10 pairs of zeros on the unit circle around $z = 1$ and $z = -1$. The pole-zero plot and the resulting magnitude plot are shown in Figure 18.9(a). To avoid zeros on the unit circles, the zeros are moved inside the unit circle to a radius 0.7 from the origin, and their mirror images with respect to the unit circle (at the same angles but at a radius $\frac{1}{0.7}$) are added. Their pole-zero plots and the resulting magnitude plots are shown in Figure 18.9(b).

To flatten the top of the passband segment of the frequency response the poles are flipped vertically. The zeros remain as before. The pole-zero plot and the resulting magnitude plot are shown in Figure 18.22(a) and (b), corresponding to Figure 18.9(a) and (b).

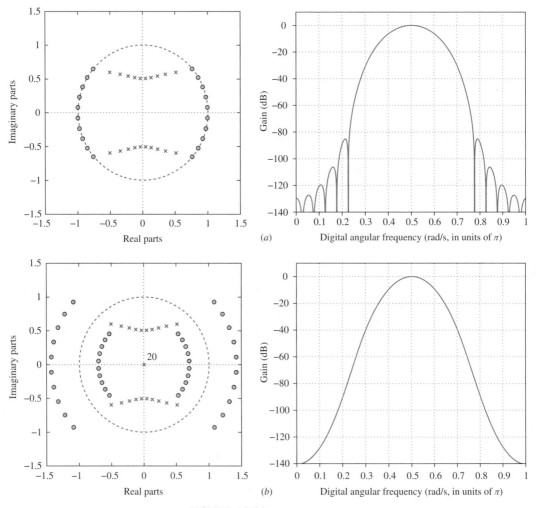

FIGURE 18.22 (For problem 15)

16. An FIR notch filter has 2 zeros at $z_{1,2} = \pm j$. We add to it two pairs of poles on both sides of the zeros at distances of $\pi/20$ rad/s, inside the unit circle at a distance of 0.1. The result is a filter with:

Two zeros at $z = \pm j$ and two pairs of poles at $p_{1,2} = 0.9e^{\pm j9\pi/20}$, $p_{3,4} = 0.9e^{\pm j11\pi/20}$

The DC gain of the filter is one. A continuous-time signal $x(t) = 1 + \cos(4{,}500\pi t) + \cos(5{,}000\pi t) + \cos(5{,}500\pi t)$ is sampled at the rate of 10 kHz and the samples are passed through the above filter. The filter's output is converted back into the continuous-time using a unity-gain A/D converter. Find the resulting signal in the form of $y(t) = A_0 + A_1 \cos(4{,}500\pi t - \theta_1) + A_2 \cos(5{,}000\pi t - \theta_2) + A_3 \cos(5{,}500\pi t - \theta_3)$.

Hint: A pair of poles at $\rho e^{\pm j\theta}$ contributes

$$\frac{1}{(1 - \rho e^{j\theta} z^{-1})(1 - \rho e^{-j\theta} z^{-1})} = \frac{1}{1 - 2\rho \cos\theta z^{-1} + \rho^2 z^{-2}}$$

to a filter's system function. See Figure 18.23.

	Filter	A_0	A_1	θ_1	A_2	θ_2	A_3	θ_3
Ans.	FIR	1	0.1564	-1.4137	0	-1.5708	0.1564	1.4137
	IIR	1.	9.5192	-0.3135	0	-1.5708	9.5192	0.3135

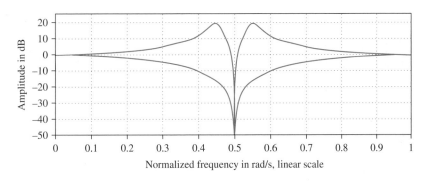

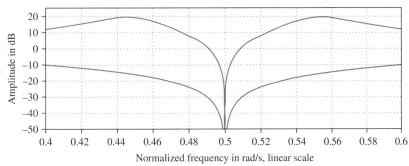

FIGURE 18.23 (For problem 16)

17. Notch filter. The *m* file for a notch filter design is given below. Run the program and measure the performance of the filter at the neighborhood of the notch.

```
r=0.96757;
k=( 2-sqrt(2)*r+r^2)/(2-sqrt(2));
num=[1 -sqrt(2) 1];
den=[1 -sqrt(2)*r r^2];
w=0:0.002*pi:pi;
H=k*freqz(num,den,w);
gain=20*log10(abs(H));
figure(1)
subplot(2,1,1)
plot(w/pi,gain); grid;
axis([0 1  -75 0])
title('magnitude plot');
xlabel('digital angular frequency (unit of pi)');
ylabel ('gain (db)');
subplot(2,1,2)
plot(w/pi,gain);grid;
axis([0.23 0.27  -3.5 -2.5])
%title('magnified plot');
xlabel('digital angular frequency (unit of pi)');
ylabel ('gain (db)');
```

Chapter Problems

All systems are discrete. Plots of $H(\omega)$ are to be done for $-\pi < \omega < \pi$ rad/s or $-0.5 < f < 0.5$ Hz.

18. A discrete signal $x(n) = \cos\frac{\pi n}{6}$ is passed through a linear filter with $h(n) = \{1, -1, 1\}$. Find the output $y(n)$.

19. The poles of a causal discrete-time LTI system are at $p_1 = 2$, $p_2 = 0.5$, and $p_3 = 0.1$. The system has no zeros. The steady-state response to a unit step is $-10/9$.

  a. Find the system function.
  b. Find the system's unit-sample response.
  c. Is the system stable? Support your answer by an appropriate argument.
  d. Find the difference equation that gives $y(n)$ for $n > 0$ and sketch a block diagram made of gains and unit delays to represent the system.
  e. Given the initial conditions $y(0) = 1$, $y(-1) = 1$ and $y(-2) = 1$, use the analytic method to solve the difference equation and find a closed-form expression for $y(n)$, $n \geq 0$, given $x(n) = 0$, $n > 0$.
  f. Repeat part e for $x(n) = 1$, $n > 0$.
  g. Repeat parts e and f using the z-transform method.

20. a. Find the system function given the following difference equation:
$$y(n) - 0.2y(n-1) = x(n)$$
  b. Find the steady-state response to $x(n) = \cos(\pi n)$.
  c. Find the magnitude and phase of the frequency response for $\omega = \pi$.
  d. Obtain b from c.

21. The unit-sample response of a digital filter is $h(n) = \{1, \underset{\uparrow}{2}, 1\}$.

    a. Find and plot $H(\omega)$.

    b. The discrete signal $x(n) = 1 + \cos\frac{n}{2}$ is passed through the above filter. Find the output $y(n)$.

22. A continuous-time analog signal $x(t) = \cos 500\pi t + \cos 200\pi t$ is sampled at the rate of 1,000 samples per second.

    a. Find $x(n)$.

    b. The resulting discrete signal $x(n)$ is passed through a filter with unit sample response $h(n) = \{1, 2, \underset{\uparrow}{3}, 2, 1\}$. Find the output $y(n)$ and its analog reconstruction $y(t)$.

23. A linear time-invariant analog filter is specified by its unit-impulse response $h_a(t) = e^{-t}u(t)$. Sample $h_a(t)$ at the rate of five samples per second and obtain $h(n)$. Find $H(z)$ and $H(\omega)$. Plot $H(\omega)$ and find the 3-dB discrete-time frequency.

24. The unit-sample response $h(n)$ of a digital filter is

$$h(n) = \begin{cases} 0.5^n, & 0 \leq n < N \\ 0, & \text{elsewhere} \end{cases}$$

Is the filter FIR? Is it linear phase? Find $H(z)$ in the form of the ratio of two polynomial functions of z^{-1} with real coefficients. How many poles and zeros does the filter have? Find them for the case of $N = 3$.

25. Determine the magnitude and phase of the frequency response of the following systems.

    a. $y(n) = x(n) + x(n+1)$

    b. $y(n) = x(n) + x(n-1)$

    c. $y(n) = x(n-1) + 2x(n) + x(n+1)$

26. Find the shortest linear-phase unit-sample response whose $H(z)$ has a zero at $0.8e^{j\frac{\pi}{4}}$.

27. a. Sample the analog signal $h(t) = \sin(1,000\pi t)/(\pi t)$ at the rate of 1,000 samples per second in a way that one sample is at $t = 0$. Find $h(n)$ and $H(\omega)$. Sketch $H(\omega)$. For what purpose, or application, may such a signal be used?

    b. Repeat part a for the rates of (i) 2,000 and (ii) 5,000 samples per second.

28. a. The unit-sample response of a digital filter is $h(n) = \{1, \underset{\uparrow}{0}, 1\}$. Find $H(z)$ and $H(\omega)$. Sketch magnitude and phase of $H(\omega)$, including scales and labels. Discuss properties of the filter. Repeat for $h(n) = \{1, \underset{\uparrow}{-1}, 1\}$.

    b. A continuous-time pulse

$$x(t) = \begin{cases} 1, & -5.1 \text{ msec} \leq t \leq 5.1 \text{ msec} \\ 0, & \text{elsewhere} \end{cases}$$

is sampled at the rate of 200 samples per second in a way that one sample is at $t = 0$. The resulting discrete-time function is $x(n)$. Find $X(z)$ and $X(\omega)$.

    c. The $x(n)$ of part b is passed through a filter with $h(n) = \{1, \underset{\uparrow}{0}, 1\}$. Using the method(s) of your choice find the output $y(n)$ of the filter. Compare $y(n)$ with $x(n)$ and discuss their differences.

    d. Repeat parts b and c for the rate of 2,000 samples per second.

29. a. A discrete-time LTI system has three zeros at $z = \pm j$ and -1. Find $H(z)$ and $H(\omega)$ (magnitude and phase). Plot $|H(\omega)|$ and find the 3-dB bandwidth.

    b. Add a zero at $z = \frac{1}{3}$ and another zero at $z = 3$. Repeat part a.

30. a. A discrete-time LTI system has 4 zeros at $z = \pm j$ and $e^{\pm j150°}$. Find $H(z)$ and $H(\omega)$ (magnitude and phase). Plot $|H(\omega)|$ and find the 3-dB bandwidth.

    b. Add a zero at $z = \frac{1}{3}$ and another zero at $z = 3$. Repeat part a.

31. A discrete-time LTI system has two poles at $0.9e^{\pm j45°}$. Find $H(z)$ and $H(\omega)$ (magnitude and phase). Plot $|H(\omega)|$ and find the 3-dB bandwidth.

32. a. An LTI system has two zeros at $z = \pm 1$ and two poles at $0.9e^{\pm j45°}$. Find $H(z)$ and $H(\omega)$ (magnitude and phase). Plot $|H(\omega)|$ and find the 3-dB bandwidth.

33. A discrete-time LTI system has two poles at $p_{1,2} = 0.9e^{\pm j\frac{\pi}{4}}$ and two zeros on the unit circle at $z_{1,2} = e^{\pm j\frac{\pi}{4}}$.
  a. Write the expression for $H(z)$ as a function of z^{-1}.
  b. Find and plot $H(\omega) = H_r(\omega)e^{j\Theta(\omega)}$.

34. The system function $H(z)$ of a discrete LTI filter has two poles at $0.8e^{\pm j\frac{\pi}{4}}$ and a double zero at $z = 0$. Find $H(z)$ in the form of the ratio of two polynomial functions of z^{-1} with real coefficients.

35. An FIR filter has 2 zeros at $z = \frac{1}{2}$ and $z = 2$. Find $H(z)$ and $|H(\omega)|$ in the form of $H_r(\omega)e^{j\Theta(\omega)}$. Plot $H_r(\omega)$ and label important points on the plot. Specify if the filter is low-pass, high-pass, bandpass, or bandstop.

36. The following three systems approximate an integrator.
  a. $y(n) = x(n) + y(n-1)$
  b. $y(n) = 0.5x(n) + 0.5x(n-1) + y(n-1)$
  c. $y(n) = 0.333x(n) + 1.333x(n-1) + 0.333x(n-2) + y(n-2)$

  In each case find $H(\omega)$. Plot its magnitude and phase. Compare with the performance of a perfect integrator.

37. The following two systems approximate a differentiator.
  a. $y(n) = x(n) - x(n-1)$
  b. $y(n) = 0.5x(n) - 0.5x(n-2)$.

  In each case find $H(\omega)$. Plot its magnitude and phase. Compare with the performance of a perfect differentiator.

38. A discrete-time LTI system has two poles at $p_{1,2} = \rho e^{\pm j\omega_0}$ and 2 zeros at $z = \pm 1$.
  a. Write the expression for $H(z)$ as a function of z^{-1}.
  b. Find $H(j\omega) = H_r(\omega)e^{j\Theta(\omega)}$.
  c. Find and plot $H_r(\omega)$ and $\Theta(\omega)$ for $\rho = 0.6, 0.8, 0.9, 0.95$ and $\omega_0 = \frac{\pi}{12}, \frac{\pi}{4}, \frac{\pi}{2}$.

  In each case specify the maximum of the magnitude and the angular frequency at which it is reached.

39. a. Find $H(z)$ as a function of z^{-1} for the following two filters.
  (i) Filter 1 has two poles at $0.95e^{\pm j\frac{\pi}{4}}$, 2 zeros at $e^{\pm j\frac{\pi}{4}}$ and a double zero at the origin.
  (ii) Filter 2 has two poles at $0.95e^{\pm j\frac{\pi}{4}}$, 2 zeros at $1.05e^{\pm j\frac{\pi}{4}}$ and a double zero at the origin.
  b. Plot $|H(\omega)|$, find the 3-dB points and determine if each filter is low-pass, high-pass, bandpass, or bandstop.

40. Filtering in time and frequency domains. Generate a 64-point Gaussian random signal with zero mean and unity variance by a random noise generator. Pass it through the 3-tap FIR filter with $h(n) = \{1, -1.5, 1\}$. Obtain filter's output by (a) linear convolution and (b) circular convolution (i.e., DFT and IDFT), and compare the two predictions.

41. Effect of zeros on the bandwidth of an IIR bandpass filter. A digital filter has a pair of complex conjugate poles at $p_{1,2} = \pm j0.95$ and a set of 4 zeros at $z_{1,2,3,4} = \pm e^{\pm j\theta}$.
  a. Find $H(z)$ and $H(\omega) = |H(\omega)|\angle H(\omega)$ as a function of θ.
  b. Specify important values [maximum value, 3-dB frequencies (ω_ℓ and ω_h), and the bandwidth $\Delta\omega = \omega_h - \omega_\ell$] for $\theta = k\pi/8$, $k = 1, 2, 3$. Plot magnitude response of the filter for the above three values of θ. Compare results with Example 18.36 in the chapter.

42. **Effect of poles on the bandwidth of an IIR bandpass filter.** A digital filter has a set of four poles at $p_{1,2,3,4} = \pm j0.95e^{\pm j\theta}$ and 2 zeros at $z = \pm 1$.

a. Find $H(z)$ and $H(\omega) = |H(\omega)|\angle H(\omega)$ as a function of θ.

b. Specify important values [maximum value, 3-dB frequencies (ω_ℓ and ω_h), and the bandwidth $\Delta\omega = \omega_h - \omega_\ell$] for $\theta = k\pi/32$, $k = 1, 2, 3$. Plot magnitude response of the filter for the above three values of θ. Compare results with Example 18.37 in the text.

18.14   Project 1: FIR Filter, Design by Windowing

Objective and Summary

In this project you will design three low-pass FIR filters (9-tap, 31-tap, and 61-tap) based on the windowing method, and measure their theoretical performance through simulation by Matlab or other software tools. You will then apply the filters to a recorded speech signal off-line and compare with theory. You will compare the filters performance with theory and with each other. If digital hardware is available, you then will run the filters in real time on a digital platform such as DSP board and measure their frequency responses. Using the microphone as the input device you will let the board filter your speech in real time and qualitatively observe its filtering effect.

Design
A Nine-Tap Low-Pass FIR Filter

Obtain the unit-sample response of a linear-phase low-pass digital FIR

$$h(n) = \{h_0, \ h_1, \ h_2, \ h_3, \ h_4, \ h_3, \ h_2, \ h_1, \ h_0\}$$
$$\uparrow$$

based on an ideal filter with a cutoff frequency corresponding to 1 kHz in the continuous-time domain. The filter is to operate on the signals listed in the filtering section of this project with $F_s = 22{,}050$ Hz. First determine the cutoff frequency ω_0 of the digital filter. Then consider the ideal low-pass filter with

$$\overline{h}(n) = \frac{\sin(\omega_0 n)}{\pi n}$$

Shift $\overline{h}(n)$ to the right by four units and multiply it by a nine-tap Hanning window ($0 \leq n \leq 8$) to obtain $h(n)$ and its frequency response $H(\omega)$. Call the filter FIR1. Plot magnitude and phase of $H(\omega)$. Save the computer code for later use in this experiment.

A 31-Tap Low-Pass FIR Filter

Using an approach similar to that of the above, design a 31-tap low-pass filter and call it FIR2. Plot the magnitude and phase of $H(\omega)$.

A 61-Tap Low-Pass FIR Filter

Repeat for a 61-tap low-pass filter according to the following steps:[8]

---

[8]The *ARRL Handbook for Radio Amateurs*, 2002, Chapter 18 and its appendix, published by ARRL-the National Association for Amateur Radio.

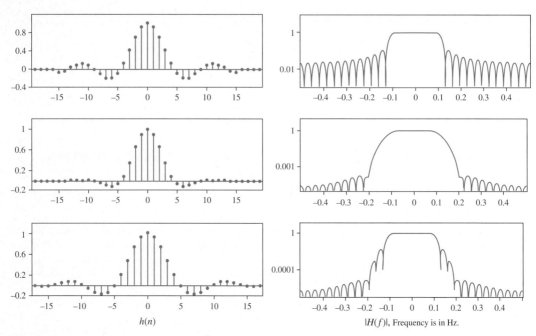

FIGURE 18.24 Constructing the 61-tap FIR3 filter to approximate an ideal low-pass filter with cutoff frequency at 0.125 Hz. Top row is a 31-tap FIR filter under rectangular window. Middle row is a 31-tap FIR filter under Blackman window. Bottom row is a 61-tap FIR filter created by the convolution of the two filters above. Unit-sample responses are shown in the left column and frequency responses in the right column.

1. Consider the ideal low-pass filter

$$\overline{h}(n) = \frac{\sin(\omega_0 n)}{\pi n}$$

2. Multiply $\overline{h}(n)$ by a 31-tap uniform window to obtain $h_1(n)$ and its frequency response $H_1(\omega)$.
3. Multiply $\overline{h}(n)$ by a 31-tap Blackman window to obtain $h_2(n)$ and its frequency response $H_2(\omega)$.
4. Convolve $h_1(n)$ and $h_2(n)$ and normalize the result to obtain the new unit-sample and frequency responses $h_3(n)$ and $H_3(\omega)$.

$$h_3(n) = \frac{h_1(n) \star h_2(n)}{Max(h_1 \star h_2)} \iff H_3(\omega) = \frac{H_1(\omega) \times H_2(\omega)}{Max(h_1 \star h_2)}$$

$h_3(n)$ is your design result and will be used in the rest of this experiment. Call it FIR3. An example is shown in Figure 18.24.

Filter Specifications

1. Find the attenuation of filters at the cutoff frequency ω_0 in dB below their DC gain, and their phase at that frequency. In the continuous-time domain, the corresponding cutoff frequency will be $f_0 = 1$ kHz.
2. Define the passband of the low-pass filters to be $\omega < \omega_p$, where the gain is less than 1 dB below the DC gain. Define the stopband to be $\omega > \omega_s$, where the gain is more than 20 dB below the DC gain. Find ω_p and ω_s for the FIR1, FIR2, and FIR3 filters. Call them ω_{p1}, ω_{s1}, ω_{p2}, ω_{s2}, ω_{p3}, and ω_{s3}, respectively. In the continuous-time domain and at the sampling frequency of $F_s = 22{,}050$ Hz the passband and stopband frequencies will be called f_p and f_s. Enter your results in Table 18.1.

TABLE 18.1 Specifications for the Three Linear-Phase Low-Pass Filters

	f_p, Hz	f_0, Hz	f_s, Hz
FIR1			
FIR2			
FIR3			

Filtering

Filtering Sinusoids

Sample the sinusoidal signal $x(t) = \cos(2\pi f t)$ at the rate of $F_s = 22{,}050$ Hz and pass it through the nine-tap filter. For the frequency values given in Table 18.2 find magnitude A, phase θ, and time delay τ in the filter's output $y(t) = A\cos(2\pi f t - \theta) = Ax(t - \tau)$. Repeat for the 31-tap and 61-tap filters. Enter your results in Table 18.2.

TABLE 18.2 Output Values for Three Linear-Phase Low-Pass Filters at Various Frequencies

Frequency f, Hz	100	f_{p1}	f_{p2}	f_0	f_{s2}	f_{s1}	9,000
FIR1 A, volts							
FIR1 θ, radian							
FIR1 τ, msec							
FIR2 A, volts							
FIR2 θ, radian							
FIR2 τ, msec							
FIR3 A, volts							
FIR3 θ, radian							
FIR3 τ, msec							

Filtering a Speech Signal

Pass the speech file *a cup of hot tea* (*cht5.wav*) through the nine-tap low-pass FIR filter you designed previously [convolving it with $h(n)$] and save it as a *.wav file. Compare the spectra of the input and output files. Play both of them and compare with each other.

You will notice that for the high sampling rate of 22 kHz used in *cht5.wav*, the nine-tap filter is inadequate. Apply the 31-tap filter (as shown in Figure 18.25) to the speech file *cht5.wav* and compare with the output obtained from the nine-tap filter. Repeat using the 61-tap filter. Obtain the input and output spectra and their power. Compare with theory.

Real-Time Filtering

In this section you will use a digital platform such as a DSP board for testing the real-time operation of the filters. First apply a sinusoidal input to the board to measure its frequency response. Then program the filters and load them onto the digital platform. Now apply the sinusoid to measure the frequency response of filters. Finally, apply the microphone as the input and let the board filter your speech. Qualitatively observe the effect.

Discussion and Conclusions

Design Methodology

Suggest a methodology for designing a low-pass FIR filter for which
    Passband attenuation $\leq A_p$ within the band $\omega \leq \omega_p$,
    Stopband attenuation $\geq A_s$ within the band $\omega \geq \omega_s$,
where A_p, ω_p, A_s, and ω_s are known.

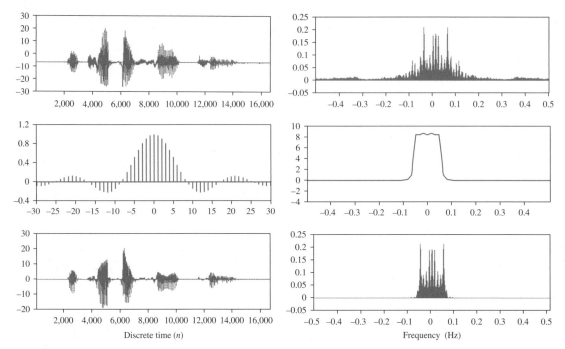

FIGURE 18.25 Low-pass filtering of the speech file $cht5.wav$ by the 31-tap FIR2 filter. Time sequences are on the left. The magnitude spectra are on the right.

Is Filtered Speech Intelligible?

How intelligible is filtered speech? Compare the outputs of FIR1, FIR2, and FIR3.

Speaker Recognition in Filtered Speech

Can you recognize the speaker by listening to filtered speech used in this experiment? Compare the outputs of FIR1, FIR2, and FIR3 in terms of speaker recognition.

Overall Conclusions

Summarize your observations and methods to design other classes of FIR filters (high-pass, bandpass, bandstop, $\cdots$) close to a desired characteristic.

18.15 Project 2: Discrete-Time Modeling and Control of a Dynamic System

Objectives and Summary

In this project you will explore digital control of a physical plant modeled by a linear time-invariant system. The plant is specified by measurements of its time and frequency responses. The project consists of two procedures. Procedure 1 is concerned with the open-loop system. First, you will model the system by a continuous-time second-order model and transform the model from the continuous- to the discrete-time domain. You will verify the validity of the discrete-time system as an operational model for the actual physical plant by comparing the step and frequency responses in both domains. This is carried out off-line and all responses are obtained by simulations. The comparison will be done mathematically. You will then implement the open-loop discrete-time system on a digital device (such as a computer or a digital signal processing board) and confirm the validity of the model by running it in real time with an analog input and output.

In procedure 2 you add feedback to the system to reduce the overshoot in its step response from a maximum of 60% to below 10%. First, you will design and simulate an analog closed-loop control system made of the second-order continuous-time system and an analog PID controller. You will then convert it to a closed-loop digital control system and test its performance both off-line, by simulation, and in real time on a digital platform. Finally, you will construct a hybrid system made of an analog plant (an op-amp electronic circuit) and a digital controller (on the digital device), all working together in real time. Again, the controller's task is to reduce the overshoot in the step-response from a maximum of 60% to below 10%. You will then run the hybrid system in real time with analog inputs and outputs and test the performance of the controller.

The tools needed for the project consist of software and a digital platform for real-time operation of the digital controller. These can be a DSP board embedded in a PC (or existing on its own) or a digital computer. (For this purpose a Texas Instruments's DSP board was used.) The dynamic system that is to be controlled is an electronic op-amp circuit.

Procedure 1 Open-Loop System

In this procedure you construct a discrete-time model of a continuous-time system.

Modeling the Plant

Consider a physical plant (Figure 18.26; e.g., a simple pendulum with small displacements) to be modeled as an LTI system. The response of the plant to a unit-step input is recorded and shown in Figure 18.26(*a*). Note the 60% overshoot. The frequency response is also given in Figure 18.26(*b*). (The quantitative values of the step and frequency responses may be found from the Matlab program given in "Simulation of the Discrete-Time System" below.)

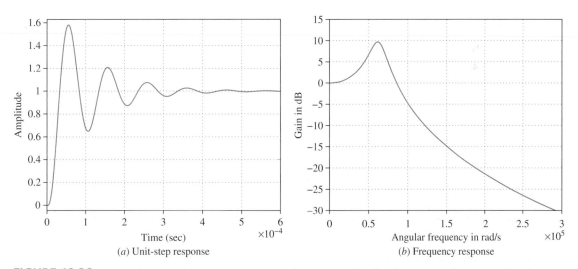

(*a*) Unit-step response   (*b*) Frequency response

FIGURE 18.26 Measured time and frequency responses of the plant. (The plant is a third-order system. The above plots were generated by the Matlab program given in the next section of this project. This information, however, is not known by the project.)

Model the plant by a second-order linear time-invariant system:

$$H(s) = \frac{\omega_0^2}{s^2 + 2\zeta\omega_0 s + \omega_0^2}$$

In the above equation, ω_0 is the undamped natural frequency and ζ is the damping ratio. From the step response of Figure 18.26(a), find ω_0 and ζ. (Determine ζ from the overshoot in the unit-step response. See the next section for a refresher on second-order systems.) Compare your findings with the following values: $\omega_0 = 62,832$ and $\zeta = 0.16$. (From here on, for the sake of uniformity, use the above model parameters.) Plot the unit-step response and the frequency response of the model. Compare, qualitatively and quantitatively, the model's responses with the measured plots in Figure 18.26. For a more detailed mathematical analysis of the model and its responses, refer to Chapter 9.

Transformation from the Continuous- to Discrete-Time Domain

The differential equation relating a continuous-time system's input $x(t)$ to its output $y(t)$ is

$$y'' + 2\zeta\omega_0 y' + \omega_0^2 y = x(t)$$

where

$$y'' = \frac{d^2\, y(t)}{dt^2} \quad \text{and} \quad y' = \frac{d\, y(t)}{dt}$$

The continuous-time system may be transformed to an equivalent discrete-time system by sampling it at intervals of T. The above differential equation is then converted to a difference equation. The transformation is done in one of several ways.

1. Forward rectangular rule using the forward-difference formula:

$$\frac{dy(t)}{dt}\Big|_{t=nT} \approx \frac{y(nT+T) - y(nT)}{T} \Rightarrow s \approx \frac{z-1}{T}$$

2. Backward rectangular rule using the backward-difference formula:

$$\frac{dy(t)}{dt}\Big|_{t=nT} \approx \frac{y(nT) - y(nT-T)}{T} \Rightarrow s \approx \frac{z-1}{Tz}$$

3. Impulse-invariant transformation:

$$h(n) = h(t)|_{t=nT} \Rightarrow H(z) = \sum \left[\text{residues of } \frac{H(s)}{1 - e^{sT}z^{-1}} \text{ at poles of } H(s) \right]$$

4. Bilinear transformation using the trapezoidal integration formula:

$$g(t) = \int_{-\infty}^{t} h(\tau)d\tau$$

$$g(nT+T) = g(nT) + \frac{T}{2}[h(nT+T) + h(nT)] \Rightarrow s \approx \frac{2}{T}\frac{z-1}{z+1}$$

The aim is to obtain an $H(z)$ that would emulate the continuous-time system when used along with A/D and D/A converters. Discuss the choice of transformation method and sampling rate, and summarize the advantages and disadvantages of each of the above methods. Then use the bilinear rule to transform $H(s)$ to $H(z)$.

Simulation of the Discrete-Time System

Use a mathematical software package to simulate the second-order discrete-time model and plot its step and frequency responses. Compare the unit-step response of the discrete-time model with that of the plant and then comment. Examine the following Matlab program. You may use this program or write one of your own.

```
zeta=0.16; f0=10000; w0=2*pi*f0
num1=[w0^2]; den1=[1 2*zeta*w0 w0^2];
r1=roots(den1); r2=[-4*w0    r1'];
num2=[4*w0^3]; den2=poly(r2);
Ts= 0.0002; Fs= 1/Ts;
%
printsys(num1,den1); printsys(num2,den2);
figure(1); step(num1,den1); grid; print -dpsc PID_V5_Figxa.eps;
figure(2); step(num2,den2); grid; print -dpsc PID_V5_Fig2a.eps;
%
[H1,w]=freqs(num1,den1);[H2,w]=freqs(num2,den2);
figure(3); plot(w,20*log10(abs(H1))); grid
axis([0,300000,-30,15]); title('simulated frequency response of the model')
ylabel('Gain in dB'); xlabel('Angular frequency in rad/s')
print -dpsc PID_V5_Figxb.eps;
figure(4); plot(w,20*log10(abs(H2))); grid
axis([0,300000,-30,15]); title('Measured frequency response of the plant')
ylabel('Gain in dB'); xlabel('Angular frequency in rad/s')
print -dpsc PID_V5_Fig2b.eps;
[num1d, den1d] = BILINEAR(num1, den1, Fs); printsys(num1d, den1d,'z');
[num2d, den2d] = BILINEAR(num2, den2, Fs); printsys(num2d, den2d,'z');
```

The above program also prints out system functions of the plant and the model.

The model: $H(s) = \dfrac{\omega_0^2}{s^2 + 2\zeta\omega_0 s + \omega_0^2}$ $H(z) = \dfrac{0.92914z^2 + 1.8583z + 0.92914}{z^2 + 1.8112z + 0.90536}$

The plant: $H(s) = \dfrac{4\omega_0^3}{(s + 4\omega_0)(s^2 + 2\zeta\omega_0 s + \omega_0^2)}$ $H(z) = \dfrac{0.89359z^3 + 2.6808z^2 + 2.6808z + 0.89359}{z^3 + 2.7347z^2 + 2.578z + 0.83607}$

Real-Time Operation of the Discrete-Time System

Implement the discrete-time model on a digital platform and run it in real time. Discuss your choice of sampling rate in A/D and D/A converters and its effect on the real-time performance of the discrete-time model.

Procedure 2 Closed-Loop System

In this procedure you produce digital control of the dynamic system.

Design of a PID Controller[9]

Use the control loop of Figure 18.27. The PID controller has three parallel paths (a proportional path k_p, an integral path k_i, and a derivative path k_d) and an adder. Its transfer function is

$$G_c(s) = k_p + \frac{k_i}{s} + k_d s$$

---

[9]Another design example for a PID control system is found in Example 9.48 of Chapter 9.

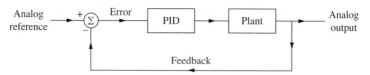

FIGURE 18.27 Simulation of analog control of the analog system.

Design of the controller requires determination of k_p, k_i, and k_d. This can be done by the root locus method, placing the poles of the closed-loop system at acceptable locations, or by writing a computer program that iteratively adjusts the coefficients in a search for an answer.[10] You also need to pay attention to the steady-state error produced by the controller. Simulate the analog (continuous-time) closed-loop control system. See Figure 18.27.

Digital Control of the Digital System

1. Transform the analog control loop to the discrete-time (digital) domain by bilinear transformation [see Figure 18.28(a)] and simulate it. In doing this, the whole control loop is digital.
2. Implement the closed-loop digital control system of Figure 18.28(b) on a digital platform (e.g., a digital signal processing board; Texas Instruments DSK board) and run it in real time. Compare to the continuous-time control system.

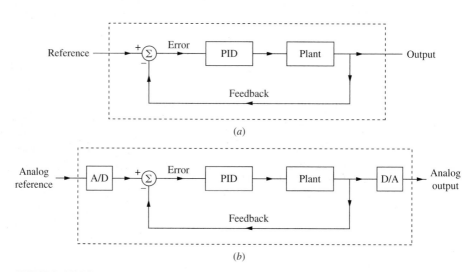

FIGURE 18.28 *(a)* Simulation of digital control of the digital system. *(b)* Real-time implementation of digital control of digital system.

Digital Control of the Analog System

In this part of the project the plant to be controlled is an analog electronic low-pass system and the controller is digital. See Figure 18.29(a). To construct the analog plant we express it in terms of its quality factor Q and undamped natural

---

[10]The coefficients will not be unique. For more, see a book on control systems design.

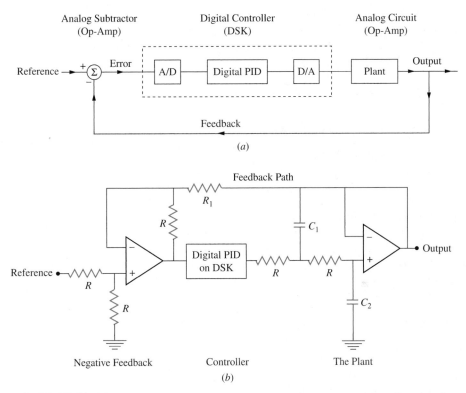

FIGURE 18.29 *(a)* The closed-loop control system with a digital PID controller and analog plant that is made of op-amp circuits. *(b)* Digital control (with external feedback) of the analog plant. Feedback is provided by the external voltage path and the summing op-amp circuit. The digital controller is implemented on the DSK board. The analog plant is modeled by the second-order low-pass 741 op-amp circuit with $R = 100\ K\Omega$, $C_1 = 4.7\ \mu F$, and $C_2 = 120$ nF. Feedback coefficient can be adjusted through R_1.

frequency ω_0:

$$H(s) = \frac{1}{\left(\frac{s}{\omega_0}\right)^2 + \frac{1}{Q}\left(\frac{s}{\omega_0}\right) + 1}$$

where $Q = 1/(2\zeta) = 3.125$. Use a unity-gain Sallen-Key low-pass filter to realize the above system. An example is shown in Figure 18.29(*b*). Simulate the analog electronic circuit using a circuit simulation package such as Spice. Test the unit-step response of the physical circuit to verify its compliance with the desired specifications.

Load the digital controller onto the digital platform and place it in the control loop. Use an op-amp circuit to subtract the output (feedback signal) from the reference signal as shown in Figure 18.29(*b*). Run the system in real time. Record its unit-step response. Measure its rise time, percentage overshoot, natural frequency, and damping ratio. Compare with those of the plant in open-loop state. Compare with theory.

Comparison and Conclusions

Find the three poles of the third-order LTI system used in this project as the plant (see "Real-Time Operation of the Discrete-Time System" in this project). Compare them with the poles of the second-order model. Why is the effect of the third pole (necessarily being a negative real pole) on the unit-step response small? At what frequency range do you expect the effect to be pronounced? Illustrate the effect of the third pole on the frequency response.

Draw your conclusions from this project and include them in the report.

18.16   Project 3: Modeling a Random Trace Generator

Summary

Figure 18.30 shows consecutive traces drawn by a human subject on paper using a pen. The abscissa is time and the ordinate is the trace generated by the motion of the pen. The subject was instructed to generate the traces as randomly as possible. The six traces shown in the figure are consecutive in time. This project is concerned with characterizing such time series and producing a mathematical model for their generation. The project consists of two procedures.

Procedure 1

In this procedure you will first sample the traces in Figure 18.30 and convert each to a discrete-time signal. For normalization purposes, let Δt in the figure be 1 msec (not the actual value). You will then compute the DTFT of each signal. You will

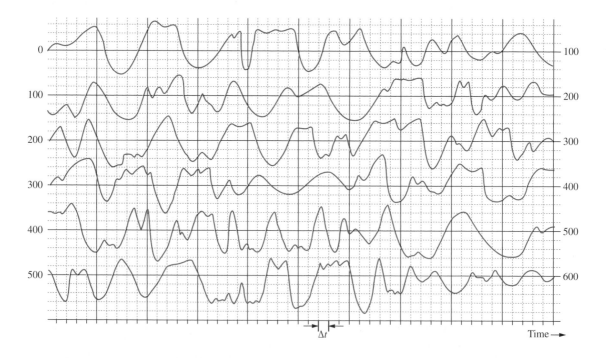

FIGURE 18.30 Randomly generated traces by a human subject. See procedure 1 in this project.

investigate the transform to see if the spectrum contains peaks (indicating the presence of resonators) and determine the complex frequencies associated with such possible peaks. A model for the generation of the traces is then constructed, which includes a random generator and a number of digital resonators. The performance of the model will be compared with the given data and a measure of error (such as the average rms of differences) will evaluate the model.

Procedure 2

In this procedure 2 you will generate your own traces following the same instructions and repeat the steps described in procedure 1.

Appendix

Electric Circuits

A.1 Circuit Elements, Laws, and Formulation of Equations

Circuit Elements

An electric circuit is made of interconnected elements or devices (resistors, capacitors, inductors, switches, diodes, transistors, amplifiers, sensors, transducers, signal, and energy sources, etc.). In this book we consider circuits in which elements and devices are interconnected through their terminals. If a device with more than two terminals is embedded in a circuit, we model it by an interconnection of two-terminal elements. With a two-terminal element are associated two algebraic quantities: current i and voltage v, called the terminal variables. By convention, the terminal where the current enters the element is labeled with a positive voltage sign with respect to the terminal where the current leaves the element. See Figure 1(a). The element is defined by its $i - v$ relationship, also called terminal characteristic. The voltage and current generally vary with time. Either i or v could constitute the input, while the other would be the output. The electric power entering an element is defined as $p(t) = v(t)i(t)$, also a function of time.

Circuit Laws

The interconnection of elements creates nodes and loops. Nodes are connection points where two or more elements come together as in Figure 1(b). Loops are closed paths formed by several elements as in Figure 1(c). Elements' currents and voltages constitute the circuit variables. In addition to the elements' terminal characteristics, the circuits are constrained by two additional laws called Kirchhoff's current law (KCL) and Kirchhoff's voltage law (KVL). These laws apply regardless of elements types. They are

KCL: Algebraic sum of currents arriving at a node is zero. $\sum_k i_k = 0$. See Figure 1(b).

KVL: Algebraic sum of voltages around a loop is zero. $\sum_\ell v_\ell = 0$. See Figure 1(c).

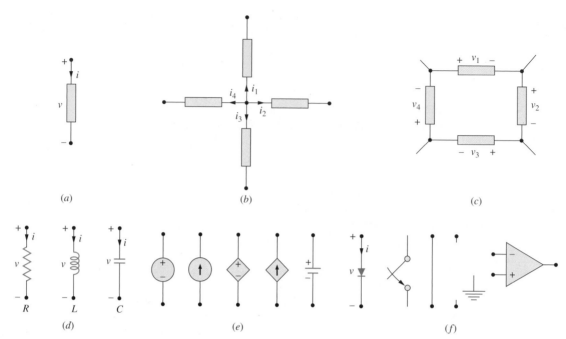

FIGURE 1 *(a)* A two-terminal circuit element. *(b)* An illustration of Kirchhoff's current law (KCL), $i_1 + i_2 + i_3 + i_4 = 0$. *(c)* An illustration of Kirchhoff's voltage law (KVL), $v_1 + v_2 + v_3 + v_4 = 0$. *(d)* Three linear models R, L, and C of passive elements. *(e)* Signal and energy sources. From left: an independent voltage source, an independent current source, a dependent voltage source, a dependent current source, and a battery. *(f)* More circuit elements and symbols. From left: a diode, a switch, a short circuit, an open circuit, ground (reference point for node voltages), and an operational amplifier.

Passive Elements

A passive element dissipates electrical energy (in the form of heat, light, mechanical work, chemical transformation, etc.) or stores it (in the form of electric and magnetic fields). On average, and on a net basis, it does not supply energy to the rest of the circuit. Resistors, capacitors, inductors, and diodes are examples of passive elements.

Linear Elements

In a two-terminal element (or device), the terminal voltage and current constitute the input-output pair. If their relationship is a linear one, the element is called linear. Three elements are modeled as linear. They are as follows: R (resistor), L (inductor), and C (capacitor). Their terminal i-v characteristics are

Resistor R in units of ohm, Ω　　　$v = Ri$

Inductor L in units of henry, H　　　$v = L\dfrac{di}{dt}$

Capacitor C in units of farad, F　　　$v = \dfrac{1}{C}\displaystyle\int_{-\infty}^{t} i(\tau)d\tau$

In this textbook we mainly consider linear circuits. The symbols for R, L, and C are shown in Figure 1(d).

Signal and Energy Sources

Sources of electric signals and energy such as batteries, generators, sensors, and transducers are complex physical devices and systems. They are modeled by electric circuits that contain, among other elements, two idealized model elements called a voltage source and a current source. Ideal voltage and current sources are two-terminal elements characterized by their voltage or current. In an ideal voltage source, the voltage is set independently from the current that is drawn from it (or passes through it). The voltage may be constant (DC) or may vary with time according to a time function. In an ideal current source, the current is set independently of the voltage across it. Again, the current may be constant (DC) or time varying.

If the voltage between the terminals of a voltage source is controlled by the voltage or current in another element, the source is called a dependent (or controlled) voltage source. Similarly then, the current through a source may be under the control of another circuit variable, in which case it is called a dependent current source. Amplifiers are examples of controlled sources. The voltage and current sources defined above are called active elements as opposed to passive elements. The circuit symbols for voltage and current sources are shown in Figure 1(e).

Switch, Short Circuit, and Open Circuit

A switch is a two-terminal device with only two states: *on* (closed) and *off* (open). In the *on* state, $v = 0$. In the *off* state, $i = 0$. The i-v characteristic of the switch is, therefore, shown by $v = 0$ (on) and $i = 0$ (off). The switch may be considered a nonlinear resistor with two values, $R = 0$ (on state) and $R = \infty$ (off state). In the *on* state, the switch is said to produce a "short circuit" between its two terminals. In the *off* state, the switch produces an "open circuit" between its terminals. The "short circuit" and "open circuit" states are also shown by a "direct connection" and "no connection," respectively. See Figure 1(f).

Formulation of Equations

Given the element values, voltages and currents throughout the circuit can be found by applying the following three sets of equations: (i) the elements' i-v characteristics, (ii) KCL, and (iii) KVL. In general, the result will be in the form of a set of differential equations containing the elements variables. The number of equations may be reduced, limiting them to the variables of interest only. See Example A.1.

E*xample*
A.1

In the circuit of Figure 2, formulate the equation from which the capacitor voltage $v_2(t)$ may be found.

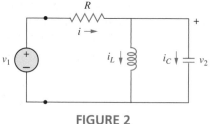

FIGURE 2

Solution

Let i, i_C, and i_L be the currents in the resistor, capacitor, and inductor, respectively. Then,

KVL around the loop: $\qquad\qquad\qquad\qquad v_1 = Ri + v_2$

KCL at the RLC node: $\qquad\qquad\qquad\qquad i = i_C + i_L$

Capacitor (i, v) characteristic: $\qquad\qquad\qquad i_C = C\dfrac{dv_2}{dt}$

Inductor (i, v) characteristic: $\qquad\qquad\qquad i_L = \dfrac{1}{L}\displaystyle\int_{-\infty}^{t} v_2 d\tau$

Eliminate currents from the above equations: $\quad RC\dfrac{dv_2}{dt} + \displaystyle\int_{-\infty}^{t} \dfrac{R}{L} v_2 dt + v_2 = v_1$

Differentiate both sides: $\qquad\qquad\qquad \dfrac{d^2 v_2}{dt^2} + \dfrac{1}{RC}\dfrac{dv_2}{dt} + \dfrac{1}{LC} v_2 = \dfrac{1}{RC}\dfrac{dv_1}{dt}$

Solution methods for differential equations have been discussed in Chapters 5 and 6. However, use of the phasor and impedance concepts provide simplifying approaches to the formulation of equations and their solutions. These are discussed below.

A.2 AC Steady State

What Is the AC Steady-State?

AC is an abreviation for alternating current.[1] In an AC (also called AC circuit) regime, the circuit variables (i.e., element currents and voltages) are periodic functions of time. Since any periodic function of time can be expanded as the sum of its harmonics (sinusoids), and since the circuit is linear, the superposition effect of harmonics applies. Therefore, without loss of generality, the analysis in the AC steady state can be reduced to an analysis of the circuit in the sinusoidal steady state. A sinusoidal steady state develops when a sinusoidal source is applied to a linear circuit. After passage of enough time, all voltages

---

[1] At the dawn of the electric century (i.e., at the turn of the 20th century), the usability of alternating current was under question. Wouldn't a current in one direction potentially undo what it had done in the opposite direction?

and currents in the circuit eventually become sinusoidal with the same frequency but possibly different amplitudes and phase angles. This is because the circuit equations are linear differential equations with constant coefficients and, as we know, the response of such equations to a sinusoid is a sinusoid. We start with response of R, L, and C to a sinusoidal voltage.

Response of R, L, and C to a Sinusoid

Let a sinusoidal voltage $v(t) = V \cos(\omega t)$ feed a circuit element. The current in the element is

For a resistor: $i = \dfrac{v}{R} = \dfrac{V}{R} \cos(\omega t) = I \cos(\omega t)$,   where $I = \frac{V}{R}$

For an inductor: $i = \dfrac{1}{L} \displaystyle\int_{-\infty}^{t} v \, dt = \dfrac{1}{L} \int_{-\infty}^{t} V \cos(\omega t) dt$

$\qquad\qquad = I \cos(\omega t - 90°)$,   where $I = \frac{V}{L\omega}$

For a capacitor: $i = C\dfrac{dv}{dt} = C\dfrac{d}{dt}[V \cos(\omega t)] = I \cos(\omega t + 90°)$, where $I = C\omega V$

AC Analysis

All voltages and currents in an AC circuit have the same functional form. They are all sinusoidal with the same frequency. The analysis of the circuit will become simpler if the sinusoidal variation with time is left out until the last stage. The analysis concerns itself with the amplitude and phase of the currents and voltages only. This is done by noting that a real-valued sinusoid is the real part of a complex exponential function:

$$v(t) = V \cos(\omega t + \theta) = \mathcal{RE}\{\overline{V}e^{j\omega t}\}$$

The complex amplitude $\overline{V} = Ve^{j\theta}$ is called the phasor representation of $v(t)$ and is also shown by $V \angle \theta$. The phasor notation can represent the terminal characteristics of R, L, and C in the AC steady-state condition, thus, by eliminating time as a variable, it reduces the differential equations to algebraic operations on amplitudes and phases. This is called the *phasor method*. The phasor method combines the amplitude and phase of a sinusoid in the form of a vector. Graphical methods and phasor diagrams would then perform the algebraic operations without resorting to any complex number algebra. This method was successfully used for the analysis and design of electric motors, transformers, and power transmission lines for several decades before Charles Steinmetz introduced the use of complex numbers in AC analysis in the early part of the 20th century.[2] Sinusoidal voltages and currents represented by phasors may be added and subtracted using the familiar vector addition and subtraction. Kirchhoff's laws may be applied to the vectors

---

[2]Charles P. Steinmetz, *Theoretical Elements of Electrical Engineering*, 2nd. ed. (New York: McGraw-Hill Publishing, 1905). See also C. P. Steinmetz, *Lectures on Electrical Engineering* in three volumes, P. L. Alger ed., (Dover Publications, 1971).

numerically or graphically. The diagram obtained by this method is called the phasor diagram and can be used to solve circuit equations or even as a tool for design. Before the advent of electronic calculators and computers, phasor diagrams were used extensively as effective tools for the analysis and design of AC circuits, devices, and systems. Phasor diagrams still provide valuable visualization tools and can illustrate the functioning and operation of AC circuits.

The relationship between the phasor representations of the terminal variables are given below.

Resistor: $v(t) = V \cos(\omega t) = Re\{\overline{V}e^{j\omega t}\}$, $i(t) = \dfrac{V}{R} \cos(\omega t)$

$$= Re\{\overline{I}e^{j\omega t}\} \Longrightarrow \overline{I} = \frac{V}{R}$$

Inductor: $v(t) = V \cos(\omega t) = Re\{\overline{V}e^{j\omega t}\}$, $i(t) = \dfrac{V}{L\omega} \cos(\omega t - 90^o)$

$$= Re\{\overline{I}e^{j\omega t}\} \Longrightarrow \overline{I} = \frac{\overline{V}}{jL\omega}$$

Capacitor: $v(t) = V \cos(\omega t) = Re\{\overline{V}e^{j\omega t}\}$, $i(t) = V C\omega \cos(\omega t + 90^o)$

$$= Re\{\overline{I}e^{j\omega t}\} \Longrightarrow \overline{I} = jC\omega \overline{V}$$

Impedance and Admittance

The impedance and admittance of a two-terminal circuit (given by Z and Y, respectively) are defined by

$$Z = \frac{\overline{V}}{\overline{I}} \quad \text{and} \quad Y = \frac{\overline{I}}{\overline{V}}$$

Admittance is the inverse of impedance. In general, both are complex numbers and frequency-dependent.

The standard unit of impedance is the Ohm (shown by Ω) and the standard unit of admittance is the Siemens (shown by S, where $1\,S = 1\,\Omega^{-1}$). The concept of impedance applies to any linear circuit (containing RLC and dependent sources) operating in the sinusoidal steady-state. The i-v relationship for linear circuits in the time and phasor domains are summarized below.

$$v(t) = V \cos(\omega t) = \mathcal{RE}[\overline{V}e^{j\omega t}] \qquad \text{where } \overline{V} = V\angle 0$$

$$i(t) = I \cos(\omega t - \theta) = \mathcal{RE}[\overline{I}e^{j\omega t}] \quad \text{where } \overline{I} = I\angle -\theta$$

$$\overline{V} = Z\overline{I}, \quad \overline{I} = Y\overline{V}, \quad Z = \overline{V}/\overline{I} \qquad \text{where } Z = |Z|\angle\theta$$

Note that the angle θ represents the phase of the impedance. Impedances combine like resistors. Two impedances in series are equivalent to an impedance whose value is the sum of the two. Two admittances in parallel are equivalent to an admittance whose value is the sum of the two. A phasor voltage applied to a series combination of two impedances is divided between them proportionally. A phasor current applied to two parallel admittances is divided between them proportionally.

Names and Notations

The impedance $Z = R + jX$ and admittance $Y = G + jW$ are complex numbers, with R, X, G and W being real functions of frequency. At a given frequency, R, X, G and W are constants. R is called resistance and X is called reactance. If X is a positive number, the circuit is said to be inductive (X being inductance). If X is a negative number, the circuit is referred to as capacitive (X being capacitance). Similarly, the real part G of an admittance is called conductance and its imaginary part W is susceptance. The unit of impedance (Z, R, and X) is the Ohm (Ω). The unit of admittance (Y, G, and W) is the Siemens.

Kirchhoff's Laws in the Phasor Domain

In the AC steady-state Kirchhoff's laws apply to phasor currents and voltages.

$$\sum_{k=1}^{N} i_k = \sum_{k=1}^{N} \mathcal{RE}[\overline{I}_k e^{j\omega t}] = \mathcal{RE}\left[e^{j\omega t} \sum_{k=1}^{N} \overline{I}_k\right] = 0 \rightarrow \sum_{k=1}^{N} \overline{I}_k = 0$$

$$\sum_{k=1}^{N} v_k = \sum_{k=1}^{N} \mathcal{RE}[\overline{V}_k e^{j\omega t}] = \mathcal{RE}\left[e^{j\omega t} \sum_{k=1}^{N} \overline{V}_k\right] = 0 \rightarrow \sum_{k=1}^{N} \overline{V}_k = 0$$

Circuit Analysis in the Phasor Domain

Consequent to the application of KVL, KCL, and the i-v relationship in the phasor domain, branch voltages and currents are found by solving a set of simultaneous equations of first order. Solution techniques are similar to those of resistive circuits, except for the fact that currents, voltages, and element values are phasors (i.e., complex numbers). The following three examples illustrate the application of the loop current and node voltage methods.

Example
A.2

In the circuit of Figure 2, let $R = 50\ \Omega$, $L = 1$ H, $C = 10^{-2}$ F, and $v_1(t) = \cos \omega t$. Find $v_2(t)$ for

a. $\omega = 9$

b. $\omega = 10$

c. $\omega = 11$

Solution
From voltage division in the phasor domain we have

$$V_2 = V_1 \frac{Z_{LC}}{R + Z_{LC}}, \quad \text{where } Z_{LC} = \frac{Z_L Z_C}{Z_L + Z_C}, \quad Z_L = j\omega L, \text{ and } Z_C = \frac{1}{j\omega C}$$

The following Matlab program implements the above for three values of ω.

```
R=50;  L=1;  C=0.01;  V1=1;  w=[9 10.0000001 11];
z1=j*L*w;  z2=1./(j*C*w);  z=(z1.*z2)./(z1+z2);
```

```
V2=V1*z./(R+z);
abs(V2)
angle(V2)*180/pi
```

The capacitor voltage v_2 is

a. $0.6877\cos(9t + 46.5°)$

b. $\cos(10t)$

c. $0.7234\cos(9t - 43.67°)$

Find the capacitor voltage $v(t)$ and inductor current $i(t)$ in the circuit of Figure 3. The current source is $i_s(l) = 10\cos(1{,}000t)$. Element values are $R_1 = 3k\Omega$, $R_2 = 2k\Omega$, $L = 1$ H, and $C = 1\ \mu$F.

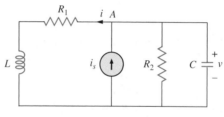

FIGURE 3

Solution

The phasor representation of the current source is $\overline{I}_s = 10\angle 0°$ A. The impedances of the inductor and capacitor at $\omega = 1{,}000$ rad/s are $j10^3$ and $-j10^3$, respectively. The source current $\overline{I}_s$ is divided between the series RL and parallel RC combinations, where

$$Z_{RL} = (3 + j)10^3\ \Omega \text{ and } Z_{RC} = \frac{2{,}000}{1 + j2} = 400(1 - j2)\ \Omega$$

Using current division,

$$\overline{I} = \overline{I}_s \frac{Z_{RC}}{Z_{RC} + Z_{RL}} = 10\frac{400(1 - j2)}{400(1 - j2) + (3 + j)10^3} = 2.626\angle - 66.8°\ A$$

$$\overline{V} = I Z_{RL} = (2.626\angle - 66.8°) \times (3 + j)10^3 = 8.304\angle - 48.4°\ V$$

The time expressions for the inductor current and capacitor voltage are

$$i(t) = 2.626\cos(1{,}000t - 66.8°) \text{ and } v(t) = 8.304\cos(1{,}000t - 48.4°)$$

Quality Factor Q

Capacitors and inductors store energy. Resistors consume energy. The *quality factor* of a passive linear circuit or a linear device that both stores and dissipates energy is defined by

$$Q \equiv 2\pi \frac{\text{Maximum energy stored}}{\text{Total energy dissipated per cycle}}$$

A quality factor generally depends on the operating frequency.

A.3 Complex Frequency and Generalized Impedance

The Complex Frequency

The phasor analysis of the sinusoidal steady state may be extended to broader classes of signals (e.g., sinusoids with exponentially growing or decaying amplitudes). This is done by employing the exponential function $\overline{V} e^{st}$, where both $\overline{V}$ and s may be complex numbers. $\overline{V}$ is called the complex amplitude (or generalized phasor) and s is called the complex frequency. Parallels between derivations related to the complex frequency (such as the generalized impedance and phasor analyses) and those related to the sinusoidal steady state can be seen throughout this section. This is expected, as the sinusoidal steady state is a special case of the complex frequency. The generalization from $j\omega$ to $s = \sigma + j\omega$ is valuable not only in providing a mathematical model for broader classes of real signals, but, more importantly, because *complex exponentials are eigenfunctions of linear circuits* (i.e., they elicit responses with the same complex frequency).

Sources that are complex functions of the complex frequency are in general not realizable. It may appear at first that the s-plane analysis has limited application. This is not the case, however. A real source may be represented by the real part of a complex source or as a sum of complex sources. Also, the response of a circuit made of real-valued linear elements is equal to the real part of its response to the complex source.

Generalized Phasor and R, L, C Responses

Let $v(t) = V e^{\sigma t} \cos(\omega t + \theta)$ be a real function of time. Consider the complex exponential function

$$\overline{V} e^{st}, \quad \text{where } \overline{V} = V e^{j\theta} \quad \text{and} \quad s = \sigma + j\omega$$

Note that

$$\overline{V} e^{st} = V e^{j\theta} e^{(\sigma + j\omega)t} = V e^{\sigma t} e^{j(\omega t + \theta)} = V e^{\sigma t} [\cos(\omega t + \theta) + j \sin(\omega t + \theta)]$$

from which

$$v(t) = Re\{\overline{V} e^{st}\}$$

The function $\overline{V} e^{st}$, often called the complex exponential signal, is completely specified by its complex amplitude, or phasor, $\overline{V}$ and the complex frequency s. Responses of R,

L, and C to such a voltage are

Resistor:	$v(t) = Re\{\overline{V}e^{st}\}, \quad i(t) = \dfrac{v(t)}{R} = Re\{\overline{I}e^{st}\}$	$\implies \quad \overline{I} = \dfrac{\overline{V}}{R}$
Inductor:	$v(t) = Re\{\overline{V}e^{st}\}, \quad i(t) = \dfrac{1}{L}\displaystyle\int_{-\infty}^{t} v\,dt = Re\{\overline{I}e^{st}\}$	$\implies \quad \overline{I} = \dfrac{\overline{V}}{Ls}$
Capacitor:	$v(t) = Re\{\overline{V}e^{st}\}, \quad i(t) = C\dfrac{dv}{dt} = Re\{\overline{I}e^{st}\}$	$\implies \quad \overline{I} = Cs\overline{V}$

Generalized Impedance and Admittance

The $\overline{i}$-$\overline{v}$ relationships for R,L,C in the phasor domain may be summarized by $\overline{V} = Z\overline{I}$, where Z is called the generalized impedance of the element. Conversely, the relation between $\overline{I}$ and $\overline{V}$ may be shown by $\overline{I} = Y\overline{V}$, where Y is called the generalized admittance. These are generalized forms of the impedance and admittance previously defined for the sinusoidal steady state. The concept of generalized impedance can be extended to any linear circuit. The concepts of series and parallel combinations (as well as voltage and current division) apply equally to the generalized impedance. The latter simplifies the analysis of circuits. More importantly, it leads to a circuit's dynamical equations being expressed in terms of s as a substitute for the time domain.

Example
A.4

a.   Use the concept of generalized impedance in the circuit of Figure 2 to find the differential equation relating v_2 to v_1.

b.   Let $R = 50\ \Omega$, $L = 1$ H, $C = 10^{-2}$ F, and $v_1 = \cos\omega t$. Find v_2.

Solution

a. Phasor-domain voltage division: $H(s) \equiv \dfrac{V_2}{V_1} = \dfrac{z_{LC}}{R + z_{LC}} = \dfrac{\frac{Ls}{1+LCs^2}}{R + \frac{Ls}{1+LCs^2}}$

$$= \dfrac{1}{RC}\dfrac{s}{s^2 + \frac{1}{RC}s + \frac{1}{LC}}$$

Reverting to the time domain: $\dfrac{d^2 v_2}{dt^2} + \dfrac{1}{RC}\dfrac{dv_2}{dt} + \dfrac{1}{LC}v_2 = \dfrac{1}{RC}\dfrac{dv_1}{dt}$

b. $H(s) = \dfrac{2s}{s^2 + 2s + 100}$, $H(\omega) = H(s)|_{s=j\omega} = \dfrac{j2\omega}{100 - \omega^2 + j2\omega}$

$v_2 = |H|\cos(\omega t + \phi)$ $|H| = \dfrac{2\omega}{\sqrt{(100 - \omega^2)^2 + 4\omega^2}}$,   and

$$\phi = 90° - \tan^{-1}\dfrac{2\omega}{100 - \omega^2}$$

Observe parallels between the above $H(s)$ and the system function defined for the input-output pair (v_1, v_2). In fact, the $H(s)$ obtained by applying the circuit laws to generalized impedances becomes the Laplace transform of the unit-impulse response of the system if we assign to it a pole-free region of convergence in the RHP of the s-plane.

A.4 Amplifiers and Op-Amp Circuits[3]

An amplifier is a two-port device, one port being the input and the other the output. As the name implies, it amplifies a signal or the power it can deliver. The signal can be in the form of a voltage or a current. A simple model of a voltage amplifier is shown in Figure 4(a) with its input-output connections. The amplifier is the segment of the circuit within the dashed box in the center. It receives input from the signal source such as the voltage source $v_1(t)$ in series with resistor R_ℓ shown in the dashed box on the left. It feeds the load such as the resistor R_ℓ shown in the dashed box on the right. The dependent source amplifies the input signal and is a required element. The input and output impedances [shown in Figure 4(a) by resistors R_i and R_o, respectively] are ideally infinity and zero, respectively. The open-loop gain of the amplifier is k. The voltage gain in the circuit of Figure 4(a) is

$$\frac{v_2}{v_1} = \frac{R_i}{R_i + R_1} \times \frac{R_\ell}{R_\ell + R_o} k$$

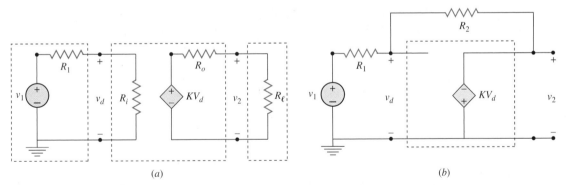

(a) $\qquad\qquad\qquad\qquad\qquad\qquad$ (b)

FIGURE 4 *(a)* The amplifier circuit shows a voltage source in series with a resistor interfacing with an amplifier (enclosed within the dashed box in the center), which feeds the load R_ℓ. The overall gain is affected by the gain of the amplifier and the interfacing circuits. *(b)* The negative feedback path across a high-gain inverting amplifier (i.e., k very large) sets the gain of the circuit at nearly $-R_2/R_1$.

---

[3]For a more detailed treatment of this subject the reader is referred to Nahvi and Edminister, *Electric Circuits*, 5th ed., Schaum's Outlines, (New York: McGraw-Hill, 2011).

The gain is less than k and depends on the connection with the rest of the circuit. A negative feedback path such as the one shown in the circuit of Figure 4(b) can reduce the sensitivity of the gain to the above factors. More importantly, under certain conditions, it can set the overall gain to a desired value.

Feedback Amplifier

In Figure 4(b) a feedback path is established from the output of the amplifier to its input. For simplicity, we set $R_i = R_\ell = \infty$ and $R_o = 0$. We will show that for a high-gain amplifier, the external circuit sets the overall gain. From the amplifier we have $v_2 = -kv_d$. From the output-input voltage divison we have $v_d = (v_1 R_2 + v_2 R_1)/(R_1 + R_2)$. Substitute v_d from the first equation into the second to find

$$\frac{v_2}{v_1} = -\frac{R_2}{R_1}\frac{k}{k + (1 + \frac{R_2}{R_1})} \approx -\frac{R_2}{R_1}, \ \text{if } k >> \left(1 + \frac{R_2}{R_1}\right)$$

Operational Amplifiers (Op-Amps)

Electronic devices called op-amps have the amplifier property described above. An op-amp is an active device, most often in the form of an integrated circuit chip, with two input terminals and one output. These constitute the input and output signal ports, respectively. The input terminals are labeled inverting and noninverting. The op-amp has connections to a DC power supply. It also has a pair of terminals for adjustment. It doesn't have a terminal for ground connection. The ground is provided by the power supply and signal sources as a common ground for the whole circuit. See Figure 5(a).

Op-amps normally have a low-output impedance. At low frequencies, their input impedance and open-loop gain are very high. As a consequence of the high-input impedance, the current through the inverting and noninverting terminals are very low. They are normally considered zero. When the op-amp is working in the linear range the voltage at the output terminal is proportional to the differential input, $v_o = k(v^+ - v^-) \equiv kv_d$. k is the open-loop gain of the op-amp. All voltages are referenced with respect to ground. In the linear range, output voltage doesn't deviate from a window defined by voltage supply. A consequence of k being high is that when the op-amp is working in the linear range $v_d \approx 0$. In that state, the circuit elements that are external to the op-amp

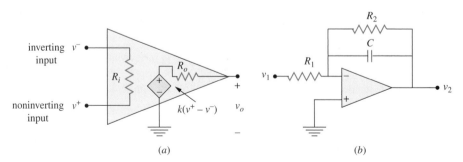

FIGURE 5 *(a)* An op-amp model. *(b)* A leaky integrator circuit using an op-amp.

force v_d near zero. A high v_d would send the output voltage into high or low levels set by the power supply. The op-amp is said to go into high or low saturation and, therefore, would not be working as a linear amplifier anymore.

Assuming the circuit is operating linearly, the analysis of a circuit containing one or more op-amps becomes simple by the following approximations: (i) the op-amp has zero input voltage, $v_d = 0$, and (ii) the op-amp terminals don't draw any current, $i^+ = i^- = 0$. Such assumptions have often been used throughout this book. Example A.5 reviews the method. At higher frequencies when the input impedance cannot be considered high (due to capacitive effects), a basic model such as that of Figure 4(a) may be used.

In the circuit of Figure 5(b) find $H(s) = V_2/V_1$.

Solution

The noninverting terminal is at ground. Assuming a linear operation, the inverting terminal is, therefore, at a virtual ground. By applying KCL at the inverting terminal and noting that the op-amp doesn't draw any current, we obtain

$$\frac{V_1}{R_1} + V_2 \left(CS + \frac{1}{R_2} \right) = 0,$$

$$\frac{V_2}{V_1} = -\frac{R_2}{R_1} \frac{1}{1 + R_2 Cs}$$

INDEX

U